P9-DNT-858

Discovering Geometry
An Inductive Approach

Michael Serra

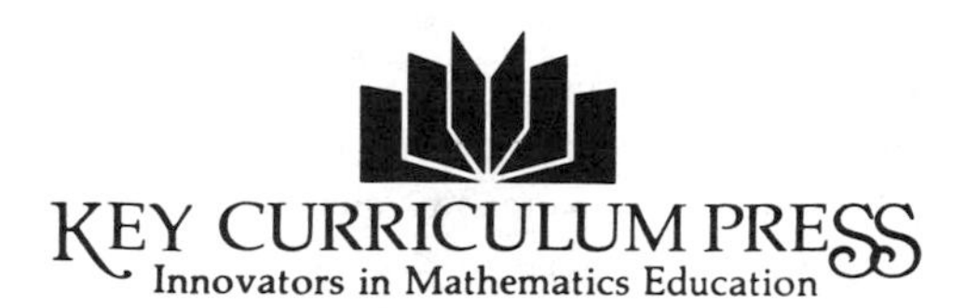

2512 Martin Luther King Jr. Way
P.O. Box 2304
Berkeley, California 94702

Author

Michael Serra
George Washington High School
San Francisco, California

Editor

Steven Rasmussen
Key Curriculum Press
Berkeley, California

Advisors

Tom Swartz
George Washington High School
San Francisco, California

David Rasmussen
Cabot High School
Nova Scotia, Canada

Dr. John Olive
University of Georgia
Athens, Georgia

Dr. J. Michael Shaughnessy
Oregon State University
Corvallis, Oregon

Katie Jurney
Catlin Gabel School
Portland, Oregon

Richard Wertheimer
School District of Pittsburgh
Pittsburgh, Pennsylvania

Teacher Consultants and Field Testers

Catlin Gabel School: Bob Eckland, Katie Jurney
Portland Public School District #1: Dave Damcke, John P. Oppedisano
Reynolds School District: Debbie Lindow, Charlene Trachsel
San Francisco Unified School District: Dr. Sam Butscher, Theresa Hernandez-Heinz, Robert Knapp, Jeff Salisbury, Tom Swartz, Edward Van Pelt, Li Oi Yu
San Lorenzo Valley Unified School District: Dianne Borchardt, Sandie Gilliam, Dennis Olson
Vancouver School District: Connie Callos

Logo Computer Activities

Dr. John Olive, University of Georgia
Dr. Robert Jensen, Emory University
Steven Rasmussen, Key Curriculum Press

20 19 98 97 96 95

The 1993 revision contains an expanded hints section.

Printed in the United States of America ISBN 0-913684-08-2

Foreword

Michael Serra has written a genuinely exciting geometry book. This book is unique in that the students actually "create" geometry for themselves as they proceed through the activities and problems. Concepts are first introduced visually, then analytically, then inductively, and finally, deductively. The spirit of this text is remarkably consistent with recent research on the development of geometric thinking in adolescents, particularly the levels of thinking in the van Hiele theory.

From the beginning, students participate in the construction of definitions. The author makes excellent use of a "non-examples and examples" approach to encourage students to build their own definitions. As new geometric figures are introduced, activities are structured so that students can discover properties of these figures. Throughout the book, students are asked to make conjectures about figures, and about relationships among figures. These conjectures are brought up again in the chapters on proof for deductive consideration.

In addition to a thorough treatment of all the topics anyone has ever wished to cover in a geometry course, Serra provides a number of extra "nuggets" and superimposes a creative developmental sequence to his topics. Measurement topics such as area and volume appear earlier than in many geometry texts, prior to formal proof. There is an extensive chapter on transformations and tessellations. The tessellations provide ample opportunity to concretely explore the symmetry patterns that make transformations so powerful. Where some texts pay lip service to these topics, Serra provides us with a full and varied menu. Excellent Logo computer activities are sprinkled throughout the text in strategic places, providing an additional environment for student discoveries. Coordinate topics are woven in wherever appropriate. Most of all, however, this book has *super* problems, written by a master story teller. Each chapter has one or more cooperative problem solving activities, which encourage interaction and communication between students. Students will have an opportunity to write and talk about their mathematics, and thus a better chance of understanding their mathematics.

Formal proof does not appear in this book until the last two chapters. However, by the time students are asked to write proofs in Serra's book, they have already made conjectures for months; formed and tested their own definitions; solved logic problems; developed visualization skills through drawings, constructions, and Logo; and studied logic, reasoning, and the nature of proof. *They are ready for proofs*!

In the past we have erred in pushing proofs on students too soon, before they had a handle on shapes and their properties. In the past we have asked many of our students to do two things simultaneously — learn geometric concepts, and learn deductive reasoning. I applaud Michael Serra's move to delay proof in his book until students have seen the whole spectrum of geometric concepts. Serra's book should gives students a better chance of learning geometry, and of learning about proof.

Finally, this is a book for "do-ers." Students constantly *do* things in this book, both alone, and with other students. If you want your students and yourself to become actively involved in the process of learning and creating geometry, then this book is for you.

Dr. J. Michael Shaughnessy
Department of Mathematics
Oregon State University

A Note to Students from the Author

What Makes *Discovering Geometry* Different?

Features of *Discovering Geometry*

This book was designed so that you and your teacher can have fun with geometry. In *Discovering Geometry* you "learn by doing." You will learn to use the tools of geometry and to perform geometric **investigations** with them. Your investigations will lead to geometric discoveries. Many of the geometric investigations are carried out in small **cooperative groups** where you jointly plan and find solutions with other students. I think you will enjoy the **humorous word problems with their cartoons**. You will help archaeologist Ertha Diggs determine the height of a Mayan jungle temple, help pirate Captain Coldhart bury his treasure, and help Hemlock Bones solve the case of the Belgian Stamp Murder. I created these problems in the hope of reducing your anxiety about word problems. In the **special projects** you will build geometric solids, make kaleidoscopes, design a racetrack, find the height of your school building, and create a mural. Every chapter closes with a special **cooperative problem solving** lesson designed to be worked in small groups and set at a lunar colony of the 21st century. I think you'll enjoy the extra challenges in the Improving Visual Thinking Skills and Improving Reasoning Skills **puzzles** that I have sprinkled throughout the book. There are also five **Logo computer activities** in the text. In one Logo activity you discover the world of a strange class of geometric shapes called *fractals*! The Logo procedures that you'll need for the computer activities are in Appendix B at the back of the book.

I have designed the special projects, puzzles, and computer activities so that you can do them independently, whether or not your class tackles them as a group. Read through them as you proceed through the book.

Chapter Sequence

I begin with a chapter on geometric art to show you that geometry is found in the art of cultures throughout the world. In Chapter 1 you will learn how to reason inductively. Inductive reasoning is the process that you will use to make geometric discoveries in this book. In Chapter 2 you will use inductive reasoning to create definitions of geometric terms. In Chapter 3 you will use a compass and straightedge to construct geometric figures. The compass and straightedge are two of the primary tools that you will use to make your geometric discoveries in the remainder of the book. In Chapter 7 you will learn to create tiling designs with geometric shapes similar to some of the M. C. Escher works used on the chapter opening pages. Finally, in the last three chapters, you will learn about the other type of reasoning which is called *deductive reasoning* or *proof*.

Please do not feel overwhelmed by the number of chapters in *Discovering Geometry*. It is not possible to discover *all* of the chapters in one school year. Your teacher will guide you through the book to create one of several different types of geometry courses possible using this text.

Suggestions For Success

It is important to be organized when working with *Discovering Geometry*. You should have a notebook with a section for definitions, a section for new geometric discoveries, and a section for daily notes and exercises. You should study your notebook regularly. You will need four tools of geometry for the investigations: compass, protractor, straightedge, and ruler. You should have a calculator handy.

You will find hints and solutions for some key exercises in Appendix A at the back of the book. Exercises that have hints are identified by an asterisk (*). Try to solve the problems on your own without looking at the hints. Refer to the hints section to check your method or as a last resort if you can't solve a problem.

Unlike most texts, *Discovering Geometry* will ask you to work cooperatively with your fellow students. This means you should pull your desks together and get to know one another. When you are working together cooperatively, you should always be willing to listen to each other, be an active participant, ask each other questions when you don't understand, and help each other when asked. When working cooperatively, you can accomplish much more than the sum total of what you can accomplish individually. And best of all, you'll have less anxiety and a whole lot more fun.

Michael Serra
George Washington High School
San Francisco, California

Contents

0 Geometric Art

Print Gallery, M. C. Escher, 1956

In this chapter you will see how geometry appears in nature. You will discover geometry's role in the art of many cultures. In this chapter you will see that geometry is not just a college preparatory math class but is a way of thinking and seeing the world. You will discover that geometry is alive in cultures and art forms around the world. In addition, you will become familiar with the use of two tools of geometry, the compass and the straightedge.

Lesson 0.1

Geometry in Nature and Art

There is one art, no more no less,
To do all things with artlessness.

– Piet Hein

Nature displays an infinite array of geometric shapes from the small atom to the greatest of the spiral galaxies. Crystalline solids, the honeycomb of the bee, snowflakes, the arrangement of seeds on sunflowers and pinecones, the spiral of the nautilus shell, the spider's web, and the regular polygons found in the basic shapes of many flowers are just a few of nature's geometric masterpieces.

Circle

Hexagon

Pentagon

Geometry includes the study of the properties of shapes such as circles, hexagons, and pentagons. Outlines of the sun and moon appear as circular shapes. A stone tossed into a still pond and the growth rings in the cross section of a tree exhibit families of circles (concentric circles). The hexagonal shape appears often in inorganic objects of nature. Snowflakes and other crystalline structures have the hexagon as their basic geometric form. The pentagonal shape, which rarely appears in inorganic objects, is one of the primary geometric forms for organic shapes (living organisms). The starfish is an organism with a pentagonal shape.

People have observed the geometric patterns in nature and have incorporated the patterns in a variety of art forms. Mandalas (Lesson 0.5) and knot designs (Lesson 0.7) are geometric art forms that appear in many cultures all over the world from early history to the present. The circular mandala design, used by the Hindus of India for meditation, is also found in the Aztec calendar stone, in the rose windows of the cathedrals of Europe, and in the intricate Islamic designs of the Moslems of North Africa. Knot designs are found in Africa, northern Scotland (Celtic art), and in Japanese and Chinese lattice designs.

During the Renaissance there was a revival in realistic drawing. Great artists such as Leonardo da Vinci, Albrecht Dürer, and Raphael of Urbino sought to portray spatial relationships on their canvases as if their scenes were being viewed through a window. Renaissance artists turned to geometry for insight in developing their techniques. You will learn about perspective in Lesson 0.8 and get a chance to create your own perspective drawings.

Celtic Art
The Celtic Design Book,
Rebecca McKillip,
Stemmer House, 1981

Islamic Art
Geometric Concepts in Islamic Art,
El-Said and Parman, Dale Seymour Publications, 1976

Modern art includes op art, which is very geometrical. Victor Vasarely is a famous op artist whose work reflects a strong interest in geometry. In Lesson 0.4 you will learn more about op art and you will get a chance to create your own op art design.

Dutch artist M. C. Escher is another 20th-century artist whose work reflects a strong interest in geometry. You will see Escher's works on each of the chapter opening pages in this text. In Chapter 7 you will make your own Escher-like creations.

Modern Art
KAT–TUZ, 1972–1975, Victor Vasarely (1908–present), Courtesy of the artist

EXERCISE SET

In this lesson your goal is to become aware of geometry in nature and geometric art found in different cultures throughout the world.

1. List six natural objects which have geometric shapes. Name the shapes.
2. Bring an object to class which exhibits natural geometry.
3.* Bring an object to school or wear an article of clothing which displays a form of hand-made or manufactured geometric art, perhaps a traditional folk art design.

Improving Visual Thinking Skills

Pickup Sticks

Pickup sticks, a good game for developing motor skills, can be turned into a challenging visual puzzle. In what order are the sticks to be picked up so that you are always removing the top stick?

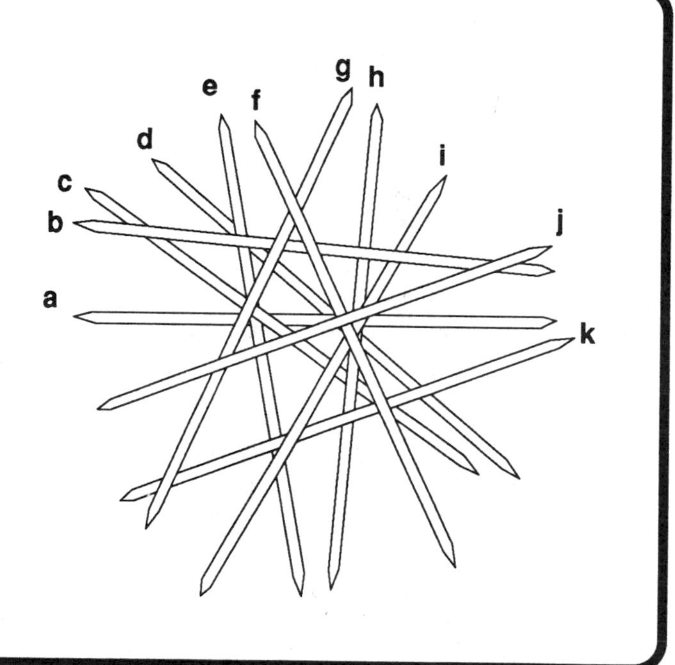

Lesson 0.2

Line Designs

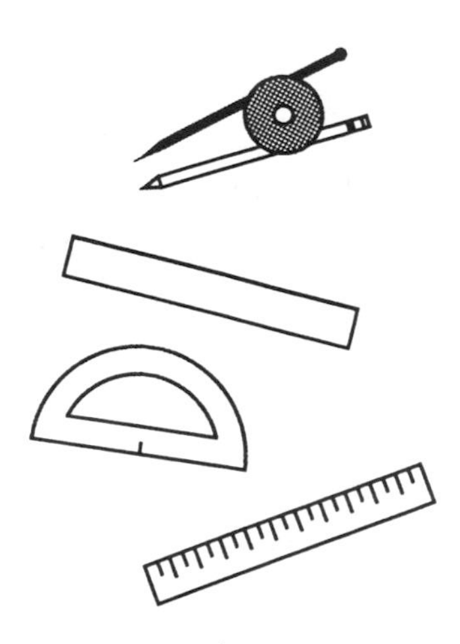

The symmetry and proportion in geometric designs make them very appealing. Geometric designs are also easy to make — provided you have the tools of geometry. There are four basic tools of geometry: compass, straightedge, ruler, and protractor.

A **compass** is a geometric tool used to construct circles. A **straightedge** is a tool used to construct straight lines. The compass and the straightedge are the classical tools of geometry. All the geometry of the ancient Greeks can be produced with just these two tools. The ancient Greeks laid the foundations of the geometry that you will study in this text.

The ruler and protractor are later inventions. A **ruler** is a tool used to measure the length of a line segment. A ruler has marks used for measurement. A straightedge has no marks. The edge of a ruler can serve as a straightedge. A **protractor** is a tool used to measure the size of an angle in degrees. In the next few lessons on geometric art, you will become familiar with these tools of geometry.

Example A

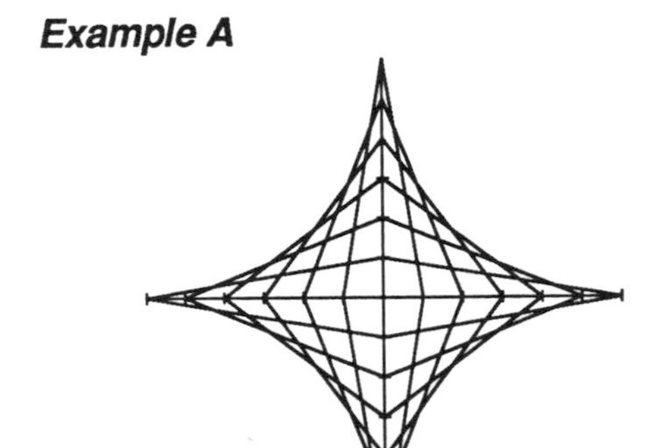

Example B

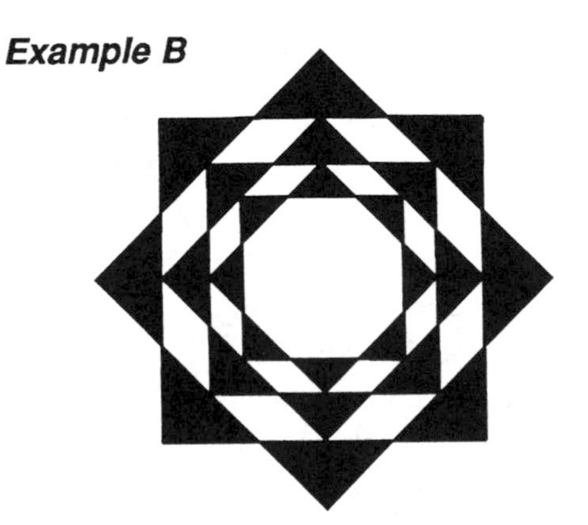

Many types of designs can be created with only straight lines. The steps for creating the two line designs above are demonstrated below. Follow along with the steps in the making of the designs before you begin the exercise set at the end of the lesson.

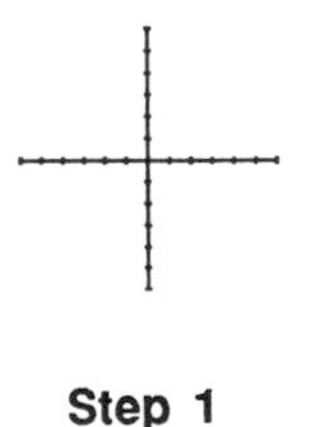

Step 1

Step 2

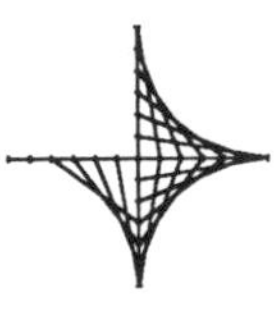

Step 3

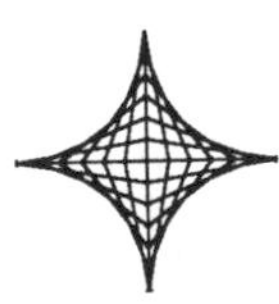

Step 4

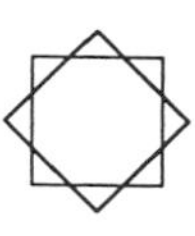

Step 1 **Step 2** **Step 3** **Step 4**

EXERCISE SET

1. Each of the linear designs below was drawn with straight lines only. Select one design and recreate it on a separate sheet of paper. Use the steps demonstrated for example A or example B to help you.

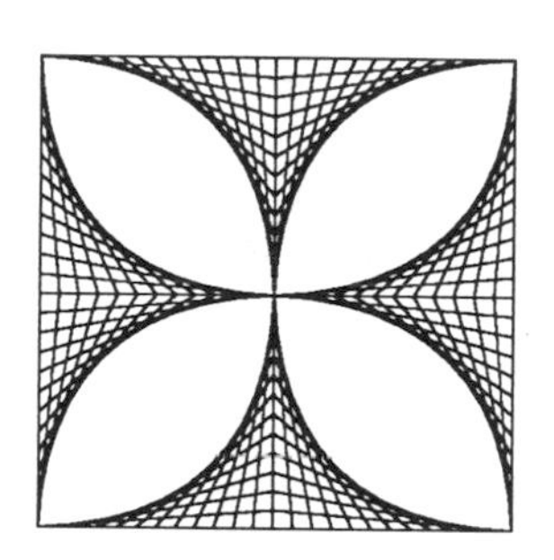

2. Design and make your own drawing using straight lines. Be creative. Look ahead to Lessons 0.4, 0.6, and 0.7 for helpful ideas.

Improving Reasoning Skills

Who Can't Count?

If there are five more consonants than vowels in the phrase, "Verbal reasoning problems are weird," then print "I can't count." Otherwise print "Count Dracula can't count either."

Lesson 0.3

Daisy Designs

The compass is a geometric tool used to construct circles. You can make very nice designs with only a compass. A daisy is one such simple design. The diagrams below give you the necessary instructions to make a daisy. Read through the steps for making the design before you begin the exercise set at the end of the lesson.

Step 1

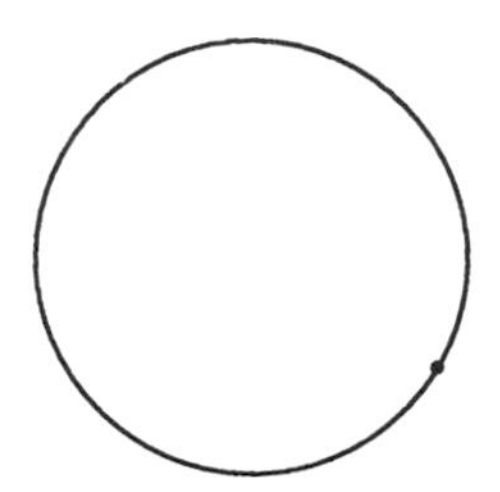

Construct a circle. Then select any point on the circle.

Step 2

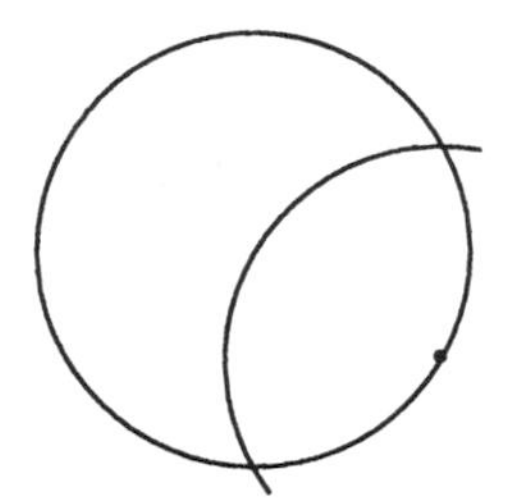

Without changing your compass setting, swing an arc centered at the point selected.

Step 3

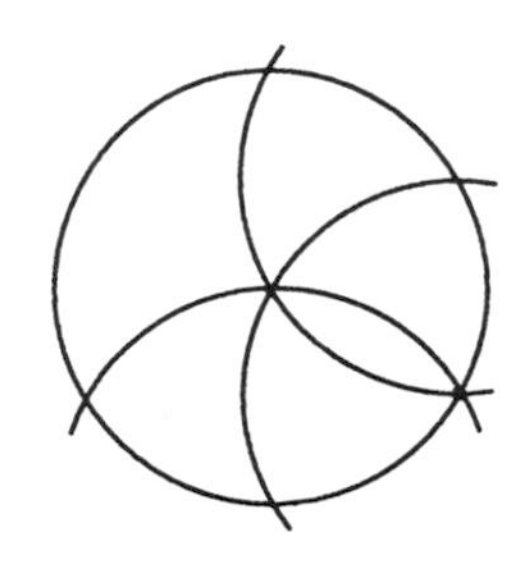

Swing an arc from each of the two new points of intersection created.

Step 4

Again, swing an arc from each of the two new point of intersection.

Step 5

Swing an arc connecting the last two points of intersection.

Step 6

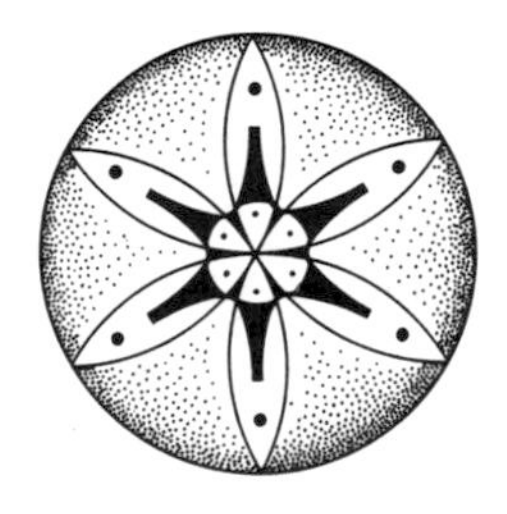

Decorate your daisy design.

The steps for making a daisy are also the steps used to construct a regular hexagon. A regular hexagon is a six-sided figure with all the sides the same length and all the angles the same size. There are six petals equally spaced on each daisy. The tips of the daisy touch the circle at points equally spaced, one compass setting apart. If you connect the tips, you will have a regular hexagon. The compass setting (called the radius of the circle) can be marked off exactly six times around the circle.

The six-pointed daisy can be turned into a twelve-pointed daisy by making another six-pointed daisy between the petals of the first daisy as shown in example A.

If, instead of stopping at the perimeter of the first circle, you continue by swinging full circles, you end up with a "field of daisies" as shown in example B.

Example A

Example B

EXERCISE SET

1. Construct your own daisy design on a full sheet of paper.
2. Using 1 inch for the compass setting (the radius), construct a central regular hexagon and six regular hexagons that each share one side with the original hexagon. It should look similar to (but larger than) the figure on the right.

Lesson 0.4

Op Art

The most beautiful thing we can experience is the mysterious. It is the true source of all art and science.

– Albert Einstein

Op art (or optical art) is a form of abstract art that uses straight lines or geometric patterns to create a special visual effect. The contrasting dark and light regions can, at times, appear to be in motion or to represent a change in surface, direction, and dimension. Victor Vasarely (1908–present) is an artist who can create misleading perceptions with his geometric optical art.

Victor Vasarely (1908–present), Courtesy of the artist

In this lesson you will create your own optical art. But first look at some examples of op art.

Example A

Example B

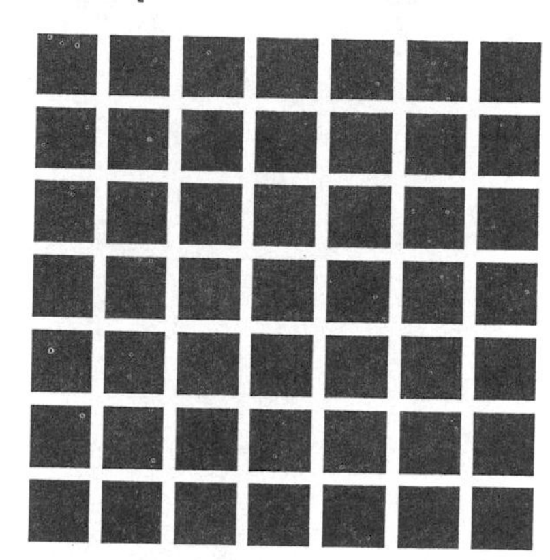

Example C

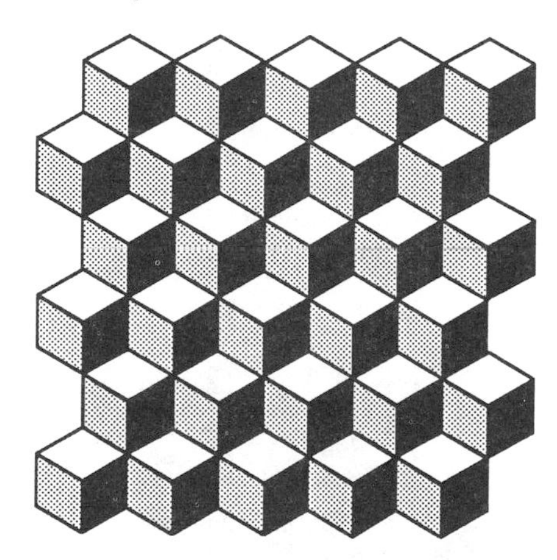

Example D

Op art designs are fun to create and not difficult to make. The steps below show how to create one kind of op art design. First, make a design in outline. Run vertical and horizontal parallel lines on your drawing space, varying the spacing to create visual hills and valleys. Finally, shade in alternate spaces.

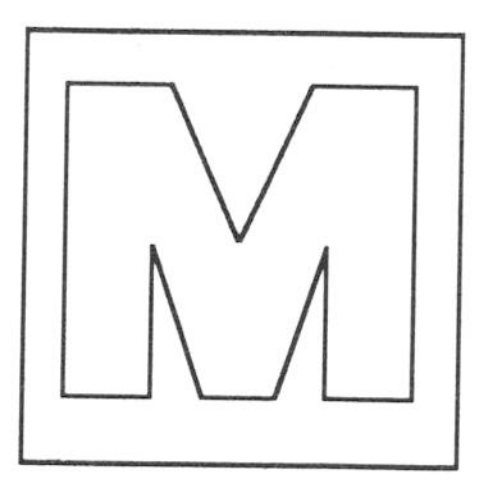

Step 1

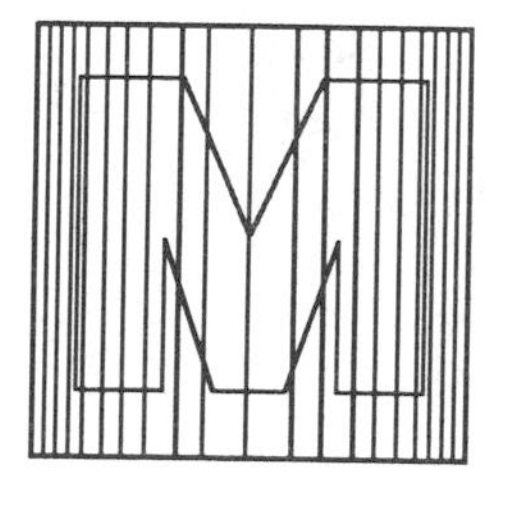

Step 2

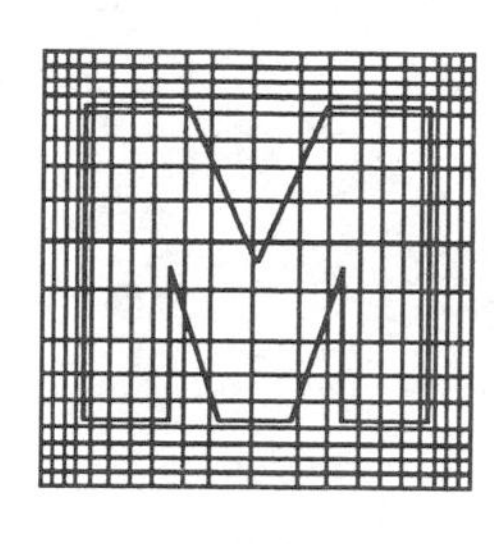

Step 3

Step 4

The steps for creating a design like example D are demonstrated for you below. First, locate a point on each of the four sides of a square the same distance from a corner. Your compass is a good tool for measuring equal lengths. Connect these four points to create another square within the first. Repeat until the squares appear to converge on the center. Be careful that you don't fall in!

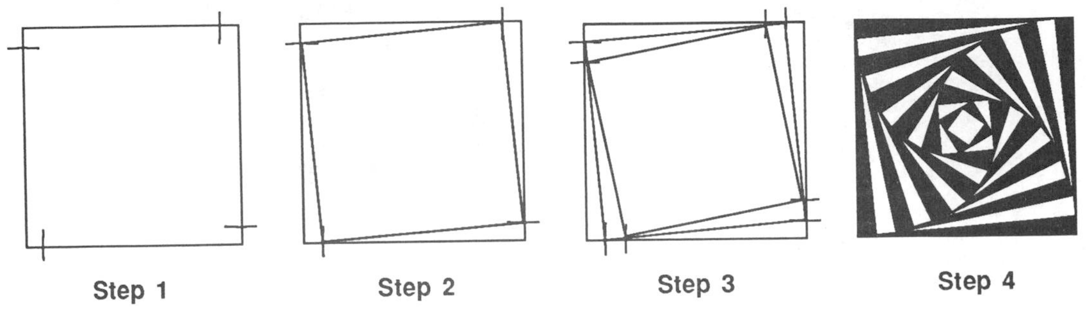

EXERCISE SET

1. What is the optical effect in example A?
2.* What is the optical effect in example B?
3.* What is the optical effect in example C?
4. What is the optical effect in example D?
5. Select one of the op art design types from this lesson and create your own version of that type of op art.

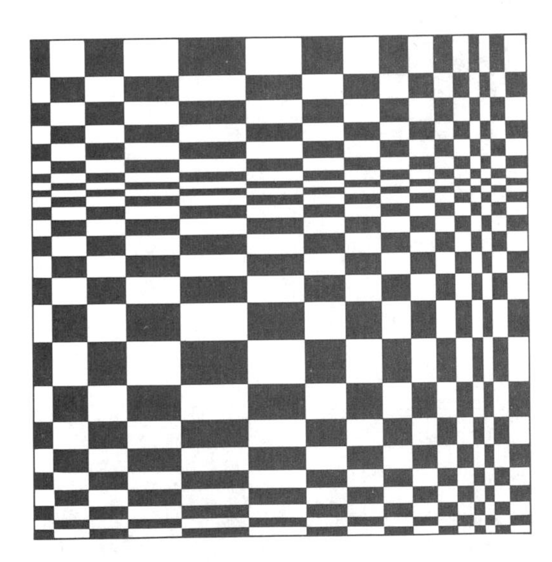

Carmen Apodaca
Geometry student

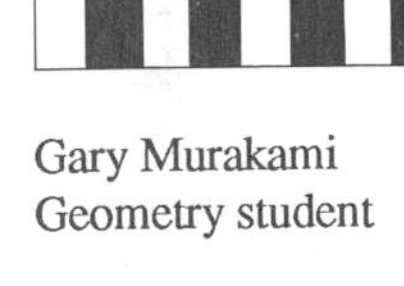

Gary Murakami
Geometry student

Special Project

Drawing the Impossible

You've seen many kinds of optical illusions. Some optical illusions at first appear to be drawings of real objects, but actually they are impossible to make (except on paper). Your task in this special project is to draw the four impossible objects below on full sheets of paper.

Three prongs from two?

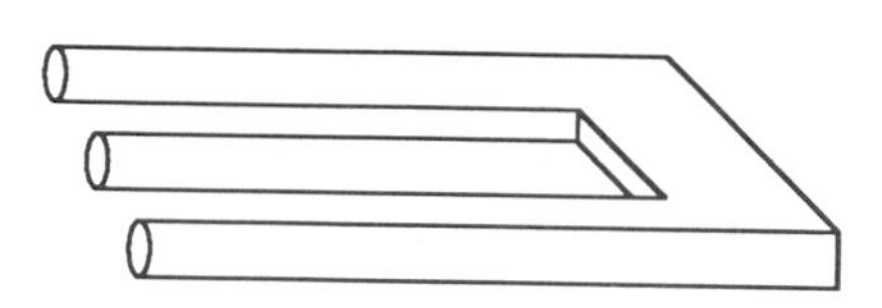

To create this drawing start with the six parallel lines. Then complete both ends.
Try not to look at one end while working on the other.

Penrose Triangle

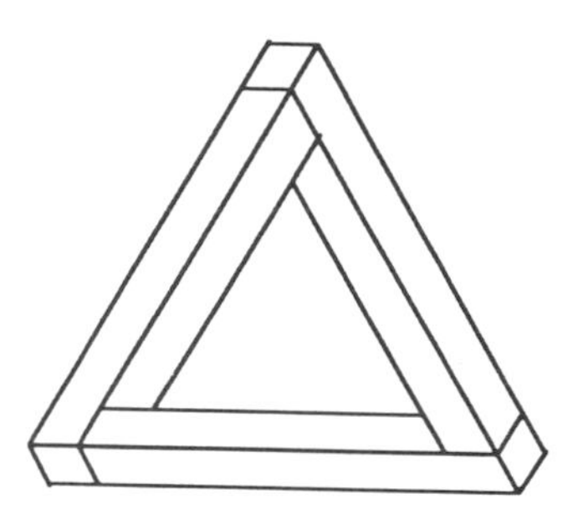

To create the Penrose Triangle (named after mathematician and avid puzzle enthusiast Roger Penrose) begin with three equal sided triangles nested one within the other.

Three towers from four?

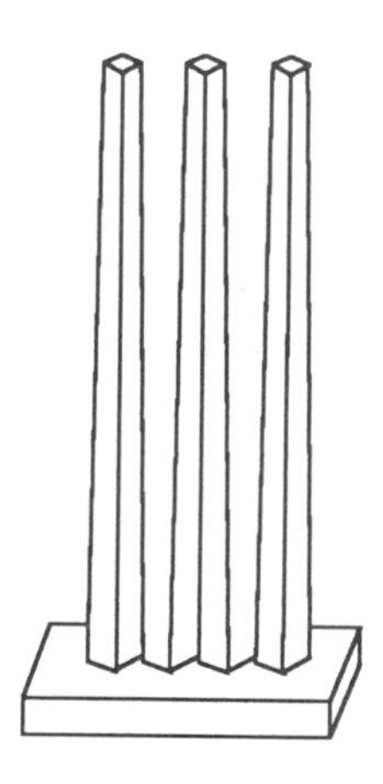

Note that the nine vertical edges of the towers in this drawing are not parallel. If you extend the edges upward, you will find that each set of three meets at a point.

Strange Shish Kabob?

This strange shish kabob is left for you to figure out on your own.

Lesson 0.5
Mandalas

Section of an Aztec Calendar Stone
Aztec & Other Mexican Indian Designs,
Caren Caraway, Stemmer House, 1984

A mandala is a circular design arranged in layers radiating from the center. The word comes from Hindu Sanskrit, the classical language of India, and means circle or center. The Hindus used mandala designs for meditation. The mandala also appears in many cultures other than those of the Eastern world. Mandalas can be found in the rose windows of medieval cathedrals of Europe. The Aztecs of Mexico created a magnificent stone mandala for their calendar. At its center was a mask of the sun god. Surrounding the mask were symbols depicting an earthquake that the Aztecs thought would end the world. Around that was a band with the signs for the days of the Aztec year. The Navahos of the American Southwest have a healing ritual in which multicolored sands are used to create circular patterns. The patient is placed in the center of the mandala design and is encircled by a ring of prayers.

The Swiss psychiatrist C. G. Jung (1875–1961) made psychological studies of the mandala and used it in treating patients. Jung and his patient would draw a mandala, starting the design at the center and continuing outward. Taking turns, the patient drew, then doctor, then patient, until they felt the mandala was complete.

The steps demonstrated below show one way of creating a mandala design. Look at these steps before you begin the exercise set at the end of the lesson.

Step 1

Step 2

Step 3

Step 4

Begin with a center design. Add details within. Add more designs, symmetrically filling in open spaces. Finally, add shading to highlight special areas.

EXERCISE SET

1. What are some examples of mandala designs that you've seen or know about?
2. What company designs or logos are also mandalas?
3. Find a quiet moment and create your own mandala. Color it.

Frances Wong
Geometry student

Scott Shanks
Geometry student

Design by the author

Donna Chang
Geometry student

Lesson 0.6
Islamic Art

Islamic art is rich in geometric forms. Islamic artists were familiar with geometry through the works of Euclid, Pythagoras, and other Greek mathematicians, and they used geometric patterns extensively in their art and architecture.

For hundreds of years, western civilization has looked on Islamic art as merely decorative. However, recently we have come to learn that Islamic art is not only beautiful and geometric, but it is also filled with religious meaning.

Islam's great prophet Muhammed (622 A.D.) preached against idolatry. Many of his followers interpreted his teachings as forbidding the representation of humans or animals in art. Therefore, instead of using human or animal forms for decorations, Islamic artists used intricate geometric patterns. Using geometry, the artists created beautiful artworks while expressing their religious beliefs in a universe ordered by mathematics and reason.

One of the most striking examples of Islamic architecture is the Alhambra, a palace in Granada, Spain. Standing for over 600 years as a tribute to Islamic artisans, the Alhambra is filled from floor to ceiling with marvelous geometric patterns. The designs in the detail on this page are but a few of the hundreds of intricate geometric patterns found in tile work, inlaid wood ceilings, hand-tooled bronze plates, and carpets in Islamic buildings like the Alhambra.

Left: Alcove in the Hall of Ambassadors, Alhambra, Granada, Spain

Below: Detail of Mosaic in Belvedere of the Sultana Aixa, Alhambra, Granada, Spain

The geometric patterns are often elaborations of basic grids of regular hexagons, equilateral triangles, or squares. The complex Islamic patterns were constructed with no more than a compass and straightedge. Being made with a compass, the patterns tend to radiate out from a central point. Their designs, therefore, are intricate arrangements of mandalas.

Creating an Islamic style mandala is not difficult. The steps below show one way. Read through the steps for making an Islamic style mandala design before you begin the exercise set at the end of the lesson.

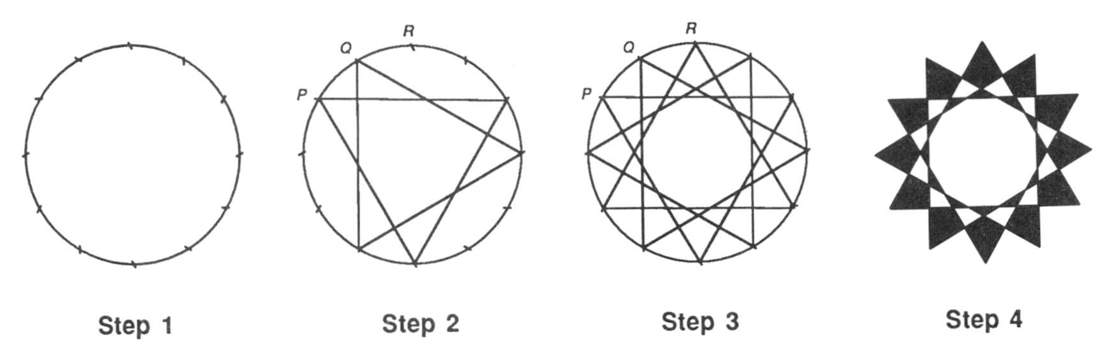

Step 1: Divide a circle into six equal parts by the "daisy construction." Locate the middle point of one of these arcs and, with the same compass setting, go around the circle marking off six new points. This divides the circle into 12 equal arcs.

Step 2: Select a point (call it *P*) and connect every fourth point. You should get a triangle. Then move clockwise to point *Q* and connect every fourth point, giving you a second triangle. Then move clockwise again to point *R* and connect every fourth point, once again giving you a triangle.

Step 3: Do this one more time.

Step 4: Decorate your design by shading in sets of regions.

The Islamic style mandala designs below make use of 8, 12, and 16-sided polygons.

EXERCISE SET

1. With a compass, draw a circle and divide it into six equal parts. Connect every second point. You should get an equilateral triangle, a triangle with all sides of equal length.

2.* Draw a circle and divide it into six equal arcs. Locate the middle point of one arc and use your compass to divide the circle into twelfths. Connect every third point. You should get a square.

3. Draw a circle and use a similar procedure to divide it into 24 equal pieces. Connect every third point. You should get a regular 8-sided polygon called an octagon.

4. Divide a large circle into 24 equal parts. Using squares, equilateral triangles, regular octagons, or some combination of them, create an Islamic style mandala.

LeDana Andrews, Geometry student

Hae Ju Suh, Geometry student

Ingrid Vida, Geometry student

David Fukuda, Geometry student

Lesson 0.7

Knot Designs

Knot designs are geometric designs that appear to weave or interlace (like a knot). The earliest known examples are found in Celtic art (pronounced Keltic). Celtic art blossomed in the northern regions of England and Scotland in the centuries following the Roman conquest. In early Celtic culture you can find fine examples of carved stone rich in geometric designs. In their carved knot designs, artists imitated on a flat surface the same patterns that appeared in three-dimensional crafts such as weaving and basketry. The *Book of Kells* is the most famous collection of Celtic drawings.

Celtic Knot
The Celtic Design Book,
Rebecca McKillip, Stemmer House, 1981

During the Renaissance Leonardo da Vinci, Albrecht Dürer, and Michelangelo created their own knot designs. In one of the ballrooms of the Sforza castle in Milan, Italy, you can still see the painting of a tree by Leonardo da Vinci. The branches of the tree intertwine as they climb up the walls and over the ceiling. Michelangelo created a simple knot design for a plaza near the Forum in Rome. It is still there for people to enjoy.

Knot Engraving
Leonardo da Vinci (1452–1519)

You can find interlacing knot designs in the art of most Mediterranean cultures. Egyptians, Greeks, Romans, Byzantines, Moors, Persians, Turks, Arabs, and Africans have all used interlocking designs in their folk art. Intricate knot designs are also found in Chinese and Japanese lattice decorations.

Carved Knot Pattern from Nigeria
African Designs, Geoffrey Williams, Dover, 1971

The last woodcut by the twentieth century Dutch artist M.C. Escher was a knot design called *Snakes*. You will see more art by Escher later in Chapter 7. You will even learn to make Escher-like drawings of your own.

Snakes, Woodcut by M. C. Escher, 1969

A knot design looks three dimensional. It appears to come out of the page. The design seems to weave in and out or over and under itself. This is a very dramatic effect and yet a knot design is not difficult to create. The steps for creating two examples of knot designs are illustrated below. Read through the step before you begin the exercise set at the end of this lesson.

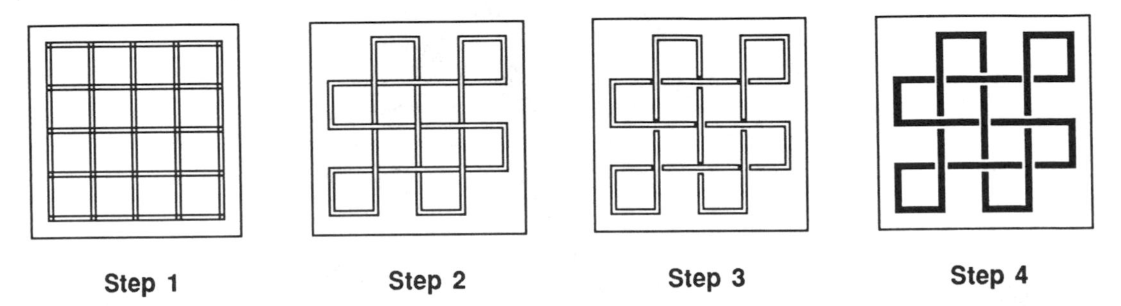

Step 1: Draw a basic design of double lines. You may find it useful to use graph paper. Use a pencil because you will need to erase.

Step 2: Erase sections where the lines overlap to give the appearance of a band or ribbon weaving over and under itself.

This is enough to create a knot design. If you would like to shade in the band, however, you must complete steps 3 and 4.

Step 3: Erase a small portion of the lines where each pair of lines intersects with another pair. This allows you to shade without losing the weaving effect.

Step 4: Shade your knot design.

A similar approach can be used to create knot designs with rings.

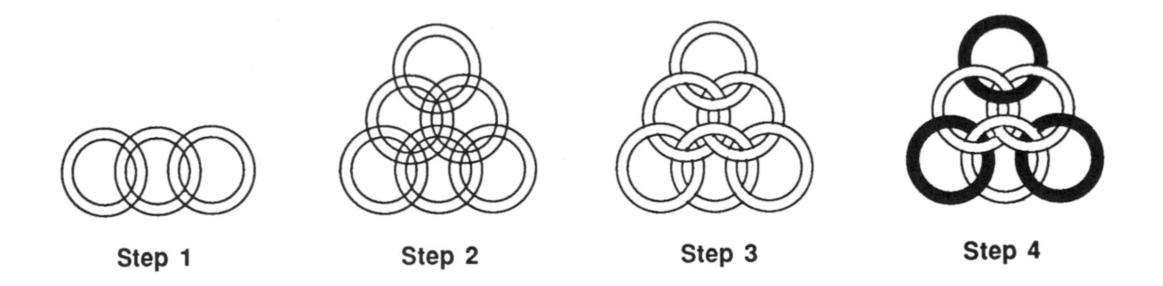

EXERCISE SET

1. With the aid of graph paper, create a knot design of your own using only straight lines.
2. With the aid of a compass or circle template, create a knot design of your own with circles.

3. One very interesting knot design is known as the Borromean Rings. This design appears on the coat of arms of the Italian Renaissance family Borromeo. The knot design consists of three rings. No two are connected, yet the three together cannot be separated. They are linked together in such a way that if any one ring is removed from the set of three, the remaining two rings are no longer connected. Got that? Good. Sketch the Borromean Rings.

Celtic Knot
Celtic Art, G. Bain
Dover Publications, 1973

Tiger Tail
Diane Cassell
(Parent of a geometry student)

Scott Shanks
Geometry student

Japanese Knot Design
The Patterns of Japan
K•D•C Co., 1987

Lesson 0.8

Perspective

Art is a lie that makes us realize the truth.

– Picasso

Many of the paintings created by European artists during the Middle Ages were commissioned by the Church. The art was symbolic; that is, people and objects in the paintings were symbols representing religious ideas. Artists were more interested in creating a symbolic scene than in accurately representing people and objects. The background in these religious paintings was usually a solid color and people were sized according to importance rather than distance from the viewer. There was no attempt to put people or objects in perspective.

The symbolic nature of art began to change in the Renaissance. Renaissance artists delighted in nature and the human form. They began to search for ways to represent the three-dimensional world more naturally on flat, two-dimensional surfaces.

Many of the Renaissance artists were engineers or architects well versed in mathematics. It is no surprise then that these artists turned to geometry to solve the

The School of Athens, Vatican fresco by Raphael of Urbino (1483–1520)

problems of perspective. **Perspective** is the technique of portraying solid objects and spatial relationships on a flat surface so that they appear true-to-life. Raphael's *School of Athens* is a perspective painting paying homage to science and philosophy. The two central figures are Plato and Aristotle. The kneeling figure in the lower left corner is Pythagoras. The figure drawing on a slate in the lower right corner is Euclid, the founder of our geometric tradition. The size of each figure is determined by the distance from the viewer to the figure. The receding arches enhance the realism of the scene.

In their writings on the subject of perspective painting, Renaissance artists were very systematic and mathematical. Leonardo da Vinci even began one of his writings on painting by saying, "Let no one who is not a mathematician read my works."

Perhaps the most influential student of perspective was the great German artist Albrecht Dürer. Dürer traveled to Renaissance Italy to study the works of the earlier masters. Dürer left many woodcuts detailing his method of perspective drawing.

Woodcut from *The Artist's Treatise on Geometry* by Albrecht Dürer (1471–1528)

From experience you have noticed that objects closer to you appear larger than similar sized objects farther away. You have probably noticed that, when looking down railroad tracks toward the horizon, the parallel tracks appear to meet at some point on the horizon line. In a perspective drawing, parallel lines running directly away from the viewer are drawn so that they come together at a point called the **vanishing point**. The vanishing point is located on the horizon line.

Can you find the vanishing point in this perspective study by Jan Vredeman de Vries?

Perspective study by Jan Vredeman de Vries (1527–1604)

A rectangular solid, or box, is one of the simplest objects to draw in perspective. Follow the steps below to draw a rectangular solid with one face viewed straight on.

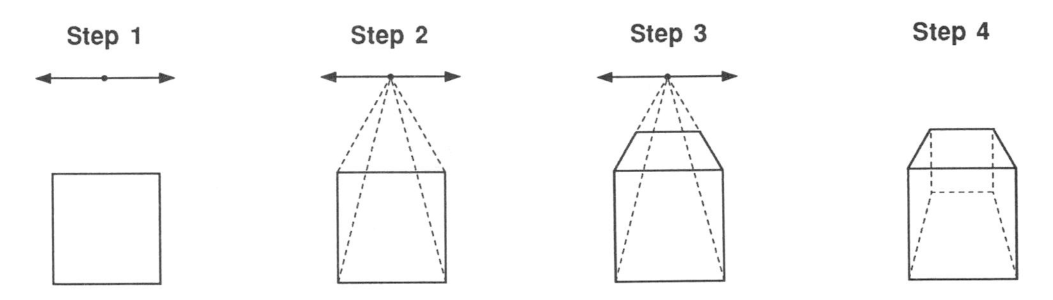

Step 1: Begin by drawing a rectangle for the front face. Draw a horizon line parallel to the horizontal edges of your rectangle and select a vanishing point on it.

Step 2: To create the edges of the box that recede from view, draw lines, called **vanishing lines**, from the corners of the rectangle to the vanishing point.

Step 3: To create the visible back horizontal edge, draw a line parallel to the horizon line. Use this line to determine the position of the back vertical edges.

Step 4: To complete the figure, draw hidden back vertical and horizontal edges. Erase the horizon line and the unnecessary portions of the vanishing lines.

Notice that only the edges of the box that appear to move away from the viewer are drawn to meet at a vanishing point. The horizontal lines that are parallel to the picture's plane are drawn parallel in the picture and are not drawn to the vanishing point. The same is true for the vertical lines that are parallel to the picture's plane.

In a perspective drawing the location of the horizon line tells you something about the position of the viewer. If the horizon line is high in the picture, then the viewer is looking from above, perhaps from a hill. If the horizon line is low, the viewer is low to the ground.

One doesn't always view an object straight on. However, as long as the front surface of the object is parallel to the picture's plane, only the lines that appear to move directly away from the viewer are drawn to a vanishing point. If an object is viewed from the right, then the vanishing point is placed to the right. If an object is viewed from the left, then the vanishing point is placed to the left.

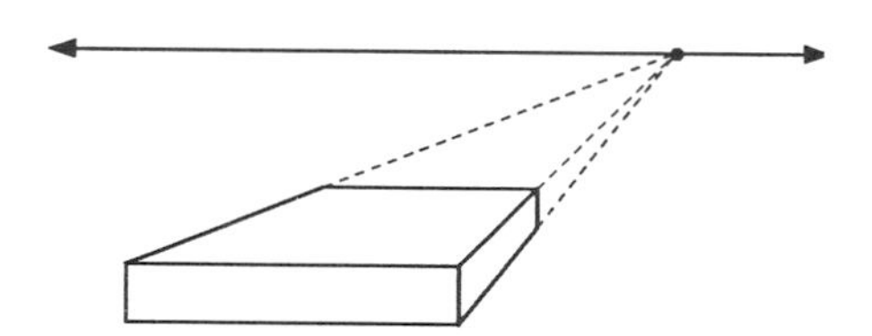

Here the viewer is looking at the rectangular box from above and to the right. Notice that the horizon line is located high in the picture and the vanishing point is placed to the right.

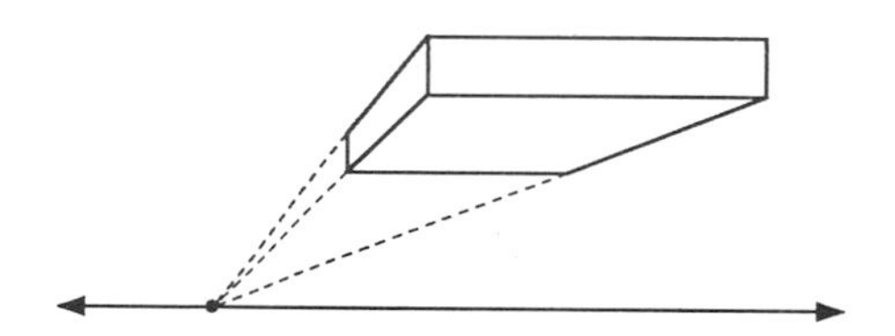

Here the viewer is looking at the rectangular box from below and to the left. Notice that the horizon line is located low in the picture and the vanishing point is placed to the left.

EXERCISE SET A

1. Use your straightedge and neatly copy the figure on the right onto an unlined piece of paper. The easiest, most accurate way to copy a simple figure made with straight lines is to place your paper over the figure and put a dot on your paper at each corner of the figure. Then take the paper off the figure and use your straightedge to connect the dots. After you have copied the figure onto your own paper, locate the vanishing point and draw the horizon line.

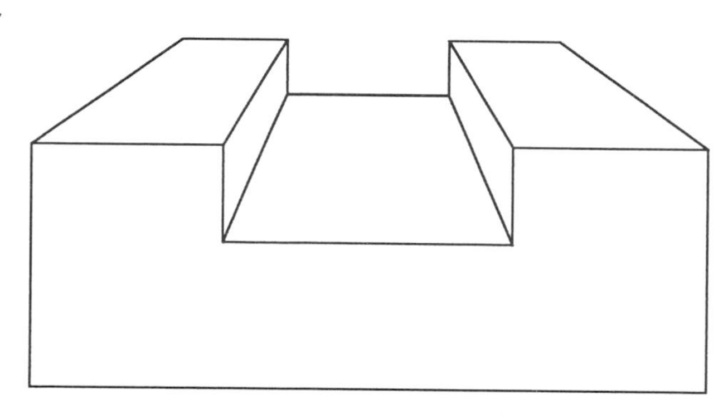

2. Draw a perspective view of a cube. Show all 12 edges. Use dashed lines for the hidden edges.

3. Draw a perspective view of a rectangular box that is viewed from above and to the right.

4.* Draw a perspective view of a rectangular box with a rectangular box removed through the center. (Your drawing will look like a rectangular shaped donut.)

In the first part of this lesson you learned to draw a perspective view of a rectangular object with the front surface parallel to the picture's plane. Drawings of this type use **one-point perspective**. If the front surface of a rectangular solid is not parallel to the picture's plane, then one needs two vanishing points. This is called **two-point perspective**. To demonstrate how to draw a figure with two point-perspective, let's look at a rectangular solid with one edge viewed straight on.

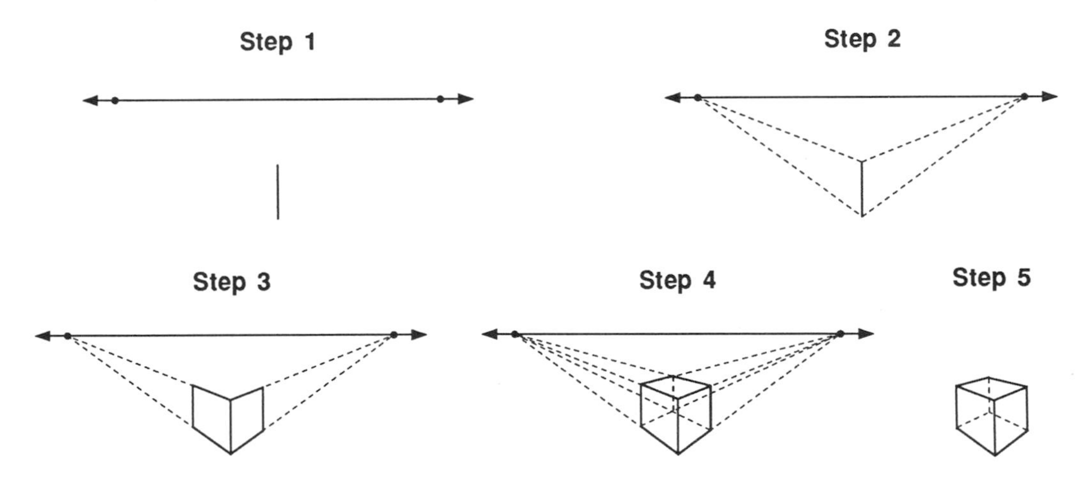

Step 1: Begin by drawing a vertical line segment for the front edge. Draw a horizon line and select two vanishing points on it.

Step 2: To create the edges of the box that recede from view, draw four lines (**vanishing lines**) from the endpoints of the vertical line segment back to the two vanishing points.

Step 3: To create the length and width of the box, draw vertical line segments intersecting the vanishing lines. The endpoints of these line segments determine the position of the back edges that recede from view.

Step 4: Draw the four remaining vanishing lines and the back hidden horizontal edge.

Step 5: Erase the unnecessary portions of the vanishing lines.

In the previous drawings, the lines that recede to the left meet at a vanishing point on the left of the horizon line. The lines that recede to the right meet at a second vanishing point on the right. All the vertical edges in the rectangular solids are parallel to the picture's plane and therefore they do not meet at a vanishing point but are drawn parallel in the picture plane.

Sometimes an object is drawn below the horizon line (viewer is above the object), and occasionally an object is drawn above the horizon line (the viewer is on the ground and the object is in the air). Many times, however, the object is so large or the viewer so close that the object is drawn both above and below the horizon line. Observe the location of the horizon line in the figure below.

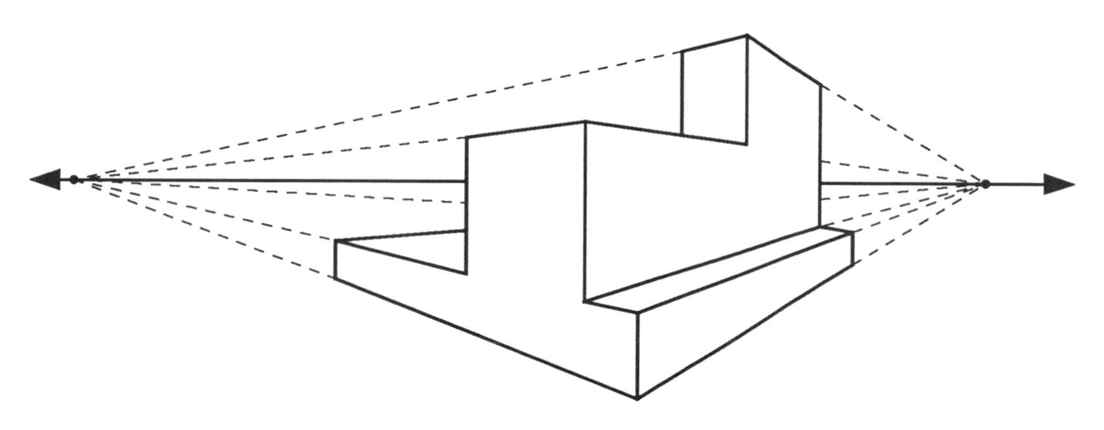

EXERCISE SET B

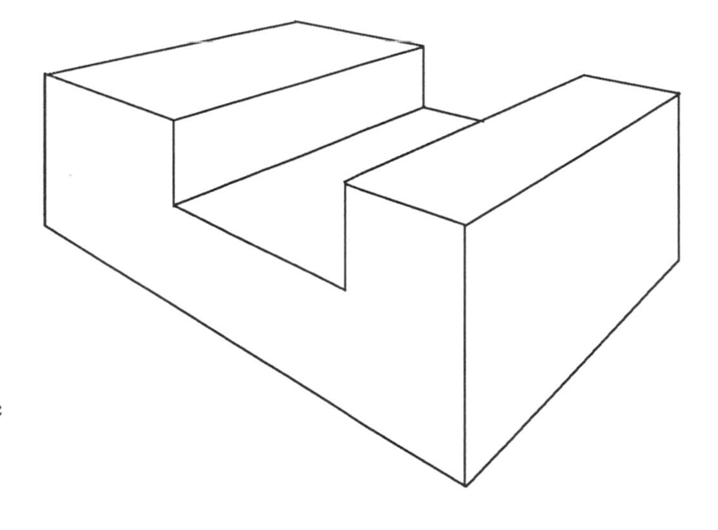

1. Use your straightedge and neatly copy the figure onto the right on an unlined piece of paper. Then locate the horizon line by finding the two vanishing points.
2. Draw a two-point perspective view of a rectangular box viewed from below and to the right.
3. Draw a two-point perspective view of a rectangular box set on top of another rectangular box.
4. Draw a two-point perspective view of a rectangular box with a rectangular window cut out of one of the faces. Show that the box has thickness.

Special Project

Block Lettering in Perspective

Perspective drawing techniques can be used to create letters or words that appear to be solid – useful for giving emphasis to an element of a design. Your task in this special project is to draw letters or words with one-point and two-point perspectives.

Drawing Block Letters in One-Point Perspective

Step 1

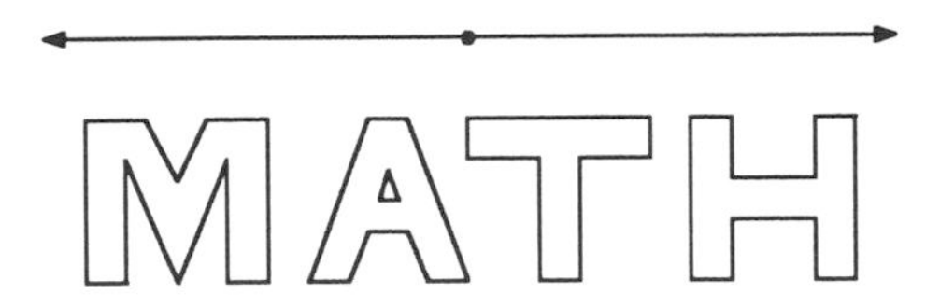

Write a word in block letters. Draw a horizon line parallel to the bottom edge of your word. Select a vanishing point on the horizon line.

Step 2

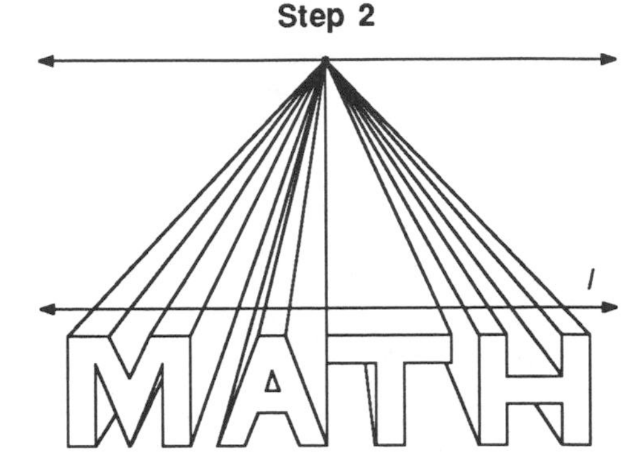

Draw vanishing lines from each corner point of the block letters back to the vanishing point. Select a thickness for your block letters and draw a line (line *l*) parallel to the horizon line.

Step 3

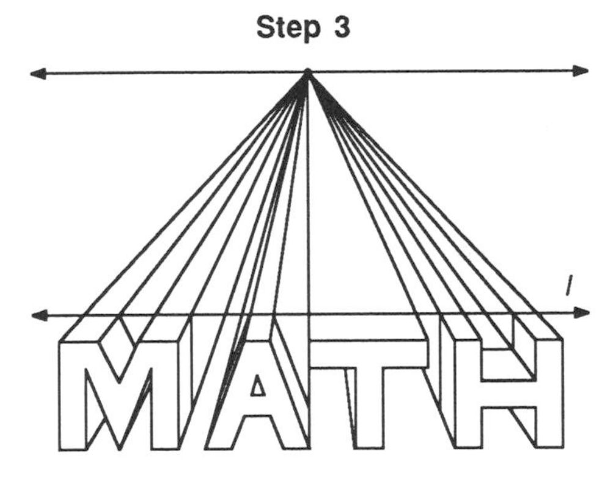

To create the back edges of your letters, draw lines parallel to the front edges, starting and ending on the points where line *l* intersects the vanishing lines.

Step 4

Erase all the vanishing lines and shade in all the sides and tops of the letters.

Now you try making block letters. Draw a perspective view of your name or initials in block letters. You can change the perspective by making the horizon line high or low and by placing the vanishing point left or right.

Drawing Block Letters in Two-Point Perspective

Step 1

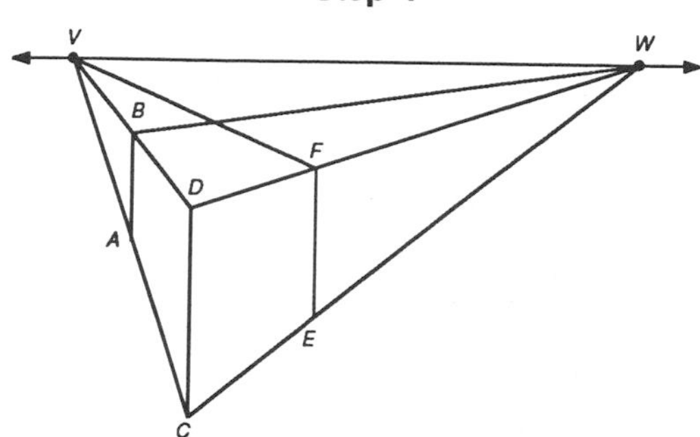

Draw a box in two-point perspective. Label the points as shown. Make height *CD* and width *CE* of your box about the same. For example, $CD = CE = 6$ cm. Do not erase the vanishing lines. Your first letter will fill the front face of this box. Select a distance between the first and second letters by drawing a vertical line *GH*. If you used 6 cm for the width *CE*, then use about 1 cm for *EG*.

Step 2

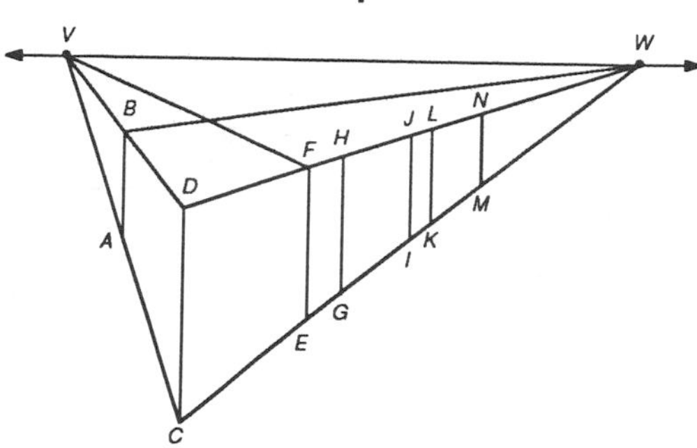

Select a width for your second box by drawing a vertical segment *IJ*. This box will eventually house your second letter. If you used 6 cm for *CE*, then use about 3 cm for *GI*. Next, select a width for the space between the second and third boxes by drawing in a vertical segment *KL*. If you used 1 cm for *EG* then use .5 cm for *IK*. Repeat this procedure for the third box.

Step 3

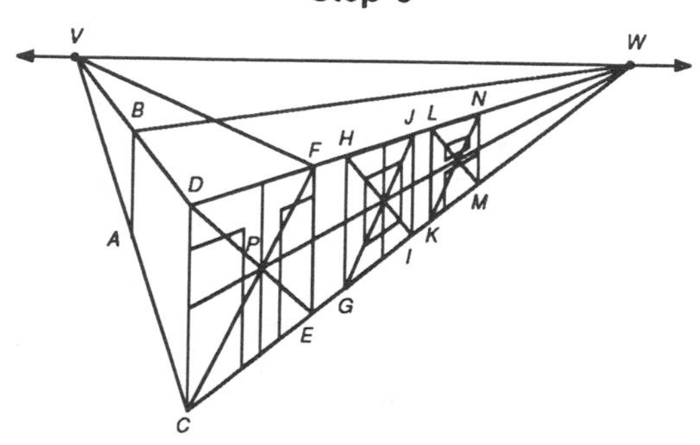

Now design a letter on the front face of each box. Draw in diagonal lines *CF*, *DE*, *HI*, *GJ*, *LM*, and *KN*. The points where these diagonals intersect are the perspective centers for each front face. Draw vertical lines through these centers. Label the center in the first box *P*. Draw line *PW*. Use this line to center each block letter on its front face.

Step 4

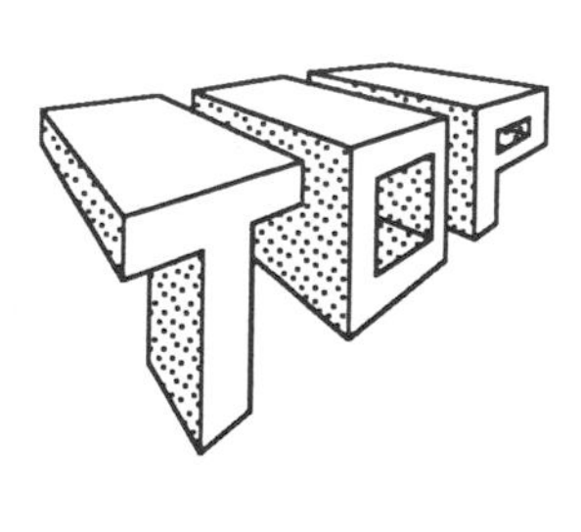

Draw all the top vanishing lines from the top front corners to the back edges of the solid letters. Draw all the vertical edges at the back of the solid letters. Draw all the remaining vanishing lines. With a pen or felt tip marker outline all the edges of the solid letters. Erase all other lines. Decorate.

Draw a two-point perspective view of your name or initials in solid letters.

Special Project
Drawing Skyscrapers

One of the more challenging perspective drawing problems is that of drawing skyscrapers. Drawing skyscrapers are challenging because there are many different rectangular solids and thus many different vanishing lines. Your task in this special project is to design a skyscraper complex with at least three towers in two-point perspective. Drawing a skyscraper in two-point perspective is demonstrated for you below. Read through the steps before you begin your own work.

Drawing a High Rise Complex

Step 1

Begin with a horizon line and two vanishing points. Draw the front vertical edge of your first building with all the vanishing lines.

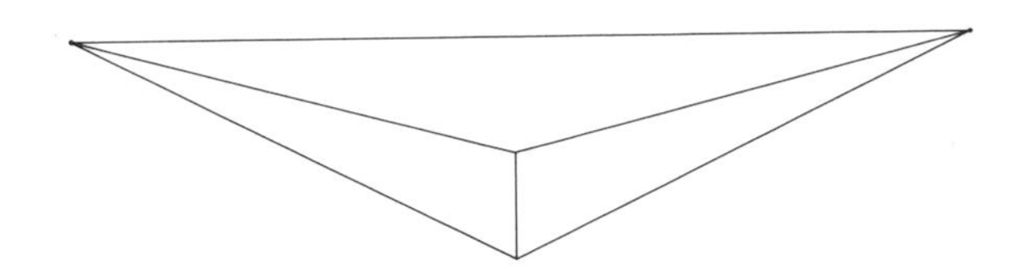

Step 2

Complete the two-point perspective view of the first building.

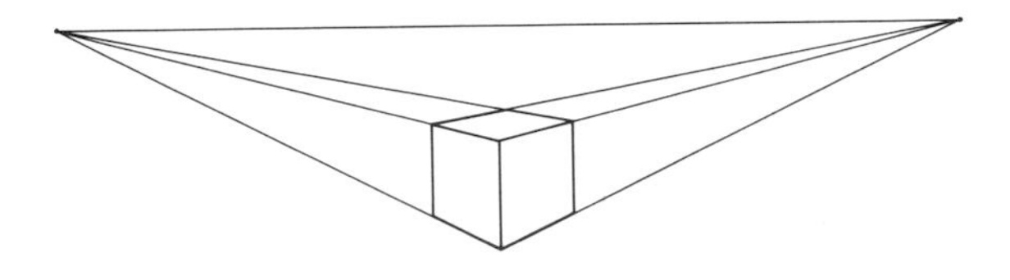

Step 3

Next, draw in a couple of the taller buildings. Start with the front vertical edge of each building and draw the vanishing lines. Complete the perspective view.

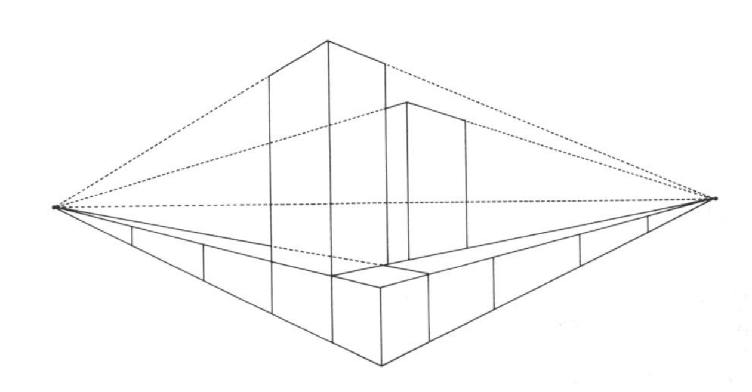

Step 4

Create additional buildings and use vanishing lines to add architectural details.

Step 5

Erase all unnecessary lines and add other details.

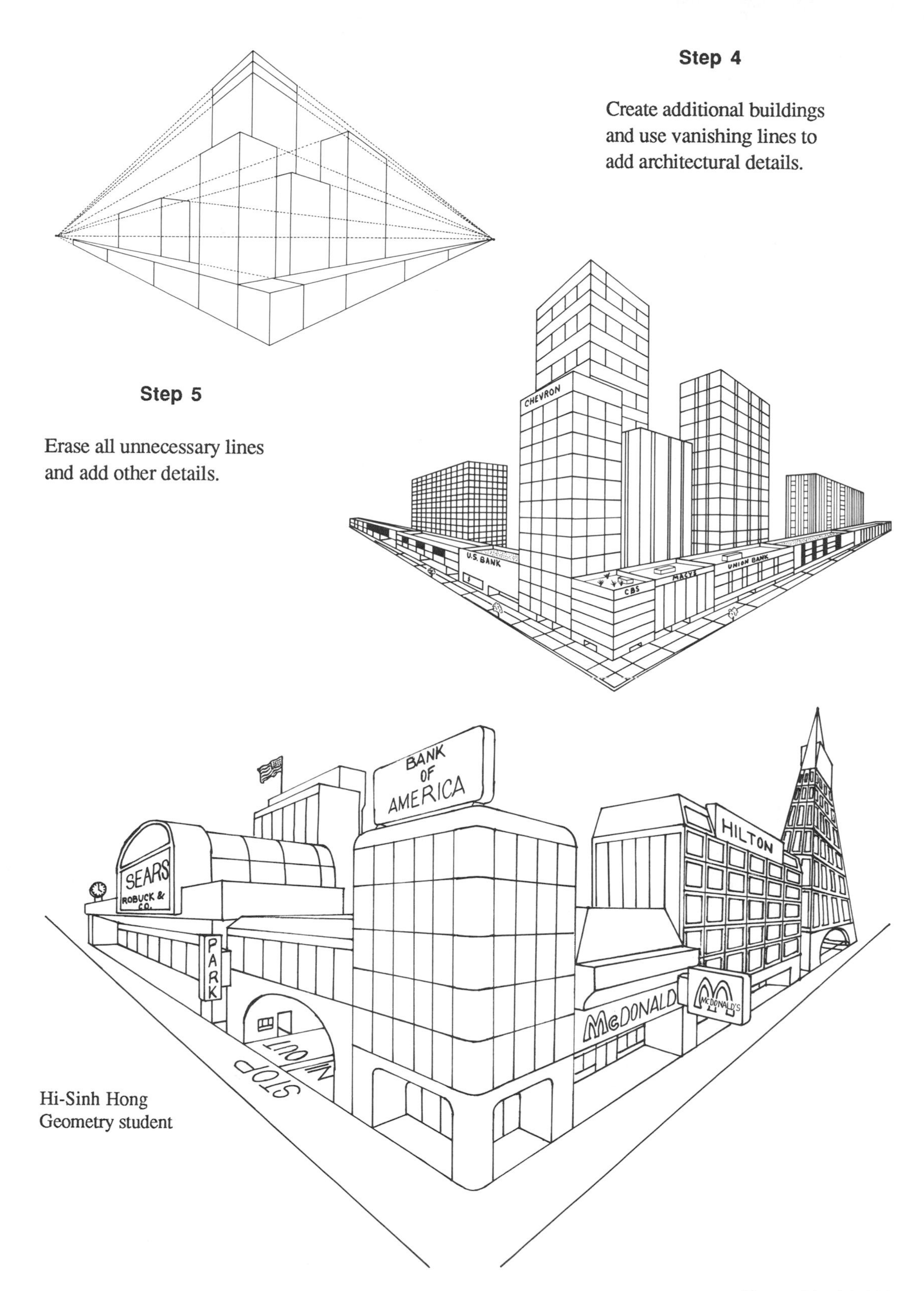

Hi-Sinh Hong
Geometry student

Special Project

Advanced Perspective

Many perspective drawings rely on diagonals. You use diagonals to create a perspective view of a peaked-roof house, to create a perspective view of a checkerboard or floor tile pattern, or to draw a series of equally-spaced trees, telephone poles, or fenceposts that recede from view.

Your task in this special project is to design a scene using at least two of the items mentioned above. The technique for drawing each of the items is outlined below. Read through the steps before you begin your own work. You need not limit yourself to the objects demonstrated. Instead, incorporate the techniques into your own design.

Drawing a Peaked-Roof House

Step 1

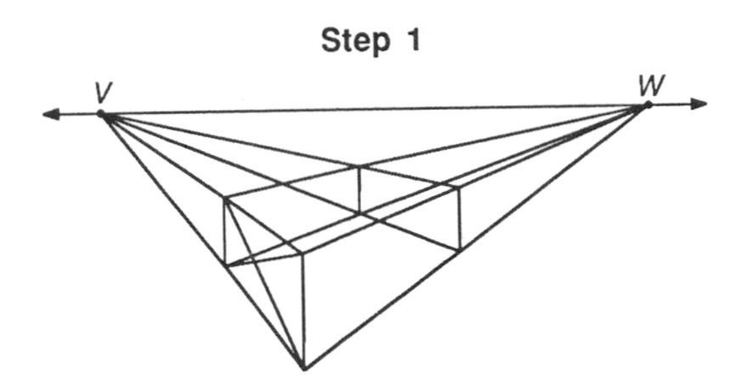

Draw the first floor with two-point perspective including hidden lines. Label your vanishing points V and W as shown. Draw diagonals connecting the opposite corners of the front end wall.

Step 2

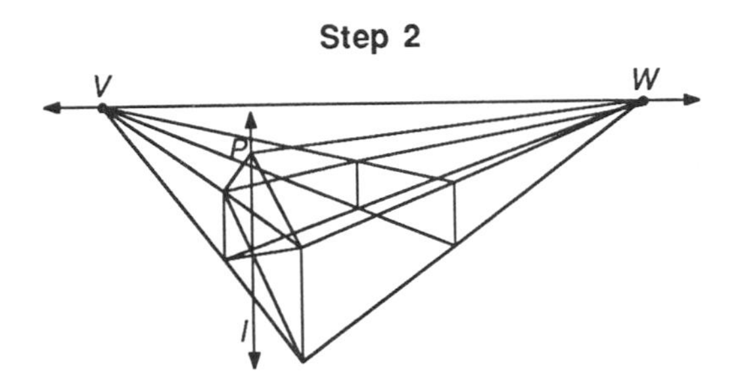

Draw vertical line l passing through the intersection of the two diagonals. Select a point P on line l as the peak of your roof and draw a vanishing line from P to W. This creates the roof line.

Step 3

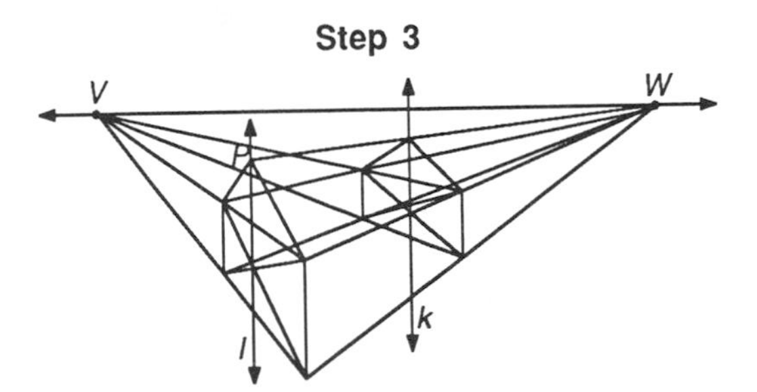

Draw the diagonals connecting the opposite corners of the back end wall. Draw vertical line k passing through the intersection point of the two diagonals. The point where line k crosses the roof line is the back end of the roof.

Step 4

Draw a door and a window. Make sure that all lines that are parallel recede to the same vanishing point. Erase unnecessary portions of vanishing lines and add details where necessary.

Perspective View of a Tiled Floor

Step 1

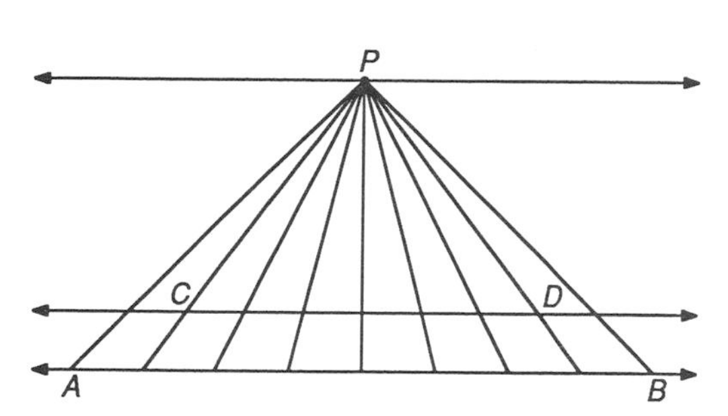

Draw a pair of horizontal parallel lines. The upper line is your horizon line. The lower line will become the front edge of your first row of squares. On the lower line, mark off eight equal lengths. Label the endpoints *A* and *B*. On the top line (your horizon line) select a vanishing point *P*. Draw all nine vanishing lines. Draw a line parallel to line *AB* and the horizon line. This line determines the back edge of your first row of squares. Label points *C* and *D* as shown in the diagram.

Step 2

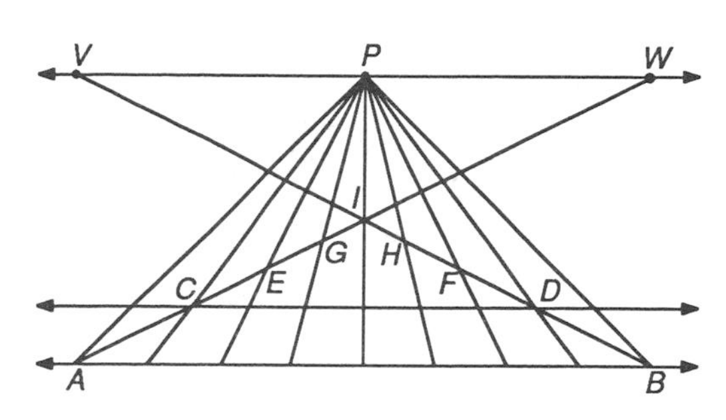

Draw diagonal segments *BD* and *AC* and extend them to the horizon line. Label the points where these diagonals intersect the horizon line as *V* and *W*. Diagonal segments *BV* and *AW* should both cross through the center vanishing line at the same point *I*. Mark off points where diagonal segments *BV* and *AW* intersect the nine vanishing lines. Label the intersections as shown.

Step 3

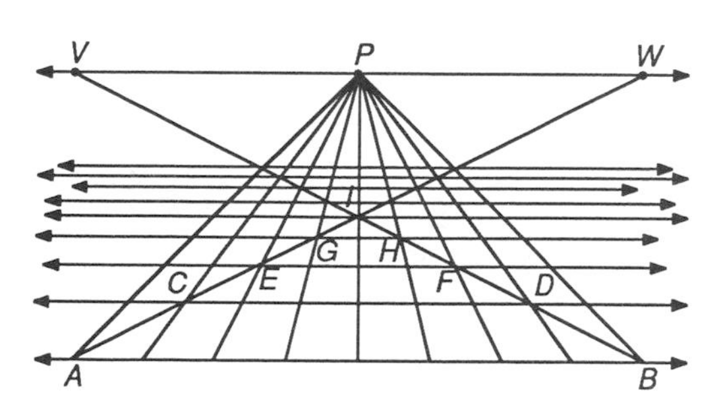

Draw a line through points *E* and *F*. This is the back line for the second row of squares. Draw a line through points *G* and *H*. This is the back edge for the third row of squares. Draw a line through point *I* parallel to the other horizontal lines. This is the back edge of the fourth row. Continue in this fashion, drawing lines through pairs of points until all nine horizontal lines have been drawn.

Step 4

Shade in alternating squares. Erase all unnecessary portions of horizontal and vanishing lines. For special effect you may add a slim rectangle to the front of the tile pattern to give it thickness.

Spacing Fenceposts in Perspective

Step 1

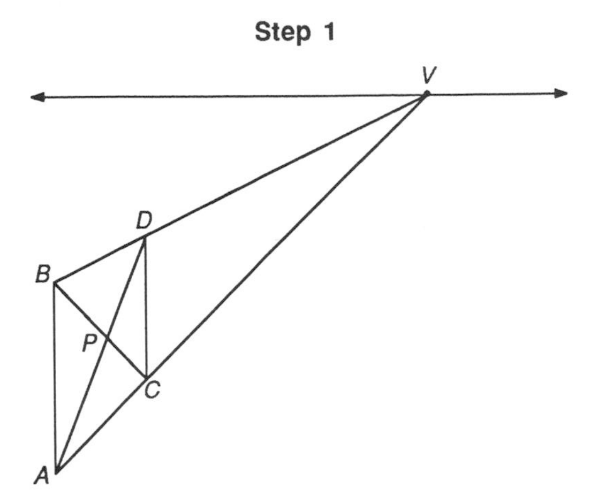

Start with a horizon line and vanishing point V. Draw the fencepost nearest the viewer and label it segment AB. Draw vanishing lines from the top and bottom of the post. Draw the second post parallel to the first at a distance from it that looks pleasing. Label it segment CD. Draw the diagonal segments AD and BC. The diagonals intersect at a point. Label it P.

Step 2

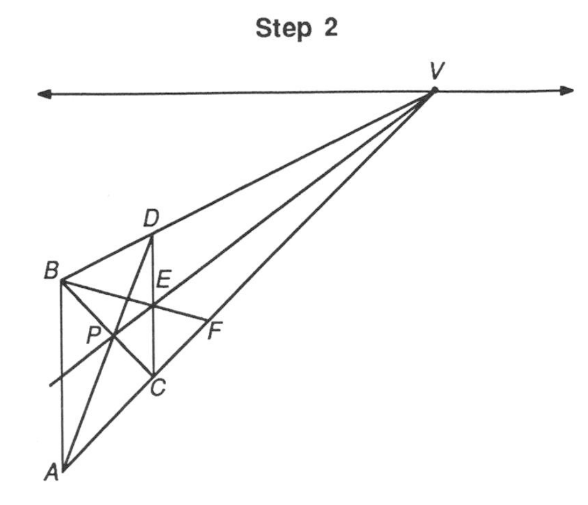

Draw a line connecting point P and the vanishing point. This line will pass through the centers of all the posts. Draw a line from point B through point E, the center of the second post. Extend it until it meets the bottom vanishing line AV. Call this point F. Point F is the base of the third post.

Step 3

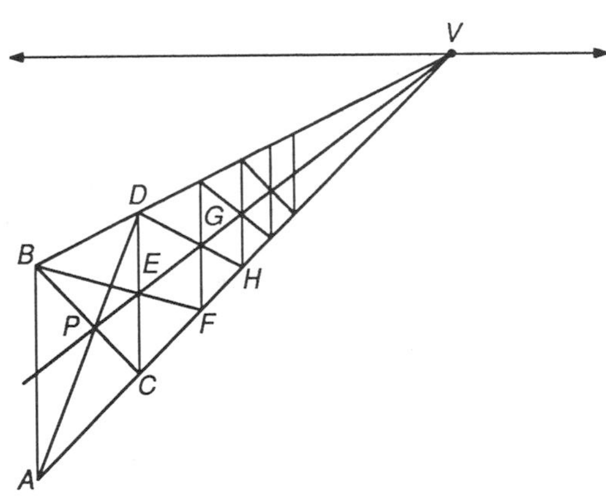

Continue in this way, drawing a line through points D and G. Point G is the middle point of the third post. Extend this line until it intersects the vanishing line AV at a point called H. This new point is the base of the fourth post. Continue until you have all the posts you want.

Step 4

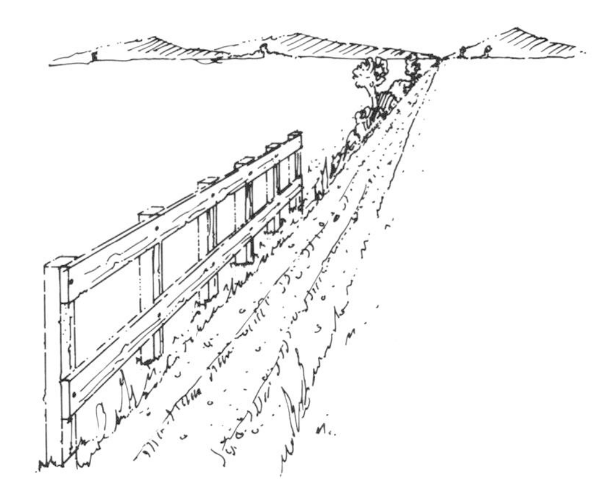

Use these lines as a guide to draw in the complete fence.

Lesson 0.9

Chapter Review

EXERCISE SET

1. List three things in nature that exhibit geometrical shapes.
2. List three cultures that use geometry in their art.
3. Name and define the four basic tools of geometry.
4. With a compass, draw a 12-petal daisy.
5. What is the optical effect of the op art design below?

6. Each of the cultures below made use of a mandala design. What was the mandala design? What purpose did each serve?

   a.* Aztecs b. Navahos c. Medieval Europeans

7. Why are geometric designs used in Islamic art?
8. Draw an original knot design.
9. Draw a rectangular box with two-point perspective.

Cooperative Problem Solving

Designing a Lunar Colony Park

It is the year 2065 on the lunar space port Galileo. The space port has been in continuous operation for the past 25 years. The space port includes the research facility, the recreational and living quarters, the agricultural and life support divisions, and the lunar landing and surface transportation department.

After years of very successful operation, many residents at retirement have opted to remain at the space port. The executive council of the space port has recently agreed to turn over to the local affiliate of the Senior Citizens Association an undeveloped agricultural pod for development as a retired astronauts park.

The agricultural pod is a hemispherical reinforced glass dome 70 meters in diameter and has been set up to reproduce many of the fruits and vegetables native to planet earth. The temperature and climate is computer controlled to create a continuous year-round growing season. This includes a light rain each Tuesday and Thursday evening.

Your job is to submit a proposal for the park design before the Galileo Planning Commission. The budget limit is $NL10,000. The Commission will make its selection of a park proposal from the oral presentations at the next commission meeting based on the following criteria:

1. Safe design in accordance with accepted Galileo traditions
2. Aesthetics, innovation, and optimum use of the area
3. Cost effectiveness

Your presentation before the Commission must include:

1. A park design (scaled map where 1 cm = 1 m) showing all improvements
2. An itemized list of costs
3. Explanations of how your design satisfies each of the three criteria

A list of materials and their costs have been itemized for you in the table below. The materials have all been reproduced to appear authentic or have been imported from the home planet at great expense. The costs have been listed in new lunar dollars ($NL).

Materials	**Cost**	
Park bench (1.5 m long)	$NL 125.00	each
Trash receptacle	40.00	each
Picnic table (2 m long)	150.00	each
Bike rack	100.00	each
Stainless steel drinking fountain	300.00	each
Plumbing for drinking fountain	25.00	per m from bathroom plumbing
Iron barbecue pit	380.00	each
Park overhead lights	500.00	each
Redwood Gazebo (5 m diameter)	1500.00	each
Bathroom (men/women 3 by 4 m)	2800.00	each
Clubhouse (redwood 5 m square)		
Prefab with electricity	9000.00	each
Prefab without electricity	6500.00	each
Putting green	20.00	per m^2
Community garden	2.00	per m^2
Asphalt (4 cm thickness)	45.00	per m^2
Trees		
Fruit	150.00	each
Shade	100.00	each
Shrubs	25.00	per m^2
Building materials		
Lumber (5 by 10 cm)	4.50	per m
Lumber (10 by 10 cm)	12.00	per m
Siding (1 by 2 m sheet)	16.00	per sheet
Asphalt shingles	6.00	per m^2
Paint	12.00	per liter (covers 15 m^2)
Railroad ties (4 m long)	15.00	each
Cement bricks	1.50	each

1 Inductive Reasoning

Liberation, M. C. Escher, 1955

This chapter introduces you to inductive reasoning. Inductive reasoning is the process of observing patterns and making generalizations about those patterns. It is the basis of the scientific method. Mathematicians use inductive reasoning to make discoveries; then they attempt to verify their discoveries logically. This is the type of reasoning that you will be using throughout this textbook. In the chapters that follow, you will make many discoveries. You will then use your discoveries to solve problems.

Lesson 1.1

What is Inductive Reasoning?

> *We have to reinvent the wheel every once in a while, not because we need a lot of wheels; but because we need a lot of inventors.*
>
> – *Bruce Joyce*

As a child you learned by experimenting with the natural world about you. You learned how to walk, talk, and hold your spoon, all by trial and error. After many tries, you eventually learned how to ride your first bicycle, discovering it was easier to keep your balance if you kept the bike moving. You learned by experience that a counterclockwise motion turns on a water faucet and a clockwise motion turns it off. Most of your learning has been by a process called inductive reasoning.

> ***Inductive reasoning*** *is the process of observing data, recognizing patterns, and making generalizations from your observations.*

Geometry is also rooted in inductive reasoning. The study of geometry began long ago in Egypt and Babylonia. The geometry of the ancients was a collection of measurements and simple procedures that seemed to give reasonably correct answers to practical problems. Those measurements and procedures were generated over a great period of time as a result of experience and observation. They were used to calculate land areas and to re-establish boundaries of agricultural fields after the yearly flooding of the Nile River. They were used to build canals, reservoirs, and the great pyramids. As you work through this book, you too will learn geometry inductively.

Inductive reasoning is not only important for learning geometry. Most scientific inquiry, including mathematical inquiry, begins with inductive reasoning. When a mathematician uses inductive reasoning, the generalization is called a **conjecture**.

Example

A scientist takes a piece of salt, turns it over a burning candle, and observes that it burns with a yellow flame. She does this with other pieces of salt, finding they all burn with a yellow flame. She therefore makes the conjecture: "All salt burns with a yellow flame."

EXERCISE SET A

Use inductive reasoning to make a conjecture.

1. Caveperson Stony Grok picks up a rock, drops it into a lake, and watches it sink. He picks up a second rock, drops it into the lake, and it also sinks. He does this five more times, and each time the rock heads straight to the bottom of the lake. Stony conjectures: "Ura nok seblu," which translates to: —?—.

2. A mathematician lands at the airport of the kingdom of Moravia. He desperately needs to use the bathroom, but he is very shy, and the social customs of the kingdom would not permit him to use the wrong bathroom. He locates the doors to what appear to be two bathrooms. He observes men enter the door marked "Warvan" and women enter the door marked "Cupore." He is finally ready to make his conjecture. How does he spell relief?

3. Salesperson Henrietta Cluck is selling square-egg makers door to door. At the first house she tells a joke about robot chicken eggs, gets a laugh, and sells a square-egg maker. At the second house she uses a "creative cooking technique" approach. Her approach is informative, but she doesn't sell her square-egg maker. At the third, fourth, sixth, and eighth houses she tries the robot egg joke, and each time she sells the square-egg maker. At the fifth and seventh houses she tries her "creative cooking techniques" approach and is unsuccessful each time. Henrietta conjectures: —?—.

Henrietta Cluck Sells Square Egg Makers

4. Juan went to a restaurant and ate sushi for the first time. A few hours later he broke out in a terrible rash. Two days later he went to another restaurant, ate sushi, and broke out in a rash. A week later Juan decided to give sushi one more try, but, unfortunately, he developed the same irritating rash. Juan conjectures: —?—.

Inductive reasoning starts with careful observation. It's necessary to sift through the observations, sorting out useful information from material that is not useful. Important information can be right in front of you but may be difficult to see. Often it is helpful to organize the information into a table. It is easier to recognize patterns in organized data than in unorganized data.

EXERCISE SET B

In Exercise 1, make careful observations. Look at the sequence of pictures in different ways. Look around the circle clockwise. Counterclockwise. Look from the top down. From the bottom up. Organization is the key in Exercise 2.

1.* A rebus is a visual riddle made up of objects or symbols representing words or phrases. The symbols in the rebus below represent words that can be arranged into a meaningful phrase. What is the rebus saying?

2. Six students attended a class party and ate a variety of foods. Something caused them to become ill. John Q. Public ate pizza, chow mein, and tacos and became ill. Homer T. Odyssey ate chow mein and tacos but not pizza. He became ill. L. J. Horner ate pizza but neither chow mein nor tacos and felt fine.

L. B. Peep didn't eat anything and also felt fine. J. Spratt ate pizza and tacos but no chow mein and became ill. James T. Kirk ate chow mein and tacos but stayed away from the pizza. He also got sick.

Make a table to organize your information. You might use the headings below for your table. Then determine the food that probably caused the illness.

Name	Ate Chow Mein?	Ate Pizza?	Ate Tacos?	Became Ill?

The story that follows, from *In Mathematical Circles* by Howard Eves, illustrates how an incorrect conjecture can result from careful observation. It might be titled *The Inductive Thinking of a Real Flea Brain.*

> A scientist had two large jars before him on the laboratory table. The jar on his left contained a hundred fleas; the jar on the right was empty. The scientist carefully lifted a flea from the jar on the left, placed the flea on the table between the two jars, stepped back, and in a loud voice said, "Jump." The flea jumped and was put in the jar on the right. A second flea was carefully lifted from the jar on the left and placed on the table between the two jars. Again the scientist stepped back and in a loud voice said, "Jump." The flea jumped and was put in the jar on the right. In the same manner, the scientist treated each of the rest of the hundred fleas and each flea jumped as ordered. The two jars were then interchanged and the experiment continued with a slight difference. This time the scientist carefully lifted a flea from the jar on the left, yanked off its hind legs, placed the flea on the table, stepped back, and in a loud voice said, "Jump." The flea did not jump, and was put in the jar on the right. A second flea was carefully lifted from the jar on the left, its hind legs yanked off, and then placed on the table between the two jars. Again the scientist stepped back and in a loud voice said, "Jump." The flea did not jump and was put in the jar on the right. The scientist treated each of the hundred fleas in the jar on the left in this manner, and in no case did a flea jump when ordered. So the scientist recorded in his notebook the following induction: "A flea, if its hind legs are yanked off, cannot hear."

You probably wouldn't make the conjecture that the "scientist" in the story made. You undoubtedly wouldn't undertake his experiment. There are, however, many common mistakes that you can make when forming conjectures. You too can jump to an incorrect conclusion even after making very good observations. This might happen because you have not studied enough data. It might happen because preconceived ideas affect your judgement. It might be because you mistake coincidence with cause and effect.

EXERCISE SET C

1. Julio, sitting in the last row of a bus, notices that six girls in a row get on the bus wearing jeans, and each girl sits down on the left side of the bus. He conjectures that each girl that gets on the bus wearing jeans will sit on the left side of the bus. What is wrong with this conjecture?

2. Vera Smart suddenly wakes up at her desk, having slept through the last fifteen minutes of her philosophy class. She notices that the first six students that enter the room for the next class are women. Vera predicts that the next student that enters will also be a woman. What is wrong with her reasoning?

   If Vera had noticed the sign on the door announcing that the next class was a graduate class in women's studies, would this make her prediction more likely to be true? Would it guarantee that the next student would be a woman?

Vera Smart Wakes Up

3. Ralph is four feet five inches tall. Ralph is in the fourth grade. He claims that he has grown three inches each year for as long as he can remember (the past three years). With the aid of his calculator, Ralph conjectures, "By the time I'm 45 years old, I'll be 161 inches tall!" What is wrong with Ralph's thinking?

4. In Mark Twain's *Life on the Mississippi,* he mentions that the Mississippi River is very crooked, and that over the years, as the bends and turns get straightened out, the river gets shorter and shorter. He lists some numerical data about the length of the river. In 1700 the river was over 1200 miles long, yet by 1875 it was only 973 miles long. Noticing the pattern, he writes, "Any calm person who is not blind or idiotic can see that 742 years from now the lower Mississippi will be only a mile and three-quarters long." What is wrong with this inductive reasoning?

EXERCISE SET D

1.* Give an example of a situation outside of school where inductive reasoning is used correctly.

2. Give an example of a situation outside of school where inductive reasoning is used incorrectly.

Improving Visual Thinking Skills

Dissecting a Hexagon I

Trace the hexagon on the right twice.

1. Divide one hexagon into three identical parts. Each part will be a rhombus.
2. Divide the other hexagon into six identical parts so that each part is a kite.

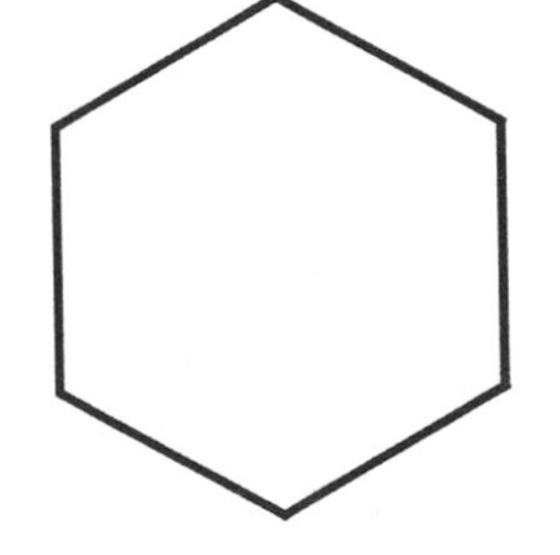

Lesson 1.2
Number Patterns

Let us teach guessing.
– George Polya

In this exercise set you're going to use inductive reasoning to look for the next number or letter in a sequence. "How will I find it?" you might ask. You'll have to guess. Your guess, however, should be a good guess — not a wild one. "How will I make a good guess?" If you make careful observations, find a pattern, and make a conjecture for the next term based on the pattern, you can be confident that your guess is a good one. It will be based on solid inductive reasoning.

EXERCISE SET A

Use inductive reasoning to find the next term of each sequence.

1. 20, 18, 16, 14, –?–
2. 1, 2, 4, 8, 16, 32, –?–
3. 1, 3, 6, 10, 15, 21, –?–
4.* 1, 4, 9, 16, 25, 36, –?–
5.* 2, 6, 15, 31, 56, 92, –?–
6. T, Q, N, K, H, E, –?–
7.* 1/6, 1/3, 1/2, 2/3, –?–
8.* a, 6, c, 12, e, 18, –?–
9. 1/2, 9, 2/3, 10, 3/4, 11, –?–
10. A, C, F, J, O, –?–
11.* 1, 1, 2, 3, 5, 8, 13, –?–
12. 1, 3, 4, 7, 11, 18, –?–
13.* 1, 3, 7, 15, 31, 63, –?–
14. 1, 3, 9, 27, 81, 243, –?–
15. 1, 5, 14, 30, 55, 91, –?–
16. 1, 10, 100, 1000, –?–
17. 0, 10, 21, 33, 46, 60, –?–
18.* 3, 5, 11, 29, 83, 245, –?–
19. 0, 3, 8, 15, 24, 35, –?–
20. 3, -12, 48, -192, 768, –?–
21.* 1, 2, 5, 14, 41, 122, –?–
22. 1, 5, 17, 53, 161, 485, –?–
23. 1/2, 1/4, 1/8, 1/16, –?–
24. 1, 3/2, 9/4, 27/8, –?–

The next few problems are very different. Watch out!

25. 18, 46, 94, 63, 52, 61, –?–
26. O, T, T, F, F, S, S, E, N, –?–
27. 4, 8, 61, 221, 244, 884, –?–
28. 6, 8, 5, 10, 3, 14, 1, –?–
29. B, 0, C, 2, D, 0, E, 3, F, 3, G, –?–
30. A E F H I K L M N T V W
B C D G J O P Q R S U
Where do the X, Y, and Z go?

In the last lesson you made conjectures about situations in the real world. How did you arrive at your conjectures? You probably used simple trial and error to arrive at a conjecture that seemed to fit the situation. Unlike the "ways of the world," mathematics is a subject governed by clear-cut rules. The rules produce patterns. Because they are based on rules, patterns in mathematics are usually more recognizable than those in real life. It is generally easier to make conjectures about mathematical patterns than about real life situations. Take heart! If you can find patterns in life, you should have no problem making conjectures about the number patterns in this lesson.

EXERCISE SET B

Study the given information until you discover a relationship that holds true for all the data. Then state your finding by completing the conjecture.

1.* $3 + 5 = 8$ $13 + 27 = 40$

$-3 + 5 = 2$ $51 + 85 = 136$

$-1 + 1 = 0$ $-141 + -85 = -226$

Conjecture: *The sum of two odd numbers is always an —?—.*

2. $24 + 57 = 81$ $147 + 534 = 681$

$2 + -7 = -5$ $-4 + -9 = -13$

$0 + 9 = 9$ $80 + -91 = -11$

Conjecture: *The sum of an odd and an even number is always —?—.*

3. $24 + 46 = 70$ $0 + 26 = 26$

$-12 + 8 = -4$ $-26 + -18 = -44$

$-18 + 28 = 10$ $22 + 58 = 80$

Conjecture: *The sum of —?—.*

4. $3 \cdot 4 = 12$ $-24 \cdot -3 = 72$

$12 \cdot 5 = 60$ $-7 \cdot 8 = -56$

$11 \cdot -4 = -44$ $-71 \cdot 0 = 0$

Conjecture: *The product of —?—.*

5. $25 \cdot 3 = 75$ $19 \cdot 7 = 133$

$-17 \cdot 31 = -527$ $-21 \cdot -7 = 147$

$-1 \cdot 41 = -41$ $103 \cdot -7 = -721$

Conjecture: *The product of —?—.*

6. $3 \cdot 4 = 12$ $8 \cdot 9 = 72$

$23 \cdot 24 = 552$ $-3 \cdot -4 = 12$

$-1 \cdot 0 = 0$ $-15 \cdot -14 = 210$

Conjecture: *The product of two consecutive integers is always —?—.*

EXERCISE SET C

Study the pattern and then complete the problem based on your observations. (Please, no calculators here. You would be missing the point. You are looking for patterns, not mere numerical answers.)

1.
$$1 \cdot 1 = 1$$
$$11 \cdot 11 = 121$$
$$111 \cdot 111 = 12321$$
$$1111 \cdot 1111 = 1234321$$
$$11111 \cdot 11111 = \text{—?—}$$

2.
$$6 \cdot 7 = 42$$
$$66 \cdot 67 = 4422$$
$$666 \cdot 667 = 444222$$
$$6666 \cdot 6667 = 44442222$$
$$66666 \cdot 66667 = \text{—?—}$$

3.
$$12345679 \cdot 9 = 111111111$$
$$12345679 \cdot 18 = 222222222$$
$$12345679 \cdot 27 = 333333333$$
$$12345679 \cdot 36 = \text{—?—}$$
$$12345679 \cdot \text{–?–} = 555555555$$

4.
$$9 \cdot 0 + 1 = 1$$
$$9 \cdot 1 + 2 = 11$$
$$9 \cdot 2 + 3 = 21$$
$$9 \cdot 3 + 4 = \text{–?–}$$
$$\text{—?—} = 41$$

5.
$$1 + (9 \cdot 0) = 1$$
$$2 + (9 \cdot 1) = 11$$
$$3 + (9 \cdot 12) = 111$$
$$4 + (9 \cdot 123) = \text{–?–}$$
$$\text{—?—} = 11111$$

6.
$$8 + (9 \cdot 0) = 8$$
$$7 + (9 \cdot 9) = 88$$
$$6 + (9 \cdot 98) = \text{–?–}$$
$$5 + (9 \cdot 987) = \text{–?–}$$
$$\text{—?—} = 88888$$

Improving Reasoning Skills

Reasonable 'rithmetic I

Each letter in these problems represents a different digit.

1. What is the value of *B*?
2. What is the value of *J*?

```
  3 7 2
  3 8 4
+ 9 B 4
-------
C 7 C A
```

```
    E F 6
    x D 7
  -------
  D D F D
  J E D
  -------
  H G E D
```

Lesson 1.3

Picture Patterns

In this lesson you will combine your inductive reasoning skills with your artistic skills. First you will use inductive reasoning to picture what the next shape will be in each pattern. Then you will draw what you visualize. Like an artist, you will be translating your thoughts into pictures.

EXERCISE SET A

Draw the next shape in each picture pattern.

1. 

2. 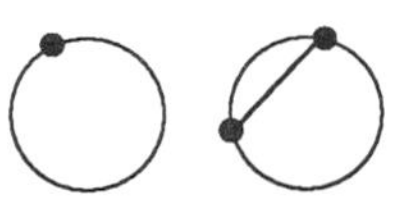  

3. 

4. 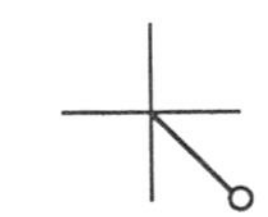  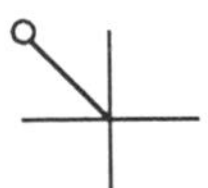

5.  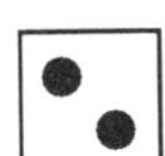 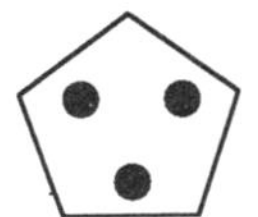

6.* 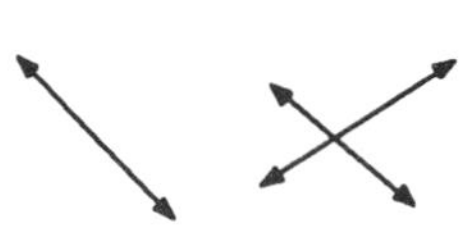 

7. 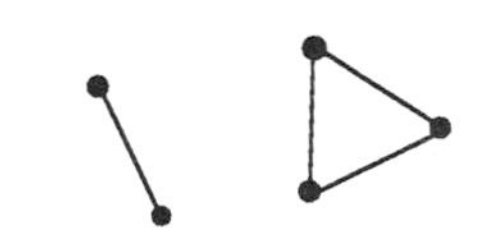

8. 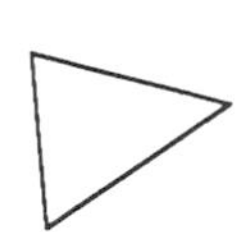 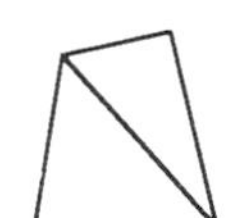 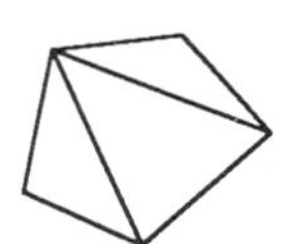

9. 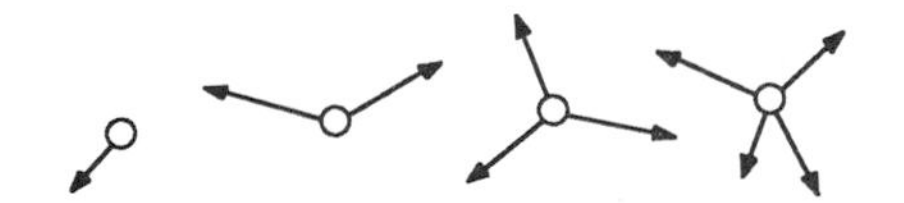

10.   

EXERCISE SET B

For this exercise set you will need graph paper. Each picture pattern below was created on a grid of squares. Follow these steps for each pattern:

Step 1: Copy the pattern (2 squares high by 16 squares long) onto graph paper.

Step 2: Continue the pattern in a 2 by 6 region at the end of the pattern.

Example

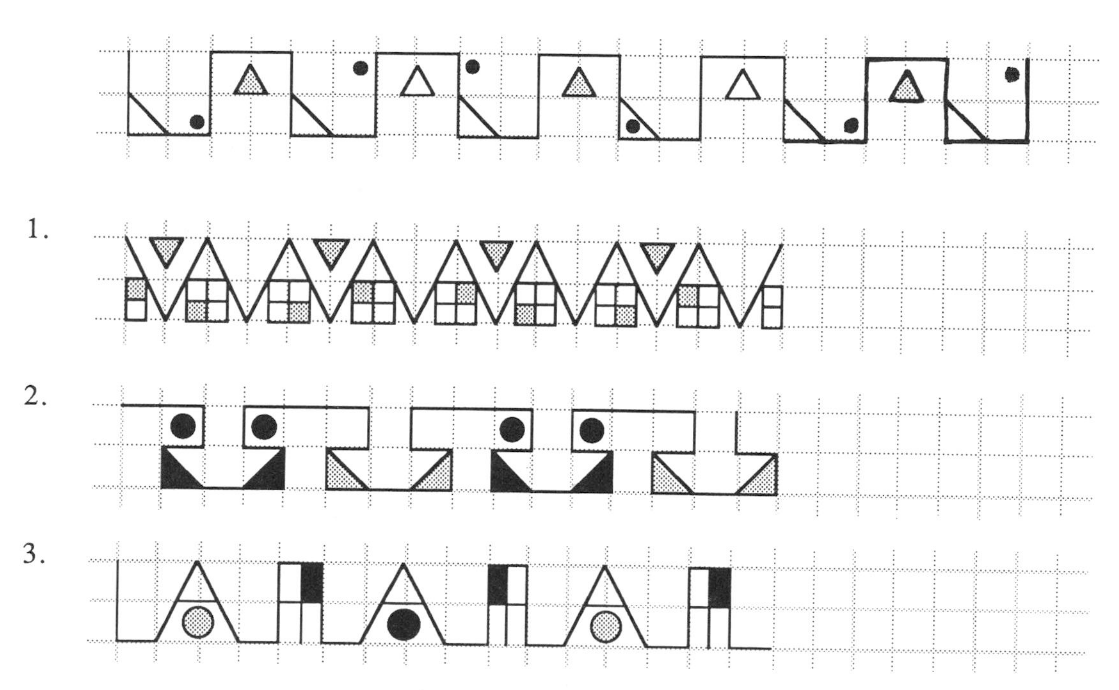

4. Now it's your turn. On graph paper, create a repeating pattern of your own.

EXERCISE SET C

Draw what the next solid will look like in each pattern.

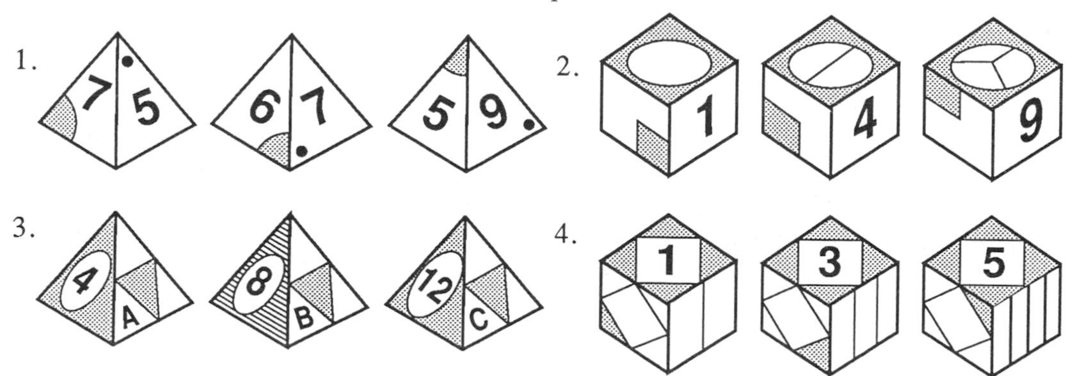

Lesson 1.4

Finding the *n*th Term

God made the integers, all else is the work of man.
– Leopold Kronecker

The pattern of numbers 3, 4, 5, 6, 7, 8, 9, . . . is an example of a **sequence** of numbers. The first term of the sequence has a value of three, the second term has a value of four, the third term has a value of five, and so on. Let's show this as a table.

Term	1	2	3	4	5	6	. . .	20	. . .	200	. . .	n
Value	3	4	5	6	7	8	. . .	22	. . .	202	. . .	$n+2$

Each number in the top row represents a **term** or position number in the sequence. The number beneath each term number is the **value** of that term of the sequence. The n and the $n + 2$ tell us the rule that connects the top row with the bottom row. This particular rule states that to find the value of any particular term in this sequence, you simply add two to the term number. Therefore, if you wish to know the value of the twentieth term of this sequence, you add 20 + 2 to get 22. The value of the twentieth term of this sequence is 22.

In previous lessons you found the next term of many sequences. What if you needed to know the value of the 20th, 200th, or 2000th term of a number sequence but you didn't know the rule? You certainly don't want to calculate all the values in between just to get one answer. What you need is a formula or rule that describes the sequence of values. Let's look at a few examples to see how you find these rules. The simplest of the formulas involve simple addition, subtraction, or multiplication by an integer.

Example A

Term	1	2	3	4	5	6	. . .	20	. . .	200	. . .	n
Value	-2	-1	0	1	2	3	. . .	–?–	. . .	–?–	. . .	–?–

To find the rule or formula that describes the sequence above, notice that the value of each term is three less than the number of the term. Therefore, the 20th term of the sequence will be 20 – 3 or 17. The value of the 200th term will be 197. In general, the value of the *n*th term is $n - 3$.

Example B

Term	1	2	3	4	5	6	...	20	...	200	...	n
Value	2	4	6	8	10	12	...	–?–	...	–?–	...	–?–

To find the rule or formula that describes the sequence above, notice that the value of each term is twice the value of the number of the term. The value of the 20th term of the sequence is 2(20) or 40. The value of the 200th term is 400. In general, the value of the nth term is $2n$. This pattern is the sequence of **even numbers**.

Sometimes the patterns are created by a combination of addition or subtraction with multiplication. The pattern below is the familiar sequence of **odd numbers.**

Example C

Term	1	2	3	4	5	6	...	20	...	200	...	n
Value	1	3	5	7	9	11	...	–?–	...	–?–	...	–?–

Every odd number is one less than an even number. 1, 3, 5, 7, 9, . . . are each one less than a term in the even sequence 2, 4, 6, 8, 10, ... Therefore, the value of each term of the sequence of odd numbers is one less than twice the number of the term. The value of the 20th term is 2(20) – 1 or 39. The value of the 200th term is 2(200) – 1 or 399. In general, the value of the nth term is $2n - 1$.

EXERCISE SET A

Find the nth term in each sequence.

1.*

1	2	3	4	5	6	...	n
5	6	7	8	9	10	...	–?–

2.

1	2	3	4	5	6	...	n
0	1	2	3	4	5	...	–?–

3.

1	2	3	4	5	6	...	n
4	8	12	16	20	24	...	–?–

4.

1	2	3	4	5	6	...	n
4	7	10	13	16	19	...	–?–

Sometimes the rule for a pattern involves more than just addition, subtraction, or multiplication by an integer (as in the last exercise set). Sometimes the rule involves the multiplication of factors, each containing n. For example, each factor in the rule $n(n + 2)$ contains n. Let's look at some examples of this type of pattern.

Example A

The rule for the sequence below is complicated. You are looking for a relationship between the number of the term and the value of the term for each pair of these numbers. You are asking, "What is the same thing I need to do with the 1 to get a 3, with the 2 to get an 8, and with the 3 to get a 15?" If the pattern is not obvious, then you might try factoring each bottom number. Perhaps the factors will display a more obvious relationship with the top number. Of the bottom numbers, 3, 15, and 35 factor in the least number of ways, so it makes sense to start with the factored form of these numbers in the hope of finding a pattern.

Term	1	2	3	4	5	6	...	20	...	200	...	n
Value	3	8	15	24	35	48	...	–?–	...	–?–	...	–?–

Factored	1 • 3 (circled)	1 • 8	1 • 15	1 • 24	1 • 35	1 • 48
		2 • 4 (circled)	3 • 5 (circled)	2 • 12	5 • 7 (circled)	3 • 16
				4 • 6 (circled)		6 • 8 (circled)

Notice that in each pair of circled factors, the top number appears as one factor and the top number plus two appears as the other factor. Therefore, the 20th term would be (20)(22) or 440. The value of the 200th term is (200)(202) or 40,400. In general, the nth term is $n(n + 2)$.

Example B

To find the rule that describes the sequence below, first try to see if you can add or subtract the same number to each term number to get the value of the term. No luck! Can you multiply or divide each term number by the same number to get the value of each term? Still no luck. Try factoring.

Term	1	2	3	4	5	6	...	20	...	200	...	n
Value	0	5	12	21	32	45	...	–?–	...	–?–	...	–?–

Factored		1 • 5 (circled)	1 • 12	1 • 21	1 • 32	1 • 45
			2 • 6 (circled)	3 • 7 (circled)	2 • 16	3 • 15
			3 • 4		4 • 8 (circled)	5 • 9 (circled)

Perhaps the factors will display a pattern that suggests a relationship with the top numbers. Display all the possible ways to factor each bottom number (skip zero since it factors in an endless number of ways). Then look across to find a pattern. Since 5 and 21 factor in the least number of ways, it makes sense to start by examining the factored forms of these numbers. Notice that in each pair of factors, one factor is three greater than the top number and one factor is one less than the top number. See if you can find a factored form of each other number that follows this pattern. You can! Therefore, the 20th term would be $(20-1)(20+3)$ or $(19)(23)$ or 437. In general, the nth term is $(n-1)(n+3)$.

EXERCISE SET B

Find the nth term in each sequence.

1.*

1	2	3	4	5	6	...	n
-1	0	3	8	15	24	...	–?–

2.

1	2	3	4	5	6	...	n
0	3	8	15	24	35	...	–?–

3.

1	2	3	4	5	6	...	n
1	6	15	28	45	66	...	–?–

4.

1	2	3	4	5	6	...	n
12	25	42	63	88	117	...	–?–

EXERCISE SET C

Draw the next shape in each pattern.

1.

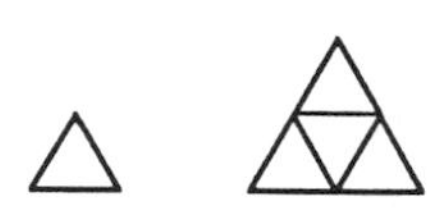

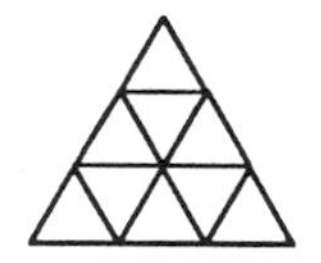

2.

3.

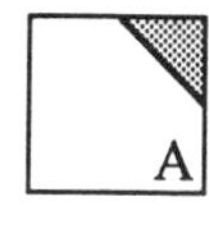

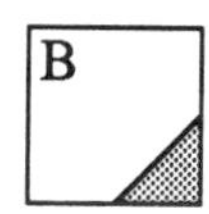

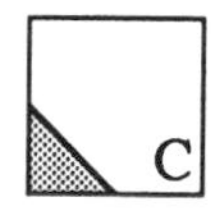

4.

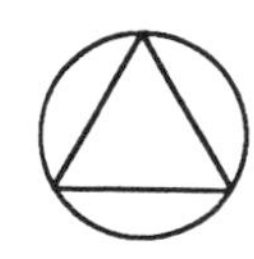

5.

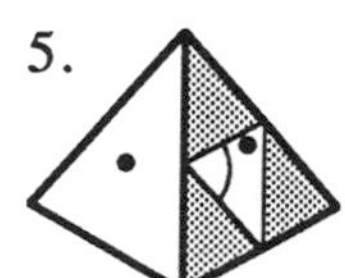

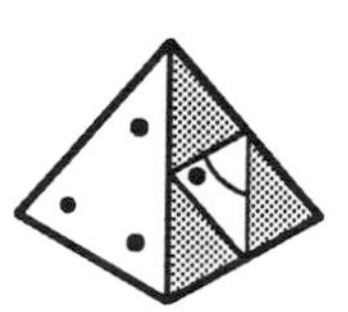

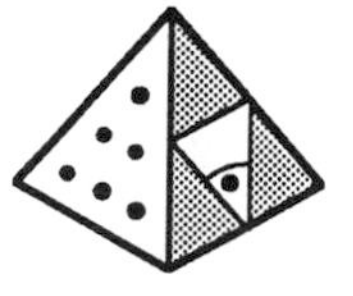

6.

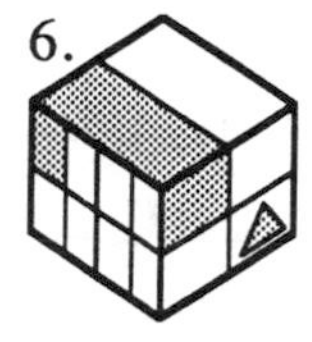

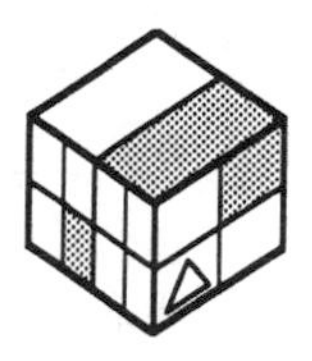

Lesson 1.5

Triangular Numbers

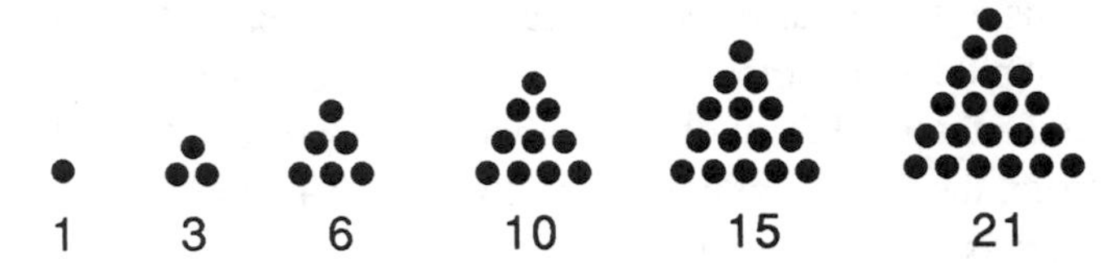

The sequence above appears in many different geometry problems. The sequence is called the **triangular numbers**. The ancient Greeks were probably the first to work with the triangular number sequence. Let's look at how you can find the rule for this sequence of triangular numbers.

Example A

In this pattern it is easy to get the next term in the sequence, but not so easy to find the rule that describes the term. You can try factoring 1, 3, 6, 10, 15, 21, hoping for a pattern to emerge that relates the terms to the factors.

Term	1	2	3	4	5	6	...	20	...	200	...	n
Value	1	3	6	10	15	21	...	–?–	...	–?–	...	–?–

Factored	1 • 1	1 • 3	1 • 6	1 • 10	1 • 15	1 • 21
			2 • 3	2 • 5	3 • 5	3 • 7

But this doesn't seem to be too helpful. There is a pattern, but it appears to alternate (one pattern with odd terms and another with even terms). If factoring doesn't help, it may be because the formula involves division as well as factoring. If you suspect that the rule involves division, try doubling or tripling the values, and then factor these numbers. If the doubled values give you a set of numbers that can be more readily put into a formula, then this formula must be divided in half to get the formula for your original sequence. This works with triangular numbers.

Term	1	2	3	4	5	6	...	20	...	200	...	n
Value	1	3	6	10	15	21	...	–?–	...	–?–	...	–?–

Doubled	2	6	12	20	30	42
Factored	(1 • 2)	1 • 6	1 • 12	1 • 20	1 • 30	1 • 42
		(2 • 3)	2 • 6	2 • 10	2 • 15	2 • 21
			(3 • 4)	(4 • 5)	3 • 10	3 • 14
					(5 • 6)	(6 • 7)

Since the nth term of the doubled sequence is $n(n + 1)$, then half of that gives the nth term of the triangular number sequence. The nth triangular number is $\frac{n(n+1)}{2}$.

The triangular number sequence doesn't always begin with a one. What would be the rule for the triangular number sequence: 6, 10, 15, 21, 28, . . . ? Let's look.

Example B

To find the rule that describes the triangular number sequence below, you would probably first look at the factors, but you would again have no luck. If you suspect that the rule involves division, you should try doubling or tripling the values, and then factor these numbers. The doubled values will again give you a set of numbers that can be more readily put into a formula. This formula must be divided in half to get the formula for your original sequence.

Term	1	2	3	4	5	6	. . .	20	. . .	200	. . .	*n*
Value	6	10	15	21	28	36	. . .	–?–	. . .	–?–	. . .	–?–

Doubled	12	20	30	42	56	72
Factored	1 • 12	1 • 20	1 • 30	1 • 42	1 • 42	1 • 72
	2 • 6	2 • 10	2 • 15	2 • 21	2 • 28	2 • 36
	(3 • 4)	(4 • 5)	3 • 10	3 • 14	4 • 14	3 • 24
			(5 • 6)	(6 • 7)	(7 • 8)	4 • 18
						6 • 12
						(8 • 9)

EXERCISE SET A

Find the *n*th term in each sequence.

1.*

1	2	3	4	5	6	. . .	*n*
0	1	3	6	10	15	. . .	–?–

2.

1	2	3	4	5	6	. . .	*n*
0	2	5	9	14	20	. . .	–?–

3.

1	2	3	4	5	6	. . .	*n*
3	6	10	15	21	28	. . .	–?–

4.

1	2	3	4	5	6	. . .	*n*
2	5	9	14	20	27	. . .	–?–

5.

1	2	3	4	5	6	. . .	*n*
-1	-1	0	2	5	9	. . .	–?–

6.*

1	2	3	4	5	6	. . .	*n*
$\frac{2}{3}$	2	4	$\frac{20}{3}$	10	14	. . .	–?–

In this lesson you have learned some new techniques for finding formulas for patterns involving division. In all the exercises we have restricted the rules to those that can be factored because most of the patterns you will encounter while making geometric discoveries in later chapters will be of this type. There are other techniques that can be used to find formulas for sequences that follow other rules.

EXERCISE SET B

Study the given information until you discover a relationship that holds true for all the data. Then state your finding by completing the conjecture.

1. $5^1 = 5$ $\quad 3^4 = 81$ $\quad 5^3 = 125$

   $1^5 = 1$ $\quad 7^2 = 49$ $\quad 9^4 = 6561$

   $13^2 = 169$ $\quad (-1)^6 = 1$ $\quad (-3)^5 = -243$

   Conjecture: *Every odd number raised to a positive integer power is —?—.*

2. $4^2 = 16$ $\quad 12^2 = 144$ $\quad 6^3 = 216$

   $(-2)^3 = -8$ $\quad 2^{10} = 1024$ $\quad 8^1 = 8$

   $2^7 = 128$ $\quad 0^4 = 0$ $\quad (-4)^3 = -64$

   Conjecture: *Every even number raised to a positive integer power is —?—.*

EXERCISE SET C

Study the pattern and then complete the exercise based on your observations.

1.*

1	= 1
1 + 3	= 4
1 + 3 + 5	= 9
1 + 3 + 5 + 7	= 16
1 + 3 + 5 + 7 + 9	= 25
1 + 3 + 5 + 7 + 9 + 11	= 36
1 + 3 + . . . + 11 + 13	= –?–
1 + 3 + . . . + 13 + 15	= –?–
1 + 3 + . . . + 15 + 17	= –?–
1 + 3 + . . . + 17 + 19	= –?–

Conjecture: *The sum of the first 30 odd numbers is –?–.*

2.

2	= 2
2 + 4	= 6
2 + 4 + 6	= 12
2 + 4 + 6 + 8	= 20
2 + 4 + 6 + 8 + 10	= 30
2 + 4 + 6 + 8 + 10 + 12	= 42
2 + 4 + . . . + 12 + 14	= –?–
2 + 4 + . . . + 14 + 16	= –?–
2 + 4 + . . . + 16 + 18	= –?–
2 + 4 + . . . + 18 + 20	= –?–

Conjecture: *The sum of the first 30 even numbers is –?–.*

EXERCISE SET D

Use inductive reasoning to find the next term of each sequence.

1. 7, 2, 5, -3, 8, -11, –?–

2.* 3, 3, 0, 3, -3, 6, -9, –?–

3. 1, 2, 6, 24, 120, –?–

4. 4, 5, 7, 11, 19, 35, 67, –?–

5.* 1, 4, 10, 20, 35, 56, –?–

6.* 1, 5, 15, 35, 70, 126, –?–

7.* 1, 6, 21, 56, 126, –?–

8. 16, 2, 8, 18, 2, 9, 20, 2, –?–

Lesson 1.6
Patterns in Geometric Shapes

Now let's apply the number pattern techniques that you've learned to some patterns in geometry. You'll be looking at lots of geometric patterns like the ones in this lesson in the chapters ahead.

EXERCISE SET A

Copy and complete each table. Make a conjecture for the value of the *n*th term.

1.* **Points Dividing a Line**

If you place 35 points on a line, into how many parts does it divide the line? Wait! Don't start placing 35 points on a line. In this problem you will discover a rule that relates the number of points placed on a line to the number of parts created by those points. Then you can use your rule to answer the question.

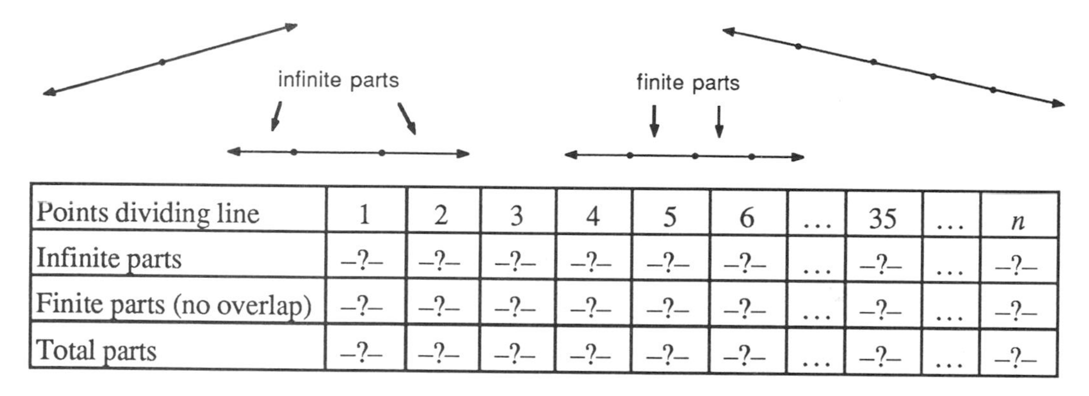

Points dividing line	1	2	3	4	5	6	...	35	...	n
Infinite parts	–?–	–?–	–?–	–?–	–?–	–?–	...	–?–	...	–?–
Finite parts (no overlap)	–?–	–?–	–?–	–?–	–?–	–?–	...	–?–	...	–?–
Total parts	–?–	–?–	–?–	–?–	–?–	–?–	...	–?–	...	–?–

2. **Line Segments Joining Random Points**

If you place 35 points on a piece of paper so that no three are in a line, how many line segments are necessary to connect each point to all the others? You don't need to draw and count all those segments. In this problem you will discover a rule relating the number of random points on a plane to the number of segments necessary to connect the points.

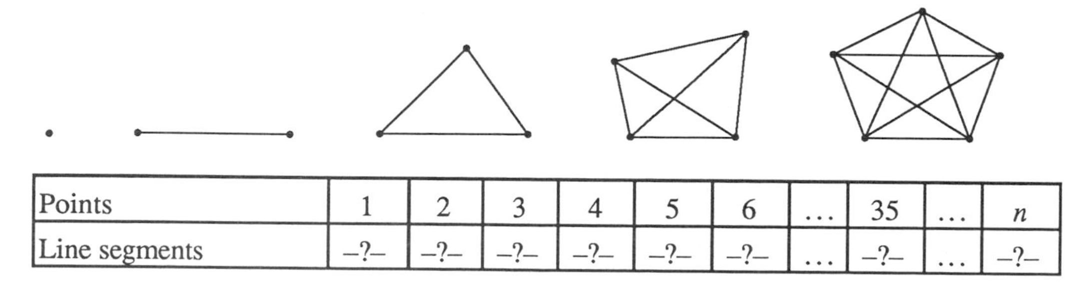

Points	1	2	3	4	5	6	...	35	...	n
Line segments	–?–	–?–	–?–	–?–	–?–	–?–	...	–?–	...	–?–

3. **Intersecting Random Lines**

If you draw 35 lines on a piece of paper so that no two lines are parallel to each other and no three lines pass through the same point, how many times will they intersect? In this problem you will discover a rule relating the number of random lines on a plane to the number of points created by their intersections.

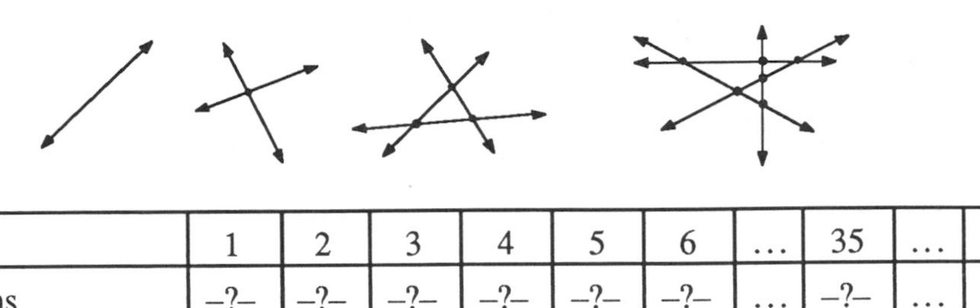

Lines	1	2	3	4	5	6	...	35	...	n
Intersections	–?–	–?–	–?–	–?–	–?–	–?–	...	–?–	...	–?–

4. **Shaded Squares in a Square Array**

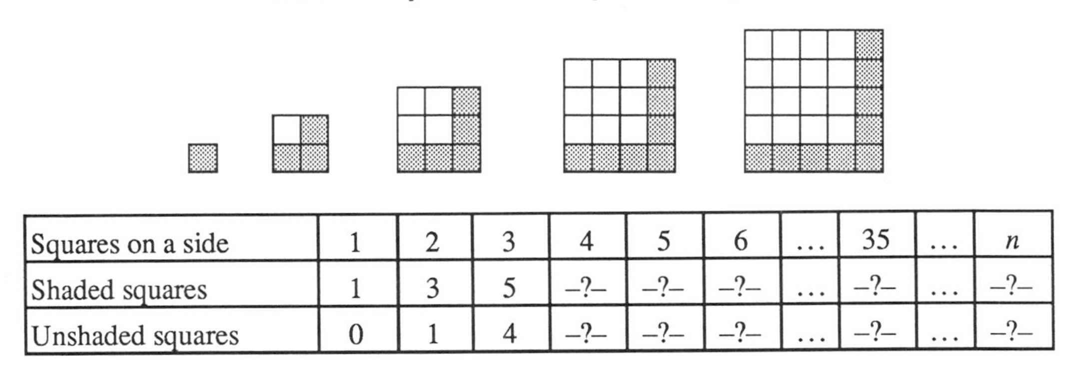

Squares on a side	1	2	3	4	5	6	...	35	...	n
Shaded squares	1	3	5	–?–	–?–	–?–	...	–?–	...	–?–
Unshaded squares	0	1	4	–?–	–?–	–?–	...	–?–	...	–?–

5. **Number of Diagonals from One Vertex in a Polygon**

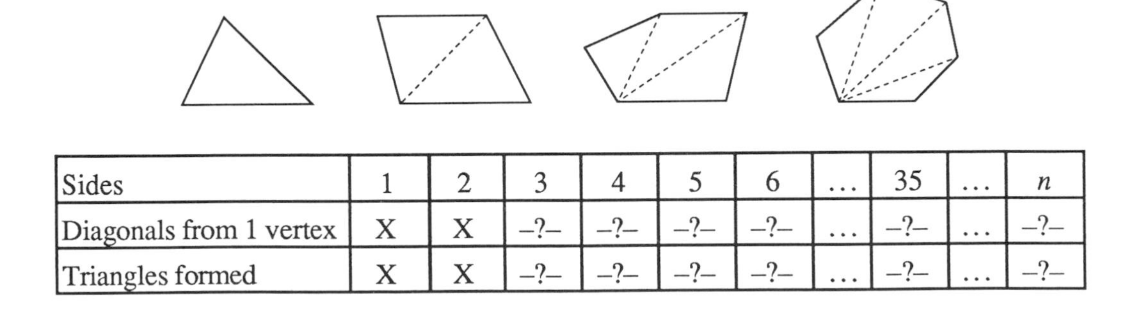

Sides	1	2	3	4	5	6	...	35	...	n
Diagonals from 1 vertex	X	X	–?–	–?–	–?–	–?–	...	–?–	...	–?–
Triangles formed	X	X	–?–	–?–	–?–	–?–	...	–?–	...	–?–

6. **Total Number of Diagonals in a Polygon**

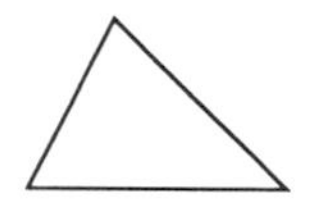
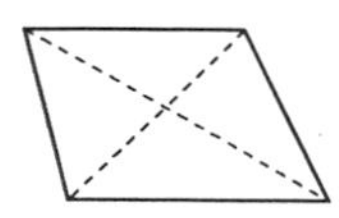
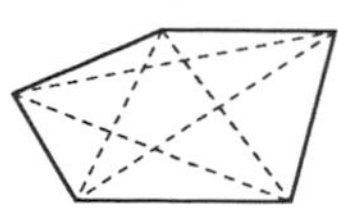
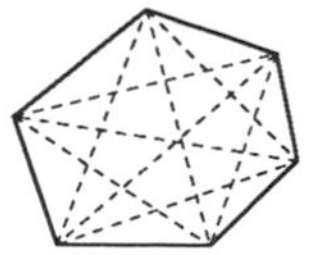

Sides	1	2	3	4	5	6	...	35	...	n
Total diagonals	X	X	–?–	–?–	–?–	–?–	...	–?–	...	–?–

EXERCISE SET B

Use what you have learned from the problems in Exercise Set A to solve each problem below.

1.* If there are 20 people sitting around a table, how many different combinations of two people can have conversations during dinner?

2. If there are 40 people at a party, how many different handshakes are possible between all the party-goers?

3.* If each team in a ten-team league plays each of the other teams four times in a season, how many league games are played during one season?

4.* Each person at a dinner table shakes hands with everyone except the two people on either side. How many handshakes will there be among eight diners?

Improving Visual Thinking Skills

Rotating Gears

In what direction will gear E rotate if gear A rotates in a counterclockwise direction?

Lesson 1.7

Solving "Big" Problems

It's amazing what one can do when one doesn't know what one can't do.

– Garfield

What is the sum of the first 5000 positive integers? What? You don't want to add up that many numbers? Well, I don't blame you. There must be an easier way! One useful question to ask in problem solving is: "Can I solve similar but easier problems?" Well, what's easier than adding up the first 5000 positive integers? Right, adding up only 10 positive integers, or 1, 2, 3, 4, 5, or 6 positive integers. Try adding smaller groups of integers and see if you find a pattern.

1	$= 1$	The sum of the first positive integer.
$1 + 2$	$= 3$	The sum of the first two positive integers.
$1 + 2 + 3$	$= 6$	The sum of the first three positive integers.
$1 + 2 + 3 + 4$	$= 10$	The sum of the first four positive integers.

Hey, they're familiar!

$1 + 2 + 3 + 4 + 5$	$= 15$	Right, they're the triangular numbers!
$1 + 2 + 3 + 4 + 5 + 6$	$= 21$	Here again, the sum of the first six positive integers is the sixth triangular number.

EXERCISE SET A

1. Since $\frac{n(n+1)}{2}$ is the *n*th triangular number, then the formula for the sum of the first *n* positive integers is also —?—.
2. What is the sum of the first 5000 positive integers?

In the series of problems above, you solved a "big" problem by first solving easier but similar problems, noticing a pattern and making a conjecture, and then using the conjecture to solve the original, difficult problem. Try solving these "big" problems.

3.* What is the sum of the first 4000 odd numbers?

4.* What is the sum of the first 7000 even numbers?

5. What is the 1000th term in the sequence: 4, 12, 24, 40, 60, 84, . . . ?

6. If the pattern below continues, what will be the total number of blocks in the one hundredth stack of blocks?

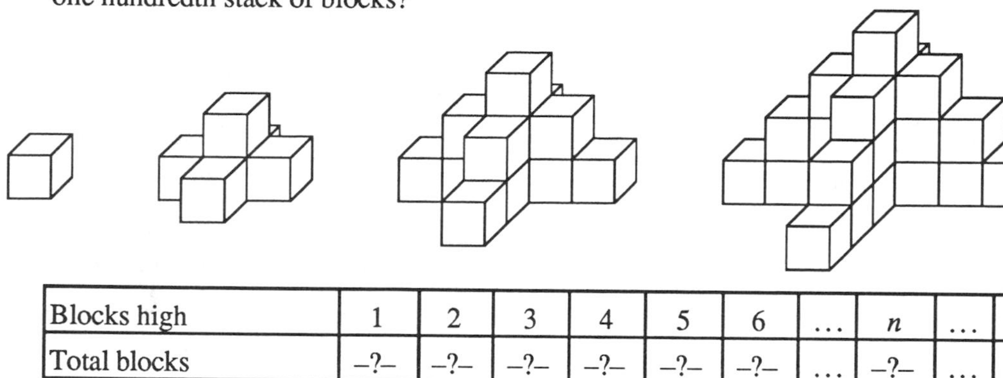

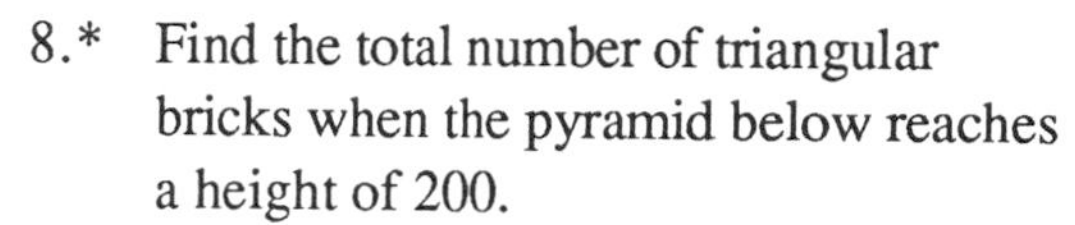

Blocks high	1	2	3	4	5	6	. . .	n	. . .	100
Total blocks	–?–	–?–	–?–	–?–	–?–	–?–	. . .	–?–	. . .	–?–

7.* Find the total number of bricks when the pyramid of bricks below reaches 100 bricks high.

8.* Find the total number of triangular bricks when the pyramid below reaches a height of 200.

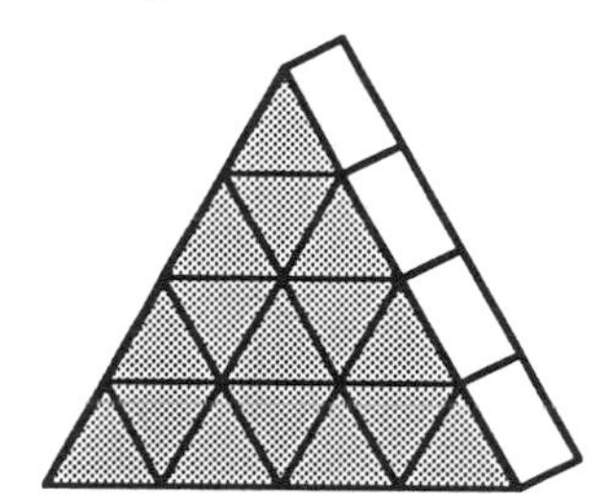

9.* How many squares of all sizes are in a 6 by 6 grid of squares? Count all the 1 by 1's, all the 2 by 2's and so on. Don't forget to count the entire 6 by 6 square. Look for patterns to help.

10.* Picture a wire grid of cubes, 6 cubes on an edge. How many different cubes of all sizes are in this 6 by 6 by 6 grid of cubes? Count all the 1 by 1 by 1 cubes, all the 2 by 2 by 2 cubes, and so on. Don't forget to count the entire 6 by 6 by 6 cube. Look for patterns to help.

EXERCISE SET B

Use what you have learned from this chapter to solve each problem below.

1. If everyone in the town of Skunk's Crossing (population 84) has a telephone, how many different two-way conversations can take place?

Find the *n*th term in each sequence.

2.

1	2	3	4	5	6	...	*n*
5	10	15	20	25	30	...	–?–

3.

1	2	3	4	5	6	...	*n*
2	12	30	56	90	132	...	–?–

4.*

1	2	3	4	5	6	...	*n*
0	2	5	9	14	20	...	–?–

5.*

1	2	3	4	5	6	...	*n*
-5	0	15	40	75	120	...	–?–

Improving Reasoning Skills

Bagels I

In this puzzle you are to determine a 3-digit number (no digit repeated) by making "educated guesses." After each guess, you will be given a clue about your guess. The clues:

bagels: no digit is correct
pico: one digit is correct but in the wrong position
fermi: one digit is correct and in the correct position

In each of the problems below, a number of guesses have been made with the clue for each guess shown to the right. From the given set of clues, determine the 3-digit number. If there is more than one solution, find them all.

1.
1 2 3 *bagels*
4 5 6 *pico*
7 8 9 *pico*
0 7 5 *pico fermi*
0 8 7 *pico*
? ? ?

2.
9 0 8 *bagels*
1 3 4 *pico*
3 8 7 *pico fermi*
2 5 6 *fermi*
2 3 7 *pico pico*
? ? ?

Special Project

Three Peg Puzzle

The three peg puzzle first appeared as a toy in 1883 in France. Shortly after it was introduced, the following story, as told by W. W. R. Ball in *Mathematical Recreations and Essays,* was associated with it. As a result, the game became known as the Tower of Brahma. It is also widely known as the Tower of Hanoi.

> In the great temple at Benares, beneath the dome which marks the center of the world, rests a brass plate in which are fixed three diamond needles, each a cubit high and as thick as the body of a bee. On one of these needles, at the creation, God placed sixty-four disks of pure gold, the largest disk resting on the brass plate, and the others getting smaller and smaller up to the top one. This is the Tower of Brahma. Day and night unceasingly the priests transfer the disks from one diamond needle to another according to the fixed and immutable laws of Brahma, which require that the priest on duty must not move more than one disk at a time and that he must place this disk on a needle so that there is no smaller disk beneath it. When the sixty-four disks shall have been thus transferred from the needle on which at the creation God placed them to one of the other needles, tower, temple, and Brahmins alike will crumble into dust, and with a thunderclap the world will vanish.

What is the smallest number of moves by which the priests can successfully transfer all 64 rings from one needle to another according to the given rules? Remember, in solving big problems, it is often useful to solve smaller, similar problems to discover a pattern that will make your difficult problem an easy one.

1.* Copy the table below. Then cut out 6 circular disks of different sizes from a sturdy piece of cardboard and physically solve this problem for 0 through 6 rings.

Rings	0	1	2	3	4	5	6	. . .	n	. . .	64
Necessary moves	–?–	–?–	–?–	–?–	–?–	–?–	–?–	. . .	–?–	. . .	–?–

2. When you have completed the table for 0 through 6 rings, find a pattern and make a conjecture about the number of moves necessary for n rings. What is the number of moves necessary for n rings?

3. Use your conjecture to find how many moves it will take the priests to transfer all 64 rings? You may express your answer using exponents.

Lesson 1.8

Patterns in Pascal's Triangle

In earlier lessons you looked for patterns in sequences of numbers. These sequences were arranged in a straight line. In this lesson you are going to look for patterns in a special triangular arrangement of numbers called **Pascal's Triangle**. Pascal's Triangle is named after Blaise Pascal, a very gifted French mathematician of the seventeenth century because he developed so many of the triangle's properties. This triangular arrangement of numbers, however, was known by the Arabian poet and mathematician Omar Khayyam (c. 1044–1123) as well as by Chinese mathematicians some 250 years before Pascal. Pascal's Triangle is shown below.

Blaise Pascal

Pascal's Triangle

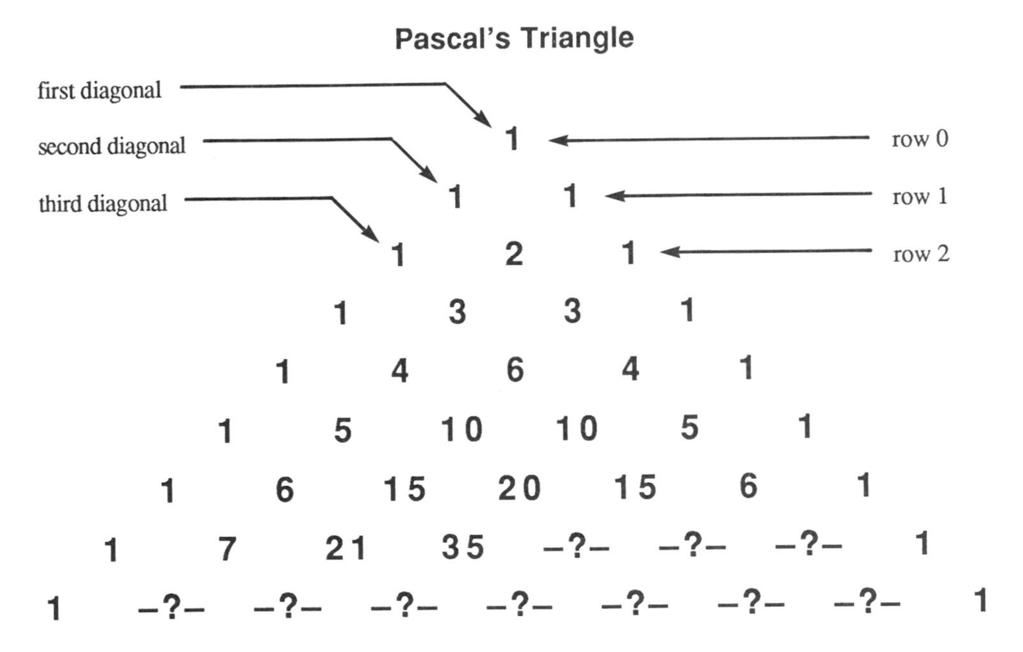

EXERCISE SET A

1.* Copy and then fill in the blanks in Pascal's Triangle.

2. The sequence in the first diagonal is: 1, 1, 1, 1, ... What is the nth term?

3. The sequence in the second diagonal is: 1, 2, 3, 4, ... What is the nth term?

4. The sequence in the third diagonal is: 1, 3, 6, 10, ... What is the nth term?

Copy and complete each table below.

5.

Number of the row	0	1	2	3	4	5	6	...	n
Number of numbers in row	–?–	–?–	–?–	–?–	–?–	–?–	–?–	...	–?–

6.*

Number of the row	0	1	2	3	4	5	6	...	n
Sum of the numbers in row	–?–	–?–	–?–	–?–	–?–	–?–	–?–	...	–?–

EXERCISE SET B

There are many other triangular arrays of numbers similar to Pascal's Triangle that have interesting patterns in them.

1.* Find the terms in the next row of the triangular array below.

```
                1
              1   1
            1   3   1
          1   5   5   1
        1   7  13   7   1
      1   9  25  25   9   1
   -?- -?- -?- -?- -?- -?- -?-
```

2.* What is the sum of the numbers in the tenth row of the triangular array below?

```
row 1                0
row 2              2   4
row 3            6   8   10
              12  14  16  18
            20  22  24  26  28
          30  32  34  36  38  40
```

3.* What is the sum of the numbers in the 500th row of the triangular array below.

```
row 1               1
row 2             1 2 1
                1 2 3 2 1
              1 2 3 4 3 2 1
            1 2 3 4 5 4 3 2 1
          1 2 3 4 5 6 5 4 3 2 1
```

4. What is the sum of the numbers in the 800th row of the triangular array below.

```
row 1                1
row 2              3   5
                 7   9   11
              13  15  17  19
            21  23  25  27  29
          31  33  35  37  39  41
```

Lesson 1.9

Chapter Review

Creative minds have always been known to survive any kind of bad training.

– Anna Freud

EXERCISE SET

1. What is inductive reasoning?

2.* What is the sum of the first 8000 odd numbers?

Find the next term of each sequence.

3. 7, 21, 35, 49, 63, 77, –?–

4.* Z, 1, Y, 2, X, 4, W, 8, –?–

Draw the next shape in each pattern.

5.

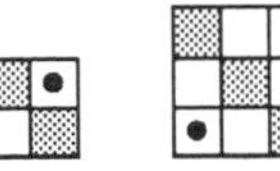

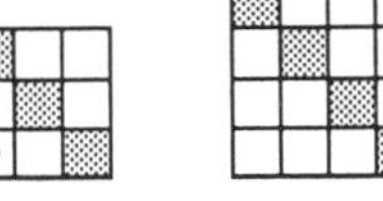

6.*

7.

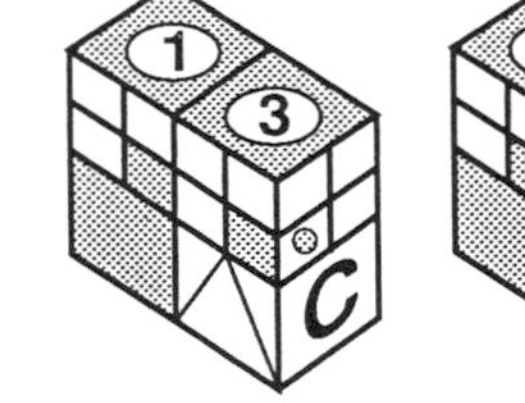

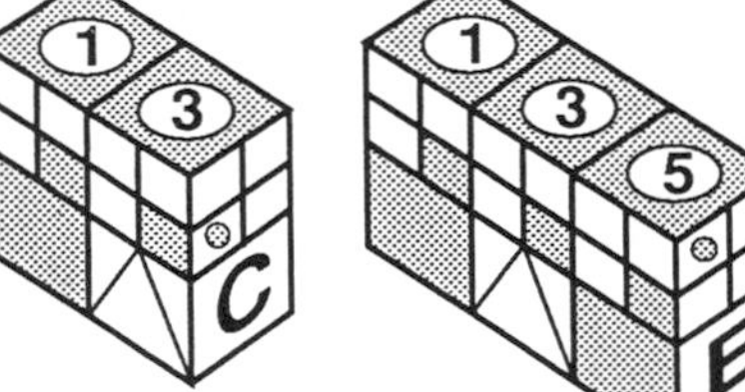

8.*

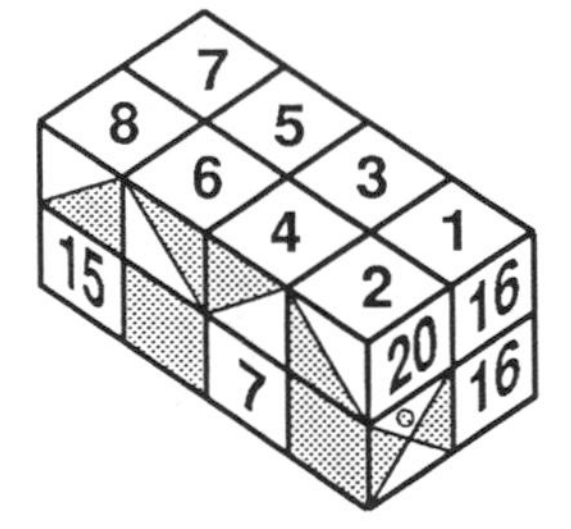

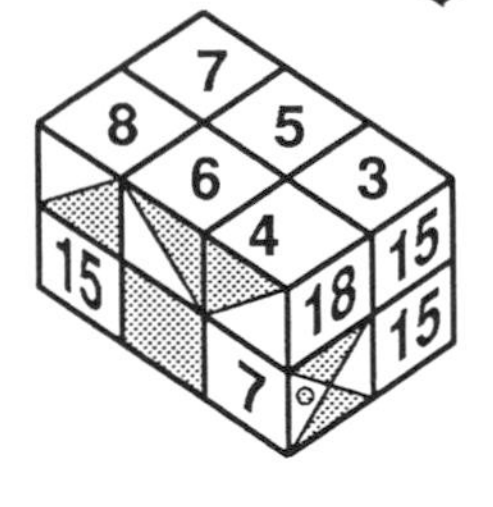

For Exercise 9 you will need graph paper. Each picture pattern below was created on a grid of squares (2 squares high by 16 squares long). Copy the pattern onto graph paper and then continue the pattern in a 2 x 8 grid at the end of the pattern. (For more help, see Lesson 1.3 Exercise Set B)

9.*

Find the *n*th term in each sequence.

10.

1	2	3	4	5	6	...	*n*
-1	1	3	5	7	9	...	–?–

11.*

1	2	3	4	5	6	...	*n*
0	3	10	21	36	55	...	–?–

12.

1	2	3	4	5	6	...	*n*
0	$\frac{4}{3}$	$\frac{10}{3}$	6	$\frac{28}{3}$	$\frac{40}{3}$	...	–?–

13.*

1	2	3	4	5	6	...	*n*
-4	0	6	14	24	36	...	–?–

Copy and complete each table. Make a conjecture for the value of the *n*th term.

14. **Triangles in a Square Array**

Squares on a side	1	2	3	4	5	6	...	35	...	*n*
Shaded triangles	1	–?–	–?–	–?–	–?–	–?–	...	–?–	...	–?–
Unshaded triangles	3	–?–	–?–	–?–	–?–	–?–	...	–?–	...	–?–

15.* **Building Block Pattern**

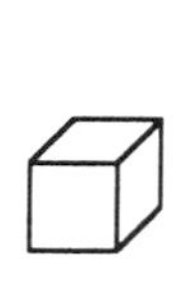
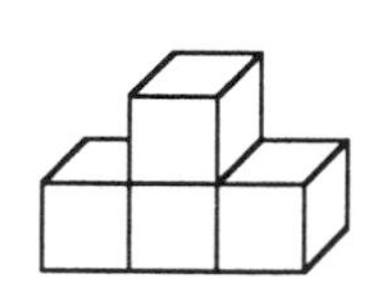
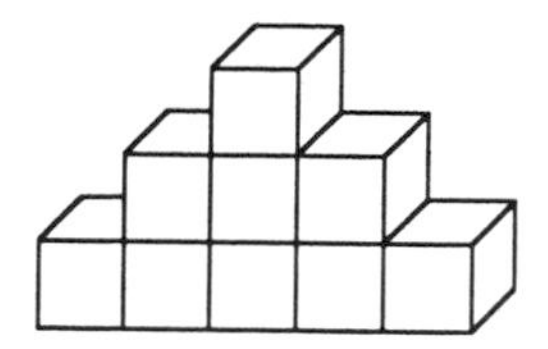
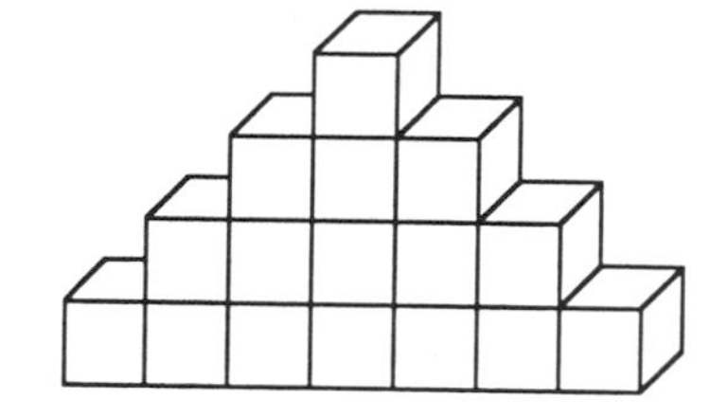

Blocks high	1	2	3	4	5	6	...	35	...	*n*
Total blocks	–?–	–?–	–?–	–?–	–?–	–?–	...	–?–	...	–?–

16. How many two-person conversations are possible at a party of 100 people?

17.* If 28 lines are drawn on a plane, what is the maximum number of points of intersection possible?

18. If a polygon has 42 sides, how many diagonals will it have?

19.* On the right is a triangle of letters. You can spell GEOMETRY by bouncing down the letters like a pinball from the G down to the Y. How many ways can you spell GEOMETRY? (You may only bounce to one of the two letters below the letter you last hit. One possible path has been marked.)

Pinball Spelling

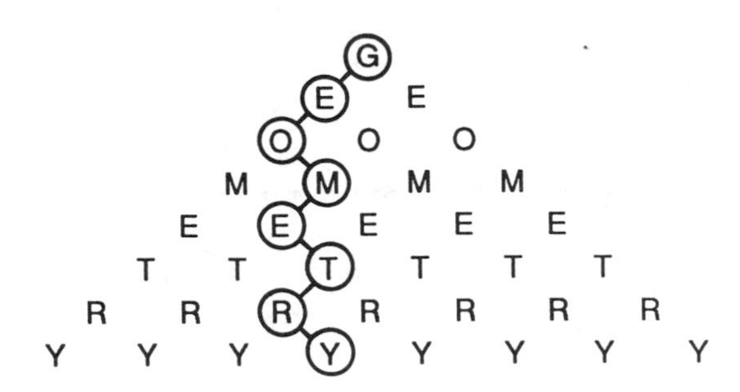

20. Below is the poem *The Blind Men and the Elephant* by the American poet John Godfrey Saxe (1816–1887). Read the poem. How might the six men of Indostan improve their inductive reasoning techniques?

The Blind Men and the Elephant

It was six men of Indostan
To learning much inclined,
Who went to see the Elephant
(Though all of them were blind).
That each by observation
Might satisfy his mind.

The First approached the Elephant,
And happening to fall
Against his broad and sturdy side,
At once began to bawl:
"God bless me! but the Elephant
Is very like a wall!"

The Second, feeling of the tusk,
Cried, "Ho! what have we here
So very round and smooth and sharp?
To me 'tis mighty clear
This wonder of an Elephant
Is very like a spear!"

The Third approached the animal,
And happening to take
The squirming trunk within his hands,
Thus boldly up and spake:
"I see," quoth he," the Elephant
Is very like a snake!"

The Fourth reached out an eager hand,
And felt about the knee.
"What most this wondrous beast is like
Is mighty plain," quoth he;
"'Tis clear enough the Elephant
Is very like a tree!"

The Fifth who chanced to touch the ear,
Said: "E'en the blindest man
Can tell what this resembles most:
Deny the fact who can,
This marvel of an Elephant
Is very like a fan!"

The Sixth no sooner had begun
About the beast to grope,
Than, seizing on the swinging tail
That fell within his scope,
"I see," quoth he, "the Elephant
Is very like a rope!"

And so these men of Indostan
Disputed loud and long,
Each in his own opinion
Exceeding stiff and strong,
Though each was partly in the right
And all were in the wrong!

Cooperative Problem Solving

Patterns at the Lunar Colony

The residential section of the lunar space port Galileo is laid out in a rectangular pattern of streets. To avoid the need for personal transport vehicles within the residential section, the engineers have developed a moving sidewalk system. Each of the streets A through K (A is the northernmost street, K is the southernmost) has sidewalks moving east and west, while the avenues 1 through 11 (1st Avenue is the westernmost avenue, 11th Avenue is the easternmost) have sidewalks moving north and south. Professor Osgood Farkle lives at the corner of A Street and Third Avenue. His friend, astrobiologist Melissa Pratt, lives at the corner of Tenth and J. Osgood would like to make each trip to his friend's house a new experience. He would like to figure out how many different ways he can travel from his house to hers if he always travels east or south. How many paths are there between the two homes if he takes only eastbound and southbound sidewalks on his way to visit his heartthrob? There is a hint for this problem in the *Hints* section.

2 Introducing Geometry

Three Worlds
M. C. Escher, 1955

This chapter introduces you to the terms and symbols of geometry. In this chapter you will write your own definitions of many geometric figures and terms. You will do this by inductive reasoning. Keep a notebook with a list of all the terms and their definitions. As each new term is defined, add the definition to your list. Draw examples next to your definitions.

Lesson 2.1

Building Blocks of Geometry

The three building blocks of geometry are points, lines, and planes. Ancient mathematicians tried to define these terms. The ancient Greeks said, "A point is that which has no part. A line is breadthless length." The Mohist philosophers in ancient China said, "The line is divided into parts and that part which has no remaining part is a point." Those definitions don't help much, do they?

A definition is a statement that clarifies or explains the meaning of a word or phrase. In order to define point, line, and plane, however, you must use words or phrases that themselves need defining or further explanation.

Have you ever encountered a word that you didn't know? Perhaps you looked it up in a dictionary, only to find another unknown word in the definition. So you looked that word up and found the word that you started with. Frustrating, isn't it?

Early mathematicians were similarly frustrated trying to define point, line, and plane. Point, line, and plane remain undefined. To help you gain an intuitive understanding of these terms, however, we offer general descriptions of them. While not precise definitions, these descriptions will give you a sense of what is meant by point, line, and plane.

A **point** is the basic unit of geometry. It has no size. It is infinitely small. It has only location. A very sharp pencil tip is a physical model of a point. A point however, is smaller than the smallest dot you can make with your pencil.

A **line** is a straight arrangement of points. There are infinitely many points in a line. It has length but no thickness. It extends forever in two directions. A taut string is a model of a line. A line, however, is thinner then any string ever made.

A **plane** has length and width but no thickness. It is a flat surface that extends forever. A wall, a floor, or a ceiling is a model of a plane. A slice of cheese is a model of a plane, but a plane is thinner than any piece of cheese you could ever slice.

Picture of a Point

P

You use a dot to represent a point. You name a point with a capital letter. This point is called point *P*.

Picture of a Line

A B

You name a line using two points on the line. $\leftrightarrow$ is the symbol for line. This line is called line *AB* or line *BA*. You can also write it $\overleftrightarrow{AB}$ or $\overleftrightarrow{BA}$.

Picture of a Plane

EXERCISE SET A

Name each line below two different ways.

1. 

2.* A R T

3. 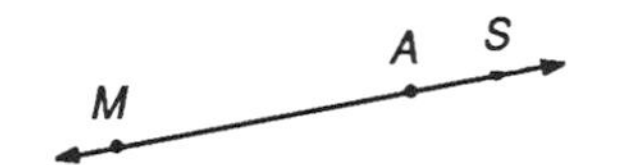

Use a ruler to draw each line below. Place two dots on each line (to represent points) and label them. Don't forget to put arrowheads on the ends of each line to show that it goes on forever.

4. $\overleftrightarrow{AB}$ 5. $\overleftrightarrow{KL}$ 6. $\overleftrightarrow{PU}$

Using the undefined terms, point, line, and plane, one can define many new geometric figures and terms. Many will be defined for you in this chapter and in those that follow. Others you will define. Each time you encounter a new geometric definition, put the definition on a definition list kept in your notebook. You should illustrate the definition with a simple picture or diagram. Whenever you define a new geometric term, add it to your list also. Read the definition list daily.

Here are your first definitions. Begin your list (drawing a picture of space could prove to be difficult).

***Space** is the set of all points.*

***A line segment** consists of two points and all the points between them that lie on the line containing the two points.*

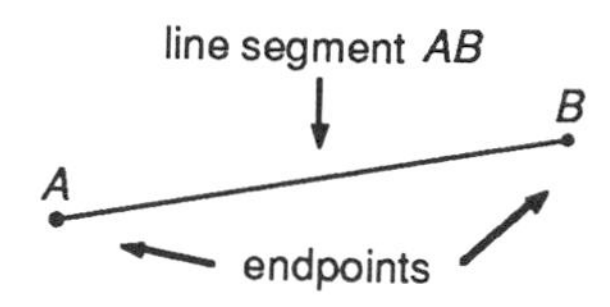

The two points are called the **endpoints** of the line segment.
The symbol for line segment is $\overline{\quad}$. The symbol is placed above the letters that name the endpoints of the line segment. Line segment AB may be written $\overline{AB}$ or $\overline{BA}$.

EXERCISE SET B

Name each line segment.

1. A C

2. 

3. X Y

Draw and label each line segment.

4. $\overline{AB}$

5. $\overline{PQ}$

6. $\overline{TS}$

Here is another very important definition. Add it to your definition list. Draw and label a picture of a ray next to your definition.

> ***Ray** AB is the part of line AB that contains point A and all the points on AB that are on the same side of A as B.*

A ray begins at a point and goes on forever in one direction. Point A is the **endpoint** of ray AB. You name a ray using two letters. The first letter is the endpoint and the second letter is any other point that the ray passes through. $\overrightarrow{\quad}$ is the symbol for ray. Ray AB may be written $\overrightarrow{AB}$. If Y is another point that the ray passes through, the ray may also be called $\overrightarrow{AY}$.

You can think of a ray as half of a line. If you cut everything to the left of point A off line AB, you would be left with ray AB.

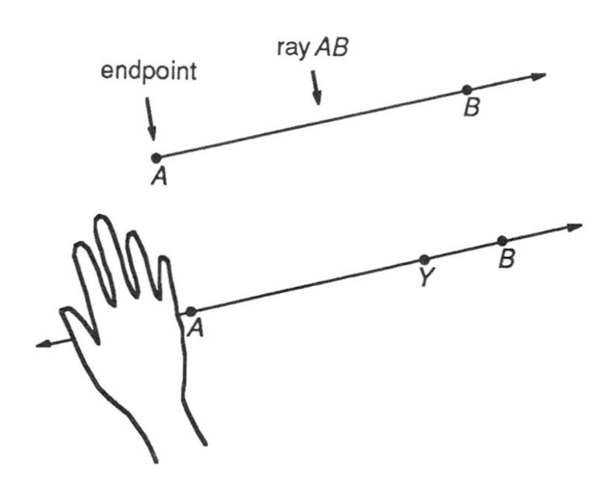

EXERCISE SET C

Name each ray two different ways.

1.* 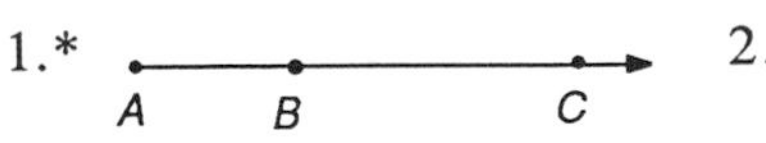

2. 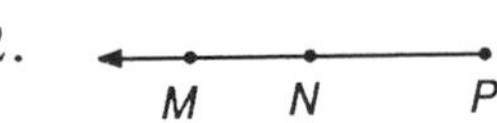

3. Z Y X

Draw and label each ray.

4. $\overrightarrow{AB}$ 5. $\overrightarrow{YX}$ 6. $\overrightarrow{MN}$

Here is a definition that is based upon the definition of a ray. Add it to your definition list. Draw and label a picture of an angle next to your definition.

*An **angle** is two rays that share a common endpoint, provided that the two rays do not lie on the same line.*

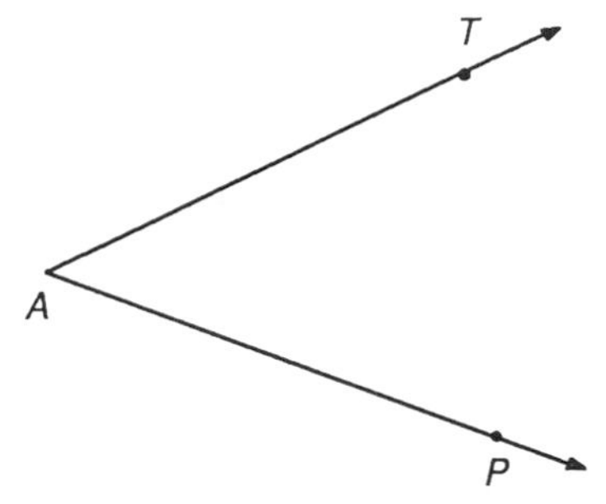

The common endpoint of the two rays that make an angle is called the **vertex** of the angle. The two rays are called the **sides** of the angle. $\angle$ is the symbol for angle. You name angles with three letters. The middle letter must be the vertex of the angle. The other two letters are points from each of the sides of the angle.

Above is the angle formed by ray *AT* and ray *AP*. Point *A*, the common endpoint of the two rays, is the vertex of the angle. $\overrightarrow{AT}$ and $\overrightarrow{AP}$ are the sides of the angle. This angle is named $\angle TAP$ or $\angle PAT$. Why do you think the definition for angle includes the expression *provided that the two rays do not lie on the same line*? What would an angle look like if the two rays that formed it lay on the same line?

EXERCISE SET D

Name each angle two different ways.

1.* 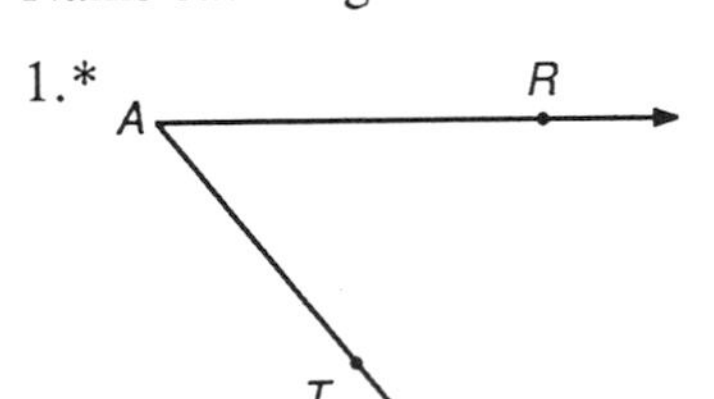

2. 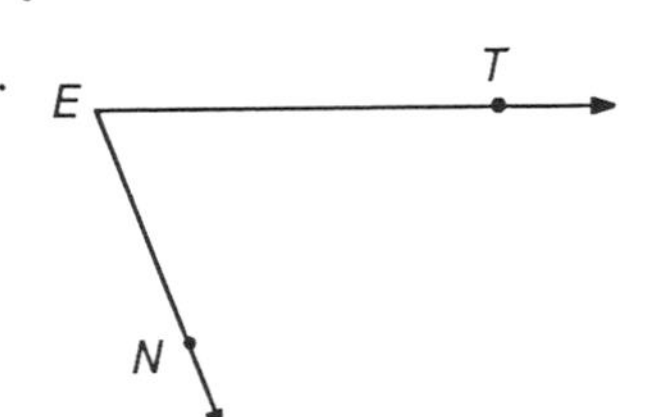

3. 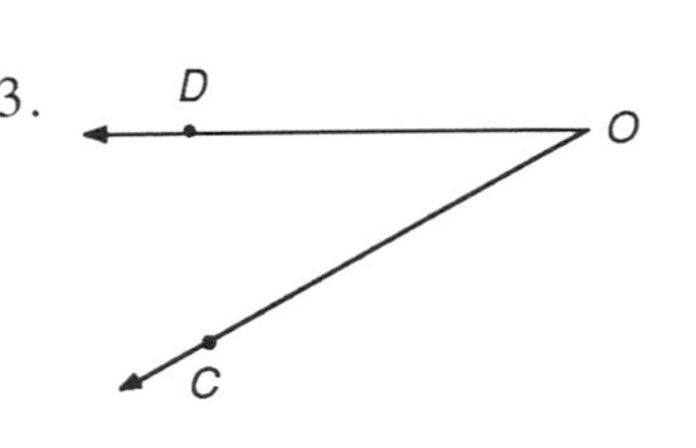

Draw and label each angle.

4. $\angle TAN$ 5. $\angle BIG$ 6. $\angle SML$

On occasion, when no confusion results, an angle can be named with just the vertex letter instead of the usual three letters.

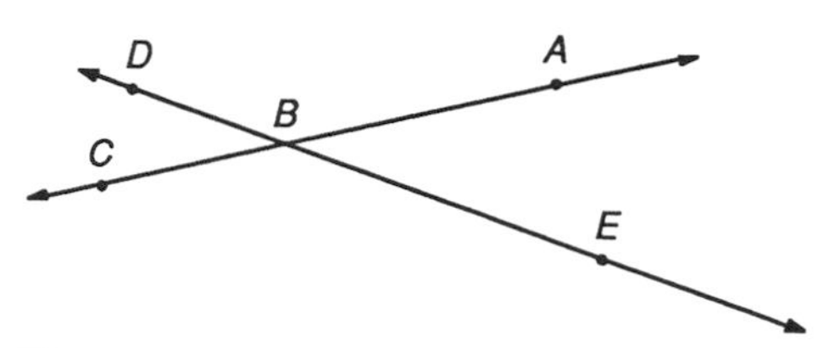

There are four angles with vertex B. You *must* use three letters to identify any of the four angles around B.

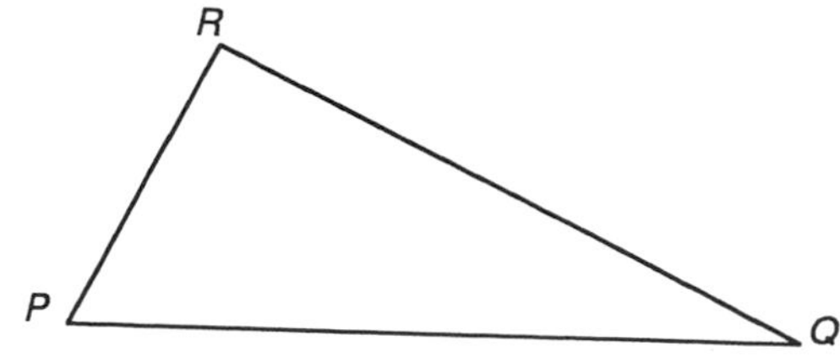

There is only one angle with vertex P. So, you *may* write $\angle P$ instead of $\angle RPQ$.

For each diagram, list the angles that can be named using only one vertex letter.

7.*

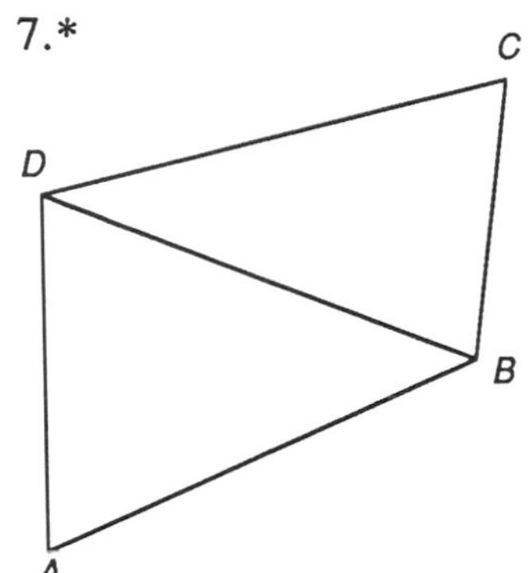

8.

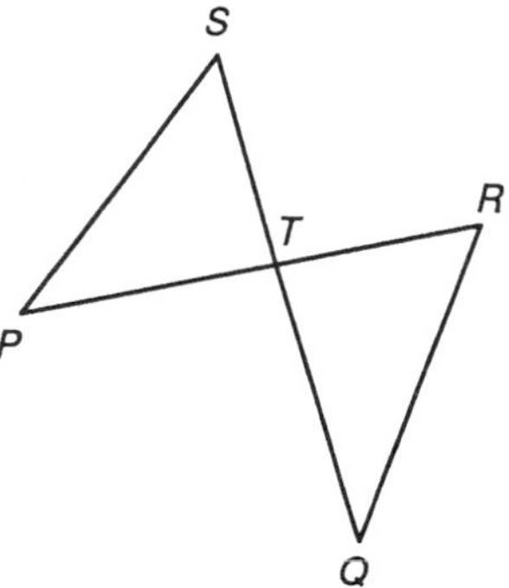

9.

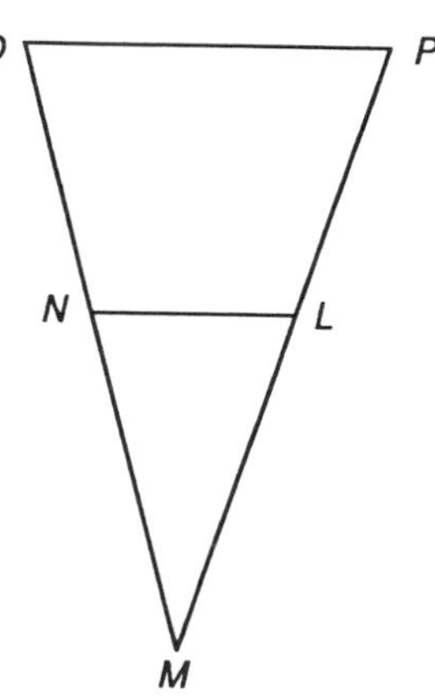

Improving Visual Thinking Skills

Picture Patterns I

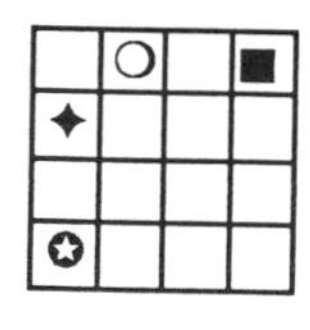
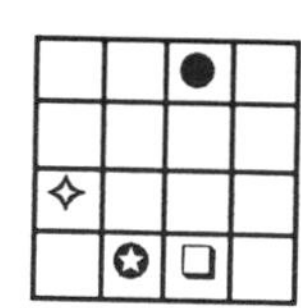
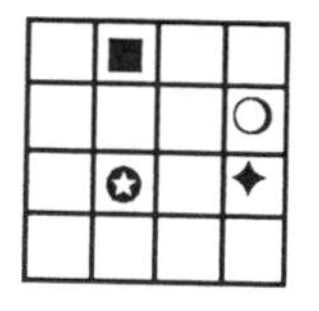
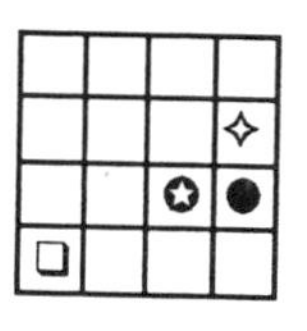
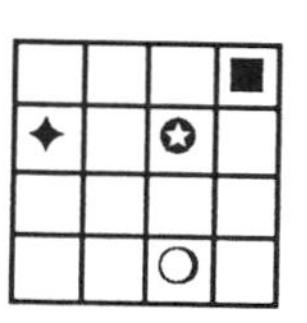

Which comes next?

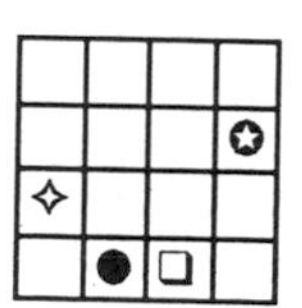

a.

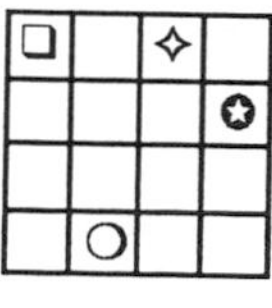

b.

c.

Lesson 2.2

Poolroom Math

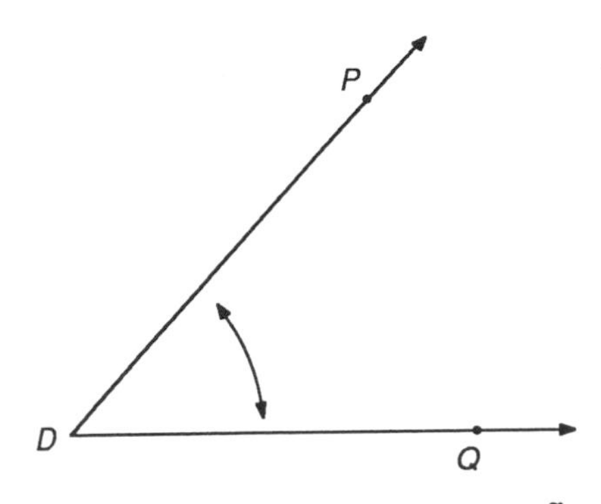

Recall that the arrowhead on the end of each ray means that the ray goes on forever. If you were to redraw $\angle PDQ$ with longer or shorter looking rays, would the size of the new angle be different? No, the lengths of rays as they appear on paper have nothing to do with the size of an angle. The actual rays go on forever whether they're drawn short or drawn long.

The size of an angle is the smallest amount of turning necessary to rotate one ray to fit over the other ray. You measure this rotation in units called **degrees**. One full circle of rotation is 360 degrees (usually written 360°). One-half circle of rotation is 180°; one-fourth is 90°; and so on. According to the definition of an angle, the size of an angle can be anything between 0° and 180°. When you wish to indicate the size or **degree measure** of an angle, write a lowercase *m* in front of the angle symbol. For example, $m\angle ZAP = 43$ means the measure of angle *ZAP* is 43 degrees. It is understood that the units for measuring angles are always degrees.

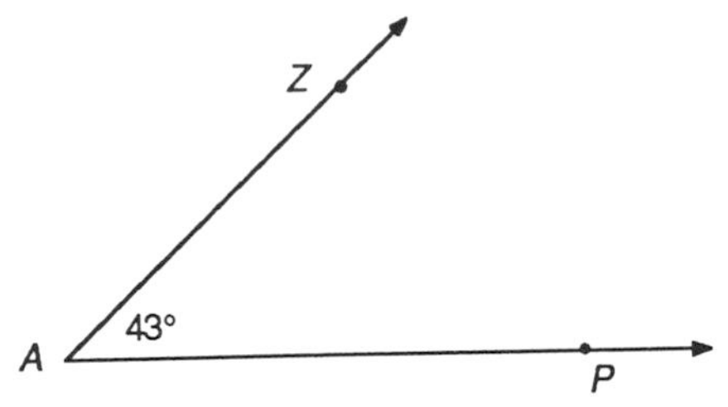

The geometric tool used to measure the number of degrees in an angle is a **protractor**. To measure angles with a protractor, follow the three steps below.

Step 1

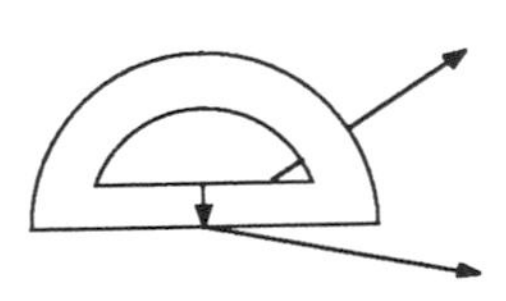

Place the center mark of the protractor on the vertex of the angle.

Step 2

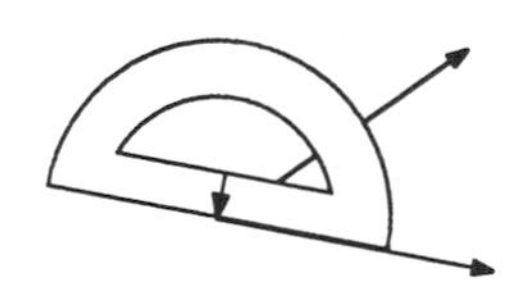

Rotate the "zero-edge" of the protractor to line up with one side of the angle.

Step 3

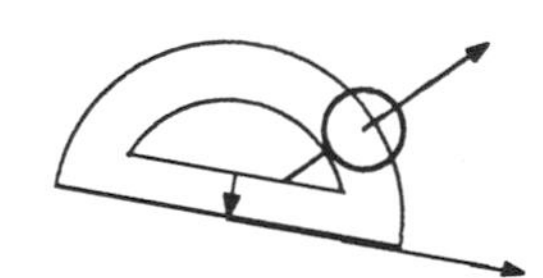

Read the size of the angle on the protractor's scale at the point where the other side crosses.

Your protractor has two scales around its edge, so you must be careful to read the correct scale. First, note whether the angle you are measuring is smaller or larger than a ninety degree angle. If the angle is smaller, use the smaller of the two numbers on the protractor. If the angle is larger than ninety, use the larger of the two numbers.

EXERCISE SET A

Using the protractor pictured, figure out the measure of each angle named below. Keep in mind that whenever you use a protractor to measure an angle, or a ruler to measure a segment, the measurement you get is only an approximation of the actual size of the angle or segment. It is impossible to measure an angle or segment exactly. You should, however, measure as accurately as you can. The symbol ≈ means *is about equal to.*

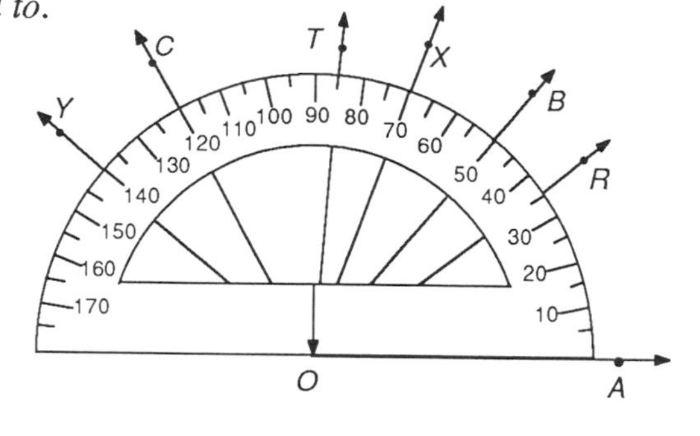

1. $m\angle AOB \approx$ –?–
2. $m\angle AOC \approx$ –?–
3. $m\angle XOA \approx$ –?–
4. $m\angle AOY \approx$ –?–
5. $m\angle ROA \approx$ –?–
6. $m\angle TOA \approx$ –?–

Use subtraction to find the size of each angle.

7.* $m\angle COB \approx$ –?–
8. $m\angle YOX \approx$ –?–
9. $m\angle XOT \approx$ –?–
10. $m\angle COY \approx$ –?–

Use your own protractor to find the size of each angle below (to the nearest degree).

11. $m\angle FAT \approx$ –?–
12. $m\angle IBM \approx$ –?–

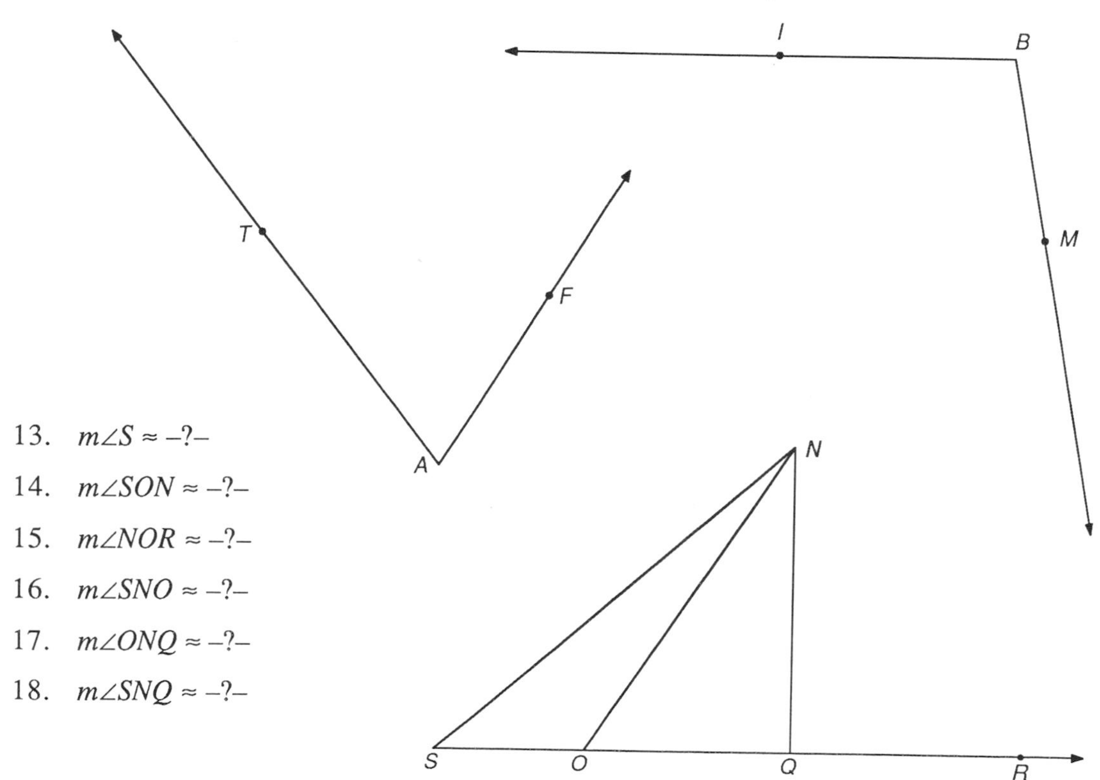

13. $m\angle S \approx$ –?–
14. $m\angle SON \approx$ –?–
15. $m\angle NOR \approx$ –?–
16. $m\angle SNO \approx$ –?–
17. $m\angle ONQ \approx$ –?–
18. $m\angle SNQ \approx$ –?–

19. Which angle is larger, $\angle SML$ or $\angle BIG$?

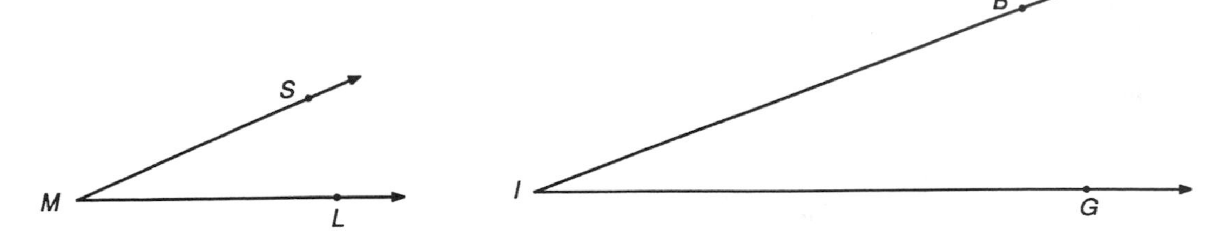

Now use your protractor to draw an angle with each size below. While it is impossible to draw an angle of a specified size exactly, draw as accurately as you can.

20. $m\angle A = 40$

21. $m\angle B = 90$

22. $m\angle C = 135$

Those who play pool know the importance of studying the angles, for pocket billiards is a game of angles. The incoming angle is the smaller angle formed by the cushion and the path of the ball approaching the cushion. The outgoing angle is the smaller angle formed by the cushion and the path of the ball leaving the cushion. If a ball is hit into a cushion without any English (spin), the ball will bounce out at the same angle at which it entered. The incoming angle will always equal the outgoing angle. The same principle, by the way, is true for a ray of light reflecting off of a mirror.

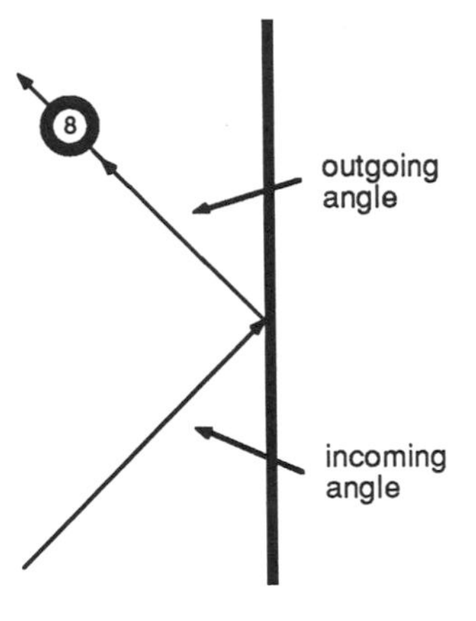

EXERCISE SET B

What is the measure of the incoming angle in each problem below? Which point will the ball hit? Use your protractor to find out.

1.

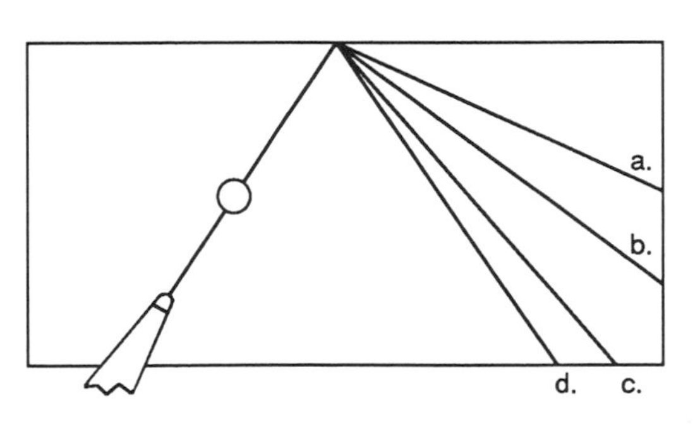

2.

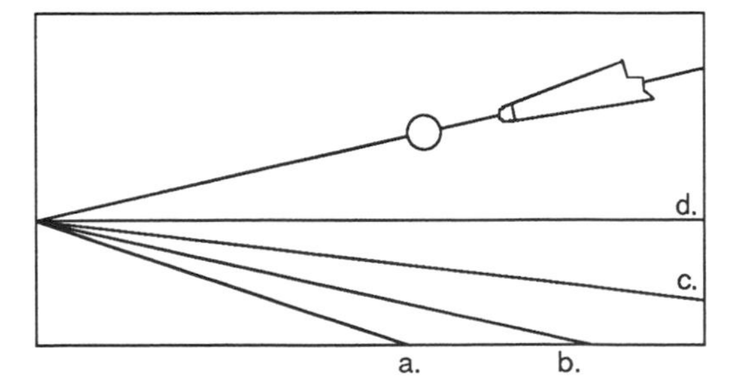

3.* If the four ball is hit as indicated, will it go in the corner pocket? Please do not draw in this book. If you're clever, you can determine the path of the ball using just your protractor and straightedge.

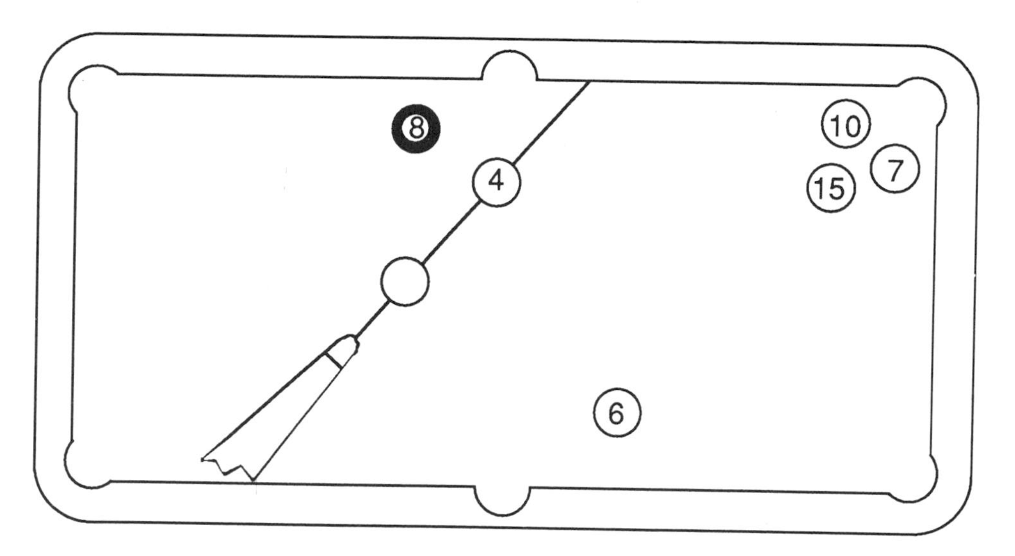

The length or measure of a line segment is the distance between its endpoints. There are two ways to indicate the length of a segment. One way is to list the two endpoints with no symbol above. For example, $MN = 2''$ means that the measure or length of segment MN is two inches. You can also indicate the measure of a segment with a lower case m in front of the line segment symbol as in $m\overline{PQ} = 4.5$ cm. If no units are used when indicating the length of a segment in a diagram, it is understood that the units are arbitrary and that the choice of units is not important in understanding the diagram.

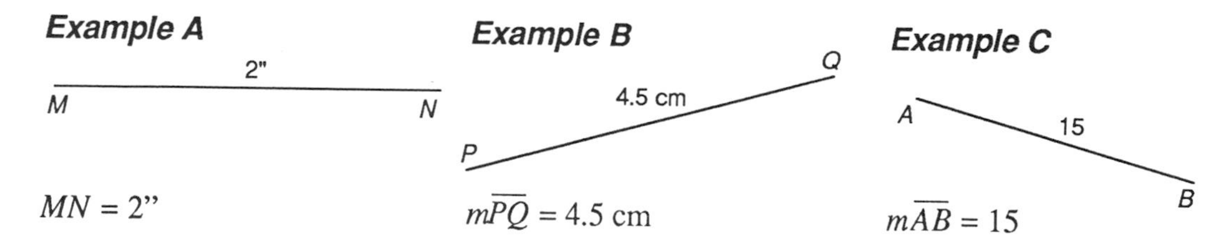

EXERCISE SET C

Use your ruler to determine the measure (to the nearest tenth of a centimeter) of each line segment below. Write your answer in the form $m\overline{AB} \approx$ –?–.

1. A ——————————————————————————— B

2. C ——————————— D

3. E ———————————————————— F

Use your ruler to draw each segment. While it is impossible to draw a segment of a certain length exactly, you should draw each as accurately as you can. Label each segment.

4. $AB = 4.5$ cm 5. $CD = 3$ inches 6. $EF = 24.8$ cm

When you label geometric figures, mark angles and segments that have the same measure with similar markings. For example, in the drawing below, $\angle DOG$ has the same measure as $\angle CAT$. Both angles are marked with an arc and one slash. Also, $m\overline{AB}$ has the same length as $m\overline{AC}$. Both segments are marked with two slashes. To indicate the measure of an angle, place the measure in the interior of the angle. To indicate the length of a segment, you place its measures near the segment.

Example A

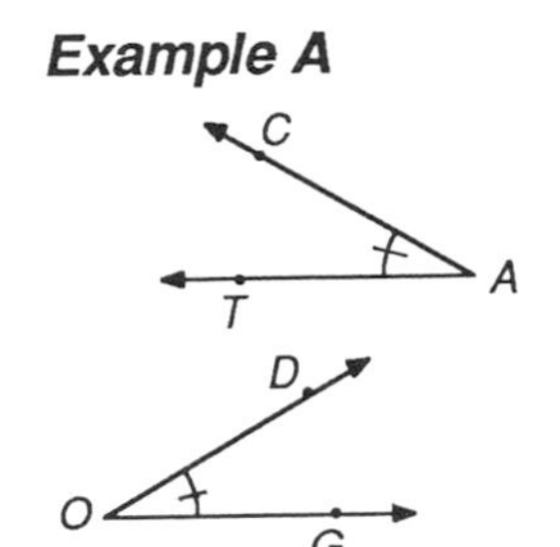

$m\angle DOG = m\angle CAT$

Example B

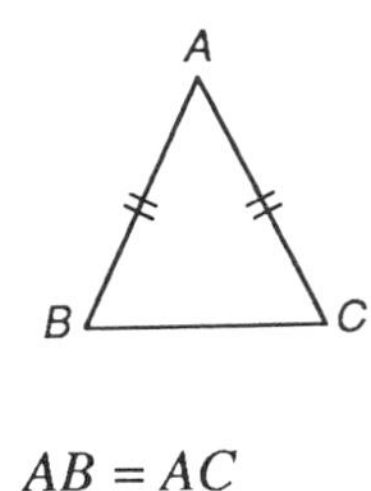

$AB = AC$

Example C

F
S
40°
H
12 cm

$m\angle FSH = 40$
$SH = 12$ cm

Example D

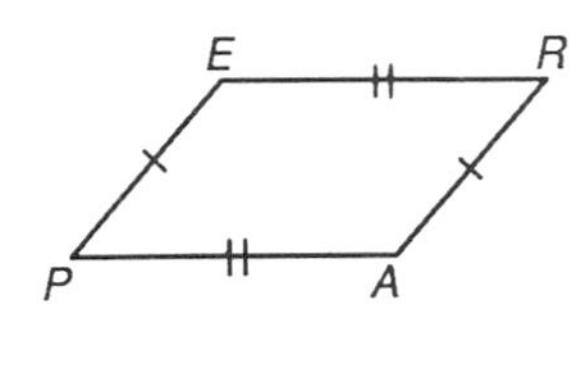

$PA = ER$
$PE = AR$

EXERCISE SET D

Copy each diagram onto a clean sheet of paper. Mark it with the information given.

1.* $m\overline{TH} = 6$ $m\angle THO = 90$
$m\overline{OH} = 8$

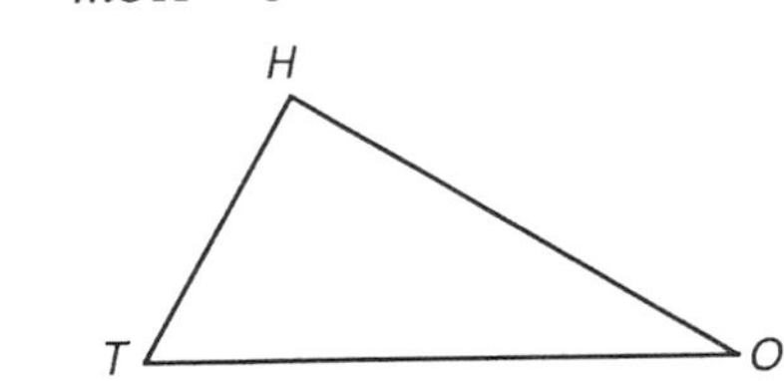

2. $RA = SA$ $m\angle T = m\angle H$
$RT = SH$

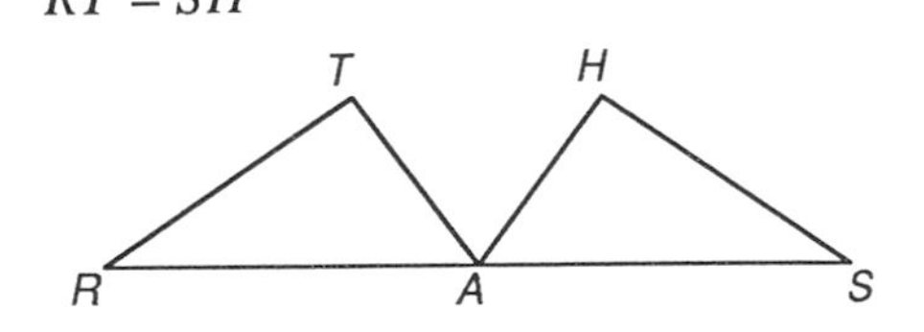

3. $AT = AG$ $m\angle AGT = m\angle ATG$
$AI = AN$ $GI = TN$

G
T
A
I
N

4. $BW = TY$ $m\angle WBT = m\angle YTB$
$WO = YO$ $m\angle BWO = m\angle TYO$

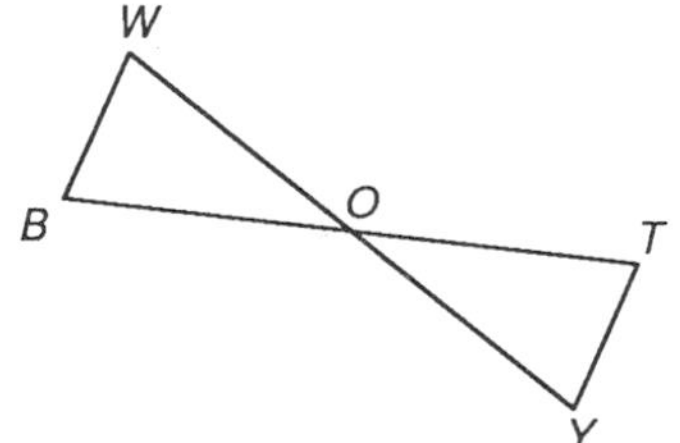

Find the missing information. Use the markings on the drawings. Do not use your protractor or ruler.

5.* $m\overline{AK} = $ –?–
$m\angle A = $ –?–

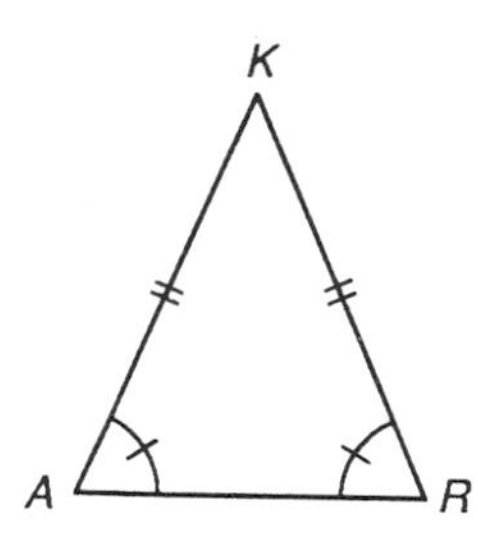

6. $MI = $ –?–
$IC = $ –?–
$m\angle M = $ –?–

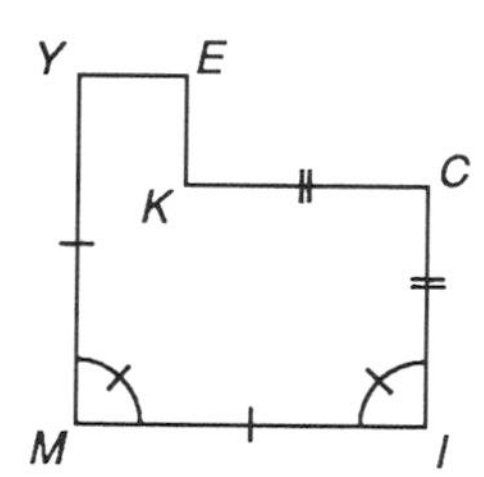

7. $m\angle MEO = $ –?–
$m\angle SUE = $ –?–
$OU = $ –?–

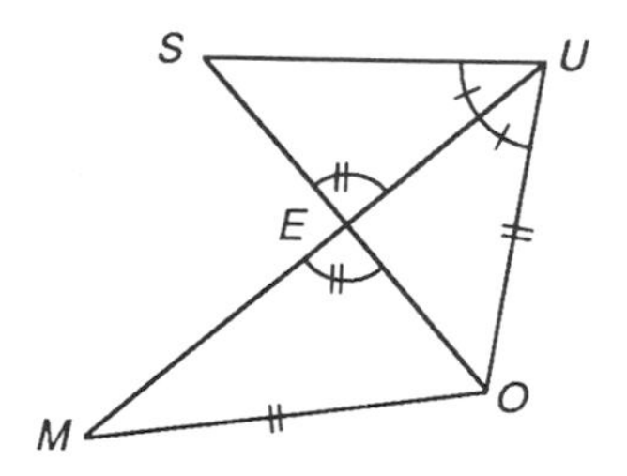

8. $m\overline{LO} = $ –?–
$m\overline{CO} = $ –?–
$m\angle C = $ –?–
$m\angle COD = $ –?–

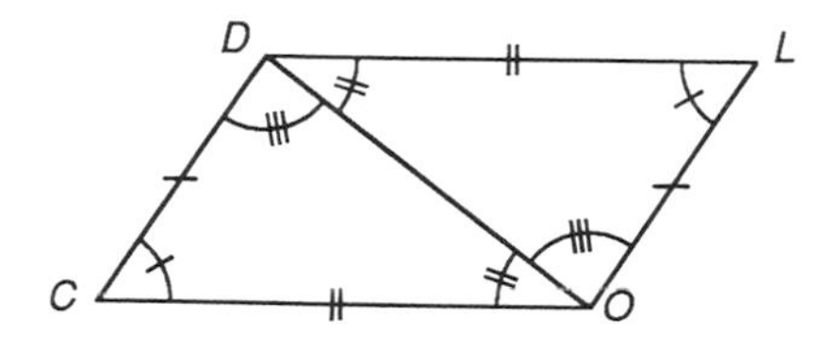

9. If $TA = TK + KN$, what is the sum of the lengths of the sides of quadrilateral $TANK$?

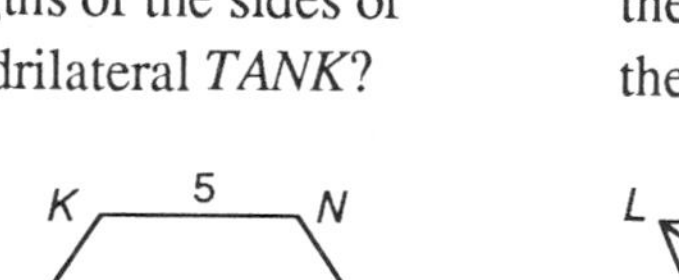

10. If $RO = 12$, $LO = 20$, and $LR = 18$, what is the sum of the lengths of the sides of ΔBOA?

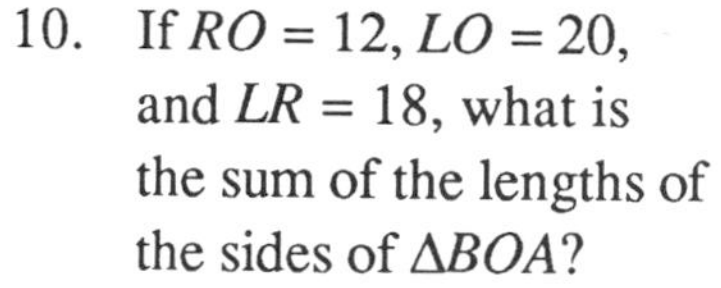

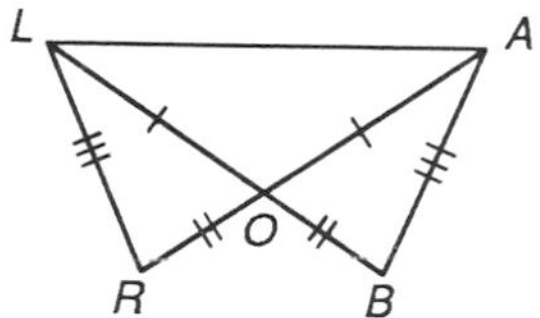

Improving Reasoning Skills

How Is This Possible?

1. The scientist shouted across the lab, "Great! I have found the antidote to the poison my son swallowed." Yet, the poisoned boy's father died three years ago. How is this possible?

2. When the corporation president arrived late to the board meeting, it made the number of men and women attending the meeting equal. When the president's secretary arrived shortly after with the president's spouse, the ratio of men to women became 3 to 2. How is this possible?

Special Project

The Daffynition Game

In this special project you will play a game. The game will sharpen your definition writing skills. It's called the Daffynition Game. Here are the rules:

1. The game is played in groups of four or five.
2. Each group must select a scorekeeper. Then choose a person to start the game. Call this person the *selector*.
3. To begin a round, the selector finds a strange new word in the dictionary. It must be a word that nobody in the group knows. (If you know the word, you should say so. The selector should then pick a new word.)
4. Each person then writes down a make-believe definition that sounds real.
5. The selector collects all the different bogus definitions and mixes them up along with the real definition (copied from the dictionary by the selector).
6. The selector then reads all the definitions out loud, trying to make them all sound as real as possible.
7. Everyone (except the selector) writes on a "guess" card the number of the definition he or she thinks is the real one. You cannot choose your own definition. All players show their guesses at the same time.
8. After scoring, the person to the selector's right becomes the selector for the next round. The game continues until everyone has had a chance to be the selector.
9. The table below explains how to score each round.

If you're the *selector*:	You get 1 point for each person who guessed the wrong definition.
If you're *not* the *selector*:	You get 1 point if you guessed correctly. You get 2 points for each person who guessed your made-up definition.

Sample Scorecard

	Round 1	Round 2	Round 3	Round 4	Total
Player A					
Player B					
Player C					
Player D					

Lesson 2.3
What's a Widget?

"When I use a word," Humpty replied in a scornful tone, "it means just what I choose it to mean – neither more nor less." "The question is," said Alice, "whether you can make a word mean so many different things." "The question is," said Humpty, "which is to be master, that's all."

– Alice's Adventures in Wonderland by Lewis Carroll

In this lesson you will write your own geometric definitions. Before you start, however, you need to understand a little about what makes a *good* definition. Good definitions are very important in geometry.

Example

Study the information, and then identify which creatures in the last group are Orks.

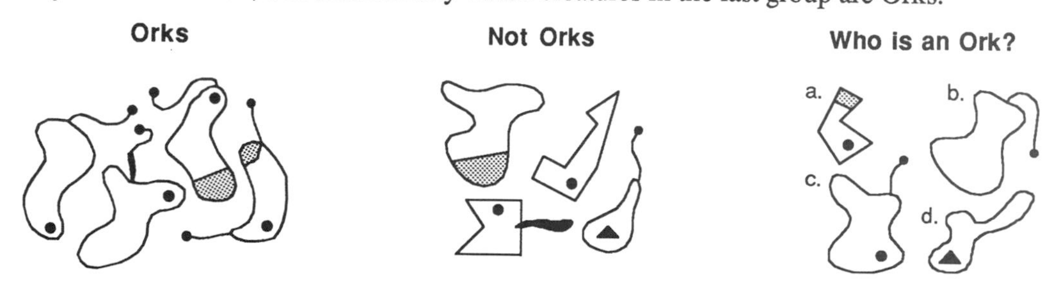

In puzzling over the groups above, you mentally sorted and compared the objects' characteristics. You might have asked yourself, "What things do all the Orks have in common and what things do Orks have that others do not have?" In other words, what characteristics make an Ork an Ork? They all have curved bodies, at least one tail, and a black dot inside. Based on this definition, you should have selected creature c as the only Ork in the last group. Before you go on to define geometric figures, try defining a few more creatures.

EXERCISE SET A

1.* Which creatures in the last group are Widgets? What makes a Widget a Widget?

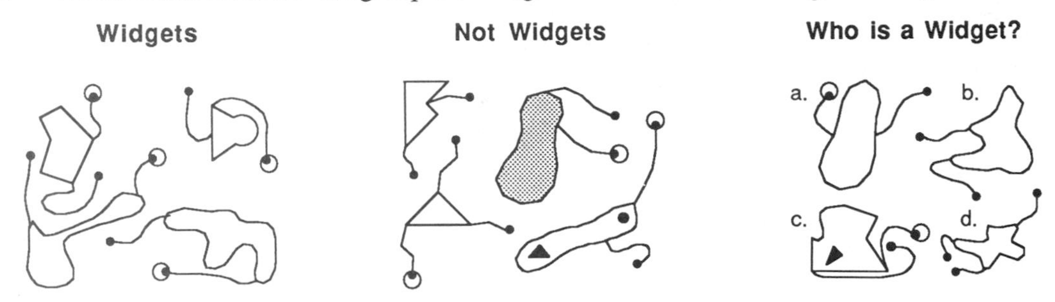

2. Which creatures in the last group are Gorfids? What makes a Gorfid a Gorfid?

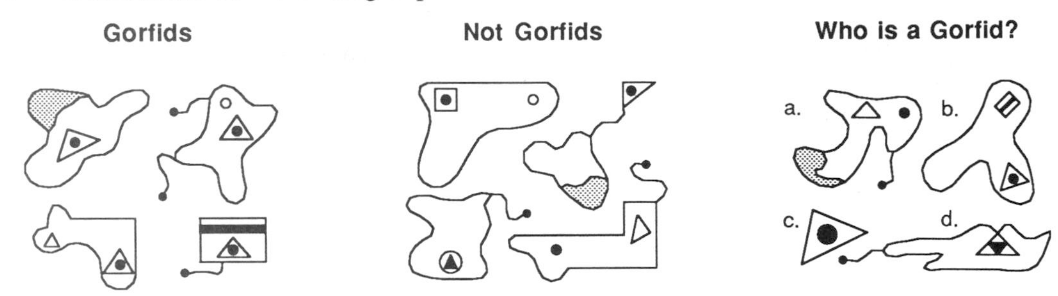

3. Which creatures in the last group are Trebleys? What makes a Trebley a Trebley?

By observing what an Ork is and what an Ork isn't, you identified those characteristics that distinguish an Ork from a non-Ork. This same process can help you write good definitions of geometric figures. When defining a geometric figure, you look for those qualities or characteristics that classify it. Then you look for those qualities that distinguish the figure from other figures in that first classification.

What makes a *good* definition? A good definition must be precise. You should avoid vague or non-mathematical terms such as "some," "about," "small," or "rounded." A good definition places an object to be defined into a class of well defined similar things and then states how it differs from other things in that class.

For example, you can define a *hexagon* as a *polygon* (you have thus classified it) *with exactly six sides* (you say how it differs from other polygons).

Once you've written a definition of a geometric figure, you should test it. That is, you should try to create a figure that meets the criteria of your definition but *isn't* what you're trying to define. You are looking for a counter-example. A **counter-example** is an example that proves a statement wrong. If you can't come up with a counter-example for your definition, that's a sign that you've defined the figure well.

If your definition is written in a form that is reversible, that is also a sign that it is written well. For example, *a hexagon is a polygon with exactly six sides* defines a hexagon. The reverse statement works just as well; *a polygon with exactly six sides is a hexagon.* (Logically every definition is reversible, but sometimes it is difficult to write it in a reversible form.)

EXERCISE SET B

You should be able to write a good definition of each geometric figure below using the information provided. Once you are satisfied with the definition that you have written, discuss it with others near you. See if they can find counter-examples that will reveal a problem with your definition. If they do, refine your definition until you are satisfied with it. Compare your definitions with theirs. Try to arrive at one common definition that your class can agree on. Add this common definition to the definition list that you have started in your notebook. Draw and label a picture to illustrate each definition in your notebook. You and your classmates are responsible for the definitions for this course — there is no glossary of terms in this text. Therefore it is very important that you keep an accurate and updated notebook listing the class accepted definitions.

1.* Define *right angle.*

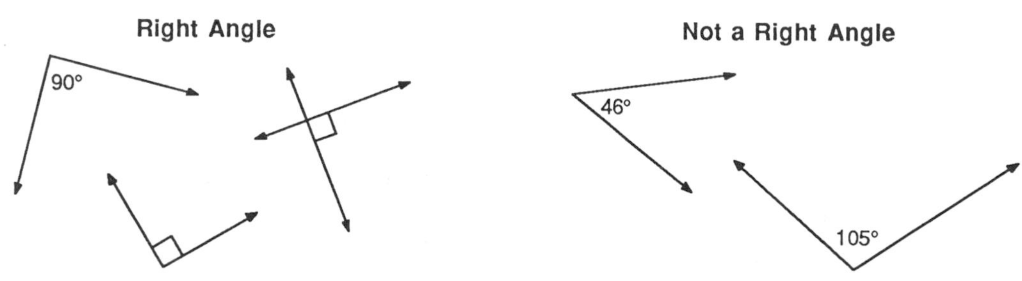

Note: A small square in the corner of an angle indicates that it measures 90°.

2.* Define *acute angle.*

Acute Angle

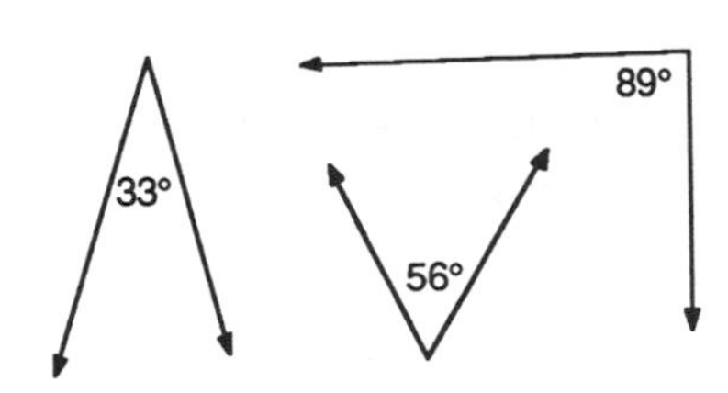

Not an Acute Angle

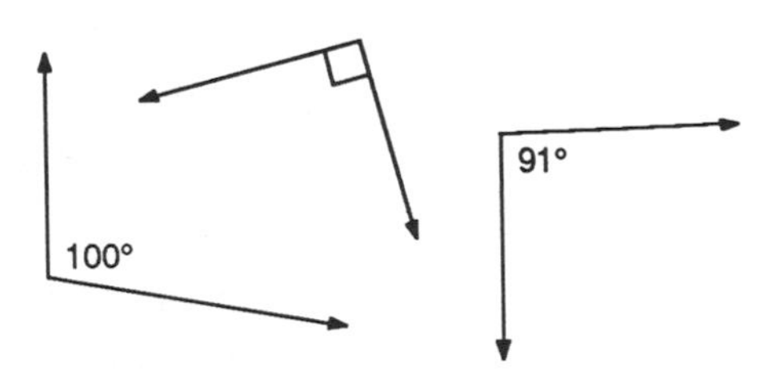

3. Define *obtuse angle.*

Obtuse Angle

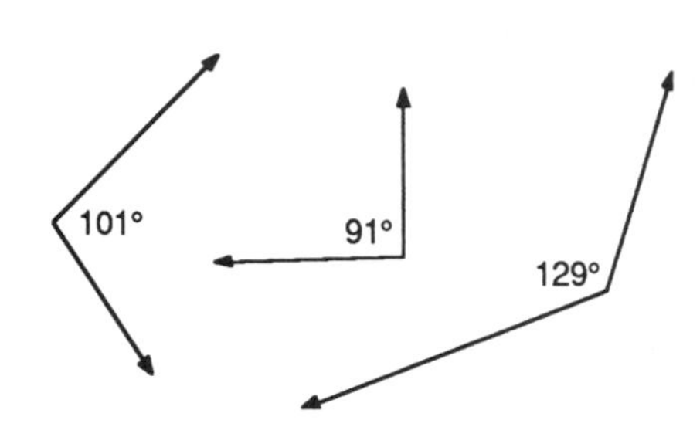

Not an Obtuse Angle

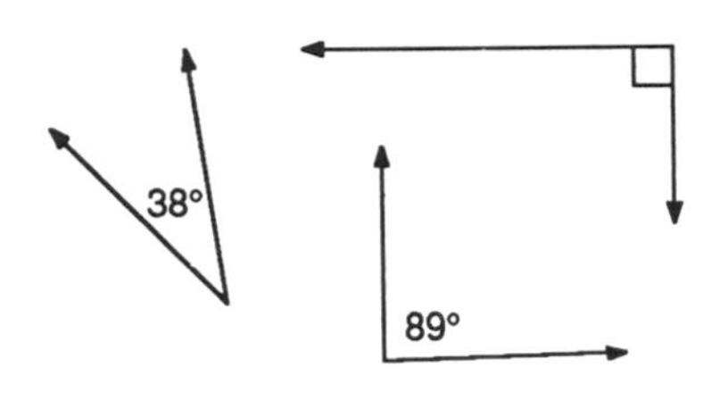

4. Define *midpoint of a segment.*

Midpoint of a Segment

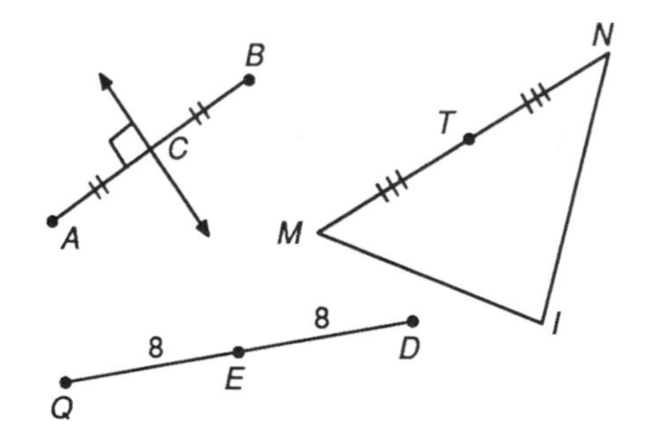

Not a Midpoint of a Segment

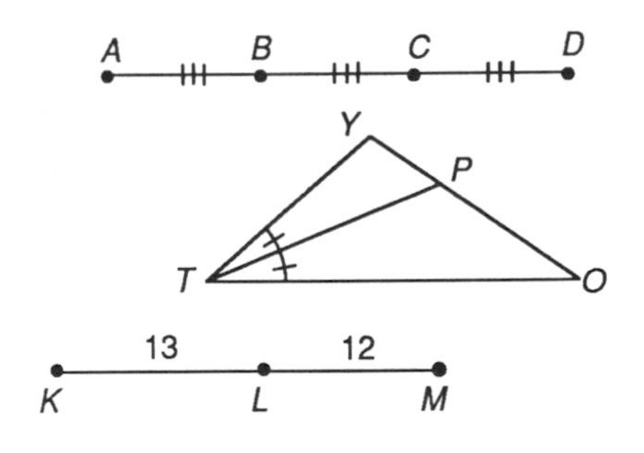

C is a midpoint of segment *AB*.
T is a midpoint of segment *MN*.
E is a midpoint of segment *QD*.

B and *C* are not midpoints of segment *AD*.
P is not a midpoint of segment *OY*.
L is not a midpoint of segment *KM*.

5. Define *angle bisector*.

Angle Bisector

Not an Angle Bisector

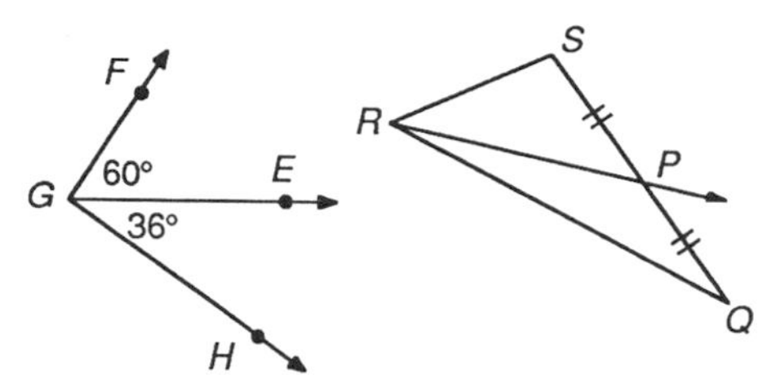

Ray *CD*, ray *OF*, and ray *MN* are angle bisectors.

Ray *GE* and ray *RP* are not angle bisectors.

EXERCISE SET C

Draw and carefully label each of the figures below. Use the special marks to indicate right angles, and segments and angles equal in measure. Use a protractor and ruler when necessary.

1. Acute angle *DOG* with a measure of 45 degrees
2. Right angle *RTE*
3. Obtuse angle *BIG* with angle bisector *IE*
4. Line segment *OY* with midpoint *L*

Improving Visual Thinking Skills

Design I

Use your geometry tools to create a larger version of the design on the right.

Then decorate your design.

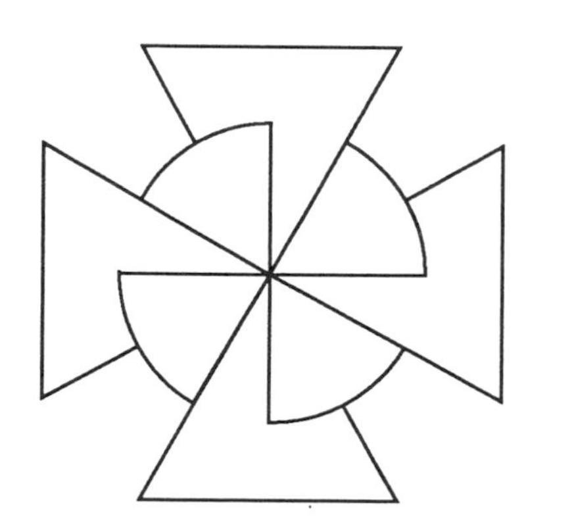

Lesson 2.4

Defining Line and Angle Relationships

> *There are two kinds of people in this world: those who divide everything into two groups, and those who don't.*
>
> – *Kenneth Boulding*

EXERCISE SET A

Write a good definition for each geometric term or figure below. Once you are satisfied with your definitions, discuss them with others. Try to arrive at one common definition that you can agree on for each term. Add these definitions to the definition list in your notebook. Draw and label a picture to illustrate each definition.

1. Define *collinear points.*

Collinear Points

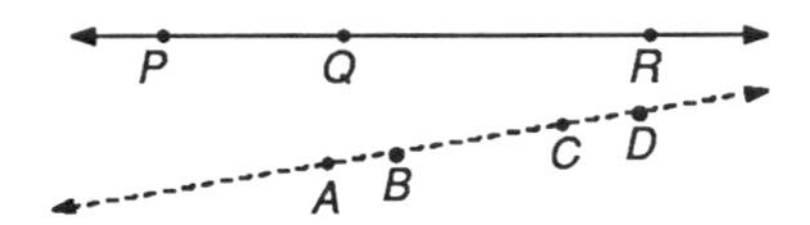

P, *Q*, and *R* are collinear.
A, *B*, *C*, and *D* are collinear.

Not Collinear Points

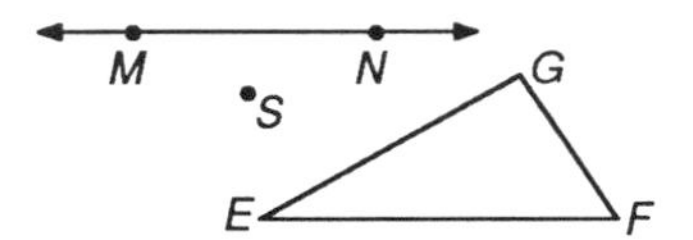

M, *N*, and *S* are not collinear.
E, *F*, and *G* are not collinear.

2. Define *coplanar points.*

Coplanar Points

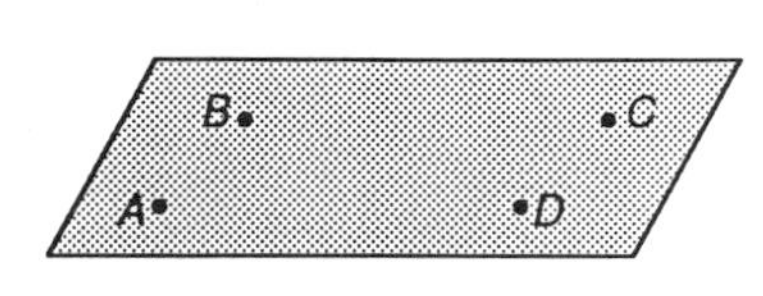

A, *B*, *C*, and *D* are coplanar.

Not Coplanar Points

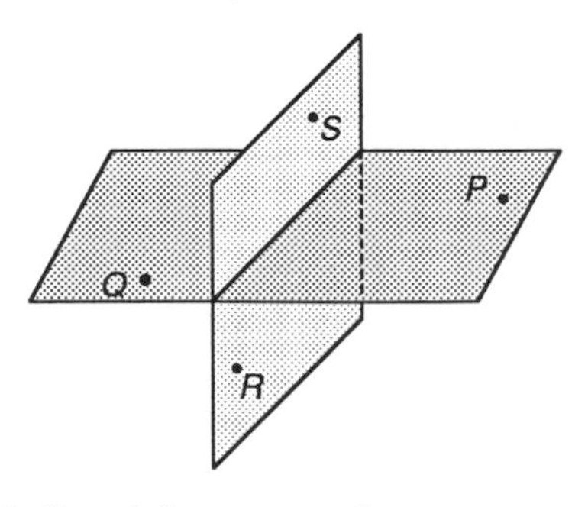

P, *Q*, *R*, and *S* are not coplanar.

3. Define *parallel lines.*

Parallel Lines

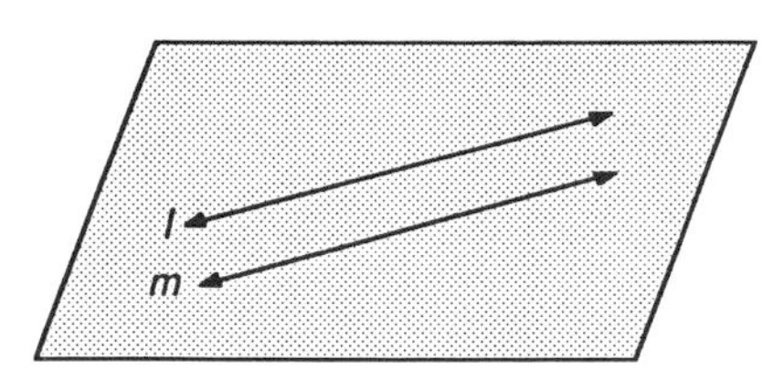

$l \parallel m$

Not Parallel Lines

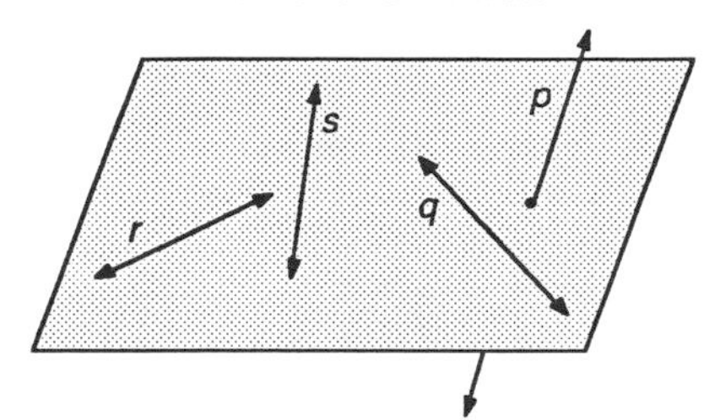

Line *r* is not parallel to line *s*.
Line *p* is not parallel to line *q*.

Note: // means *is parallel to.*

4. Define *perpendicular lines.*

Perpendicular Lines

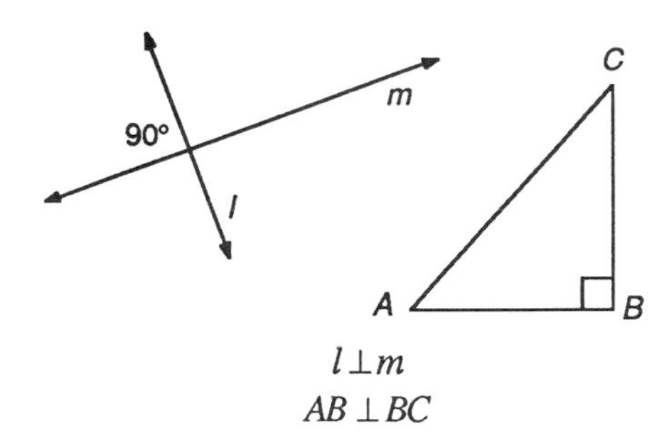

$l \perp m$
$AB \perp BC$

Not Perpendicular Lines

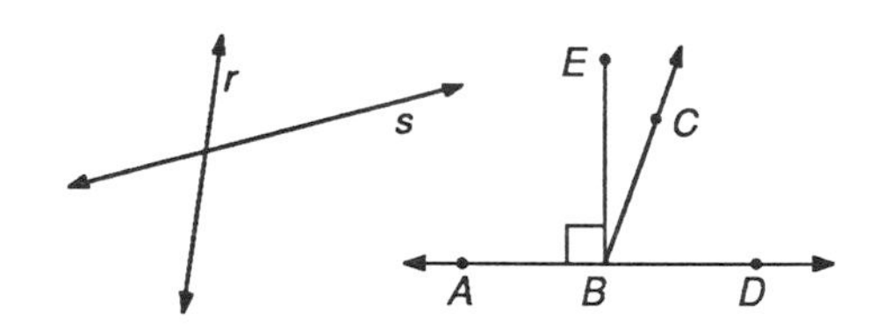

Line *r* is not perpendicular to line *s*.
Ray *BC* is not perpendicular to line *AD*.

Note: ⊥ means *is perpendicular to.*

5. Define *pair of complementary angles.*

Pair of Complementary Angles

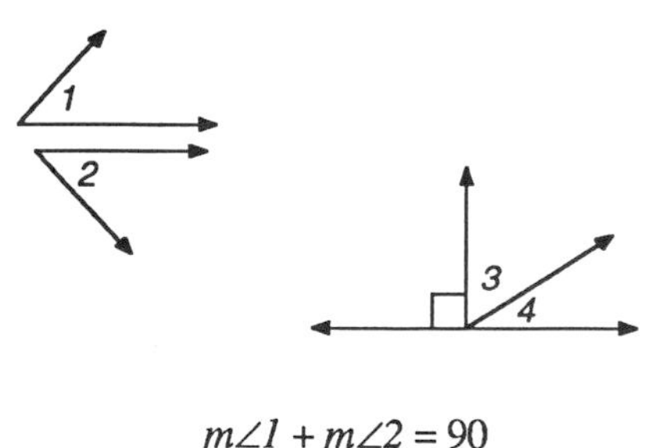

$m\angle 1 + m\angle 2 = 90$
$m\angle 3 + m\angle 4 = 90$

Not a Pair of Complementary Angles

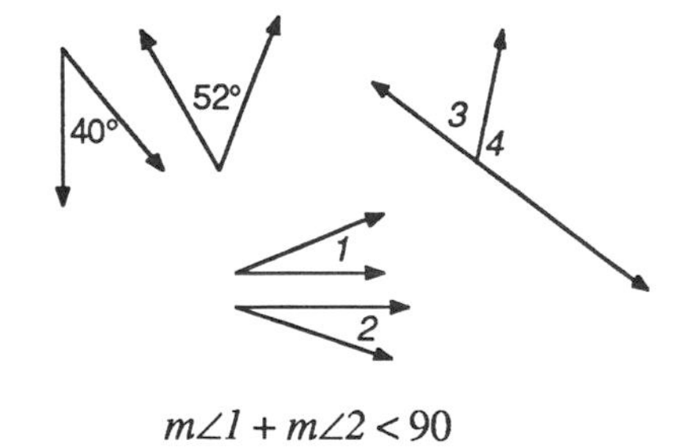

$m\angle 1 + m\angle 2 < 90$

Note: There is a third way to identify angles. Sometimes, when you wish to identify angles in a diagram, you name them by number.

6. Define *pair of supplementary angles.*

Pair of Supplementary Angles

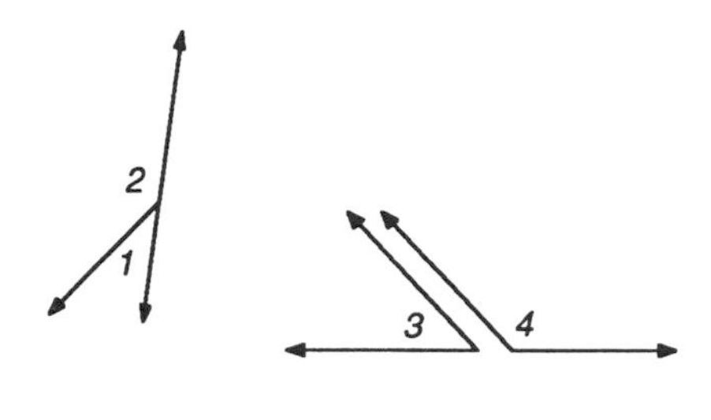

$m\angle 1 + m\angle 2 = 180$
$m\angle 3 + m\angle 4 = 180$

Not a Pair of Supplementary Angles

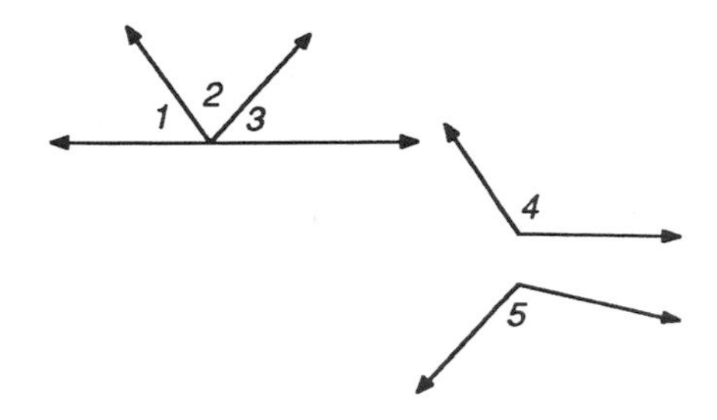

$m\angle 1 + m\angle 2 < 180$
$m\angle 4 + m\angle 5 > 180$

Quite often geometric definitions are easier to write if you refer to named figures. For example, you can define a line segment by saying: *$\overline{AB}$ consists of points A and B and all points on $\overleftrightarrow{AB}$ that are between A and B.* You can define a ray by saying: *$\overrightarrow{AB}$ consists of $\overline{AB}$ and all other points P on $\overleftrightarrow{AB}$ such that B is between A and P.* You can define the midpoint of a line segment by saying: *M is a midpoint of $\overline{AB}$ if M is a point on $\overline{AB}$ and AM = MB.* The definitions for the next two terms are easier to write if you use this technique.

7.* Define *pair of vertical angles.*

Pair of Vertical Angles

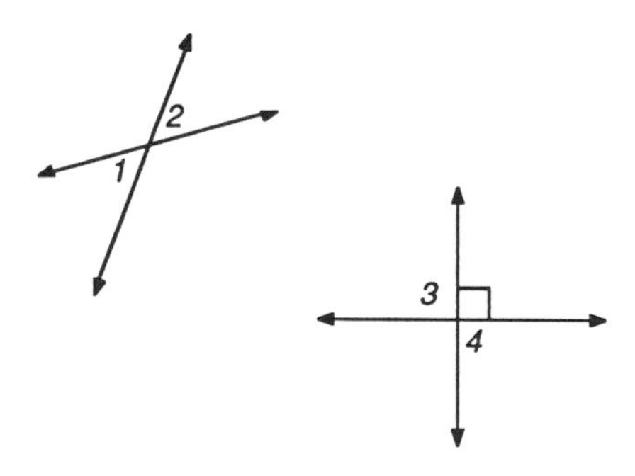

$\angle 1$ and $\angle 2$ are a pair of vertical angles.
$\angle 3$ and $\angle 4$ are also vertical angles.

Not a Pair of Vertical Angles

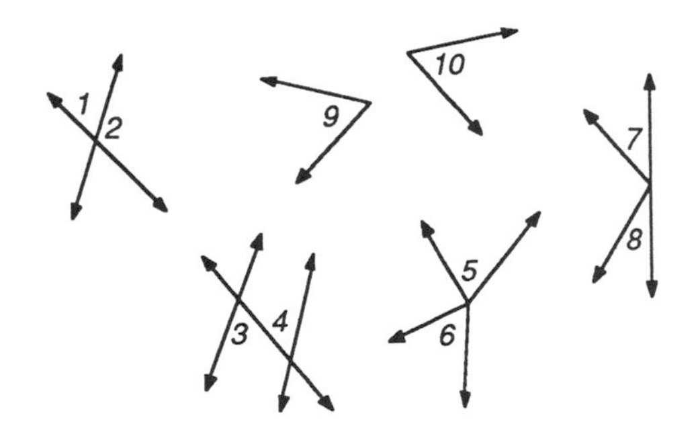

$\angle 1$ and $\angle 2$ are not vertical angles.
$\angle 3$ and $\angle 4$ are not vertical angles.
$\angle 5$ and $\angle 6$ are not vertical angles.
$\angle 7$ and $\angle 8$ are not vertical angles.
$\angle 9$ and $\angle 10$ are also not vertical angles.

8.* Define *linear pair of angles.*

Linear Pair of Angles

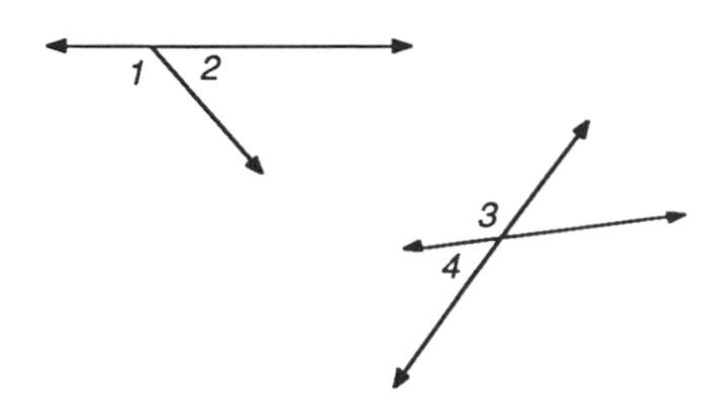

∠*1* and ∠*2* are a linear pair of angles.
∠*3* and ∠*4* are a linear pair of angles.

Not a Linear Pair of Angles

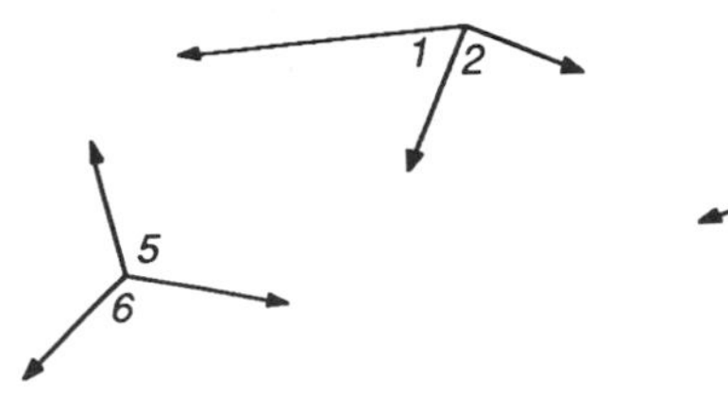

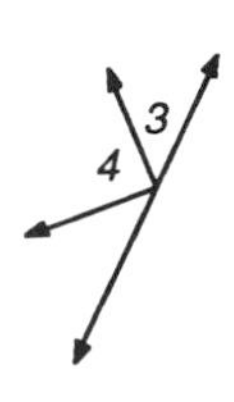

∠*1* and ∠*2* are not a linear pair of angles.
∠*3* and ∠*4* are not a linear pair of angles.
∠*5* and ∠*6* are not a linear pair of angles.

EXERCISE SET B

This exercise set will help you visualize relationships between geometric figures in the plane and in space. All the statements below except three are true. Make a sketch or use physical objects to demonstrate each true statement. For the three false statements, produce a counter-example demonstrating that each is false. Pencil tips and thumb tacks can represent points. Rulers, pencils, or stiff wires can represent lines.

1.* For every line segment there is one and only one midpoint.

2. For every angle there is one and only one angle bisector.

3. If two different lines intersect, then they intersect at one and only one point.

4. If two different circles intersect, then they intersect at one and only one point.

5.* There is one and only one line perpendicular to a given line through a given point on the given line.

6. In a plane, there is one and only one line perpendicular to a given line through a given point on the given line.

7. There is one and only one line perpendicular to a given line through a given point not on the given line.

8. For every triangle there is one and only one right angle.

9. Through a given point not on a given line there is one and only one line that can be constructed parallel to the given line.

EXERCISE SET C

Sketch and carefully label each of the figures below. Use a protractor and ruler when necessary.

1. Collinear points P, T, and S
2. Parallel lines $\overleftrightarrow{PA}$ and $\overleftrightarrow{LE}$
3. $\overleftrightarrow{PE}$ perpendicular to $\overline{AR}$
4. Complementary angles $\angle A$ and $\angle B$ with $m\angle A = 40$
5. Supplementary angles $\angle C$ and $\angle D$ with $m\angle D = 40$

Improving Reasoning Skills

Spelling Card Trick

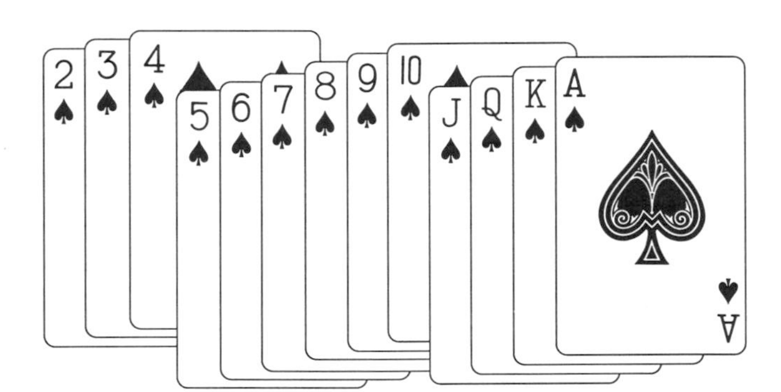

The card trick below uses one complete suit from a regular deck of playing cards. How must you arrange the cards so that you can successfully complete the trick? Here is what your audience should see and hear as you perform.

1. As you take the top card and place it on the bottom of the pile, say "*A*."
2. Then take the second card, place it on the bottom of the pile, and say "*C*."
3. Take the third card, place it on the bottom, and say "*E*."
4. You've just spelled *ACE*. Now take the fourth card and turn it face up on the table. It should be an ace.
5. Continue in this fashion, saying "*T*," "*W*," and "*O*" for the next three cards. Then turn the next card face up. It should be a two.
6. Continue spelling *THREE, FOUR, . . . , JACK, QUEEN, KING*. Each time you spell a card, the next card turned face up should be that card.

Lesson 2.5

Defining Polygons

> *A* ***polygon*** *is a closed geometric figure in a plane formed by connecting line segments endpoint to endpoint with each segment intersecting exactly two others.*

Each line segment is called a **side** of the polygon. Each endpoint where the sides meet is called a **vertex** of the polygon.

Polygons

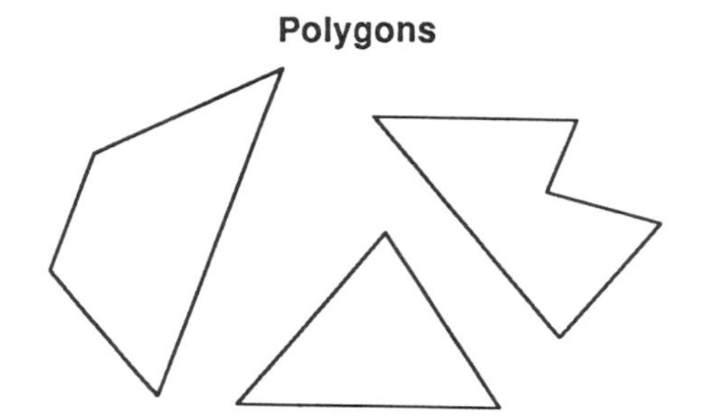

Not Polygons

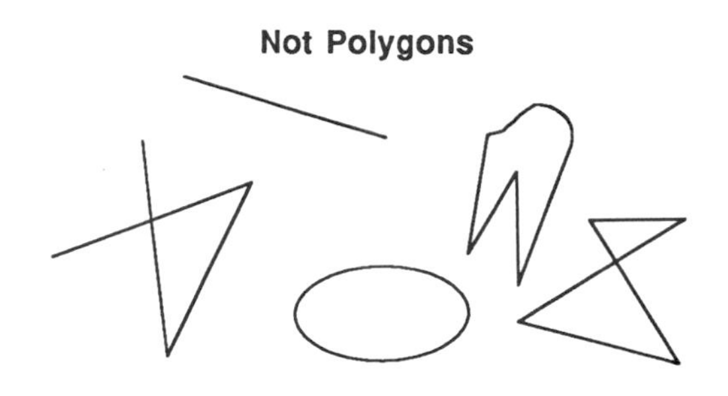

A **convex polygon** is a polygon in which no segment connecting two vertices is outside the polygon. A **concave polygon** is a polygon in which at least one segment connecting two vertices is outside the polygon. From here on in this book when we speak of a *polygon,* we will always mean a *convex polygon.*

Convex Polygons

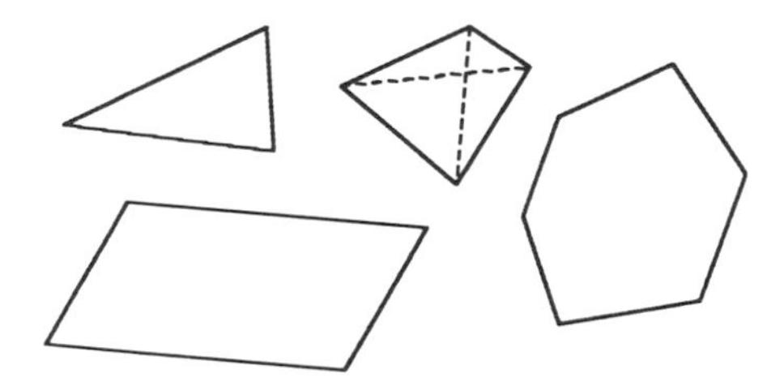

Concave Polygons

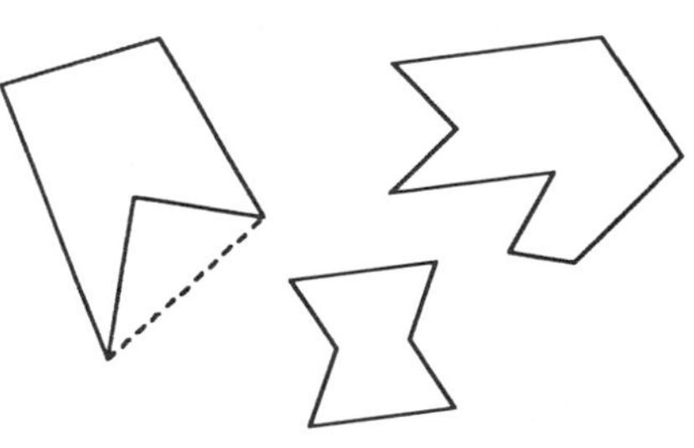

You classify polygons by the number of sides they have.

Sides	Classification
3	**Triangle**
4	**Quadrilateral**
5	**Pentagon**
6	**Hexagon**

Sides	Classification
7	**Heptagon**
8	**Octagon**
9	**Nonagon**
10	**Decagon**

Sides	Classification
11	**Undecagon**
12	**Dodecagon**
n	***n*-gon**

When referring to a specific polygon, you list in succession the capital letters representing consecutive vertices. For example, the hexagon on the right can be referred to as hexagon *ABCDEF*. You could also call it hexagon *DCBAFE*.

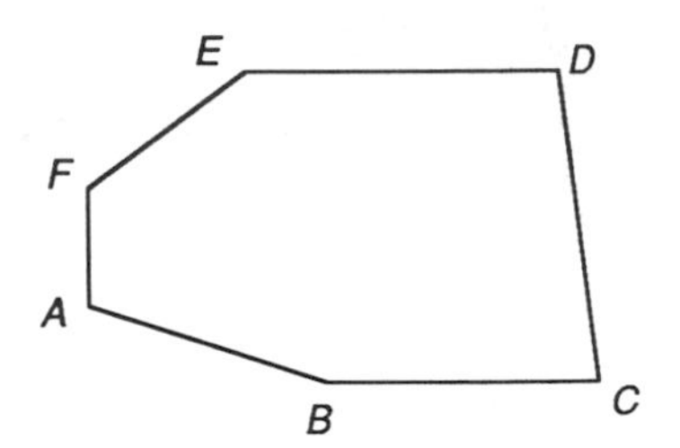

EXERCISE SET A

Classify each polygon.

1.* 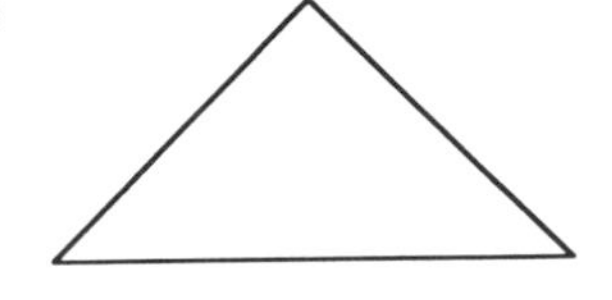

2. 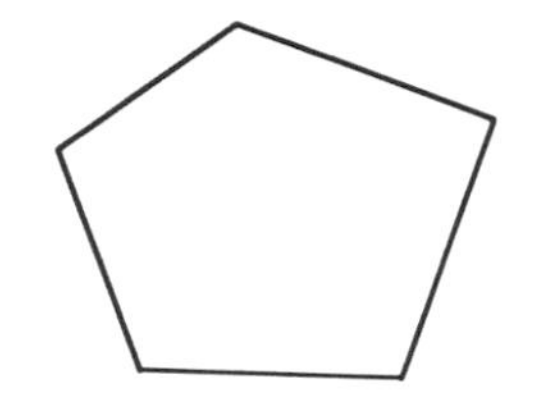

3. 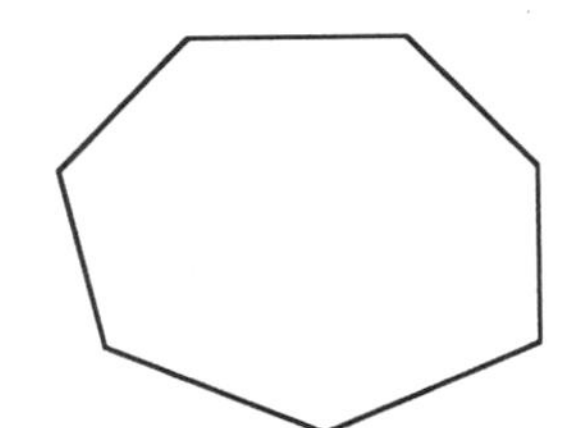

4. 

5. 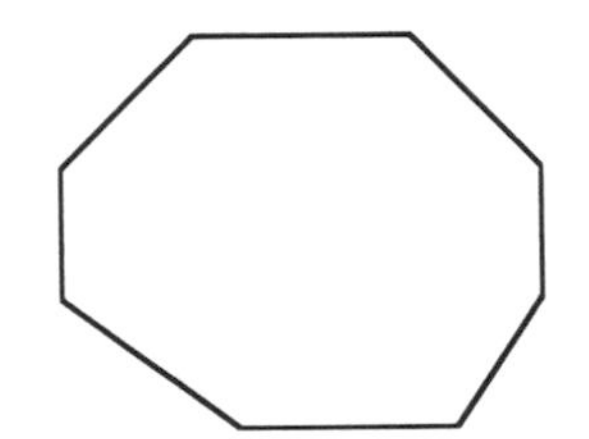

6. 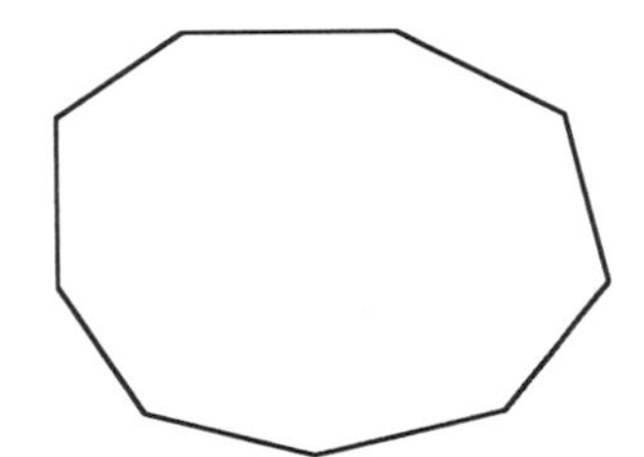

Drawing Tip

Drawing a Polygon

Do you remember your very first attempt at drawing a polygon with eight or more sides? You might have ended up with a spiral like the one shown at right. To draw polygons easily, first, lightly sketch a circle. Next, place the points you need on the circle. Then, connect the points to form your polygon. Finally, erase the circle.

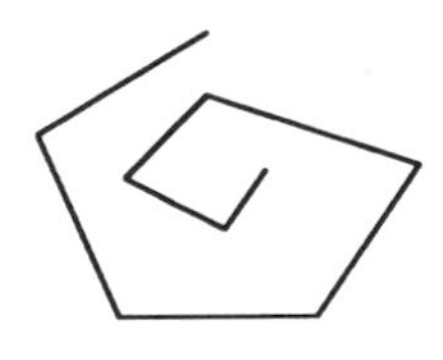

Yep!

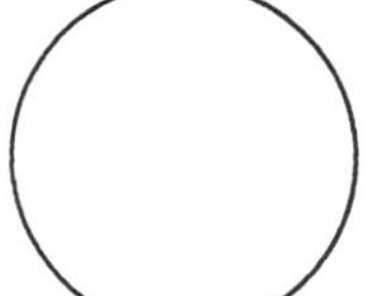

Step 1

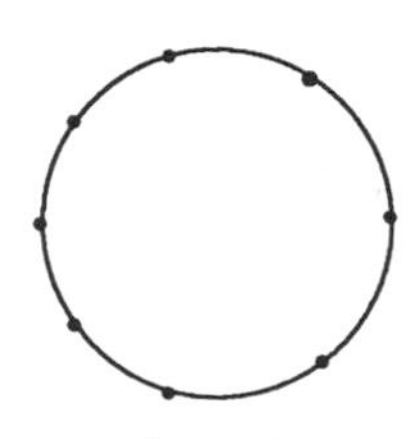

Step 2

Step 3

Step 4

Draw an example of each polygon.

7. Quadrilateral 8. Decagon 9. Dodecagon 10. Octagon

Give one possible name for each polygon.

11.*

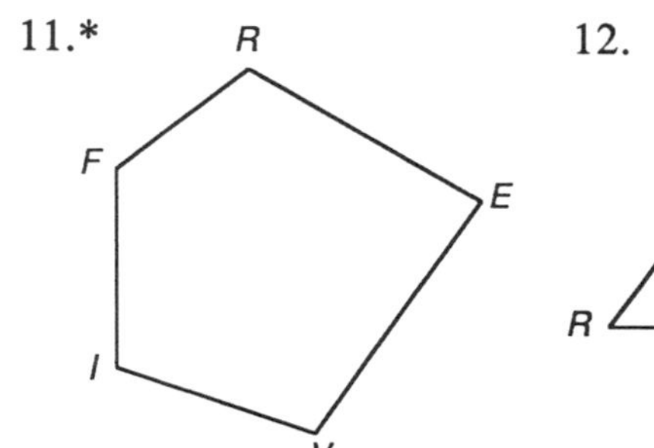

12.

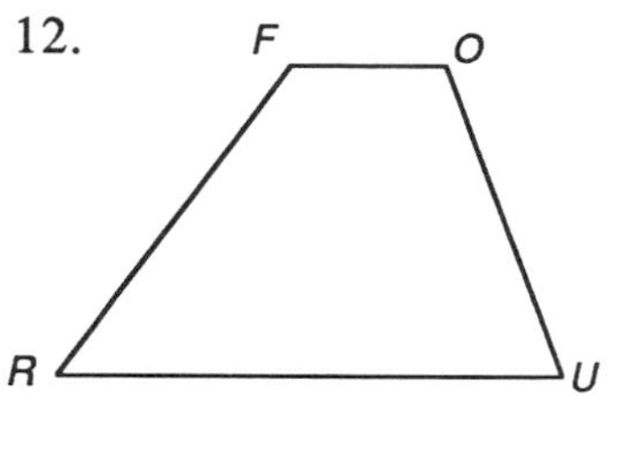

13.

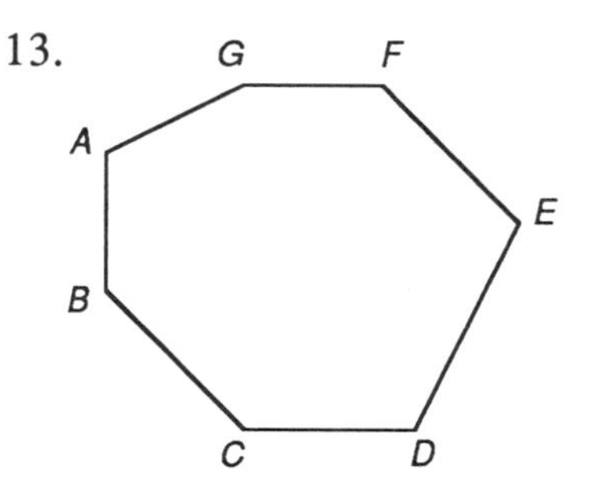

If two vertices of a polygon are connected by a side, then they are **consecutive vertices** in the polygon. If two sides share a common vertex, then they are **consecutive sides**. If two angles share a common side, then they are **consecutive angles**.

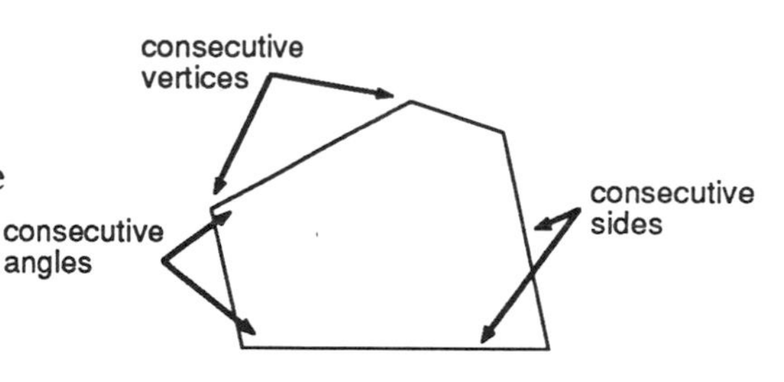

EXERCISE SET B

Write a good definition of each geometric term or figure below. Once you are satisfied with the definitions that you have written, discuss them with others near you. Try to arrive at one common definition that your class can agree on for each term. Add these definitions to the definition list in your notebook. Draw and label a picture to illustrate each definition.

1. Define *diagonal of a polygon.*

Diagonal of a Polygon

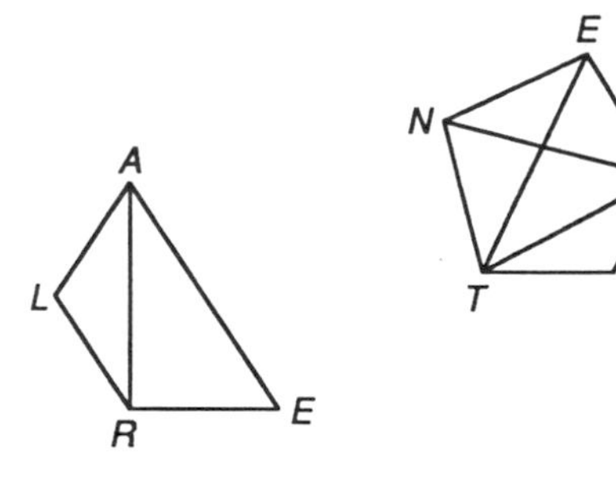

Segments *AR*, *PN*, *TE*, and *PT* are diagonals.

Not a Diagonal of a Polygon

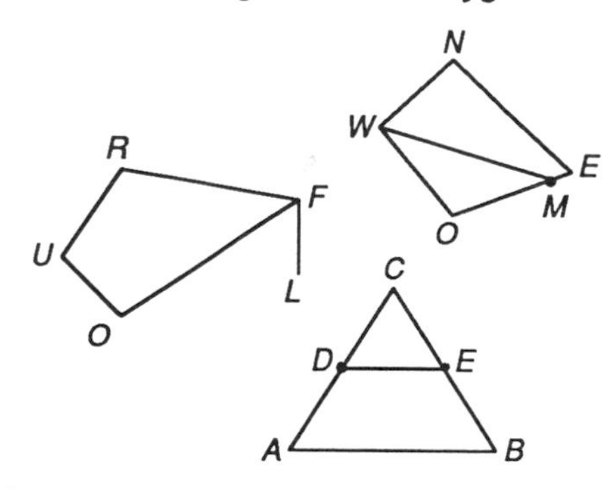

Segments *FL*, *FO*, *WM*, and *DE* are not diagonals.

2. Define *equilateral polygon.*

Equilateral Polygon | **Not an Equilateral Polygon**

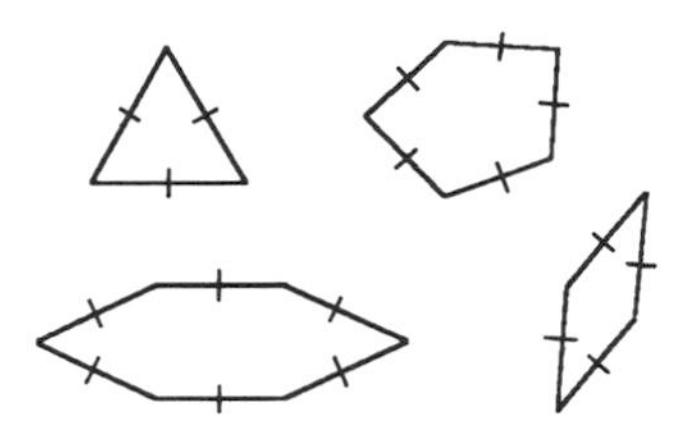

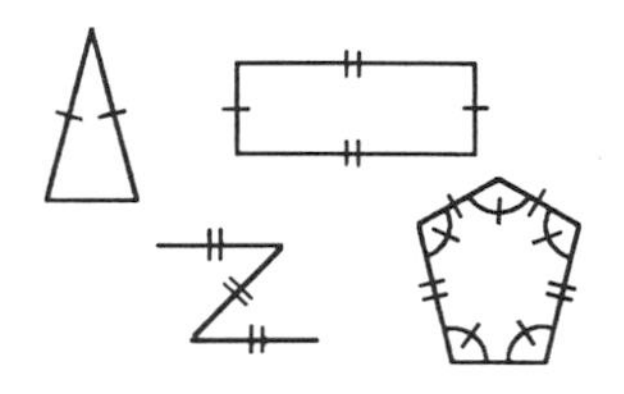

3. Define *equiangular polygon.*

Equiangular Polygon | **Not an Equiangular Polygon**

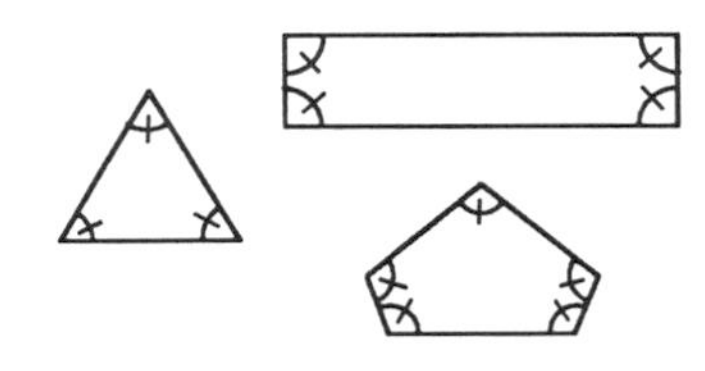

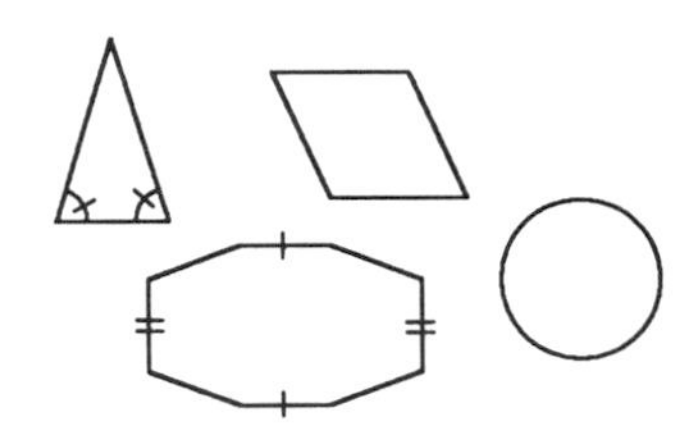

4.* Define *regular polygon.*

Regular Polygon | **Not a Regular Polygon**

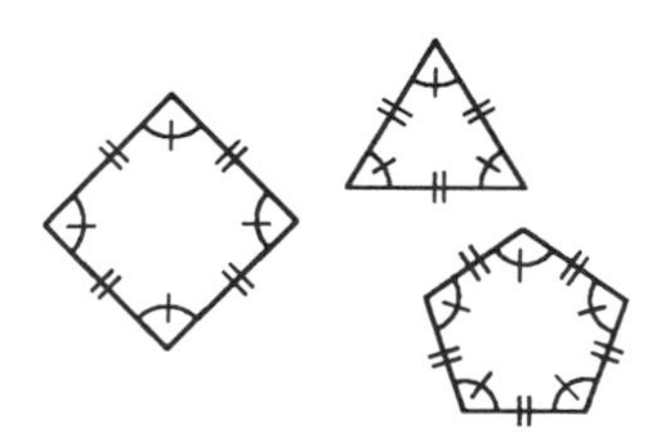

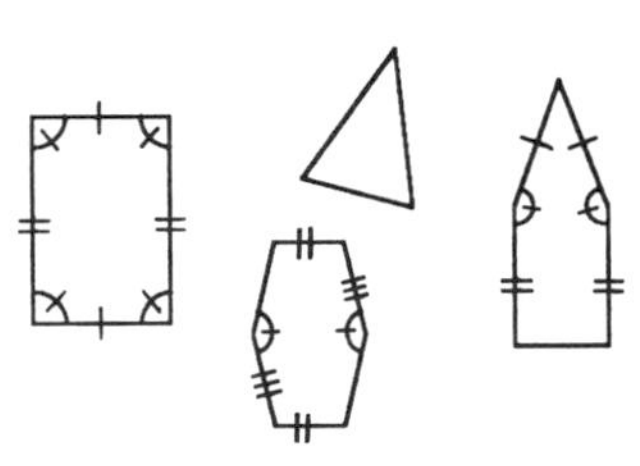

EXERCISE SET C

Sketch and carefully label each figure.

1. Pentagon $PENTA$ with $PE = EN$
2. Hexagon $NGAXEH$ with $m\angle HEX = m\angle EXA$
3. Complementary angles $\angle ABC$ and $\angle DEF$ with $m\angle ABC = 75$
4. Supplementary angles $\angle RAT$ and $\angle MSE$ with $m\angle RAT = 75$
5. $\overleftrightarrow{AB}$, $\overleftrightarrow{CD}$, and $\overline{EF}$ with $\overleftrightarrow{AB} \parallel \overleftrightarrow{CD}$ and $\overleftrightarrow{CD} \perp \overline{EF}$
6. Equiangular quadrilateral $QUAD$ with $QU \neq QD$

Lesson 2.6

Defining Triangles

You have already seen shortcut symbols for *parallel* and *perpendicular*. In this lesson you will learn another shortcut symbol. The Δ symbol represents the word *triangle*. ΔABC means the same as triangle ABC.

EXERCISE SET A

Write a good definition of each geometric figure or term. Discuss them with others in your class. Agree on a common set of definitions for your class and add them to your definition list. Draw and label a picture to illustrate each definition.

1.* Define *right triangle.*

Right Triangle

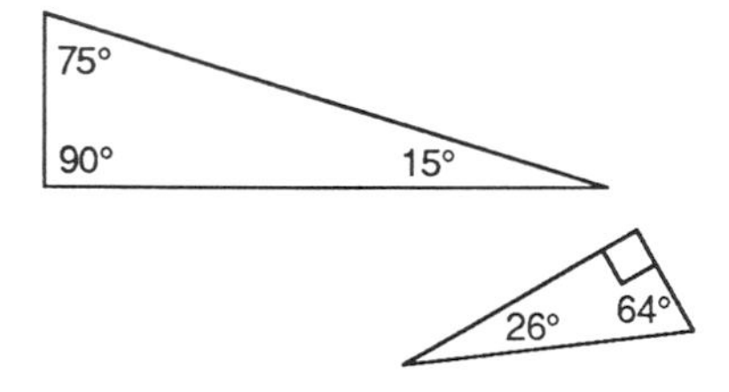

Not a Right Triangle

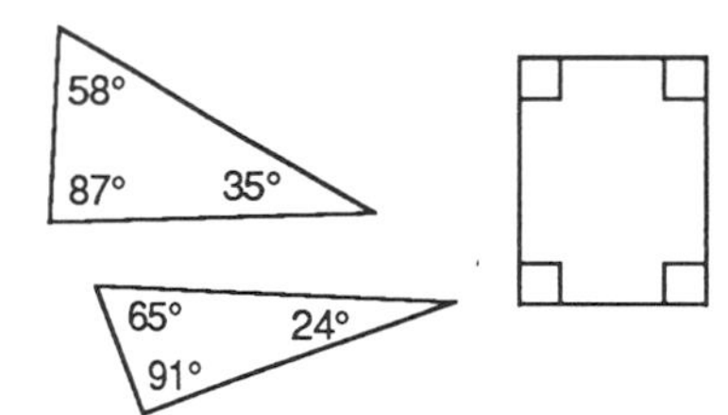

2. Define *acute triangle.*

Acute Triangle

Not an Acute Triangle

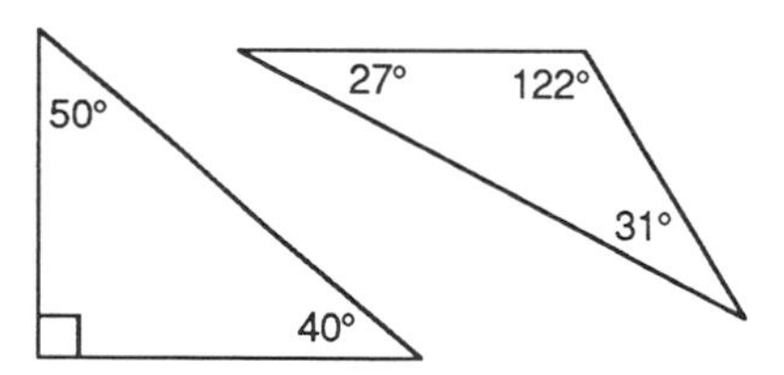

3. Define *obtuse triangle.*

Obtuse Triangle

Not an Obtuse Triangle

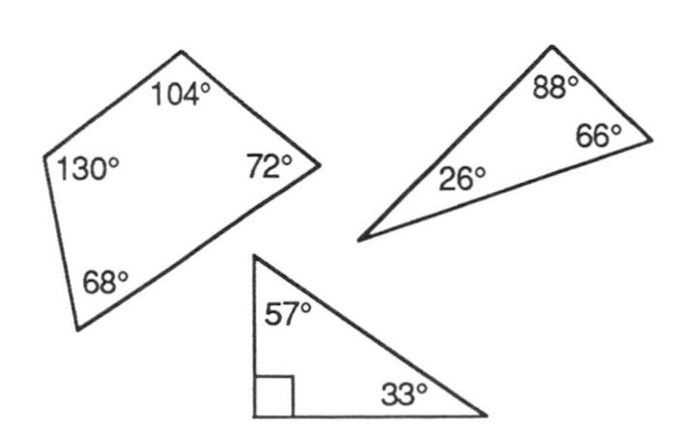

4. Define *scalene triangle.*

Scalene Triangle

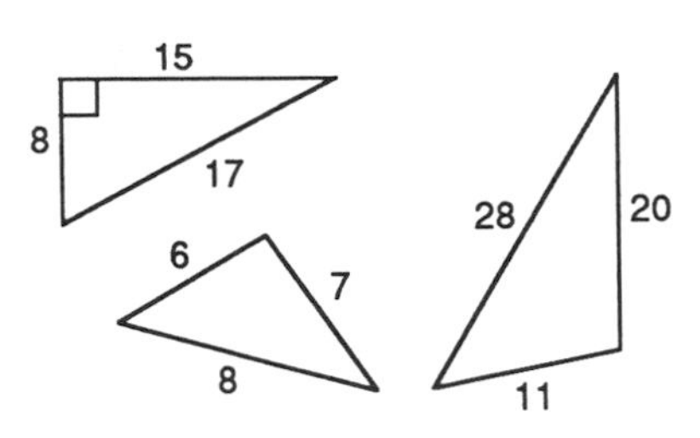

Not a Scalene Triangle

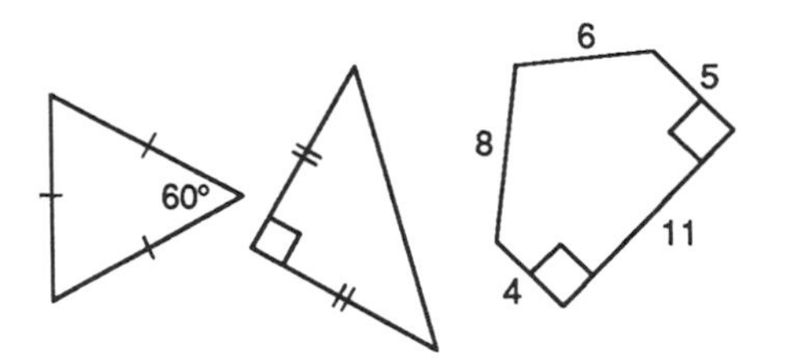

5. Define *isosceles triangle.*

Isosceles Triangle

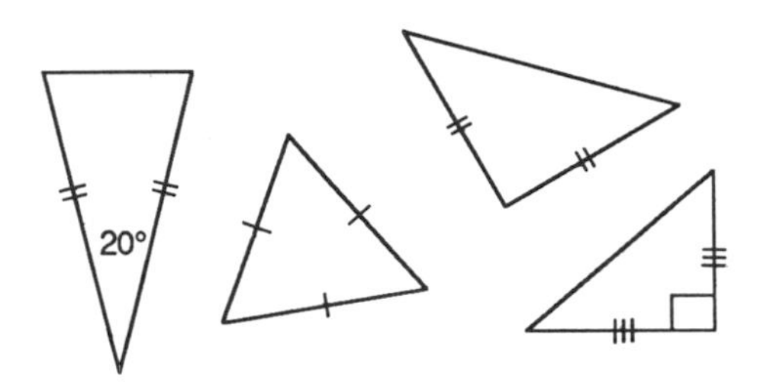

Not an Isosceles Triangle

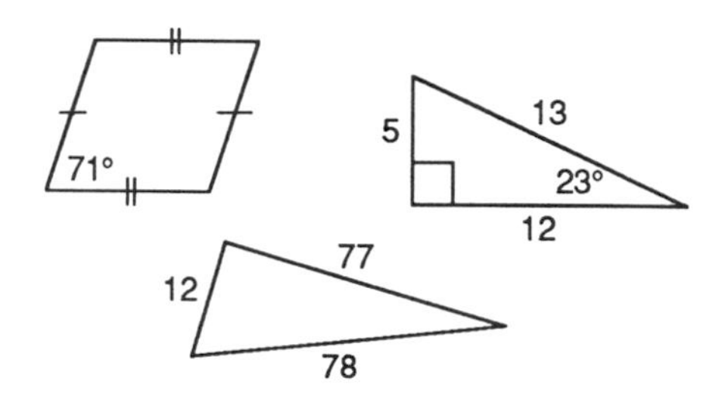

In an isosceles triangle, the angle between the two sides of equal length is called the **vertex angle**. The side opposite the vertex angle is called the **base** of the isosceles triangle. The two angles opposite the two sides of equal length are called the **base angles** of the isosceles triangle.

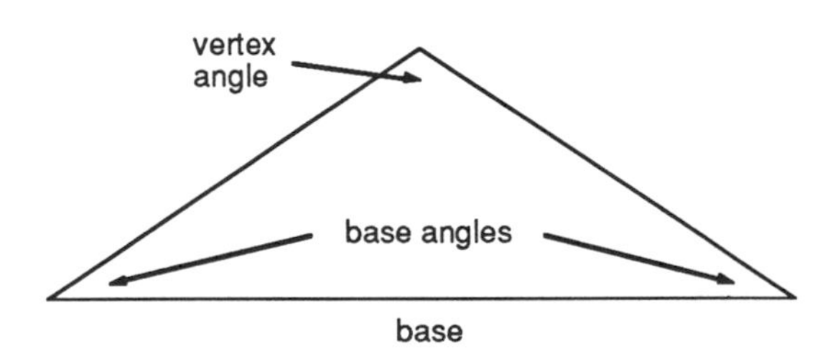

6. Define *equilateral triangle.*

Equilateral Triangle

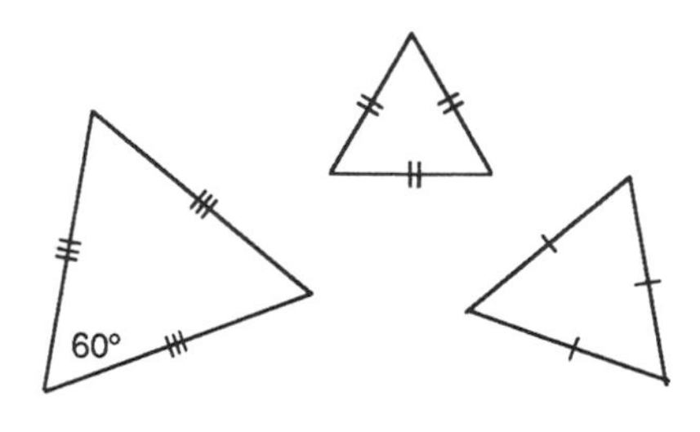

Not an Equilateral Triangle

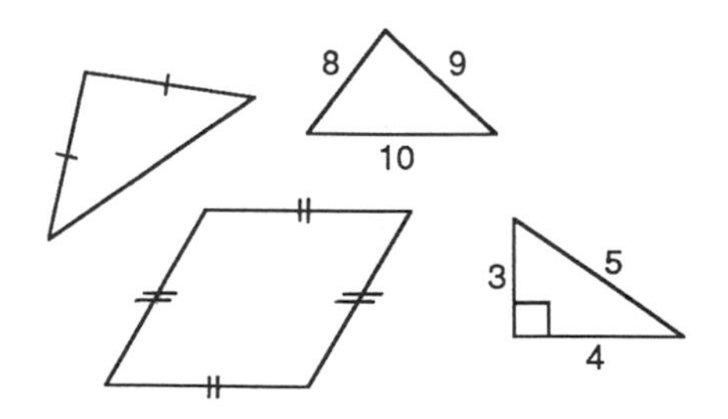

7. Define *median of a triangle.*

Median of a Triangle

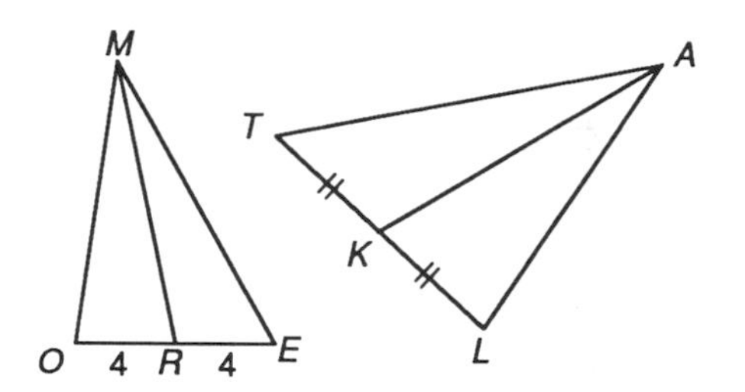

Segments *MR* and *AK* are medians.

Not a Median of a Triangle

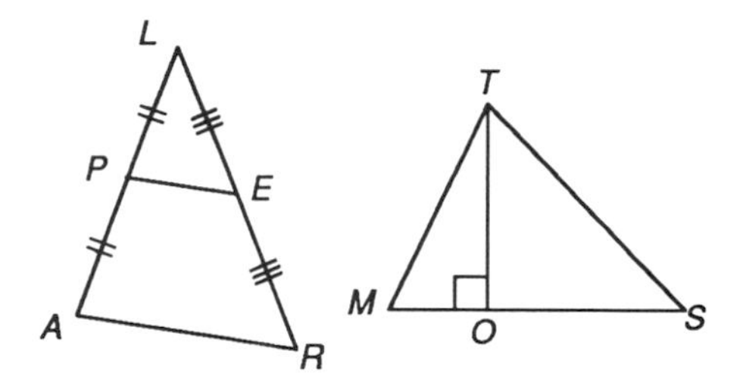

Segments *PE* and *TO* are not medians.

8. Define *altitude of a triangle.*

Altitude of a Triangle

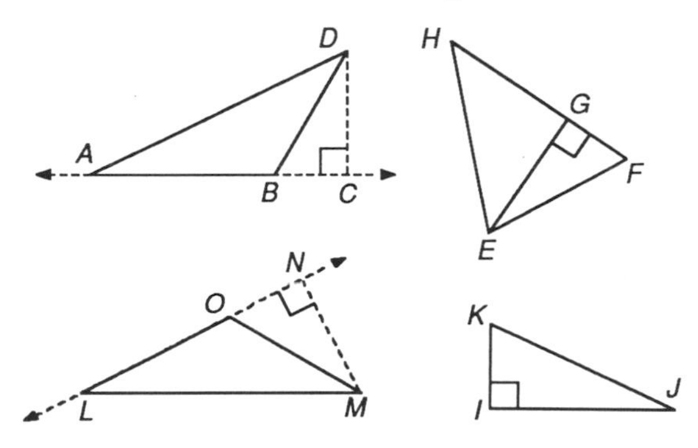

Segments *CD*, *EG*, *IK*, and *MN* are altitudes.

Not an Altitude of a Triangle

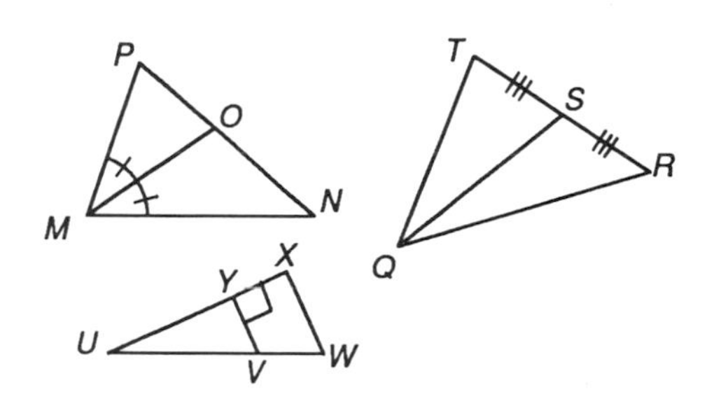

Segments *MO*, *QS*, and *VY* are not altitudes.

In a triangle, the length of the altitude is called the **height**. A triangle has three different altitudes and, therefore, it has three different heights.

EXERCISE SET B

Sketch and carefully label each figure.

1. Acute isosceles ΔACT with $AC = CT$
2. Obtuse isosceles ΔOBT with $OB = OT$
3. Right isosceles ΔRGT with $RT = GT$, $m\angle RTG = 90$
4. Scalene ΔSCL with median $\overline{CM}$
5. Equilateral ΔEQL with altitude $\overline{QT}$
6. Equilateral octagon *OCTAPUSH* with $m\angle CTA = 90$
7. Regular hexagon *SPIDER* with equilateral triangle *PDR*

EXERCISE SET C

Match.

1. $\overrightarrow{AB}$
2. AB
3. $\overline{AB}$
4. $\overleftrightarrow{AB}$
5. $\angle ABC$
6. Isosceles ΔABC
7. Right ΔABC
8. Equilateral ΔABC
9. $\overleftrightarrow{AB} \perp \overleftrightarrow{CD}$
10. $\overleftrightarrow{AB} \parallel \overleftrightarrow{CD}$
11. Pair of vertical angles
12. Pair of complementary angles
13. Pair of supplementary angles
14. Hexagon
15. Octagon
16. Median of a triangle
17. Altitude of a triangle
18. Angle bisector in a triangle

a. 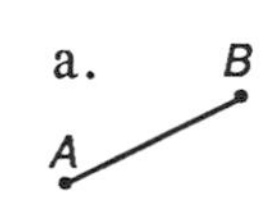

b. 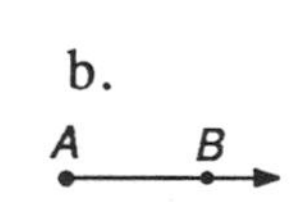

c. 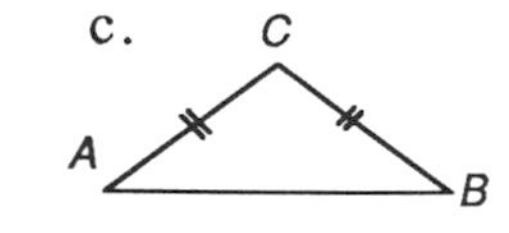

d. 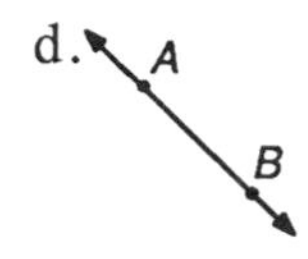

e. 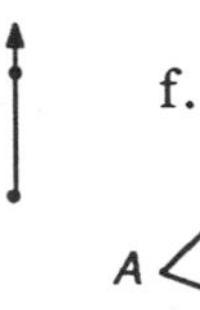

f. 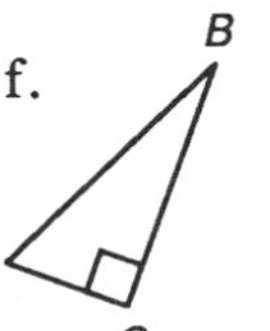

g. 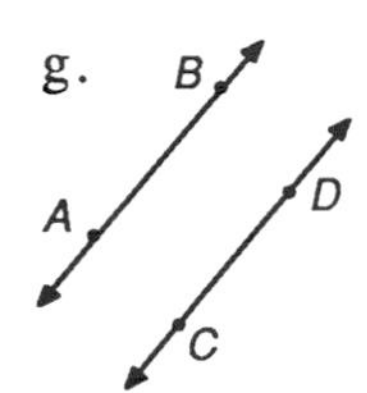

h. 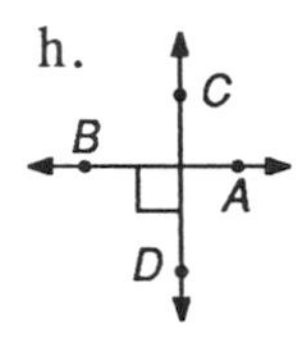

i. 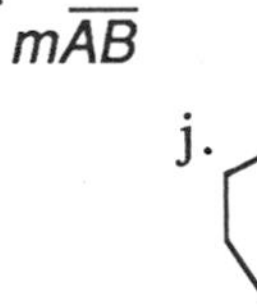

j.

k. 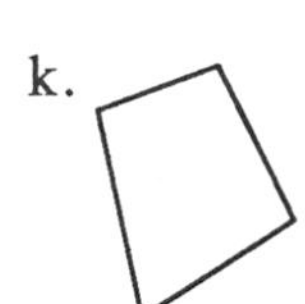

l. 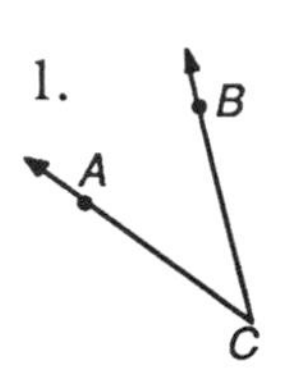

m. 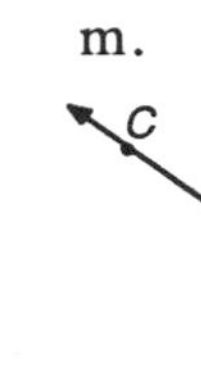

n. 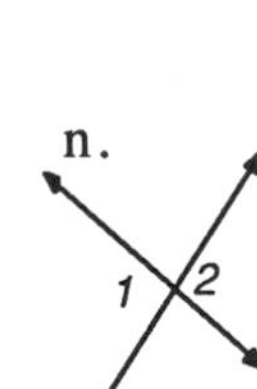

o. 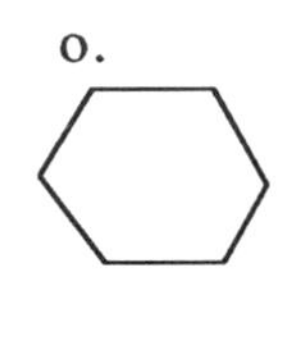

p. 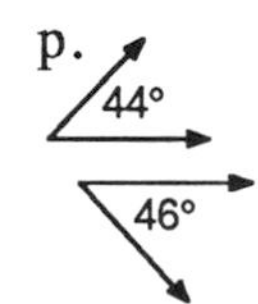

q. 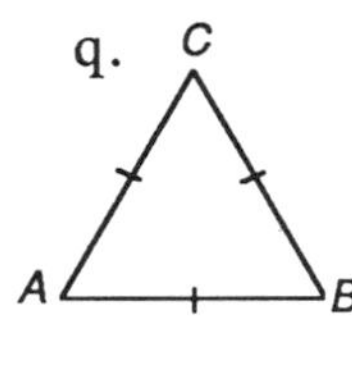

r. 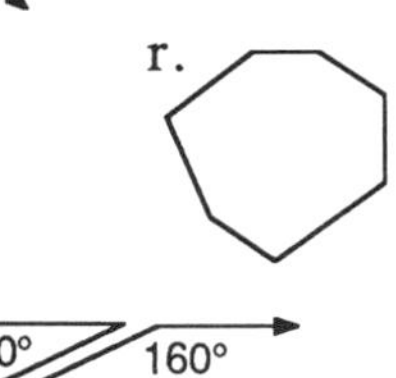

s.

t. 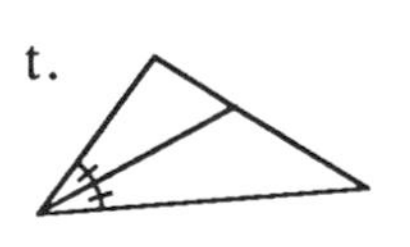

u. 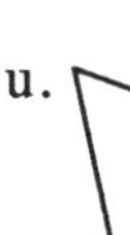

v. 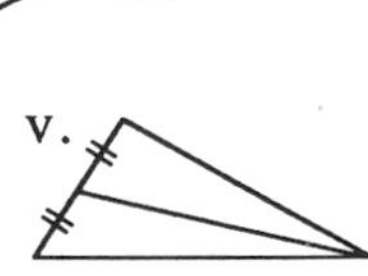

Lesson 2.7

Defining Special Quadrilaterals

A mind that is stretched by a new idea can never go back to its original dimensions.

– Oliver Wendell Holmes

EXERCISE SET A

Write a good definition of each geometric term or figure. Discuss them with others in your class. Agree on a common set of definitions for your class and add them to your definition list. Draw and label a picture to illustrate each definition.

1.* Define *trapezoid.*

Trapezoid **Not a Trapezoid**

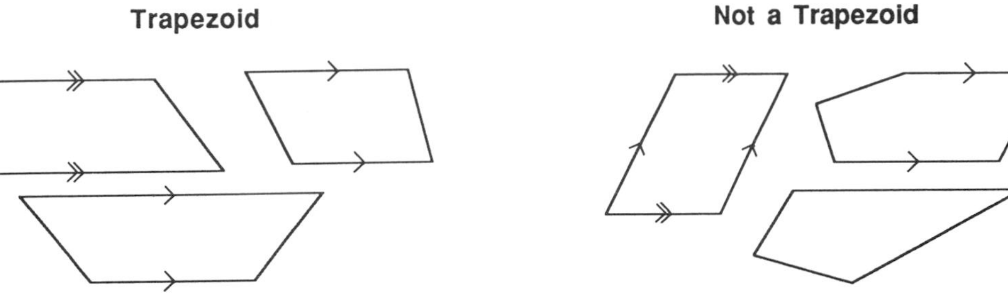

Note: Matching arrows in a diagram indicate parallel segments.

2. Define *parallelogram.*

Parallelogram

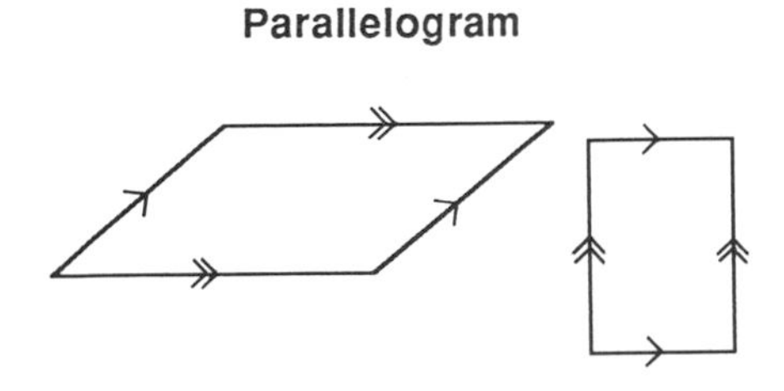

Not a Parallelogram

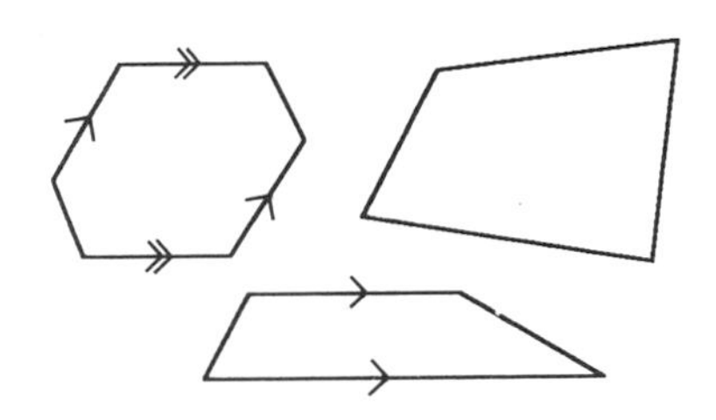

3. Define *rhombus.*

Rhombus

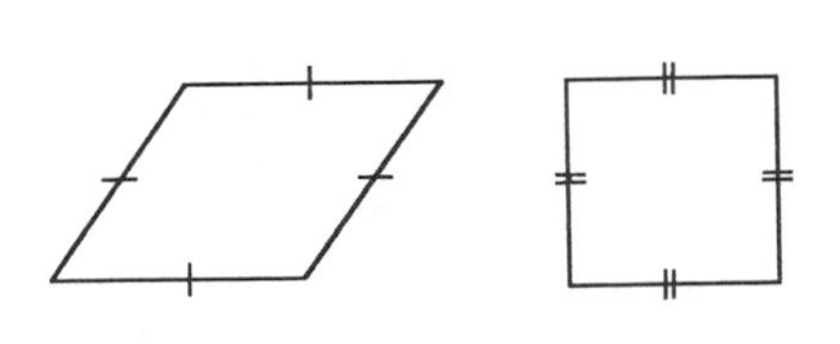

Not a Rhombus

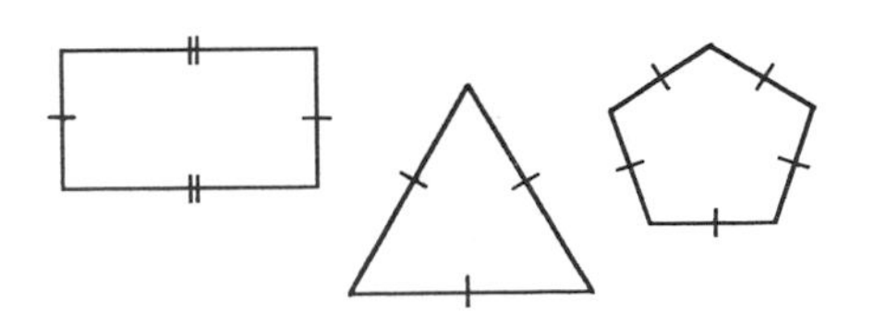

Both polygons are parallelograms.

4.* Define *rectangle.*

Rectangle

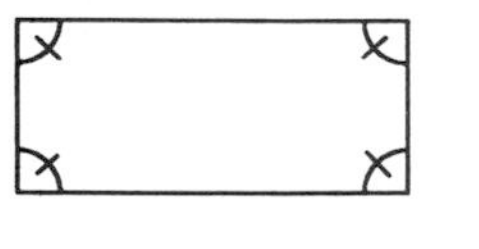

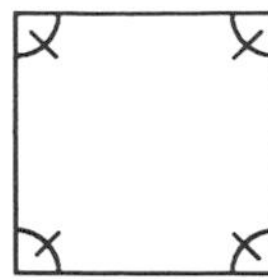

Not a Rectangle

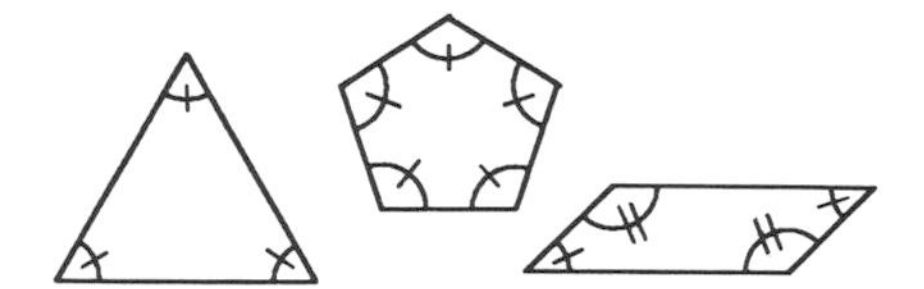

Both polygons are parallelograms.

5.* Define *square.*

Square

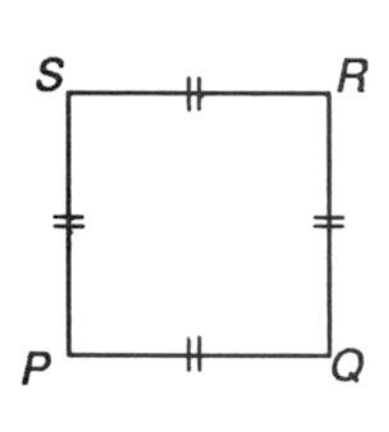

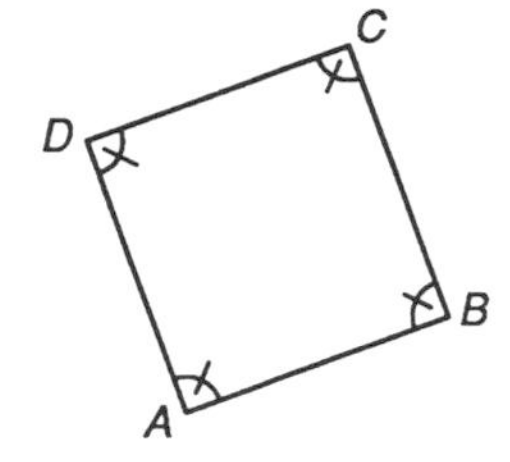

Quadrilateral *ABCD* is a rhombus.
Quadrilateral *PQRS* is a rectangle.

Not a Square

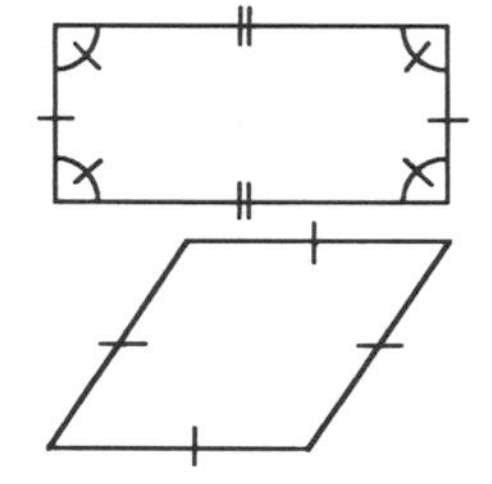

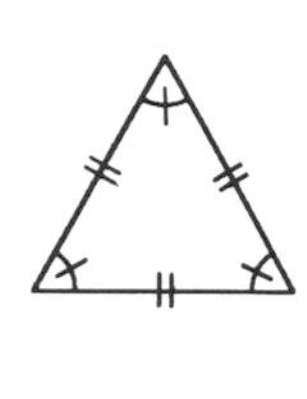

EXERCISE SET B

1. Many of the geometric figures that you have defined are closely related to each other. A diagram can help you see the relationships between them. At right is a concept map showing the relationships between members of the triangle family. This type of concept map is known as a tree diagram because the relationships are shown as branches of a tree. Copy and fill in the missing branches of the tree diagram for triangles.

Tree Diagram for Triangles

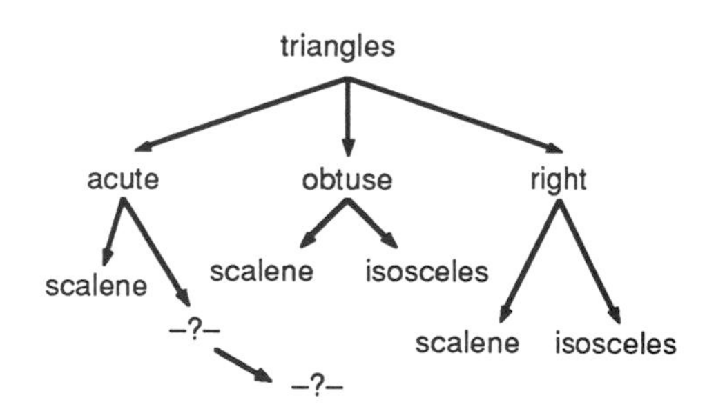

2. At right is a concept map showing the relationships between some members of the parallelogram family. This type of concept map is known as a Venn diagram. Copy the diagram and fill in the missing names in the blank regions.

Venn Diagram for Parallelograms

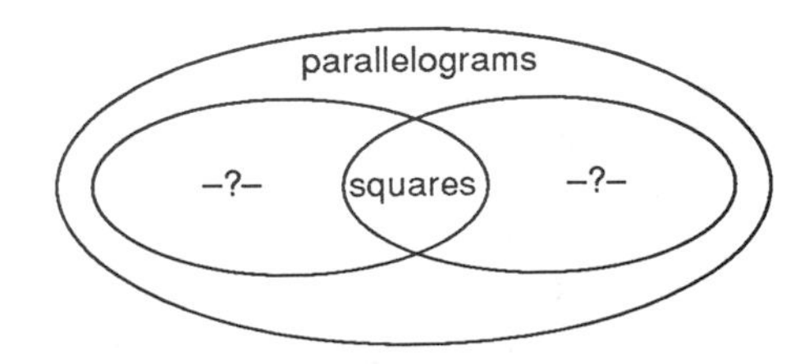

EXERCISE SET C

In Exercises 1 to 10, write the words or symbols that makes each statement true.

1. The three undefined elements of geometry are —?—, —?—, and —?—.
2. *Line AB* is written in symbolic form as —?—.
3. *Line segment AB* is written in symbolic form as —?—.
4. *Ray AB* is written in symbolic form as —?—.
5. *Angle ABC* is written in symbolic form as —?—.
6. The point where the two rays of an angle meet is called the —?—.
7. The geometric tool used to measure the size of an angle is called a —?—.
8. *Segment AB is parallel to segment CD* is written in symbolic form as —?—.
9. *Line AB is perpendicular to line CD* is written in symbolic form as —?—.
10. The angle formed by a light ray coming into a mirror is —?— the angle formed by a light ray leaving the mirror.

In Exercises 11 to 20, identify each statement as true or false.

11. An angle is measured in degrees.
12. An acute angle is an angle whose measure is less than 90.
13. An obtuse triangle has exactly one angle whose measure is greater than 90.
14. A diagonal is a line segment that connects any two vertices of a polygon.
15. If the measures of two angles add up to 90, then they are supplementary.
16. If two lines intersect forming a right angle, then the lines are perpendicular.
17. An angle bisector is a ray or line segment that divides an angle into two angles of equal measure.
18. A median of a triangle is a line segment from a vertex to the midpoint of the opposite side.
19. A trapezoid is a device used to capture zoids.
20. A rhombus is a parallelogram with all the sides the same measure.

EXERCISE SET D

Draw a triangle that fits each name. If impossible, write *not possible*. Use your geometric tools to make your drawings as accurate as possible.

1. Scalene triangle
2. Scalene acute triangle
3. Obtuse isosceles triangle
4. Isosceles right triangle
5. Equilateral right triangle
6. Scalene isosceles triangle

Lesson 2.8

Space Geometry

Space. The final Frontier.
– Star Trek

In the beginning lessons of this chapter you were introduced to point, line, and plane and you used those terms to define a wide range of other geometric figures – from rays to polygons. All of your work, however, was done on a single flat surface, a single plane. In this lesson you will learn a little about space geometry — the geometry of objects that are not restricted to a flat surface (not necessarily coplanar). Let's start by visualizing relationships between points, lines, and planes in space.

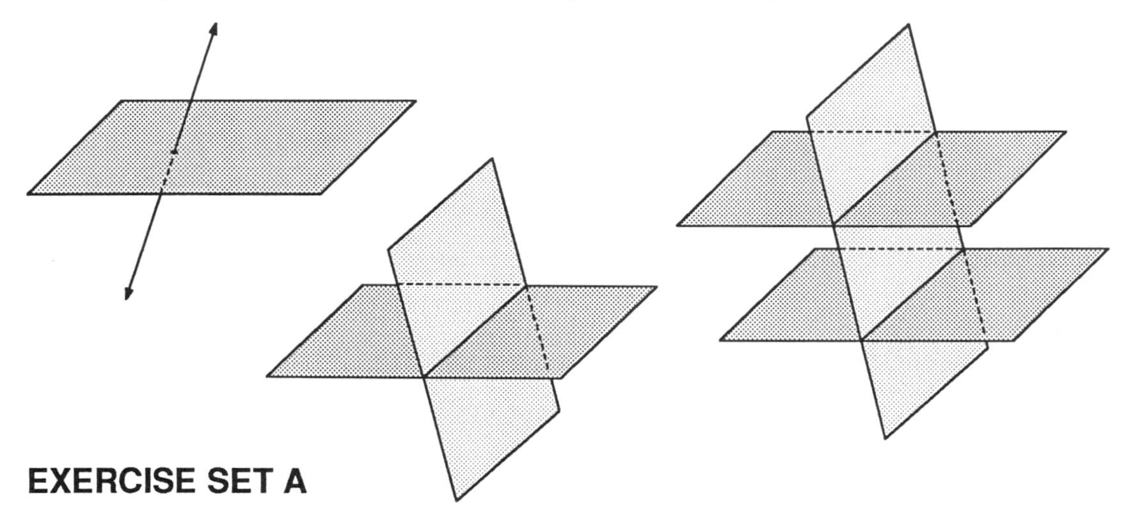

EXERCISE SET A

This exercise set will help you visualize relationships between geometric figures in the plane and in space. All the statements below except two are true. Make a sketch or use physical objects to demonstrate each true statement. For the two false statements, produce a counter-example demonstrating that each is false. Pencil tips and thumb tacks can represent points. Rulers, pencils, or stiff wires can represent lines. Sheets of heavy paper or cardboard can represent planes.

1. One and only one distinct line can be drawn through two different points.
2. One and only one distinct plane can be made to pass through three noncollinear points.
3. One and only one distinct plane can be made to pass through one line and a point not on the line.
4. If a line intersects a plane not containing it, then the intersection is exactly one point.
5.* If two lines are perpendicular to the same line, then they are parallel.
6. If two different planes intersect, then their intersection is a line.

7. If a line and a plane have no points in common, then they are parallel.

8. If two different planes do not intersect, then they are parallel.

9. If a plane intersects two parallel planes, then the lines of intersection are parallel.

10. If three random planes intersect (no two parallel and no three through the same line), then they divide space into six parts.

11. If a line is perpendicular to two lines in a plane but the line is not contained in the plane, then the line is perpendicular to the plane.

12.* If two lines are perpendicular to the same plane, then they are parallel to each other.

You've already seen a wide range of flat or plane geometric figures in this text. You will also encounter a variety of space figures (called solids). Solids of revolution are one group of space figures. The exercises below will help you visualize these solids.

EXERCISE SET B

Each figure on the left represents a card with a wire attached. Each figure on the right represents what a figure on the left would look like if it were revolved by spinning the wire between your fingers. Match each two dimensional figure on the left with a solid of revolution on the right

1. 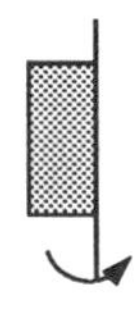

a. 
b. 
c. 
d. 

2. 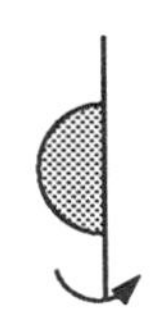

a. 
b. 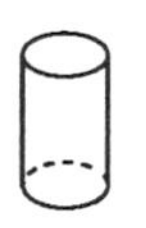
c. 
d. 

3. 

a. 
b. 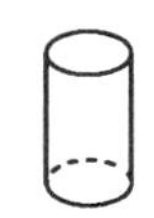
c. 
d. 

In the last exercise set you used plane figures to generate solids of revolution. In the next exercise set you will make plane figures by slicing a solid with a plane. When a solid is cut by a plane, the resulting plane figure is called a **section**.

EXERCISE SET C

Match each solid sliced by a plane with its section.

1. 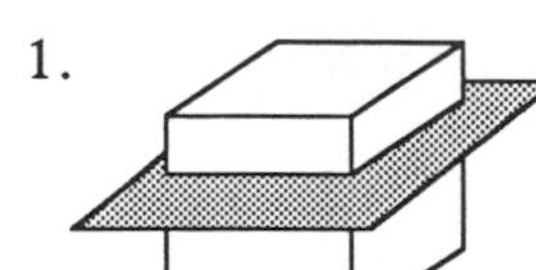

a. 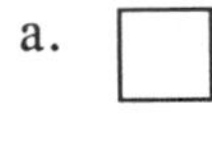

b. 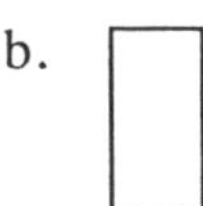

c. 

d. 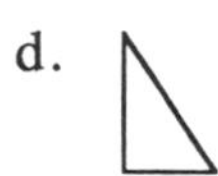

2. 

a.

b.

c.

d.

3. 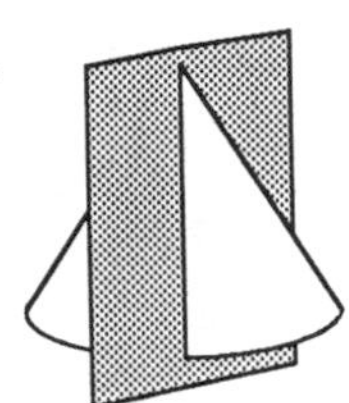

a. 

b. 

c. 

d. 

Examples of five geometric solids that you will be working with in this book are shown below. You probably recognize that three of these are solids of revolution. Formal definitions of these solids will come in a later chapter. We're sure that the shapes of the solids are already familiar to you even if you are not familiar with their formal names. The dashed lines in the drawings (called hidden lines) represent parts of the drawing that would not be visible if the object were actually solid.

Prism **Pyramid** **Cylinder** **Cone** **Sphere**

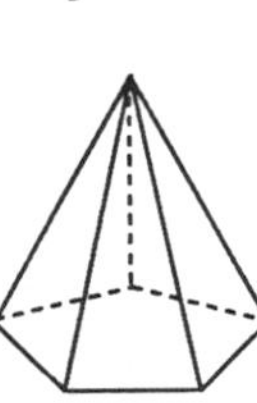

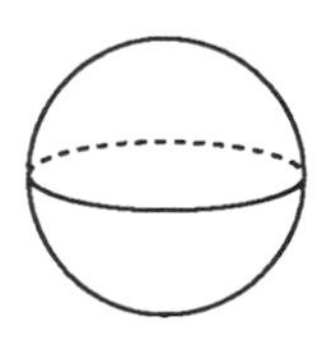

EXERCISE SET D

Draw each figure below. Study the drawing tips on the next page before you start.

1. Cylinder
2. Cone
3. Prism with a hexagonal base
4. Pyramid with a heptagonal base
5. Sphere

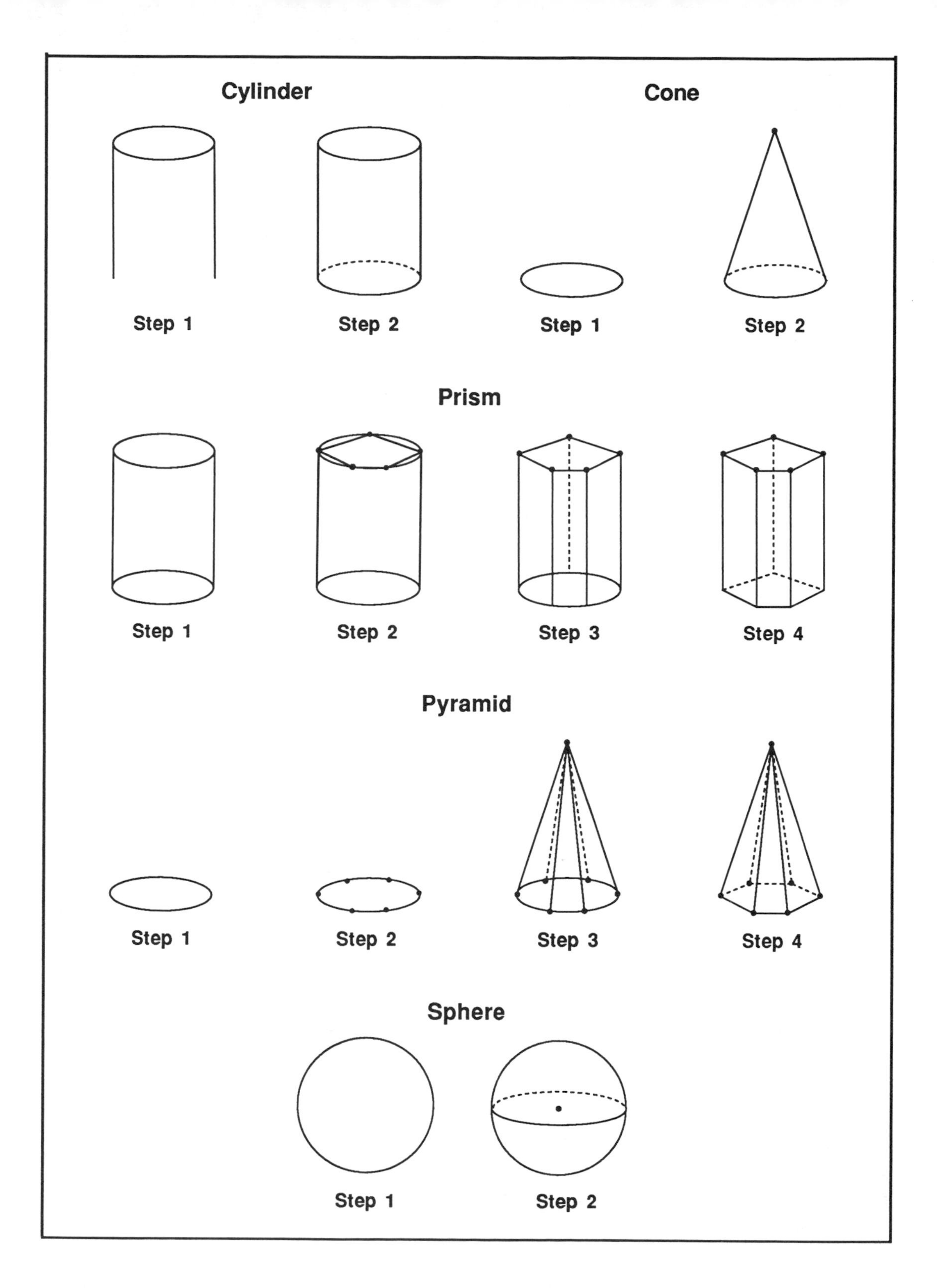
Cylinder
Cone
Step 1
Step 2
Step 1
Step 2
Prism
Step 1
Step 2
Step 3
Step 4
Pyramid
Step 1
Step 2
Step 3
Step 4
Sphere
Step 1
Step 2

Lesson 2.9

A Picture is Worth a Thousand Words

You can observe a lot just by watching.

– Yogi Berra

The ability to draw geometric solids is an important visual thinking skill. In Chapter 0, you learned to draw rectangular solids in perspective. While perspective views are nice, they take time to make. You can draw solids more easily if you don't worry about perspective. Below are drawing tips for rectangular solids. These steps result in drawings that are not perspective views. Study the steps below before attempting to draw rectangular prisms in the next exercise set.

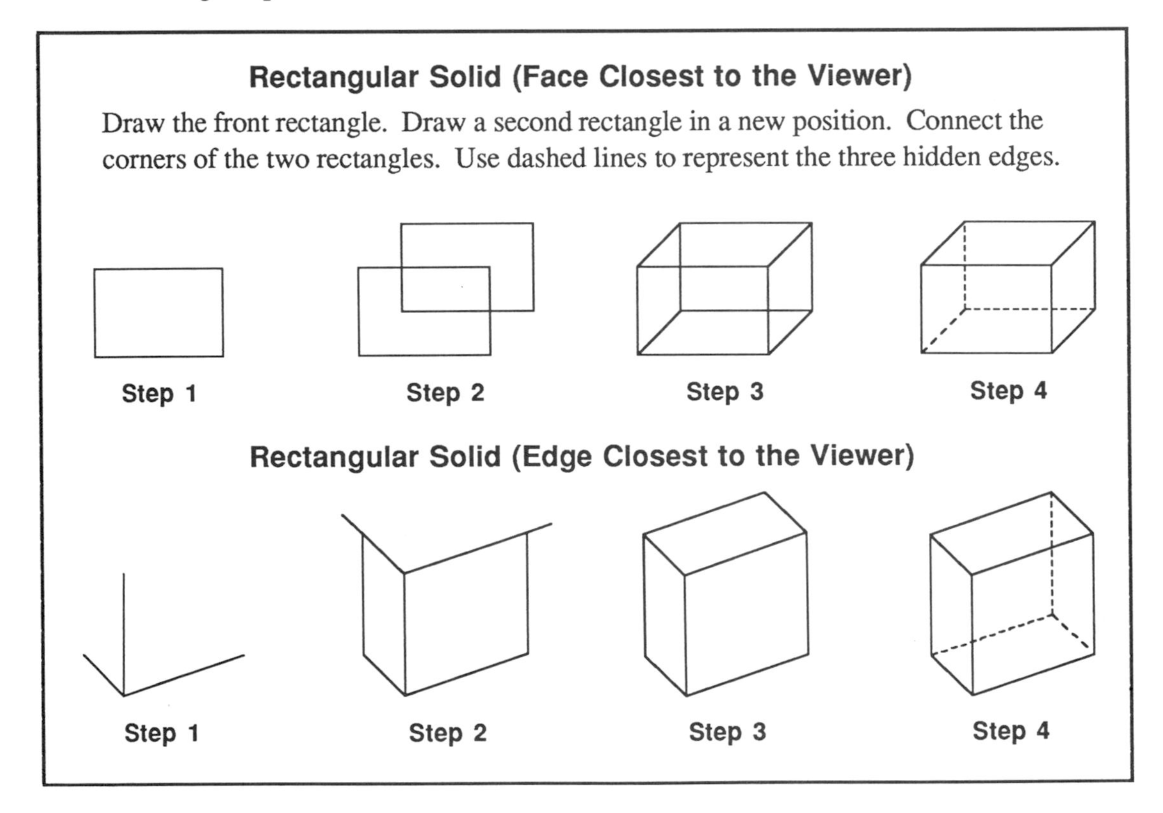

EXERCISE SET A

Draw and label each solid figure described below. For Exercises 1 to 3 you should make scaled drawings. For example, in Exercise 2 you must draw a rectangular solid 2 m by 3 m by 4 m. There is no way that you're going to draw a solid that is actually that big. You should draw the solid at a smaller scale so that the dimensions look about 2 units by 3 units by 4 units and then label the figure with meters.

1. A rectangular solid with a base of 6 cm by 12 cm and a height of 3 cm with a face closest to the viewer.

2. A rectangular solid 2 m by 3 m by 4 m, sitting on its biggest face, and with an edge closest to the viewer.

3.* A rectangular solid 3 inches by 4 inches by 5 inches resting on its smallest face. Draw lines on the three visible surfaces showing how the solid can be divided into 60 cubic inch boxes.

4. A cube (a prism with each face a square). Draw lines on the three visible faces to show how it can be divided into 27 equal smaller cubes.

5.* A cube with a plane passing through it. The section is equilateral triangle *CUT*.

6. A cube with a plane passing through it. The section is trapezoid *ZOID*

Find the lengths x and y in Exercises 7 and 8 (each angle on each polygonal side of the block is a right angle).

7.

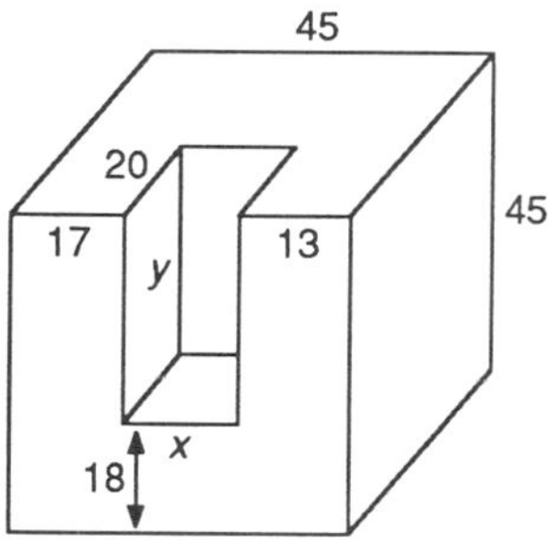

8.

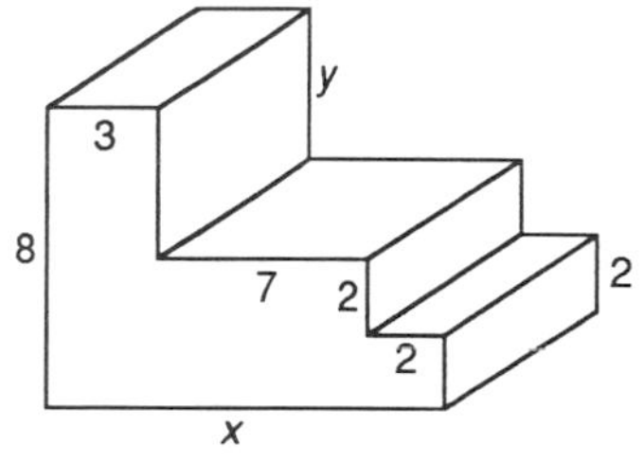

Have you ever heard the expression, "A picture is worth a thousand words?" That expression certainly applies to geometry. A drawing of an object often conveys more information in a quicker way than a long description in words. It is often useful to translate a long written description into a drawing or labeled diagram. This is an important visual thinking skill that you can master with practice.

Example

What is the object being described in the paragraph below?

> I am examining an object that is essentially a cylinder 23 cm in height. From the base to a point 16 cm above the base, it is straight-sided with a diameter of 8 cm. From 16 cm above the base to 18 cm above the base, the diameter is 7 cm. Evenly spaced about the surface of this portion is a screw thread. Between a height of 18 cm and the top, the diameter is uniformly reduced from 7 cm to 4 cm in an arch, ending at an opening at the top. The opening reveals a highly reflective interior. The main body of the cylinder is cool and very hard to the touch. It has no perceptible odor and, when struck with a metal object, gives off a "thunking" sound.

To determine what the object is that is being described, read through the paragraph again, stopping occasionally to draw what is being described, redrawing or adding details to your drawing as new information is revealed.

Initial Drawing

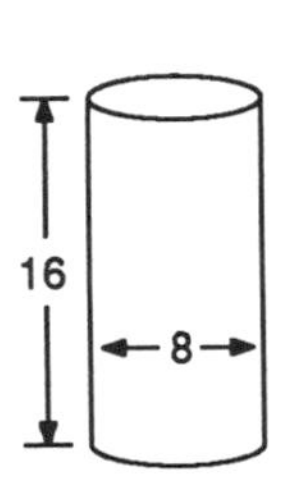

From the base to a point 16 cm above the base, it is straight-sided with a diameter of 8 cm.

Improved Drawing

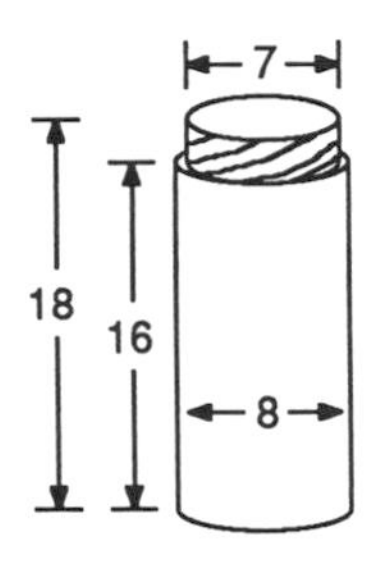

From 16 cm above the base to 18 cm above the base, the diameter is 7 cm. Evenly spaced about the surface of this portion is a screw thread.

Final Drawing

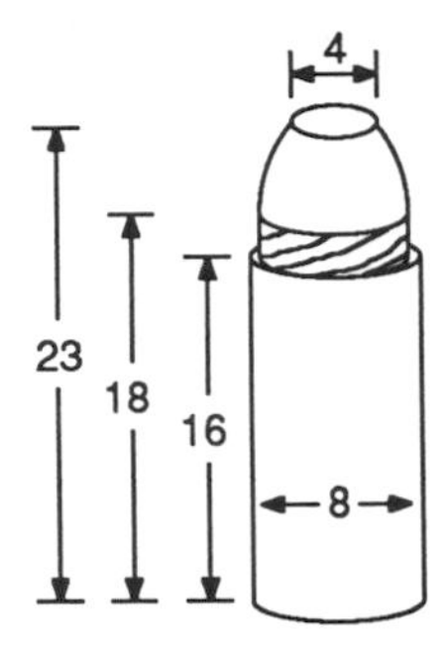

Between a height of 18 cm and the top, the diameter is uniformly reduced from 7 cm to 4 cm in an arch, ending at an opening at the top.

The object has a highly reflective interior. The main body of the cylinder is cool and very hard to the touch. It has no perceptible odor and, when struck with a metal object, gives off a "thunking" sound. Combine this information with the drawing and you should be able to conclude that the object is a thermos.

EXERCISE SET B

Draw the object being described in each paragraph below using a centimeter ruler. What is each object?

1. I am examining an object that is primarily a hexagonal prism with a cone at one end and a cylinder at the other end. The total length is 17 cm. The diameters of both the cone and cylinder are approximately 7 mm. The hexagon base of the prism fits exactly within the circular base cylinder and the circular base of the cone fits exactly within the hexagon . The height of the cylinder is 15 mm and the height of the cone is about 18 mm. The tip of the cone (top 4 mm) is black and the opposite end of the object (bottom 4 mm of the cylinder) is pink. The hexagonal prism is painted yellow with occasional dents that appear to be teeth marks.

2.* I am examining an object that is primarily a rectangular prism measuring 7.6 cm by 13.5 cm by 1.4 cm thick. There are 38 small rectangular prisms protruding from the top face of the object. The 38 prisms are arranged on the top surface in two different rectangular arrays. The main body of the object appears to be black plastic while the top surface appears to be a silvery metal. The small raised prisms on the top surface appear to be made of a rubber substance. In the top third of the top surface there is a window 1 cm by 4.5 cm, centered, with the longer side running parallel to the 7.6 cm edge. This window displays a grey-green surface beneath.

One set of the raised rectangular prisms are black and are arranged in a four by five rectangular array with five running parallel to the 7.6 cm width and four running parallel to the 13.5 cm edge. Each small prism measures 7 mm by 8 mm (running parallel to the width) by 1 mm high. Each prism is positioned approximately 5 mm from its vertical and horizontal neighbors. The entire set of 20 is positioned in a third of the top surface approximately 8 mm from each of three edges. There are symbols on the top face of each raised prism.

The second set of raised rectangular prisms are grey and are arranged in a three by six rectangular array with six running parallel to the width of the top surface and three running parallel to the 13.5 cm length. Each small prism measures 4 mm (running parallel to the length) by 6 mm by 1 mm high. Each prism is positioned approximately 4 mm from its horizontal neighbors and approximately 6 mm from its vertical neighbors. The array of 18 rectangular prisms is centered on the top surface between the window and the set of black raised prisms. There are symbols on the top face of each raised prism.

EXERCISE SET C

You should be thoroughly convinced that a picture, if not necessarily "worth" a thousand words, can clearly require that many words to describe it accurately. Below are two familiar items. Describe them accurately in a thousand words or less.

1. Peanut butter and jelly sandwich
2. Carton of milk to wash it down

Lesson 2.10

Translating Word Problems into Drawings

We talk too much; we should talk less and draw more.

– Goethe

What is seven times eight? Fifty-six, of course. Do you realize that you just solved a word problem? Not much of a word problem, but a word problem nonetheless. Unfortunately, not all word problems are that easy.

There is no single way to solve a word problem. There are, however, some general techniques that may help you.

Suggestion 1: Read and reread the problem. Take special care to understand the relationships between the given facts. Make sure you understand what it is that you are being asked to find.

Suggestion 2: Organize the information in the problem. A diagram is one of the best ways to organize information. A picture is worth a thousand words!

Let's look at some examples showing how to use these suggestions.

Example A

Amy is older than Bryce. Carole is younger than Bryce. Dotty is older than Amy. Who is oldest?

Did you read the problem several times? You're given information about the relative ages of Amy, Bryce, Carole, and Dotty. You're trying to find out who is oldest.

Now you should make a diagram based on the given information. You can start your diagram with the information from the first sentence. Then you can add to the diagram as new information is revealed.

Diagram from First Sentence	Diagram after Second Sentence	Diagram after Third Sentence
		D
A	A	A
B	B	B
	C	C

From the diagram the answer is apparent. Dotty is the oldest.

Example B

In Reasonville, streets that begin with a vowel run north-south unless they end with the letter *d*, in which case they run east-west. All other streets run in either direction. Euclid Street runs perpendicular to Pascal Street. Fermat Street runs parallel to Pascal Street. In which direction does Fermat Street run?

In rereading you learn that streets beginning with vowels run north-south unless they end with the letter *d*, in which case they run east-west. Euclid begins with a vowel but ends with the letter *d*, so it runs east-west. You are trying to find the direction of Fermat Street.

Now draw a diagram.

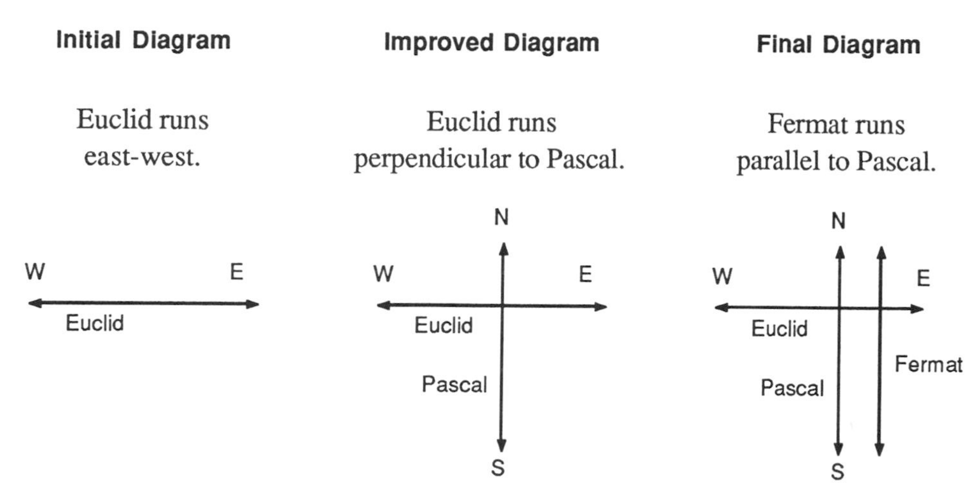

The final diagram reveals the answer. Fermat Street runs north-south.

Example C

Volumes 1 and 2 of a two volume set of math books are next to one another on a shelf in their proper order (Volume 1 on the left; Volume 2 on the right). Each front and back cover is one-eighth inch thick and the pages portion of each book is one inch thick. If a bookworm starts at page one of Volume 1 and burrows all the way through to the last page of Volume 2, how far will she travel?

Take a moment and try solving the problem in your head.

Did you get two and one-fourth inches? It seems reasonable, doesn't it? Guess what? That's not the answer. Let's get organized.

Reread the problem to identify what information you are given and what you are trying to find.

You are given the thickness of each cover, the thickness of the page portion, and the positions of the books on the shelf. You are trying to find how far it is from the first page of Volume 1 to the last page of Volume 2.

Draw a picture and locate the position of the pages referred to in the problem.

Now "look" how easy it is to solve the problem.

She just traveled one-fourth of an inch through the two covers!

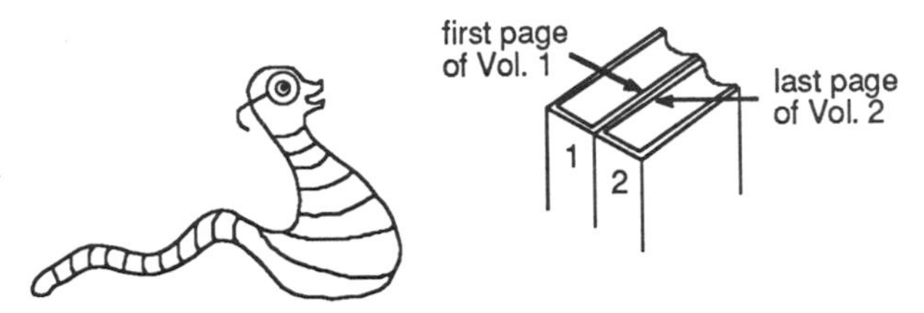

EXERCISE SET A

Now it is your turn. Read and reread each problem carefully, determining what information you are given and what it is that you are trying to find. Then draw and label a diagram. Finally, solve the problem.

1. Arturo is taller than Ben. Carlita is taller than Donna. Carlita is shorter than Ben. Which one is the tallest?

2. Moe, Larry, and Curlie got different scores in a geometry test. Their last names are Wright, Wong, and Gray, but not necessarily in that order. Moe's geometry score was 8 points higher than Larry's but only 2 points higher than Curlie's. Wong had the lowest score in the class and Gray had the highest of the three. What is Larry's last name?

3. Mary Ann has been contracted to design and build a fence around the outer edge of a rectangular garden plot 25 feet by 45 feet. The posts are to be set five feet apart. How many posts will she need?

4.* F. Freddie Frog is at the bottom of a 30 foot well. Each day he jumps up three feet, but then, during the night, he slides back down two feet. How many days will it take Fast Freddie to get to the top and out?

5. Midway through a 2000 meter race, a photo was taken showing the positions of all five runners. The picture shows Meg 20 meters behind Edith. Edith is 50 meters ahead of Wanda who is 20 meters behind Olivia. Olivia is 40 meters behind Nadine. At this point in the race, who is ahead? Who is second? Who is third? Who is fourth? Who is holding up the rear? (In your diagram, use *M* for Meg, *E* for Edith, and so on.)

6. Egbert Throckmorton III decided to run away from home. So, being only eight years old and not permitted to cross the street alone, he instructed Jeeves, the family chauffeur, to drive him to Disneyland. After traveling awhile at 20 miles per hour, Egbert realized that he forgot his credit cards and so returned home at three times the speed at which he left. As he passed through the front gate of Throckmorton Manor, he noticed that he had been gone exactly one hour. How far away from home did he travel before turning back?

Lost in Birnam Woods

7. Beth Mack is lost in the woods 15 km east of the north-south Birnam Woods Road. She begins walking in a zig-zag pattern: 1 km south, 1 km west, 1 km south, 2 km west, 1 km south, 3 km west, and so on. Beth walks at the rate of 4 km/h (kilometers per hour). If it is 3 pm now and the sun sets at 7:30, will there still be sunlight when she reaches Birnam Woods Road?

8.* A silver 1963 Cobra followed with familiar ease the winding Bernal Heights Road to its conclusion, the home of professor Lance Michaels. Climbing out of his car, Lance is met at his doorstep by ATF agent Kim Lee. "Professor Michaels, we need your help again." "Hi Kim. Yes, I know. I just spoke to Stone on the way up. Come in. Would you like some espresso?" "Lance, we intercepted this coded message which we believe contains the date and time of an assassination attempt at a foreign embassy. It is a series of 1909 x's and o's. Any ideas?"

"Lets see? 1909 is 23 x 83. Both factors are prime, so there are only two ways of arranging these x's and o's into rectangles. Perhaps, on rearranging them into a rectangle, the x's or o's will make a picture or spell out a message? Let's type them into the computer and see what comes up. Here it comes."

SUN. EVE. FIRST TIME BIG AND SMALL HANDS ARE 70° APART

"Well it looks like these fellas like to play games. Remember what I said in my introductory lecture on problem solving? Solving big problems is often just solving a series of little problems. Well, one down and, hopefully, one to go. When is the first time after 6 P.M. that the hour and minute hands are 70° apart? You were always the quickest in class, so I'll let you solve this one, Kim, but here is a hint. The fraction that the big hand has moved since six in its complete hour cycle is equal to the fraction that the hour hand has moved between the six and seven. Good Luck!"

Kim was able to solve the problem and capture the terrorists near the embassy. Could you? What was the time set for the assassination attempt?

Agent Kim Lee Stops the Terrorists

EXERCISE SET B

Match each term on the left with a figure on the right.

1. Angle bisector in a triangle
2. Median in a triangle
3. Altitude in a triangle
4. Diagonal in a polygon
5. Pair of complementary angles
6. Pair of supplementary angles
7. Pair of vertical angles
8. Obtuse angle
9. Acute angle

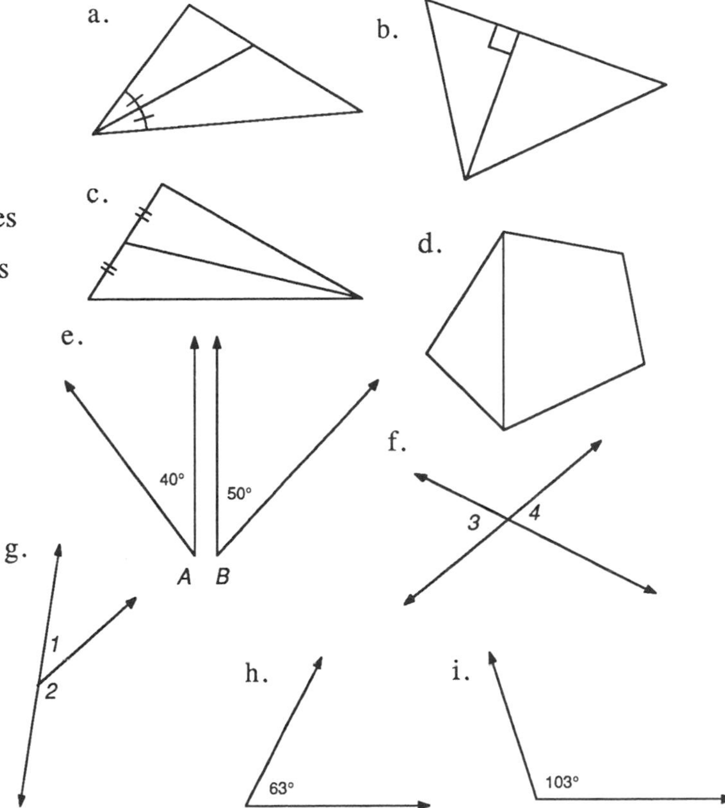

Improving Visual Thinking Skills

Coin Swap I

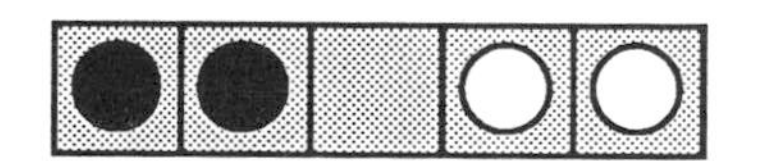

Find two light coins (dimes) and two dark coins (pennies) and arrange them on a grid of five squares as shown. Your task is to switch the position of the two light and two dark coins in exactly 8 moves. A coin can slide into an empty square next to it or it can jump over one coin into an empty space. Record your solution by drawing eight diagrams showing the moves.

Lesson 2.11

Chapter Review

EXERCISE SET A

Identify each statement as true or false.

1. The three basic building blocks of geometry are point, line, and plane.
2. *The ray from P through Q* is written in symbolic form as $\overrightarrow{PQ}$.
3. *The line segment from P to Q* is written in symbolic form as $\overline{PQ}$.
4. *The length of line segment PQ* is written in symbolic form as PQ.
5. The vertex of angle PDQ is point P.
6. An acute angle is an angle whose measure is less than 90.
7. A scalene triangle is a triangle with no two sides the same length.
8. A diagonal is a line segment in a polygon connecting any two non-consecutive vertices.
9. If $\overleftrightarrow{AB}$ intersects $\overleftrightarrow{CD}$ at point P, then $\angle APD$ and $\angle BPC$ are a pair of vertical angles.
10. If the sum of the measures of two angles is 180, then the two angles are complementary.
11. If two lines lie in the same plane and do not intersect, then they are parallel.
12. The symbol for perpendicular is $\perp$.
13. An altitude in an acute triangle is a perpendicular line segment connecting a vertex with the opposite side.
14. A trapezoid is a quadrilateral having exactly one pair of parallel sides.
15. A rhombus is a parallelogram with all the angles equal in measure.
16. A square is a rectangle with all the sides equal in length.
17. To show that an angle is a right angle, you mark it with a little box.
18. The vertex angle of an isosceles triangle is between the two sides of equal length.
19. A polygon with ten sides is called a decagon.
20. Exactly four statements in this exercise set are false.

EXERCISE SET B

Match each term with its drawing

1. Acute isosceles triangle
2. Isosceles right triangle
3. Rhombus
4. Trapezoid
5. Octagon
6. Prism
7. Pyramid
8. Cylinder

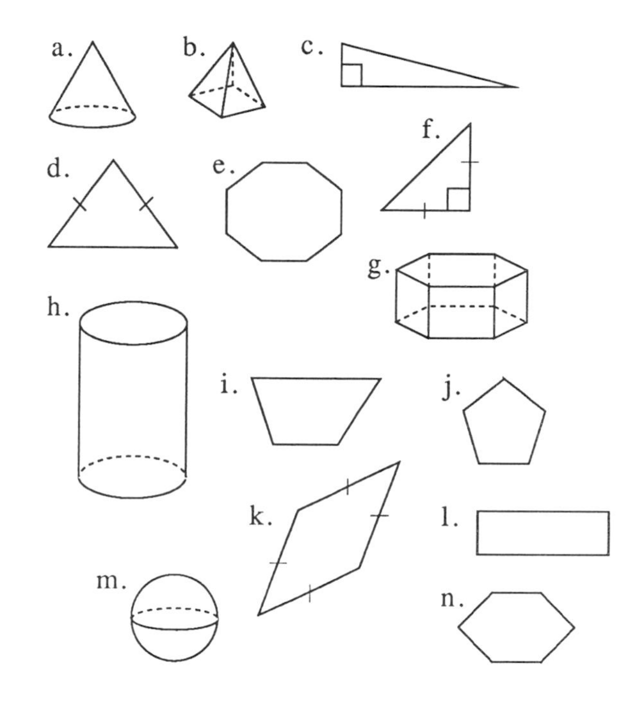

EXERCISE SET C

Sketch and label each figure.

1. Trapezoid *TRAP* with $\overline{TR} \parallel \overline{AP}$ and $\overline{TR} \perp \overline{AR}$
2. Isosceles right ΔABC with $AB = BC$
3. Scalene ΔPTS with $PS = 3$, $ST = 5$, $PT = 7$, and median $\overline{SO}$
4. Hexagon *REGINA* with diagonal $\overline{AG}$ parallel to sides $\overline{RE}$ and $\overline{NI}$
5. Trapezoid *TRAP* with $\overline{AR}$ and $\overline{PT}$ the non-parallel sides. Let E be the midpoint of $\overline{PT}$ and Y be the midpoint of $\overline{AR}$. Draw $\overline{EY}$.
6. Square *ABCD* with each side trisected by two points. The order of the points is: $A, E, F, B, G, H, C, I, J, D, K, L$. Draw $\overline{FG}$, $\overline{HI}$, $\overline{JK}$, $\overline{LE}$.
7. A dodecagon with every other vertex connected by diagonals to form a hexagon. Shade in the triangular regions between the hexagon and dodecagon.
8. Obtuse isosceles triangle *OLY* with $OL = YL$ and median $\overline{LM}$
9. A rectangular solid 2 by 3 by 5 inches resting on its largest face. Draw lines on the three visible faces showing how it can be divided into 30 equal smaller cubes.
10. A cube with a plane passing through it. The section is rectangle *RECT*.

11.* A cube with a plane passing through it. The section is hexagon *SPIDER*.

EXERCISE SET D

1. Which creatures in the last group are Brewsters?

Brewsters

Not Brewsters

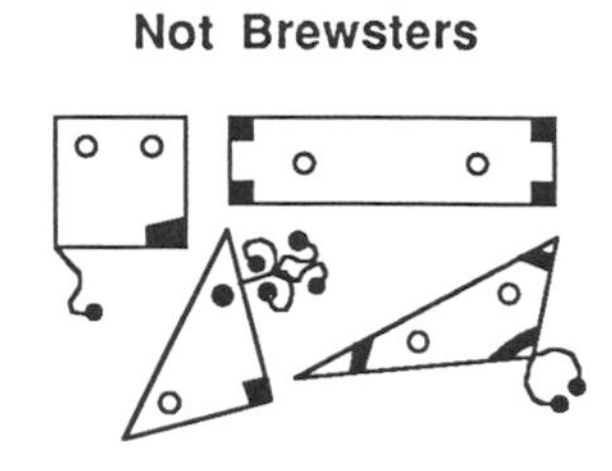

Who is a Brewster?

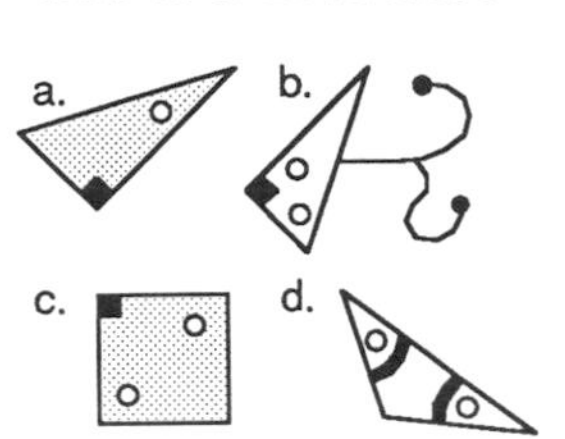

2. What characterizes a good definition?

3. What is the measure of $\angle A$? Use your protractor.

Find the lengths x and y in Exercises 4 to 6.

4.

20

18

x y 12

In Exercises 5 and 6, each angle on each polygonal side of the block is a right angle.

5.

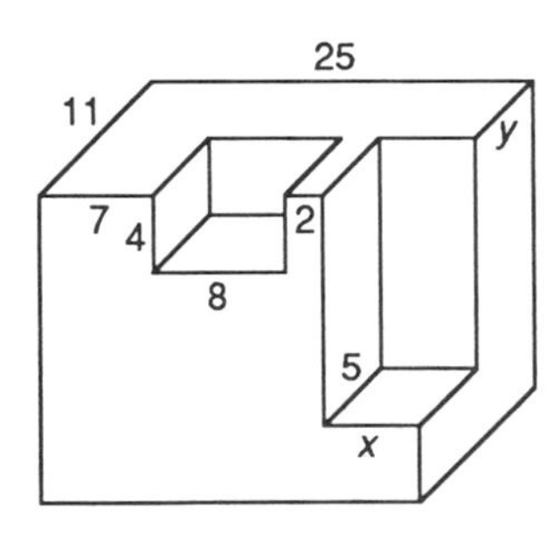

6.

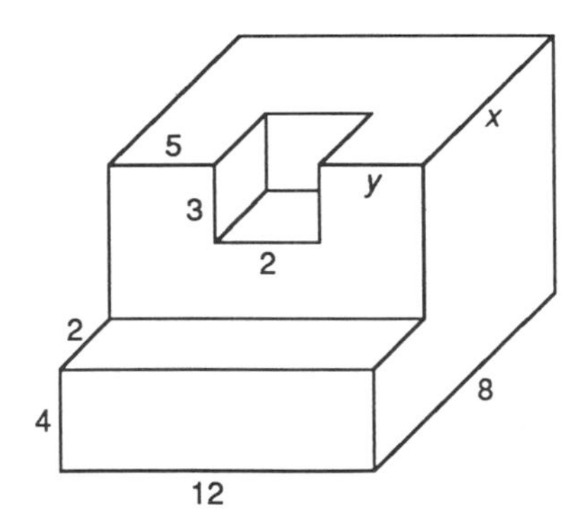

7. The box on the right is wrapped with two strips of ribbon as shown. What length of ribbon was needed to decorate the box?

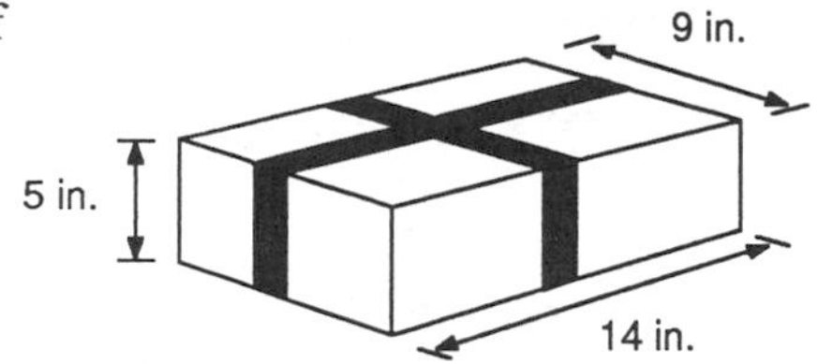

8. Make a concept map (either a tree diagram or a Venn diagram) for quadrilaterals.

9. A pair of parallel interstate gas and power lines run 10 meters apart and are equally distant from relay station A. The power company needs to locate a gas monitoring point on one of the lines which is exactly 12 meters from relay station A. Draw a diagram showing possible locations.

10.* A motion efficiency expert in an electronics assembly plant needs to locate a point that is equally distant to two major work-inspection stations and yet remain at least 4 meters, and at most 20 meters, from a heat sensitive wall. The work stations are 30 meters apart and are each 12 meters from the heat sensitive walls. Draw a diagram showing possible locations.

11. A large aluminum ladder with wheels was resting vertically against the research shed at midnight when it began to slide down the vertical edge of the shed. A burglar was clinging to the ladder's midpoint holding a pencil flashlight (which was visible in the dark). Witness Jane Seymour claimed to see the ladder slide. What did she see? That is, what was the path taken by the bulb of the flashlight? Draw a diagram showing the path. (Devise a physical test to check your visual thinking. You might try sliding a meter stick against a wall or you might plot points on graph paper.)

12. At one point in a race, Ringo was 15 feet behind Paul and 18 feet ahead of John. John was trailing George by 30 feet. Paul was ahead of George by how many feet?

13.* The outer measurement across the grooves of a record album is nine inches. The inner portion of the record with no grooves is three inches across. The record has 20 grooves per inch. It plays at $33\frac{1}{3}$ revolutions per minute. How far does the needle travel (linear tracking turntable) while the record plays from the beginning (outer groove) to the end (inner groove)?

14. Edith is a TV cameraperson; Stella is a textbook editor; Janet is a ceramic artist; and Mary is an auto mechanic. The editor has a higher income than Mary. The artist earns less than the mechanic. Edith earns more than the mechanic but less than Stella. Who earns the most?

15. Juanita is at the front of the Zephyr Express train heading east at 40 miles/hour. Starting at the front of the train, Juanita walks towards the rear of the train at .5 miles/hour. If the train is a quarter mile long, how far forward (east) does she travel while walking towards the rear of the train?

Cooperative Problem Solving

Geometrivia

The geometry students at the lunar space port Galileo have just completed a geometry puzzle called Geometrivia for their earth friends at the NASA Academy in Atlanta. The students here at the space port have been trained to learn in cooperative small groups. Geometrivia has been designed with that in mind. Geometrivia is a multistep problem solving puzzle. In this puzzle you are to answer each lettered question, convert the letters of the answer to numbers according to the table, add the numbers, and, finally, substitute the sum for that letter into the formula. When all letter variables have been replaced by numbers, calculate the value of x. It is a whole number. Some questions require information gained from the chapter. Some questions require library research. Still other questions require that you find a particular word in a sentence from the chapter. Geometrivia demands team effort.

Example

The first question reads as follows:

a The line segment in a triangle connecting a vertex to the midpoint of the opposite side is called a —?—.

Step 1: Find the answer. The answer is *median.*

Step 2: Substitute numbers for letters according to the table:
12 for m; 4 for e; 3 for d; 8 for i; 0 for a; 12 for n.

Step 3: Add the numbers found in the word:
$12 + 4 + 3 + 8 + 0 + 12 = 39$.

Step 4: Your sum is the value for the letter a, so $\boldsymbol{a} = 39$.

Questions

a The line segment in a triangle connecting a vertex to the midpoint of the opposite side is called a —?—.

b The famous set of math books called the *Elements* was written by —?—.

c The symbol $\overrightarrow{AB}$ represents the —?— *AB*.

d "Based on this definition, you should have selected —?— C as the only Ork in the last group." (This sentence can be found in one of the first five lessons of this chapter.)

e A —?— is a ten sided polygon.

f Albert Einstein was born in the month of —?—.

g An angle whose measure is less than 90 is —?—.

h A, B, D, G, K, P, –?–

i A triangle with two sides the same length is —?—.

j Archimedes was a native of the Greek city of Syracuse on the island of —?—.

k A —?— is a quadrilateral with exactly two sides parallel.

l The symbol for —?— is //.

m —?— was third midway in the race between Meg, Edith, Wanda, Olivia, and Nadine.

n The vertex of angle *SUN* is point —?—.

p Sir Isaac Newton was born on the holiday called —?—.

Table

a	*b*	*c*	*d*	*e*	*f*	*g*	*h*	*i*	*j*	*k*	*l*	*m*	*n*	*o*	*p*	*q*	*r*	*s*	*t*	*u*	*v*	*w*	*x*	*y*	*z*
0	1	2	3	4	5	6	7	8	9	10	11	12	12	11	10	9	8	7	6	5	4	3	2	1	0

Formula

$$X = \frac{\frac{n(j-g)}{k}\left[\frac{(a-d)(i-p)}{h}\right]^{\frac{mn}{l}}}{\sqrt{\frac{c(l-m)+b(e-f)}{bc}+\frac{h(c-n)}{g-c}}}$$

3 Geometric Constructions

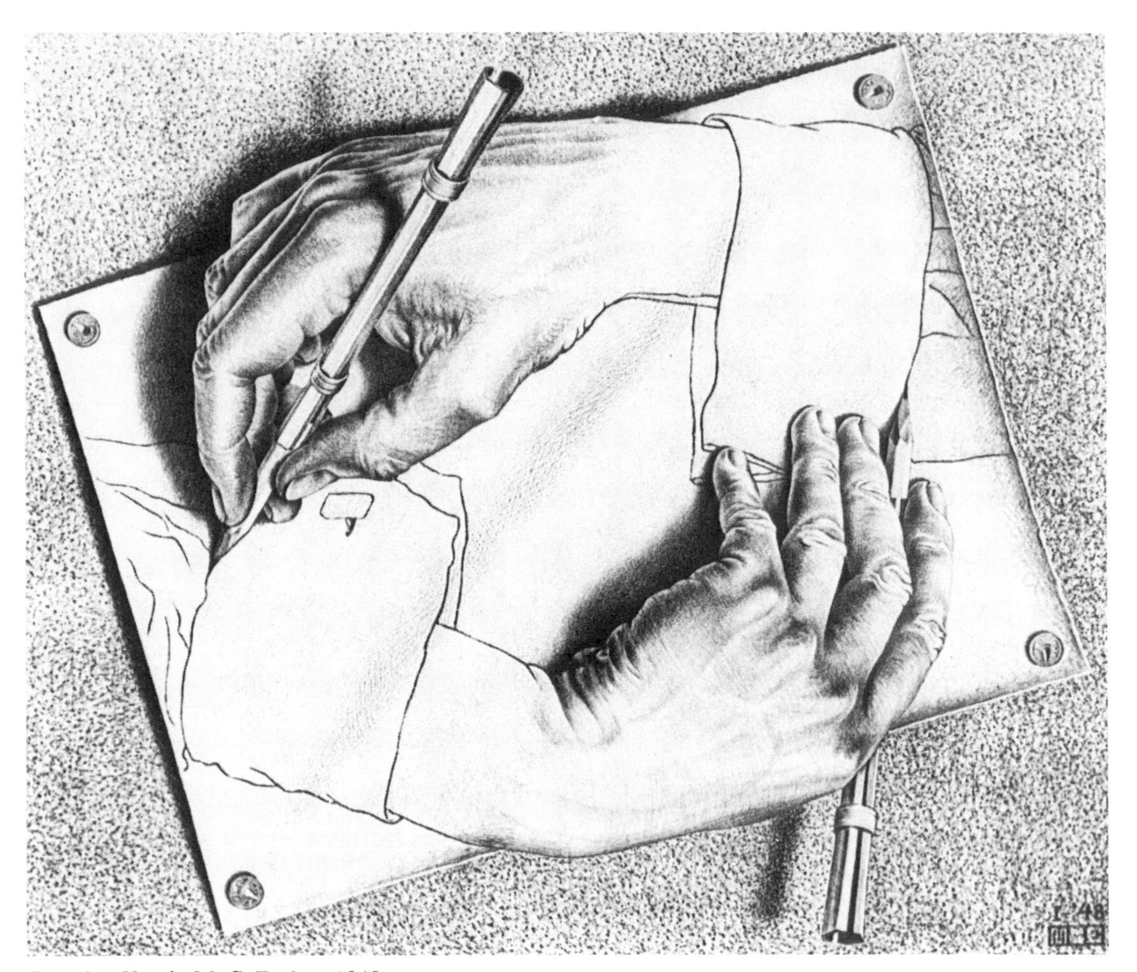

Drawing Hands, M. C. Escher, 1948

In this chapter you will learn how to perform the basic geometric constructions. Geometric shapes that can be created using only the compass and straightedge are called geometric constructions. In this chapter you will perform your first geometric investigations and make conjectures at the conclusion of those investigations. Your work will go smoother and be more enjoyable if you work cooperatively in groups, sharing the tasks and comparing results. You should keep a record of your investigations and a separate list of your conjectures in your notebook.

Lesson 3.1

Duplicating Segments and Angles

Geometric figures that are created with only a compass and straightedge are called **geometric constructions**. Since the Golden Age of Greece, people have made a game of geometric constructions. Over the centuries, people have enjoyed using their compass and straightedge, to discover just what they could and could not construct. In this lesson you will get your chance to play the game of geometric construction.

Neither a ruler nor a protractor is ever used to create geometric constructions. Rulers and protractors are measuring tools, not construction tools. (You may use a ruler as a straightedge in constructions provided you do not use its marks for measuring.)

Construction Tips

1. You begin a construction by selecting a starting point. From that point you usually make a line (or ray or segment) with your straightedge, or a circle (or arc) with your compass. A line used to begin a construction is often referred to as a construction line.
2. Once your construction is underway, you generally position your construction tools using clearly identified points. These are usually points where two lines, two circles, or a circle and a line intersect.
3. Don't erase your construction marks (segments, arcs, circles) when you finish a construction. Your construction marks show your method. They are the footprints that allow someone to track your progress. In a construction, the method of construction is as important as the finished product.
4. For a complicated construction, make a sketch before starting. Then label it. This may help clarify your task.

How would you duplicate a line segment using only your compass and straightedge?

Investigation 3.1.1

Construct a line segment. Experiment and discover a method to duplicate the segment. Your new segment must be the same length as the original. Discuss your method with others near you.

This is the first of many geometric investigations that you will make as you work through this text. You should keep a record of your geometric investigations in your notebook. Clearly label each investigation for future reference. Write a statement summarizing the results of your investigation underneath your work. Some investigations will lead to conjectures.

You've just discovered how to duplicate segments. Using only a compass and straightedge, how would you duplicate an angle? In other words, given an angle, how would you construct a second angle that is the same size? Remember, you may not use your protractor. It is a measuring tool, not a construction tool.

Investigation 3.1.2 

Construct an angle. Experiment and devise a method for duplicating it using only a compass and straightedge. Share your ideas and methods with others near you.

Here are a couple of suggestions that might help you in this task. Begin by constructing an angle on the top half of a clean sheet of paper. Then construct a ray at the bottom of the paper. The ray will be one side of the angle. The endpoint of the ray will be the vertex of the duplicate angle. You're half finished! Now construct an arc on the original angle using the vertex as center. This arc will help you locate the other side of the angle. You're on your own from here!

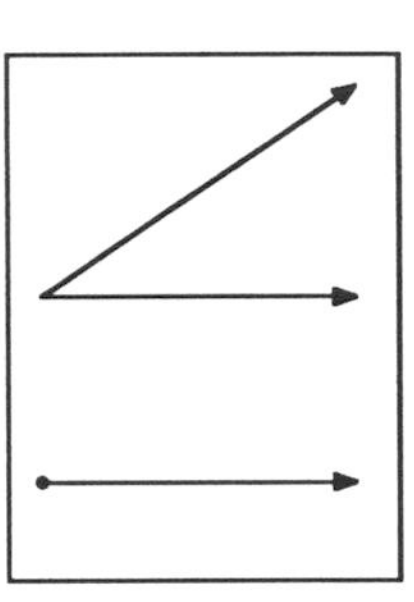

EXERCISE SET

1. Duplicate by construction the three line segments below onto your own paper. Label them as they're labeled in the book. (Please don't write in this book.)

A————————————*B* *C*——————*D* *E*————————————*F*

The ancient Greeks developed their algebra using geometric constructions. Numbers represented the lengths of line segments. Addition of two numbers was the sum of the lengths of two segments. Subtraction was the difference of the lengths. Use your copies of the segments above to construct the segments in Exercises 2 and 3.

2.* Construct a line segment with length $AB + CD$.

3. Construct a line segment with length $AB + 2EF - CD$.

4. Construct an acute angle. Label it $\angle TNY$. Then duplicate it by the method you've devised using your compass and straightedge.

5. Construct an obtuse angle. Label it $\angle LGE$. Then duplicate it.

Duplicate each angle below onto your own paper. We've placed an arc in each angle so that you may duplicate it without writing in this book. If you work carefully, you'll leave no clues (holes, arcs) behind to help the student who uses the book next year.

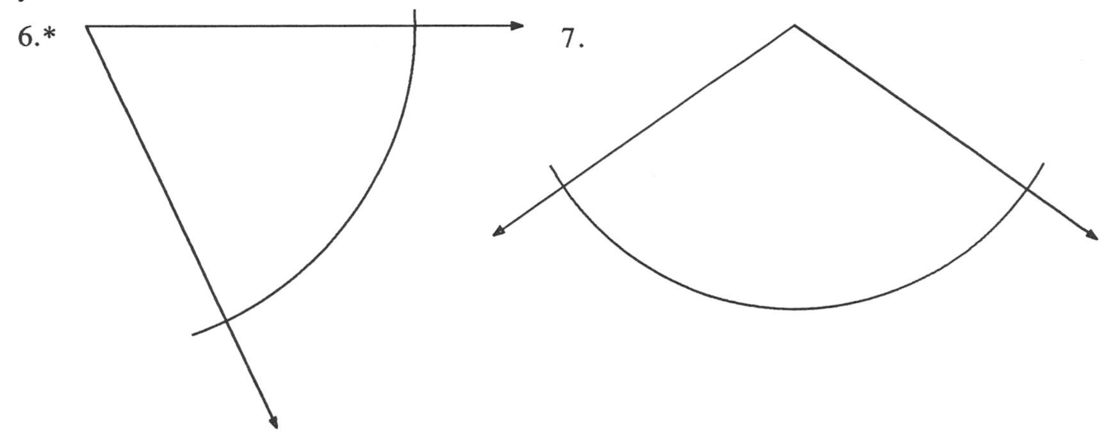

8. Construct two acute angles on your paper. Construct a third angle with a measure equal to the sum of the measures of the first two angles. Remember you cannot use a protractor.

Now, if you can duplicate line segments and angles, you can duplicate polygons.

9. Construct an acute triangle as large as you can make it on the top half of your paper. Then duplicate it onto the bottom half using your compass and straightedge. Make sure to leave all of your construction marks behind to show your method.

10. Construct an equilateral triangle with each side the length of the segment below.

11.* Construct a quadrilateral. Label it quadrilateral *FOUR*. Then duplicate it using your compass and straightedge. Label the vertices of your copy F_1, O_1, U_1, and R_1 so that vertex F matches vertex F_1, O matches O_1, etc.

Note: F_1, O_1, U_1, and R_1 (read "*F* sub one, *O* sub one, *U* sub one, and *R* sub one") are examples of points labeled with subscripts. The subscripts are not funny exponents! We occasionally use subscripts to label points or sets of points that are somehow related to another set of points.

Lesson 3.2
Constructing Perpendicular Bisectors

*A **segment bisector** is a line which passes through the midpoint of a segment.*

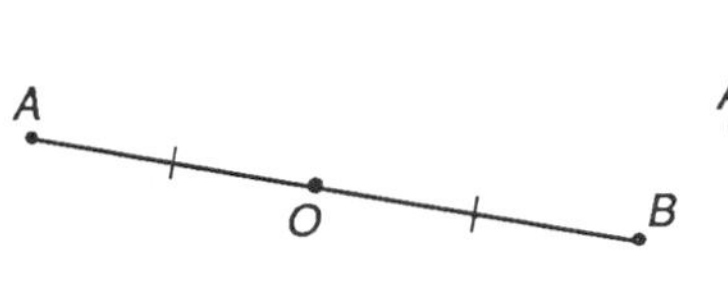

Segment AB has midpoint O.

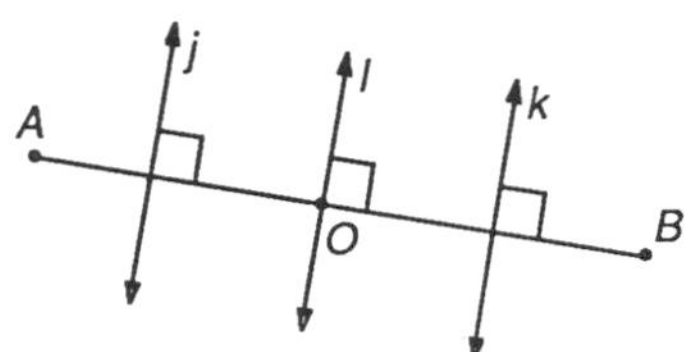

j, k, and l are perpendicular to $\overline{AB}$.

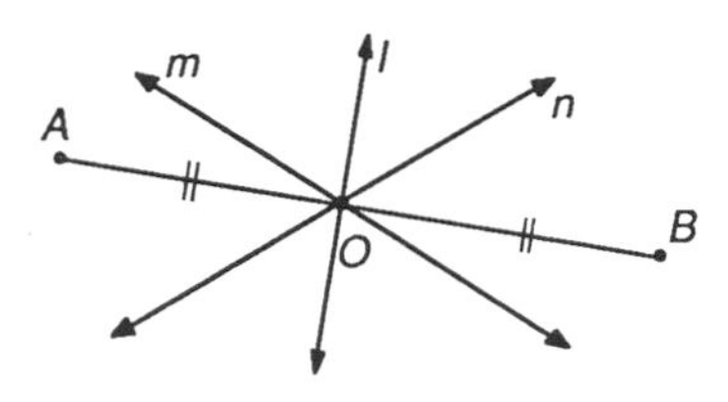

l, m, and n bisect $\overline{AB}$.

*A **perpendicular bisector** of a line segment is a line which divides the line segment into two equal parts (bisects it) and which is also perpendicular to the line segment.*

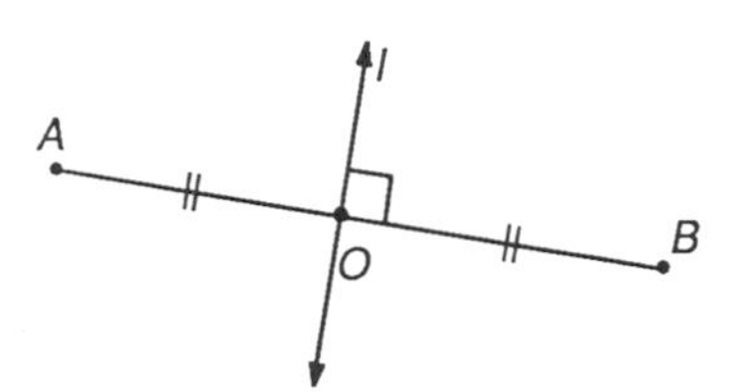

Line l is a perpendicular bisector of $\overline{AB}$.

While by definition segment bisectors and perpendicular bisectors are lines, segments and rays that are contained in the lines are also often referred to as segment or perpendicular bisectors.

There is one and only one midpoint on a line segment. In a plane, there is one and only one line perpendicular to a segment through each point on the segment. Combining these two statements from the last chapter, it seems reasonable to conclude that in a plane a segment has one and only one line as a perpendicular bisector. Let's investigate.

Investigation 3.2.1

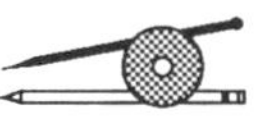

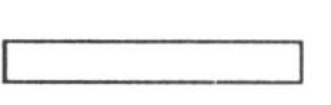

Step 1

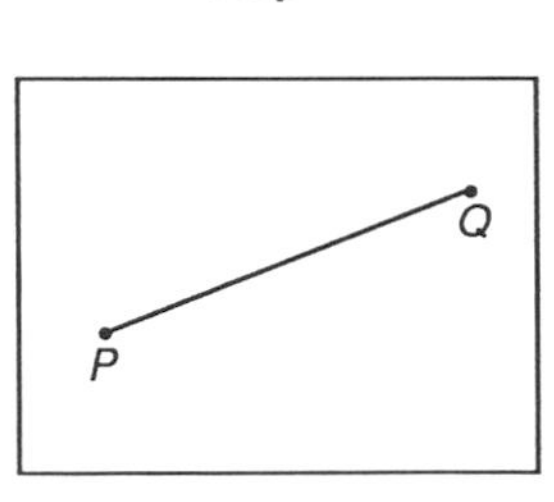

Step 2

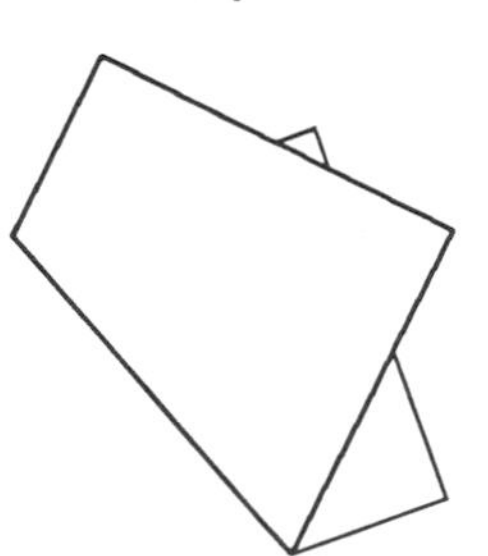

Step 3

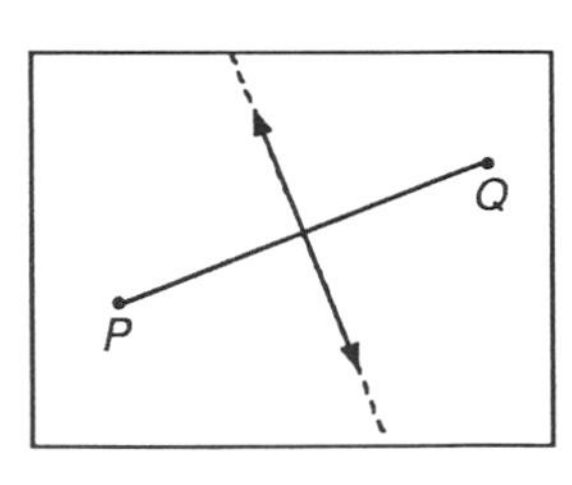

Step 1: On a clean sheet of paper, construct a line segment. Label it $\overline{PQ}$.

Step 2: Fold your paper over so that endpoints P and Q coincide (land exactly on top of each other). Crease your paper along the fold.

Step 3: Unfold your paper. Draw a line in the crease. The line appears to be a perpendicular bisector of $\overline{PQ}$. Check with your ruler and protractor to verify that the line in the crease is indeed a perpendicular bisector of $\overline{PQ}$.

You can demonstrate that the line is the perpendicular bisector with a little logic. The two segments created by the fold are equal since they fit over each other (coincide). Therefore the crease bisects the segment. The two angles are equal since they coincide. Since the two angles form a linear pair of angles, their sum is 180. Thus each measures 90. If each measures 90, then the crease is perpendicular to the segment. Therefore the line in the crease is a perpendicular bisector of the segment.

Is the line the only perpendicular bisector of the segment? Are there other lines that are also perpendicular bisectors of the segment? Can you find another? Try. Compare your results with the results of others around you. Does a segment have one and only one perpendicular bisector?

What else is true about a perpendicular bisector? How would you describe the relationship between the points on the perpendicular bisector with the endpoints of the segment being bisected? Let's perform one more step in our investigation.

Step 4: Place three points on your perpendicular bisector. Label them A, B, and C. With your compass, measure the distances PA and QA. Measure PB and QB. Measure PC and QC.

Step 4

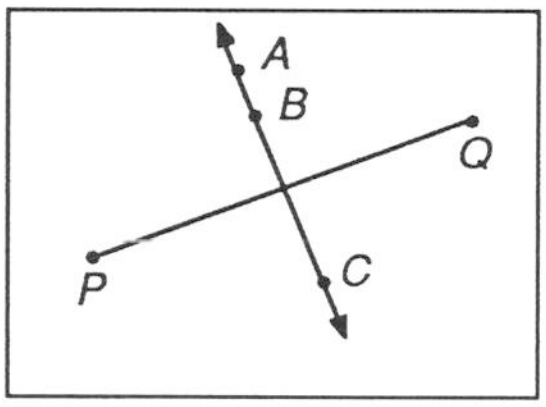

What do you notice about the two distances from each point on the perpendicular bisector to the endpoints of the segment? Compare your results with the results of others near you. You should now be ready to state a conjecture.

Conjecture: *If a point is on the perpendicular bisector of a segment, then it is —?— from the endpoints.*

Complete the conjecture and write it underneath your investigative work as part of the statement summarizing your investigation.

In the last investigation you discovered one way to find the perpendicular bisector of a segment — by paper folding. But this chapter is about geometric constructions, and paper folding is not allowed in a geometric construction. How can you find the perpendicular bisector of a segment by geometric construction, that is, using only a compass and straightedge? In the last investigation you discovered that a point on the perpendicular bisector is equally distant from the two endpoints. What if you turn this around and locate a point that is equally distant from the endpoints of a segment. Will it be on the perpendicular bisector? What if you locate two points that are each equally distant from the endpoints of a segment. Can those points help you construct the perpendicular bisector? Maybe we're on to something. Let's investigate.

Investigation 3.2.2 

Construct a line segment. Start with the following steps:

Step 1

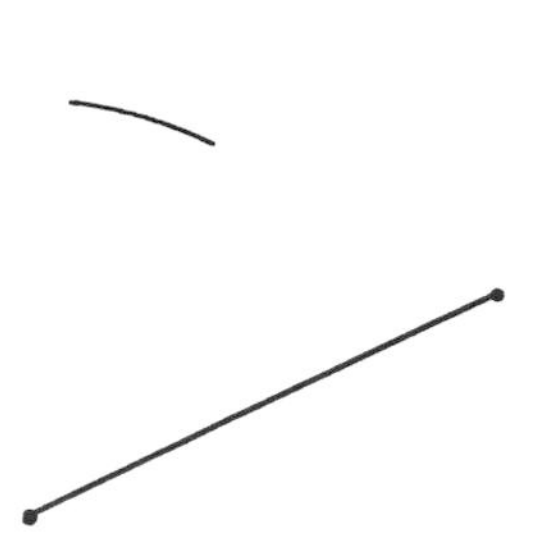

Using one endpoint as center, swing an arc on one side of the segment.

Step 2

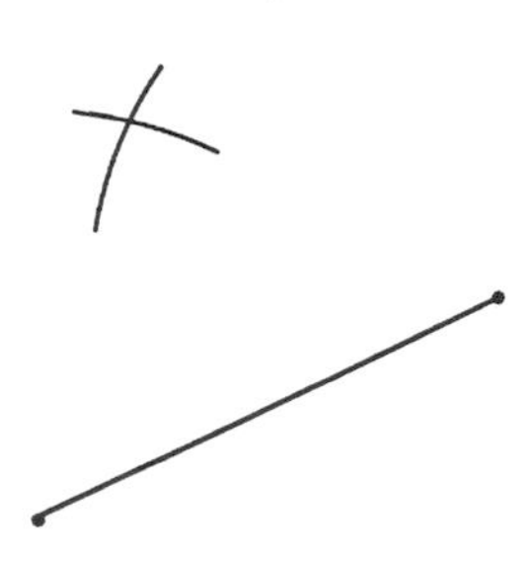

Using the same compass setting but using the other endpoint as center, swing a second arc intersecting the first.

Step 3

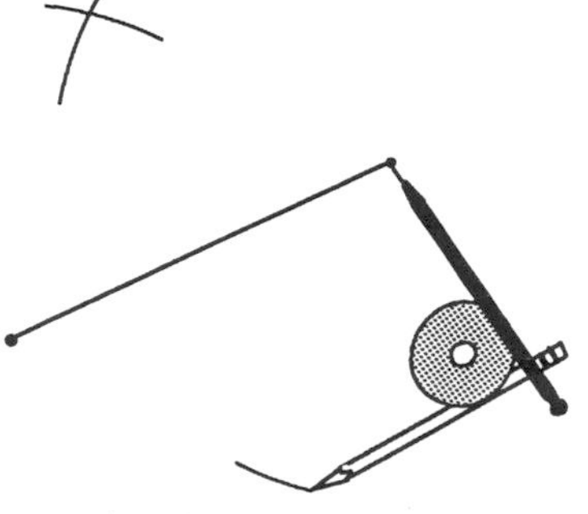

Using the same compass setting and endpoint, swing a third arc on the other side of the segment.

Now you're on your own. You constructed one point equally distant from the endpoints of the segment. Use your compass to find another point equally distant from the endpoints. Use these points to construct a line. Is the line the perpendicular bisector of the segment?

While you couldn't use the paper folding technique of the first investigation in constructing the perpendicular bisector, you can use it to check to see if the line you constructed with your compass and straightedge is, in fact, the perpendicular bisector. Try it. Is the line you constructed the perpendicular bisector of the segment? Write a summary of what you did in this investigation beneath your construction work.

EXERCISE SET A

Match each term with its drawing.

1. Acute scalene triangle
2. Obtuse isosceles triangle
3. Right isosceles triangle
4. Acute isosceles triangle
5. Obtuse scalene triangle
6. Right scalene triangle

a. 50° 8 8 65° 65°

b. 4 4

c. 3 5 4

d.

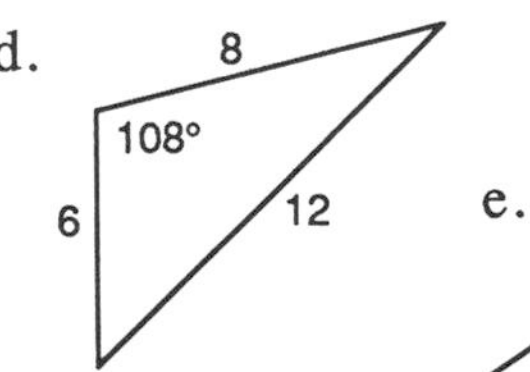

e. 6 120° 6

f.

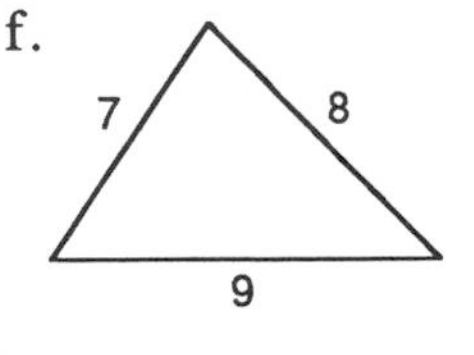

EXERCISE SET B

1. Construct and label $\overline{AB}$. Construct the perpendicular bisector of $\overline{AB}$.

2.* Construct and label $\overline{QD}$. Construct perpendicular bisectors to divide $\overline{QD}$ into four equal parts.

3.* Construct a line segment so close to the edge of a piece of paper that you can only swing arcs on one side of the segment. Then construct the perpendicular bisector of the segment.

4.* Construct $\overline{MN}$ with length equal to the average length of $\overline{AB}$ and $\overline{CD}$ below.

A ———————————— B C ———————————— D

5.* Using $\overline{AB}$ and $\overline{CD}$ above, construct the distance $2AB - \frac{1}{2}CD$.

Improving Visual Thinking Skills

Jigsaw Puzzle

Which of the five pieces cannot be found in the jigsaw puzzle? (Do not flip over any piece.)

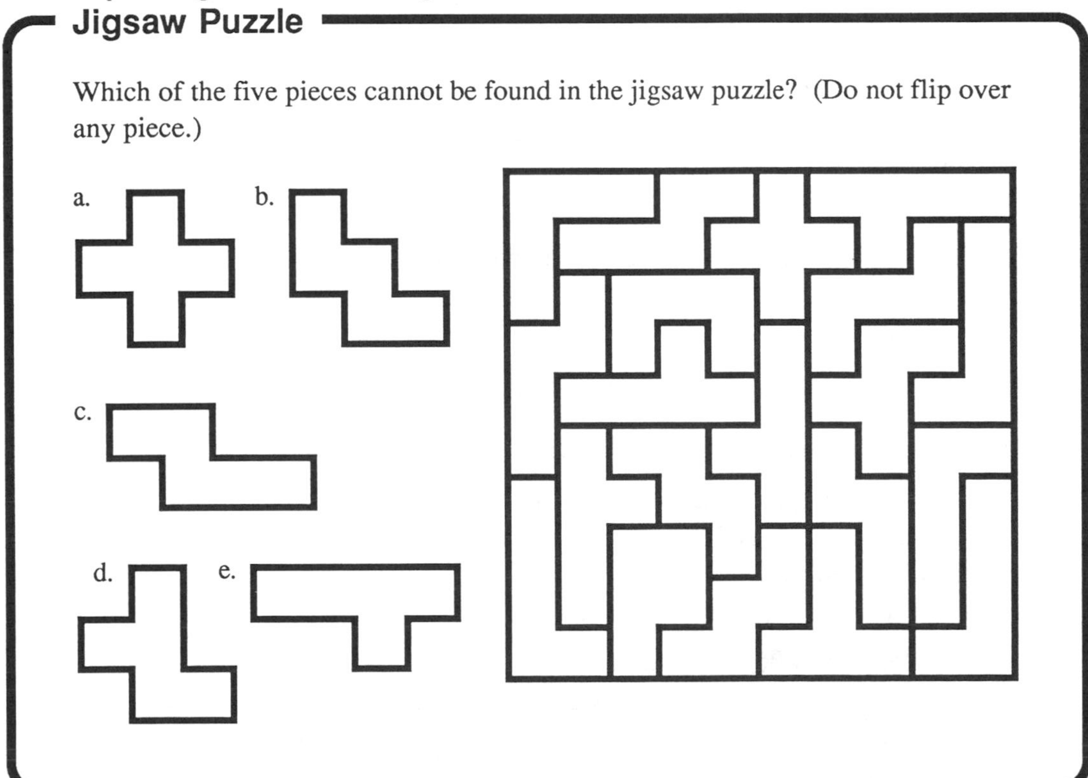

Lesson 3.3

Constructing Perpendiculars

The perpendicular bisector construction that you learned in the last lesson is very important because it locates the midpoint of a segment and it creates a right angle. The construction of a perpendicular segment from a point to a line (with the point not on the line) is another very important construction. You already know how to construct perpendicular bisectors. You can use that know-how to construct a perpendicular from a point to a line.

Investigation 3.3.1

Construct a line. Place a point near but not on the line.

Starting with the steps below, devise a method to construct a perpendicular from the point to the line.

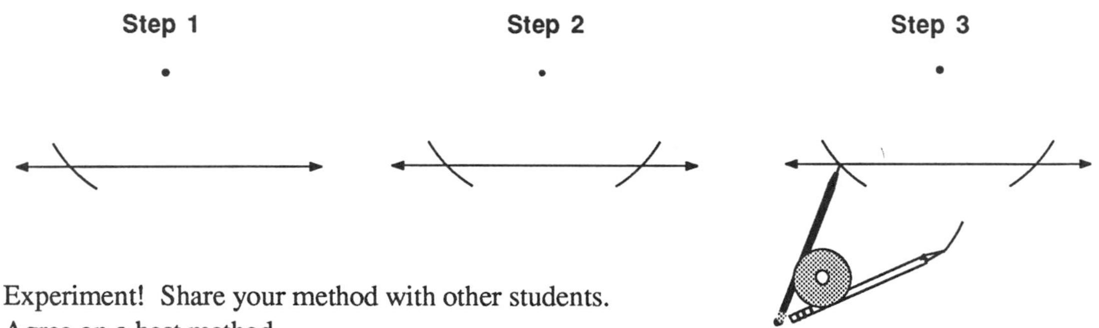

Experiment! Share your method with other students. Agree on a best method.

You have just discovered a method of constructing a perpendicular from a given point to a given line. Let's see what you can discover about this perpendicular segment.

Investigation 3.3.2

On a clean sheet of paper perform the following steps.

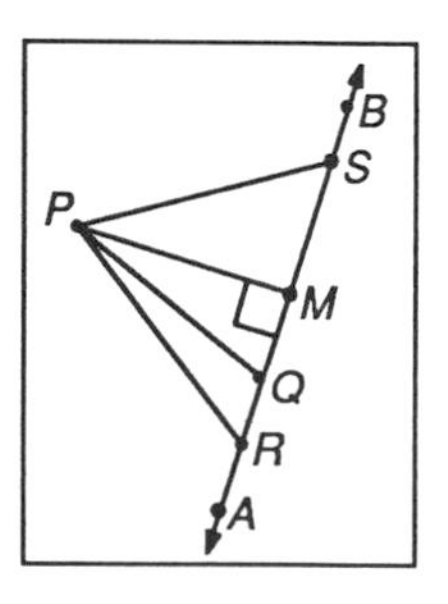

Step 1: Construct and label line *AB* and a point *P* not on $\overleftrightarrow{AB}$.

Step 2: Construct a perpendicular from *P* to $\overleftrightarrow{AB}$ and label the point of intersection of the perpendicular and the given line, point *M*.

Step 3: Select three points on $\overleftrightarrow{AB}$ and label them *Q*, *R*, and *S*. Draw segments *PQ*, *PR*, and *PS*.

Step 4: With your compass compare the lengths *PM*, *PQ*, *PR*, and *PS*.

How do the lengths compare? Can you find a segment from P to $\overleftrightarrow{AB}$ that is shorter than $\overline{PM}$? Compare your results with the results of others near you. You should be ready to state your observations as a conjecture.

Conjecture: *The shortest distance from a point to a line is measured along the —?— from the point to the line.*

Write the conjecture as your summary statement beneath your investigative work.

The construction of a perpendicular from a point to a line lets you find *the* distance from a point to a line. The definition below is based on this construction.

> *The* ***distance from a point to a line*** *is the length of the perpendicular segment from the point to the line.*

Add the definition to your definition list.

EXERCISE SET A

1. Construct an obtuse angle. Place a point inside the angle. Now construct perpendiculars from the point to both sides of the angle.

2. Construct an acute triangle. Label it ΔABC. Construct altitude $\overline{CD}$. (We didn't forget point D. Your job is to locate it!)

3.* Construct obtuse ΔOBT with $\angle O$ obtuse. Construct altitude $\overline{BU}$. In an obtuse triangle, an altitude can fall outside of the triangle. To construct an altitude from point B in your triangle, you must extend the side $\overline{OT}$.

4. Rondo, the distant cousin of Frodo, is at point R near the juncture of the rivers Loudwater and Hoarwell. He has just sensed the approach of a band of angry Orcs. Help him determine the shortest path to a river and save him from becoming the main course at an Orc banquet. Toward which river should he run? Trace the map and then use a construction to demonstrate that you have found the shortest path.

Loudwater

Hoarwell

R

5.* How can you construct a perpendicular to a line through a point that is on the line? Construct a line. Mark a point on your line. Now experiment. Devise a method to construct a perpendicular to your line at the point. This is another important construction.

6. Construct a line. Mark two points on the line and label them Q and R. Now construct a square $SQRE$ with $\overline{QR}$ as a side.

In this chapter you've had practice constructing geometric figures. In the previous chapters you drew and sketched many figures. As you go on in this text you will make many more geometric figures. You will see the words *sketch*, *draw*, and *construct* often. Each has a specific meaning.

When you ***sketch an equilateral triangle***, you may make a free hand sketch of a triangle that looks equilateral. You need not use of any geometric tools.

When you ***draw an equilateral triangle***, you should draw it carefully and accurately. Use your geometric tools. You may use a protractor to measure angles and a ruler to measure the sides in order to make them appear equal in measure.

When you ***construct an equilateral triangle***, you must use only compass and straightedge. Your construction guarantees that your triangle is equilateral.

EXERCISE SET B

Sketch, draw, or construct each figure. Then label the vertices with the appropriate letters. When you sketch or draw, use the special marks to indicate right angles, parallel segments, and segments and angles equal in measure.

1. Sketch obtuse triangle FAT with $m\angle A > 90$ and median $\overline{AY}$.
2. Draw isosceles right triangle RGT with $RG = GT$ and angle bisectors $\overline{RH}$ and $\overline{TI}$.
3. Sketch pentagon $CINKO$ with $CI = IN$ and $CN = NO$.
4.* Draw octagon $EFGHIJKL$ with diagonals $\overline{FH}$, $\overline{HJ}$, $\overline{JL}$, and $\overline{LF}$ forming a square.
5. Construct isosceles ΔABC with $AB = BC$.

Improving Reasoning Skills
Chew on This for Awhile

If the third letter before the second consonant after the third vowel in the alphabet is in the twenty-sixth word of this problem, then print the fortieth word of this problem and then print the twenty-second letter of the alphabet after this word. Otherwise list three uses for chewing gum.

Lesson 3.4

Constructing Angle Bisectors

*An **angle bisector** is a ray which has an endpoint on the vertex of the angle and divides the angle into two angles of equal measure.*

While the definition states that the bisector of an angle is a ray, you may also refer to a segment contained in the ray with an endpoint at the vertex as an angle bisector.

In the investigation to follow you will take a closer look at the bisector of an angle.

Investigation 3.4.1

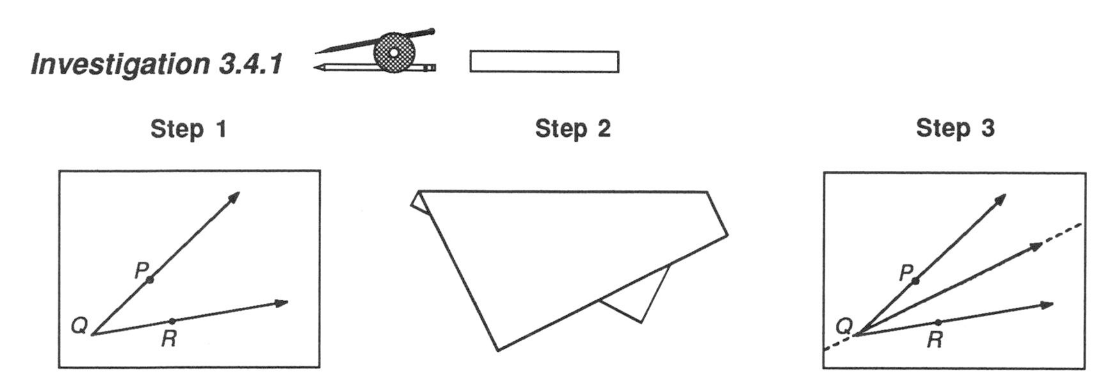

Step 1: On a clean sheet of paper, construct a large acute angle. Label it $\angle PQR$.

Step 2: Fold your paper over so that $\overrightarrow{QP}$ and $\overrightarrow{QR}$ coincide. Crease your paper along the fold.

Step 3: Unfold your paper. Draw a ray with endpoint Q along the crease. The ray appears to be the angle bisector of $\angle PQR$. (A quick check with your protractor can convince you that the ray in the crease is indeed the angle bisector. However, it is not necessary to use your protractor since the fold formed two angles that coincided.)

Clearly, an angle has a bisector. Does an angle have one and only one bisector? Can you find another ray that also bisects the angle?

What else is true about an angle bisector? How would you describe the relationship between the points on the angle bisector with the sides of the angle being bisected? Let's perform one more step in our investigation.

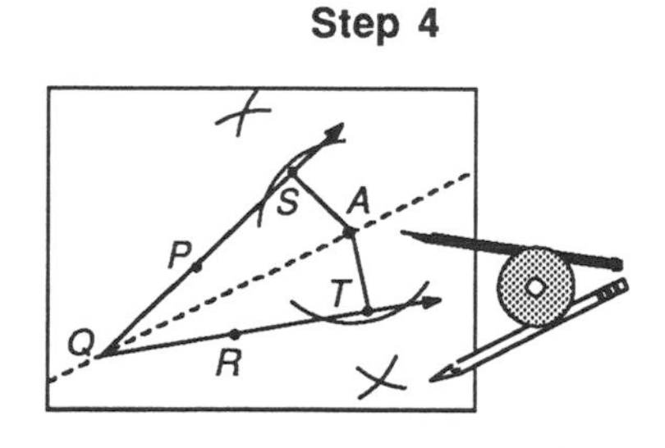

Step 4: Place a point on your angle bisector. Label it A. Construct a perpendicular from A to $\overrightarrow{QP}$ and label the point as S. Construct a perpendicular from A to $\overrightarrow{QR}$ and label the point as T. With your compass, measure the distances AS and AT.

What do you notice about the two distances from a point on the angle bisector to the sides of the angle? Compare your results with the results of others near you. You should now be ready to state a conjecture.

Conjecture: *If a point is on the bisector of an angle, then it is —?— from the sides of the angle.*

Complete the conjecture and write it underneath your investigative work as part of the statement summarizing your investigation.

You've found the bisector of an angle by paper folding. Now let's see if you can devise a method to construct the angle bisector.

Investigation 3.4.2

Construct an acute angle.

Starting with the steps below, devise a method to construct the bisector of the angle.

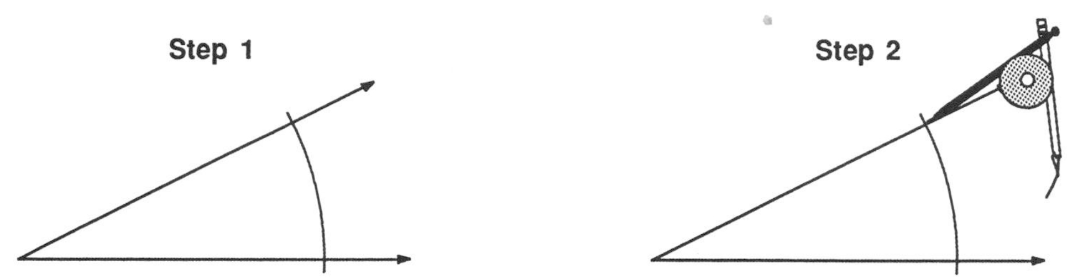

Experiment! Once you think you have constructed the angle bisector, you can fold your paper to see if the ray you constructed actually is the bisector. Share your method with other students. Agree on a best method.

One more investigation. In the first lesson of this chapter you constructed an equilateral triangle. What can you discover about equilateral triangles? Let's investigate.

Investigation 3.4.3

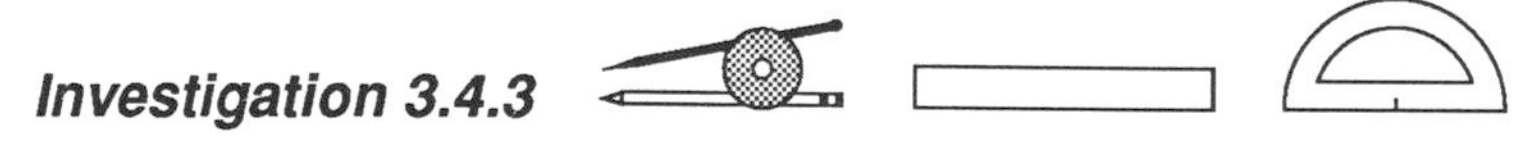

Step 1: On a clean sheet of paper, construct a large equilateral triangle.

Step 2: With your protractor measure the three angles of the triangle.

What is the measure of each angle of the equilateral triangle? Try this again. Construct a second equilateral triangle. What is the measure of each angle? Compare your results with the results of others. You should be ready to state a conjecture.

Conjecture: *The measure of each angle of an equilateral triangle is —?— .*

Complete the conjecture and write it underneath your investigative work as part of the statement summarizing your investigation.

EXERCISE SET A

Match each construction with its diagram.

1. Construction of angle bisector
2. Construction of median
3. Construction of perpendicular bisector
4. Construction of altitude
5.* Construction of a 60 degree angle

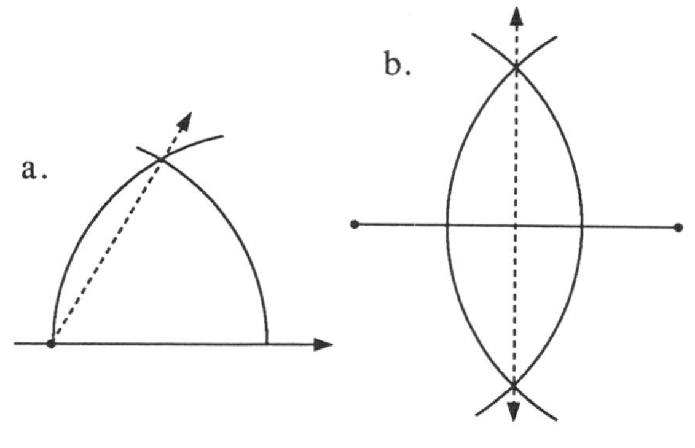

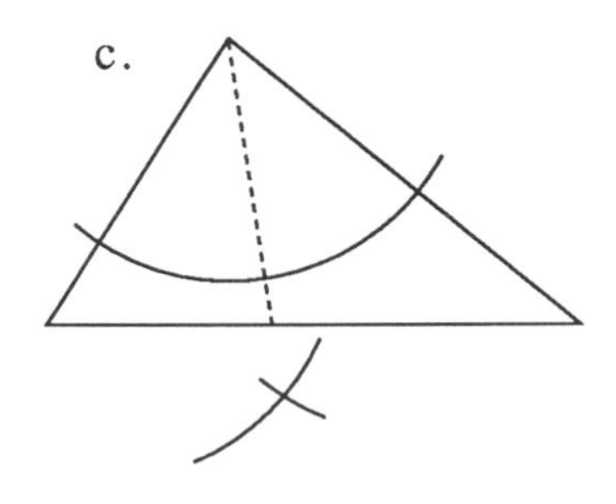

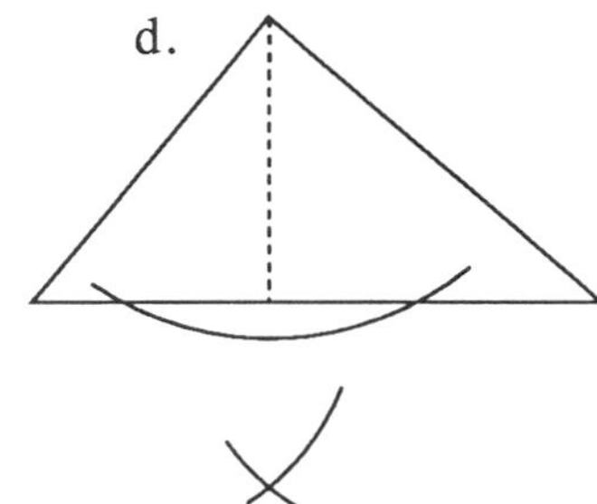

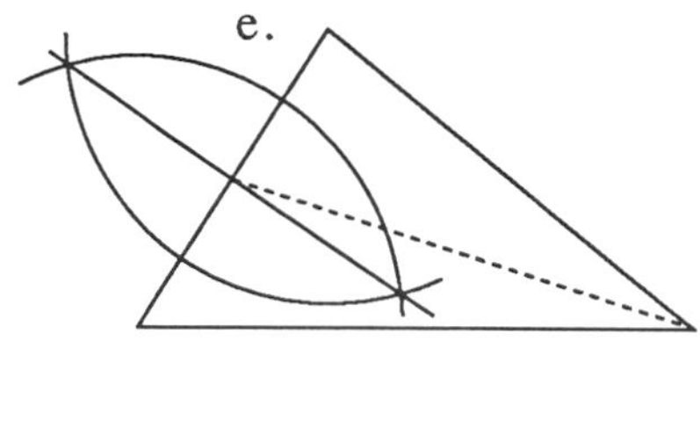

EXERCISE SET B

You now know how to duplicate angles and bisect angles. You can also construct angles of 60 and 90 degrees. With this as a start, you can construct angles of many other sizes. In Exercises 1 to 8, describe how you would construct an angle with each of the measures below. (Don't actually perform the construction.)

1.* 30   2. 45   3. 15   4. 120

5. $22\frac{1}{2}$ 6. 75   7.* $52\frac{1}{2}$ 8. 135

Now for the hard part! In Exercises 9 to 11, construct an angle with each given measure. Identify clearly which angle in your construction has the desired measure. Remember, you may use only your compass and straightedge. No protractor!

9.* 135   10. 105   11. $7\frac{1}{2}$

12. Construct an obtuse triangle. Bisect each angle.

13. Construct a large scalene obtuse ΔABC (filling your paper) with $\angle B$ the obtuse angle. Construct the median $\overline{BM}$, the angle bisector $\overline{BR}$, and the altitude $\overline{BS}$. What is the order of the points on $\overline{AC}$? Complete the conjecture: The foot of the —?— is always between the foot of the —?— and the foot of the —?— . Compare your results with the results of others near you. Is the order of points always the same for the foot of the median, foot of the altitude, and foot of the angle bisector?

Computer Activity

Parallelograms with Logo

In this computer activity you will use the Logo computer language to explore parallelograms. Explorations with Logo can take time. You should not feel you have to complete every task. To undertake this computer activity you do not need previous experience with Logo. You will, however, need the following:

- Computer and disk with Logo computer language
- Resource person familiar enough with Logo to help you get started
- Disk with the procedure called PARA (The listing of the PARA procedure is in *Appendix B: Logo Procedures.*)

Start up your computer using the disk with Logo on it. You're ready to go!

1. The Logo turtle will respond to your commands. Your first goal is to discover the action that the turtle takes when you give it some common Logo commands. Experiment with the commands below. Do they all work in your version of Logo? Try different numbers with any command which has a number. You need to press the return key after typing each command. Typing errors can be corrected by pressing return and retyping the command.

Command	Short Form	Command	Short Form	Command	Short Form
FORWARD 50	FD 50	CLEARSCREEN	CS	SHOWTURTLE	ST
BACK 60	BK 60	DRAW		HIDETURTLE	HT
LEFT 90	LT 90	CLEARGRAPHICS	CG	FULLSCREEN	Ctrl-L or F
RIGHT 45	RT 45	PENUP	PU	TEXTSCREEN	Ctrl-T
HOME		PENDOWN	PD	SPLITSCREEN	Ctrl-S

2. Use the commands above (with different numbers) to make a rectangle. List the commands that you use. You can fix mistakes by "undoing" a command with its "opposite," or by clearing the screen and starting over. Now make a square. Make a parallelogram that is not a rectangle. Make a rhombus that is not a square. Make a list of the commands that you use for each quadrilateral.

Now load the PARA procedure from your procedures disk.

3. The procedure PARA needs three inputs: :SIDE1, :ANGLE, and :SIDE2. For example, typing PARA 20 75 30 and pressing return will start the turtle working. Experiment with PARA using inputs of your own. What does PARA do? What do the inputs control?

4. How many parallelograms do you see in the figure on the right? Using only the PARA procedure and the simple turtle commands listed above (use them to shift the turtle's position), create a box like the one you see. List the commands that you use.

5. Make the box using only PARA and the turn commands LEFT and RIGHT.

6. Make a taller box, a wider box, a longer box, the same box as if viewed from a different angle. What inputs to the PARA procedure did you have to change? Did you have to change any of your turtle moves?

7. The designs below were made using PARA together with simple Logo commands. What designs can you make with PARA? Try some simple designs to start. If you create something you would like to keep and have access to a printer, ask your Logo resource person how to print your picture in your version of Logo.

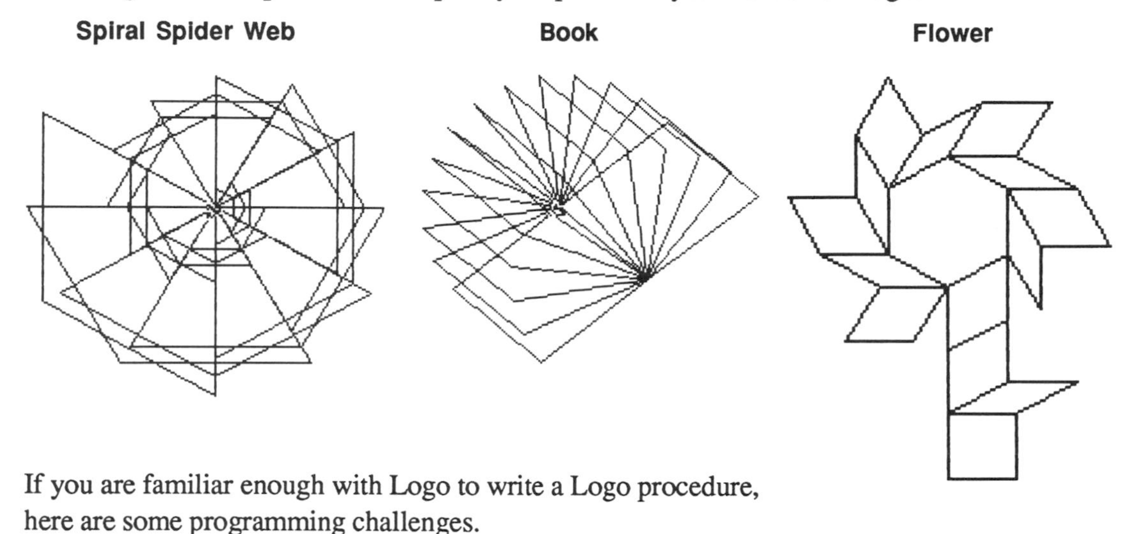

If you are familiar enough with Logo to write a Logo procedure, here are some programming challenges.

8. Use the PARA procedure to create a new procedure called BOX which takes four inputs: :HEIGHT, :WIDTH, :LENGTH, and :ANGLE. Write BOX so that when the turtle has finished BOXing, it is in exactly the same position and has the same heading as it did before it BOXed.

9. Will your BOX procedure draw any size or shape of BOX (within the limits of your screen)? What is the geometric name for the solid represented by your BOX.

10. Make a STACK of BOXes like you see at right.

11. Make three STACKs of BOXes in an ARRAY.

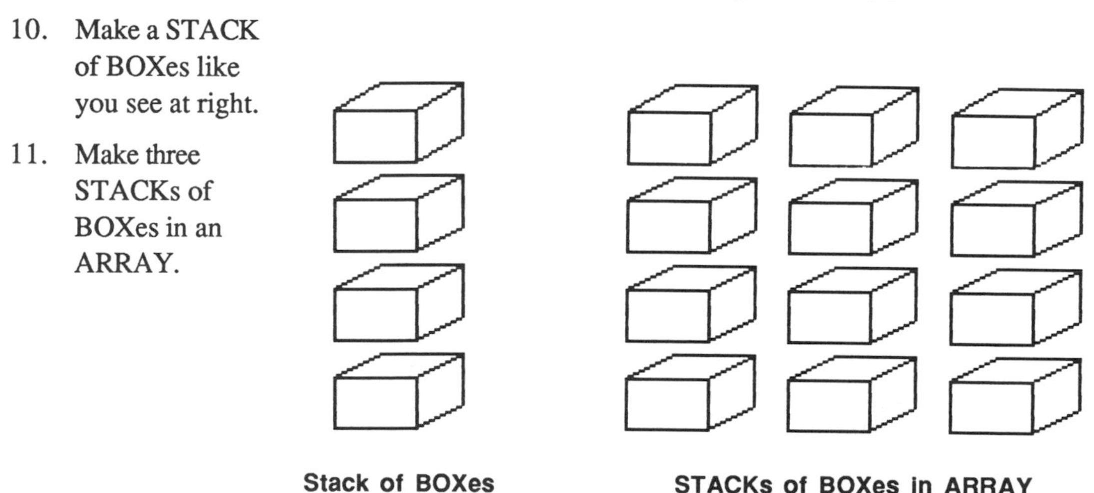

Lesson 3.5

Constructing Triangles and Quadrilaterals

If you can combine a given set of triangle parts (arranged in a given way) to create a triangle, and if all the triangles created with those parts turn out to be identical, then the given set of parts **determines** a triangle. (Two triangles are identical if they can be made to fit over each other by sliding, rotating, or flipping.)

EXERCISE SET

In each exercise below, first sketch and label the figure that you are going to construct. (This helps organize your construction.) Then perform the construction. The arc drawn in each angle will help you copy that angle accurately without writing in the book. In Exercises 1 to 4, the given parts determine a triangle.

1.* Given: Construct: ΔMAS

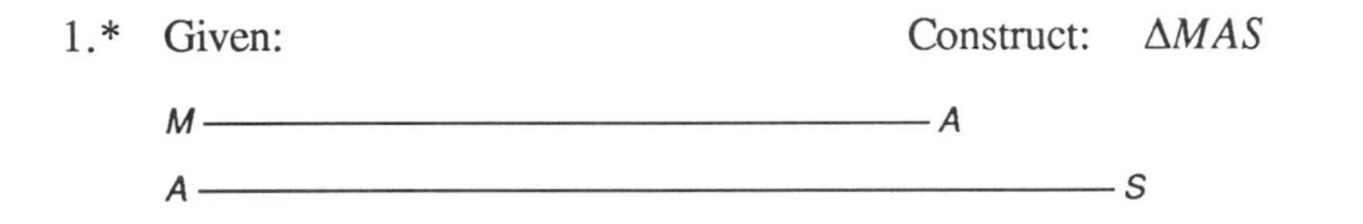

2. Given: Construct: ΔDOG

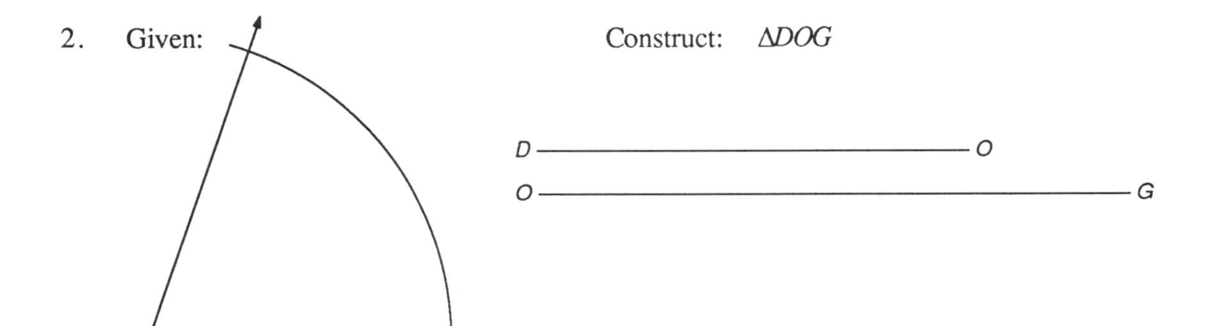

3. Given: Construct: ΔIGS

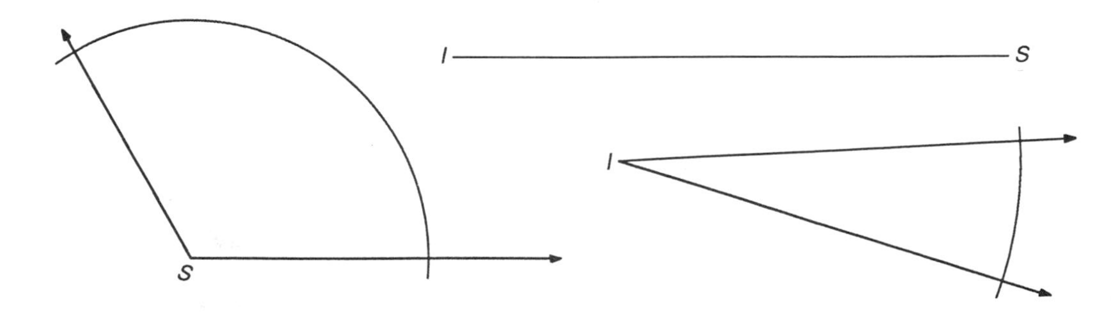

4.* The two segments and angle below do not determine a triangle.

Given:

Construct: Two different triangles named ΔABC having the three given parts.

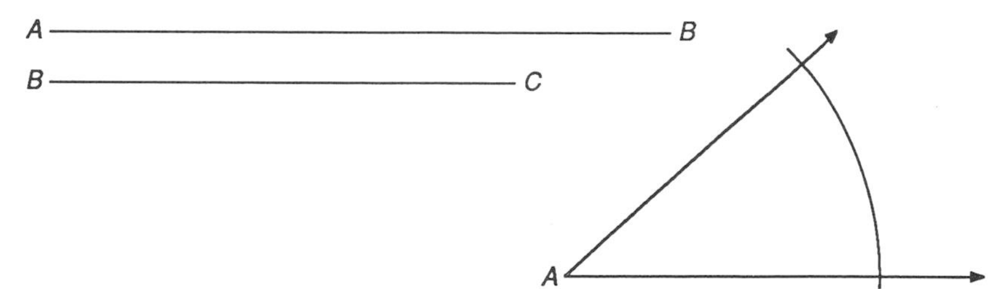

5. Construct a quadrilateral with one pair of sides of equal length.

6. Construct a quadrilateral with two pairs of consecutive sides of equal length. A quadrilateral with two pairs of consecutive sides of equal length is called a **kite**.

7. Construct a quadrilateral with two pairs of opposite sides of equal length.

8. Construct a quadrilateral with exactly three sides of equal length.

9. Construct a quadrilateral with all four sides of equal length.

Improving Reasoning Skills

The Harmonic Triangle

The pattern at right is called the harmonic triangle. What is the next row in the harmonic triangle?

$$1$$

$$\frac{1}{2} \quad \frac{1}{2}$$

$$\frac{1}{3} \quad \frac{1}{6} \quad \frac{1}{3}$$

$$\frac{1}{4} \quad \frac{1}{12} \quad \frac{1}{12} \quad \frac{1}{4}$$

$$\frac{1}{5} \quad \frac{1}{20} \quad \frac{1}{30} \quad \frac{1}{20} \quad \frac{1}{5}$$

$$\frac{1}{6} \quad \frac{1}{30} \quad \frac{1}{60} \quad \frac{1}{60} \quad \frac{1}{30} \quad \frac{1}{6}$$

? ? ? ? ? ? ?

Lesson 3.6
Constructing Parallel Lines

Parallel lines *are lines which lie in the same plane and do not intersect.*

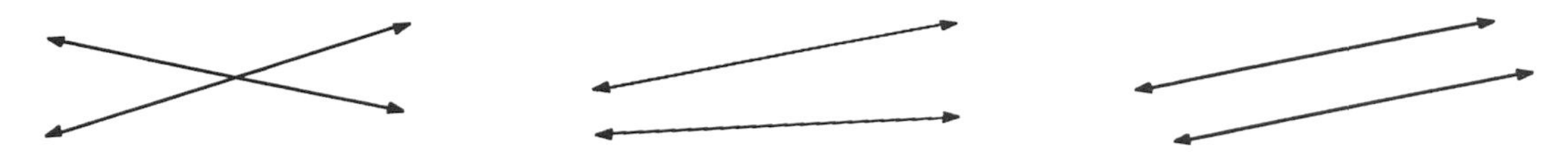

The first pair of lines intersect. They are clearly not parallel. The second pair of lines do not meet as drawn. However, if they are extended, they will intersect. Therefore they are not parallel. But what about the third pair? Hard to tell, right? You could extend them, but you might need to draw them on a roll of toilet paper to extend them far enough. If two lines in a plane are parallel, then they must always be the same distance apart. If the distances are not the same, then they are not parallel.

You can also say that if two lines are always the same distance apart, then they are parallel. This gives you a way of constructing parallel lines (called the *equidistant method*). The steps below show how you can construct a line parallel to a given line through a given point. The given line is m and the given point is P.

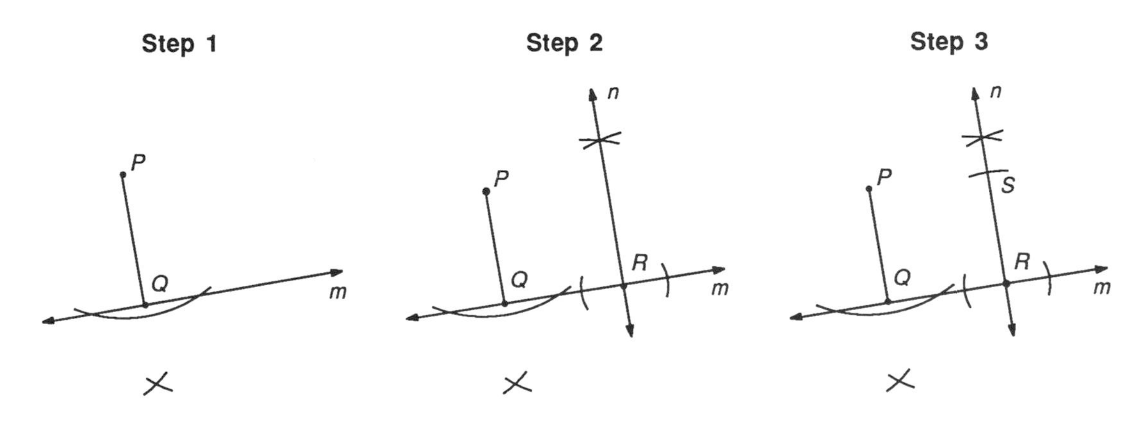

Step 1: Construct a perpendicular from the given point P to the given line m. Label the point of intersection as Q (PQ is the distance from P to m).

Step 2: Select a point R on line m (as far away from P as possible) and construct a line through R perpendicular to m. Label the perpendicular n.

Step 3: With your compass, locate a point S on n (P and S are on the same side of m) so that $PQ = RS$.

Step 4: Construct $\overleftrightarrow{PS}$. Since $PQ = RS$, then $\overleftrightarrow{PS} \parallel m$.

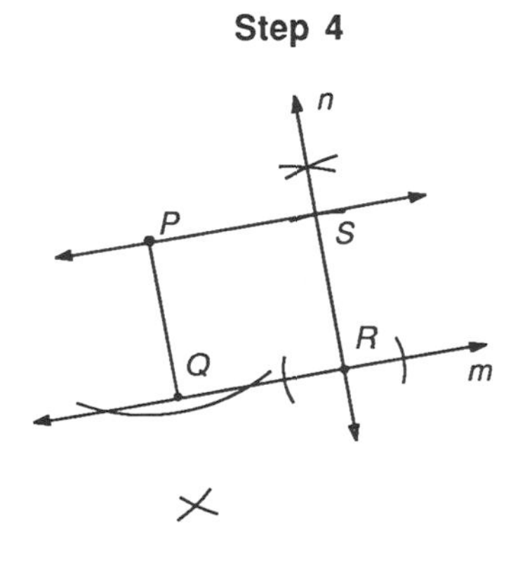

Here is another method for constructing parallel lines. It is a little easier and is based on the rhombus.

> ***A rhombus** is a parallelogram with all sides equal in measure.*

In the last lesson you constructed a quadrilateral with all four sides equal in measure. Did you notice that it was a rhombus? In other words, did you notice that the opposite sides turned out parallel even though you didn't purposely construct the quadrilateral to be a parallelogram. It turns out that if a quadrilateral has four equal length sides, then the opposite sides must be parallel and the quadrilateral must be a rhombus. You can use this to construct parallel lines. Constructing a rhombus is quicker and easier than constructing two perpendiculars. Therefore, if you wish to construct a pair of parallel lines, construct a rhombus. Voilà! You get parallel lines — FREE!

This method of constructing parallel lines is called the *rhombus method.* Start with a line, l. Select a point not on line l and label it P. Then follow the steps.

Step 1

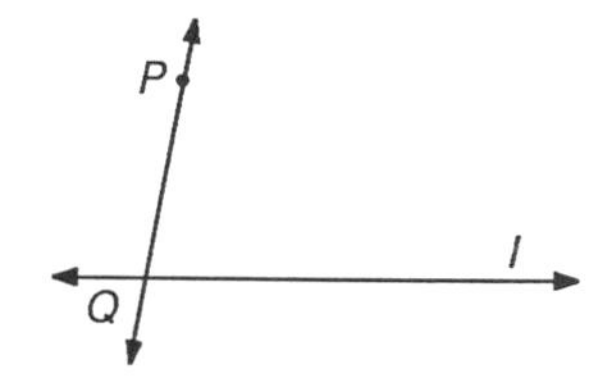

Construct any line from point P that intersects line l. Label the point of intersection Q.

Step 2

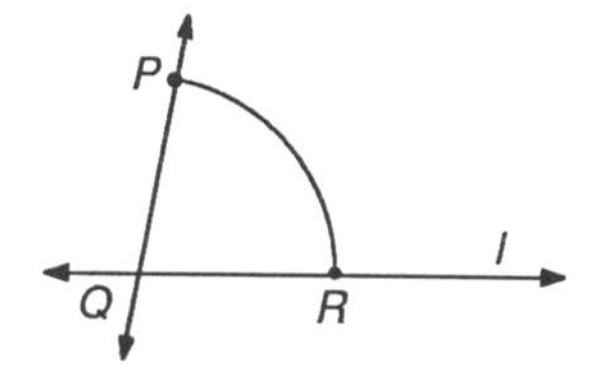

Swing an arc with center Q of radius PQ. Label as R the point where the arc crosses line l.

Step 3

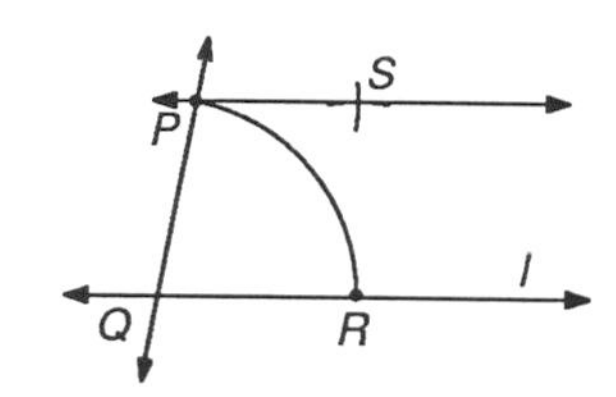

With P and R as centers, swing arcs of radius PQ. Label the intersection S. Construct $\overleftrightarrow{PS}$. $PQRS$ is a rhombus, so $\overleftrightarrow{QR} \parallel \overleftrightarrow{PS}$.

EXERCISE SET A

1.* Sketch trapezoid $ZOID$ with $\overline{ZO} \parallel \overline{ID}$, T the midpoint of $\overline{OI}$, and R the midpoint of $\overline{ZD}$. Put $\overline{TR}$ in your sketch.

2. Draw rhombus $ROMB$ with $m\angle R = 60$ and diagonal $\overline{OB}$.

3. Draw rectangle $RECK$ with diagonals $RC = EK = 8$ cm intersecting at point W.

4. Sketch $\overleftrightarrow{AB}$, $\overleftrightarrow{CD}$, $\overleftrightarrow{EF}$, and $\overleftrightarrow{GH}$ with $\overleftrightarrow{AB} \parallel \overleftrightarrow{CD}$, $\overleftrightarrow{CD} \perp \overleftrightarrow{EF}$, and $\overleftrightarrow{EF} \parallel \overleftrightarrow{GH}$.

5. Draw octagon $ALTOSIGN$ with $\overline{AS} \perp \overline{GT}$.

EXERCISE SET B

1. Construct a line. Select a point not on the line. Construct a second line parallel to the first line that passes through the point (using the equidistant method).

2. Construct a line. Select a point not on the line. Construct a second line parallel to the first line that passes through the point (using the rhombus method). Which method is easier, the equidistant method or the rhombus method?

3. A rectangle can also be defined as a parallelogram with one right angle. Construct rectangle *RECT* with $\overline{RE}$ and $\overline{EC}$ as a pair of consecutive sides. Which method is better here?

   R ——————————————————— E

   E ——————————— C

4. Construct trapezoid *TRAP* (using the equidistant method) with $\overline{TR}$ and $\overline{AP}$ as the two parallel sides and with *AP* as the distance between them. (Many solutions!)

   T ——————————————————— R

   A ——————————— P

5. Construct parallelogram *GRAM* (using the equidistant method) with $\overline{RG}$ and $\overline{RA}$ as two consecutive sides and *ML* as the distance between $\overline{RG}$ and $\overline{AM}$. (How many solutions?)

   G ——————————————————— R

   R ———————————————— A

   M ——————————— L

6. Construct a large acute scalene triangle and label it ΔABC. Locate the midpoint of side $\overline{AB}$ and label it *M*. Locate the midpoint of side $\overline{BC}$ and label it *N*. Using the rhombus method, construct lines through points *M* and *N* parallel to the sides $\overline{AC}$ and $\overline{AB}$ respectively.

Improving Visual Thinking Skills

Painted Faces I

Some unit cubes are assembled to form a larger cube, and then some of the faces of this larger cube are painted. After the paint dries, the larger cube is disassembled into the unit cubes and it is found that 24 of these have no paint on any of their faces. How many faces of the larger cube were painted?

Lesson 3.7
Construction Problems

People who are only good with hammers see every problem as a nail.

– Abraham Maslow

The sum of the lengths of the sides of a polygon is its ***perimeter****.*

EXERCISE SET

The lower case letter above some segments represents the length of the segment.

1.* Given: x, y

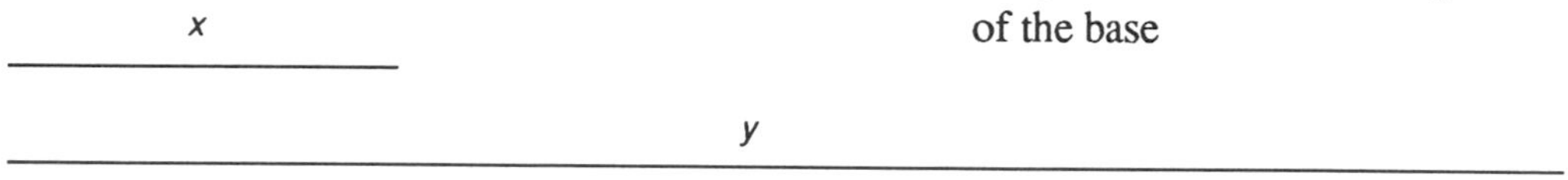

Construct: Isosceles triangle CAT with y the perimeter and x the length of the base

2. Given: z

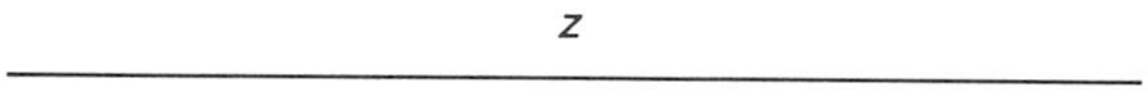

Construct: An isosceles right triangle wit the length of each of the two equal sides

3. Given: R———A, A———T, R———T

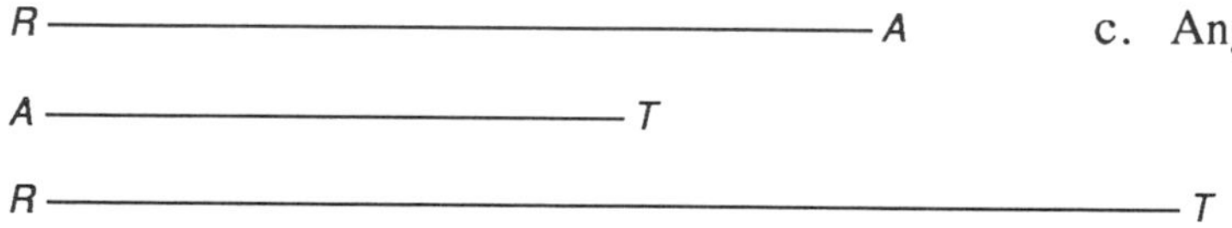

Construct:
a. ΔRAT
b. Median $\overline{TM}$
c. Angle bisector $\overline{RB}$

4. Given: M———S, M———E, M

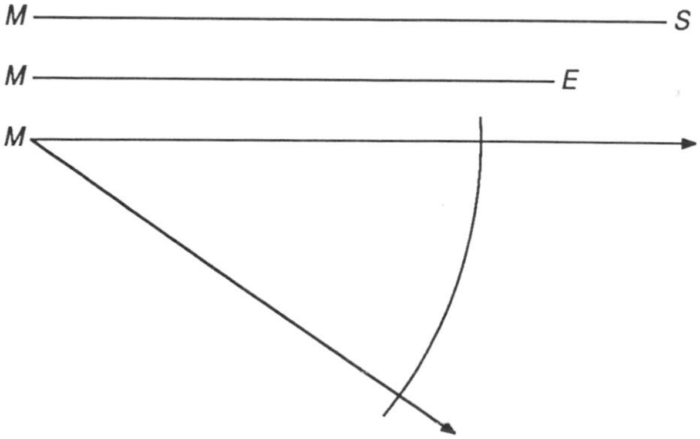

Construct:
a. ΔMSE
b. Perpendicular bisector of $\overline{MS}$
c. $\overline{OU}$ where O is the midpoint of $\overline{MS}$ and U is the midpoint of $\overline{SE}$

5. Construct and label ΔREG (your choice). Then construct another triangle that has a perimeter equal to half the perimeter of ΔREG.

6. Given: Construct: A square with z the perimeter

z

7. Given: Construct: A rhombus with x the length of each side and $\angle A$ one of the acute angles

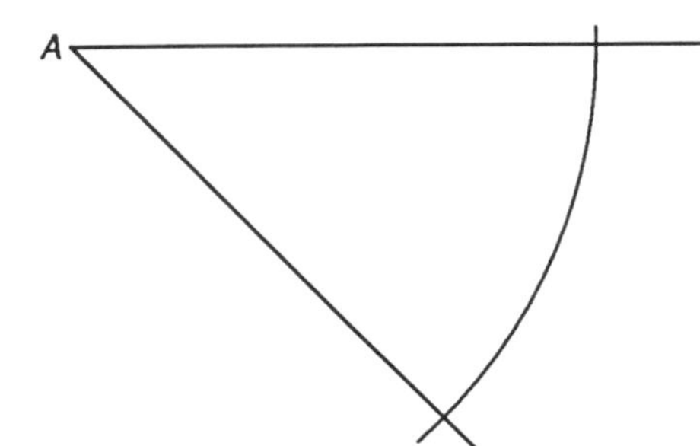

x

Improving Reasoning Skills

Murder at the Socratic Liars Club

Hemlock Stones, the not yet famous consulting detective, has just been called in to solve the mysterious murder at the Socratic Liar's Club. There are five suspects each of whom has sworn by the club oath to make two true statements and one false one whenever speaking to someone on the club's premises. From their recorded statements below, Hemlock Stones was able to determine "who dunit." Can you?

Professor: I did not kill Henley. I never owned a knife. Lance did it.

Ethel: I didn't kill him. I don't own a knife. The others are crazy.

Phoebe: I'm innocent. Lance is the killer. I don't even know Dutch.

Lance: I'm innocent. Dutch is guilty. The Professor lied when he said I did it.

Dutch: I didn't kill him. Ethel is the murderer. Phoebe and I are old friends.

Lesson 3.8

Constructing Points of Concurrency

You now can construct angle bisectors, perpendicular bisectors of the sides, medians, and altitudes in triangles. In this lesson and the next lesson you will discover special properties of these lines and segments.

Investigation 3.8.1

Step 1: Construct a large acute triangle on the top half of a clean sheet of unlined paper. Construct a large obtuse triangle on the bottom half of the paper.

Step 2: Construct the three angle bisectors in each triangle.

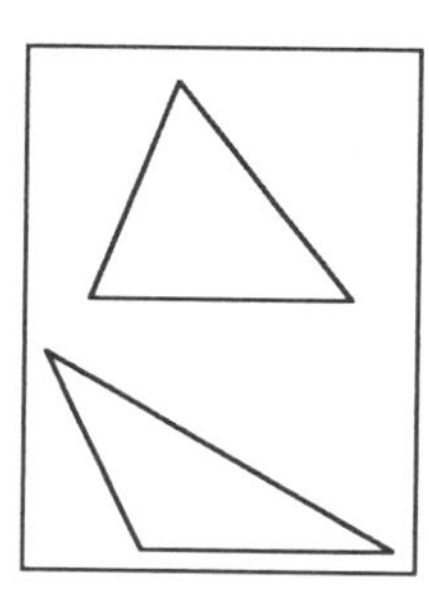

What do you notice? Compare your results with the results of others near you. State your observations as a conjecture.

Conjecture: *The three angle bisectors of a triangle —?—.*

Investigation 3.8.2

Step 1: Construct a large acute triangle on the top half of a clean sheet of unlined paper. Construct a large obtuse triangle on the bottom half of the paper.

Step 2: Construct the three perpendicular bisectors of the sides in each triangle.

What do you notice? Compare your results with the results of others near you. State your observations as your next conjecture.

Conjecture: *The three perpendicular bisectors of the sides of a triangle —?—.*

Investigation 3.8.3

Step 1: Construct a large acute triangle on the top half of a clean sheet of unlined paper. Construct a large obtuse triangle on the bottom half of the paper.

Step 2: Construct the three altitudes of each triangle.

Step 3: Construct lines through the altitudes of the obtuse triangle.

What do you notice? Compare your results with the results of others near you. State your observations as your next conjecture.

Conjecture: *The three altitudes (or the lines through the altitudes) of a triangle —?—.*

Were you surprised by the results of your investigations? The definition below will help you describe your findings.

__Concurrent lines__ (segments or rays) are lines which lie in the same plane and intersect in a single point.

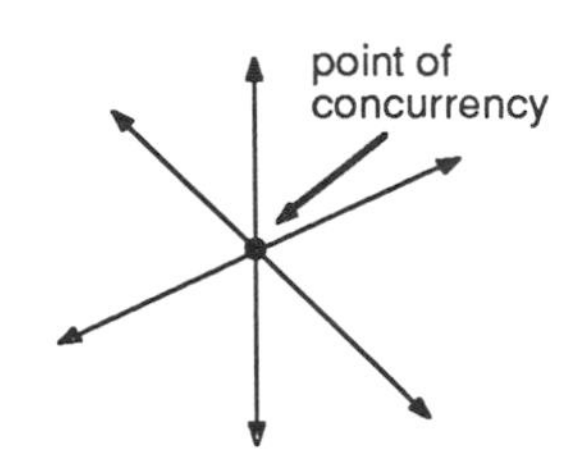

The point of intersection is the **point of concurrency**. It's no big deal for two lines to be concurrent. In a plane, any two lines that aren't parallel are concurrent. It is special, however, for three or more lines to be concurrent.

Each of the points of concurrency that you found in your investigations has a name. The point of concurrency of the three angle bisectors in a triangle is the **incenter**. The point of concurrency of the three perpendicular bisectors of the sides of a triangle is the **circumcenter**. The point of concurrency of the three altitudes or the lines through the altitudes of a triangle is the **orthocenter**.

The incenter and circumcenter are useful in geometric constructions. They help you neatly construct circles in and about triangles.

A circle is __inscribed in a polygon__ when it touches each side of the polygon at exactly one point. (The polygon is circumscribed about the circle.)

A circle is __circumscribed about a polygon__ when it passes through each vertex of the polygon. (The polygon is inscribed in the circle.)

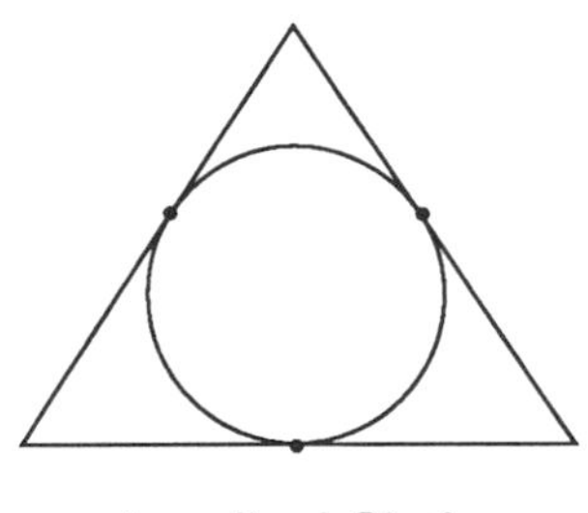

Inscribed Circle
(Circumscribed Triangle)

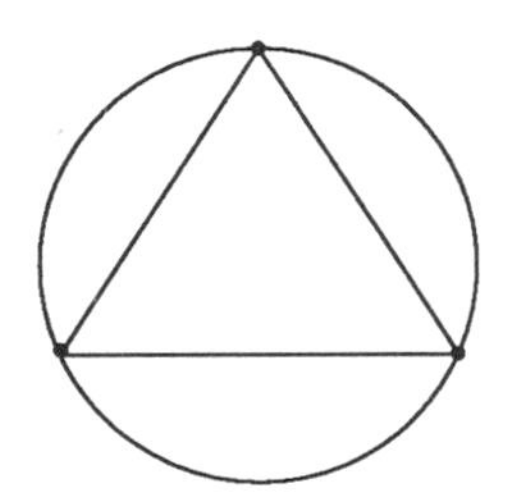

Circumscribed Circle
(Inscribed Triangle)

From the conjecture you made earlier about angle bisectors, you know that each point on an angle bisector is equally distant from the sides of the angle. Since the incenter of a triangle is on all three angle bisectors, then it is equally distant from all three sides. Since it is equally distant from all three sides, the incenter is the center of a circle inscribed in the triangle.

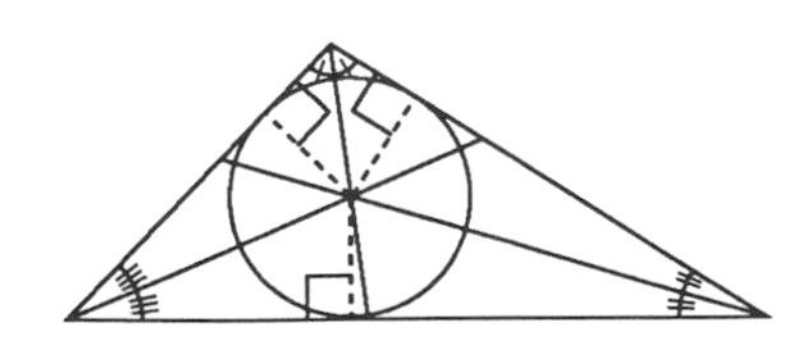

From the conjecture you made earlier about perpendicular bisectors, you know that each point on a perpendicular bisector of a segment is equally distant from the ends. If the circumcenter of a triangle is on all three perpendicular bisectors, then it is equally distant from all three vertices of the triangle. Since it is equally distant from all three vertices of the triangle, the circumcenter is the center of a circle circumscribed about the triangle.

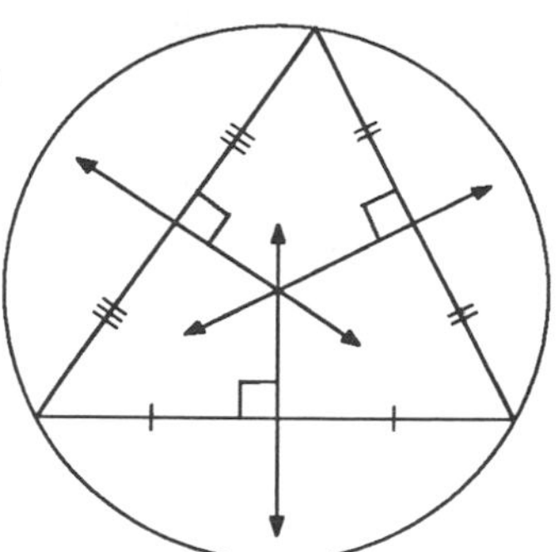

EXERCISE SET A

For each exercise below, make a sketch and answer the question.

1.* Hubert Honda is the dirt bike officer of Buford T. Pettibone State Park. He wishes to position himself at a point that is the same distance from each of three, straight, intersecting (not concurrent) bike paths. Help Hubert locate this point so that in an emergency he will be able to get to any one of the paths by the shortest route possible. Which point of concurrency does Hubert need to locate?

2. Stained glass artist Susie Sunshine wishes to inscribe a circle in a triangular portion of her latest abstract design. Which point of concurrency in the triangular section of her design does Susie need to locate?

3. Rosita is installing a round sink in her new kitchen countertop. She has marked three points on the countertop indicating points through which a circle must pass to install the sink. Which point of concurrency of the triangle connecting the three points must she locate to construct the circle?

4. It is the year 2060 on the lunar colony Galileo. The Director of Food Production for Galileo, Ima Gourmet, has been instructed to locate the food production facility equally distant from the three residential sections of the lunar colony. Which point of concurrency does Ima need to locate?

EXERCISE SET B

Once you know that in a triangle the three angle bisectors, three medians, three perpendicular bisectors, and three altitudes are concurrent, you no longer need to construct all three segments to locate a point of concurrency. Any two will locate the point of concurrency. You should always construct the third segment, however, to check the accuracy of your work.

1. Construct a large triangle. Follow the steps below to construct a circle inscribed in the triangle.

   Step 1: Construct the incenter (I) of the given triangle.

   Step 2: Construct a perpendicular from the incenter to one of the sides. This perpendicular distance (d) is the radius of the inscribed circle.

   Step 3: Construct the inscribed circle with center at I and radius d.

2. Construct a triangle. Follow the steps below to construct a circle circumscribed about the triangle.

   Step 1: Construct the circumcenter (C) of the given triangle.

   Step 2: Measure with your compass the distance (r) from the circumcenter to one of the vertices.

   Step 3: Construct the circumscribed circle with center at C and radius of r.

EXERCISE SET C

In the investigations you should have noticed that the incenter was always in the interior of the triangle. However, the circumcenter and orthocenter were sometimes in the interior and sometimes in the exterior of the triangle.

1. Notice that the circumcenter is in the interior of acute triangles and on the exterior of obtuse triangles. Use graph paper to draw a right triangle with even length sides. Where is the circumcenter of a right triangle?

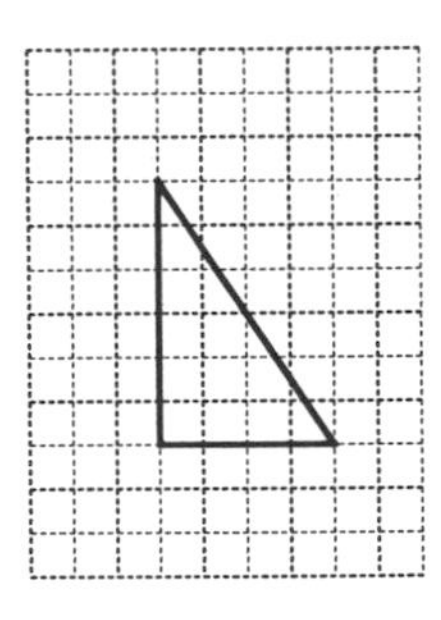

2. Notice that the orthocenter is in the interior of acute triangles and on the exterior of obtuse triangles. Use graph paper to draw a right triangle. Where is the orthocenter of a right triangle?

Lesson 3.9
The Centroid

The Universe may
be as great as they say.
But it wouldn't be missed
if it didn't exist.

– Piet Hein

Investigation 3.9.1  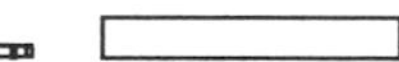

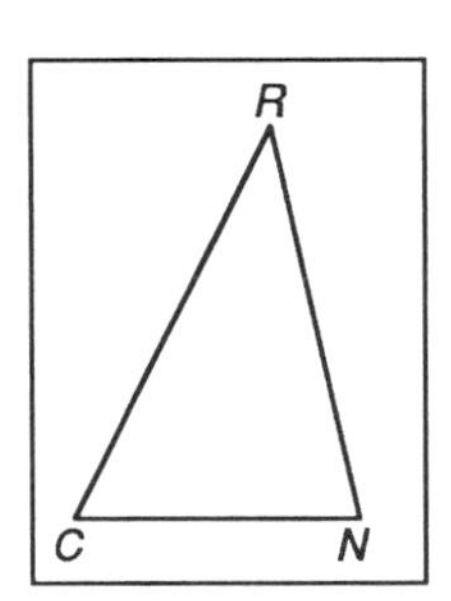

Step 1: On a clean sheet of unlined paper, construct a large scalene acute triangle. Make the triangle as large as you can.

Step 2: Construct the three medians in each triangle.

Step 3: On another clean sheet of unlined paper, construct a large obtuse triangle. Make the triangle as large as you can.

Step 4: Construct the three medians in each triangle.

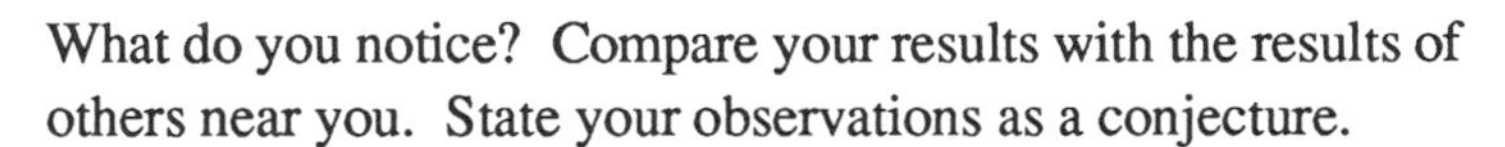

What do you notice? Compare your results with the results of others near you. State your observations as a conjecture.

Conjecture: *The three medians of a triangle —?—.*

The point of concurrency of the three medians of a triangle is the **centroid.**

The three medians of a triangle are not only concurrent at the centroid but the centroid also divides each of the medians into the same ratio. Let's investigate.

Investigation 3.9.2  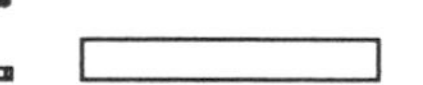

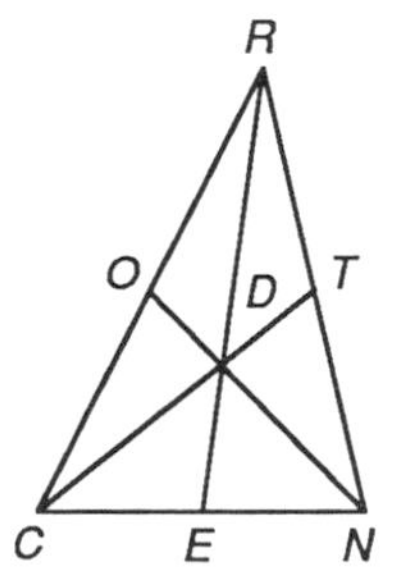

Step 1: Take the scalene acute triangle you constructed in the investigation above and label it ΔCNR as shown on the right. Don't worry if your triangle is a different shape.

Step 2: Label the three medians $\overline{CT}$, $\overline{NO}$, and $\overline{RE}$. Label the centroid D.

Step 3: With your compass, measure $\overline{DT}$. Compare $\overline{DT}$ with $\overline{CD}$ by seeing how many times you can mark off the distance DT on $\overline{CD}$. (If, for example, you can mark off DT three times on $\overline{CD}$, then $\overline{CD}$ is three times longer than $\overline{DT}$.) How many times longer than $\overline{DT}$ is $\overline{CD}$?

Step 4: Compare the lengths of the parts of the other two medians. ($\overline{RD}$ is how many times longer than $\overline{DE}$? $\overline{ND}$ is how many times longer than $\overline{DO}$?)

Step 5: Complete these equations:

$CD = (–?–)\ DT$ $\qquad$ $RD = (–?–)\ DE$ $\qquad$ $ND = (–?–)\ DO$

Did you get the same number each time? State your discovery as a conjecture.

Conjecture: *The centroid of a triangle divides each median into two parts so that the distance from the centroid to the vertex is —?— the distance from the centroid to the midpoint.*

The centroid is perhaps the most useful point of concurrency, especially in physics. If you had a thin solid object of uniform density in the shape of a triangle, the centroid would be the center of mass of the triangle. This is sometimes called the center of gravity. You could balance the triangle at the centroid.

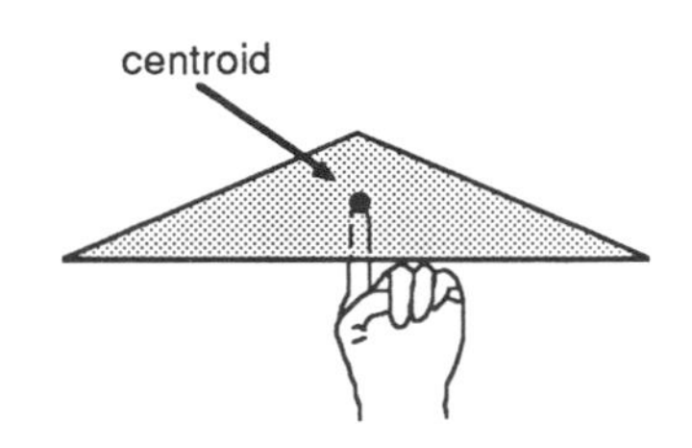

You've learned how to locate four points of concurrency associated with triangles.

Incenter

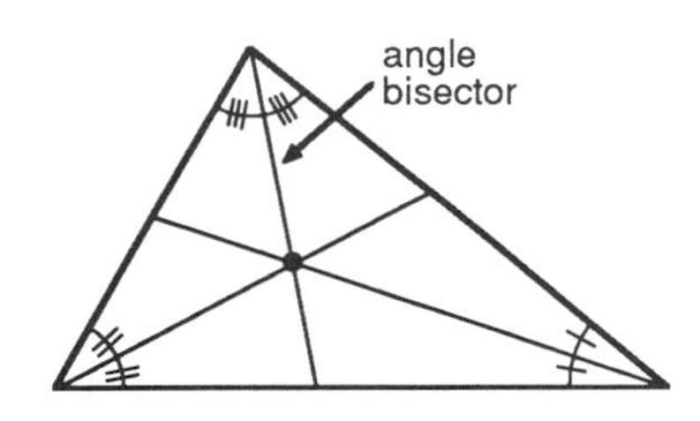

Circumcenter

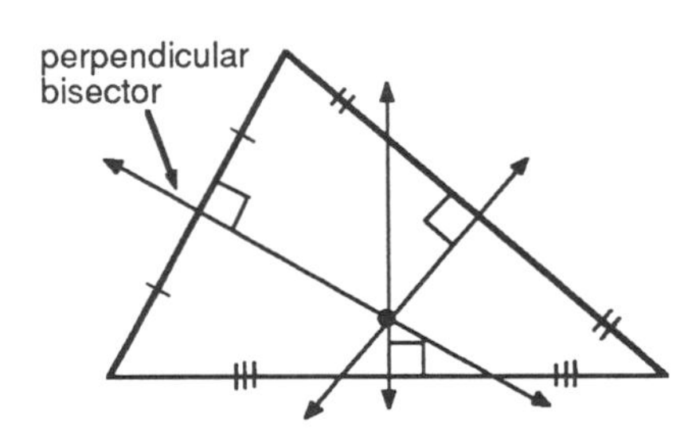

Orthocenter

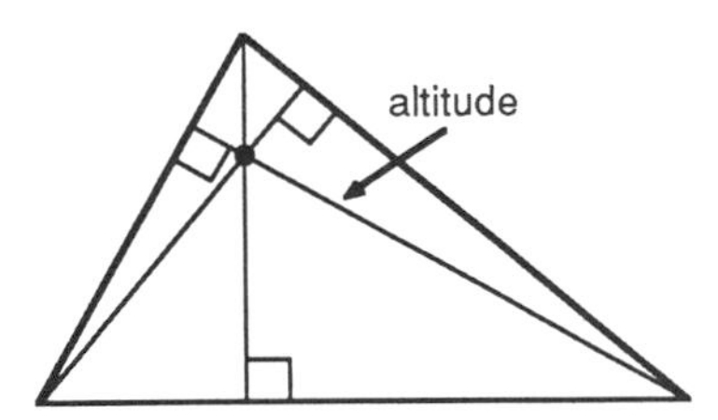

Centroid

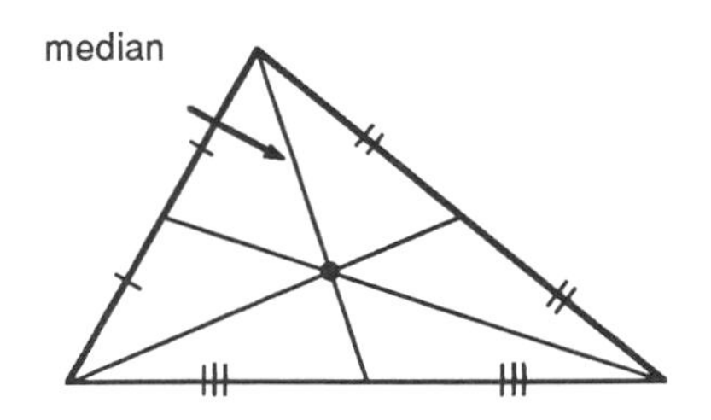

EXERCISE SET A

Use your new conjecture to find each length.

1.* M is the centroid.
$CM = 16$
$MO = 10$
$TS = 21$
$AM =$ –?–
$SM =$ –?–
$TM =$ –?–
$UM =$ –?–

2. Y is the centroid.
$PY = 8$
$TY = 18$
$FY = AY + 4$
$AY =$ –?–
$GY =$ –?–
$IY =$ –?–

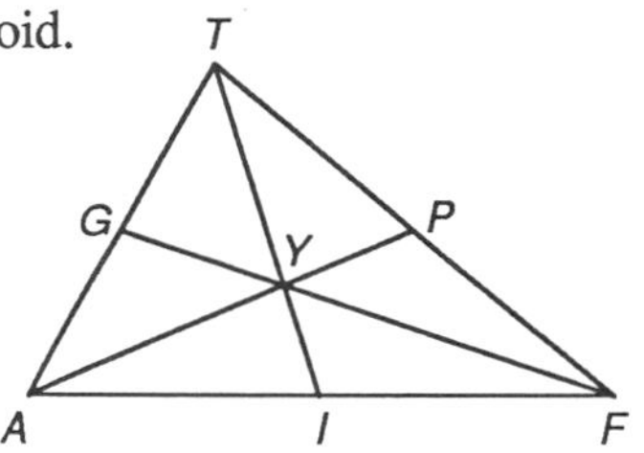

3. Z is the centroid.
$CZ = 14$
$TZ = 30$
$RZ = AZ$
$RH =$ –?–
$TE =$ –?–

4. G is the centroid.
$GI = GR = GN$
$PR = 36$
$DG =$ –?–
$IG =$ –?–

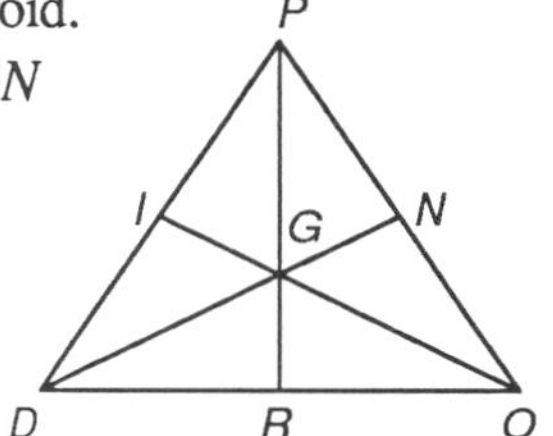

EXERCISE SET B

1. Birdy McFly is designing a large triangular hang glider. She needs to locate the center of gravity for her glider. Which point does she need to locate?

2. Construct a triangle (as large as possible) on a sheet of matboard or heavy cardboard. Construct the centroid. Cut out the triangle. Place the eraser of your pencil beneath the centroid and try to balance your triangle on the tip of your eraser. Does it balance? Can you balance it on your pencil tip?

3. Construct a scalene triangle (as large as possible). Construct the centroid, incenter, circumcenter, and orthocenter. What do you notice about the four points of concurrency (are any the same or are they all different)? Three of them are collinear. Which three?

   The line through them is called the **Euler Line**. The Euler Line is named after the Swiss mathematician Leonard Euler (1707–1783) who proved that the three points of concurrency were collinear. Euler's mathematical output was so tremendous that forty volumes of his writings, representing just a small part of his work, have been published.

4. Construct a large isosceles triangle. Construct the centroid, incenter, circumcenter, and orthocenter. What do you notice about the four points of concurrency?

5. Construct a large equilateral triangle. Construct the centroid, incenter, circumcenter, and orthocenter. What do you notice?

Special Project

Center of Mass for a Cardboard Triangle

In Lesson 3.9 you learned that the centroid of a triangle is the center of mass. In Lesson 3.9, Exercise B2, you constructed a triangle on cardboard, located the centroid, cut out the triangle, and, hopefully, balanced it on its centroid.

There is another way to locate the center of mass of a triangle by using a plumb line instead of a geometric construction. You will need the following materials:

- Cardboard triangle
- Pins
- String
- Paper clips

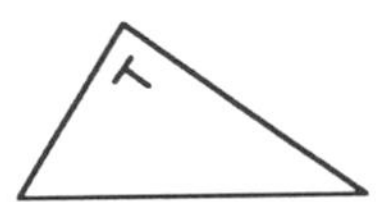

Locate the materials above and then follow these steps:

Step 1: Stick a pin anywhere through the interior of the cardboard triangle. Pin the triangle loosely to a bulletin board. The triangle should be able to hang freely by its own weight.

Step 2

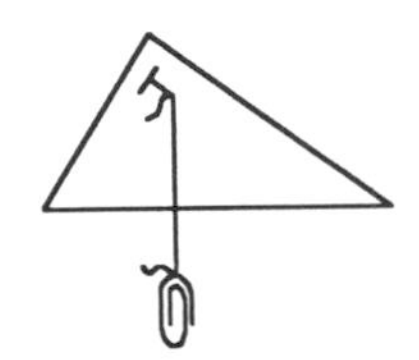

Step 2: Tie one or two paper clips to a string and loop the string over the pin so that the weighted string hangs over the interior of the triangle. This string with paper clip weights is a plumb line. Draw a line on the cardboard triangle showing the path of the string.

Step 3

Step 3: Stick the pin through another point in the interior of the triangle and hang the plumb line. Draw another line on the cardboard triangle showing the new path of the string.

Step 4

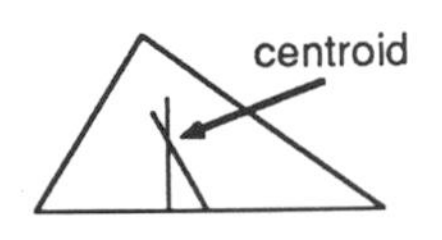

Step 4: The center of mass of the cardboard triangle is where the two lines cross. Check to see if your cardboard triangle balances on that point.

While the geometric construction method of finding the center of mass you learned in Lesson 3.9 worked only for triangles, this physical procedure works for any cardboard polygon. Can you discover a geometric construction method that locates the center of mass of a cardboard quadrilateral? (You could then use the plumb line method to check to see if your geometric construction method works.)

Special Project

Constructing the Nine-Point Circle

In 1820 French mathematicians Charles Brianchon and Jean Victor Poncelet published a paper that contained the proof of the following statement:

The circle which passes through the feet of the perpendiculars, dropped from the vertices of any triangle on the sides opposite them, passes also through the midpoints of these sides as well as through the midpoints of the segments which join the vertices to the point of intersection of the perpendiculars.

Did you catch all that? This circle is called the **nine-point circle**. Agreed, it is a very difficult statement to follow, but perhaps it will have more meaning after you have completed this special project. Constructing the nine-point circle is a good test of your construction skills and your ability to follow written instructions.

Step 1

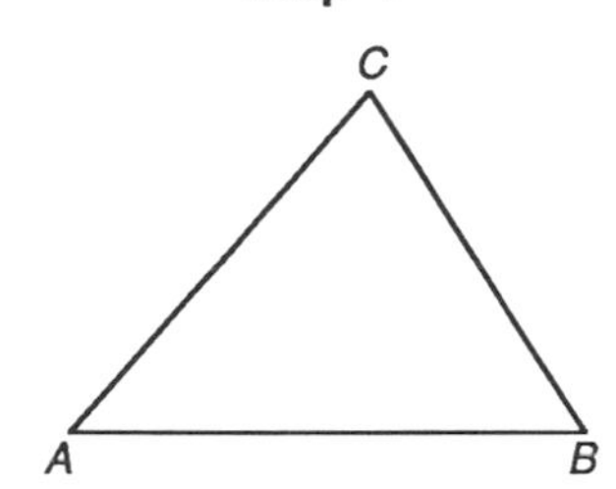

Construct a large scalene triangle on a clean sheet of unlined paper. Label it ΔABC.

Step 2

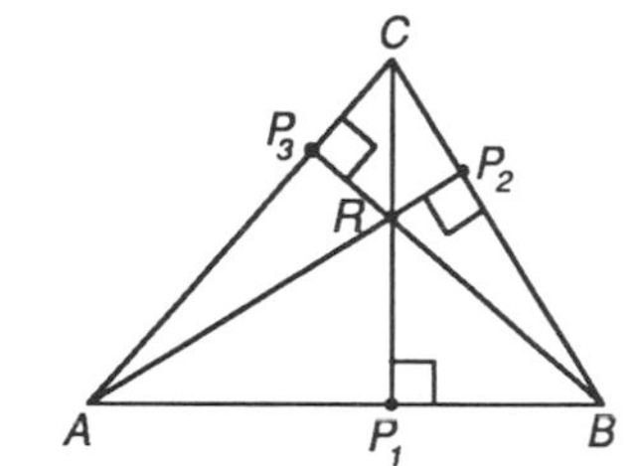

Construct the altitude to each side of the triangle and label the points of intersection with the sides P_1, P_2, and P_3. Label the orthocenter R.

Step 3

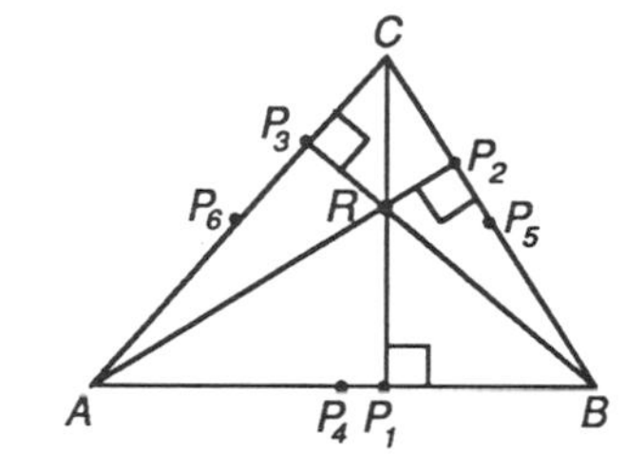

Construct the midpoint of each side of the triangle. Label the points P_4, P_5, and P_6 so that P_4 is the midpoint of $\overline{AB}$ and P_5 is the midpoint of $\overline{BC}$.

Step 4

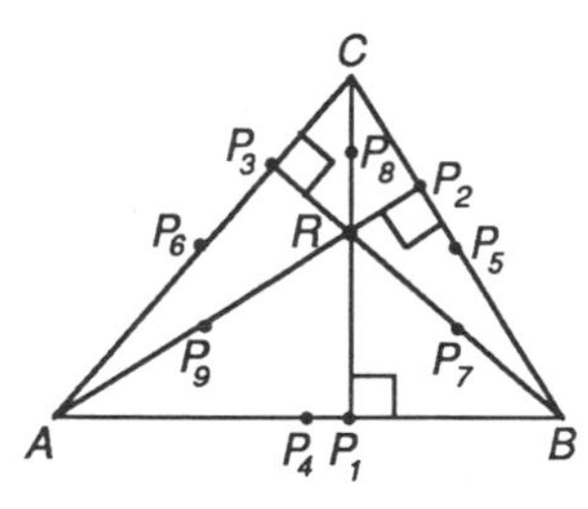

Construct the midpoints of $\overline{BR}$, $\overline{CR}$, and $\overline{AR}$ and label the points P_7, P_8, and P_9 so that P_7 is the midpoint of $\overline{BR}$ and P_8 is the midpoint of $\overline{CR}$.

Step 5

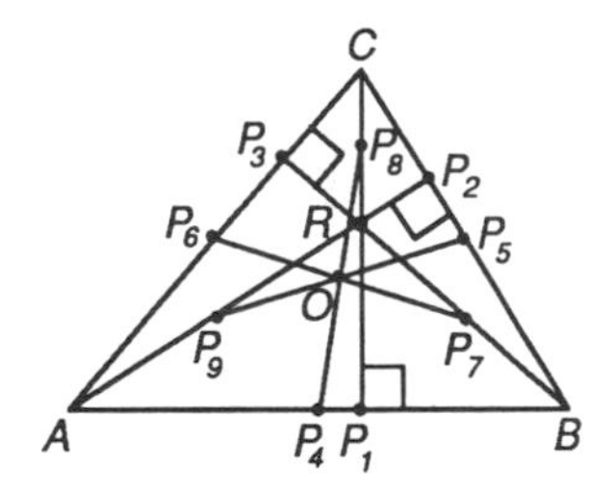

Construct the line segments connecting points P_4 to P_8, P_5 to P_9, and P_6 to P_7. They should all intersect in one point. Label that point O.

Step 6

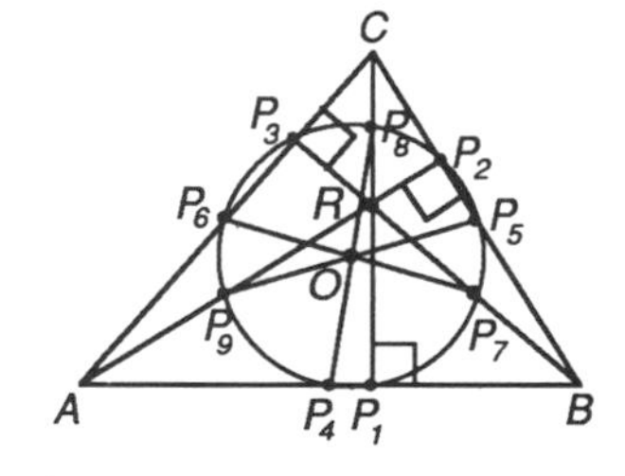

Construct a circle with radius OP_1 and center at point O. It should pass through all nine points: $P_1, P_2, P_3, P_4, P_5, P_6, P_7, P_8, P_9$.

Lesson 3.10
Chapter Review

EXERCISE SET A

Identify each statement as true or false.

1. A geometric construction uses a protractor and ruler.
2. An acute angle is an angle whose measure is greater than 90.
3. An isosceles triangle is a triangle with no sides the same length.
4. A diagonal is a line segment in a polygon connecting any two vertices.
5. If the sum of the measures of two angles is 180, then the two angles are complementary.
6. If two lines do not intersect, then they are parallel.
7. A trapezoid is a quadrilateral having exactly one pair of parallel sides.
8. A rhombus is a parallelogram with all sides equal in measure.
9. A square is a rhombus with all angles equal in measure.
10. A polygon with ten sides is called a decagon.
11. If a point is equally distant from the endpoints of a segment, then it must be the midpoint of the segment.
12. The shortest distance from a point to a line segment is the length of the shortest segment from the point to each endpoint.
13. It is not possible to construct a 15° angle.
14. If $\overline{CD}$ is a median of ΔABC and M is the centroid, then $CM = 3MD$.
15. The set of all the points in the plane that are a given distance from a line segment is a pair of lines parallel to the given segment.
16. The incenter of a triangle is the point of intersection of the three medians.
17. The centroid of a triangle is the point of intersection of the three altitudes.
18. The circumcenter of a triangle is the point of intersection of the three angle bisectors.
19. The orthocenter of a triangle is the point of intersection of the three perpendicular bisectors.
20. Exactly four statements in this exercise set are true.

EXERCISE SET B

Match each construction with its drawing.

1. Construction of a perpendicular bisector
2. Construction of an angle bisector
3. Construction of a perpendicular from a point to a line
4. Construction of a perpendicular through a point on a line
5. Construction of a centroid
6. Construction of a circumcenter
7. Construction of an incenter
8. Construction of an orthocenter

a.

b.

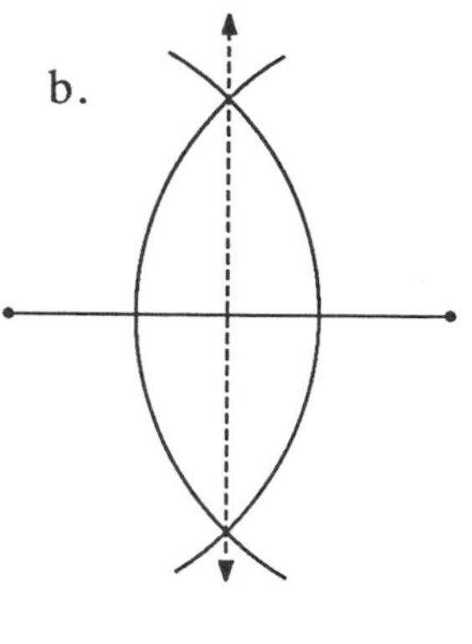

c.

d.

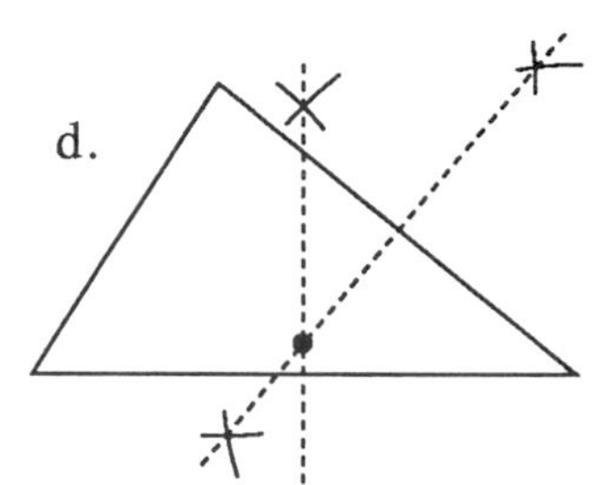

e.

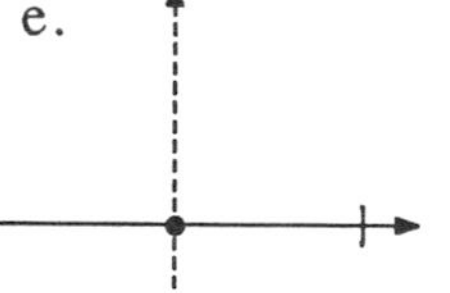

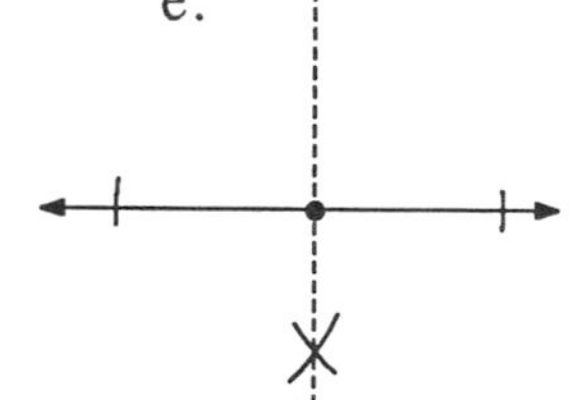

f.

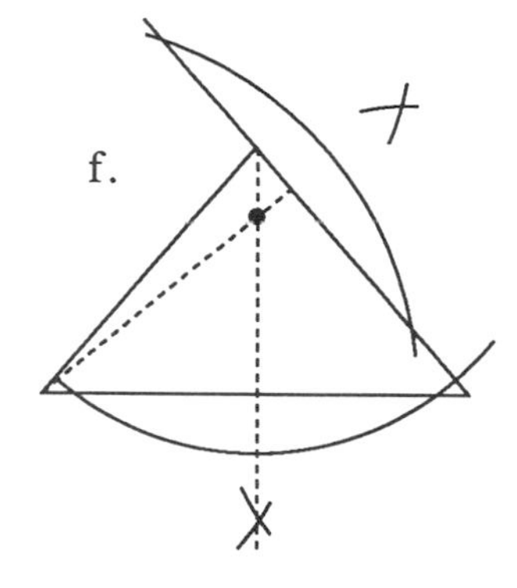

g.

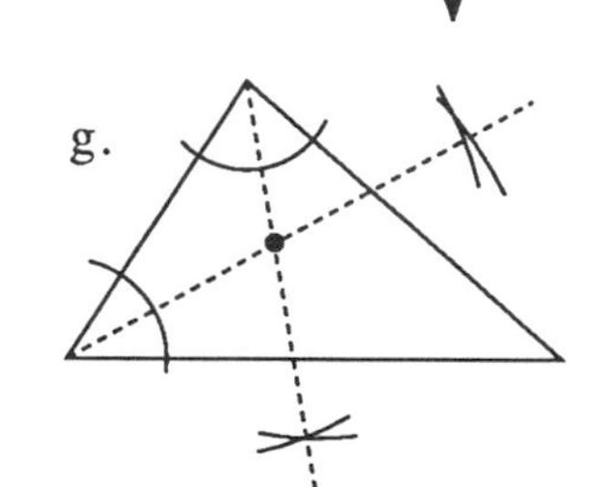

h.

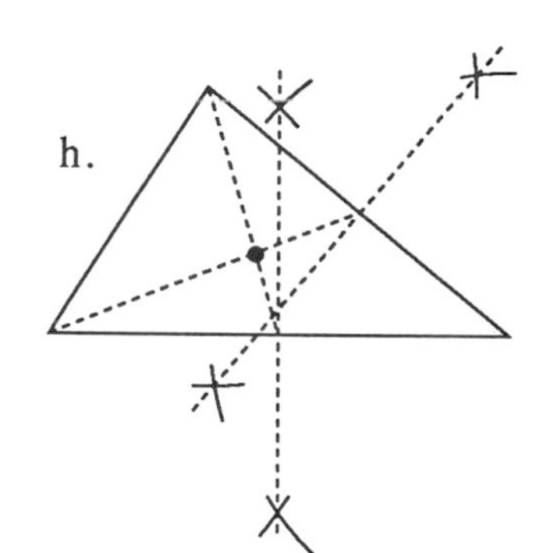

EXERCISE SET C

1. Construct an angle. Then duplicate it using your compass and straightedge.
2. Construct a line segment. Then construct its perpendicular bisector.
3. Construct a line and a point not on the line. Construct a perpendicular to the line through the point.
4. Construct an angle. Bisect it.
5. Construct an angle of 22.5°.
6. Construct a line and a point not on the line. Construct a second line parallel to first line that passes through the point.

Use the segments and angles below to construct each figure in Exercises 7 to 14. The lower case letter above each segment represents the length of the segment in units.

x

y

z

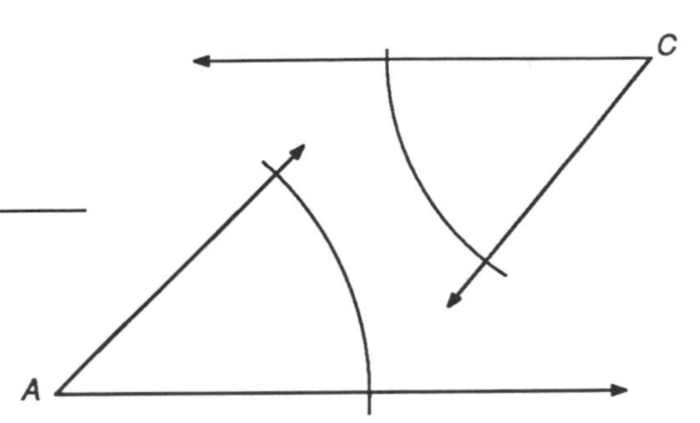

7. ΔABC given $\angle A$, $\angle C$, and $AC = z$

8.* A segment with length $2y + x - \frac{1}{2}z$

9. ΔPQR with $PQ = 3x$, $QR = 4x$, and $PR = 5x$

10.* ΔABD with $m\angle A = m\angle B$ and $AB = 3x$ (Make sure that you use $\angle A$ above.)

11. Isosceles triangle ABD with $AB = BD = 2y$ (Use $\angle A$ above.)

12. Triangle FUN with $m\angle F = m\angle U = 45$ and $FU = 2y - x$

13. Quadrilateral $ABFD$ with $m\angle A = m\angle B$, $AD = BF = y$, and $AB = 4x$

14.* Quadrilateral $RHOM$ with $RH = \frac{7}{2}x$, $RO = 2y$, and $HO = z$.

Improving Visual Thinking Skills

Paint By Numbers

Each puzzle below has one piece with a missing number. What number is missing in each puzzle?

1.

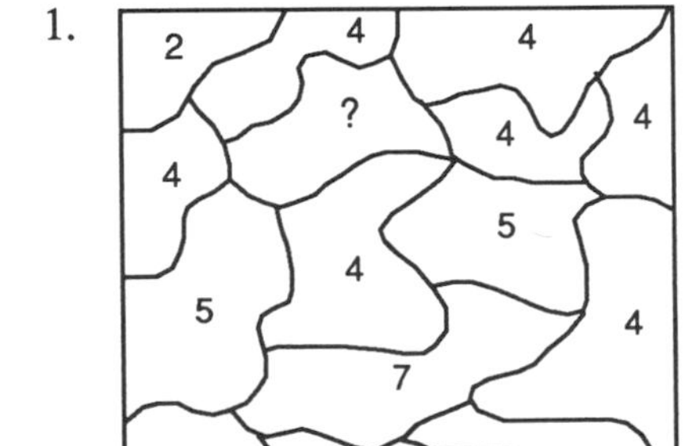

2.

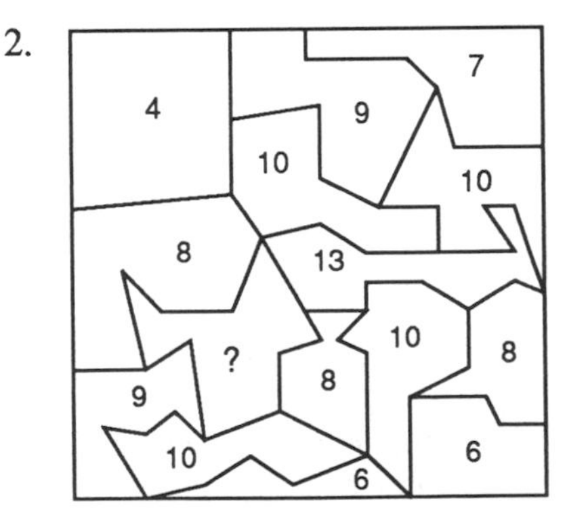

Cooperative Problem Solving

Lunar Survival

You have recently joined the lunar research team aboard the space station *Entropy*. On the way to the uninhabited Lunar Research Outpost your space station transport vehicle malfunctioned. This forced you and the crew to make an emergency landing on the lunar surface 150 kilometers from your intended destination. All of your communication devices are inoperable. After the emergency the crew was able to retrieve from the transport vehicle a number of items (listed below). Since you are at least one thousand kilometers from the lunar space port Galileo, survival depends on the crew reaching the outpost. Your first survival task is to rank the listed items in terms of their importance to the crew's ability to reach the Lunar Research Outpost.

Copy the list of items below onto a clean sheet of paper. Without discussion with fellow crew members, rank the items on your list. Then discuss with your crew the relative importance of each salvaged item. After a thorough group discussion, rank the items by placing the number 1 by the most important item; number 2 by the second most important item; and so on through number 14, the least important. Finally, to the right of each item, give a reason for your ranking. According to NASA scientists, there is a correct ranking.

- Solar-powered FM receiver-transmitter
- Signal flares
- Box of matches
- 20 meters of nylon rope
- First-aid kit containing injection needles
- Two 50-kilogram tanks of oxygen
- Parachute silk
- Food concentrate
- Portable heating unit
- One case dehydrated milk
- 20 liters of water
- Magnetic compass
- Life raft
- Star map of moon's constellations

4 Discovering Geometry

Cycle, M. C. Escher, 1938

Each lesson in this chapter will present a topic for you to explore, often using your geometric tools. Through your explorations you will make some important conjectures. It is important to remember these conjectures along with the definitions you encounter because you will build on them in the lessons to come. You should be keeping a definition list in your notebook. Keep a list of conjectures there also.

Lesson 4.1

Discovering Angle Relationships

Discovery consists of looking at the same thing as everyone else and thinking something different.

– A. Szent-Gyorgyi

Before you begin this lesson, review the definitions of the following pairs of special angles. The wording of these definitions may not be exactly the same as the wording you used in Chapter 2; however, the idea should be the same.

Two angles are ***complementary angles*** *if their measures add to 90.*

Two angles are ***supplementary angles*** *if their measures add to 180.*

If $\overleftrightarrow{PQ}$ intersects $\overleftrightarrow{RS}$ at point T, then ∠PTS and ∠RTQ are ***vertical angles****. ∠PTR and ∠STQ are also a pair of vertical angles.*

If $\overleftrightarrow{PQ}$ intersects $\overleftrightarrow{RS}$ at point T, then ∠PTS and ∠STQ form a ***linear pair of angles****. Three other linear pairs of angles are also formed.*

Investigation 4.1.1

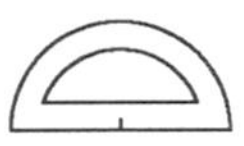

On a clean sheet of paper draw two lines, $\overleftrightarrow{PQ}$ and $\overleftrightarrow{RS}$, intersecting at point T (as shown on the right). With a protractor carefully measure all four angles and record the measures.

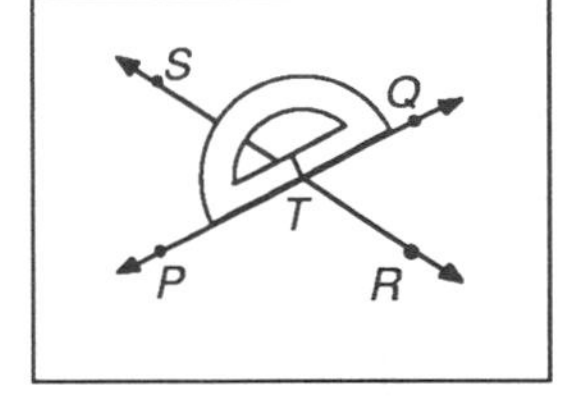

$m\angle PTS \approx$ –?– $m\angle RTQ \approx$ –?–

$m\angle PTR \approx$ –?– $m\angle STQ \approx$ –?–

Do you notice anything special about the measures of the pair of vertical angles ∠*PTS* and ∠*RTQ*? What about vertical angles ∠*PTR* and ∠*STQ*?

Try this again. On a sheet of paper draw two lines, $\overleftrightarrow{JK}$ and $\overleftrightarrow{LM}$, intersecting at point *N*. Measure all four angles. Compare your results with the results of others near you. State your findings as a conjecture.

If two angles are vertical angles, then —?—.
(Vertical Angles Conjecture)

Did you complete the conjecture with the same words as others in your class? Discuss it and agree on a common wording. Write the completed conjecture beneath your investigative work.

The Vertical Angles Conjecture is a conjecture that will be very useful to you later in this text. You will make many more conjectures as you proceed through this book. The conjectures that are important to keep track of have been numbered and, in most cases, named. Start a list of these numbered conjectures in your notebook. (You should already be keeping a list of definitions there.) This will be your only list of completed conjectures and you will need to refer to the conjectures often. The Vertical Angles Conjecture should be the first conjecture on your conjecture list. From now on, each time you complete a conjecture, add it to your conjecture list. Review your conjecture list from time to time and study it before tests.

What about a pair of linear angles? Are they equal in measure? Clearly not. Are they related at all? What about their sum?

Investigation 4.1.2

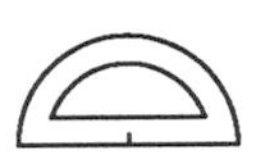

Find the sum of each pair of linear angles formed by intersecting lines *PQ* and *RS*.

$m\angle PTS + m\angle STQ \approx$ –?– $\qquad$ $m\angle PTR + m\angle PTS \approx$ –?–

$m\angle PTR + m\angle RTQ \approx$ –?– $\qquad$ $m\angle RTQ + m\angle STQ \approx$ –?–

Find the sum of each pair of linear angles formed by intersecting lines *JK* and *LM*.

$m\angle JNM + m\angle MNK \approx$ –?– $\qquad$ $m\angle JNL + m\angle JNM \approx$ –?–

$m\angle JNL + m\angle LNK \approx$ –?– $\qquad$ $m\angle LNK + m\angle MNK \approx$ –?–

Compare your results with those of others near you. State a conjecture about angles that form a pair of linear angles. Add it to your conjecture list.

C-2 *If two angles are a linear pair of angles, then —?—.*
(Linear Pair Conjecture)

At the conclusion of each investigation you should discuss your results with others in your class and agree on a common way to complete the conjecture (if there is one). Write the conjecture or a short statement summarizing your investigation beneath your investigative work and add the conjecture to your conjecture list.

The next conjecture is an immediate result of the Linear Pair Conjecture. If two angles are supplementary and they are equal in measure, what must be true of the two angles? State your observations as your next conjecture.

If two angles are both equal in measure and supplementary, then each angle measures —?—.
(Equal Supplements Conjecture)

Let's see how this conjecture can be supported with the help of algebra and a little logical reasoning.

A convincing logical argument of a conjecture is called a **proof** of the conjecture. The algebraic argument on the left proves the Equal Supplements Conjecture. It's called a paragraph proof because the proof is given in paragraph form.

Paragraph Proof:

If x and y are the measures of two angles that are supplementary, then:

$x + y = 180$

If x and y are equal, then:

$x + x = 180$

$2x = 180$

$x = 90$

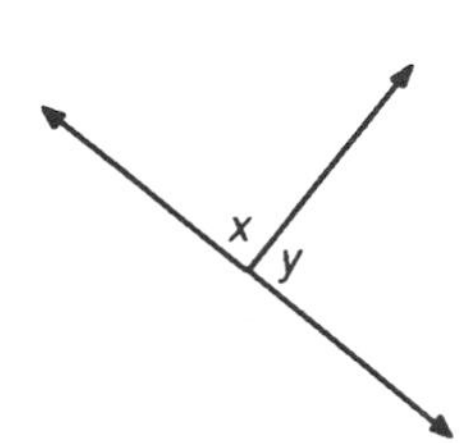

EXERCISE SET

Without using a protractor, but with the aid of your two new conjectures, determine the measure of each lettered angle below. You may find it helpful to trace the diagrams onto a clean sheet of unlined paper so you can write on them. You do not need to find the measure of the angles in alphabetical order, but you should list your answers in alphabetical order to make them easier to check.

1.

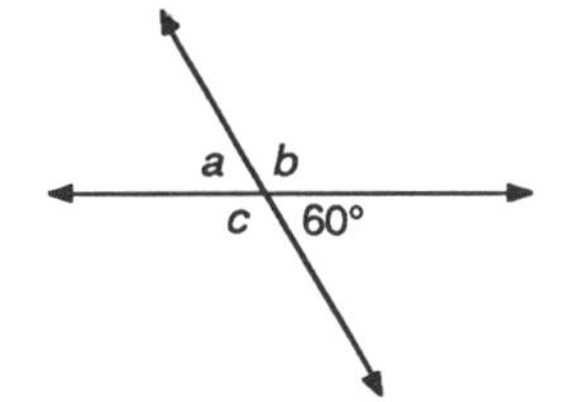

2.

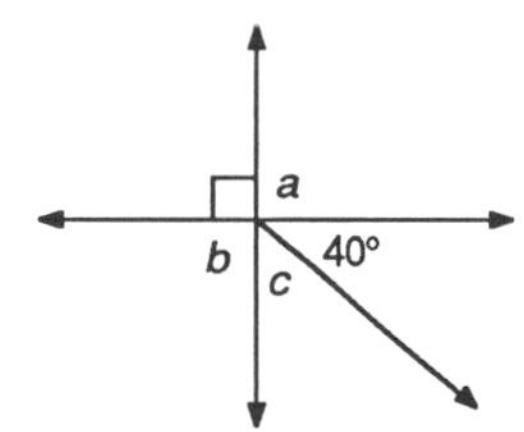

3.*

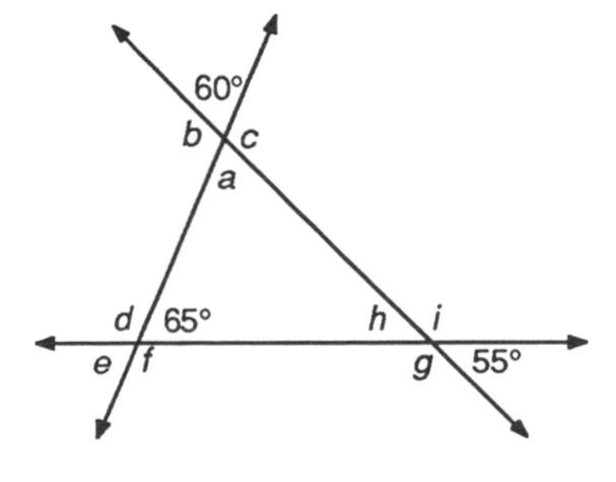

4.

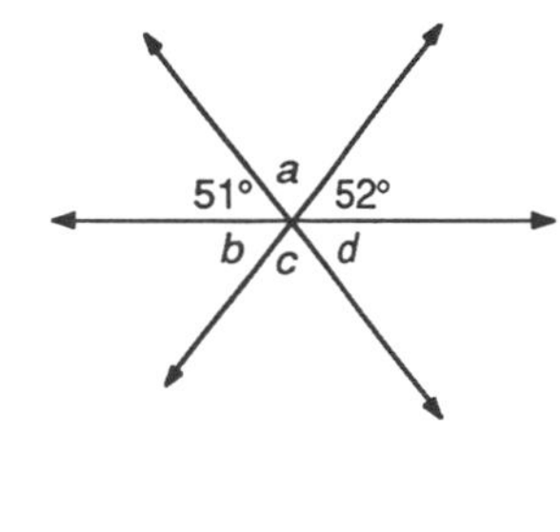

5.

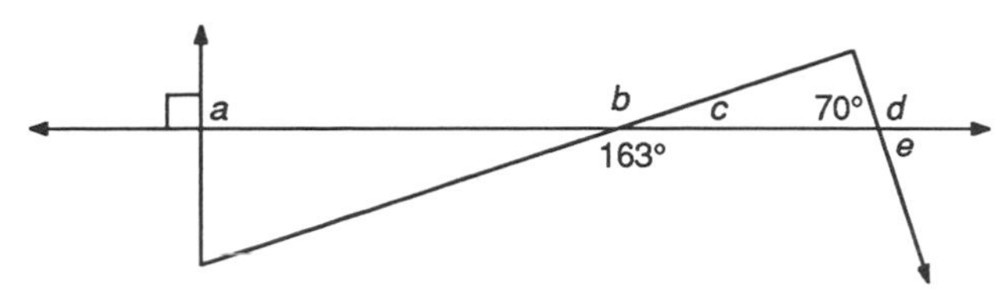

Lesson 4.2

Triangle Sum Conjecture

In this lesson you will discover a very important fact about the sum of the measures of the three angles in every triangle.

Investigation 4.2.1  

Step 1: Construct two large acute triangles and two large obtuse triangles on a clean sheet of paper.

Step 2: Measure the three angles of each triangle as accurately as possible with your protractor.

Step 3: Find the sum of the measures of the three angles in each triangle.

Compare your results with the results of others near you. Do you always seem to get close to the same result? What appears to be the sum of the three angles in every triangle? Before you make a conjecture, check this another way.

Step 1

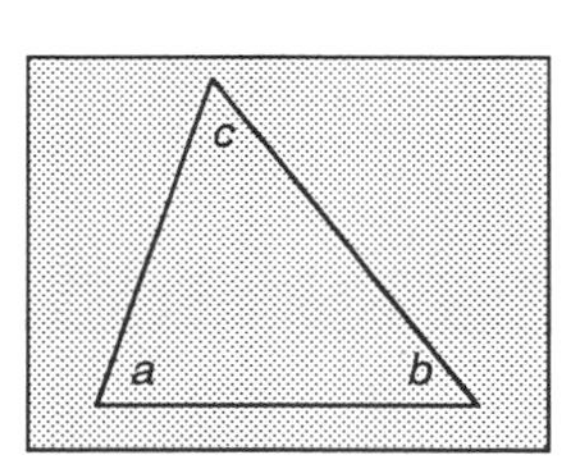

Write a, b, and c in the interiors of the three angles of one of the acute triangles.

Step 2

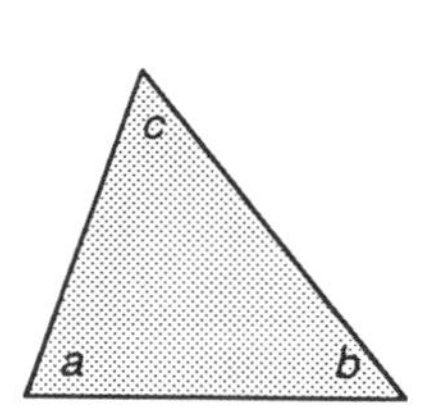

Carefully cut out the triangle.

Step 3

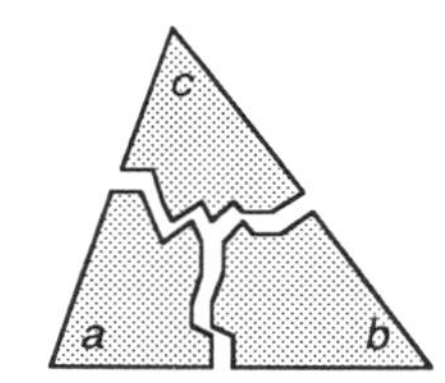

Tear off the three angles.

Step 4

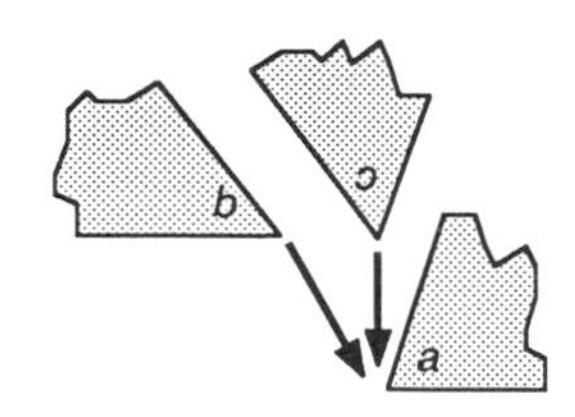

Arrange the three angles so that their vertices meet at a point to show their sum.

What is the sum of the measures of the three angles of any triangle? Make a conjecture and add it to your conjecture list.

The sum of the measures of the three angles of every triangle is —?—. ***(Triangle Sum Conjecture)***

The next conjecture is an immediate result of the Triangle Sum Conjecture.

Investigation 4.2.2

Calculate the measure of the third angle in each triangle below.

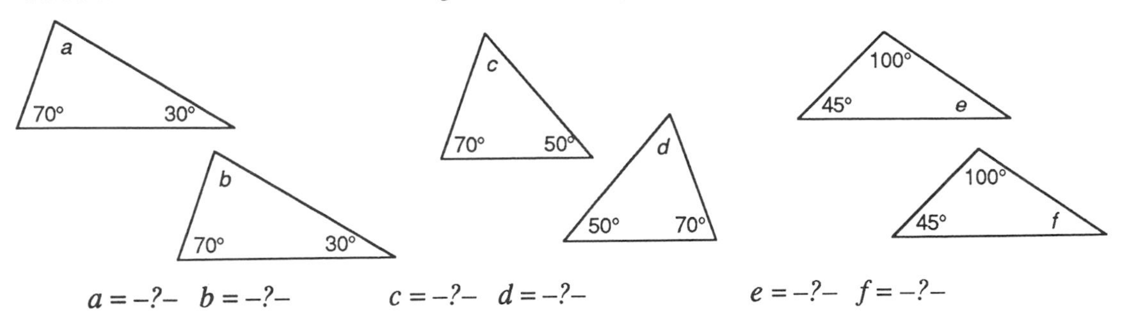

$a = –?–$ $b = –?–$ $c = –?–$ $d = –?–$ $e = –?–$ $f = –?–$

In each pair of triangles, two angles of one triangle were equal in measure to two angles of the other triangle. What did you discover about the measures of the remaining two angles? How do a and b, c and d, e and f compare?

Compare your results with the results of others near you. Can you support your conjecture with a logical algebraic argument? State a conjecture.

If two angles of one triangle are equal in measure to two angles of another triangle, then the remaining two angles —?—. ***(Third Angle Conjecture)***

EXERCISE SET A

Use the Triangle Sum Conjecture to determine the measure of each lettered angle. You might find it helpful to trace the more complicated diagrams.

1.* $x = –?–$

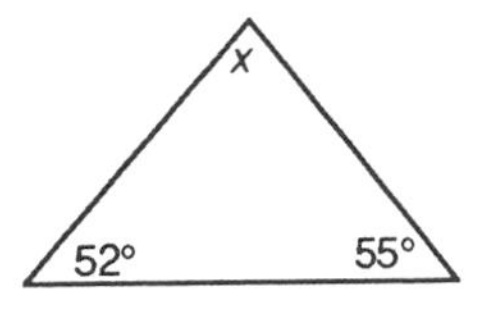

2. $y = –?–$

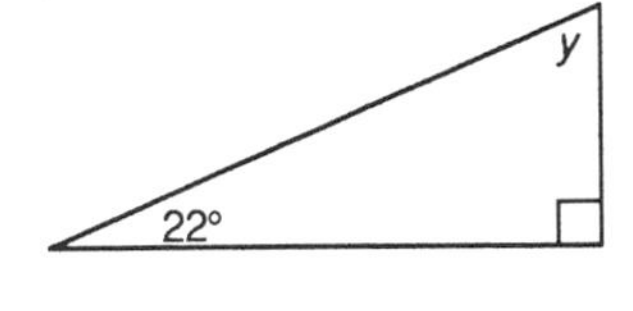

3.* $z = –?–$

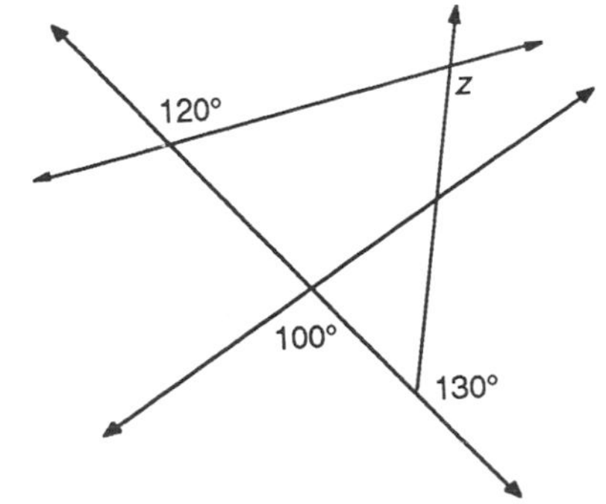

4. $w = –?–$

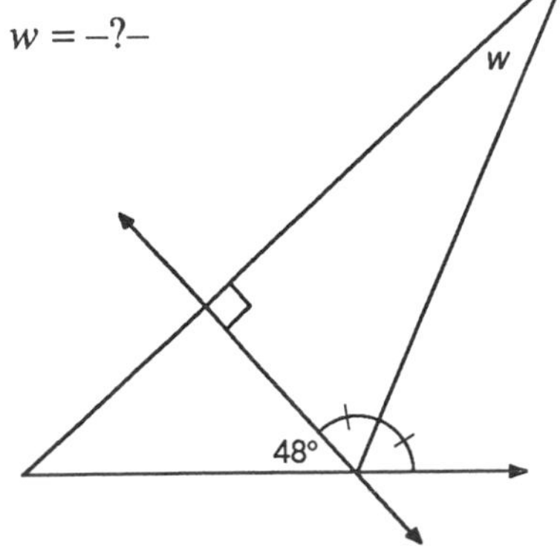

5.* $a = \text{–?–}$
$b = \text{–?–}$
$c = \text{–?–}$
$d = \text{–?–}$
$e = \text{–?–}$

6. $m = \text{–?–}$
$n = \text{–?–}$
$p = \text{–?–}$
$q = \text{–?–}$
$r = \text{–?–}$
$s = \text{–?–}$
$t = \text{–?–}$
$u = \text{–?–}$

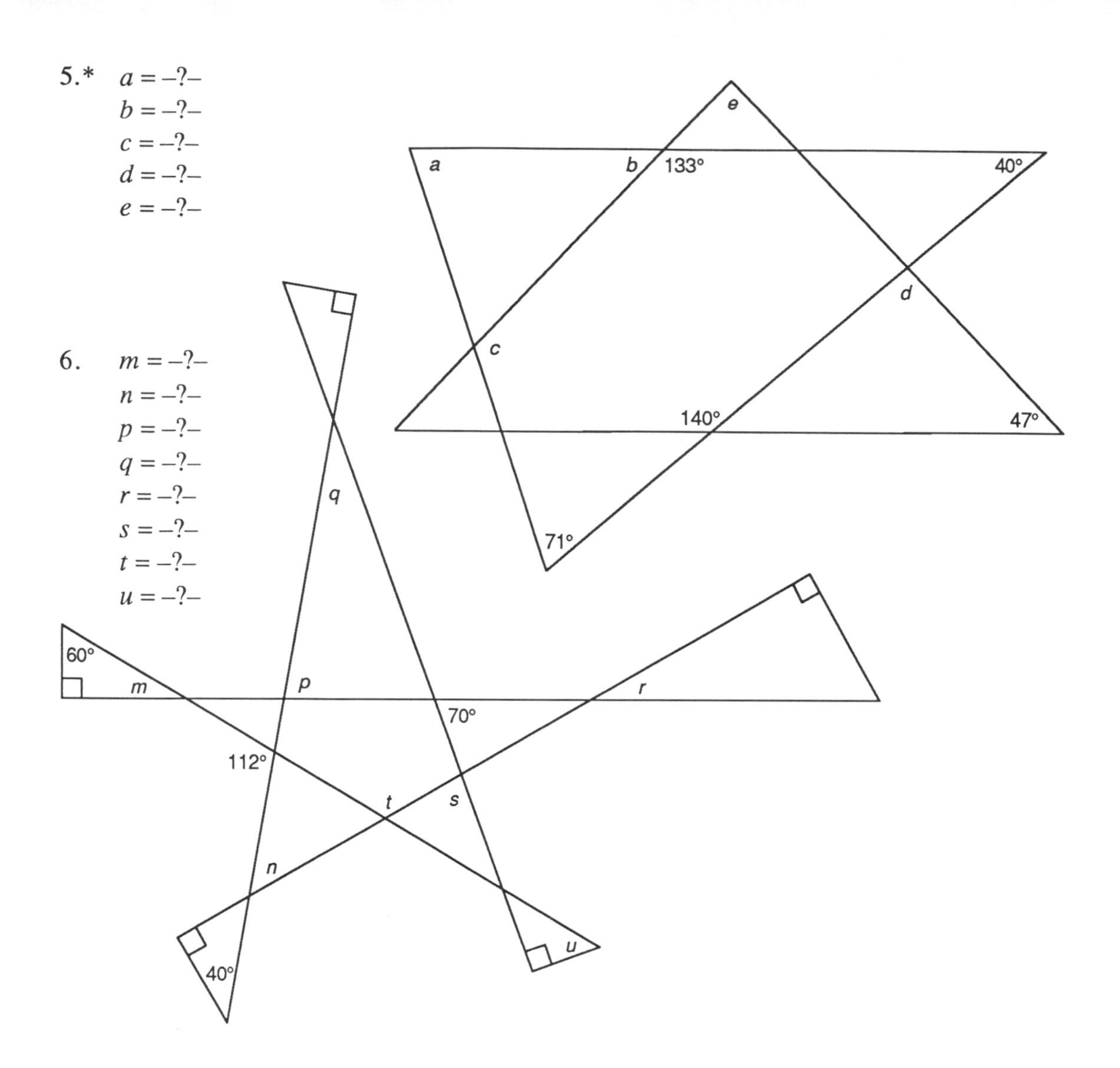

EXERCISE SET B

1. Find x.

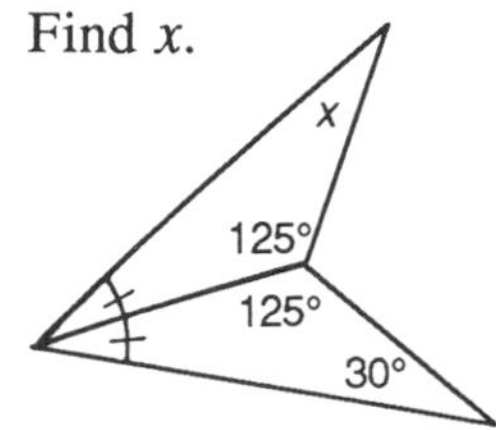

2. Find y.

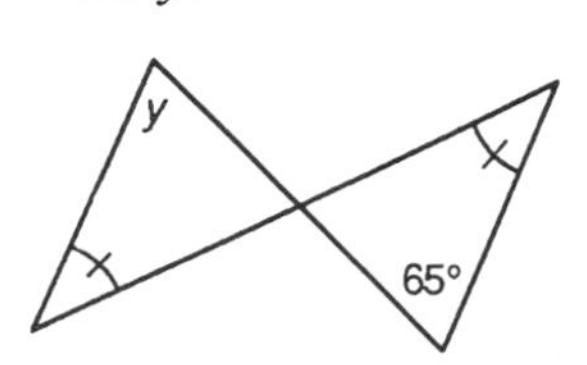

3. Let a, b, and c represent the measures of the three angles of ΔABC and let a, b, and d represent the measures of the three angles of ΔDEF. Use algebra to prove that the Third Angle Conjecture follows from the Triangle Sum Conjecture. (That is, show $c = d$.)

EXERCISE SET C

Perform each construction. Remember, in constructions you are not permitted to use protractor or ruler, only compass and straightedge. Arcs have already been constructed through the angles. You can use the arcs to duplicate the given angles without having to write in this book.

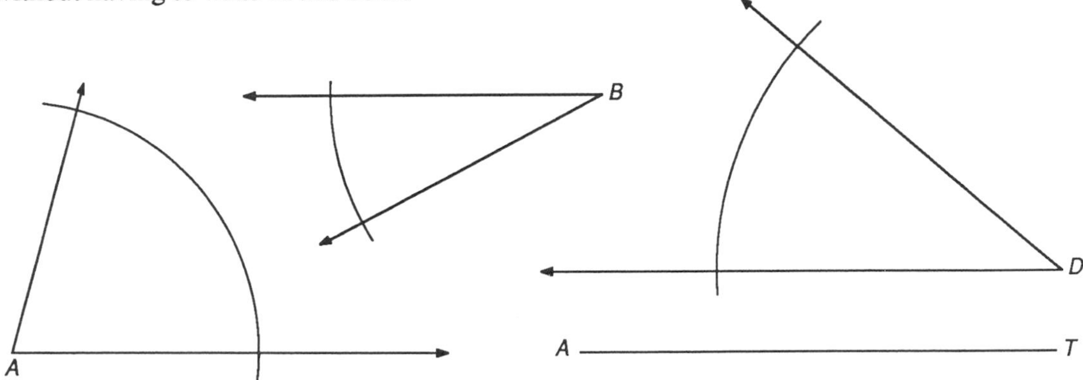

1.* Given angles A and B of ΔABC, construct angle C.

2.* In ΔDEF, $\angle E$ and $\angle F$ are equal in measure. Given $\angle D$, construct $\angle F$.

3. Given $\angle A$, $\angle B$, and side $\overline{AT}$, construct ΔABT.

Improving Reasoning Skills

Mudville Monsters

The starting 11 members of the Mudville Monsters football team and their coach Osgood Gipper have been invited to compete in the Smallville Punt, Pass, and Kick Competition. Upon arriving at the outskirts of the town, they find they must get across the deep Smallville River. The only available way across is with a small boat owned by two very small Smallville football players. The boat holds just one Monster visitor or the two Smallville players. The Smallville players agree to help the Mudville players across if the visitors agree to pay $5 each time the boat crosses the river. If the Monsters have a total of $200 between them, do they have enough money to get all twelve to the other side of the river?

Lesson 4.3

Polygon Sum Conjecture

In the last lesson you learned that the sum of the measures of the three angles of every triangle is 180. Is there a similar conjecture for the sum of the measures of the four angles of a quadrilateral? Is there a conjecture for the sum of the measures of the five angles of a pentagon? Is there a conjecture for the sum of the measures of the angles of *any* polygon? Let's first look at quadrilaterals.

Investigation 4.3.1

On a clean sheet of paper, construct two large quadrilaterals. Using a protractor, find the sum of the measures of the four angles in each quadrilateral.

Did you get close to the same answer both times? As in the last lesson, this result can be demonstrated by cutting out a quadrilateral from a piece of heavy paper and arranging all four angles about a point. Try it. Cut out one of your quadrilaterals. Tear off the four angles and arrange them around a point to determine their sum. Try this with the other quadrilateral. What appears to be the sum each time? State your next conjecture.

The sum of the measures of the four angles of a quadrilateral is —?—.

Now let's prove your conjecture using the Triangle Sum Conjecture. The diagonal $\overline{OR}$ in the quadrilateral *FOUR* has divided *FOUR* into two triangles.

Paragraph Proof:

Since $a + b + c = 180$ and since $d + e + g = 180$ (Triangle Sum Conjecture), then it follows that $a + b + c + d + e + g = 360$ (by addition).

But, $m\angle FOU = b + g$ and $m\angle FRU = c + e$ (from the drawing).

Since $a + (b + g) + d + (c + e) = 360$, then $m\angle F + m\angle O + m\angle U + m\angle R = 360$ in quadrilateral *FOUR* .

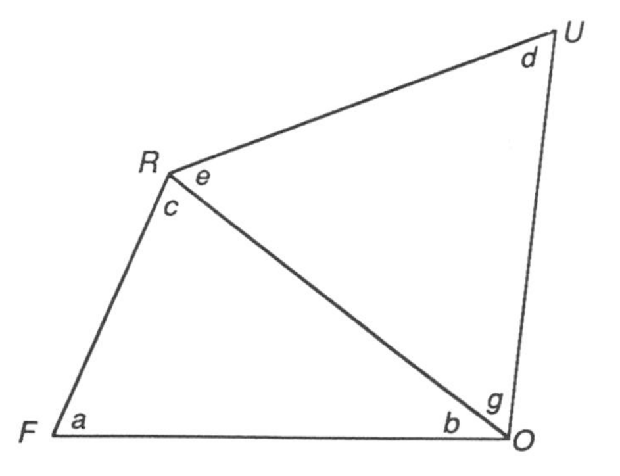

What about pentagons and all the other polygons? Is there a formula for the sum of the measures of the angles of any polygon? In other words, is there a relationship between the number of sides of a polygon and the total number of degrees in all its angles? It turns out there is.

Investigation 4.3.2

Copy and complete the table below to find the formula that relates the number of sides of a polygon and the total number of degrees in all its angles.

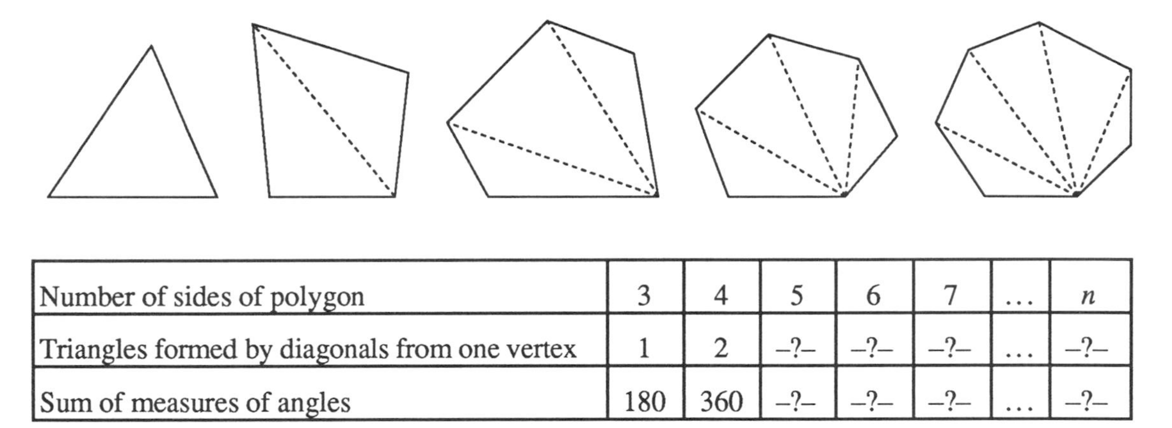

Number of sides of polygon	3	4	5	6	7	...	n
Triangles formed by diagonals from one vertex	1	2	–?–	–?–	–?–	...	–?–
Sum of measures of angles	180	360	–?–	–?–	–?–	...	–?–

State your next conjecture.

C-7 *The sum of the measures of the n angles of an n-gon is —?—.* ***(Polygon Sum Conjecture)***

Here are three more definitions from Chapter 2. (Again, don't worry if the wording differs slightly from the wording you used in your definitions.)

*An **equiangular polygon** is a polygon with all angles equal in measure.*

*An **equilateral polygon** is a polygon with all sides equal in length.*

*A **regular polygon** is a polygon with all sides equal in length and all angles equal in measure. It is equiangular and equilateral.*

Can you predict the measure of each angle in an equiangular polygon?

Investigation 4.3.3 

Copy and complete the table below to determine the measure of each angle of an equiangular *n*-gon.

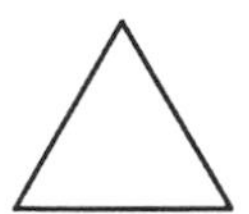  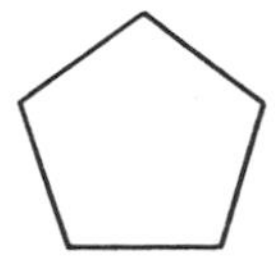  

Sides of equiangular polygon	3	4	5	6	7	. . .	n
Sum of measures of angles	(1)(180)	(2)(180)	(3)(180)	–?–	–?–	. . .	–?–
Measure of each angle	$\frac{(1)(180)}{3}$	$\frac{(2)(180)}{4}$	$\frac{(3)(180)}{5}$	–?–	–?–	. . .	–?–

State your next conjecture.

The measure of each angle of an equiangular n-gon is —?—.

EXERCISE SET A

Trace the diagrams below. Calculate the measure of each lettered angle.

1.* $x =$ –?–

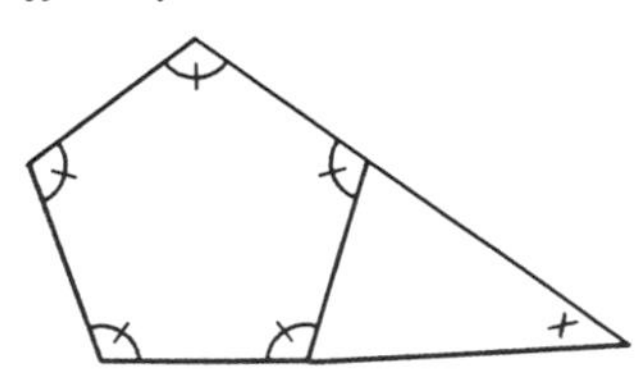

2. $y =$ –?–

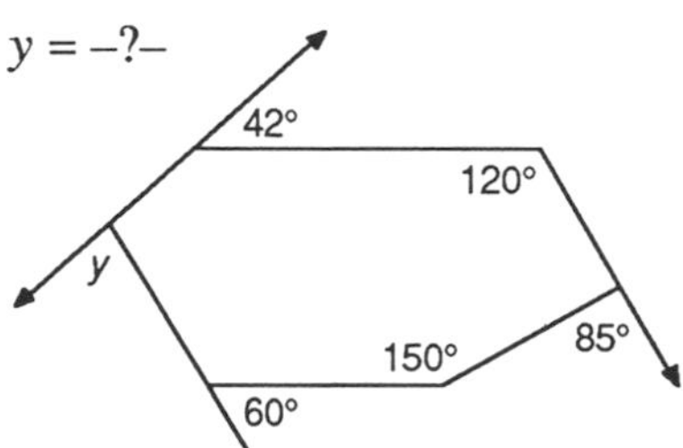

3.* $z =$ –?–

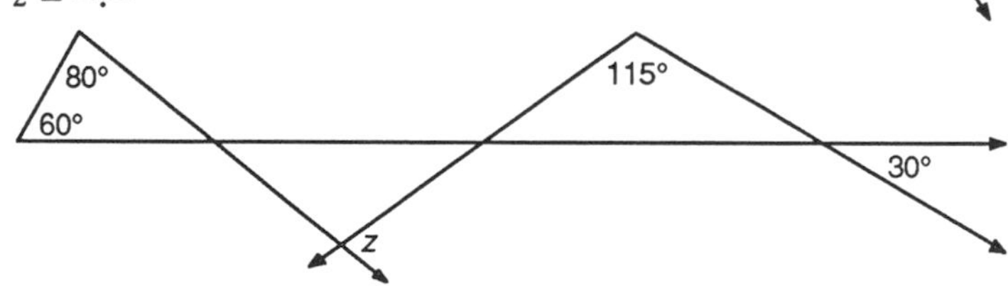

EXERCISE SET B

Use your conjectures from this lesson to answer each question.

1. What is the sum of the measures of the angles of a decagon?
2. What is the sum of the measures of the angles of a 25-gon?
3.* How many sides does a polygon have if the sum of the angle measures is 2700?
4. What is the measure of each angle of an equiangular decagon?

5. What is the measure of each angle of a regular hexagon?

6.* How many sides does an equiangular polygon have if each angle measures 170?

7. How many sides does an equiangular polygon have if an angle measures 174?

8. Archaeologist Ertha Diggs has uncovered a piece of ceramic plate. The original plate appears to have been in the shape of a regular polygon. If the original plate was a regular 16-gon, it was probably a ceremonial dish from the 3rd Century Ho Ping dynasty. If it was a regular 18-gon, it was probably a palace dinner plate from the 12th century Wai Ting dynasty. Ertha measures each of the sides of her piece and finds that they are all the same size. She conjectures that all the sides of the original whole plate were the same size. She measures each of the angles on the piece and finds that they are all the same size. She conjectures that all the angles on the original whole plate were equal. If each angle measures 160, what dynasty did the plate come from?

EXERCISE SET C

1. Six equilateral triangles can fit about a point ($6 \cdot 60 = 360$) without gaps or overlapping. Can you do this with any triangle? Try it. Draw a triangle. Make five more copies. Label the interiors of the three angles *a*, *b*, and *c* correspondingly in each triangle. Cut them out. Try arranging them about a point.

2. Four rectangles can fit about a point ($4 \cdot 90 = 360$) without gaps or overlapping. Can you do this with any quadrilateral? Try it. Draw a quadrilateral. Make three more copies. Label the interiors of the four angles *a*, *b*, *c*, and *d* correspondingly in each quadrilateral. Cut them out. Try arranging them about a point.

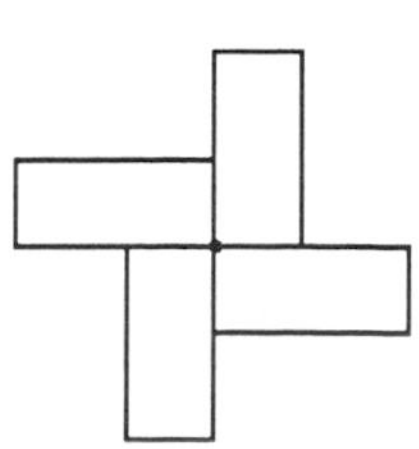

Lesson 4.4

Exterior Angles of a Polygon

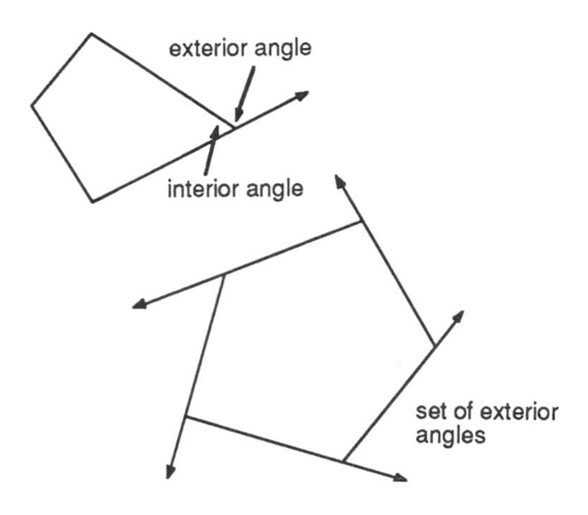

If you extend one side of a polygon from one endpoint, then you have constructed an **exterior angle** at that vertex. (To avoid confusion, we refer to the angle of the polygon at that vertex as the interior angle.) If you extend each side of a polygon to form one exterior angle at each vertex, you have a set of exterior angles for the polygon.

In the last lesson you discovered a formula for the sum of the measures of the interior angles of every polygon. In this lesson you will discover a formula for the sum of the measures of a set of exterior angles of a polygon.

Let's investigate the sum of the measures of a set of exterior angles of five different polygons. This investigation will be easier if you work in small groups. Form groups of four or five members. Each group will need five sheets of paper.

Investigation 4.4.1

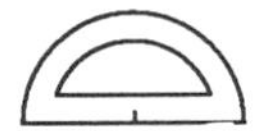

Construct a large triangle on one sheet of paper, a large quadrilateral on the second, a large pentagon on the third, a large hexagon on the fourth, and a large heptagon on the fifth piece of paper. Distribute the five polygons, one to each of the group members. (If you have four members in your group, one person should take the triangle and quadrilateral.) Each person should perform the steps below.

Step 1: Construct a set of exterior angles.

Step 2: With a protractor, measure all the interior angles of the polygon except one.

Step 3: Use the Polygon Sum Conjecture to calculate the measure of the remaining interior angle.

Step 4: Use the Linear Pair Conjecture to calculate the measure of each exterior angle.

Step 5: Calculate the sum of the measures of the set of exterior angles.

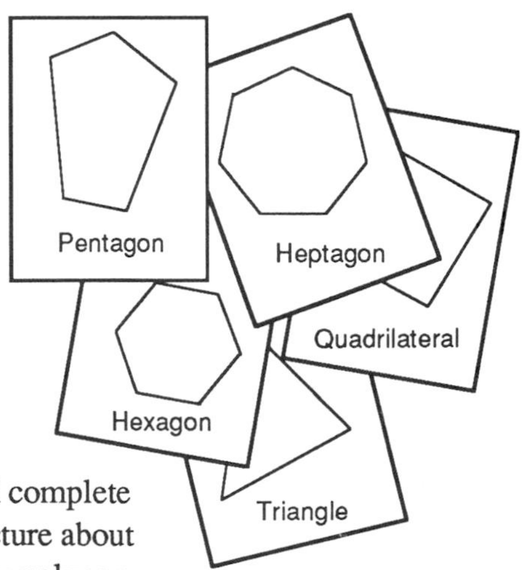

Share your results with your group members. Copy and complete the table on the next page. When finished, state a conjecture about the sum of the measures of a set of exterior angles of any polygon.

Number of sides of polygon	3	4	5	6	7	...	n
Sum of measures of exterior angles	–?–	–?–	–?–	–?–	–?–	...	–?–

State your conjecture.

The sum of the measures of one set of exterior angles —?—.

For each exterior angle of a triangle there corresponds an adjacent interior angle and a pair of remote interior angles. The **remote interior angles** are the two angles that do not have the same vertex as the exterior angle.

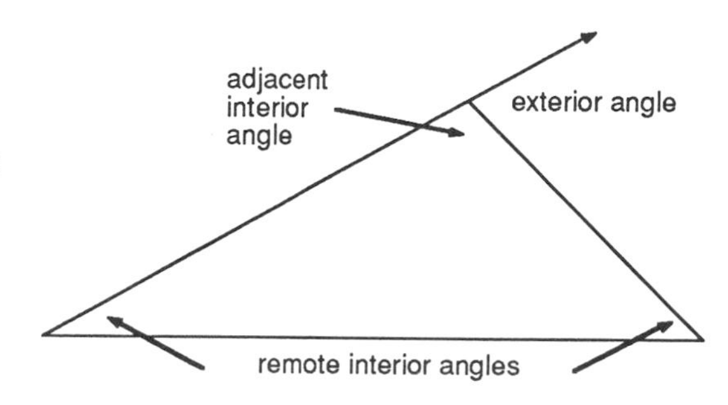

In the next investigation, you are going to discover the relationship between an exterior angle and its two remote interior angles.

Investigation 4.4.2

Draw a large acute scalene triangle ABC on your paper.

Step 1

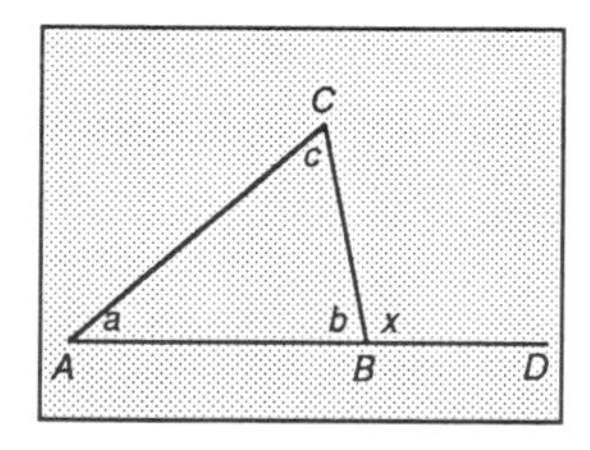

Extend segment AB through point B and label a point D beyond point B. Put an a in the interior of $\angle A$, a b in the interior of $\angle ABC$, a c in the interior of $\angle C$, and an x in exterior angle CBD.

Step 2

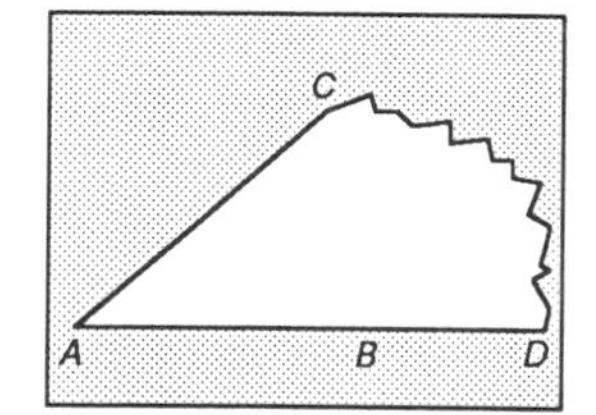

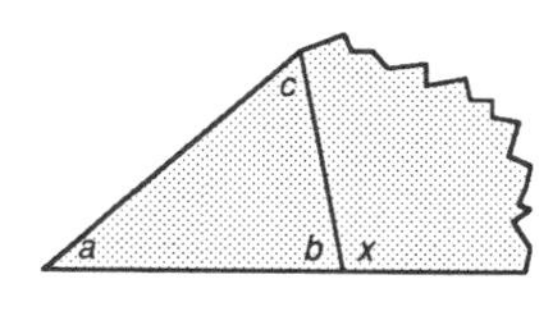

Cut out the triangle with its exterior angle attached.

Step 3

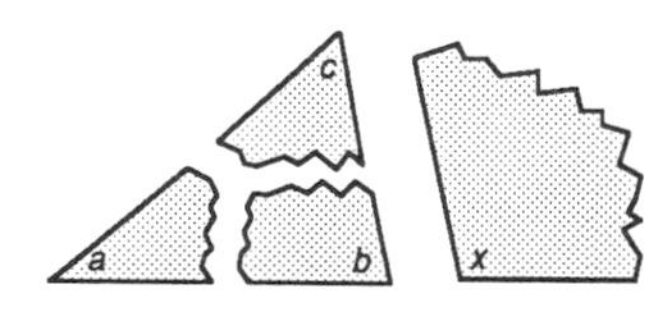

Cut off the exterior angle (the angle with the x). Tear off the two remote interior angles (the angles with a and c).

Step 4

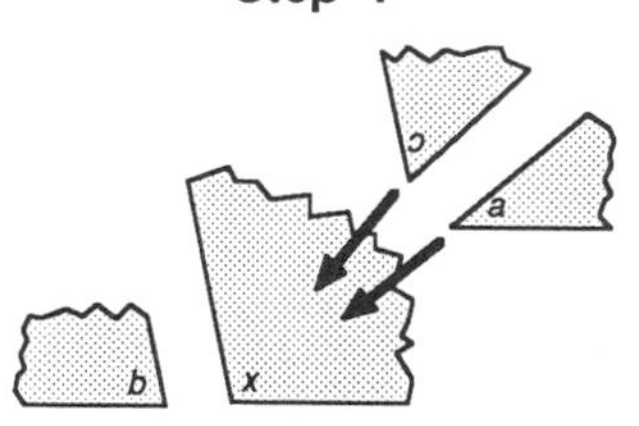

Place the two remote interior angles over the exterior angle, positioning the vertex points together and arranging the two remote interior angles without overlap to show their sum.

What do you notice about the sum of the measures of the two remote interior angles and the measure of the exterior angle? State your observations as a conjecture.

C-10 *The measure of an exterior angle of a triangle —?—.*
(Exterior Angle Conjecture)

The Exterior Angle Conjecture can be verified with algebra and logical reasoning. In the diagram of ΔCAT below, $\overline{CA}$ has been extended to form exterior angle RAT.

Paragraph Proof:

Since $\angle CAT$ and $\angle RAT$ form a linear pair of angles, $y + z = 180$.

Since the sum of the measures of the three angles of a triangle equals 180, you know that $x + y + w = 180$ also.

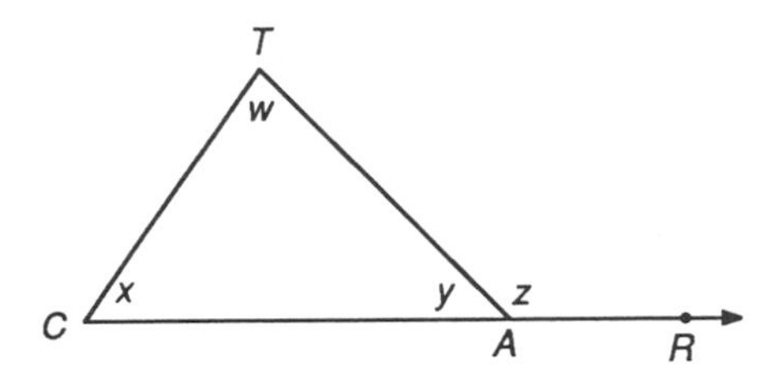

Therefore $x + y + w = y + z$.

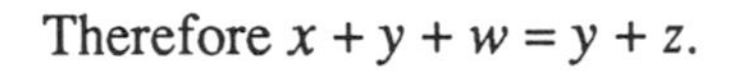

Subtracting y from both sides of this equation gives $x + w = z$, proving your conjecture.

EXERCISE SET A

1. What is the sum of the measures of a set of exterior angles of a decagon?
2. Four exterior angles of a pentagon measure 63, 67, 58, and 64. What is the measure of the remaining exterior angle?
3.* What is the measure of each exterior angle of a regular octagon?
4.* How many sides does a regular polygon have if each exterior angle has a measure of 24?
5. What is the sum of the measures of the interior angles of a dodecagon?
6. How many sides does a polygon have if the sum of the interior angle measures is 7380?
7. What is the measure of each interior angle of a regular octagon?
8.* How many sides does a regular polygon have if each interior angle measures 165?

EXERCISE SET B

Use the Exterior Angle Conjecture to solve each problem below.

1. $t + p =$ –?–

135°

2. $m + n =$ –?–

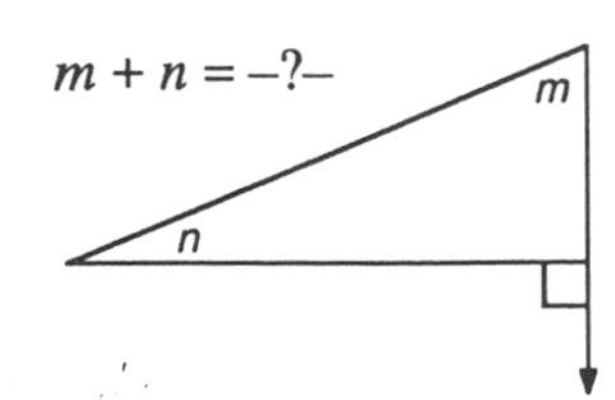

3. $r = $ –?–

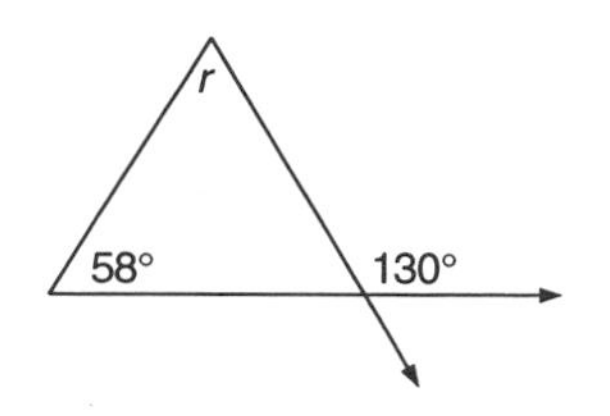

4. $x = $ –?–

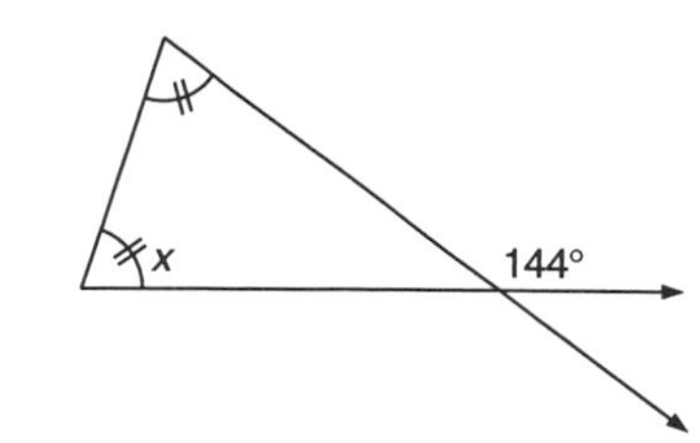

5. $y = $ –?–

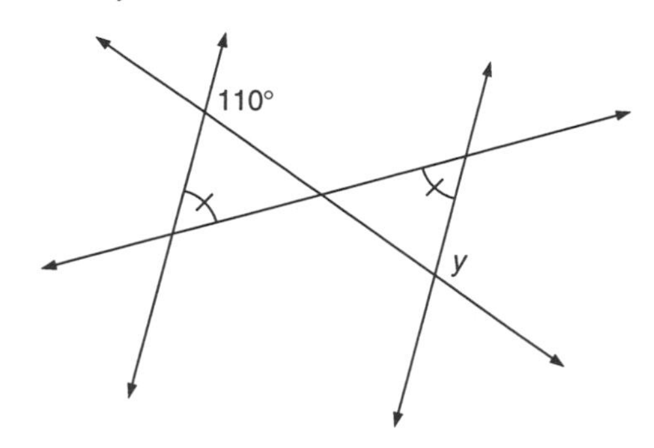

6.* $z = $ –?–

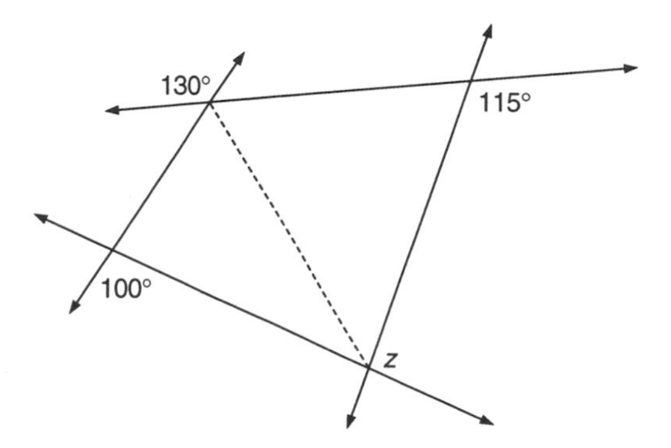

Improving Visual Thinking Skills

Match This!

Perhaps you have seen matchstick puzzles before. Here is a classic from British puzzle enthusiast Henry E. Dudeney (1857–1930). The matches represent a farmer's hurdles (portable fence pieces) placed so that they form six pens of equal size. After one hurdle is stolen, the farmer must rearrange the remaining twelve to form another six pens of equal size. How can it be done?

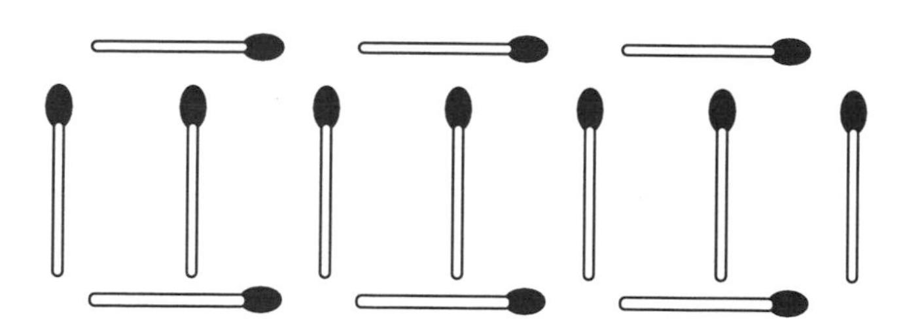

Special Project

Traveling Networks

The River Pregel runs through the university town of Konigsberg (now Kaliningrad in the Soviet Union). In the middle of the river are two islands connected to each other and the rest of the city by seven bridges.

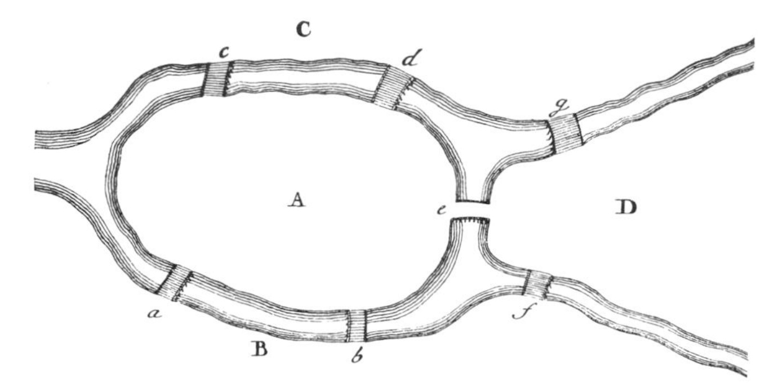

The Seven Bridges of Konigsberg

Over the years a tradition developed among the college students of Konigsberg. They challenged each other to make a round trip over all seven bridges, walking over each bridge once and only once before returning to the starting point. For a long time no one was able to do it, and yet no one was able to show that it couldn't be done. In 1735, students finally wrote to the great Swiss mathematician Leonard Euler asking for his help on the problem. Euler was able to show that the feat cannot be done. (The illustration above is from Euler's manuscript.) The walking is good exercise, but it is impossible to cross all the bridges without crossing at least one of them more than once.

Euler reduced the problem to a network of paths connecting the two sides of the river C and B, and the two islands A and D, as shown in the network at right. Then Euler demonstrated that the task was impossible. Euler's solution began the branch of geometry called topology. In this special project you will work with a variety of networks to see if you can come up with a rule to determine whether a network can or cannot be "traveled."

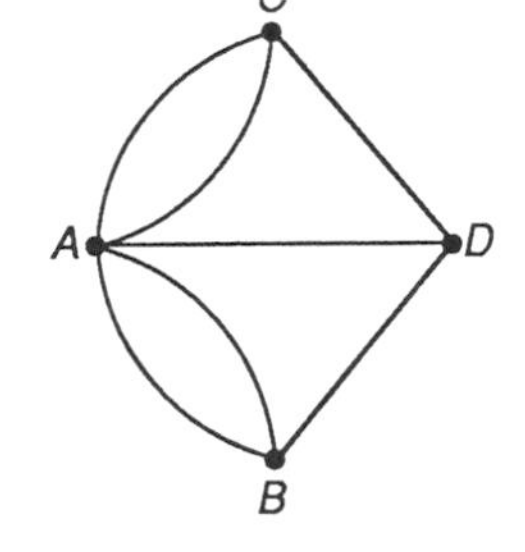

A collection of points connected by paths is called a **network**. When we say a network can be traveled, we mean that the network can be drawn with a pencil without lifting the pencil off the paper and without retracing any paths. (Points can be passed over more than once.) Try the networks on the next page to see which ones can be traveled and which are impossible to travel. It will make things easier if you know that the number of paths is not important in determining whether or not a network can be traveled. This leaves only the points to be investigated. There are two types of points: odd points and even points.

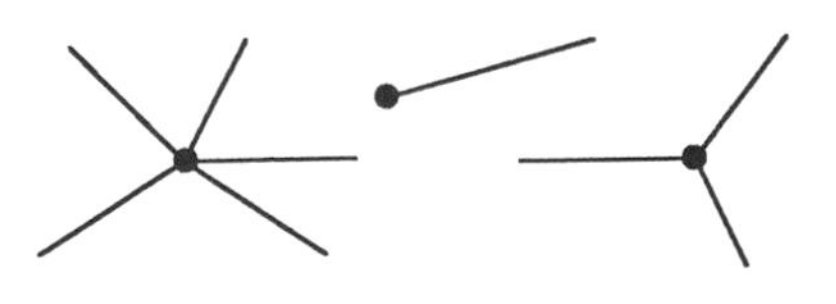

Odd Points

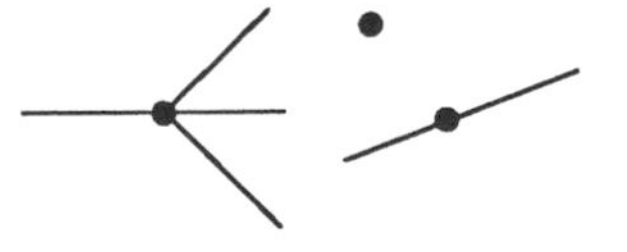

Even Points

Copy and complete the table. When you have completed the table, you should find that seven networks cannot be traveled.

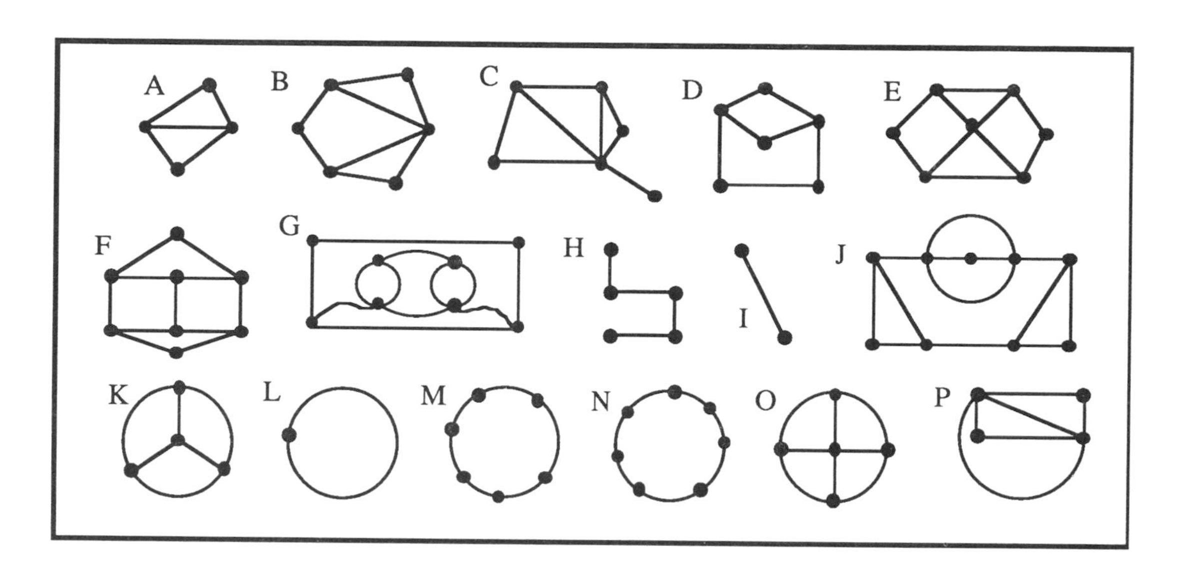

Network	A	B	C	D	E	F	G	H	I	J	K	L	M	N	O	P
Number of odd points	-?-	-?-	-?-	-?-	-?-	-?-	-?-	-?-	-?-	-?-	-?-	-?-	-?-	-?-	-?-	-?-
Number of even points	-?-	-?-	-?-	-?-	-?-	-?-	-?-	-?-	-?-	-?-	-?-	-?-	-?-	-?-	-?-	-?-
Can it be traveled? (y/n)	-?-	-?-	-?-	-?-	-?-	-?-	-?-	-?-	-?-	-?-	-?-	-?-	-?-	-?-	-?-	-?-

Look carefully at networks L, M, N, and P. These four networks have only even points. Notice that the network can be traveled whether you have one even point or seven even points — as long as there are no odd points. It appears that you can have as many even points as you desire and the network can still be traveled. Therefore it is the odd points that can cause trouble for network travelers! Study the table to see how the number of odd points determines whether or not a network can be traveled. When you see the relationship, you will be able to complete the conjecture below.

Conjecture: *A network can be traveled whenever —?—.*

By successfully completing the conjecture, you have solved the network problem using inductive reasoning. After you have solved a problem this way, it's a good idea to test your solution on some more examples, and then, if possible, to verify it using logical reasoning. Let's take a closer look at the conjecture above using a bit of logical reasoning.

Let's agree that any network that has all even points can be traveled. Let's look at some networks with odd points that we were able to travel. If a network can be traveled and it has odd points, where do you have to start and where do you end? Look carefully at networks A, B, D, and H. How would you travel them?

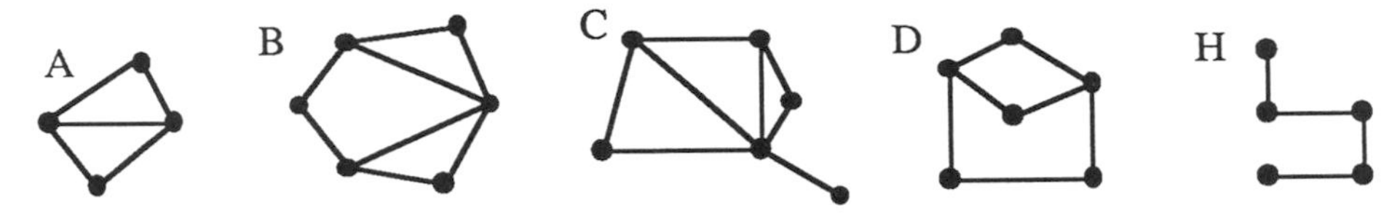

Did you notice that if you start at one odd point, you *must* end at the other? If you start at one of the even points you cannot cover all the paths. If there are more than two odd points, select one as your starting point (for example, one with three segments leaving it). If you start at this odd point, you must leave it, return to it, and leave it again in order to cover all lines leaving this point. Meanwhile, since you didn't start at the other odd points, you must also go to each of the other odd points, leave them, and eventually return to each of them in order to cover all the paths at each of the odd points. But how can you possibly end at more than one point? This leaves only one logical conclusion.

Use your conjecture to solve each problem.

1. Draw the River Pregel and the two islands shown on the first page of this special project. Draw an eighth bridge so that the students can travel over all the bridges exactly once if they start at point C and end at point B.

2. Draw the River Pregel and the two islands. Can you draw an eighth bridge so that students can travel over all the bridges exactly once, starting and finishing at the same point? How many solutions can you find?

3.* Secret Agent Samantha Spade must search Baron Von Evol's home to find the microfilm stolen from Laserlite Industries. The microfilm, containing the plans for a laser satellite, is to be sold by the Baron to foreign agents tomorrow. The floor plan of the house is shown on the right. Agent Samantha Spade must find a quick route through all the rooms. If she enters the house through the door to room A and leaves through the door at room G, can she find a path through all the rooms, passing through each and every door exactly once? If possible, show the route. If impossible, explain why it cannot be done. Draw a network in your explanation. Can it be done if she omits one door from her path?

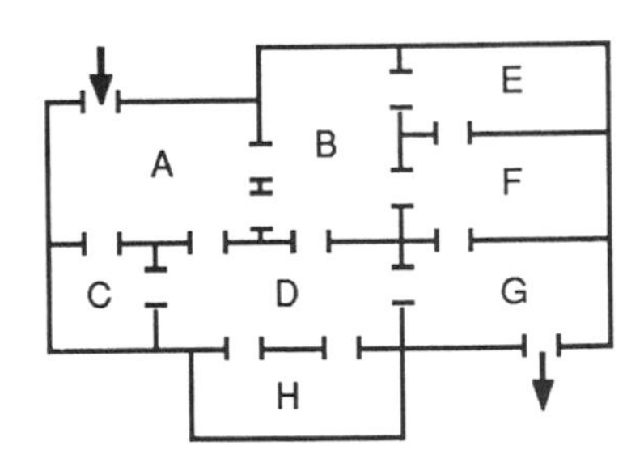

Von Evol Home

4.* There are eight major roads in Podunk County. A county road map is shown above. Elmer Rhodes, the county highway engineer, has a truck designed to insert "dummy bumps" (raised plastic squares) into the middle of the road. (They're called "dummy bumps" because you hear the sound "dummy, dummy, dummy" when you drive over them) Elmer's truck is not very fuel efficient, however, and he wants to find the shortest possible route covering the eight county roads. The truck is kept at point B on the map and must return to the same point. Elmer suspects he may have to go over at least one road more than once. From what we have learned, he is correct. What is the shortest distance he must travel to cover all eight county roads? Distances shown are measured in kilometers.

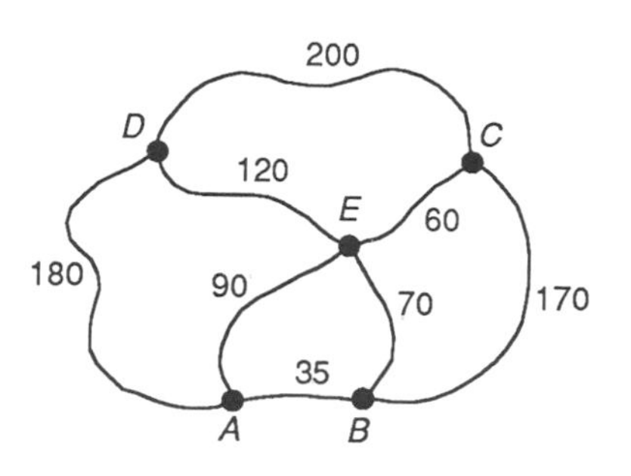

Podunk County

Improving Visual Thinking Skills

Counting Cubes

How many cubes were used to build each solid?

1.

2.

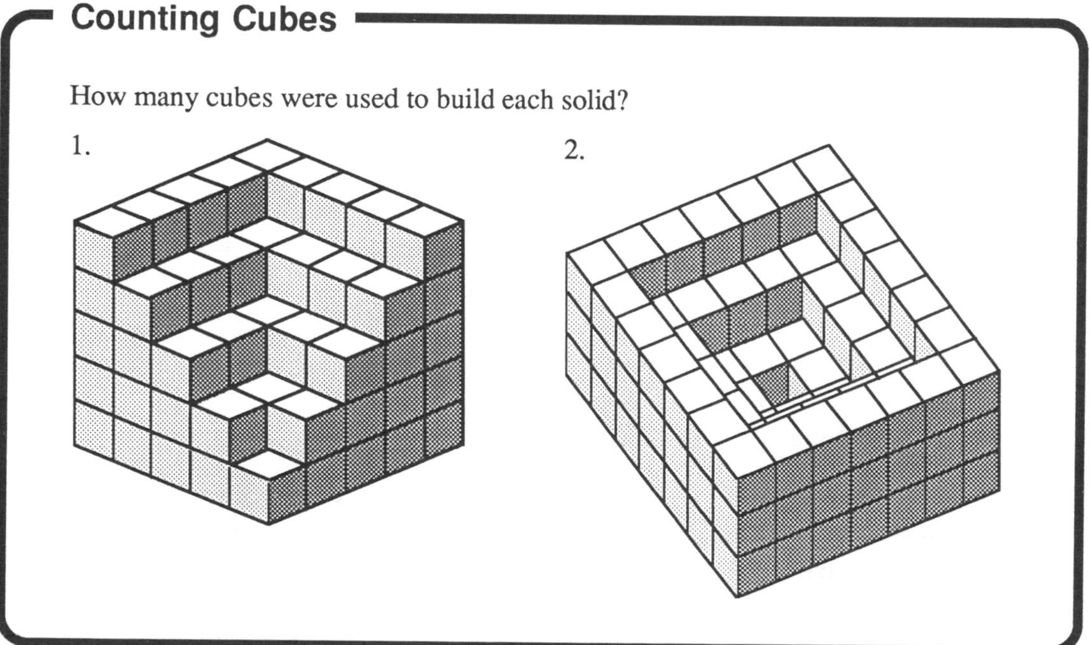

Lesson 4.5
Triangle Inequalities

Education is that which allows you to get into more intelligent trouble.

– Anonymous

In this lesson you will discover two very basic geometric inequalities.

Investigation 4.5.1

Use your compass to copy the segments below onto a clean sheet of paper. Construct a triangle with each set of segments below as sides.

Given: Construct: ΔDOG

D ———————————— O

O ———————————————— G

D ———————————————————— G

Given: Construct: ΔCAT

C ———————— A

A ———————————— T

C ———————————————————————— T

Given: Construct: ΔFSH

F ————————————————————————— S

S ——————————— H

F ———————— H

Were you able to construct a triangle each time? Why or why not? What does the statement: *The shortest path between two points is a straight line* have to do with this investigation? Discuss your results with others near you. State your observations as your next conjecture.

C-11 *The sum of the lengths of any two sides of a triangle is —?— the length of the third side.*
(Triangle Inequality Conjecture)

The Triangle Inequality Conjecture relates the lengths of the three sides of a triangle. In the next investigation you will discover the relationship between the size of angles in triangles and the size of the sides opposite those angles.

Investigation 4.5.2 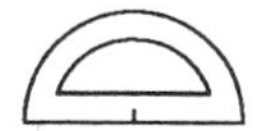

On a clean sheet of paper construct a large acute triangle on the top half of your paper and a large obtuse triangle on the bottom half. For each triangle:

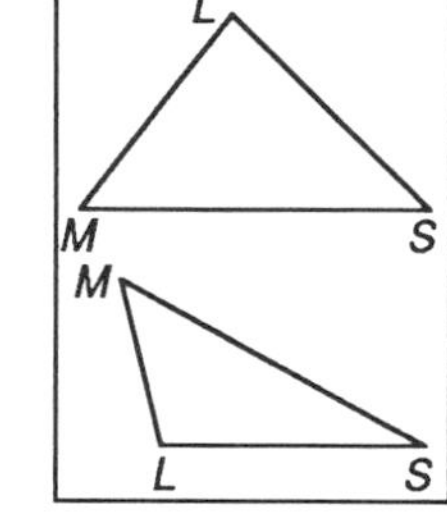

Step 1: Measure the angles in each triangle. Label the largest, $\angle L$; the second largest, $\angle M$; and the smallest, $\angle S$.

Step 2: Measure the three sides. Which side is the largest? Label it by placing the lower case l near the middle of the side. Which side is the second largest? Label it m in the same way. Which side is the smallest? Label it s.

Which side, l, m, or s, is opposite the largest angle? Which side is opposite the second largest angle? Which side is opposite the smallest angle?

Discuss your results with others near you. State your observations as a conjecture.

In a triangle, —?—.

EXERCISE SET A

For each set of lengths, determine whether it is possible to draw a triangle with sides of the given measures. If possible, write *yes*. If not possible, write *no*.

1. 3 cm, 4 cm, 5 cm
2. 4 m, 5 m, 9 m
3 5 ft., 6 ft., 12 ft.
4. 3.5 cm, 4.5 cm, 7 cm
5. 4", 5", $8\frac{1}{2}$"
6. .5 m, .6 m, 12 cm

EXERCISE SET B

In Exercises 1 and 2, the letter on each side of the triangles indicates the size of that side. In Exercises 3 and 4, the letter indicates the size of an angle. Use your new conjecture to arrange the letters in order from large to small.

1.*

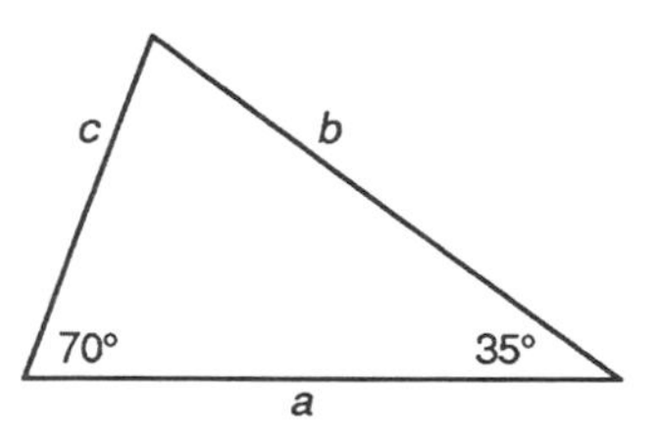

2.

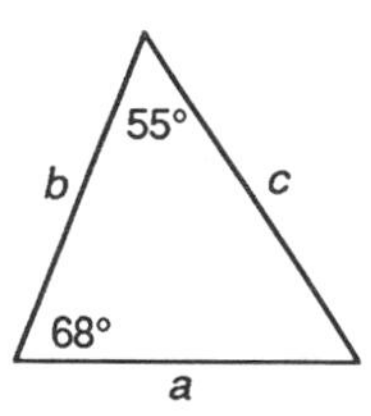

3.

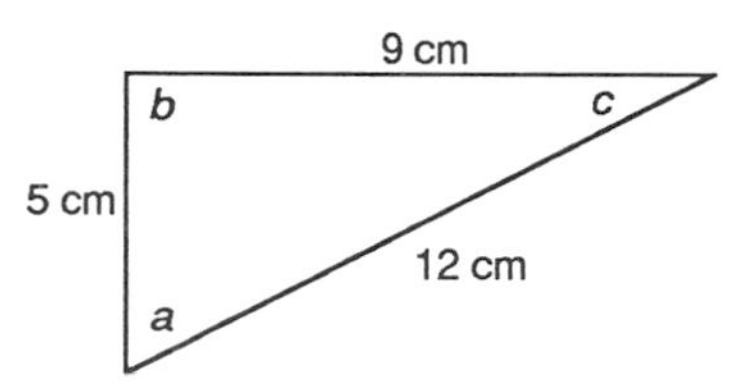

4.

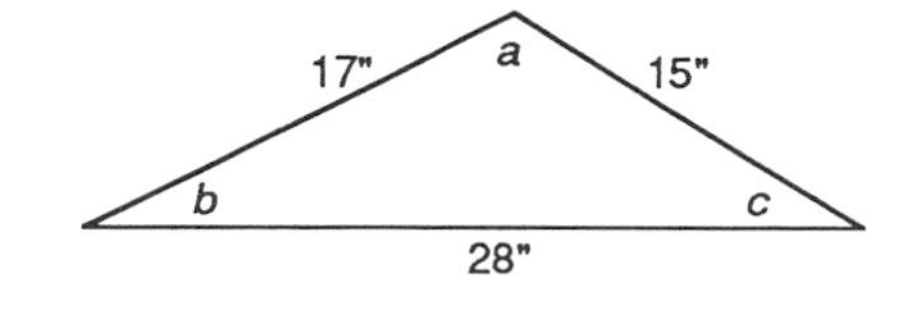

Lesson 4.6
Discovering Properties of Isosceles Triangles

*An **isosceles triangle** is a triangle with at least two sides the same length.*

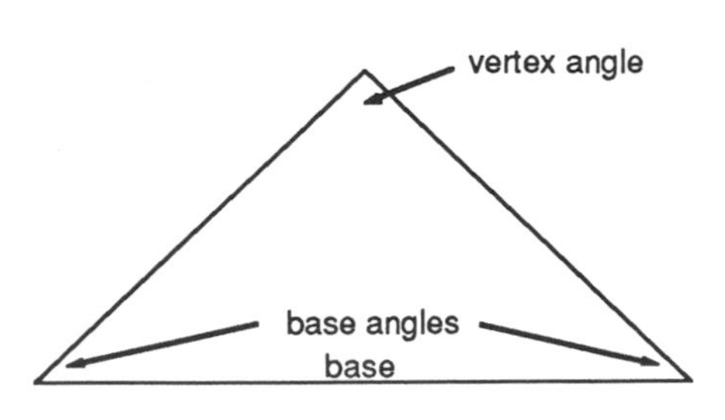

In an isosceles triangle, the angle between the two sides of equal length is called the **vertex angle** and the other two angles are called the **base angles**. The side between the two base angles is called the **base** of the isosceles triangle.

In this lesson you are going to construct isosceles triangles and investigate the measures of the base angles. This investigation works best in small cooperative groups of four or five members. You, or you and members of your group, will do this four times, twice using isosceles triangles that have acute vertex angles, and twice using isosceles triangles with obtuse vertex angles.

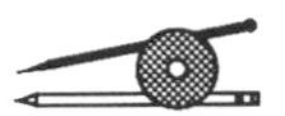
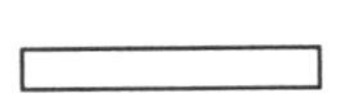

Step 1

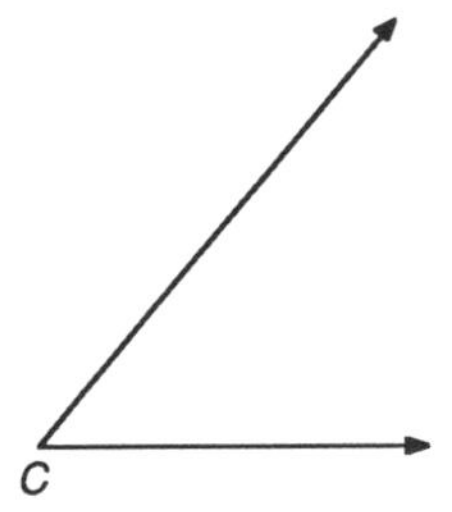

Construct an angle. Label the vertex C. This angle will be the vertex angle of your isosceles triangle.

Step 2

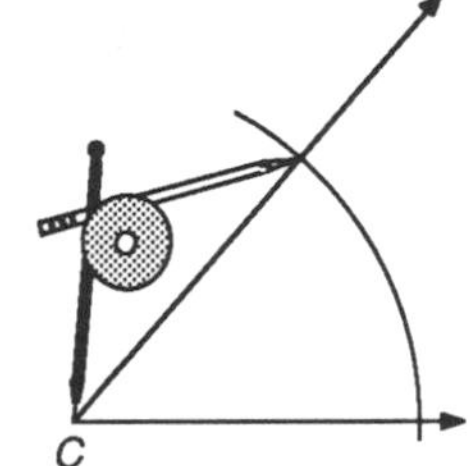

Place the pointer of your compass on point C and swing an arc passing through the two sides of $\angle C$.

Step 3

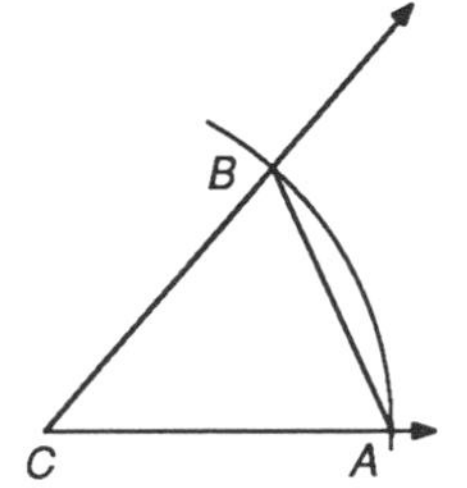

Label the two points A and B. Construct $\overline{AB}$. You have constructed isosceles ΔABC.

Step 4

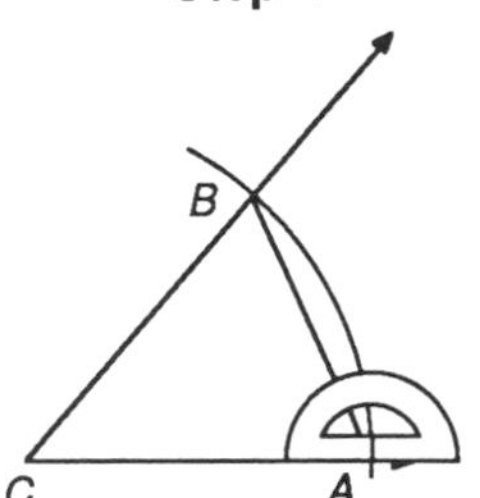

Use your protractor to measure the base angles of isosceles ΔABC.

Repeat these steps three more times to construct four isosceles triangles, two with obtuse vertex angles and two with acute vertex angles. Compare your results with the results of others near you.

What relationship do you notice between the base angles of each isosceles triangle? State your observations as your next conjecture.

If a triangle is isosceles, then —?—.
(Isosceles Triangle Conjecture)

Is the reverse (called the converse) of the conjecture also true? Conjecture 13 says: *If a triangle is isosceles, then the bases angles are equal in measure.* This is an "if-then" or conditional statement. If you switch the two parts of the conditional statement you create the converse of the conditional statement. The converse of Conjecture 13 says: *If a triangle has two angles of the same size, then the triangle is isosceles.* Just because a conditional statement is true does not mean that its converse must be true. Is the converse of Conjecture 13 true? Let's investigate.

In the next investigation you will construct triangles that have two angles equal in measure. Then you will measure the sides of the triangles to see if the triangles are isosceles. The investigation also works best if you work in groups of four or five.

Investigation 4.6.2

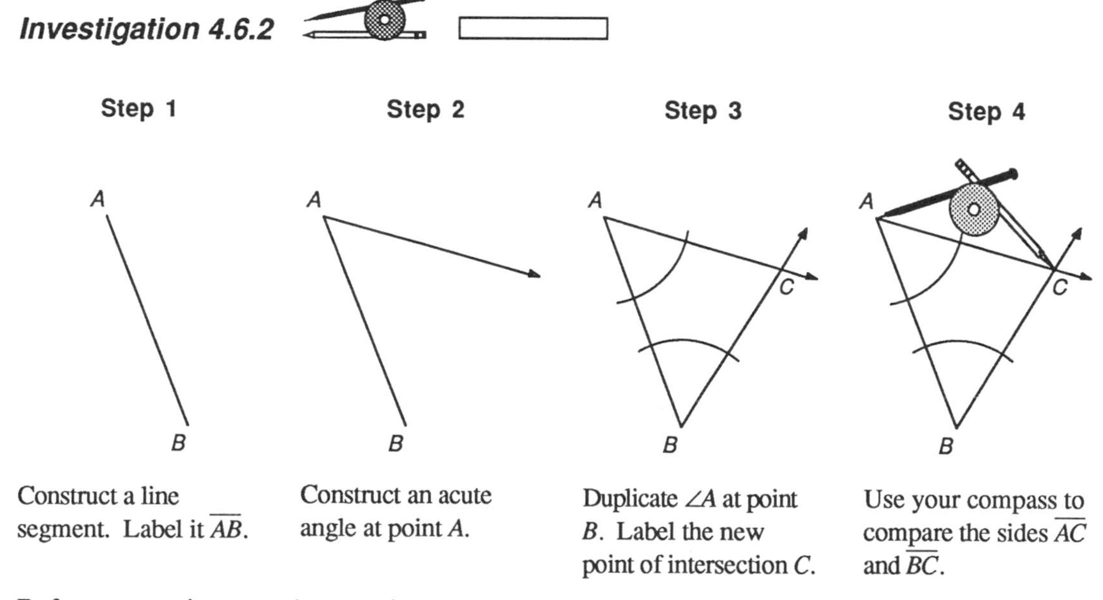

Step 1: Construct a line segment. Label it $\overline{AB}$.

Step 2: Construct an acute angle at point A.

Step 3: Duplicate $\angle A$ at point B. Label the new point of intersection C.

Step 4: Use your compass to compare the sides $\overline{AC}$ and $\overline{BC}$.

Before attempting to make a conjecture, you should repeat these steps three more times. When finished, you will have constructed four triangles, each having two angles, $\angle A$ and $\angle B$, equal in measure.

What relationship do you notice between the sides $\overline{AC}$ and $\overline{BC}$ in each of the triangles you constructed? State your observation as your next conjecture.

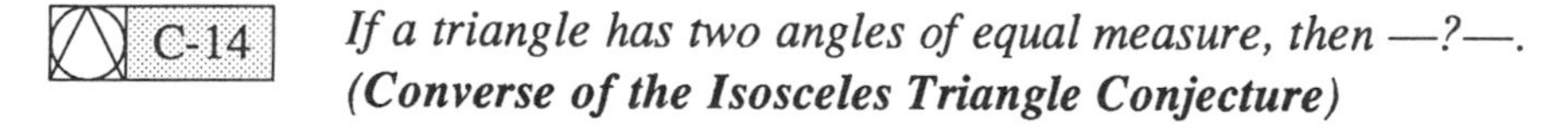

If a triangle has two angles of equal measure, then —?—.
(Converse of the Isosceles Triangle Conjecture)

Once you have discovered the converse of the Isosceles Triangle Conjecture, you can use it with a little logical reasoning to arrive at the next conjecture. In Chapter 3 you discovered that if a triangle is equilateral, then each angle measures 60. If each angle measures 60, then all the angles have the same measure. Therefore, *if a triangle is equilateral, then it is equiangular.* Is the converse also true? The converse says: *If a triangle is equiangular, then it is equilateral.*

The converse of the Isosceles Triangle Conjecture can be used to show that this is true. Read the paragraph proof below. It proves that if the converse of the Isosceles Triangle Conjecture is true, then the conjecture: *If a triangle is equiangular, then it is equilateral* is also true.

Paragraph Proof:

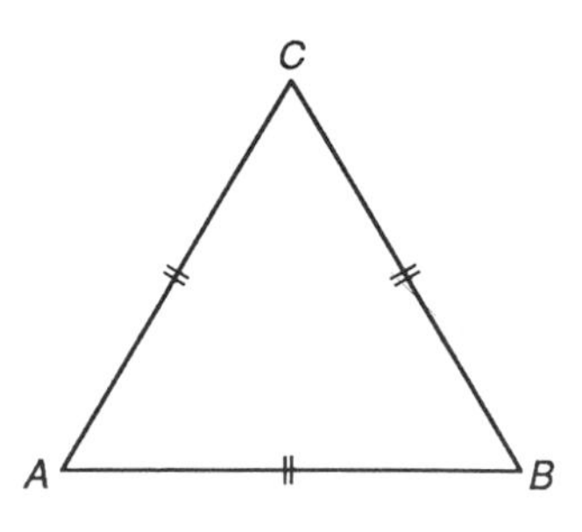

If ΔABC is equiangular, then $m\angle B = m\angle C$ and $m\angle A = m\angle C$
If $m\angle B = m\angle C$, then $AB = AC$.
If $m\angle A = m\angle C$, then $AB = BC$.
But, if $AB = AC$ and $AB = BC$, then $AB = AC = BC$ and thus ΔABC is equilateral.
Therefore, *if a triangle is equiangular, then it is equilateral.*

Copy and complete the conjecture below and add it to your conjecture list.

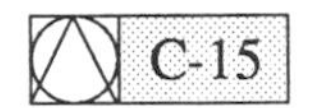

An equilateral triangle is —?— and, conversely, an equiangular triangle is —?—.
(Equilateral Triangle Conjecture)

If a triangle is equilateral, then it is equiangular, and therefore it is a regular triangle. In this text we will usually refer to such a triangle as equilateral, but it will be understood that it is also equiangular and thus regular.

When a conditional statement and its converse are both true, they can be combined into one statement called an "if and only if" or **biconditional** statement. The two conjectures: *If a triangle is equilateral, then it is equiangular* and the converse: *If a triangle is equiangular, then it is equilateral* can be combined into one "if and only if" statement. The "if and only if" statement says: *A triangle is equilateral if and only if it is equiangular.*

EXERCISE SET

Use your new conjectures to find the missing measures in each problem. You should not use a protractor or ruler.

1.* $m\angle H = $ –?–

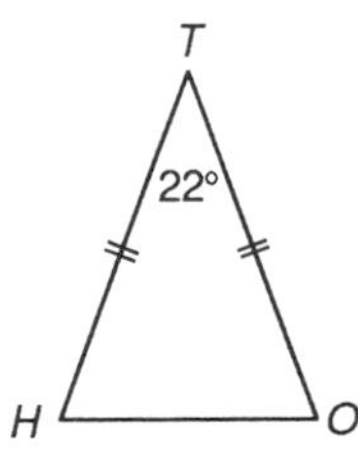

2. $m\angle G = $ –?–

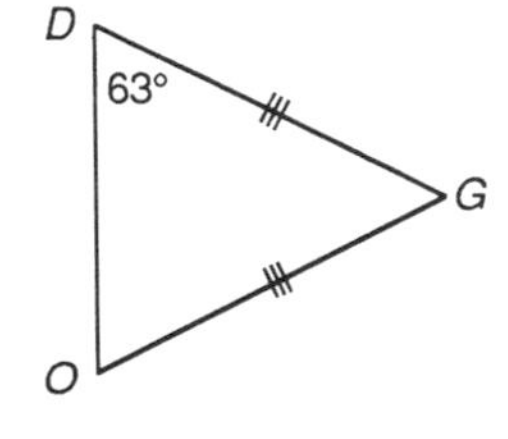

3. $m\angle OLE = $ –?–

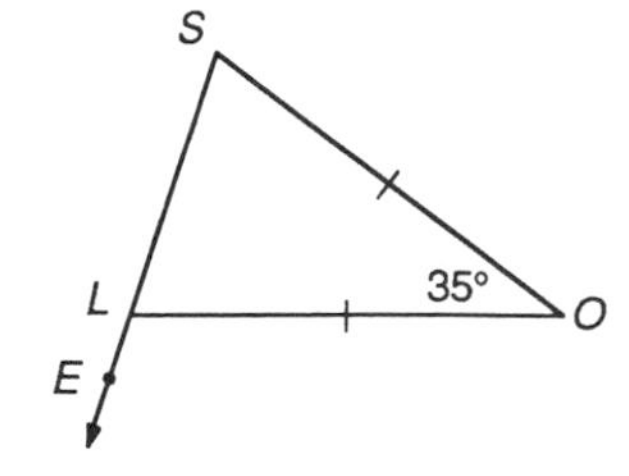

4. $m\angle R =$ –?–
$m\overline{RE} =$ –?–

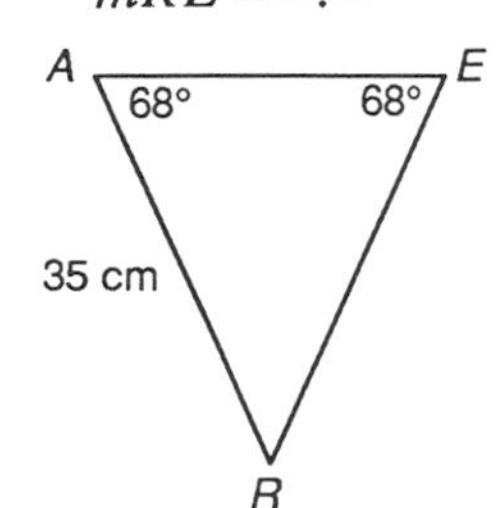

5. $m\angle Y =$ –?–
$m\overline{UO} =$ –?–

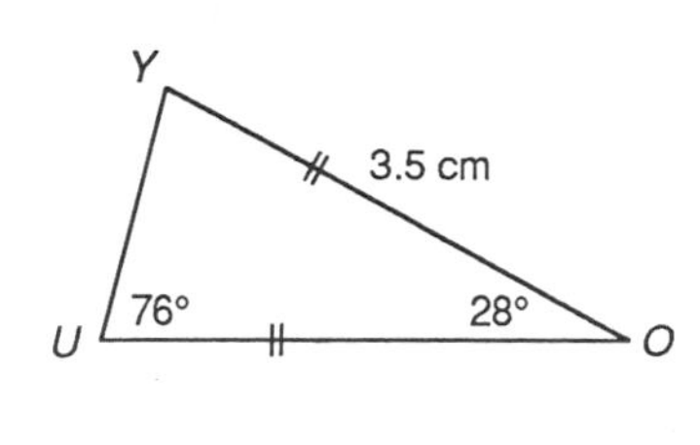

6. The perimeter of ΔMUD is 38 cm.
$m\angle D =$ –?–
$m\overline{MD} =$ –?–

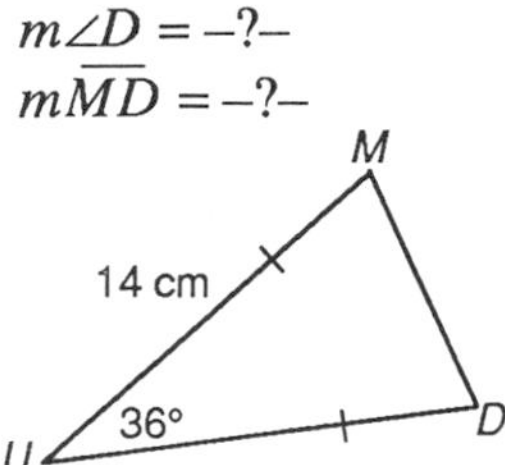

7.* Trace the diagram below. Calculate the measure of each lettered angle.

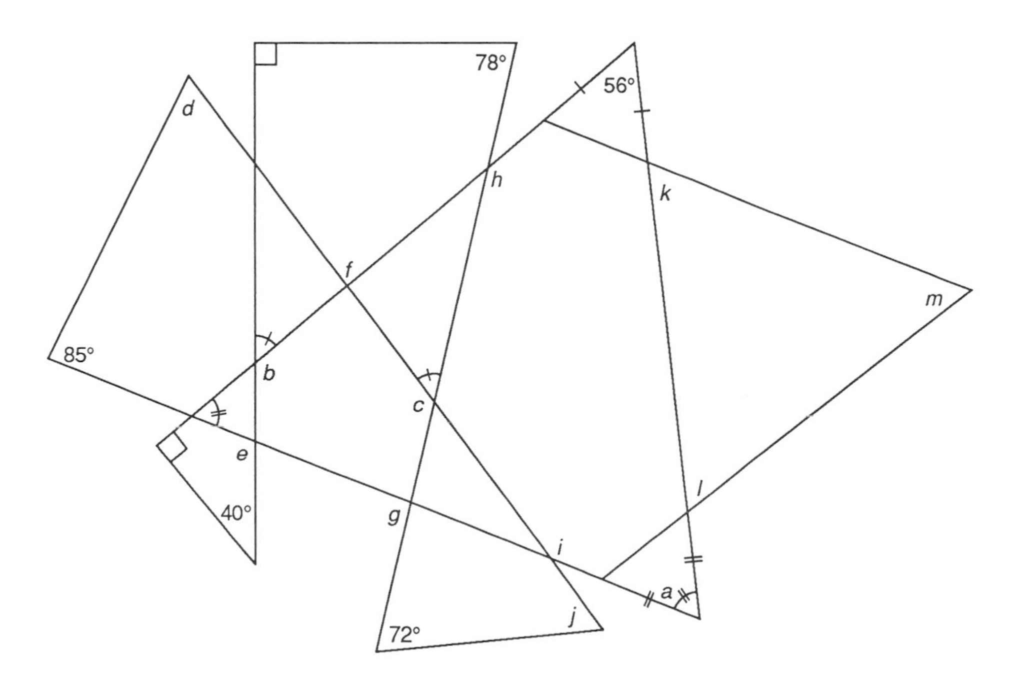

Improving Visual Thinking Skills

Coin Swap II

Find three light coins (dimes) and three dark coins (pennies) and arrange them on a grid of seven squares as shown.
Your task is to switch the position of the three light and three dark coins in exactly 15 moves. A coin can slide into an empty square next to it or it can jump over one coin into an empty space. Record your solution by listing in order which color coin is moved. For example, your list might begin DLDLDLDL . . . .

Lesson 4.7

Discovering Properties of Parallel Lines

Two or more lines are ***parallel*** *if and only if they are in the same plane and do not intersect.*

A line passing through two or more other lines in a plane is called a **transversal.** A transversal intersecting two lines creates three different types of angle pairs.

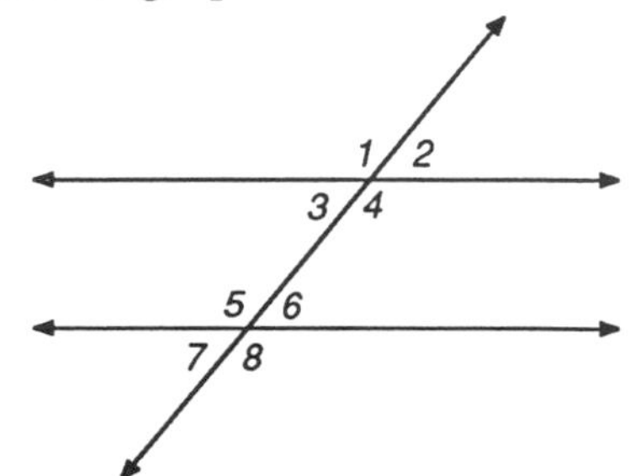

$\angle 1$ and $\angle 5$ are one pair of **corresponding angles.** Can you find three more pairs of corresponding angles?

$\angle 3$ and $\angle 6$ are one pair of **alternate interior angles**. Do you see another pair of alternate interior angles?

$\angle 2$ and $\angle 7$ are one pair of **alternate exterior angles**. Do you see another pair of alternate exterior angles?

When a pair of parallel lines are cut by a transversal, there is a special relationship among the angles. Let's investigate.

Investigation 4.7.1

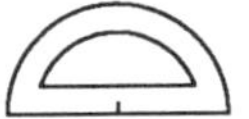

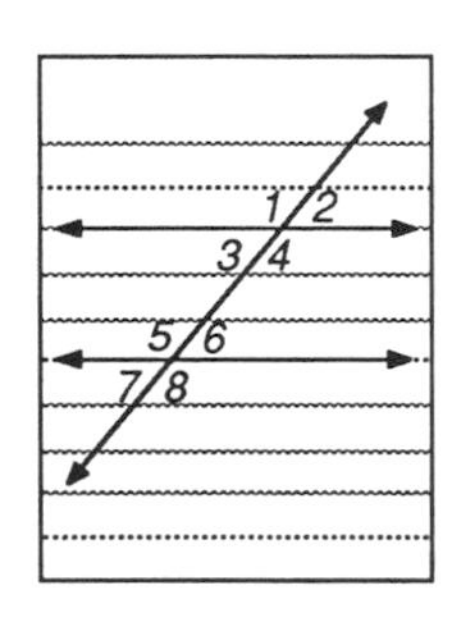

Using the lines on a piece of graph paper or a piece of lined paper as a guide, draw a pair of parallel lines. (We will assume that the lines on the paper have been created carefully by the paper's manufacturer.) If you don't have ruled paper, use both edges of your ruler or straightedge to create parallel lines. Now draw a transversal intersecting the parallel lines. Label the angles with numbers as in the diagram above.

Measure $\angle 1$. Then calculate the measures of the other three angles that share the same vertex. Next measure $\angle 7$ and calculate the measures of its three neighbors. Organize your findings into a table like the one on the right.

Angle	Measure	Angle	Measure
$\angle 1$	–?–	$\angle 5$	–?–
$\angle 2$	–?–	$\angle 6$	–?–
$\angle 3$	–?–	$\angle 7$	–?–
$\angle 4$	–?–	$\angle 8$	–?–

Do you notice a pattern between the measures of pairs of corresponding angles, alternate interior angles, or alternate exterior angles? You should be ready to state at least three conjectures.

Let's start by focusing on the relationship between corresponding angles. Beneath your table, copy and complete the conjecture below.

Conjecture: *If two parallel lines are cut by a transversal, then the corresponding angles are —?—.*

What happens if the lines you start with are not parallel? A quick sketch ought to convince you that the conjecture won't work if the lines are not parallel.

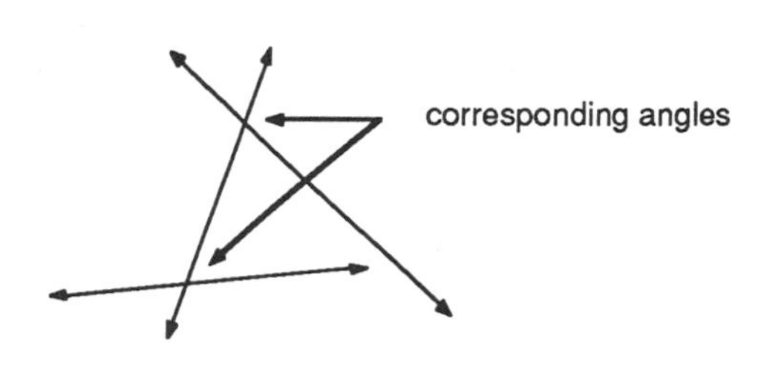

Let's turn the last conjecture around and look at its converse. Suppose you start with two lines cut by a transversal such that pairs of corresponding angles are equal in measure. What does that tell you about the two lines? Will they be parallel? Is it possible for the angles to be equal in measure but the lines not parallel? Let's investigate.

Investigation 4.7.2

Step 1

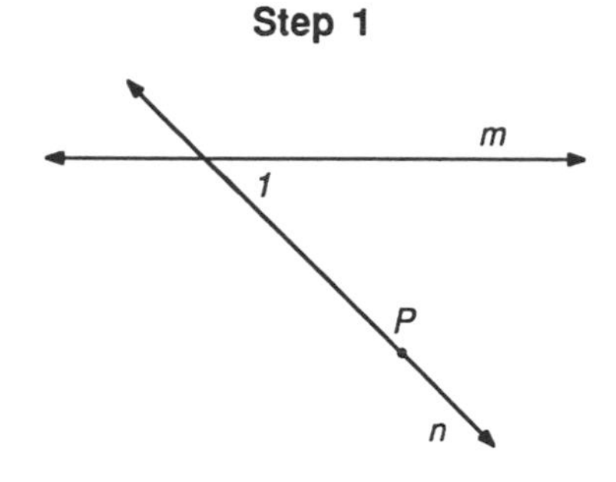

On a clean unlined sheet of paper, construct two lines, *m* and *n*, intersecting to form an angle as shown above. Label it $\angle 1$. Select a point on line *n* and label it *P*.

Step 2

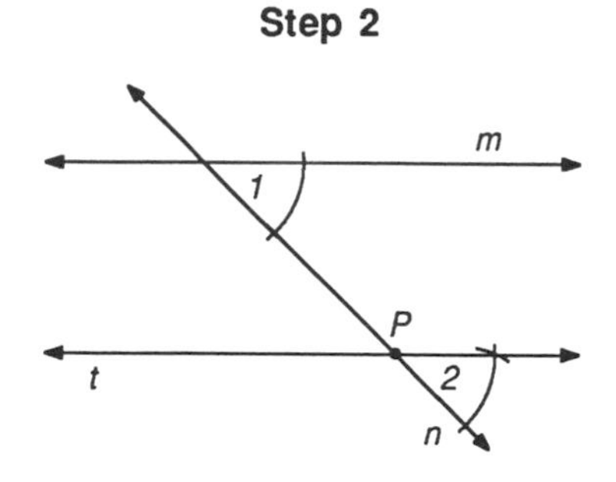

Using line *n* as one side, construct $\angle 2$ with vertex *P* such that $m\angle 2 = m\angle 1$. Extend the new side of $\angle 2$ and label it line *t*.

Do the two lines *m* and *t* appear to be parallel? Compare your results with the results of others near you. Then, beneath your construction, copy and complete the conjecture below.

Conjecture: *If two lines are cut by a transversal forming pairs of equal corresponding angles, then the lines are —?—.*

The last two conjectures you made are converses of each other. (It is not always true that a statement and its converse are both true, but in this case they are both true.) Let's combine the two conjectures into one.

If two parallel lines are cut by a transversal, then —?—. Conversely, if two lines are cut by a transversal forming pairs of corresponding angles equal in measure, then —?—.
(CA Conjecture)

You probably found patterns in the first investigation involving pairs of alternate interior angles and pairs of alternate exterior angles. Let's apply logical reasoning to make conjectures about them also.

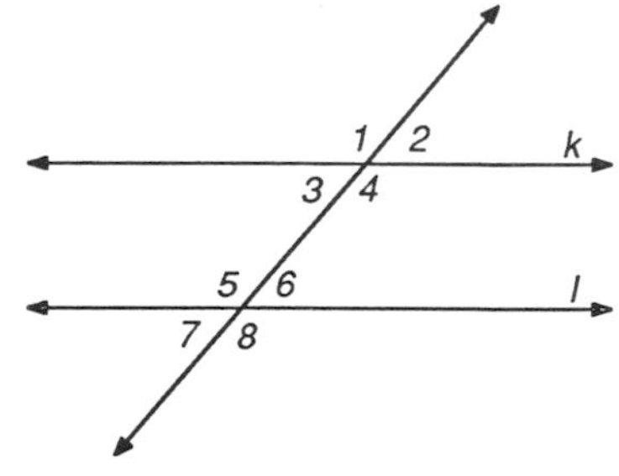

You know from the CA Conjecture that if lines k and l are parallel, then $m\angle 3 = m\angle 7$ ($\angle 3$ and $\angle 7$ are corresponding angles).

You also know that $m\angle 7 = m\angle 6$ (they're a pair of vertical angles).

Therefore, if $m\angle 3 = m\angle 7 = m\angle 6$, then $m\angle 3 = m\angle 6$ (one pair of alternate interior angles are equal in measure).

Similar logic shows that $m\angle 4 = m\angle 5$ (the other pair of alternate interior angles are equal in measure).

How about the converse? Can you use logical reasoning to show that if $\angle 3$ and $\angle 6$ are equal in measure, the lines will be parallel? That's not difficult either. If $m\angle 3 = m\angle 6$, then $m\angle 3 = m\angle 7$ since $\angle 7$ and $\angle 6$ are a pair of vertical angles. If $m\angle 3 = m\angle 7$, then $k \parallel l$ by the CA Conjecture. You have just logically demonstrated the next conjecture. Based on the reasoning above, combine the two conjectures about alternate interior angles into one conjecture.

C-17 *If two parallel lines are cut by a transversal, then —?—. Conversely, if two lines are cut by a transversal —?—.* ***(AIA Conjecture)***

Similar logical arguments can be applied to alternate exterior angles. The arguments are left for you as an exercise. The conjecture will read as follows:

C-18 *If two parallel lines —?—. Conversely, —?—.* ***(AEA Conjecture)***

How many lines can you construct parallel to line l that also pass through point P? One? Two? You probably guessed correctly; there's one. There are other types of geometry where the correct answer is not one! How can that be?

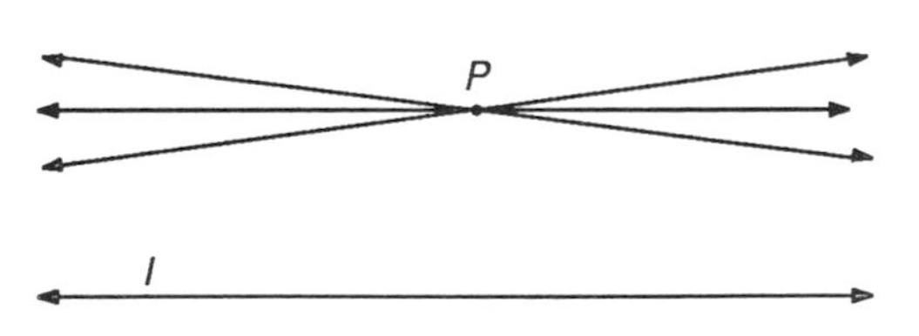

The geometry you are discovering in this text is called Euclidean geometry. It is named after the famous Greek mathematician Euclid, who first organized geometry into a logical structure. Euclid lived around 300 B.C. and wrote a series of books, *The Elements*. One of the basic propositions of Euclidean geometry is that through a point not on a line there is *exactly one* line that can be drawn parallel to a given line.

While it seems logical to conclude, as Euclid did, that exactly one line can be constructed through a point parallel to a given line, it is in fact possible to make other assumptions. Perhaps no parallel lines pass through the point. Perhaps more than one parallel line can be constructed. These two other possible assumptions have led to the development of non-Euclidean geometries. Hyperbolic geometry assumes through a given point not on a given line there passes *more than one line* parallel to the given line. Elliptic geometry assumes through a given point not on a given line there passes *no lines* parallel to the given line. Although they seem to go against our intuitive notion of parallel in the real world, non-Euclidean geometries actually describe some realities better than Euclidean geometry. Albert Einstein, for example, used non-Euclidean geometries in developing his theory of relativity. You will learn more about non-Euclidean geometries at the end of Chapter 15.

Using only compass and straightedge, how do you construct one line through point P parallel to line l? The technique below is based on the AIA Conjecture.

Investigation 4.7.3 

Start with a line l and a point P not on the line. Then follow the steps below.

Step 1

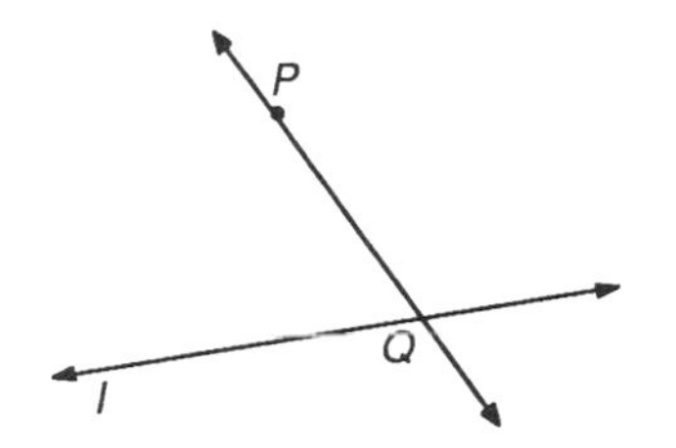

Construct any line from point P that intersects line l. Label the point of intersection as Q.

Step 2

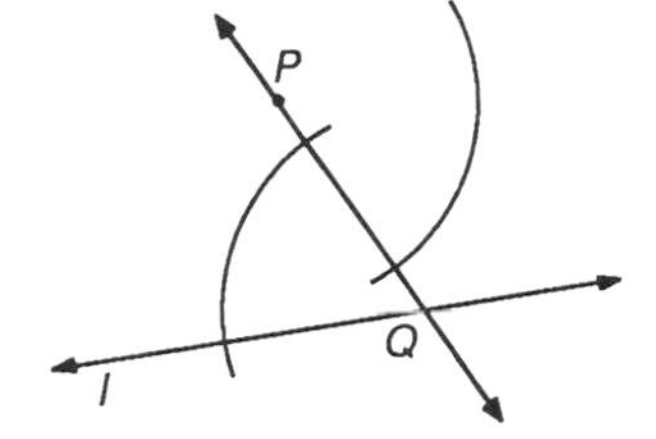

With points P and Q as centers, swing arcs on opposite sides of the transversal.

Step 3

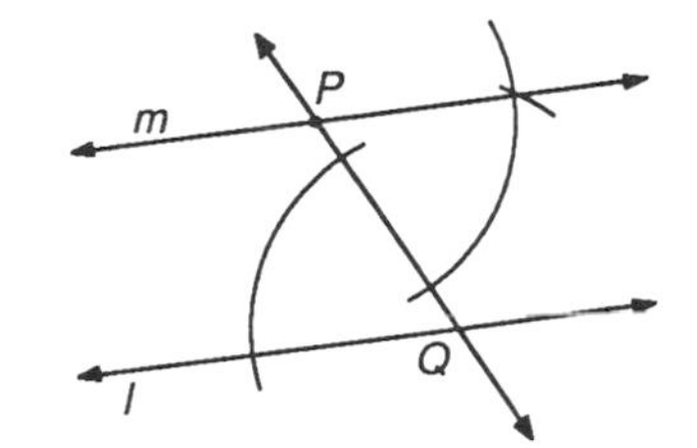

Duplicate angle Q at point P. Construct line m. Line l is parallel to line m.

EXERCISE SET A

Use your new conjectures in the problems below. Remember that matching arrows on pairs of lines indicate that they are parallel.

1. $r \parallel s$
   $w = $ –?–

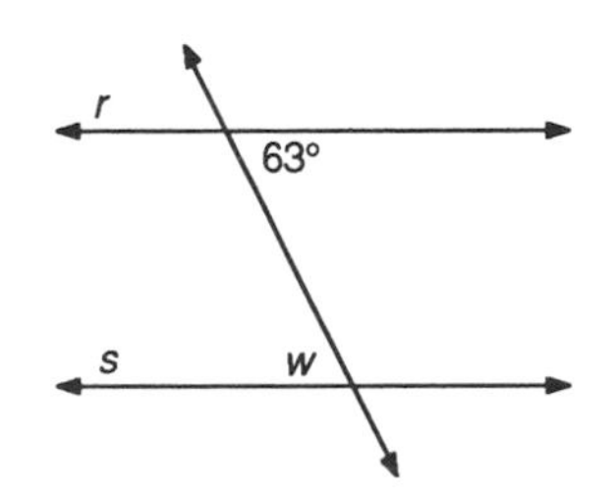

2. $p \parallel q$
   $x = $ –?–

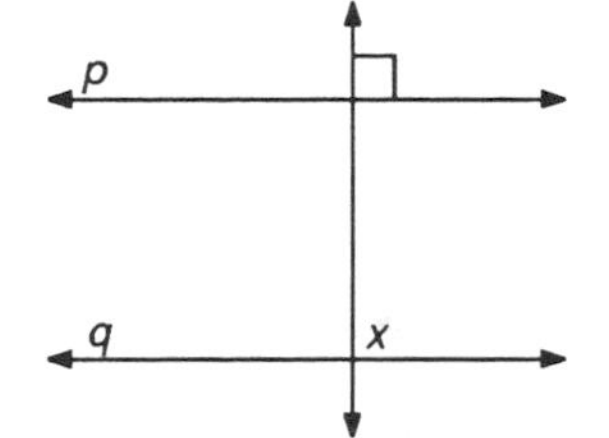

3. Is line k parallel to line l?

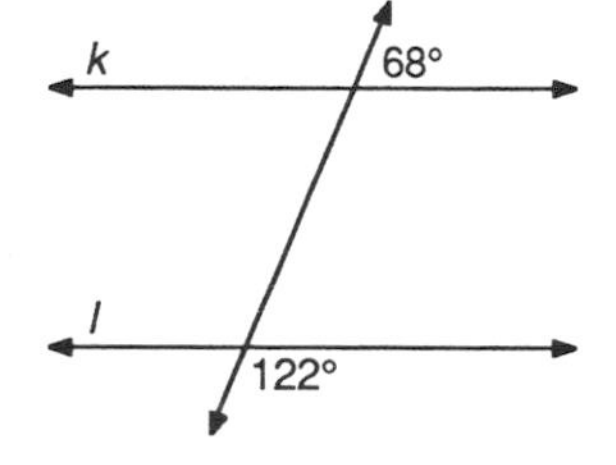

4. Quadrilateral *HELP* is a parallelogram. $y = -?-$

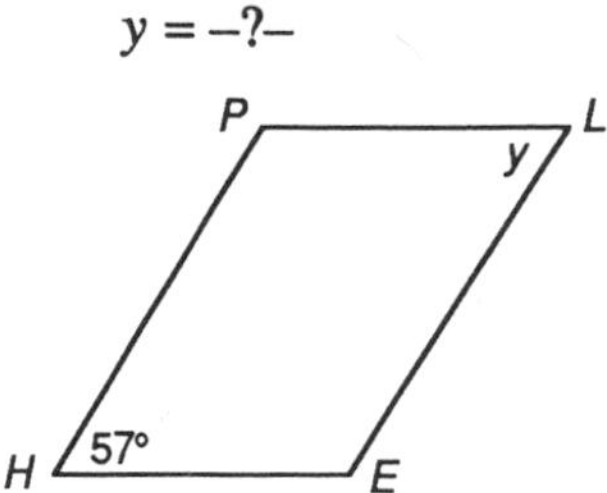

5. Is quadrilateral *ABCD* a parallelogram?

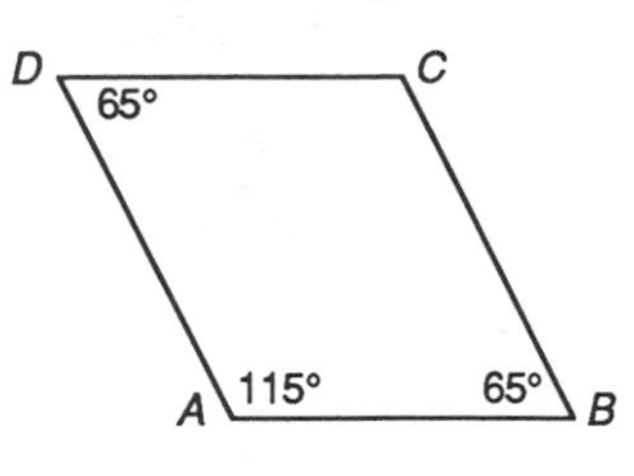

6.* $m \parallel n$
$z = -?-$

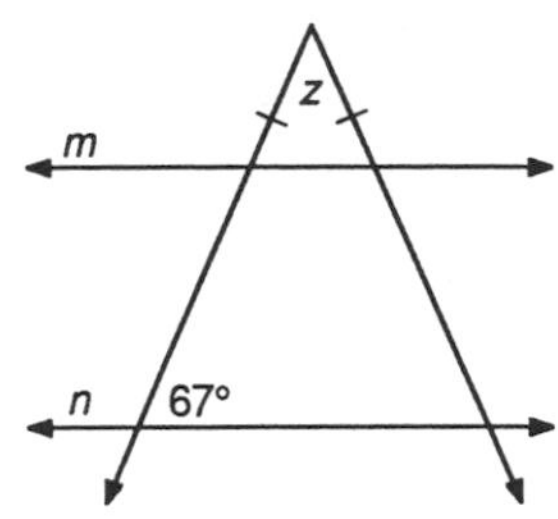

7. Trace the diagram below. Calculate the measure of each lettered angle.

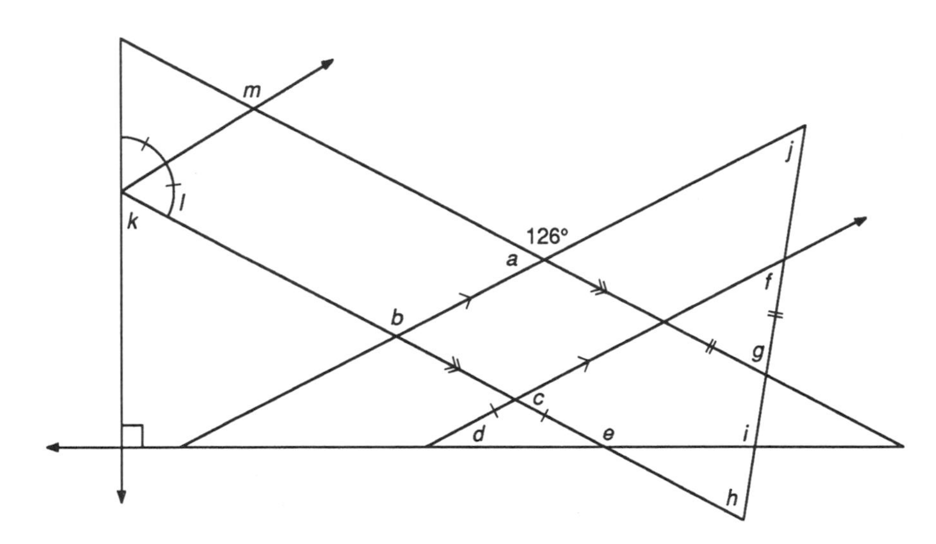

EXERCISE SET B

Perform each construction below on a separate sheet of unlined paper. Make your constructions as large as possible.

1. Construct a line on your paper. Label it *l*. Select a point not on the line and label it *M*. Construct a line through the point *M* parallel to *l*.

2. Construct an acute angle on your paper. Label it $\angle PAR$. Place point *K* in the interior of $\angle PAR$. Construct a line *m* through *K* parallel to $\overrightarrow{AP}$. Construct a line *n* through *K* parallel to $\overrightarrow{AR}$.

3. The method for constructing parallel lines in Investigation 4.7.3 is based on the AIA Conjecture. Could you have based it on another conjecture? Construct a pair of parallel lines basing your method on the AEA Conjecture.

4. Draw a diagram and write out the logical argument to support the AEA Conjecture. You may use the argument for the CA Conjecture as a guide.

Special Project

Euler's Formula for Networks

Leonard Euler

Networks consist of points and paths (curved lines). Between the paths are the regions formed by connecting the points with the paths. There is a relationship between these three parts of a network that is true for all networks that can be drawn on a piece of paper. This formula is called Euler's Formula for Networks. Copy and complete the table below using the networks from *Special Project: Traveling Networks*. When counting regions, include the region that is left on the outside of the network. (The reason for counting the outer region will become clear when you find Euler's Formula for Solids in a later chapter.)

Network	A	B	C	D	E	F	G	H	I	J	K	L	M	N	O	P
Number of points (P)	-?-	-?-	-?-	-?-	-?-	-?-	-?-	-?-	-?-	-?-	-?-	-?-	-?-	-?-	-?-	-?-
Number of regions (R)	-?-	-?-	-?-	-?-	-?-	-?-	-?-	-?-	-?-	-?-	-?-	-?-	-?-	-?-	-?-	-?-
No. of paths or lines (L)	-?-	-?-	-?-	-?-	-?-	-?-	-?-	-?-	-?-	-?-	-?-	-?-	-?-	-?-	-?-	-?-

Once you have completed your table, search for a pattern that will lead to a conjecture. You are looking for a relationship between the three parts of a network. In other words, given the number of points (P) and the number of regions (R), you want a formula to calculate the number of paths or lines (L). Perhaps by adding, subtracting, multiplying, or some combination of the three, you will find a formula for all networks. When you've found your formula, complete the conjecture below.

Conjecture: *If P stands for the number of points in a network, L stands for the number of lines in the network, and R stands for the the number of regions in the network, then an equation that combines all three for any network is —?—.* ***(Euler's Formula for Networks)***

Use your conjecture to solve each problem.

1. If a network has 26 points and 41 paths, then how many regions does the network create?
2. If a network has 36 points and 19 regions, then how many paths does the network have?
3. Draw a network that has eight points and nine enclosed regions. How many paths will you have to draw between points?

Lesson 4.8

Discovering Properties of Trapezoids

Recall the definition of a trapezoid from Chapter 2.

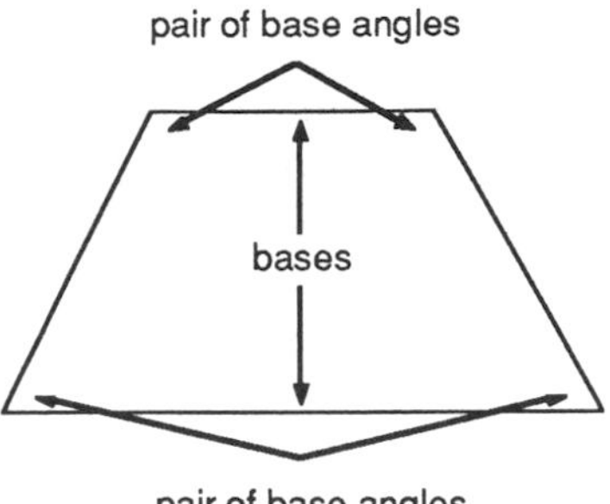

> *A **trapezoid** is a quadrilateral with exactly one pair of parallel sides.*

In a trapezoid, the parallel sides are called **bases**. A pair of angles that share a base as a common side are called a pair of **base angles**.

> *A trapezoid with the two non-parallel sides the same length is called an **isosceles trapezoid**.*

In the next investigation you are going to discover a property of isosceles trapezoids.

Investigation 4.8.1

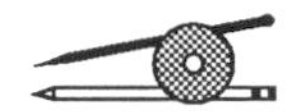
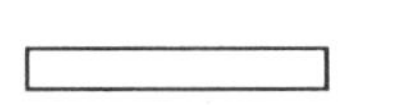
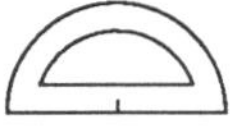

Step 1: Using the lines on a piece of graph paper or a piece of lined paper as a guide, draw a pair of parallel lines. (If you don't have ruled paper, use both edges of your straightedge to create parallel lines.)

Step 2: Using your compass, construct an isosceles trapezoid.

Step 3: Measure each pair of base angles.

Repeat steps 1 to 3. What do you notice about each pair of base angles in each trapezoid? Share your observations with others near you. State your observations as your next conjecture.

The base angles of an isosceles trapezoid are —?—. ***(Isosceles Trapezoid Conjecture)***

EXERCISE SET

Now that you know how to construct parallel lines, you can construct trapezoids.

1. Draw and label trapezoid *TRAP* with one base $\overline{TR}$. What is the other base? Name the two pairs of base angles.

2. On a clean sheet of unlined paper, using the construction for parallel lines, construct a trapezoid.

3.* Given $\angle A$, $\angle B$, base $\overline{AB}$, and non-parallel side $\overline{BC}$ below, construct trapezoid $ABCD$.

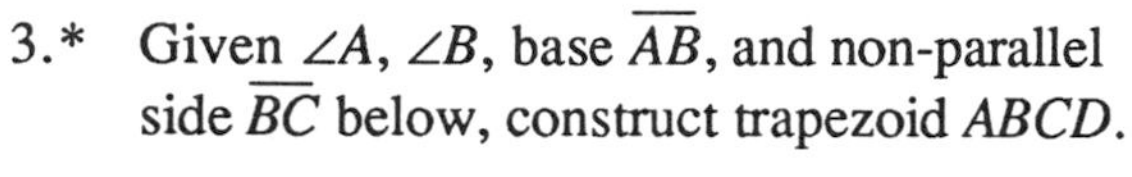

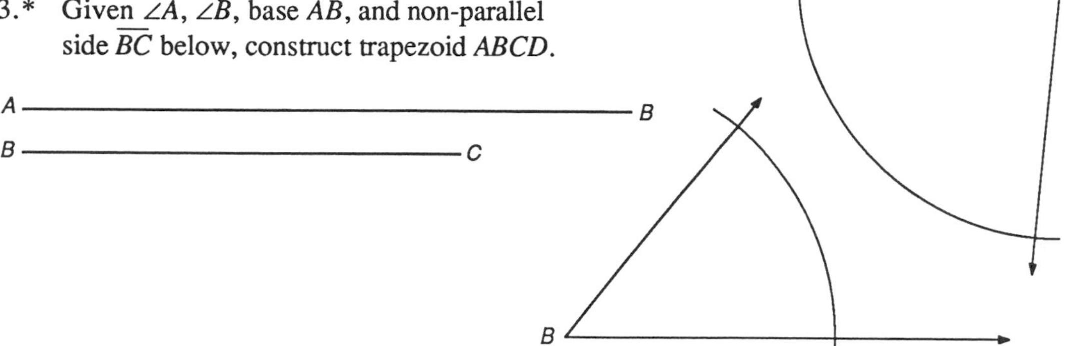

4.* Given bases $\overline{EF}$ and $\overline{GH}$, non-parallel side $\overline{EH}$, and $\angle E$, construct trapezoid $EFGH$.

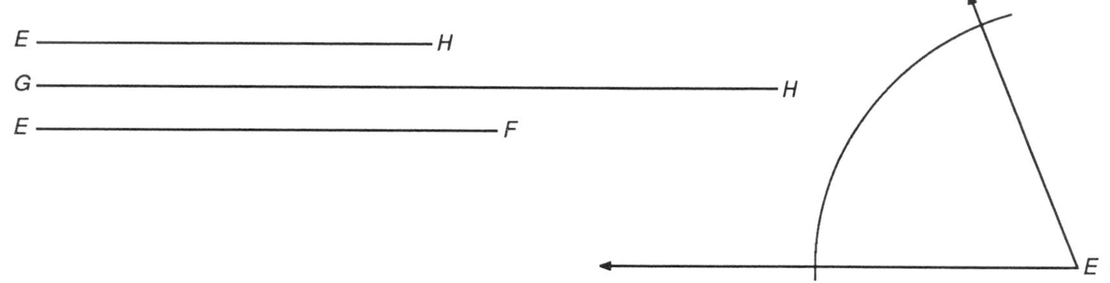

5. Draw and label isosceles trapezoid $ZOID$ with $\overline{ZO}$ as one base. What is the other base? Name the two pairs of base angles. Name the two sides of equal length.

6. Given base $\overline{MO}$, a non-parallel side $\overline{ON}$ with $ON = MK$, and $\angle M$, construct the isosceles trapezoid $MONK$.

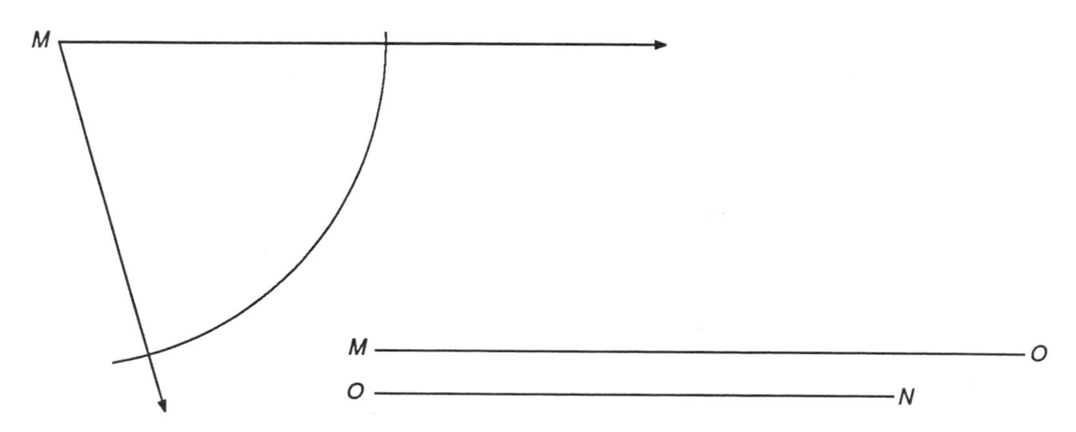

Lesson 4.9

Discovering Properties of Midsegments

In this lesson you will discover special properties of segments called midsegments. In the first investigation you will discover two properties about a segment that connects the midpoints of two sides of a triangle. The segment connecting the midpoints of two sides of a triangle is called a **midsegment** of a triangle.

Investigation 4.9.1 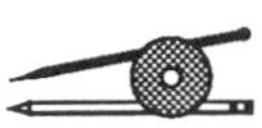 

Step 1: Construct scalene ΔSCL.

Step 2: Construct the midpoint of $\overline{LS}$. Label it E.

Step 3: Construct the midpoint of $\overline{LC}$. Label it A.

Step 4: Construct midsegment $\overline{EA}$.

Step 5: With your compass, compare EA and SC. EA is what fraction of SC?

Step 6: With a protractor, compare $\angle LEA$ and $\angle LSC$. Does it appear that $\overline{EA} \parallel \overline{SC}$?

Repeat steps 1 to 6. Compare your results with the results of others near you. State your observations as your next conjecture.

*A midsegment of a triangle is —?— to the third side and —?— the length of —?—. **(Triangle Midsegment Conjecture)***

In the second investigation you will discover two properties about a line segment that connects the midpoints of the non-parallel sides of a trapezoid. The line segment connecting the midpoints of the two non-parallel sides of a trapezoid is called the **midsegment** of the trapezoid.

Investigation 4.9.2  

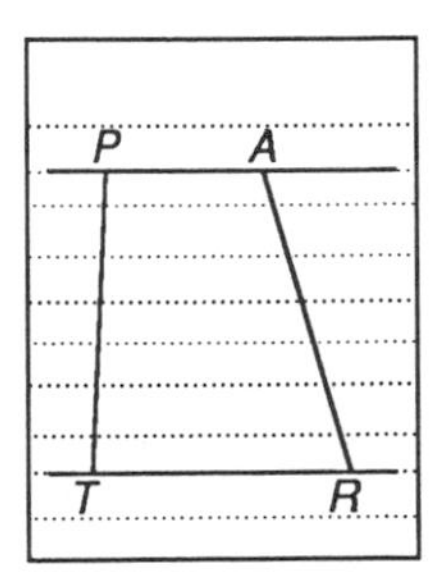

Step 1: Using the lines on a piece of lined paper as a guide, draw a pair of parallel lines. (Draw the lines at least 6 cm apart.)

Step 2: Make a trapezoid. Label it *TRAP* with $\overline{TR}$ and $\overline{AP}$ as the bases.

Step 3: Construct the midpoint of non-parallel side $\overline{RA}$. Label it E.

Step 4: Construct the midpoint of non-parallel side $\overline{PT}$. Label it Z.

Step 5: Construct $\overline{EZ}$.

Step 6: With a protractor, compare $\angle PTR$ and $\angle PZE$. Does it appear that $\overline{ZE} \parallel \overline{TR}$? With a protractor, compare $\angle APT$ and $\angle EZT$. Does it appear that $\overline{ZE} \parallel \overline{PA}$?

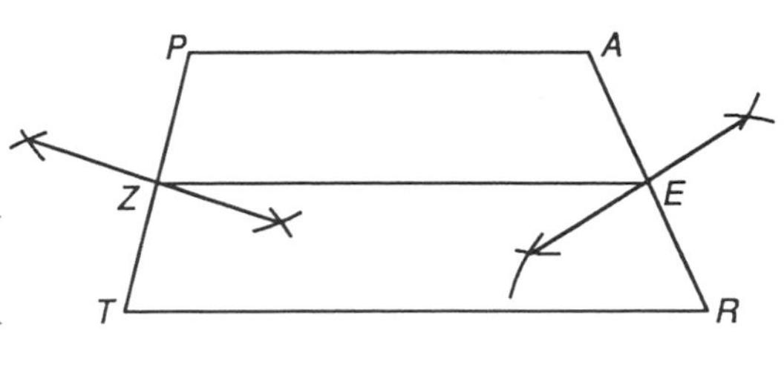

Repeat steps 1 to 6. Compare your results with the results of others near you. Then, beneath your construction, copy and complete the conjecture below.

Conjecture: *The midsegment of a trapezoid is —?—.*

What else can you discover about the midsegment of a trapezoid?

Investigation 4.9.3

Make a trapezoid as you did in the previous investigation. Label it *ABCD*.

Step 1

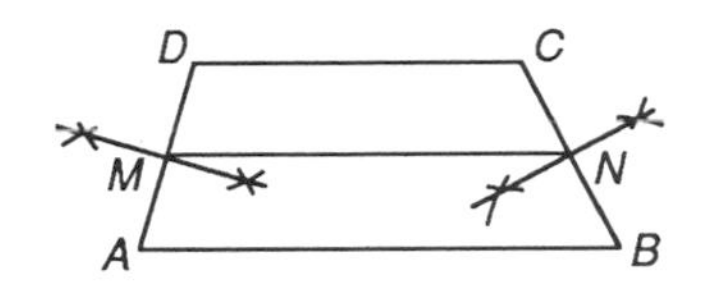

Construct the midsegment. Label it $\overline{MN}$.

Step 2

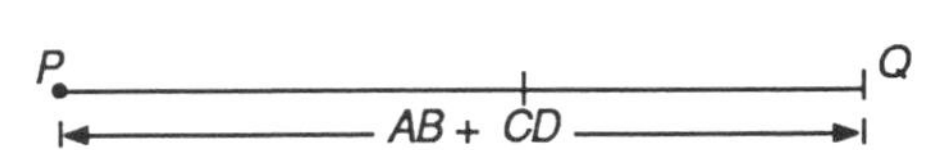

Draw a construction line below the trapezoid. On this line, construct segment *PQ* with length *AB* + *CD* (the sum of the lengths of the bases).

Step 3

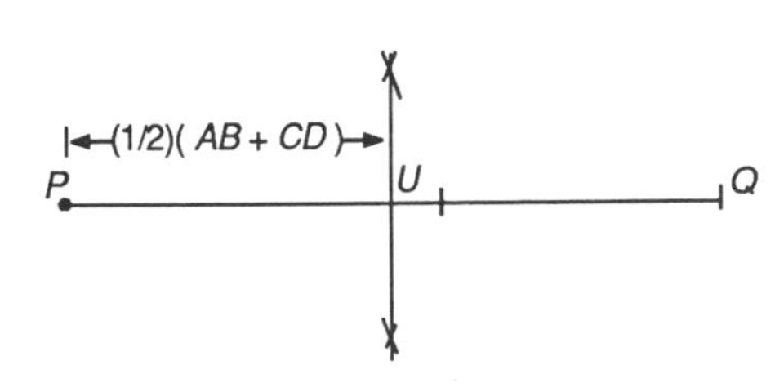

Find the midpoint of $\overline{PQ}$. Label it *U*. Since the average of the lengths of the bases is half their sum, *PU* is the average of the lengths of the bases.

Step 4

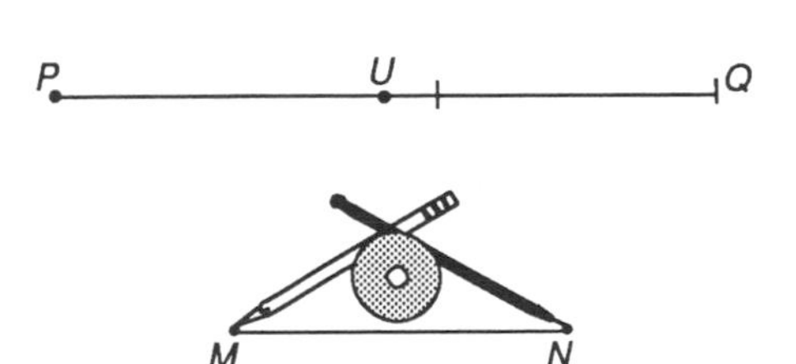

With your compass, compare *PU*, the average of the lengths of the two bases, with *MN*, the length of the midsegment.

Perform the investigation with another trapezoid. What do you notice about the length of the midsegment and the average of the two bases? Compare your results with the results of others. Beneath your construction, copy and complete the conjecture.

Conjecture: *The midsegment of a trapezoid is equal in length to —?—.*

Now combine the two trapezoid midsegment conjectures into one and add it to your conjecture list.

*The midsegment of a trapezoid is —?— to the bases and is equal in length to —?—. (**Trapezoid Midsegment Conjecture**)*

EXERCISE SET A

Use your new conjectures to solve each problem.

1.* What is the perimeter of ΔPOT?

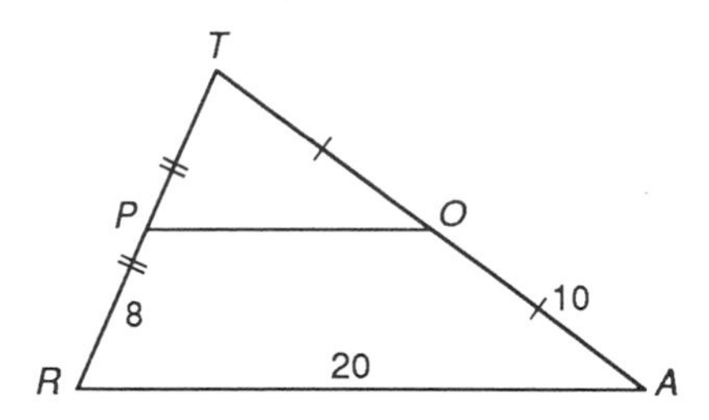

2. $x = $ –?– $y = $ –?–

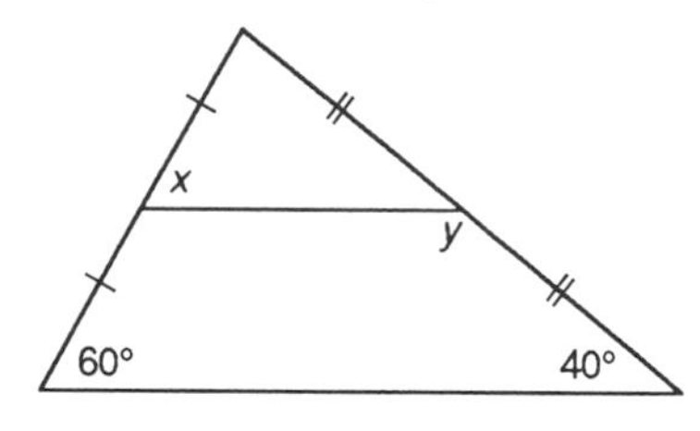

3.* $z = $ –?–

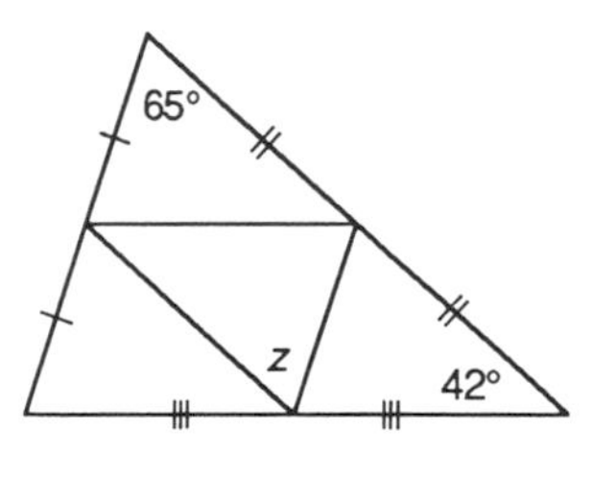

4. What is the perimeter of ΔBFD?

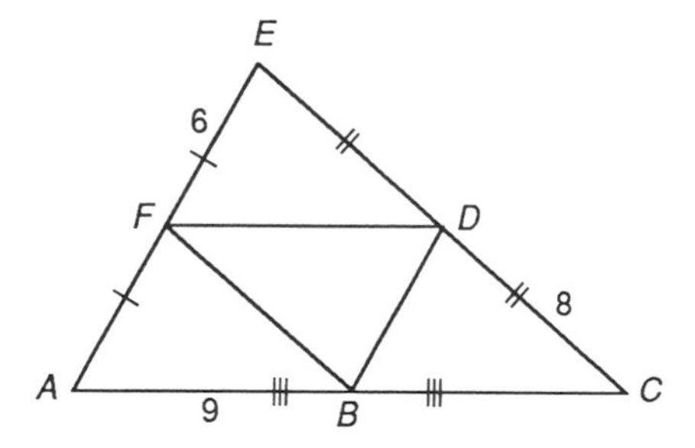

In Exercises 5 and 6, both figures are trapezoids.

5. $m = $ –?– $n = $ –?–

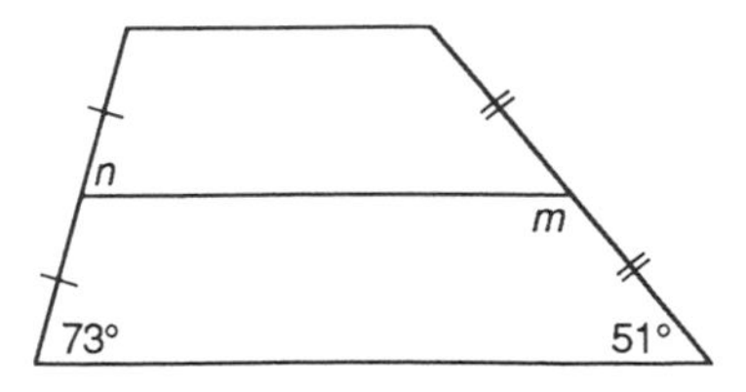

6. $q = $ –?–

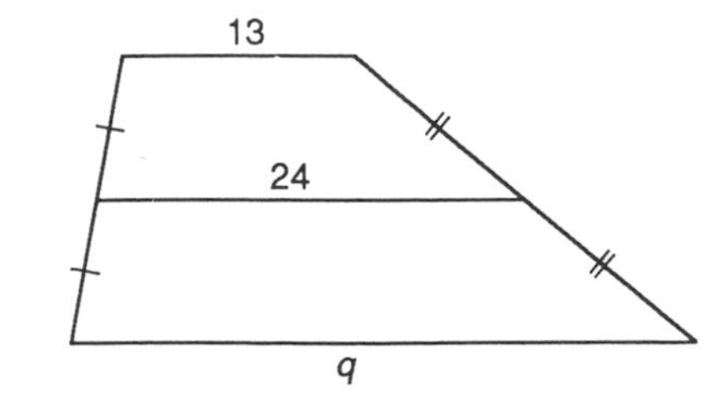

EXERCISE SET B

1. How many midsegments are there in a triangle? How many are there in a trapezoid?

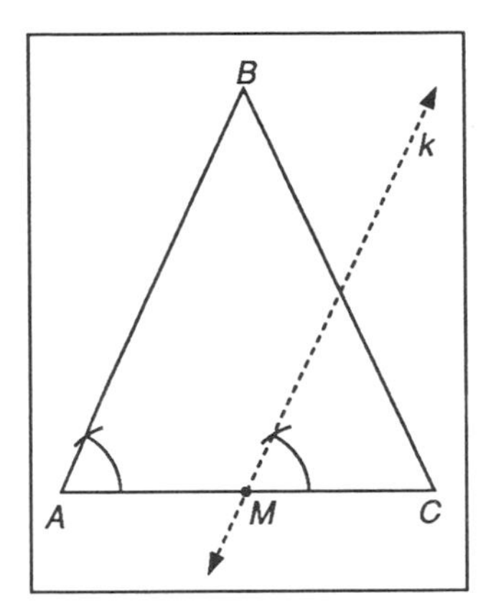

2. Is the converse of the Triangle Midsegment Conjecture true? If you construct a line through a midpoint of one side of a triangle that is parallel to a second side of the triangle, must it pass through the midpoint of the third side? Investigate. Construct and label a triangle like the one at right (but larger). Find the midpoint M of side $\overline{AC}$ in triangle ABC. (The construction marks for the midpoint are not shown.) Construct a line k through point M parallel to the side $\overline{AB}$.

   Did you notice where the line passed through the third side? With your compass, compare the lengths of the two segments on that third side. How do they compare? Compare your observations with the observations of others near you. Does the line pass through the midpoint of the third side?

3. On a clean sheet of paper construct a large quadrilateral. Label it $PQRS$. Construct the midpoints of the four sides. Label as A the midpoint of $\overline{PQ}$. Label as B the midpoint of $\overline{QR}$. Label as C the midpoint of $\overline{RS}$. Label as D the midpoint of $\overline{SP}$. Construct quadrilateral $ABCD$. Are the opposite sides of $ABCD$ parallel? What type of polygon does $ABCD$ appear to be?

Improving Reasoning Skills

Logical Vocabulary

Here is a logical vocabulary challenge. It is sometimes possible to change one word to another of equal length by changing one letter at a time. This gives you a new word at each stage. For example, DOG can be changed to CAT in exactly three moves:

DOG ⇒ DOT ⇒ COT ⇒ CAT

Change MATH to each of the following words in exactly four moves.

1. MATH ⇒ –?– ⇒ –?– ⇒ –?– ⇒ ROSE
2. MATH ⇒ –?– ⇒ –?– ⇒ –?– ⇒ LIVE
3. MATH ⇒ –?– ⇒ –?– ⇒ –?– ⇒ HOST
4. MATH ⇒ –?– ⇒ –?– ⇒ –?– ⇒ BUST
5. MATH ⇒ –?– ⇒ –?– ⇒ –?– ⇒ COPS

Lesson 4.10
Discovering Properties of Parallelograms

*A **parallelogram** is a quadrilateral with the opposite sides parallel.*

In this lesson you will discover some special properties of parallelograms.

Investigation 4.10.1

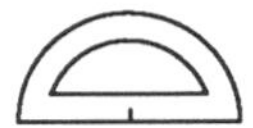

Step 1

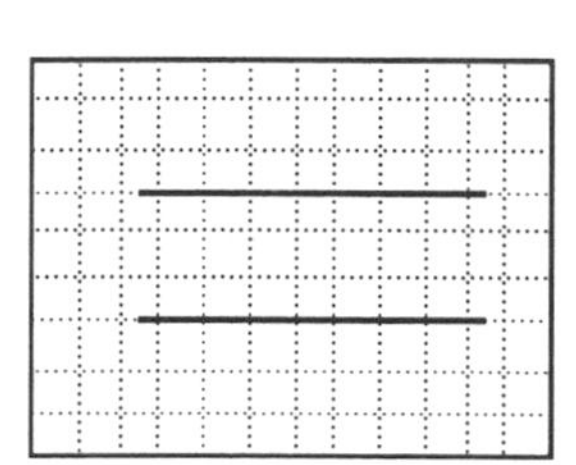

Using the lines on a piece of graph paper as a guide, draw a pair of parallel lines at least 6 cm apart.

Step 2

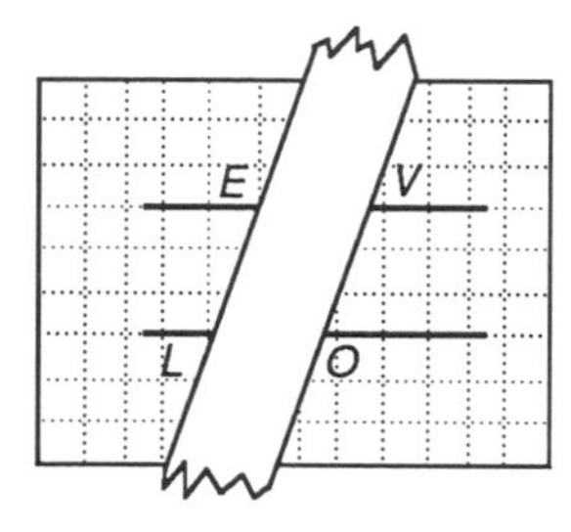

Make a parallelogram using the parallel edges of your straightedge or use the points on the graph paper. Label your parallelogram *LOVE.*

Step 3

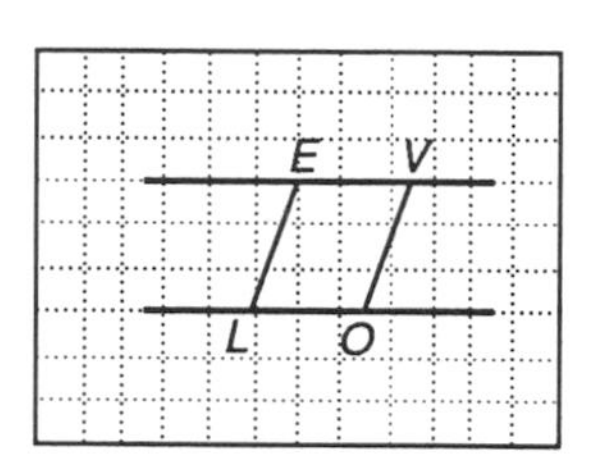

Measure $\angle ELO$ with a protractor. Then, using what you know about parallel lines, calculate the remaining angles. Write the measures in the angles.

Repeat steps 1 to 3. Compare your results with the results of others near you. State your observations as your next conjecture.

C-22 *The —?— of a parallelogram are —?—.*

In Chapter 2 you learned that two angles in a polygon that share a common side are called consecutive angles. In parallelogram *LOVE*, $\angle LOV$ and $\angle EVO$ are a pair of consecutive angles. The consecutive angles of a parallelogram are not equal in measure, but there is a relationship between them. Compare your observations with the observations of others near you. State a conjecture about the consecutive angles in a parallelogram.

C-23 *The —?— of a parallelogram are —?—.*

Investigation 4.10.2 

With your compass, compare the lengths of the opposite sides of the parallelograms you made.

Compare your results with the results of others near you. State a conjecture about the opposite sides of a parallelogram.

C-24 *The —?— of a parallelogram are —?—.*

Investigation 4.10.3 

Construct the diagonals $\overline{LV}$ and $\overline{EO}$ in your parallelograms as shown below. Label the point of intersection of the two diagonals point M.

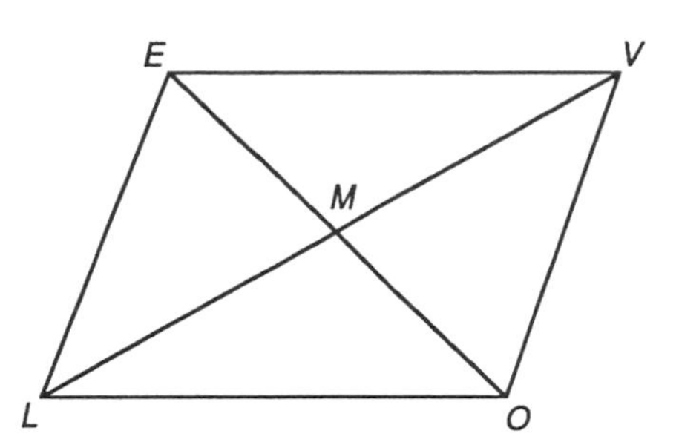

Does $LM = VM$?

Does $MO = ME$?

If $LM = MV$, then M is the —?— of LV.

If $MO = ME$, then M is the —?— of OE.

Does $\overline{EO}$ bisect $\overline{LV}$?

Does $\overline{LV}$ bisect $\overline{EO}$?

Compare your observations with the observations of others near you. State a conjecture about the diagonals of a parallelogram.

C-25 *The —?— of a parallelogram —?—.*

EXERCISE SET A

Use your new conjectures to solve the problems below. All figures are parallelograms.

1. $a = $ –?– $b = $ –?–

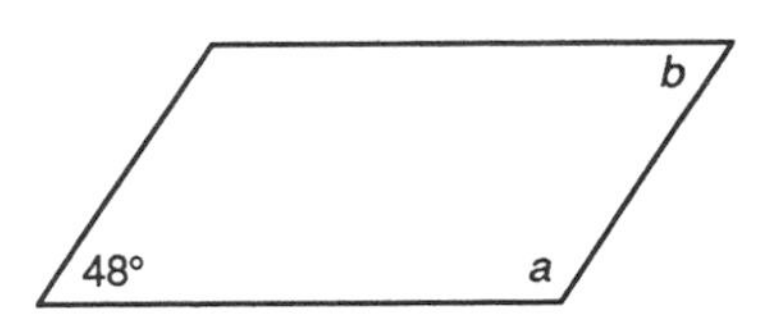

2. $c = $ –?– $d = $ –?–

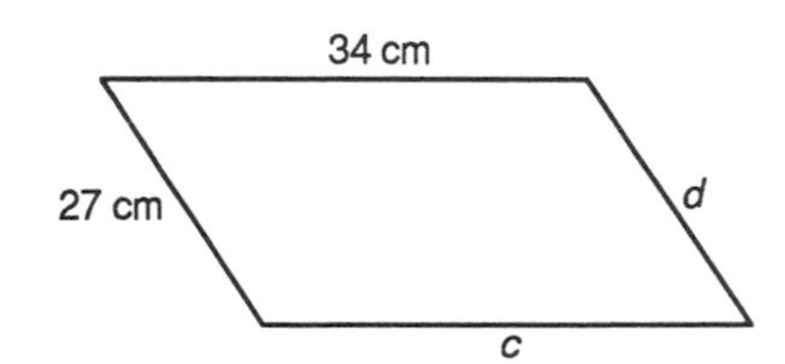

3. $e = $ –?– $f = $ –?–

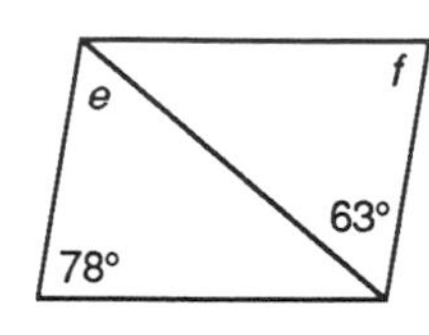

4.* What is the perimeter of the parallelogram?

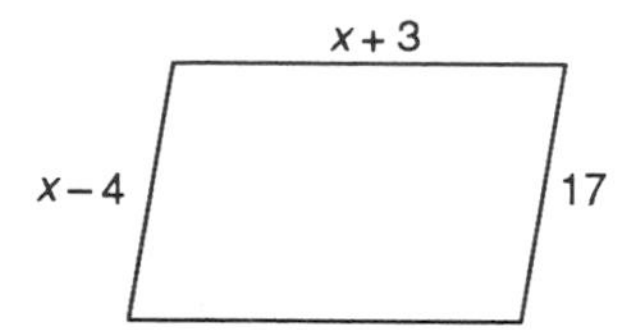

5. $g = $ –?– $h = $ –?–

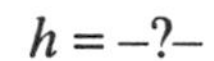

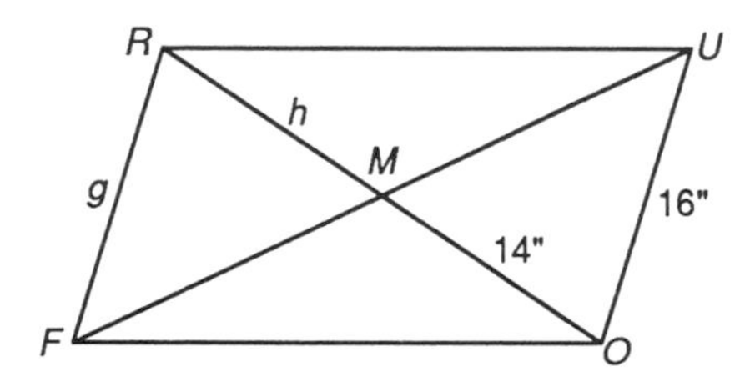

6.* $VF = 36$ m $EF = 24$ m $EI = 42$ m
What is the perimeter of ΔNVI?

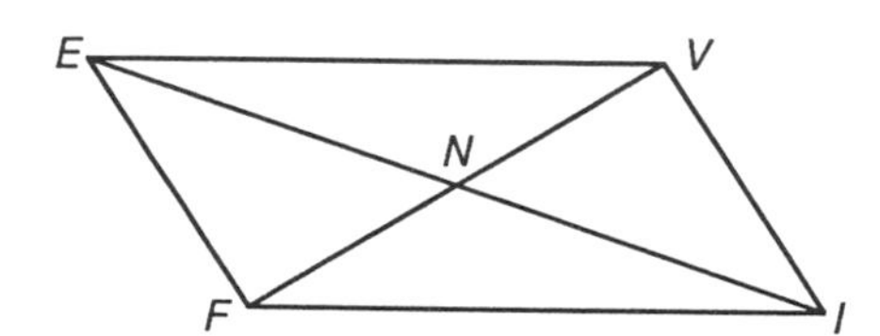

EXERCISE SET B

1. Given side $\overline{AB}$, side $\overline{BC}$, and angle A, construct parallelogram $ABCD$.

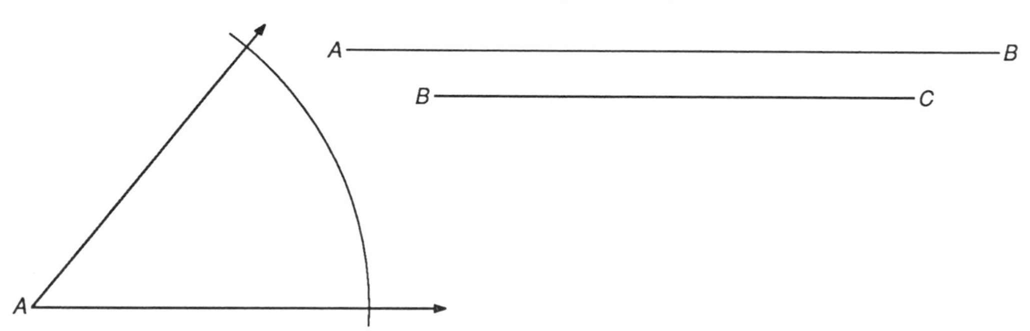

2.* Given side $\overline{MO}$ and two diagonals, $\overline{MN}$ and $\overline{EO}$, construct parallelogram $MONE$.

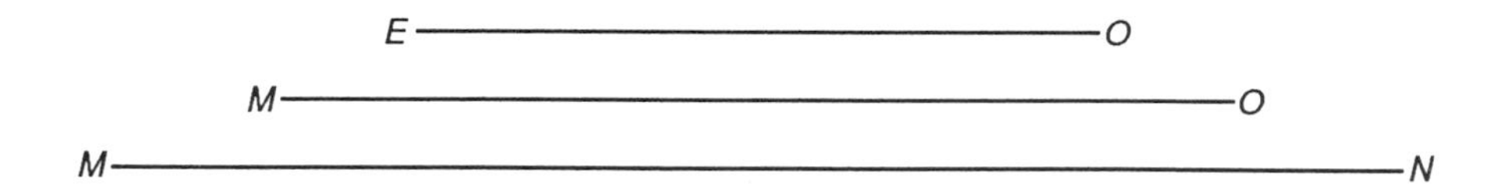

Lesson 4.11

Discovering Properties of Special Parallelograms

In this lesson you will discover some properties of rhombuses, rectangles, and squares. Let's start by looking at a rhombus.

*A **rhombus** is an equilateral parallelogram.*

A rhombus is a parallelogram. Conjecture 25 tells us that the diagonals of a parallelogram bisect each other. Therefore, by Conjecture 25, the diagonals of a rhombus must also bisect each other. But there's something else true about the diagonals of a rhombus. Let's investigate.

Investigation 4.11.1

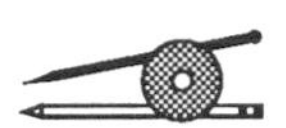
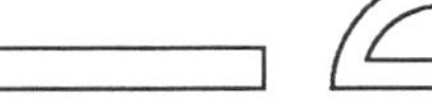

Step 1: Construct a rhombus.

Step 2: Draw in the diagonals.

Step 3: Using your protractor, measure the angles formed at the intersection of the diagonals. Write the measures in the angles.

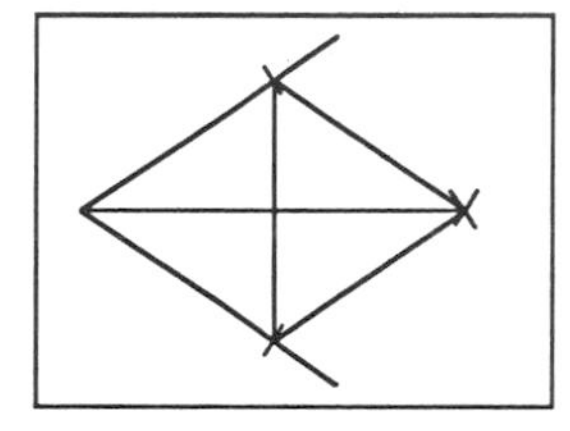

Repeat steps 1 to 3. Compare your results with the results of others near you. Then state a conjecture about the intersection of the diagonals in a rhombus.

C-26 *The —?— of a rhombus are —?— of each other.*

Measure the angles formed by the diagonals at each vertex of your rhombus and write the measures in the angles. What do you observe? Compare your results with the results of others. State a conjecture about the pair of angles at each vertex.

C-27 *The —?— of a rhombus —?— the angles of the rhombus.*

Now let's turn our attention to rectangles. What conjectures can you make?

*A **rectangle** is an equiangular parallelogram.*

What is the measure of each angle of a rectangle? A little logical reasoning can give you the answer. What is the sum of the measures of the four angles of a rectangle? If all four angles have the same measure, then what is the measure of each angle? Make a conjecture.

The measure of each angle of a rectangle is —?—.

Investigation 4.11.2 

Step 1

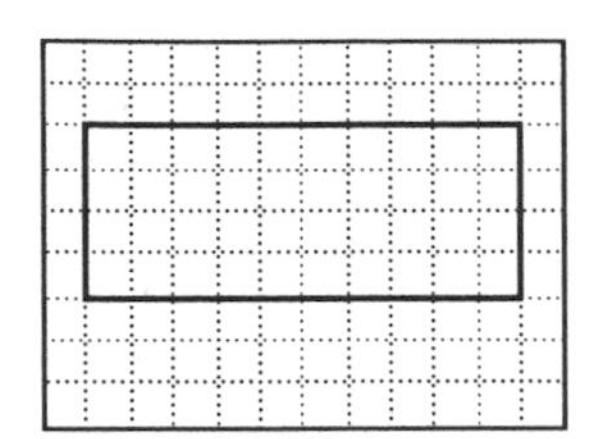

Using the grid of lines on a piece of graph paper as a guide, draw a rectangle with opposite sides at least 6 cm apart.

Step 2

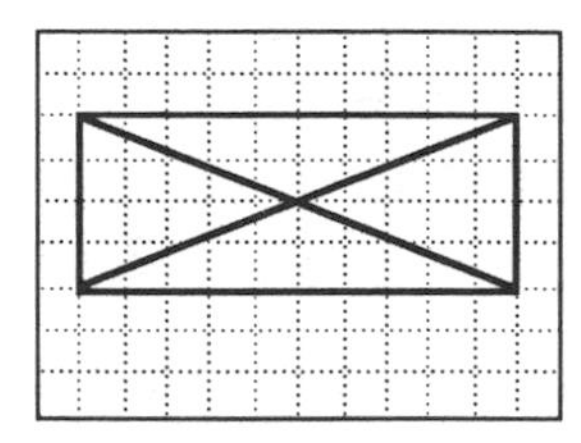

Draw in both diagonals. With your compass, compare the lengths of the two diagonals.

Repeat the investigation with two more rectangles. Compare your results with the results of others near you. State a conjecture about the diagonals of a rectangle.

The —?— of a rectangle are —?—.

A square is both a rectangle and a rhombus. Therefore it can have two definitions.

*A **square** is an equiangular rhombus.*

*A **square** is an equilateral rectangle.*

Squares have all the properties of rectangles and rhombuses. Since rectangles and rhombuses are both parallelograms, they have all the properties of parallelograms.

EXERCISE SET A

Using what you learned in this lesson, identify each statement as true or false.

1.* The diagonals of a parallelogram are equal in measure.

2. The diagonals of a rhombus are equal in measure.

3. The consecutive angles of a rectangle are equal in measure.

4. The diagonals of a rectangle bisect each other.

5. The diagonals of a rectangle bisect the angles.

6. The diagonals of a square are perpendicular bisectors of each other.

7. The diagonals of a parallelogram are perpendicular to each other.

8. Every rhombus is a square.

9. Every square is a rectangle.

10. A diagonal divides a square into two isosceles right triangles.

EXERCISE SET B

1. *RHOM* is a rhombus.
   $a = $ –?– $b = $ –?– $c = $ –?–

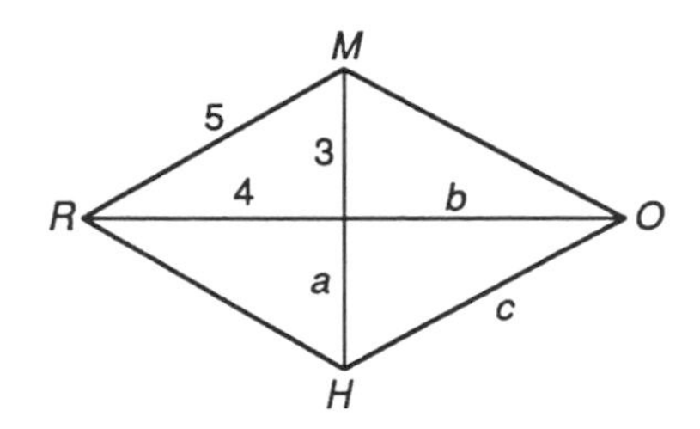

2. *WREK* is a rectangle.
   $WE = $ –?–

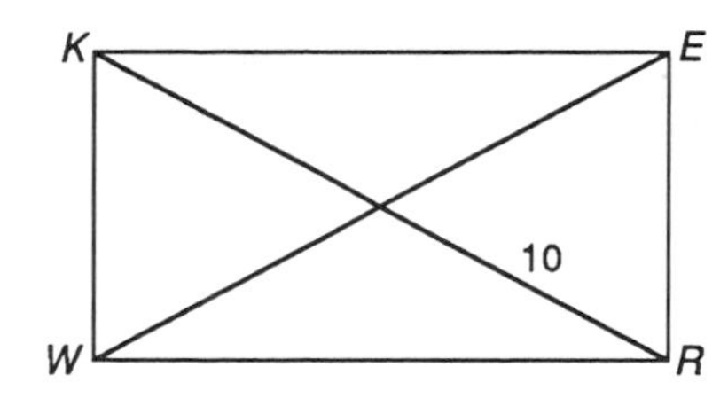

3.* *PEAR* is a parallelogram.
   $y = $ –?–

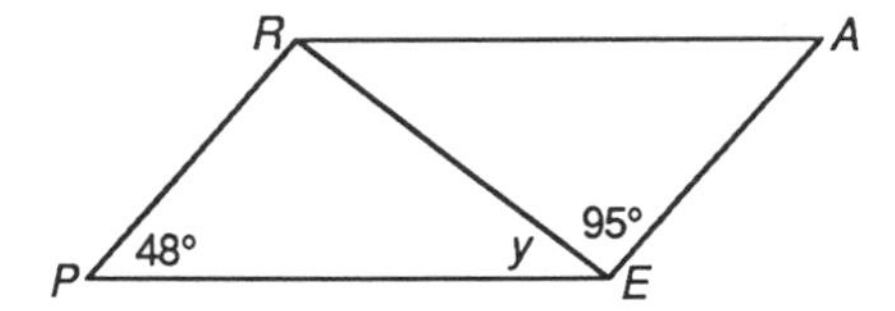

4. *SQRE* is a square.
   $x = $ –?–

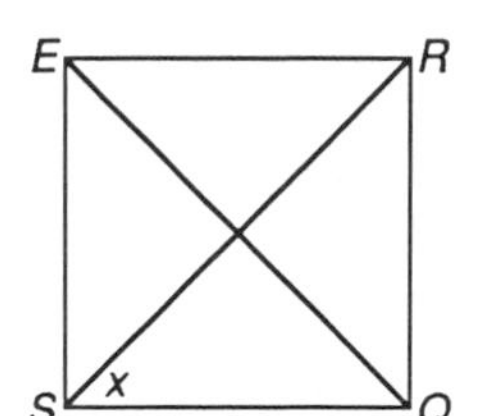

EXERCISE SET C

1.* Given $\angle B$ and diagonal $\overline{BK}$, construct rhombus *BAKE*.

B ———————————————— K

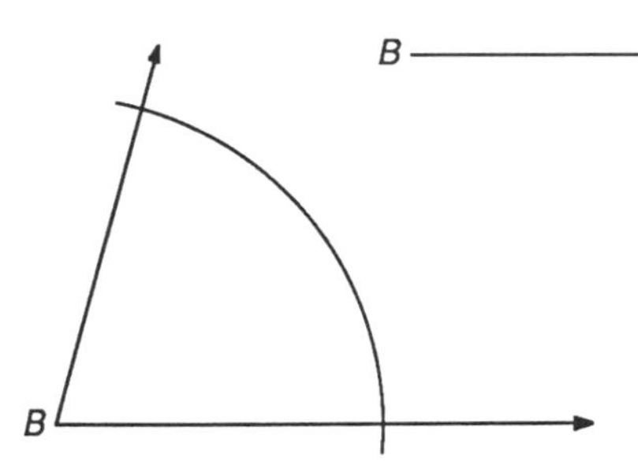

2.* Given side $\overline{RT}$ and diagonal $\overline{RC}$, construct rectangle *RECT*.

R ———————————————— C

R ———————— T

3.* Given the diagonal $\overline{LV}$, construct square *LOVE*.

L ———————— V

Lesson 4.12

Coordinate Geometry

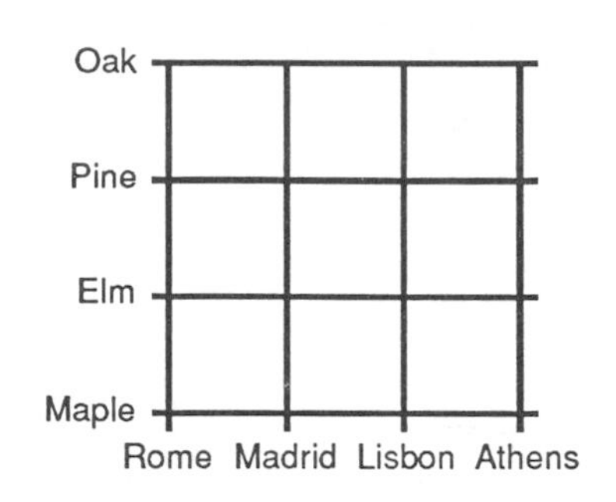

Scott and Viki have agreed to meet downtown to go shopping. Scott suggested they meet on Pine Street because that's where his favorite shops are located. Viki pointed out that Pine is a long street. She asked, "Where on Pine?" Scott replied, "I'll meet you at the intersection of Pine and Lisbon." Now the location is clear. Viki and Scott used coordinate geometry to locate a point on a map as the intersection of two lines.

René Descartes

Ancient Egyptians, Chinese, Greeks, and Romans used various coordinate systems to locate points in surveying and map making. By the seventeenth century, the age of European exploration, the need for accurate maps and the development of easy to use algebraic symbols combined to give birth to modern coordinate geometry. The French philosopher and mathematician René Descartes (1596–1650) is credited with the development of this new method called Cartesian coordinate geometry. (*Cartesian* comes from the Latin form of Descartes' name.)

In algebra you graphed points on a coordinate system labeled with an x-axis and a y-axis. Let's review. If you wish to locate a point A at (3, 4), you start at the origin (0, 0) and move three units to the right, then four units up. (3, 4) are called the coordinates of point A. The x-coordinate of A is 3 and the y-coordinate is 4.

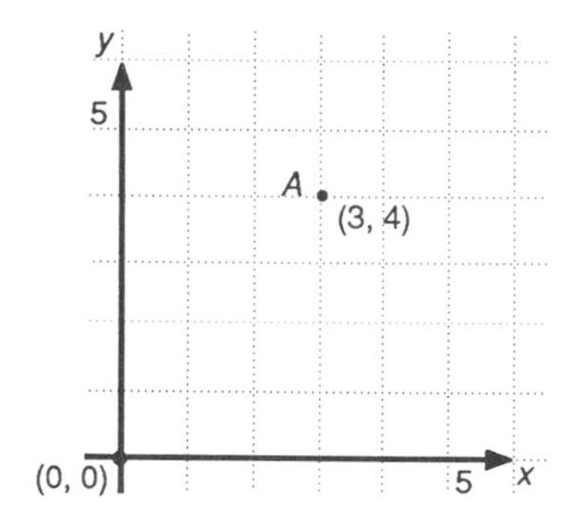

EXERCISE SET A

Name the coordinates of a point or pair of points necessary to complete each polygon.

1.* Isosceles ΔABC

2. Right ΔABC

3.* Isosceles right ΔABC

4. Square $ABCD$

5. Rectangle $ABCD$ with $AD = \frac{1}{2}AB$

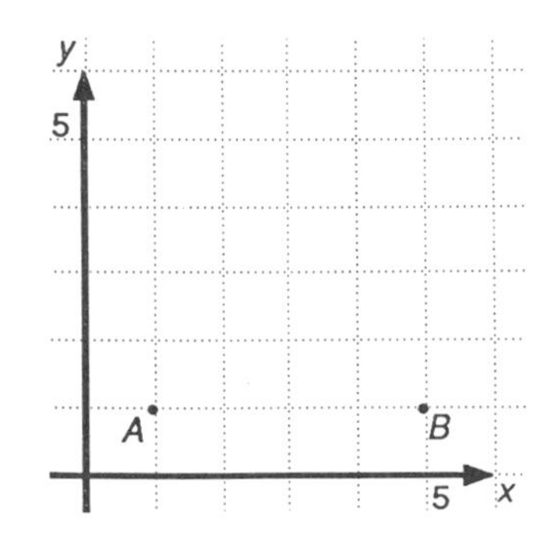

6.* Parallelogram $ABCD$

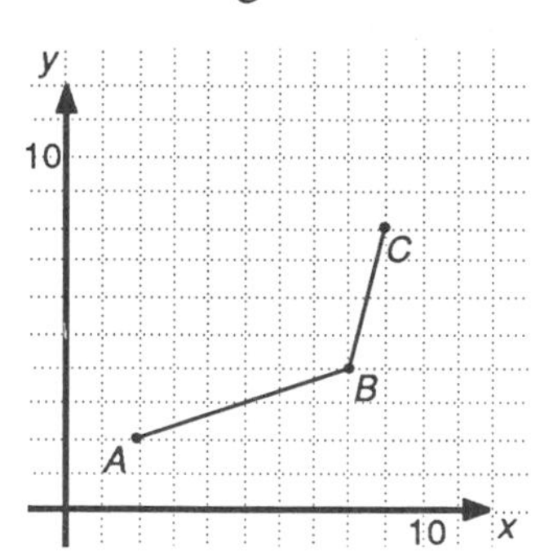

7. Rhombus $ABCD$

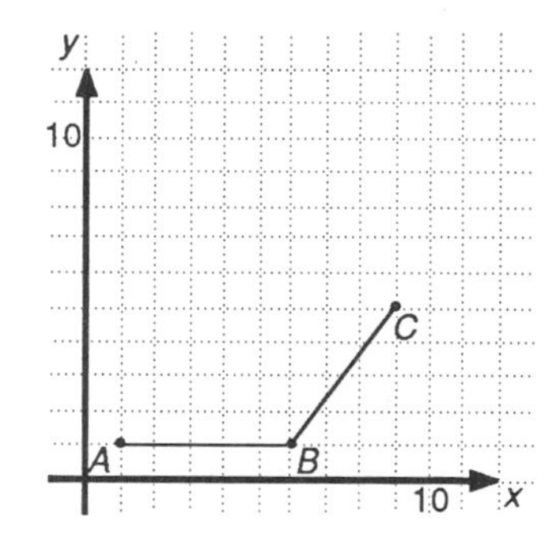

8. Rectangle $ABCD$

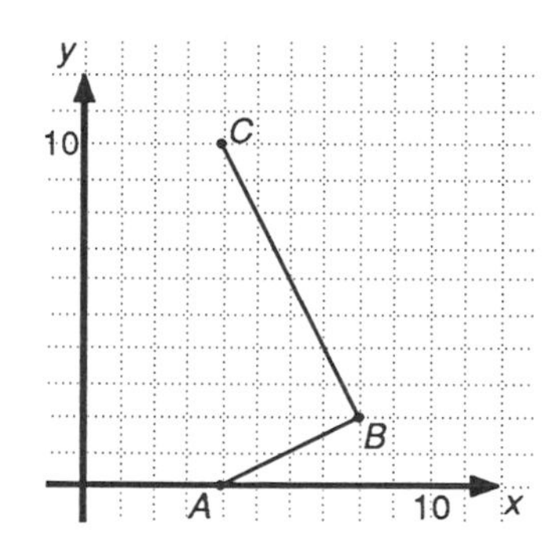

9. What are the midpoints of $\overline{AB}$ and $\overline{CD}$?

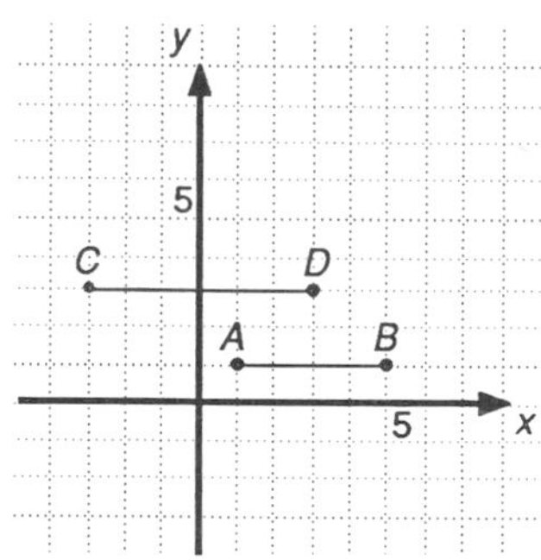

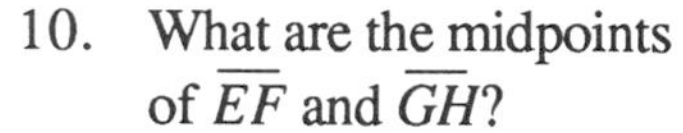

10. What are the midpoints of $\overline{EF}$ and $\overline{GH}$?

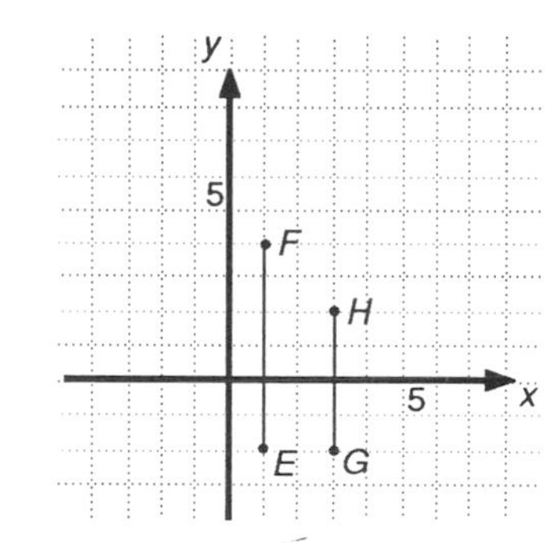

11. What are the midpoints of $\overline{IJ}$ and $\overline{KL}$?

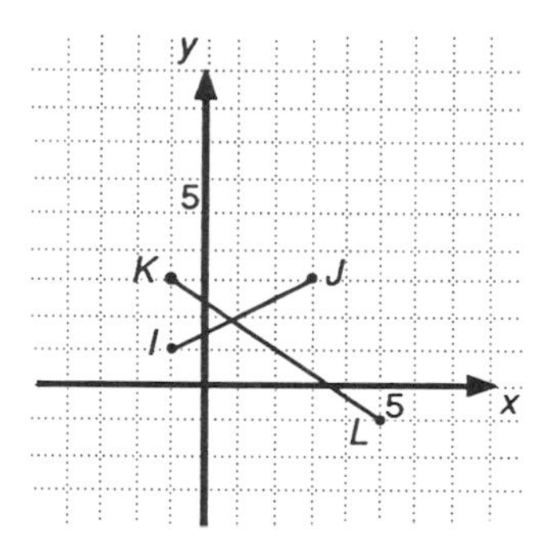

In Exercises 9 and 10, the lines were either horizontal or vertical. What if a segment is slanted at an arbitrary angle and you know only the coordinates of its endpoints? How did you find the coordinates of the midpoints in Exercise 11? Let's take a closer look.

Investigation 4.12.1

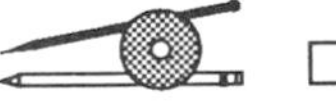

Look back at Exercises 9 to 11 and complete these statements:

The endpoints of $\overline{AB}$ are (1, 1) and (5, 1). The midpoint of $\overline{AB}$ is (–?–, –?–).

The endpoints of $\overline{CD}$ are (3, 3) and (-3, 3). The midpoint of $\overline{CD}$ is (–?–, –?–).

The endpoints of $\overline{EF}$ are (1, -2) and (1, 4). The midpoint of $\overline{EF}$ is (–?–, –?–).

The endpoints of $\overline{GH}$ are (3, 2) and (3, -2). The midpoint of $\overline{GH}$ is (–?–, –?–).

The endpoints of $\overline{IJ}$ are (-1, 1) and (3, 3). The midpoint of $\overline{IJ}$ is (–?–, –?–).

The endpoints of $\overline{KL}$ are (-1, 3) and (5, -1). The midpoint of $\overline{KL}$ is (–?–, –?–).

How does the x-coordinate of the midpoint compare with the x-coordinates of the two endpoints in each segment? How does the y-coordinate of the midpoint compare with the y-coordinates of the two endpoints in each segment? Are you ready to make a conjecture? Let's test your conjecture.

Try this with another segment. On graph paper select two points and construct the segment connecting them. Construct the perpendicular bisector to locate the midpoint. Find the coordinates of the midpoint. How do the x- and y-coordinates of the midpoint compare with the x- and y-coordinates of the two endpoints? Compare your results with the results of others near you. State a conjecture and add it to your conjecture list.

C-30 *If (x_1, y_1) and (x_2, y_2) are the coordinates of the endpoints of a segment, then the coordinates of the midpoint are —?—.* ***(Coordinate Midpoint Conjecture)***

Let's look at two examples of how you can use the Coordinate Midpoint Conjecture.

Example A

Find the midpoint of $\overline{AB}$ with endpoints (-8, 13) and (14, 7).

$$x = \frac{(-8 + 14)}{2} = \frac{6}{2} = 3$$

$$y = \frac{(13 + 7)}{2} = \frac{20}{2} = 10$$

The midpoint of $\overline{AB}$ is (3, 10).

Example B

Find the missing endpoint of $\overline{CD}$ if one endpoint is (3, 9) and the midpoint is (0, 5).

$$0 = \frac{(x + 3)}{2} \text{ or } x + 3 = 0 \text{ or } x = -3$$

$$5 = \frac{(y + 9)}{2} \text{ or } y + 9 = 10 \text{ or } y = 1$$

The missing endpoint of $\overline{CD}$ is (-3, 1).

EXERCISE SET B

In Exercises 1 to 3, determine the coordinates of the midpoint of the segment joining each pair of points.

1. (12, -7) and (-6, 15)
2. (-17, -8) and (-1, 11)
3. (14, -7) and (-3, 18)
4. One endpoint of a segment is (12, -8). The midpoint is (3, 18). Find the coordinates of the other endpoint.
5. Parallelogram *ABCD* has vertices *A* (0, 0); *B* (6, 0); *C* (12, 8); and *D* (6, 8). Find the coordinates of the midpoints of both diagonals.

6.* Find the two trisection points of $\overline{AB}$ if *A* has coordinates (0, 0) and *B* has coordinates (9, 6).

Lesson 4.13

The Slope of a Line

The slope of a line is a measure of its incline or steepness. The *grade* of a road is a measure of its steepness or slope. The *pitch* of a roof is a measure of its slope. The *incline* of a ramp is a measure of its slope.

If a road rises a vertical distance of 50 meters for every horizontal distance of 300 meters, then the grade or slope of the road is calculated by:

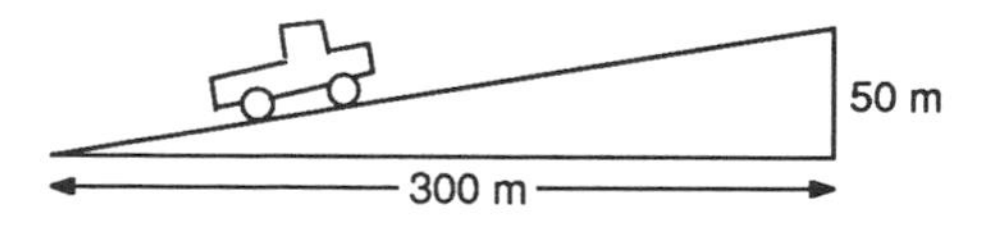

$$\text{slope} = \frac{\text{rise}}{\text{run}} = \frac{\text{vertical change}}{\text{horizontal change}} = \frac{50}{300} = \frac{1}{6}$$

Coordinate geometry can be used to measure the slope of lines. Informally, the slope of a line in a coordinate plane is the quotient of the change in the vertical direction (rise) divided by the change in the horizontal direction (run) that is necessary to get from one point on the line to another point on the line. Formally:

> *The **slope m of a line** (or segment) through P_1 and P_2 with coordinates (x_1, y_1) and (x_2, y_2) where $x_1 \neq x_2$ is $m = \dfrac{y_2 - y_1}{x_2 - x_1}$.*

Let's look at a few examples that show how you can calculate the slope of a line.

Example A

Find the slope of $\overleftrightarrow{AB}$.

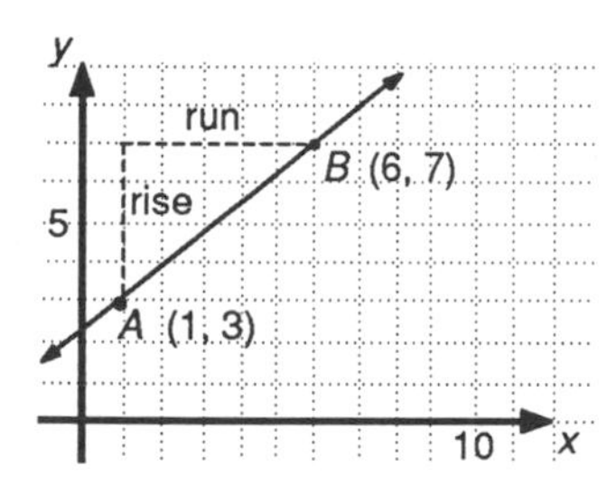

The slope of AB is $\dfrac{7-3}{6-1}$ or $\dfrac{4}{5}$.

Example B

Find the slope of $\overleftrightarrow{CD}$.

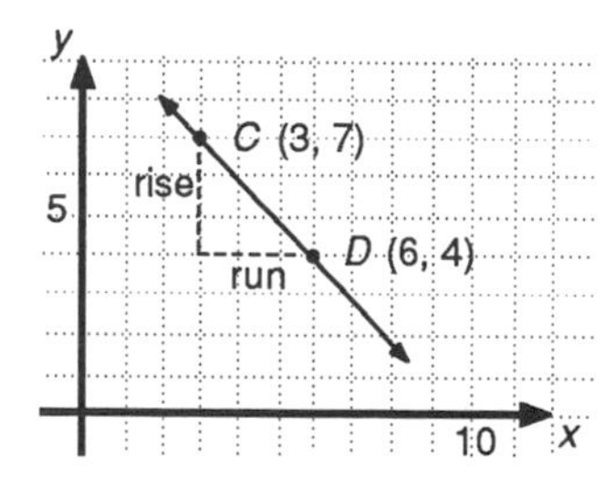

The slope of CD is $\dfrac{7-4}{3-6}$ or $\dfrac{3}{-3}$ or -1.

Notice from the examples that the slope can be positive or negative. If the line rises from the left to the right as in example A, then the slope is positive. If the line falls from the left to the right as in example B, then the slope is negative.

The slope can also be zero or it can be undefined.

Example C

Find the slope of $\overleftrightarrow{EF}$.

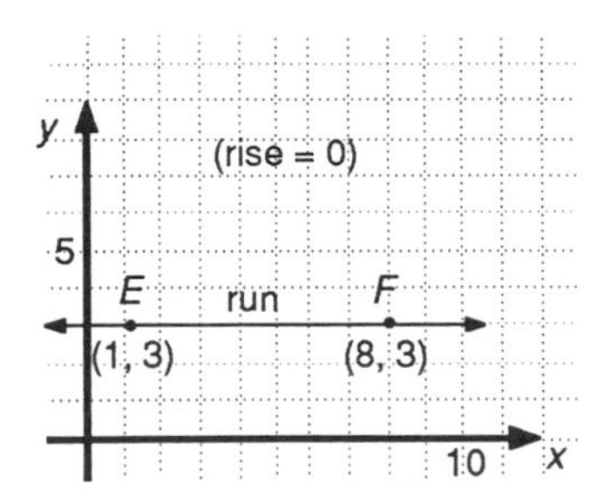

The slope of EF is $\frac{3-3}{8-1}$ or $\frac{0}{7}$ or 0.

Example D

Find the slope of $\overleftrightarrow{GH}$.

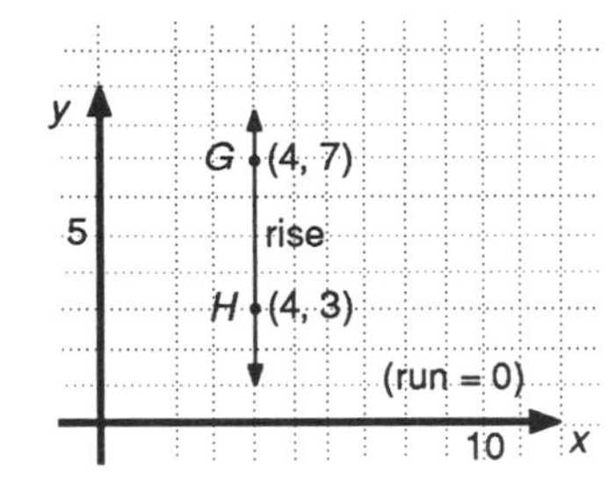

The slope of GH is $\frac{7-3}{4-4}$ or $\frac{4}{0}$ which is undefined.

Examples C and D illustrate two special cases in the application of the slope formula. If the line is horizontal, then the slope is zero. A zero slope means the line doesn't slope up or down. If the line is vertical, then the slope is undefined. That is why the slope definition says $x_1 \neq x_2$.

EXERCISE SET

Calculate the slope of each segment.

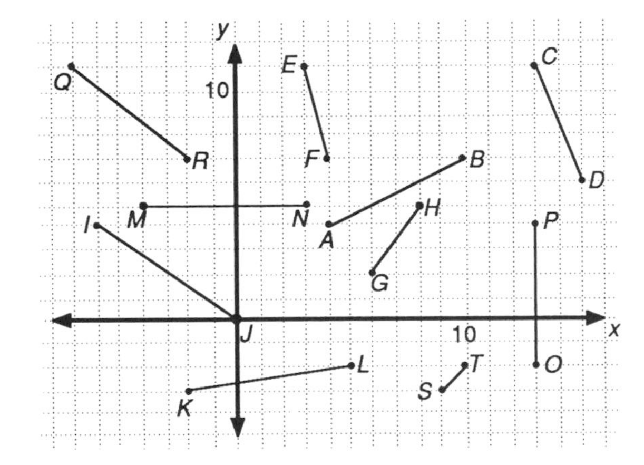

1. $\overline{AB}$
2. $\overline{CD}$
3. $\overline{EF}$
4. $\overline{GH}$
5. $\overline{IJ}$
6. $\overline{KL}$
7. $\overline{MN}$
8. $\overline{OP}$
9. $\overline{QR}$
10. $\overline{ST}$

11. Find the slope of the line through the points (16, 0) and (12, 8).
12. Find the slope of the line through the points (-3, -4) and (-16, 8).
13. Find the slope of the line through the point (6, 8) and the midpoint of the segment with endpoints (-7, 20) and (15, -10).
14. Quadrilateral $ABCD$ has vertices A (0, 0); B (6, 0); C (14, 8); and D (3, 8). Find the slope of each of the four sides.
15. Quadrilateral $EFGH$ has vertices E (0, 0); F (13, 0); G (18, 12); and H (5, 12). Find the slopes of the two diagonals.

Lesson 4.14

Slopes of Parallel and Perpendicular Lines

If two lines are parallel, what will be true of their slopes? If two lines are perpendicular, what will be true of their slopes? In this lesson you will make conjectures about the properties of the slopes of parallel and perpendicular lines.

Investigation 4.14.1

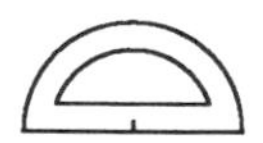

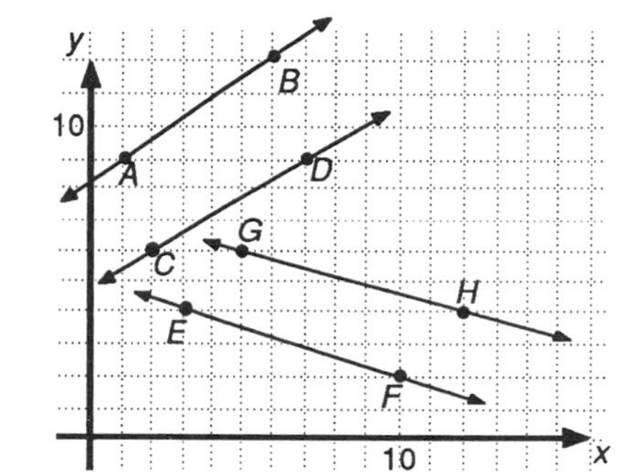

Step 1: Copy the pairs of lines shown on the right onto your own graph paper.

Step 2: Find the slope of each line.

Step 3: With your straightedge, draw a transversal through each pair of lines that look parallel.

Step 4: With your protractor, measure a pair of corresponding angles formed by each transversal.

If the slopes are equal, do the lines appear to be parallel? Compare your results with the results of others near you.

Let's try this another way.

Investigation 4.14.2

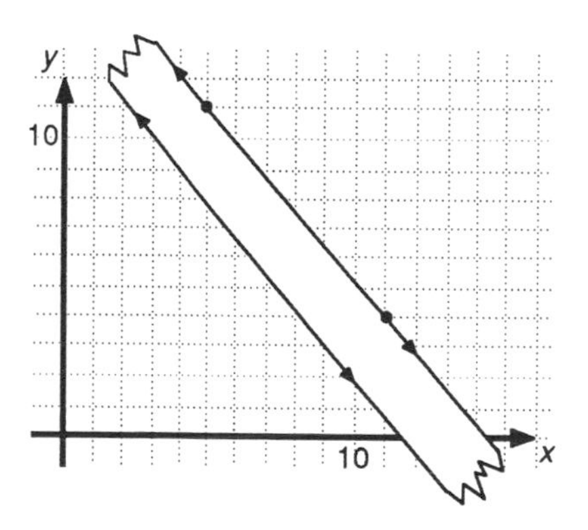

Step 1: Position your straightedge on your graph paper so that one edge passes through two points of the graph paper.

Step 2: Draw a line through the two points. Draw a line along the other edge of your straightedge.

Step 3: Form a "rise/run" right triangle with your two points. Find the slope of the line.

Step 4: Find a point that your other line passes through. From this point , will the same "rise" and "run" put you back on this line?

If the two lines are parallel, must their slopes be equal? Compare your results with the results of others near you. You should be ready to state your observations as your next conjecture.

C-31 *In a coordinate plane, two lines are —?— if and only if their slopes are —?—. **(Parallel Slope Conjecture)***

What about perpendicular lines? Is there a relationship between the slopes of perpendicular lines? (Since we are asking, there must be. Right?) What is the relationship? Let's investigate.

Investigation 4.14.3

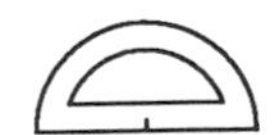

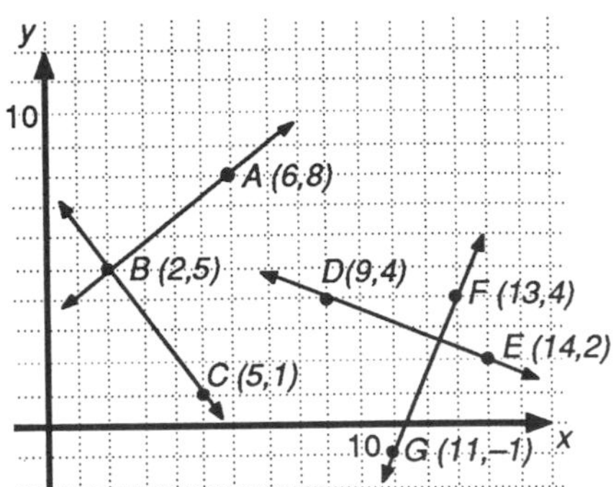

Step 1: Copy the pairs of lines shown on the right onto your own graph paper.

Step 2: Find the slope of each line.

Step 3: With your protractor, measure the angle formed at the intersection of each pair of lines.

If the lines are perpendicular, what is the relationship between their slopes? Compare your results with the results of others.

Let's try this another way.

Investigation 4.14.4

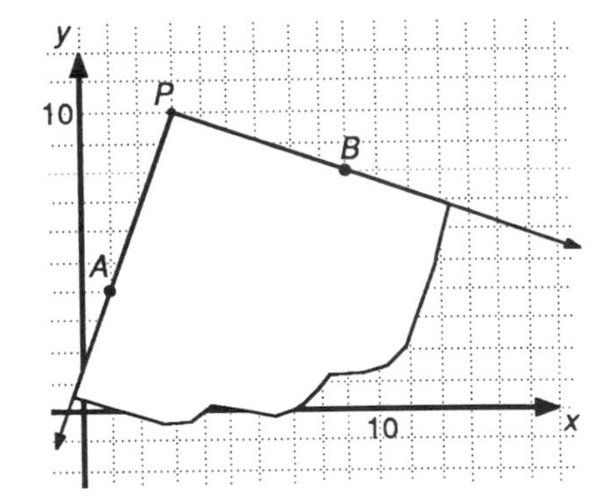

Step 1: Place the right-angled corner of a sheet of paper on a point of the graph paper. Label this point P.

Step 2: Rotate the right-angled corner on P until each edge crosses over a point of the graph paper. Label the two points A and B.

Step 3: Find the slope of $\overrightarrow{PA}$ and $\overrightarrow{PB}$.

Since the edges of the paper are perpendicular, $\overrightarrow{PA}$ and $\overrightarrow{PB}$ are perpendicular. What are the slopes of $\overrightarrow{PA}$ and $\overrightarrow{PB}$? Compare your results with the results of others near you. What is the relationship between the slopes of perpendicular lines? State your observations as your next conjecture.

C-32 *In a coordinate plane, two lines are —?— if and only if their slopes are —?—.* ***(Perpendicular Slope Conjecture)***

EXERCISE SET A

Determine whether each pair of lines through the points below is parallel, perpendicular, or neither.

A (1, 2); B (3, 4); C (5, 2); D (8, 3); E (3, 8); F (6, 9); G (-2, -3); H (10, 3)

1. $\overleftrightarrow{AB}$ and $\overleftrightarrow{BC}$	2. $\overleftrightarrow{AB}$ and $\overleftrightarrow{CD}$	3. $\overleftrightarrow{AB}$ and $\overleftrightarrow{DE}$	4. $\overleftrightarrow{CD}$ and $\overleftrightarrow{EF}$
5. $\overleftrightarrow{BC}$ and $\overleftrightarrow{CD}$	6. $\overleftrightarrow{BC}$ and $\overleftrightarrow{DE}$	7. $\overleftrightarrow{CD}$ and $\overleftrightarrow{GH}$	8. $\overleftrightarrow{DE}$ and $\overleftrightarrow{GH}$

EXERCISE SET B

Determine whether each figure below is a trapezoid, parallelogram, rectangle, or just an ordinary quadrilateral.

1.

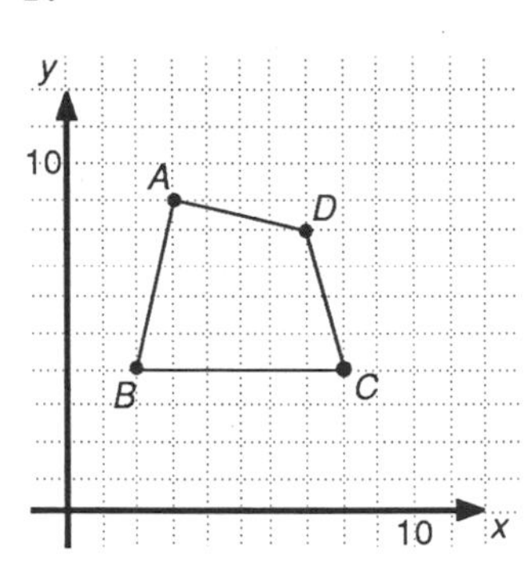

2.

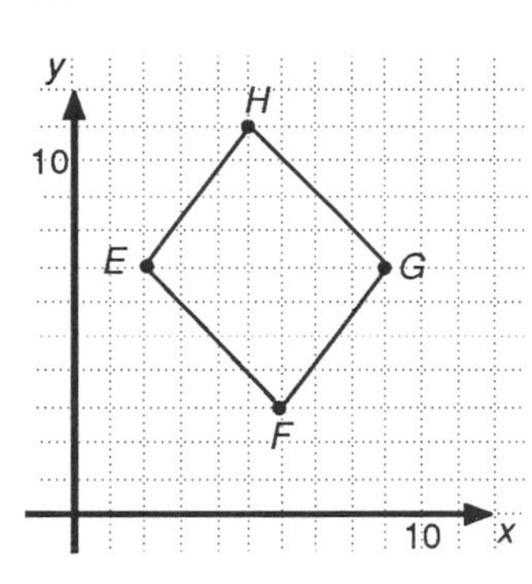

3.

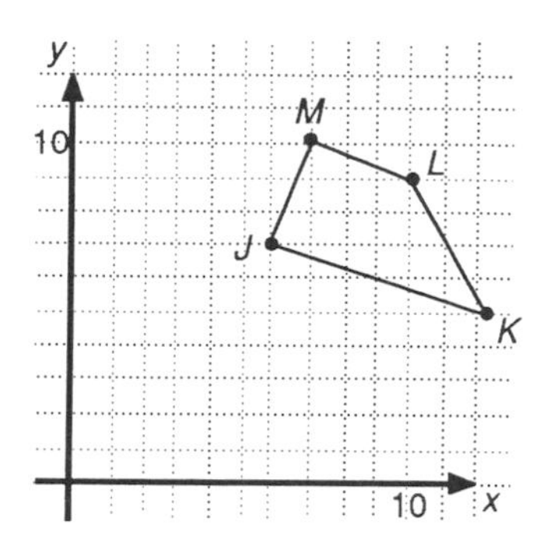

4.

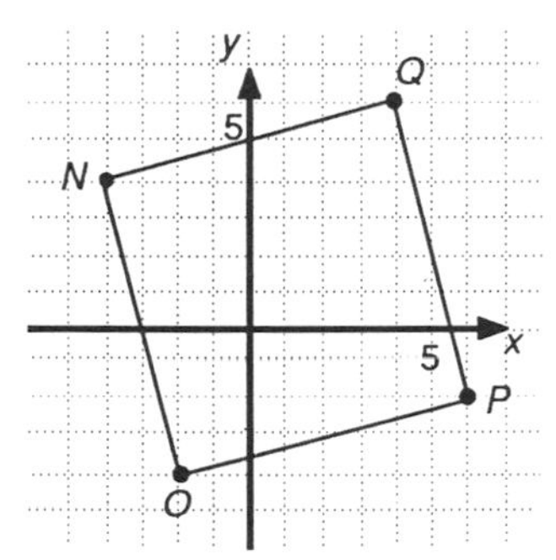

5.

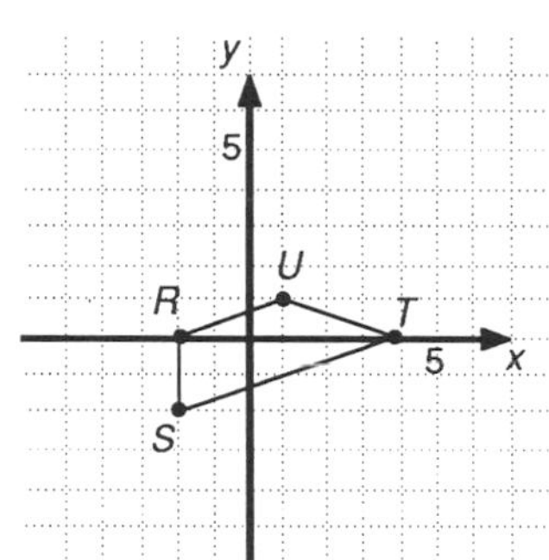

6.

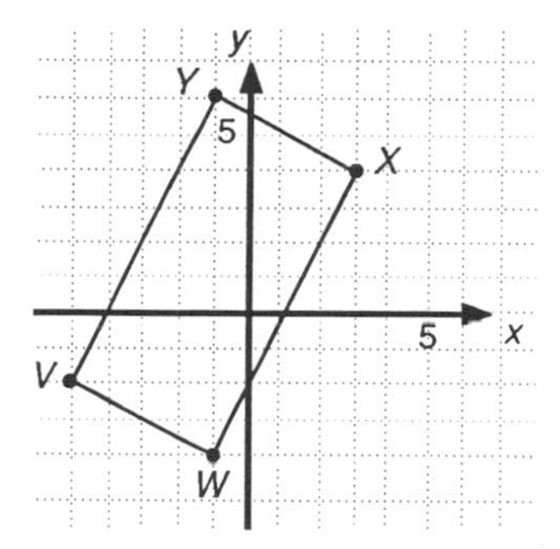

Improving Visual Thinking Skills

Dissecting a Hexagon II

Trace the hexagon on the right twice.

1. Divide one hexagon into four identical parts. Each part will be a trapezoid.
2. Divide the other hexagon into eight identical parts.

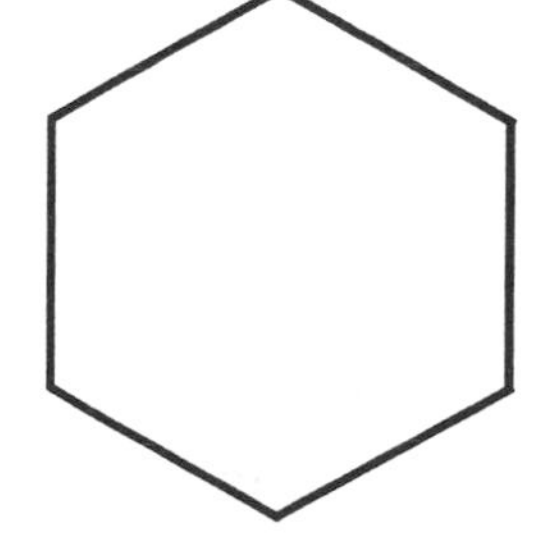

Lesson 4.15

Chapter Review

If it takes one girl 36 seconds to blow up a balloon, how long will it take four girls trying to blow up the same balloon?

– Anonymous

EXERCISE SET A

Complete each conjecture.

1. If two angles are vertical angles, then they are —?—. (C-1 or **Vertical Angle Conjecture**)
2. If two angles are a linear pair of angles, then they are —?— . (C-2 or **Linear Pair Conjecture**)
3. If two angles are both equal in measure and supplementary, then each angle measures —?—. (C-3 or **Equal Supplements Conjecture**)
4. The sum of the measures of the three angles of every triangle is —?—. (C-4 or **Triangle Sum Conjecture**)
5. If two angles of one triangle are equal in measure to two angles of another triangle, then the remaining two angles are —?—. (C-5 or **Third Angle Conjecture**)
6. The sum of the measures of the four angles of a quadrilateral is —?—. (C-6)
7. The sum of the measures of the n angles of an n-gon is —?—. (C-7 or **Polygon Sum Conjecture**)
8. The measure of each angle of an equiangular n-gon is —?—. (C-8)
9. The sum of the measures of one set of exterior angles —?—. (C-9)
10. The measure of an exterior angle of a triangle equals the sum of the measures of the two —?—. (C-10 or **Exterior Angle Conjecture**)
11. The sum of the lengths of any two sides of a triangle is —?— the length of the third side. (C-11 or **Triangle Inequality Conjecture**)
12. In a triangle, the longest side is opposite —?—. (C-12)
13. If a triangle is isosceles, then the base angles are —?—. (C-13 or **Isosceles Triangle Conjecture**)
14. If a triangle has two angles of equal measure, then the triangle is —?—. (C-14 or **Converse of the Isosceles Triangle Conjecture**)
15. An equilateral triangle is —?— and, conversely, an equiangular triangle is —?—. (C-15 or **Equilateral Triangle Conjecture**)

16. If two parallel lines are cut by a transversal, then the —?— are equal in measure. Conversely, —?—. (C-16 or **CA Conjecture**)

17. If two parallel lines are cut by a transversal, then the —?— are equal in measure. Conversely, —?—. (C-17 or **AIA Conjecture**)

18. If two parallel lines are cut by a transversal, then the alternate exterior angles are equal in measure. Conversely, —?—. (C-18 or **AEA Conjecture**)

19. The base angles of an isosceles trapezoid are —?—. (C-19 or **Isosceles Trapezoid Conjecture**)

20. A midsegment of a triangle is —?— to the third side and —?— the length of the third side. (C-20 or **Triangle Midsegment Conjecture**)

21. The midsegment of a trapezoid —?— to the bases and equal in length to the —?— of the lengths of the bases. (C-21 or **Trapezoid Midsegment Conjecture**)

22. The opposite angles of a parallelogram are —?—. (C-22)

23. The consecutive angles of a parallelogram are —?—. (C-23)

24. The opposite sides of a parallelogram are —?—. (C-24)

25. The diagonals of a parallelogram —?—. (C-25)

26. The —?— of a rhombus are perpendicular bisectors of each other. (C-26)

27. The diagonals of a rhombus —?— the angles of the rhombus. (C-27)

28. The measure of each angle of a rectangle is —?—. (C-28)

29. The diagonals of a rectangle are —?—. (C-29)

30. If (x_1, y_1) and (x_2, y_2) are the coordinates of the endpoints of a segment, then the coordinates of the midpoint are —?—. (C-30 or **Coordinate Midpoint Conjecture**)

31. In a coordinate plane, two lines are —?— if and only if their slopes are —?—. (C-31 or **Parallel Slope Conjecture**)

32. In a coordinate plane, two lines are —?— if and only if their slopes are —?—. (C-32 or **Perpendicular Slope Conjecture**)

EXERCISE SET B

Identify each statement as true or false.

1. When you make a conjecture from observations, you use inductive reasoning.

2. The total number of diagonals in a polygon of n sides is $n(n-2)$.

3. Point, line, and plane are undefined terms in our geometry.

4. A quadrilateral is a triangle with all the sides equal in measure.

5. The base angles of an isosceles triangle are complementary.

6. The opposite angles of an isosceles trapezoid are supplementary.

7. Every square is a rhombus.

8. The diagonals of a rectangle are perpendicular to each other.

9. The compass is a tool used to measure the size of an angle in degrees.

10. The three angle bisectors of a triangle intersect at the centroid.

11. The sum of the measures of the five angles of a pentagon is 540.

12. The largest angle of a triangle is opposite the smallest side.

13. The midsegment of a trapezoid is parallel to the bases.

14. The measure of each angle of a regular dodecagon is 150 degrees.

15. If the base angles of an isosceles triangle each measure 48, then the vertex angle has a measure of 132.

16. The diagonals of a square are perpendicular bisectors of each other.

17. The diagonals of a rhombus divide the rhombus into four right triangles.

18. If (0, 0); (5, 0); and (5, 4) are three vertices of a rectangle, the fourth is (0, 4).

19. If A is (0, 0); B is (3, 2); C is (6, 8); and D is (10, 3), then $\overleftrightarrow{AB} \perp \overleftrightarrow{CD}$.

20. There are just as many true statements as false statements in this exercise set.

EXERCISE SET C

1. Find the measure of each lettered angle in the diagram below.

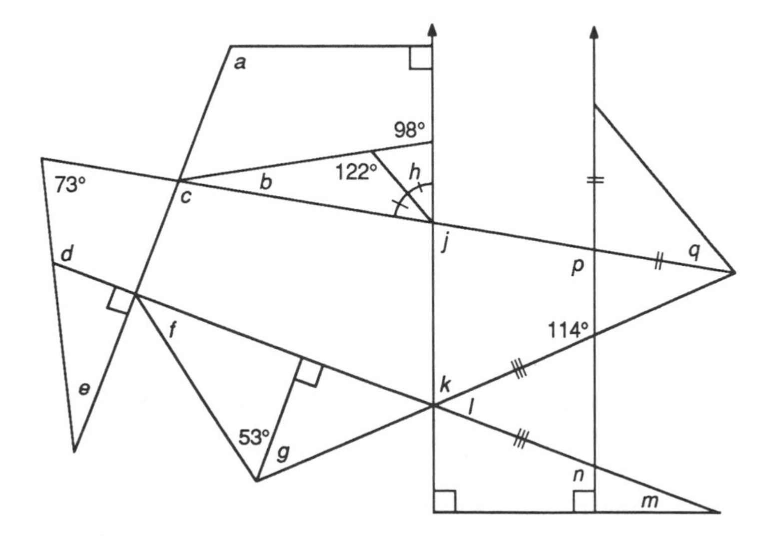

EXERCISE SET D

In Exercises 1 to 4, find the midpoint of the segment with the given points as endpoints.

1. (6, 4); (12, -2) 2. (5, -8); (-3, -8) 3. (15, 0); (7, 5) 4. (-3, 4); (6, 7)

In Exercises 5 to 8, find the slope of the line through each pair of points.

5. (0, 1); (1, 0) 6. (2, 3); (-2, -3) 7. (4, 5); (3, 5) 8. (-1, -2); (-1, -1)

In Exercises 9 to 12, determine whether the lines are parallel, perpendicular, or neither.

A (1, 0); B (3, 2); C (5, 1); D (8, 4); E (3, 4)

9. $\overleftrightarrow{AB}$ and $\overleftrightarrow{BC}$ 10. $\overleftrightarrow{AB}$ and $\overleftrightarrow{CD}$ 11. $\overleftrightarrow{AE}$ and $\overleftrightarrow{BC}$ 12. $\overleftrightarrow{AE}$ and $\overleftrightarrow{ED}$

EXERCISE SET E

Use the segments and angles below to construct each figure. The lower case letter above each segment represents the length of the segment in units.

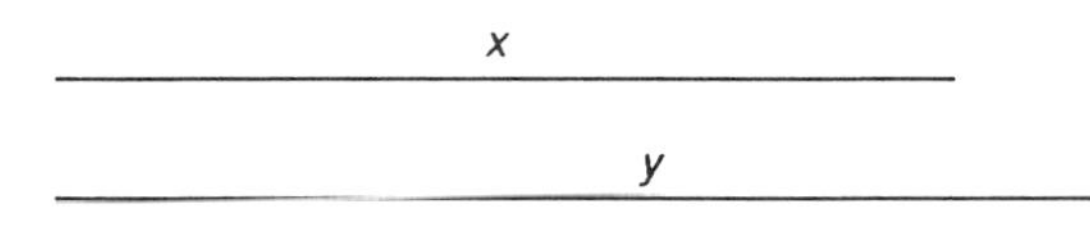

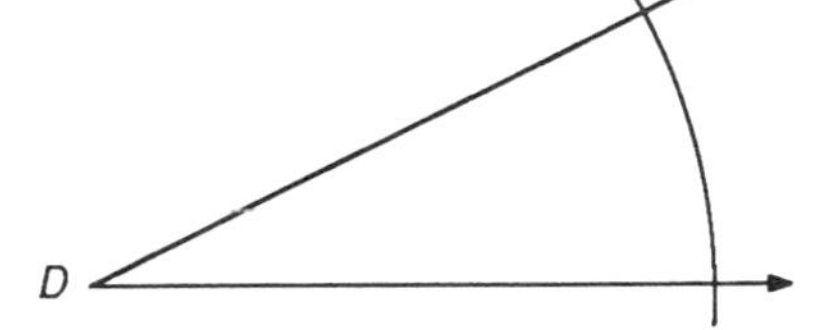

1. Construct ΔDOG given $\angle D$, $DO = x$, and $DG = y$.
2. Construct median $\overline{DT}$, angle bisector $\overline{OA}$, and altitude $\overline{GC}$ in ΔDOG.
3. Construct rhombus $RHOM$ with x and y the lengths of the two diagonals.

Improving Visual Thinking Skills

Patchwork Cubes

The large cube on the right is built from 13 double cubes like the one shown plus one single cube. What color must the single cube be and where must it be positioned?

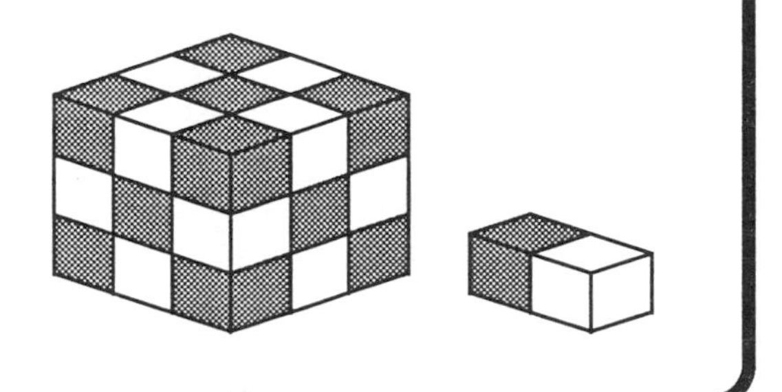

Cooperative Problem Solving

Geometrivia II

The geometry students at the lunar space port Galileo have just received a geometry puzzle called Geometrivia II from their earth friends at the NASA Academy in Atlanta. The students at the NASA Academy enjoyed the Geometrivia puzzle they received from their lunar friends at Galileo a few months ago and they have reciprocated. The students at the NASA Academy have also been trained to learn in cooperative small groups. Geometrivia II has been designed with that in mind. Geometrivia II is another multistep problem solving puzzle. In the Network Puzzle, find the measure for each lettered angle. Substitute the values into the formula. The answers to the Search and Solve problems are in this chapter. Find the answers to the Library Research problems in the math section of your library. When all the variables in the formula have been replaced, calculate x. It is a whole number.

Network Puzzle

a = –?–

b = –?–

c = –?–

d = –?–

e = –?–

$l_2 \parallel l_3$

$l_1 \perp l_3$

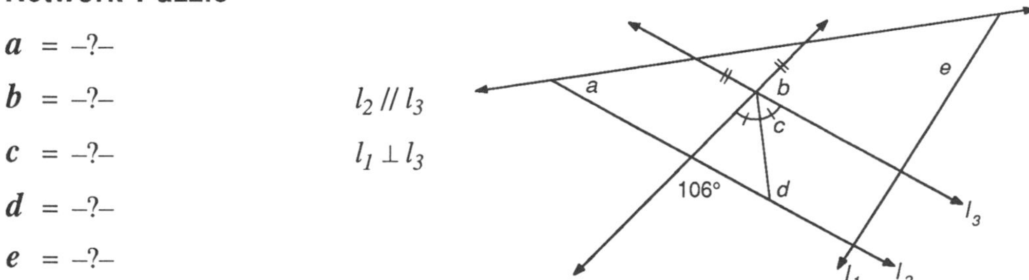

Search and Solve

f Answer from 4.3 B7

g Answer from 4.4 A4

h Answer from 4.9 A6

i Answer from 4.10 A4

j Answer from 4.11 B4

Library Research

k The number of books in the original set of Euclid's *Elements*

l The numerical value for the Egyptian hieroglyphic symbol: ∩ (heel bone)

m Age of an octogenarian in years divided by number of diagonals in a pentagon

n The year of Johan Kepler's birth minus the year of Galileo Galilei's birth

p Goldbach's Conjecture claims that every even integer greater than or equal to four can be represented as the sum of —?— prime numbers.

Formula

$$X = \frac{\left[\sqrt{\dfrac{lm-(k+d)}{\dfrac{f}{g}+\dfrac{c}{e}}}\,\right]^{\left(\frac{f}{h-g}\right)}}{\sqrt{\dfrac{(m-g)(b-f)}{(np-k)}+\dfrac{(d-i)}{j}+\dfrac{(m-np)}{(a-h)}}}$$

5 Congruence

Depth, M. C. Escher, 1955

Two geometric figures are congruent if they have the same shape and same measurements. In this chapter you are going to discover what conditions guarantee that two triangles are congruent. Congruence is very important in our technological society. Modern assembly line production relies on identical or congruent parts. In the assembly of an automobile, for example, parts must be produced so that any particular part can be used in the same position in any car coming down the assembly line.

Lesson 5.1

Congruence

Some painters transform the sun into a yellow spot, others transform a yellow spot into the sun.

– Picasso

EXERCISE SET A

In each of the puzzles below, determine which two figures are identical. If necessary, trace the figures and flip, turn, or slide the tracings over the others to make your decision.

1.* a.  b.  c.  d. 

2. a.  b.  c.  d. 

3. a. 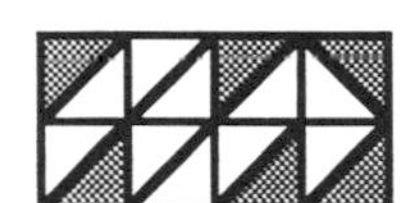 b.  c.  d. 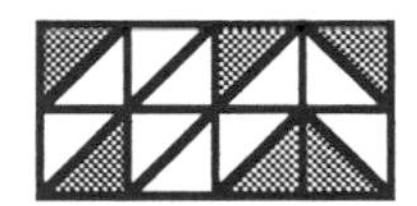

Two figures that have the exact same size and shape are **congruent**.

> *If two segments have the same measure, they are* ***congruent segments****.*
>
> *If two angles have the same measure, they are* ***congruent angles****.*

The relationship between congruent figures is like the relationship between equal numbers. The symbol for congruence is ≅ and is read *is congruent to*. You use the = symbol between equal numbers and the ≅ symbol between congruent geometric figures.

C

A D

$AC = 3.2$ cm
$DC = 3.2$ cm

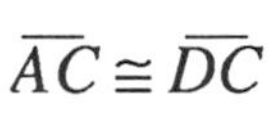

$\overline{AC} \cong \overline{DC}$

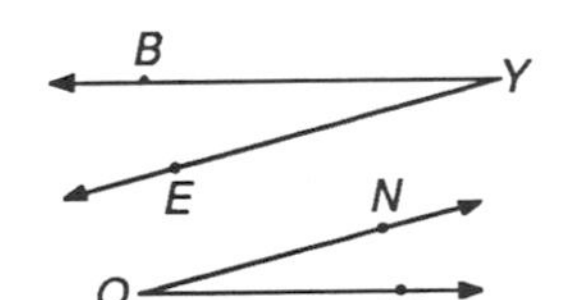

$m\angle BYE = 14$
$m\angle NOW = 14$

$\angle BYE \cong \angle NOW$

You can establish a correspondence between any two polygons with the same number of sides. The order of the letters in a statement of correspondence tells you which segments and which angles in the two polygons correspond. The symbol for *corresponds to* is $\leftrightarrow$. If a correspondence is set up between quadrilaterals $FOUR$ and $XYZW$, then $\overline{FO}$ corresponds to $\overline{XY}$ ($\overline{FO} \leftrightarrow \overline{XY}$), $\overline{OU}$ corresponds to $\overline{YZ}$, and so on. Also, angle F corresponds to angle X ($\angle F \leftrightarrow \angle X$), angle O corresponds to angle Y, and so on. One special correspondence between polygons is a congruence correspondence.

> *Two polygons are* ***congruent polygons*** *if and only if they have all their corresponding angles congruent and all their corresponding sides congruent.*

The definition above says two things. It says that if the angles and sides of one polygon are congruent to the corresponding angles and sides of another polygon, then the two polygons are congruent. For example, if the three angles of ΔWHY are congruent to the three corresponding angles of ΔNOT, and if the three sides of ΔWHY are congruent to the corresponding three sides of ΔNOT, then ΔWHY is congruent to ΔNOT. Notice that when you write the symbolic statement of congruence of the two triangles, the corresponding letters of the congruent angles are in the same position.

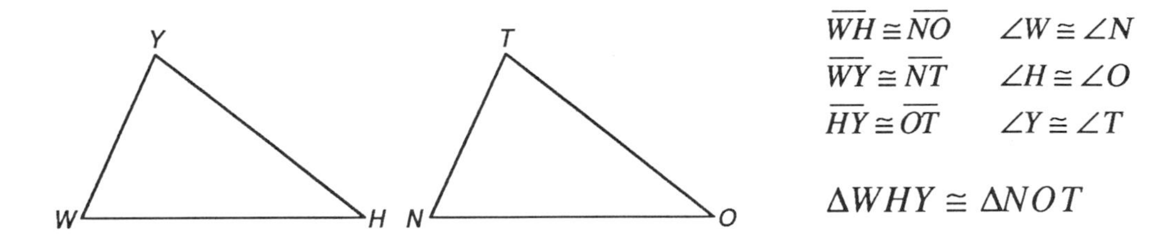

The definition also says that if two polygons are congruent, then the corresponding angles and sides are congruent. (All definitions are reversible, remember?) For example, if you are given that quadrilateral $LETS$ is congruent to quadrilateral $BOGY$, it follows that the four pairs of corresponding angles and the four pairs of corresponding sides are also congruent.

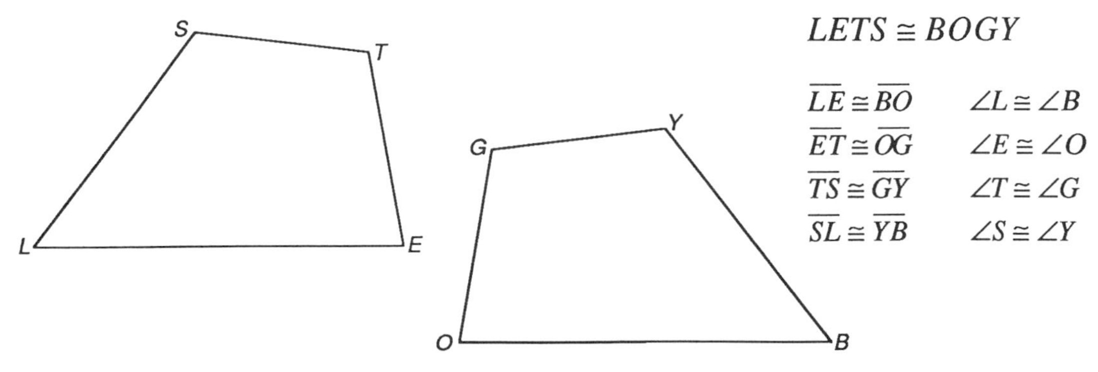

EXERCISE SET B

In Exercises 1 and 2, plot on graph paper the given triangle and the given segment. Locate a point F so that it appears that $\Delta ABC \cong \Delta DEF$. Can you find more than one point?

1.

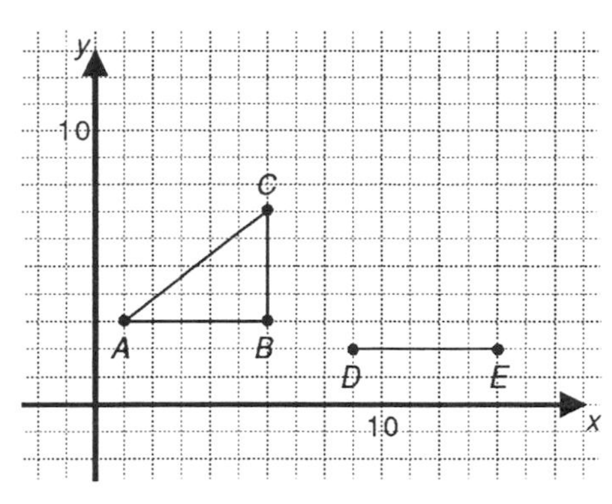

2.

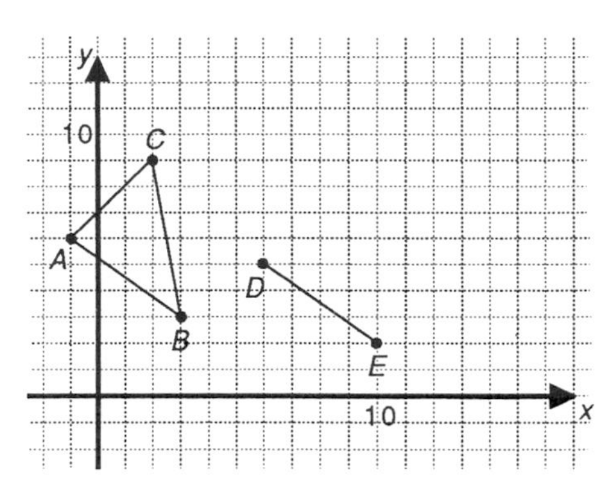

In Exercises 3 and 4, plot on graph paper the given quadrilateral and the given segment. Locate two points F and G so that it appears that $ABCD \cong EFGH$.

3.

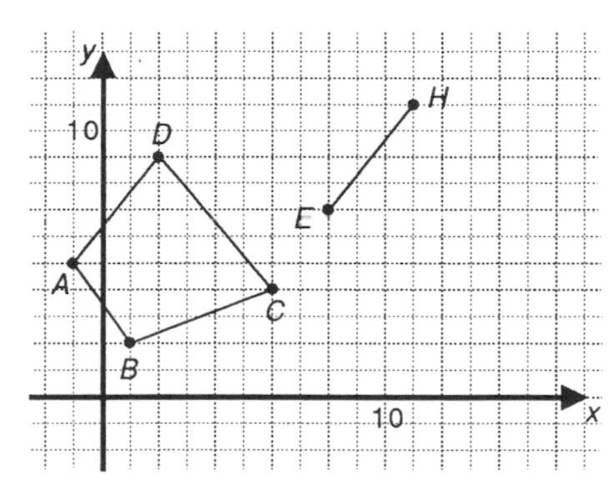

4.

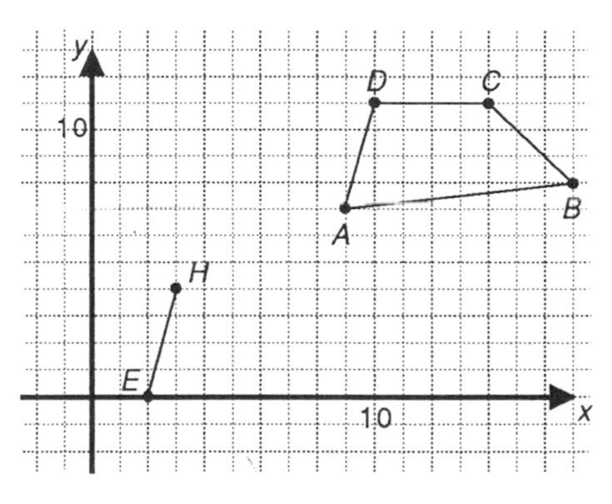

EXERCISE SET C

From the information given, determine the correct congruence statement.

1.* $\Delta DOG \cong \Delta$ —?—

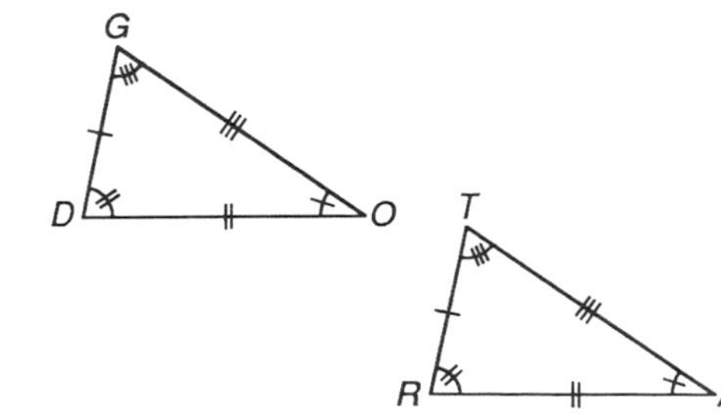

2. $\Delta COM \cong \Delta$ —?—

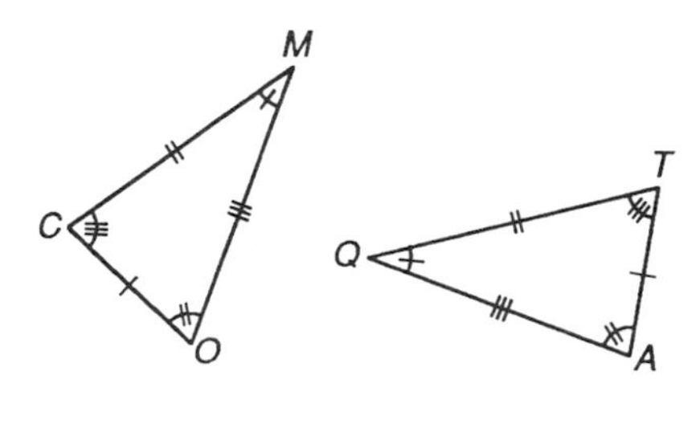

3. *HOSE* is a parallelogram.
$\Delta HOE \cong \Delta$ —?—

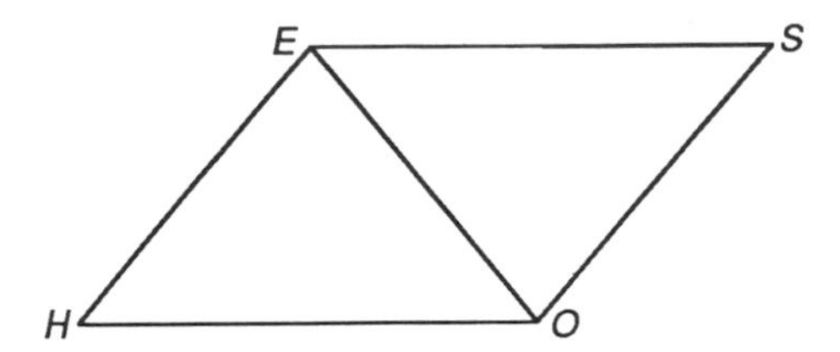

4. $\overline{MO} \parallel \overline{SE}$
$\Delta MOU \cong \Delta$ —?—

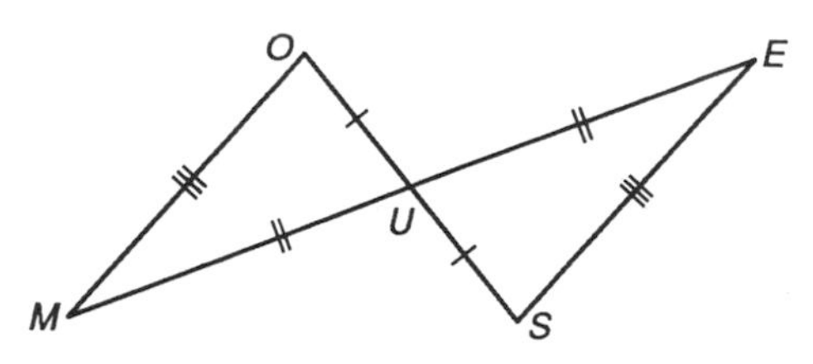

Find the missing measures in each pair of congruent polygons below.

5. $\Delta HAT \cong \Delta CEK$

$m\angle H =$ –?– $\quad CE =$ –?–

$m\angle A =$ –?– $\quad EK =$ –?–

$m\angle T =$ –?– $\quad CK =$ –?–

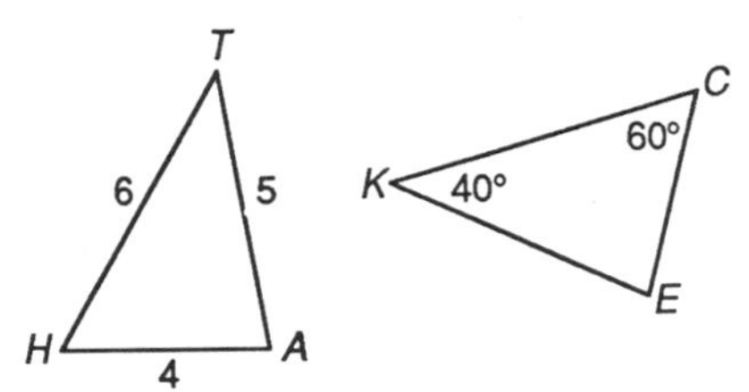

6.* $THINK \cong POWER$

$PR =$ –?– $\quad RE =$ –?–

$EW =$ –?– $\quad WO =$ –?–

$PO =$ –?–

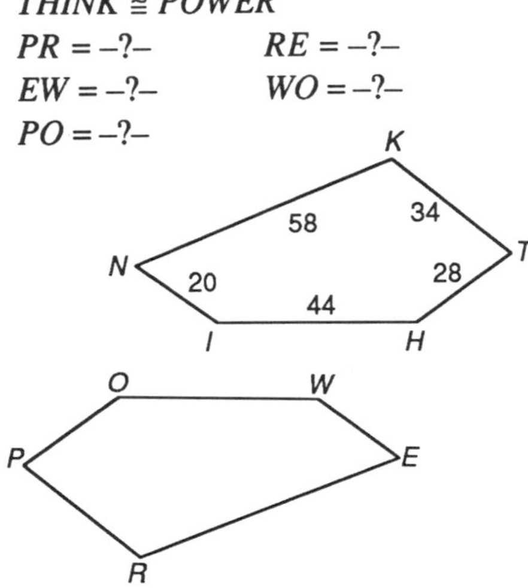

EXERCISE SET D

Sketch and label each figure below.

1. Isosceles triangle *MAN* with angle *A* the vertex angle and $\overline{NY}$ a median
2. Trapezoid *TRAP* with $\overline{TR} \parallel \overline{AP}$ and with diagonals $\overline{TA}$ and $\overline{PR}$ intersecting at *E*
3. Rhombus *DIAM* with $m\angle D = 60$ and diagonal $\overline{MI}$

Improving Visual Thinking Skills

Painted Faces II

Unit cubes are assembled to form a larger cube, and then some of the faces of this larger cube are painted. After the paint dries, the larger cube is disassembled into the unit cubes and it is found that 32 of these have no paint on any of their faces. How many faces of the larger cube were painted?

Lesson 5.2

SSS Shortcut? AAA Shortcut?

A building contractor has just assembled two massive triangular trusses that will support the roof in a recreation hall. Before the crane hoists them into place, she wants to verify that the two triangular trusses are identical. According to the definition of congruent polygons, if the three sides and three angles of one triangle are congruent to the corresponding three sides and three angles of another, then the two triangles are congruent. Does that mean the contractor must measure and compare all six parts of both triangles? There must be a shortcut. It would seem that only a few of the measurements are needed — but how many and which ones? Let's see if we can answer this question.

Wouldn't it be nice if the contractor could measure only one side in each triangle, find that they're congruent, and know that the triangles themselves were congruent? In other words, if she knew that one side of a triangle is congruent to one side of another triangle, could she state that the triangles must be congruent? Clearly they might be. But must they be? It only takes one counter-example to demonstrate that they don't have to be. Below are two triangles with one pair of congruent sides. The triangles are clearly not congruent. It is clear from this counter-example that it takes more than one pair of congruent sides to make two triangles congruent. The contractor will have to measure more than one pair of sides.

Side

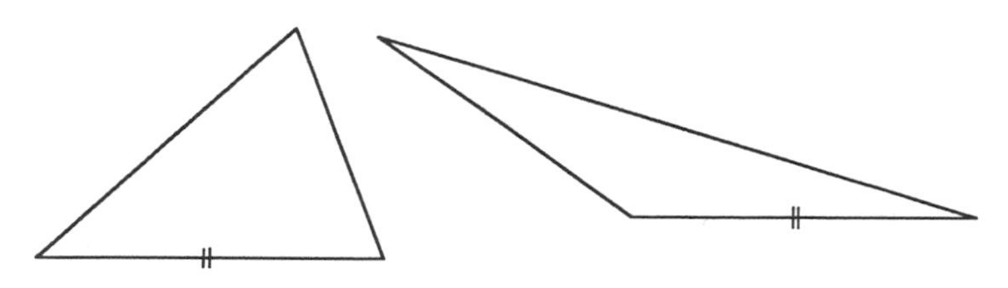

One pair of congruent sides. Triangles not congruent.

It is easy to show that two triangles need not be congruent simply because one angle in each is congruent. Below are two triangles with a pair of congruent angles. Clearly the triangles are not congruent to each other. The contractor will have to measure more than one pair of angles.

Angle

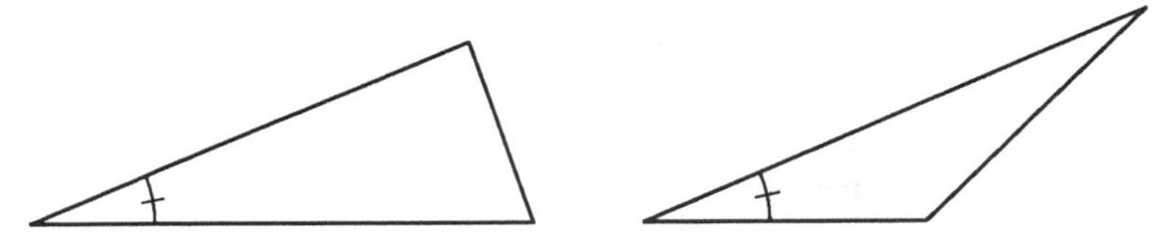

One pair of congruent angles. Triangles not congruent.

If two different triangles have two pairs of congruent parts (two pairs of congruent sides, or two pairs of congruent angles, or a pair of congruent sides and a pair of congruent angles), must the two triangles be congruent? A few counter-examples will demonstrate that they need not be.

Side-Side

Two pairs of congruent sides. Triangles not congruent.

Angle-Angle

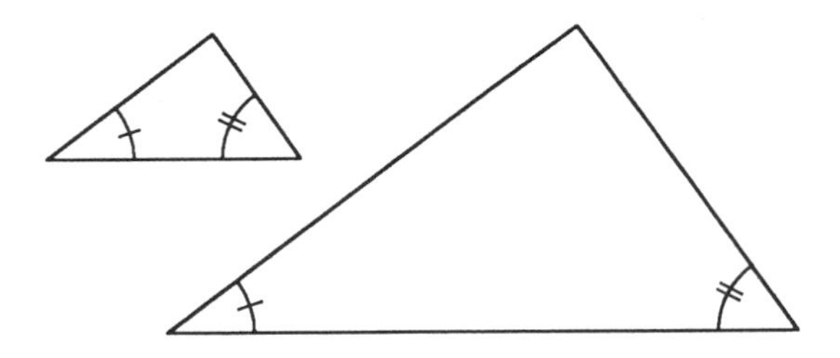

Two pairs of congruent angles. Triangles not congruent.

Side-Angle

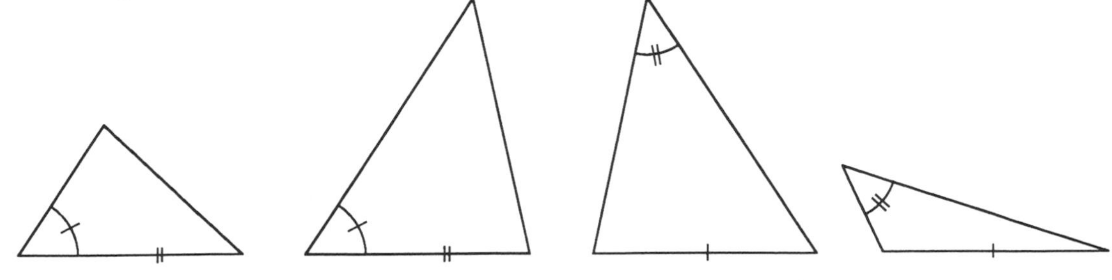

Pair of congruent sides and pair of congruent angles. Triangles not congruent.

So, if there is a shortcut that the contractor can use to show that two triangles are congruent, she must measure and compare at least three parts of each triangle. What are the combinations of three parts of a triangle that she could measure? The six different combinations of congruent pairs of three parts are diagrammed below.

Side-Side-Side

Three pairs of congruent sides.

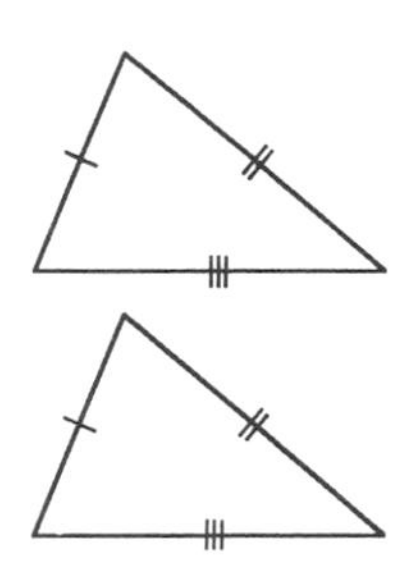

Angle-Angle-Angle

Three pairs of congruent angles.

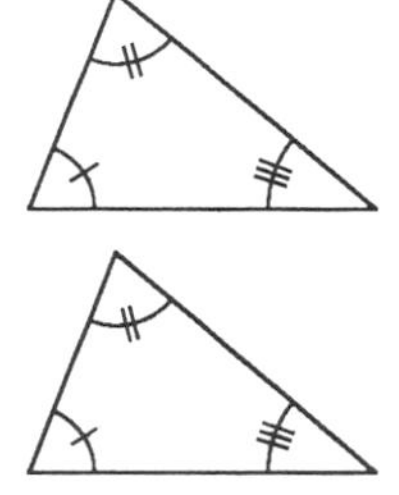

Side-Angle-Side

Two pairs of congruent sides and one pair of congruent angles (angle between sides).

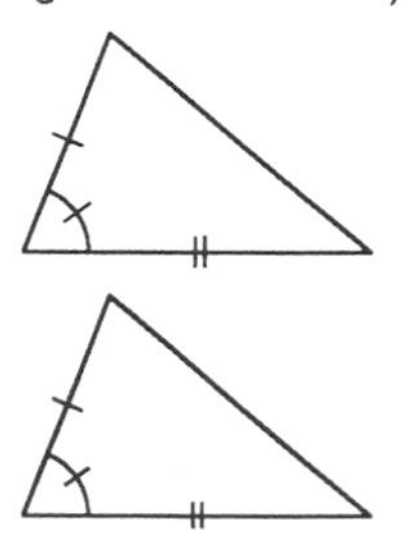

Side-Side-Angle

Two pairs of congruent sides and one pair of congruent angles (angle not between sides).

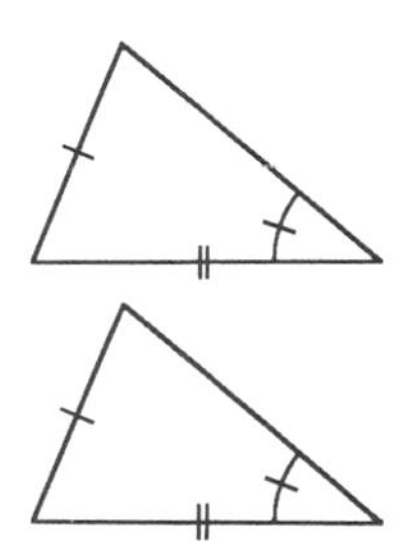

Angle-Side-Angle

Two pairs of congruent angles and one pair of congruent sides (side between angles).

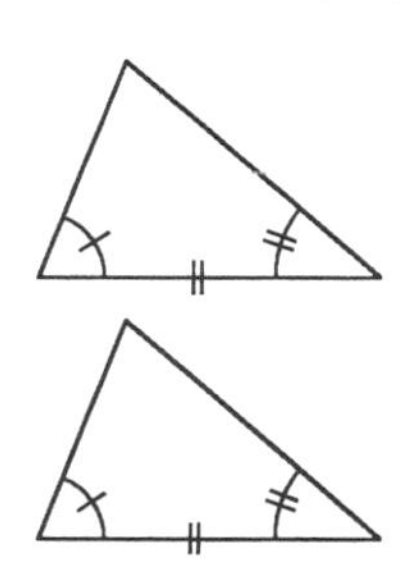

Side-Angle-Angle

Two pairs of congruent angles and one pair of congruent sides (side not between angles).

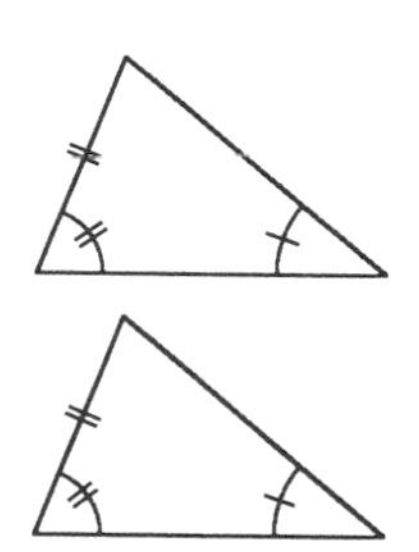

In the rest of this lesson and the two to follow, you are going to investigate the six possible combinations of pairs of three parts of a triangle to see if they lead to shortcuts for establishing two triangles congruent. Unfortunately, it's too late to help the contractor. She has become impatient and gone ahead and measured all six pairs of triangle parts and found them to be congruent. She's concluded that the triangular trusses are congruent based on the definition of congruent polygons. The crane is already lifting them into position. Later in this text you will be asked to determine whether or not pairs of triangles are congruent. Once you've discovered the congruence shortcuts, you will be able to use them to shorten your work.

If three sides in one triangle are congruent to the three sides in another triangle (SSS), must the two triangles be congruent? Let's find out. Work the next two investigations in groups of four or five.

Investigation 5.2.1

The three segments below are the sides of acute ΔABC.

Step 1: Use the three segments to construct a triangle on your own paper.

Step 2: Compare your triangle with ΔABC. Is it congruent or is it different? (One way to see if the triangles are congruent is to place them on top of each other and hold them up to a window or light. You may have to flip one of them over to get them to match.)

Step 3: Can you construct a triangle using the three segments above that is not congruent to ΔABC? Try it. Construct a second triangle using the three segments that is *not* congruent to ΔABC. Could anyone in your group construct a different triangle?

Step 4: Let's try this again. The three segments below are the sides of obtuse ΔCBS. Construct a triangle using these three segments as sides.

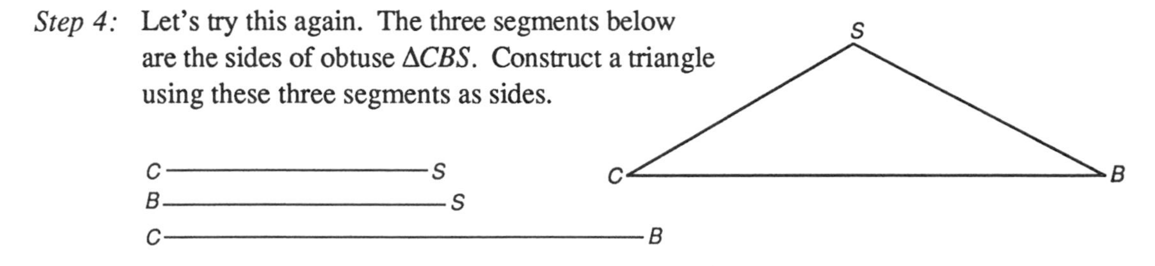

Step 5: Compare your triangle with ΔCBS and with those constructed by others in your group. Are they all congruent?

Step 6: Can you construct a triangle using the three segments above that is not congruent to ΔCBS? Try it. Construct a second triangle using the three segments that is *not* congruent to ΔCBS.

If three sides of a triangle are congruent to three sides of another, must the triangles be congruent? Make a conjecture. We will call it the SSS Congruence Conjecture.

C-33 *If the three sides of one triangle are congruent to the three sides of another triangle, then —?—.*
(SSS Congruence Conjecture)

This is an important conjecture. Write it below your investigative work. Add it to your conjecture list. Beneath your conjecture, copy and complete the "picture statement" below. It's a graphic way to illustrate the SSS Congruence Conjecture.

If you know this:

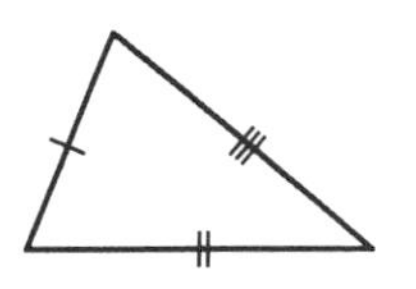

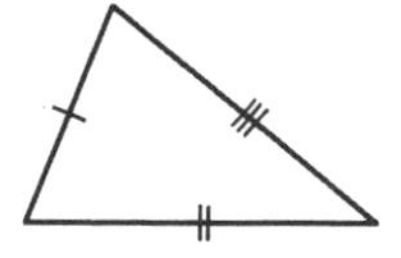

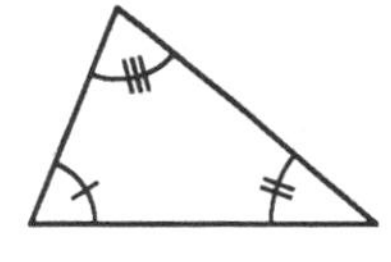

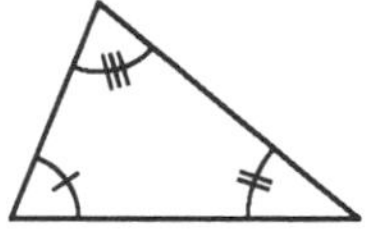

In the next investigation you will determine whether three pairs of congruent angles (AAA) force two triangles to be congruent? (Of course, the measures of the three angles must add to 180 or else no triangle would be possible.)

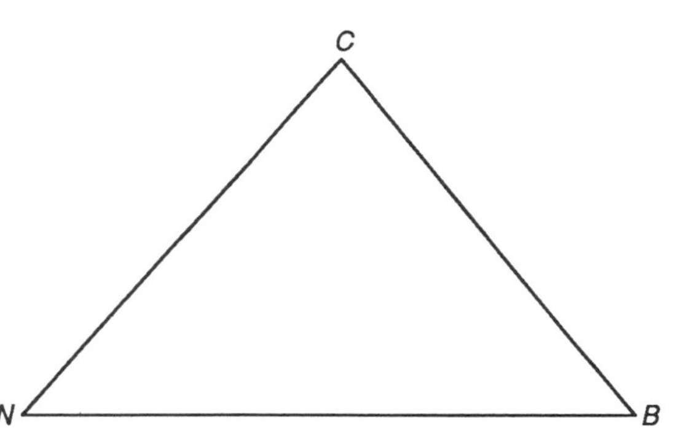

Investigation 5.2.2

The three angles below are the angles of acute ΔNBC.

Step 1: Use the three angles to construct a triangle on your own paper.

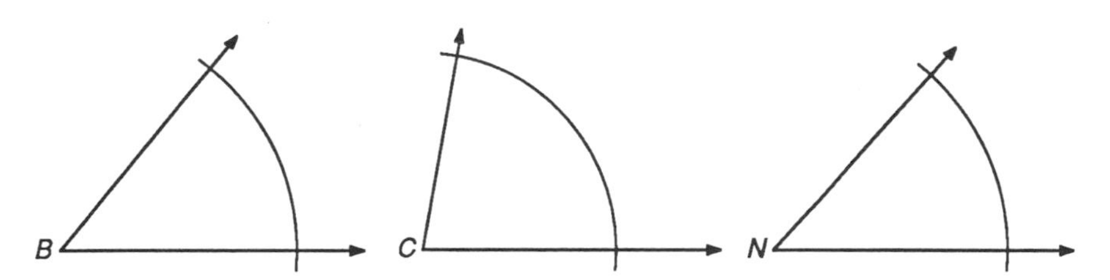

Step 2: Compare your triangle with ΔNBC and the triangles constructed by others in your group. Are they all congruent or are some different? Can you construct a triangle *not* congruent to ΔNBC that uses the three angles above?

Based on your observations, complete the conjecture below.

Conjecture: *If the three angles of one triangle are congruent to the three angles of another triangle then the two triangles (are/are not) necessarily congruent.*

Write the conjecture beneath your investigative work as your summary statement.

EXERCISE SET

From the information given, complete each statement. If the triangles cannot be shown to be congruent from the information given, write *cannot be determined.* Make sure that you base your answer on the given information. Do not assume that segments or angles are congruent just because they appear congruent.

1.* $\Delta TAN \cong \Delta$ —?—

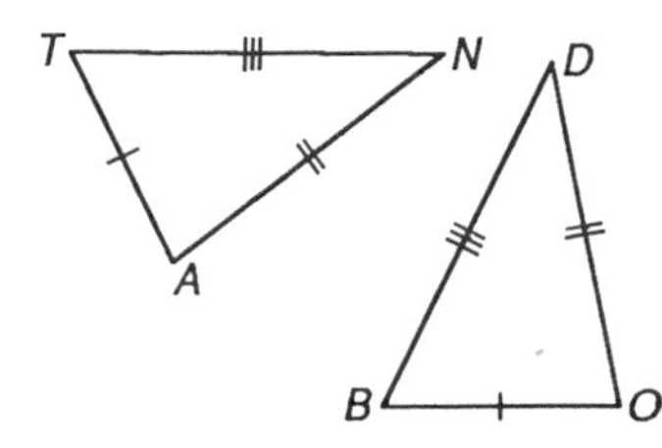

2. $\Delta GIT \cong \Delta$ —?—

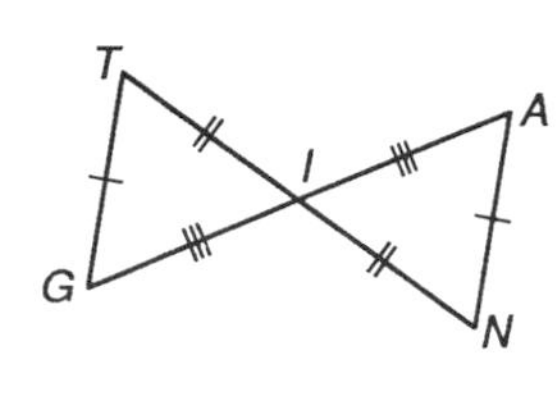

3. $\Delta DOG \cong \Delta$ —?—
Why?

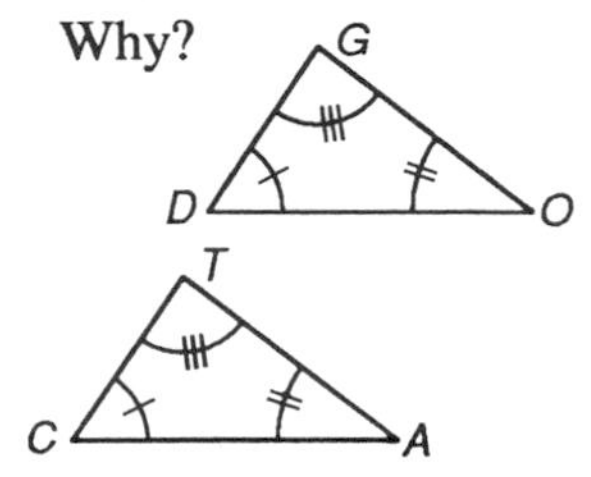

4. $\Delta ARE \cong \Delta$ —?—

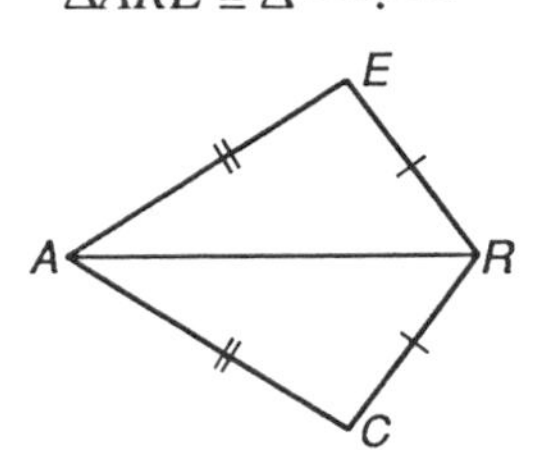

5. $\Delta BUY \cong \Delta$ —?—
Why?

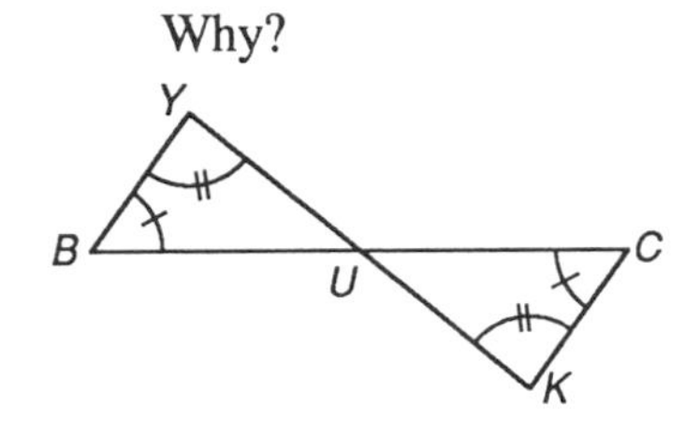

6. $\overleftrightarrow{AB} \parallel \overleftrightarrow{CD}$
$\Delta ABE \cong \Delta$ —?— Why?

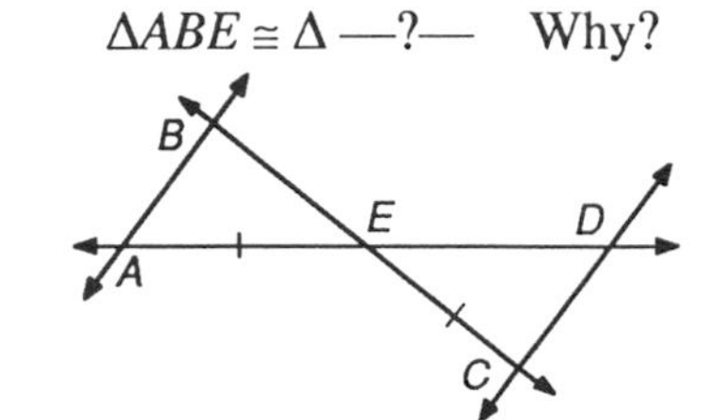

7. $\Delta MSE \cong \Delta$ —?—

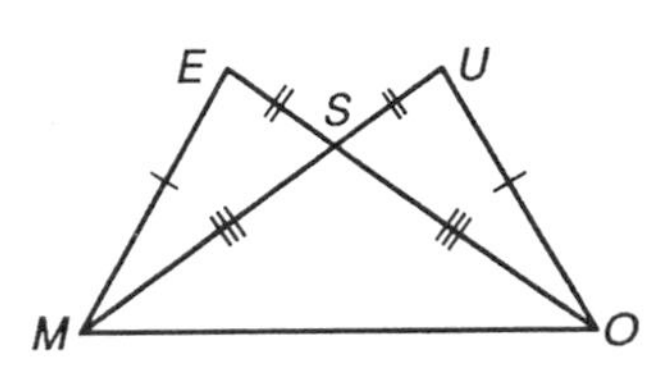

8. *HALF* is a parallelogram.
$\Delta HAF \cong \Delta$ —?—

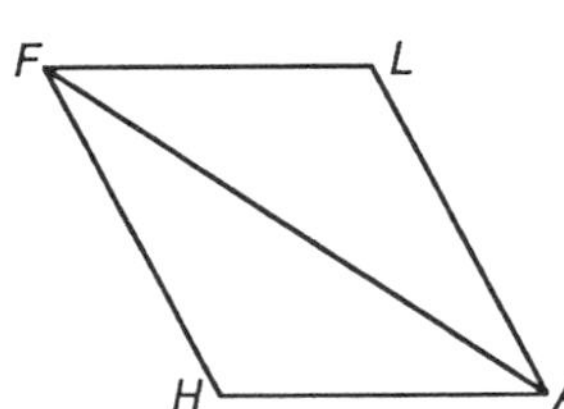

9. *SAEN* is a rectangle.
M is a midpoint.
$\Delta AMS \cong \Delta$ —?—
Why?

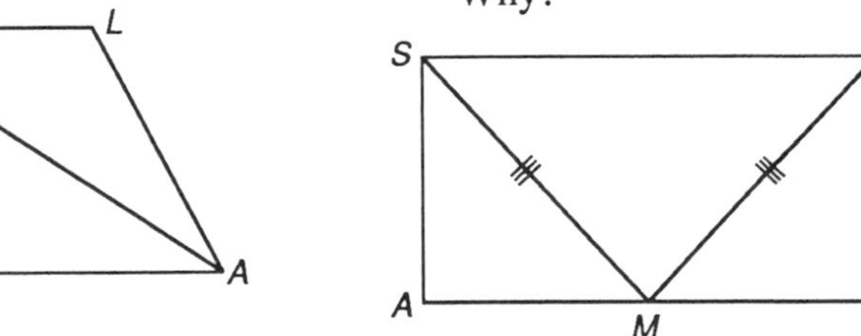

10.* $\Delta FSH \cong \Delta$ —?—
Why?

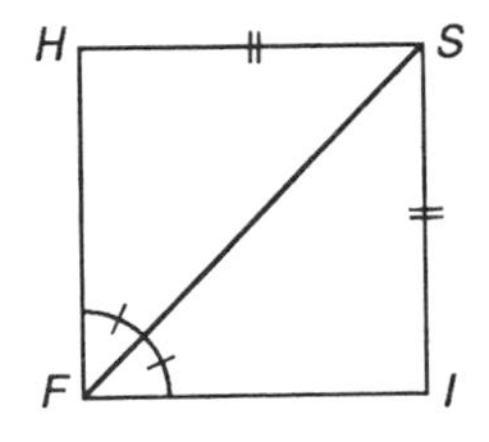

11. ΔARC is isosceles with $\angle C$ the vertex angle.
$\overline{CK}$ is a median.
$\Delta AKC \cong \Delta$ —?—
Why?

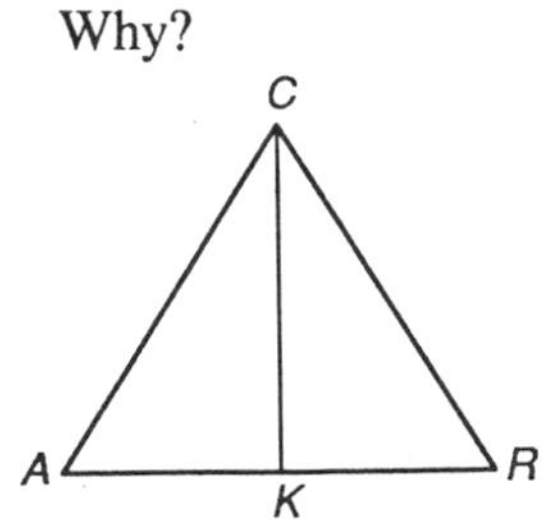

12. *PQRS* is a parallelogram.
$\Delta PQS \cong \Delta$ —?—
$\Delta PRQ \cong \Delta$ —?—
$\Delta PQM \cong \Delta$ —?—

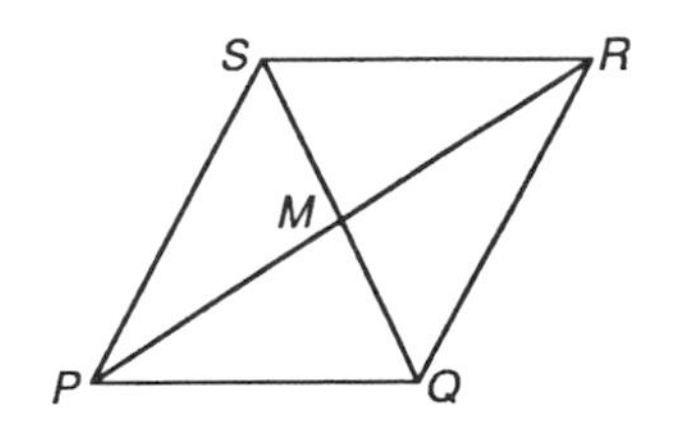

13.* $\Delta DOG \cong \Delta$ —?—
Why?

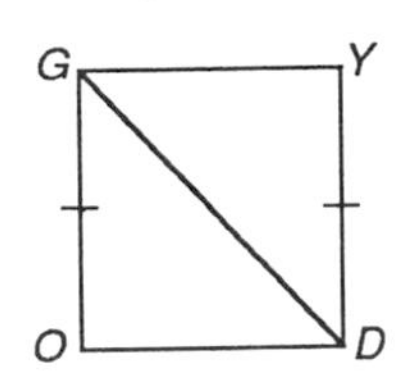

14. $\Delta CAT \cong \Delta$ —?—
Why?

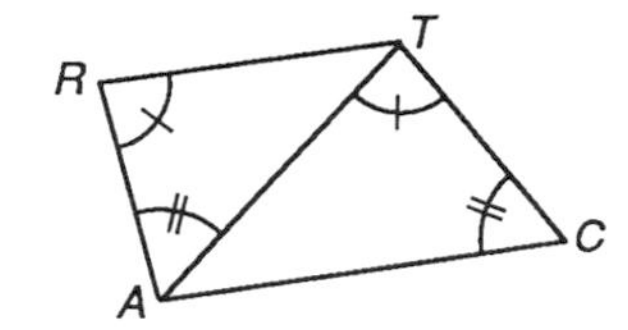

15. *CARD* is a parallelogram.
$\Delta RAM \cong \Delta$ —?—
$\Delta RAC \cong \Delta$ —?—
$\Delta RAD \cong \Delta$ —?—

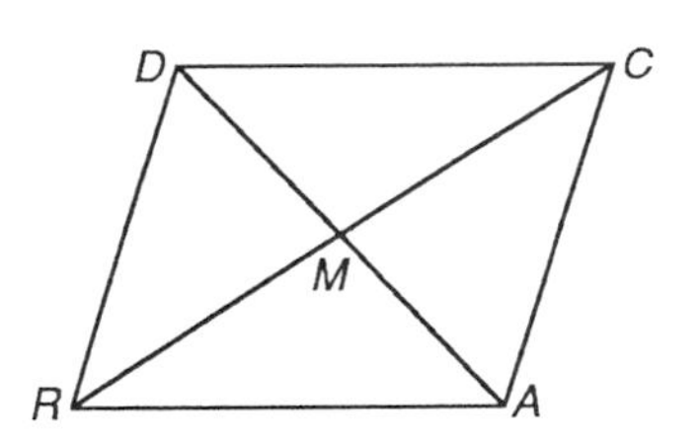

Improving Visual Thinking Skills

Folding Cubes I

In each of the three problems below, the figure on the left represents an unfolded cube. When the figure on the left is folded, which cube on the right will it become?

1. 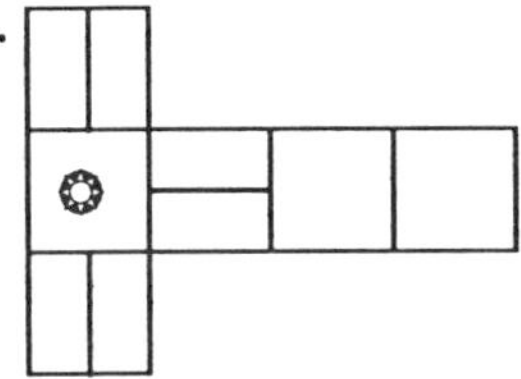

a. b. c.

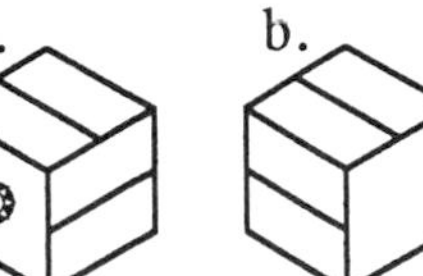

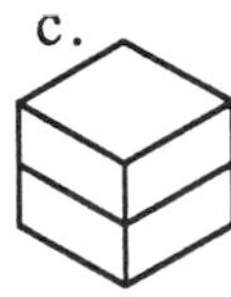

2. 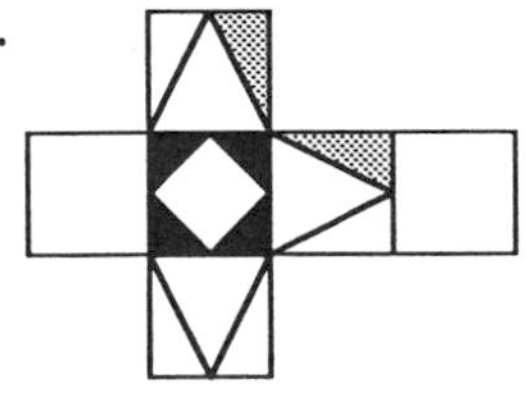

a. b. c.

3. 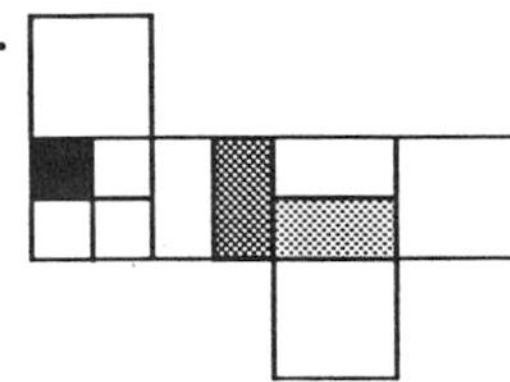

a. b. c.

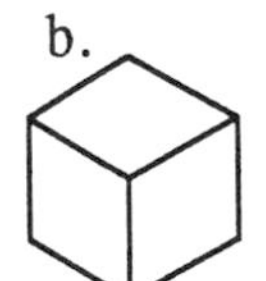

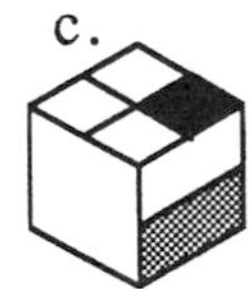

Lesson 5.3

SAS Shortcut? SSA Shortcut?

In the last lesson you discovered that if three sides of one triangle are congruent to the three sides of another triangle, then the two triangles are congruent (SSS Congruence Conjecture). You also found out that if three angles of one triangle are congruent to the three corresponding angles of another triangle, then the two triangles need not be congruent. In this lesson you will continue your search for triangle congruence shortcuts. You should continue to do the investigations in groups of four or five.

If two sides and an angle in one triangle are congruent to two sides and an angle in another triangle, must the two triangles be congruent? There are two possible arrangements of the sides and the angle that you have to consider. The angle might be between the sides, or the angle might not be between the sides.

Side-Angle-Side

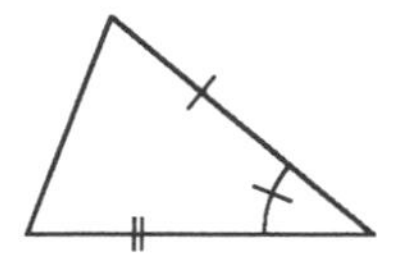

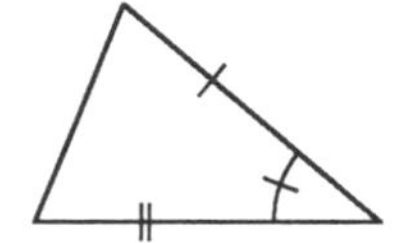

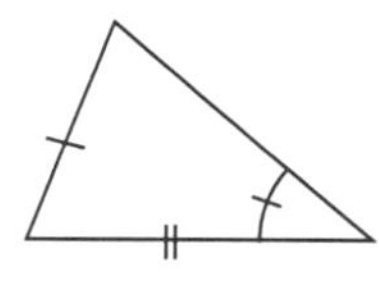

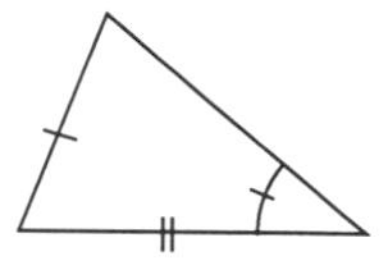

If two sides and the angle between them in one triangle are congruent to two sides and the angle between them in another triangle (SAS), must the two triangles be congruent?

Investigation 5.3.1

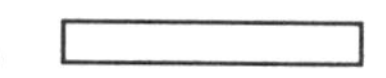

The two segments below are two sides of acute ΔNFL. The acute angle below is the angle between them.

Step 1: Construct a triangle on your own paper using these two segments as sides and the angle as the angle between them.

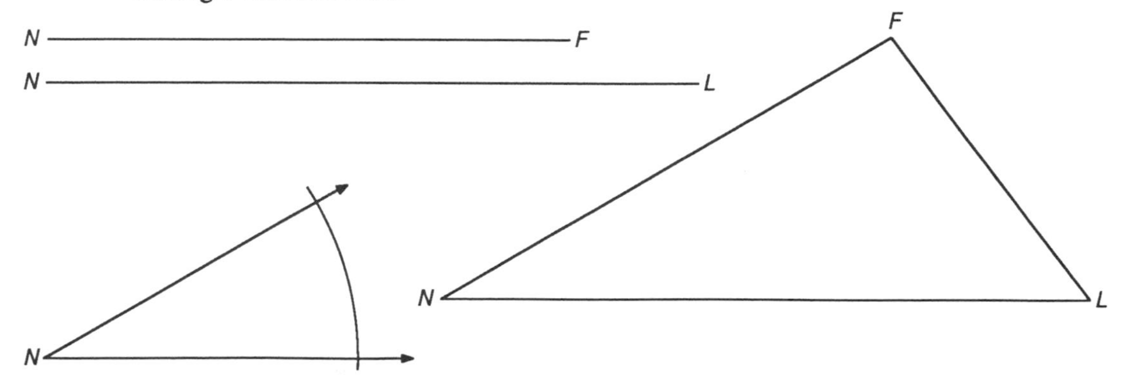

Step 2: Compare your triangle with ΔNFL and the triangles constructed by others in your group. Are they all congruent?

Step 3: Can you construct a triangle using the two segments and angle given above that is not congruent to ΔNFL? Try it. Construct a second triangle using these pieces that is *not* congruent to ΔNFL.

Step 4: Let's try this again. The two segments below are two sides of obtuse ΔAFC and the obtuse angle is the angle between them. Construct a triangle using these two segments as sides and the angle as the angle between them.

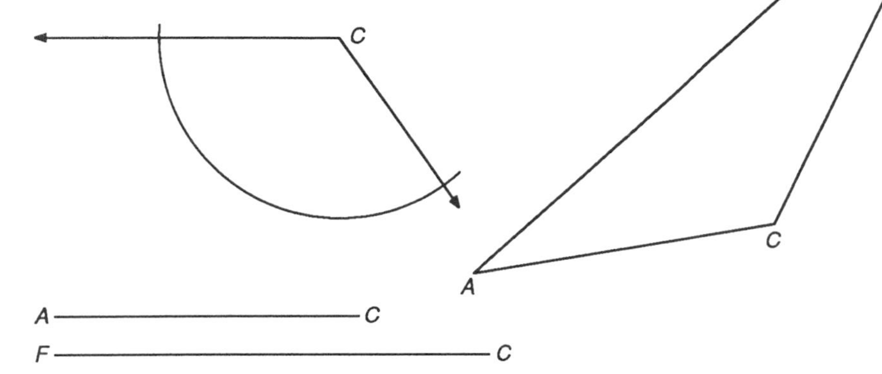

Step 5: Compare your triangle with ΔAFC and the triangles constructed by others in your group. Are they all congruent?

Step 6: Can you construct a triangle that uses the two sides and angle above that is not congruent to ΔAFC? Try it. Construct a second triangle using these pieces that is *not* congruent to ΔAFC.

You should be ready to state a conjecture.

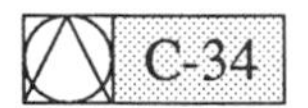

If two sides and the angle between them in one triangle are congruent to two sides and the angle between them in another triangle, then —?—. ***(SAS Congruence Conjecture)***

This is an important conjecture. Add it to your conjecture list. Beneath your conjecture, copy and complete the "picture statement" below.

If you know this:

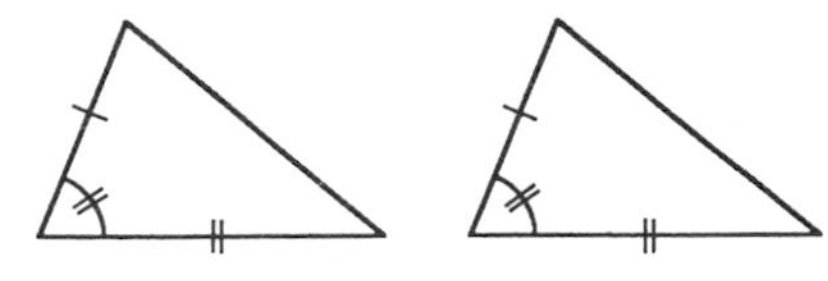

Then you also know this:

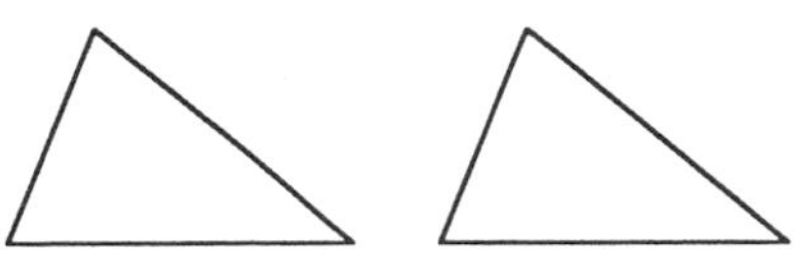

(You mark the drawing on your paper.)

If two sides and an angle not between them in one triangle are congruent to the corresponding two sides and an angle not between them in another triangle (SSA), must the two triangles be congruent?

Investigation 5.3.2

The two segments below are two sides of ΔTWO and the angle is an angle not between them.

Step 1: Construct a triangle using these two segments as sides and the angle as an angle that is not between the given sides.

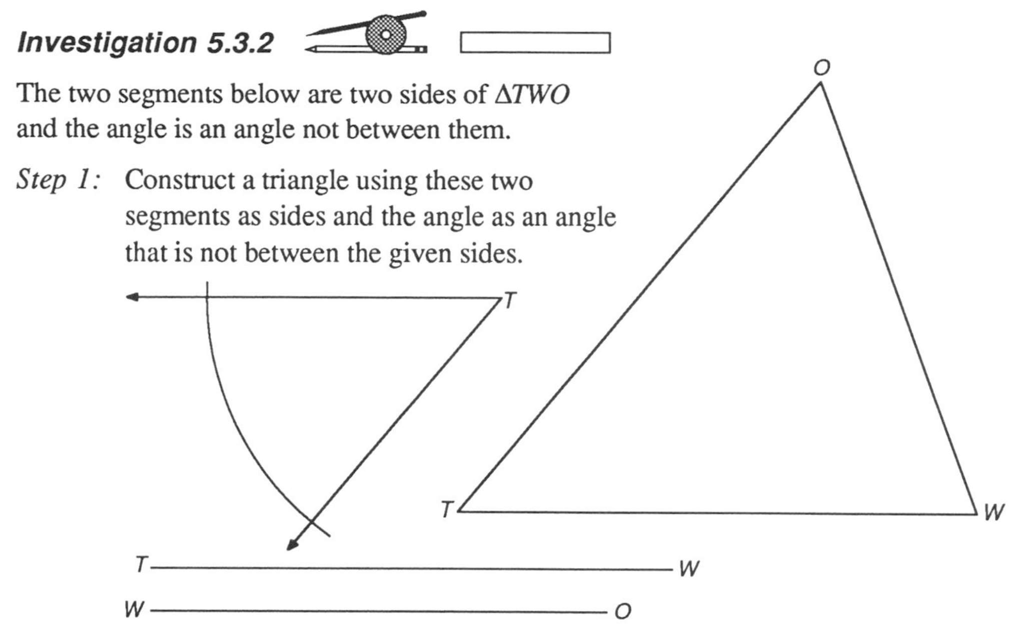

Step 2: Compare your triangle with the triangles created by others in your group. Are they all congruent? Are some different? Beneath the construction you made for this investigation, complete and copy the conjecture below.

Conjecture: *If two sides and an angle that is not between the two sides in one triangle are congruent to the corresponding two sides and an angle that is not between the two sides in another triangle, then the two triangles (are/are not) necessarily congruent.*

EXERCISE SET

From the information given, complete each statement. If the triangles cannot be shown to be congruent from the information given, write *cannot be determined.*

1.* What conjecture tells you that ΔLUZ is congruent to ΔIDA?

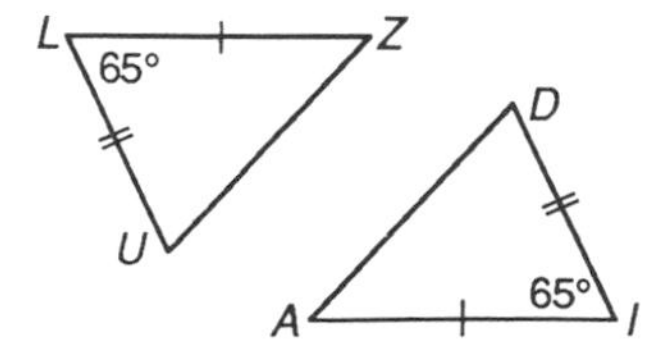

2. What conjecture tells you that ΔCAV is congruent to ΔCEV?

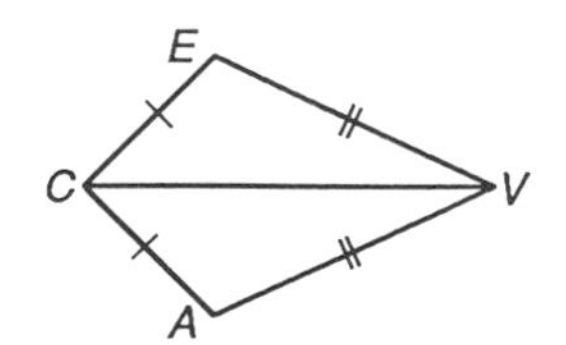

3. What conjecture tells you that ΔARC is congruent to ΔERN?

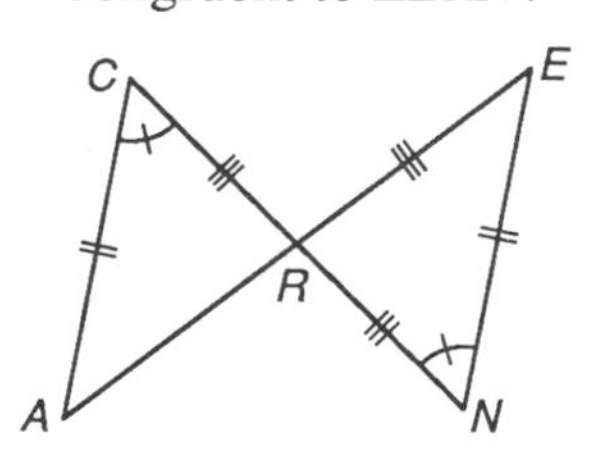

4. $\Delta COT \cong \Delta$ —?—

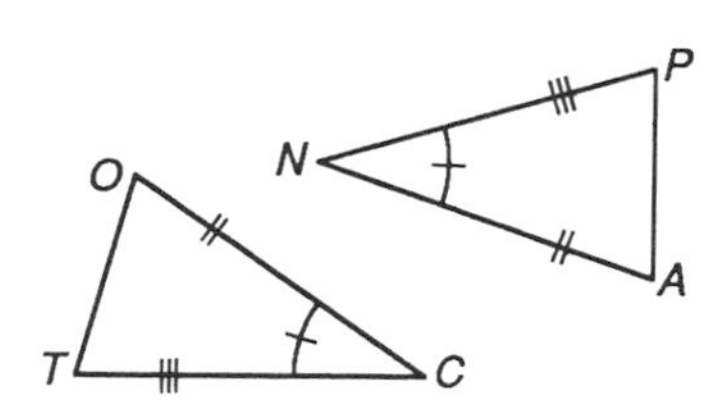

5. $\Delta SAT \cong \Delta$ —?—

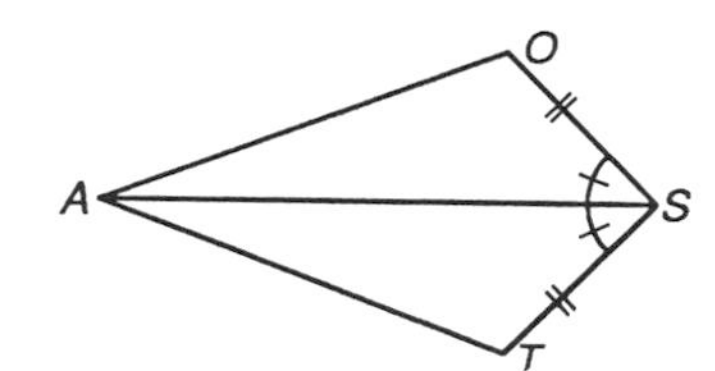

6. $\overline{AT}$ is an angle bisector. $\Delta LAT \cong \Delta$ —?—

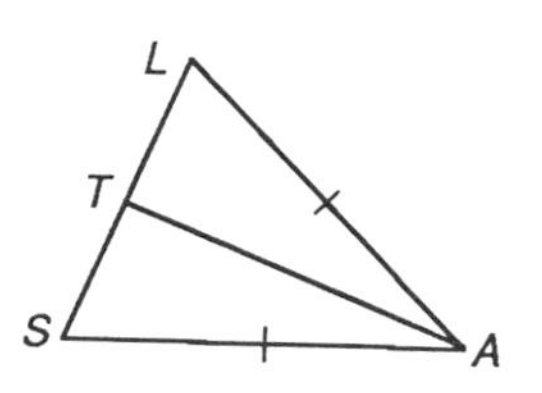

7. M is the midpoint of both $\overline{WN}$ and $\overline{OE}$. $\Delta MEN \cong \Delta$ —?—

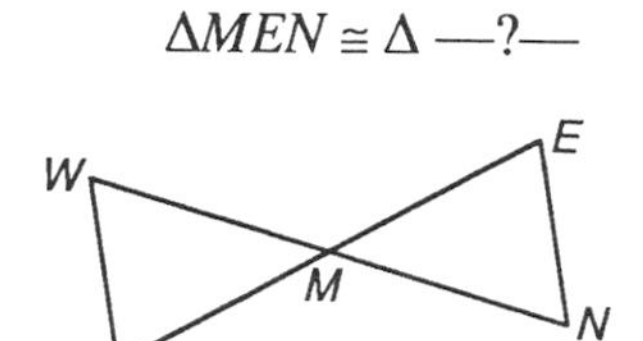

8. $\Delta PDQ \cong \Delta$ —?—

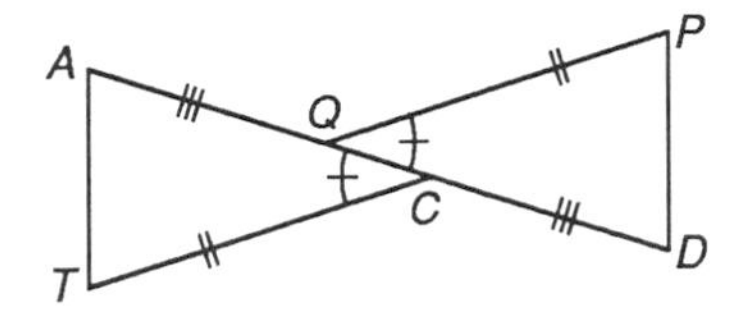

9. $\overline{AM}$ is a median. $\Delta CAM \cong \Delta$ —?—

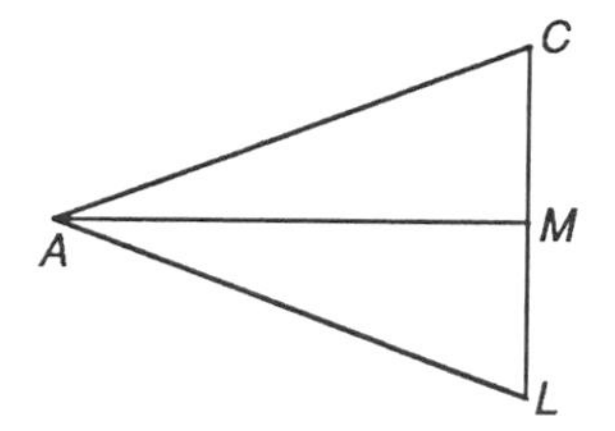

10. $\Delta RAT \cong \Delta$ —?—
Why?

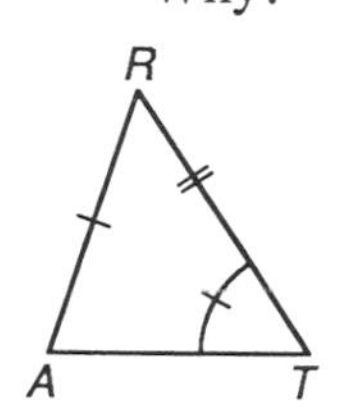

11. $\Delta TRP \cong \Delta$ —?—
Why?

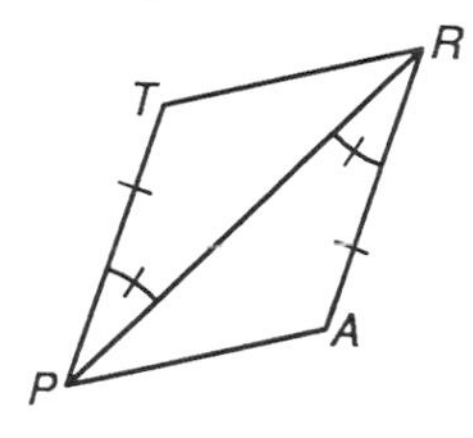

12. $\Delta CHZ \cong \Delta$ —?—
Why?

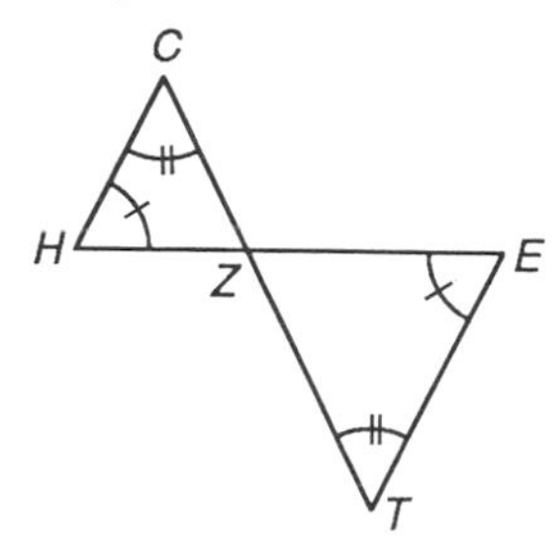

13.* Why is $\angle CGH$ congruent to $\angle NGI$? Why is $\overline{HG}$ congruent to $\overline{IG}$? Why is ΔCGH congruent to ΔNGI?

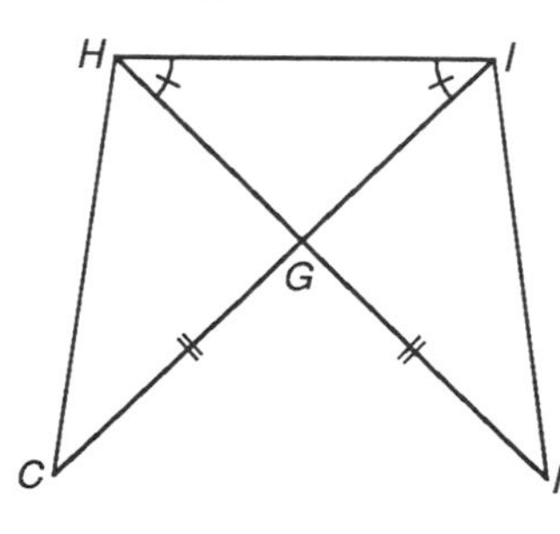

14. *CARBON* is a regular hexagon. $\Delta ACN \cong \Delta$ —?—

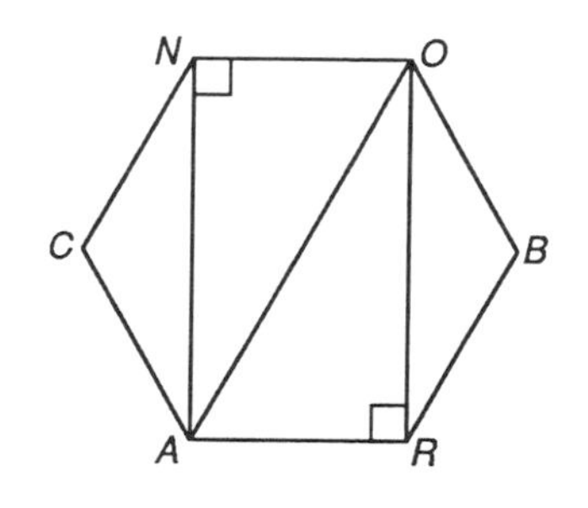

15. M is the midpoint of $\overline{JA}$. *JADE* is an isosceles trapezoid. $\Delta MAD \cong \Delta$ —?—

Lesson 5.4

ASA Shortcut? SAA Shortcut?

In this lesson you will investigate the remaining possibilities for triangle congruence shortcuts. You should continue to do the investigations in groups of four or five. You will work with members of your group to see if two angles and a side lead to a shortcut. Again, there are two cases. You could be given two angles and the side between them, or you could be given two angles and a side not between them.

Angle-Side-Angle ***Side-Angle-Angle***

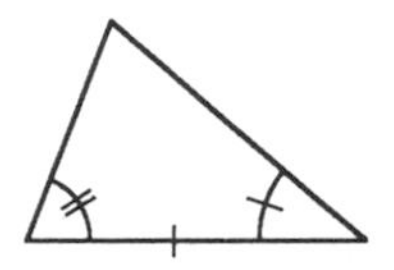
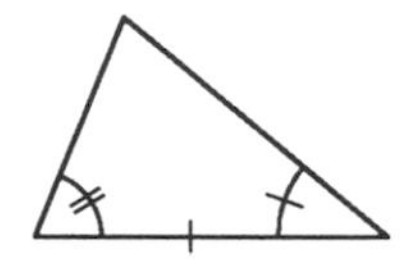
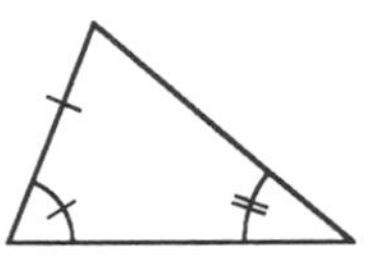
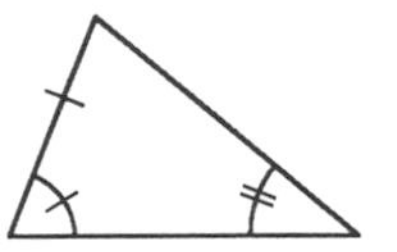

If two angles and the side between them in one triangle are congruent to two angles and the side between them in another triangle (ASA), must the two triangles be congruent? Let's investigate.

Investigation 5.4.1

The two angles below are two angles of acute ΔMTV and the segment is the side between them.

Step 1: Construct a triangle using these two angles and the segment as the side between them.

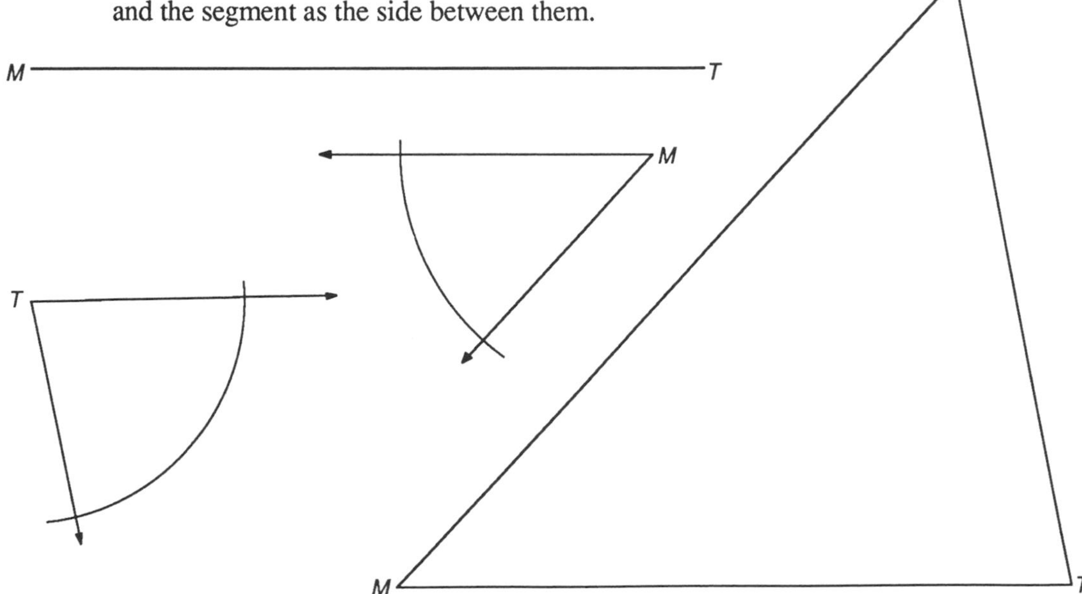

Step 2: Compare your triangle with the triangles constructed by others in your group. Are they congruent?

Step 3: Can you construct a triangle using the two angles and segment that is not congruent to ΔMTV? Try it. Construct a second triangle using these pieces that is *not* congruent to ΔMTV. Could anyone in your group construct a different triangle?

Step 4: Try again with the parts of obtuse ΔVHI. The two angles below are two angles of obtuse ΔVHI and the segment below is the side between them. Construct a triangle using these two angles and the segment as the side between them.

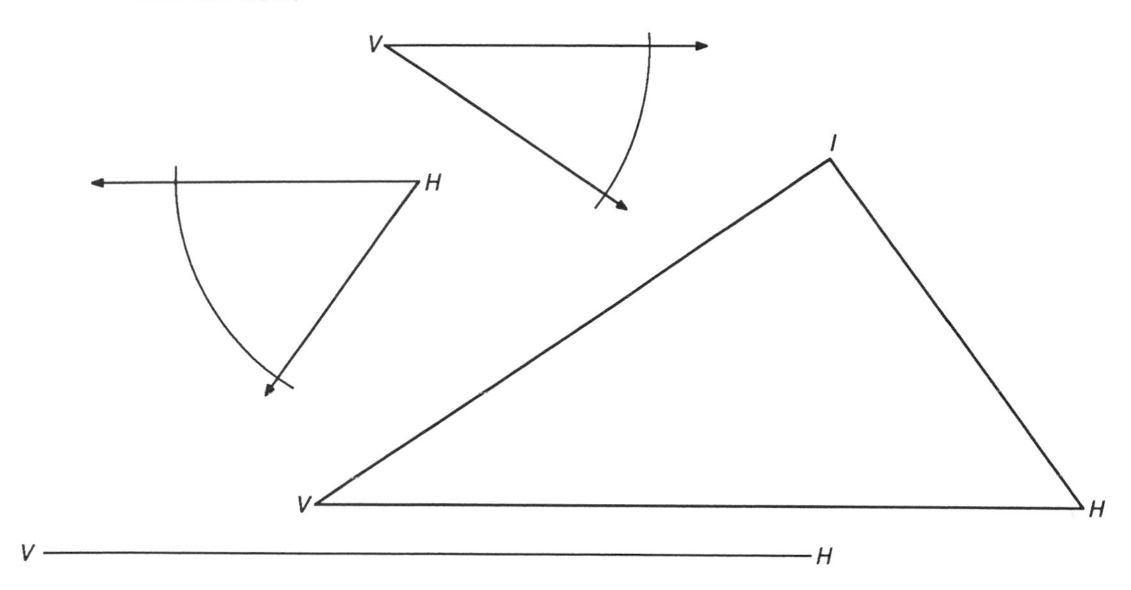

Step 5: Compare your triangle with ΔVHI and the triangles constructed by others in your group. Are they all congruent?

Step 6: Can you construct a triangle that uses the two angles and side above that is not congruent to ΔVHI? Try it. Construct a second triangle using these pieces that is *not* congruent to ΔVHI.

You should be ready to state a conjecture.

C-35 *If two angles and the side between them in one triangle are congruent to —?—. **(ASA Congruence Conjecture)***

Add this new conjecture to your conjecture list. Beneath your conjecture, copy and complete the "picture statement" below.

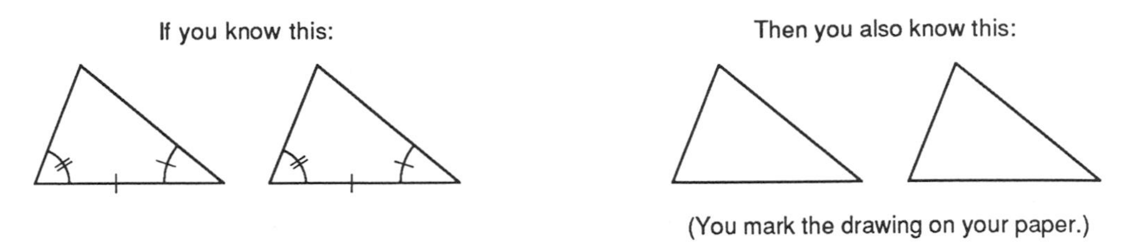

What if the given side is not between the two given angles? If two angles and a side not between them of one triangle are congruent to the corresponding two angles and side of another triangle (SAA), must the two triangles be congruent?

Investigation 5.4.2

The two angles below are two angles of acute ΔTBS and the segment is a side not between them.

Step 1: Construct a triangle using these two angles and the segment as a side not between them. (If you need the third angle, construct the sum of the two given angles and what remains of 180 is the third angle.)

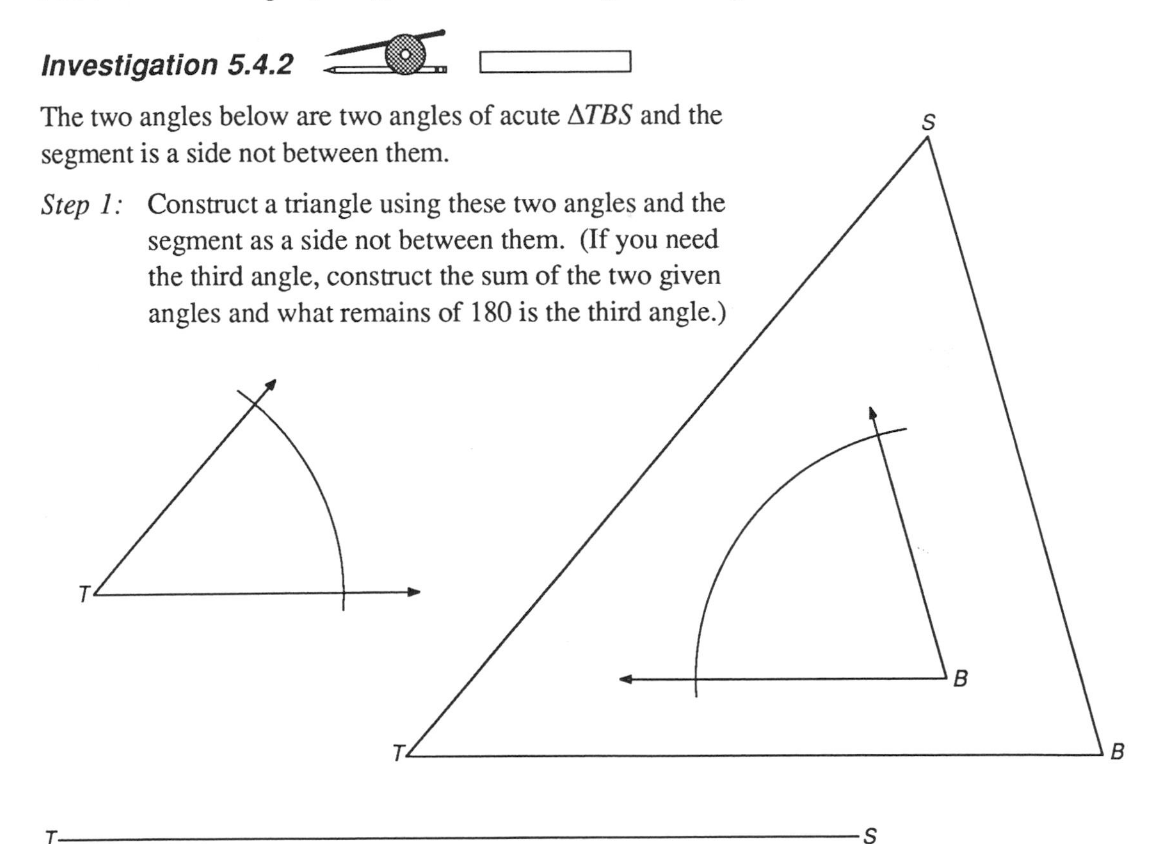

Step 2: Compare your triangle with ΔTBS and with the triangles constructed by others in your group. Are they all congruent?

State a conjecture.

C-36 *If two angles and a side that is not between them in one triangle are congruent to the corresponding —?—. **(SAA Congruence Conjecture)***

The SAA Congruence Conjecture can be demonstrated with a little logical reasoning. In the two triangles below, two angles and a side of one triangle are congruent to two angles and a side in the other triangle. What about the third pair of angles?

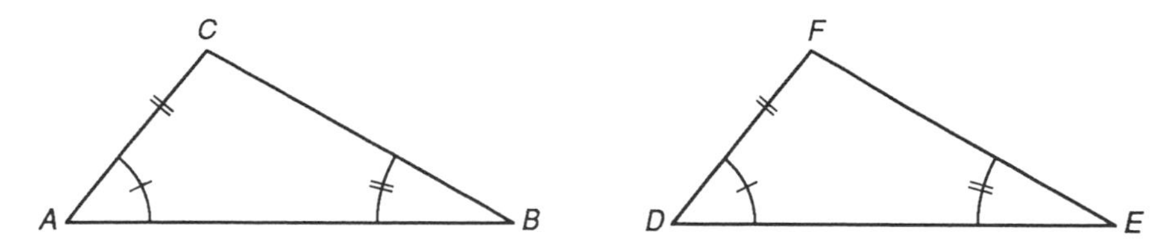

If $\angle A \cong \angle D$ and $\angle B \cong \angle E$, then $\angle C \cong \angle F$.

The Third Angle Conjecture tells you that the third pair of angles are also congruent. If the third pair of angles are congruent, then the two triangles are congruent by ASA. Therefore, every SAA congruence is really just a special case of ASA. However, for convenience, we'll consider them as two separate conjectures.

Add this new conjecture to your conjecture list. Beneath your conjecture, copy and complete the "picture statement" below.

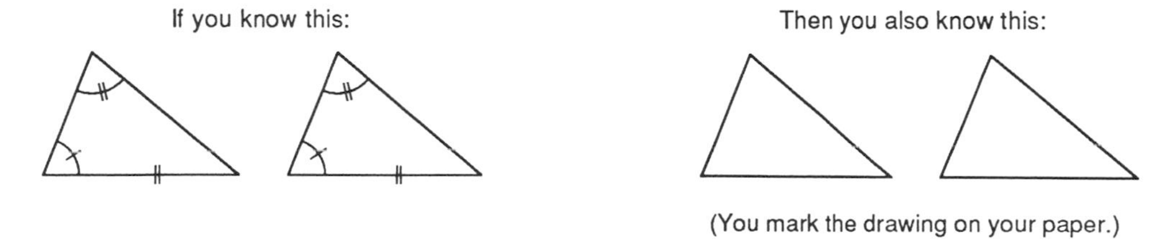

You now have SSS, SAS, ASA, and SAA as shortcuts to demonstrate that two triangles are congruent.

EXERCISE SET A

From the information given, complete each statement. If the triangles cannot be shown to be congruent from the information given, write *cannot be determined.*

1. Which conjecture tells you that ΔBOX is congruent to ΔCAR?

2. Which conjecture tells you that ΔKOP is congruent to ΔOKR?

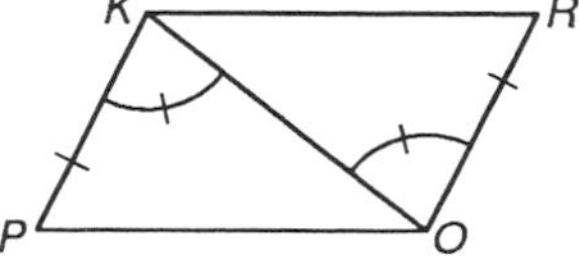

3. Which conjecture tells you that ΔFAD is congruent to ΔFED?

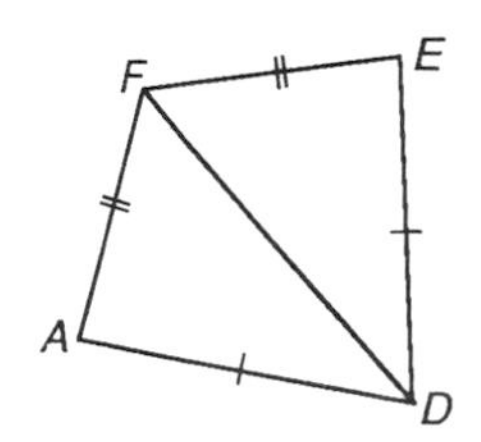

4.* Which conjecture tells you that ΔGAS is congruent to ΔIOL?

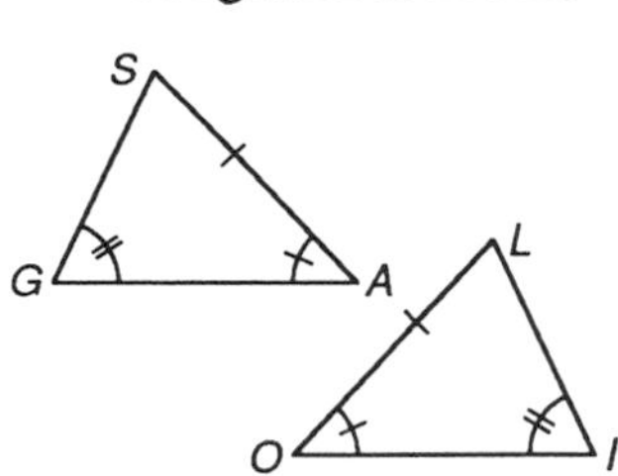

5. $\Delta WOM \cong \Delta$ —?—
Why?

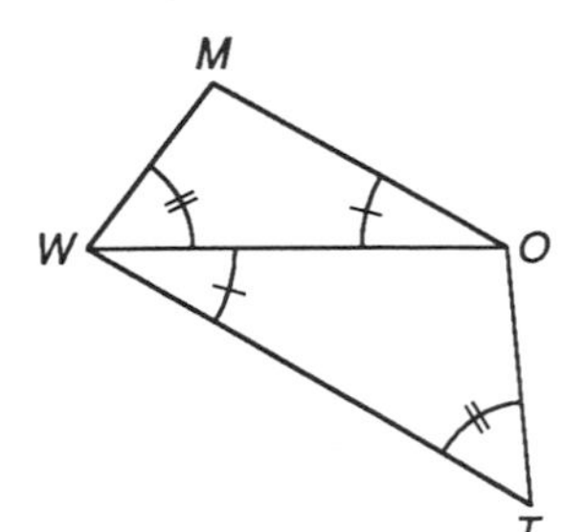

6. $\Delta BLK \cong \Delta$ —?—
Why?

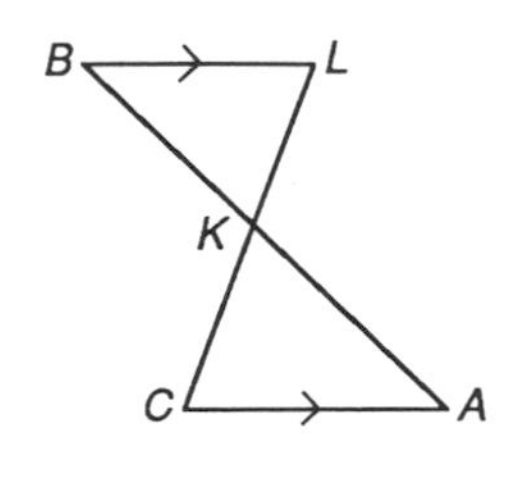

7.* $\Delta MAN \cong \Delta$ —?—
Why?

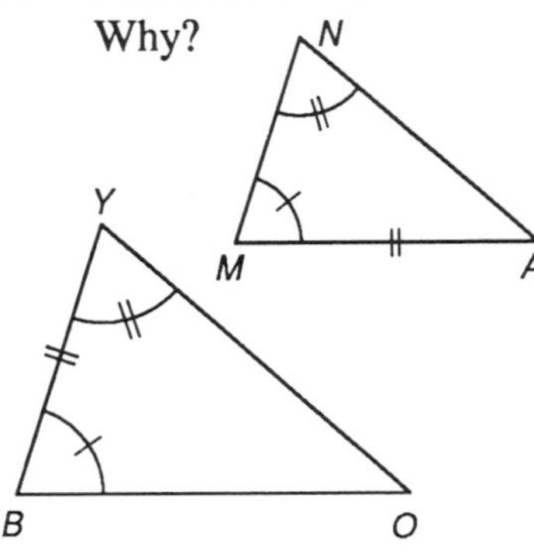

8. $\overline{AW} \perp \overline{WL}$, $\overline{WL} \perp \overline{KL}$
$\Delta LAW \cong \Delta$ —?—
Why?

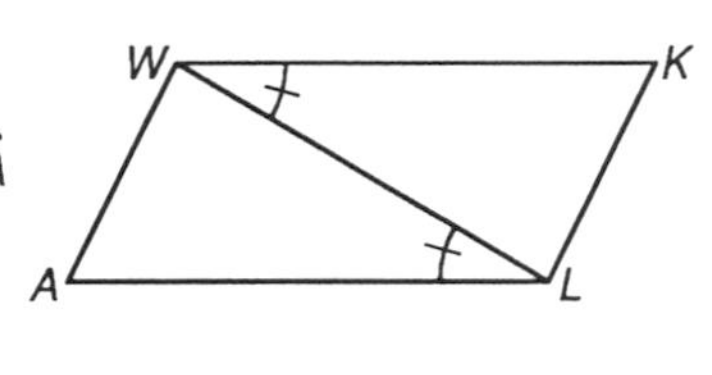

9.* $\overline{PO} \perp \overline{OI}$
$\Delta POL \cong \Delta$ —?—
Why?

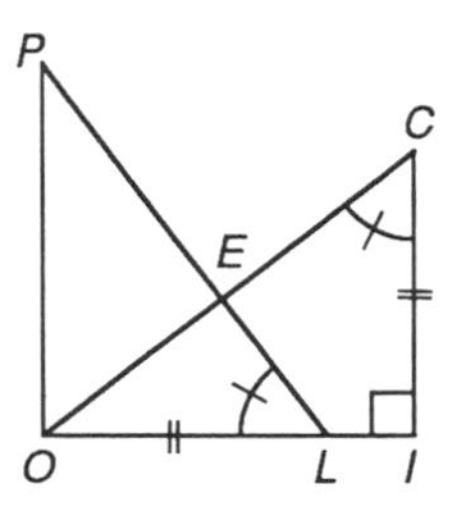

10. $\overline{HO} \parallel \overline{YT}$
$\Delta HOW \cong \Delta$ —?—
Why?

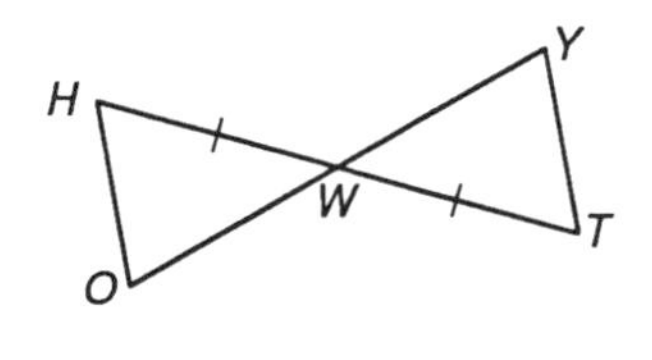

11. $\Delta CAE \cong \Delta$ —?—
Why?

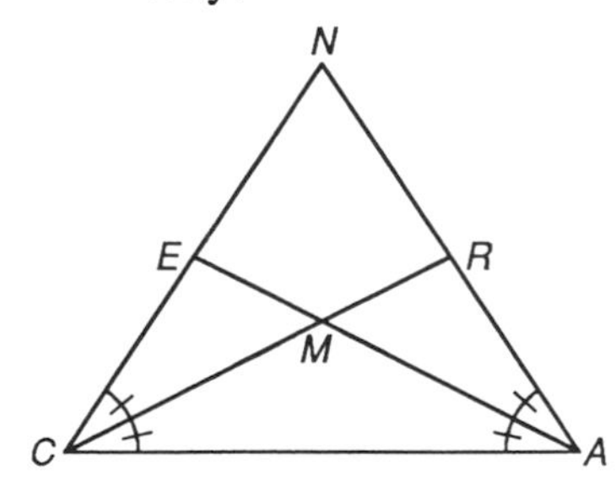

12. $\Delta PTR \cong \Delta$ —?—
Why?

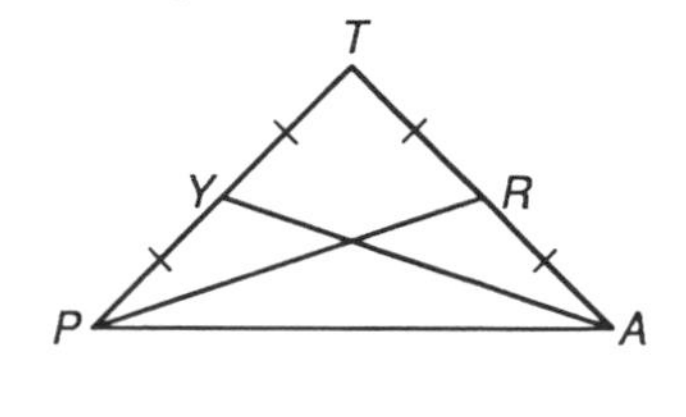

13. $PO = PR$
$\Delta POE \cong \Delta$ —?—
$\Delta SON \cong \Delta$ —?—

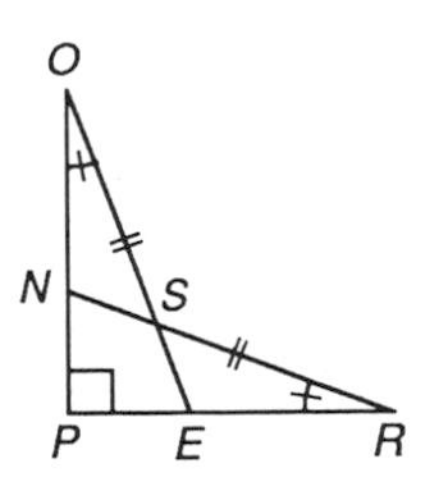

14. ΔSLN is equilateral.
Is ΔTIE equilateral?

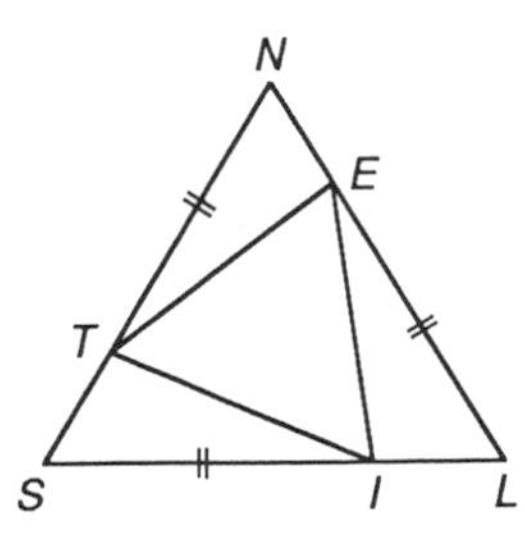

15. Δ —?— $\cong \Delta$ —?—
Why?

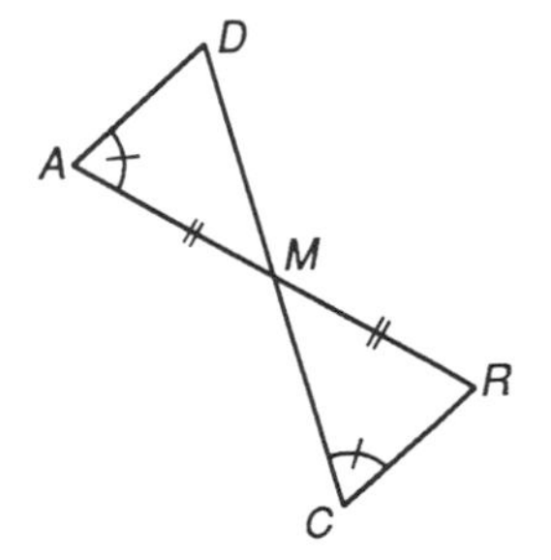

EXERCISE SET B

1. Construct a triangle on the top half of a sheet of paper. On the bottom half of your paper, use the SSS shortcut to construct a second triangle congruent to the first.

2. On another page, use the ASA shortcut to construct two congruent triangles.

3. Use the SAS shortcut to construct two congruent triangles.

4.* Construct two triangles that are *not* congruent even though the three angles of one triangle are congruent to the three angles of the other triangle.

5.* Construct two triangles that are *not* congruent even though two sides and an angle not between them in one triangle are congruent to two sides and an angle not between them in the other triangle.

Improving Reasoning Skills

Reasonable ’rithmetic II

Each letter in these problems represents a different digit.

1. What is the value of *C*?
2. What is the value of *D*?
3. What is the value of *K*?
4. What is the value of *N*?

$$\begin{array}{rrrrr} & 8 & 7 & 8 & 9 \\ & 3 & B & A & 7 \\ & 4 & 8 & 2 & A \\ + & 7 & A & B & 5 \\ \hline 2 & C & 2 & 8 & 7 \end{array}$$

$$\begin{array}{rrrrr} & D & E & F & F \\ - & E & 2 & F & 6 \\ \hline & 1 & 9 & 9 & 7 \end{array}$$

$$\begin{array}{r|rrr} & & G & J \\ \hline 7 & H & G & K \\ & 2 & 1 & \\ \hline & & H & K \\ & & H & K \\ \hline \end{array}$$

$$\begin{array}{r|rrr} & & 5 & 2 \\ \hline L\,Q & N & M & 2 \\ & N & P & \\ \hline & & M & 2 \\ & & M & 2 \\ \hline \end{array}$$

Special Project

Buried Treasure

Only the most foolish of mice would hide in a cat's ear.
But only the wisest of cats would think to look there.

– Andrew Mercer

Trace the outline of the island that appears on the treasure map and the points A, B, D, H, M, and S onto a clean sheet of paper. On your tracing, find the locations of the buried treasures.

1.* Pirate Alphonse is standing at the edge of Westend Bay (A), and his cohort, pirate Beaumont, is 60 paces to the north at point B. Each pirate can see Captain Coldhart off in an easterly direction burying a treasure. With his sextant, Alphonse measures the angle between the Captain and Beaumont and finds that it's 85°. With his sextant, Beaumont measures the angle between the treasure and Alphonse and finds it to be 48°. Alphonse and Beaumont mark their positions with large boulders and return to their ship, confident that they have enough information to return later and recover the treasure.

   Can they recover the treasure? Which conjecture (SSS, SAS, ASA, or SAA) guarantees they'll be able to find it? If it is possible, use your geometric tools to locate the position of the treasure on the map. Mark it with an X.

2. Captain Coldhart, convinced someone in his crew stole his last treasure, has decided to be more careful about burying his latest booty. He gives his trusted first mate, Dexter, two ropes, the lengths of which only the Captain knows. The Captain instructs Dexter to nail one end of the shorter rope to Hangman's Tree (H) and secure the longer rope through the eyes of Skull Rock (S). The Captain, with the ends of the two ropes in one hand and the treasure chest tucked under the other arm, walks away from the shore to the point where the two ropes become taut. The Captain buries the treasure at the point where the two ropes come together, collects his ropes, and returns confidently to the ship.

   Has the Captain given himself enough information to recover the treasure? Which conjecture (SSS, SAS, ASA, or SAA) insures the uniqueness of the location? Locate the position of this second treasure on the map if the two ropes are 100 paces and 150 paces in length. Mark the treasure location on your map.

3. After the theft of two very important ropes from his locker and the subsequent disappearance of his trusted first mate, Dexter, Captain Coldhart is determined that no one shall find the location of his latest buried treasure. The Captain, with his new first mate, Endersby, walks out to Deadman's Point (D). The Captain instructs his first mate to walk inland along a straight path for a distance of 130 paces. There Endersby is to drive his sword into the ground for a marker (M), turn and face in the direction of the captain, turn at an arbitrary

angle to the left, continuing to walk for another 60 paces, stop, and wait for the Captain. The Captain measures the angle formed by the lines from Endersby to himself and from himself to the sword. The angle measures 26°. The Captain places a boulder where he is standing (*D*), walks around the bay to Endersby, and instructs him to bury the treasure at this point. Alas, poor Endersby is buried with the treasure. Has the Captain given himself enough information to locate the treasure? If he has, determine the unique location of the treasure? If he does not have enough information, explain why not? How many possible locations for the treasure are there? Find them on the map.

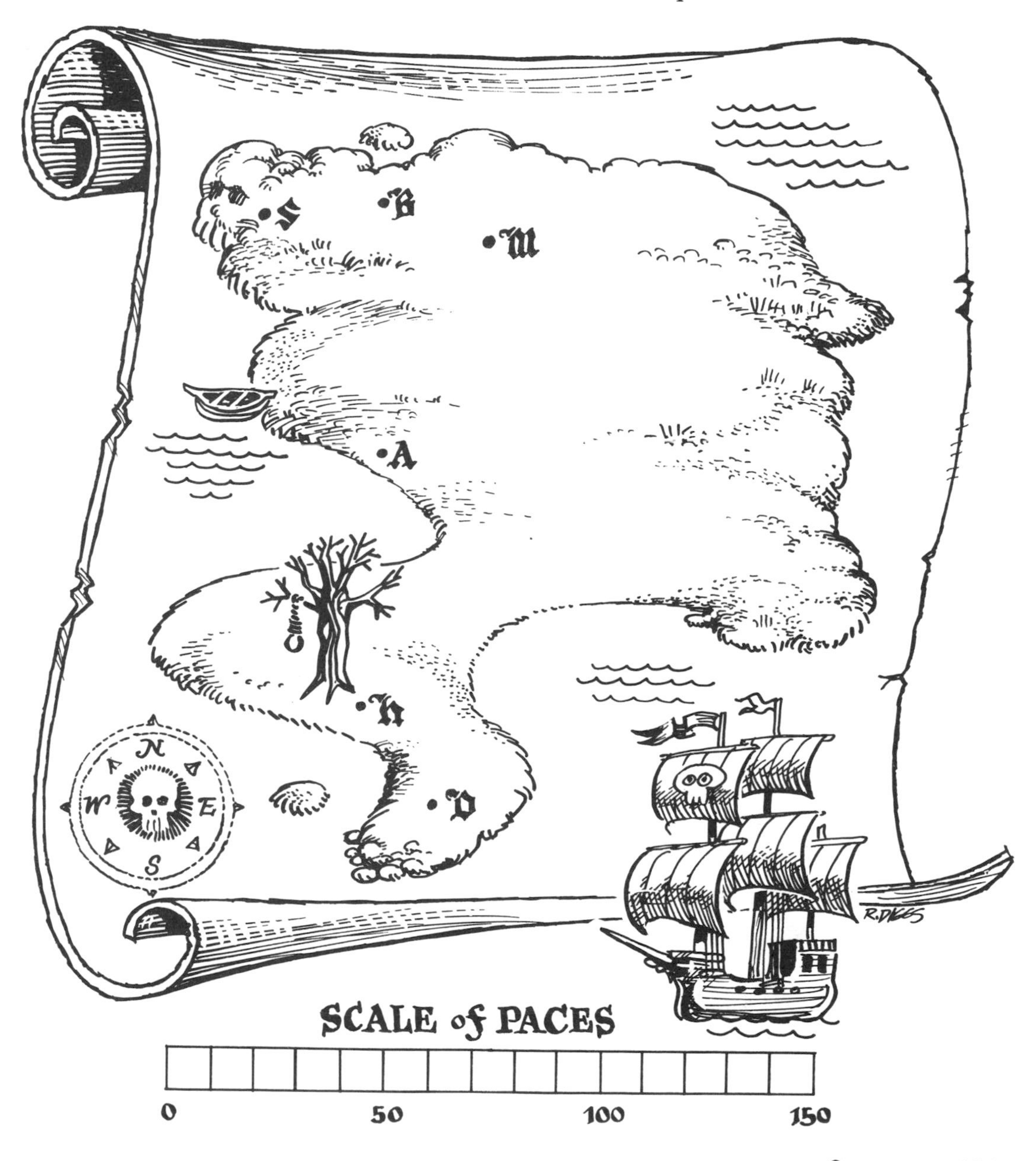

Lesson 5.5
CPCTC

Nature's Great Book is written in mathematical symbols.

– Galileo

In the previous lessons you discovered four shortcuts (SSS, SAS, ASA, and SAA) for showing that two triangles are congruent. A common technique for showing that two segments or two angles are congruent is first to show that they are corresponding parts of congruent triangles.

The definition of congruent polygons states: If two triangles are congruent, then the **c**orresponding **p**arts of those **c**ongruent **t**riangles are **c**ongruent. We'll use the letters CPCTC to refer to this part of the definition. Let's see how you can use congruent triangles and CPCTC to show two segments are congruent.

Example

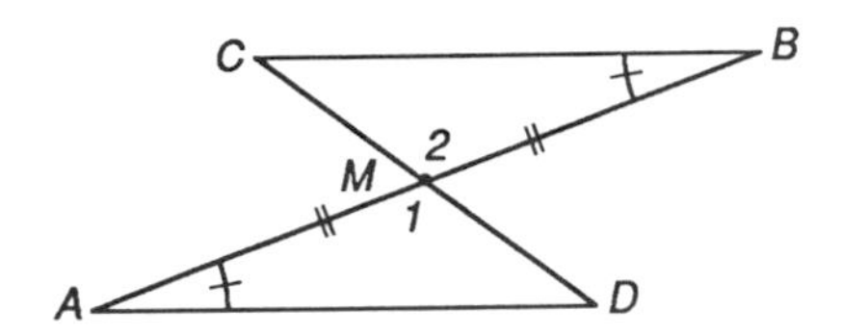

Is $\overline{AD} \cong \overline{BC}$? If you think so, you are correct. But can you give a logical argument showing that they are congruent? Here is one possible logical argument:

Since $\angle A \cong \angle B$, $\overline{AM} \cong \overline{BM}$, and $\angle 1 \cong \angle 2$, then $\Delta AMD \cong \Delta BMC$ by ASA Congruence Conjecture. If $\Delta AMD \cong \Delta BMC$, then $\overline{AD} \cong \overline{BC}$ by CPCTC.

EXERCISE SET

Copy the diagrams onto your paper and mark them with the given information. Provide a short argument that demonstrates whether or not the segments or angles indicated are congruent. If it is necessary to show that two triangles are congruent, then state which triangles are congruent and which shortcut conjecture (SSS, SAS, ASA, or SAA) proves them congruent.

1.* $\overline{AR} \cong \overline{ER}$
$\overline{EC} \cong \overline{CA}$
Is $\angle E \cong \angle A$?
Why?

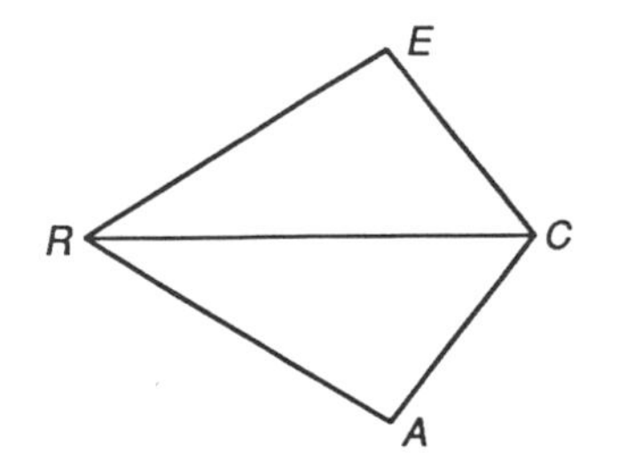

2. $\overline{SE} \cong \overline{SU}$
$\angle E \cong \angle U$
Is $\overline{MS} \cong \overline{OS}$?
Why?

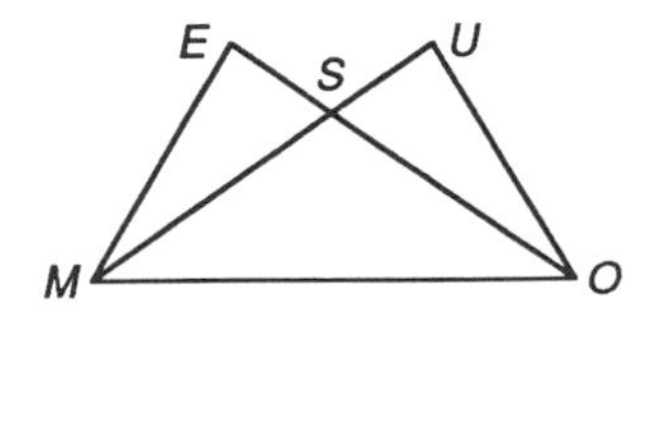

3.* $\overline{SA} \parallel \overline{NE}$
$\overline{SE} \parallel \overline{NA}$
Is $\overline{SA} \cong \overline{NE}$?
Why?

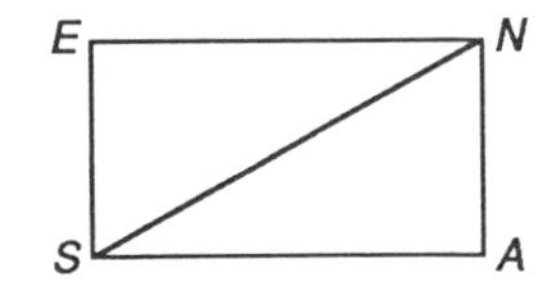

4. $\angle E \cong \angle W$
M is the midpoint of $\overline{WE}$.
Is $\overline{MO} \cong \overline{MN}$? Why?

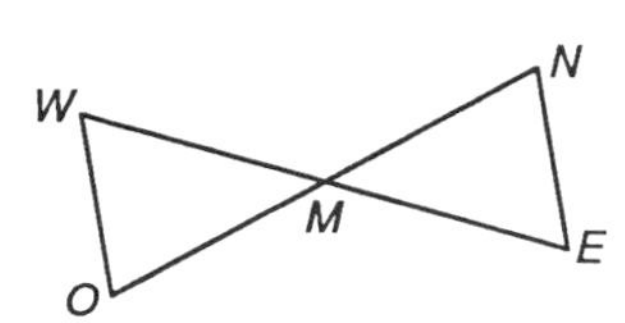

5.* $\overline{CS} \cong \overline{HR}$
$\angle 1 \cong \angle 2$
Is $\overline{CR} \cong \overline{HS}$? Why?

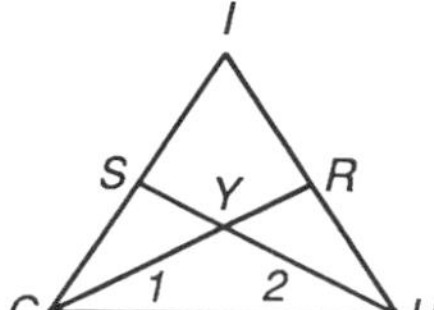

6. $\overline{MN} \cong \overline{MA}$
$\overline{ME} \cong \overline{MR}$
Is $\angle E \cong \angle R$? Why?

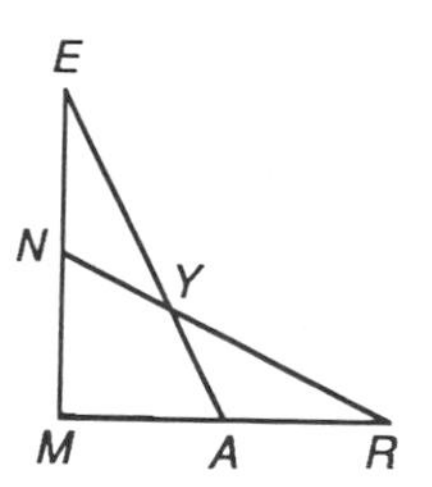

7. $\angle S \cong \angle T$
$\angle R \cong \angle A$
$\overline{RE} \cong \overline{AE}$
Is $\overline{RT} \cong \overline{SA}$? Why?

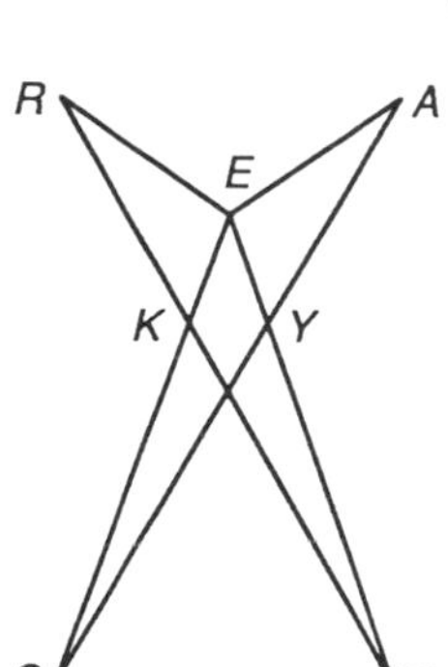

8.* $\overline{FO} \cong \overline{FR}$
$\overline{OU} \cong \overline{UR}$
Is $\angle O \cong \angle R$? Why?

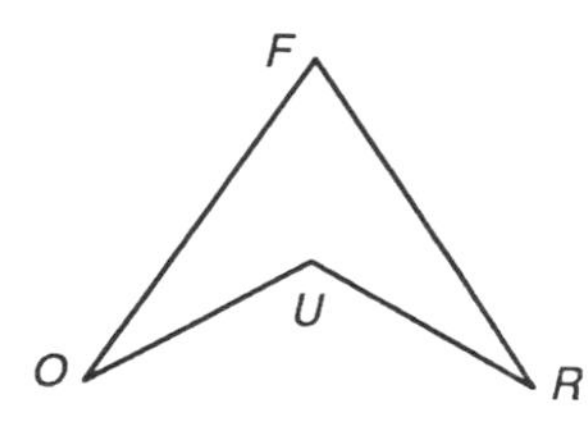

9.* $\overline{BT} \cong \overline{UE}$
$\overline{BU} \cong \overline{TE}$
Is $\angle B \cong \angle E$? Why?

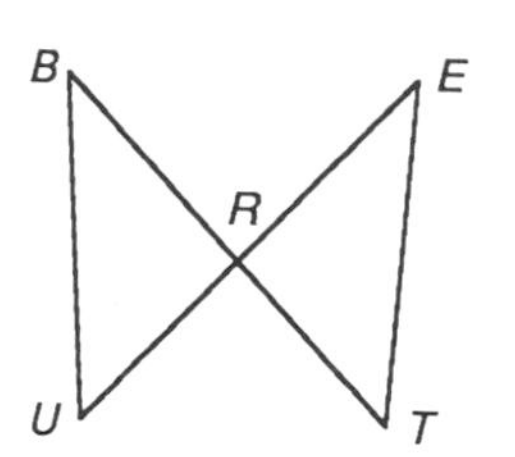

Improving Visual Thinking Skills

Design II

Use your geometry tools to create a larger version of the design on the right.

Then decorate your design.

Lesson 5.6

Flowchart Thinking

In Chapter 2 you used concept maps to visualize the relationships between sets of polygons. A concept map can also help to plan and visualize logical thinking. A **flowchart** is a concept map that shows a step-by-step procedure through a complicated system. Actions are represented in boxes. Arrows connect the boxes to show the flow of the action. Computer programmers use flowcharts to plan the logic in programs.

You will use flowcharts to demonstrate that what you have discovered through inductive reasoning is logically true. Flowcharts can make your logical thinking visible and help others to follow your reasoning. The example below is a flowchart showing a logical argument for Exercise 5.5 A1. A logical argument presented in the form of a flowchart is called a **flowchart proof**.

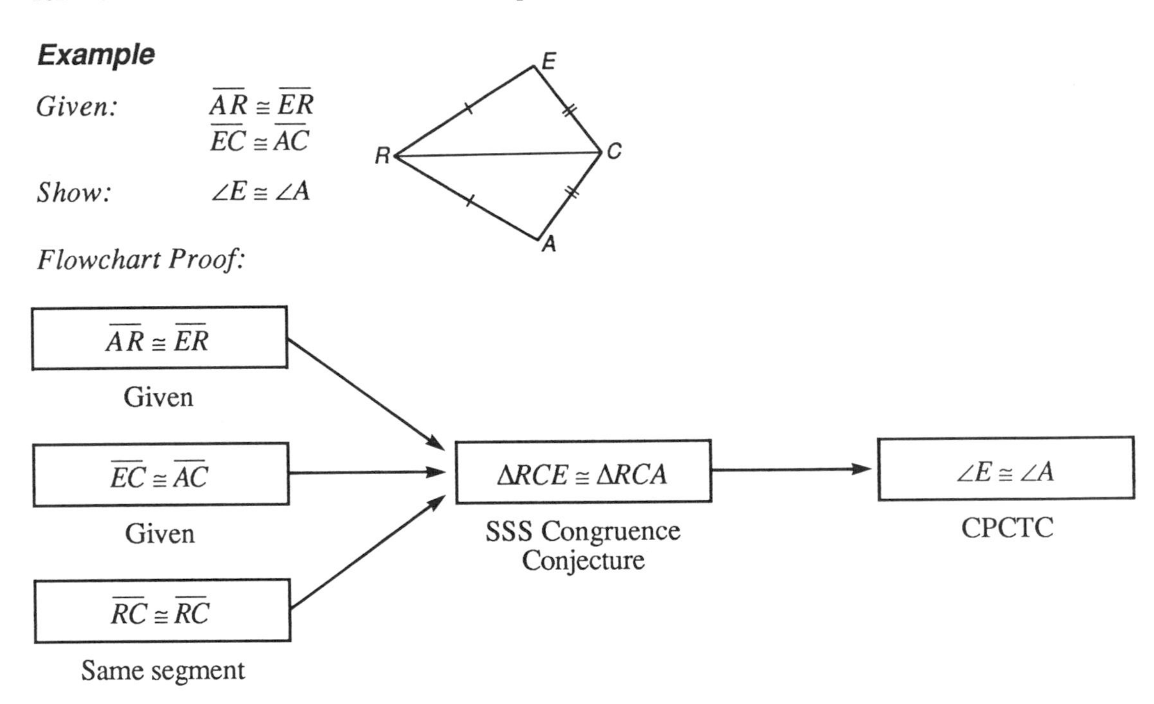

In the example above, you saw that the logical argument flows according to the arrows from the information that you are given to the conclusion you are trying to demonstrate. The logical reason supporting each statement is written beneath its box.

Of course, in an argument such as this, your proof is only as good as the conjectures you use to make your argument. The proof demonstrates that if you accept the given information and if you accept the conjectures in the argument, then the conclusion logically follows. You will learn more about writing proofs in Chapters 13 to 15.

EXERCISE SET A

Copy the flowchart. Provide each missing reason or statement in the proof.

Given: $\overline{SE} \cong \overline{SU}$

$\angle E \cong \angle U$

Show: $\overline{MS} \cong \overline{SO}$

Flowchart Proof:

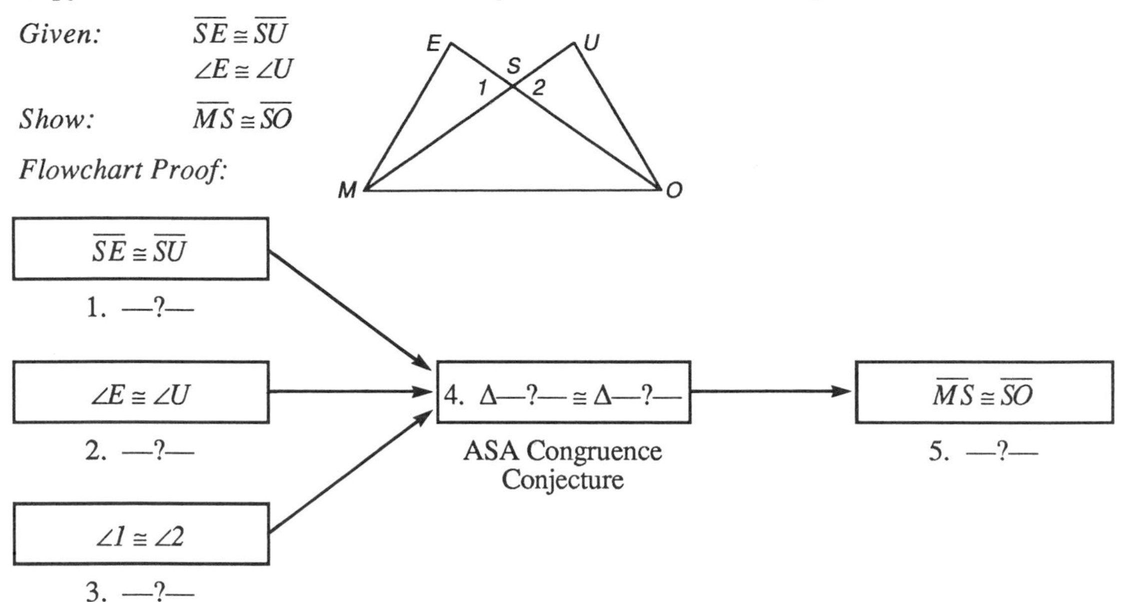

EXERCISE SET B

Copy the flowchart. Provide each missing reason or statement in the proof.

Given: M is the midpoint of $\overline{WE}$

M is the midpoint of $\overline{NO}$

Show: $\overline{WO} \cong \overline{EN}$

Flowchart Proof:

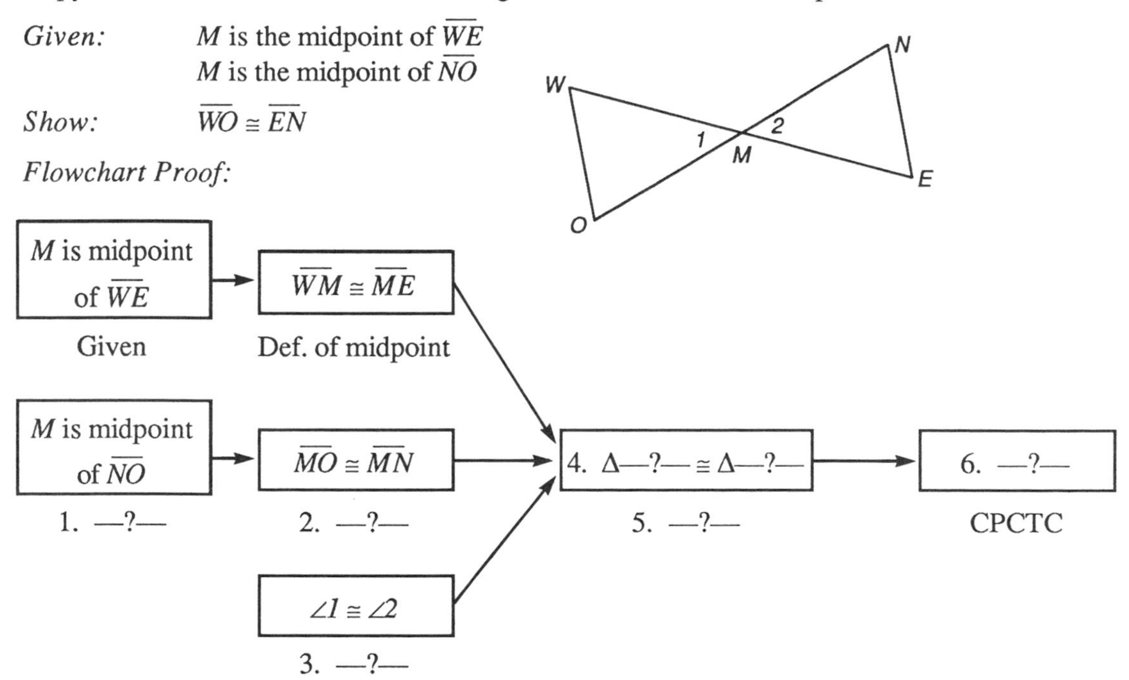

EXERCISE SET C

Complete the flowchart proof below to logically demonstrate the argument. There is always more than one flowchart proof that can be written for a logical argument. One possible proof for this argument has been started for you.

Given: $\overline{SA} \parallel \overline{NE}$
$\overline{SE} \parallel \overline{NA}$

Show: $\overline{SA} \cong \overline{NE}$

Flowchart Proof:

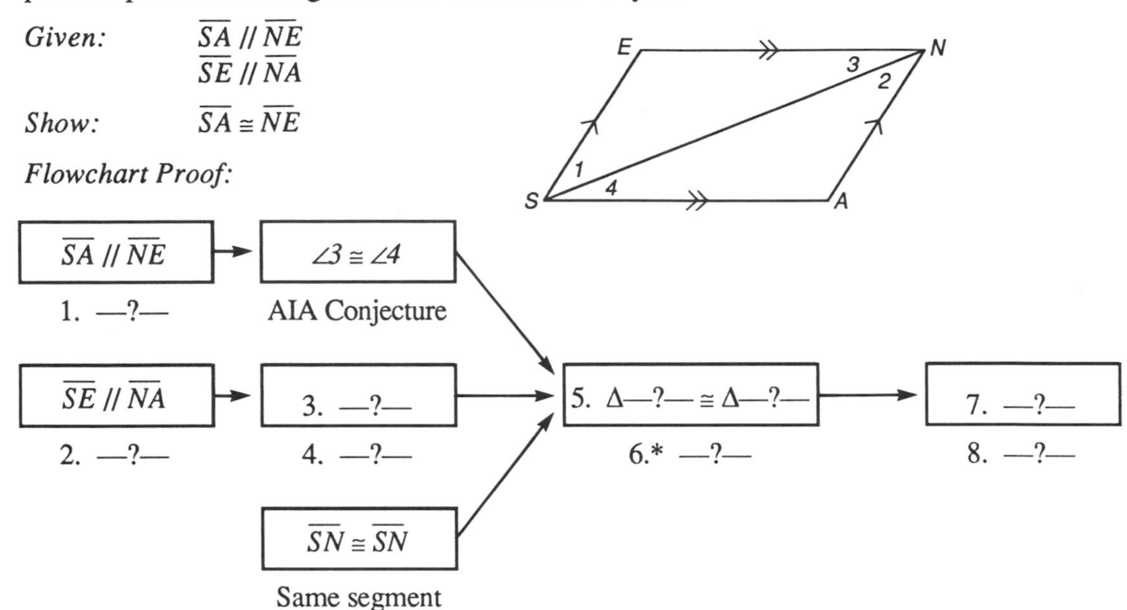

EXERCISE SET D

1.* Now it's your turn. Write a flowchart proof to demonstrate the argument below. Writing a proof can be difficult. You may find it very helpful to work with others on the proof, sharing your ideas.

Given: $\angle S \cong \angle T$
$\angle R \cong \angle A$
$\overline{RE} \cong \overline{AE}$

Show: $\overline{RT} \cong \overline{SA}$

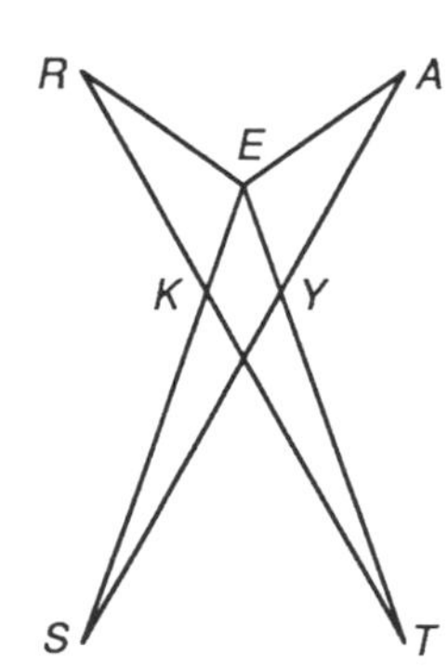

Computer Activity

Regular Polygons and Star Polygons with Logo

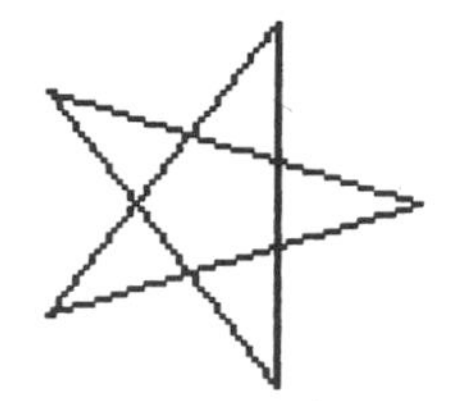

Regular Star Polygon

In this computer activity you will use the Logo computer language to explore regular polygons and star polygons. Regular star polygons, like regular polygons, are closed figures in a plane with straight line sides equal in length that meet to form angles equal in measure. Unlike the sides of polygons though, the sides of star polygons can intersect. You do not need prior experience with Logo for this activity. You will need:

- Computer and disk with Logo computer language
- Resource person familiar enough with Logo to help you get started
- Disk with the file of procedures called POLYGONS (The listings of the procedures are in *Appendix B: Logo Procedures*.)

Start up your computer using the disk with Logo on it. You're ready to go! If you've undertaken an earlier special project with Logo, you can start with task 2.

1. The Logo turtle will respond to your commands. Your first goal is to discover the action that the turtle takes when you give it some common Logo commands. Experiment with the commands below. Do they all work in your version of Logo? Try different numbers with any command which has a number. You need to press the return key after typing each command. Typing errors can be corrected by pressing return and retyping the command.

Command	Short Form	Command	Short Form	Command	Short Form
FORWARD 50	FD 50	CLEARSCREEN	CS	SHOWTURTLE	ST
BACK 60	BK 60	DRAW		HIDETURTLE	HT
LEFT 90	LT 90	CLEARGRAPHICS	CG	FULLSCREEN	Ctrl-L or F
RIGHT 45	RT 45	PENUP	PU	TEXTSCREEN	Ctrl-T
HOME		PENDOWN	PD	SPLITSCREEN	Ctrl-S

2. Use the commands above (with different numbers) to make a polygon. List the commands that you use. Now make a regular triangle. What commands did you use? Make a square. What other regular polygons can you make? Make a list of the commands that you use for each.

Now load the file POLYGONS from your procedures disk. The POLYGONS file contains four procedures: SIDEPOLY, REGPOLY, ANYPOLY, and STARPOLY.

3. The procedure SIDEPOLY needs only one input. For example, typing SIDEPOLY 5 will set the turtle to work. Experiment with SIDEPOLY using inputs of your own. What does SIDEPOLY do? What does the input control? To fit polygons with more than 12 sides on the screen, you will need to PENUP, move the turtle to a new position, and then PENDOWN. Create a polygon with 20 sides so that it appears unbroken on the screen (without wrapping).

4. The procedure REGPOLY needs two inputs. Try REGPOLY 5 25. What does REGPOLY do? What does the first input control? What does the second input control?

5. Draw a circle using REGPOLY. Explain your answer.

6.* The procedure ANYPOLY needs three inputs. The first two are related to the inputs used by SIDEPOLY and REGPOLY The third is the measure of an angle. Unlike SIDEPOLY and REGPOLY, ANYPOLY will not always create a polygon. What inputs will make ANYPOLY draw regular polygons? See if you can find inputs that will make ANYPOLY draw each of the regular polygons in the table. (Note: Logo knows how to do arithmetic, so inputs may be in the form of an arithmetic expression such as 15 * 3 or 180 * 3 or 180 / 2.)

Polygon	Triangle	Square	Pentagon	Hexagon	Heptagon	Octagon	Nonagon
1st input	–?–	–?–	–?–	–?–	–?–	–?–	–?–
2nd input	–?–	–?–	–?–	–?–	–?–	–?–	–?–
3rd input	–?–	–?–	–?–	–?–	–?–	–?–	–?–

7. Find a formula for inputs that make a regular polygon with *n* sides.

8. ANYPOLY can be used to create star polygons. Experiment. What inputs to ANYPOLY will create a star pentagon? Can you create a star hexagon and a star dodecagon? Find a formula for inputs that will make a star polygon with *n* sides.

9. The procedure STARPOLY needs three inputs also. The first two are like the first two inputs used by ANYPOLY. The third input is a special number related to the first input. The third input should always be less than the first. STARPOLY can surprise you. Would you have guessed that STARPOLY 6 50 2 draws a triangle? Experiment with STARPOLY. What kinds of star polygons can you make?

10. Make a star table like the one below for STARPOLY showing the first and third inputs. (Remember to keep the third input less than the first.) Continue the table until you can make a conjecture allowing you to predict future results. Predict which numbers for the third input to STARPOLY 12 30 –?– will create star polygons. Explain how you know.

First input	Third input	Star? (yes/no)	Regular polygon?	Number of points/sides
3	1	no	yes	3
3	2	–?–	–?–	–?–
4	1	–?–	–?–	–?–
4	2	–?–	–?–	–?–
4	3	–?–	–?–	–?–
5	1	–?–	–?–	–?–
⋮	⋮	⋮	⋮	⋮

Improving Reasoning Skills

Magic Squares

A magic square is an arrangement of numbers in a square array such that the numbers in every row and every column have the same total. In some magic squares the two diagonals have the same totals as the rows and columns. For example, in the magic square to the right, the sum of each row is 18, the sum of each column is 18, and the sum of each diagonal is also 18.

5	10	3
4	6	8
9	2	7

1. Complete the 5 x 5 magic square below using the numbers 6, 7, 9, 13, 17, 21, 23, 24, 27, and 28.

20			8	14
	19	25	26	
	12	18		30
29	10	11		
22			15	16

2. Find x in the magic square at right. Hint: Since each row and column must add to the same total, set the algebraic sum of one row or column equal to another and solve for x.

x	$x - 1$	8
$x + 5$	5	$x - 3$
$x - 2$	$x + 3$	$x + 2$

Improving Reasoning Skills

Euler's Magic Square

The mathematician Leonard Euler (1707–1783) created a unique magic square. He designed a magic square in which the entire 8 x 8 array was not only a magic square, but each corner 4 x 4 array was also a magic square. Euler added a twist for chess fans. He arranged the numbers so that it would be possible to move from the 1 to the 2, to the 3, and so on, all the way to the 64 by making the L-shaped move of the chess knight. However, the sum of each diagonal of this magic square is not the same as the row and column sum. Copy and complete his magic square shown at right.

1		31		33		63	
	51		3		19		
		49		15			
			45		61	36	13
		25				21	
28		8	41	24			
43	6	55			10		
				58		38	

Lesson 5.7

Isosceles Triangles Revisited

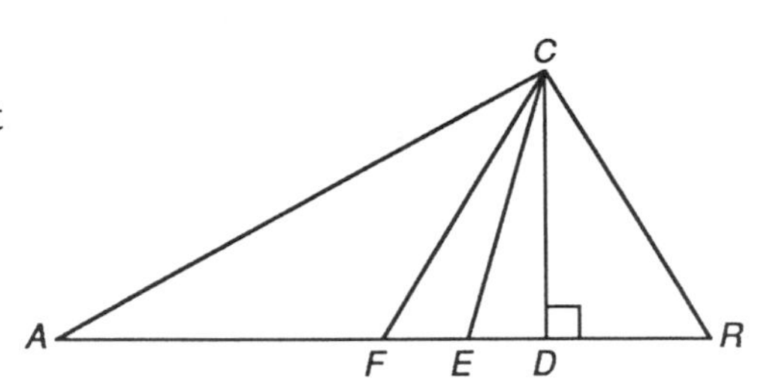

In this lesson you will make a conjecture about a special segment in isosceles triangles. Then you will use logical reasoning to support your conjecture.

In ΔARC, $\overline{CD}$ is the altitude to the base $\overline{AR}$. $\overline{CE}$ is the angle bisector of the vertex angle ACR. $\overline{CF}$ is the median to the base $\overline{AR}$. From the diagram it is clear that the angle bisector, altitude, and median can all be different line segments. But is this true for all triangles? Let's investigate.

Investigation 5.7.1

Step 1: Construct a large isosceles triangle on a sheet of unlined paper. Label it ΔARK with K the vertex angle.

Step 2: Construct angle bisector $\overline{KD}$ (with D on $\overline{AR}$).

Step 3: With your protractor, measure $\angle ADK$ and $\angle RDK$.

How do $\angle ADK$ and $\angle RDK$ compare? Are they equal? Beneath your construction, copy and complete the conjecture below.

Conjecture: *The bisector of the vertex angle of an isosceles triangle is also the —?— to the base.*

Step 4: With your compass, measure $\overline{AD}$ and $\overline{RD}$.

Is D the midpoint of $\overline{AR}$? Beneath your construction, copy and complete the conjecture below based on your observations from step 4.

Conjecture: *The bisector of the vertex angle of an isosceles triangle is also the —?— to the base.*

Compare your results (both conjectures above) with the results of others near you. You should be ready to state your next conjecture.

C-37 *In an isosceles triangle, the bisector of the vertex angle is also the —?— to the base and the —?— to the base.* ***(Vertex Angle Conjecture)***

EXERCISE SET A

In each exercise, ΔABC is isosceles with $AC = BC$.

1.* Perimeter $\Delta ABC = 48$
$AC = 18$
$AD = $ –?–

2. $m\angle ABC = 72$
$m\angle ADC = $ –?–

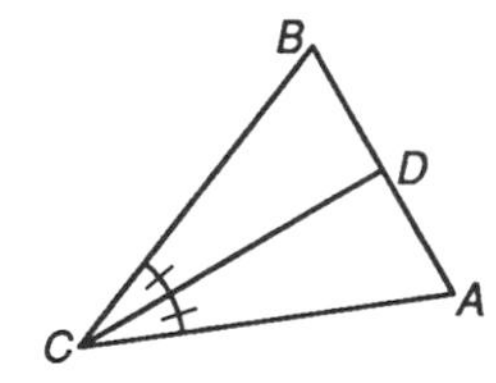

3. $m\angle CAB = 45$
$m\angle ACD = $ –?–

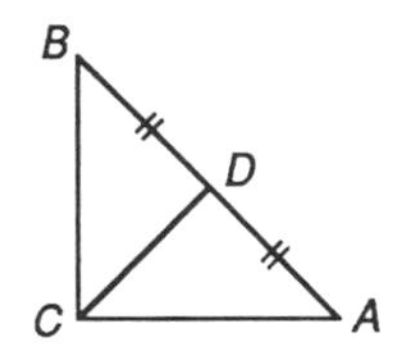

Examples A and B show how you can use a flowchart to logically demonstrate the first part of the Vertex Angle Conjecture. The second part of the conjecture is left for you to complete in Exercise Set B.

The conjecture in example A is used to prove the conjecture in example B. You will also need to use it to prove the conjecture in Exercise Set B.

Example A

Show by logical reasoning that the conjecture is true.

Conjecture: *The bisector of the vertex angle in an isosceles triangle divides the isosceles triangle into two congruent triangles.*

Given: ΔABC is isosceles with $\overline{AC} \cong \overline{BC}$
$\overline{CD}$ is the bisector of $\angle C$

Show: $\Delta ADC \cong \Delta BDC$

Flowchart Proof:

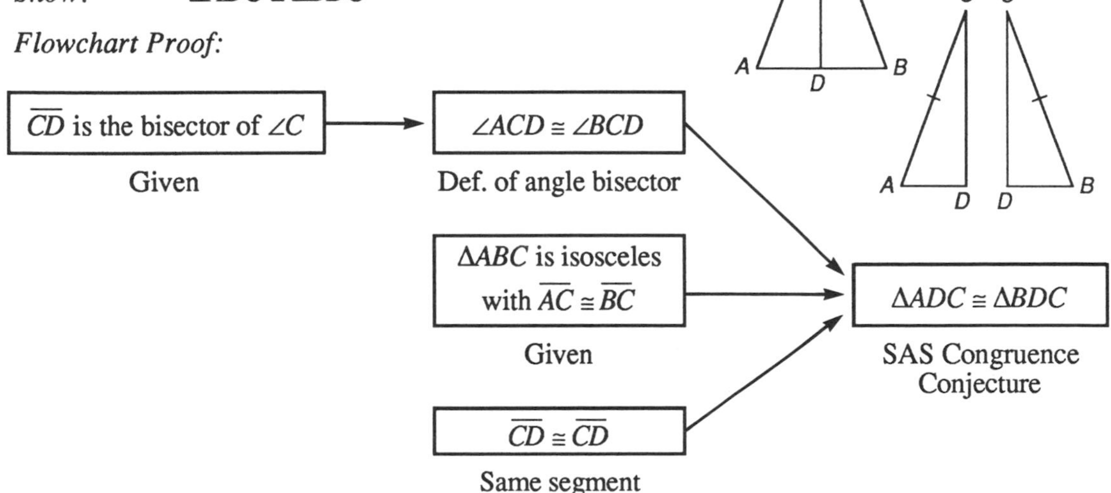

Example B

Show by logical reasoning that the bisector of the vertex angle in an isosceles triangle is also the altitude to the base.

Conjecture: *The bisector of the vertex angle in an isosceles triangle is also the altitude to the base.*

Given: ΔABC is isosceles with $\overline{AC} \cong \overline{BC}$
$\overline{CD}$ is the bisector of $\angle C$

Show: $\overline{CD}$ is an altitude

Flowchart Proof:

- ΔABC is isosceles with $\overline{AC} \cong \overline{BC}$ and $\overline{CD}$ is the bisector of $\angle C$ — Given
- → $\Delta ADC \cong \Delta BDC$ — By example A
- → $\angle 1 \cong \angle 2$ — CPCTC
- $\angle 1$ and $\angle 2$ form a linear pair — Def. of linear pair
- → $\angle 1$ and $\angle 2$ are supplementary — Linear Pair Conjecture
- → $\angle 1$ and $\angle 2$ are right angles — Congruent Supplements Conjecture
- → $\overline{CD}$ is an altitude — Def. of altitude

One part of the Vertex Angle Conjecture has now been proven. Now it's your turn to prove the other part.

EXERCISE SET B

The flowchart below logically demonstrates that the bisector of the vertex angle in an isosceles triangle is also the median to the base. Copy the flowchart proof. Provide each missing statement or reason.

Conjecture: *The angle bisector of the vertex angle in an isosceles triangle is also the median to the base.*

Given: ΔABC is isosceles with $\overline{AC} \cong \overline{BC}$
$\overline{CD}$ is the bisector of $\angle C$

Show: $\overline{CD}$ is a median

Flowchart Proof:

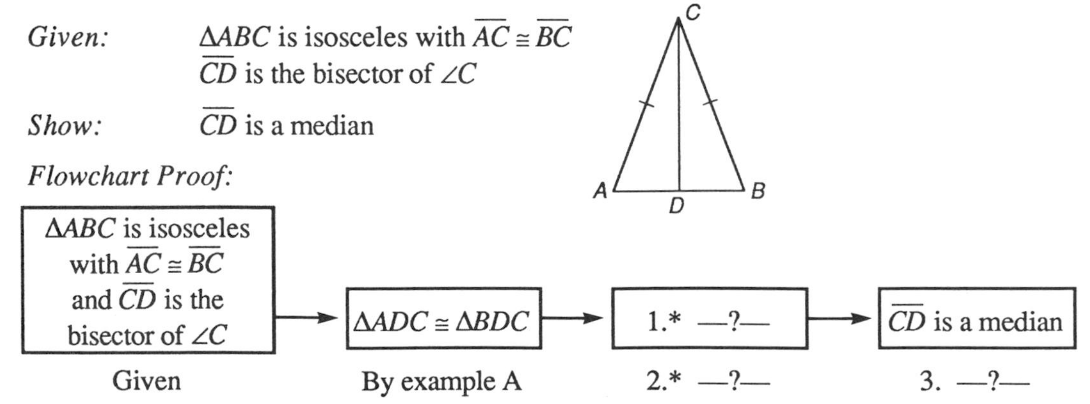

Lesson 5.8

Geometric Probability

We cannot leave the haphazard to chance.

– N. F. Simson

You've just been visiting a friend in jail. Now your only hope of success is buying property on St. James Place. To do that, you'll need a roll of six on your next turn! (That's right, we're on the *Monopoly* board.) Will you succeed? You ask yourself, "What's the probability of rolling a six on my next roll?"

What are my chances? What are the odds? What's the probability? You've heard these expressions many times and have *probably* used them yourself. Probability theory, an important branch of mathematics, attempts to answer questions like these. Probability is important to many occupations — from insurance agents to professional backgammon players. Probability is a measure of the likelihood that an event will happen. The **probability** that a particular outcome will occur out of a number of equally likely possible outcomes is the ratio of the number of ways that particular outcome can occur (successful outcomes) to the total number of possible outcomes.

$$\text{probability} = \frac{\text{number of successful outcomes}}{\text{number of possible outcomes}}$$

Probability is a ratio between zero and one. The ratio tells how likely it is that an event will happen. A probability of zero means that an event will not happen. A probability of one means that it is certain to happen.

Let's look at a few examples of probability problems.

Example A

What is the probability of a "heads" facing up on one toss of a coin?

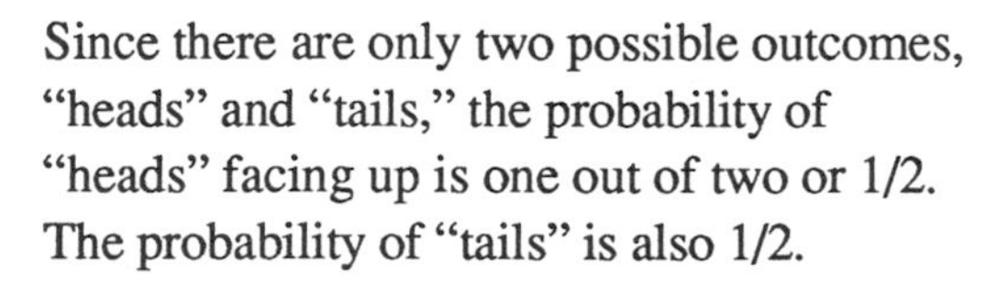

Since there are only two possible outcomes, "heads" and "tails," the probability of "heads" facing up is one out of two or 1/2. The probability of "tails" is also 1/2.

Example B

What is the probability of rolling a 6 with one toss of a single die?

Since there are six possible outcomes (1, 2, 3, 4, 5, and 6) and only one 6, the probability of rolling a 6 is 1/6. In fact, each of the numbers 1 through 6 has an equal probability of 1/6.

Example C

What is the probability of rolling a sum of 6 with a pair of dice?

Let's organize our counting by setting up a table.

Roll (Total on Both Dice)	Ways of Getting Total	Number of Ways to Get Total
2	(1, 1)	1
3	(1, 2) (2, 1)	2
4	(1, 3) (2, 2) (3, 1)	3
5	(1, 4) (2, 3) (3, 2) (4, 1)	4
6	(1, 5) (2, 4) (3, 3) (4, 2) (5, 1)	5
7	(1, 6) (2, 5) (3, 4) (4, 3) (5, 2) (6, 1)	6
8	(2, 6) (3, 5) (4, 4) (5, 3) (6, 2)	5
9	(3, 6) (4, 5) (5, 4) (6, 3)	4
10	(4, 6) (5, 5) (6, 4)	3
11	(5, 6) (6, 5)	2
12	(6, 6)	1

By organized counting we see that there are five ways of rolling a total of six and 36 total rolls possible. Therefore, the probability of rolling a six is 5/36.

Example D

What is the probability of randomly selecting one of the longest diagonals from among all the diagonals of a regular hexagon?

By organized counting you can see that there are three long diagonals and six shorter diagonals for a total of nine diagonals. Therefore the probability of selecting one of the longest diagonals is 3/9 or 1/3.

In many geometric probability problems you will not be able to count the events. The next example demonstrates a problem which involves more than organized counting.

Example E

Given $\overline{AB}$ with midpoint M, what is the probability of randomly selecting a point on the segment closer to point A or point B than to point M?

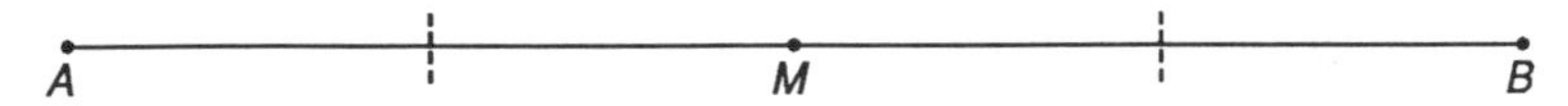

All the points to the left of the midpoint of $\overline{AM}$ are closer to A than to M. This is one-fourth of the entire segment. All the points to the right of the midpoint of $\overline{MB}$ are closer to B than to M. This is also one-fourth of the entire segment. Therefore, the probability of randomly selecting a point closer to either A or B than to M is 1/2.

EXERCISE SET

Use organized counting when possible to solve these probability problems. When it is not possible to use organized counting, draw diagrams for help.

1. What is the probability of randomly selecting one of the longest diagonals from among all the diagonals of a regular octagon?

2. What is the probability of randomly selecting one of the shortest diagonals from among all the diagonals of a regular octagon?

   A, B, and C represent sets containing lengths of segments as indicated.

   A = {3, 4} B = {3, 6, 9} C = {3, 12}

3.* If one length is randomly selected from set A, one length from set B, and one length from set C, what is the probability that a triangle can be formed with segments having the three selected lengths?

4. If one length is randomly selected from set A, one length from set B, and one length from set C, what is the probability that you can form an isosceles triangle whose sides have the three selected lengths?

5. If one length is randomly selected from set A, one length from set B, and one length from set C, what is the probability that you can form an equilateral triangle whose sides have the three selected lengths?

6. Given $\overline{AB}$, what is the probability of randomly selecting a point on the segment closer to point A than to point B?

7. Given $\overline{AB}$ with midpoint M, what is the probability of randomly selecting a point on the segment closer to point M than to point A?

8.* Given $\overline{AB}$ with midpoint M, what is the probability of randomly selecting a point on the segment closer to point M than to points A or B?

9.* If three different points are selected randomly from six equally spaced points on a circle, what is the probability that the triangle formed by connecting these three points will be equilateral?

Improving Reasoning Skills

Pick a Card

Nine cards are arranged in a 3 by 3 array. Every jack borders (horizontally or vertically but not diagonally) on a king and on a queen. Every king borders on an ace. Every queen borders on a king and an ace. There are at least two aces, two kings, two queens, and two jacks. What kind of card is in the center position of the 3 by 3 array.

Lesson 5.9

Chapter Review

If it takes 12 musicians one hour to play Beethoven's Third Symphony, how long will it take 24 musicians to play the same symphony?

– Anonymous

EXERCISE SET A

Identify each statement as true or false.

1. The measure of each interior angle of a regular hexagon is 120.
2. The base angles of an isosceles triangle are supplementary.
3. If a triangle has two angles of equal measure, then the third angle is acute.
4. The capital letters CPCTC are an abbreviation for the phrase: *Corresponding Parts of Constructed Triangles are Concurrent.*
5. A logical argument presented in the form of flowchart is a paragraph proof.
6. The probability of randomly selecting a point closer to either endpoint than to the midpoint of a segment is 2/3.
7. The intersection of the altitudes of a triangle is called the centroid.
8. If ΔDOG is congruent to ΔRAT, then $\overline{DO}$ is congruent to $\overline{TR}$.
9. Exercise 1 is the only true statement in this exercise set.

EXERCISE SET B

Use the segments and angles below to construct each figure. The lower case letter above each segment represents the length of the segment in units.

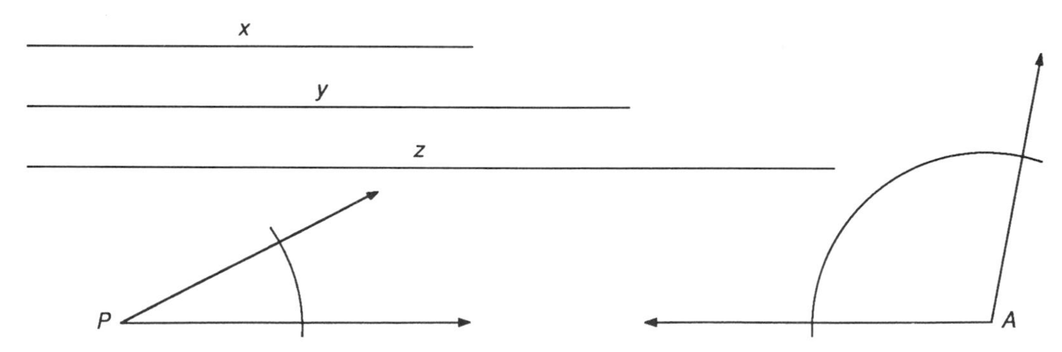

1. Construct ΔPAL given $\angle P$, $\angle A$, and $AL = x$.
2. Construct two triangles ΔPBS that are not congruent to each other, given $\angle P$, $PB = z$, and $BS = x$.
3. Construct square $SQRE$ with y the lengths of the diagonals.

EXERCISE SET C

From the information given, determine which triangles, if any, are congruent. State the congruence conjecture that supports the congruence statement. If the triangles cannot be shown to be congruent from the information given, write *cannot be determined.*

1.*

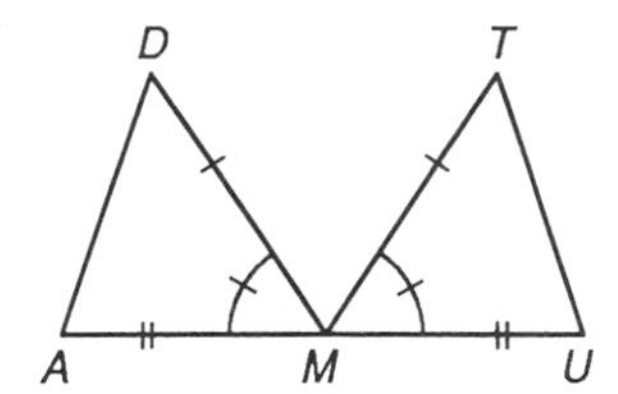

2.

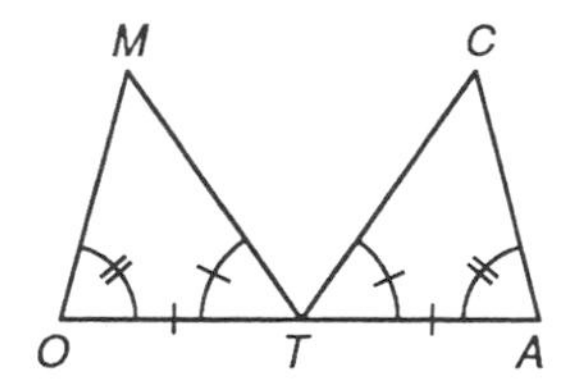

3.

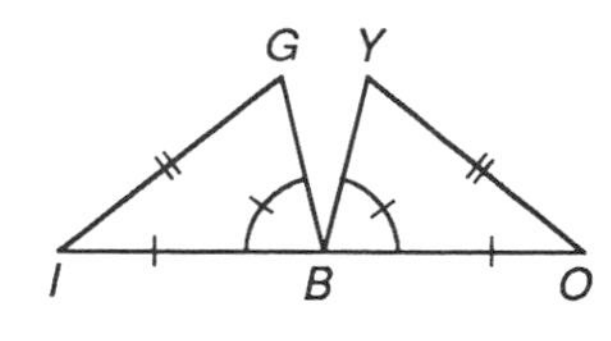

4.*

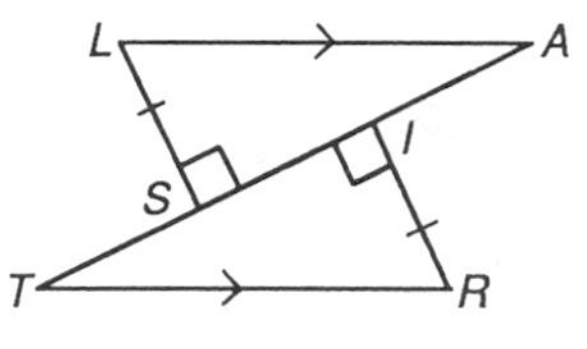

5.

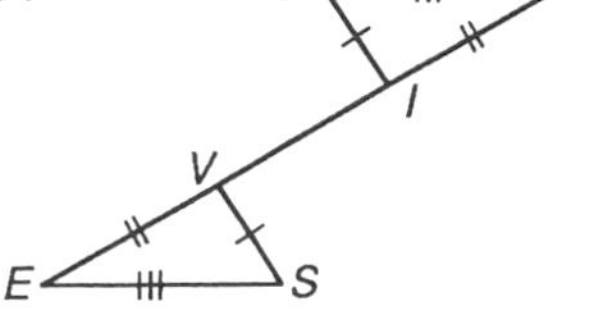

6.

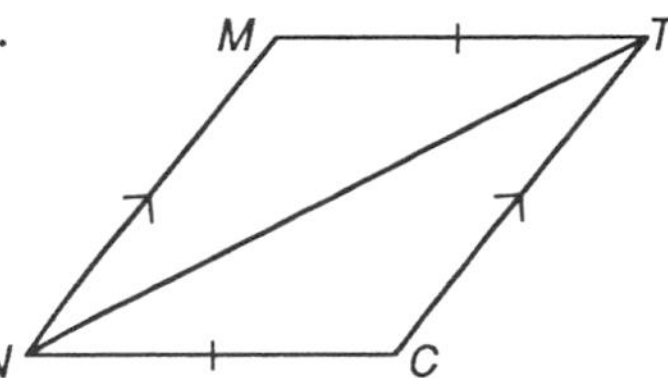

7.*

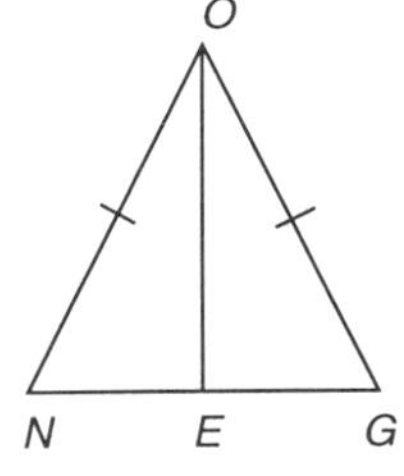

8.

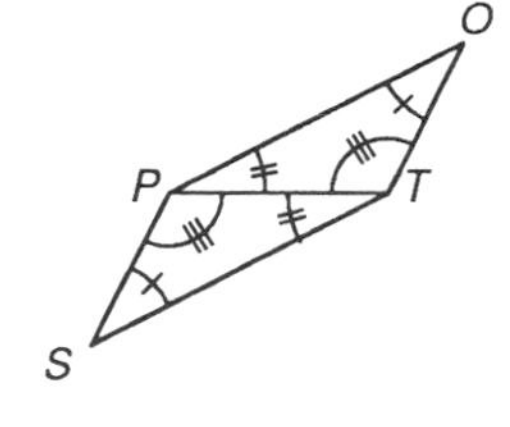

9.* ΔLAI is isosceles with $LA = IA$.

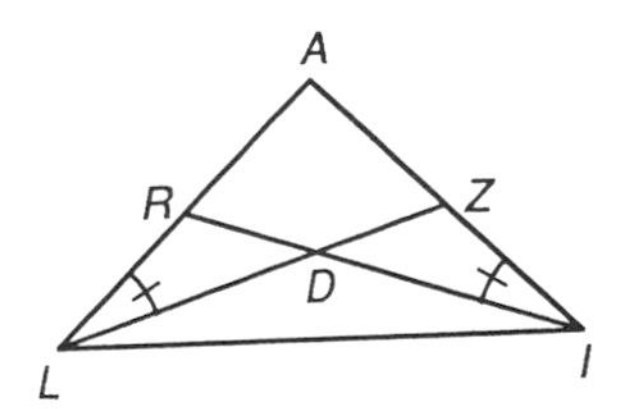

EXERCISE SET D

1. Find the measure of each lettered angle in the diagram below. $l \parallel k$

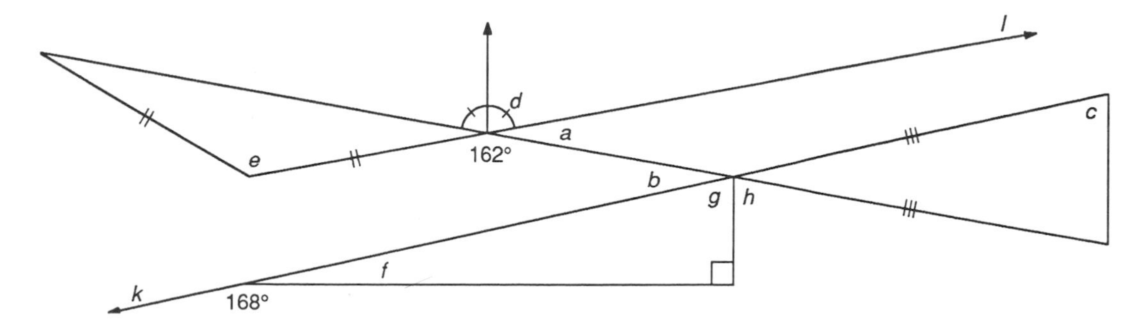

Cooperative Problem Solving

Construction Games from Centauri

The geometry students of the lunar colony have received a communication from the Centauri star-system. The Centaurian geometry students have sent two of their classic games of geometric construction. The geometry developed by Centaurian mathematicians is almost identical to that of the geometry studied on Earth, however, they did develop alternative geometric constructions. Like the ancient Greeks of Earth, the early Centaurian geometers played with a compass and straightedge, as if it were a game, to see what geometric figures they could create. They also went on to develop other construction games. In one construction game, they tried to see what they could create using just a right triangle. In a another construction game, they tried to see what they could create using just a pair of parallel lines.

You and your CPS team, like the students at the lunar colony, are going to play the two games of construction from Centauri. In each game you must play by the rules and "construct" different geometric figures. With compass and straightedge constructions, you were able to duplicate segments and angles, construct angle bisectors and perpendicular bisectors, construct a perpendicular from a point to a line and a perpendicular through a point on a line. How many of these will you be able to create with a different set of tools?

When you play the Centauri construction games, the term *construct* does not mean construct with a compass and straightedge. Instead, it means create with only the tool or tools allowed by the game. Once you have figured out a method to construct each figure, use your geometric definitions and conjectures to support your method. Explain why your construction works. Good Luck.

Game One: Constructions with a Right Triangle

In this first construction game you are permitted to use only a *right triangle*. With your right triangle tool you can construct straight lines, construct perpendiculars, duplicate the acute angles of the right triangle, and duplicate the lengths of the legs of the right triangle. (You may not assume to know the measures of the angles or the relationship between the lengths of the sides.) If you do not have a plastic drafting right triangle, make a right triangle for yourself out of poster-board or heavy cardboard.

1. Construct a perpendicular from a given point to a given line.
2. Construct a perpendicular through a given point on a given line.
3. Construct a rectangle.
4. Construct a rhombus.
5. Construct a perpendicular bisector of a segment.
6. Construct an angle bisector.

Can you develop any new right triangle constructions to send as a challenge back to the geometry students on Centauri?

Game Two: Constructions with Parallel Lines

In this second construction game you are permitted to use only the two parallel edges of a straightedge or a ruler. If you use a ruler, you cannot use the markings on it. With your parallel line tool you can construct parallel lines a fixed distance apart. (With parallel lines come all the properties of parallels and parallelograms!)

1. Construct a rhombus.
2. Construct a rectangle.
3. Construct the angle bisector of a given angle.
4. Construct a perpendicular bisector of a given segment (where the length of the segment is greater than the distance between your parallel lines).
5. Construct a perpendicular through a given point on a given line.
6. Construct a perpendicular bisector of a given segment (where the length of the segment is less than the distance between your parallel lines).
7. Construct a line parallel to a given line through a point not on the given line.
8. Construct a perpendicular to a given line through a point not on the given line.

Can you develop any new parallel line constructions to send as a challenge back to the geometry students on Centauri?

6 Circles

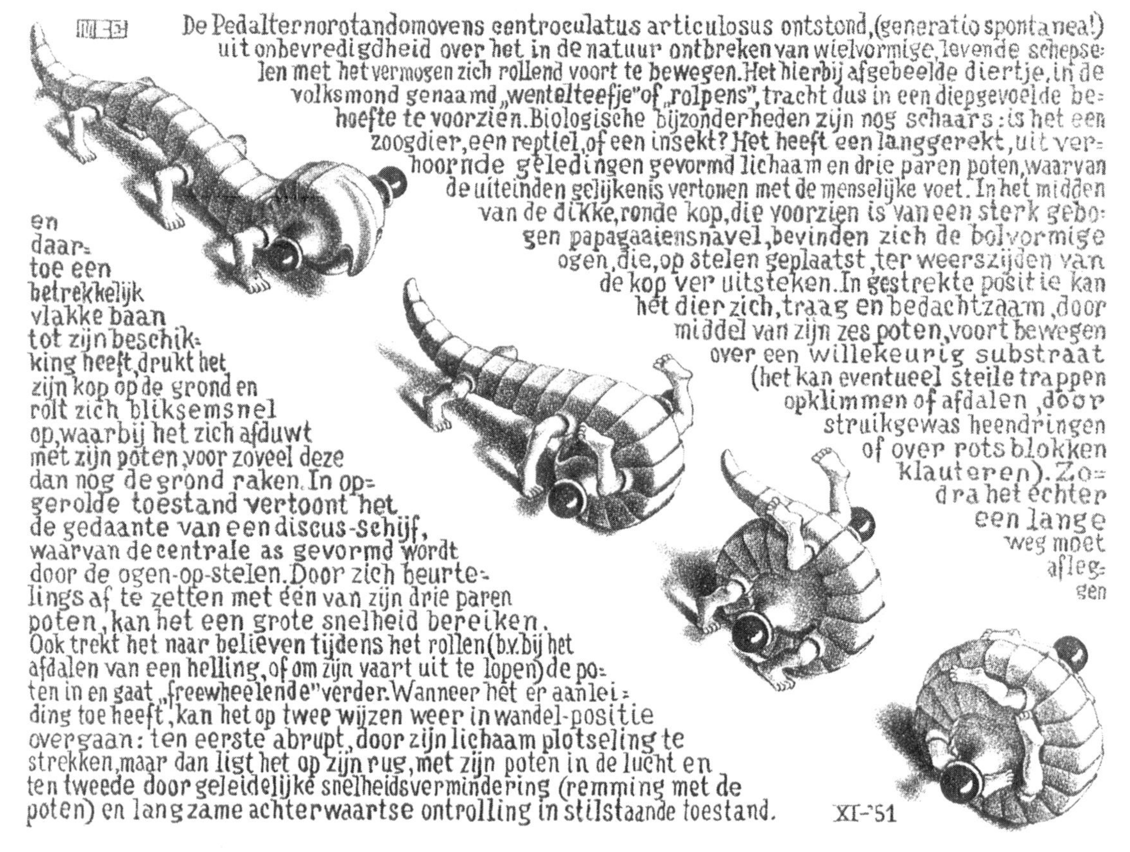

Curl-Up, M. C. Escher, 1951

This chapter introduces you to the properties of the circle. Using your geometric tools, you will discover many relationships between the angles and line segments in and around circles. One important property of the circle is the relationship between the circumference (perimeter of the circle) and the diameter of the circle (the longest distance across the circle). Circular wheels and gears are two of the most important applications of the properties of circles.

Lesson 6.1

Defining Circles

Tracing the Stars

The giant Anglo-Australian Observatory at Siding Springs, New South Wales, was silhouetted against a time exposure of star trails. The wide-angle telescope plays a central role in the search for a 10th planet, believed by astronomers to lie at the boundaries of the solar system. The six small circles at the left are the pointers of the Southern Cross.

The circle is a geometric shape that is all around you. Unless you walked to school this morning, you arrived on a vehicle with circular wheels. Civilizations have rolled forward on wheels for thousands of years. Wheels gave rise to circular gears which ushered in the industrial revolution. Potter's wheels, clocks, and windmills were great advances to civilization based on applications of wheels.

*A **circle** is the set of all points in a plane at a given distance from a given point in the plane.*

center

O

radius

The given distance is the **radius** of the circle. A segment from a point of the circle to the center is also called a **radius**.

The given point is the **center** of the circle. You name a circle by its center. The circle on the right, with center O, is called circle O. In this text, when you see a dot at the center of a circle, you may assume that it represents the center point.

By the definition of a circle, any two radii (plural of radius) of the same circle are congruent. If two circles have the same radius, then they are **congruent circles**. If two or more circles share the same center, then they are **concentric circles**.

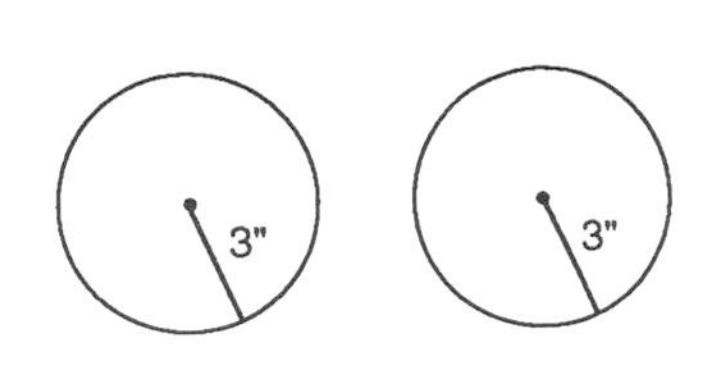

Congruent Circles

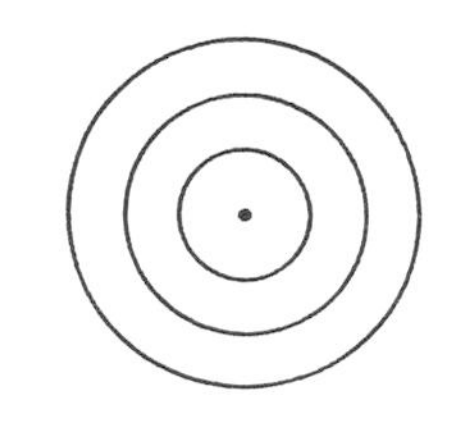

Concentric Circles

EXERCISE SET A

Write a definition of each geometric term. Discuss your definitions with others in your class. Agree on a common set of definitions and add them to your definition list. Draw and label a picture to illustrate each definition.

1. Define *chord.*

Chord

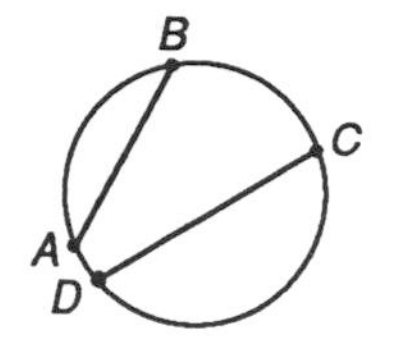

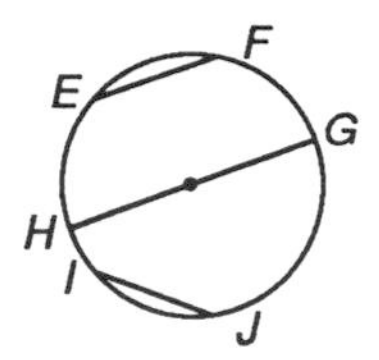

$\overline{AB}$, $\overline{CD}$, $\overline{EF}$, $\overline{GH}$, and $\overline{IJ}$ are chords.

Not a Chord

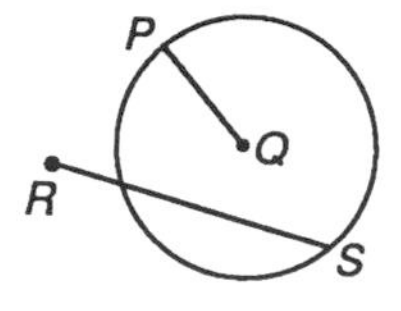

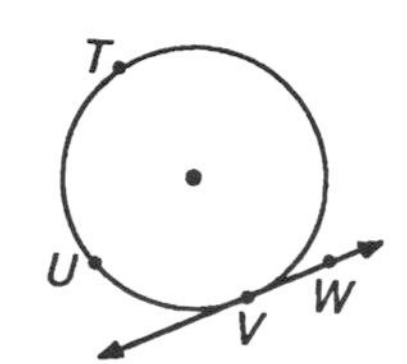

$\overline{PQ}$, $\overline{RS}$, TU, and $\overleftrightarrow{VW}$ are not chords.

2. Define *diameter*.

Diameter

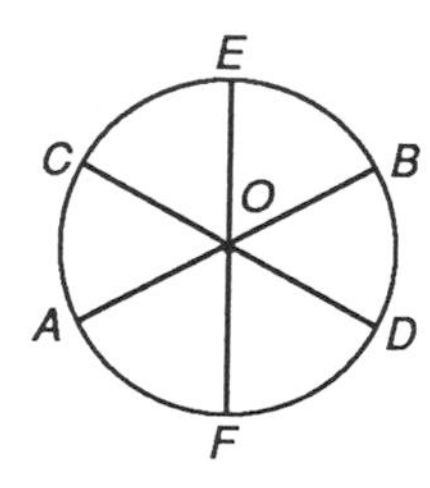

$\overline{AB}$, $\overline{CD}$, and $\overline{EF}$ are diameters of circle O.

Not a Diameter

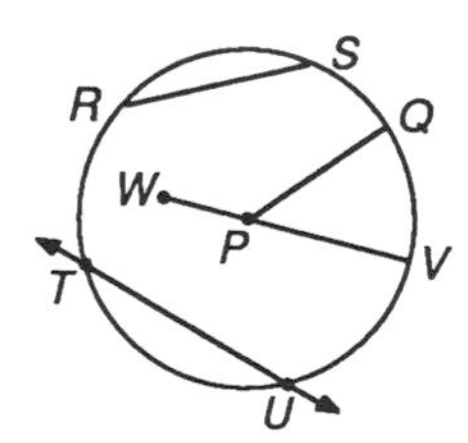

$\overline{PQ}$, $\overline{RS}$, $\overleftrightarrow{TU}$, and $\overline{VW}$ are not diameters of circle P.

Note: Like radius, diameter can refer to a line segment or to the length of a segment. You may say a diameter has a length of 5 cm or that the diameter is 5 cm.

3. Define *secant.*

Secant

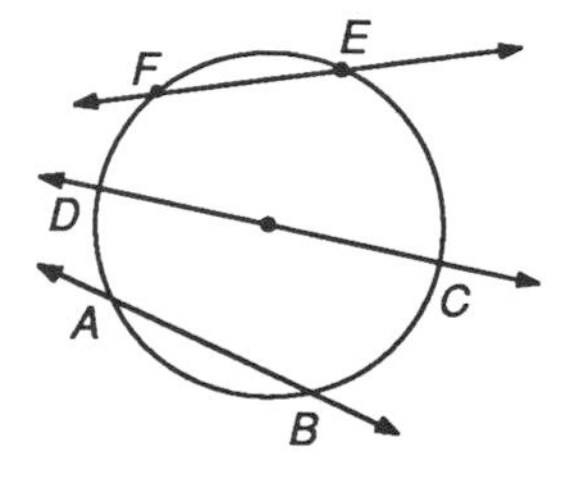

$\overleftrightarrow{AB}$, $\overleftrightarrow{CD}$, and $\overleftrightarrow{EF}$ are secants.

Not a Secant

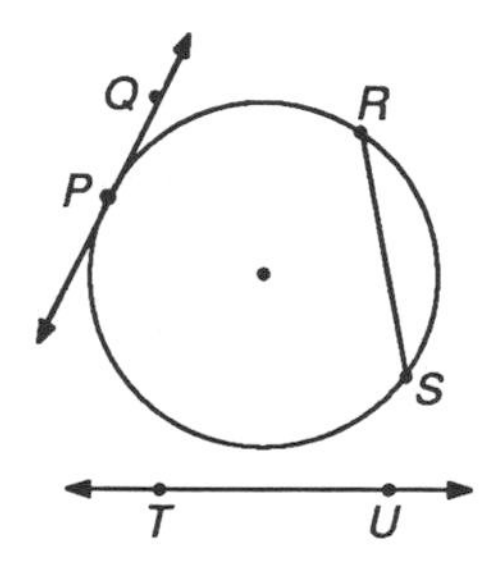

$\overleftrightarrow{PQ}$, $\overline{RS}$, $\overleftrightarrow{TU}$ are not secants.

4. Define *tangent.*

Tangent

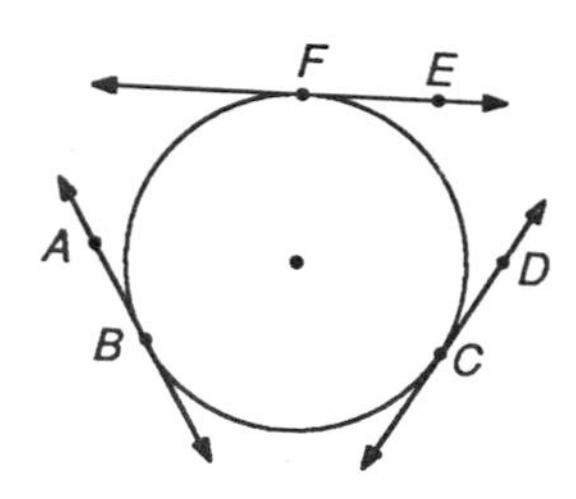

$\overleftrightarrow{AB}$, $\overleftrightarrow{CD}$, and $\overleftrightarrow{EF}$ are tangents.

Not a Tangent

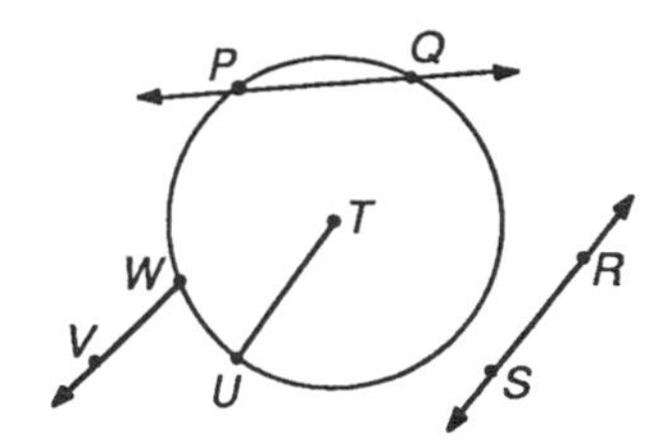

$\overleftrightarrow{PQ}$, $\overleftrightarrow{RS}$, $\overline{TU}$, and $\overrightarrow{VW}$ are not tangents.

Note: The term tangent is also used in more than one way. You may say line *AB* is a tangent, or you may say line *AB* is tangent to circle *O*. The point where the tangent touches the circle is called the **point of tangency.**

5.* Define *inscribed angle.*

Inscribed Angle

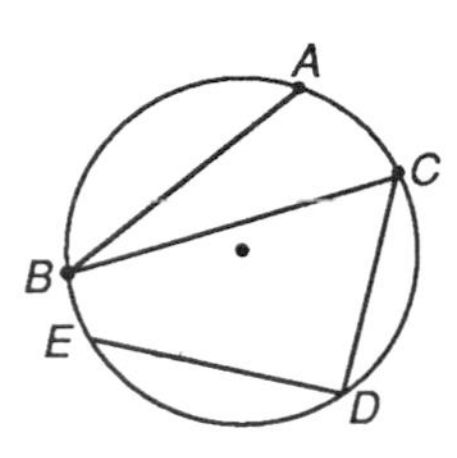

$\angle ABC$, $\angle BCD$, and $\angle CDE$ are inscribed angles.

Not an Inscribed Angle

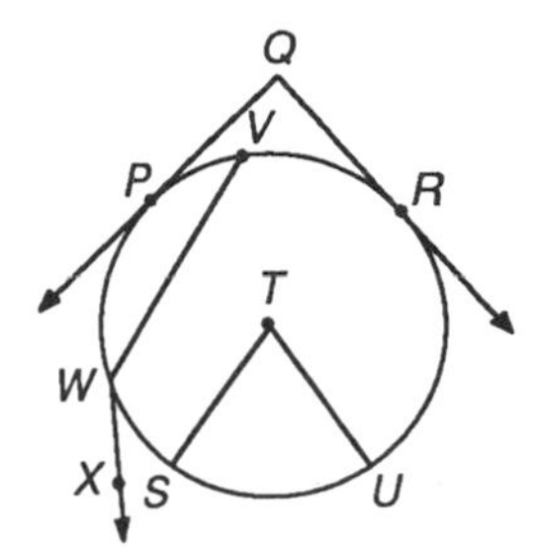

$\angle PQR$, $\angle STU$, and $\angle VWX$ are not inscribed angles.

6. Define *central angle.*

Central Angle

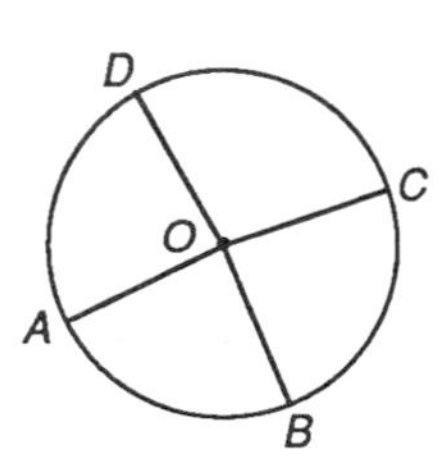

$\angle AOB$, $\angle BOC$, $\angle BOD$, and $\angle DOA$ are central angles of circle *O*.

Not a Central Angle

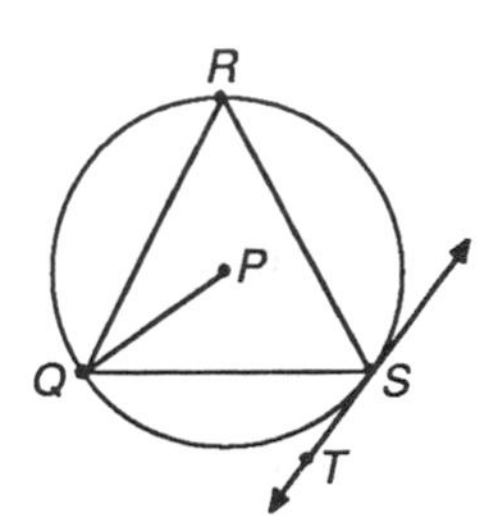

$\angle PQR$, $\angle PQS$, $\angle RQS$, and $\angle QST$ are not central angles of circle *P*.

*An **arc of a circle** is two points on the circle and the continuous (unbroken) part of the circle between the two points.*

The two points are called the **endpoints** of the arc. The symbol for an arc is ⌒. The symbol is placed above the letters that name the endpoints of the arc. Arc AB is written $\overset{\frown}{AB}$ or $\overset{\frown}{BA}$. Just as you classify angles into three types, you divide arcs into three types: minor arcs, semicircles, and major arcs.

*A **semicircle** is an arc of a circle whose endpoints are the endpoints of a diameter.*

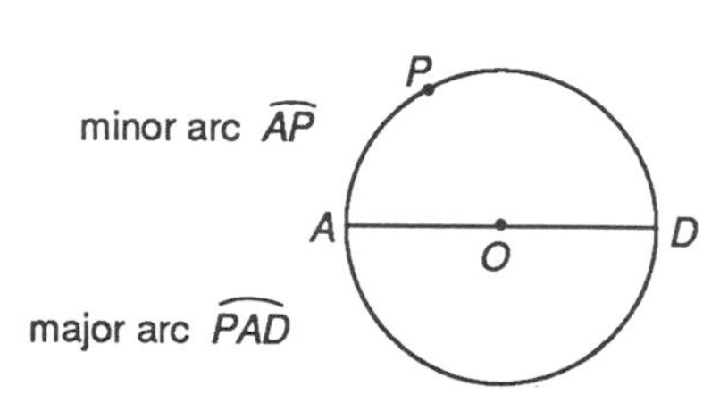

*A **minor arc** is an arc of a circle that is smaller than a semicircle of the circle.*

*A **major arc** is an arc of a circle that is larger than a semicircle of the circle.*

You name minor arcs with the letters of the two endpoints of the arc. You name semicircles and major arcs with the letters of three points — the first and last letters are the endpoints and the middle letter is any other point on the arc.

EXERCISE SET B

Use the diagram on the right for Exercises 1 to 7.

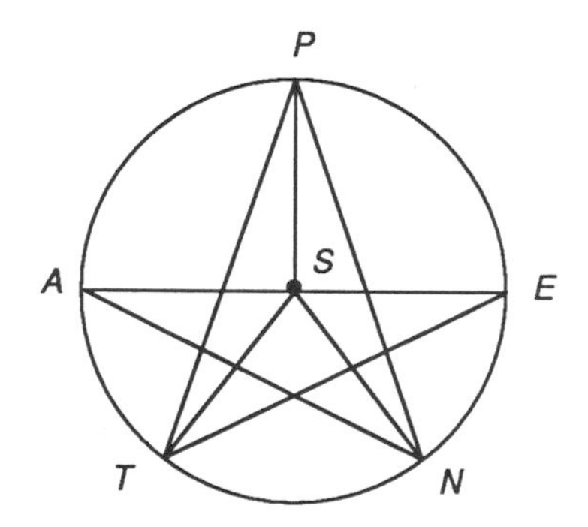

1. Name three chords.
2. Name five central angles.
3. Name one diameter.
4. Name two inscribed angles.
5. Name five radii.
6. Name five minor arcs.
7. Name two semicircles.

Use the diagram on the right for Exercises 8 to 14.

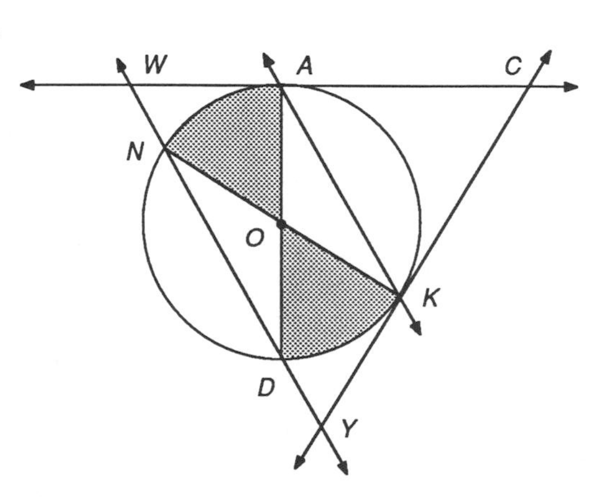

8. Name two tangents.
9. Name two secants.
10. Name three minor arcs.
11. Name two major arcs.
12. Name one central angle.
13. Name one inscribed angle.
14. Name two chords.

EXERCISE SET C

Use the segments below with lengths r and s in Exercises 1 to 4.

r ______________________ s ______________________

1. Construct two circles with radius r intersecting at two points. Label the centers P and Q. Label the points of intersection of the two circles A and B. Construct quadrilateral $PAQB$. What type of quadrilateral is $PAQB$?

2. Construct two circles with radius s intersecting in exactly one point. What is the distance between the two centers?

3. Construct a pair of concentric circles with radii r and s.

4.* Construct two circles with radius r so that each circle passes through the center of the other circle. Label the centers P and Q. Construct $\overline{PQ}$ connecting the centers. Label the points of intersection of the two circles A and B. Construct chord $\overline{AB}$. What is the relationship between $\overline{AB}$ and $\overline{PQ}$?

5. Do you remember the daisy construction from Chapter 0? Construct a circle with radius s. With the same compass setting, see how many times the radius fits about the circle. If you are careful, you should be able to divide the circle into six congruent arcs. Construct the chords to form a regular hexagon inscribed in the circle. Construct radii to each of the six points on the circle. What type of triangles are formed?

EXERCISE SET D

1. Name two types of vehicles that use wheels, two household appliances that use wheels, and two uses of the wheel in the world of entertainment.

2. Name two places or objects where concentric circles appear.

3. Collect two different circular objects (coffee can, saucer, etc.) that can be used to trace circles (the larger the better) onto a sheet of notebook paper. Bring them to class. (Remember show and tell?) You'll need them again in Lessons 6.2 and 6.5. (Make sure that you bring objects that won't be missed at home for a few days.)

Lesson 6.2

Discovering Circle Properties

In this lesson you will discover some properties of chords, arcs, and central angles. The first investigation is about chords and central angles.

Investigation 6.2.1 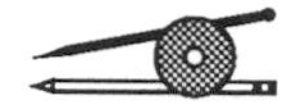 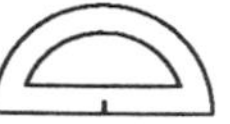

Step 1: Construct a large circle. Label the center O.

Step 2: Construct two congruent chords in your circle. (Use your compass to guarantee that they are congruent.) Label the chords $\overline{AB}$ and $\overline{CD}$.

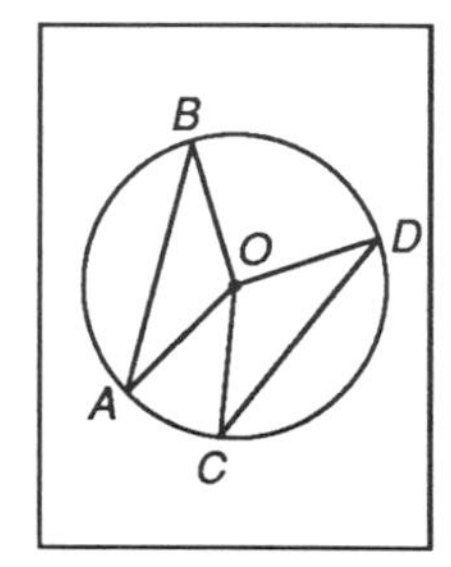

Step 3: Construct radii $\overline{OA}$, $\overline{OB}$, $\overline{OC}$, and $\overline{OD}$.

Step 4: Now, measure $\angle BOA$ and $\angle COD$ with your protractor.

Compare your results with the results of others near you. State your observations as your next conjecture.

If two chords in a circle are congruent, then they determine two central angles that are —?—.

It is not difficult to present an argument demonstrating that this conjecture fits logically with your previous conjectures.

Paragraph Proof:

$OA = OB = OC = OD$ since by definition all radii of the same circle are congruent. It is given that $AB = CD$. Therefore $\Delta BOA \cong \Delta COD$ by the SSS Congruence Conjecture. Therefore, by CPCTC, all the corresponding angles are congruent. In particular, $\angle BOA \cong \angle COD$.

You measure a minor arc by the number of degrees in its central angle. For example, the central angle BOA on the right has a measure of 40 and, therefore, the measure of the intercepted arc AB is 40 ($m\overset{\frown}{AB} = 40$). A semicircle has a measure of 180. A circle has a measure of 360. The measure of a major arc is 360 minus the measure of the minor arc making up the remainder of the circle. For example, the measure of major arc BCA is $360 - 40$ or 320 ($m\overset{\frown}{BCA} = 320$).

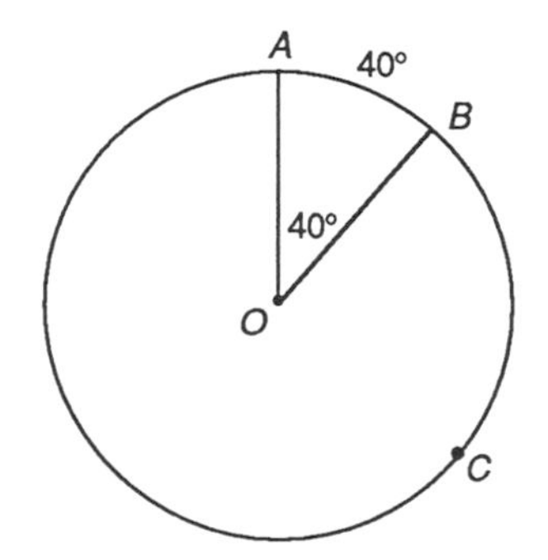

If two central angles are congruent, it follows that their two intercepted arcs must be congruent.

Your next conjecture follows almost immediately from Conjecture 38 and the definition of arc measure. Conjecture 38 states: *If two chords in a circle are congruent, then they determine two central angles that are congruent.* However, the measure of an arc is equal to the measure of its central angle. Therefore, if two central angles are congruent, their intercepted arcs must be congruent. These two statements can be linked together.

If two chords in a circle are congruent, then their —?— are congruent.

In the next two investigations you will discover relationships about chords that are congruent and chords that are not congruent.

Investigation 6.2.2

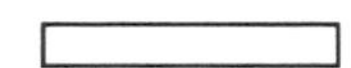

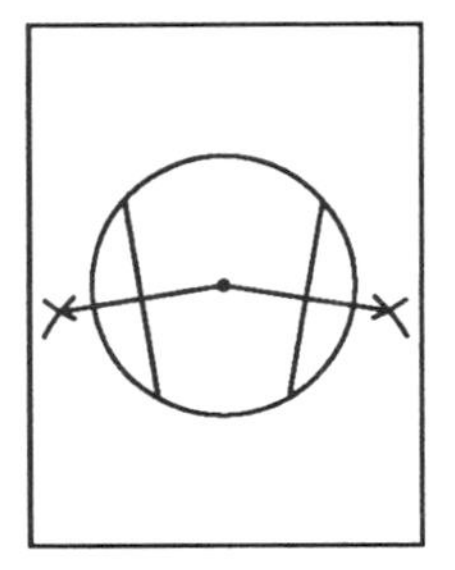

Step 1: Construct a large circle. Mark the center.

Step 2: Construct two non-parallel congruent chords that are not diameters.

Step 3: Construct the perpendiculars from the center to each chord.

Step 4: With your compass, compare the distances from the center to the chords (measure along the perpendicular from the center to the chords).

State your observations as your next conjecture.

Two congruent chords in a circle are —?— from the center of the circle.

In the next investigation you will discover a property of perpendicular bisectors of chords in a circle.

Investigation 6.2.3

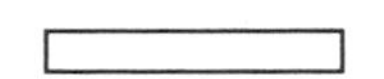

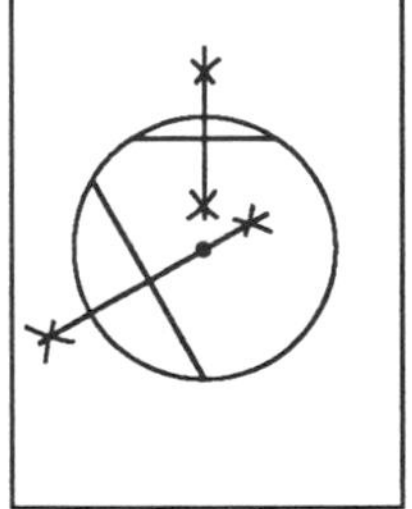

Step 1: Construct a large circle. Mark the center.

Step 2: Construct two non-parallel chords that are not diameters.

Step 3: Construct the perpendicular bisector of each chord and extend them until they intersect.

What is special about the point of intersection? Compare your results with the results of others near you. State your observations as a conjecture.

The perpendicular bisector of a chord —?—.

EXERCISE SET A

Use your new conjectures to solve each problem below. Which conjecture supports your conclusion?

1. $x = $ –?–

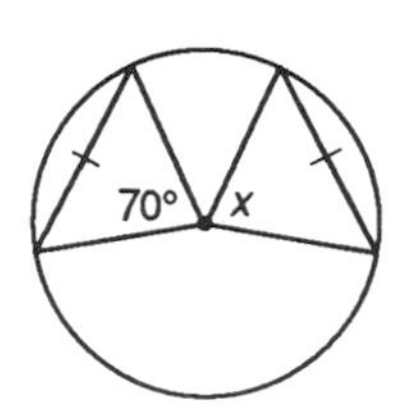

2. $x = $ –?–

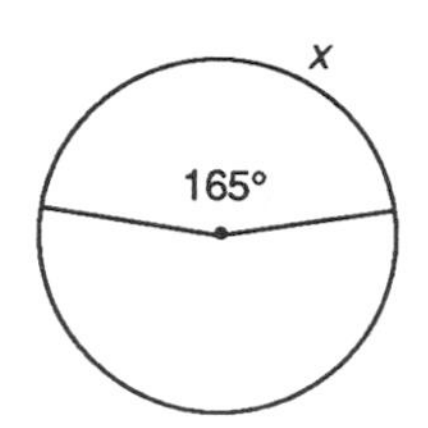

3.* $x = $ –?–

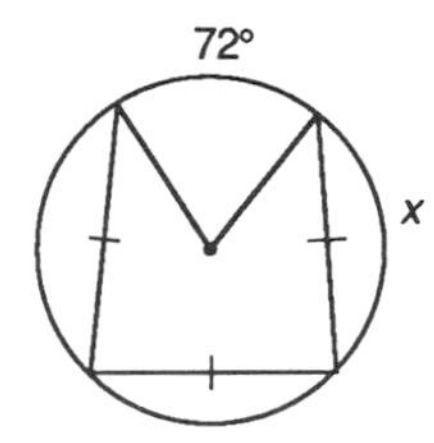

4. $x = $ –?–

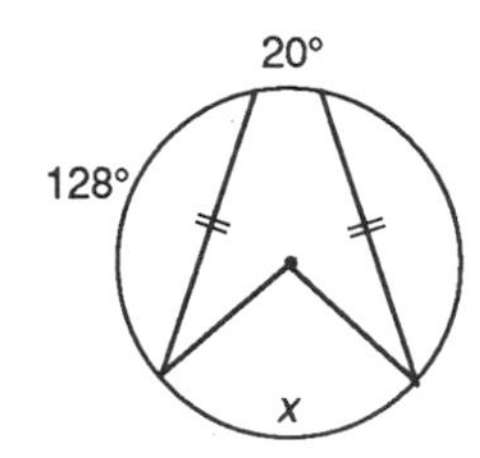

5. $AB = CD$
$PO = 8$ cm
$OQ = $ –?–

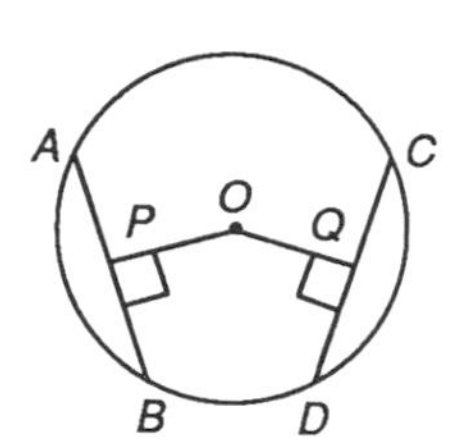

6. $AB = 6$ cm $\quad OP = 4$ cm
$CD = 8$ cm $\quad OQ = 3$ cm
$BD = 6$ cm
What is the perimeter of *OPBDQ*?

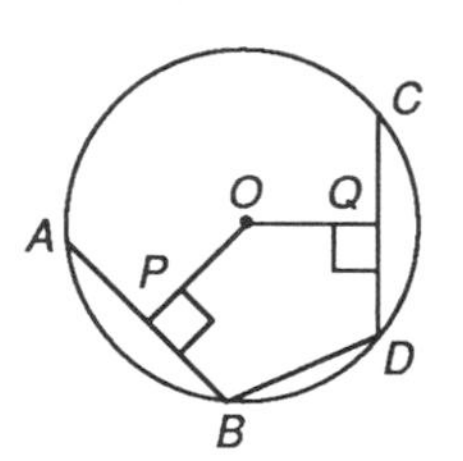

EXERCISE SET B

Conjecture 41 is useful if you need to find the unmarked center of a circle.

1. Use one of the circular objects you collected in Lesson 6.1 to trace a circle onto a clean sheet of paper. Don't use your compass (because then you'll know where the center is). Construct two chords. Construct the perpendicular bisector of each. Locate the center of the circle. Label it *O*.
2. Use another circular object to trace a large minor arc. Locate by construction a point on the arc equally distant from the arc's endpoints. Label it *P*.
3. Construct a triangle. Construct a circle passing through all three vertices. (The sides of the triangle become chords of the circle.) Why does this seem familiar?
4. Draw a circle. Draw two chords of the circle of unequal length. Which is closer to the center of the circle, the larger chord or smaller chord?
5. Draw two circles with different length radii. Draw a chord in each circle with the same length. Each chord determines a central angle. Draw the central angles. Which central angle is larger?

6. A piece of ceramic plate was recently dug up on the island of Samos. The Greek government gives rewards for all antiquities discovered. It is known by archaeologists that classical plates with this particular design have a diameter of 24 cm. Elton Notle, the discoverer of the piece of plate, wishes to calculate the diameter to see if it might be from a classical Greek plate. Trace the outer edge of the plate onto a sheet of paper. Determine the diameter for Elton Notle. Perhaps he will share the reward if it is part of a classic Greek plate.

7. A large object recently crashed into the lunar surface leaving a gigantic circular crater (between marks 24, 25, 34, and 35). Scientists at NASA have reason to believe the object may have been launched by creatures from another galaxy. Unfortunately, the single photo received from the satellite shows only a portion of the new crater. NASA cartographers must locate the center of the crater and determine its radius in order to program the Lunar Landrover to retrieve the mysterious space object. Trace the outer edge of the crater in the diagram onto a clean unlined sheet of paper. Locate the crater's center. Using the scale shown, find the radius in kilometers.

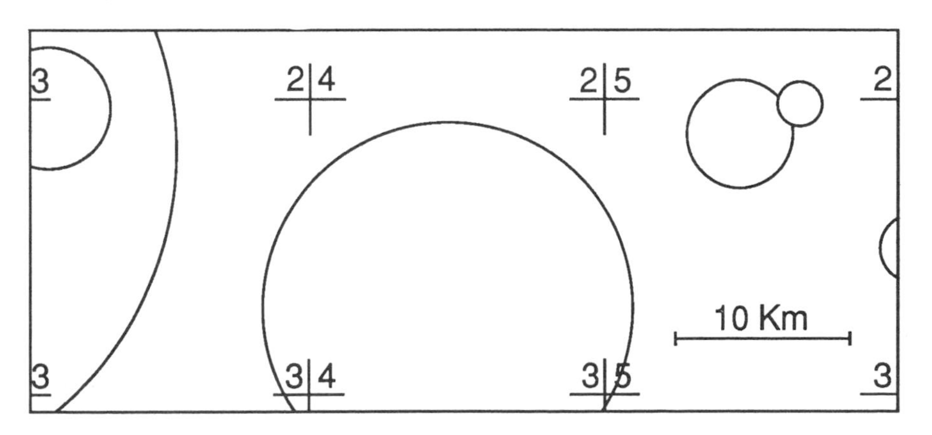

Lesson 6.3
Discovering Tangent Properties

We are, all of us, alone
Though not uncommon
In our singularity.

Touching,
We become tangent to
Circles of common experience,
Co-incident,
Defining in collective tangency
Circles
Reciprocal in their subtle
Redefinition of us.

In tangency
We are never less alone,
But no longer

Only.

– Gene Mattingly

In this lesson you will discover some properties of tangents. In the first investigation you will discover something about the angle formed by a tangent and the radius drawn to the point of tangency.

Investigation 6.3.1  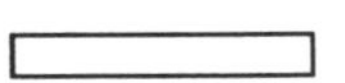 

Step 1: Construct a large circle. Label the center O.

Step 2: Using your straightedge, draw a line which appears to touch the circle at only one point. Label the point T. Construct $\overline{OT}$.

Step 3: Use your protractor to measure the angles at T.

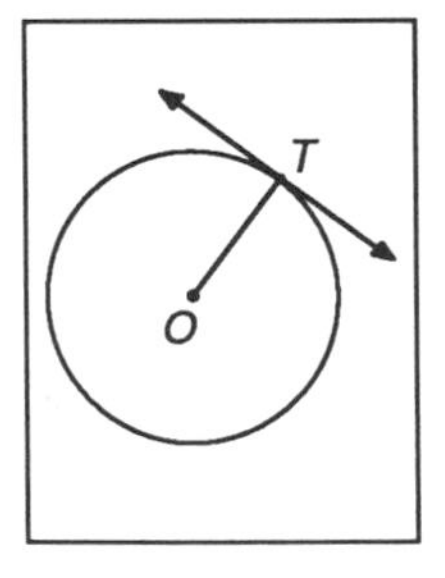

Compare your results with the results of others near you. State your observations as a conjecture.

A tangent to a circle is —?— to the radius drawn to the point of tangency. ***(Tangent Conjecture)***

In the next investigation you will discover something about the lengths of segments tangent to a circle from a point outside the circle.

Investigation 6.3.2  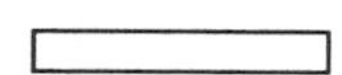

Step 1: Construct a circle. Label the center E.

Step 2: Choose a point outside the circle and label it N.

Step 3: Draw two lines through N which appear to be tangent to the circle. Mark and label the points where these lines appear to touch as A and G.

Step 4: With your compass, compare NA and NG. ($\overline{NA}$ and $\overline{NG}$ are called **tangent segments.**)

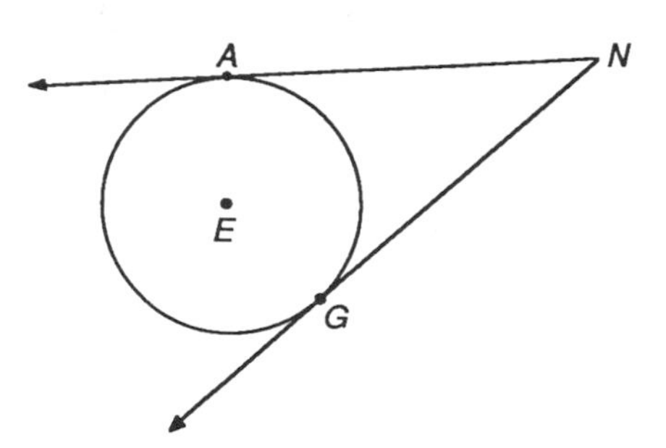

Compare your results with the results of others near you. State your observations as your next conjecture.

 C-43 *Tangent segments to a circle from a point outside the circle are —?—. **(Tangent Segments Conjecture)***

EXERCISE SET A

1.* m and n are tangents.
$w = $ –?–

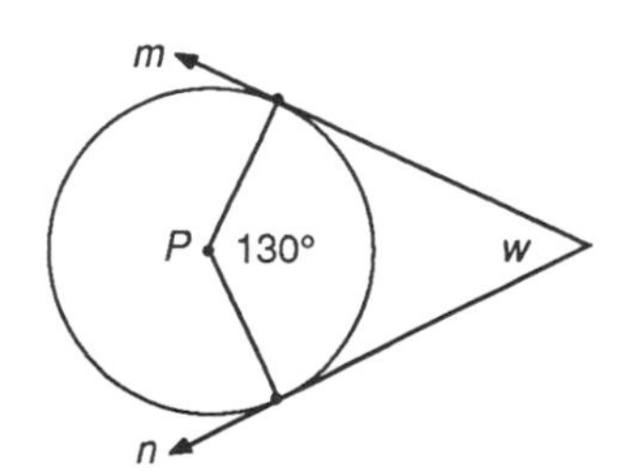

2.* r and s are tangents.
$x = $ –?–

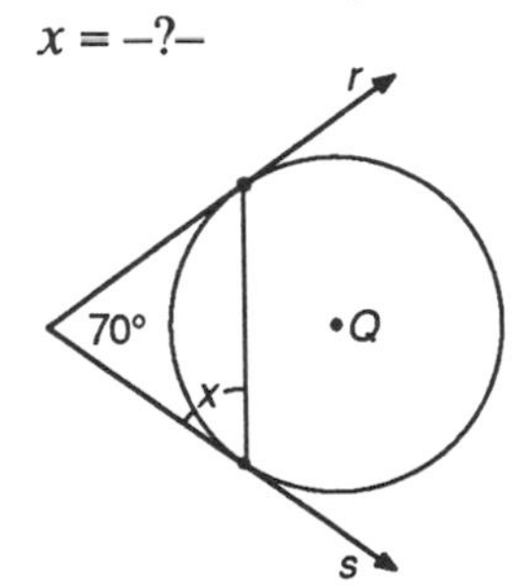

3. k is a tangent.
$y = $ –?–

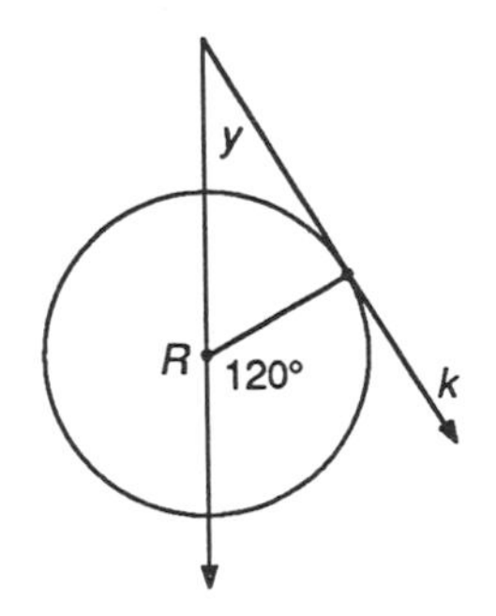

4. t is a tangent to both circles.
$z = $ –?–

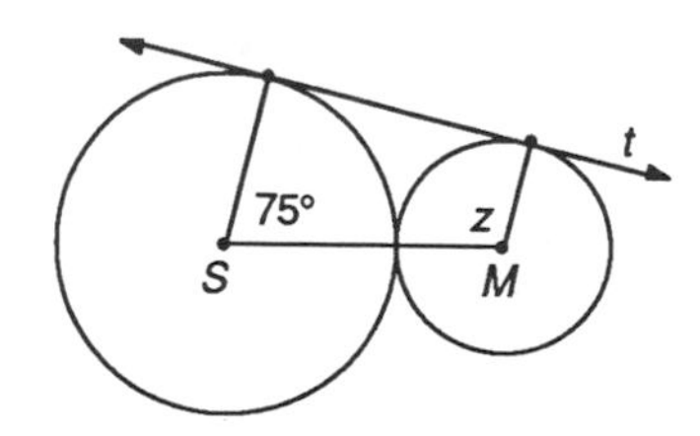

5. Quadrilateral $ABCD$ is circumscribed about circle O.
$CD = 14$ $BR = 4$
$AD = 11$ $AM = 5$
What is the perimeter of $ABCD$?

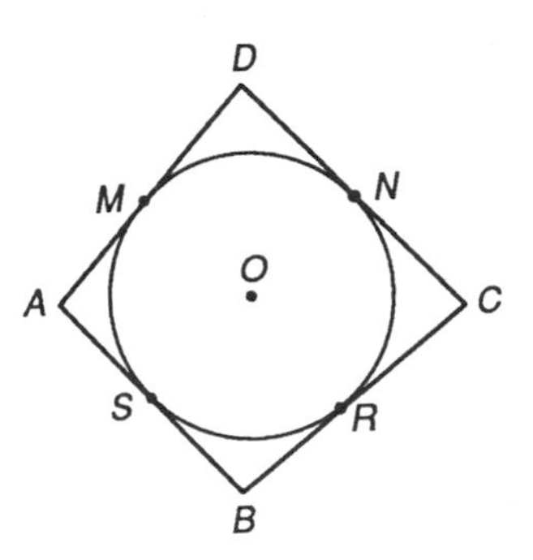

6. Quadrilateral $ABCD$ is circumscribed about circle O.
$BX = 13$ $CD = 12$
What is the perimeter of $ABCD$?

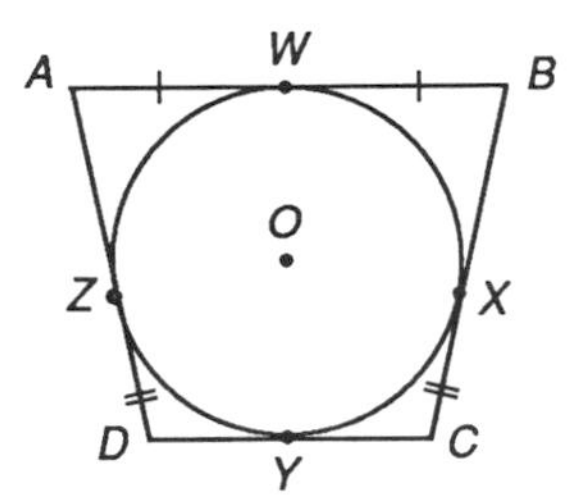

A line or segment that is tangent to two circles is called a ***common tangent****.*

The diagram below shows two types of common tangents. A **common external tangent** *does not* intersect the line segment connecting the centers of the circles. A **common internal tangent** *does* intersect the line segment connecting the centers of the two circles.

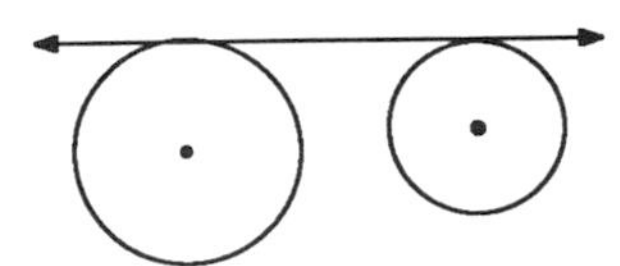

Common External Tangent

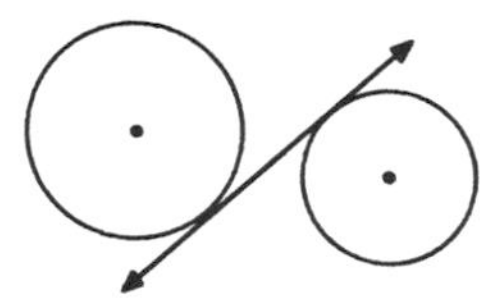

Common Internal Tangent

Tangent circles *are two circles that are tangent to the same line at the same point.*

They can be **internally tangent** or **externally tangent** as shown in the diagrams.

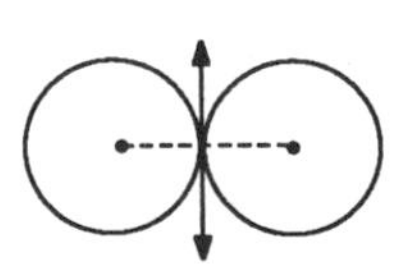

Externally Tangent Circles

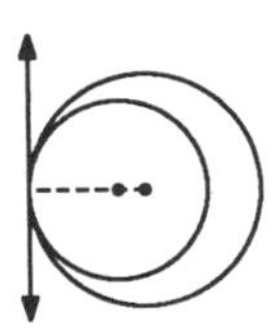

Internally Tangent Circles

EXERCISE SET B

In each construction, make a sketch of what you are trying to construct and label it. This helps clarify your task. Use the segments below with lengths r, s, and t.

r ________________ s ____________ t __________

1.* Construct a circle with radius r. Mark a point on the circle. Construct a tangent through this point.

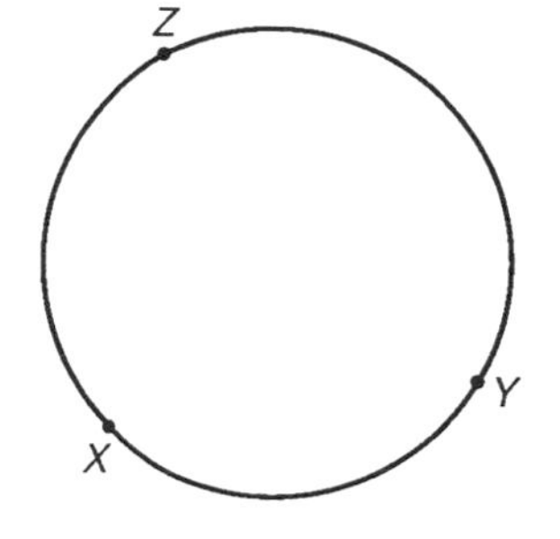

2. Construct a circle with radius t. Choose three points on the circle and label them X, Y, and Z (see diagram at right). Construct a triangle which is circumscribed about the circle and tangent at points X, Y, and Z.

3. Construct two congruent externally tangent circles with radius s.

4. Construct a third circle that is both congruent and externally tangent to the two circles of Exercise 3.

5. Construct two internally tangent circles with radii r and t.

6. Construct a third circle with radius s that is externally tangent to both circles of Exercise 5.

7. In Chinese cosmology, all things are divided into two natural principles, Yin and Yang. Yin is the feminine principle characterized by darkness, cold, or wetness. Yang is the masculine principle characterized by light, heat, or dryness. The two principles combine to produce the harmony of nature. The symbol for Yin-Yang is shown at right. Construct your own Yin-Yang symbol. Start with one large circle. Then construct two circles with half the diameter that are internally tangent to the large circle and externally tangent to each other. Finally, construct small circles that are concentric to the two inside circles. Shade or color your Yin-Yang symbol.

Lesson 6.4

Arcs and Angles

Recall that you measure a minor arc by the number of degrees in its central angle. The measure of a major arc is 360 minus the measure of the minor arc making up the remainder of the circle. How does the measure of an inscribed angle compare with the measure of its intercepted arc? Let's investigate.

Investigation 6.4.1

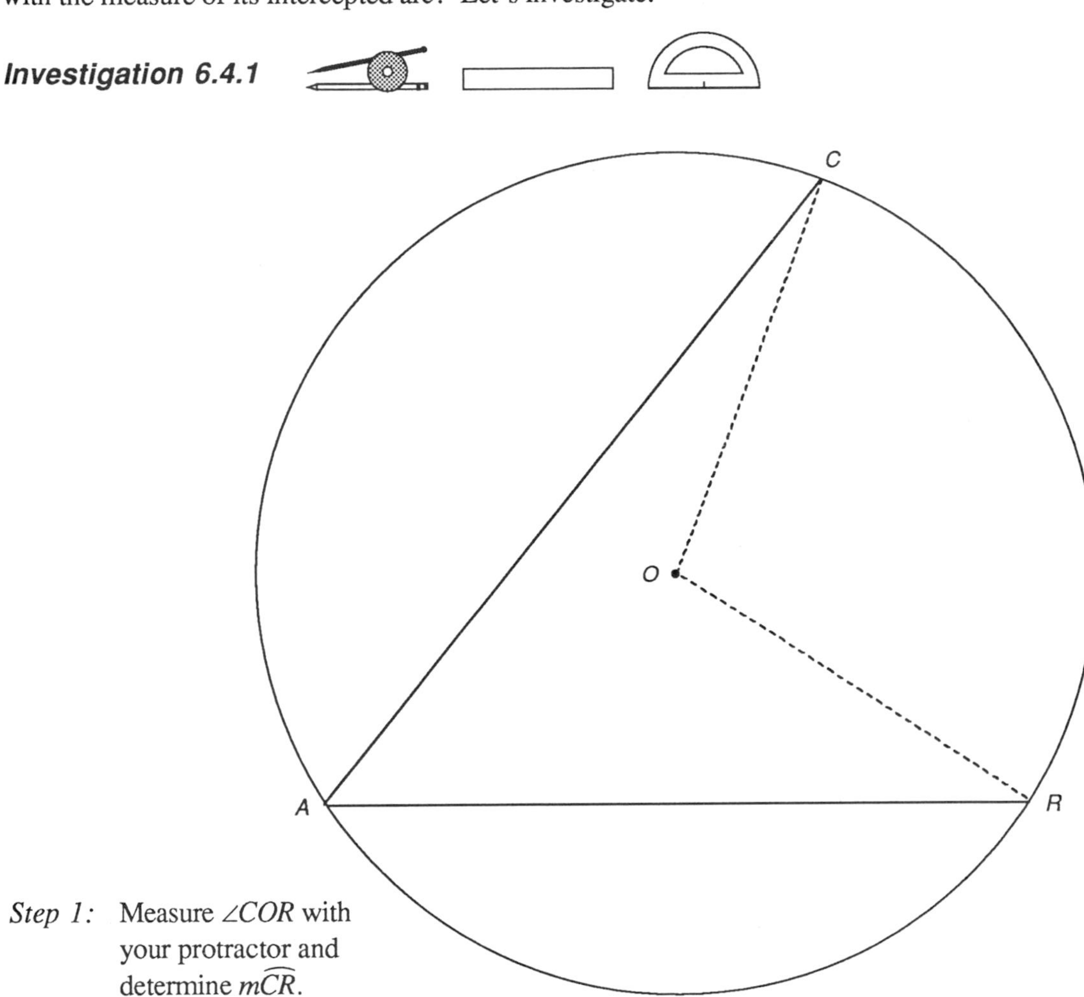

Step 1: Measure $\angle COR$ with your protractor and determine $m\widehat{CR}$.

Step 2: Measure $\angle CAR$. How does $m\angle CAR$ compare with $m\widehat{CR}$?

Step 3: Construct a circle of your own with an inscribed angle and its corresponding central angle.

Step 4: Measure the central angle. What is the measure of the intercepted arc?

Step 5: Measure the inscribed angle. How does the measure of the inscribed angle compare with the measure of its intercepted arc?

Compare your results with the results of others. State a conjecture.

 C-44 *The measure of an inscribed angle in a circle —?—.* ***(Inscribed Angle Conjecture)***

In the drawing on the right, $\angle AQB$ and $\angle APB$ both intercept $\overset{\frown}{AB}$. $\angle AQB$ and $\angle APB$ are both inscribed in $\overset{\frown}{APB}$. $\angle AQB$ and $\angle APB$ appear to be congruent. Can you find angles inscribed in the same arc that are not congruent? Let's investigate.

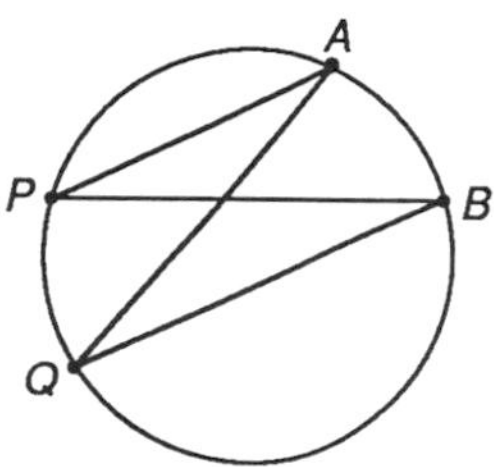

Investigation 6.4.2  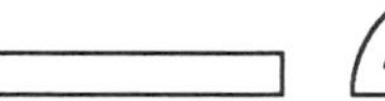 

Step 1: Construct a large circle.

Step 2: Select two points on the circle Label them A and B.

Step 3: Select a point P on the major arc and construct inscribed $\angle APB$.

Step 4: Measure $\angle APB$ with your protractor.

Step 5: Select another point Q on major arc APB and construct inscribed $\angle AQB$.

Step 6: Measure $\angle AQB$. How does the measure of $\angle AQB$ compare with the measure of $\angle APB$?

Repeat steps 1 to 6 with points P and Q selected on the minor arc AB. Compare the measure of $\angle AQB$ with the measure of $\angle APB$? Compare your results with the results of others near you. Do you think you can find an angle inscribed in APB that is not congruent to $\angle APB$? State your observations as a conjecture.

 C-45 *Angles inscribed in the same arc are —?—.*

Next you will discover a property of angles inscribed in semicircles.

Investigation 6.4.3  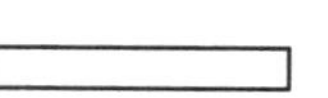

Step 1: Construct a large circle.

Step 2: Construct a diameter.

Step 3: Inscribe three angles in the same semicircle.

Step 4: Measure each angle with your protractor.

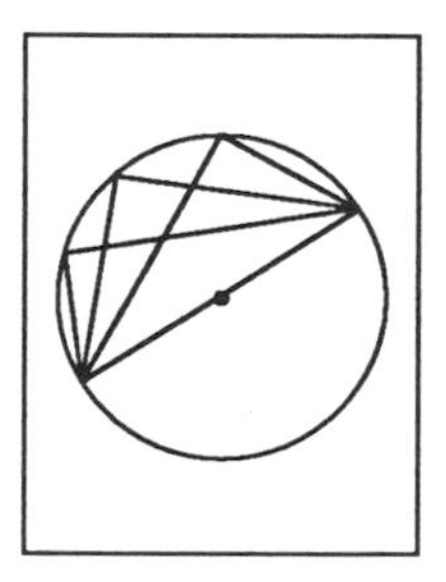

Compare your results with the results of others and make a conjecture.

 C-46 *Every angle inscribed in a semicircle is —?—.*

Now you will discover a property of the angles of a quadrilateral inscribed in a circle.

Investigation 6.4.4  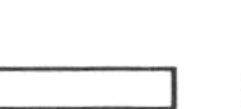 

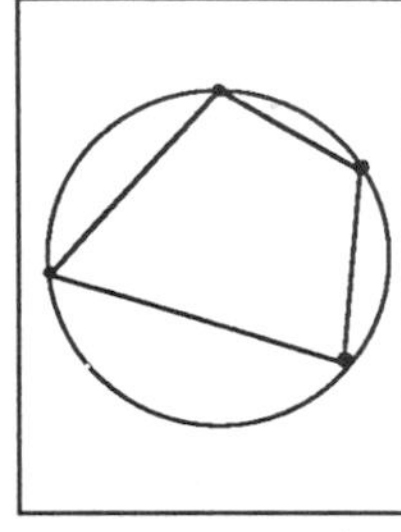

Step 1: Construct a large circle.

Step 2: Construct an inscribed quadrilateral.

Step 3: Measure each of the four inscribed angles. Write the measure in each angle.

It is unlikely that any of the angles are congruent, but there is a special relationship between some pairs of angles. Compare your observations with the observations of those near you. State your findings as your next conjecture.

The —?— angles of quadrilateral inscribed in a circle are —?—.

Next you will discover a property of arcs formed by parallel lines intersecting a circle.

Investigation 6.4.5  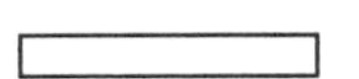

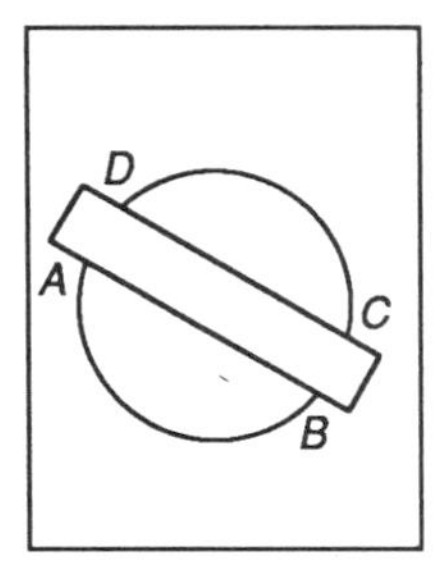

Step 1: Construct a large circle.

Step 2: The top and bottom edges of your straightedge should be parallel. Lay your straightedge across the circle so that both edges pass through the circle.

Step 3: Draw lines along both edges of the straightedge. Label one chord AB and the other CD as shown.

Step 4: With your compass, compare the distances AD and BC.

What does this tell you about the arcs AD and BC? Repeat these steps with something else having parallel edges (or use lined paper). Compare your results with the results of others near you. State your observations as a conjecture.

Parallel lines intercept —?— arcs on a circle.

EXERCISE SET

Use your new conjectures to solve each problem.

1. $a = $ –?–

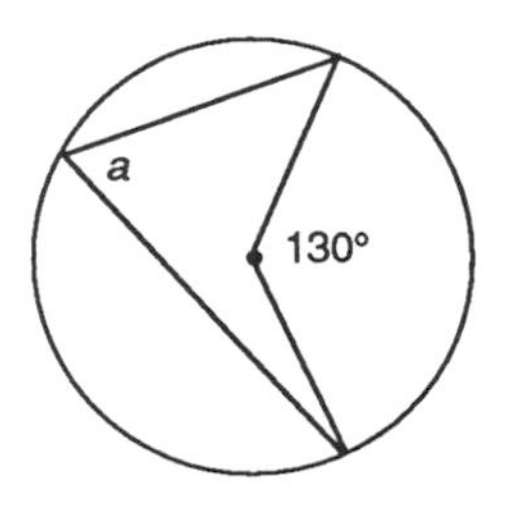

2. $b = $ –?–

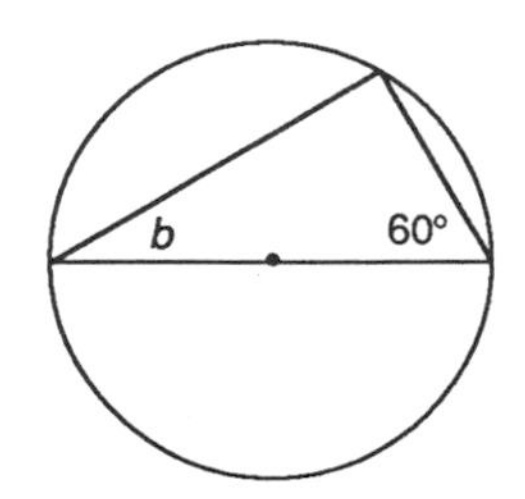

3.* $c = $ –?–

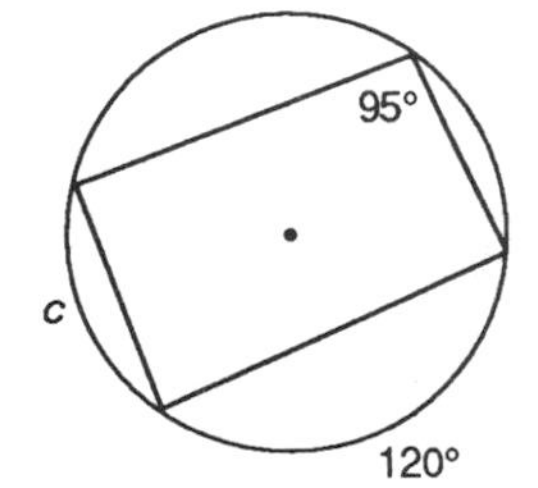

4. $d = -?-$
$e = -?-$

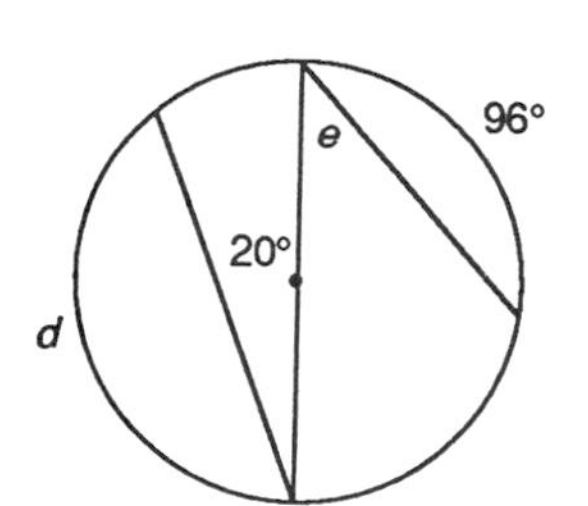

5. $f = -?-$
$g = -?-$

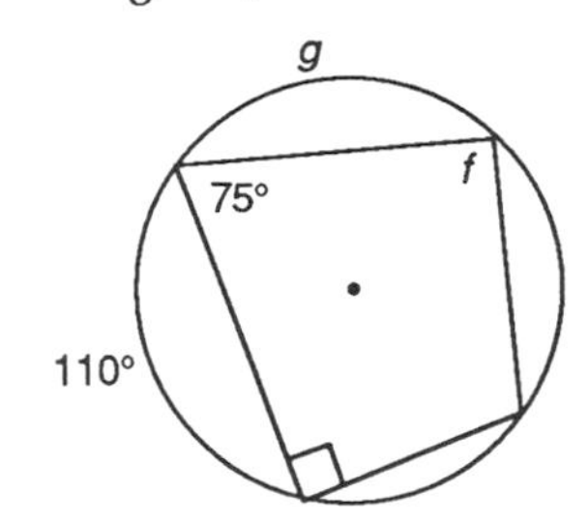

6.* $h = -?-$

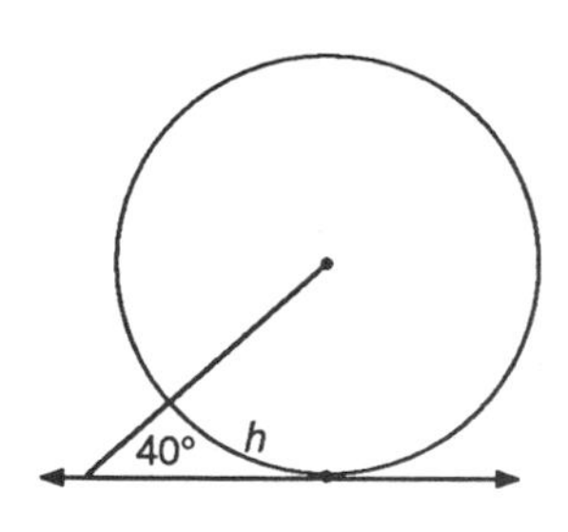

7. $k = -?-$

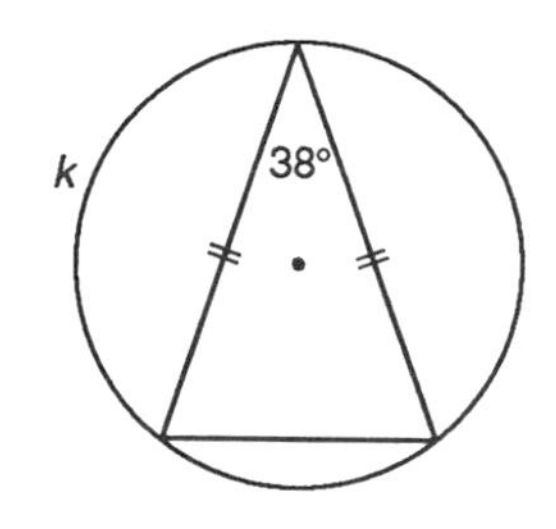

8. $m = -?-$
$n = -?-$

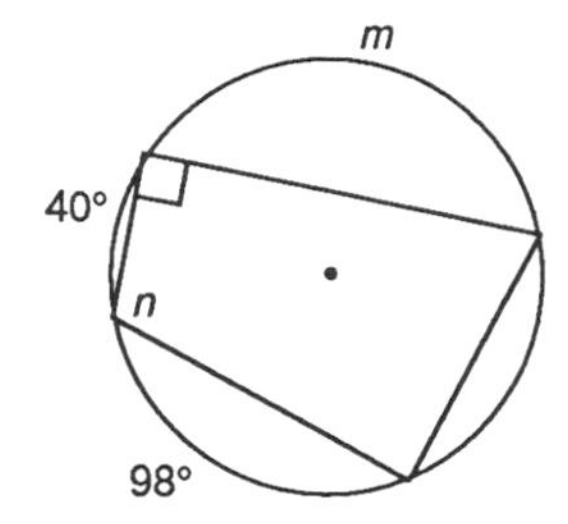

9.* $\overline{AB} \parallel \overline{CD}$
$p = -?-$
$q = -?-$

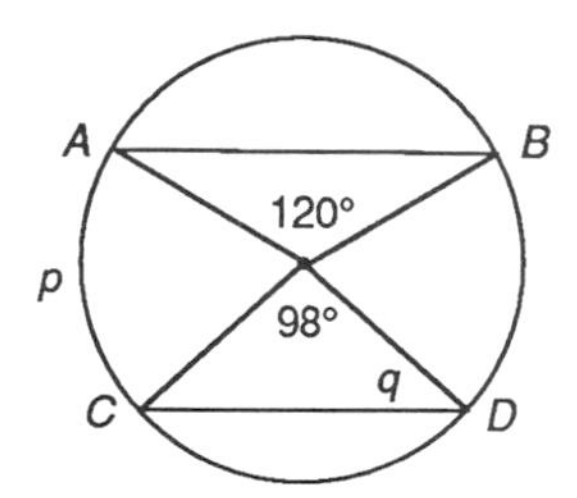

10. $r = -?-$
$s = -?-$

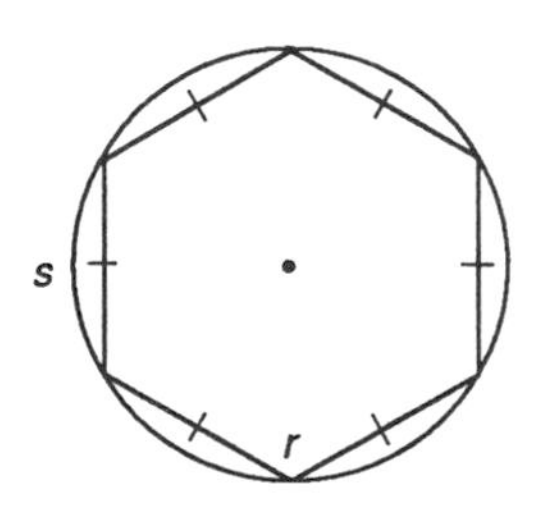

11.* What is the sum of
$a + b + c + d + e$?

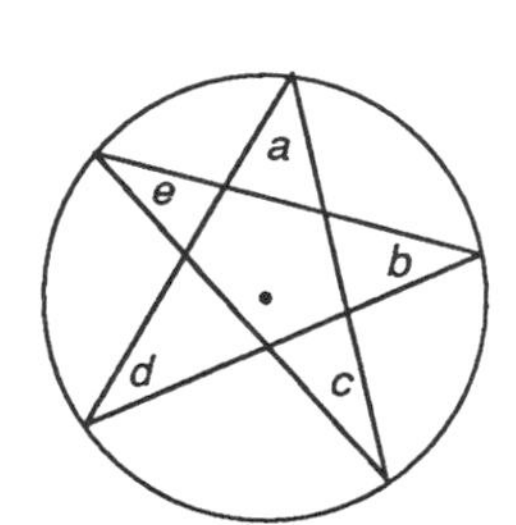

12.* $y = -?-$

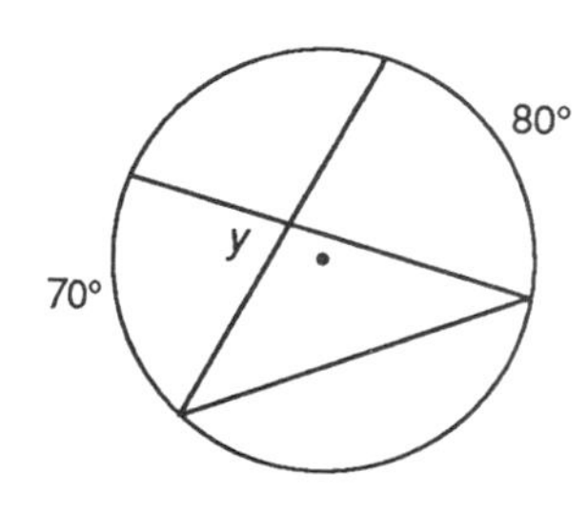

Improving Reasoning Skills

Container Problem I

You have two unmarked containers, a 9-liter container and a 4-liter container, and an unlimited supply of water. In table, symbol, or paragraph form describe the process necessary to end up with exactly 3 liters in one of the containers.

Lesson 6.5

The Circumference/Diameter Ratio

Here's a nice puzzle.

One of two quarters remains motionless while the other rotates around it, never slipping but always tangent to it. When the rotating quarter has completed a turn around the stationary quarter, how many turns has it made around its own center point?

Did you guess one? Two? Three? The solution to the puzzle is not as obvious as it first appears. The best way of seeing the solution is actually to roll one coin about the other. Mark both coins with a felt-tip pen or pencil and try it.

In a polygon, the "distance around the figure" is called the perimeter. In circles, the "distance around the figure" is called the **circumference**. In the puzzle above, one quarter is rolling along the circumference of the other quarter. In first thinking about the puzzle, you probably thought that since the circumferences of the two coins were the same, one coin would rotate about the other in one revolution. This was not the case, right?

In this lesson you are going to discover (or perhaps rediscover) the relationship between the diameter and the circumference of every circle. Once you know this relationship, you can measure a circle's diameter and calculate its circumference.

You probably already know this relationship. In every circle the circumference is slightly more than three times the diameter. If you measure the circumference and diameter of a circle and divide the circumference by the diameter, you get a number slightly larger than three. The more accurate your measurements, the closer your ratio will come to a special number called π (pi), pronounced like one of your favorite desserts.

In the next investigation you will experimentally determine an approximate value of π by measuring circular objects and calculating the circumference/diameter ratio. You should work in groups of four or five, sharing the tasks of measuring, recording, and calculating. Let's see how close you come to the actual value of π.

Investigation 6.5.1 

For this investigation, you will need the following special materials:

- Round objects you collected in Lesson 6.1 (the larger the objects the better)
- Meter stick or metric sewing tape

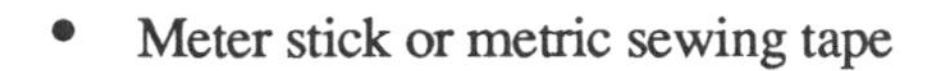

- Sewing thread or thin string to measure the circumference of each round object

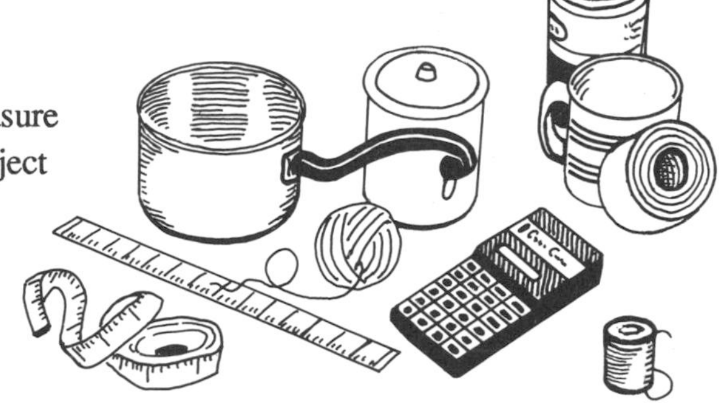

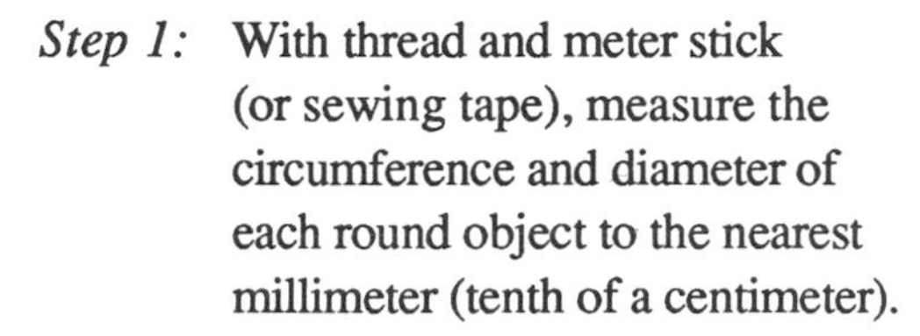

Step 1: With thread and meter stick (or sewing tape), measure the circumference and diameter of each round object to the nearest millimeter (tenth of a centimeter).

Step 2: Make a table similar to the one below and record the circumference (C) and diameter (D) measurements for each round object.

Name of object	–?–	–?–	–?–	–?–
Circumference (C)	–?–	–?–	–?–	–?–
Diameter (D)	–?–	–?–	–?–	–?–
C/D	–?–	–?–	–?–	–?–

Step 3: Calculate C/D, and place the answers in your table.

Step 4: Calculate the average of your C/D results.

Compare the average of your C/D results with the C/D averages of other groups. Notice the closeness of the C/D answers? You should now be convinced that C/D is very close to 3 for every circle. We define the ratio: $C/D = \pi$. If you solve this formula for C, you get a formula for finding the circumference of a circle in terms of the diameter. Since the diameter is twice the radius ($D = 2r$), you also can get a formula for finding the circumference in terms of the radius. State your conjecture.

If C is the circumference and D is the diameter of a circle, then there is a number π such that C = —?—. Since D = 2r where r is the radius, then C = —?—.
(Circumference Conjecture)

The number π is an irrational number. Its decimal form never ends and never repeats. The symbol π is a letter from the ancient Greek alphabet. Mathematicians began using it in the eighteenth century to represent the exact value of the circumference/diameter ratio. When you work a problem involving π, you too must use this symbol if you want to give your answer exactly.

What do you think a carpenter would say if you asked her to cut a board so that it was 3π feet long? In practical situations, approximations for π are more useful than a funny looking symbol. 3.14 is a rough decimal approximation of π and 22/7 is a rough fractional approximation. In this text, you'll use these approximations in solving some problems. In some situations, however, much more exact approximations are appropriate.

π has fascinated and intrigued mathematicians for millenniums. The ancient Egyptians used 4/3 to the fourth power (about 3.1605) as their approximation of π. The Chinese by 480 AD were using 355/113 as their approximation of π (which is accurate to six decimal places). Today, computers have calculated π accurate to millions of decimal places.

How do you use the Circumference Conjecture. Let's look at two examples.

Example A

If a circle has a circumference of 12π meters, what is the radius?

$$\begin{aligned} C &= 2\pi r \\ 12\pi &= 2\pi r \\ r &= 6 \end{aligned}$$

The radius is 6 meters.

Example B

If a circle has a diameter of 3 meters, what is the circumference? Use 3.14 for π.

$$\begin{aligned} C &= \pi D \\ &\approx (3.14)(3) \\ &\approx 9.42 \end{aligned}$$

The circumference is about 9.42 meters.

EXERCISE SET

Use the Circumference Conjecture to solve each problem below. Do not use an approximation for π in Exercises 1 to 8.

1. If $r = 5$ cm, find C.
2. If $C = 5\pi$ cm, find D.
3. If $C = 24$ m, find r.
4.* If $D = 5\pi$ m, find C.
5. If a circle has a diameter of 12 cm, what is its circumference?
6. If a circle has a circumference of 46π meters, what is its diameter?
7.* If a circle is inscribed in a square with a perimeter of 24 cm, what is the circumference of the circle?
8. If a circle with a circumference of 16π inches is circumscribed about a square, what is the length of a diagonal of the square?

In Exercises 9 to 12, use the symbol $\approx$ to show that your answer is an approximation.

9. If $D = 5$ cm, find C. Use 3.14 for π.
10. If $r = 4$ cm, find C. Use 3.14 for π.
11. If $r = 7$ m, find C. Use 22/7 for π.
12.* If $C = 44$ m, find r. Used 22/7 for π.

Lesson 6.6

Around the World

EXERCISE SET

1. Around the world in eighty days? If the diameter of the earth is 8,000 miles, calculate your speed in miles per hour if you were to take 80 days to circumnavigate the earth about the equator. Use 3.14 for π.

$$\text{speed} = \frac{\text{distance}}{\text{time}} = \frac{\text{circumference}}{(80 \text{ days} \times 24 \text{ hours})}$$

2. A wheel has a diameter of 35 cm and travels 66 meters. How many revolutions did it make in the trip? (In other words, how many circumferences of the circle are in 66 meters?) Use 22/7 for π.

3.* A satellite has recently been placed in a nearly circular orbit 2000 kilometers above the earth's surface. Given that the radius of the earth is approximately 6400 kilometers and that the satellite completes its orbit in 12 hours, calculate the speed of the satellite in kilometers per hour. Use 22/7 for π.

4. Patty's Pizza Palace is known throughout the city for its delicious pizza with extra thick crust around the edge. Their small "Mama" pizza has a 6 inch radius and sells for $9.75. The medium size "Papa" sells for $12.00 and is a savory 8 inches in radius. The large pizza, "Uncle Guiseppi," is a hefty, mouth-watering 20 inches in diameter and sells for $16.50. Since the edge is the thickest and best part of a Patty's pizza, calculate which size gives the most pizza edge per dollar. Find the circumference of the pizza.

5. Forest Ranger Felicia Ragamuffin gives school tours through the Majestic Redwoods National Park. Someone in every tour undoubtedly asks, "What is the diameter of the giant redwood tree near the park entrance." Felicia has worked out a creative response. The redwood has a circular trunk. She asks the group to find students who can arrange themselves around the base of the tree so that by hugging the tree with arms outstretched, they can just touch fingertips to fingertips. Assuming that the arm span of each student is roughly equivalent to his or her height, Felicia asks the group to calculate the diameter of the tree. In one group, four students with heights of 138 cm, 136 cm, 128 cm, and 126 cm were able to ring the tree. Using 22/7 for π, what is the approximate diameter of the redwood?

6. Wilbur Wrong is flying his model plane at the end of a wire 28 meters long. He is holding it so that the wire is level with the ground. His brother, Orville Wrong, clocks the plane at 16 seconds per revolution. What is the speed of their plane (m/sec)? They may be Wrong but you could be right! Use 22/7 for π.

Lesson 6.7
Arc Length

Here is a short three problem investigation.

Investigation 6.7.1

What fraction of its circle is each arc?

1. $\overset{\frown}{AB}$ is what fraction of circle P?
2. $\overset{\frown}{CED}$ is what fraction of circle O?
3. $\overset{\frown}{EF}$ is what fraction of circle T?

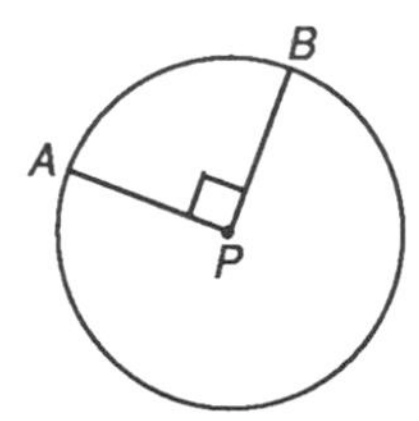

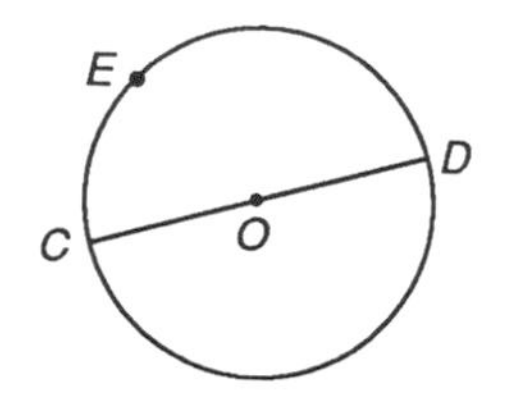

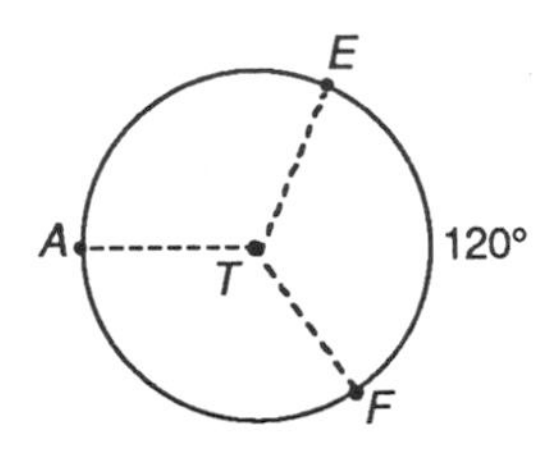

How did you solve the three problems? In Problem 1, you probably "just knew" that it was one-fourth of the circle because you have seen one-fourth of a circle so many times. But why is it one-fourth? It is one-fourth because the arc measures 90, a full circle measures 360, and 90 is one-fourth of 360. As in Problem 1, you probably "just knew" that the arc in Problem 2 was one-half the circle. But it is half because the arc measures 180 and 180 is one-half of 360. In Problem 3, you may or may not have recognized right away that the arc was one-third of the circle. But the arc is one-third of the circle because 120 is one-third of 360.

What fraction of circle O is $\overset{\frown}{GH}$ in the diagram on the right? Here the task is a little more difficult. The fraction is not obvious. But by dividing the arc measure by 360 (80 divided by 360) and then simplifying, you can calculate that the arc is two-ninths of the circle.

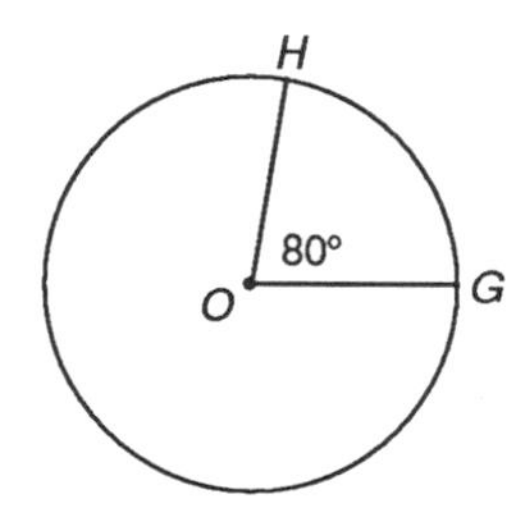

This lesson is about arc length. What is arc length?

> *The **length of an arc** (arc length) is some fraction of the circumference of its circle.*

Arcs are measured in degrees. Arc measure is some fraction of 360 degrees. Arc length is measured, like other lengths, in some unit of length.

Many people confuse arc measure with arc length. Arc measure and arc length are not the same. Perhaps the illustration on the right will help you avoid this confusion. The angle intercepts two different arcs. Both arcs have the same degree measure (40), but clearly, the lengths of the two arcs are not equal.

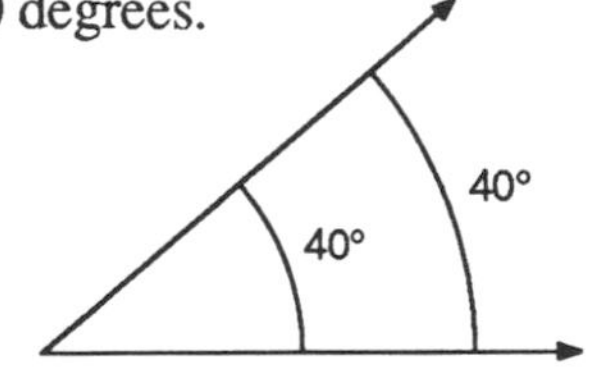

Here is another short investigation. This time you must calculate arc length. You'll need to know what fraction each arc is of its circle and you'll need to know the circle's circumference. With these two pieces of information, you can calculate the arc length.

Investigation 6.7.2

What is the length of each arc?

1. Length of $\overset{\frown}{AB}$ is –?–.
   Radius is 12 meters.

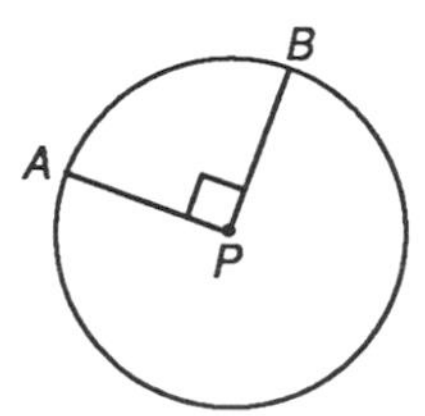

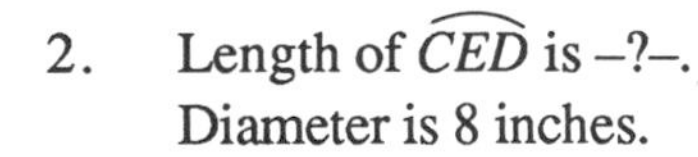

2. Length of $\overset{\frown}{CED}$ is –?–.
   Diameter is 8 inches.

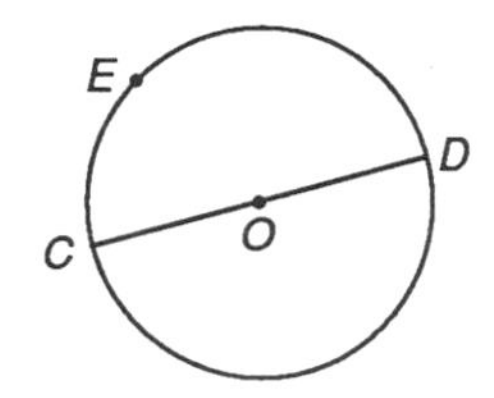

3. Length of $\overset{\frown}{EF}$ is –?–.
   Radius is 36 feet.

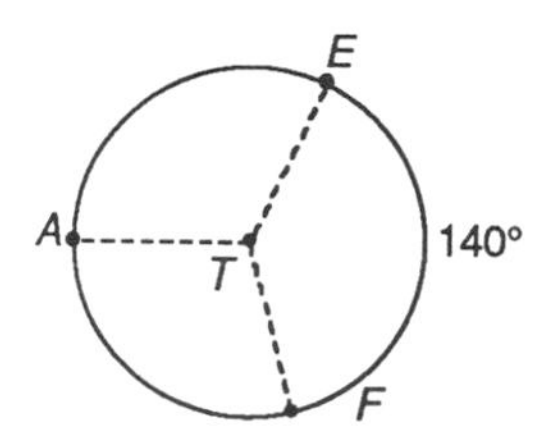

How did you solve these three problems? Perhaps your thinking went like this:

Problem 1: $\frac{90}{360} = \frac{1}{4}$ so $\overset{\frown}{AB}$ is $\frac{1}{4}$ of a circle. (Perhaps you "just knew" it was $\frac{1}{4}$.)
The radius is 12 meters, so the circumference is 24π meters.
$\frac{1}{4}$ of 24π is 6π, so the arc length of $\overset{\frown}{AB}$ is 6π meters.

Problem 2: $\frac{180}{360} = \frac{1}{2}$ so $\overset{\frown}{CED}$ is $\frac{1}{2}$ of a circle. (Perhaps you "just knew" it was $\frac{1}{2}$.)
The diameter is 8 inches, so the circumference is 8π inches.
$\frac{1}{2}$ of 8π is 4π, so the arc length of $\overset{\frown}{CED}$ is 4π inches.

Problem 3: $\frac{140}{360} = \frac{7}{18}$ so $\overset{\frown}{EF}$ is $\frac{7}{18}$ of a circle. (Not obvious from the picture.)
The radius is 36 feet so the circumference is 72π feet.
$\frac{7}{18}$ of 72π is 28π, so the arc length of $\overset{\frown}{EF}$ is 28π feet.

Was this your method? Generalize this method and state it as a conjecture for finding the length of an arc. Add it to your conjecture list.

C-50 *The arc length equals the —?— divided by —?—, times —?—.*
(Arc Length Conjecture)

How do you use this new conjecture? Let's look at a few sample problems.

Example A

If the radius is 24 cm and $m\angle BTA = 60$, then what is the length of $\widehat{AB}$?

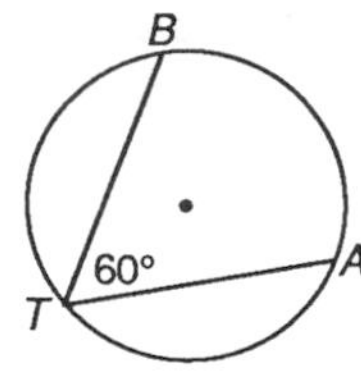

$m\angle BTA = 60$, so $m\widehat{AB} = 120$. $\frac{120}{360} = \frac{1}{3}$.

So the arc length is $\frac{1}{3}$ of the circumference.

Therefore: $\quad$ Arc length $= \frac{1}{3}C$

$$= \frac{1}{3}(48\pi)$$

$$= 16\pi$$

The arc length is 16π cm.

Example B

If the length of $\widehat{ROT}$ is 116π meters, what is the radius of the circle?

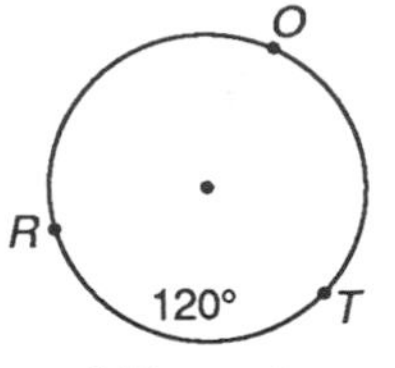

$m\widehat{ROT} = 240$, so $\widehat{ROT}$ is $\frac{240}{360}$ or $\frac{2}{3}$ of the circumference. But the arc length is 116π.

Therefore:

$$116\pi = \frac{2}{3}C$$

$$116\pi = \frac{2}{3}(2\pi r)$$

$$348\pi = 4\pi r$$

$$87 = r$$

The radius is 87 meters.

EXERCISE SET A

Use the Arc Length Conjecture to solve each problem. State each answer in terms of π.

1. Length of $\widehat{CD}$ is –?–.

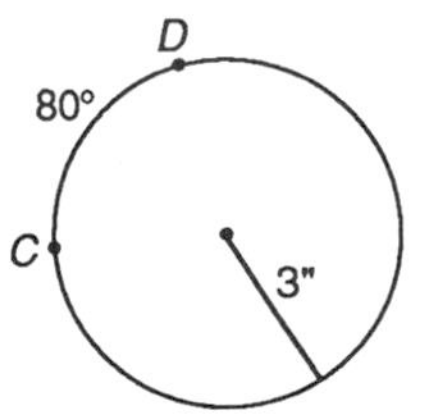

2. Length of $\widehat{EF}$ is –?–.

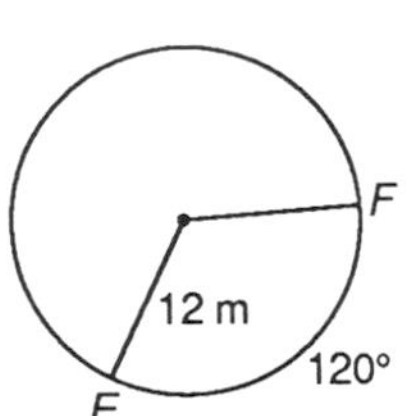

3.* Length of $\widehat{BIG}$ is –?–.

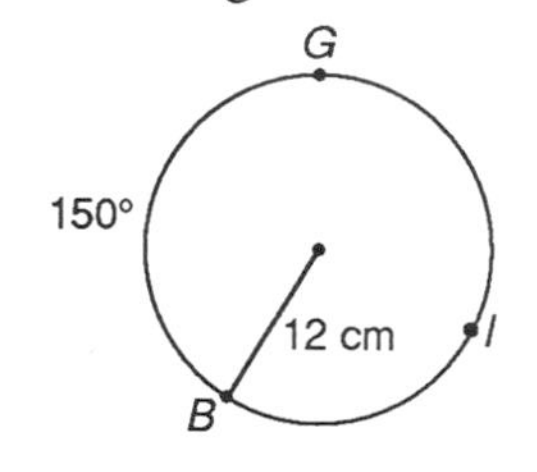

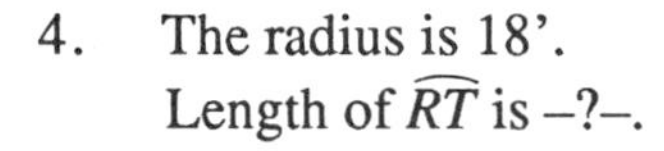

4. The radius is 18'. Length of $\widehat{RT}$ is –?–.

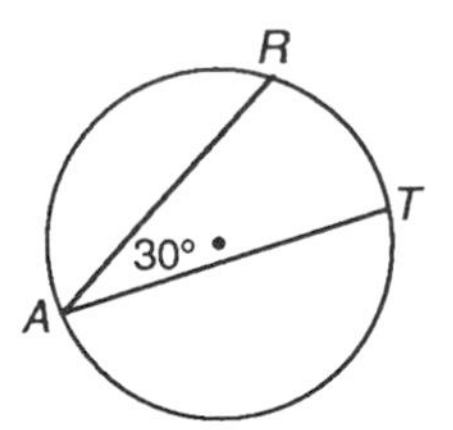

5. The diameter is 48 cm. Length of $\widehat{AC}$ is –?–.

6.* The radius is 9 m. Length of $\widehat{SO}$ is –?–.

7. Length of $\widehat{AB}$ is 6π m. The radius is –?–.

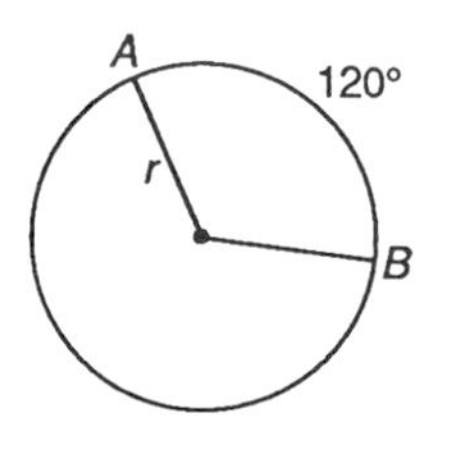

8. Length of $\widehat{TV}$ is 12π in. The diameter is –?–.

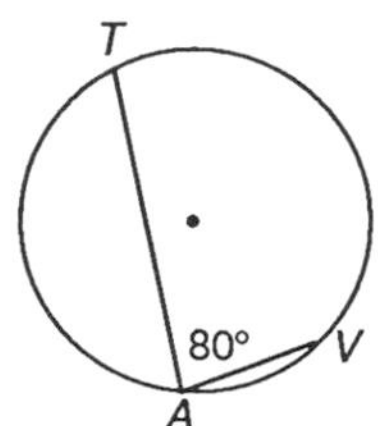

9.* Length of $\widehat{AR}$ is 40π cm. The radius is –?–. $\overleftrightarrow{PA} \parallel \overleftrightarrow{RE}$

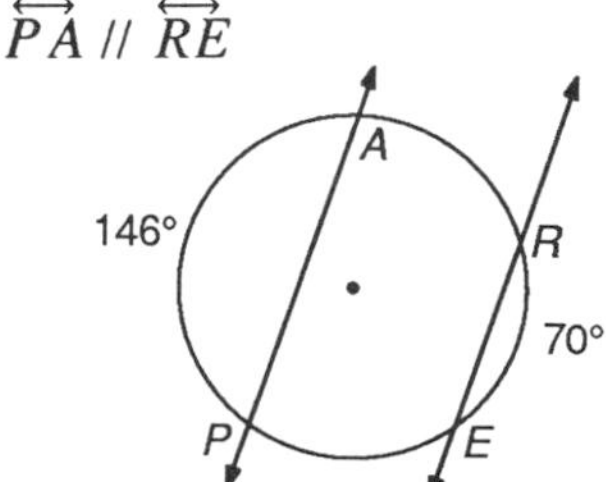

EXERCISE SET B

1. What is the diameter (to the nearest meter) of a circular track if one-fourth of the inside circumference is 100 meters?

2. What is the measure of the angle formed by the hands of a clock at 10:20? It is not 180. What fraction of a complete cycle has the big hand moved since 10? Hasn't the little hand moved that same fraction of the way from 10 to 11?

3.* Design Engineer Bridgette Dobar is in the process of designing the new bridge that will span the Round River Valley. To her it seems fitting that the arch of the bridge be the arc of a circle since it will be spanning the Round River Valley. For aesthetic reasons the arc should be 60°. Bridgette now needs to calculate the length of the arc. The bridge will span 180 m across the valley. Using 3.14 for π, help Bridgette check her work by calculating the arc length of the bridge.

60°

180m

4.* FBI agent Dudley Sharp has just received a coded message stating that a bomb has been planted somewhere near the clock tower of Big Bert. The note gives clues to the location of the bomb and the time it will go off. After decoding, the clues state:

Location: nO eht elttil dnah fo giB treB.

Time: When the tip of the little hand has traveled 40π cm past 10:00 — KABOOM!

Dudley quickly called Colonel Karl Kolone, the clock repairman, and found that the length of the little hand of Big Bert was 60 cm. At what time will the bomb go off? Where is the bomb located?

Special Project

Racetrack Geometry

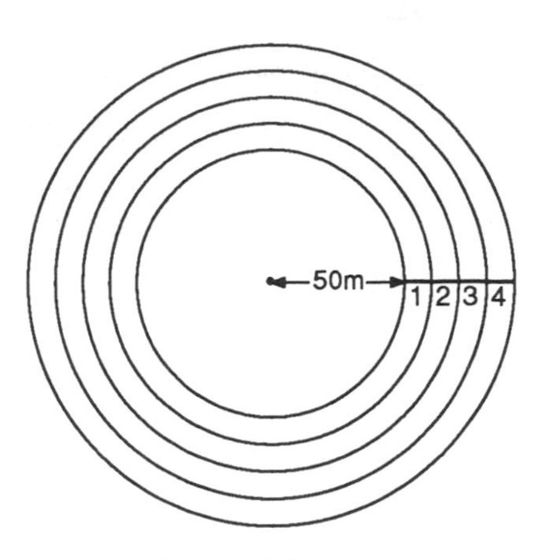

If you had to start and finish at the same line, which lane of the racetrack on the right would you choose to run in? Sure, the inside lane. If the runners in the four lanes were to start and finish at the line shown, the runner in the inside lane would have an obvious advantage because that lane is the shortest. For a race to be fair, runners in the outside lanes must be given head starts.

Your task in this special project is to design a 4-lane oval track with straightaways and semicircular ends. The semicircular ends must have inner diameters of 50 meters so that the distance of one lap in the inner lane is 800 meters. Draw starting and stopping segments in each lane so that an 800 meter race can be run in all four lanes

What do you need to know to design such a track? You will need to determine the length of the straightaways. You will also need to determine the head start for each of the runners in the outer lanes so that each has 800 meters to the finish line. (Use 3.14 for π.) Before you begin creating your racetrack, you will need to determine:

- Does the radius of the circle play a part in determining the head start?
- Does the width of the lane play a part in determining the head start?
- Does the length of the straightaways play a part in determining the head start?

To answer these questions, try calculating the lengths of a few sample racetracks. For example, if the inner radius of the circular track pictured above is 50 meters and each lane is 1 meter wide, you can calculate the distance each runner must travel in one lap if each runner must stay in his or her own lane.

Copy and complete the table. r is the radius of the circle that defines the inside edge of each lane. C is the circumference. All distances are in meters.

Lane 1	Lane 2	Lane 3	Lane 4
$r = 50$	$r = 51$	$r = $ –?–	$r = $ –?–
$C = 100\pi$	$C = $ –?–	$C = $ –?–	$C = $ –?–

To make a race fair, you can look in the table above to determine how much of a head start each runner in the outer lanes must have. For the circular track above, it turns out that the runner in lane 2 must have a head start of 2π meters over the runner in lane 1; the runner in lane 3 must have a 2π meters head start over the runner in lane 2; and, the runner in lane 4 must also have a 2π meters head start over the runner in lane 3. With these head starts, each runner will travel 100π meters.

Is the head start always 2π meters? Try other tracks to find out. In the table on the left, the lane distance is kept the same but the inner radius of the track is changed from the first example. In the table on the right, the lane width is changed but the inner radius is the same as the radius of the track on the left. Copy and complete the tables below or make up some of your own. All distances are in meters.

Circular track with inner radius of 65 m and lane width of 1 m

Lane 1	Lane 2	Lane 3	Lane 4
$r = 65$	$r = 66$	$r = $–?–	$r = $–?–
$C = 130\pi$	$C = $–?–	$C = $–?–	$C = $–?–

Circular track with inner radius of 65 m and lane width of 1.5 m

Lane 1	Lane 2	Lane 3	Lane 4
$r = 65$	$r = 66.5$	$r = $–?–	$r = $–?–
$C = 130\pi$	$C = $–?–	$C = $–?–	$C = $–?–

From these examples you may be able to answer the first two questions. Most tracks go around a playing field and have straightaways. What about the length of the straightaways (S)? Copy and complete the tables below to calculate the total distance (T) for each lane for the type of track shown on the right. Clearly $T = 2S + 2\pi r$.

Lane 1	Lane 2	Lane 3	Lane 4
$r = 30$	$r = 31$	$r = 32$	$r = 33$
$S = 100$	$S = 100$	$S = 100$	$S = 100$
$T = 200 + 60\pi$	$T = $–?–	$T = $–?–	$T = $–?–

Lane 1	Lane 2	Lane 3	Lane 4
$r = 30$	$r = 31$	$r = 32$	$r = 33$
$S = 200$	$S = 200$	$S = 200$	$S = 200$
$T = 400 + 60\pi$	$T = $–?–	$T = $–?–	$T = $–?–

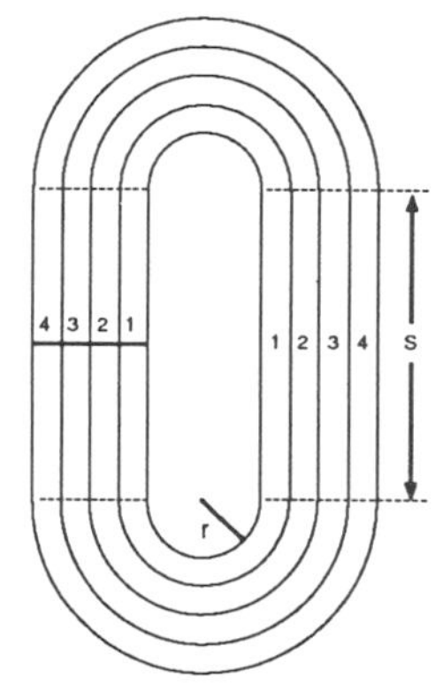

From these tables you may be able to answer the third question. If you can answer all three questions, you are ready to start designing your racetrack.

Again, your task in this special project is to design a 4-lane oval track with straightaways and semicircular ends. The semicircular ends must have inner diameters of 50 meters so that the distance of one lap in the inner lane is 800 meters. You determine a width for the lanes. Draw starting and stopping segments in each lane so that an 800 meter race can be run in all four lanes.

On your mark, get set, GO!

Lesson 6.8
Chapter Review

Euclid

Old Euclid drew a circle
On a sand-beach long ago.
He bounded and enclosed it
With angles thus and so.
His set of solemn graybeards
Nodded and argued much
Of arc and of circumference,
Diameters and such.
A silent child stood by them
From morning until noon
Because they drew such charming
Round pictures of the moon.

– Vachel Lindsay

EXERCISE SET A

Identify each statement as true or false.

1. A circle is the set of all points in space at a given distance from a given point.
2. A chord is a segment connecting the center of a circle to any point of the circle.
3. A diameter is a line segment connecting any two points of a circle.
4. A radius is a chord passing through the center of a circle.
5. A secant is a line in the plane of a circle passing through one point of the circle.
6. A tangent is a line in the plane of a circle containing two points of the circle.
7. The center of a circle is a point of the circle.
8. The measure of an arc is equal to one-half the measure of its central angle.
9. Two circles are congruent if they have the same center.
10. The area of a circle is the perimeter or distance "around the circle."
11. The ratio of the diameter to the circumference of a circle is π.
12. Every statement in this exercise set is false.

EXERCISE SET B

1. Construct an acute scalene triangle. Construct the circumscribed circle.
2. Construct an acute scalene triangle. Construct the inscribed circle.
3. Construct a rectangle. Construct the circumscribed circle.
4. Construct a rhombus. Construct the inscribed circle.

EXERCISE SET C

1. $a = –?–$

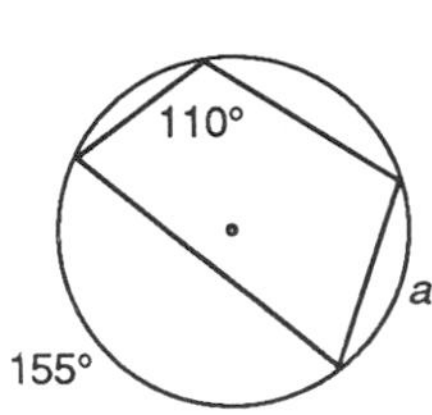

2.* $b = –?–$

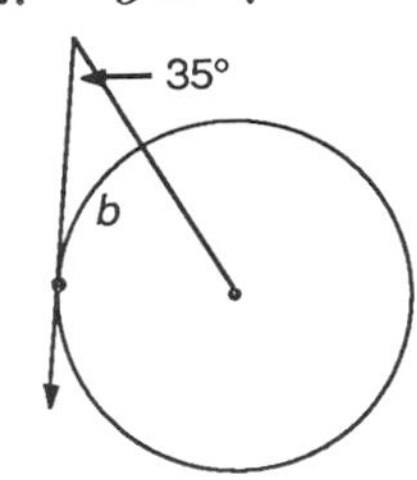

3. $c = –?–$

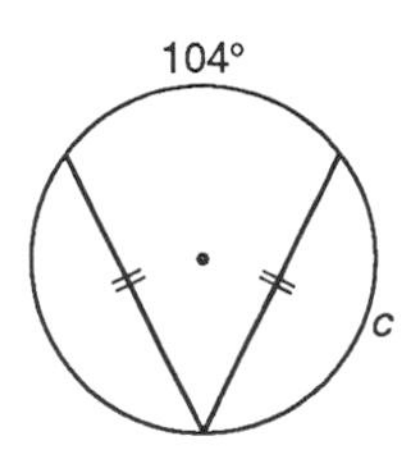

4. $d = –?–$

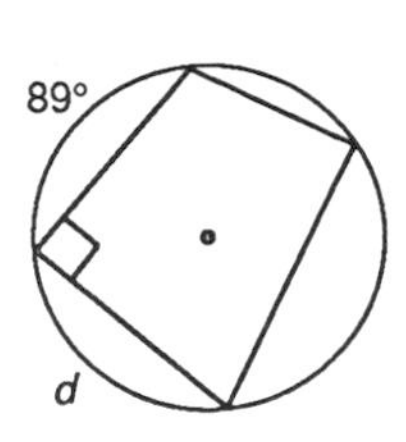

5.* $e = –?–$

60°
e
64°

6.* $f = –?–$

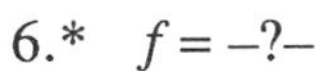

88°
f
118°

7. Use 22/7 for π.
Circumference?

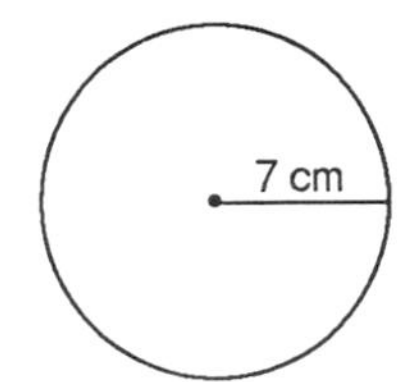

8. Use 3.14 for π.
Circumference?

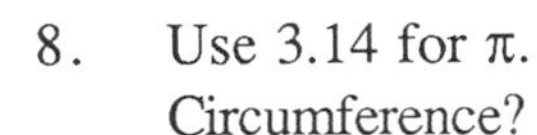
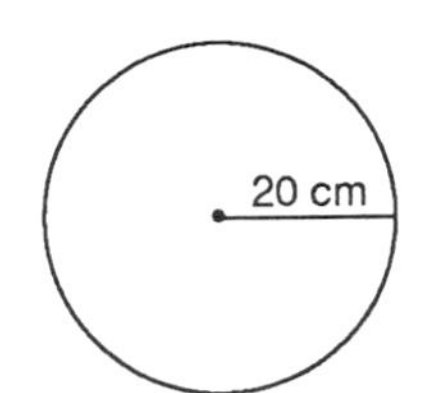

9. Circumference is 42π m.
$r = –?–$

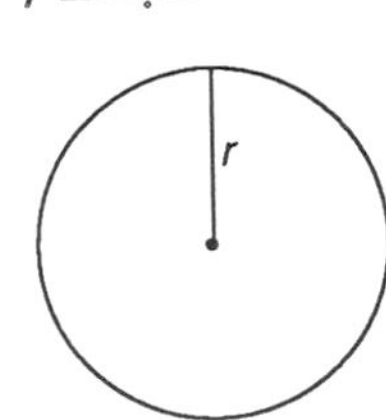

10.* Use 22/7 for π.
Circumference is 132 cm.
$D = –?–$

D

11.* $r = 27$ cm
The arc length of $\widehat{AB}$ is –?–.

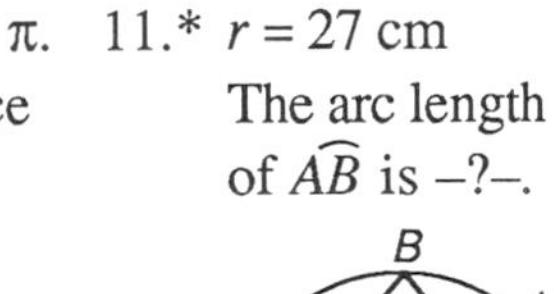

B
100°
A
T
r

12.* $r = 36'$
The arc length of $\widehat{CD}$ is –?–.

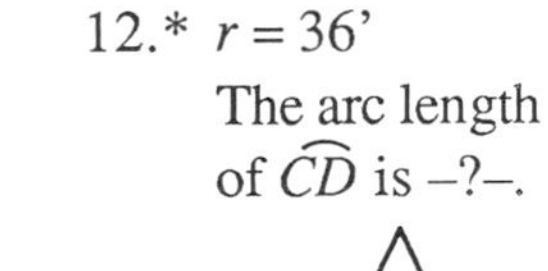
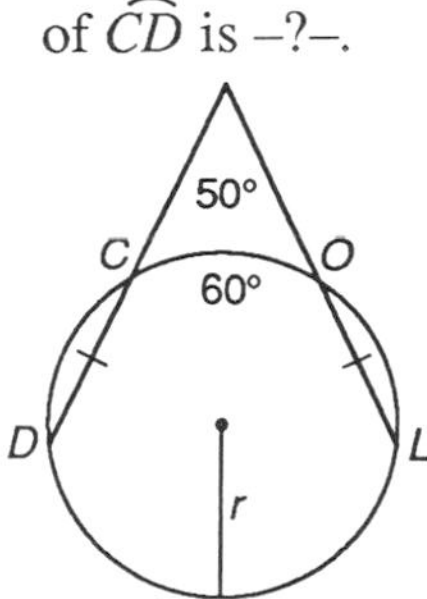

EXERCISE SET D

1.* Stacie tosses her boomerang in a perfect circle so that it just passes around a flagpole 50 meters from where she is standing. If Stacie and the flagpole are at opposite ends of a diameter of the circular path, what is the distance (to the nearest meter) that her boomerang traveled in its roundtrip? (Use 22/7 for π.)

2.* Captain Mattingly wishes to determine the length of his anchor chain. He counts 42 turns of anchor chain on the cylindrical anchor spool. What is the length of the anchor chain if the diameter of the anchor spool is one meter? (Use 22/7 for π.)

3. Elvira has just caught her pet ocelot Gato about to eat traveling salesperson Harms Nobody. Elvira must now build a pen for Gato out behind the castle to keep him out of Harms way. To give her ocelot maximum area with the minimum of fencing, Elvira figures it must be a circular region. If the radius of the circle is to be 42 feet, and 6-foot-high chain-link fencing costs 10 kopeks per foot, how many kopeks will it cost to build the fence around poor confused and hungry Gato? Use 22/7 for π.

The Round Table

4.* King Arthur has just finished celebrating the grand opening of Camelot and has come up with a marvelous new idea. Instead of the King at the head of a long rectangular table and all his knights stretched out along its length, how about a round table so that everyone is seen as an equal! Arthur has sent instructions down to Merlin in his magic chamber to design and create just such a round table. It is to be an oak table large enough to seat 150 knights, including the King. Each knight is to have 2 feet along the edge of the table. In addition, Arthur wishes to have inlaid strips of rosewood emanating from the center, dividing the table into 150 sectors. Normally, this is an easy task for Merlin. However, today he is still asleep from all the merriment of last night. The task is now up to his poor assistant Willie. Can you help him out? Using 22/7 for π, calculate the diameter of the table.

5.* One nautical mile was originally defined to be the length of one minute of arc of a great circle of the earth. (A great circle on a sphere is the intersection of the sphere and a plane that cuts through the sphere's center.) However, scientists discovered that the earth is not a perfect sphere. It is fatter through the equator than through the poles. Defined as one minute of arc, therefore, one nautical mile could take on a range of values. It would be shortest along a great circle through the poles and longest along the great circle called the equator. To remedy this, an international unit of one nautical mile was established. The international nautical mile is defined as 1.852 kilometers (about 1.15 miles). Given that the polar radius of the earth is 6357 kilometers and that the equatorial radius of the earth is 6378 kilometers, calculate one nautical mile by the original definition near a pole and one nautical mile by the original definition near the equator. Show that the international nautical mile is between both values. (There are 60 minutes of arc in each degree.)

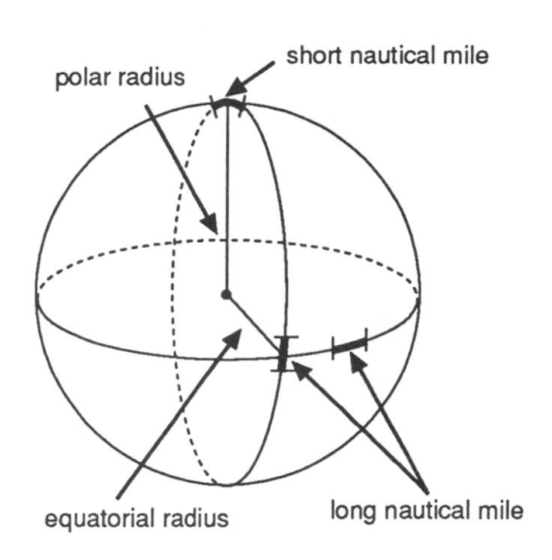

Improving Reasoning Skills

Bagels II

In this puzzle you are to determine a 3-digit number (no digit repeated) by making "educated guesses." After each guess, you will be given a clue about your guess. The clues:

bagels:	no digit is correct
pico:	one digit is correct but in the wrong position
fermi:	one digit is correct and in the correct position

In each of the problems below, a number of guesses have been made with the clue for each guess shown to the right. From the given set of clues, determine the 3-digit number. If there is more than one solution, find them all.

1.

1 2 3	*pico*
4 5 6	*pico*
7 8 9	*pico*
9 4 1	*bagels*
3 7 5	*pico*
6 3 8	*pico*
? ? ?	

2.

1 9 8	*pico fermi*
7 6 5	*bagels*
4 3 2	*pico*
1 2 9	*pico fermi*
? ? ?	

Cooperative Problem Solving

Designing a Theater for Galileo

You and your architectural engineering team are competing for the contract to design the new circular theater. The theater is to be built beneath the great dome *R-3* at the lunar space port Galileo. The Arts Director of Galileo has asked each potential engineering team to submit its design and its calculations for the new circular theater with revolving center stage. Although overall design is important, the job will go to the team that demonstrates the design with the greatest seating capacity. The director has given you the following restrictions and guidelines:

1. The theater must be only one level and seat at least a thousand people.
2. The stage should be at least ten meters in diameter.
3. The outer diameter of the theater interior should be at most 42 meters.
4. The seating should be divided into sections by equally spaced aisles radiating from center stage. There should be no fewer than four radial aisles and no more than eight radial aisles. Each radial aisle should be at least one meter in width.

5. There should be two concentric aisles. The innermost concentric aisle about the stage should be at least one meter wide and at most two meters wide. The outer concentric aisle should be at least two meters wide and at most four meters wide and ring the perimeter of the theater.

6. Safety codes at the lunar colony require that each seat be at least 60 centimeters wide and that each seating position be at least 90 centimeters in depth.

7. Safety codes require that there be no more than 30 seats in any row.

Your job is to draw a scaled plan of the theater (including seating, aisles, and stage). Your plan should maximize seating capacity. You should be able to support your design with calculations verifying your seating capacity. You will need to calculate:

1. The number of rows and the number of seats in each row.

2. The width at the stage end of each radial aisle and the width at the back end of each radial aisle. The width of the concentric aisles.

3. Total seating capacity for your plan.

There is a hint leading to a possible solution for this problem in the *Hints* section.

7 Transformations and Tessellations

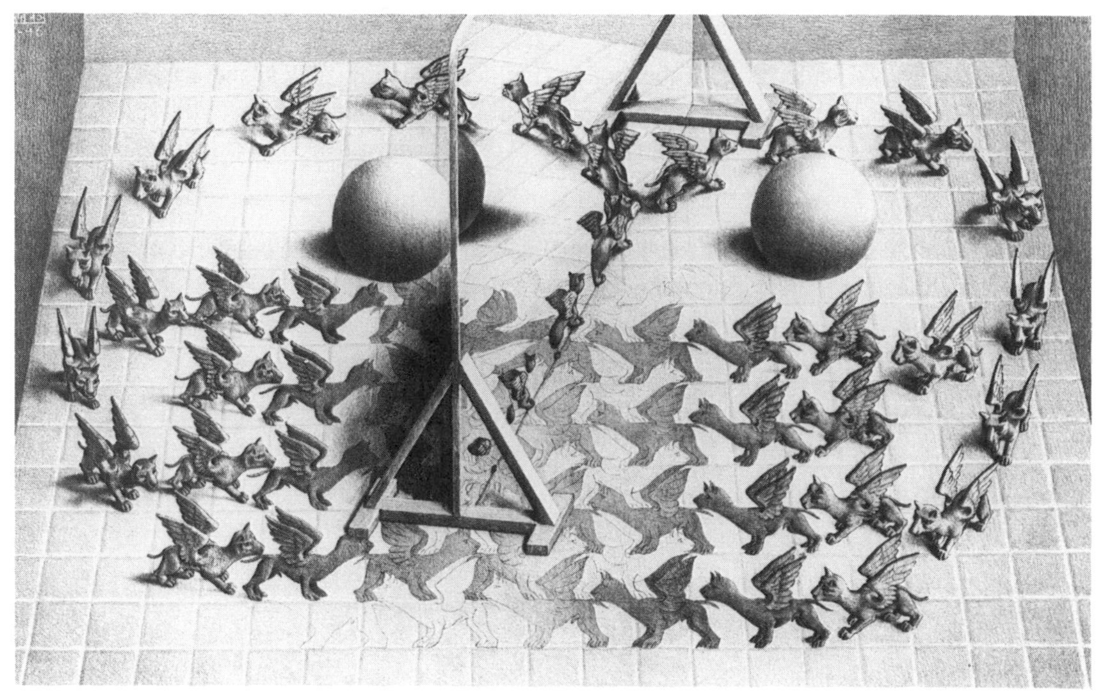

Magic Mirror, M. C. Escher, 1946

Symmetry is found in art, architecture, dance, music, poetry, and of course mathematics. In this chapter you will discover some basic properties of symmetry. You will use the properties of symmetry to create special tiling patterns called tessellations. The hexagon pattern in the honeycomb of the bee is a tessellation of regular hexagons. Tessellations in the world are found in Islamic designs. The famous Dutch artist, Maurits Cornelis Escher, after observing the Islamic designs in the Alhambra in Granada, Spain, began creating tessellations using recognizable shapes such as birds, fish, reptiles, and humans. In this chapter you will learn a few of his techniques and create your own designs.

Lesson 7.1
Transformations

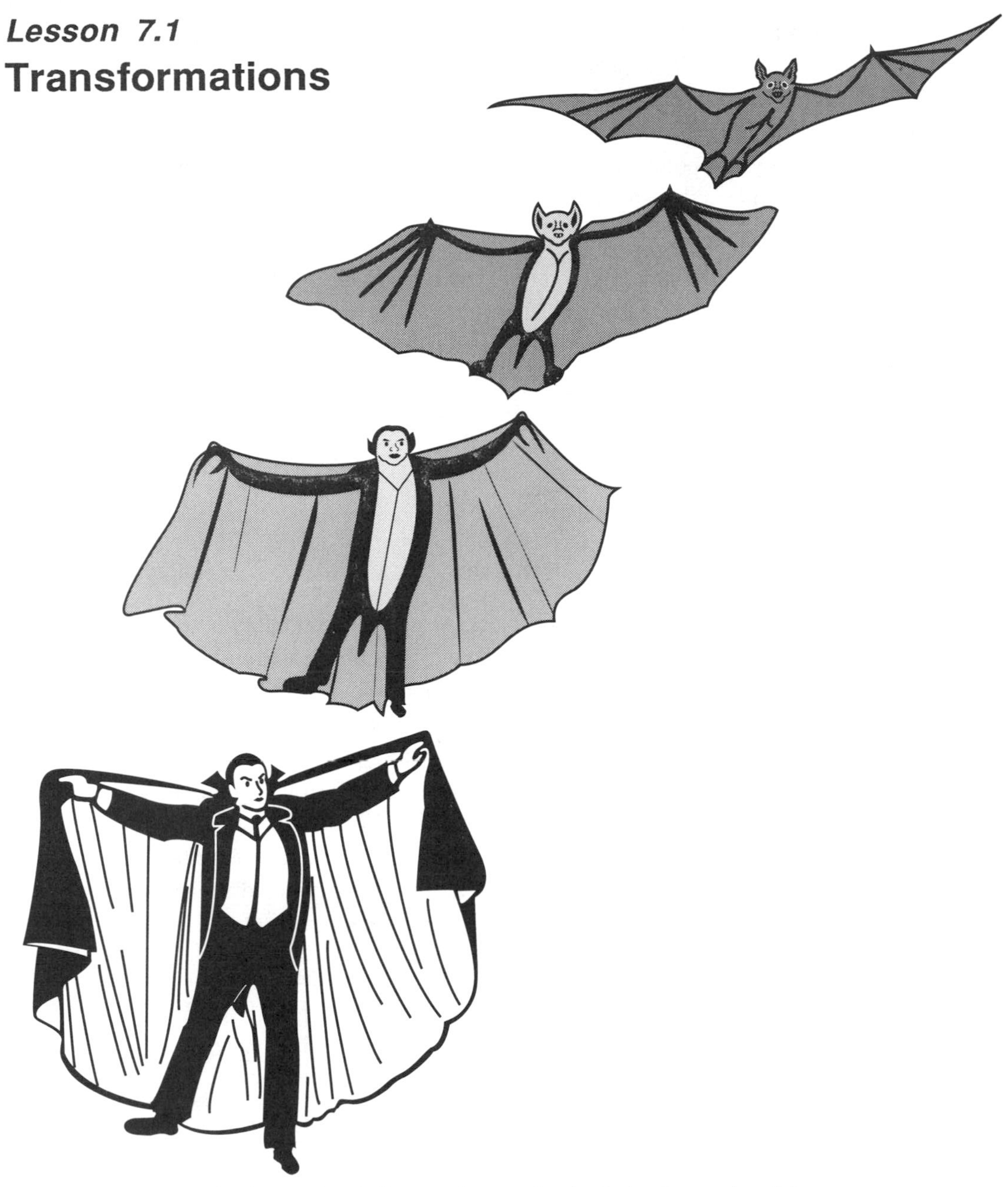

Geometry is not only the study of figures, it can also include the study of the movement of those figures. If you move all the points of a geometric figure according to set rules, you can create a new geometric figure. Geometers describe such a motion by establishing a correspondence between the points of the original figure and the points of the new figure (called the **image**). If each point of a plane figure can be paired with exactly one point of its image on the plane and each point of the image can be paired with exactly one point of the original figure, then the correspondence is called a **transformation**.

When Count Dracula transforms himself into a bat, he changes his size and shape. In this lesson you are going to look at special transformations that do not change the size or shape of figures but may change the location of figures in a plane. A transformation that preserves size and shape is called an **isometry**. In an isometry, the distance between any two points in the image of a figure is equal to the distance between the original points in the original figure. The image is always congruent to the original in an isometry.

Three types of isometries are translation, rotation, and reflection. You probably know these as the motions: slide, turn, and flip.

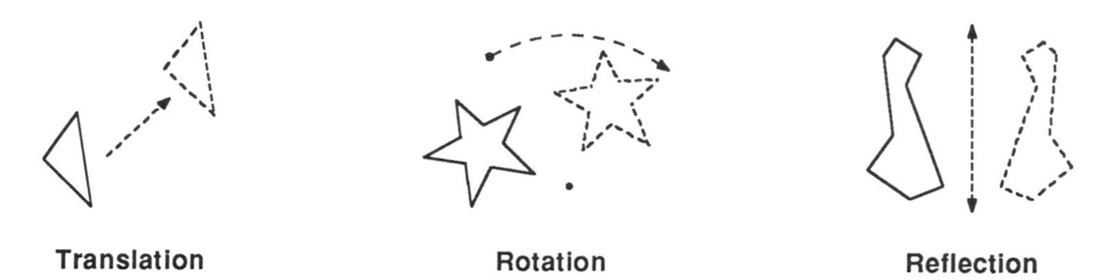

A **translation** is the simplest type of isometry. If you draw a figure on a piece of paper and then slide the paper on your desk along a straight path, your slide motion models a translation isometry. In a slide, points in the original figure move an identical distance along parallel paths to the image. In a translation, the distance between a point and its image is always the same. That is, all points in the original figure are the same distance from each of their images. A distance and a direction together define a translation. You can use an arrow, called a **translation vector**, to show distance and direction. The length of the translation vector from starting point to tip represents the distance, and the direction that the arrow is pointing represents the direction of the translation.

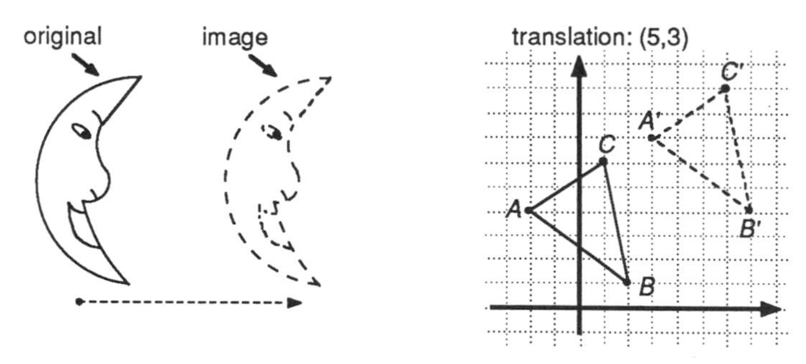

A translation can also be represented on a coordinate grid using an ordered pair of numbers. The first number represents the distance a point of the original figure moves in the x-direction to its position in the image and the second number represents the distance the point moves in the y-direction.

A **rotation** is a second type of isometry. If you draw a figure on a piece of paper, place a dot on the paper, put your pencil point on the dot, and rotate the page about the point, your turn motion models a rotation isometry. In a turn motion, points in the original figure rotate or turn an identical number of degrees about a fixed center point. A center of rotation together with an amount and direction of rotation define a rotation. You can use a **rotation vector** to define the amount and direction of rotation. The rotation vector is an arc of a circle (with an arrowhead at one end) whose center is the center of rotation. The degree measure of the arc represents the degrees of rotation. If no direction is given, the direction of rotation is assumed to be counter-clockwise.

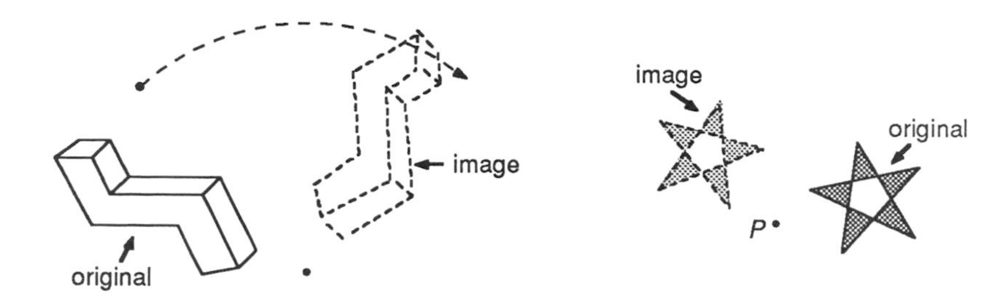

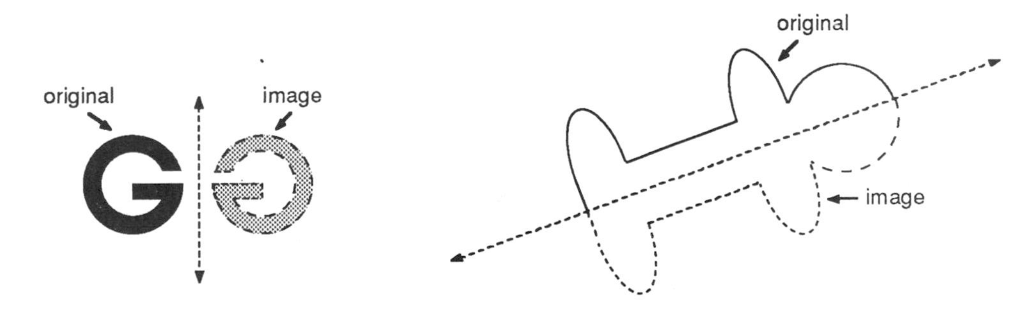

A **reflection** is a third type of isometry. If you draw a figure on a piece of paper, place the edge of a mirror on your paper, and look at the figure in the mirror, you will see the figure reflected. A reflection produces a figure's "mirror image." A **line of reflection** defines a reflection. The line of reflection is the perpendicular bisector of every segment joining a point in the figure with the image of the point. If a point in the figure is on the line of reflection, then the point is its own image. Each point in a figure is reflected or flipped about the reflection line as if the reflection line were a mirror.

Any transformation in which the image coincides with the original figure is called an **identity** transformation. The translation identity on a coordinate grid would be the ordered pair (0, 0). The rotation identity would be a rotation of 0° (or any multiple of 360°). For figures other than points, segments, and lines, there is no reflection identity. It takes two consecutive reflections about the same reflection line for the reflected image of a plane figure to coincide with the original figure.

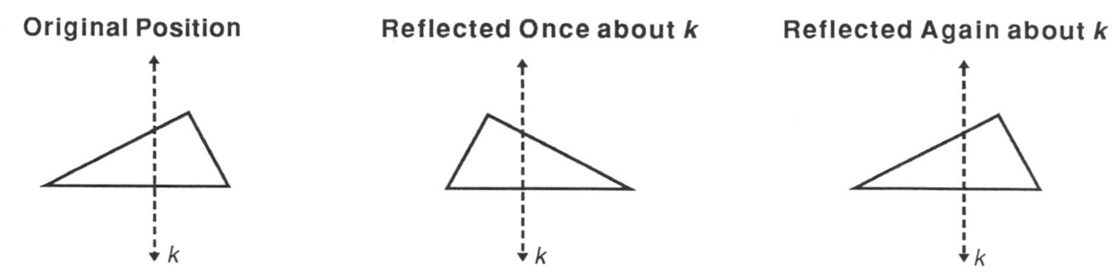

Geometers have demonstrated that every possible isometry of plane figures from one position in the plane to another position in the plane can be completed by one of the three single motion isometries or by a special combination of two of them: a translation and reflection. This two-step isometry is called a **glide reflection**. A pair of foot steps is the most common example of a glide reflection.

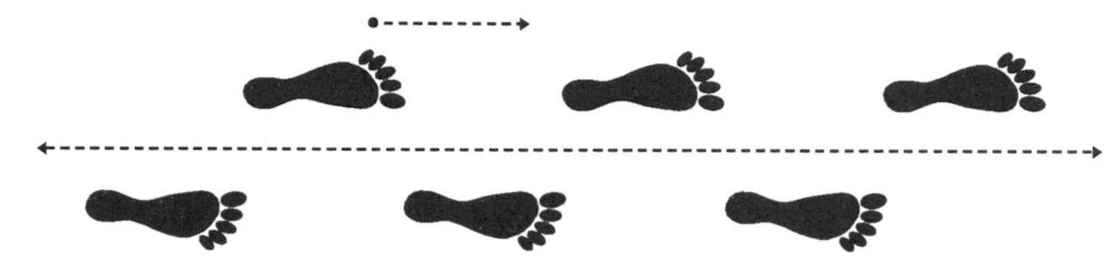

EXERCISE SET

In Exercises 1 and 2, identify which figure on the right is a translated image of a figure on the left. Trace figures if necessary.

1. a. b.

2. c.

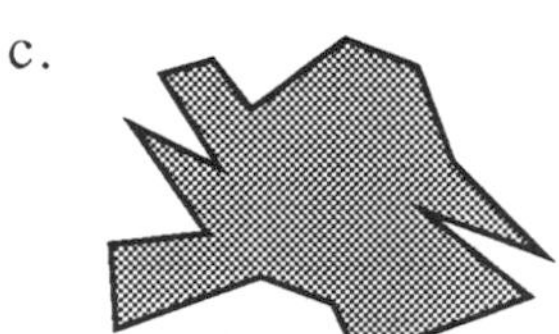

d.

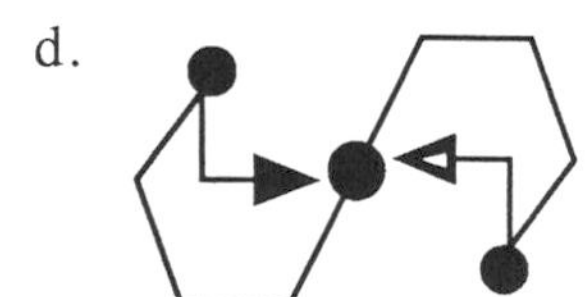

3. Trace the figure and point of rotation *P* onto a sheet of paper. With your copy over the original, rotate the design clockwise 90° about *P*. Trace the original figure again.

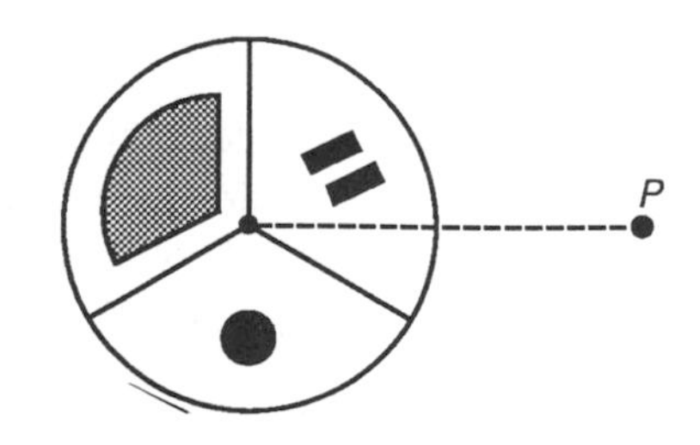

4.* Each figure below is the rotated image of the other. Copy both figures onto a sheet of paper. Use your compass to locate the center of rotation.

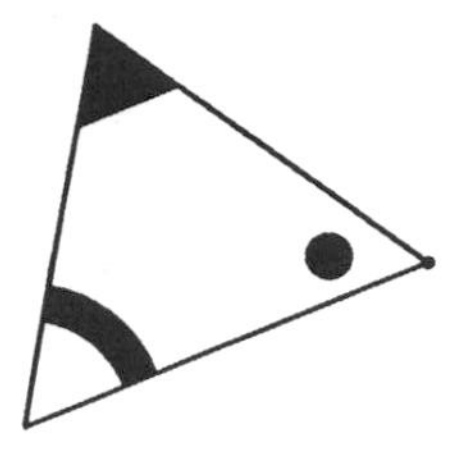

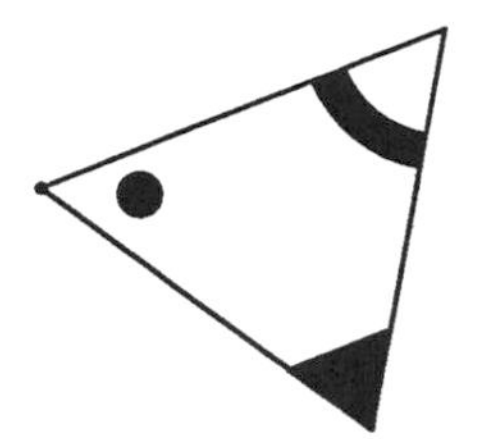

5. Copy the figure and reflection line onto a sheet of paper. Draw the reflected image.

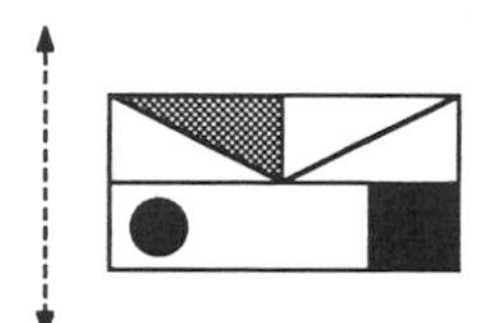

6. Each figure below is the reflection of the other. Copy both figures onto a sheet of paper. Locate the line of reflection.

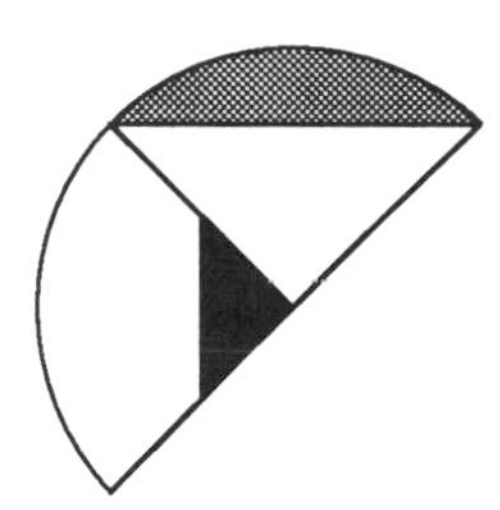

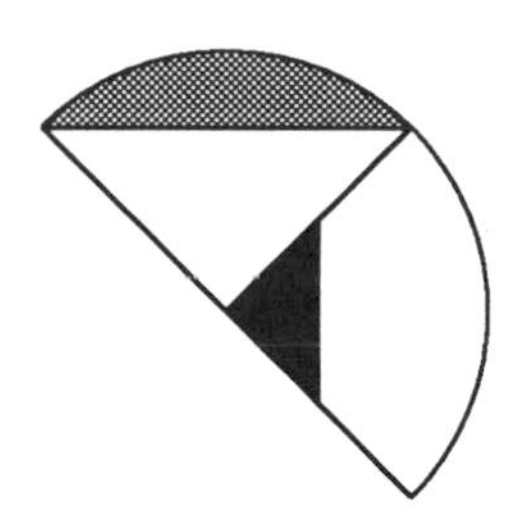

7. Copy the two figures below onto graph paper. Each figure is the glide reflected image of the other. Sketch two more glide reflected figures.

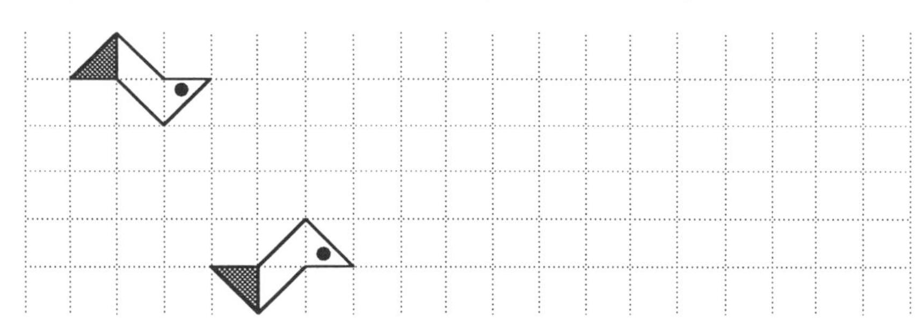

8. Copy the figure, reflection line, and translation vector onto a sheet of paper. Sketch two more glide reflected figures.

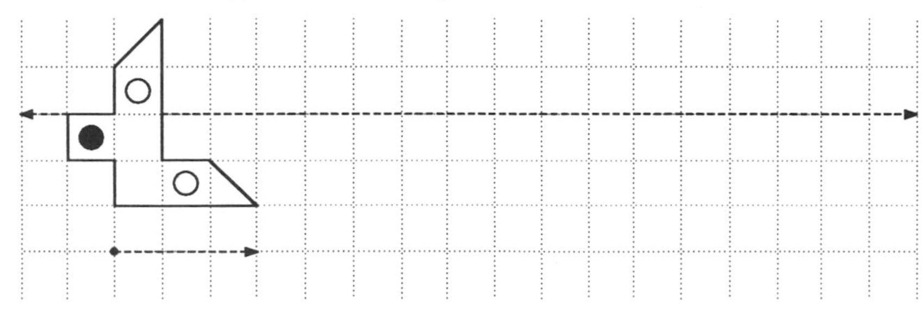

Lesson 7.2

Properties of Isometries

She puzzled over this for some time, but at last a bright thought struck her. "Why it's a Looking-glass book, of course! And, if I hold it up to the glass, the words will all go the right way again."

– Alice in Through the Looking-Glass by Lewis Carroll

In the first lesson of this chapter you were introduced to a number of new concepts and a lot of new terms. In this lesson you will get a chance to use what you learned to discover some properties of isometries.

EXERCISE SET

1. Trace the figure and the translation vectors onto a piece of paper and then create the translated image. (Only one vector is needed to define the translation, but two will prevent you from accidentally rotating the image during the translation.)

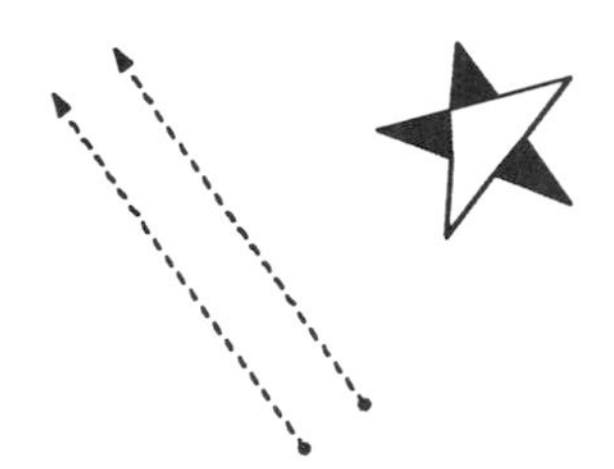

2. Copy the figure onto graph paper. Locate the vertices of the image after a translation of (-6, -5) followed by a translation of (14, 3). Create the translated image by connecting the vertices.

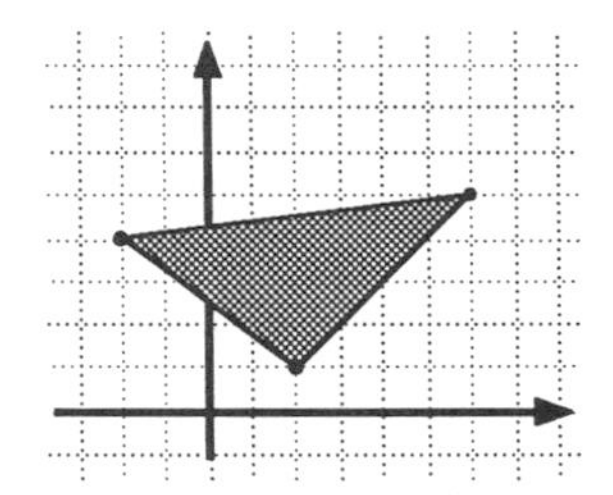

3. What single translation is equivalent to the two translations in Exercise 2?

4. What single translation can replace the three translations: (2, 3) followed by (-5, 7) followed by (13, 0)?

5. Trace the figure, center of rotation, and the rotation vector onto a piece of paper. Create the rotated image.

6. Trace the figure and center of rotation onto tracing paper. Perform a rotation of 80° followed by a rotation of 50° about Q.

7. What single rotation is equivalent to the two rotations in Exercise 6?

8. What is one rotation that can replace the three rotations: 45° followed by 50° followed by 85° all about the same center of rotation? Is this true if the centers of rotation are different?

9.* Trace the figure and reflection line onto a piece of paper. Create the reflected image.

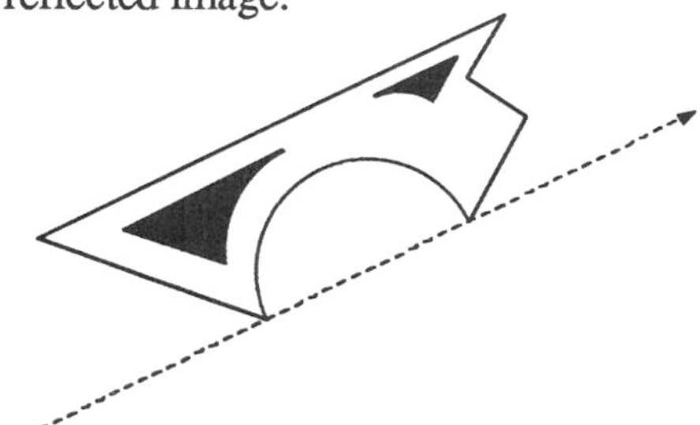

10. Trace the figure and reflection lines onto a piece of paper. Create the reflected image over the first reflection line and then the reflection of the image over the second reflection line.

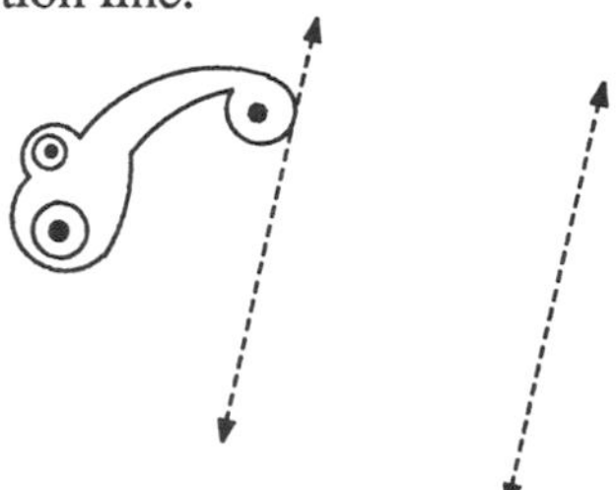

11.* After completing Exercise 10, you should notice that consecutive reflections over two parallel reflection lines results in a final image that is equivalent to the image which would be produced in a translation isometry. How does the distance between parallel reflection lines compare to the translation distance that produces the equivalent image?

12. If point A on a figure is 6 cm from reflection line m and 16 cm from reflection line n (reflection lines m and n are parallel and 10 cm apart) and the figure is then reflected about each reflection line, how far has the figure been translated?

13. Trace the figure and reflection line onto a piece of paper. Create the reflected image over the first reflection line and then the reflection of the image over the second reflection line.

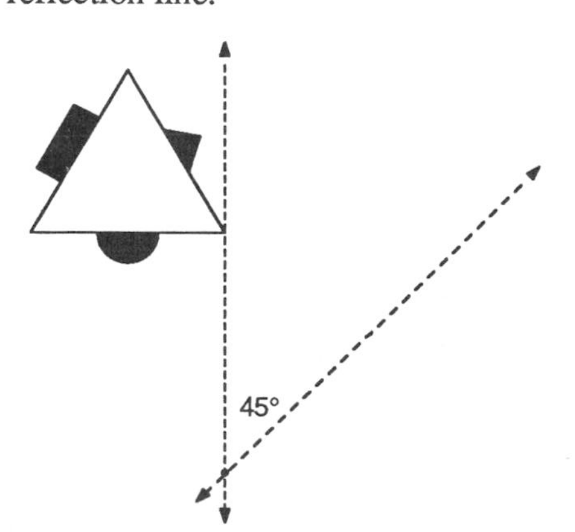

14. Perform the rotation indicated.

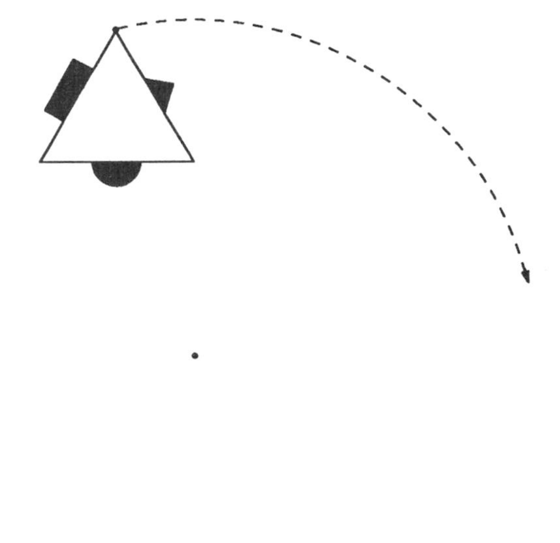

15. You should notice that consecutive reflections over two intersecting reflection lines results in an image that is equivalent to the image which would be produced in a rotation isometry. How does the acute angle between the intersecting reflection lines compare to the angle of rotation that produces the equivalent image? Where is the center of rotation?

16. Two reflection lines m and n intersect forming a 40° angle. A figure is on the exterior of the 40° angle. If the figure is reflected about each reflection line, how many degrees has the figure been rotated?

17. Trace the figure and reflection line onto a piece of paper. Reflect the figure twice over the reflection line.

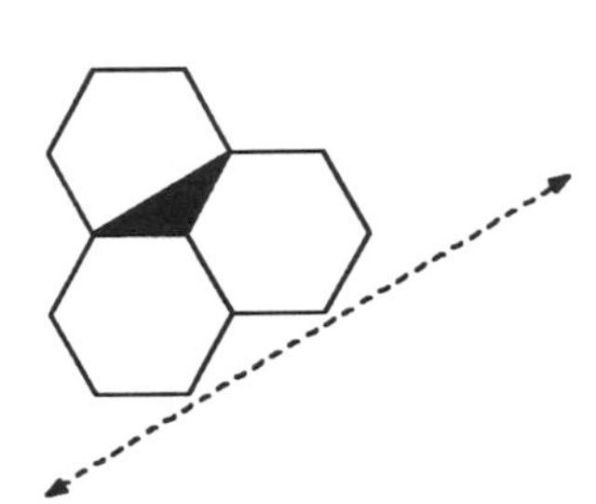

18. Perform a 180° rotation twice (two half-turns) about the center of rotation.

19. You should notice that if you reflect a design about an axis twice, then the image coincides with —?—. If you rotate a design 180° twice the image coincides with —?—.

Improving Visual Thinking Skills

Picture Patterns II

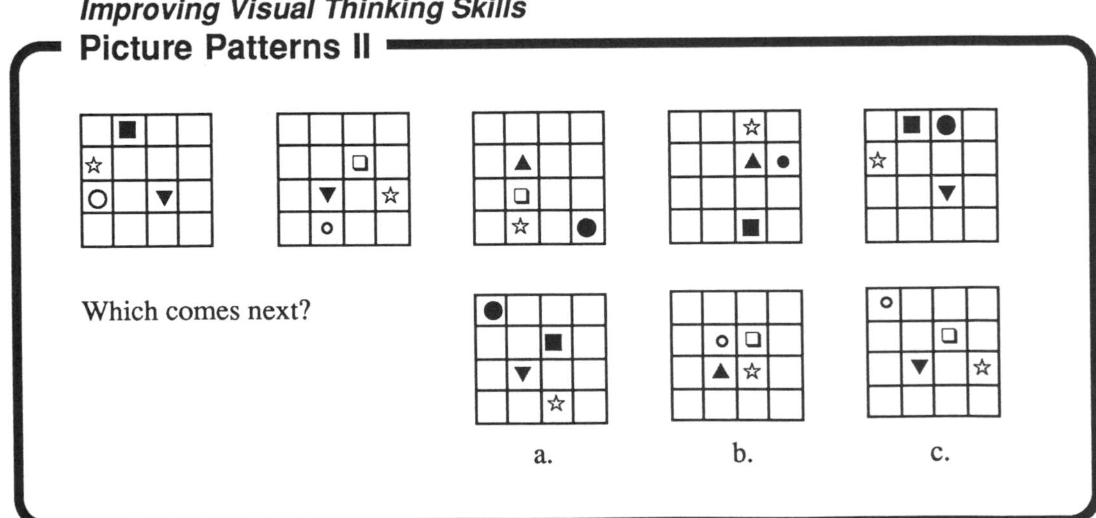

Special Project

Creating a Frieze

Copy the grid and design below onto a sheet of tracing paper. Reflect the design by folding it under and along the first vertical reflection line. Trace the back side of the design to create the first reflected image. Repeat this process across the grid to create a repeating pattern called a strip pattern or **frieze**. (Or, if you prefer, create your own design on graph paper and then create a pattern by repeated reflections of it.)

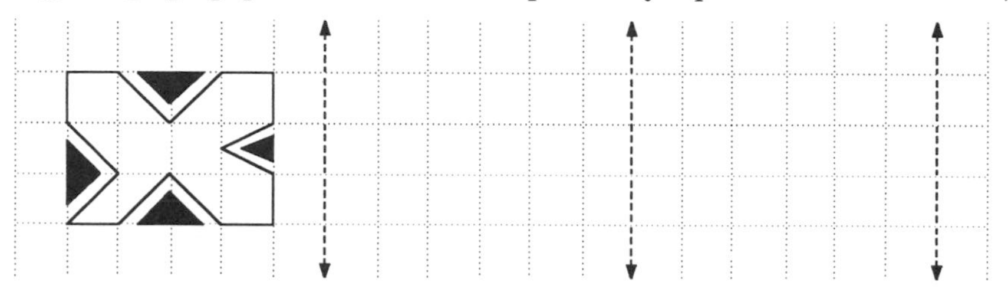

Copy the grid and design below onto a sheet of tracing paper. Translate the design on the tracing paper according to the length of the translation vector. Trace to create the first translated image. Repeat this process across the grid to create a strip pattern. (Again, if you prefer, create your own design on graph paper and then create a pattern by repeated translations of it.)

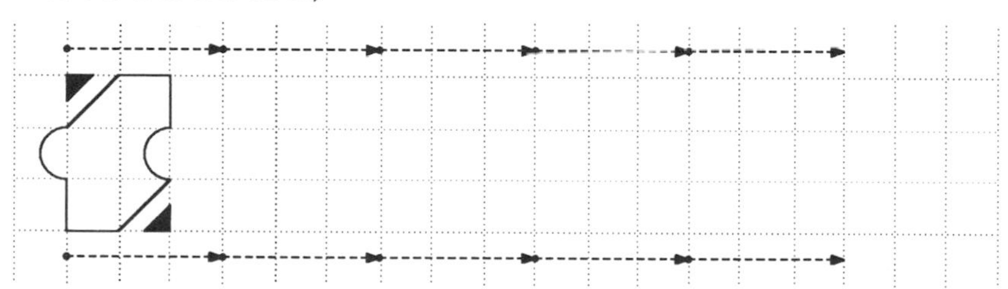

Copy the grid, the design, the translation vector, and the horizontal reflection line onto tracing paper. Translate the design on the tracing paper according to the length of the translation vector. Reflect the design by lifting the tracing paper up and flipping it over the reflection line. Trace the image to create the first glide reflected image. Repeat this process across the grid to create a frieze.

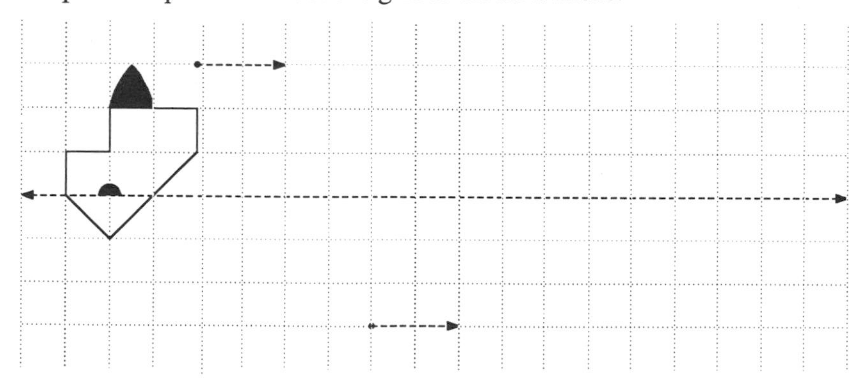

Lesson 7.3
Symmetry

Symmetry is one idea by which man through the ages has tried to comprehend and create order, beauty, and perfection.

– H. Weyl

The word *symmetry* may bring to the mind of an artist other words like *balance, harmony*, and *equally proportioned.* Flowers, insects, fish, birds, and many other natural objects are symmetric. The human body is symmetric. The chambered nautilus and crystals grow with the aid of symmetry. Since symmetry appears so abundantly in nature — in plants, animals, and crystalline structures — it is not surprising that artists throughout history have taken pleasure in symmetric designs.

Symmetry in Nature

Because symmetric designs are so naturally pleasing, symmetric symbols are very popular. As a young child you probably made symmetric designs in school by folding or cutting paper, or using ink blots. Many companies use symmetric designs (called logos) as their corporate symbol. Many countries throughout the world use symmetry in their national flags.

Symmetry in Flags

Jamaica

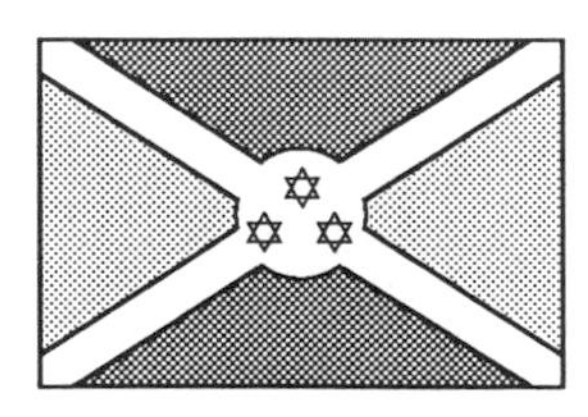

Burundi

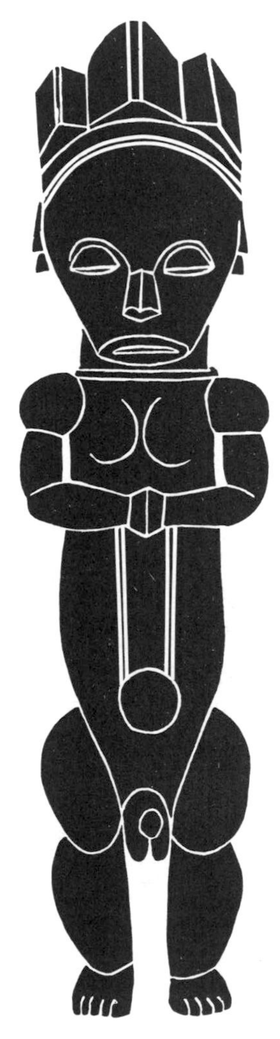

Symmetry in Art
Wooden Guardian Figure, Gabon
African Designs, Williams, Dover, 1971

In geometry, symmetry is defined in terms of isometries. When a figure undergoes an isometry (other than the identity) and the resulting image coincides with the original, then the figure has a transformational symmetry. There is a symmetry for each of the four isometries discussed.

Reflectional Symmetry

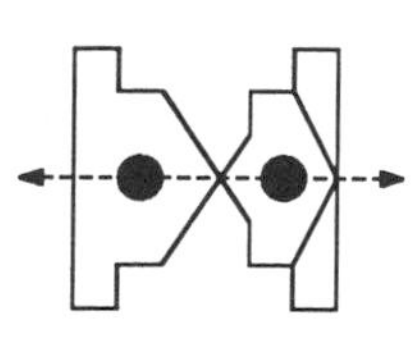

If a figure can be reflected about some line in such a way that the resulting image coincides with the original, then the figure has a **reflectional symmetry**. Reflectional symmetry is also called **line symmetry**. The line is called the **axis of symmetry** or **line of symmetry**.

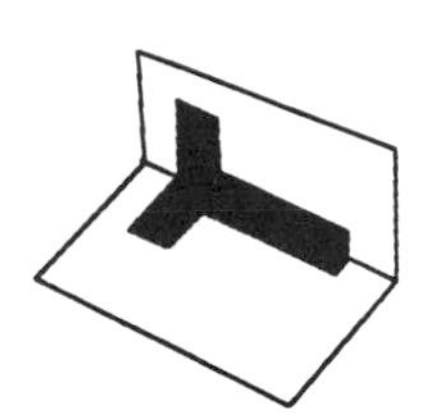

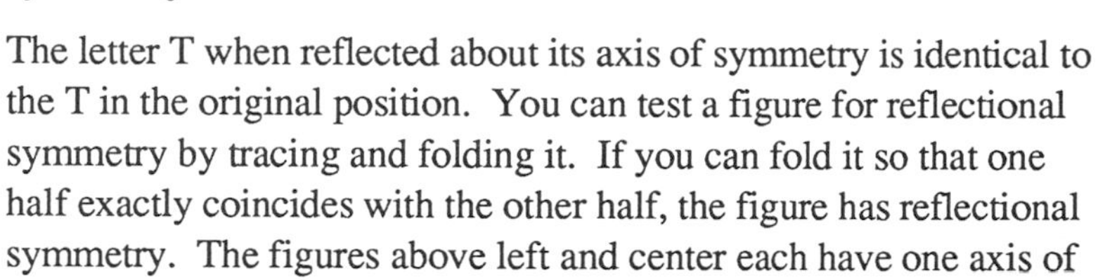

The letter T when reflected about its axis of symmetry is identical to the T in the original position. You can test a figure for reflectional symmetry by tracing and folding it. If you can fold it so that one half exactly coincides with the other half, the figure has reflectional symmetry. The figures above left and center each have one axis of symmetry. The figure above right has two lines of symmetry.

Reflectional symmetry is also called **mirror symmetry** because half a figure with reflectional symmetry is a mirror image of the other half. If you place a mirror on a figure's line of symmetry, perpendicular to the plane of the figure, the half-figure and its image produce a complete figure.

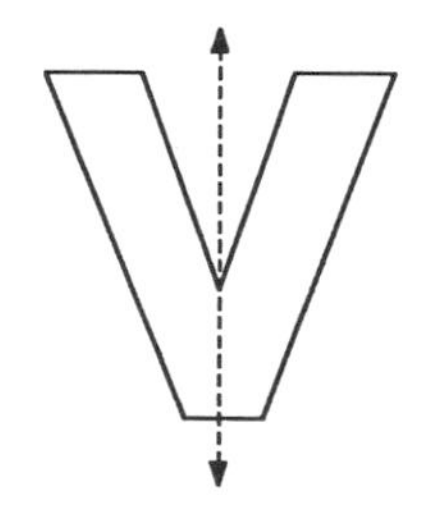

Vertical Symmetry

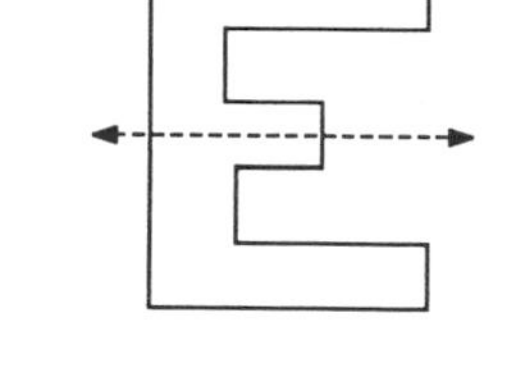

Horizontal Symmetry

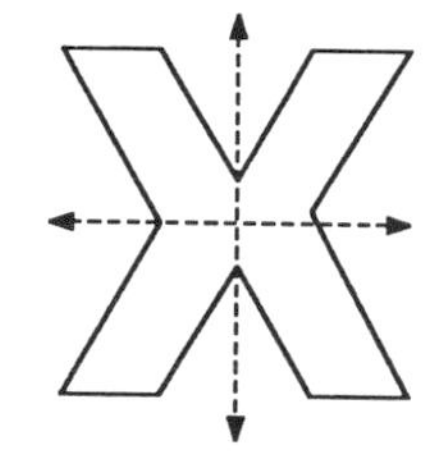

Horizontal and Vertical Symmetry

If a figure's line of symmetry is vertical, the figure has a vertical symmetry. The letter V has a vertical axis of symmetry. Likewise, if the line of symmetry is horizontal, the figure has a horizontal symmetry. The letter E has horizontal symmetry. Some figures, like the letter X, have both horizontal and vertical symmetries.

Rotational Symmetry

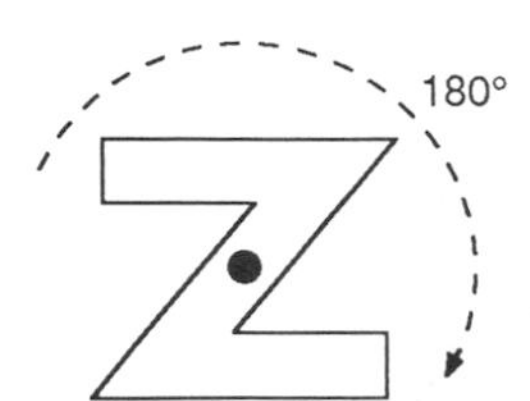

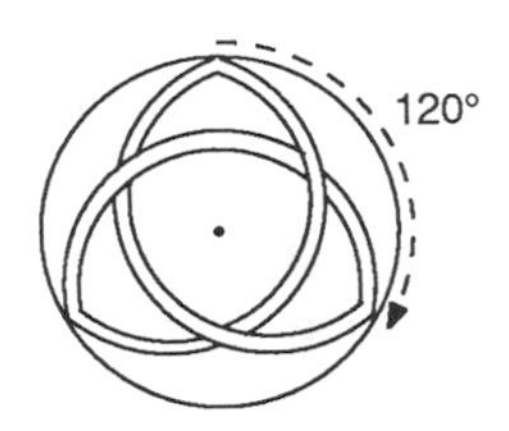

If a figure can be rotated n degrees about a point (where $n < 360$) in such a way that the resulting image coincides with the original figure, then the figure has a **rotational symmetry** of n degrees. You can trace a figure and test it for rotational symmetry. Place the copy exactly over the original, put your pen or pencil point on the center to hold it down, rotate the copy, and try to make the copy and original coincide. The letter Z has a rotational symmetry of 180° because when it is rotated 180° about a point, the image coincides with the original figure. A rotational symmetry of 180° is called **point symmetry**. A figure can have more than one rotational symmetry. The figure above center has rotational symmetries of 120° and 240°. The figure above right has rotational symmetries of 90°, 180°, and 270°. Of course every image is identical to the original figure after a rotation of some multiple of 360° (the rotation identity). However, we don't call a figure symmetric if the rotation identity is the only symmetry that it has.

EXERCISE SET A

Identify the type (or types) of symmetry in each design.

1.

2.

3.

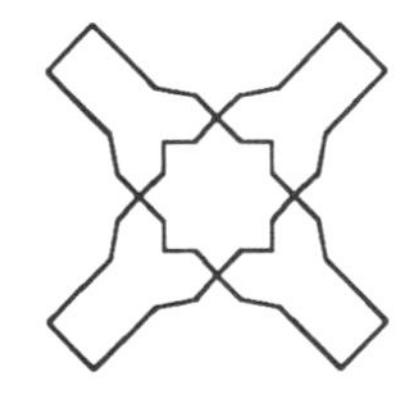

4.

5.

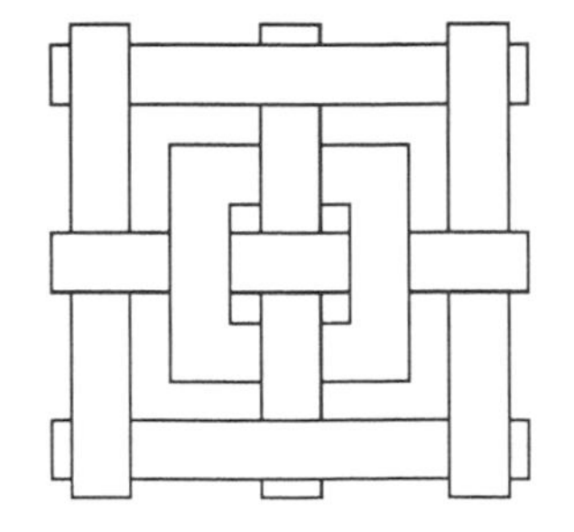

6.

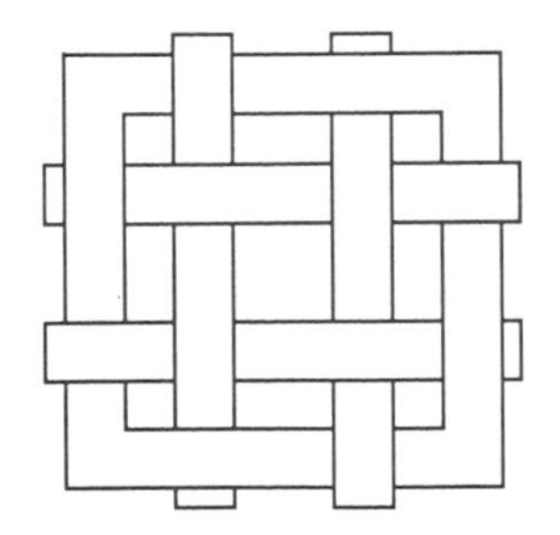

EXERCISE SET B

1. The three of diamonds has point symmetry because when rotated 180°, it is identical to the original. The three of clubs, however, does not have point symmetry. What are the other playing cards, 2 through 10 (non-face cards), that have point symmetry?

2. Miriam-the-Magician placed four cards face-up on her magic table (the four cards shown below left). Blind-folded, she instructed someone from her audience to come up on the stage and turn exactly one card upside-down.

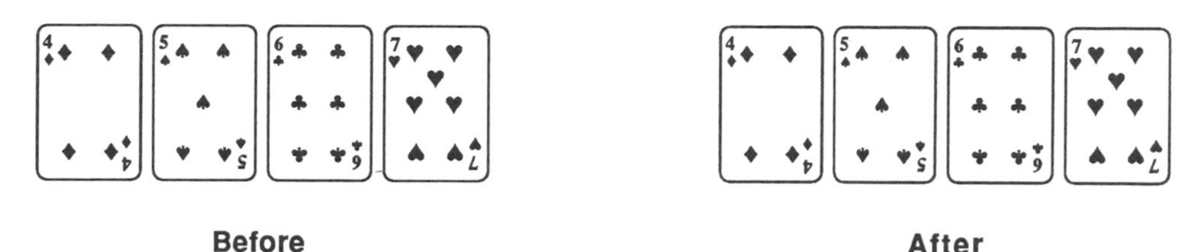

Before **After**

Miriam removed her blindfold and was able to determine which card was turned 180° by "magically reading the psychic energy emanating from the cards." Of course, it may be her magic, but it is possible to figure out which card was turned over without magic. Can you figure out which card was turned?

3. The alphabet printed below contains letters which are point symmetric. List them. If you need help, turn the page upside down.

A B C D E F G H I J K L M N O P Q R S T U V W X Y Z

4. List the letters from the alphabet above that have a horizontal axis of symmetry.

5. List the letters from the alphabet above that have vertical symmetry.

6. The word DECODE when flipped about its horizontal axis of symmetry remains unchanged. Find another such word.

7. The word TOMATO when written in column form has a vertical axis of symmetry. Thus, when a mirror is placed on or parallel to the axis of symmetry, TOMATO is seen in the mirror. Find another such word.

T
O
M
A
T
O

8. Design a logo with rotational symmetry for "Happy Time" Ice Cream Company. (Or make up a company name, perhaps your own name, and design a logo for it.)

9. Your colonizing spaceship has just landed on *Tao*, the small earth-like planet of the binary star system Gemini in the Andromeda galaxy. You have been instructed by Captain James Kirk IV to design a flag for the new colony. Design the flag with reflectional symmetry.

10. The design above comes from *Inversions,* a book by Scott Kim. Not only does the design spell the word *mirror*, it does it in mirror symmetry! Scott is able to turn *m*'s into *r*'s and other magic, but you can do almost the same thing. Using the vertically symmetric letters of the the alphabet from Exercise 3, find a word or words that have vertical symmetry. This can be done with some palindromic words, words that spell the same forward and back. If all the letters in the palindrome have vertical lines of symmetry, then the palindromic word will have mirror symmetry. Words like MOM, WOW, TOOT, and OTTO all look like their images when seen in a mirror.

11. Can you make out the fancy word above? It's *symmetry*. But it not only spells symmetry, it *is* symmetric. Rotate the design 180° and it is identical to the original. This design also comes from *Inversions* by Scott Kim. Scott is able to write many words so that they possess rotational or reflectional symmetry. Everyone can write OTTO so that it is point symmetric. If you can design an R so that when rotated 180°, it can pass for a *y*, then you can write the name Roy so that it has point symmetry. Try it! Another word Scott suggests we can all write with point symmetry is *chump*. Try writing *chump* (in script) so that when rotated 180°, it reads the same.

12. If you try writing your name with the opposite hand that you normally write with, you will probably not be very successful. Try it. Scott Kim suggests that if you write with both hands simultaneously, your "opposite" hand improves. Try it. Holding a pen in each hand, place the pen points together in the middle of a blank sheet of paper. Write you name with both hands simultaneously — the right hand moving left to right and left hand moving right to left. Hold a mirror vertical to the backward writing. Did your "opposite" handwriting improve? (If you are left-handed, start with your pens at the far left and right of the paper and write with your left hand moving left to right and your right hand moving right to left.

13.* Did you notice that some letters have both horizontal and vertical symmetries? Did you also notice that all the letters that had both horizontal and vertical symmetries also had point symmetry? Is this a coincidence? Not at all. Use what you have learned in this chapter to explain why this is no coincidence.

14. Scott Kim suggests that symmetry can also be found in handshakes. There is a nice rotational symmetry to a handshake between two people. What about handshakes between three people? The illustrations below show three different symmetric handshakes between three people. Now it is your turn. Find two partners and create a symmetric three-person handshake different from the three shown. Make a sketch of your handshake.

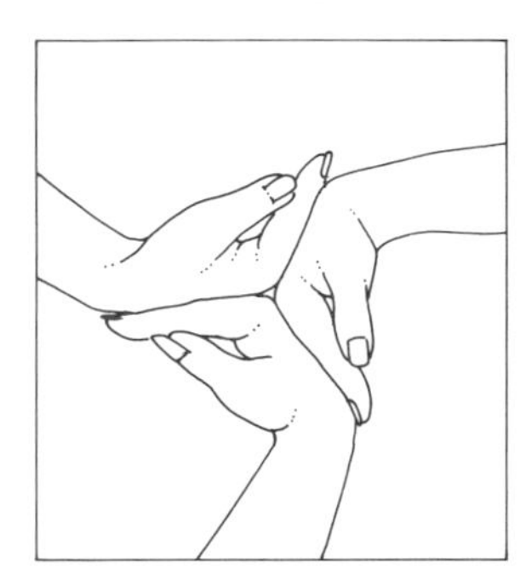

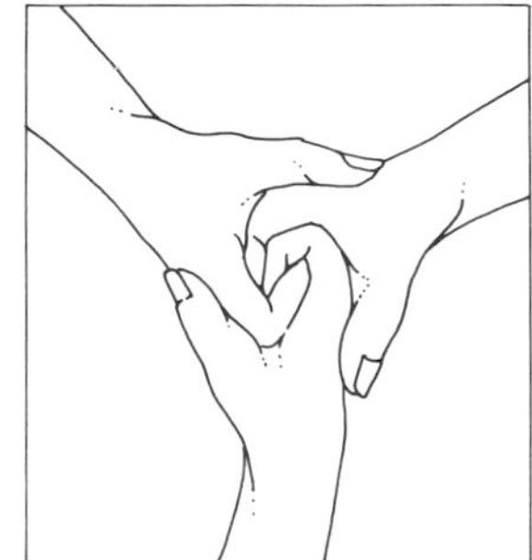

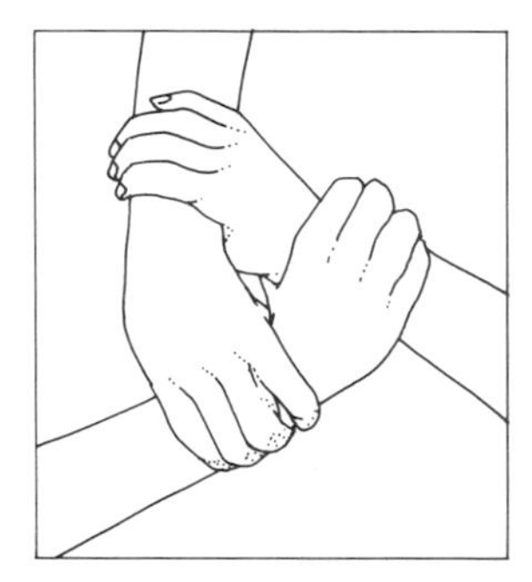

EXERCISE SET C

Did you know that the flag section is the most used section of the *World Book Encyclopedia*? This exercise set requires a little library research.

1. Is the flag of Puerto Rico symmetric?
2. Does the flag of Kenya have rotational symmetry?
3. Find three flags that have rotational symmetry.
4. Find a flag that has a horizontal axis of symmetry but not a vertical axis of symmetry.

The complete alphabet in reflectional symmetry from *Inversions* by Scott Kim

Lesson 7.4

Properties of Symmetries

Some polygons have no symmetry. Some polygons have a single symmetry. Regular polygons, however, have many symmetries. A square, for example, has four reflectional symmetries — one vertical, one horizontal, and one through each diagonal.

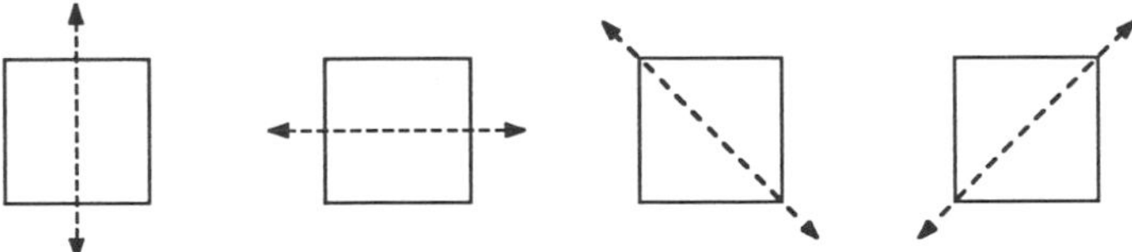

The Four Reflectional Symmetries of a Square

A square can be rotated 90°, 180°, 270°, and 0° (the identity rotation) and it will coincide with itself. A square has four rotational symmetries.

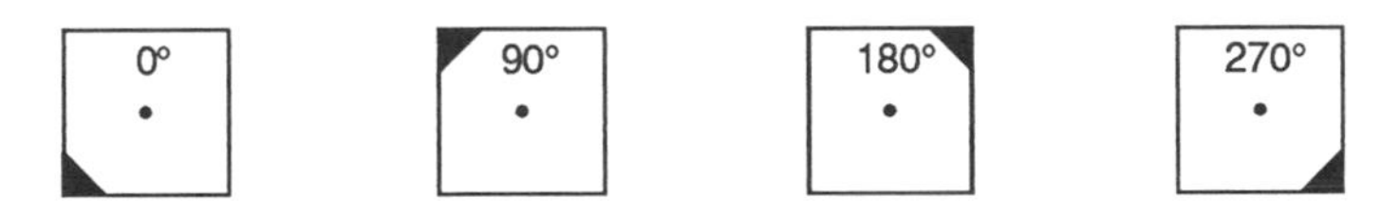

The Four Rotational Symmetries of a Square

Does the number of sides of a regular polygon determine the number of symmetries of each type? Let's investigate.

Investigation 7.4.1

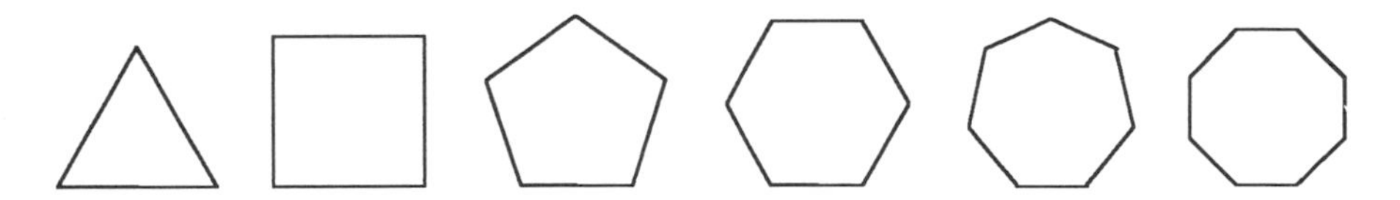

Find the number of reflectional and rotational symmetries for each of the regular polygons listed in the table below. If necessary, use a mirror to locate the axes of symmetry for each of the regular polygons. To find the number of rotational symmetries, you may wish to trace each regular polygon above onto tracing paper. Place the tracing over the original and rotate until the tracing coincides with the original. In listing the rotational symmetries, include the 0° identity rotation.

Number of sides of regular polygon	3	4	5	6	7	8	...	n
Number of reflectional symmetries	–?–	4	–?–	–?–	–?–	–?–	...	–?–
Number of rotational symmetries (< 360°)	–?–	4	–?–	–?–	–?–	–?–	...	–?–

When you've completed the table, make a conjecture about the number of reflectional and rotational symmetries in a regular n-gon.

C-51 *A regular polygon of n sides has —?— reflectional symmetries and —?— rotational symmetries*

You've just worked with two types of symmetries — reflectional symmetry and rotational symmetry. The other two symmetries — translational symmetry and glide reflectional symmetry — really exist only for infinite patterns. If a finite design is said to have translational or glide reflectional symmetry, it is understood that this is true only if the particular design were to continue indefinitely.

Translational Symmetry

Glide Reflectional Symmetry

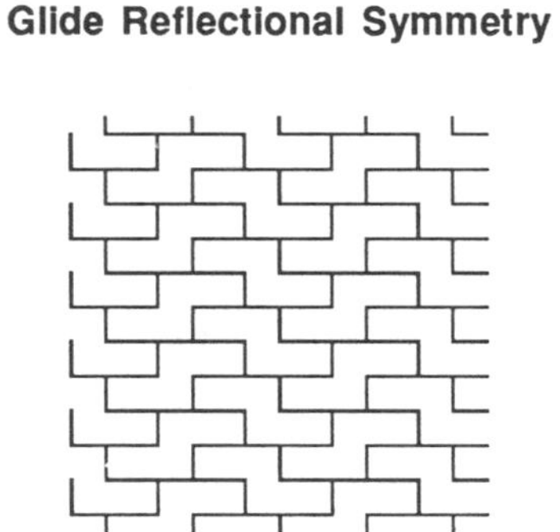

If a pattern can be translated a given distance in a given direction in such a way that the image coincides with the original, then the pattern has **translational symmetry**. The given distance and given direction are identified by the translation vector. In the pattern above left, any vector with the same length as the vector shown (or a multiple of the length) that is parallel to it can be a translation vector. If a design can undergo a glide reflection isometry in such a way that the image coincides with the original, then the design has **glide reflectional symmetry**.

EXERCISE SET

1. How many reflectional symmetries does an isosceles triangle have?
2. How many reflectional symmetries does a rhombus have?
3. Sketch a figure that has point symmetry but not line symmetry.
4. Name a polygon that has a 36° rotational symmetry.
5. Is it possible for a triangle to have exactly one line of symmetry? Exactly two? Exactly three? Support your answers with sketches.
6. Is it possible for a quadrilateral to have exactly one line of symmetry? Exactly two? Exactly three? Exactly four? Support your answers with sketches.

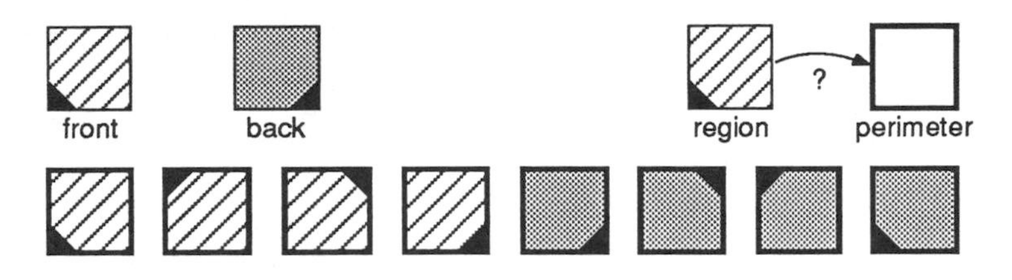

Eight Ways a Square Card Can Fit into its Perimeter

7. The square region above has lines on its front side and shading on its back side. It can be placed in its square perimeter in eight different ways. How many different ways can a regular polygonal region fit in its regular polygonal perimeter? To help you answer this question complete the table below.

Number of sides of regular polygon	3	4	5	6	7	...	n
Ways polygonal region can fit	–?–	8	–?–	–?–	–?–	...	–?–

8. A regular tetrahedron is a pyramid in which each of the four faces is an equilateral triangle. Imagine a regular tetrahedron and its tetrahedron case like those shown on the right in which the regular tetrahedron fits perfectly in its case. How many different ways can it be placed in its case?

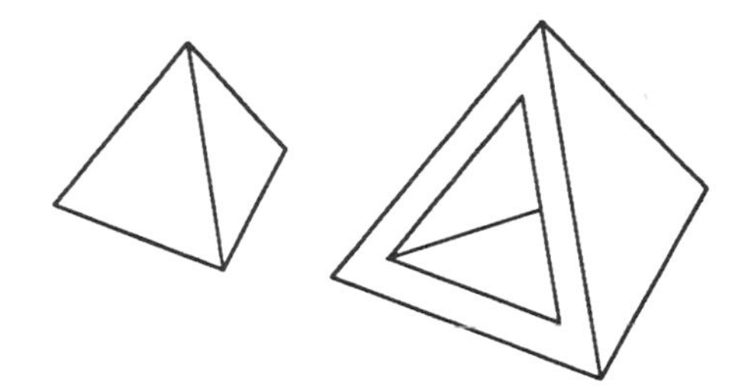

9. How many ways can a cube be placed in its case?

10. Assume the design below continues infinitely. Does it have translational symmetry? Rotational symmetry? Reflectional symmetry? Glide reflectional symmetry?

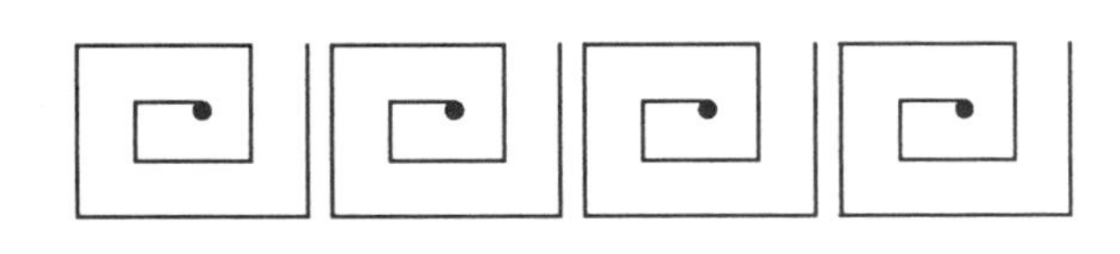

11. Assume the design below continues infinitely. Does it have translational symmetry? Rotational symmetry? Reflectional symmetry? Glide reflectional symmetry?

12. Assume the design below continues infinitely. What type or types of symmetry does it have?

13.* Assume the design below continues infinitely. There are two different types of centers of rotational symmetry in it. One set of points are the centers of 120° and 240° rotational symmetries. Another set of points are the centers of 60°, 120°, 180°, 240°, and 300° rotational symmetries. Copy the figure, and then locate and label the two different types.

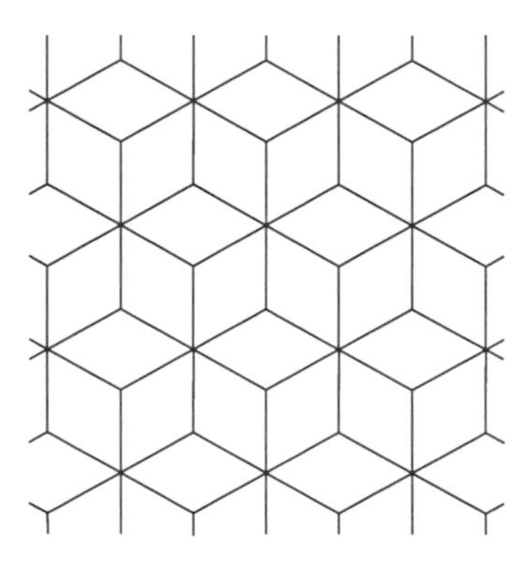

Improving Reasoning Skills

Logical Liars

Five students have just completed a logic contest. To confuse the school reporter, Lois Lovely, each student agreed to make one true and one false statement to her when she interviewed them. Lois was clever enough to figure out the winner — are you? See if you can determine the results of the contest from their statements.

Frances:	Kai was second. I was fourth.
Leyton:	I was third. Charles was last.
Denise:	Kai won. I was second.
Kai:	Leyton had the best score. I came in last.
Charles:	I came in second. Kai was third.

Special Project

Building Kaleidoscopes

Everyone has looked through a kaleidoscope. But how many people know how they are made or how they work their magic? The kaleidoscope was invented by Sir David Brewster, a physicist, in 1816. There are many varieties of kaleidoscopes on the market today. Some scopes have colored glass or plastic that tumbles about the end section creating an infinite series of symmetric patterns. Some have nothing in the end section but a lens and the designs that are created depend on what the kaleidoscope is aimed at. Some have a marble on the end. Some modern scopes have colored non-mixing liquids in the end section that create flowing symmetric designs. Some handcrafted kaleidoscopes sell for well over $100.

Your project as a group is to:

1. Design a kaleidoscope.
4. Make a diagram showing your kaleidoscope plan.
2. Collect the raw materials and tools necessary to make your kaleidoscope.
3. Build the kaleidoscope.

The diagram on the right is an exploded view of a typical kaleidoscope. This entire scope must be rotated for the colored pieces to tumble into the different designs. To create a kaleidoscope with an end cap that turns while the body remains stationary is a little trickier. This involves a slightly larger piece of tube for the rotating end cap. The material for a kaleidoscope can vary, depending on what the group can find and/or afford. The tube can be a cardboard cylinder found around the home. For a kaleidoscope to last your lifetime, use plastic plumbing pipe (PVC). The round clear pieces can be glass, lenses, clear plastic, or clear acetate. The three reflecting surfaces can be mirror glass, glass painted black on one side, or clear vinyl painted black on one side. The colored pieces that go into the end chamber can be almost anything translucent. Don't put too many colored pieces in between the spacers! If you use too many, they don't tumble well and they may block out too much light.

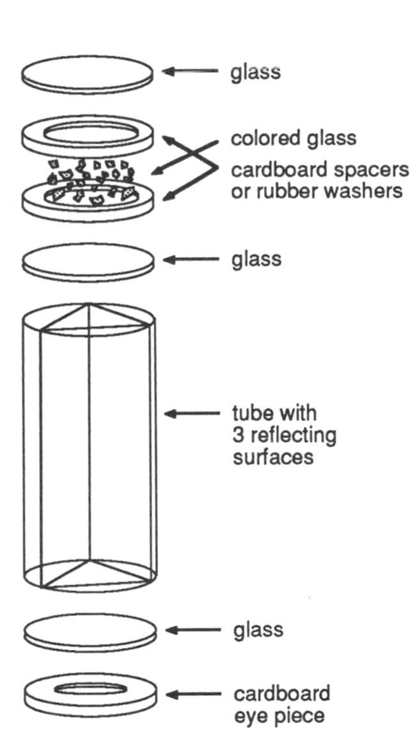

Lesson 7.5

Tessellations with Regular Polygons

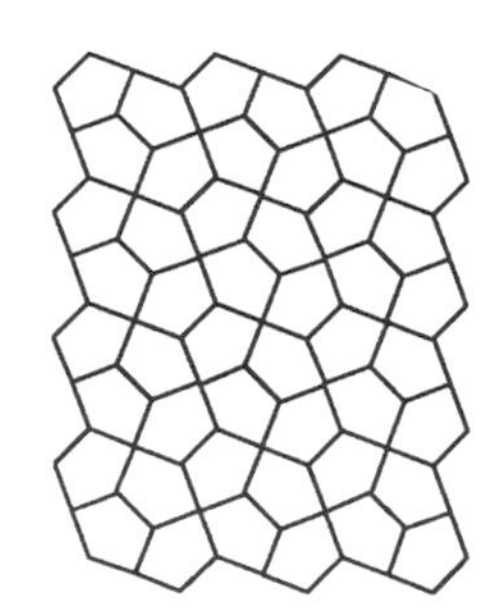

You can find mosaic tile patterns in many places. The tile pattern on the right is made of identical pentagons (non-regular) that could completely cover a floor without gaps or overlaps. Geometers call such a tiling pattern a tessellation or simply a tiling. A **tessellation** or **tiling** is an arrangement of closed shapes that completely covers the plane without overlapping and without leaving gaps. The pentagon tessellation can be found in the street tiling of portions of Cairo and many other ancient cities in the Islamic world. When a tessellation uses only one shape, it's called a **pure tessellation**.

But you don't have to go around the world in search of tessellations. You can find them in every home. Floor tiles and brick fireplaces have tessellating patterns. Where are other tessellations in your home?

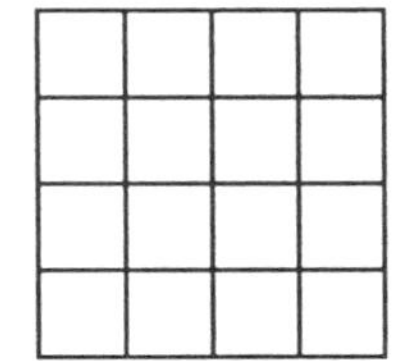

Tessellations are found in nature. The molecular structures of many crystals show tessellating patterns. The honeycomb of the bee is a tessellation using regular hexagons. Where are other tessellations in nature?

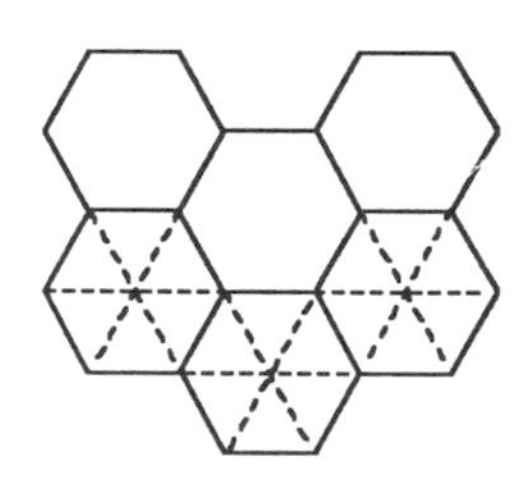

Which of the regular polygons can be used to create a pure tessellation? We have already mentioned that squares (floor tile) and regular hexagons (honeycomb) create pure tessellations. Since each regular hexagon can be divided into six equilateral triangles, it follows logically that equilateral triangles can also be used to create a pure tessellation. Will other polygons tessellate?

In the next investigation you'll need to trace regular polygons. On the next page you'll find a series of eight regular polygons — all with sides of the same length. Unless you have a plastic template that has outlines of all (or most) of the regular polygons pictured, you'll need to make a set of poster-board or heavy cardboard polygons to use as templates. One set of templates can be shared by a group of four or five students. To make each regular polygonal template:

Step 1: Place an unlined sheet of paper over the polygons. Select a polygon and put a dot over each vertex of the polygon.

Step 2: Put the paper on top of your poster-board. Using a sharp pencil point or the point of your compass, make a mark through your paper onto the poster-board. Use your straightedge to connect the marks.

Step 3: Cut out the polygonal shape. You have a template for the regular polygon.

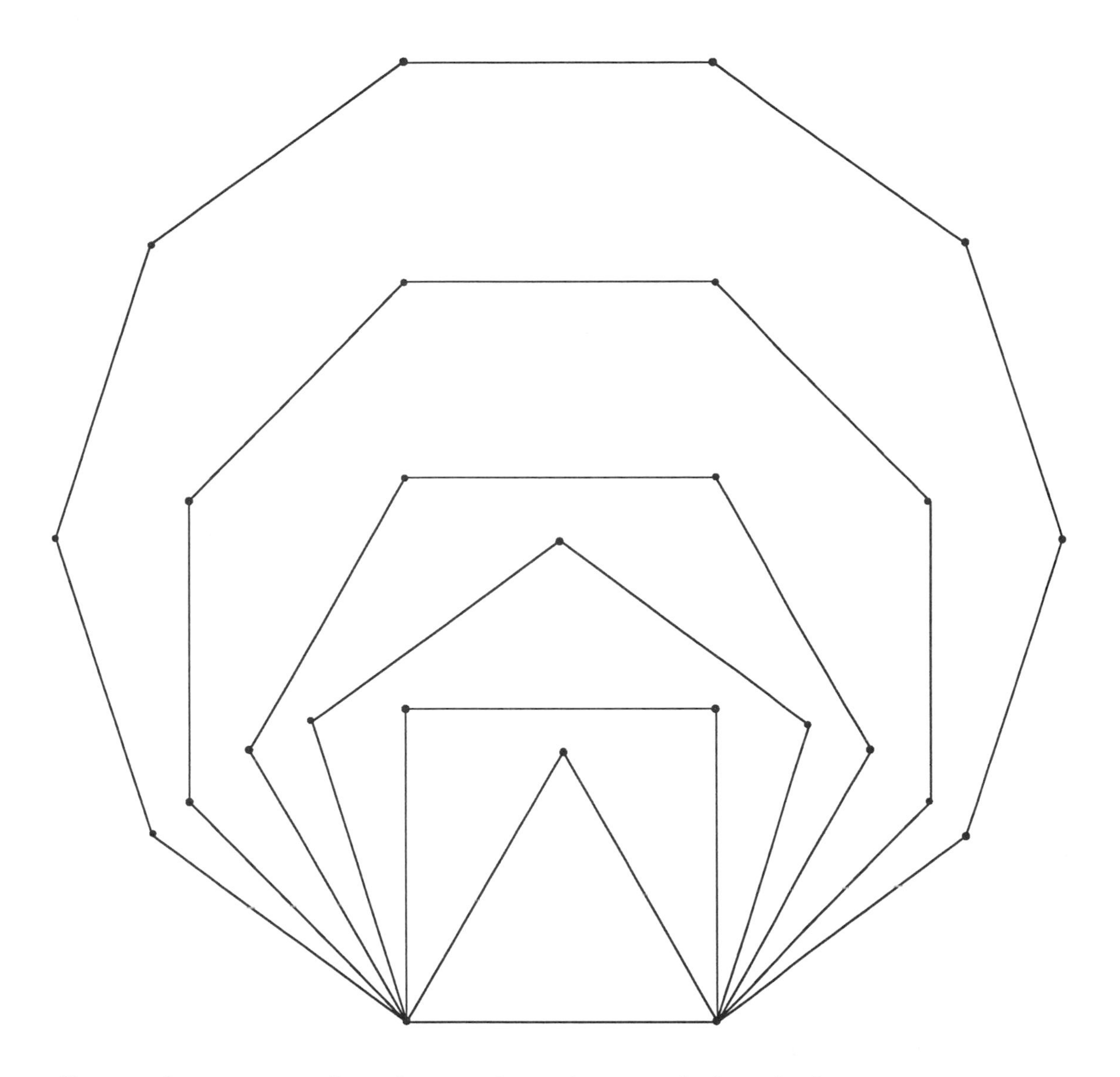

Once you have your set of templates, you're ready to start the investigation.

Investigation 7.5.1

Which regular polygons tessellate? If you're not sure whether a regular polygon can create a tessellation, use your templates and find out.

Before you make a conjecture let's look at this logically. In order for a set of regular polygons to create a tessellation, the measures of all the angles that come together at a point must add up to exactly 360°. And since all the angles are the same, the measure of each interior angle must be a factor of 360°. Since the measure of each angle of a regular polygon is less than 180°, it follows that there must be at least three regular polygons that meet at each point. What is the measure of each angle of a regular pentagon? Can three or more fit exactly about a point? How many angles from a polygon having more than six sides can meet at a point? This information is left for you to organize. You may want to make sketches. Copy and complete the table.

Number of sides of regular polygon	3	4	5	6	7	8	10	12
Degrees in each interior angle	–?–	–?–	–?–	–?–	–?–	–?–	–?–	–?–
Angles that fit about a point	–?–	–?–	–?–	–?–	–?–	–?–	–?–	–?–

Based on your investigation and reasoning, you should be ready to state a conjecture about the regular polygons that can be used to create a pure tessellation.

C-52 *The —?— and only —?— regular polygons that create pure tessellations are —?—.*

Tessellations that involve more than one type of shape are called **semi-pure tessellations**. The octagon-square combination on the right is an example of a semi-pure tessellation. You can find this type of tiling in both Islamic and Christian art. The pattern appears in the Moorish Alhambra in Granada, Spain, as well as the Renaissance floor mosaics of St. Mark's Basilica in Venice, Italy.

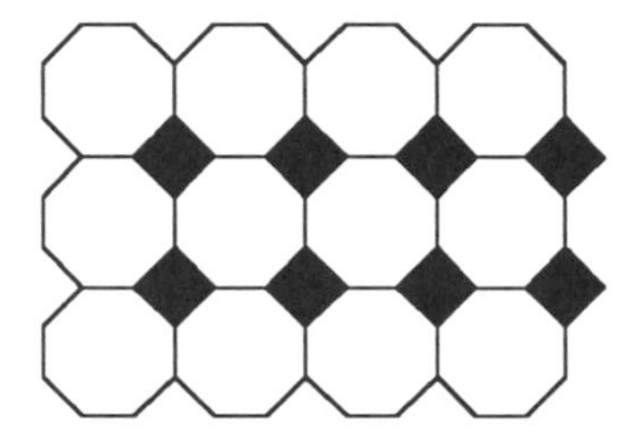

In the octagon-square tessellation, two regular octagons and a square meet at each vertex point. When the same combination of regular polygons meet in the same order at each vertex in a semi-pure tessellation, the tessellation is called a **semiregular tessellation**. There are eight different semiregular tessellations. Semiregular tessellations are identified by listing the number of sides of the polygons about each vertex. You start with the polygon with the least number of sides and then list the number of sides of each polygon as you move (clockwise or counterclockwise) about the vertex. The octagon-square tessellation shown above is identified as 4.8.8. Below are two more examples of semiregular tessellations.

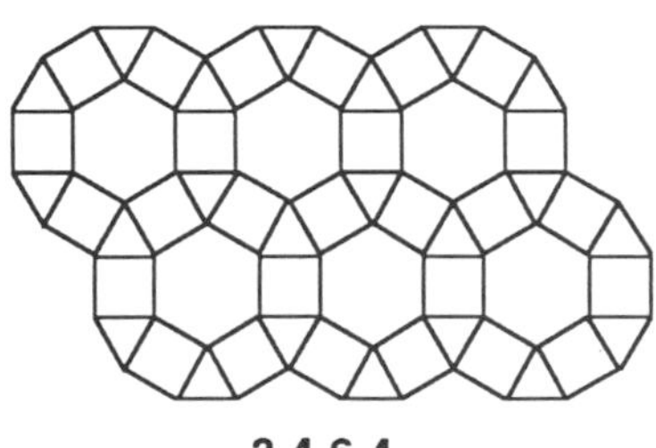

3.4.6.4

The same order of polygons appears at each vertex. The sum of the measures of the angles at a vertex is 360 (60 + 90 + 120 + 90 = 360).

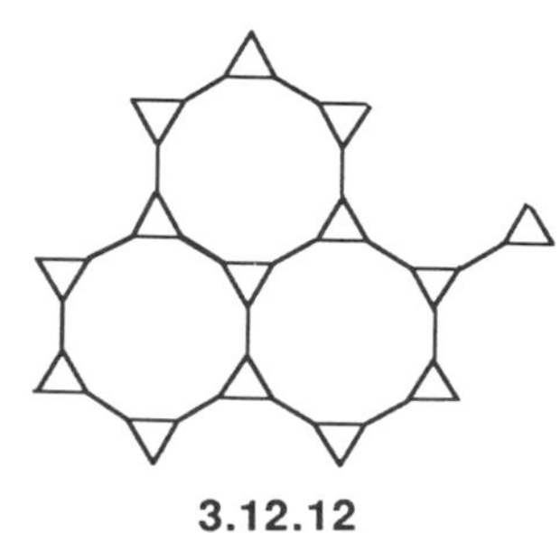

3.12.12

The same order of polygons appears at each vertex. The sum of the measures of the angles at a vertex is 360 (60 + 150 + 150 = 360).

If the arrangement at each vertex in a semi-pure tessellation of regular polygons is not the same, then the tessellation is called a **demiregular tessellation**. Two demiregular tessellations are pictured below.

In the 2-uniform tessellation $3.4.3.12/3.12^2$, two orders of polygons appear at vertices.

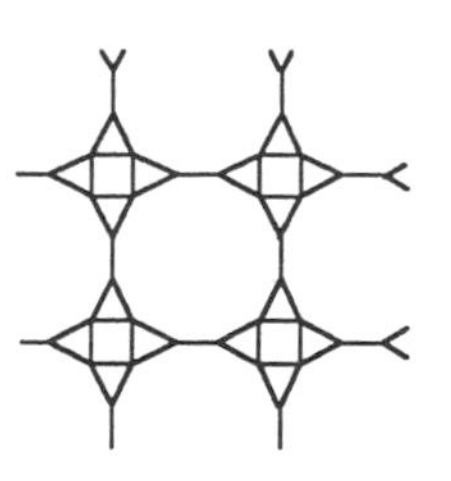

$3.4.3.12/3.12^2$ (2-uniform)

In the 3-uniform tessellation $3^3.4^2/3^2.4.3.4/4^4$, three orders of polygons appear at vertices.

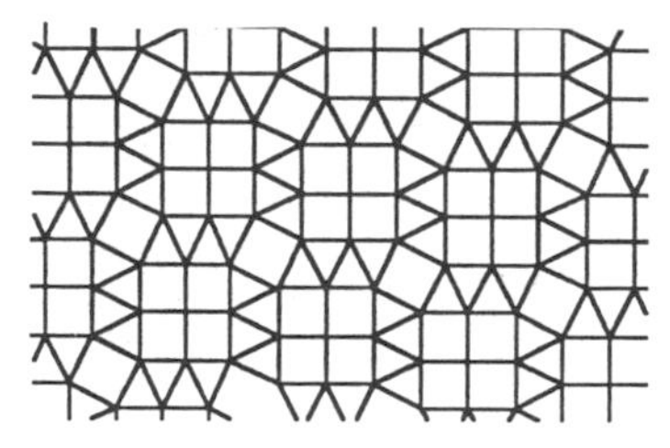

$3^3.4^2/3^2.4.3.4/4^4$ (3-uniform)

Like semiregular tessellations, demiregular tessellations are identified by listing the number of sides of the polygons about the vertices. Unlike semiregular tessellations, however, more than one possible combination occurs. You separate each vertex numbering with / and abbreviate repetitions with exponents. There are 20 different demiregular tessellations that have two different types of vertices (called 2-uniform). There are also demiregular tessellations that have three or more different types of vertices. In fact the number of 4-uniform tessellations is still an unsolved problem. Let's look at two more examples.

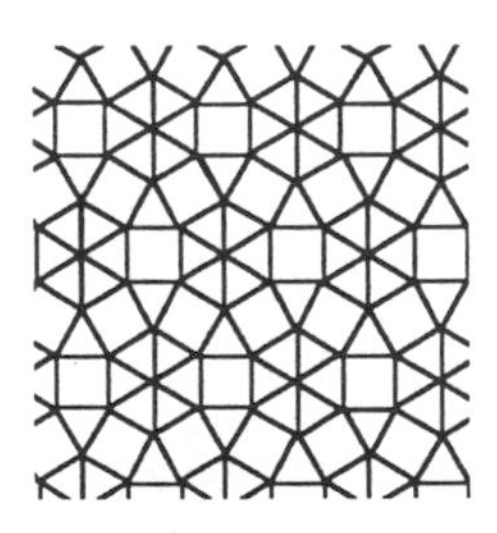

$3^6/3^2.4.3.4$

In the 2-uniform tessellation $3^6/3^2.4.3.4$, two orders of polygons appear at the vertices.

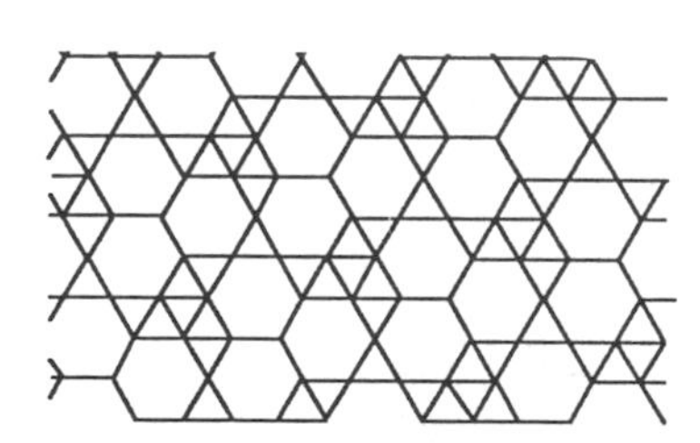

$3^4.6/3^2.6^2/3.6.3.6/6^3$

In the 4-uniform $3^4.6/3^2.6^2/3.6.3.6/6^3$ tessellation, four orders of polygons appear.

EXERCISE SET

Give the numerical name for each of these semiregular tessellations.

1.

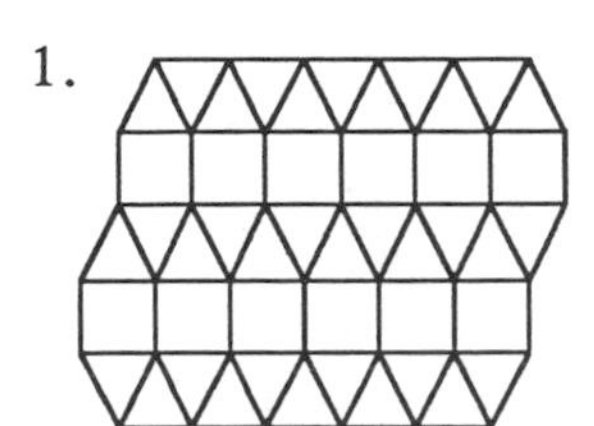

2.

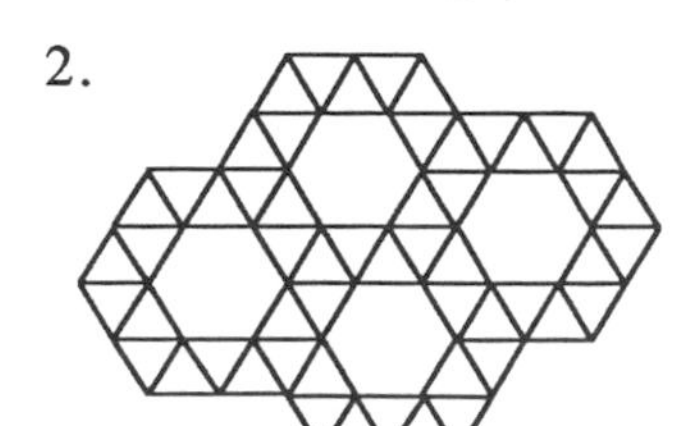

3.

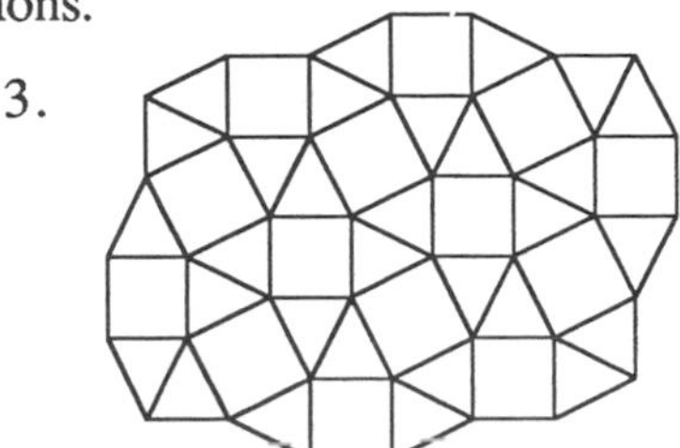

Use the templates of regular polygons you made earlier in Exercises 4 and 5.

4. Sketch a 3.3.3.3.6 tessellation different from the tessellation in Exercise 2.

5. Sketch the 3.6.3.6 tessellation.

6. Using your templates, show that two regular pentagons and a regular decagon fit about a point, but that 5.5.10 does not create a semiregular tessellation.

Give the numerical name for each of these demiregular tessellations.

7.

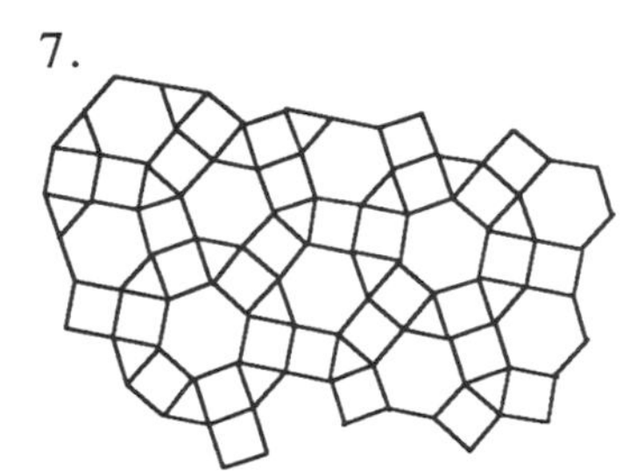

8.

9.

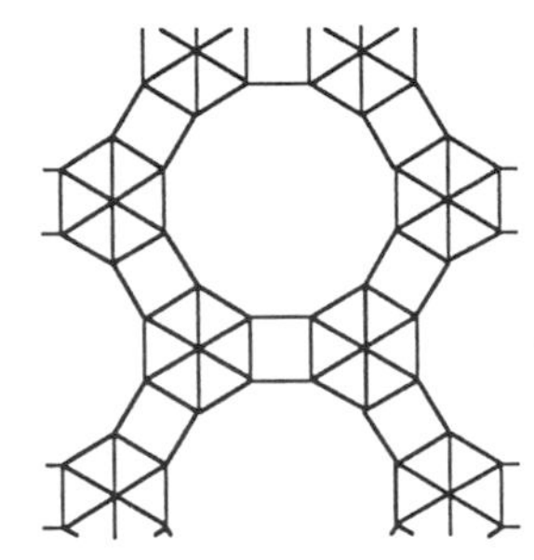

10. Using your templates, create the demiregular tessellation $3.12^2/3.4.3.12$. Draw the design on a full sheet of paper. Decorate and color the design.

11. Using your templates, create the demiregular tessellation $3^3.4^2/4^4$ on the top half of a full sheet of paper. There are two distinct forms for $3^3.4^2/4^4$. Can you find the other? Create the second on the bottom half of your paper. Color the top tessellation design so that it retains all its symmetries. Color the bottom tessellation design so that if you take color into account, it retains a reflectional symmetry but not a rotational symmetry.

12. A tessellation of equilateral triangles is shown on the right. When you connect the center of each triangle across the common sides of the tessellating triangles, you get another tessellation. This new tessellation is called the **dual** of the original tessellation. Every tessellation of regular polygons has a dual. The dual of the equilateral triangle tessellation is the regular hexagon tessellation. Copy the square and regular hexagon tessellations pictured on the first page of this lesson. Create the dual of each.

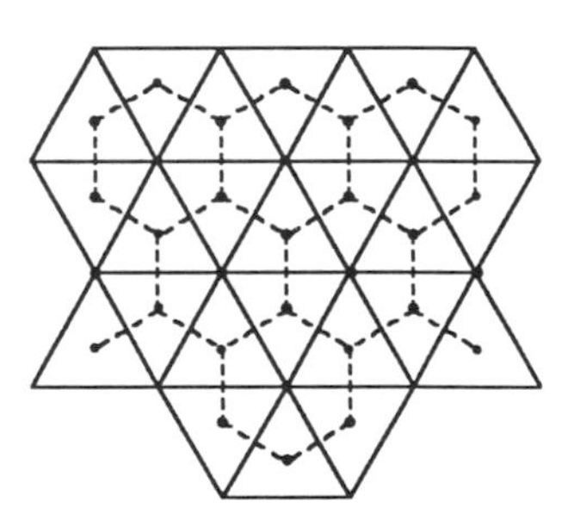

13. You can create dual tessellations from semi-pure tessellations of regular polygons (but they may not be tessellations of regular polygons). Try it. Sketch the dual of 4.8^2.

Lesson 7.6
Tessellations with Non-Regular Polygons

In the last lesson, you tessellated with regular polygons. You learned that there are only three pure tessellations of regular polygons and eight semiregular tessellations. What about tessellations of non-regular polygons? Will any scalene triangles tessellate? Let's investigate.

Investigation 7.6.1

Stack three pieces of paper together and fold them in half. Construct a scalene triangle on one half of the top half-sheet and cut it out, cutting through all six half-sheets of paper. You now have six congruent scalene triangles. Use one triangle as a template and trace the triangle onto what's left of the top half-sheet. Cut again. You now have twelve congruent scalene triangles. Label the interior of the angles in each triangle as a, b, and c, making sure that all twelve angles labeled as a are congruent, all twelve labeled as b are congruent, and all twelve labeled as c are congruent. Using your twelve congruent scalene triangles, try to create a tessellation.

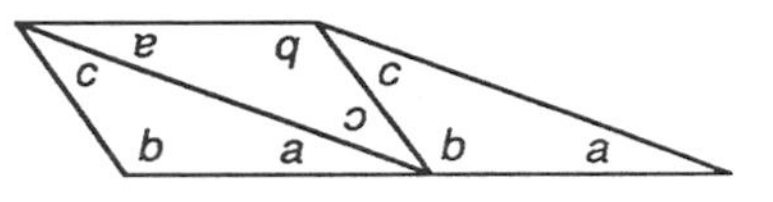

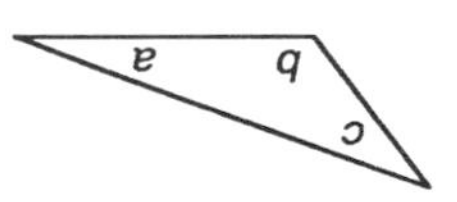

Observe the angles about each point. How many times did each angle of the triangle fit about each point? What is the sum of the measures of the three angles of a triangle? Compare your results with the results of others. State your next conjecture.

C-53 *—?— triangle will create a pure tessellation.*

You have seen squares and rectangles tile the plane. You can probably visualize tiling with parallelograms. Will any quadrilateral tessellate? Let's investigate.

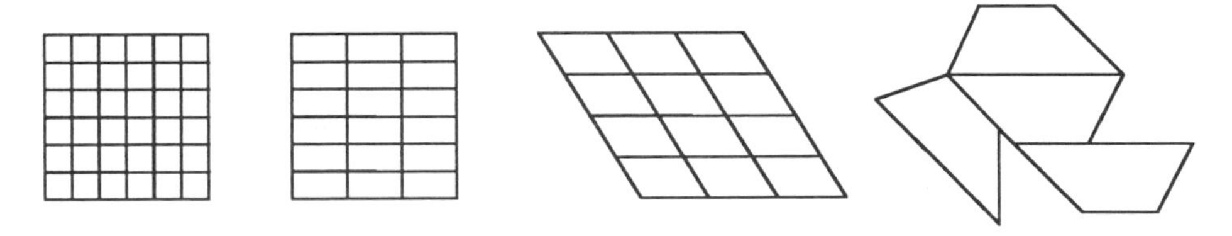

Investigation 7.6.2

Cut out twelve congruent quadrilaterals (not parallelograms). Label the interior of the angles in each quadrilateral as a, b, c, and d. Using your twelve congruent quadrilaterals, try to create a tessellation.

Observe the angles about each point. How many times did each angle of your quadrilateral fit about each point? What is the sum of the measures of the angles of a quadrilateral? Compare your results with the results of others. State a conjecture.

C-54 *—?— quadrilateral will create a pure tessellation.*

Since the regular pentagon does not create a tessellation, you know that *not every* pentagon will tessellate. Is there at least one pentagon that will tessellate? Yes, there are many. What types of pentagons tessellate? Prior to 1967, it was thought that all tessellating pentagons could be classified into five types. But that year, Richard B. Kershner of John Hopkins University found three more. Most everyone thought that the problem had been solved — until 1975 when Marjorie Rice, a San Diego mother of five, entered the picture. After reading an article in *Scientific American* by Martin Gardner on Kershner's discovery of the new types of tessellating pentagons, Marjorie Rice began her own investigations. With no formal training in mathematics beyond high school general math, she discovered within a few months a ninth type of pentagon that tessellates. By 1977, Marjorie Rice had discovered four more types of tessellating pentagons. As of 1988, no one has found other types of tessellating pentagons or proved that there were no others. What are *all* the types of pentagons that tessellate? The tessellating pentagon problem remains unsolved.

Two of the Pentagonal Tessellations Discovered by Marjorie Rice

Type 9 discovered in February 1976

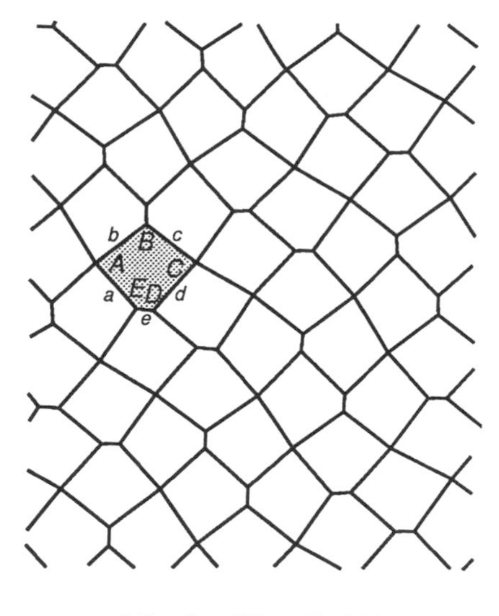

$2E + B = 2D + C = 360$
$a = b = c = d$

Type 13 discovered in December 1977

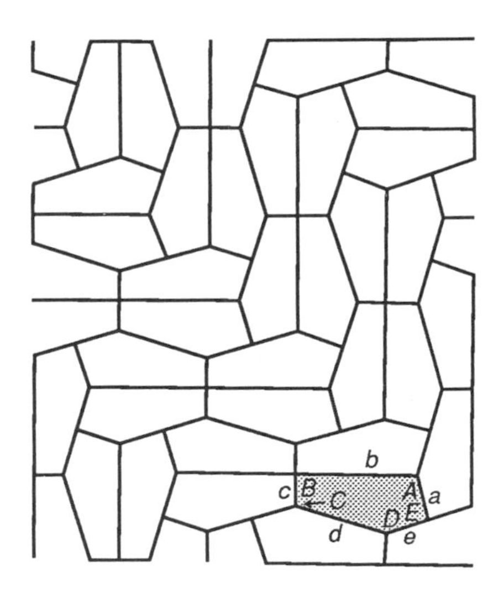

$B = E = 90 \quad 2A + D = 360 \quad 2C + D = 360$
$a = e \quad a + e = d$

Capital letters represent angle measures in the shaded pentagon. Lower case letters represent lengths of sides.

EXERCISE SET

1. Another very beautiful pentagon tessellation uses equilateral pentagons (the sides are congruent but not the angles). An example is the Cairo street tiling shown at the beginning of Lesson 7.5. The construction of an equilateral pentagon is shown below. On poster-board or heavy cardboard, construct an equilateral pentagon and use it as a template to recreate the Cairo street tiling.

Step 1

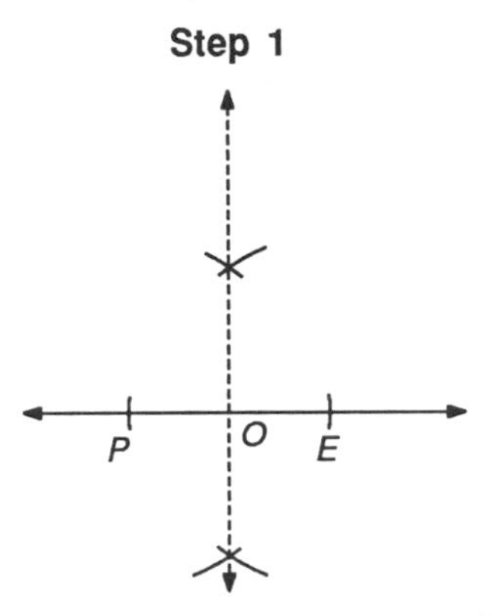

Construct a segment (a side of your equilateral pentagon) and label it $\overline{PE}$. Construct the perpendicular bisector of $\overline{PE}$. Label the midpoint as point O.

Step 2

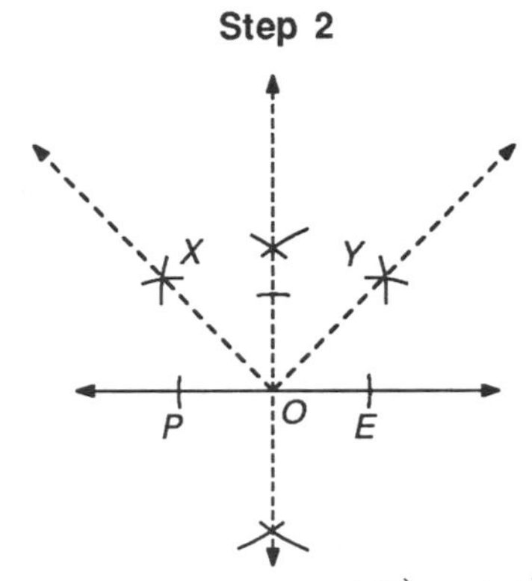

Bisect both right angles with $\overrightarrow{OX}$ and $\overrightarrow{OY}$.

Step 3

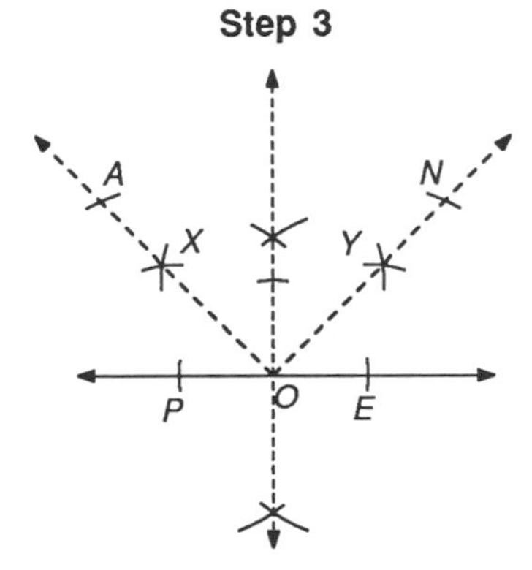

From P swing an arc with radius equal to PE through $\overrightarrow{OX}$. Label the intersection as point A. From E swing an arc with radius equal to PE through $\overrightarrow{OY}$. Label the intersection as point N.

Step 4

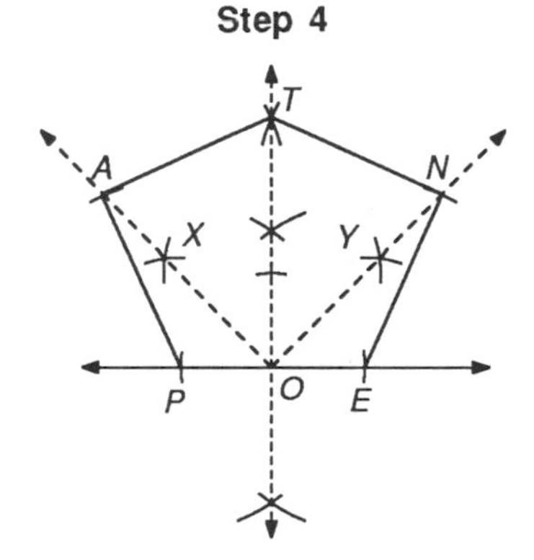

From points N and A swing arcs with radius equal to PE. Label their intersection as T. Construct pentagon $PENTA$.

2. Create a full-page color tessellation of triangles or quadrilaterals.

3. A non-convex (concave) quadrilateral is shown on the right. Can any non-convex quadrilateral tile the plane? Try it. Create your own non-convex quadrilateral and try to create a tessellation with it. Decorate your attempt.

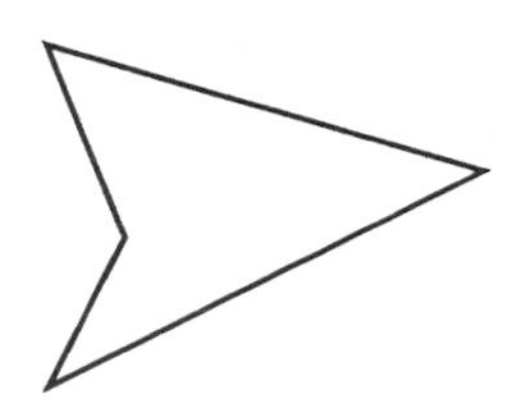

Lesson 7.7
Translation Tessellations

Islamic artists are the masters of tessellation. The Greek Pythagoreans, who felt that "numbers ruled the universe," were an important influence on Islamic thought. The Islamic artists, therefore, were not only trained in geometry, but their philosophy saw mathematics as essential in understanding the universe. The artists expressed this philosophy in their art. The Alhambra, a thirteenth century Moorish palace in Granada, Spain, is one of the finest examples of the precise mathematical art of Islam that exists today.

In 1936, the Dutch artist M. C. Escher traveled to Spain and became fascinated with the tile patterns of the Alhambra. Escher spent days in the Alhambra sketching the tessellations on the walls and ceilings. One of his sketches is shown above. Escher writes on tessellations, "This is the richest source of inspiration that I have ever tapped. . . ." But Escher did not limit himself to pure geometric tessellations as did the Islamic artists. Escher writes:

> What a pity it was that Islam forbade the making of "images". . . . I find this restriction all the more unacceptable because it is the recognizability of the components of my own patterns that is the reason for my never ceasing interest in this domain.

Escher spent many years learning how to use translations, rotations, and reflections on a grid of equilateral triangles, regular hexagons, or parallelograms to create tessellations of birds, fish, reptiles, and humans. One striking example is Escher's translation tessellation design used in a tile mural for the Liberal Christian Lyceum, in The Hague, Netherlands. In this lesson you will learn to create your own tessellations of recognizable shapes.

Untitled work by M. C. Escher, 1960

The four steps demonstrate how Escher may have created his "Pegasus-type" tessellation.

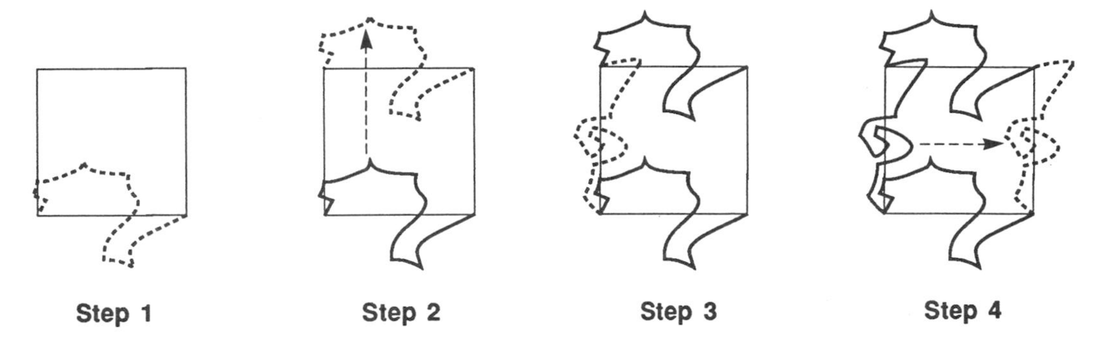

The simplest way to create a non-polygonal tessellation is by changing the opposite sides of a square or parallelogram tessellation. The steps below show how Robert Canete, a geometry student, created *Leap Frog*.

Leap Frog, Robert Canete, Geometry student

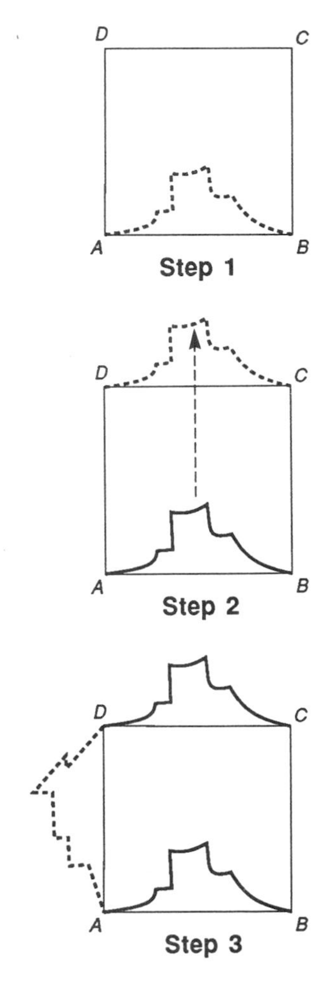

Step 1: Start with one square from a tessellation of squares (although any parallelogram will work with this method). Connect one side $\overline{AB}$ of the square with a curve, call it $\widetilde{AB}$ (curve AB).

Step 2: Place tracing paper or clear plastic over $\widetilde{AB}$ and copy it with a felt tip pen onto the tracing paper or clear plastic. Place the copy beneath the original and slide it so that the endpoints of $\widetilde{AB}$ line up with the endpoint of $\overline{CD}$. Retrace the curve on the original so that it now connects with the endpoint of $\overline{CD}$.

Step 3: Repeat this process with a curve connecting points A and D. That is, connect one side $\overline{AD}$ of the square with a curve, call it $\widetilde{AD}$.

Step 4: Copy $\widetilde{AD}$ onto tracing paper or clear plastic and transfer it across to the opposite side $\overline{BC}$.

Step 5: When completed, trace the entire figure onto the tracing paper or clear plastic and move it to the next square. Trace the entire figure onto the next square. Fill the grid of squares with your figure. You have created a non-polygonal translation tessellation.

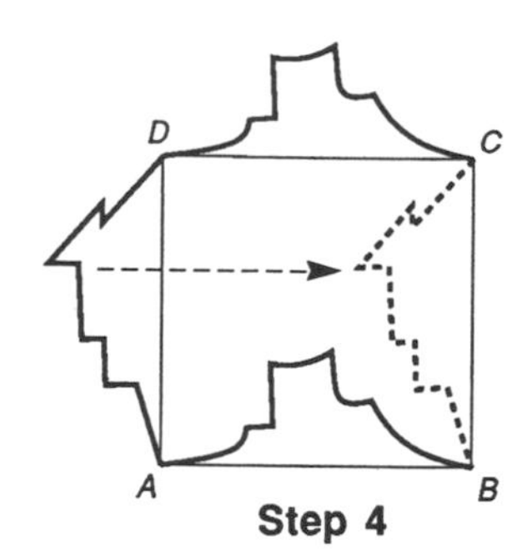

Step 4

Step 5

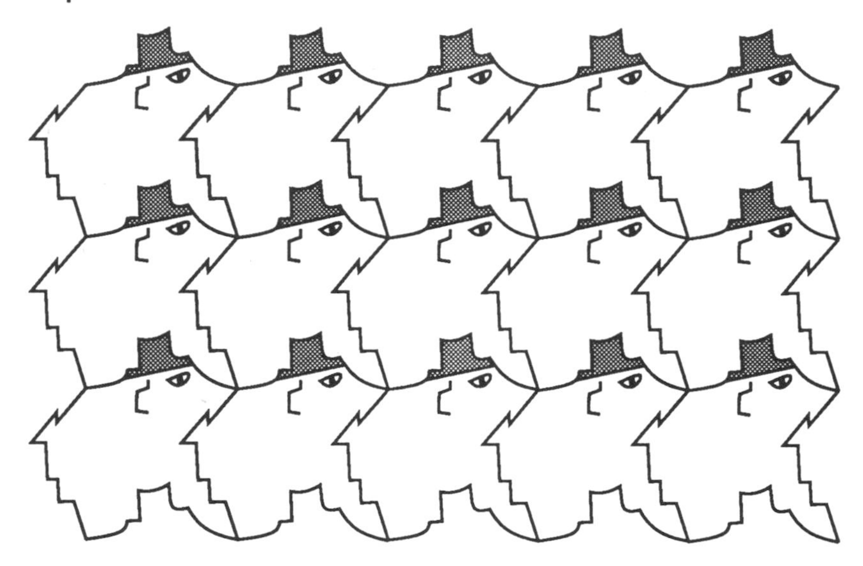

Below is another translation tessellation which uses squares as the basic structure.

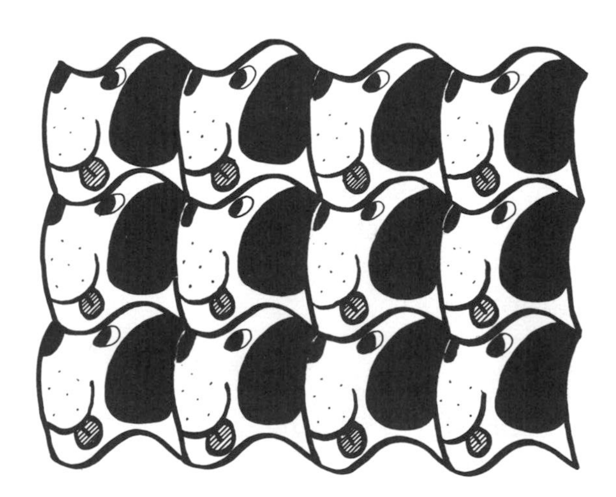

Dog Prints
Gary Murakami
Geometry student

The translation technique outlined on the previous page can be used with regular hexagons. The only difference is that with hexagons there are three sets of opposite sides. Therefore, you need to make three sets of opposite curves. The six diagrams below demonstrate the steps necessary in creating the example on the right.

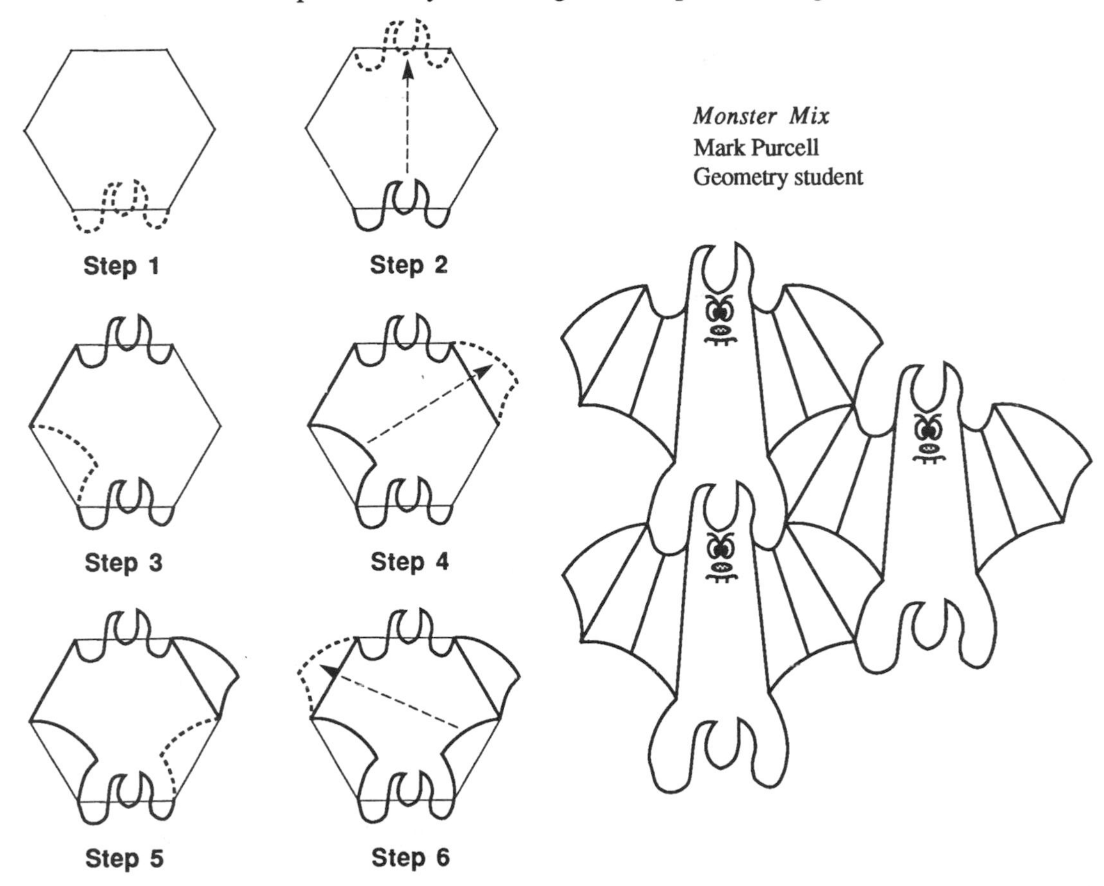

Monster Mix
Mark Purcell
Geometry student

EXERCISE SET

In Exercises 1 to 3, copy the tessellating shapes and fill them in so that they become recognizable figures. When you make your own tessellating designs of recognizable shapes, you'll appreciate this practice!

1.

2.

3.

Identify the basic tessellation grid (squares, parallelograms, or regular hexagons) that was used in creating the translation tessellations below.

4.

Cat Pack
Renee Chan, Geometry student

5.

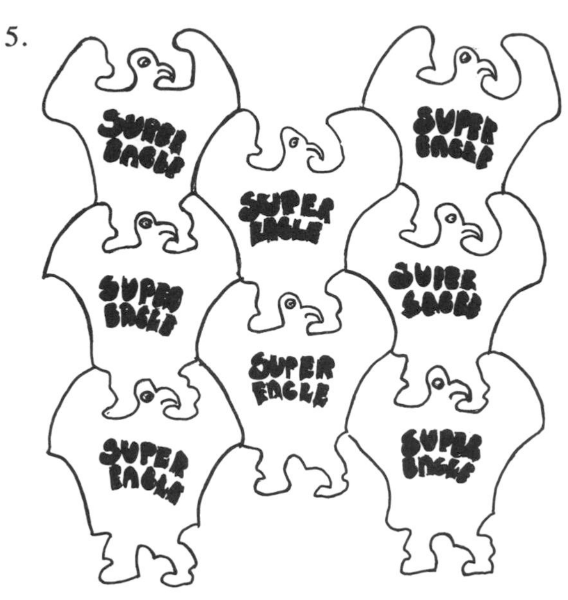

Super Eagle
Carol Witter, Geometry student

6.

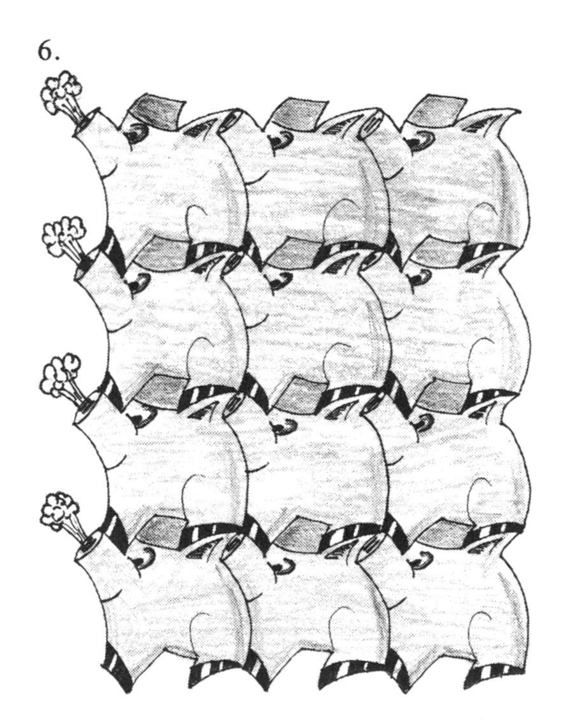

Snorty the Pig
Jonathan Benton, Geometry student

7.

Old Wise One
Serene Tam, Geometry student

8. Now it's your turn is to create a tessellation of "recognizable shapes" using squares as the basic structure and using the translation method demonstrated on the previous pages. At first, you will probably end up with shapes that look like amoebas or spilled milk, but with practice you can begin to see recognizable images within your amoeba-like tessellating shapes. Decorate your design. Give it a title. You will need the following materials:

   - Tracing paper or clear plastic
   - Paper with a grid of large squares
   - Colored pencils, felt-tip markers, and ball point pens

9. Now try this with a grid of regular hexagons as the basic structure. Create a tessellation of "recognizable shapes" using the translation method. Decorate your design. You will also need:

   - Paper with a grid of large hexagons

Improving Visual Thinking Skills

Mystic Octagram

On the right is the mystic Octagram of the Knights of Daze. This mysterious pattern is a lock which must be solved like a puzzle to gain entrance to the forbidden Hall of Daze. To solve the puzzle, you must make eight moves in the proper sequence. To make each move (except the last), you place a gold coin on an empty circle and then slide it along a diagonal to another empty circle. You must place the first coin on circle 1 and then slide it to either circle 4 or circle 6. You must place the last coin on circle 5. You do not slide the last coin. Once the puzzle is solved, you will gain entrance to the mysterious and forbidden Hall of Daze.

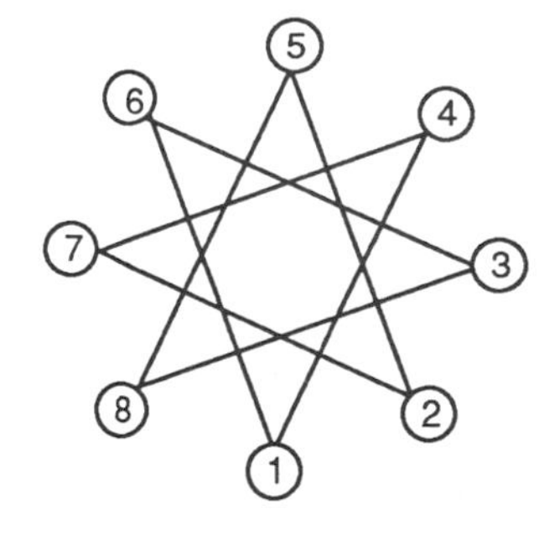

Coin:	Placed on:		Slid to:
First	1	→	–?–
Second	–?–	→	–?–
Third	–?–	→	–?–
Fourth	–?–	→	–?–
Fifth	–?–	→	–?–
Sixth	–?–	→	–?–
Seventh	–?–	→	–?–
Eighth	5		

Solve the puzzle. If you can't find gold coins, use pennies. Copy and complete the chart to show your solution.

Lesson 7.8
Rotation Tessellations

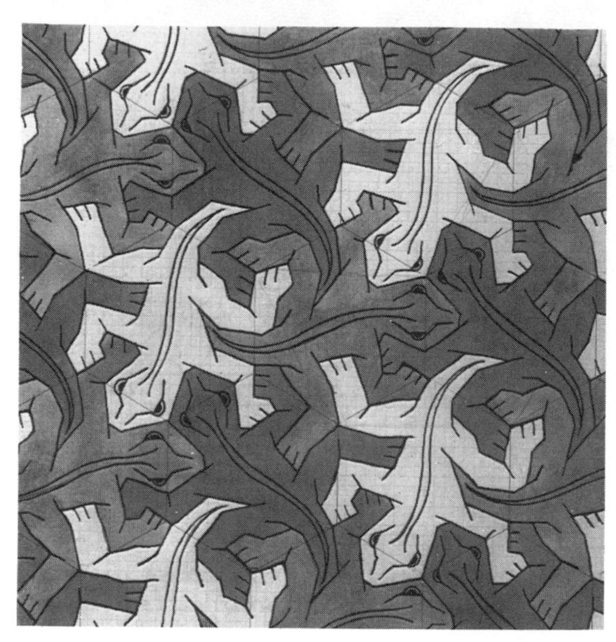

Study of Regular Division of the Plane with Reptiles,
M. C. Escher, 1939

In the previous lesson you made recognizable shapes by translating curves from opposite sides of the regular hexagon or square. In a translation tessellation, all the figures face in the same direction. In this lesson you will use rotations of curves on a grid of equilateral triangles or regular hexagons. The designs formed will revolve about points in the tessellation.

The six steps outlined below demonstrate how you might create a tessellating reptile like those used by Escher in *Study of Regular Division of the Plane with Reptiles*. Each reptile uses rotations of three different curves about three alternating points of a regular hexagon.

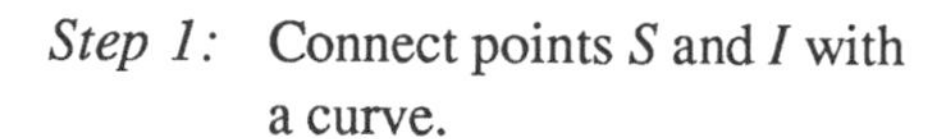

Step 1: Connect points S and I with a curve.

Step 2: Rotate $\widetilde{SI}$ about point I so that point S rotates to coincide with point X.

Step 3: Connect points G and X with a curve.

Step 4: Rotate $\widetilde{GX}$ about point G so that point X rotates to coincide with point O.

Step 5: Create $\widetilde{NO}$.

Step 6: Rotate $\widetilde{NO}$ about point N so that point O rotates to coincide with point S.

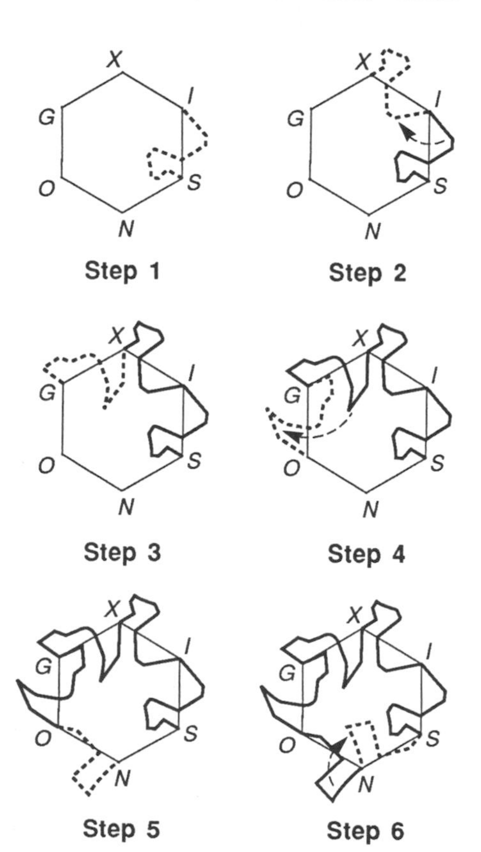

Remember that Escher worked long and hard to perfect each of the three curves in his drawing. He adjusted each curve until he got what he recognized as a reptile. When you are working on your own design, keep in mind that you should probably redraw your curves many times until something that you recognize appears.

The Study of the Regular Division of the Plane with Reptiles was a study for a number of Escher's works. It appeared in *Metamorphosis II* — a woodcut that is 13 feet long! The reptile study was probably best used in his famous lithograph *Reptiles*. In *Reptiles* the little creatures become three-dimensional and leave the two-dimensional study, crawl over numerous objects and then re-enter the two-dimensional study of reptiles. Escher loved to play with reality!

Reptiles, Lithograph by M. C. Escher, 1943

Another method used by Escher utilizes rotations on an equilateral triangle grid. Two sides of each equilateral triangle will have the same curve, rotated about their common point. The third side will be a curve with point symmetry. The four steps below demonstrate how you might make a tessellating flying fish like one used by Escher.

Study of the Regular Division of the Plane, M. C. Escher

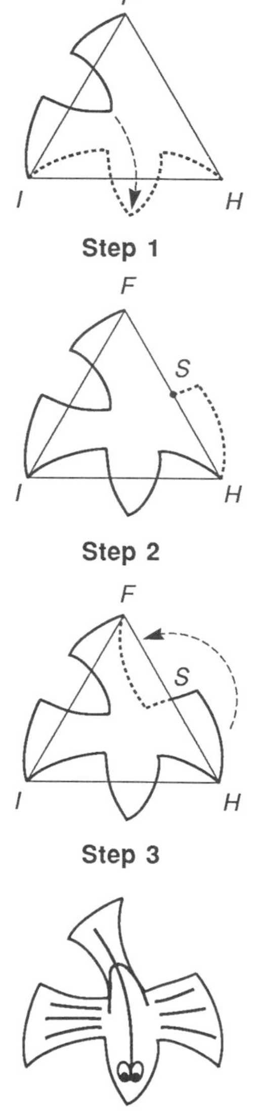

Step 1: Connect points F and I with a curve. Then rotate the curve 60° about point I so that it becomes $\widetilde{IH}$.

Step 2: Find the midpoint S of $\overline{FH}$ and draw $\widetilde{SH}$.

Step 3: Rotate $\widetilde{SH}$ 180° to produce $\widetilde{FS}$. Together $\widetilde{FS}$ and $\widetilde{SH}$ become the point symmetric $\widetilde{FH}$.

Step 4: With a little added detail, the design becomes a flying fish.

With just a slight variation in the curves, the resulting shape will appear more like a bird than a flying fish. The steps are outlined below.

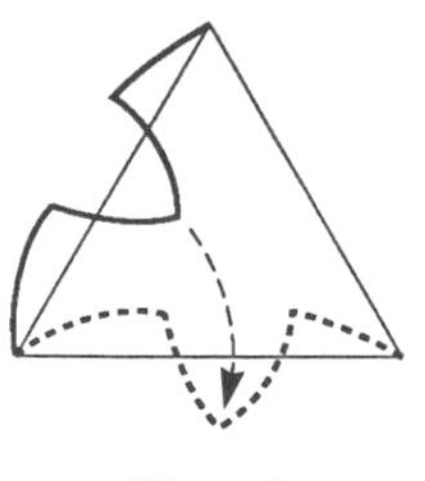

Step 1

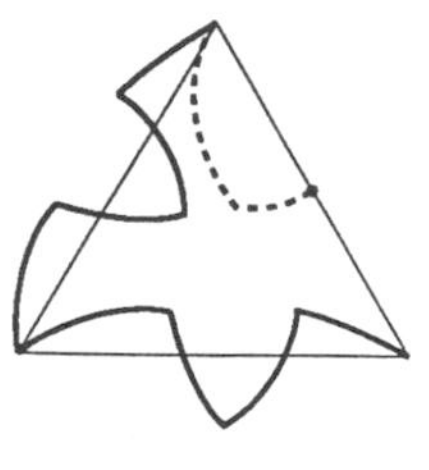

Step 2

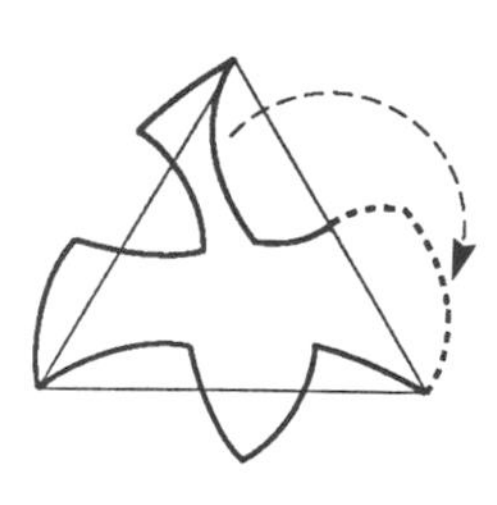

Step 3

Step 4

EXERCISE SET

1. Student Garret Lum started with a grid of regular hexagons and subdivided each hexagon into three rhombuses to create the design below. Copy a portion of the design and locate the vertices of the regular hexagonal grid. Then divide one hexagon into its three rhombuses.

The Wizard's Pets
Garret Lum, Geometry student

2. Copy the finished bird (step 4) from the previous page. Use it to make a tessellating design.

In the exercises to follow, you are to create tessellation designs using rotations. You will need tracing paper or clear plastic, a grid of regular hexagons, and a grid of equilateral triangles.

3. Create a tessellating design of recognizable shapes using a grid of regular hexagons. Decorate and color your art.

4. Create a tessellating design of recognizable shapes using a grid of equilateral triangles. Decorate and color your art.

Lesson 7.9
Reflection Tessellations

Imagination is the highest kite we can fly.
– Lauren Bacall

In this lesson you will use reflections of curves to create shapes that together with their mirror images will create a tessellation. Half the shapes in a reflection tessellation face in one direction and half in the opposite direction.

Escher's woodcut *Day and Night* uses reflection in its design. In this woodcut, Escher gradually changes the shape of the patches of farm land into black and white birds. The birds are going in opposite directions. Escher's method is not as simple as it first appears — notice that the tails of the white birds curve down while the tails of the black birds curve up.

Day and Night, Woodcut by M. C. Escher, 1938

Two methods of creating a reflection tessellation are demonstrated in this lesson. One method uses a grid of rhombuses; the other uses a grid of isosceles triangles. The rhombus method uses reflections and rotations with only one curve to create a tessellation. This is the technique used by Escher in his *Study of Regular Division of the Plane with Human Figures*. The four steps outlined below demonstrate how you might create a tessellating human figure like the one used by Escher in his design.

Step 1: Create $\widetilde{PI}$.

Step 2: Reflect $\widetilde{PI}$ about the vertical axis of symmetry $\overleftrightarrow{PL}$ to get $\widetilde{PF}$.

Step 3: Rotate $\widetilde{PF}$ 120° about point F creating $\widetilde{FL}$.

Step 4: Reflect $\widetilde{FL}$ about the vertical axis of symmetry $\overleftrightarrow{PL}$ to get $\widetilde{LI}$.

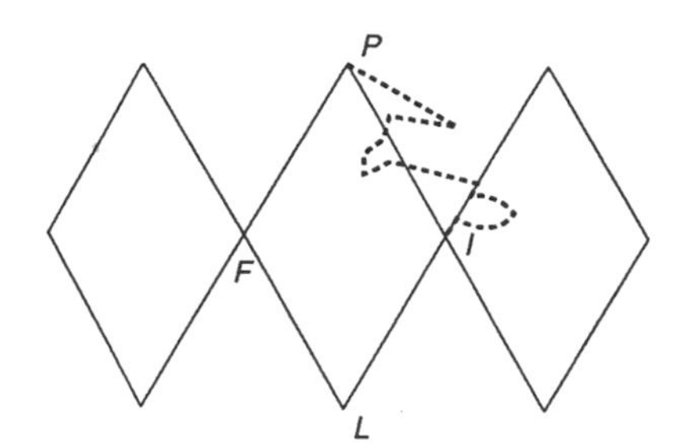

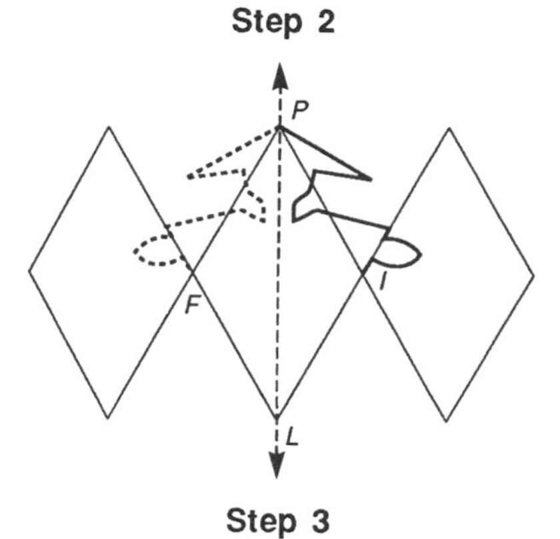

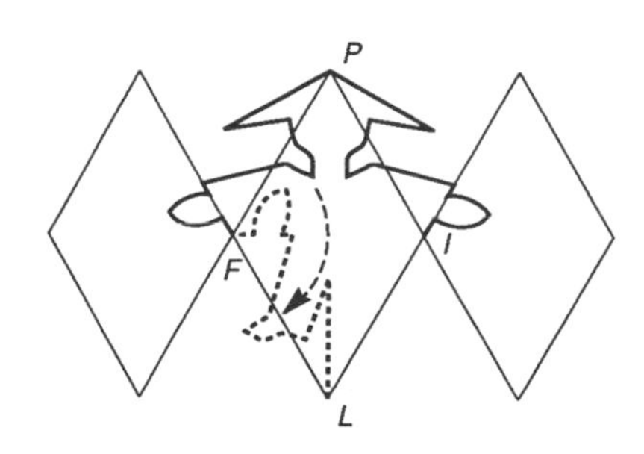

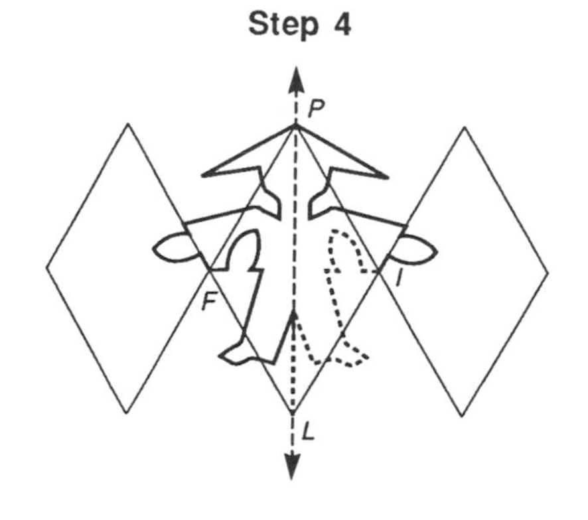

Study of Regular Division of the Plane with Human Figures,
Pencil and watercolor by M. C. Escher, 1936

Another method used by Escher is the isosceles triangle method. Escher used this basic approach with a variety of twists. The steps below show how you can make a tessellating bird like the one used by Garret Lum in his *South Pole or Bust* tessellation. If you make a copy of Garret Lum's tessellating bird, you can use it to fill a page with a glide reflection tessellation.

Step 1: Connect the endpoints of one of the equal length sides $\overline{PE}$ with a curve $\widetilde{PE}$.

Step 2: Reflect $\widetilde{PE}$ about a line parallel to the base $\overline{PL}$ midway between the base and vertex E. Then translate the reflected curve to the opposite equal length side $\overline{LE}$. ($\widetilde{PE}$ undergoes a glide reflection.) This results in point P moving to coincide with point E and E moving to coincide with point L.

Step 3: Connect the endpoints of the third side with a curve that has point symmetry. Connect the midpoint O of $\overline{PL}$ to point P with $\widetilde{PO}$. Then rotate $\widetilde{PO}$ about point O so that point P rotates to coincide with point L. This creates $\widetilde{OL}$. $\widetilde{PO}$ and $\widetilde{OL}$ together create the point symmetric $\widetilde{PL}$.

Step 4: Add detail to finish the bird.

Step 1

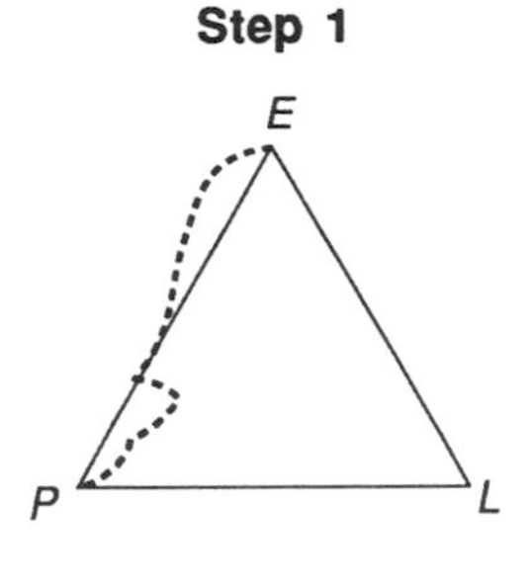

Step 2

E

P

L

Step 3

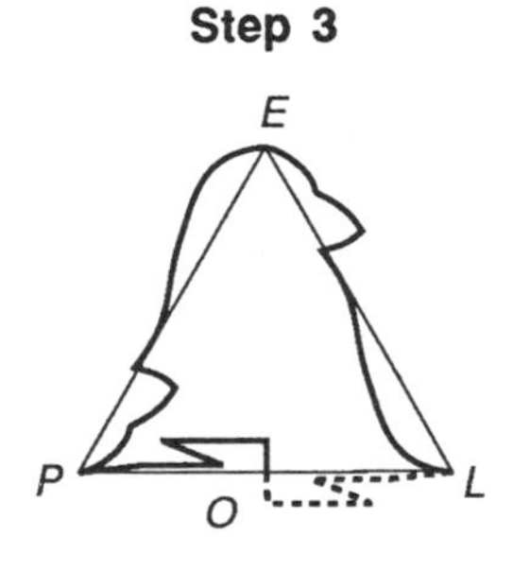

Step 4

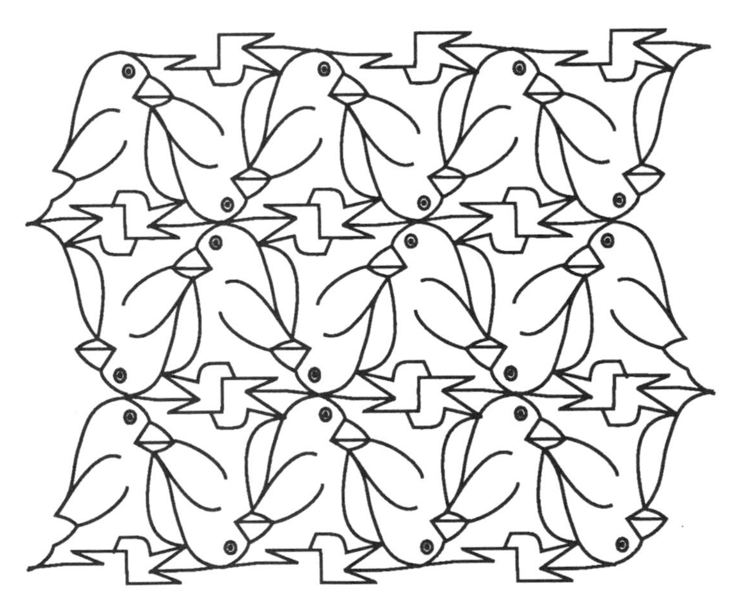

South Pole or Bust
Garret Lum, Geometry student

EXERCISE SET

In this exercise set, you are to create tessellation designs using reflections. You will need tracing paper or clear plastic, a grid of rhombuses, or a grid of isosceles triangles. Once you have completed a tessellating figure, you will have to duplicate it many times. You might use tracing paper, retracing your original copy many times into a tessellation design. If your design has a lot of art work within the shape, it would probably be easier to use a copy machine to make multiple copies. Cut out the copies and paste them together into a tessellation. Another student-made reflection tessellation design is shown below.

1. Create a tessellating design of recognizable shapes using a grid of rhombuses and using reflections and rotations. Decorate and color your art.

2. Create a tessellating design of recognizable shapes using a grid of isosceles triangles and using glide reflections. Decorate and color your art.

Pyramid of Prayers
Anita Lim
Geometry student

Improving Reasoning Skills

Millionaire Jetsetters

The three millionaire jetsetters, Lance, Sean, and Farrah, own a total of 16 sportscars. Included among these are 7 Corvettes, 4 Alfa Romeos, half as many Porsches as Alfas, and the remainder Jaguar XKE's. Lance prefers the styling of the Corvette and Alfa Romeo, owning two and one respectively. Sean loves variety so he owns one Corvette, one Porsche, one Jag, and two Alfas. How many of each kind of car does Farrah own?

Lesson 7.10
Chapter Review

EXERCISE SET A

Complete each conjecture from Chapters 5, 6, and 7.

1. If the three sides of one triangle are congruent to the three sides of another triangle, then —?—. (C-33 or **SSS Congruence Conjecture**)
2. If two sides and the angle between them in one triangle are congruent to two sides and the angle between them in another triangle, then —?—. (C-34 or **SAS Congruence Conjecture**)
3. If two angles and the side between them in one triangle are congruent to —?—. (C-35 or **ASA Congruence Conjecture**)
4. If two angles and a side that is not between them in one triangle are congruent to —?—. (C-36 or **SAA Congruence Conjecture**)
5. In an isosceles triangle, the bisector of the vertex angle is also the —?— to the base and the —?— to the base. (C-37 or **Vertex Angle Conjecture**)
6. If two chords in a circle are congruent, then they determine two central angles that are —?—. (C-38)
7. If two chords in a circle are congruent, then their —?— are congruent. (C-39)
8. Two congruent chords in a circle are —?— from the center of the circle. (C-40)
9. The perpendicular bisector of a chord passes —?— of the circle. (C-41)
10. A tangent to a circle is —?— to the radius drawn to the point of tangency. (C-42 or **Tangent Conjecture**)
11. Tangent segments to a circle from a point outside the circle are —?—. (C-43 or **Tangent Segments Conjecture**)
12. The measure of an inscribed angle equals —?— the measure of the intercepted arc. (C-44 or **Inscribed Angle Conjecture**)
13. Angles inscribed in the same arc are —?—. (C-45)
14. Every angle inscribed in a semicircle is a —?—. (C-46)
15. The opposite angles of quadrilateral inscribed in a circle are —?—. (C-47)
16. Parallel lines intercept —?— arcs on a circle. (C-48)
17. If C is the circumference of a circle and D is the diameter of the circle, then there is a number π such that $C =$ —?—. Since $D = 2r$ where r is the radius, then $C =$ —?—. (C-49 or **Circumference Conjecture**)

18. The arc length equals the degree measure of the arc divided by 360, times —?—. (C-50 or **Arc Length Conjecture**)

19. A regular polygon of n sides has —?— reflectional symmetries and —?— rotational symmetries. (C-51)

20. The —?— and only —?— regular polygons that create pure tessellations are —?—. (C-52)

21. —?— triangle will create a pure tessellation. (C-53)

22. —?— quadrilateral will create a pure tessellation. (C-54)

EXERCISE SET B

1. What is the isometry in which the image has the same orientation (faces the same way) as the original?

2. What are the two isometries in which the image has the opposite orientation as the original?

3. A translation of (5, 12) followed by a translation of (-8, -6) is equivalent to what single translation?

4. A rotation of 140° followed by a rotation of 260° about the same point is equivalent to what single rotation about the point?

5. Two consecutive reflections over a pair of parallel reflection lines 12 cm apart is equivalent to a translation. What is the length of that translation vector?

6. Identify the type or types of symmetry in each of the designs below.

a. 

b. 

c. 

d. 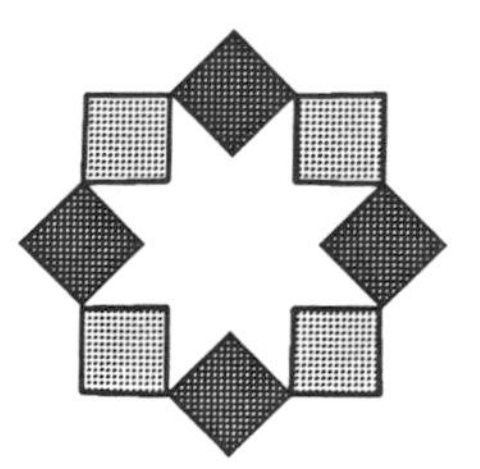

e. 

f. 

7. Identify each of the tessellations below using its numerical name.

a.

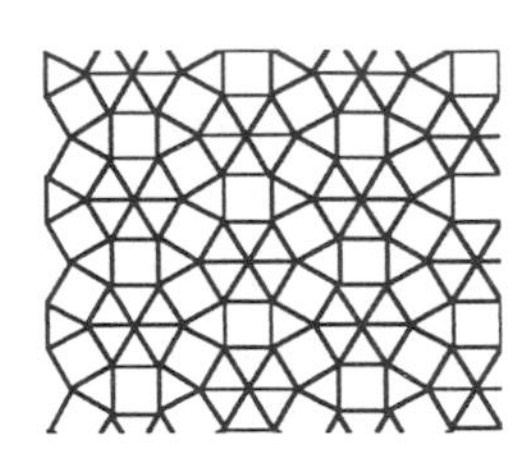

b.

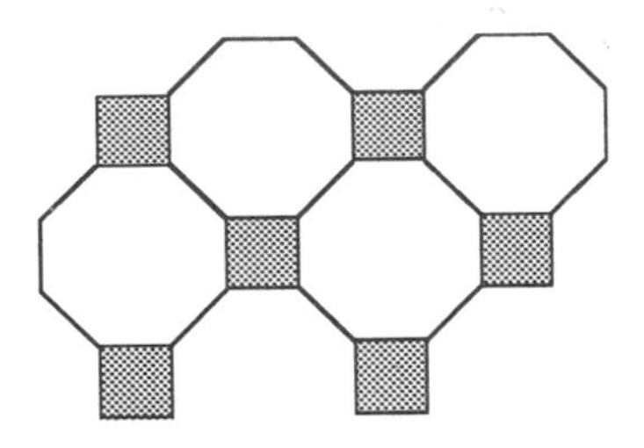

8. Identify the method used in each of the tessellations below.

a.

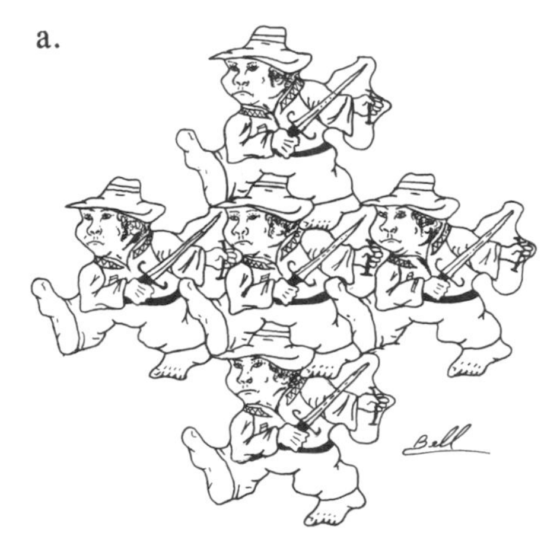

b.

c.

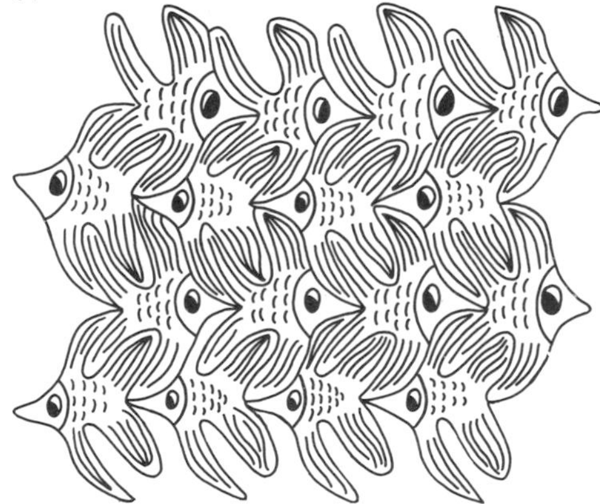

a. Robert Bell, Geometry student

b. Geometry student

c. Serene Tam, Geometry student

9. Your friendly librarian, or the local news stand may help with this one! Most professional football teams wear their team logo on their helmet. The logo on the Houston Oilers helmet is an oil derrick. It, like most logos, has symmetry. What other professional football team has a logo that possesses symmetry? What type of symmetry? Sketch it.

10. The logo for the Portland Trailblazers basketball team has rotational symmetry. What other professional sports team has a logo that possesses rotational symmetry? Sketch it.

Cooperative Problem Solving

Games in Space

It is the year 2066 aboard the recreational space station *Empyrean.* The Board of Directors has just sent a directive to the station's Education and Research Council asking that it begin development of new games. The games are for the entertainment and possible education of the visiting students from Earth.

You are part of a group on the space station's Education and Research Council. Your group's task is to design and build a game. After designing and testing the game, you should collect the materials for the game, assemble them, print rules, replay and retest the game, and, finally, package it. Develop the game with the following criteria in mind:

- The game is intended for students in the eighth grade and up.
- The game must be educational and entertaining. (It might give drill and practice in learning basic geometric facts, or it might develop reasoning skills, problem solving skills, or communication skills.)
- The game should have a clear set of rules.
- It is recommended but not required that the game have a set of playing pieces and a playing board larger than the size of a standard chess board.
- The game's average playing time should be less than one hour.

8 Area

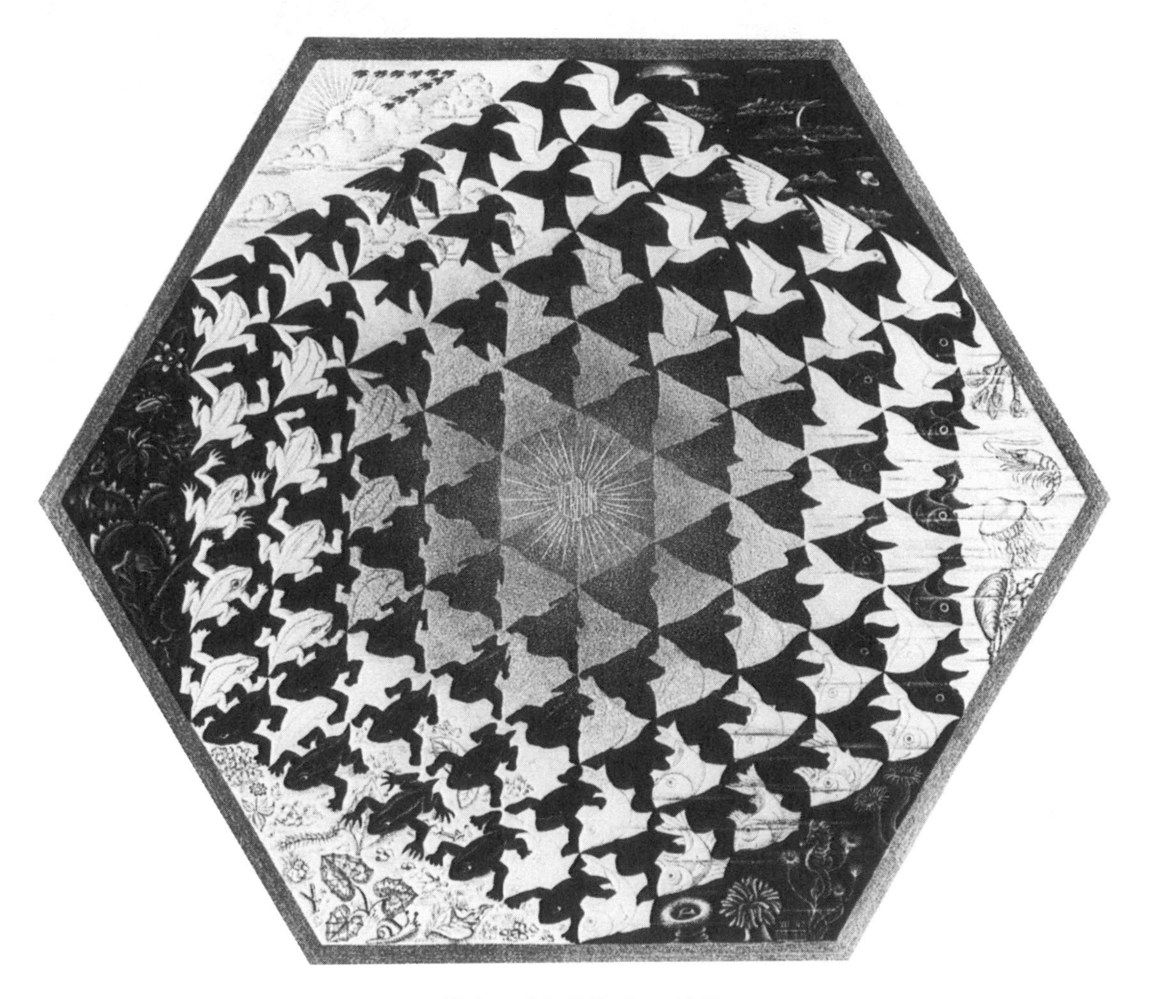

Verbum, M. C. Escher, 1942

In this chapter you will discover formulas (by using your geometry tools) that can be used to find the areas of rectangles, parallelograms, triangles, regular polygons, circles, and combinations of them. Some of the formulas you may already know. Your investigations will be physical proofs of these formulas. You will use these area formulas to solve problems in this chapter and in the chapters to follow.

Lesson 8.1
Areas of Rectangles and Parallelograms

*The **area** of a plane figure is the measure of the region enclosed by the figure.*

People in many occupations work with areas. Carpenters calculate the areas of walls, floors, and roofs to order materials for construction. Painters calculate the area of surfaces to be painted to know how much paint will be needed for a job. Decorators need to know the areas of carpeting and drapery materials to install in homes. In this chapter you will discover formulas for finding the areas of the regions within triangles, parallelograms, trapezoids, regular polygons, and circles.

The area of a figure is measured by the number of squares of a unit length (square units) that can be arranged to completely fill the figure. (For some figures, like the figure below, the squares may have to be cut up and rearranged.) Triangles, hexagons, or any other shape that tessellates could be used to measure area, but the square is used because of its simplicity.

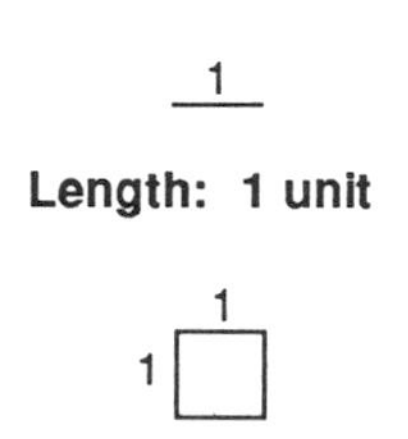

Area: 1 square unit

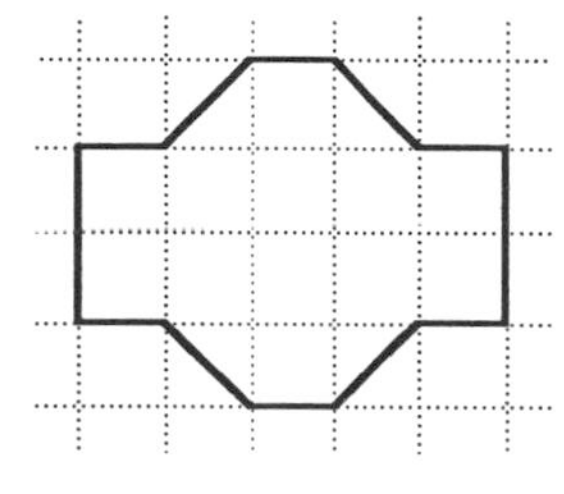

Area: 14 square units

The area of the plane figure above is 14 square units. There are 12 unit squares that fit completely inside the figure. The remainder of the figure can be filled by cutting up and rearranging 2 additional squares.

EXERCISE SET A

Estimate the area of each figure. The area of each square is one square unit.

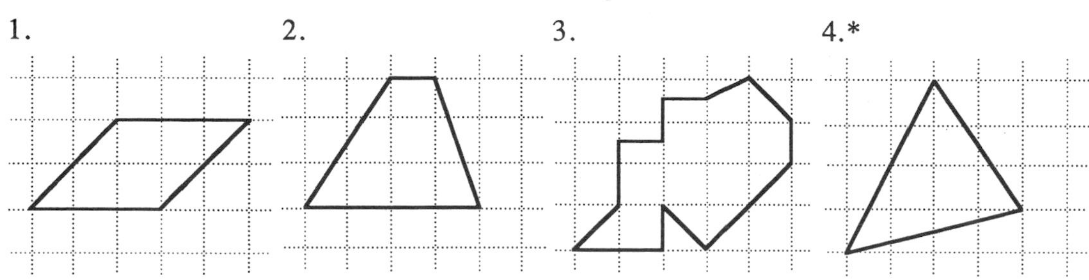

Investigation 8.1.1

It's easy to find the area of rectangles. Find the area of each rectangle in square units.

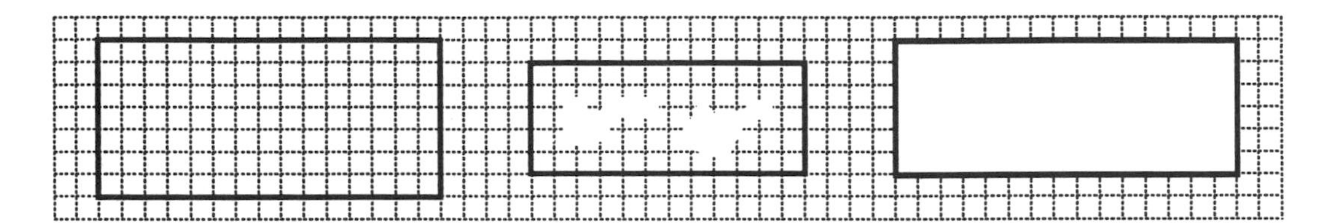

To find the area of the first rectangle you can simply count squares. You can do the same for the second, but it's a little harder because portions of some squares are blanked out. To find the area of the third rectangle you could copy the rectangle, draw in the lines, and count the squares; but, there's an easier method.

Any side of a rectangle (or parallelogram) may be called a **base**. A segment perpendicular to the base with one endpoint on it and the other endpoint on the opposite side is the **altitude** to that base. The length of the altitude is the **height**.

Since the length of the base indicates the number of squares in each row and the height indicates the number of rows, you can use these terms to state a formula for the area. Do you see how? State your next conjecture.

C-55 *The area of a rectangle is given by the formula —?— where A is the area, b is the length of the base, and h is the height of the rectangle.* ***(Rectangle Area Conjecture)***

How do you use the conjecture? Here are some tips. Write down the formula. Substitute the given or known values into the formula. Solve for the remaining variable. If the dimensions are measured in inches, feet, or yards, the area is measured in square inches (sq. in.), square feet (sq. ft.), or square yards (sq. yd.). If the dimensions are in centimeters or meters, the area is in square centimeters (cm^2) or square meters (m^2). Let's look at a few sample problems.

Example A

What is the area of a rectangle with a base of 88 feet and a height of 17 feet?

$$\begin{aligned} A &= bh \\ &= (88)(17) \\ &= 1496 \end{aligned}$$

88'

17'

The area is 1496 square feet.

Example B

What is the height of a rectangle that has an area of 7.13 m^2 and a base 2.3 m long?

$$\begin{aligned} A &= bh \\ 7.13 &= (2.3)h \\ 7.13/2.3 &= h \\ h &= 3.1 \end{aligned}$$

2.3 m

$A = 7.13\ m^2$

The height is 3.1 meters.

EXERCISE SET B

Use the Rectangle Area Conjecture to solve each problem. Each quadrilateral is a rectangle. Use the appropriate unit in each answer.

1. $A = -?-$

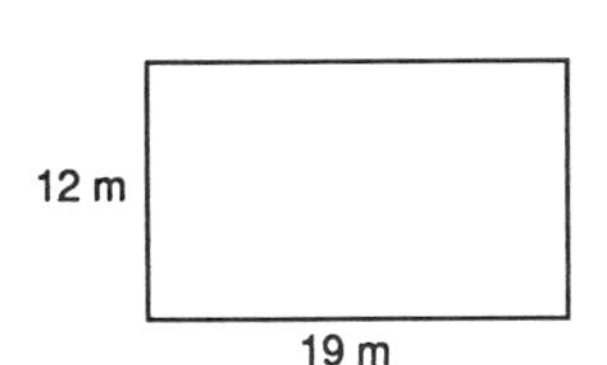

2. $A = -?-$

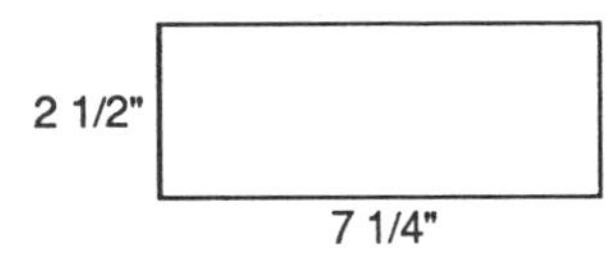

3. $A = -?-$

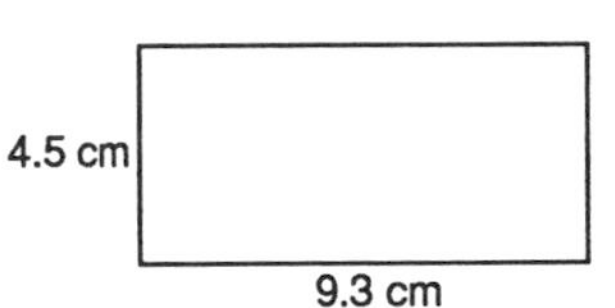

4. $A = -?-$

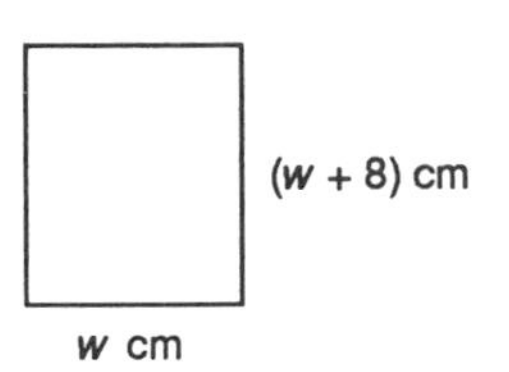

5.* $A = 96$ sq. yd.
$b = -?-$

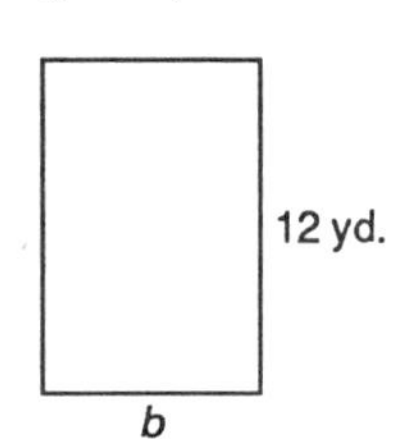

6. $A = 273\ \text{cm}^2$
$h = -?-$

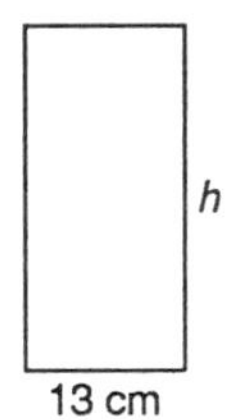

7. $A = 375$ sq. ft.
$h = -?-$

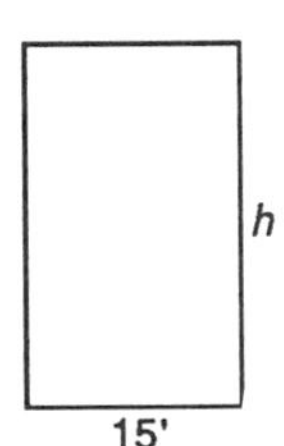

8. The perimeter is 40 ft.
$A = -?-$

7'

9. $A = 264$ sq. ft.
What is the perimeter?

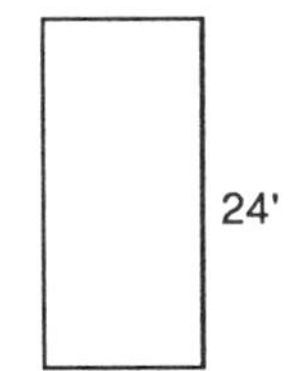

You probably already know many area formulas. The investigations leading to area formulas in this lesson should be thought of as physical demonstrations of these formulas. Physical demonstrations may help you remember the formulas longer.

Investigation 8.1.2

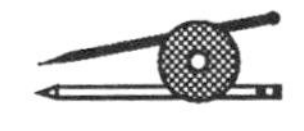

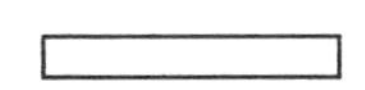

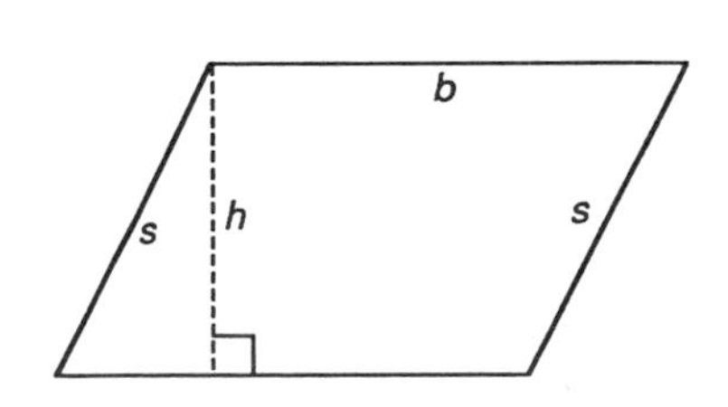

On heavy paper or cardboard, construct a parallelogram. Label the parallelogram as shown in the diagram. Construct an altitude from the vertex of the upper obtuse angle to the lower base. Label it as shown. Cut out the parallelogram and then cut along the altitude. You will have two pieces, a triangle and a trapezoid. Try arranging the two pieces into other shapes without overlapping them. Is the area of each of these new shapes the same as the original parallelogram? You should find a rectangle as one of your new shapes. What is the area of this rectangle? What is the area of the original parallelogram? State your next conjecture.

The area of a parallelogram is given by the formula —?— where A is the area, b is the length of the base, and h is the height of the parallelogram. ***(Parallelogram Area Conjecture)***

EXERCISE SET C

Use the Parallelogram Area Conjecture to solve each problem. Each quadrilateral is a parallelogram. Label each answer with the appropriate unit.

1. $A = $ –?–

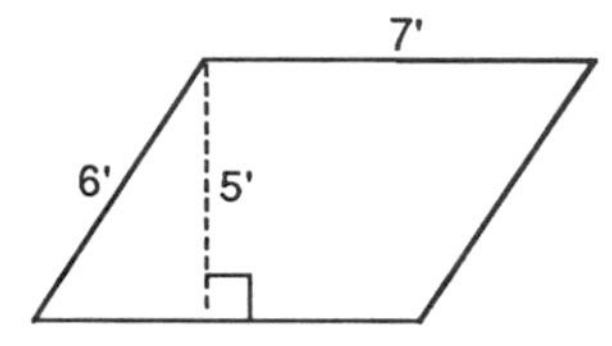

2. $A = $ –?–

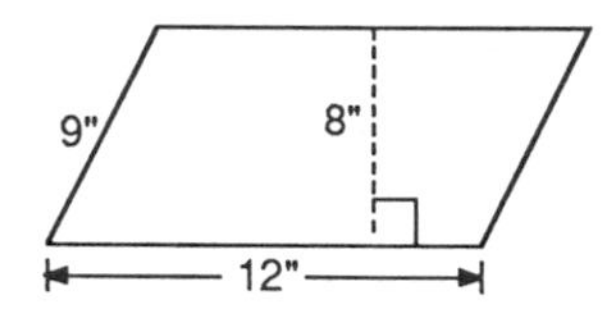

3. $A = $ –?–

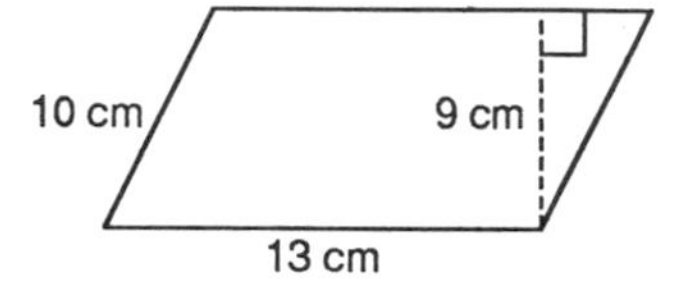

4. $A = $ –?–

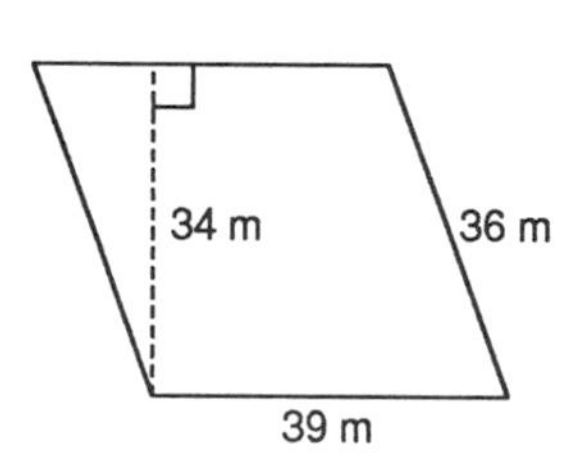

5. $A = $ –?–

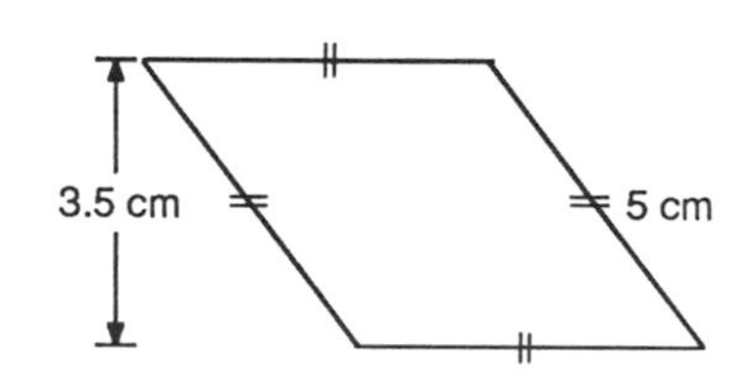

6. $A = 176$ sq. yd.
   $h = $ –?–

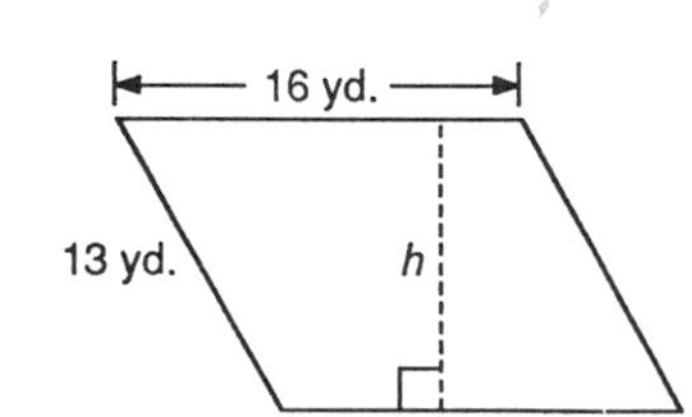

7. $A = 48x^2$ sq. in.
   $b = $ –?–

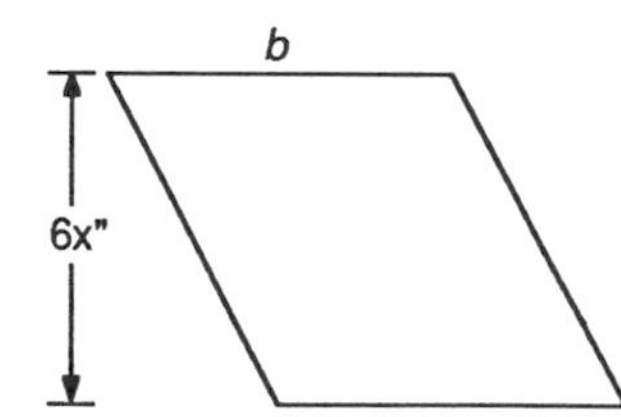

8.* $A = 2508\ \text{cm}^2$
   The perimeter is –?–.

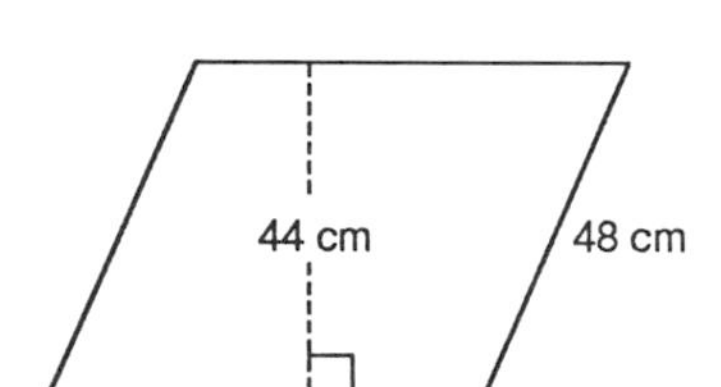

9. The shaded area is –?–.

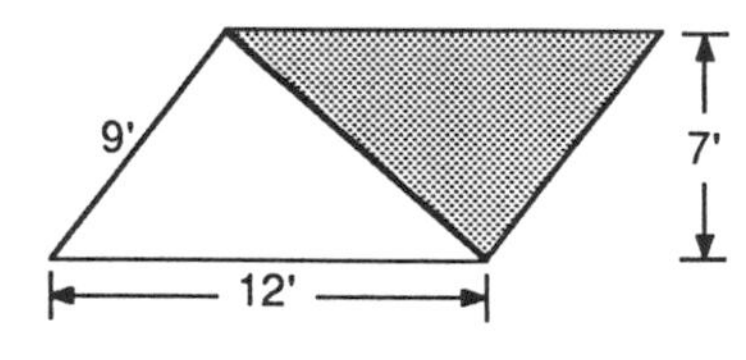

Lesson 8.2

Areas of Triangles and Trapezoids

In this lesson you will physically demonstrate the formula for the area of triangles and the formula for the area of trapezoids.

Investigation 8.2.1

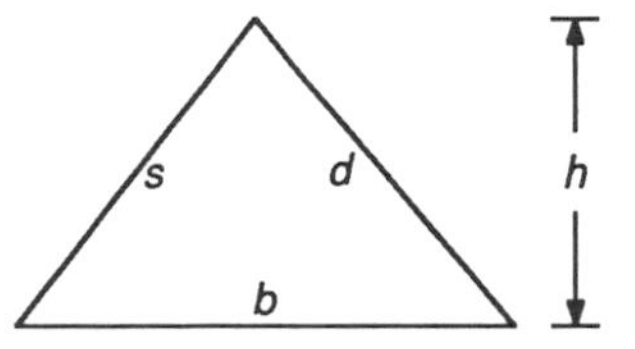

On heavy paper or cardboard, construct a triangle about 6 cm in height by 10 cm at the base. Label the triangle as shown in the diagram. Cut out the triangle and make a copy of it. Cut out the copy. Arrange the two triangles to form a figure for which you already have an area formula. Is the area of this figure equal to the sum of the areas of the two triangles? If two polygons are congruent, their areas are equal. What is the area of one of the triangles? Make a conjecture.

The area of a triangle is given by the formula —?— where A is the area, b is the length of the base, and h is the height of the triangle. (***Triangle Area Conjecture***)

EXERCISE SET A

Use your new area conjecture to solve each problem.

1. $A = $ –?–

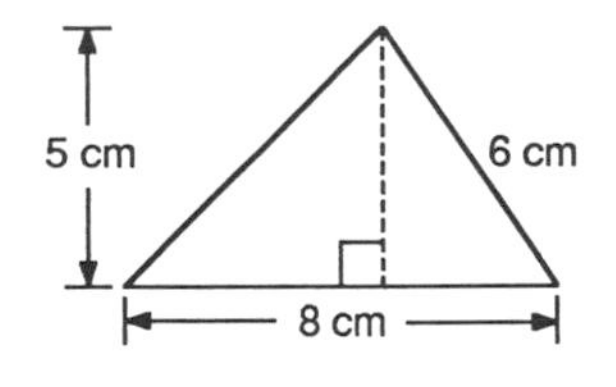

2. $A = $ –?–

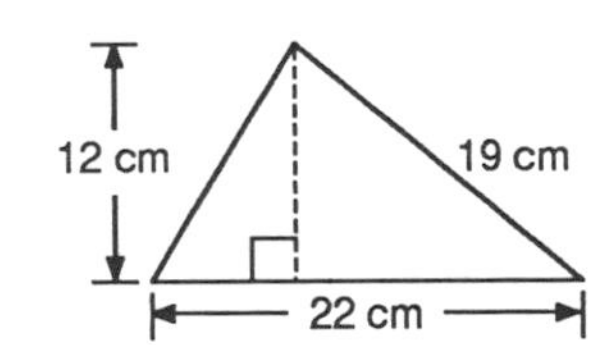

3.* $A = $ –?–

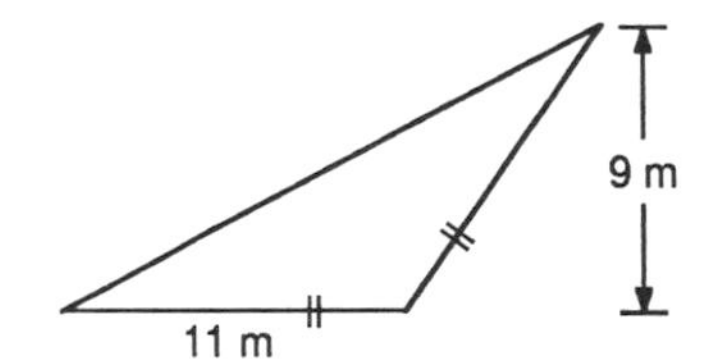

4. $A = 39\text{ cm}^2$
$h = $ –?–

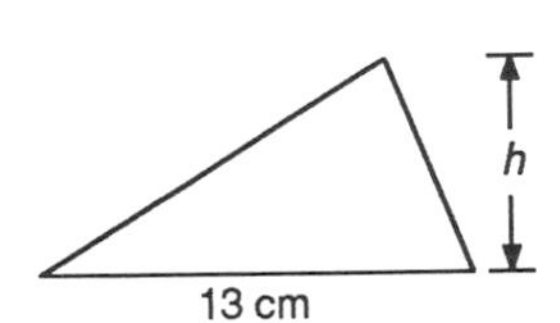

5.* $A = 31.5$ sq. ft.
$h = $ –?–

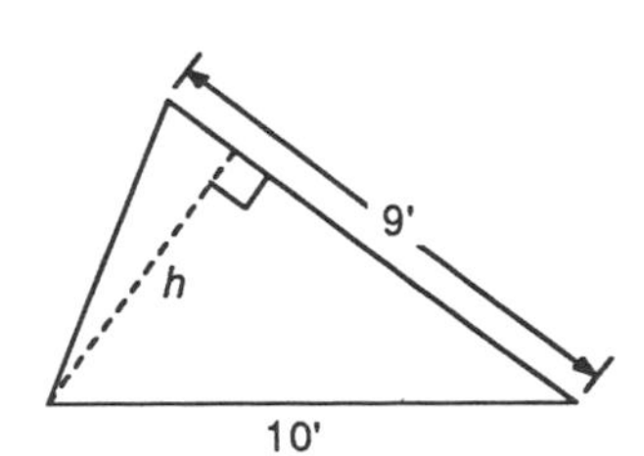

6. $A = $ –?–

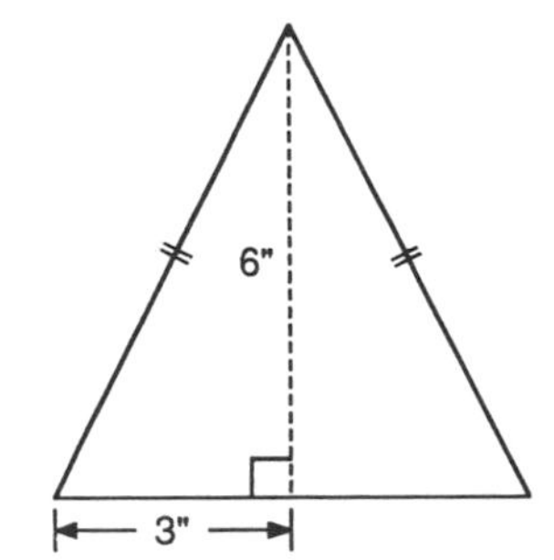

7. $h = $ –?–

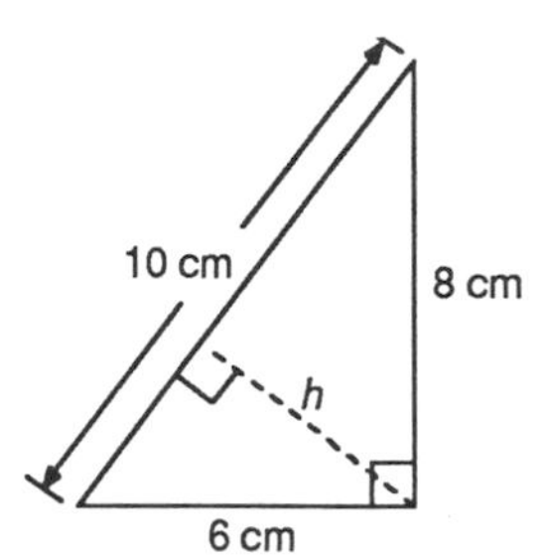

8. If the area is 924 cm^2, what is the perimeter?

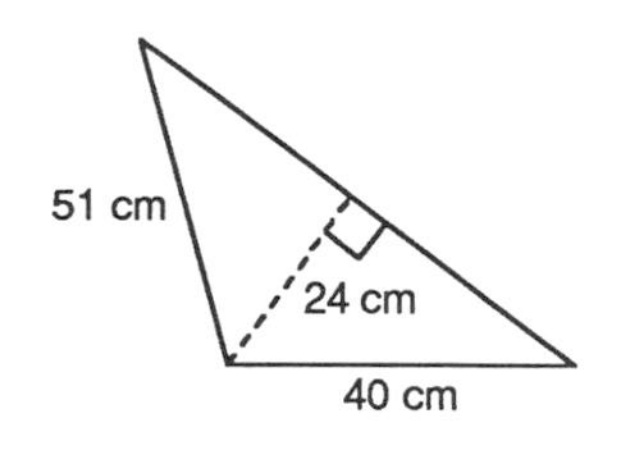

9.* $x = $ –?–
$y = $ –?–

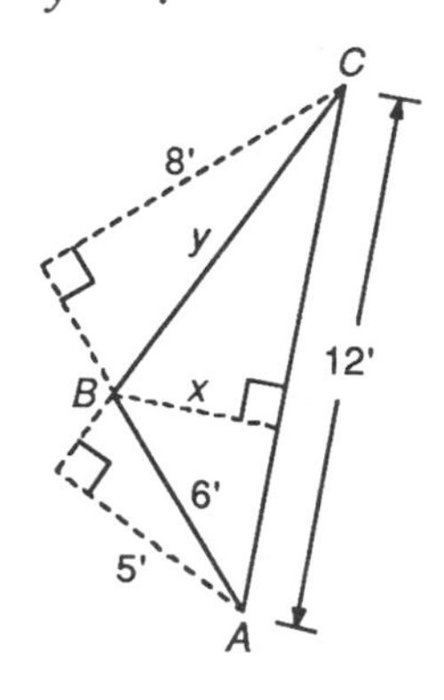

Investigation 8.2.2  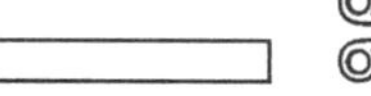 

As you remember from Chapter 3 (yeah, sure!), the parallel sides of the trapezoid are called the bases and the other two sides are called the non-parallel sides. The **altitude** of a trapezoid is a segment from one base to the other that is perpendicular to both. The **height** of a trapezoid is the length of an altitude.

Construct a trapezoid about 6 cm in height by 10 cm at the base on heavy paper or cardboard. Construct an altitude from one of the endpoints of the smaller base to the larger base. Label the trapezoid as shown in the diagram. Cut out the trapezoid and trace it onto another piece of paper. Cut out the copy. Arrange the two trapezoids to form a figure for which you already have an area formula.

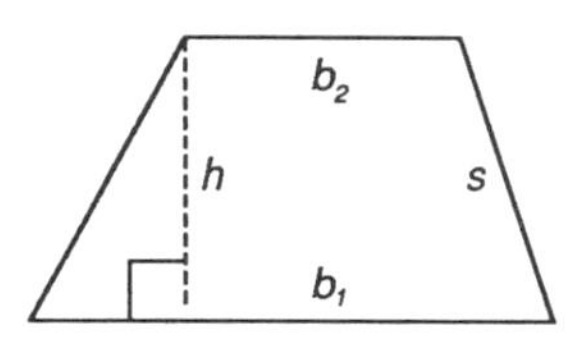

What type of polygon is created by the two trapezoids placed next to each other? What is the height of this figure? What is the length of the base? What is the area? What is the area of one of the trapezoids? State a conjecture.

The area of a trapezoid is given by the formula —?— where A is the area, b_1 and b_2 are the lengths of the two bases, and h is the height of the trapezoid. ***(Trapezoid Area Conjecture)***

EXERCISE SET B

Use the Trapezoid Area Conjecture to solve. Each quadrilateral is a trapezoid.

1. $A = $ –?–

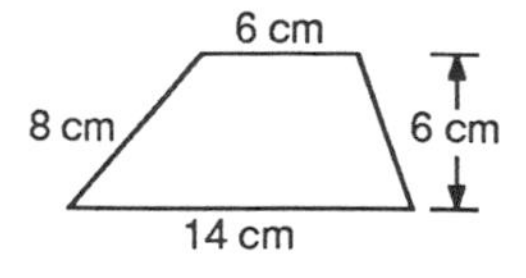

2. $A = 50\ cm^2$
$h = $ –?–

3.* $A = $ –?–

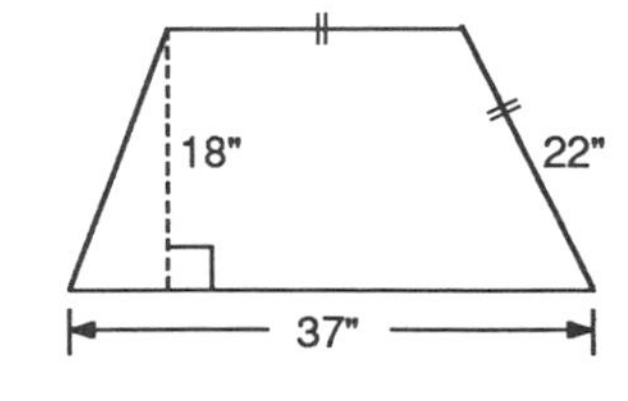

4.* $A = 180 \text{ m}^2$
$b = $ –?–

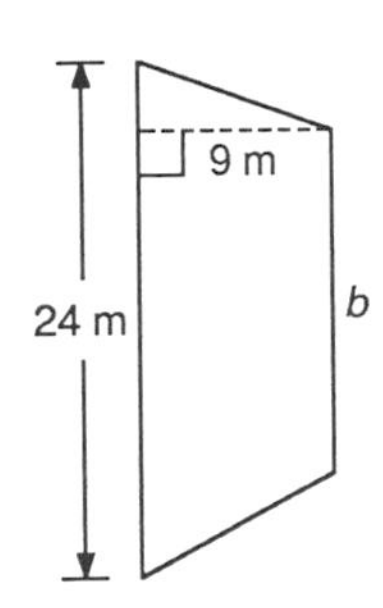

5. If the area is 204 cm^2 and the perimeter is 62 cm, then h is –?–.

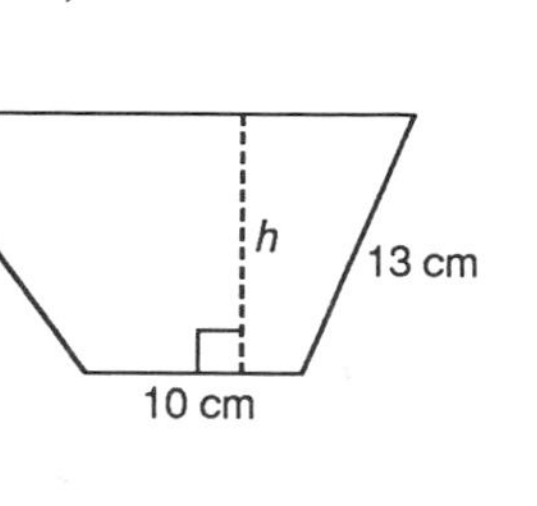

6. $A = 84$ sq. ft.
The perimeter is –?–.

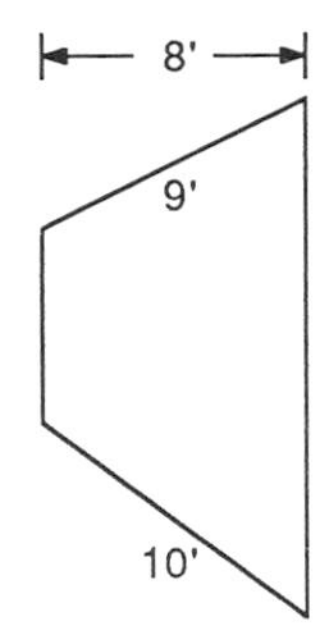

EXERCISE SET C

1.* Show how you can arrive at the formula for the area of a trapezoid by dividing the trapezoid into two triangles.

2.* Show how you can arrive at the formula for the area of a trapezoid by dividing the trapezoid into a parallelogram and a triangle.

3.* Find another method for arriving at the formula for the area of a trapezoid. Consider ways of dividing the trapezoid into pieces and rearranging them.

Improving Visual Thinking Skills

Knot or Not?

Which of the following ropes will tie into a knot when pulled from the ends?

a.

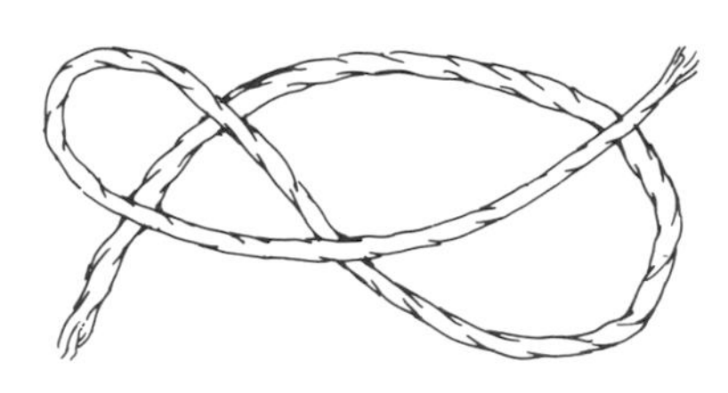

b.

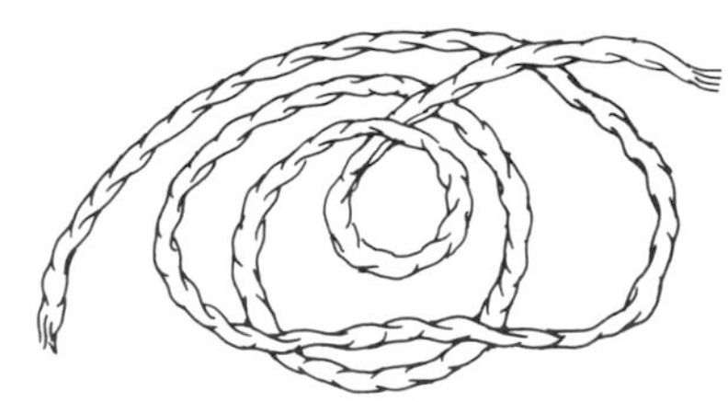

c.

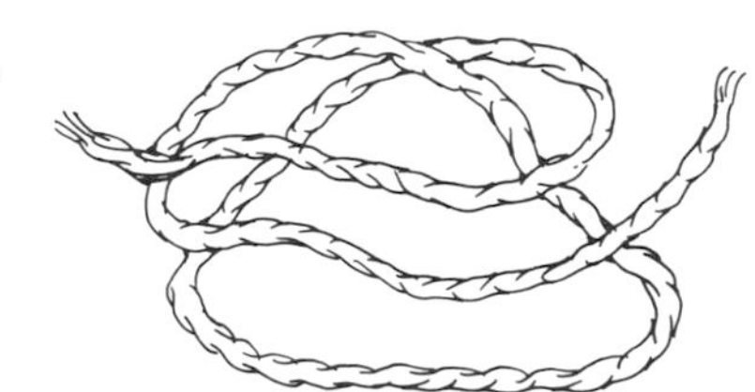

d.

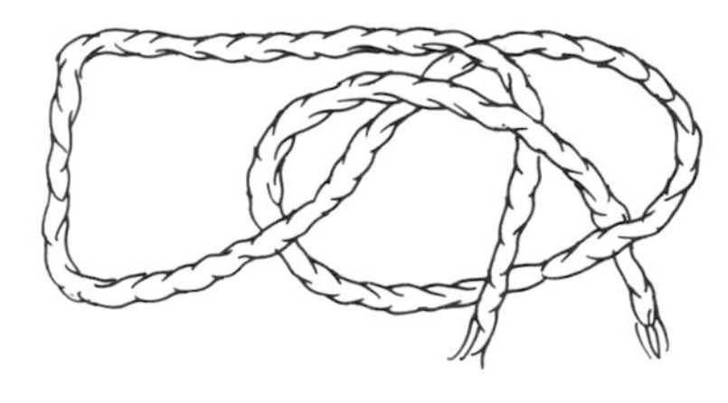

Lesson 8.3

Measuring Area

In the first two lessons of this chapter you discovered formulas for finding areas of rectangles, parallelograms, triangles, and trapezoids. Now let's see if you can use them! In this lesson you will find the approximate areas of irregularly shaped figures using area formulas that you already know. This lesson works best in groups, sharing the tasks. Before you begin to measure, discuss in your group the best strategy for each exercise. Group members should get different (but close) results in each exercise. You might average your results to arrive at one group answer.

EXERCISE SET

1.* Use your straightedge and neatly copy the figure below onto a piece of paper. The easiest and most accurate way to copy a figure made with straight lines is to place your paper over the figure and put a dot on your paper at each vertex of the figure. Then take the paper off the figure and use your straightedge to connect the dots. Use your centimeter ruler to approximate the area of the figure.

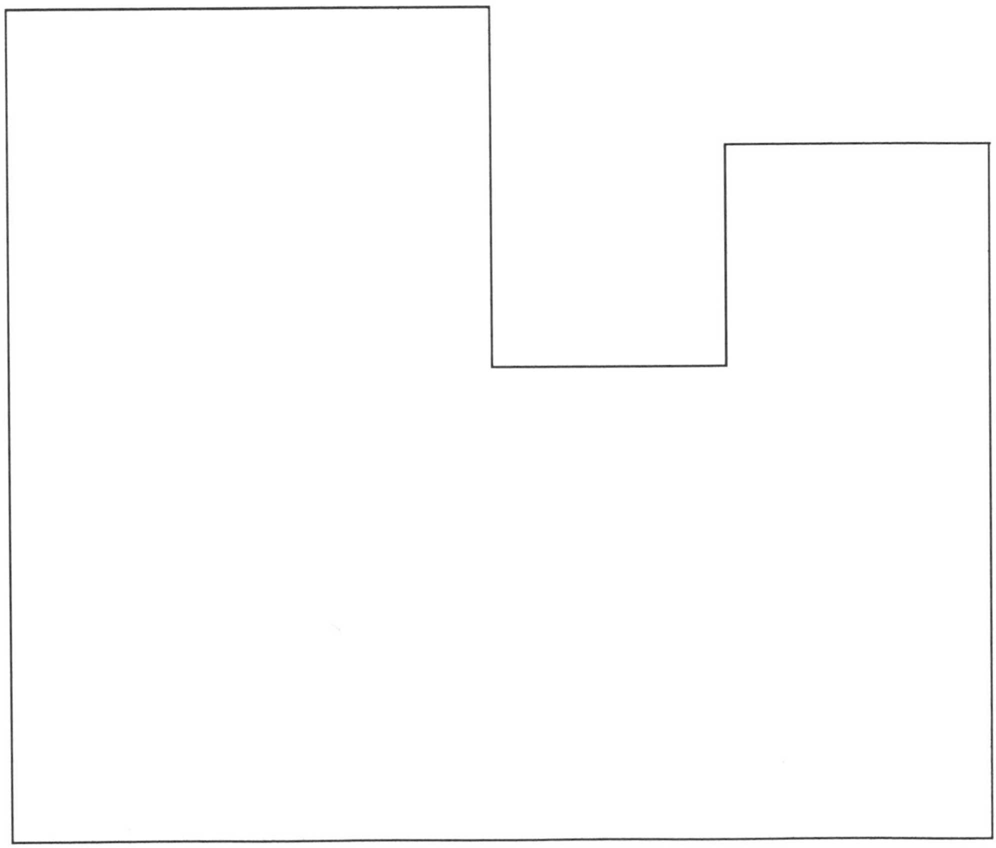

2.* Which parallelogram has the greater perimeter? What is its approximate perimeter in cm? Which parallelogram has the greater area? What is its approximate area in cm²?

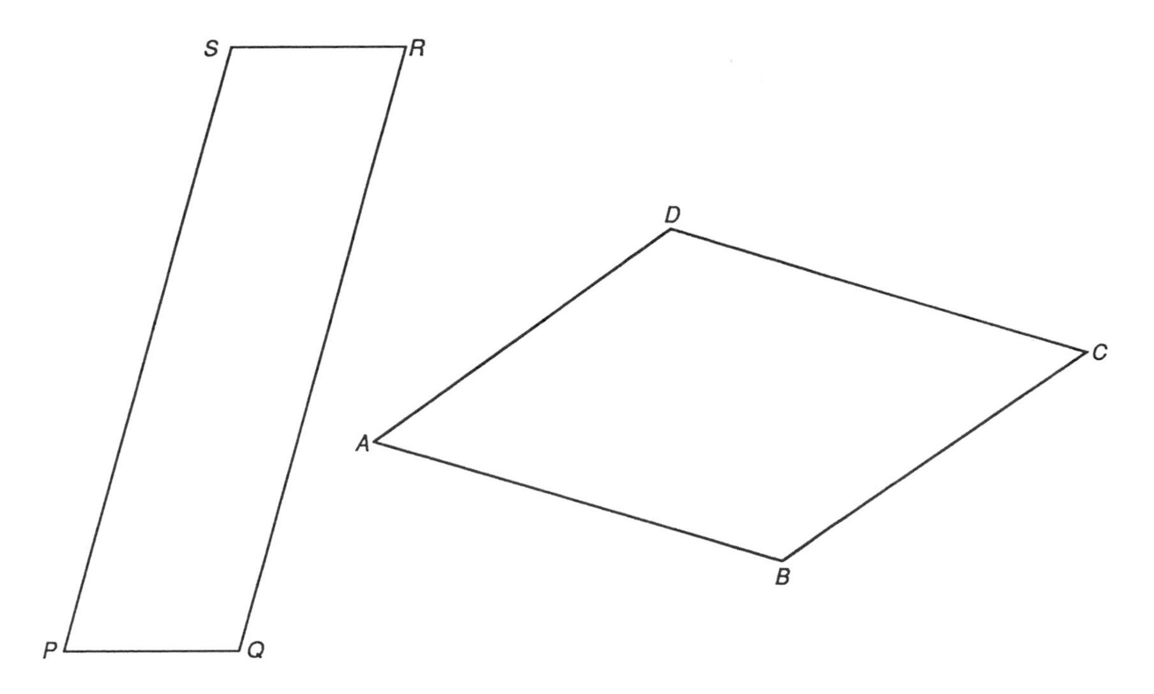

3.* Use your straightedge and neatly copy the figure below onto a piece of paper. Use your centimeter ruler to approximate its area.

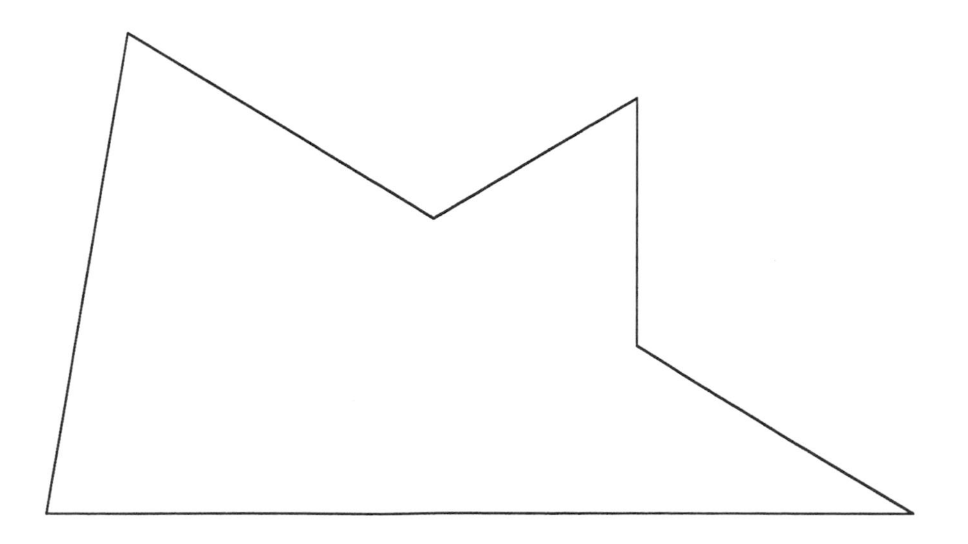

4. Which trapezoid has the greater perimeter? What is its approximate perimeter in cm? Which trapezoid has the greater area? What is its approximate area in cm^2?

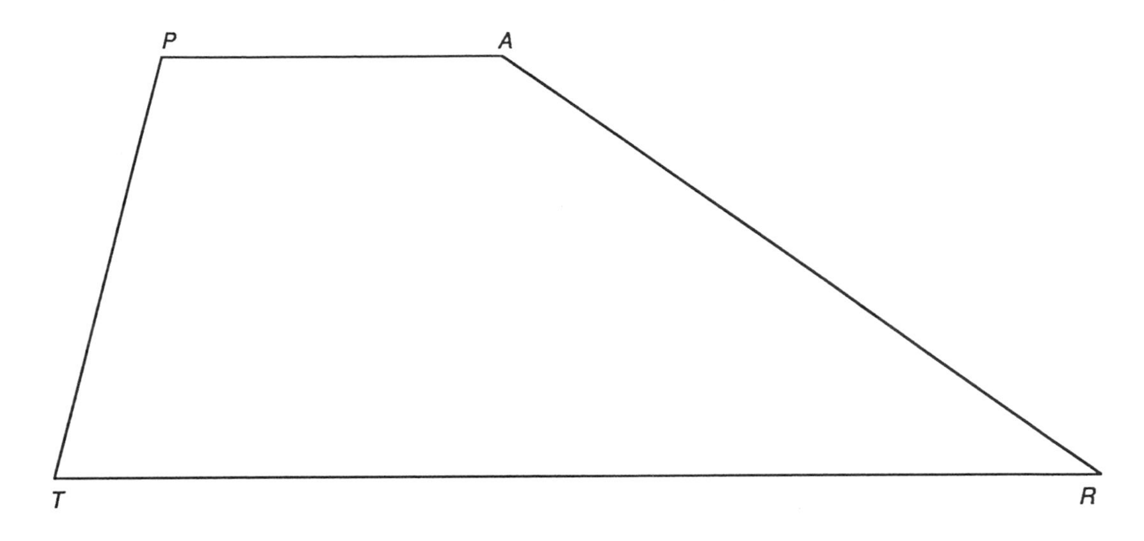

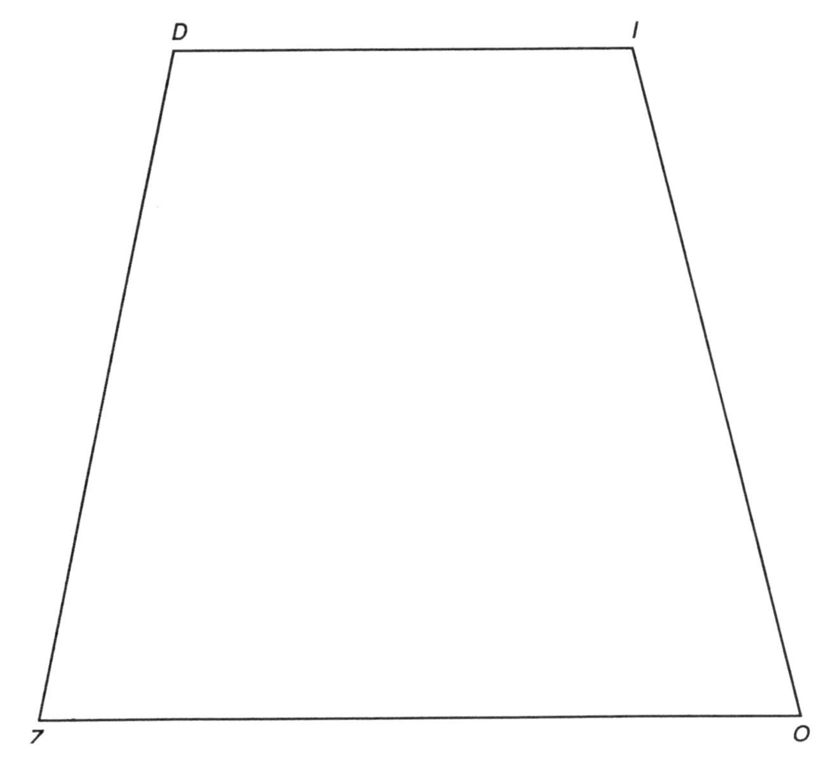

5.* Use your straightedge and neatly copy the regular dodecagon below onto a piece of paper. Use your centimeter ruler to approximate its area. Try doing it in the least number of measurements. How about two measurements?

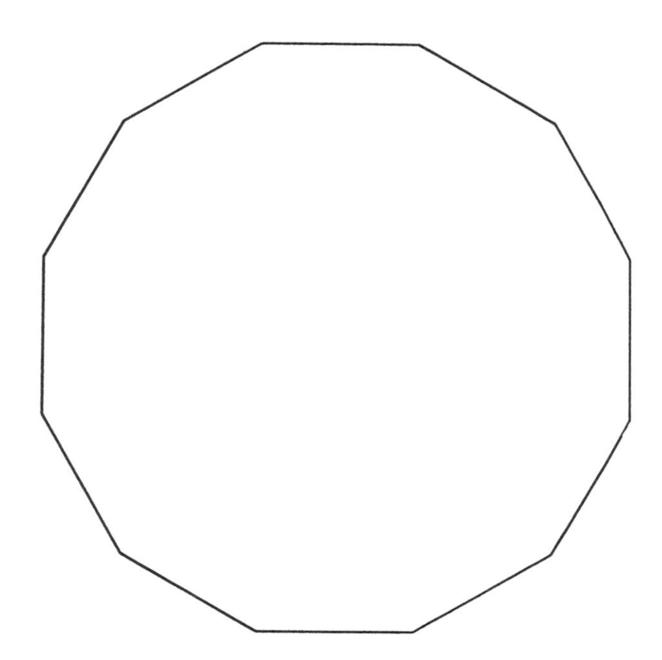

6.* Place a sheet of lined paper over the region below aligning one of the lines on the lined paper over the vertical segment of the figure. Make a copy of the region by tracing the figure onto the lined paper. Use your centimeter ruler to approximate its area.

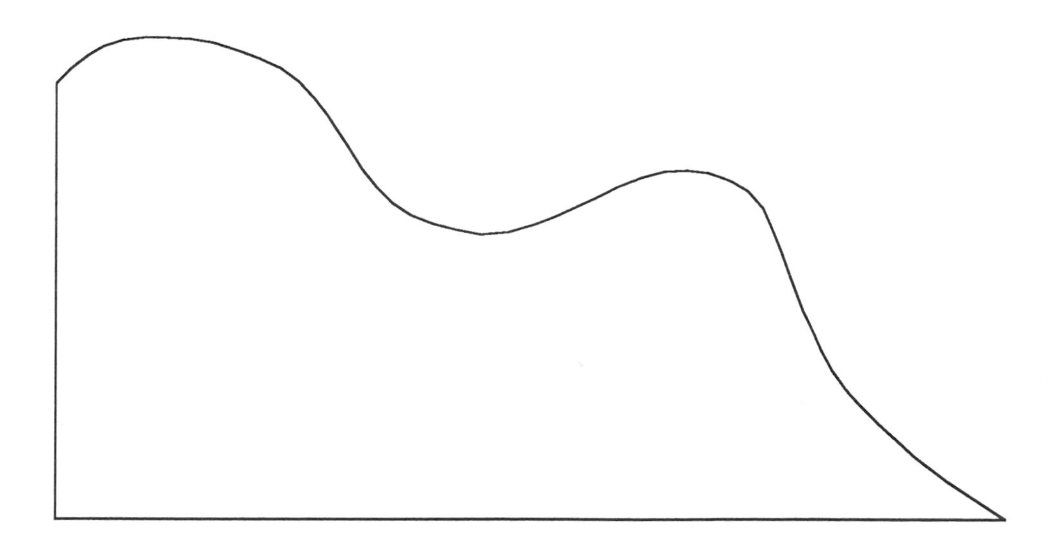

Lesson 8.4

Area Problems

The Great Architect of the universe now begins to appear as a pure mathematician.

– J.H. Jeans

You can't buy 12 and 11/16 gallons of paint at a paint store; you must buy 13 gallons. If your calculations tell you need 5.25 bundles of shingles to shingle a house, you have to buy 6 bundles. In this type of "rounding off," you must always "round up." The problems in this lesson involve buying gallons of paint, bundles of shingles, square yards of carpet, and square feet of tile. Round up.

EXERCISE SET

1. Bernice Kravitts is an apprentice painter. Her boss, Rene P. Casso, has the flu. It is up to Bernice to make an estimate on painting 148 rooms in a new motel with one coat of base paint and one coat of finishing paint. The four walls and ceiling of each room must be painted. Each room is 14' by 16' by 10' high.

   a. Calculate the total area of all the surfaces to be painted with one coat. (Ignore doors and windows.)

   b. One gallon of base paint covers 500 square feet. How many gallons of base paint are required for the job?

   c. One gallon of finish paint covers 250 square feet. How many gallons of finish paint will it take to complete the job?

2.* Barbara's Bigtime Bakery baked the world's largest chocolate cake. (It was also the world's worst cake as 343 people got sick after eating it.) The length was 600 cm, the width 400 cm, and the height was 180 cm. Barbara and her two assistants, Boris and Bernie, applied green peppermint frosting on the four sides and top. How many liters of frosting did they need for this dieter's nightmare? One liter of green peppermint frosting covers about 1200 square centimeters.

Barbara, Bernie, and Boris Bake a Big Cake

3.* Aunt Teak is ready to have wall-to-wall carpeting installed. The carpeting she has selected costs $14 per square yard, the padding is $3 per square yard, and the installation is $3 per square yard. What will it cost Aunt Teak to carpet the three bedrooms and hallway shown in the diagram on the right? (Ignore the thickness of the walls.)

4. Aunt Teak wishes to have 1 foot square Italian Buff ceramic tile in the entry way and kitchen. Italian tiles cost $5 each. Aunt Teak wishes to have 4 inch by 4 inch square ceramic tile on each bathroom floor. The bathroom tile she has selected costs $.45 per tile. What will it cost Aunt Teak to tile her home?

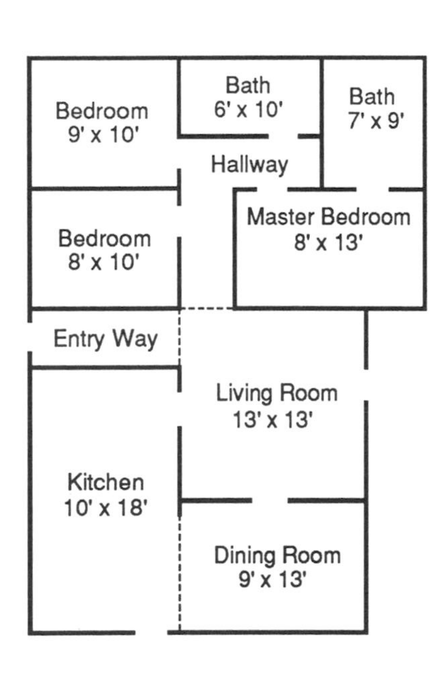

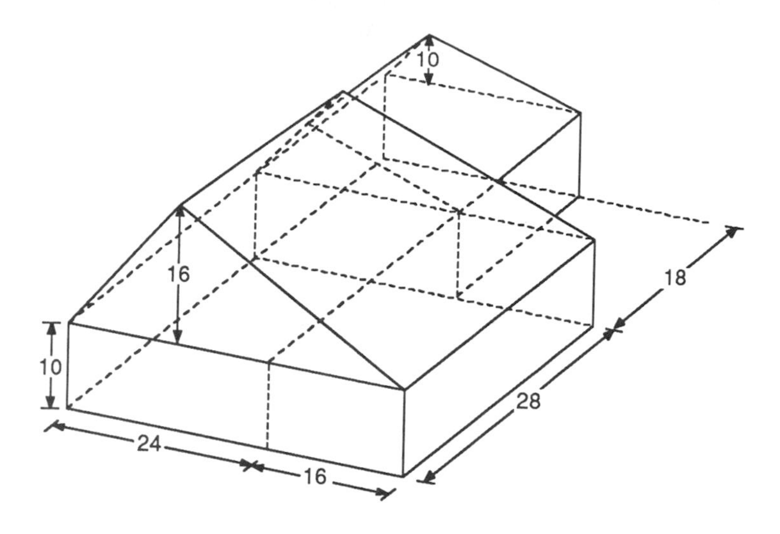

5.* Tom and Betty are planning to paint the exterior walls of their cabin. The paint they have selected costs $24 per gallon and covers 150 to 300 square feet per gallon depending on the condition of the surface to be painted. Tom and Betty decide to calculate the amount of paint needed based on 150 square feet per gallon because the cabin exterior appears to be very dry. How much will it cost to paint the exterior walls of their cabin (all vertical surfaces)? All measurements shown above are in feet.

6.* Claudette and Henri are about to paint the exterior walls of their country farm home (all four exterior vertical surfaces) and put new cedar shingles on the roof. Their home has a gambrel roof typical of homes of rural gentry back in their home country, France. The paint selected costs $25 per gallon and covers 250 square feet per gallon. The wood shingles cost $65 per bundle and each bundle covers 100 square feet. How much will this home improvement cost Claudette and Henri? All measurements shown below are in feet.

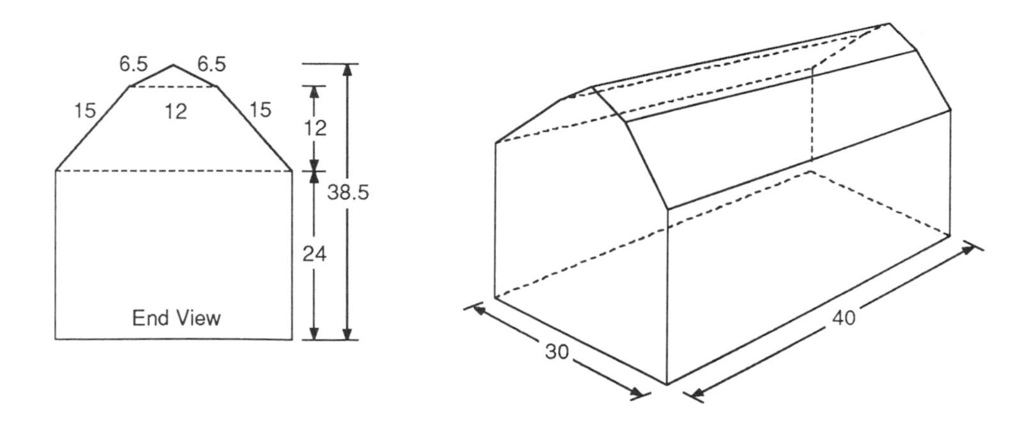

Lesson 8.5

Areas of Regular Polygons

In this lesson you will discover a formula for finding the area of any regular polygon. A regular pentagon, a regular hexagon, and a regular octagon are shown below inscribed in circles. You can divide a regular polygon of n sides into n congruent isosceles triangles by drawing in the radii from the center of the circumscribed circle to each vertex of the polygon. Then you can find the area of one isosceles triangle and multiply by the number of isosceles triangles in the polygon (n) to find the area of the regular polygon of n sides.

*An **apothem of a regular polygon** is a perpendicular segment from the center of the polygon's circumscribed circle to a side of the polygon.*

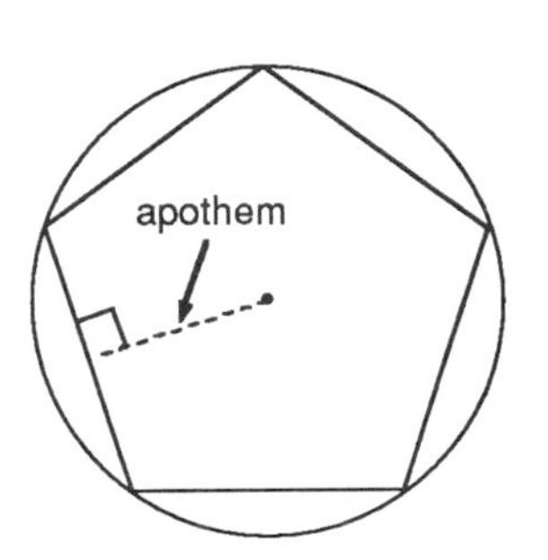

You may also refer to the length of the segment as the apothem.

In the following investigation, you will calculate the area of regular polygons. Then you will write a formula for the area of a regular polygon.

Investigation 8.5.1

Let the length of the apothem of each of the regular polygons be equal to a and let the length of each side of the regular polygons be equal to s.

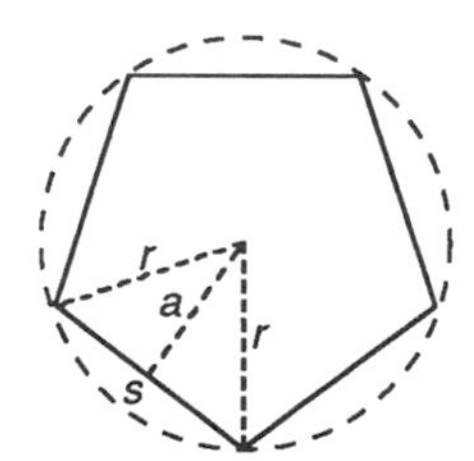

Area of a Regular Pentagon
$A = (1/2)as(5)$

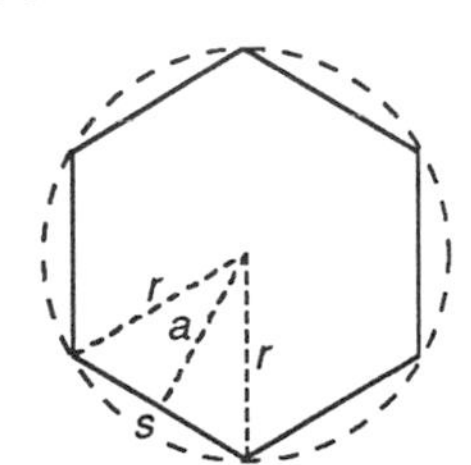

Area of a Regular Hexagon
$A = (1/2)as(6)$

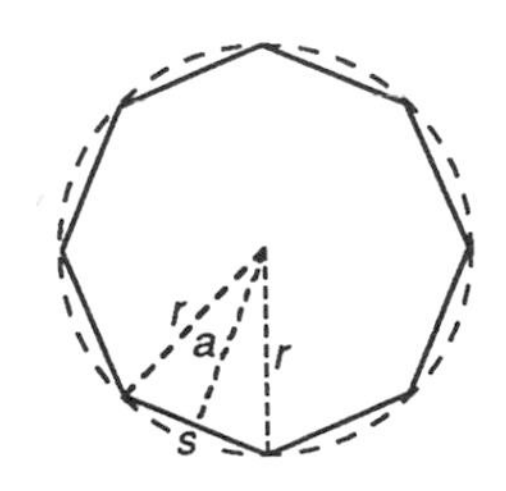

Area of a Regular Octagon
$A = (1/2)as(8)$

Copy the table below. Then find the area of each regular polygon in the table in terms of a and s.

Number of sides	3	4	7	9	10	11	12	...	n
Area of polygon	–?–	–?–	–?–	–?–	–?–	–?–	–?–	...	–?–

Your last entry in the table can be restated as your next conjecture.

C-59 *The area of a regular polygon is given by the formula —?— where A is the area, a is the apothem, s is the length of each side, and n is the number of sides of the regular polygon. Since the length of each side times the number of sides is the perimeter (sn = p). The formula can also be written as A = (1/2)a –?–.* ***(Regular Polygon Area Conjecture)***

EXERCISE SET

1. Use your compass and straightedge to construct a regular hexagon with sides of 8 cm. Use your new conjecture and a centimeter ruler to approximate its area.
2. Use your geometric tools to draw a regular octagon with sides of 8 cm. Use your new conjecture and a centimeter ruler to approximate its area.

In Exercises 3 to 6, the length of each side and apothem have been measured for you. Use your new conjecture to find the area of each regular polygon accurate to the nearest square centimeter. The apothem is a, s is the length of a side, and p is the perimeter.

3. Pentagon; $a \approx 3$ cm and $s \approx 4.4$ cm
4. Decagon; $a \approx 9.7$ cm and $s \approx 14.1$ cm
5. Octagon; $a \approx 12.1$ cm and $p \approx 80$ cm
6. Nonagon; $p \approx 63$ cm and $a \approx 9.6$ cm

In Exercises 7, 10, and 11, express each answer accurate to the nearest square centimeter. The apothem is a, p is the perimeter, and n is the number of sides of the regular polygon.

7. Find the area of a regular polygon with $a \approx 12$ cm and $p \approx 81.6$ cm.
8. Find the perimeter of a regular polygon to the nearest tenth of a meter if $a \approx 9$ m and $A \approx 259.2\ m^2$.
9. Find the length of each side of a regular polygon to the nearest foot if $a \approx 80'$, $n = 20$, and $A \approx 20{,}000$ sq. ft.

10. Find the shaded area of the regular octagon *ROADSIGN*. The apothem is about 20 cm. *GI* is about 16.6 cm.

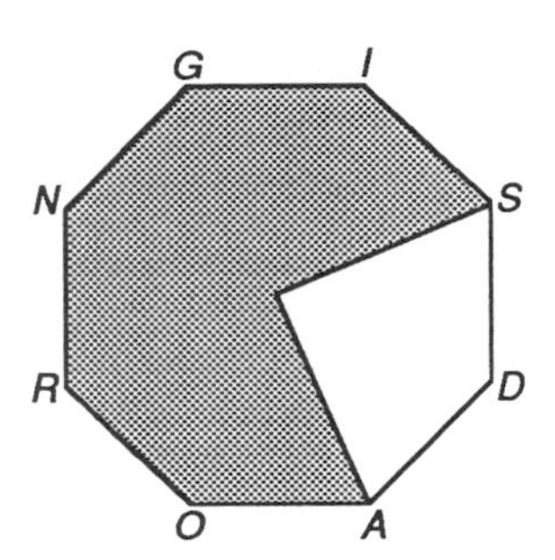

11.* Find the shaded area of this regular hexagonal donut. The smaller apothem and smaller sides are half as long as the longer apothem and longer sides.

$a \approx 6.9$ cm
$r \approx 8$ cm

12.* As the number of sides of a regular polygon inscribed in a circle gets larger and larger, the perimeter of the polygon gets closer and closer to the —?— of the circle. As the number of sides of the regular polygon increases, the apothem gets closer and closer to the —?— of the circle. As the number of sides increases, the area of the polygon gets closer and closer to the area of the —?—. Use this reasoning to make a conjecture about the formula for the area of a circle.

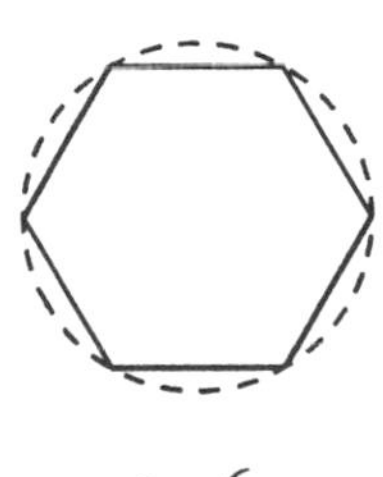

$n = 6$

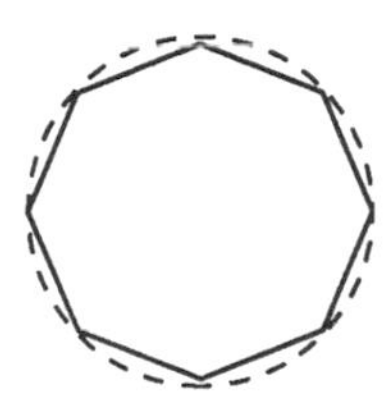

$n = 8$

$n = 12$

Improving Reasoning Skills

Scrambled Arithmetic

In the addition statement 65 + 28 = 43, all the digits are correct, but they are in the wrong places! The correct equation should be 23 + 45 = 68. In each of the three equations below the operations are correct and all the digits are correct, but some of the digits are in the wrong places. Find the correct equations.

1. 11 + 66 = 457
2. 39 x 11 = 75
3. $\frac{783}{52} = 31$

Lesson 8.6

Areas of Circles

In this lesson you will discover a formula for calculating the area of a circle.

Investigation 8.6.1 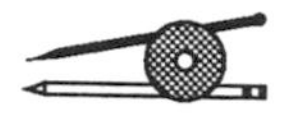 

Step 1: With your compass, make a circle with a radius of approximately three inches. Cut out the circular region.

Step 2: Fold the circular region in half. Fold it in half a second time. Fold it in half a third time. Fold it in half one last time.

Step 3: Unfold your circular region and cut it along the folds into 16 wedges.

Step 4: Arrange the 16 wedges in a row, alternating the tips up and down to form a shape that resembles a parallelogram.

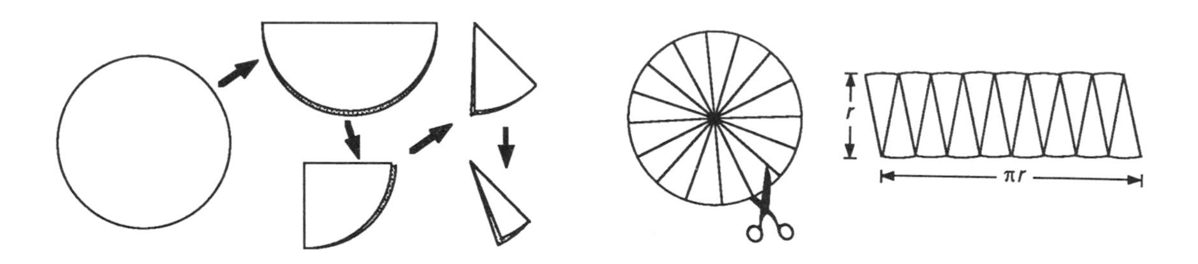

You have just taken apart a circular region and rearranged it into a shape resembling a parallelogram. If you cut it into more wedges, you can rearrange it to look like a rectangle. No area has been lost or gained in this change from circle to "rectangle" so the area of this new "rectangle" is the same as the area of the original circle. You know how to find the area of a rectangle. If the radius of the original circle is r, and the circumference of the original circle is $2\pi r$, then you can determine the base and height of the "rectangle" and thus its area. State your next conjecture.

The area of a circle is given by the formula —?— where A is the area and r is the radius of the circle. ***(Circle Area Conjecture)***

How do you use this new conjecture? Let's look at a few sample problems.

Example A

If a circle has a diameter of 6 inches, what is the area accurate to the nearest tenth of a square inch? Use 3.14 for π.

$$\begin{aligned} A &= \pi r^2 \\ &\approx (3.14)(3^2) \\ &\approx (3.14)(9) \\ &\approx 28.26 \end{aligned}$$

The area is about 28.3 square inches.

Example B

If a circle has an area of 144π m^2, what is the radius?

$$\begin{aligned} A &= \pi r^2 \\ 144\pi &= \pi r^2 \\ 144 &= r^2 \\ r &= 12 \end{aligned}$$

The radius is 12 meters.

EXERCISE SET A

Use your new conjecture to solve each problem. State each answer in terms of π unless told to use an approximation.

1. If $r = 3$", then $A =$ –?–.
2. If $r = 7$ cm, then $A =$ –?–.
3.* If $r = 1/2$ yard and you use 22/7 for π, then $A \approx$ –?–.
4. If $C = 12\pi$ inches, then $A =$ –?–.
5. If $C = 314$ meters and you use 3.14 for π, then $A \approx$ –?–.
6. If $A = 9\pi$ cm^2, then $r =$ –?–.
7.* If $A = 3\pi$ square inches, then $r =$ –?–.
8. If $A = .785$ m^2 when you use 3.14 for π, then $r \approx$ –?–.

EXERCISE SET B

1. A small college TV station can broadcast its programming a radius of 60 km. How many square kilometers of viewing audience does the station have? Use 3.14 for π.
2. Juanita's dog Cecil has been secured to a post by a chain 7 meters long. In how many square meters can Cecil play? Use 22/7 for π.
3.* The strength of a bone or a muscle is proportional to its cross sectional area. If a cross section of one muscle is a circular region with a radius of 3 cm and a second identical type of muscle has a cross section that is a circular region with a radius of 6 cm, how many times stronger is the second muscle?

Lesson 8.7

Any Way You Slice It

In the previous lesson you discovered a formula for calculating the area of a circle. In this lesson you will discover how to calculate the area of different portions of a circle.

If you were told to cut a slice of your favorite pie, your slice would probably be in the shape of a sector of a circle. If for some reason you were permitted only one straight cut with your knife, your slice would probably be in the shape of a segment of a circle. If your brother or sister didn't like the pie crust he or she might cut out a circular slice from the center of the pie. The shape that remains is called an annulus.

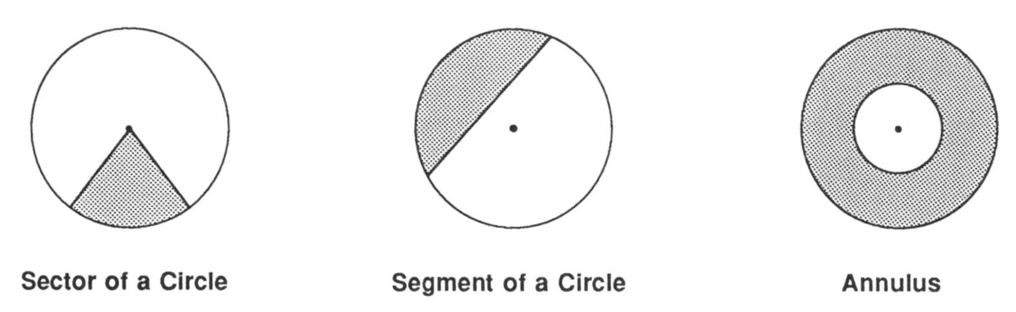

Sector of a Circle **Segment of a Circle** **Annulus**

*A **sector of a circle** is the region between an arc and two radii of a circle.*

*A **segment of a circle** is the region between an arc and a chord of a circle.*

*An **annulus** is the region between two concentric circles.*

In this lesson you will solve problems that include finding the areas of sectors of circles, segments of circles, and annuluses. "Picture equations" are helpful when you are trying to visualize the area of some of these regions. The picture equations below show you how to find the area of a sector of a circle, the area of a segment of a circle, and the area of an annulus.

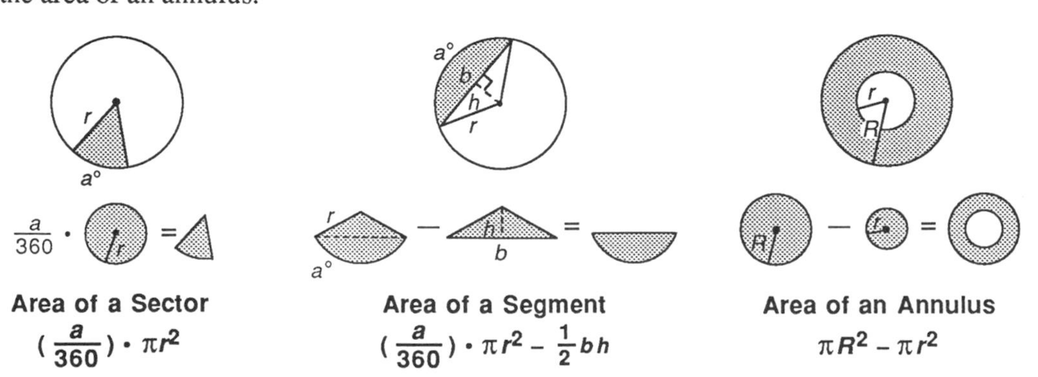

Area of a Sector
$\left(\frac{a}{360}\right) \cdot \pi r^2$

Area of a Segment
$\left(\frac{a}{360}\right) \cdot \pi r^2 - \frac{1}{2} bh$

Area of an Annulus
$\pi R^2 - \pi r^2$

Example A

Find the area of the sector.

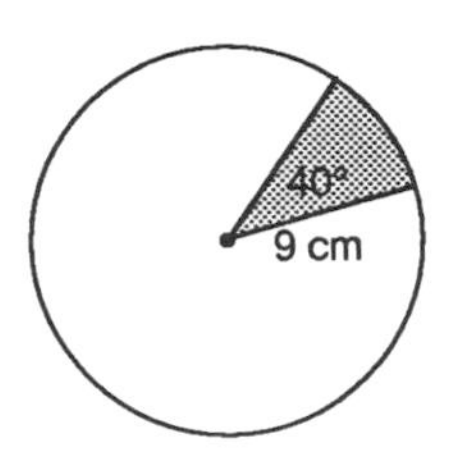

Since $r = 9$ cm, then the area of the circle is $\pi 9^2$ or 81π cm^2. Since $40/360 = 1/9$, then the area of the sector is $(1/9)(81\pi$ cm$^2)$ or 9π cm^2.

Example B

Find the area of the segment.

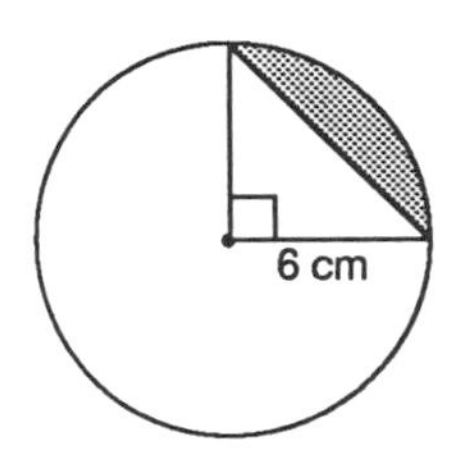

Since $r = 6$ cm, then the area of the circle is $\pi 6^2$ or 36π cm^2. Since the sector is 1/4 of the circle, then the area of the sector is $(1/4)(36\pi$ cm$^2)$ or 9π cm^2. Since the triangle is a right triangle, then its area is $(1/2)(6)(6)$ or 18 cm^2. Therefore, the area of the segment is the area of the sector (9π cm^2) less the area of the triangle (18 cm^2) or $(9\pi - 18)$ cm^2.

Example C

Find the area of the annulus.
$R = 12$ cm and $r = 8$ cm.

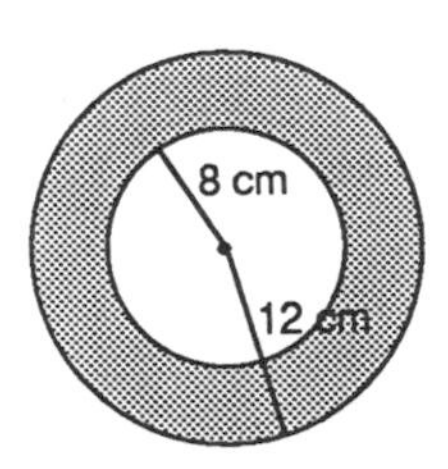

The area of the annulus is equal to the area of the larger circle (144π cm^2) less the area of the smaller circle (64π cm^2). Therefore, the area of the annulus is $144\pi - 64\pi$ or 80π cm^2.

Example D

Find x. The shaded area is 14π cm^2 and the radius is 6 cm.

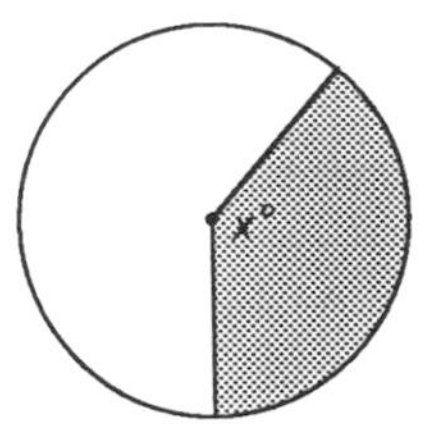

The sector's area is $\frac{x}{360}$ of the circle's.

Therefore:

$$14\pi = \left(\frac{x}{360}\right)(36\pi)$$

$$\frac{(360)(14\pi)}{(36\pi)} = x$$

$$x = 140$$

The central angle measures 140.

EXERCISE SET

What is the shaded area in each problem below? r is the radius of the circle. If two circles are shown, r is the radius of the smaller and R is the radius of the larger. All given measurements are in centimeters.

1. $r = 20$

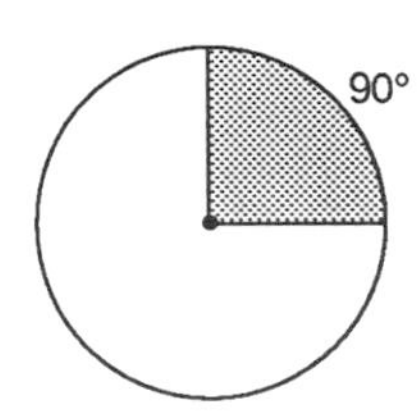

2. $r = 6$

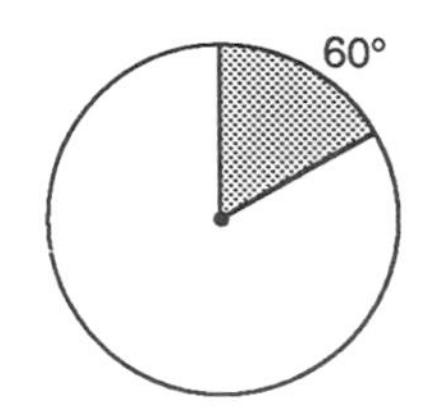

3.* $r = 8$

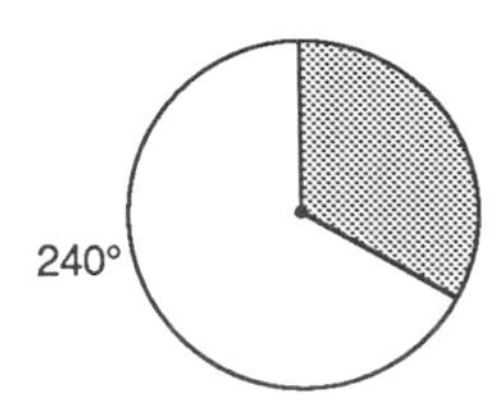

4. $r = 16$

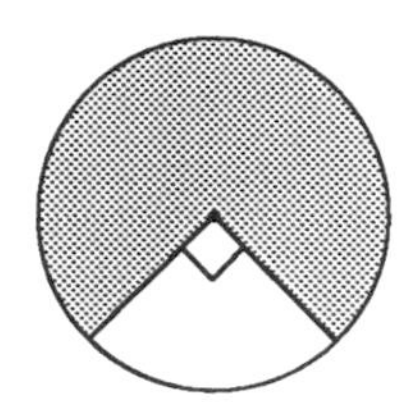

5. $r = 2$

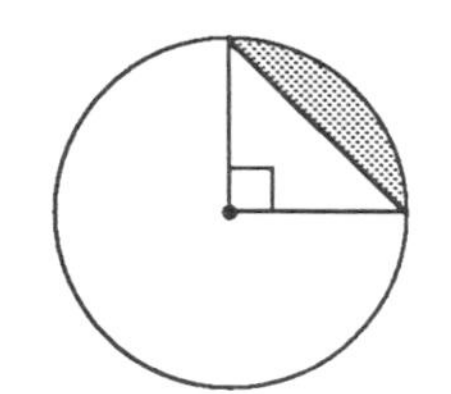

6.* $r = 8$

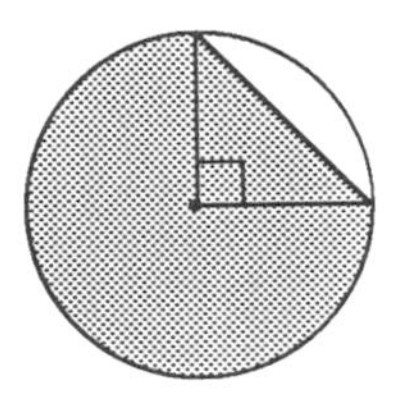

7. $R = 7$
$r = 4$

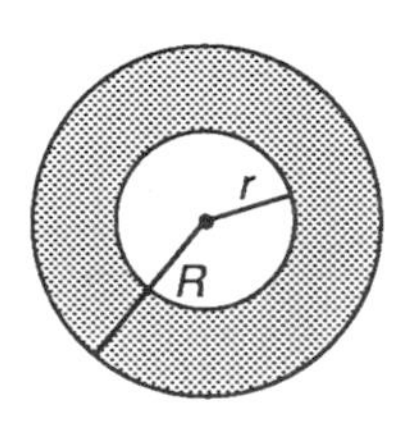

8. $r = 2$

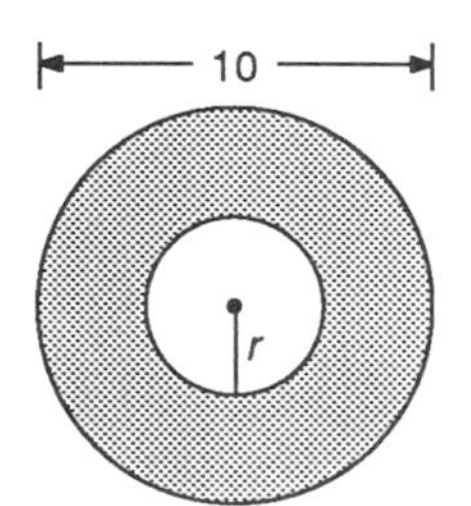

9.* $R = 12$
$r = 9$

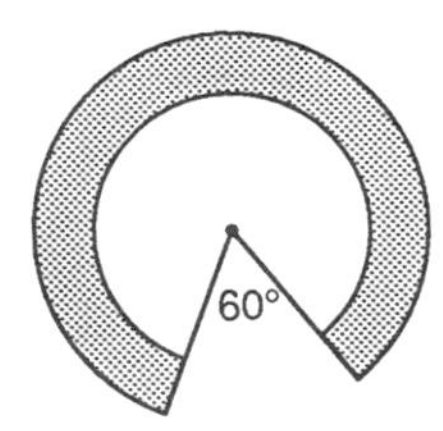

Find the radius in each problem below.

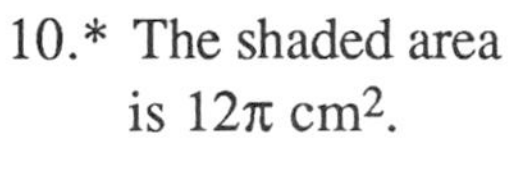

10.* The shaded area is 12π cm^2.

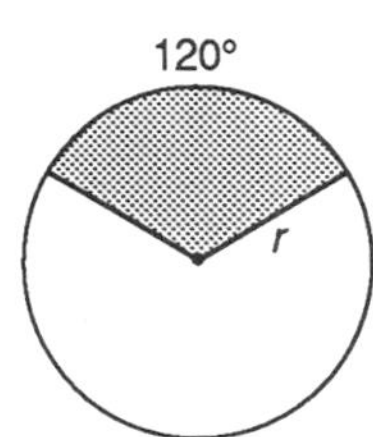

11.* The shaded area is $(49\pi - 98)$ cm^2.

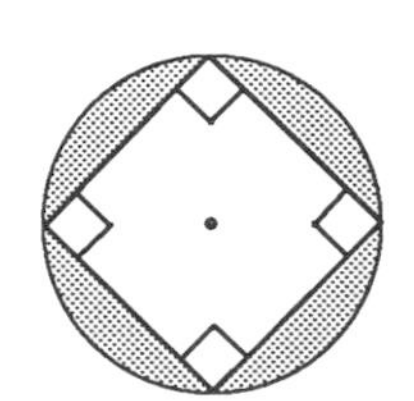

12.* The area of the annulus is 32π cm^2.

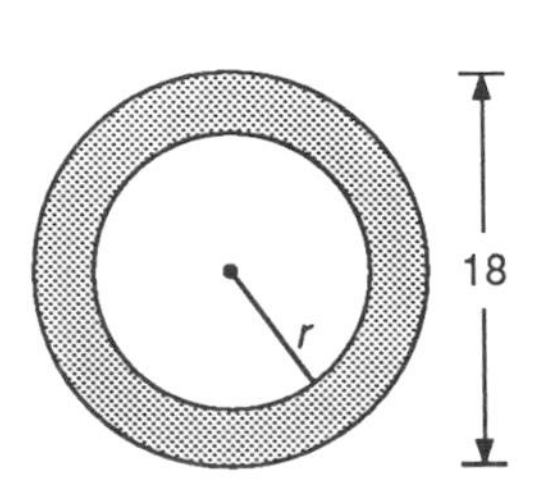

Find the measure of $\angle ABC$ in each problem below.

13.* The shaded area is 120π cm^2.
$r = 24$ cm

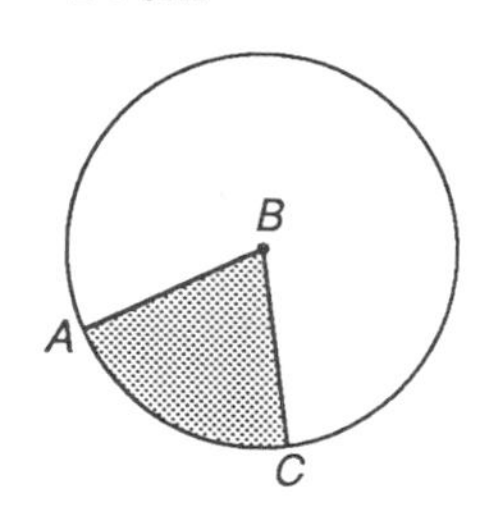

14.* The shaded area is 10π cm^2.
$R = 10$ $r = 8$

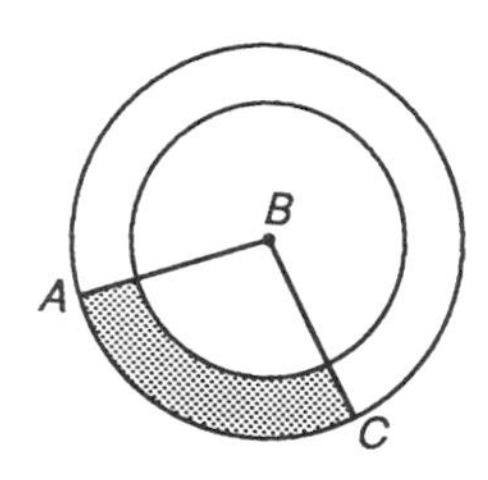

What is the shaded area in each problem below. Each arc is an arc from a circle. In Exercises 16 to 20, circles are externally tangent. In Exercise 22, the smaller circles are externally tangent to each other and are each internally tangent to the larger circle. All given measurements are in centimeters.

15.

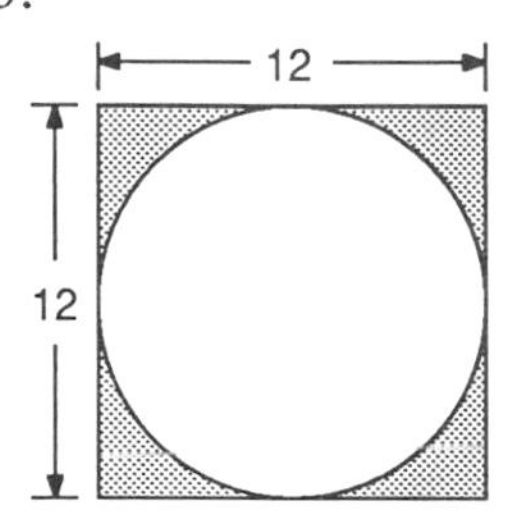

16.

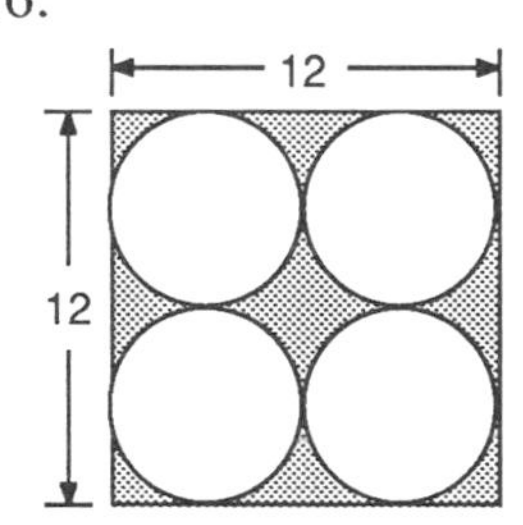

17.

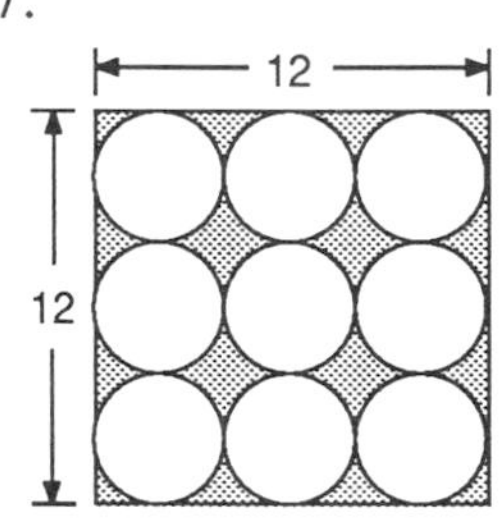

18.

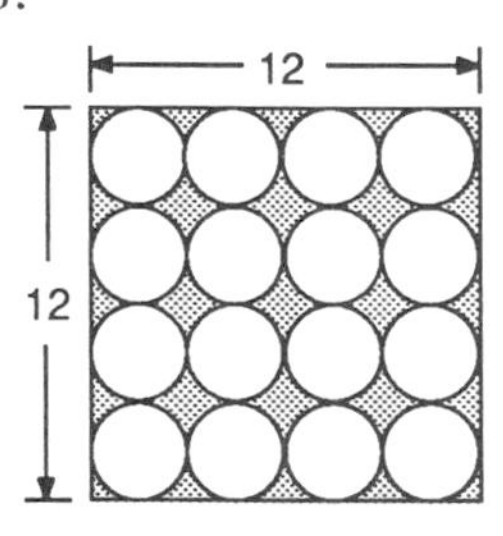

19.

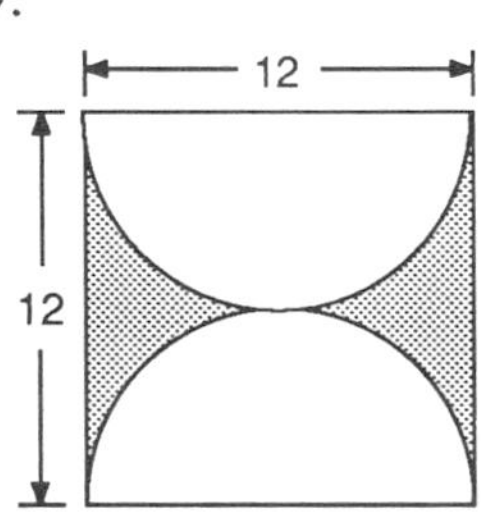

20.

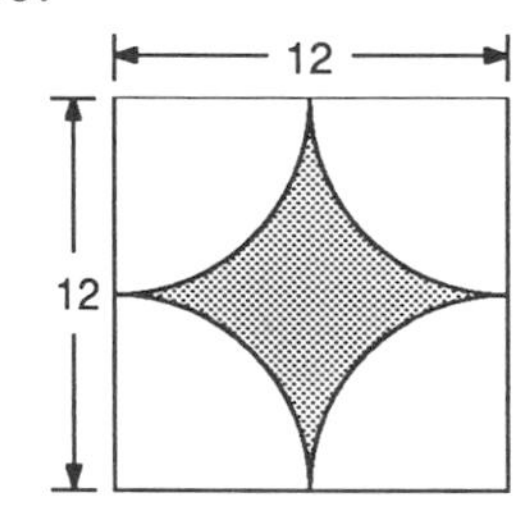

21.*

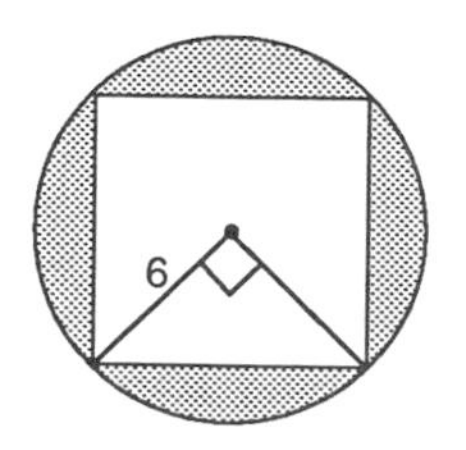

22.

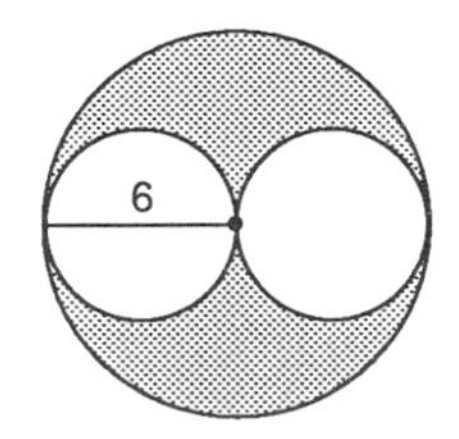

23.*

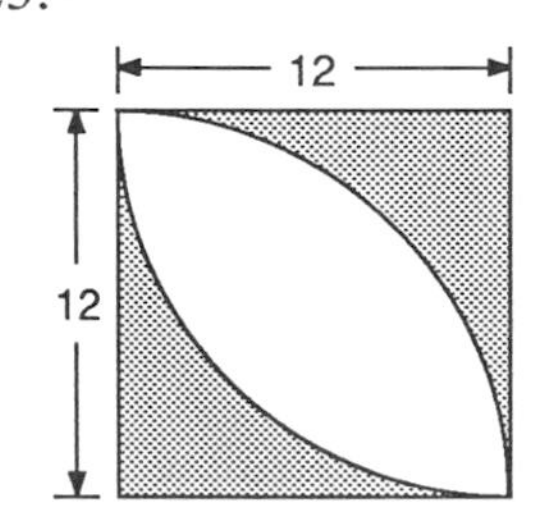

Computer Activity

Logo Spirolaterals

In this computer activity you will use the Logo computer language to explore spirolaterals. As you can see from the pictures below, spirolaterals come in a variety of shapes. Spirolaterals can be quite beautiful. Some are based on right angles. These are called right spirolaterals. Other spirolaterals are of a more general type and are based on other angles.

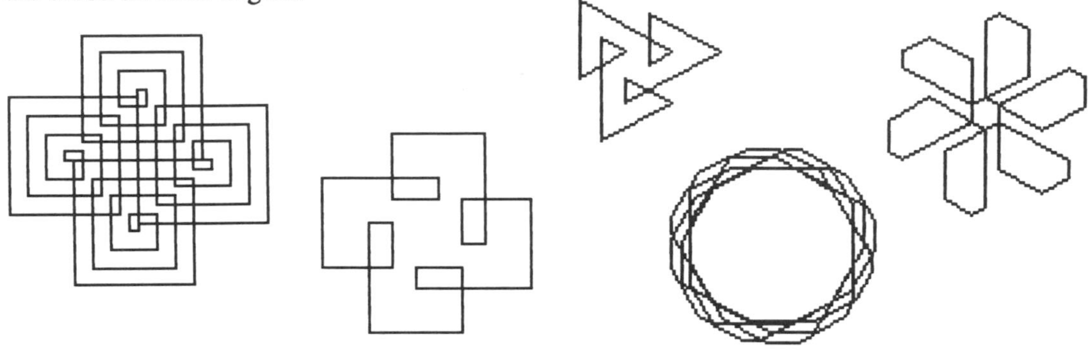

Right Spirolaterals **Spirolaterals**

You do not need previous experience with Logo to try this activity. You may want to work on it over an extended period of time. You should not feel that you have to try every experiment or answer every question. In fact, as you work with spirolaterals you may develop investigations and questions of your own that you want to pursue. Feel free to explore on your own. For this activity you will need:

- Computer and disk with Logo computer language
- Resource person familiar enough with Logo to help you get started
- Disk with the file of procedures called SPIROS
  (The listings of the procedures are in *Appendix B: Logo Procedures*.)
- Graph paper

1. On graph paper, draw a 1-2-3 right spirolateral. Select a starting point and draw a segment of unit length. Then turn right 90° and add a segment of length two. Turn 90° right again and draw a segment of length three. Turn 90° right and repeat the sequence. Continue until you return to your starting point or spiral off your graph paper. Draw 1-2-3-4 and 1-2-3-4-5 right spirolaterals.

Start up your computer with Logo and load the file called SPIROS from your procedures disk. The SPIROS file includes three right spirlolateral drawing procedures called RT.S, RT.ANY, RT.S3.

RT.S takes one number as an input: the order of the spirolateral. The **order** of a spirolateral is the number of turns made (or segments marched off) before the pattern used in the spirolateral repeats. If you input 3, you will create the same 1-2-3 right spirolateral that you created on graph paper. If you input 4, you'll get the 1-2-3-4 right spirolateral. With some inputs the spirolateral will keep spiralling forever. If this happens, halt RT.S with the Logo "break" key sequence. (Ask your Logo resource person what it is for your version of Logo.)

The SCALE procedure included in SPIROS will help you keep big spirolaterals on the screen. SCALE requires one input. It determines how many turtle steps RT.S makes when marching off one spirolateral unit. SCALE 10 will create nice size units for inputs to RT.S up to 10. SCALE 4 will work well with RT.S for inputs up to 20. You can use other inputs to SCALE as well.

2. Use RT.S to make spirolaterals of order 1, 2, 3, 4, and 5. Which spirolaterals returned to their starting positions (were closed)? How many times did the spirolateral have to go through the sequence before it closed? Which spirolaterals spun off forever (were open)? Make a table like the one on the right to keep track of your results.

Order of spirolateral	Closed or open?	If closed, repeats needed to close
1	—?—	–?–
2	—?—	–?–
3	—?—	–?–
4	—?—	–?–
⋮	⋮	⋮

3. Experiment with RT.S spirolaterals of higher orders. What is the pattern? How can you determine whether a spirolateral will be closed or open? If it is closed, how can you predict the number of times the sequence needs to repeat before the turtle returns to its starting point? This is the **degree** of the spirolateral.

RT.S3 creates order-3 spirolaterals. It takes three numbers as inputs. If the inputs are each a different number, the spirolateral will always be one of three types of closed spirolaterals. The number of units in each segment will be indicated by tiny tic-marks.

Use SCALE to keep large RT.S3 spirolaterals on your screen. If you want several RT.S3 spirolaterals on the screen at the same time, you can use the MOVETO or MOVE procedures included in SPIROS to move the turtle to a new starting position. MOVETO requires two numbers as inputs: the x- and y-coordinates of the point to which you want to move. MOVE requires one number as input and moves the turtle in the direction it is facing that number of spirolateral units.

Type A (RT.S3 2 5 3) **Type B (RT.S3 1 3 5)** 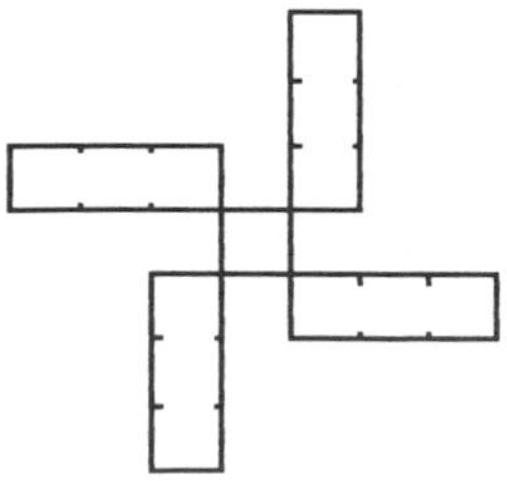 **Type C (RT.S3 3 4 5)**

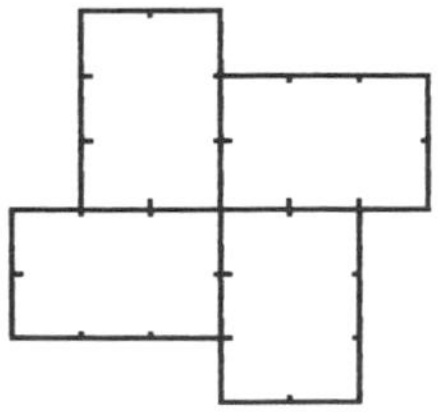

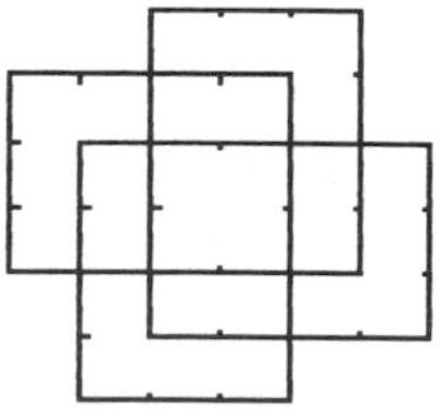

4. Experiment with RT.S3. Try many triplets of inputs and make a table to show which inputs result in which type of spirolateral. How can you predict from the three inputs which type of spirolateral will result? Does the order of the numbers make a difference?

5. How can you predict the area of the inner-most square in types B and C spirolaterals?

RT.ANY creates right spirolaterals based on a list of numbers input inside square brackets. You must leave a space between each number in the list. RT.ANY [1 1 2 3 5 8] creates a spirolateral based on the Fibonacci sequence.

6. Make a spirolateral using your own sequence of numbers. Make a spirolateral based on your telephone number. Try some negative numbers in your input list. How do negative inputs affect the movement of the turtle? Use negative inputs to create concave spirolaterals.

7.* Why do some spirolaterals return to the starting point while others don't? Can you predict which ones will close and which ones will not?

8. What happens if the sequence of numbers you use with RT.ANY consists of duplicate subsequences. Try RT.ANY [1 2 1 2]. Does this right spirolateral close? Can you use a shorter sequence of numbers to create the same spirolateral? What is the order of RT.ANY [1 2 1 2]? (Look back at the definition of the order of a spirolateral.) RT.ANY assumes that your input list contains the number of turns or segments that you want to make before repeating. If your input list consists of duplicate subsequences, you can confuse RT.ANY and cause it to lose track of the proper order of the spirolateral. To avoid confusing RT.ANY, don't use input lists with duplicate subsequences.

9. Can you create the familiar design on the right using RT.ANY? What inputs does it take? Could this design completely fill the screen without overlaps or gaps? What can you say about a shape that can do this? Can you use RT.ANY to create another polygon which will tessellate. Use MOVE, RT, and LT to fit your shapes together. You can see if you're right by typing TESS.CROSS.

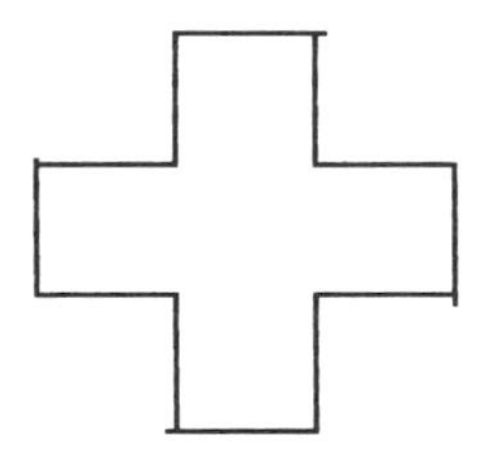

In addition to the right spirolateral procedures, the SPIROS file includes two more general spirlolateral drawing procedures called SPIRO, and ANY.SPIRO along with their various subprocedures.

SPIRO requires two inputs: the order of the spirolateral (number of turns or segments) and the angle measure of the turns. Unlike the RT.S procedure you used earlier, SPIRO will make a spirolateral based on any angle you give it. Like RT.S, SPIRO will draw segments with lengths increasing by one unit each time, starting with a segment of length one.

10. SPIRO 10 120 draws the design on the right. What polygon is this design based on? Why? Try some other inputs to SPIRO. Some angles will take many repetitions before the spirolateral closes. You can find out how many by typing PR :DEGREE after drawing your spirolateral. Non-closing spirolaterals will report a degree of 1. Their degree is actually unspecified as no amount of repetitions will cause them to close.

SPIRO 10 120

11. Make a table like the one on the right to keep track of your inputs to SPIRO. Do you see any patterns emerging from this table? Look at the product of the numbers in the first two columns. Find a way to predict whether a SPIRO spirolateral will close based on the inputs for order and angle.

Order	Angle	Closed or open?	If closed, repeats needed to close
–?–	–?–	—?—	–?–
–?–	–?–	—?—	–?–
–?–	–?–	—?—	–?–
⋮	⋮	⋮	⋮

ANY.SPIRO creates spirolaterals based on a list of numbers. ANY.SPIRO works similarly to the RT.ANY procedure you used earlier except that you need to input the angle used for turning as well as the list of numbers for your spirolateral. Enclose your input list in square brackets and leave a blank space between each number. For example, ANY.SPIRO [2 3 2 4] 72 creates a spirolateral based on a pentagon.

12. Experiment with ANY.SPIRO. Change the number of numbers in your input list (the order of the spirolateral) while keeping the angle the same. Use the information from your table to help with your designs.

13. Use some negative numbers in your input list. Make a cube like the one below right using just ANY.SPIRO.

14. Make a six-pointed star.

15. The "pinwheel" spirolateral below was created by ANY.SPIRO. Both zeros and negative numbers were used in the input list. What inputs to ANY.SPIRO could produce this pinwheel?

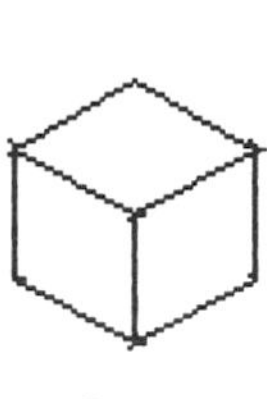

Cube

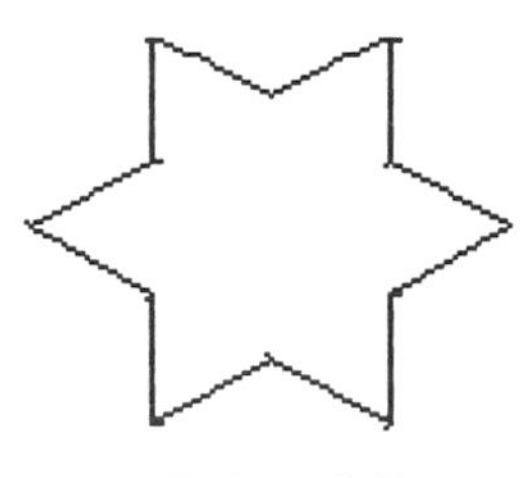

Six-Pointed Star

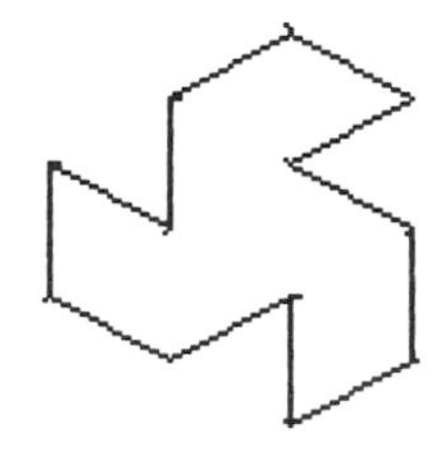

Pinwheel

Lesson 8.8
Surface Area

In this lesson you will find the surface area of prisms, pyramids, and cylinders. The **surface area** of each of these solids is the sum of the areas of all of the faces or surfaces that enclose the solid. The faces include the tops and bottoms (bases) and the remaining surfaces (lateral faces or surfaces).

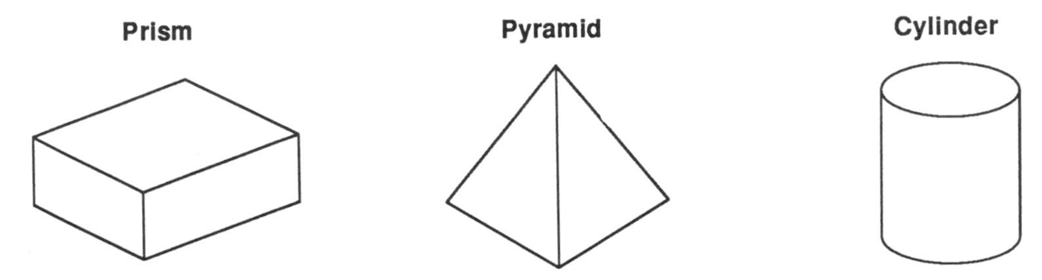

To find the surface areas of prisms and pyramids follow the steps below.

Step 1: Draw a diagram of each face of the solid as if the solid were cut apart at the edges and laid flat. Label the dimensions.

Step 2: Calculate the area of each face. If some faces are identical, you need only calculate the area of one and multiply by the number of identical faces.

Step 3: Find the total area of all the faces (bases and lateral faces).

Example A

Find the surface area of the prism on the right. Each face is a rectangle.

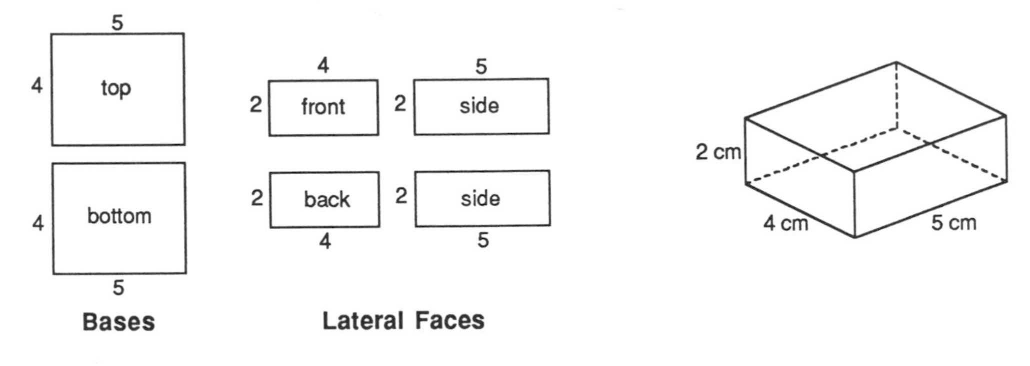

Surface area $= (2)(4)(5) + (2)(2)(4) + (2)(2)(5)$
$= 40 + 16 + 20$
$= 76$

The surface area of the prism is 76 cm^2.

Example B

Find the surface area of the square based pyramid on the right.

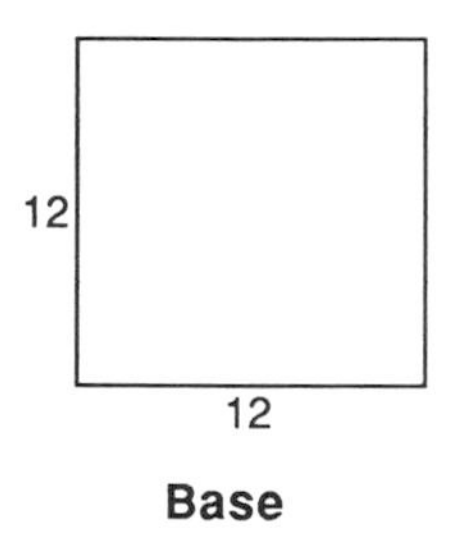

Base

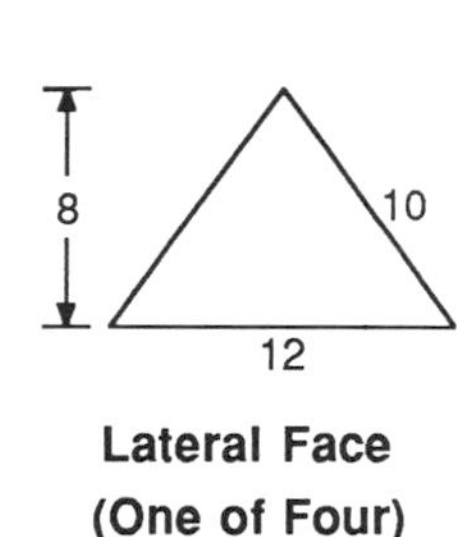

Lateral Face (One of Four)

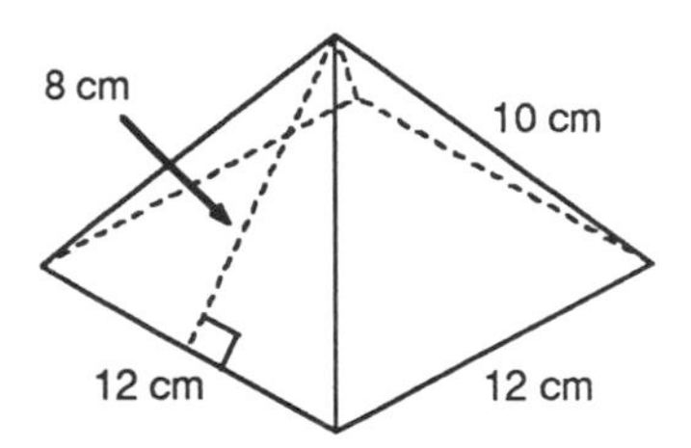

Surface area $= (4)[(1/2)(12)(8)] + (12)(12)$
$= 192 + 144$
$= 336$

The surface area of the square based pyramid is 336 cm^2.

The total surface area of a cylinder is the sum of the lateral surface area and the areas of the bases. The lateral surface is the curved surface on a cylinder. You can think of the lateral surface as a wrapper. You can slice the wrapper and lay it flat to get a rectangular region. The height of the rectangle is the height of the cylinder. The base of the rectangle is the circumference of the circular base of the cylinder. The lateral surface area is the area of the rectangular region.

Example C

Find the surface area of the cylinder on the right to the nearest square inch. Use 3.14 for π.

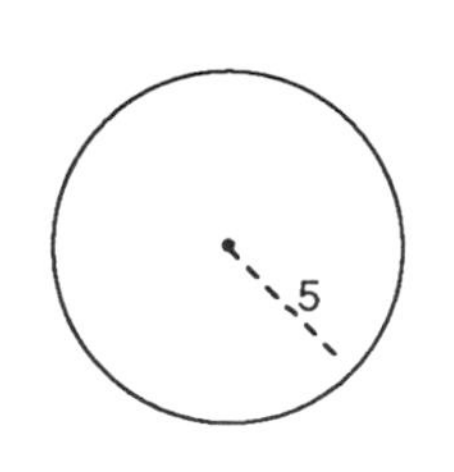

Base (One of Two)

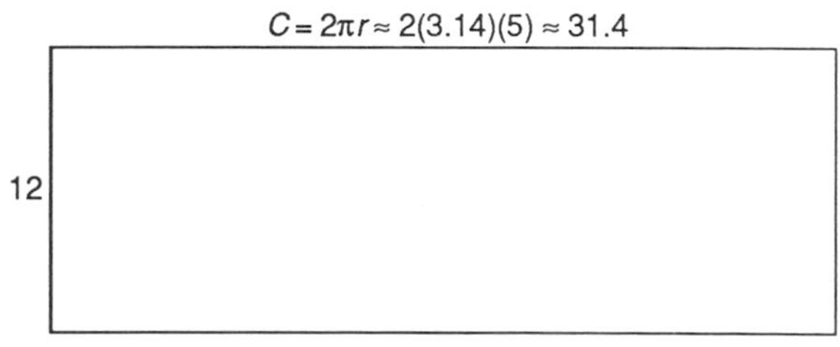

Lateral Surface

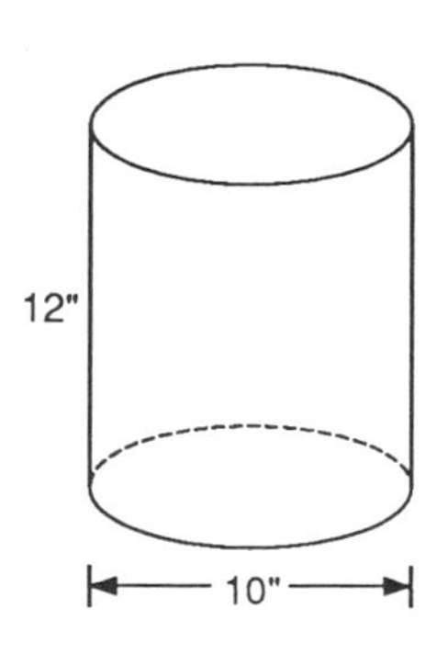

Surface area $\approx (2)(3.14)(5^2) + (2)(3.14)(5)(12)$
$\approx 157 + 376.8$
≈ 533.8

The surface area of the cylinder is about 534 square inches.

Example D

Find the surface area of the solid on the right to the nearest square centimeter.

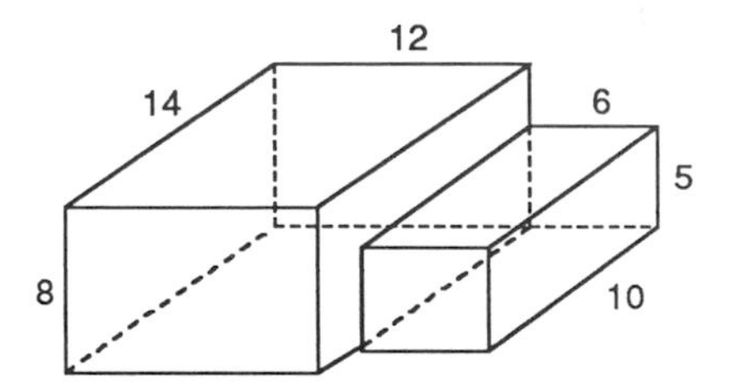

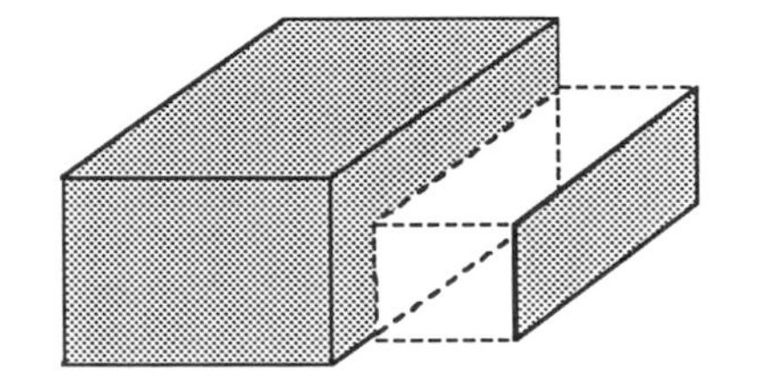

+

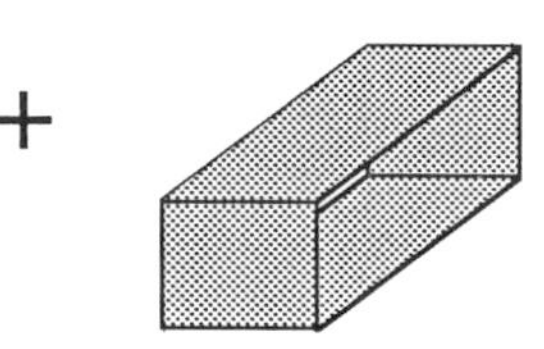

$$\begin{aligned}\text{Area} &= [(2)(12)(14) + (2)(8)(12) + (2)(8)(14)] + [(2)(6)(10) + (2)(6)(5)] \\ &= [336 + 192 + 224] + [120 + 60] \\ &= 752 + 180 \\ &= 932\end{aligned}$$

The surface area of the solid is 932 cm^2.

EXERCISE SET A

Find the surface area for each prism, pyramid, and cylinder below. All quadrilaterals are rectangles. All given measurements are in centimeters.

1.

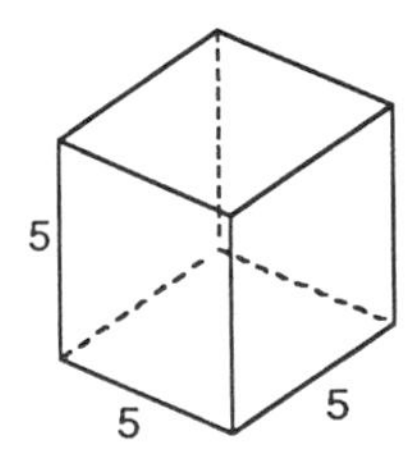

2.

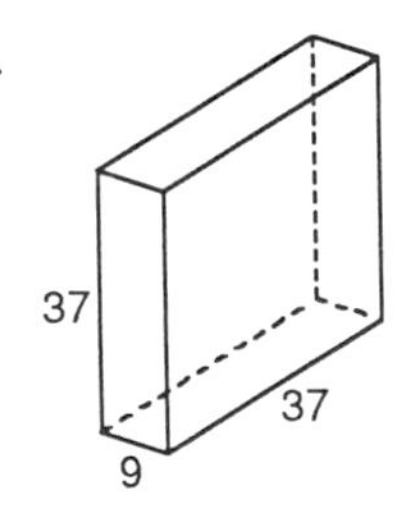

3.

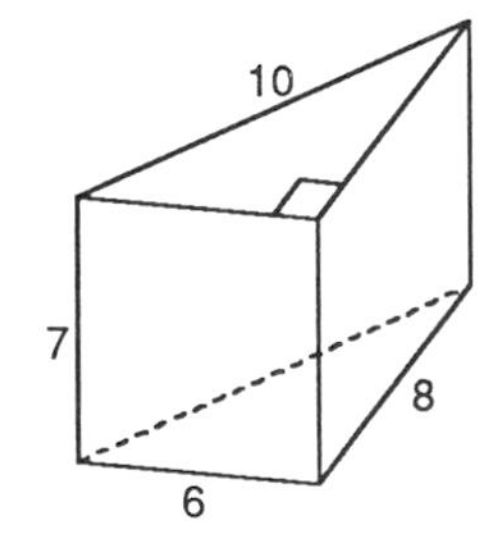

4. Use 22/7 for π.

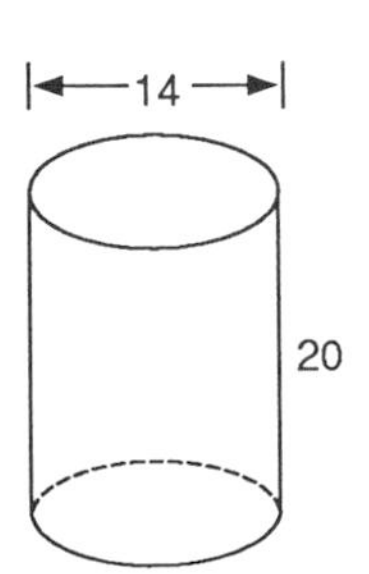

5. The base is a square.

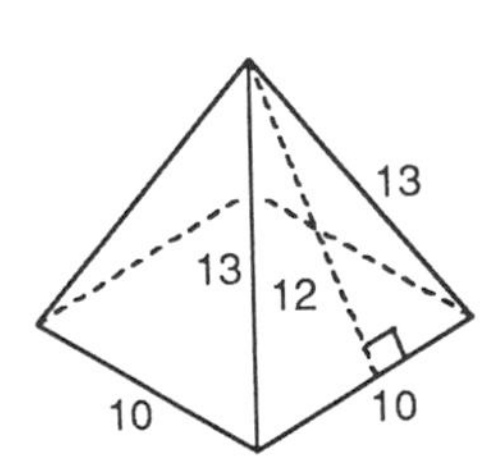

6.*

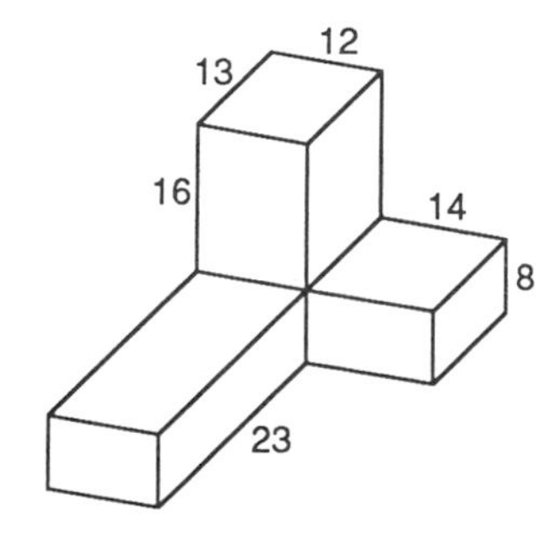

7.* The base is a regular pentagon with apothem $a \approx 11$ and side $s \approx 16$. Each lateral edge $t \approx 17$ and the height of a face $h \approx 15$. (Give answer to the nearest cm^2.)

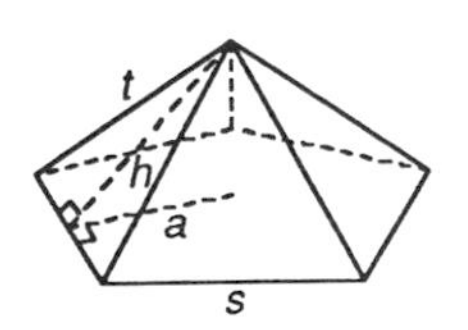

8.* $D = 8$
$d = 4$
$h = 9$

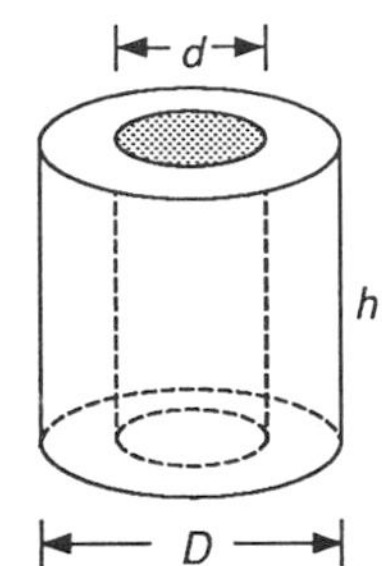

9.* Use 3.14 for π. Round your answer to the nearest cm^2.

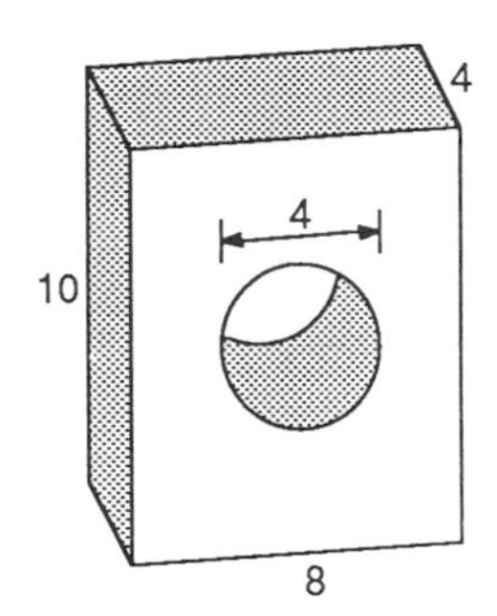

EXERCISE SET B

1. Two dart boards appear on the right. Each is one foot square. The circular targets drawn on each board are externally tangent to each other and just touch the edge of the boards. On which board do you have the best probability of landing in a circle assuming you throw darts randomly and count the throw only if it hits the board? Calculate the probability for each board. (If you make a dart board like one of these and throw enough darts at it, keeping track of your "success" rate, you can usc it to approximate the value of π.)

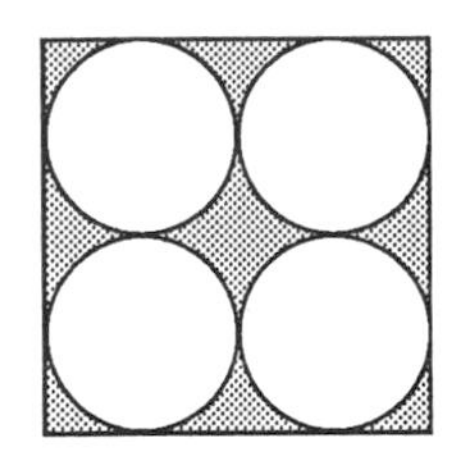

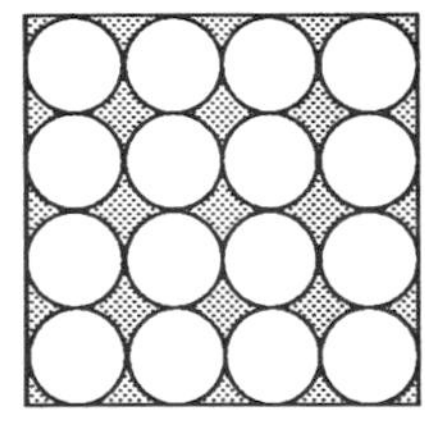

2.* What is the probability that a point randomly selected in the interior of a circle will be closer to the center of the circle than to the circle itself?

3.* The target shown on the right has a bull's eye (center circle) with a diameter of 8 cm. The width of each ring is 4 cm. Counting only the random tosses that hit the target, what is the probability of hitting the bull's eye?

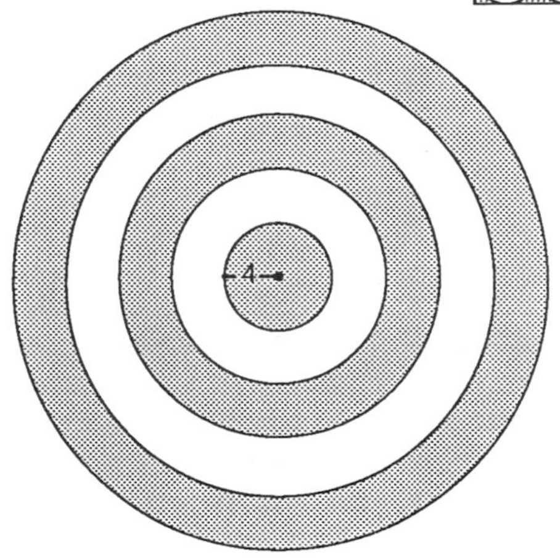

Lesson 8.9

Area Problems: The Sequel

EXERCISE SET A

1. What is the total surface area of a rectangular room with dimensions of 6 meters by 7 meters by 4 meters high? Ignore doorways and windows

2. Ernesto plans to build a pen for his pet iguana. What is the area (to the nearest m^2) of the largest pen that Ernesto can make with 100 meters of fencing?

3. The amount of water that a pipe can carry depends upon the area of the cross section of the pipe opening. How many two-inch diameter pipes are needed to carry the same amount of water as one six-inch diameter pipe?

4.* The George Washington High School Math Club held its Fibonacci Birthday Party at Pedro's Pizzeria. The 16 members present all agreed they should order 4-item pizzas. Together they had $84. Being very calculating, they agreed they should determine which purchase would give the most pizza per dollar.

Pedro's Pizzeria Price List

Type	Size	3-items*	4-items*
Small	10″	$9.00	$10.50
Medium	12″	11.25	12.75
Large	14″	14.00	15.50
Extra Large	16″	18.00	19.50

*You choose from: Salami, pepperoni, linguisa, mushrooms, onions, bell pepper, garlic.

They came up with four possibilities for spending their money. They decided to purchase either 8 small pizzas, 6 medium pizzas, 5 large pizzas, or 4 extra large pizzas (sizes listed are diameters). Determine which choice gives the best deal (most pizza per dollar). Will the best deal give them the most pizza? What is the most pizza they can get for their $84? (Try any combination of pizzas on the price list.)

Schulhoff Tam
High school student

EXERCISE SET B

Use the figures for Exercises 1 to 3. Unless the dimensions indicate otherwise, assume each quadrilateral is a rectangle.

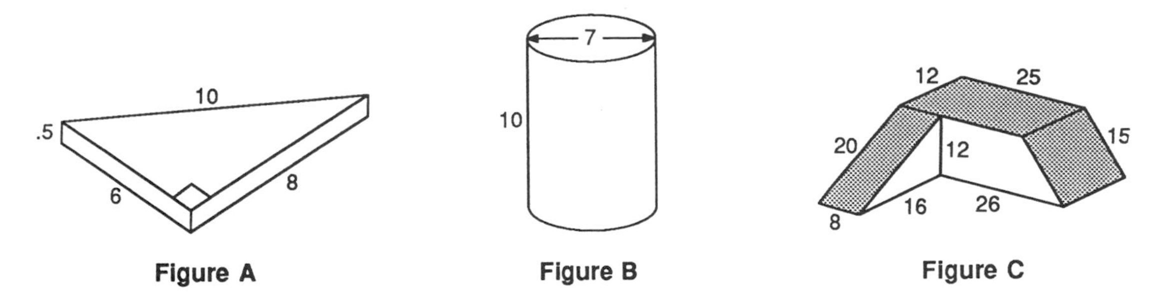

Figure A Figure B Figure C

1. Assume the measurements in figure A are in centimeters. You are to produce 10,000 of these widgets and each must be electroplated with a thin layer of high conducting silver. Find the total silver cost if silver plating costs $1 for each 200 square centimeters.
2. Assume the measurements in figure B are in meters. Find the cost of painting the exterior of nine of these large cylindrical chemical storage containers with anti-rust sealant. The sealant costs $32 per gallon. Each gallon covers 18 square meters. The exterior bottoms are not to be painted. Use 22/7 for π.

3.* Assume the measurements in figure C are in feet. The three shaded regions are to be covered with asphalt shingles which cost $35 per bundle. Each bundle contains enough shingles to cover 100 square feet. The remaining vertical surfaces (three are trapezoids and one is a right triangle) are to be covered with wood stain at $15 per gallon. Each gallon covers 150 square feet. Find the total cost of the project.

EXERCISE SET C

Graph each figure below on graph paper and then find its area.

1. Parallelogram *ABCD* with *A* (0, 0); *B* (14, 0); and *D* (6, 8)
2. Triangle *ABC* with *A* (0, 0); *B* (16, 0); and *C* (12, 9)
3. Circle *O* with points *P* (-8, 2) and *Q* (12, 2) as the endpoints of a diameter

Lesson 8.10

Chapter Review

EXERCISE SET A

Complete each conjecture.

1. The area of a rectangle is given by the formula —?— where A is the area, b is the length of the base, and h is the height of the rectangle. (C-55 or **Rectangle Area Conjecture**)
2. The area of a parallelogram is given by the formula —?— where A is the area, b is the length of the base, and h is the height of the parallelogram. (C-56 or **Parallelogram Area Conjecture**)
3. The area of a triangle is given by the formula —?— where A is the area, b is the length of the base, and h is the height of the triangle. (C-57 or **Triangle Area Conjecture**)
4. The area of a trapezoid is given by the formula —?— where A is the area, b_1 and b_2 are the lengths of the two bases, and h is the height of the trapezoid. (C-58 or **Trapezoid Area Conjecture**)
5. The area of a regular polygon is given by the formula —?— where A is the area, a is the apothem, s is the length of each side, and n is the number of sides of the regular polygon. Since the length of each side times the number of sides is the perimeter ($sn = p$). The formula can also be written as $A = (1/2)a$–?–. (C-59 or **Regular Polygon Area Conjecture**)
6. The area of a circle is given by the formula —?— where A is the area and r is the radius of the circle. (C-60 or **Circle Area Conjecture**)

EXERCISE SET B

Make a sketch to illustrate each term below.

1. Annulus
2. Prism
3. Pyramid
4. Cylinder
5. Apothem
6. Sector of a circle
7. Segment of a circle

EXERCISE SET C

Diagram and explain in a paragraph how you derived the area formula for each figure.

1.* Parallelogram
2. Trapezoid
3. Circle

EXERCISE SET D

Solve each problem below. All given measurements are in centimeters.

1. $A = 576 \text{ cm}^2$
   $h = -?-$

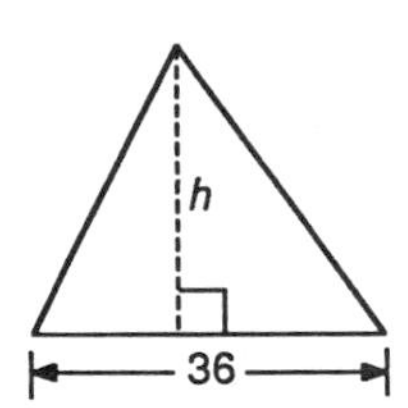

2. $A = 576 \text{ cm}^2$
   $h = -?-$

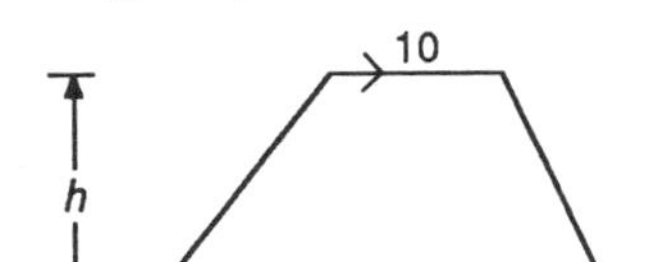

3. $A = 576\pi \text{ cm}^2$
   The circumference is –?–.

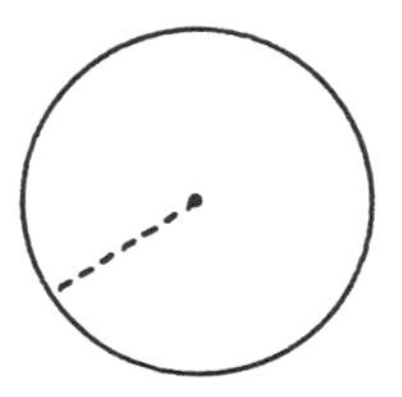

4. $C = 18\pi$ cm
   $A = -?-$

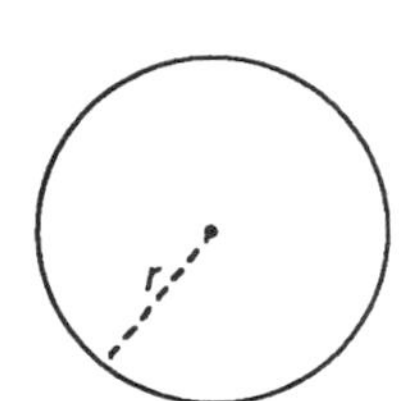

5. $A = 126 \text{ cm}^2$
   $a = 13$ cm
   $h = 9$ cm
   $b = -?-$

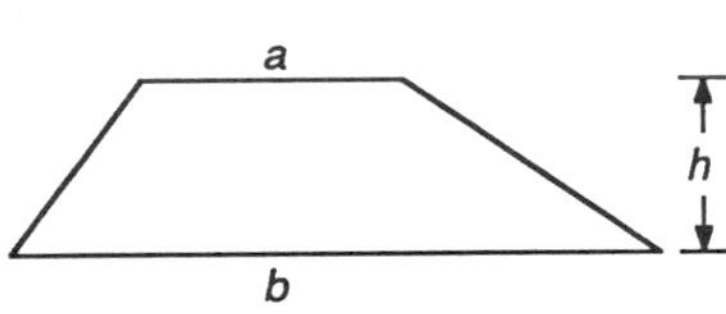

6. The area of the sector is $16\pi \text{ cm}^2$.
   $m\angle FAN = -?-$

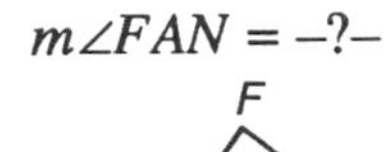
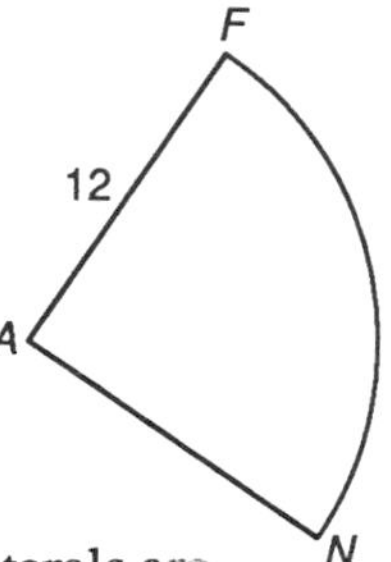

Find the surface area of each object. Unless told otherwise, quadrilaterals are rectangles. All given measurements are in centimeters.

7.

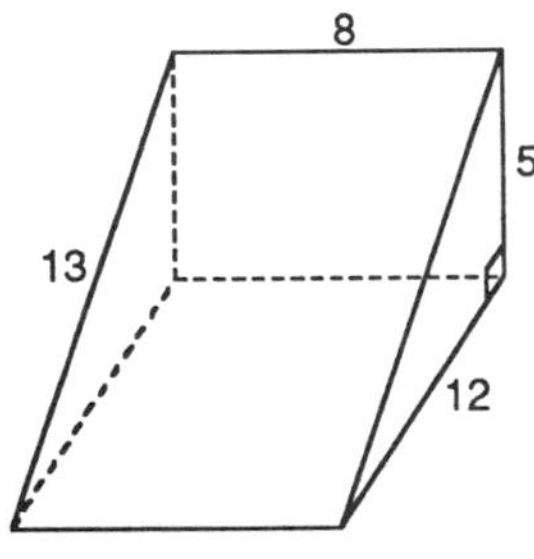

8.* The base is a trapezoid.

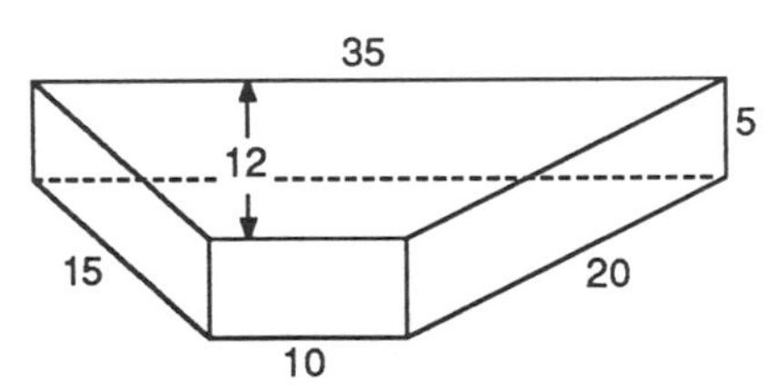

What is the shaded area in each diagram below? All given measurements are in cm.

9. Use 22/7 for π.

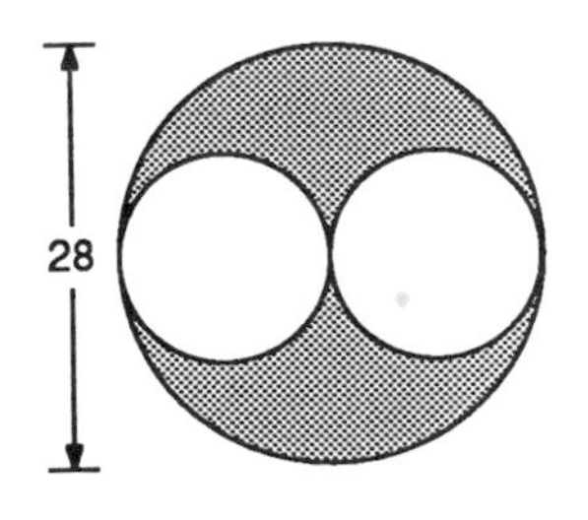

10.* *HELP* is a square. All the arcs are arcs of a circle of radius 4.

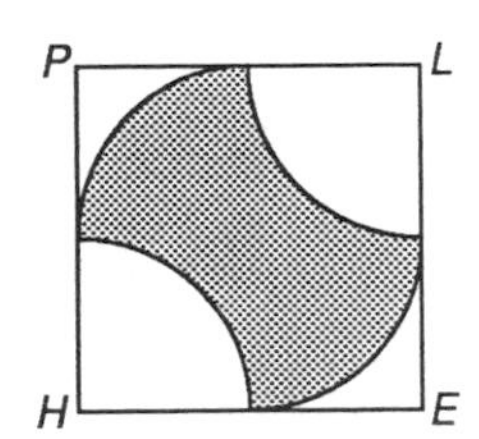

11. The sum of the lengths of the two bases of a trapezoid is 22 cm and its area is 66 cm^2. What is the height of the trapezoid?

12. Find the area of a regular pentagon to the nearest tenth of a square centimeter if the apothem is approximately 6.9 cm and each side is approximately 10 cm.

13.* Utopia Park has just installed a circular fountain 8 meters in diameter. The Senior Citizens Committee for Better Parks for All People has requested that a 1.5 m wide path be paved around the fountain. If paving costs the city $10 per m^2, find the cost to the nearest dollar of the paved ring around the fountain. (Use 3.14 for π.)

14.* What is the area of the largest rectangular pen that Igor can make for his pet Isosceles with 100 meters of fencing if one side of the castle is used for one side of the pen?

15.* Which is a better (tighter) fit: a round peg in a square hole or a square peg in a round hole?

16. Al Dente's Pizzeria sells cheese pizza by the slice according to the diagram on the right. Which slice gives the best deal (the most pizza per dollar)?

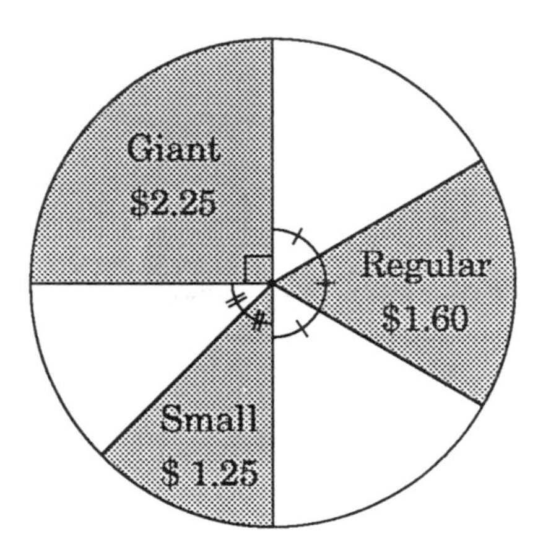

Al Dente's Pizza-by-the-Slice

Improving Visual Thinking Skills

Scrambled Vocabulary

Unscramble the letters to form familiar geometry terms.

Example: tubsoe — obtuse

1. eucat
2. legan
3. extrev
4. indema
5. dusiar
6. tompindi
7. mohsurb
8. soilsecse
9. nestgem
10. sulanun
11. nenagtt
12. megyoter

Cooperative Problem Solving

Discovering New Area Formulas

A mysterious capsule buried in an unexplored crater on the back side of the moon has recently been ~~unearthed~~ ~~unmooned~~ dug up. The space scientists have called in noted archaeologist Ertha Diggs. Ertha's speciality is language translation. After many months of careful research, Ertha and her team have concluded that the capsule is indeed from an extraterrestrial civilization. According to their investigation, the capsule was one of a series of capsules sent to young star-systems in the neighborhood of Alpha Centauri. In her translations Ertha found five area formulas that were new to her. She has asked for our help. Which, if any, of the five area formulas translated from the mysterious capsule are true? If a formula is true, show it, using algebra and what you have learned in this chapter. If a formula is not true, prove it is not true by placing numerical values into the formula and demonstrating that they do not work. You will have found a counter-example. Can you correct the incorrect formula?

Translations

1. *To find the area of a triangle, use* $A = mh$ *where* m *is the length of the midsegment of the triangle and* h *is the height of the triangle.*

2. *To find the area of a trapezoid, use* $A = mh$ *where* m *is the length of the midsegment of the trapezoid, and* h *is the height of the trapezoid.*

3. *To find the area of a rhombus, use* $A = rs$ *where* r *and* s *are the lengths of the two diagonals.*

4. *To find the area of a sector of a circle, use* $A = (1/2)rs$ *where* r *is the radius and* s *is the length of the arc.*

5.* *To find the area of a slice of a ring, use* $A = (1/2)s(m + n)$ *where* s *is the distance between two arcs, and* m *and* n *are the lengths of the arcs.*

9 Pythagorean Theorem

Waterfall,
M. C. Escher, 1961

In this chapter you will discover a property of right triangles known as the Pythagorean Theorem. The Pythagorean Theorem is one of the most important concepts in all of mathematics. It allows you to calculate the distance between two points. You will encounter the Theorem of Pythagoras and its applications in many other math classes (such as calculus). In this chapter you will also discover a number of conjectures related to the Pythagorean Theorem which you will use to solve problems.

Lesson 9.1

Square Roots

In an isosceles triangle the sum of the square roots of the two equal sides is equal to the square root of the third side.

– The Scarecrow in The Wizard of Oz by L. Frank Baum

The ancient Chinese, Babylonians, and Egyptians discovered a special relationship between the lengths of the sides of a right triangle. This relationship is perhaps the most important in all of geometry. Since working with this relationship involves squares and square roots, you will need to review some operations on square roots.

Simplifying Square Roots

$\sqrt{50} = \sqrt{25 \cdot 2} = \sqrt{25} \cdot \sqrt{2} = 5\sqrt{2}$

$\sqrt{84} = \sqrt{4 \cdot 21} = \sqrt{4} \cdot \sqrt{21} = 2\sqrt{21}$

$\sqrt{126} = \sqrt{9 \cdot 14} = \sqrt{9} \cdot \sqrt{14} = 3\sqrt{14}$

Multiplying Square Roots

$(\sqrt{3})(\sqrt{2}) = (\sqrt{6})$

$(\sqrt{5})^2 = (\sqrt{5})(\sqrt{5}) = 5$

$(2\sqrt{3})^2 = (2\sqrt{3})(2\sqrt{3}) = 4 \cdot 3 = 12$

Rationalizing the Denominator

$$\sqrt{\frac{2}{3}} = \frac{\sqrt{2}}{\sqrt{3}} = \frac{\sqrt{2} \cdot \sqrt{3}}{\sqrt{3} \cdot \sqrt{3}} = \frac{\sqrt{6}}{3}$$

$$\sqrt{\frac{5}{8}} = \frac{\sqrt{5}}{\sqrt{8}} = \frac{\sqrt{5} \cdot \sqrt{2}}{\sqrt{8} \cdot \sqrt{2}} = \frac{\sqrt{10}}{\sqrt{16}} = \frac{\sqrt{10}}{4}$$

EXERCISE SET A

Express each square root in simplest form.

1.* $\sqrt{12}$ 2. $\sqrt{18}$ 3. $\sqrt{24}$ 4.* $\sqrt{32}$ 5. $\sqrt{40}$

6. $\sqrt{48}$ 7. $\sqrt{60}$ 8. $\sqrt{75}$ 9. $\sqrt{83}$ 10. $\sqrt{85}$

11. $\sqrt{90}$ 12. $\sqrt{96}$ 13. $\sqrt{120}$ 14.* $\sqrt{185}$ 15.* $\sqrt{490}$

16. $\sqrt{576}$ 17.* $\sqrt{720}$ 18. $\sqrt{722}$ 19. $\sqrt{784}$ 20. $\sqrt{828}$

Express each product in simplest form.

21.* $(3\sqrt{2})^2$ 22. $(4\sqrt{3})^2$ 23. $(2\sqrt{3})(\sqrt{2})$ 24. $(3\sqrt{6})(2\sqrt{3})$ 25. $(7\sqrt{3})^2$

Rationalize the denominator, then simplify.

26. $\sqrt{\frac{1}{3}}$ 27.* $\sqrt{\frac{5}{12}}$ 28. $\sqrt{\frac{7}{8}}$ 29. $\sqrt{\frac{35}{72}}$ 30. $\sqrt{\frac{1}{2}}$

There is a special relationship between the areas of squares constructed on the three sides of a right triangle. The two dissection puzzles in Exercise Set B are intended to get you thinking about this special relationship.

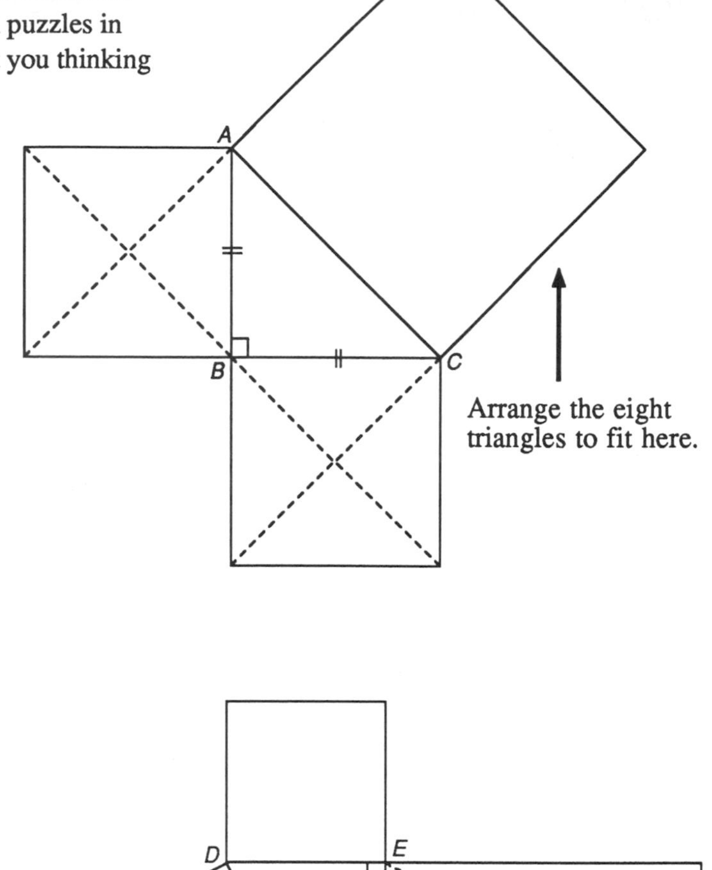

EXERCISE SET B

1. Right triangle *ABC* is isosceles. Trace the diagram to the right onto another sheet of paper. Cut out the four triangles in each of the two small equal squares and arrange them to exactly cover the larger square.

2. Side $\overline{EF}$ is twice the length of side $\overline{DE}$ in right triangle *DEF*. Trace the diagram below right onto another sheet of paper. Cut out the small square and the four triangles from the square on leg $\overline{EF}$ and arrange them to exactly cover the large square.

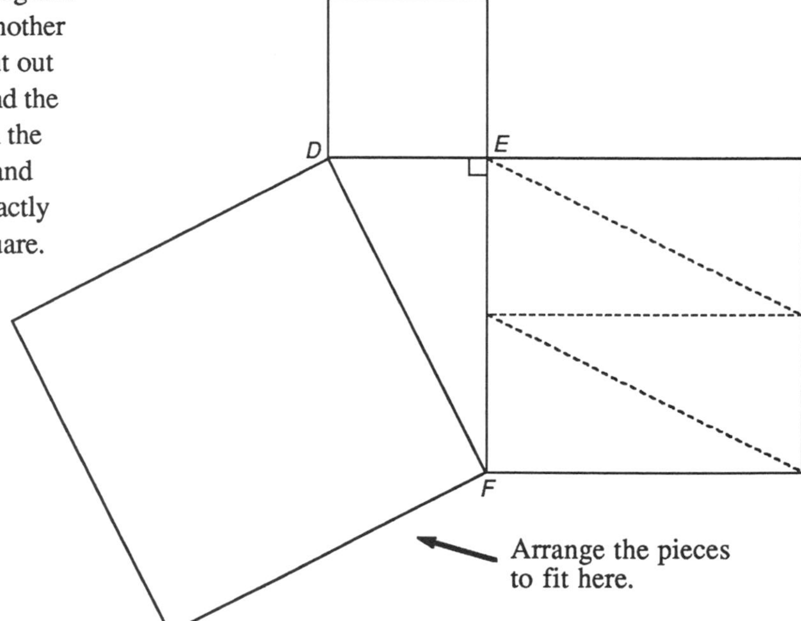

Lesson 9.2
The Theorem of Pythagoras

There is a surprising connection between the lengths of the three sides of any right triangle. It is probably the most useful property in all high school mathematics. It is useful because it helps you calculate the distance between two points. Nobody knows at what point in history this relationship was first discovered. The ancient Babylonians and Chinese saw this relationship. Some math historians believe that the Egyptians too used a special case of this property of right triangles.

In a right triangle, the side opposite the right angle is called the **hypotenuse**. The other two sides are called **legs**. In the diagram, a and b represent the lengths of the legs, and c the length of the hypotenuse.

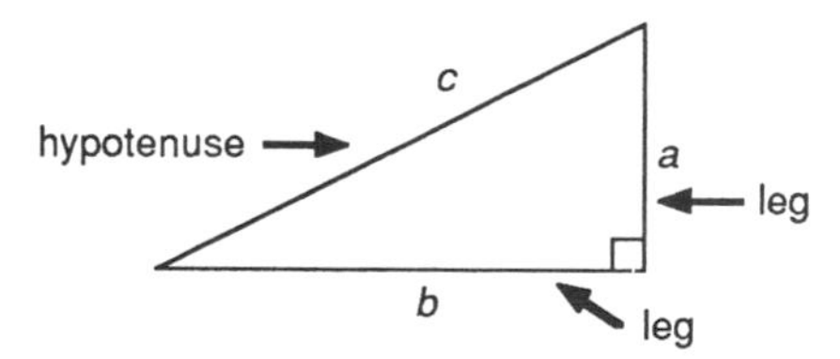

At the end of the previous lesson you worked two dissection puzzles. In each puzzle you rearranged the pieces from the two smaller squares to fit on the larger square. This demonstrated that the sum of the areas of the squares constructed on the two legs of those right triangles equalled the area of the square constructed on the hypotenuses of the triangles. The investigation to follow is a dissection puzzle that demonstrates that this property is true for any right triangle.

Investigation 9.2.1  

On a full sheet of paper, perform the steps below. Note that the arcs and segment extensions necessary to complete steps 1 and 2 are not indicated in the diagram.

Step 1: Construct a scalene right triangle in the middle of your paper (hypotenuse down). Label it so that the hypotenuse is $\overline{AB}$ and the longer leg is $\overline{BC}$.

Step 2: Construct a square on each side of the triangle. Label the square on the longer leg *BCDE*. Label the square on the smaller leg *AGFC*. Label the square on the hypotenuse *ABIH*.

Step 3: Locate the center of *BCDE* (intersection of the two diagonals). Label the point *O*.

Step 4: Through point *O*, construct line *j* perpendicular to the hypotenuse.

Step 5: Through point *O*, construct line *k* perpendicular to line *j*. Line *k* is parallel to the hypotenuse. Lines *j* and *k* divide *BCDE* into four parts.

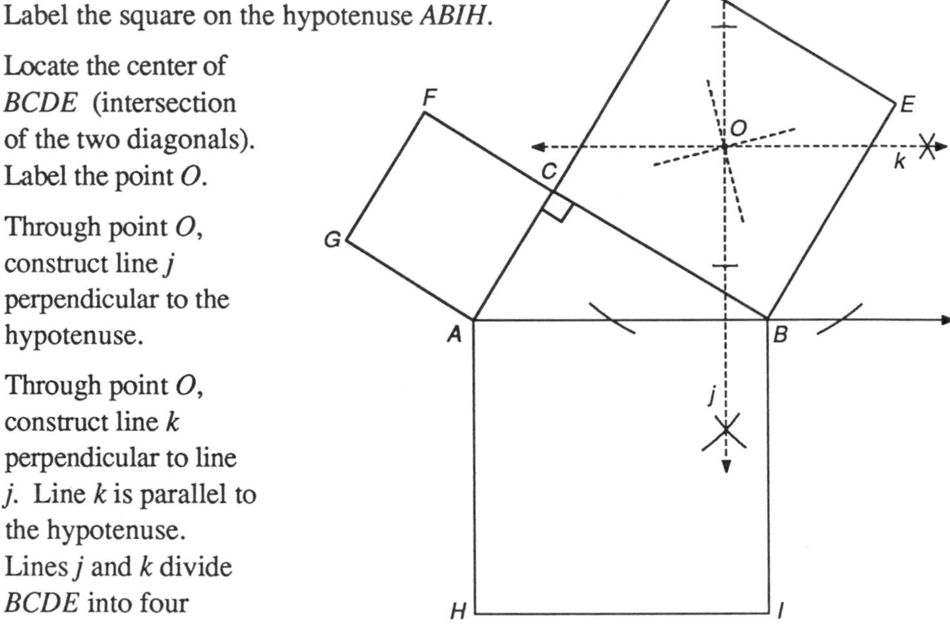

Step 6: Cut out the smaller square *AGFC* and the four parts of square *BCDE*. Arrange them to exactly cover the square *ABIH* on the hypotenuse.

If you are successful, then you have demonstrated that the area of the square on the hypotenuse is equal to the sum of the areas of the squares on the two legs of your triangle. Compare your results with the results of others near you. If the lengths of the two legs of a right triangle are a and b, then the areas of the squares on the legs would be a^2 and b^2. If the length of the hypotenuse is c, then the area of the square on the hypotenuse is c^2. State your observations as your next conjecture.

C-61 *In a right triangle, if a and b are the lengths of the legs and c is the length of the hypotenuse, then —?—.* ***(Pythagorean Theorem)***

Your last conjecture is known as the Pythagorean Theorem, named after Pythagoras, the Greek philosopher who demonstrated that it is true. Recall that a theorem is a statement that has been proved. While you have discovered the relationship between the lengths of the sides of a right triangle, you have not actually proved it. Rather than calling Conjecture 61 the Pythagorean Conjecture, however, we will call it the Pythagorean Theorem because it is well known by that name.

If the Pythagorean Theorem works for right triangles, perhaps it works for all triangles. A quick check demonstrates that it doesn't hold for other triangles.

Acute Triangle

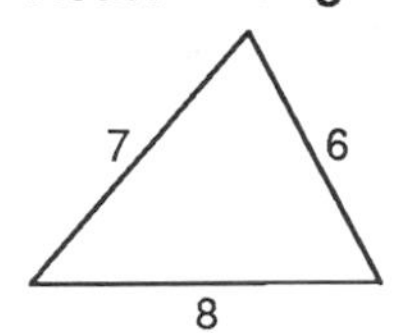

$6^2 + 7^2 > 8^2$

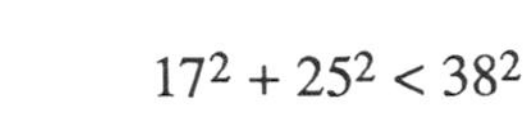

$17^2 + 25^2 < 38^2$

Obtuse Triangle

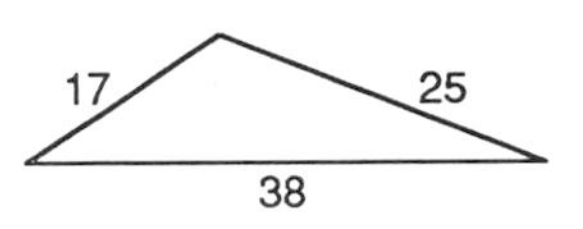

EXERCISE SET

Use the Pythagorean Theorem to find each missing length.

1.* $c = -?-$

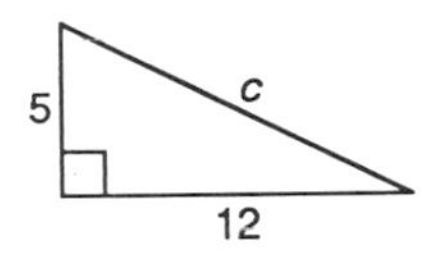

2. $c = -?-$

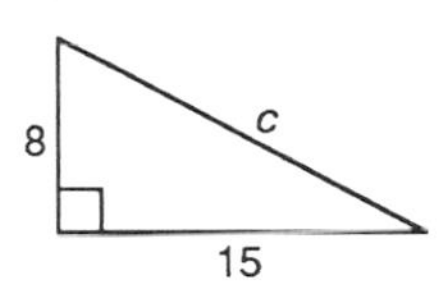

3. $c = -?-$

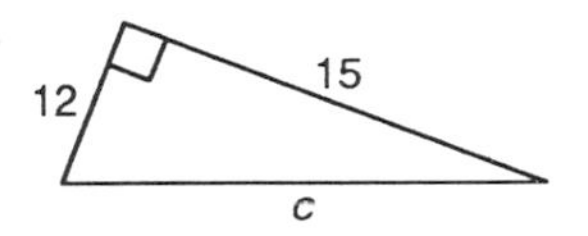

4. $c = -?-$

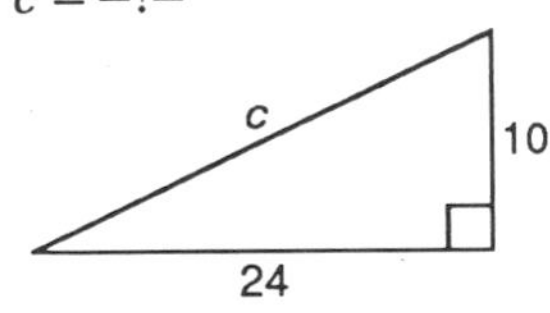

5. $d = -?-$

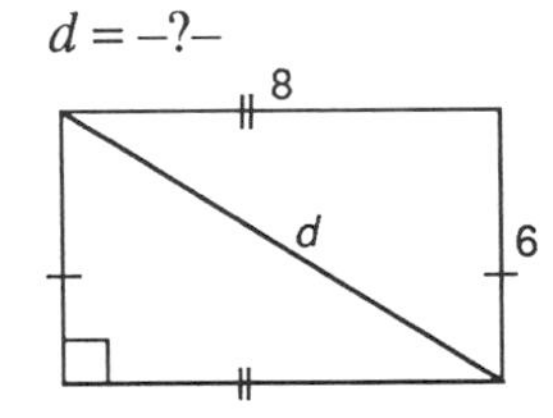

6.* $c = -?-$

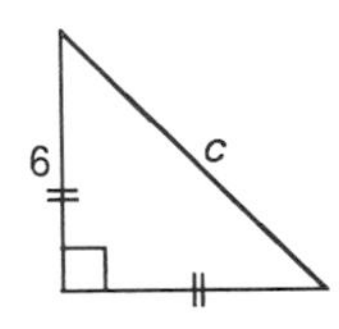

7. $r = -?-$

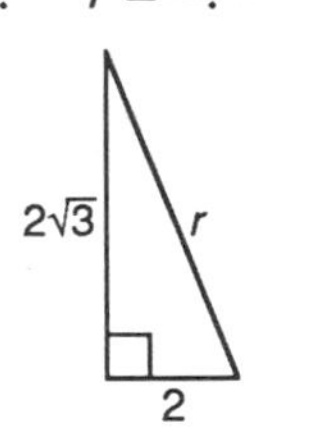

8.* $b = -?-$

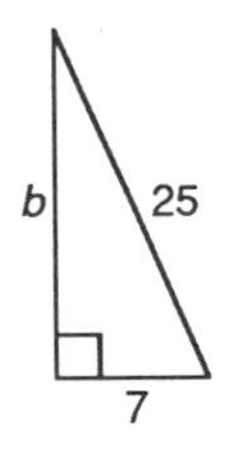

9. $a = -?-$

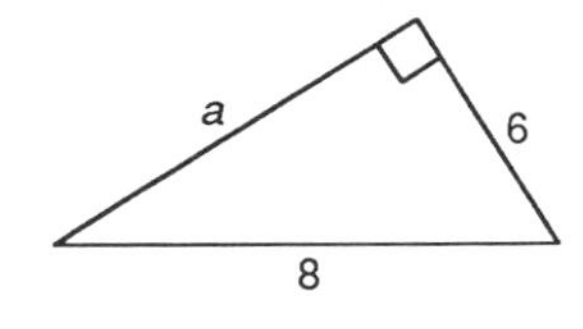

10. $h = -?-$

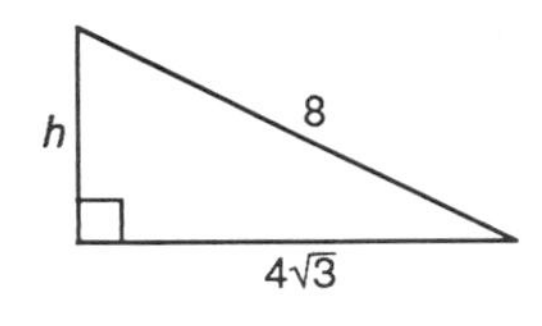

11.* $s = -?-$

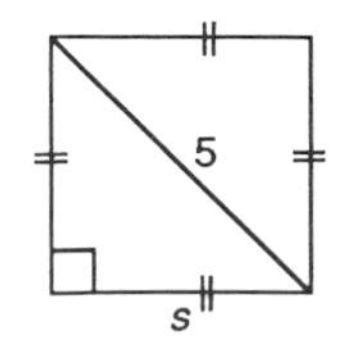

12.* $x = -?-$

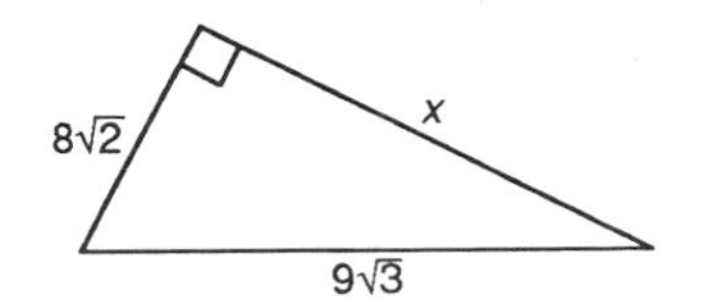

Lesson 9.3

Is the Converse True?

You have just seen in the previous lesson that if a triangle is a right triangle, then the area of the square on its hypotenuse is equal to the sum of the areas of the squares on the two legs. What about the other way around (the converse)? If x, y, and z are the lengths of the three sides of a triangle and they work in the Pythagorean formula, does it follow that the triangle is a right triangle? Let's find out.

Three positive integers that work in the Pythagorean formula are called **Pythagorean triples.** In this investigation you will construct triangles with sides that have lengths that are Pythagorean triples.

Investigation 9.3.1 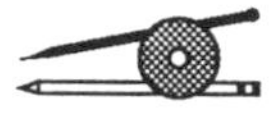 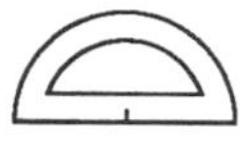

Step 1: Use a ruler and carefully draw segments of length 8 cm, 15 cm, and 17 cm. (8-15-17 is a Pythagorean triple since $8^2 + 15^2 = 17^2$.)

Step 2: With a compass, construct a triangle using the segments above.

Step 3: With a protractor, measure the largest angle.

Were you close to a right triangle? Try this again. Select a set of lengths from the list of Pythagorean triples below. Repeat the steps above with your new lengths.

Pythagorean triples: 3-4-5, 5-12-13, 6-8-10, 9-12-15, 12-16-20, 7-24-25

Did you get a right triangle again? Compare your results with the results of others. State your results as your next conjecture.

If the lengths of the three sides of a triangle work in the Pythagorean formula, then the triangle —?—.
(Converse of Pythagorean Theorem)

According to a popular story, a special case of the converse of the Pythagorean Theorem was used by Egyptian "rope stretchers." Ancient Egyptian tombs have pictures of scribes and their assistants carrying ropes with equally spaced knots on them. (See the illustration at the beginning of the lesson) Some people speculate that the ropes were used to create right triangles. Thirteen equally spaced knots would divide the rope into twelve equal lengths. If one scribe held the two ends of the rope together (knots 1 and 13), and two helpers held the rope at knots 4 and 8, stretching the rope tight, it would create a right triangle with the right angle at knot 4. It has been suggested that rope stretchers helped reestablish land boundaries after the yearly flooding of the Nile. It has also been suggested that rope stretchers helped in the construction of the pyramids.

Investigation 9.3.2

Try this yourself. With a rope or string twelve units in length (feet, inches, or meters), mark off four points, A, B, C, and D, creating lengths of 3 units, 4 units, and 5 units. Hold the two ends (points A and D) while a friend holds point B and another friend holds point C. Carefully stretch the rope tight. The rope should form a right triangle! What conjecture does this demonstrate?

EXERCISE SET A

Use the Converse of the Pythagorean Theorem to determine whether each triangle is a right triangle.

1.*

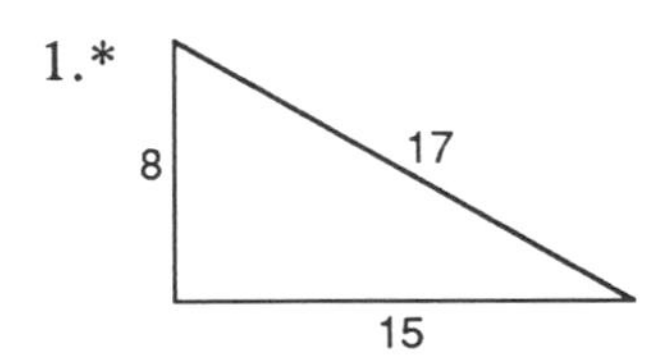

2.

13
5
12

3.

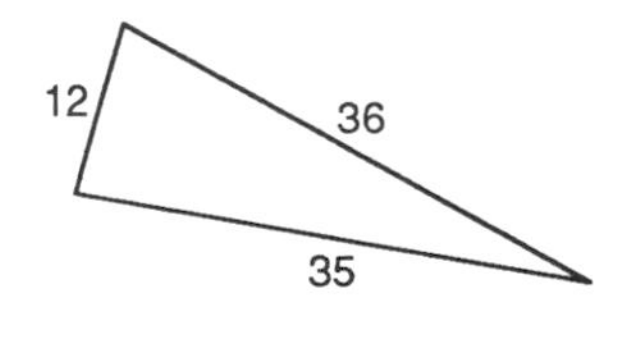

4.

22
12
18

5.

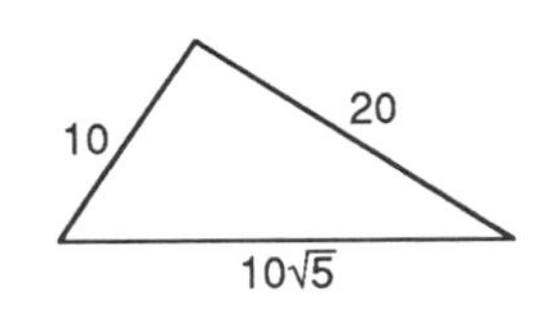

6.

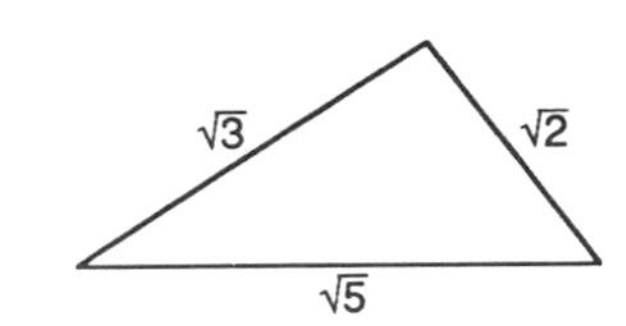

EXERCISE SET B

Use the Pythagorean Theorem or its converse to solve each problem.

1. Is a triangle with sides of 9 feet, 12 feet, and 18 feet a right triangle?

2.* Find the area of a right triangle with a hypotenuse of 13 inches and one leg of 5 inches.

3.* A parallelogram has sides of 9 cm and 12 cm and a 15 cm diagonal. Is the parallelogram a rectangle? If not, is the diagonal of 15 cm the longer or shorter diagonal?

4. Find the perimeter of a rectangle with a diagonal of 39 cm and a side of 36 cm.

5. A triangular plot of land has boundary lines 45 meters, 60 meters, and 70 meters long. The 60 meter boundary line runs north-south. Is there a boundary line for the property that runs due east-west?

6.* At Martian high noon, Dr. Rhonda Bend leaves the Martian U.S. Research Station traveling due east at 60 km/hr. One hour later Professor I.M. Bryte takes off from the station heading north straight for the polar ice cap at 50 km/hr. How far apart will the doctor and the professor be at 3 P.M. Martian time? Express your answer to the nearest kilometer.

7. Dr. Rhonda Bend is exploring the Martian landscape. She is standing at point C, 288 meters from the base of a vertical cliff (point B). To find the height of the cliff, she focuses a sonic beam at a rock on the top of the cliff (point A) as shown in the diagram on the right. The beam bounces off the rock and returns. She records the time it takes for the sonic beam to return and calculates the distance from A to C to be 480 meters. What is the height of the cliff to the nearest meter?

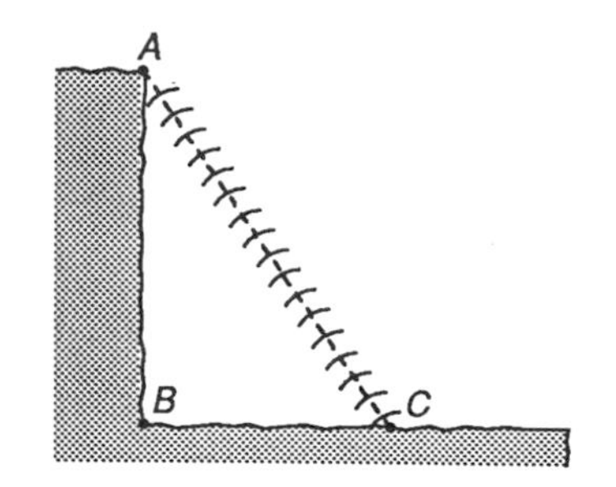

Improving Visual Thinking Skills

Dissecting a Hexagon III

Trace the hexagon on the right six times. Then divide each hexagon into twelve identical parts. Do it differently each time.

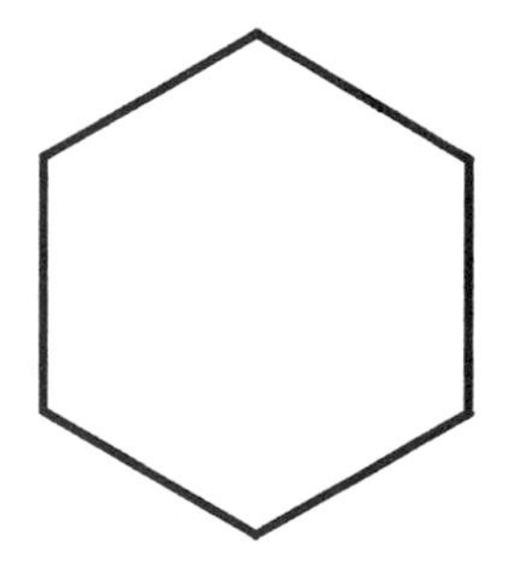

Lesson 9.4

Word Problems

You must do things you think you cannot do.

– Eleanor Roosevelt

EXERCISE SET

1. What is the length of the diagonal of a square whose sides measure 8 cm.
2. The lengths of the three sides of a right triangle are consecutive integers. Find them.
3. The lengths of the three sides of a right triangle are consecutive even integers. Find them.
4. Find the area of a right triangle with hypotenuse of 17 cm and one leg of 15 cm.
5.* The diagonal of a square measures 32 meters. What is the area of the square?
6. A rectangular closet has a perimeter of 20 feet and a width of 3 feet. Find the diagonal.
7. The legs of an isosceles triangle are 6 inches and the base is 8 inches. Find the area.
8. A rectangular garden 6 meters wide has a diagonal of 10 meters. Find the perimeter of the garden.
9. How high up on a building will a 15 foot ladder reach if the foot of the ladder is placed five feet from the building?
10. A baseball infield is a square, each side measuring 90 feet. What is the distance from home plate to second base?
11. To find the distance between two points A and B on the opposite ends of a lake, a surveyor sets a stake at point C so that angle ABC is a right angle. By measuring, she finds AC to be 160 meters and BC to be 128 meters. How far across the lake is it from point A to point B?

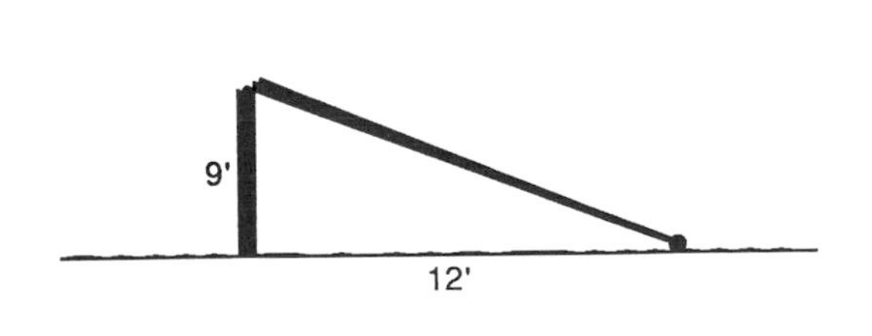

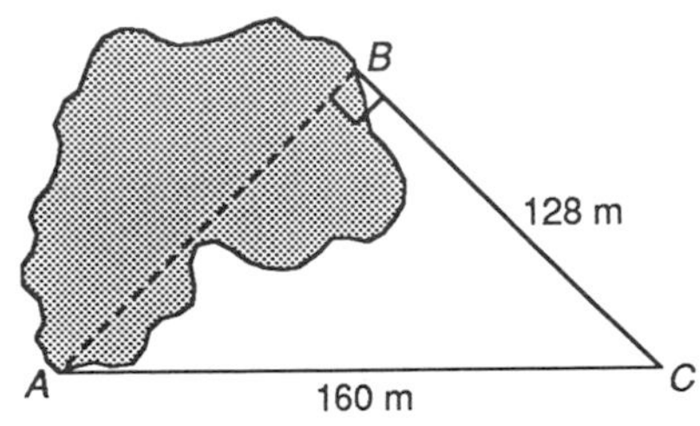

12.* A flagpole has cracked 9 feet from the ground and fallen as if hinged. The top of the flagpole hit the ground 12 feet from the base. How tall was the flagpole before it fell?

13.* A woman travels one mile due north, then two miles due east, then three miles due north again, and then once more east for 4 miles. How far is she from her starting point? (It is less than $5 + \sqrt{5}$ miles.)

David and Goliath

14.* Once upon a time there lived a boy named David Dogood, but most people called him Dave for short. Dave was the city frisbee champ.

On the other side of town lived Goliath. Most people called him "Goliath Sir." Goliath was the leader of a club called the "Thugs." Goliath worked at a circus. He was billed as the meanest man on stilts.

One day there came a showdown. Dave stood one meter to his shoulders. Dave had a strong shot with his frisbee, but it was only good within a range of 26 meters. On stilts Goliath's nose was 25 meters above the ground. His nose was his only weakness (aside from chocolate cake). Goliath's nose bled very easily, causing him to faint. Dave had to get close enough for his frisbee shot to hit the nose of Goliath. How close did he have to get?

Lesson 9.5

Multiples of Right Triangles

What happens if you double the lengths of the sides of a right triangle? Will the new lengths form a right triangle? What if all three sides are tripled? Let's find out.

Investigation 9.5.1

Step 1: Select one of the common right triangles shown below. Double the length of each side of your chosen triangle.

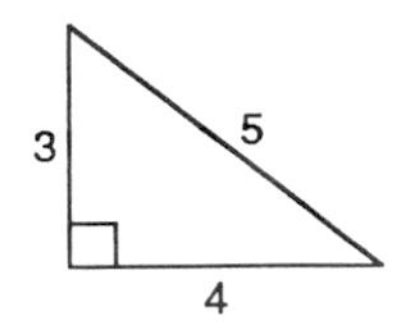

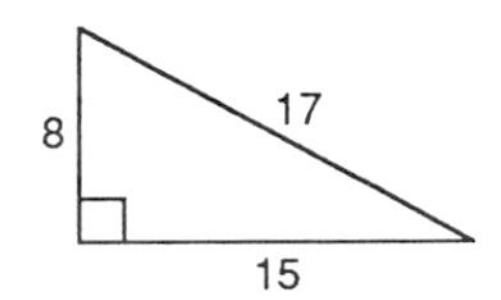

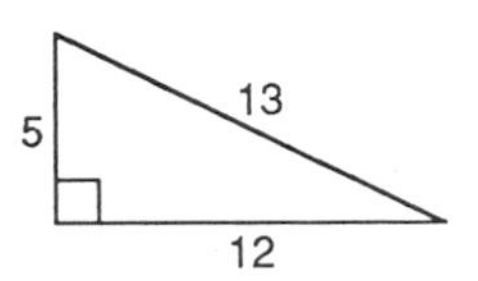

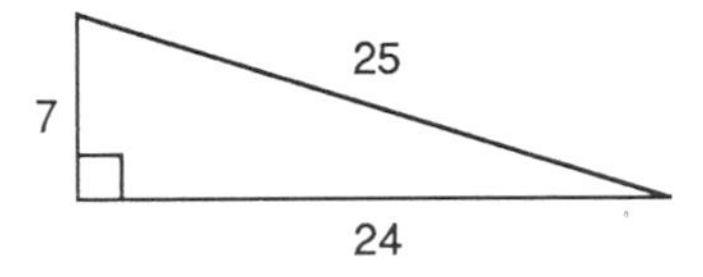

Step 2: Substitute these new lengths into the Pythagorean formula. If they work in the equation, then the new lengths will form a right triangle.

Step 3: Triple the length of each side of another triangle from the group above.

Step 4: Substitute these new lengths into the Pythagorean formula. Will they form a right triangle?

Step 5: Now multiply the length of each side of another triangle by another number. It doesn't have to be a whole number.

Step 6: Substitute these new lengths into the Pythagorean formula. Will they form a right triangle?

You should be ready to state a conjecture about multiples of right triangles.

If you multiply the lengths of all three sides of any right triangle by the same number, the resulting triangle will be a —?—.

In other words, *If* $a^2 + b^2 = c^2$, *then* $(an)^2 + (bn)^2 = (cn)^2$. This is a very useful conjecture. If you are familiar with the Pythagorean triple 3-4-5, then you will also be able to solve problems that use multiples of this triple such as 6-8-10 or 9-12-15.

The lengths of the sides of the four right triangles shown at the top of the page are examples of Pythagorean triples called **primitives**. The three numbers have no common integer factors. From these primitives, you can create new right triangles by multiplying the primitives by the same number. These new Pythagorean triples are called **multiples**.

In the investigation, you discovered that if you multiply the lengths of all three sides of a right triangle by the same number, the new lengths will form a right triangle.

Does this work if you add or subtract the same number to each side of a right triangle? A quick check shows that adding the same number to each of the numbers in a Pythagorean triple does not create a new Pythagorean triple. Adding 2 to each number in 3-4-5 gives 5-6-7, but $5^2 + 6^2 > 7^2$. Likewise, subtracting 1 from each number in 3-4-5 gives 2-3-4, but $2^2 + 3^2 < 4^2$.

Conjecture 63 raises a question worth investigating. If two sides of a right triangle have a common factor, must the third side have the same factor? Let's find out.

Investigation 9.5.2

Step 1: Select two integers that have a common factor (for example, 6 and 8). They can represent the lengths of two sides of a right triangle.

Step 2: Sketch two different right triangles. Label the two legs of one of the triangles with the two numbers you selected in step 1. Label the hypotenuse r. On the other triangle, label the hypotenuse with the larger number and one leg with the other number. Label the third side s.

Step 3: Use the Pythagorean Theorem to solve for r and s. (With 6 and 8 as the lengths of the two legs, $r = 10$. With 6 the length of a leg and 8 the length of the hypotenuse, $s = 2\sqrt{7}$. 6 and 8 have 2 as a common factor and both 10 and $2\sqrt{7}$ have 2 as a factor.) Do your r and s have the same factor as the two integers you chose?

Repeat the investigation with a second pair of integers that have a common factor. Do your r and s have the same factor as your second pair of integers? Compare your results with the results of others near you. You should be ready to state a conjecture.

C-64 *If the lengths of two sides of a right triangle have a common factor, then —?—.*

This conjecture can be quickly verified with algebra. There are two cases. The two legs could have a common factor (case 1), or the hypotenuse and a leg could have a common factor (case 2). The proof of the first case is shown below. The second case is left as an exercise.

Paragraph Proof of Case 1:

The legs have a common factor. Let ar and br be the lengths of the legs of a right triangle (two sides having a common factor r) and let x represent the length of the third side (in this case the hypotenuse). Because the triangle is right, the three values work in the Pythagorean formula.

$$\begin{aligned} (ar)^2 + (br)^2 &= x^2 \\ a^2r^2 + b^2r^2 &= x^2 \\ r^2(a^2 + b^2) &= x^2 \\ r\sqrt{a^2 + b^2} &= x \end{aligned}$$

Therefore, x (the length of the hypotenuse) also has r as a factor.

This is a very useful conjecture. If you recognize that the lengths of two sides of a right triangle have a common factor, you then know that the third side must also have that same factor. Let's look at an example of how you can use this conjecture to make solving a problem easier.

Example

Find the length of one leg of a right triangle with a hypotenuse of 35 cm and a leg of 28 cm.

Step 1

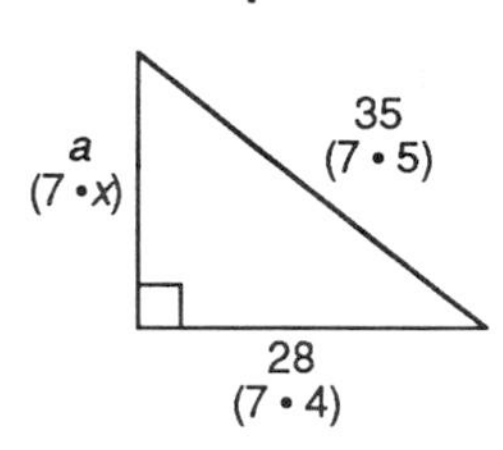

Since the lengths of two of the sides are multiples of 7, the length of the third side must be a multiple of 7.

Step 2

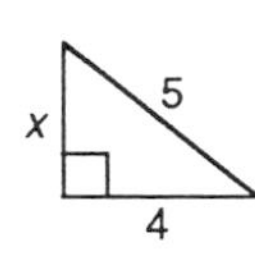

When the three sides are written without the common factor 7, the other factors (x, 4, and 5) are revealed as part of the familiar 3-4-5 triple.

Step 3

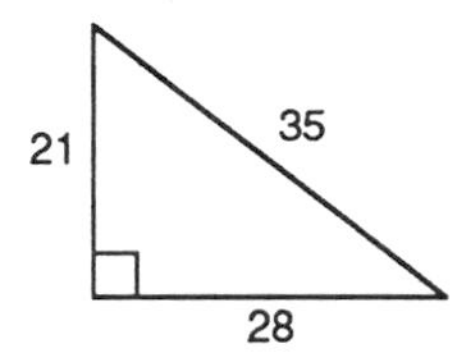

If the triangle is a multiple of a 3-4-5 right triangle, then $x = 3$. Therefore the length of the third side in the original triangle is (3)(7) or 21.

The other leg is 21 cm long.

EXERCISE SET A

Each exercise (except one of them) involves one of the four most common Pythagorean primitives. Recognize them and you can save yourself a lot of work!

1. $a = $ –?–

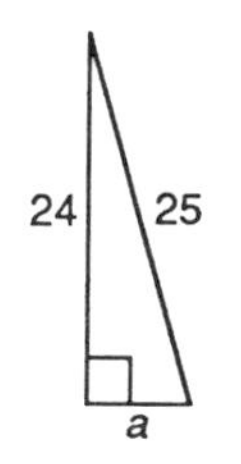

2. $b = $ –?–

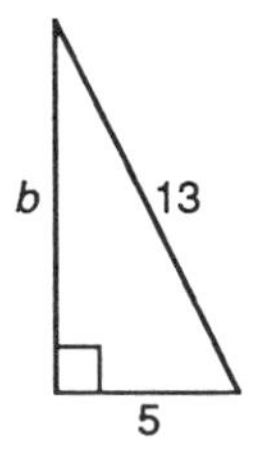

3. What is the perimeter?

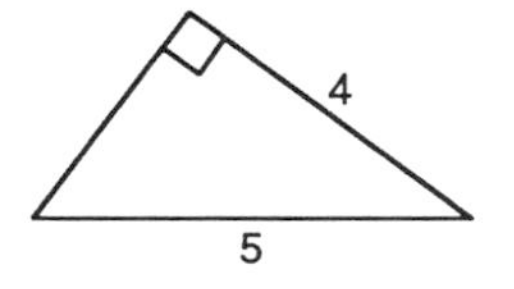

4. $c = $ –?–

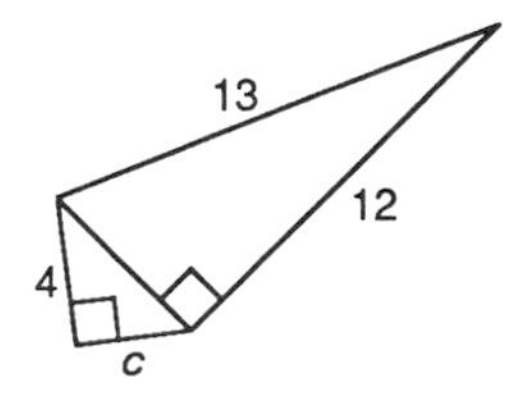

5.* The area of the rectangle is 168 sq. ft. $d = $ –?–

6. What is the area of the shaded rectangle?

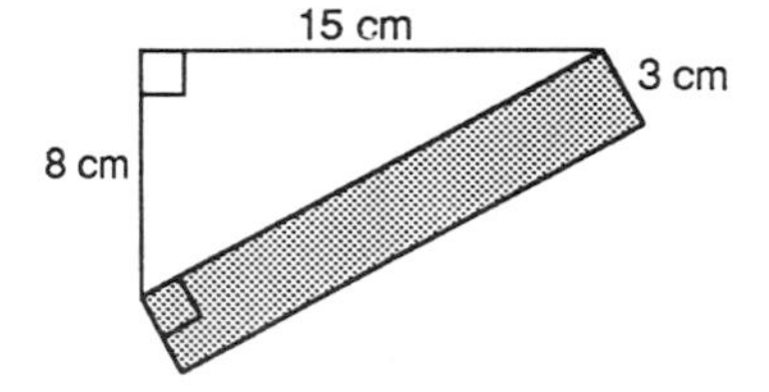

7. What is the shaded area?

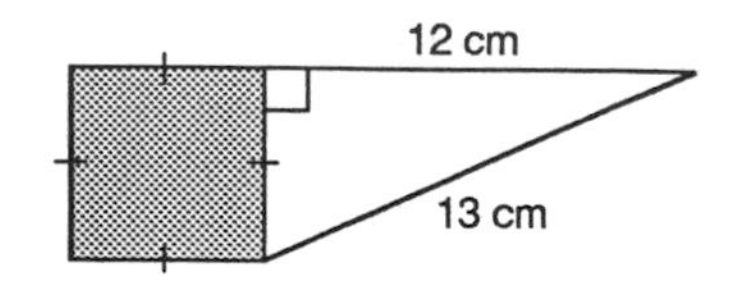

8.* The arc is a semicircle. What is the shaded area?

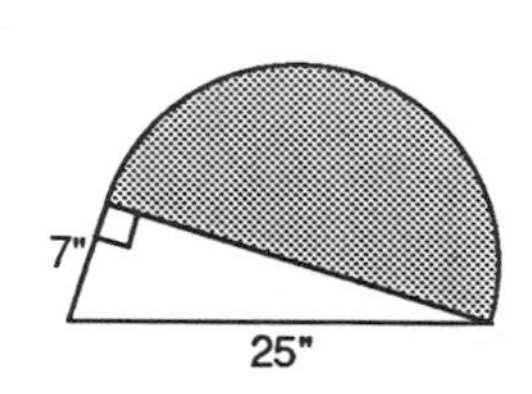

9. $m =$ –?–

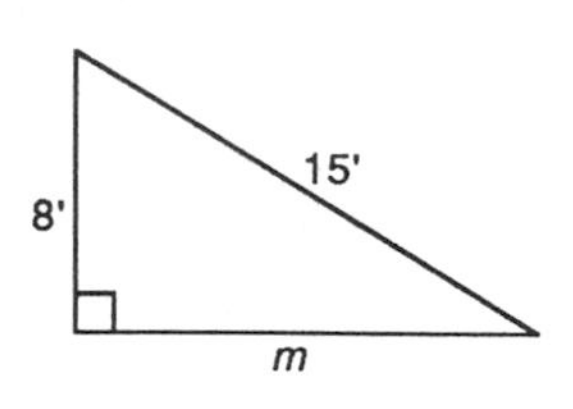

EXERCISE SET B

Copy the table below. Then complete it by creating multiples of the most common Pythagorean primitives.

Pythagorean Triples

Primitives	Doubles	Triples	4 times	10 times
3-4-5	**6-8-10**	1.* —?—	2. —?—	3. —?—
5-12-13	4. —?—	**15-36-39**	5. —?—	6. —?—
8-15-17	7. —?—	8. —?—	**32-60-68**	9. —?—
7-24-25	10. —?—	11. —?—	12. —?—	**70-240-250**

EXERCISE SET C

Each right triangle below has sides with lengths that are multiples of a Pythagorean primitive. All measurements are in centimeters. Check your completed table of multiples of Pythagorean triples (Exercise Set B) or use your new conjectures to solve for the indicated value.

1.* $a =$ –?–

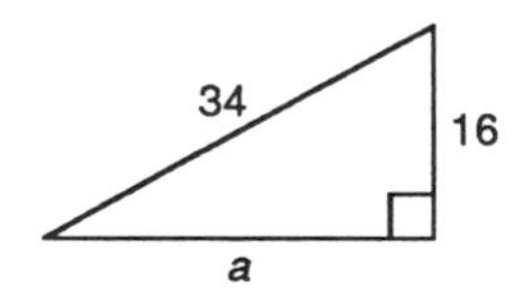

2. $b =$ –?–

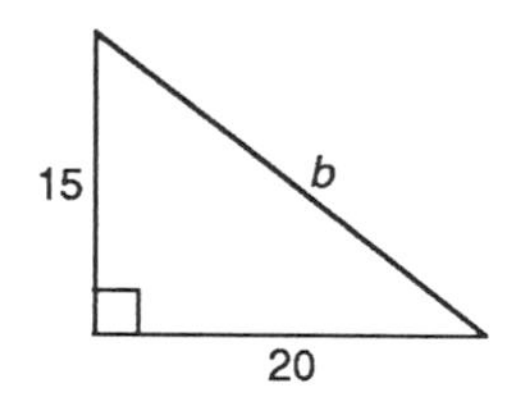

3. $c =$ –?–

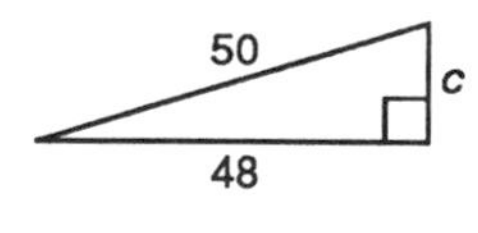

4. $d =$ –?–

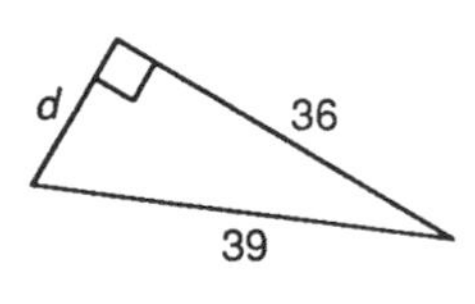

5.* $e =$ –?–

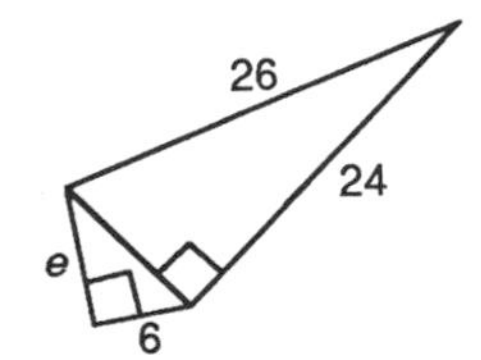

6. $f =$ –?–

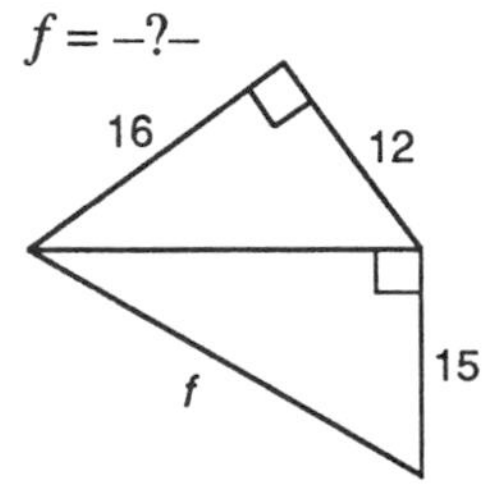

7. Use algebra to complete the proof of Conjecture 64, case 2. Use the proof of case 1 as a model.

Lesson 9.6

Return of the Word Problems

You may be disappointed if you fail,
but you are doomed if you don't try.
– Beverly Sills

EXERCISE SET

Solve each word problem. Watch for special right triangles and multiples of familiar Pythagorean triples.

1. Find the area of a right triangle with a leg of 6 feet and a hypotenuse of $3\sqrt{13}$ feet.

2. Find the length of the hypotenuse of an isosceles right triangle with legs of 3.5 meters. Express your answer accurate to the nearest tenth of a meter.

3.* The area of an isosceles right triangle is 98 square inches. What is the length accurate to the nearest inch of the hypotenuse?

4.* Meteorologist Paul Windward and his fiancee, geologist Raina Stone, are rushing to Lost Wages, Nevada, to wed at the Lost Wages Wedding Emporium. Paul lifts off in his balloon at noon from Pecos Gulch heading due east for Lost Wages. With the prevailing wind blowing from west to east, he averages a land speed of 30 km/hr. This allows him to arrive in Lost Wages in 4 hours. Meanwhile Raina is 160 km to the north of Pecos Gulch. At the moment of Paul's lift off, Raina hops into her Jeep and heads directly for Lost Wages. At what average speed must she travel to arrive at Lost Wages at the same time as Paul?

5.* A giant California redwood tree 36 meters tall cracked in a violent earthquake and fell as if hinged. The tip of the once beautiful tree hit the ground 24 meters from the base. Researchers want to investigate the crack. How many meters up from the base of the tree do the researchers have to climb? See the diagram below.

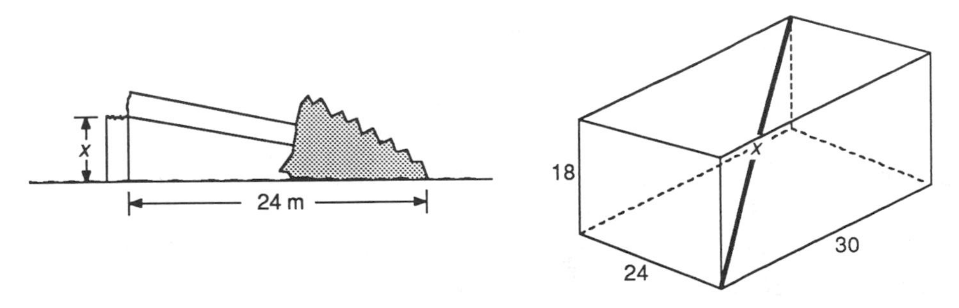

6.* What is the longest stick that can be placed inside a box with inside dimensions of 24 inches, 30 inches, and 18 inches? See the diagram above.

7.* A 25-foot ladder is placed against a building. The bottom of the ladder is 7 feet from the building. If the top of the ladder slips down 4 feet, how many feet will the bottom slide out? No, it is not 4 feet. This is a two-step problem, so draw two right triangles.

8.* The front and back walls of an A-frame cabin are shaped like isosceles triangles, each with a base of 10 meters. The equal sides of each isosceles triangle are 13 meters. The entire front of the cabin is made of double-pane insulated glass that is 1 cm thick. What is the area of one isosceles triangle? If the glass was purchased for $120 per square meter, what did the glass for the front of the cabin cost?

Romeo and Julie

9. It all began one afternoon when Romeo's mother would not let him out of the house. This caused his girlfriend Julie to wonder, "Where is he now?" So Julie decided to visit Romeo and talk to him from beneath his balcony. As Julie was calling out, "Romeo, oh Romeo, where the heck are you, Romeo?" she slipped and fell into the moat which was directly beneath the balcony. Then and there she decided that the two of them should elope! After drying herself off, she went to the local hardware store and purchased a rope so that Romeo could slide down it and escape. If Romeo's balcony is 6 meters up from the moat, what is the shortest length of rope needed to reach from Romeo's balcony to the ground on the opposite side of the 4.5 meter wide moat?

Lesson 9.7

Two Special Right Triangles

In this lesson you will use the Pythagorean Theorem to discover some relationships between the sides of two special right triangles. Problems involving these right triangles are often found on college entrance exams and achievement tests and are used often in trigonometry.

One special right triangle is an isosceles right triangle. It is also referred to as a 45-45 right triangle because those are the measures of its acute angles. If you bring the opposite corners of a square piece of paper together and fold, you get an isosceles right triangle. It is half of a square.

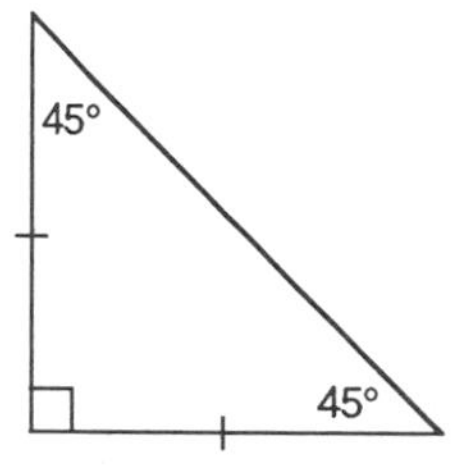

Isosceles Right Triangle

Investigation 9.7.1

Find the length of the hypotenuse of each isosceles right triangle.

1.* $a = $ –?–

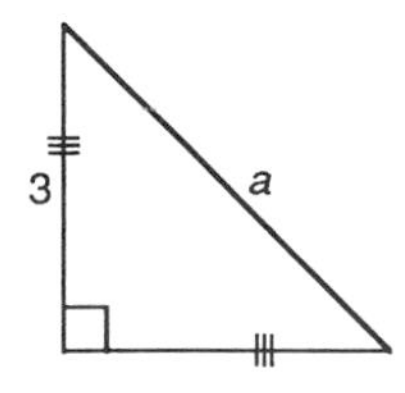

2. $b = $ –?–

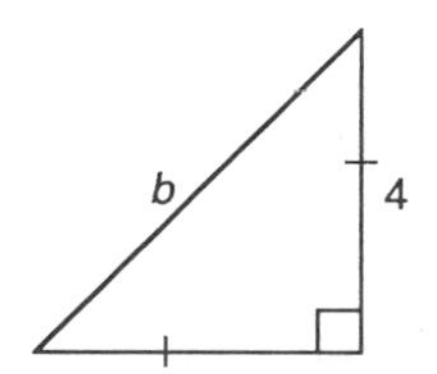

3. $c = $ –?–

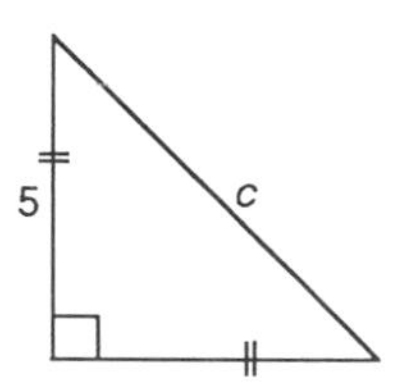

4. $d = $ –?–

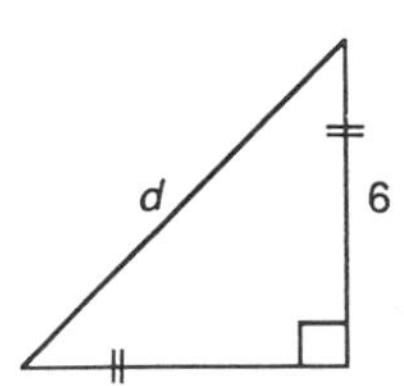

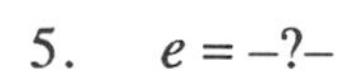

5. $e = $ –?–

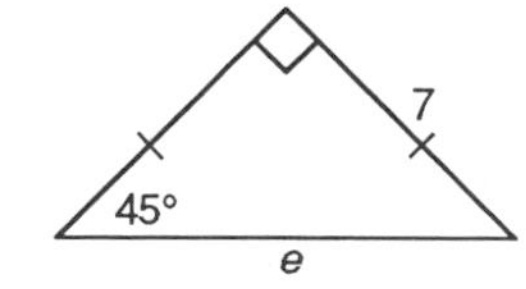

6. $f = $ –?–

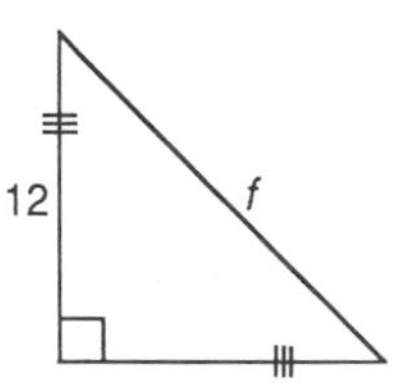

Did you notice something interesting about the relationship between the length of the hypotenuse and the length of the legs in each problem in the investigation? State your observations as your next conjecture.

In an isosceles right triangle, if the legs have length x, then the hypotenuse has length —?—.
(Isosceles Right Triangle Conjecture)

You can use algebra to verify the Isosceles Right Triangle Conjecture.

$$c^2 = x^2 + x^2$$
$$c^2 = 2x^2$$
$$c = x\sqrt{2}$$

The second special right triangle is the 30-60 right triangle. Imagine folding an equilateral triangle along one of its axes of symmetry. The right triangle that you get is a 30-60 right triangle. A 30-60 right triangle is half of an equilateral triangle.

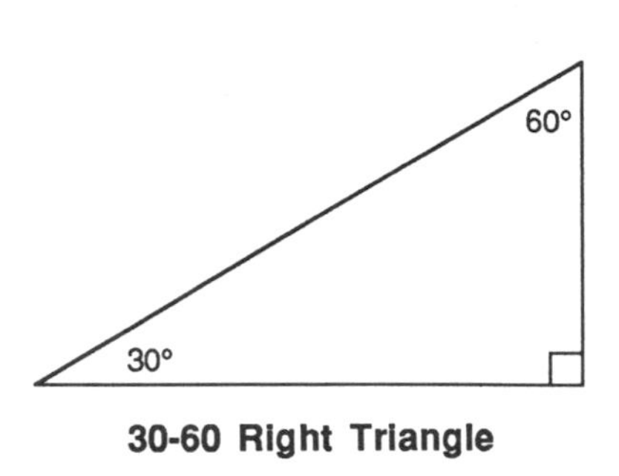

30-60 Right Triangle

Investigation 9.7.2

Let's start by using a little deductive thinking to reveal a useful relationship in 30-60 right triangles. Triangle *ABC* is equilateral. $\overline{CD}$ is an altitude.

1. What are $m\angle A$ and $m\angle B$?
2. What are $m\angle ACD$ and $m\angle BCD$?
3. What are $m\angle ADC$ and $m\angle BDC$?
4. Is $\Delta ADC \cong \Delta BDC$? Why?
5. Is $\overline{AD} \cong \overline{BD}$? Why?

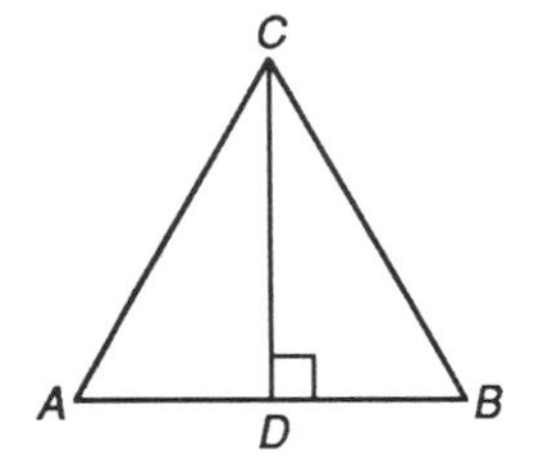

Notice that altitude $\overline{CD}$ divides the equilateral triangle into two right triangles with acute angles of 30 and 60 degrees. Look at just one of the 30-60 right triangles, say ΔADC. How do *AC* and *AD* compare? State your findings as a conjecture.

Conjecture: *In a 30-60 right triangle, if the side opposite the 30 degree angle has length x, then the hypotenuse has length —?—.*

Investigation 9.7.3

Let's see what else you can discover about 30-60 right triangles. Find the length of the indicated side in each 30-60 right triangle. All measurements are in centimeters.

1. $a = $–?–
2. $b = $–?–
3. $c = $–?–

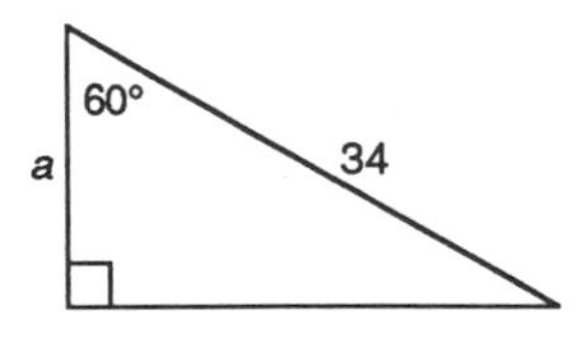

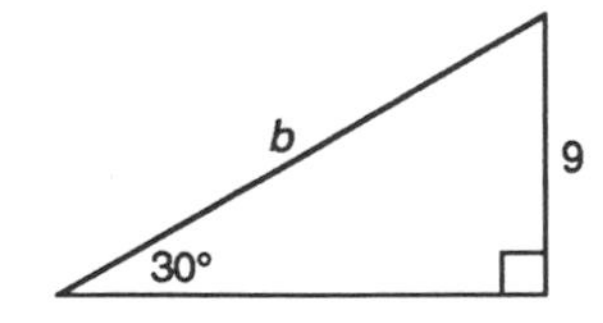

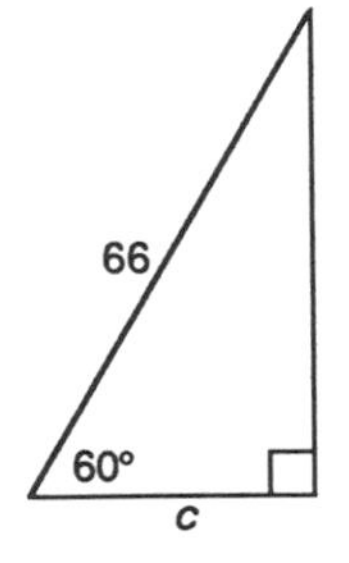

4. $d = -?-$

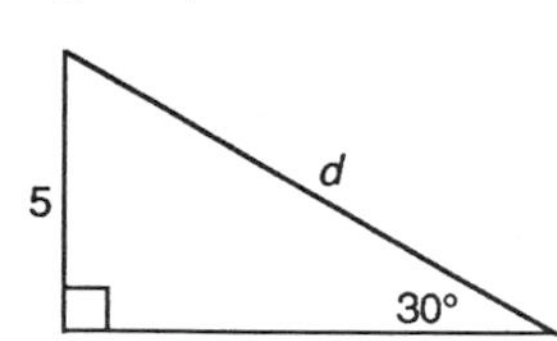

5. $e = -?-$

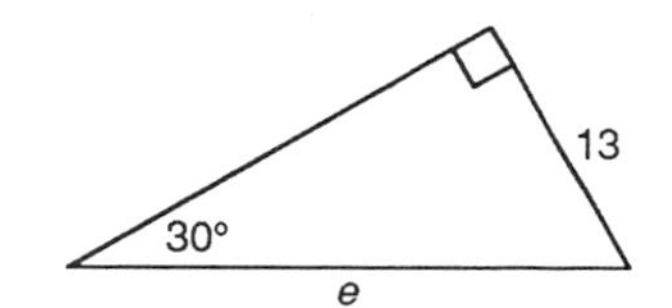

6. $f = -?-$

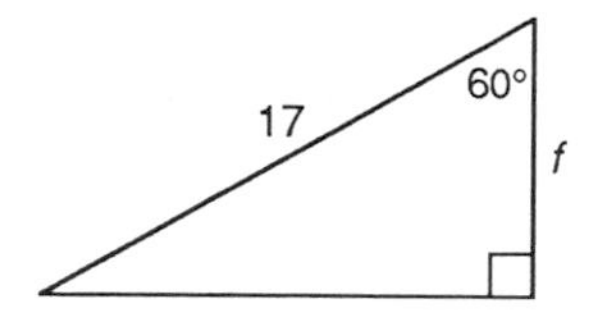

Now use the conjecture made in the previous investigation and the Pythagorean Theorem to find the length of each indicated side.

7.* $j = -?-$

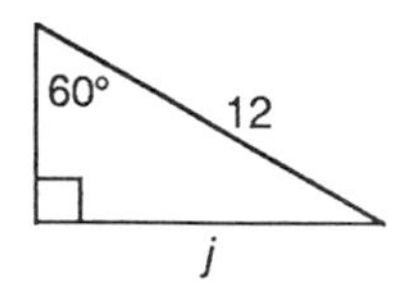

8. $k = -?-$

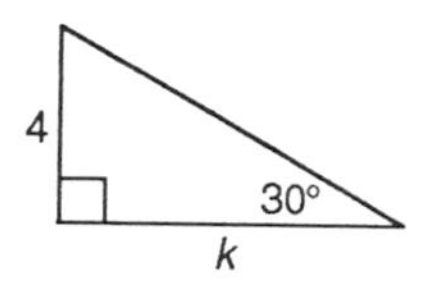

9. $m = -?-$

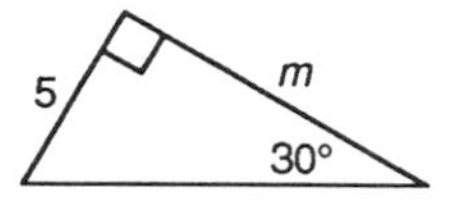

10. $n = -?-$

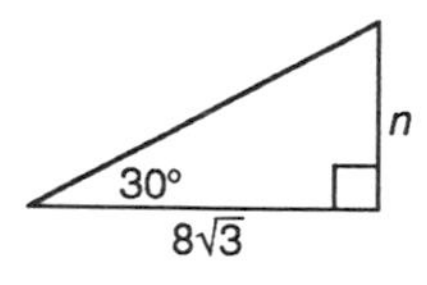

11. $p = -?-$

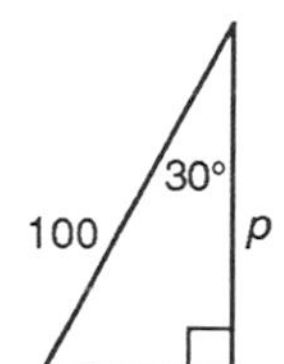

12. $s = -?-$

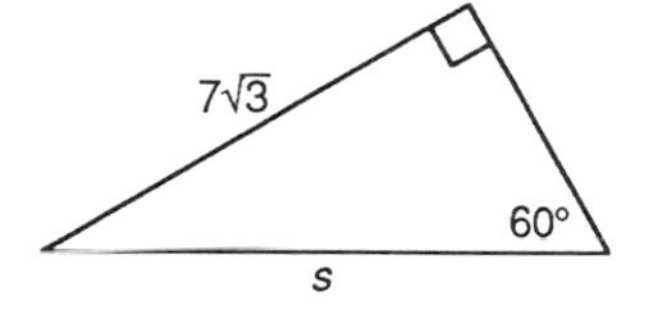

You should have noticed a pattern in your answers. Combine your observations with the previous conjecture and state your next conjecture.

C-66 *In a 30-60 right triangle, if the shorter leg has length x, then the longer leg has length —?— and the hypotenuse has length —?—.* ***(30-60 Right Triangle Conjecture)***

So far you are familiar with the right triangles shown below.

Isosceles Right

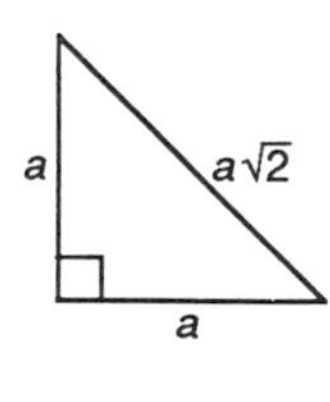

Pythagorean Primitives

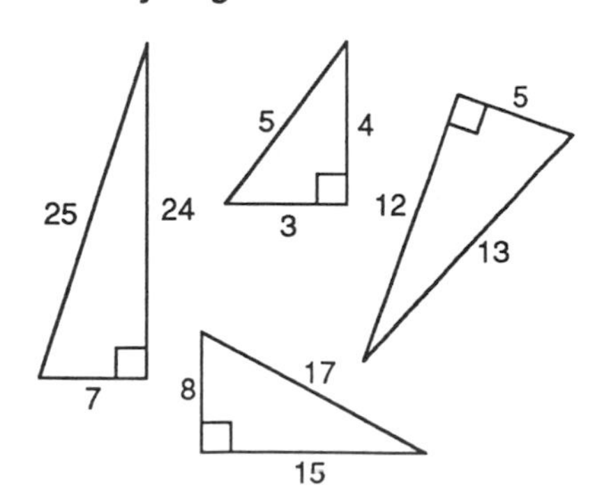

30-60 Right

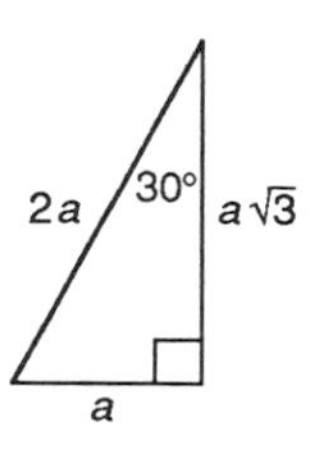

EXERCISE SET

Solve each exercise using your new conjectures. In many of the exercises, you don't need to resort to the Pythagorean formula. All measurements are in centimeters.

1. $a =$ –?–

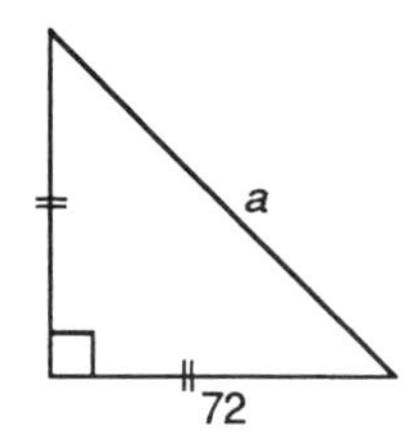

2.* $b =$ –?–

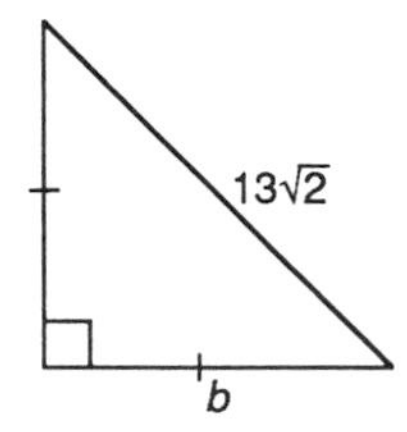

3. What is the perimeter of square *SQRE*?

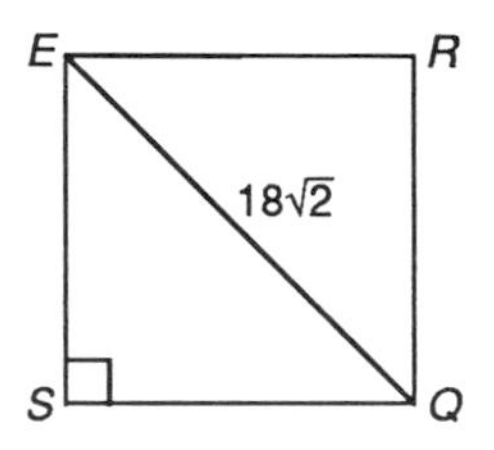

4.* What is the area of the triangle?

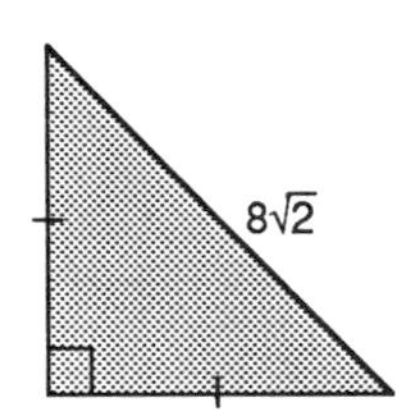

5. $c =$ –?–

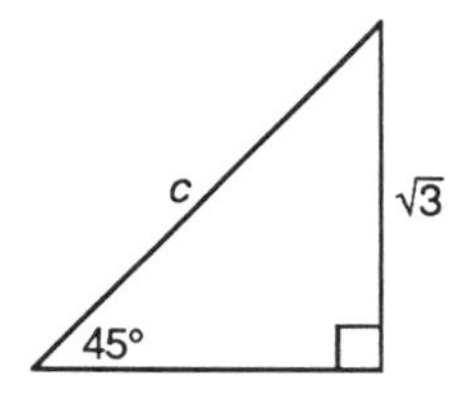

6. $d =$ –?–

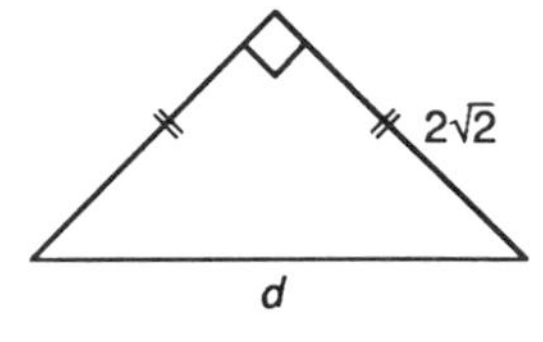

7.* $a =$ –?– $b =$ –?–

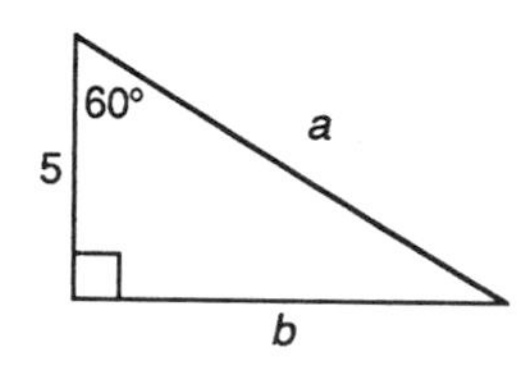

8. $c =$ –?– $d =$ –?–

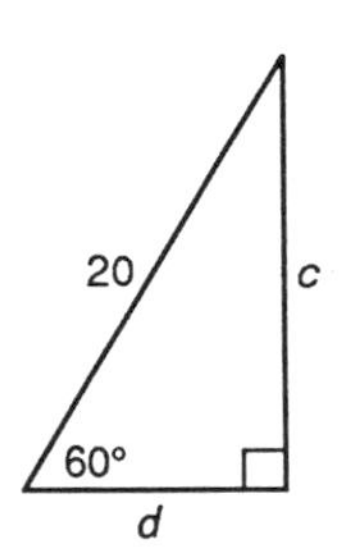

9. $e =$ –?– $f =$ –?–

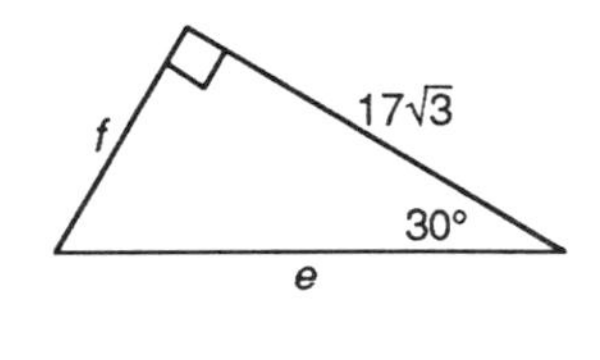

10. $g =$ –?– $h =$ –?–

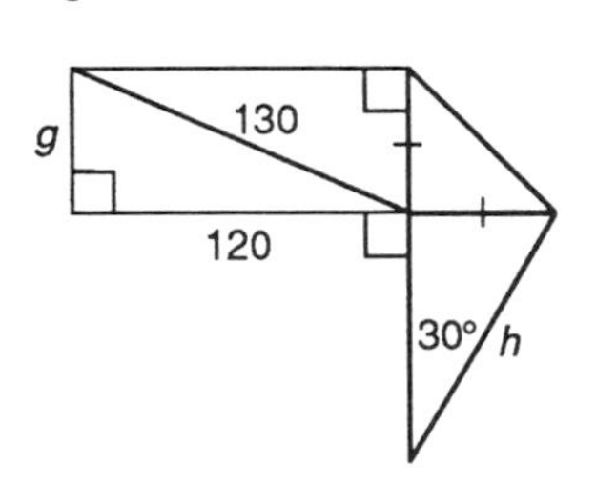

11.* $k =$ –?– $m =$ –?–

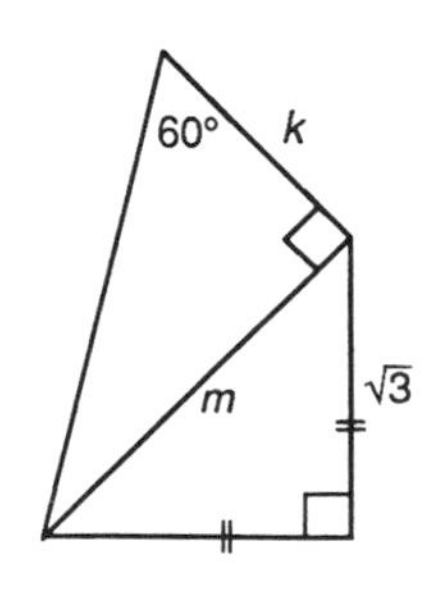

12. $n =$ –?– $p =$ –?–

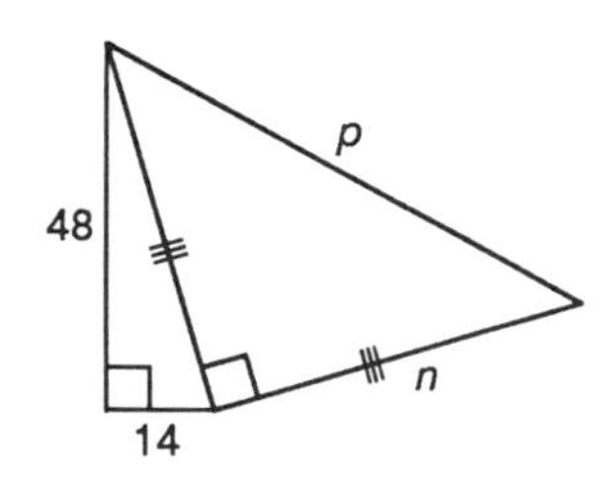

13. $r = -?-$ $s = -?-$

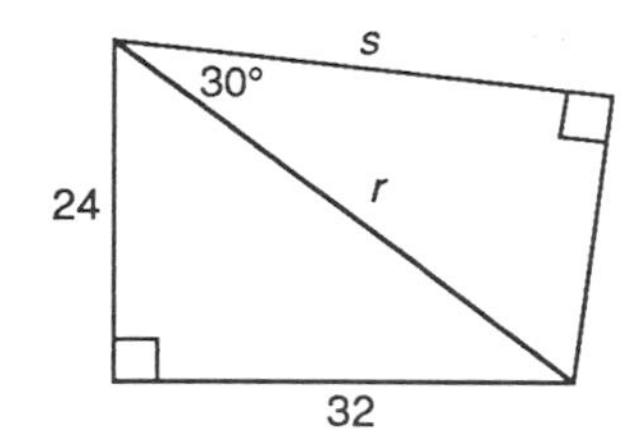

14. $t = -?-$ $v = -?-$

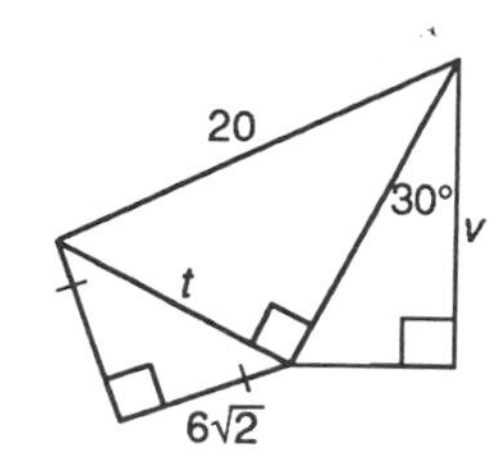

15.* $w = -?-$ $y = -?-$

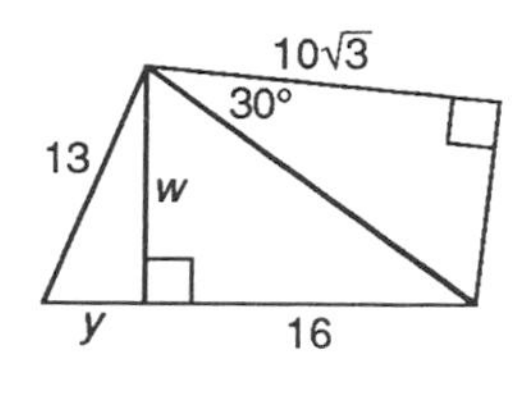

16. An oil tanker is entering the Bay of Niteall near the seaside towns of Lazyville and Sleepyton. The oil tanker is due north of Sleepyton and due east of Lazyville. In order to determine the distance to shore, the communications officer, Elmer Mudd, sends out a blast on the fog horn. The horn automatically signals a light to flash back to the ship from the light towers in each town as soon as the sound strikes the sound sensors in the towers. It took 6 seconds for the sound to reach Lazyville and 8 seconds to reach Sleepyton. Elmer was able to calculate not only the distance from the ship to each town but also the distance between the two towns. Can you? What is the distance in feet between the two towns if sound travels at 1088 feet per second at sea level?

17.* A = {3, 13, 15}; B = {4, 8, 12}; C = { 5, 9, 17}

A, B, and C represent sets that contain lengths of segments. If one length is randomly selected from set A, one length from set B, and one length from set C, what is the probability that segments with the selected lengths can form a right triangle?

Improving Visual Thinking Skills

Coin Swap III

Find four light coins (dimes) and four dark coins (pennies) and arrange them on a grid of nine squares as shown. Your task is to switch the position of the four light and four dark coins in exactly 24 moves. A coin can slide into an empty square next to it or it can jump over one coin into an empty space. Record your solution by listing in order which color coin is moved. For example, your list might begin DLDLDDLLDDD. . . .

Lesson 9.8

The Equilateral Triangle

An equilateral triangle is an isosceles triangle three times over. (Each angle is the vertex angle of an isosceles triangle.) Therefore each of the three angle bisectors is also a median and an altitude. For example, in equilateral ΔABC below, $\overline{AE}$, $\overline{BF}$, and $\overline{CD}$ are each an angle bisector, median, and altitude. These three segments divide the equilateral triangle into six overlapping 30-60 right triangles and six smaller non-overlapping 30-60 right triangles. One of the six overlapping 30-60 right triangles is ΔCDB. You should be able to find the other five. One of the six non-overlapping 30-60 right triangles is ΔADM. You should be able to find the other five.

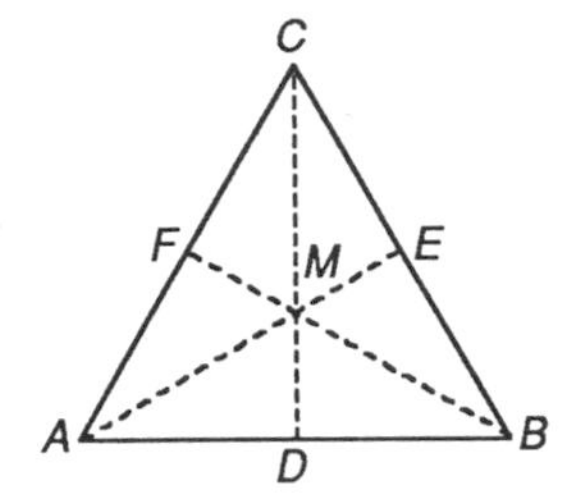

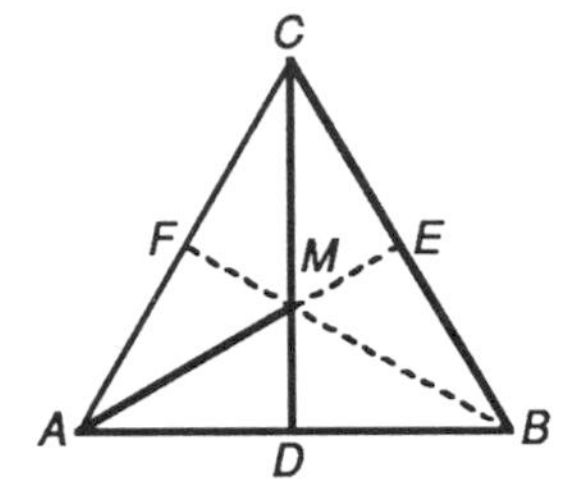

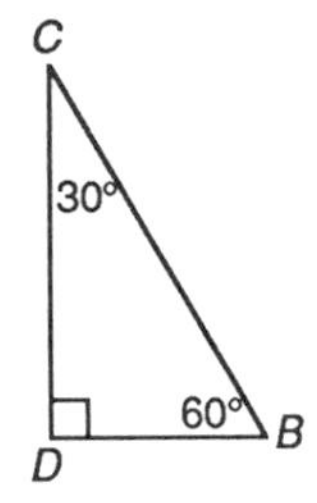

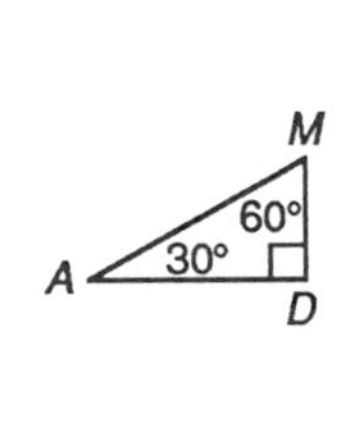

In addition, you learned in Chapter 3 that the medians of a triangle divide each other in a ratio of 1:2. Therefore: $EM = (1/2)AM$, $DM = (1/2)CM$, and $FM = (1/2)BM$.

Using these properties of equilateral triangles you can solve many difficult problems.

Example A

What are the lengths w, x, y, and z? The large triangle is equilateral with sides of 36.

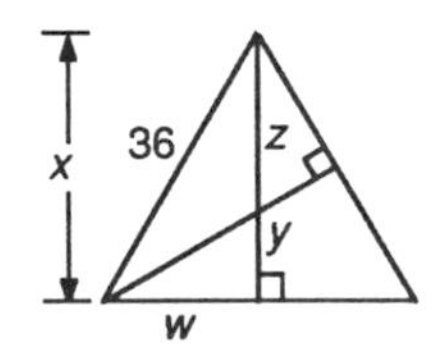

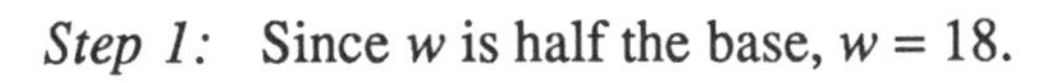

Step 1: Since w is half the base, $w = 18$.

Step 2: w is the length of the shorter leg and x is the length of the longer leg of a 30-60 right triangle. Therefore, $x = w\sqrt{3} = 18\sqrt{3}$.

Step 3: Since $z = 2y$ and $y + z = x$, $3y = x$. Therefore, $y = (1/3)x = (1/3)(18\sqrt{3}) = 6\sqrt{3}$.

Step 4: Since $z = 2y$, then $z = (2)(6\sqrt{3}) = 12\sqrt{3}$.

Example B

What are the lengths w, x, y, and z? The large triangle is equilateral and the partial length along one median is 12.

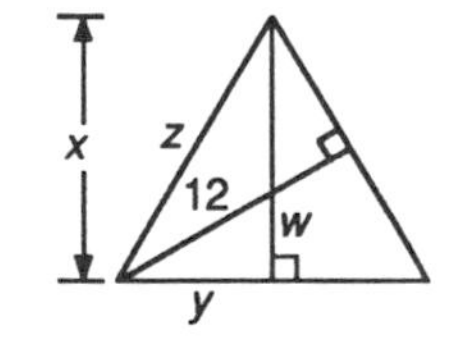

Step 1: w is the length of the shorter leg in a 30-60 right triangle that has hypotenuse 12. So $w = (1/2)(12) = 6$.

Step 2: Since y is the length of the longer leg in a 30-60 right triangle and $w = 6$ is the length of the shorter leg, $y = w\sqrt{3} = 6\sqrt{3}$.

Step 3: Since z is the length of the hypotenuse in a 30-60 right triangle with y as the length of the shorter leg, $z = 2y$. Therefore, $z = (2)(6\sqrt{3}) = 12\sqrt{3}$.

Step 4: The medians divide each other in a ratio of 1:2, therefore $x = 3w = (3)(6) = 18$.

EXERCISE SET A

In Exercises 1 to 9, each large triangle is equilateral. Use what you know about 30-60 right triangles to find x and y in each figure.

1.*

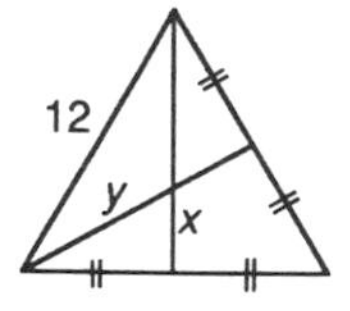

2.*

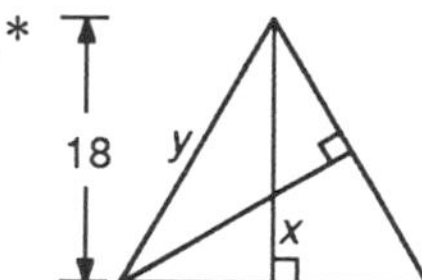

3.*

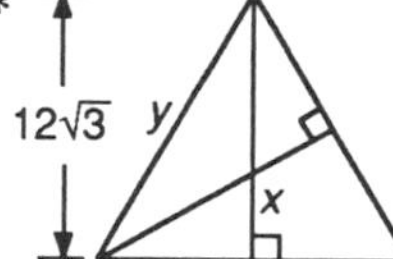

4.*

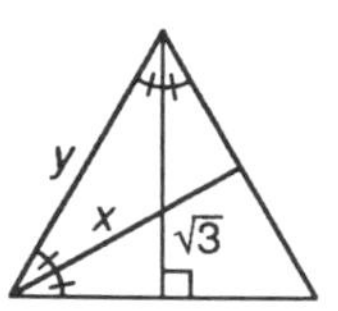

5.

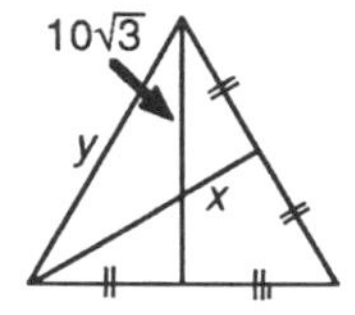

6.

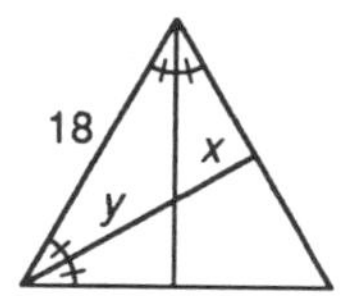

Find the area of each equilateral triangle. All measurements are given in centimeters.

7.*

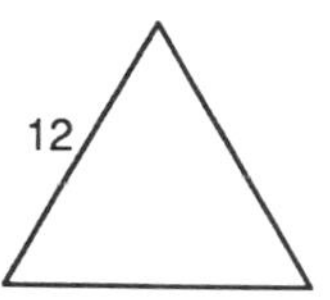

8.

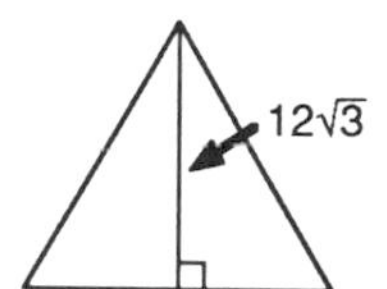

9.*

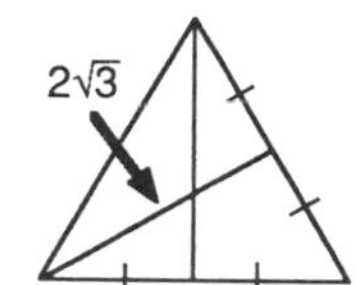

The point of concurrency of the three angle bisectors in a triangle is the incenter.

The incenter is the center of the inscribed circle. Point M is the incenter of equilateral ΔABC and segment MD is the radius of the inscribed circle.

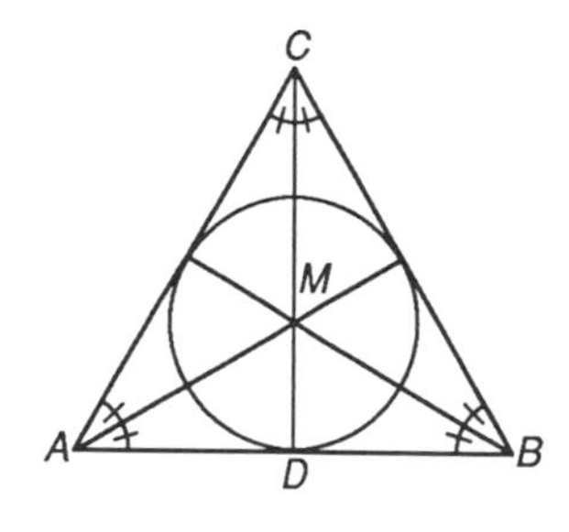

The point of concurrency of the three perpendicular bisectors in a triangle, is the circumcenter.

The circumcenter is the center of the circumscribed circle. Point M is the circumcenter of ΔABC and segment MC is the radius of the circumscribed circle.

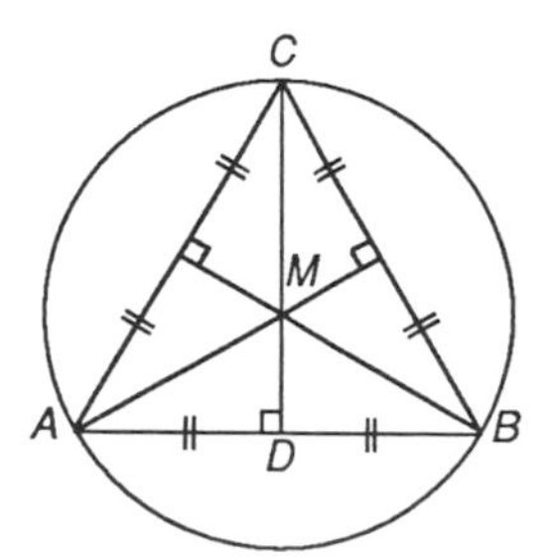

EXERCISE SET B

In Exercises 1 to 3, each triangle is equilateral. Find the area of the triangle. Find the area of the inscribed circle. Find the area of the circumscribed circle.

1.* $AB = 6$ cm

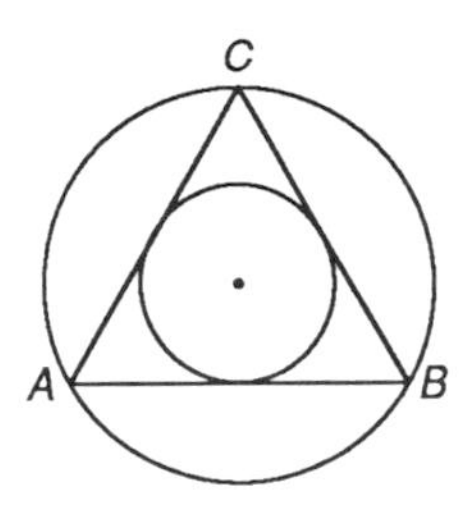

2. $DE = 2\sqrt{3}$ cm

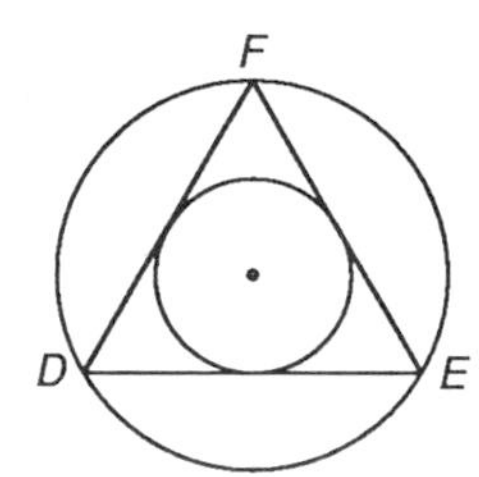

3.* $IJ = 9\sqrt{3}$ cm

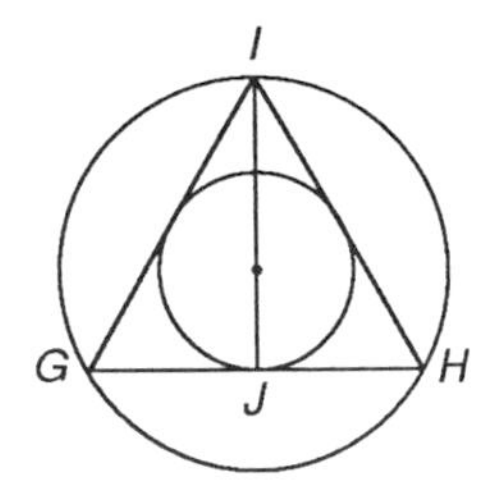

4.* The area of equilateral ΔREG is $25\sqrt{3}$ m^2. Find the perimeter of ΔREG. How many times larger is the area of the circumscribed circle than the area of the inscribed circle?

5. Find the area of a regular hexagon with each side of length 12 cm.

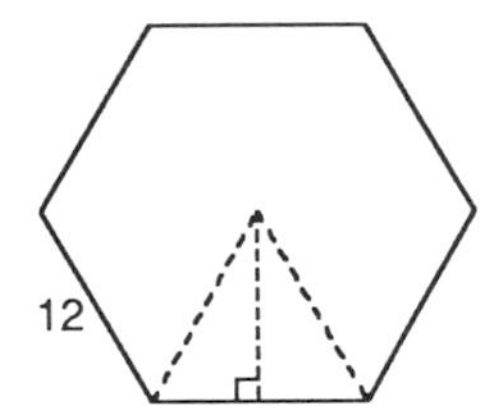

6.* Find the area of a regular hexagon circumscribed about a circle with a radius of 9 inches.

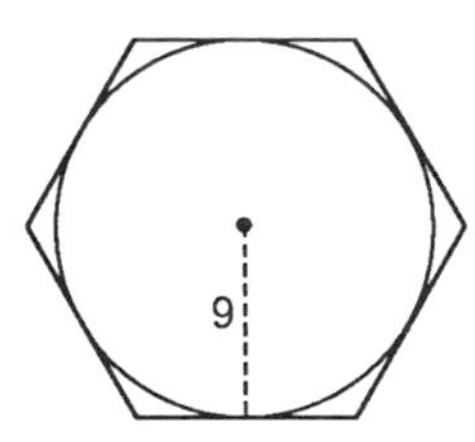

In Exercises 7 and 8, the triangles shown are equilateral. Find the shaded area in each exercise to the nearest square centimeter. Use 3.14 for π and 1.73 for $\sqrt{3}$.

7. $NP = 12$ cm

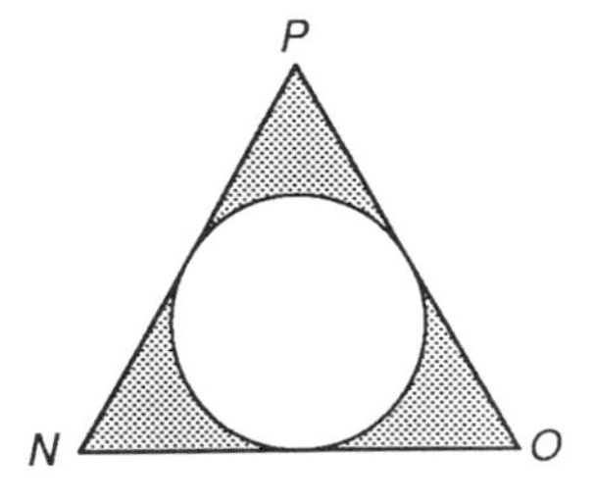

8. $QS = 12$ cm

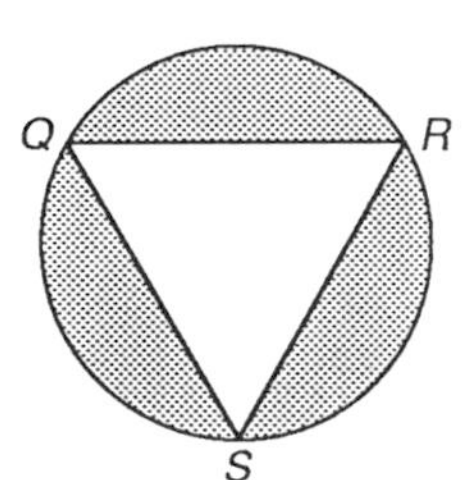

Lesson 9.9

Distance in Coordinate Geometry

Scott is located at the intersection of Second Street and Third Avenue and his sister Viki is located at the intersection of Seventh Street and Eighth Avenue. If Viki were to walk on the sidewalks along the streets and avenues between them, the shortest route would be ten blocks. It is easy to find distances along the horizontal or vertical, you simply count blocks. However, if Scott were able to fly straight to Viki, you would need the Pythagorean Theorem to calculate the distance. What is the distance to the nearest meter from Scott to Viki if each block is approximately 50 meters long?

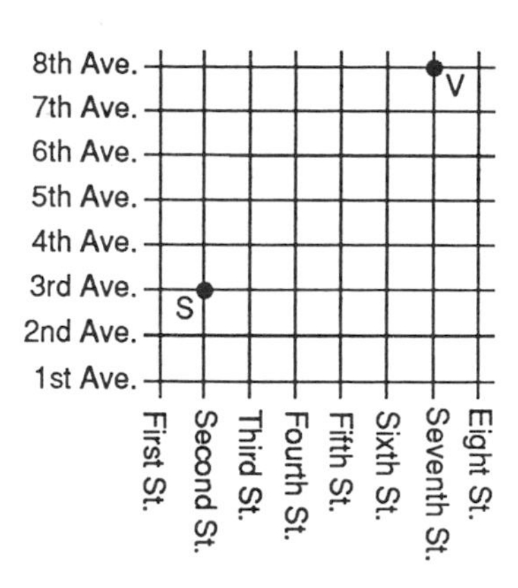

The grid of streets is like a coordinate plane. A coordinate grid is made of two sets of parallel lines, one set running perpendicular to the other set. Thus, every segment in the plane (non-vertical and non-horizontal segments) is the hypotenuse of some right triangle. You can use the Pythagorean Theorem to find the distance between any two points on a coordinate plane.

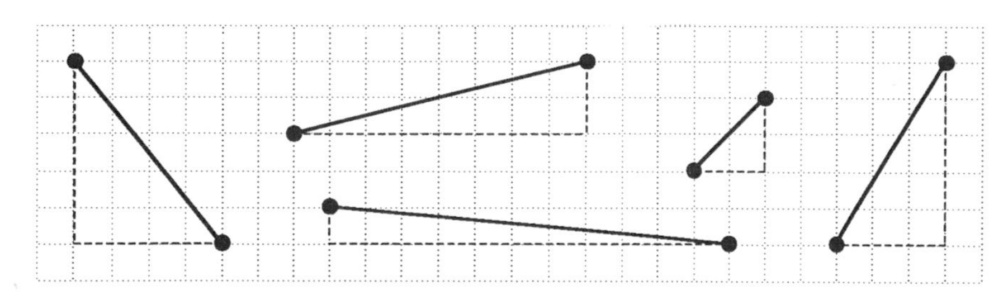

EXERCISE SET A

Copy each graph below onto your own graph paper. Turn each given segment into the hypotenuse of a right triangle using the grid lines as a guide.

1.

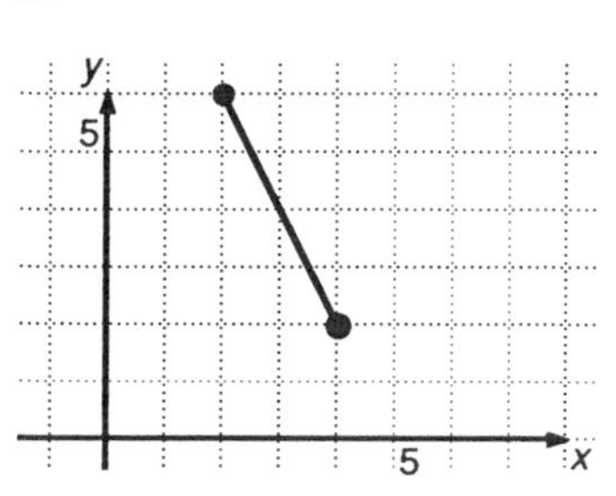

2.

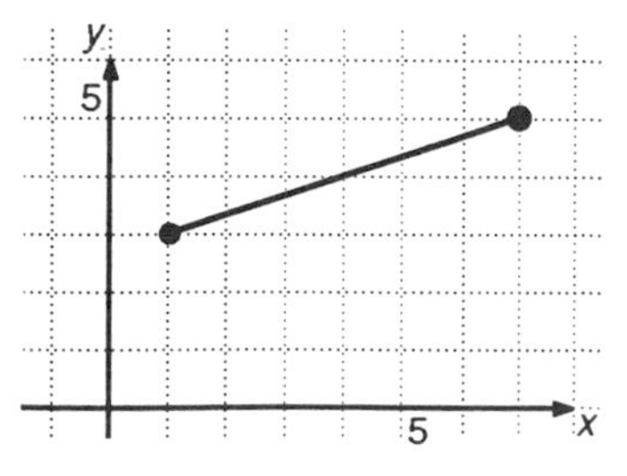

3.

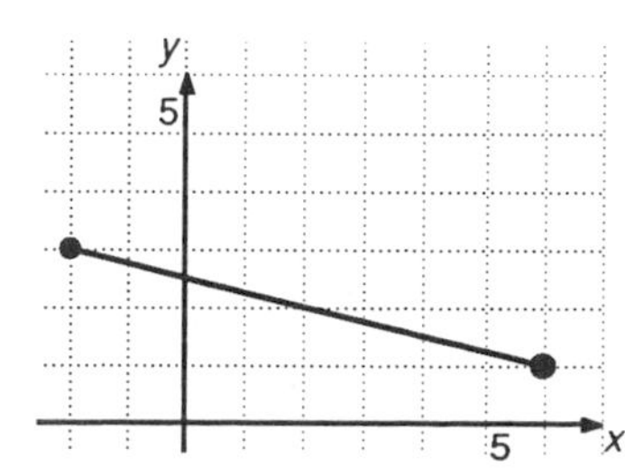

In Exercises 4 to 6, find the length of each segment.

4*.

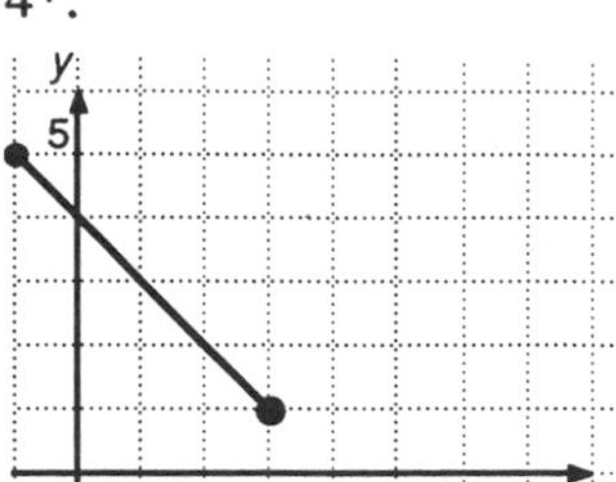

5.

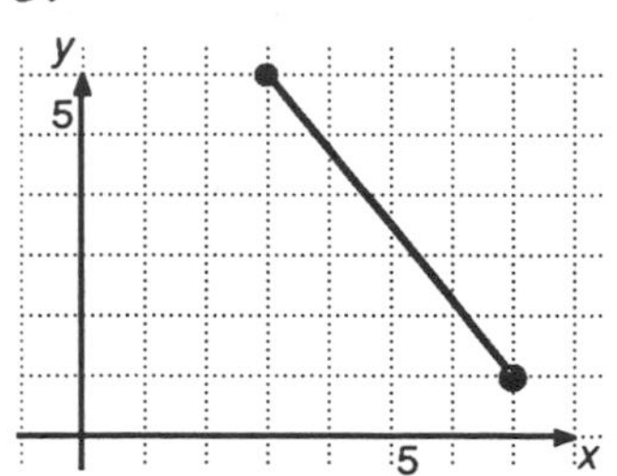

6.

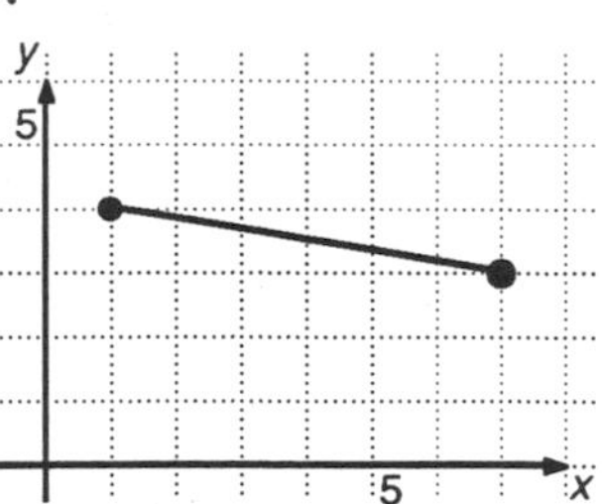

In Exercises 7 to 9, graph the points, and then find the distances between them.

7.* (1, 2); (13, 7)   8. (-5, -8); (3, 7)   9. (-9, -6); (3, 10)

10. What is the distance from Scott to Viki to the nearest meter? (You did read the introduction, didn't you?)

In the previous exercise set, you found the length of a segment by connecting it to a right triangle on the graph paper and then applying the Pythagorean Theorem. What if the points are so far apart that your graph paper isn't large enough to plot them? For example, what is the distance between the points (15, 37) and (42, 73)? Clearly, the coordinates are too large to graph on ordinary graph paper. You need a rule or formula that uses the coordinates of the two given points to calculate the distance between them.

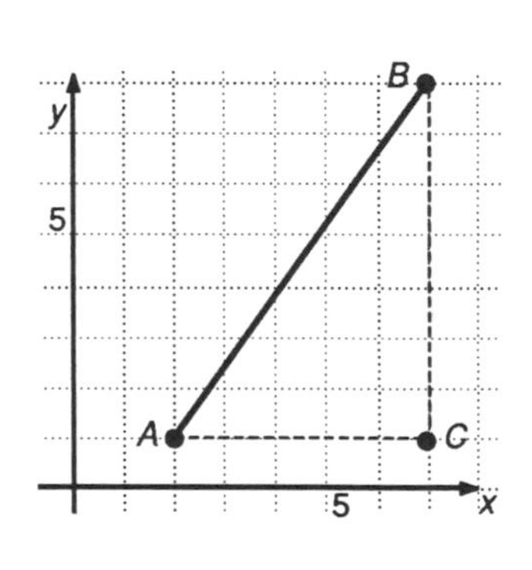

To find the distance between A and B on the right, you can simply count the squares on side $\overline{AC}$ and the squares on side $\overline{BC}$, and then use the Pythagorean Theorem to determine AB. However, when the distances are too great to count physically, there is still a nice way to find the lengths. You can find the vertical distance BC by subtracting the y-coordinates of points A and B. Since distance is never negative, subtract the smaller coordinate from the larger. $BC = 8 - 1 = 7$. You can find the horizontal distance AC by subtracting the x-coordinates of points A and B. $AC = 7 - 2 = 5$. Now you can find the length of $\overline{AB}$: $AB^2 = (7 - 2)^2 + (8 - 1)^2$, and therefore $AB = \sqrt{5^2 + 7^2} = \sqrt{74}$.

Can you generalize this result and come up with a formula for the distance between any two points in the coordinate plane? Compare the method shown above with the problems you solved in Exercise Set A. Compare your observations with those of others. Copy and complete the conjecture below and add it to your conjecture list.

C-67 *If the coordinates of points A and B are (x_1, y_1) and (x_2, y_2) respectively, then $AB^2 = (—?—)^2 + (—?—)^2$ and $AB = \sqrt{(—?—)^2 + (—?—)^2}$.* ***(Distance Formula)***

Let's look at a few examples to see how you can use the Distance Formula to find the distance between two points and how you can use the Coordinate Slope Conjecture to determine whether a triangle is a right triangle.

Example A

Find AB if the coordinates of A are (8, 15) and the coordinates of B are (-7, 23).

$$\begin{aligned} AB^2 &= (8 - -7)^2 + (23 - 15)^2 \\ &= 15^2 + 8^2 \\ &= 289 \end{aligned}$$

$$AB = \sqrt{289} = 17$$

Example B

Is ΔABC with vertices A (9, -9); B (13, -4); and C (8, 0) a right triangle?

The slope of $\overline{AB}$ is $\frac{-4 - -9}{13 - 9}$ or $\frac{5}{4}$.

The slope of $\overline{BC}$ is $\frac{-4 - 0}{13 - 8}$ or $\frac{-4}{5}$.

The slope of $\overline{AB}$ is the negative reciprocal of the slope of $\overline{BC}$, thus the two segments are perpendicular and ΔABC is a right triangle.

EXERCISE SET B

In Exercises 1 to 3, use the Distance Formula to find the distance between each pair of points. In Exercise 6, you'll need the Coordinate Slope Conjecture.

1.* (10, 20); (13, 16) 2. (-15, -18); (3, 6) 3. (-19, -16); (-3, 14)

4. Find the perimeter of ΔABC with vertices A (2, 4); B (8, 12); and C (24, 0).

5. Determine whether ΔDEF with vertices D (6, -6); E (39, -12); and F (24, 18) is scalene, isosceles, or equilateral.

6.* Determine whether ΔGHI with vertices G (2, 6); H (18, 2); and I (12, 12) is isosceles, right, isosceles right, or equilateral.

EXERCISE SET C

1.* A circle of radius 6 has a chord $\overline{AB}$ of length 6. If point C is selected randomly on the circle, what is the probability that ΔABC is obtuse?

2. A circle has a central angle AOB which measures 80. If point C is selected randomly on the circle, what is the probability that ΔABC is obtuse?

Lesson 9.10

Circles and the Pythagorean Theorem

In Chapter 6 you discovered a number of properties that involved right angles in and around circles. In this lesson you will use these conjectures with the Pythagorean Theorem to solve some challenging problems. Let's review. Two of the most useful conjectures are listed below.

> *A tangent to a circle is perpendicular to the radius drawn to the point of tangency.* ***(Tangent Conjecture)***
>
> *Every angle inscribed in a semicircle is a right angle. (C-46)*

Let's look at a few examples of how you can use these conjectures with the Pythagorean Theorem to solve problems.

Example A

$AN = 12$ cm. Find the shaded area.

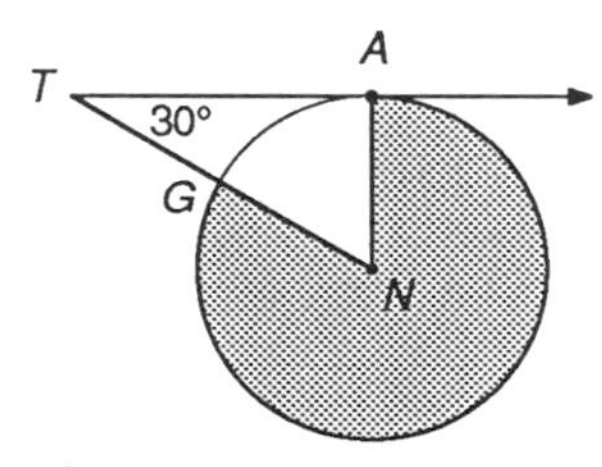

Since $\overrightarrow{TA}$ is a tangent, ΔTAN is a 30-60 right triangle. The area of the circle is $\pi 12^2$ or 144π cm^2. Since 60/360 is 1/6, then the shaded area is 5/6 of the area of the circle. Therefore the shaded area is $(5/6)(144\pi)$ or 120π cm^2.

Example B

$AB = 6$ cm and $BC = 8$ cm. Find the shaded area.

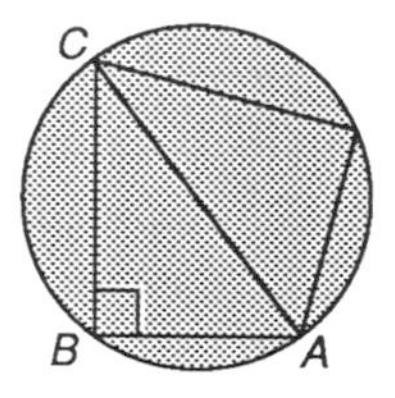

Since $\angle ABC$ is a right angle, $\widehat{ABC}$ is a semicircle and $\overline{AC}$ is a diameter. If $AB = 6$ cm and $BC = 8$ cm, then $AC = 10$ cm (by the Pythagorean Theorem) and the radius is 5 cm. If the radius is 5 cm, the area of the circle is $\pi 5^2$ or 25π cm^2.

EXERCISE SET A

Use the two conjectures above with the Pythagorean Theorem to find the shaded area in each figure. All measurements are given in centimeters.

1.* $OD = 24$

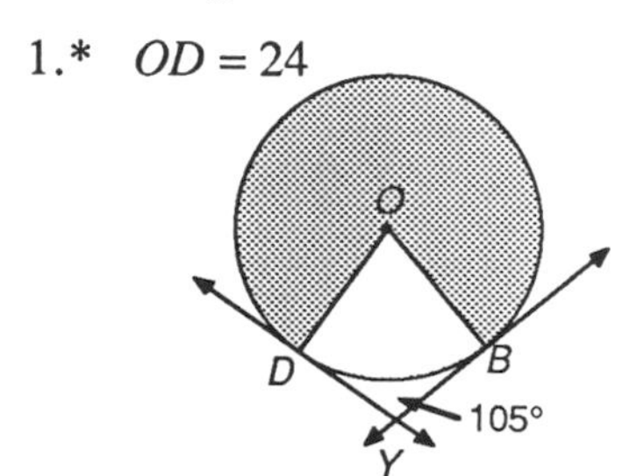

2. $RC = 9$

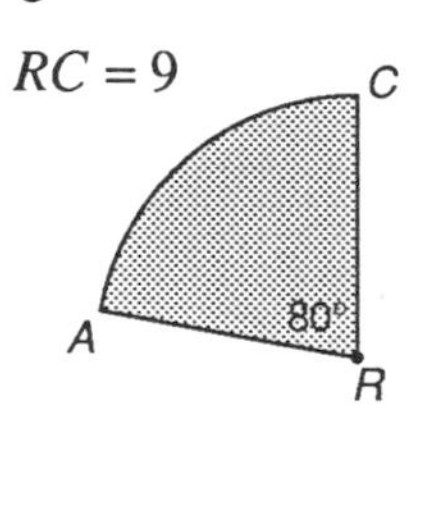

3.* $BT = 6\sqrt{3}$

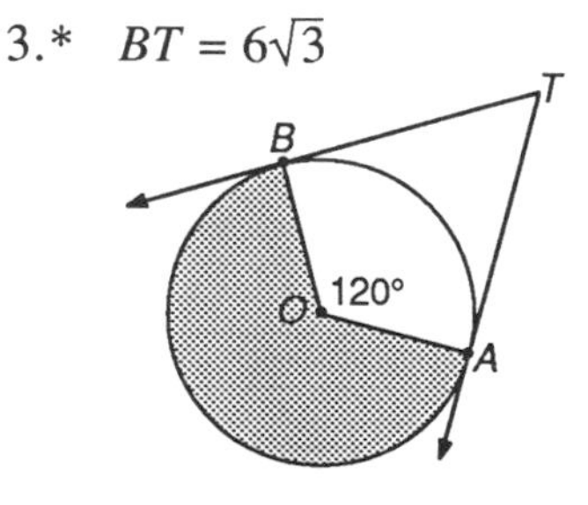

4.* $HA = 8\sqrt{3}$

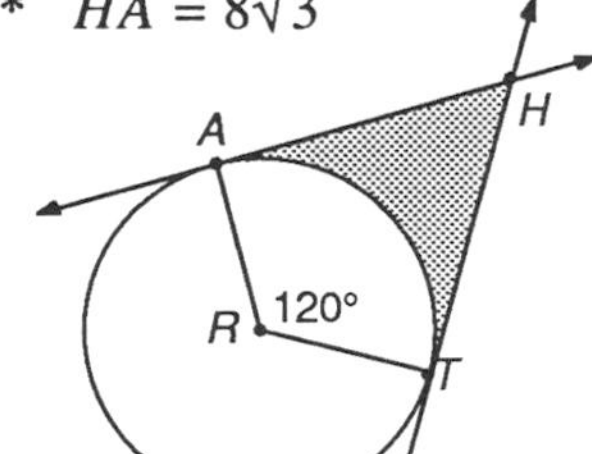

5. $HT = 8\sqrt{3}$

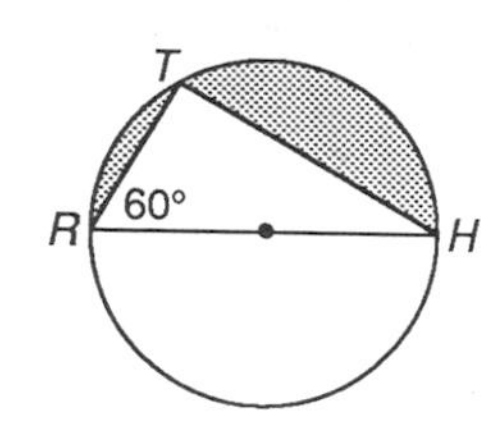

6.* $HO = 8\sqrt{3}$

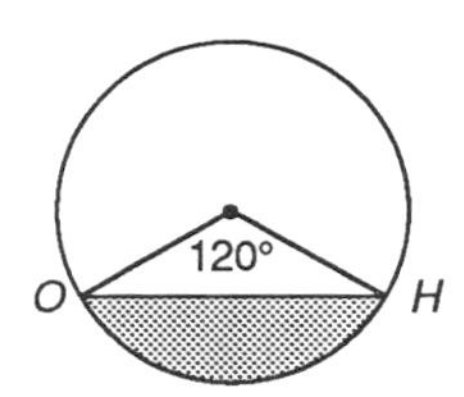

EXERCISE SET B

1.* A 3 meter wide circular track is shown on the right. The radius of the inner circle is 12 meters. What is the longest path in a straight line that stays on the track? (In other words, find AB.)

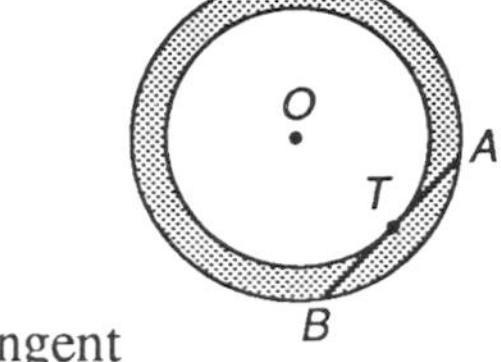

2.* An annulus has a 36 cm chord of the outer circle that is also tangent to the inner concentric circle. Find the area of the annulus.

3.* Sector ARC has a radius of 9 cm and an angle of 80°. When sector ARC is cut out and $\overline{AR}$ and $\overline{RC}$ are taped together, it forms a cone. $\overset{\frown}{AC}$ becomes the circumference of the base of the cone. What is the height of the cone?

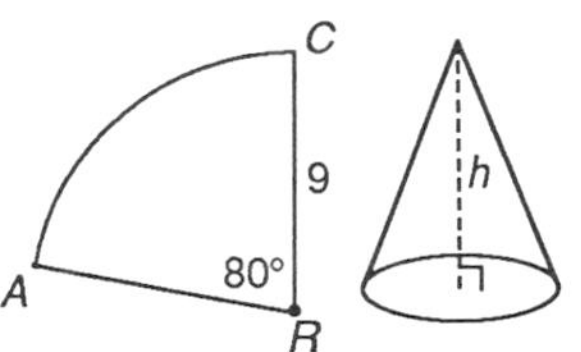

4.* Wilber Ness is a ranger for the National Forestry Service. His job is to measure the circumference of certain marked trees in his section of the forest. The marked trees have been selected to be cut in order to "thin" the forest (providing room for growth on unmarked trees). The largest marked tree has a circumference of 336 cm at a point 5 m up from its base. What is the length of a side (to the nearest centimeter) of the largest square piece of 5 m long timber that can be cut from a cross section of this tree? (In other words, calculate the length of the side of a square that can be inscribed in a circle with a circumference of 336 cm.) Use 3.14 for π and 1.41 for $\sqrt{2}$.

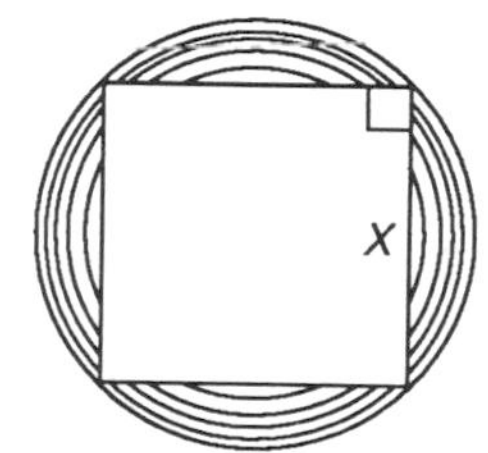

5.* A belt on a piece of machinery needs to be replaced. The belt runs around two wheels, crossing between them, so that the larger wheel turns the smaller in the opposite direction. The diameter of the larger wheel is 36 cm and the diameter of the smaller is 24 cm. The distance between the centers of the two wheels is 60 cm. The belt crosses 24 cm from the center of the smaller wheel. What is the length of the belt?

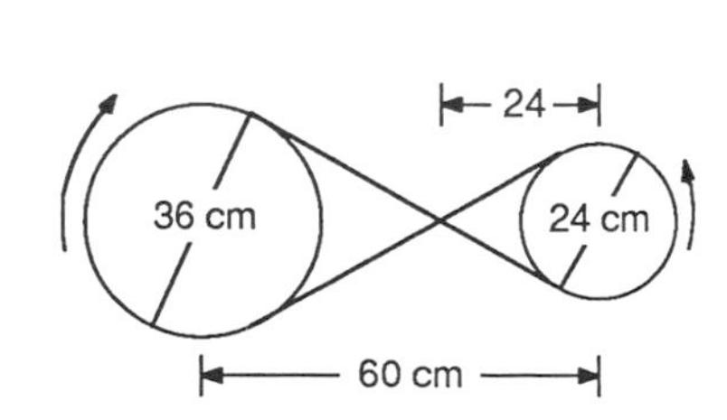

King Arthur and his Human Cannonball

6. King Arthur plans to shoot his court jester out of a cannon making him history's first human cannonball. However, Lady Guinevere's summer cottage is in the line of fire. The cannon, designed by Merlin, can fire the jester up to 3 miles. The cannon does not shoot very accurately. The jester may land up to 30 degrees to the right or 30 degrees left of the line of fire. Arthur has sent his trusted knight, Sir Lancelot, to find out where the good lady would like her castle relocated.

   "Lance darling, Rose La Fleur, my chambermaid, has spoken to Merlin and he has told her that the cannon is aimed directly at my cottage."

   "My Lady, the cottage must be moved. You must help me figure out the shortest route out of the poor unfortunate jester's landing zone," spoke Sir Lancelot. "My dear Lady G., if your cottage is 1.5 miles from the castle, what is the shortest route out of the path of Merlin's monstrous cannon?" Can you help Lady Guinevere and Sir Lancelot? What is the shortest distance out of the potential landing zone of the human cannonball?

Lesson 9.11

Constructions with the Pythagorean Theorem

In this lesson you will use the Pythagorean Theorem to solve geometric construction problems.

Investigation 9.11.1

Step 1: Construct a right triangle. Label it ΔPDQ with the hypotenuse $\overline{DQ}$.

Step 2: Construct the median to the hypotenuse. Label the median $\overline{PR}$.

Step 3: With your compass, compare the distances PR, DR, and QR.

Try this with another right triangle and then compare your results with those of others. State your observations as your next conjecture.

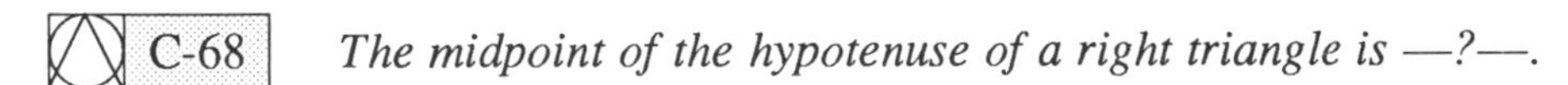

C-68 *The midpoint of the hypotenuse of a right triangle is —?—.*

Since the circumcenter of a triangle is the center of the circumscribed circle, it is the point in a triangle that is the same distance from all three vertices. Therefore the midpoint of the hypotenuse of a right triangle is the circumcenter.

EXERCISE SET A

1.* Given the segment with length a below, construct segments with lengths $a\sqrt{2}$, $a\sqrt{3}$, and $a\sqrt{5}$.

a

2. Construct right ΔDEF with $\angle D$ as an acute angle and $\overline{DE}$ the hypotenuse. Next construct a circle that circumscribes ΔDEF. What is $\overline{DE}$ in relation to the circle?

D

D —— E

3.* Given $\angle A$ and radius $\overline{OR}$, construct circle O so that it is tangent to both sides of $\angle A$.

O —— R

A

4. Construct a tangent to a circle from a given point outside the circle.

   Step 1: On a full sheet of paper, construct a circle with a radius of about 1 inch. Label the center O.

   Step 2: Locate a point about 3 inches to the left of the circle. Label it X.

   Step 3: Construct $\overline{XO}$. Construct the midpoint. Label it M.

   Step 4: With M as center, construct a semicircle with $\overline{XO}$ the diameter.

   Step 5: Label as T the point where circle O and the semicircle intersect.

   Step 6: Construct tangent $\overleftrightarrow{XT}$.

5.* Construct the right ΔTOM given the median $\overline{OA}$ to the hypotenuse $\overline{TM}$, and one leg $\overline{OT}$.

   O —————————— A   O —————————————— T

6. Construct the right ΔMAT given the median $\overline{AE}$ to the hypotenuse $\overline{MT}$, and the altitude $\overline{AG}$ to the hypotenuse.

   A ———————————————— E   A ———————————————— G

7. Construct a right triangle with sides of lengths 6 cm, 8 cm, and 10 cm. Locate the midpoint of each side. Construct a semicircle on each side with the midpoints of the sides as centers. Find the area of each semicircle.

8. Construct a right triangle with sides of lengths 6 cm, 8 cm, and 10 cm. Construct an equilateral triangle on each side of the right triangle using each side of the right triangle as a side of an equilateral triangle. Find the area of each equilateral triangle.

EXERCISE SET B

1. What is the area of a circle that circumscribes an isosceles right triangle with legs 8 cm in length?

2. What is the area of a circle that is inscribed in an isosceles right triangle with legs 8 cm in length?

3. What is the area of the square that can be placed in an isosceles right triangle with legs 8 cm in length? The two sides of the square rest on the legs of the triangle and a vertex of the square touches the hypotenuse of the triangle.

Lesson 9.12

Chapter Review

EXERCISE SET A

Complete each statement.

1. The —?— is the side opposite the right angle in a right triangle.
2. Any three positive integers that work in the Pythagorean formula are called —?—.
3. In a right triangle, if a and b are the lengths of the legs and c is the length of the hypotenuse, then —?—. (C-61 or **Pythagorean Theorem**)
4. If the lengths of the three sides of a triangle work in the Pythagorean formula, then the triangle —?—. (C-62 or **Converse of Pythagorean Theorem**)
5. If you multiply the lengths of all three sides of any right triangle by the same number, the resulting triangle will be a —?—. (C-63)
6. If the lengths of two sides of a right triangle have a common factor, then —?—. (C-64)
7. In an isosceles right triangle, if the legs have length x, then the hypotenuse has length —?—. (C-65 or **Isosceles Right Triangle Conjecture**)
8. In a 30-60 right triangle, if the shorter leg has length x, then the longer leg has length —?— and the hypotenuse has length —?—. (C-66 or **30-60 Right Triangle Conjecture**)
9. If the coordinates of points A and B are (x_1, y_1) and (x_2, y_2) respectively, then $AB^2 = (\text{—?—})^2 + (\text{—?—})^2$ and $AB = \sqrt{(\text{—?—})^2 + (\text{—?—})^2}$. (C-67 or **Distance Formula**)
10. The midpoint of the hypotenuse of a right triangle is the same distance from —?—. (C-68)

EXERCISE SET B

All measurements are given in centimeters.

1. Simplify $3\sqrt{320}$.

2. $x = $ –?–

3. Is ΔABC a right triangle?

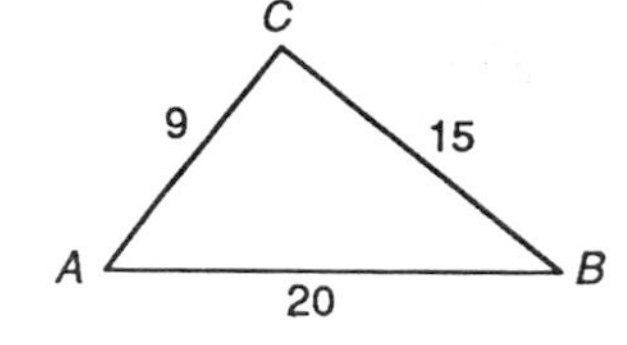

4. The area of the square is 144 cm². What is d?

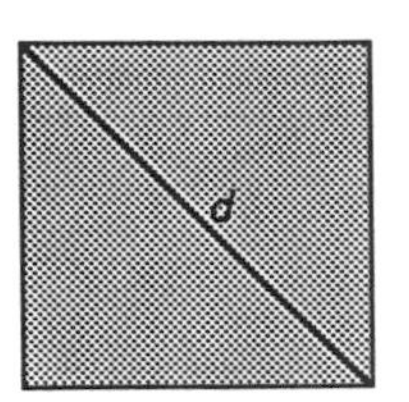

5. What is the area of the triangle?

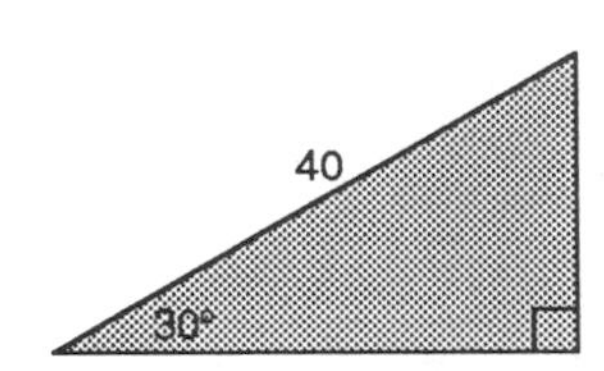

6. $AB = -?-$

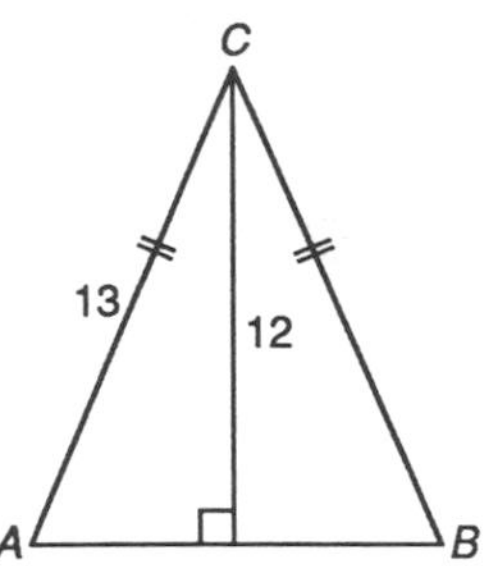

7. $AB = -?-$

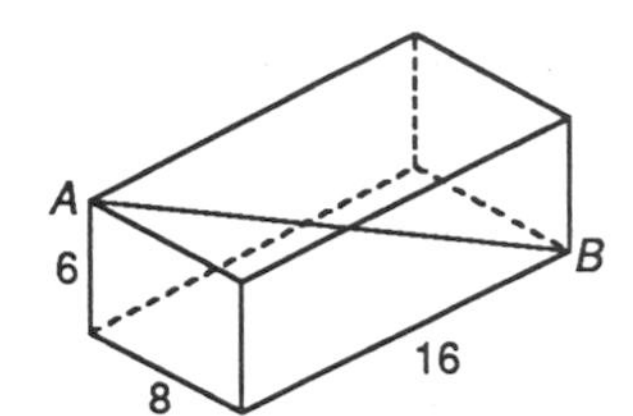

8. What is the area of trapezoid $ABCD$?

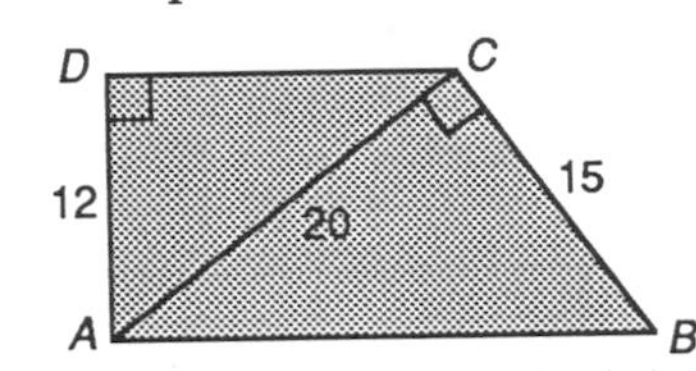

9.* $QE = 2\sqrt{2}$
What is the shaded area?

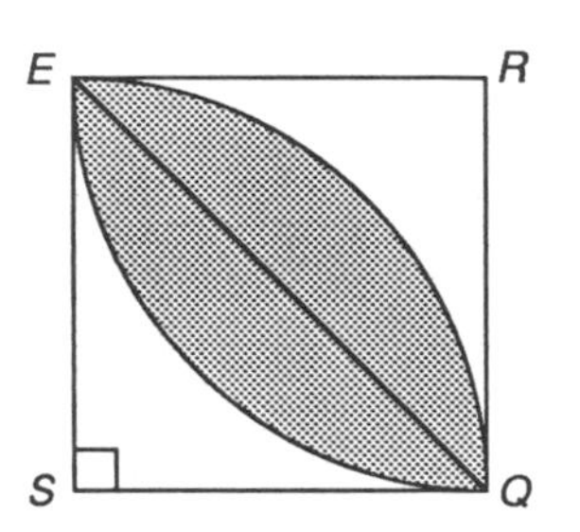

EXERCISE SET C

1. What is the length of the hypotenuse of a right triangle with legs of 300 and 400 feet?

2. What are the lengths of the two legs of a 30-60 right triangle if the length of the hypotenuse is $12\sqrt{3}$?

3. Determine whether ΔABC with vertices A (3, 5); B (11, 3); and C (8, 8) is isosceles, isosceles right, or equilateral.

4.* Two cars leave point A at noon. One car travels north at 45 mph; the other travels east at 60 mph. How far apart are the two cars after two hours?

5. Find the original height to the nearest foot of a fallen flagpole if it cracked and fell as if hinged forming an angle of 45 degrees with the ground. The tip of the pole hit the ground 12 feet from the base of the flagpole.

6. Flora Fluty is away at camp and wishes to mail her wooden flute back home to her family. The flute is 24" long. Will it fit diagonally in a box with inside dimensions of 12" by 16" by 14"? (Ignore the thickness of the flute.)

7. Find the area of an equilateral triangle with sides of 6 meters.

8.* Find the circumference of a circle inscribed in an equilateral triangle with a height of $12\sqrt{3}$.

9. Find the perimeter of an equilateral triangle if its height is $7\sqrt{3}$.

10.* Guido Palumbo is a Venetian gondolier. Guido needs to know how deep it is in front of his pier. He notices a water lily sticking straight up from the water with the blossom 18 centimeters above the surface. Guido then pulls the lily to one side, keeping the stem straight, until the blossom touches the water at a spot 36 centimeters from where the stem first broke the surface. From this data Guido is able to calculate the depth of the water. Can you? What is the depth of the water?

11. After an argument, Paul and Paula walk away from each other at what appears to be a right angle. Paul is walking 2 km/hr and Paula is walking 3 km/hr. After 1 hour Paula stops walking. After 2 hours Paul stops. If they were then 5 kilometers away from each other, did they originally walk away from each other at right angles? Now that they have decided to make up, how long will it take them to reach each other if they continue to walk at their same speeds straight towards each other?

12.* Desert prospector Sagebrush Sally hops on her dirt bike and (with a full tank of gas) leaves camp traveling 60 km/hr due east. After two hours she stops and does a little prospecting — with no luck. So she hops back on her bike and heads due north for two hours at 45 km/hr. She stops again, does a little more prospecting, and this time hits paydirt. Since the greatest distance Sagebrush Sally has ever traveled on one tank of gas is 350 kilometers, she assumes that is the maximum distance she will be able to travel this time. Does she have enough fuel to get back to camp? If not, what is the closest she can come to camp? What should she do?

13. A knight's move in chess is from one end of an L to the other as shown on the right. If a knight were on the square marked with an X in the diagram, it could move in a single move along any of the paths indicated to one of the white squares with an arrowhead. What is the greatest total distance (from center of a square to center of a square) that a knight can move in two consecutive moves on a chessboard with squares that are 6 cm on a side? A chessboard has 64 squares arranged in an 8 by 8 array.

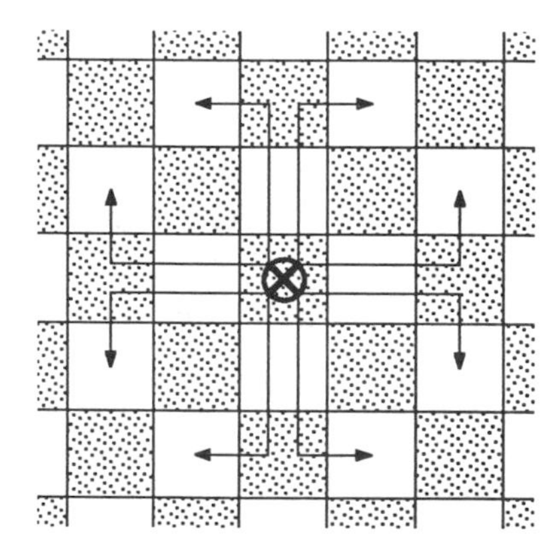

14. A chessboard has 64 squares that are 6 cm on a side. The squares alternate between red and black as shown in the diagram. What is the radius of the largest circle that can be drawn on the board in such a way that the circle lies entirely on black squares?

Cooperative Problem Solving

Pythagoras in Space

1. Astrophysicists at the lunar research facility Galileo II are planning a very large construction on the back side of the lunar surface. In the event that another intelligent life form sends robotic space probes into our solar system, our astrophysicists would like an obvious lunar surface marking that would indicate the existence of an intelligent civilization. They have decided that the Pythagorean Theorem is an important fact that must be known by intelligent life forms throughout the universe. Therefore, they have decided to construct a large right triangle with a square on each side. Bulldozers will smooth away the regions within the three squares and the right triangle. The borders of the squares will be one meter wide and made of highly reflective material. The planners are considering a right triangle with sides of 30 km, 40 km, and 50 km. What is the total area that they must smooth away? What is the total length of meter-wide reflective material needed for this monumental task? What do you think of the idea?

2. The first robot built by the Galileo II research team at the Artificial Intelligence Lab is a very simple one. The first program for the experimental robot instructs it to travel in a rectangular spiral, traveling 1 meter east, 2 meters north, 3 meters west, 4 meters south, 5 meters east again, and so on, until it runs into something. If the robot travels uninterrupted at the rate of 1 meter per minute, how far away from its starting point will it be after 2 hours?

3.* There are three large water pipes that run from the hydroponics and water recycling plant to the dormitory-recreation facility of the lunar space port Galileo. Engineers plan to enclose the three water pipes in a ceramic-coated triangular casing for protection from radiation. Each of the three water pipes has an outside diameter of 60 cm. If the cross section of the casing is an equilateral triangle, what is the smallest length possible for a side? (Use 1.7 for $\sqrt{3}$.)

4.* Machinists at the space port's machine shop have fabricated an 8 meter long pipe that is 3 centimeters in diameter. They must deliver it to a construction site down a corridor 3 meters wide by 3 meters high that has a right angled corner. The pipe must not be bent or twisted. Will it fit around the corner? What is the longest pipe that would fit around the corner?

5.* The space port's research team is planning a 30th anniversary celebration. They plan to decorate the facility by wrapping a wreath of olive branches about the microwave receiver tower. The tower is a cylindrical column 70 meters tall and 8 meters in circumference. Since it is the 30th anniversary of space port Galileo, they plan to wrap the wreath uniformly about the column exactly 30 times. How long will the spiral wreath be?

10 Volume

Cubic Space Division, M. C. Escher, 1952

Volume is the amount of space contained within an object. In this chapter you will discover three conjectures about volume. You will discover the formulas for finding the volumes of prisms, pyramids, cylinders, cones, and spheres. Volume is important in many occupations. You will also use the area formulas and the Pythagorean Theorem in this chapter. You should know them by heart by now. Of all the geometric topics, area, volume, and the Pythagorean Theorem have the most applications.

Lesson 10.1
Polyhedrons

In geometry, most of the figures that you work with are flat (plane figures). Plane figures have only two dimensions. In this chapter you will work with solids. Solids have three dimensions: length, width, and height. You see and touch solids all the time. Some solids, like rocks and plants, are very irregular, but many other solid objects have shapes that can be easily described using common geometric terms. Some of these geometric solids occur in nature: oranges, sea urchins, salt crystals. Others are manufactured: books, baseballs, soup cans, ice cream cones. In this lesson you will learn about one important group of geometric solids called polyhedrons. Later you will look more closely at two types of polyhedrons: prisms and pyramids. You will also study three types of geometric solids with curved surfaces: spheres, cylinders, and cones.

A solid formed by flat surfaces enclosed by polygons is called a ***polyhedron***.

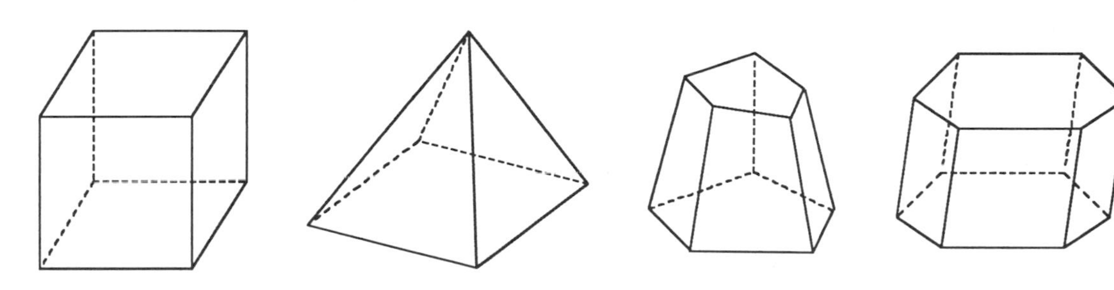

The flat polygonal surfaces of a polyhedron are called its **faces**. Although a face of a polyhedron includes the polygon and its interior region, we identify the face by naming the polygon that encloses it. When we say that the face of a polyhedron is a triangle, we really mean that the face is a triangular region. A segment where two faces intersect is an **edge**. A point of intersection of three or more edges is a **vertex**. The table identifies the parts of the polyhedron on the right.

Faces	Edges	Vertices
	$\overline{PO}$	
ΔPOL	$\overline{OL}$	P
ΔOLY	$\overline{PL}$	O
ΔPYL	$\overline{OY}$	L
ΔPOY	$\overline{PY}$	Y
	$\overline{LY}$	

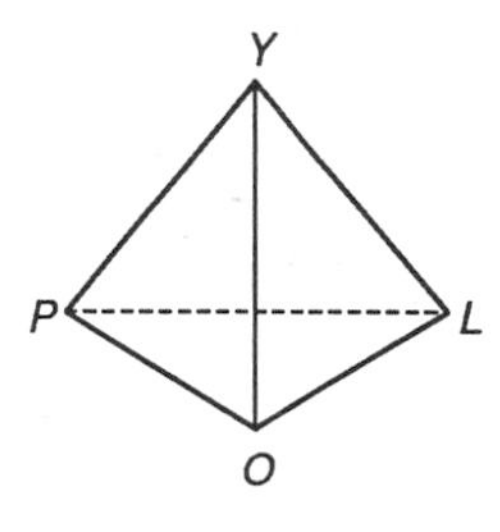

Just as a polygon is classified by its number of sides, a polyhedron is classified by its number of faces.

Faces	Classification
4	**Tetrahedron**
5	**Pentahedron**
6	**Hexahedron**

Faces	Classification
7	**Heptahedron**
8	**Octahedron**
9	**Nonahedron**

Faces	Classification
10	**Decahedron**
11	**Undecahedron**
12	**Dodecahedron**

If each face of a polyhedron is enclosed by a regular polygon and each face is congruent to the other faces and the faces meet at each vertex in exactly the same way, then the polyhedron is called a **regular polyhedron**.

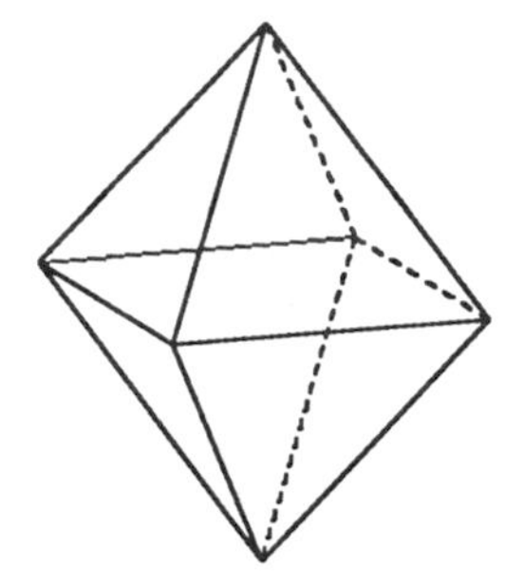

Regular Octahedron

EXERCISE SET A

Let's build some stick polyhedrons. If possible, you should work in groups of four or five on this activity. Each group will need the following materials:

- Toothpicks or wooden barbecue skewers (about 150)
- Modeling clay (a lump the size of an extra large egg) or gum drops (about 75 plus a few to eat) or dried peas (soak them overnight before using them)

Share the tasks among the group. Share your ideas and insights. Have fun! Don't eat all your gum drops before you finish.

Using toothpicks or barbecue skewers as edges and small balls of modeling clay, gum drops, or peas as connectors, build each stick polyhedron described below. You may have to cut or break some sticks. Save the polyhedrons that you make. You'll need them to complete the next exercise set.

1.

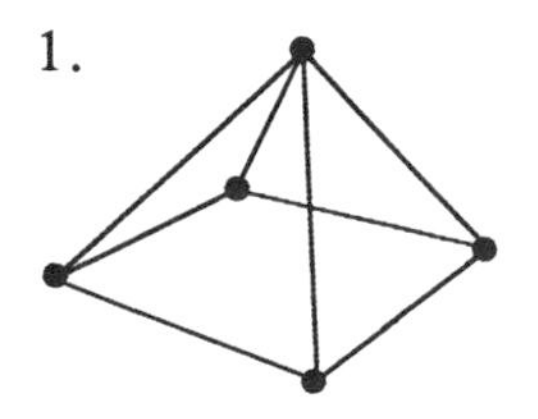

2.

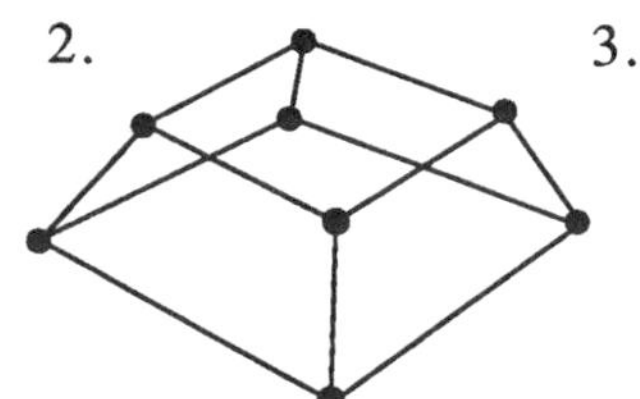

3.

4.

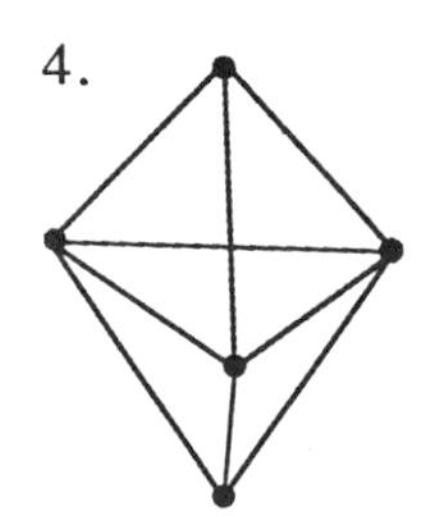

5. Build a stick tetrahedron.

6. Build a stick octahedron.

7. Build a stick decahedron.

8. Build a stick polyhedron that has four vertices and six edges. What type of polyhedron did you build? Can you build another polyhedron with a different number of faces that also has four vertices and six edges?

9. Build a stick polyhedron that has six vertices and twelve edges. What type of polyhedron did you build? Can you build another polyhedron with a different number of faces that also has six vertices and twelve edges?

10. Build another stick polyhedron of your own design.

EXERCISE SET B

1. Copy and complete the chart below for each of the stick polyhedrons that you built in Exercise Set A. Use the ten stick polyhedrons to complete the chart. You will have to count the number of vertices (V), edges (E), and faces (F) of each polyhedron.

Polyhedron	Vertices (V)	Edges (E)	Faces (F)
Exercise A1	–?–	–?–	–?–
Exercise A2	–?–	–?–	–?–
Exercise A3	–?–	–?–	–?–
⋮	⋮	⋮	⋮
Exercise A10	–?–	–?–	–?–

2. There is a special relationship relating the number of vertices (V), edges (E), and faces (F) in a polyhedron. Search the table to discover this relationship. By adding, subtracting, or multiplying (or some combination of the three) V, E, and F you can create a formula that will work for all polyhedrons. This formula was first discovered by the famous Swiss mathematician Leonard Euler and is commonly known as Euler's Formula.

Now that you have discovered the formula relating the number of vertices, edges, and faces of a polyhedron, use it to answer each question below.

3. If a solid has 8 faces and 12 vertices, how many edges will it have?

4. If a solid has 7 faces and 12 edges, how many vertices will it have?

Special Project

The Five Platonic Solids

Plato

Regular polyhedrons have intrigued mathematicians for thousands of years. They were important to ancient Greek scholars who placed a great emphasis on the study of science: chemistry, physics, mathematics, and biology. Greek philosophers saw the principles of these disciplines as the guiding forces in the universe. Empedocles studied nature and believed that all things were composed of four elements: earth, air, fire, and water. Another great philosopher, Pythagoras, conjectured that numbers ruled the universe. All forces, both natural and human, obeyed the neat precise rules of arithmetic. The most famous of all Greek philosophers, Plato, combined many of the ideas of Empedocles, Pythagoras, and others and set up a complete explanation of the nature of all things in his dialogue *Timaeus*.

Plato reasoned that since all objects are three dimensional, their smallest parts, atoms, must be in the solid shape of the regular polyhedrons which are explainable by mathematics. It turns out that there are only five regular polyhedrons. These five geometric solids are commonly called the **Platonic Solids**.

In Plato's view, all things were composed of the five different atoms. Four of them, earth, air, fire, and water, he took from Empedocles. The fifth atom was the atom that made up the cosmos, the stars and planets in the sky. Plato assigned one of the five regular solids as the shape for each of the five atoms.

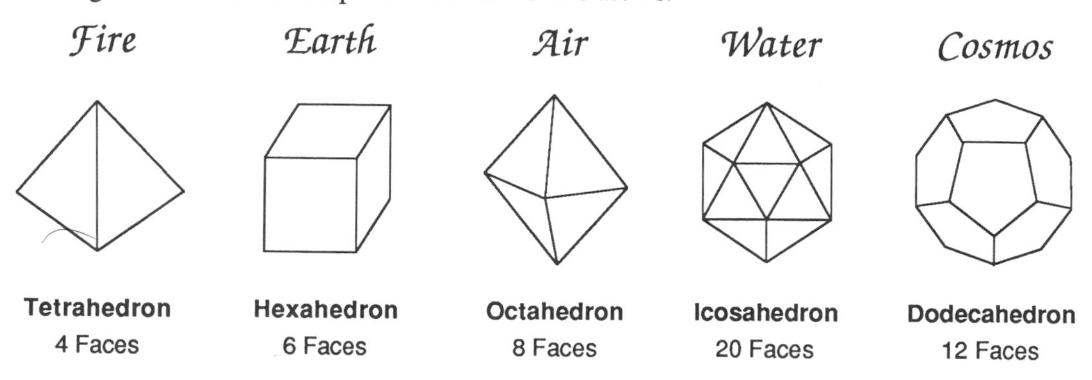

Plato argued that fire atoms were made in the shape of regular tetrahedrons since fire was the lightest atom and the tetrahedron had the least number of sides. In addition, the regular tetrahedron had the sharpest points and, therefore, it must be responsible for the sharp sting of fire. Plato reasoned further that since fire, air, and water reacted the most with each other, they must have atoms that are similar in shape. Since the octahedron, icosahedron, and tetrahedron all have equilateral triangles for faces, then air, water, and fire must have these shapes. It followed that since air was the second lightest of the three atoms, its shape must be the octahedron since the octahedron has the second least number of sides of the three. The water atom was therefore in the shape of the regular icosahedron. Plato then claimed that the earth atoms were in the shape of cubes or regular hexahedrons since the cube was very stable like earth. The fifth and remaining regular polyhedron, the dodecahedron, was so unlike the others, having pentagonal faces, that Plato argued it must be the shape of the atoms of the cosmos.

What would each of the five Platonic solids look like when unfolded? Use the pictures above to help visualize the solids. There is more than one possible way to unfold each polyhedron.

1. Complete the picture on the right to show what the tetrahedron would look like if it were cut open along the three lateral edges and unfolded in one piece. One face is missing.

2. Complete the picture of what the hexahedron would look like if it were cut open along the lateral edges and three top edges and unfolded. Two faces are missing. How many other patterns are possible for a regular hexahedron? Sketch them.

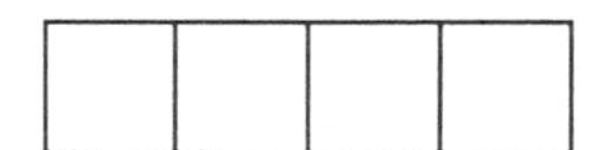

3. To the right is a picture of what the icosahedron would look like if it were cut along some edges and unfolded in one piece. When folded back together, the five top triangles all meet at one top point. The edge labeled with x will line up with which edge, a, b, or c?

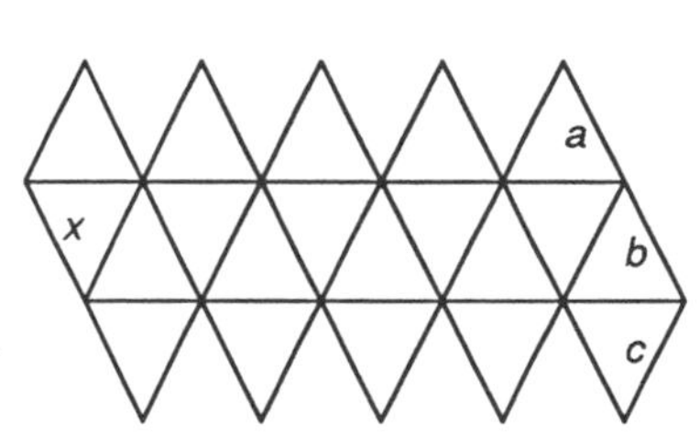

4. The octahedron is similar to the icosahedron but with only eight equilateral triangles. Complete the picture on the right to show what the octahedron would look like if it were cut along some edges and unfolded in one piece. Two faces are missing.

5. The dodecahedron is made with twelve regular pentagons. Suppose you were to cut the dodecahedron into two equal parts. They would resemble two flowers each having a pentagon center and five pentagon shaped petals around the center pentagon. If one of the halves of the dodecahedron were cut along edges connecting the petals and unfolded, what would it look like? Complete the pattern for half a dodecahedron.

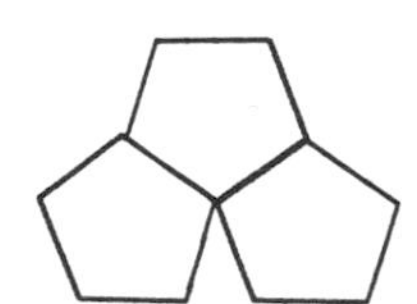

Lesson 10.2

Prisms and Pyramids

All of the solids below are prisms.

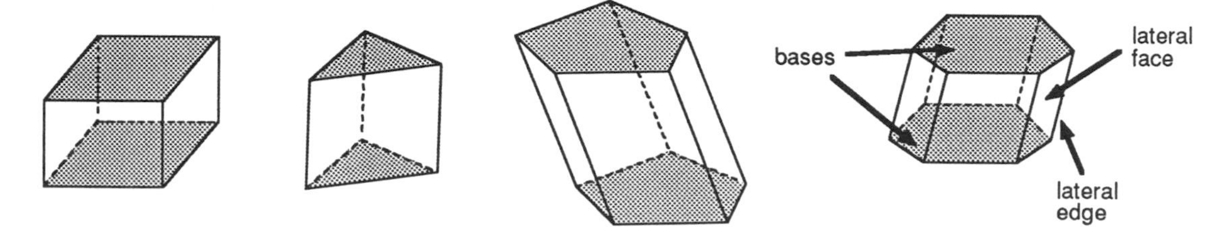

The two shaded faces of each prism are called the **bases** of the prism. The faces that are not the bases are called the **lateral faces**. The lateral faces meet to form the **lateral edges**.

How would you write a good definition of a prism? You might begin by writing down some statements based on your observations.

Statement 1: A prism is a polyhedron.

Statement 2: The bases are congruent and parallel polygons.

Statement 3: The lateral faces are parallelograms formed by segments that connect the corresponding vertices of the bases.

You can combine these three statements into a definition.

> *A **prism** is a polyhedron with two faces (bases) that are congruent and parallel polygons and whose other faces (lateral faces) are parallelograms formed by segments connecting the corresponding vertices of the bases.*

You classify prisms by their bases. A prism with triangular bases is a **triangular prism**; one with hexagonal bases is a **hexagonal prism**; etc.

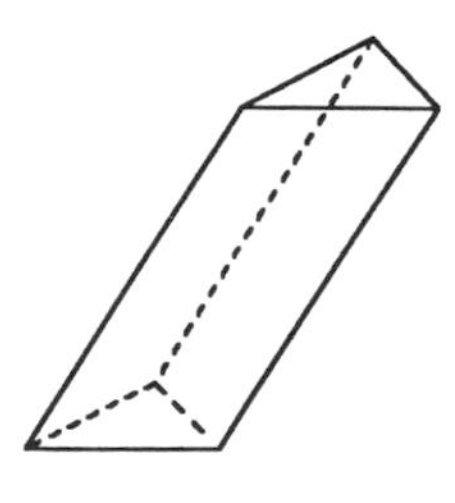

Triangular Prism

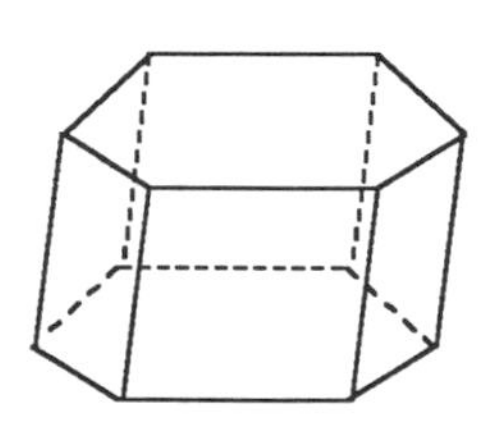

Hexagonal Prism

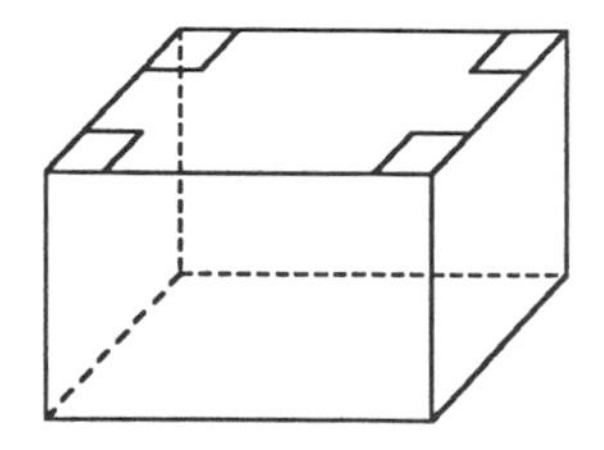

Rectangular Prism

When all the lateral faces of a prism are rectangles (when the lateral edges are perpendicular to the bases), it is a **right prism**. A prism that is not a right prism is called an **oblique prism**. An **altitude** of a prism is a perpendicular segment from one base to the plane of the other. The **height** of the prism is the length of an altitude.

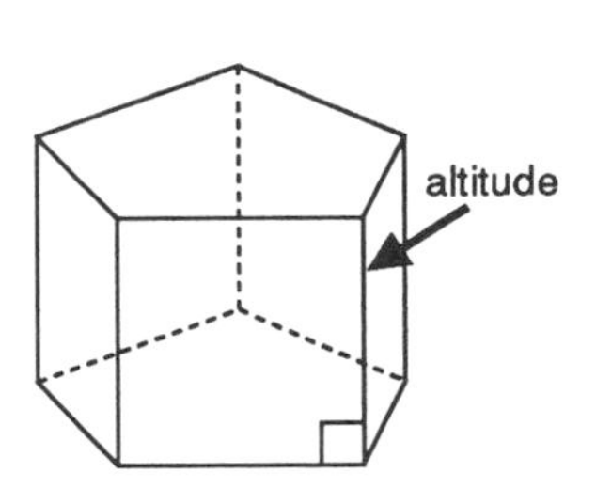

Right Pentagonal Prism

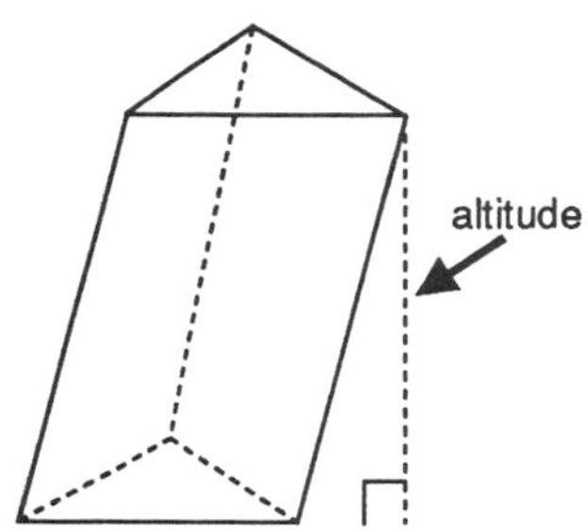

Oblique Triangular Prism

All of the solids below are pyramids.

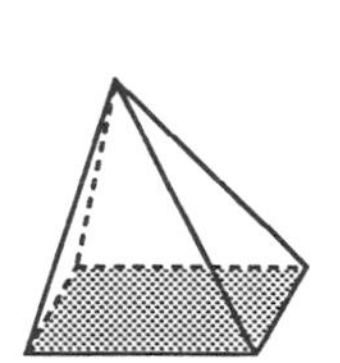

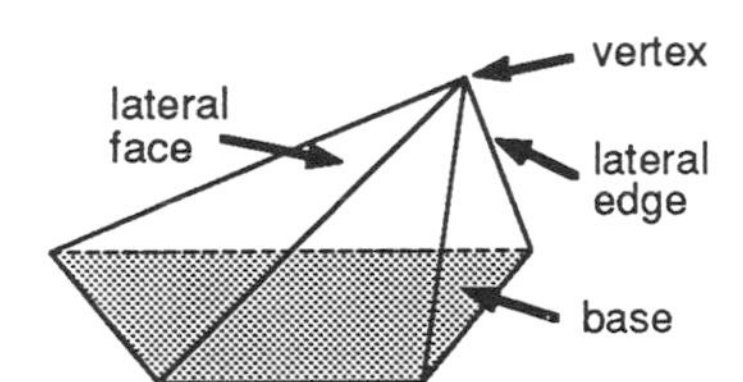

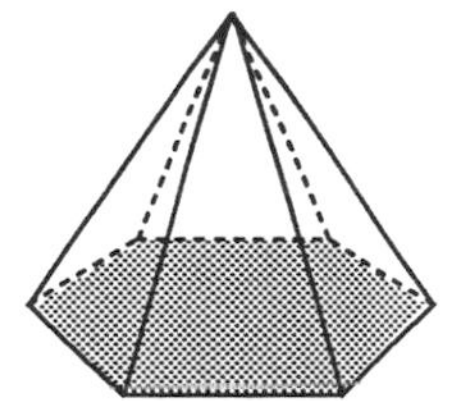

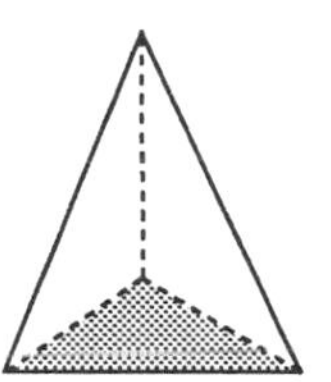

The shaded face of each pyramid is called the **base** of the pyramid. The faces that are not the bases are called the **lateral faces**. The lateral faces meet to form the **lateral edges**. The common vertex of the lateral faces is *the* **vertex** of the pyramid.

Pyramids are also classified by their bases. The pyramids of Egypt are square pyramids because they have square bases.

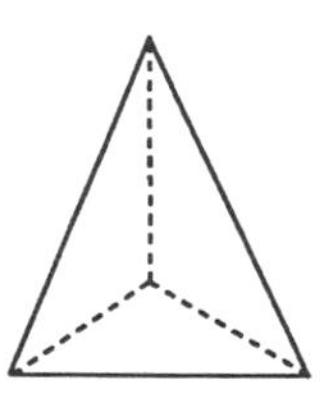

Triangular Pyramid

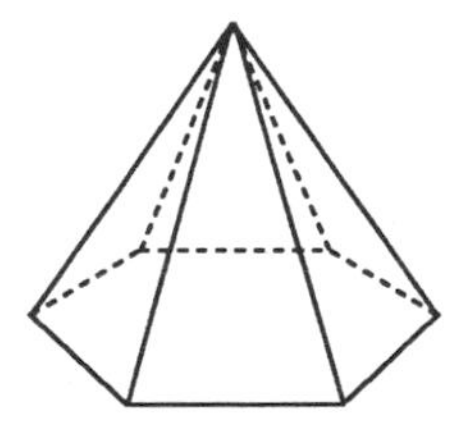

Hexagonal Pyramid

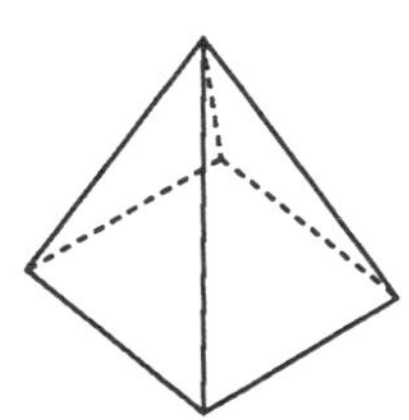

Square Pyramid

The perpendicular segment from the vertex to the plane of the base is the **altitude** of the pyramid and the **height** is the length of the altitude.

EXERCISE SET A

Now it's your turn to write a good definition of a pyramid. Begin by writing down some statements based on your observations.

1. A pyramid is a —?—.
2. The base is a —?—.
3. The lateral faces are —?— formed by segments that connect the vertices of the base to the —?—.

Now combine these three statements into a definition.

4. Define *pyramid.*

EXERCISE SET B

In the diagrams at the right, all measurements are in centimeters.

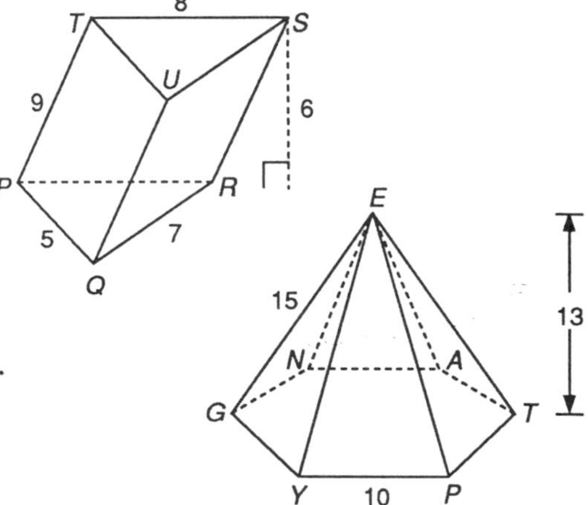

1. Name the bases of the prism.
2. Name all the lateral faces of the prism.
3. Name all the lateral edges of the prism.
4. What is the height of the prism?
5. Name the base of the pyramid.
6. Name the vertex of the pyramid.
7. Name all the lateral edges of the pyramid.
8. What is the height of the pyramid?

EXERCISE SET C

Identify each statement as true or false.

1. The lateral face of a pyramid is always a triangular region.
2. A lateral edge of a pyramid is always perpendicular to the base.
3. A lateral face of a prism can be a triangular region.

4.* Every slice of a prism cut parallel to the bases is congruent to the bases. (Remember, a section is the plane figure formed by slicing a solid with a plane. A section cut parallel to a base of a solid is called a **cross section.**)

5. Every cross section of a pyramid is congruent to the base.
6. Every cross section of a pyramid has the same shape as the base but has a different size.

EXERCISE SET D

How many cubes with one centimeter on an edge will fit into each container below?

1. A box 2 centimeters on each inside edge
2. A box 3 centimeters on each inside edge
3. A box 4 centimeters on each inside edge
4. A box measuring 3 cm by 4 cm by 5 cm on the inside edges

EXERCISE SET E

Draw and label each solid. Use dashed lines to show the hidden edges.

1. A triangular pyramid with the base an equilateral triangular region (Use the proper marks to show that the base is equilateral.)
2. A hexagonal prism with regular bases
3. A rectangular prism that is not a right prism

Improving Visual Thinking Skills

Missing Shapes

Sketch the two missing shapes.

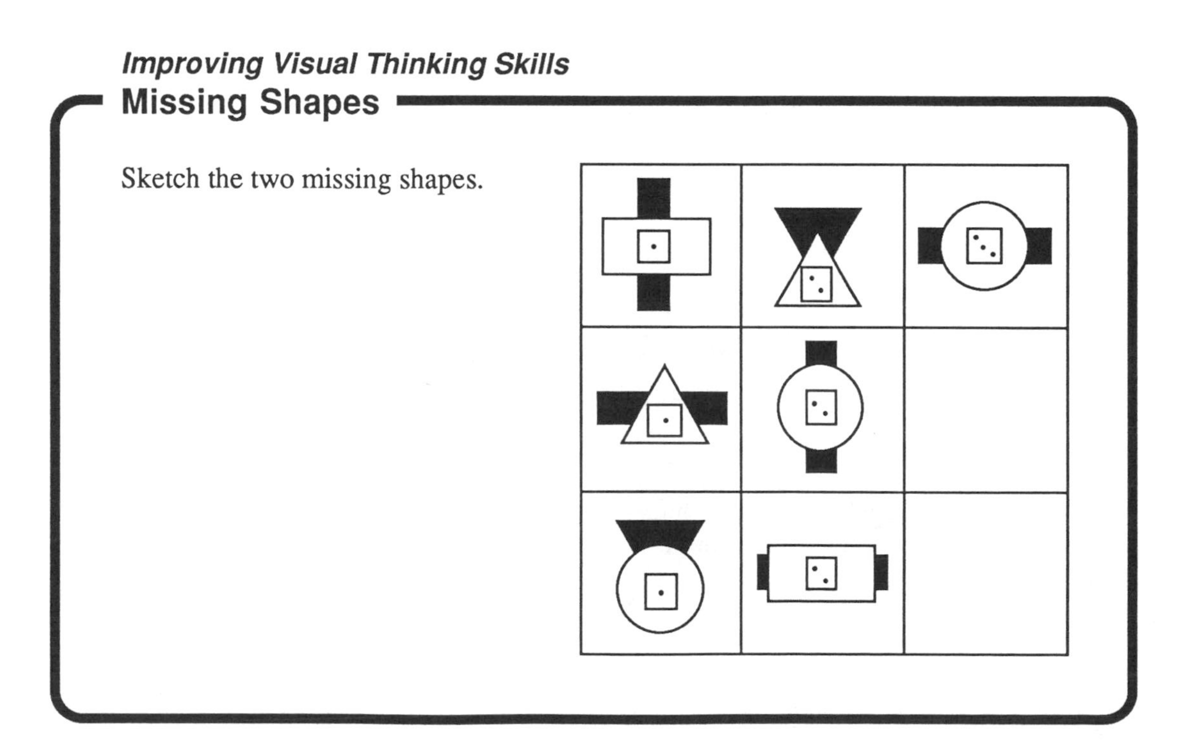

Lesson 10.3

Solids with Curved Surfaces

Polyhedrons are geometric solids with flat surfaces. This lesson introduces three important geometric solids that have curved surfaces.

The solid with a curved surface that all sports fans know so well is the ball or sphere. A sphere is a three-dimensional circle. Marbles, peas, scoops of ice cream have surfaces that are spheres.

> *A **sphere** is the set of all points in space at a given distance from a given point.*

The given distance is called the **radius** and the given point is the **center**. A **hemisphere** is half a sphere.

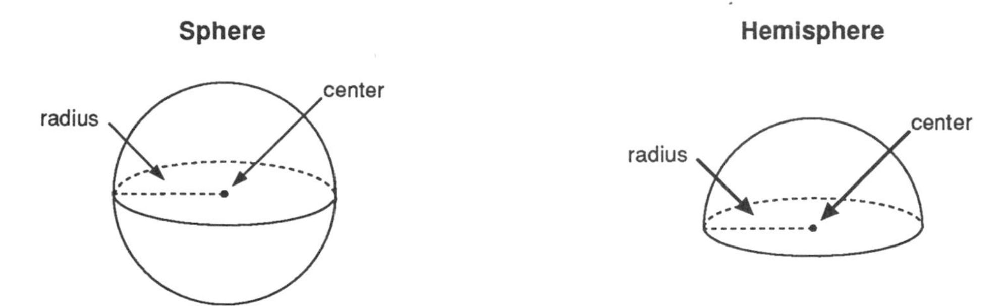

Another solid with a curved surface is a cylinder. Soup cans, rolls of paper towels, records, and plumbing pipes are shaped like cylinders. Like a prism, a cylinder has two bases that are both parallel and congruent. Instead of polygons, however, the bases of cylinders are circles. The segment connecting the centers of the circles is called the **axis** of the cylinder. A cylinder can be defined using its axis.

> *A **cylinder** is a solid composed of two congruent circles in parallel planes, their interiors, and all the line segments parallel to the axis with endpoints on the two circles.*

The circles and their interiors are the **bases**. The **radius** of the cylinder is the radius of a base.

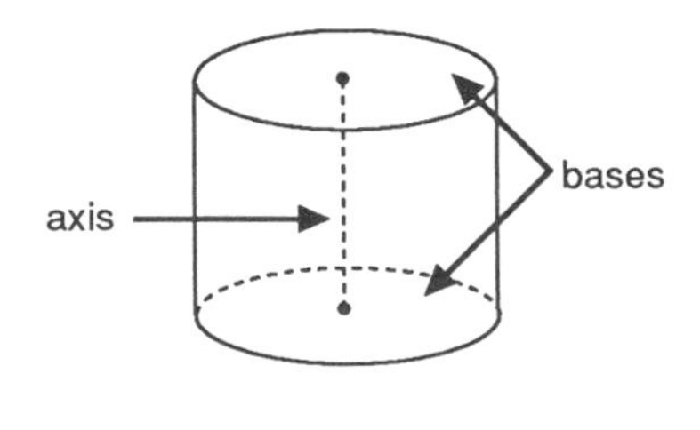

Cylinder

If the axis of a cylinder is perpendicular to the bases, then the cylinder is a **right cylinder**. A cylinder that is not right is **oblique**.

Oblique Cylinder

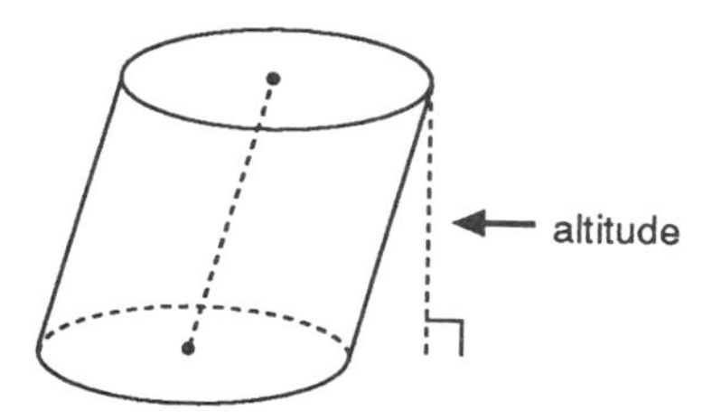

Right Cylinder

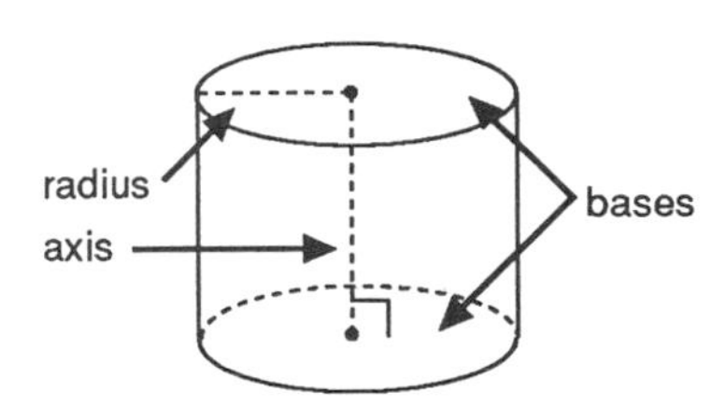

The **altitude** of a cylinder is a perpendicular segment from the plane of one base to the plane of the other. The **height** of a cylinder is the length of an altitude.

A third type of solid with a curved surface is a cone. Funnels, ice cream cones, and martini glasses are shaped like cones. Like a pyramid, a cone has a base and vertex.

> *A* ***cone*** *is a solid composed of a circle, its interior, a given point not on the plane of the circle, and all the segments from the point to the circle.*

The circle and its interior make up the **base** of the cone. The **radius** of the cone is the radius of the base. The point is the **vertex** of the cone. The **altitude** of a cone is the perpendicular segment from the vertex to the plane of the base. The **height** of a cone is the length of the altitude. If the line segment connecting the vertex of a cone with the center of its base is perpendicular to the base, then it is a **right cone**.

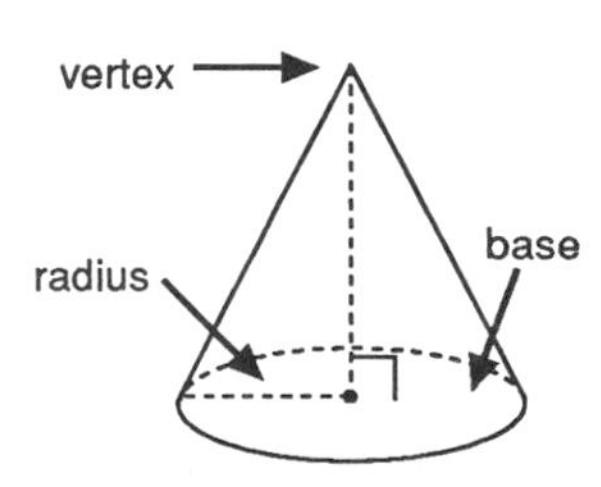

Right Cone

EXERCISE SET A

Identify each statement as true or false.

1. The lateral surface of a right cylinder, when unwrapped and laid flat, is rectangular.
2.* The lateral surface of a right circular cone, when unwrapped and laid flat, is triangular.
3. Every cross section of a cylinder is congruent to the base.
4. Every cross section of a cone is the same shape as the base but a different size.

5. The length of a segment from the vertex of a cone to the circle defining the base is the height of the cone.
6. The length of the axis of a right cylinder is the height of the cylinder.
7. Every slice of a sphere passing through the center of the sphere is congruent.
8. The longest segment connecting two points on a sphere must pass through the center of the sphere.

EXERCISE SET B

Draw and label each solid. Use dashed lines to show the hidden edges.

1. A cylinder with a height that is twice the diameter of the base (Use x and $2x$ to indicate the height and diameter.)
2. A right cone with the height half the diameter of the base
3. A sphere with a 90 degree slice to the center removed
4. A cylinder with both the radius and height r, a cone with both radius and height r resting flush on one base of the cylinder, and a hemisphere with radius r resting flush on the other base of the cylinder

Spheres, right cylinders, and right cones are solids of revolution. There are many other familiar shapes that are solids of revolution. Donuts, washers, bowls, and flower pots have shapes that are solids of revolution.

EXERCISE SET C

What solid would result if you spun each two-dimensional figure in Exercises 1 to 7 about the axis indicated? Match each two-dimensional figure on the top row with a solid of revolution on the bottom row.

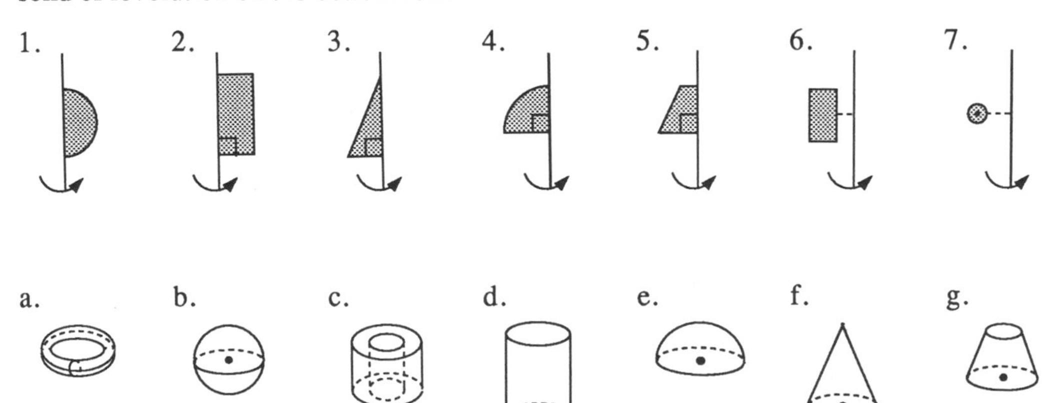

EXERCISE SET D

Match the shape of each real object on the left with a geometric name on the right. You may use a geometric name more than once or not at all.

1.*	Die	a.	Cylinder
2.	Tomb of Egyptian rulers	b.	Cone
3.	Holder for a scoop of ice cream	c.	Square prism
4.	Wedge or doorstop	d.	Square pyramid
5.	Box of breakfast cereal	e.	Sphere
6.	Bowl	f.	Triangular pyramid
7.*	Ingot of silver	g.	Octagonal prism
8.	Birthplace of a bee	h.	Pentagonal prism
9.	Stop sign	i.	Triangular prism
10.	U.S. military headquarters	j.	Trapezoidal prism
11.*	Pup tent	k.	Rectangular prism
12.	Moon	l.	Heptagonal pyramid
13.	Can of tuna fish	m.	Hexagonal prism
		n.	Hemisphere

Improving Reasoning Skills

Space Cadet Lance Marvelous

Once upon a time Space Cadet Lance Marvelous was forced to land on the mysterious planet Eslaf. The planet was inhabited by two mutant groups of people. One group, the Rail, spoke only lies to anyone of another group and only the truth to each other. The other group on the planet Eslaf, the Lepsog, always spoke the truth except to a Rail, to whom they always lied.

Upon stepping off his spacecraft, Lance was approached by three Eslafians. Lance inquired to which group they belonged. The first spoke something to Lance but the creature mumbled and Lance was unable to make out what was said. The second said to Lance, "She said she was a Rail." The third said to the second, "You lie!" From this Lance was able to determine the group of the third Eslafian. Can you? To which group did the third Eslafian belong?

Lesson 10.4

Volume of Prisms and Cylinders

Say what you know, do what you must, come what may.

– Motto of Sonia Kovalevsky

***Volume** is the measure of the amount of space contained in a solid.*

On a typical day you will usually encounter some volume problems. Whether you are shopping for a container of ice cream, buying gas for your car, or trying to fit last night's leftovers into a freezer container, you must be familiar with volume. Many occupations also require familiarity with volume. A home economist compares volume and price of different brands of a product to find the best bargain. A chef must measure the correct volume of each ingredient in a cake to be sure of a tasty success. An engineer needs to calculate volume to determine the weights of sections of a bridge in order to calculate stresses. Chemists, biologists, physicists, and geologists must all make careful volume measurements in their research.

To measure volume you use cubic units: cubic inches, cubic feet, cubic yards, cubic centimeters (cm^3), cubic meters (m^3), etc. The volume of an object is the number of unit cubes that can be arranged to completely fill the space within the object.

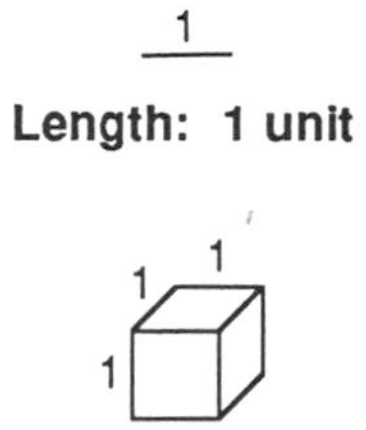

1

Length: 1 unit

1 1 1

Volume: 1 cubic unit

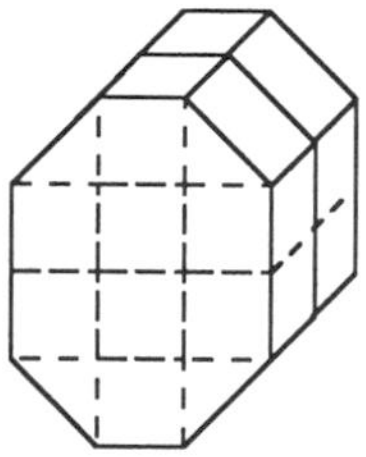

Volume: 20 cubic units

If an object is irregularly shaped, it is quite difficult to figure out how many unit cubes it contains. You'll learn a method in Lesson 10.8 which is useful in determining the volume of irregularly shaped objects. Let's begin our investigation of volume by looking at the volumes of some simple solids.

In an earlier lesson you calculated the numbers of cubes with one centimeter on an edge that would fit into different sized containers. The containers were right rectangular prisms with whole number dimensions which enabled the cubes to fit into them perfectly. It's easy to find the volume of a solid when it is a right rectangular prism with whole number dimensions.

Investigation 10.4.1

Find the volume of each right rectangular prism below. (How many cubes with one centimeter on an edge will fit into each solid?)

1. 2.

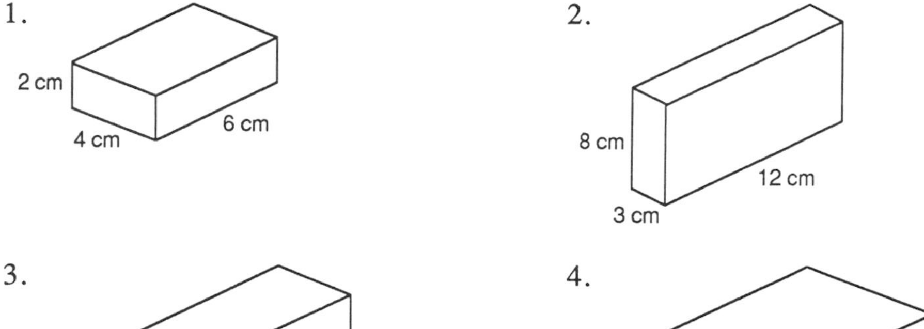

3.

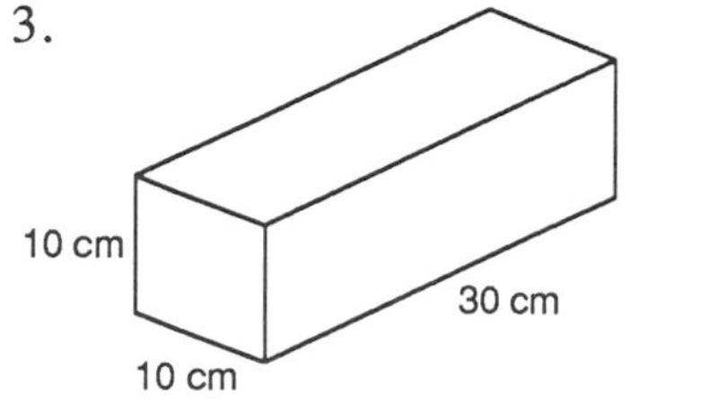

4.

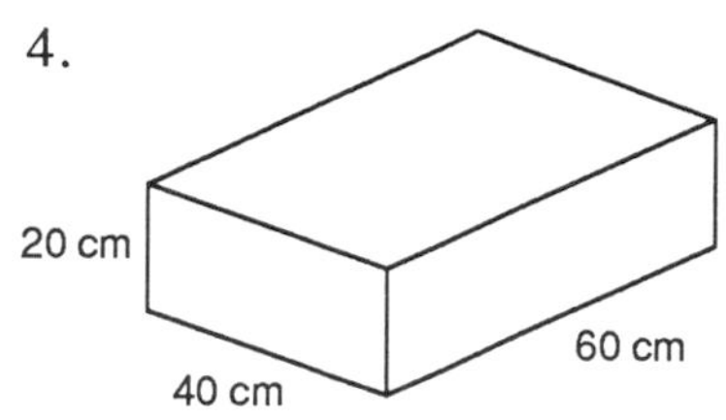

The problems above should lead you to a formula for the volume of any right rectangular prism. For example, if a right rectangular prism holds 6 layers of cubes (one inch on each edge) and each layer is 4 cubes by 5 cubes, then the volume would be (4)(5)(6) or 120 cubic inches. Notice that the number of cubes resting on the base equals the number of square units in the area of the base. The number of layers of cubes equals the number of units in the height of the prism. Therefore, you can multiply the area of the base by the height of the prism to calculate the volume. Complete the statement below.

Conjecture: *If B is the area of the base of a right rectangular prism and H is the height of the solid, then the formula for the volume is V = —?—.*

It is almost as easy to visualize the volume of a right prism that has non-rectangular bases as it is to visualize the volume of a right prism with rectangular bases. Just as you can find the area of a parallelogram by rearranging it into a rectangle, you can find the volume of a right parallelogram based prism by rearranging it into a right rectangular prism. Just slice the parallelogram based prism into two prisms as shown in the diagram below. Then move the triangular prism to form a right rectangular prism. You can calculate the volume by using the formula above.

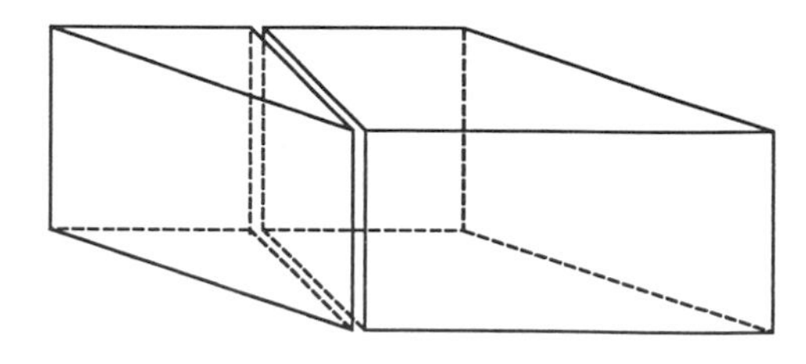

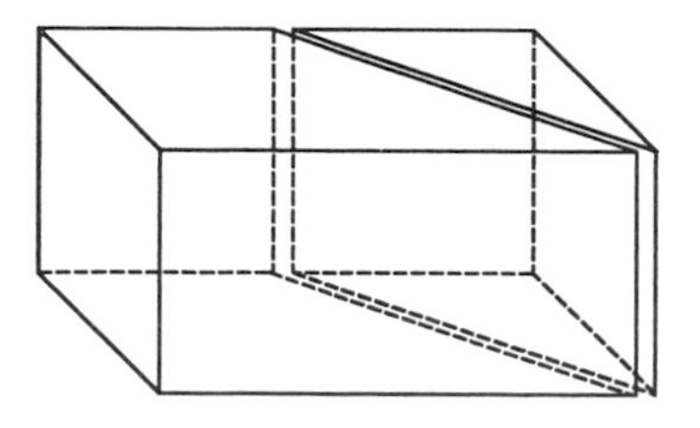

Likewise, the volume of a right triangular prism or a right trapezoidal prism can be determined. By a similar argument a right cylinder can be rearranged into a right prism with a rectangular base (just as you rearranged a circle to find its area).

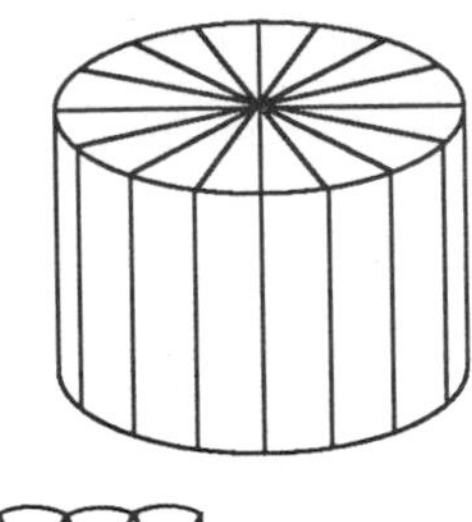

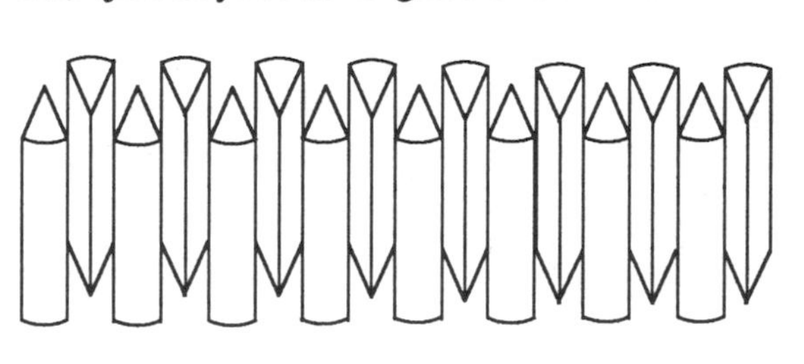

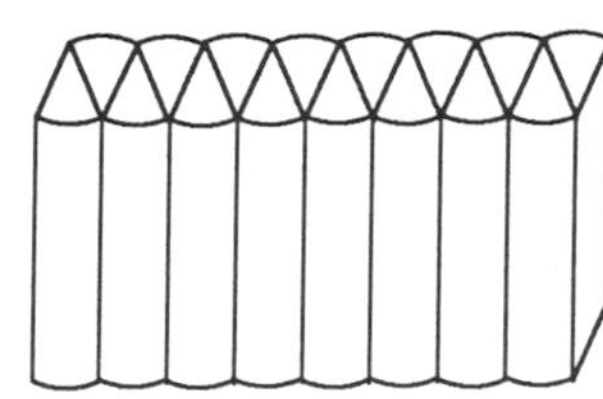

Therefore, the formula for finding the volume of right rectangular prisms can be extended to cover all right prisms and right cylinders. Complete the statement below.

Conjecture: *The volume of a right prism (or cylinder) is equal to the volume of a right rectangular prism (or right cylinder) with the same base —?— and the same —?—.*

What about the volume of a prism or cylinder when the solid is not a right prism or right cylinder? Suppose you wanted to find the volume of the oblique rectangular prism below with a base 8 inches by 11 inches and a height of 4 inches.

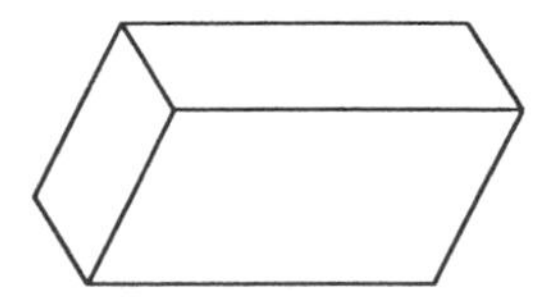

Oblique Rectangular Prism

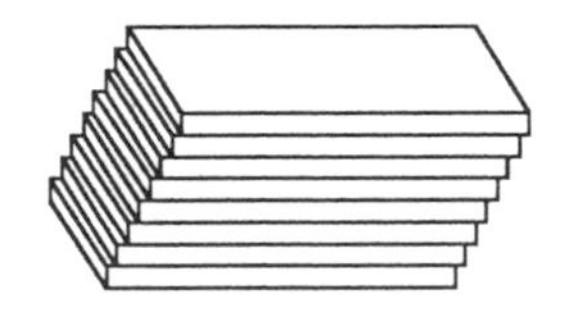

Stacked Plywood

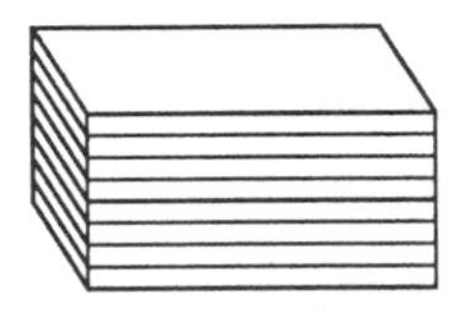

Plywood Stacked Straight

The shape of the oblique rectangular prism can be approximated by a stack of eight plywood pieces, each a right rectangular prism 8 inches by 11 inches by 1/2 inch thick. The stack can be easily rearranged into a right rectangular prism and its volume can be easily calculated using the formula $V = BH$.

$$\begin{aligned} V &= BH \\ &= [(8)(11)](4) \\ &= (88)(4) \\ &= 352 \end{aligned}$$

The volume is 352 cubic inches.

In rearranging the plywood, the shape has changed but certainly the volume of wood hasn't changed. If you replace the plywood pieces with a 4-inch high stack of thinner cardboard sheets, the cardboard stack more closely approximates the shape of the original oblique rectangular prism but still has a volume of 352 cubic inches. If you replace the cardboard sheets with manila folder paper so that the stack is still 4 inches tall, the manila stack even more closely resembles the original prism but still has a volume of 352 cubic inches.

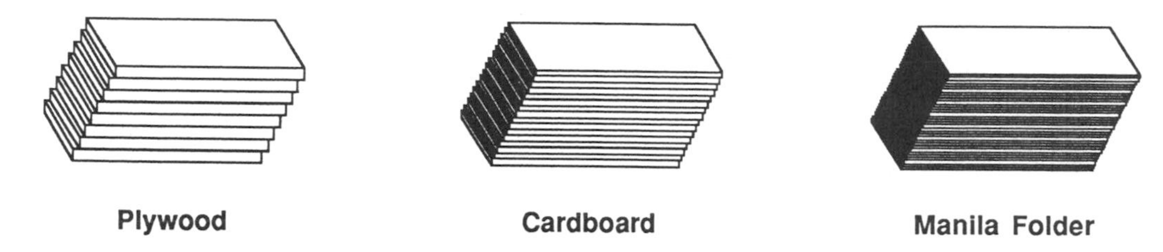

You can continue to make these approximations as close as you please to the shape of the original oblique rectangular prism by using sufficiently thin sheets. Then, by rearranging the sheets into a right rectangular prism, you can see that the volume of the oblique rectangular prism is the same as the volume of a right rectangular prism with the same base and the same height.

By a similar argument, an oblique cylinder can be shown to have the same volume as a right cylinder with the same base and height.

Let's use the stacking model to extend your formula for the volume of right prisms and cylinders to oblique prisms and cylinders. Complete the statement below.

Conjecture: *The volume of an oblique prism or cylinder is the same as the volume of a right prism which has the same base —?— and the same —?—.*

Finally, let's combine the three conjectures you've made in this lesson into a single conjecture for finding the volume of any prism or cylinder (whether it's right or oblique).

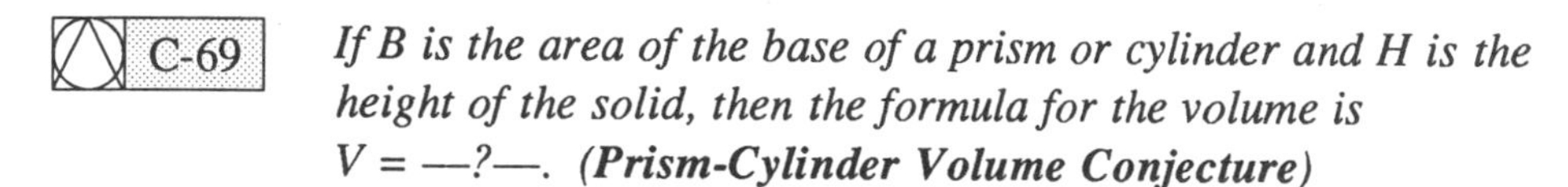

If B is the area of the base of a prism or cylinder and H is the height of the solid, then the formula for the volume is V = —?—. ***(Prism-Cylinder Volume Conjecture)***

Notice that the same volume formula applies to all prisms (regardless of the type of base they have). To calculate the volume of a prism, first identify the type of base the prism has and use the appropriate area formula to calculate its area. Then multiply the area of the base by the height of the prism. Notice that in oblique prisms, the lateral edges are no longer at right angles to the bases. You do *not* use the length of the lateral edge as the height.

Example A

Find the volume of a trapezoidal prism that has a height of 15 cm. The trapezoidal base has a height of 5 cm and the two bases of the trapezoid measure 4 and 8 cm.

$$\begin{aligned} V &= BH \\ &= [(1/2)(5)(4+8)](15) \\ &= (30)(15) \\ &= 450 \end{aligned}$$

The volume is 450 cm^3.

Example B

Find the volume of a cylinder that has a base with a radius of 6 inches and a height of 7 inches.

$$\begin{aligned} V &= BH \\ &= (\pi 6^2)(7) \\ &= 36\pi(7) \\ &= 252\pi \end{aligned}$$

The volume is 252π cubic inches.

EXERCISE SET A

Find the volume of each right prism or right cylinder named below. All measurements are in centimeters.

1. Rectangular prism

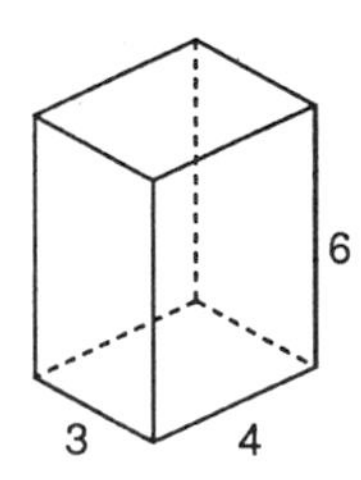

2.* Right triangular prism

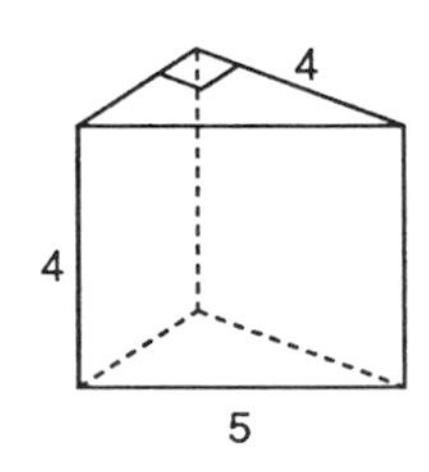

3.* Trapezoidal prism

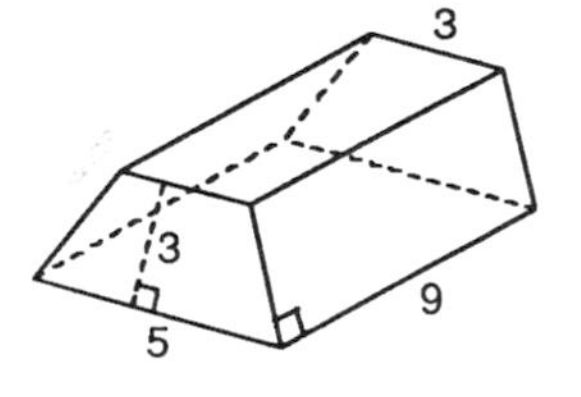

4. Cylinder

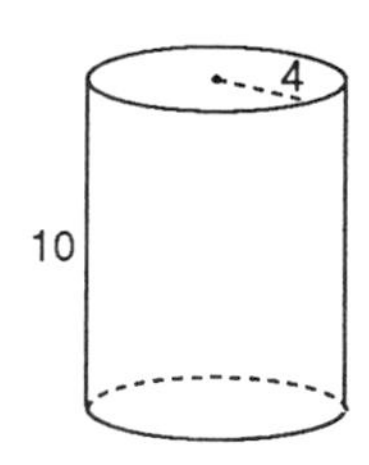

5.* Semicircular cylinder

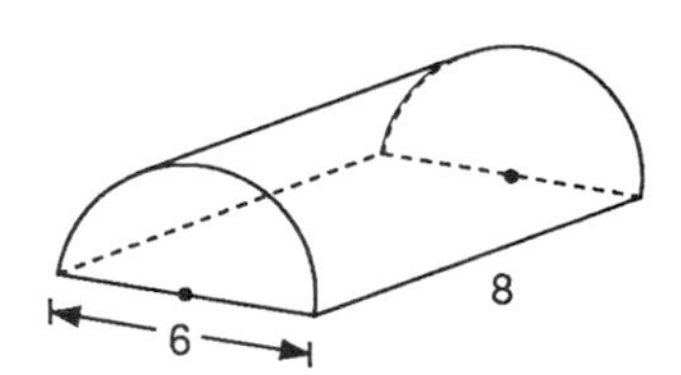

6.* Cylinder with a 90° slice removed

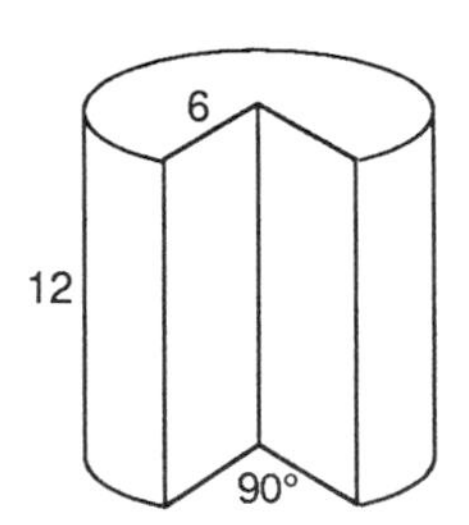

EXERCISE SET B

Use the information about the base and height of each solid to find the volume. All measurements given are in centimeters.		Triangular Prism	Rectangular Prism	Trapezoidal Prism	Cylinder
Information about base of solid	Height of solid	Use b, h, H.	Use b, h, H.	Use b, b_2, h, H.	Use r, H.
$b = 6, b_2 = 7, h = 8, r = 3$	$H = 20$	1.* $V = -?-$	4. $V = -?-$	7. $V = -?-$	10. $V = -?-$
$b = 9, b_2 = 12, h = 12, r = 6$	$H = 20$	2. $V = -?-$	5. $V = -?-$	8. $V = -?-$	11. $V = -?-$
$b = 8, b_2 = 19, h = 18, r = 8$	$H = 23$	3. $V = -?-$	6. $V = -?-$	9. $V = -?-$	12. $V = -?-$

Improving Visual Thinking Skills

Folding Cubes II

In each of the two problems below, the figure on the left represents a folded cube. When the cube on the left is unfolded, which figure on the right will it become?

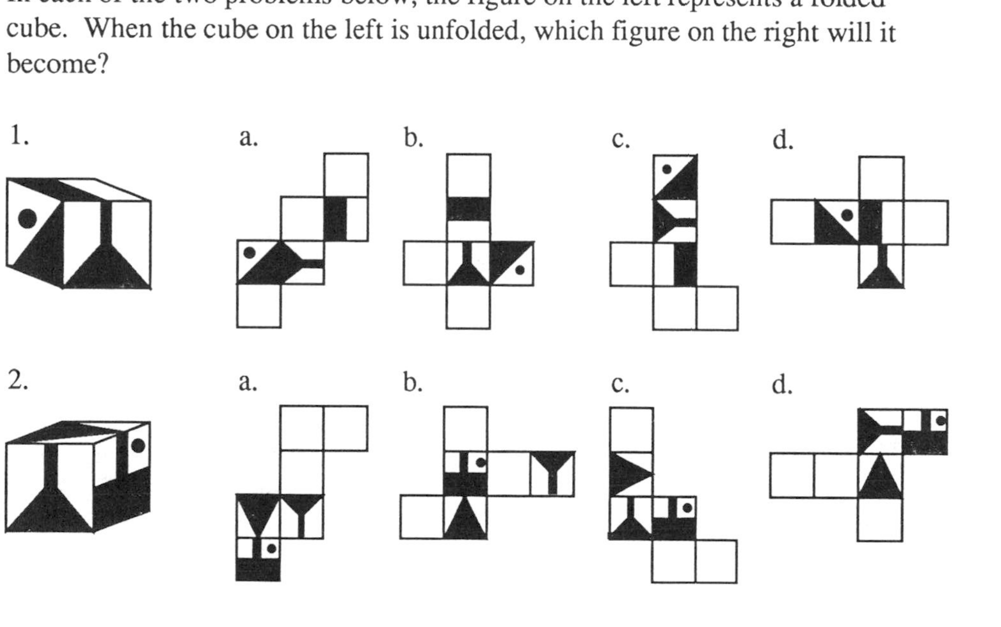

Lesson 10.5

Volume of Pyramids and Cones

In this lesson you will compare the volumes of pyramid-prism pairs and cone-cylinder pairs that have congruent bases and the same height.

If a prism and a pyramid have congruent bases and are the same height, will they have the same volume? The diagram below suggests that the pyramid will have less volume. How much less? How many times bigger is the prism than the pyramid? Is there a relationship? Does this relationship hold for cylinders and cones?

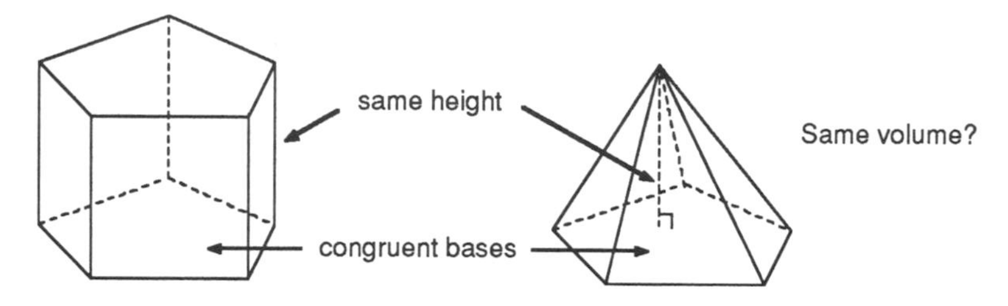

There is a simple relationship between the volumes of prisms and pyramids with congruent bases and the same height, and cylinders and cones with congruent bases and the same height. You can discover this relationship with a little investigation. You will need:

- Hollow pairs of prisms and pyramids with congruent bases and the same height
- Hollow pairs of cylinders and cones with congruent bases and the same height
- Sand or water (if your models are watertight)
- Cardboard box (sand) or plastic dishpan (sand or water)

If you already have the models you need for this investigation, follow the steps below. If models are not readily available, start by making your own. (See *Special Project: Building Solids*.)

This investigation works best if the tasks are shared. Work in groups of four or five.

Investigation 10.5.1

Step 1: Choose a prism-pyramid pair with congruent bases and the same height. (Try to use a pair with bases different from those of other groups.)

Step 2: Working over a box or dishpan, fill your pyramid with sand or water and pour the contents into your prism. Do it carefully. About what fraction of the prism is filled by the volume of one pyramid?

Step 3: Check your conjecture by repeating step 2 until the prism is filled.

Step 4: Choose a cone-cylinder pair with congruent bases and the same height and repeat the procedure in steps 2 and 3. (Try to use a pair with bases different from those of other groups.)

Compare your results with the results of others. Did you get similar results with both your pyramid-prism and cone-cylinder pairs? You should be ready to make a conjecture.

If B is the area of the base of a pyramid or cone and H is the height of the solid, then the formula for the volume is V = —?—. ***(Pyramid-Cone Volume Conjecture)***

Notice that all pyramids (regardless of the type of base they have) use the same volume formula. To calculate the volume of a pyramid, first identify the type of base the pyramid has and use the area formula for that base to calculate its area. Then take the product of the area of the base and the height of the pyramid and multiply by the fraction discovered in the investigation above.

EXERCISE SET A

Find the volume of each solid named below. All measurements are in centimeters.

1.* Trapezoidal pyramid

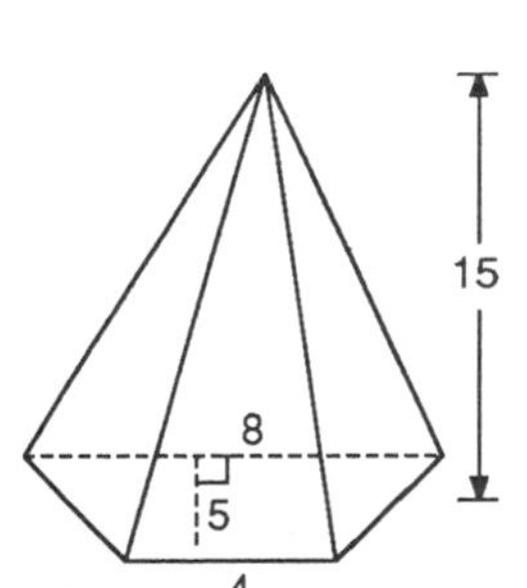

2.* Right cone

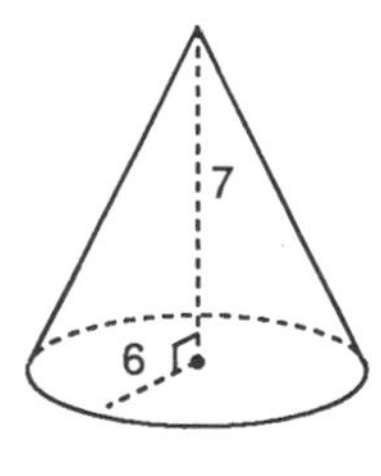

3. Square pyramid

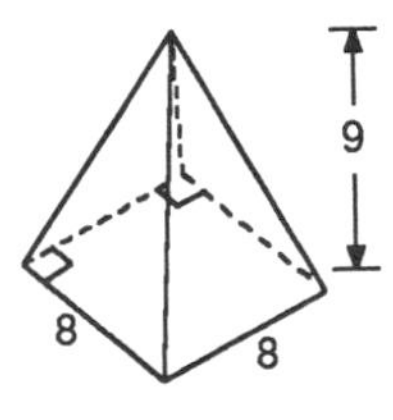

4. Semicircular cone

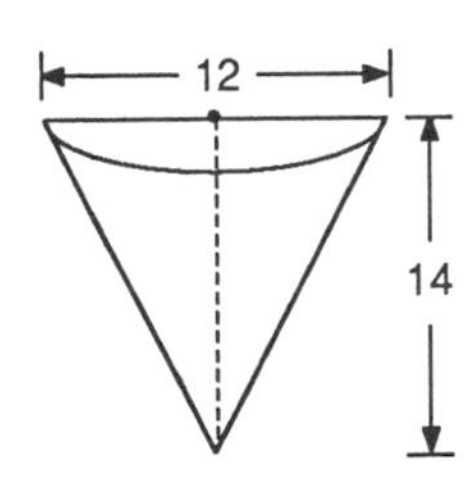

5.* Right triangular pyramid

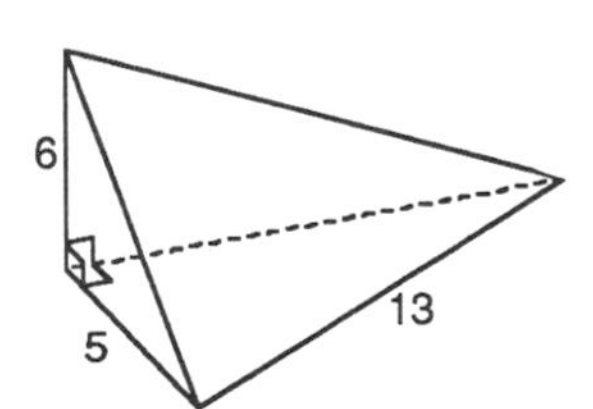

6.* Cylinder with cone removed

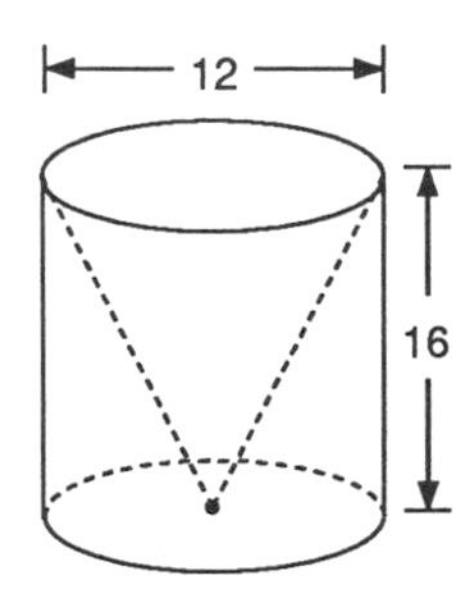

EXERCISE SET B

Use the information about the base and height of each solid to find the volume. All measurements given are in centimeters.		Triangular Pyramid	Rectangular Pyramid	Trapezoidal Pyramid	Cone
Information about base of solid	Height of solid	Use b, h, H.	Use b, h, H.	Use b, b_2, h, H.	Use r, H.
$b = 6, b_2 = 7, h = 6, r = 3$	$H = 20$	1. $V = $ –?–	4. $V = $ –?–	7. $V = $ –?–	10. $V = $ –?–
$b = 9, b_2 = 22, h = 8, r = 6$	$H = 20$	2. $V = $ –?–	5. $V = $ –?–	8. $V = $ –?–	11. $V = $ –?–
$b = 13, b_2 = 29, h = 17, r = 8$	$H = 24$	3. $V = $ –?–	6. $V = $ –?–	9. $V = $ –?–	12. $V = $ –?–

Improving Reasoning Skills

Bert's Magic Hexagram

Bert is the queen's favorite drone. When not busy with the queen, he entertains himself with puzzles. Bert was in the process of creating a magic hexagram on the front of a grid of 19 hexagons of the honeycomb when he was interrupted by a visit from the queen. When Bert's magic hexagram (like its cousin the magic square) is completed, it will have the same sum in every straight hexagonal row, column, or diagonal (whether it is three, four, or five hexagons long). For example, B + 12 + 10 is the same sum as B + 2 + 5 + 6 + 9 which is the same sum as C + 8 + 6 + 11. Bert planned to use just the first 19 positive integers (his age in days), but he only had time to place the first twelve integers before he had to see to the queen's needs. Can you complete Bert's magic hexagram? What are the values for A, B, C, D, E, F, and G?

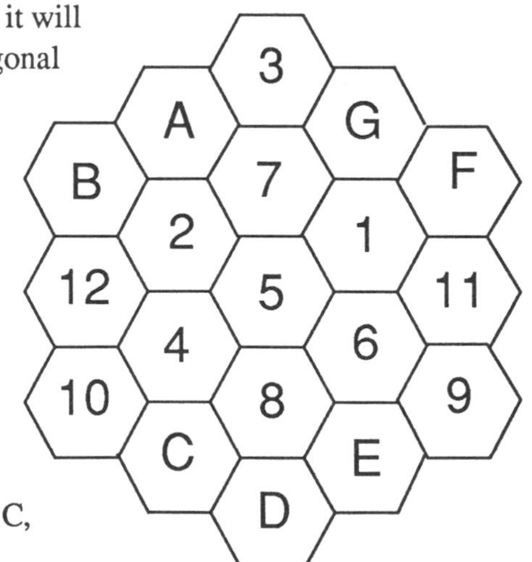

Special Project

Building Solids

There are models available commercially that are designed to show the relationship between the volumes of prisms and pyramids with congruent bases and the same height and the volumes of cylinders and cones with congruent bases and the same height. Making your own models, however, can be both instructive and enjoyable. This special project and the investigation in Lesson 10.5 work best if the tasks are shared. Work in groups of four or five.

Prisms and Pyramids

Each group should build a prism-pyramid pair with bases different from those of other groups. To build a prism-pyramid pair with congruent bases and the same height, you need:

- Matboard or very rigid cardboard (about 18" by 24")
- Two plastic drafting right triangles (or two rigid objects with right angles)
- Scissors and tape
- Compass and straightedge (at least 18" long)

Step 1: From a point P near one side of the matboard, swing an arc AE (greater than a semicircle) with a radius of about 5 inches. To build a pyramid with a quadrilateral base, select three points B, C, and D on the arc. (If you wish to build a pyramid with a triangular base, select two points on the arc; for a pentagonal base, select four points; and so on.) Draw chords $\overline{AB}$, $\overline{BC}$, $\overline{CD}$, $\overline{DE}$ and segments $\overline{PA}$, $\overline{PB}$, $\overline{PC}$, $\overline{PD}$, and $\overline{PE}$.

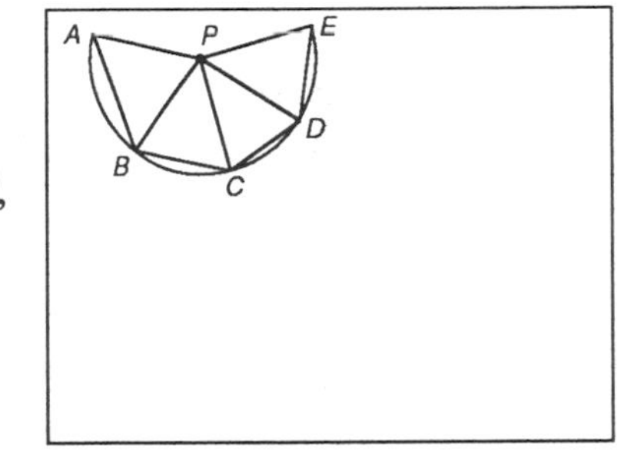

Step 2: Near one edge of the matboard, draw a segment and on it mark off lengths AB, BC, CD, and DE. (This segment will be the base of a large rectangle that will determine the lateral surface of your prism. Give yourself room to work.)

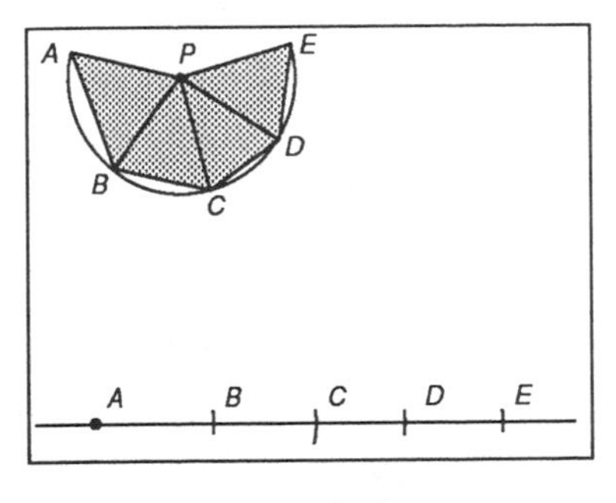

Step 3: Cut out hexagon $PABCDE$. Score the edges $\overline{PB}$, $\overline{PC}$, and $\overline{PD}$ with a ball point pen or the point of your compass. Crease these segments and bring the edges $\overline{PA}$ and $\overline{PE}$ together to form a quadrilateral based pyramid. Tape along the inside and outside of edge $\overline{PA}/\overline{PE}$.

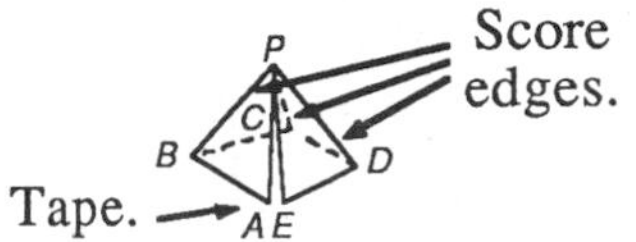

Step 4: Stand the pyramid on the matboard and gently press down on its vertex so that all four base edges lie flush on the board. Mark the vertices of the base on the matboard and connect them with a straightedge. Cut out the quadrilateral base. Use it to make a second copy for the prism. Cut off one corner of the base (to make a spout) and attach the base to the pyramid to make the pyramid rigid.

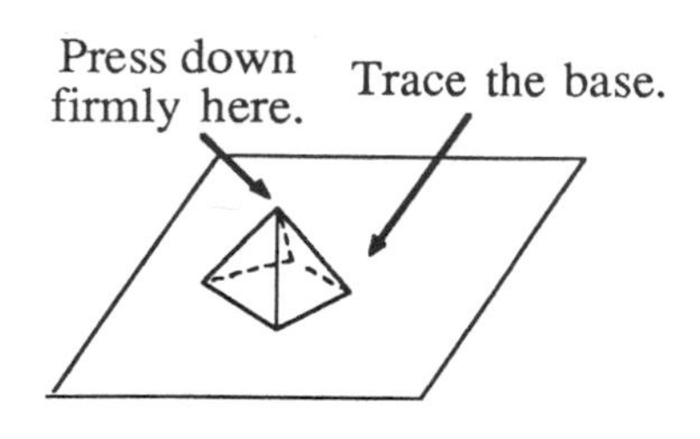

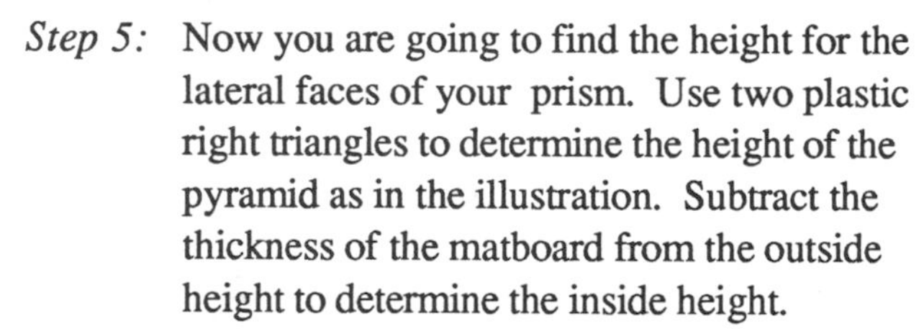

Step 5: Now you are going to find the height for the lateral faces of your prism. Use two plastic right triangles to determine the height of the pyramid as in the illustration. Subtract the thickness of the matboard from the outside height to determine the inside height.

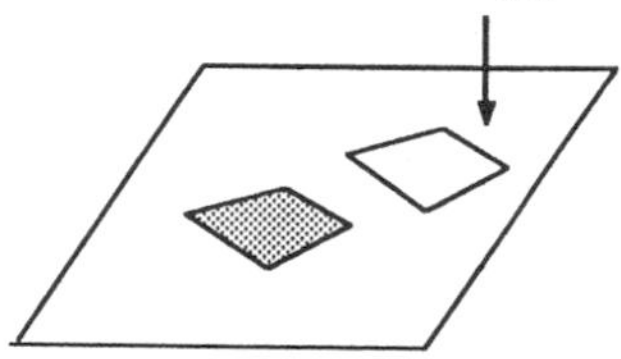

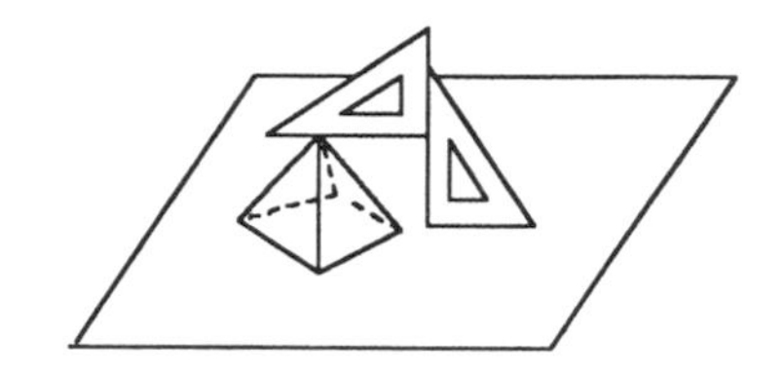

Step 6: Next, you are going to make the lateral faces of your prism. Construct perpendiculars from the endpoints of the segment you made with length $AB + BC + CD + DE$. From the endpoints of the segment, mark the inside height of your pyramid on the two perpendiculars. Complete the rectangle. The base should have length $AB + BC + CD + DE$ and the height should be the height of your pyramid. Across the top of the rectangle mark off lengths AB, BC, CD, and DE (the same lengths as on the bottom). Connect the points above each other on the two bases to make rectangles for each lateral face of your prism.

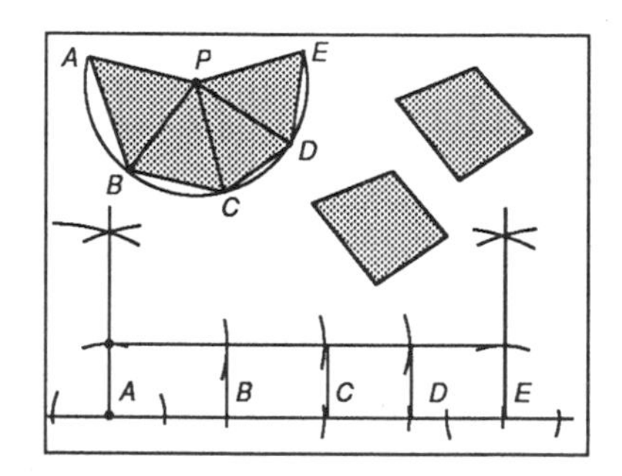

Step 7: Cut out the quadrilateral base and the rectangular lateral surface. Score the necessary lateral edges. Fold and tape edges together securely to form a quadrilateral prism.

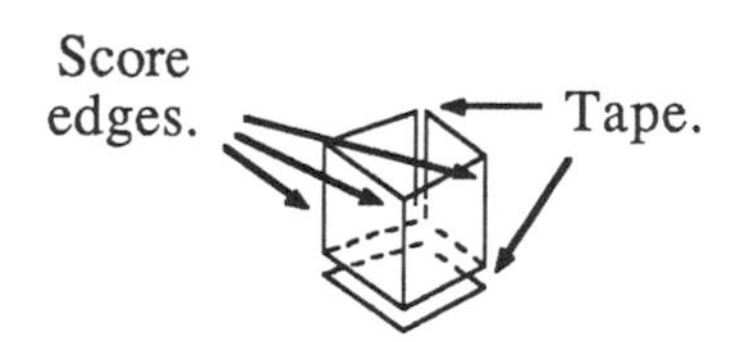

You now have a prism and pyramid with congruent bases and the same height.

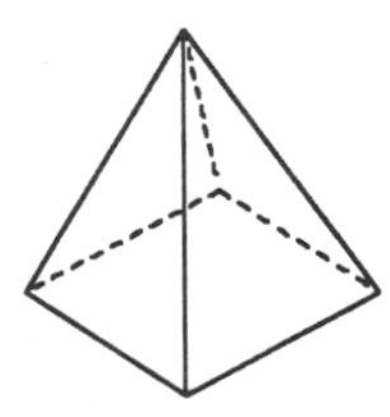

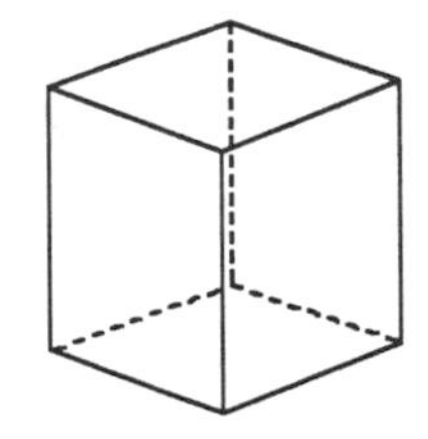

Cylinders and Cones

To build a cylinder-cone pair with congruent bases and the same height, you can start with a can. Then you only need to build a cone with the same height and a base congruent to the base of the can. Each group should perform the steps below, but start with a different size can. You will need the following materials:

- Large clean empty can (coffee can, large juice can, tomato can, etc.)
- Manila folder
- Scissors and tape

Step 1: Force a manila folder rolled into a cone into the can until the tip firmly touches the bottom of the can. With the help of a classmate, adjust the open end of the cone so that it is tight against the rim of the can. Tape the cone securely.

Step 2: Mark a circle on the cone where the rim of the can touches the cone. Cut off the excess manila folder along the circle.

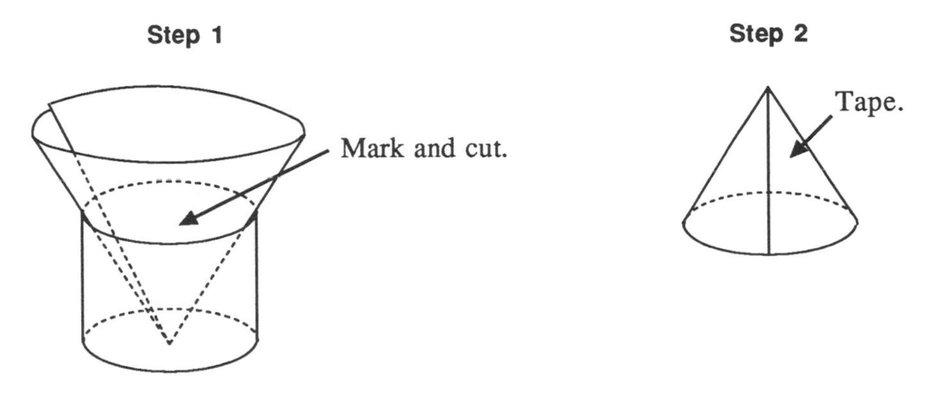

You now have a cylinder and cone with congruent bases and the same height.

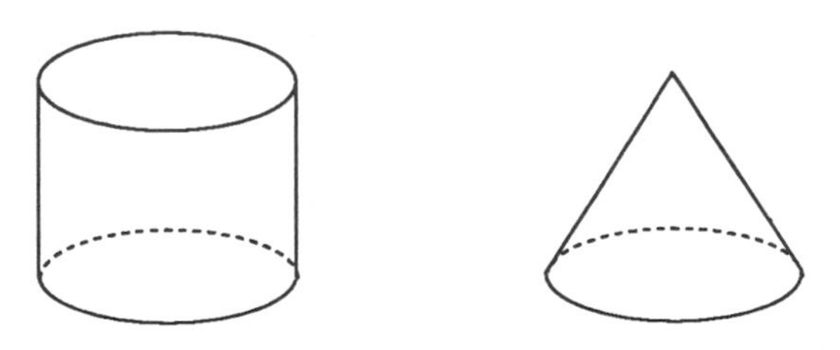

Lesson 10.6

Interpreting Diagrams

When working with the geometry of solids, it is important to be able to read and interpret diagrams of these objects. It is also important to be able to create labeled diagrams from written instructions.

EXERCISE SET A

Name each solid. Then find the volume of each. All measurements are given in centimeters. Each quadrilateral is a rectangle.

1.

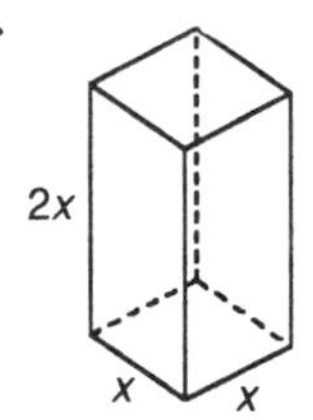

2.

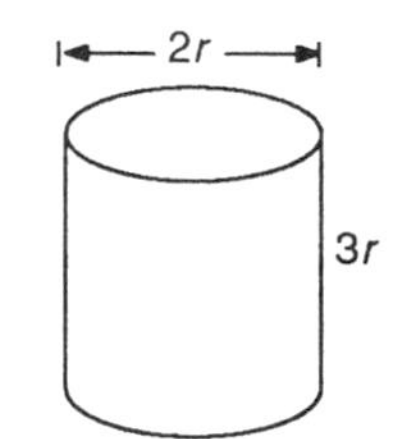

3.

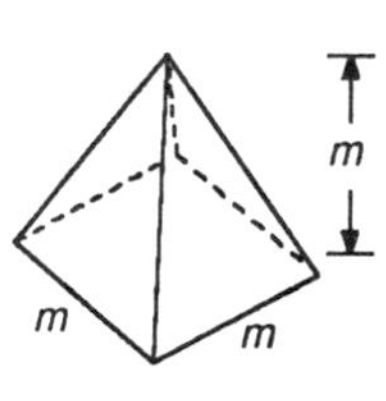

4.

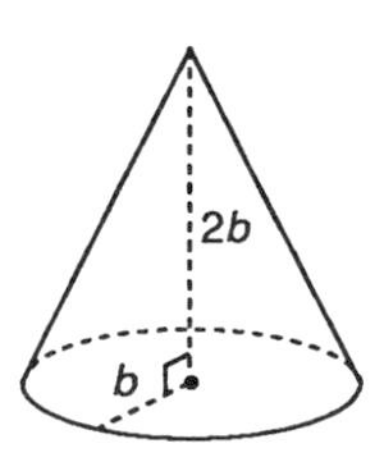

5.*

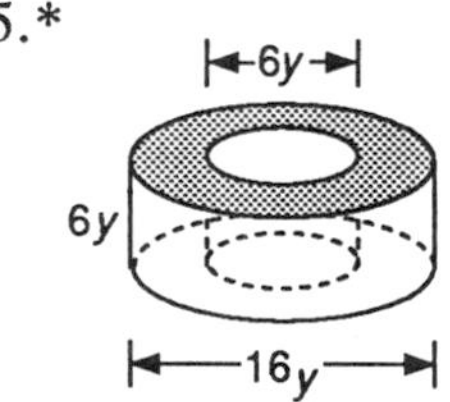

6.

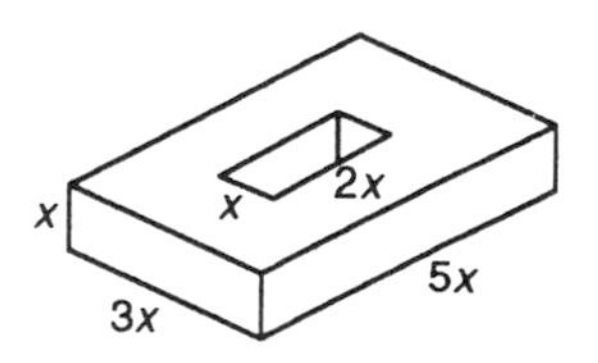

Find the volumes of the liquids in the rectangular prism and cone below. All measurements are given in centimeters.

7.

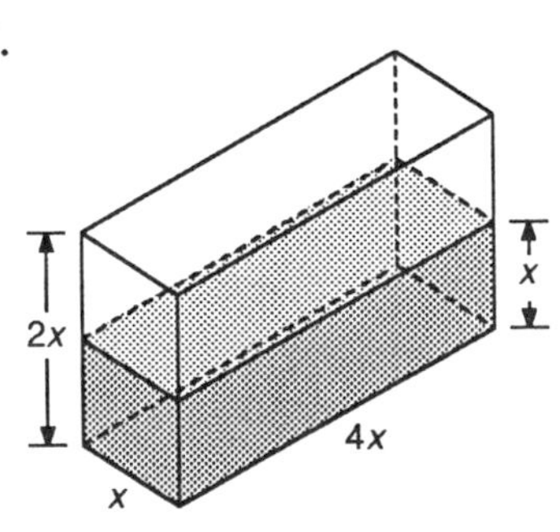

8.*

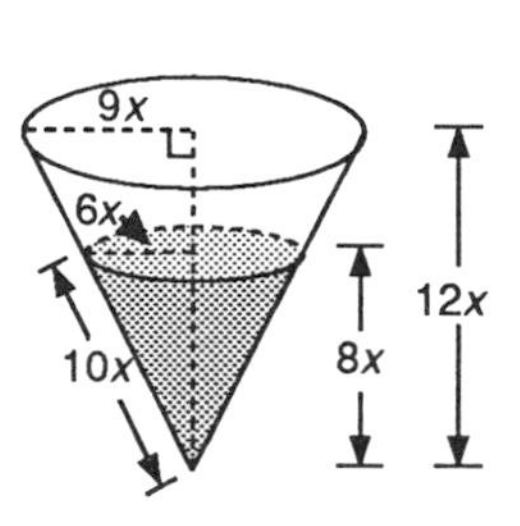

Sketch and label each solid described below. Find the volume.

9. A right rectangular prism. The base is 12 cm by 16 cm. The height is 4 cm.

10. A right trapezoidal prism. The trapezoidal base has a height of 4" and bases of 8" and 12". The height of the prism is 24".

11. A triangular pyramid. The base is equilateral with each side 6 meters long. The vertex of the pyramid is 3 meters directly over the incenter of the triangle.

12.* A right cylinder with a height of T. The radius of the base is $\sqrt{Q}$.

13. A square pyramid. The height of the pyramid is 9 feet and meets the square base at the intersection of the two diagonals. Each side of the base is 6 feet.

14. A chocolate cake of diameter 24 cm and height of 14 cm that has a slice cut out with a vertex angle of 45°.

15.* A right triangular prism with height $K + 7$. The base is determined by an isosceles right triangle with a hypotenuse of $K\sqrt{2}$.

16. A right cone. The slant height (the length of a segment from the vertex to the circumference of the base of a right cone) is $13B$. The radius of the base is $5B$. Calculate and label the height of the cone.

17. A regular hexagonal prism with a cylinder bored out from the center of the top base to the center of the bottom base. Each edge of the regular hexagon is 8 cm. The height of the prism is 16 cm. The cylinder has a radius of 6 cm.

18.* A cube with a cone bored out of the center of the top face. The radius of the cone is r and the height of both the cone and the cube is $2r$.

EXERCISE SET B

A rectangular prism 3 cm by 3 cm by 4 cm is painted red on all six faces. It is then cut into 36 cubes 1 cm on a side and all 36 cubes are placed in a bag.

1.* What is the probability of randomly selecting a cube from the bag that is painted red on exactly three faces?

2. What is the probability of randomly selecting a cube from the bag that is painted red on exactly two faces?

3. What is the probability of randomly selecting a cube from the bag that is painted red on exactly one face?

4. What is the probability of randomly selecting a cube from the bag that is painted red on no faces?

Lesson 10.7

Volume Problems

It is quite a three-pipe problem.

– Sherlock Holmes in The Red Headed League by Sir Arthur Conan Doyle

EXERCISE SET A

1. Find the volume of a rectangular prism with edges of 1.5, 2.5, and 3 cm.
2. Find the volume of a cone with a diameter of 8 inches and a height of 6 inches.
3. Find the volume of a cube with an edge of 1/2 inch.
4.* Find the volume of a rectangular prism that has dimensions twice the size of the dimensions of another rectangular prism that has a volume of 120 cm^3.

Sometimes, if you know the volume of a solid, you can calculate an unknown length of a base or the solid's height.

Example A

The volume of a triangular prism is 1440 cm^3. The base is a right triangle with legs 8 and 15 cm in length. Find the height of the prism.

$$V = BH$$
$$1440 = [(1/2)(8)(15)](h)$$
$$1440 = (60)(h)$$
$$24 = h$$

8 15 h

The height of the prism is 24 cm.

Example B

The volume of a cylinder is 2816 m^3. Find the radius of the base of the cylinder if it has a height of 14 m. Use 22/7 for π.

$$V = BH$$
$$V = (\pi r^2)H$$
$$2816 \approx [(22/7)r^2](14)$$
$$2816 \approx (44)r^2$$
$$64 \approx r^2$$
$$r \approx 8$$

r 14

The radius is about 8 m.

EXERCISE SET B

1.* Find the height of a cone with a volume of 138π cubic meters and base area of 46π square meters.
2. A triangular pyramid has a volume of 180 cubic centimeters and a height of 12 cm. Find the length of a side of the triangular base if the height to that side is 6 cm.
3.* A trapezoidal pyramid has a volume of 3168 cm^3. The height of the pyramid is 36 cm and the lengths of the two bases of the trapezoidal base are 20 cm and 28 cm. What is the height of the trapezoidal base?

4. The volume of a cylinder is 628 cm^3. Find the radius of the base if the cylinder has a height of 8 cm. Use 3.14 for π.

EXERCISE SET C

In this set of exercises, you will need the following information:

- Water weighs about 63 pounds per cubic foot.
- A cubic foot of water is about 7.5 gallons.

1.* A king-sized waterbed mattress measures 5.5 feet by 6.5 feet by 8 inches deep. How much does the water in this waterbed weigh to the nearest pound?

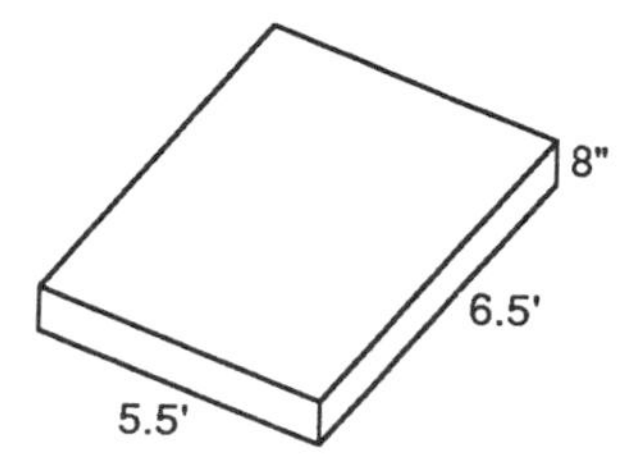

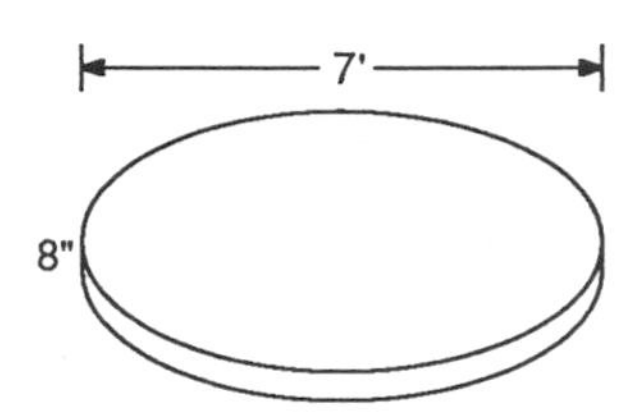

2. A round waterbed mattress measures 7 feet in diameter by 8 inches thick. How many gallons of water are in this waterbed? Use 22/7 for π.

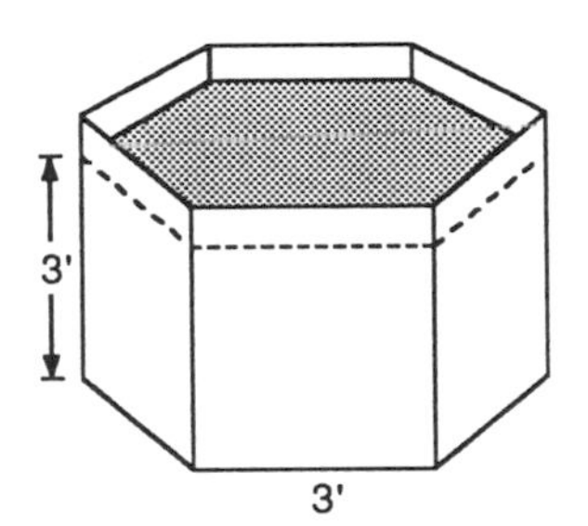

Minutes to Raise Temperature 10° F								
Gallons	350	400	450	500	550	600	650	700
Minutes	9	10	11	12	14	15	16	18

3.* Ingrid's hot tub was built in the shape of a regular hexagonal prism as shown in the diagram. The water temperature in the tub is now 93° F and Ingrid wishes to raise the water temperature to 103°. The chart on the hot tub heater tells the approximate number of minutes the heater takes to raise different amounts of water 10° F. Help Ingrid determine how long it will take to raise the temperature to 103°. Use 1.7 for $\sqrt{3}$. (First, calculate the number of cubic feet of water. Second, calculate the number of gallons. Third, use the chart to determine the approximate number of minutes to raise that many gallons 10°.)

4.* A swimming pool was built in the shape of the prism shown on the right. How many gallons of water can the pool hold?

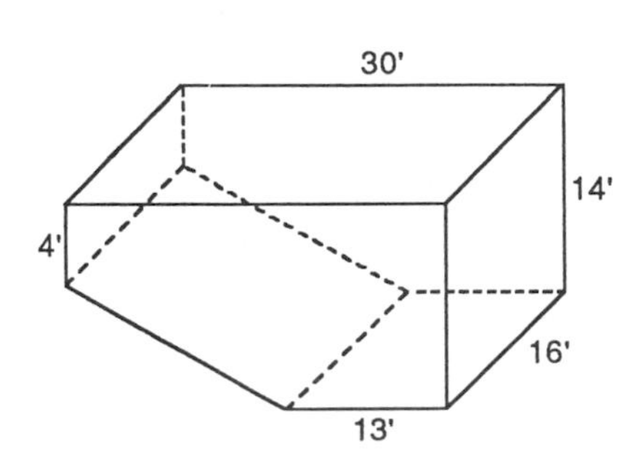

Special Project

Constructing the Platonic Solids

In the *Special Project: The Five Platonic Solids*, you discovered what each of the five Platonic solids would look like when unfolded and laid flat. Now use these patterns to construct and assemble the five Platonic solids. You will need the following materials:

- Poster-board, heavy cardboard, or manila folders
- Compass and straightedge
- Scissors
- Glue, paste, or cellophane tape
- Colored pens or pencils for decorating the solids

Before you begin, read through the tips on construction below.

Construction Tips

1. To construct the icosahedron, octahedron, and tetrahedron you need to construct networks of equilateral triangles. You should build these three solids at the same time.
2. To construct the hexahedron or cube, you will need to construct perpendiculars in order to guarantee that each side is a square.
3. To construct the dodecahedron, you will need to construct regular pentagons. To construct a regular pentagon, follow the six steps outlined below.

Step 1

Construct a circle. Construct two perpendicular diameters.

Step 2

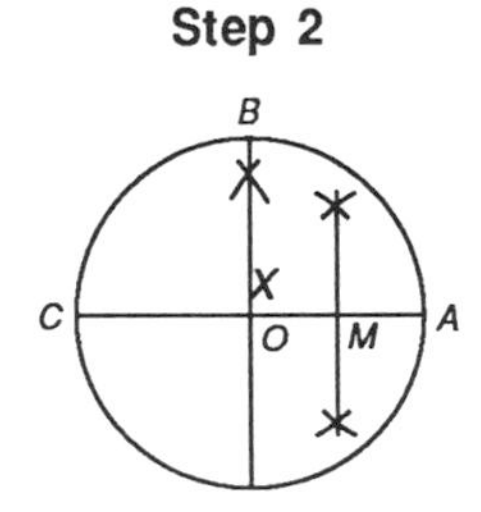

Find M, the midpoint of $\overline{OA}$.

Step 3

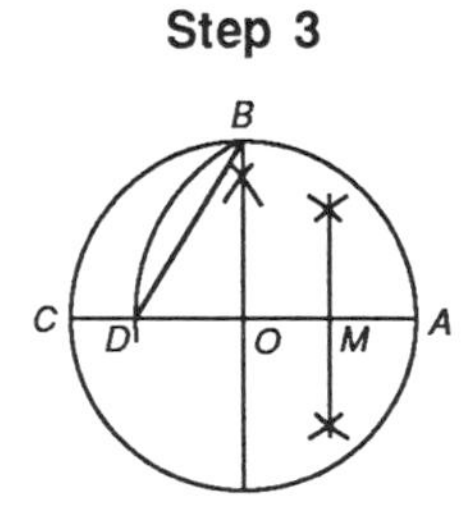

Swing an arc with radius BM intersecting $\overline{OC}$ at D.

Step 4

B

C D O M A

BD is the length of each side of the pentagon.

Step 5

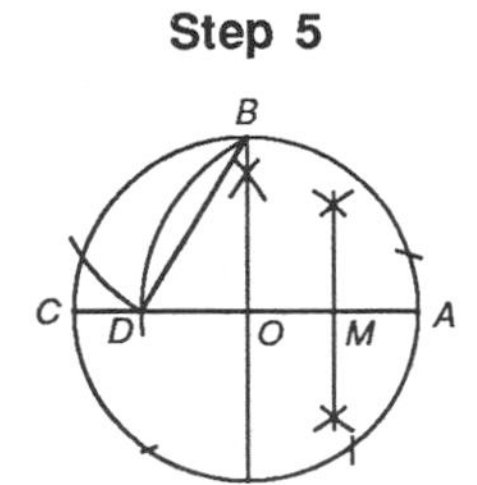

Starting at *B*, mark off *BD* on the circumference five times.

Step 6

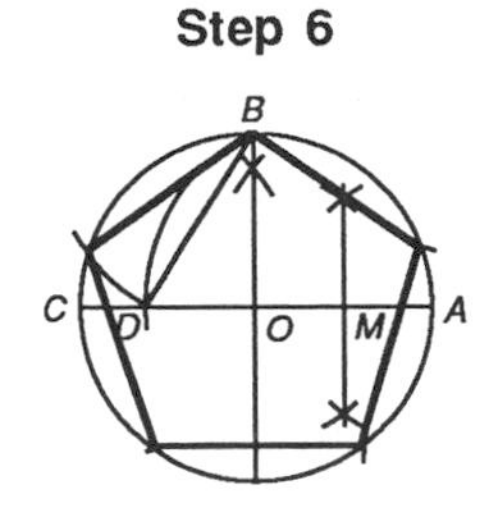

Connect the points to form a pentagon.

4. Construct the dodecahedron in two parts. Each part is composed of a regular pentagon with five regular pentagons around it. To construct one part:

   Step 1: Construct a large regular pentagon (see tip 3).

   Step 2: Lightly construct all the diagonals in the pentagon. This gives you a smaller regular pentagon which will be one of the twelve pentagonal faces of your dodecahedron.

   Step 3: Through each of the vertices of this central pentagon, construct the diagonals and extend them all the way to the sides of the larger pentagon. Find the five pentagons that encircle the central pentagon.

   Step 4: Erase the unnecessary line segments.

Step 1

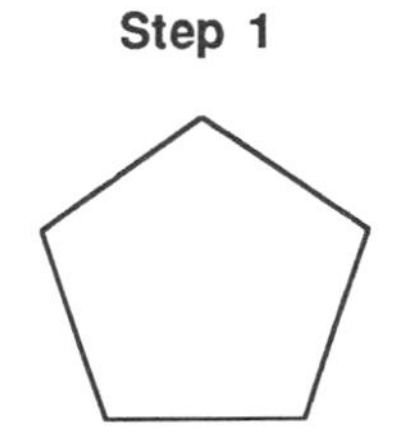

Step 2

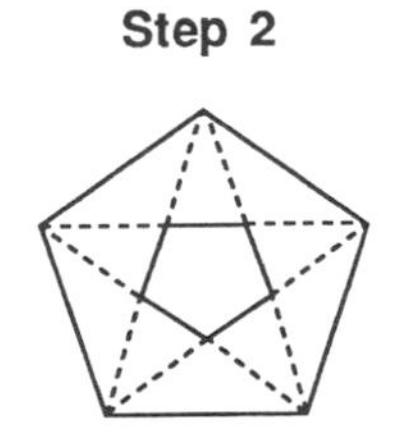

Step 3

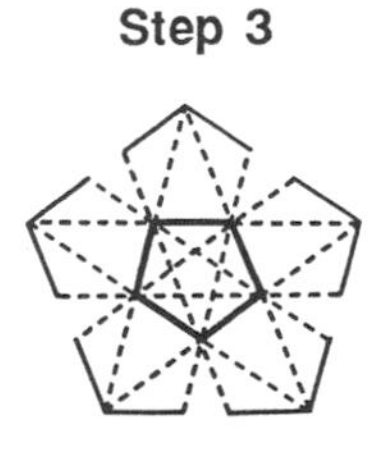

Step 4

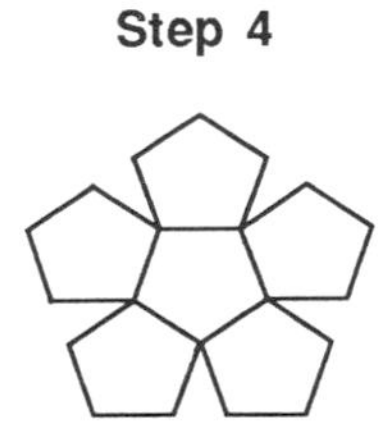

5. Decorate each Platonic solid *before* you cut it out.

6. Leave tabs on some edges for gluing your solid together.

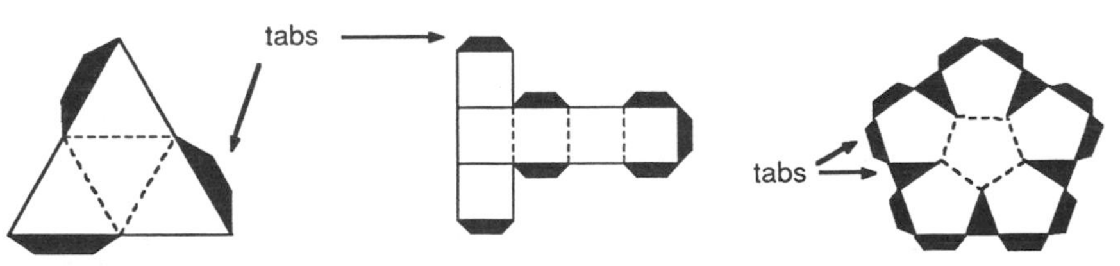

7. Score each edge that is to be folded by running a ball point pen or compass point heavily over the fold lines. Do this on both sides.

Lesson 10.8

Displacement and Density

I have a simple philosophy. Fill what's empty. Empty what's full. And scratch where it itches.

– Alice Roosevelt

If you step into a bath tub that is filled to the brim, what will happen? Right, the water will overflow. If you fill your glass to the brim with root beer and then add a scoop of ice cream, what will happen? Right, you'll have a mess! The volume that overflows in each case equals the volume of the solid that is below the liquid level. This volume is called an object's displacement. You can determine the volume of an irregularly shaped object by measuring its displacement.

Example

Geologist Crystal Stone wishes to calculate the volume of an irregularly shaped rock. She places it into a rectangular prism containing water. The base of the container is 10 cm by 15 cm. When the rock is put in the container, Crystal notices that the water level rises 2 cm due to the rock displacing its volume of water. This new "slice" of water has a volume of (2)(10)(15) or 300 cm^3. Therefore, the volume of the rock is 300 cm^3.

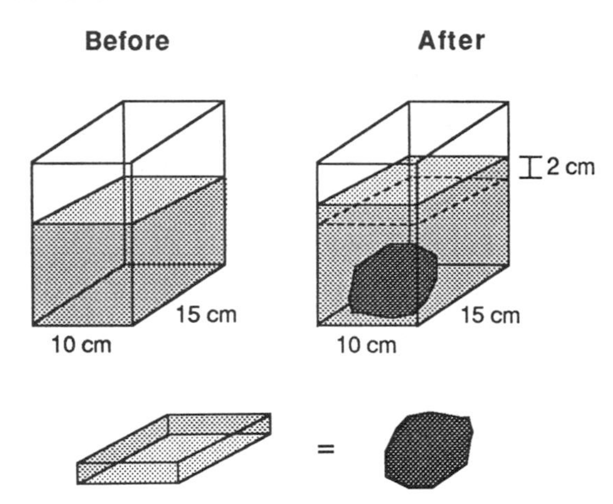

EXERCISE SET A

1. A rock is added to a rectangular prism with base dimensions of 15 cm by 15 cm and it raises the water level 3 cm. What is the volume of the rock?

2. A solid glass ball is dropped into a cylinder with a radius of 6 cm and it raises the water level 1 cm. What is the volume of the glass ball?

3. A fish tank 10" by 14" by 12" high is the home of a fat goldfish named Columbus. When he is taken out for some fresh air, the water level in the tank drops 1/3 inch. What is the volume of Columbus to the nearest cubic inch?

4.* A block of ice is placed into an ice chest containing water and it causes the water level to rise 4 cm. The base of the ice chest is 35 cm by 50 cm and it is 30 cm high. When ice floats in water, one-eighth of its volume floats above the water level and seven-eighths floats beneath the water level. What is the volume of the block of ice?

5. A piece of wood is placed in a cylindrical container and it causes the water level to rise 3 cm. The radius of the container is 5 cm. What is the volume of the piece of wood if this type of wood floats half out of the water?

Count Dragula

6. Count Dragula, the infamous Transylvanian hemoglobin expert, wishes to calculate the size of the elephant heart that he is about to transplant into the body of his ailing friend, Dr. Boris Nogoodnik. The Count places the frozen organ into a graduated cylinder with a diameter of 18 cm containing a 60/40 mix of alcohol and water at 2.5 degrees Celsius. If the liquid level rises 5 cm, what is the volume of the elephant heart? (Boris survived the transplant, but he ran away and joined the circus!)

An important property of a material is its density. Density is the amount of matter in a given amount of volume. Density is calculated by dividing the mass in grams by the volume in cubic centimeters (density = mass/volume). A chemist wishing to identify an unknown clump of metal could weigh the clump to determine its mass, determine its volume by displacement, calculate its density, and, finally, look in a chemical handbook to identify the compound from its density.

Density Table

Metal	Density	Metal	Density
Silver	10.5 g/cm^3	Lithium	.54 g/cm^3
Lead	11.3 g/cm^3	Sodium	.97 g/cm^3
Platinum	21.4 g/cm^3	Potassium	.86 g/cm^3
Gold	19.3 g/cm^3	Aluminum	2.81 g/cm^3

Example

A clump of metal weighing 351.4 grams is dropped into a cylindrical container causing the water level to rise 1.1 cm. The radius of the base of the container is 3.0 cm. What is the density of the metal? Assuming the metal is pure, what is the metal? Use 3.14 for π.

$$\begin{aligned} \text{Volume} &= [\pi(3.0)^2](1.1) \\ &\approx [(3.14)(9)](1.1) \\ &\approx 31.086 \\ &\approx 31.1 \end{aligned}$$

$$\begin{aligned} \text{Density} &\approx 351.4/31.1 \\ &\approx 11.3 \end{aligned}$$

The density is 11.3 grams per cubic centimeter. Therefore the metal is lead.

EXERCISE SET B

Use the density table above and volume displacement to solve each problem.

1. Sandy found a clump of metal. It appears to be either silver or lead. Hoping it is silver, she drops it into a rectangular container half filled with water. She observes that the water level rises 2.0 cm. Since the base of the container is a square 5 cm on an edge, she is able to calculate the volume of the clump. She weighs the metal and finds that it is 525 grams. What is the density of the metal? Is it lead or silver?

2. The crime lab has received a small piece of very reactive metal. The technician weighs the soft metal and finds that it weighs 54.3 grams. It appears to be either lithium, sodium, or potassium, all highly reactive with water. The lab technician, therefore, places the metal into a glass graduated cylinder of radius 4 cm containing a non-reactive liquid. The metal causes the liquid level to rise 2.0 cm. Which metal is it? Use 3.14 for π.

3.* A chemist is given a clump of metal and is told that it is sodium. She weighs the metal and finds that it weighs 145.5 grams. To test to see if it is indeed sodium, she places it into a container containing a non-reactive liquid that is a square prism with a base 10 cm on each edge. If it is sodium, how many centimeters will the liquid level rise?

Sherlock Holmes and the Inca Idol

4. Sherlock Holmes has just returned home excited. He rushes to his chemical lab, takes a mysterious Inca idol out of his carrying case, and weighs it. "3088 grams," Mr. Holmes says in anticipation. "Now, let's check its volume." Sherlock Holmes takes out a glass graduated container. It has a square base 10 cm on each side. He adds water to the container and records the water level. It's 3.0 cm. He places the idol into the container and reads the new water level as 4.6 cm. After a few minutes of the mental calculation that Holmes enjoys so much, he turns to Dr. Watson. "This confirms my theory about Colonel Banderson. Only he could have been to Peru in that time period. Quick Watson, the game is afoot. We are off to the train station."

   Poor Dr. Watson is still standing with the *London Daily* in his hands. "Holmes, you amaze me. Is it gold?" questions the good Doctor.

   "If it has a density of 19.3 grams per cubic centimeter, it is gold," smiles Sherlock Holmes. "If it is gold, then the Colonel is in danger and Andre Pierrot is our murderer. If it is a fake, then Lady Wellingsforth is our murderer and we are off to Wellingsforth Manor."

   Watson is still waiting for the answer. "Well?" queries the Doctor.

   Holmes, heading for the door, smiles and says, "It's elementary, my dear Watson. Elementary geometry, that is."

   What is the volume of the idol? Is the idol gold and Pierrot our murderer?

Lesson 10.9

Volume of a Sphere

Nothing in life is to be feared,
it is only to be understood.

– Marie Curie

In this lesson we are going to develop a formula for the volume of a sphere. (By volume of a sphere, we mean the amount of space contained within the sphere.) You will use a concept called Cavalieri's Principle, named after the Italian mathematician Bonaventura Cavalieri (1598–1647). Cavalieri's Principle says: *If two solids have the same cross sectional area whenever they are sliced at the same height, then the two solids must have the same volume.* Using this principle, you will demonstrate that the two solids below have the same volume.

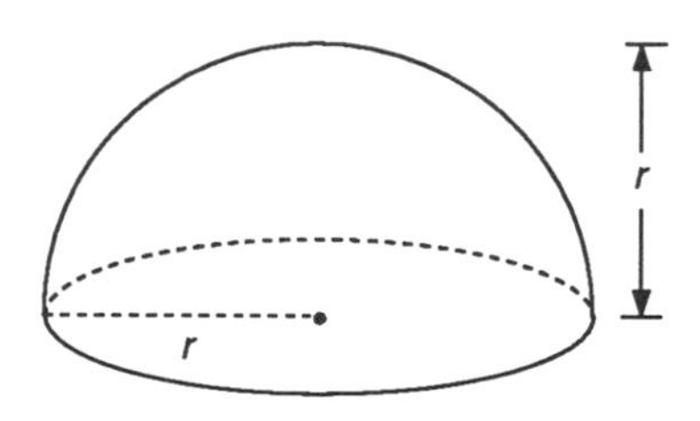

Hemisphere

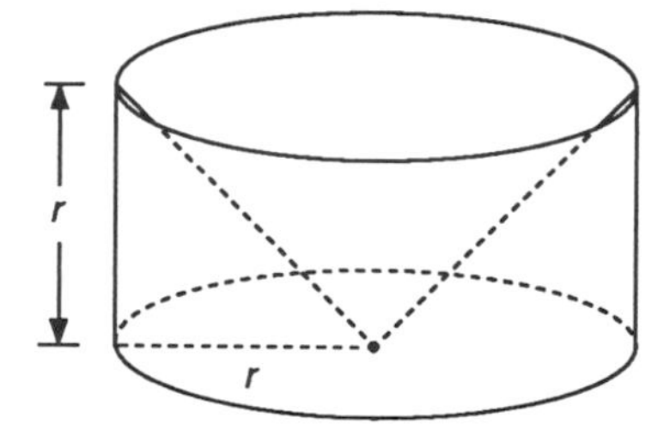

"Cylinder Less Cone"

The hemisphere has radius r. The solid on the right is a cylinder with height r and radius r which has a cone of height r and radius r bored out of it. We will demonstrate that when these two solids are sliced at the same height, their cross sectional areas are the same. Then we will apply Cavalieri's Principle to the two solids to prove that they have the same volume. Since we already know formulas for the volume of a cylinder and the volume of a cone, we can calculate the volume of the "cylinder less cone." The hemisphere must have this same volume. Confused? Let's take a closer look.

The first diagram on the next page shows a hemisphere with a radius of 15 cm. A plane is passed through it 9 cm up from the base creating a circular cross section. To the right of the hemisphere is a cylinder with a radius of 15 cm and height of 15 cm that has a cone with radius and height of 15 cm bored out. A plane is also passed through this solid 9 cm up from the base creating a ring (annulus) cross section.

Cross Section of Hemisphere

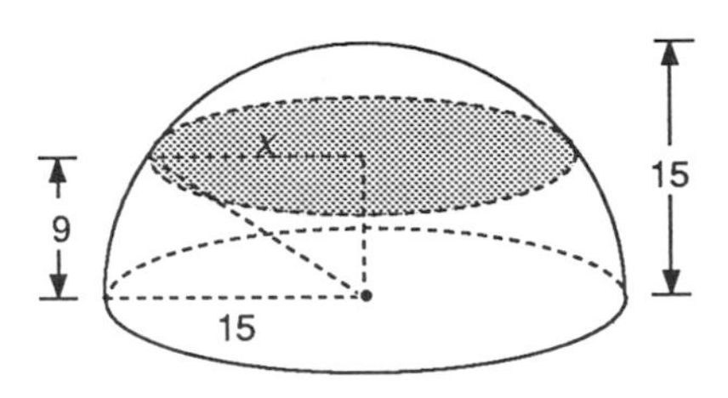

$$x^2 + 9^2 = 15^2$$
$$x^2 = 15^2 - 9^2$$
$$x^2 = 225 - 81$$
$$x^2 = 144$$
$$x = 12$$

$$\text{Area}_{\text{circle}} = \pi r^2 = \pi(12)^2 = 144\pi$$

Cross Section of "Cylinder Less Cone"

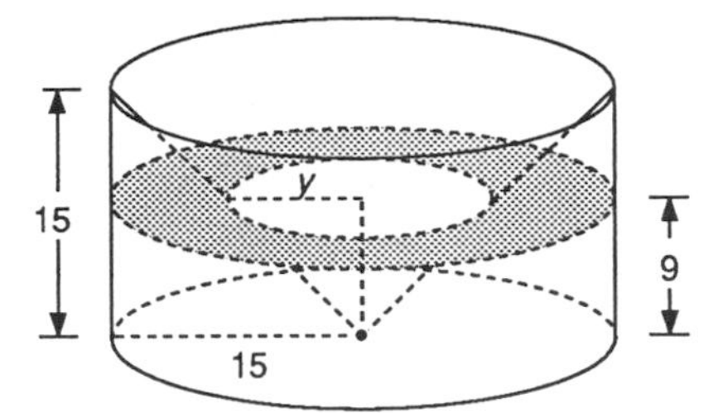

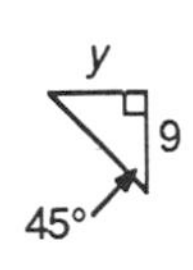

Since the height and radius of the cone are the same, the big right triangle is isosceles and thus so is the small right triangle. Therefore, $y = 9$.

$$\text{Area}_{\text{annulus}} = \pi R^2 - \pi r^2 = \pi(15^2) - \pi(9^2) = \pi(225 - 81) = 144\pi$$

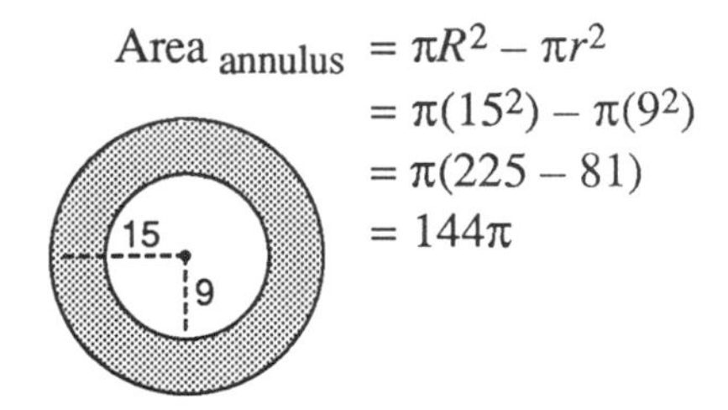

The calculations show that the two cross sectional areas are both 144π cm^2.

EXERCISE SET A

1.* A plane is passed through the hemisphere of radius 15 cm at a point 12 cm up from the base. What is the area of the circular cross section?

2. A plane is passed through the "cylinder less cone" with radius and height 15 cm at a point 12 cm up from the base. What is the area of the annulus-shaped cross section?

In the introductory discussion the two solids pictured are shown to have the same cross sectional area when they are both sliced 9 cm up from the base. In Exercise Set A you found that when the same two solids are sliced 12 cm up from the base, the two new cross sectional areas are also equal. The logical argument which follows demonstrates that this is true for any pair of hemispheres and "cylinder less cones" with the same radius and height when sliced at the same height. To demonstrate that a hemisphere with radius r and a "cylinder less cone" with radius and height r have the same cross sectional areas, we select an arbitrary height h and, with the help of algebra, find the cross sectional area of each solid at height h.

Cross Section of Hemisphere

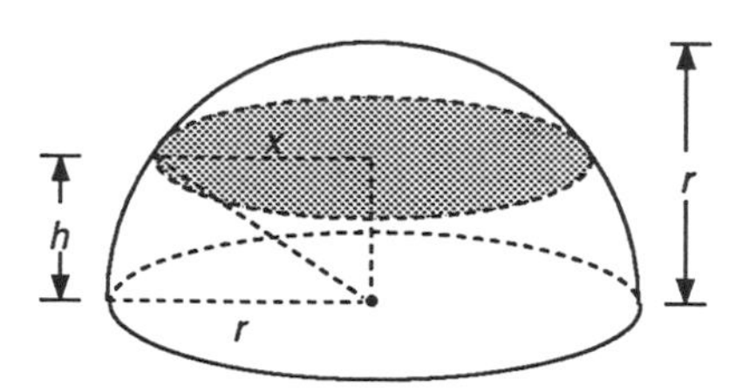

The area of the circular cross section is πx^2. But $x^2 + h^2 = r^2$ by the Pythagorean Theorem. Therefore $x^2 = (r^2 - h^2)$. Thus, with a little algebra, the cross sectional area is $\pi(r^2 - h^2)$.

Cross Section of "Cylinder Less Cone"

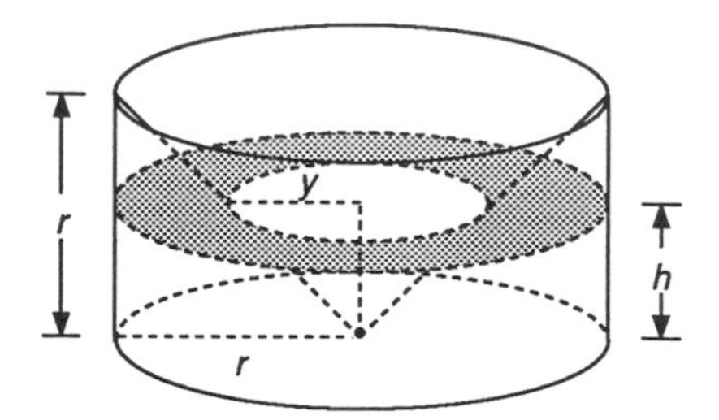

The area of the annulus is $\pi r^2 - \pi y^2$. But, since y and h are the lengths of legs of an isosceles triangle, $y = h$. Thus, the area of the annulus is $\pi r^2 - \pi h^2$. By factoring you get $\pi(r^2 - h^2)$.

Since we selected an arbitrary height, and both cross sectional areas turned out to be $\pi(r^2 - h^2)$, then the two solids must have the same volume by Cavalieri's Principle.

You can calculate the volume of the cylinder minus the volume of the cone. The volume you arrive at must also be the volume of the hemisphere.

$$
\begin{aligned}
\text{Volume}_{\text{cylinder}} &= BH \\
&= (\pi r^2)r \\
&= \pi r^3
\end{aligned}
$$

$$
\begin{aligned}
\text{Volume}_{\text{cone}} &= (1/3)BH \\
&= (1/3)(\pi r^2)r \\
&= (1/3)\pi r^3
\end{aligned}
$$

Therefore, the volume of the "cylinder less cone" is $\pi r^3 - (1/3)\pi r^3$ or $(2/3)\pi r^3$. By Cavalieri's Principle, the hemisphere must also have volume $(2/3)\pi r^3$. Finally, since the volume of a sphere is twice the volume of a hemisphere, you can arrive at a formula for the volume of a sphere. State this as your next conjecture.

The volume of a sphere with radius r is given by the formula —?—. ***(Sphere Volume Conjecture)***

Example A

Find the volume of a sphere that has a radius of 6 inches.

$$
\begin{aligned}
V &= (4/3)\pi r^3 \\
&= (4/3)\pi 6^3 \\
&= (4/3)(216)\pi \\
&= 288\pi
\end{aligned}
$$

The volume is 288π cubic inches.

Example B

Find the radius of a sphere that has a volume of 2304π cubic centimeters.

$$
\begin{aligned}
V &= (4/3)\pi r^3 \\
2304\pi &= (4/3)\pi r^3 \\
2304 &= (4/3)r^3 \\
1728 &= r^3
\end{aligned}
$$

Using a calculator, you can find that $r = 12$. The radius of the sphere is 12 cm.

EXERCISE SET B

Find the volume of each solid. All measurements are in centimeters.

1.

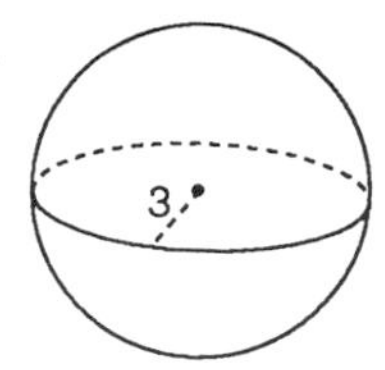

2.

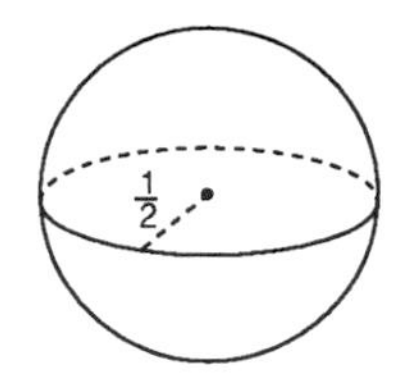

3.

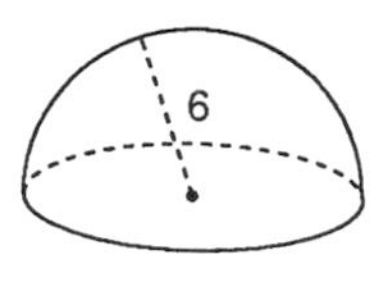

4.*

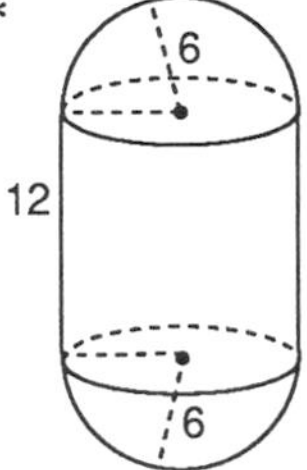

5.

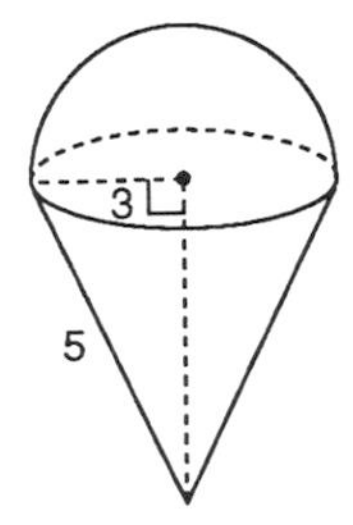

6.*

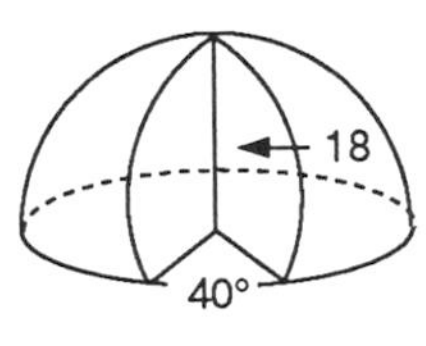

EXERCISE SET C

1. A sphere has a volume of 972π cubic inches. Find its radius.

2. A hemisphere has a volume of 18π cm^3. Find its radius.

3. What is the volume of the greatest sphere that can be carved out of a cube with edges of 6 cm?

4.* A cone of radius r and height r is bored out of a hemisphere with a radius of r as shown in the diagram. What is the remaining volume? Which is greater, the volume bored out or the volume remaining?

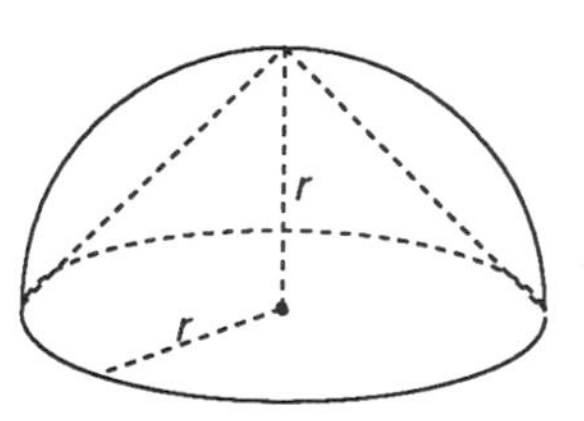

Improving Visual Thinking Skills

Mental Blocks

In the top figure on the right, every cube is lettered exactly alike. Copy and complete the two dimensional representation of one of the cubes to show how the letters are arranged on the six faces.

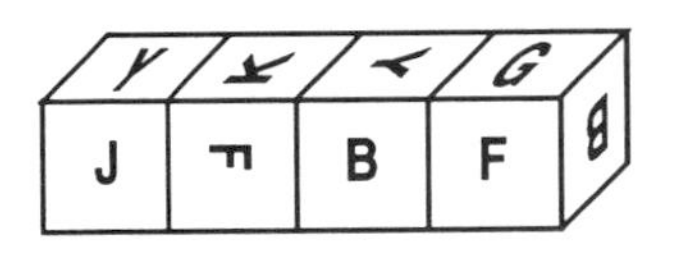

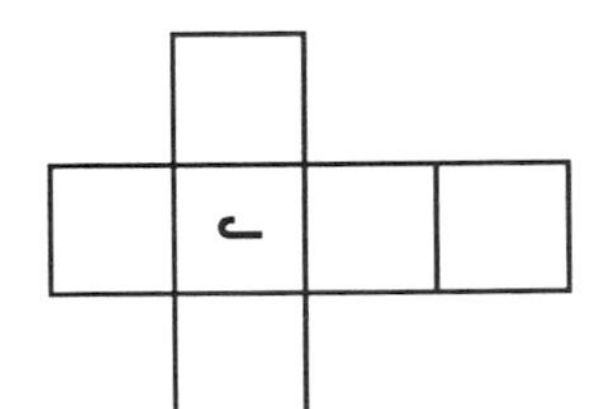

Lesson 10.10

Volume of a Sphere: The Saga Continues

EXERCISE SET

1. A sphere of ice cream is placed on your ice cream cone. Both have a diameter of 8 cm. The height of your cone is 12 cm. Will all the ice cream, if pushed with your tongue into the cone, fit?

2.* Lickety Split ice cream comes in a cylindrical container with an inside diameter of 6 inches and a height of 10 inches. The company claims to give the customer 25 scoops of ice cream in each container. The scoops of ice cream advertised are spheres with 3 inch diameters. How many scoops will each container really hold? Should Lickety Split be reported to the Better Business Bureau?

3. A 10 cm tall cylindrical glass 8 cm in diameter is filled to one cm from the top with water. If a golf ball 4 cm in diameter is dropped into the glass, will the water overflow?

4. A can of tennis balls has an inside diameter of 7 cm and a height of 20 cm. If the diameter of a tennis ball is 6 cm, how much of the space in a tennis ball can is not occupied by the three balls?

5. Marco Roni has just made two dozen meatballs. Each meatball has a 2 inch diameter. He wishes to cook them in sauce for a while, but the large pot may not be big enough to hold them plus the sauce. Right now, before the meatballs are added, the sauce is 2 inches from the top of the 14 inch diameter pot. Will the sauce spill over when the meatballs are added to the pot?

6.* Can you pick up a solid steel ball of radius 6 inches? If the steel has a density of .28 pounds per cubic inch, what is the weight of the ball to the nearest pound? Use 3.14 for π.

Schulhoff Tam
High school student

7. Find the weight to the nearest pound of a hollow steel ball with an outer diameter of 14 inches and a thickness of 2 inches. The steel has a density of .28 pounds per cubic inch. Use 22/7 for π.

8.* A hollow steel ball has a diameter of 14 inches and weighs 327.36 pounds. If the steel has a density of .28 pounds per cubic inch, find the thickness of the ball. Use 22/7 for π.

Nanu Unan

9.* Nanu Unan is a resident of Snowflake, Alaska. Nanu Unan is allergic to snow. Nanu is planning to melt down his igloo and build in its place a redwood A-frame four-unit condo. The local authorities need to know the volume of snow that will be melted down. Snowflake ordinance #S-3417-113-Q permits no more than 16 cubic meters of snow to be melted at any one time. Nanu's home is a hemispherical shell of .3 meters thickness and has an inner diameter of 5.4 meters. Ignoring his door and TV antenna hole, what is the volume of snow in Nanu Unan's igloo? Use 3.14 for π.

Lesson 10.11

Revenge of the Volume Problems

EXERCISE SET

1.* One night Tom Tow and Laurie Ranger were sitting about the campfire discussing some of the not-so important topics in the scheme of things. Just before retiring, the topic of discussion turned to tent size. Tom claimed his tent was larger than Laurie's. Laurie sleeps in a regulation Texas Ranger pup tent. The height is 5', the length is 6', and the width is 5'. Tom, on the other hand, has a traveling version of an Indian tepee. The height is 5' and the greatest diagonal across the floor is 8'. Tom's version of the tepee has only six poles. It is a regular hexagonal pyramid. Which tent has the larger floor area? Which tent has the larger volume?

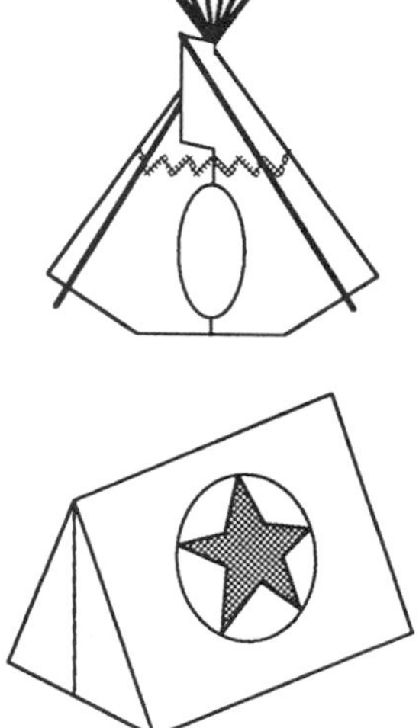

2. Sylvia Concrete discovered that someone opened the valve on her cement truck during the night of March 31 and all the contents of the truck leaked out to form a giant cone of hardened cement. For insurance purposes, Sylvia needs to figure out how much cement is in the giant cone. The circumference of the base of the cone is 44 feet and the cone has a height of 6 feet. Use 22/7 for π to calculate the volume of the cone of cement to the nearest cubic foot.

3.* An amulet was recently uncovered high in the Andes by Seymour Hills, noted adventurer and archaeologist. Professor Hills must do some calculations to determine if the charm is an authentic Inca relic or a modern tourist trinket. If the volume of the regular hexagonal ring is equal to the volume of the regular hexagonal hole in its center, then it is an ancient charm worn by a high priest. If not, then it is a modern imitation. From the dimensions shown in the diagram, determine if it is a modern fake or an authentic Inca relic.

2 cm
4 cm
6 cm

4.* Rosita Aguas is a plumbing contractor. She needs to deliver 200 lengths of steel pipe to a construction site. Each cylindrical steel pipe is 160 cm long, has an outer diameter of 6 cm, and has an inner diameter of 5 cm. Steel has a density of about 7.7 grams per cubic centimeter. Rosita needs to know if her quarter-tonne truck can handle the weight of the pipes? (One tonne is 1000 kg.) What is the weight to the nearest kilogram of these 200 pipes? How many loads will it take to deliver the 200 lengths of steel pipe? Use 22/7 or 3.14 for π.

5. General Ahmed is the engineer in charge of constructing the Pharaoh's pyramid. The pyramid will house the tomb of the reigning Pharaoh. The pyramid plans, written on the new invention paper, call for a regular square-based pyramid, 600 cubits on a side. The altitude is 400 cubits. The General has just learned that the Pharaoh is on his death bed and has at most 30 days to live. Unfortunately, the pyramid is incomplete. The pyramid stands only 300 cubits high with a square top 150 cubits on a side. If the Pharaoh dies before the pyramid is completed, the General and his family will meet an untimely end. Calculate the remaining volume that General Ahmed needs to complete within 30 days. Express your answer in cubic cubits.

Peaches del Mar

6.* Peaches del Mar is an unhappy mermaid. When she left the sea to be closer to her surfer friend Bonsai, he promised her a swimming tank with at least 2 million cubic feet of water. He built Peaches a cylindrical swimming tank, but she finds it too small. Having studied math at Pisces University by the Sea, she was able to measure the tank. In order to do this she swam from one edge of the tank due north 160 feet until she came to an edge. She turned and swam due east 120 feet until she came to the edge again. From this she was able to calculate the diameter of the tank. Next, she dove down and swam diagonally to the bottom edge on the opposite side of the tank, traveling the longest distance possible in the tank, a distance of 250 feet. With this Peaches was able to calculate the depth of the tank and, therefore, the volume. Did Bonsai keep his promise to Peaches? What is the volume of Peaches' new home to the nearest thousand gallons? Use 3.14 for π.

Lesson 10.12

Chapter Review

When you stop to think,
don't forget to start up again.

– Anonymous

EXERCISE SET A

Complete each conjecture.

1. If B is the area of the base of a prism or cylinder and H is the height of the solid, then the formula for the volume is $V =$ —?—.
   (C-69 or **Prism-Cylinder Volume Conjecture**)

2. If B is the area of the base of a pyramid or cone and H is the height of the solid, then the formula for the volume is $V =$ —?—.
   (C-70 or **Pyramid-Cone Volume Conjecture**)

3. The volume V of a sphere with a radius of r is given by the formula —?—.
   (C-71 or **Sphere Volume Conjecture**)

Complete each statement. Then draw a diagram to illustrate the statement.

4. A —?— is a polyhedron with two faces (bases) that are congruent and parallel polygons and whose other faces (lateral faces) are parallelograms formed by segments connecting the corresponding vertices of the bases.

5. A —?— is a solid composed of a circle, its interior, a given point not on the plane of the circle, and all the segments from the point to the circle.

6. The —?— of a prism determines its name.

7. The —?— of a prism are the faces that are not bases.

8. The —?— of a prism are the line segments where the lateral faces intersect.

9. A —?— prism has lateral faces that are perpendicular to the two bases.

10. A —?— is the set of all points in space at a given distance from a given point.

11. A —?— is a solid composed of two congruent circles in parallel planes, their interiors, and all the line segments parallel to the axis with endpoints on the two circles.

12. A solid formed by flat surfaces enclosed by polygons is called a —?—.

13. The —?— of a pyramid is a perpendicular segment from the vertex to the plane of the base.

EXERCISE SET B

Find the volume of each solid. Each quadrilateral is a rectangle. All given measurements are in centimeters.

1.

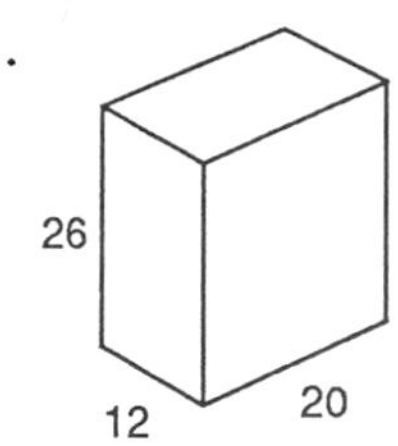

2.

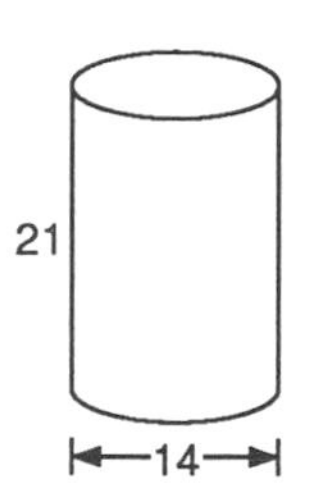

3.

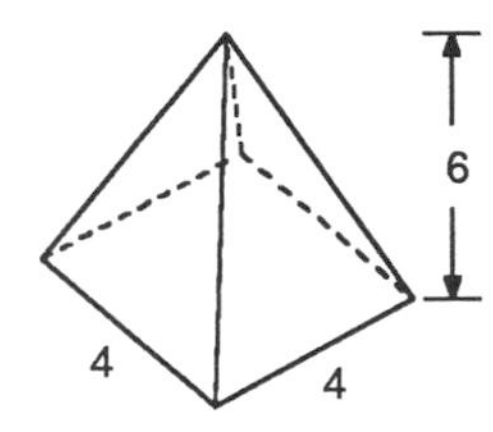

4.

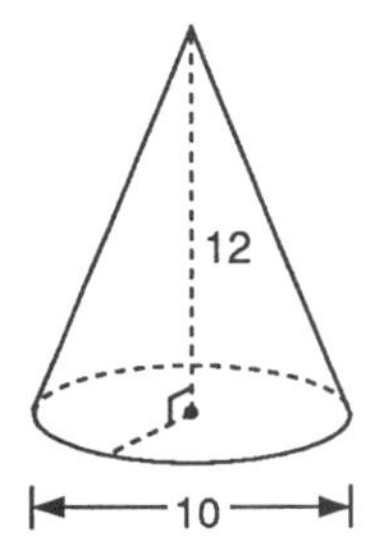

5.*

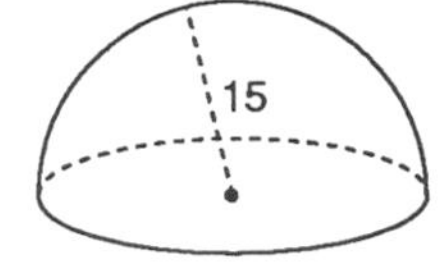

6.*

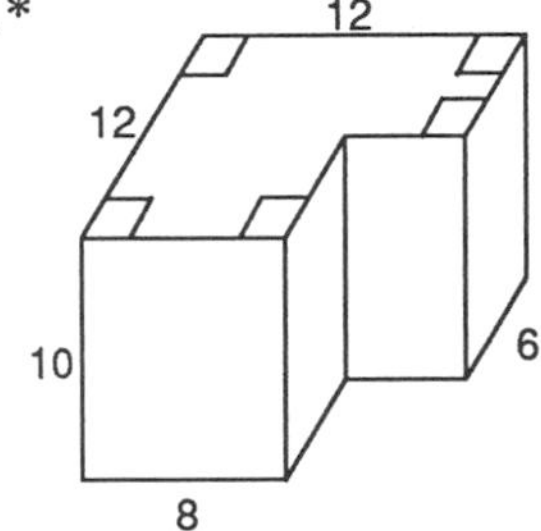

Calculate an unknown length given the volume of the solid.

7. Find h.
$V = 896 \text{ cm}^3$

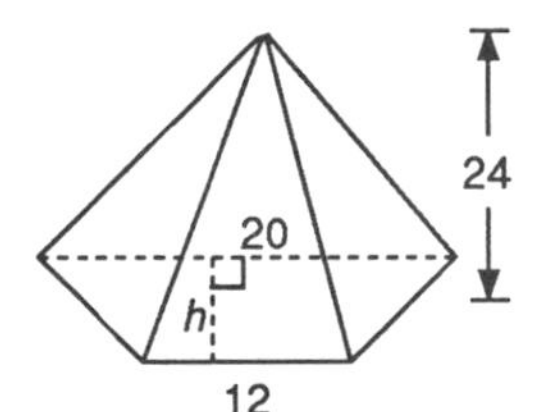

8. Find r.
$V = 1728\pi \text{ cm}^3$

9.* Find r.
$V = 256\pi \text{ cm}^3$

EXERCISE SET C

1.* Sheri Holmes, distant French cousin to the famous sleuth of Baker Street, has returned to her flat on the left bank with a small art deco piece painted a deep green. Sheri must determine the density of the heavy metal objet d'art in order to solve the crime she has been working on for weeks. She places it into a regular hexagonal glass graduated prism and finds that the water level rises 4 cm. Each edge of the hexagonal base is 5 cm. She quickly places the object on her Metler balance and finds it weighs 5457 grams. Sheri calculates the object's volume and then its density. Next she checks the density table (Lesson 10.8) to determine its composition. If it is platinum, it is from the missing Rothchild collection and Inspector Clouseau is the thief. If not, then the Baron is guilty of fraud. What is the composition of the art deco objet d'art? (Use 1.7 for $\sqrt{3}$.)

Tessie Tassahara

2.* Tessie Tassahara is planning an extravagant party to celebrate the installation of her new hot tub. She has decided to fill her tub with sparkling cider. (She loves bubbles!) Sparkling cider is sold in liter bottles. While the party is to be extravagant, Tessie doesn't want to buy more sparkling cider than she needs. Tessie decides to experiment. She fills the tub and gets in. Water flows over the top. Tessie notices that when she gets out of the tub, the water level drops 2 cm. She decides to put just enough cider in the tub so that when she and her eight guests get in, the cider rises to the brim of the tub but doesn't spill out. The tub has a diameter of 240 cm and a height of 150 cm. If you assume all eight of her guests are roughly her same size, how many liter bottles of sparkling cider must she buy? (Use 3.14 for π and assume 1000 cm^3 of cider per liter.) Can you suggest a way she can get all that cider home?

Cooperative Problem Solving

Once Upon a Time

You and your CPS team are competing in a school wide cooperative problem solving event at the lunar colony high school. All the teams have three tasks in their cooperative problem solving competition. The geometry teacher, the creative writing teacher, and the art teacher at the lunar colony have created a set of five illustrated geometry problems. The first task is to work coöperatively in small groups to solve the five problems. The problems are pure fantasy. In addition to being challenging geometry problems, they are intended to set a story telling mood. Hopefully, you will enjoy them and be inspired to create your own word problems. That is your second task. As a group you are to devise a geometric word problem. The problem should have a story or setting. The problem should demonstrate a need for the answer. It should use geometry in its solution. The third task is to illustrate your problem. It should be fun.

Secret Agent Carmen Serita Castillo

1.* Secret Agent Carmen Serita Castillo has finally located the fortress of her archenemy Evil McNasty. She scales the fortress wall, locates the power supply to the compound, and sets a timer to shut down all power at midnight. Carmen Serita locates a rooftop door and drops lithely through the door into a darkened room. She finds herself, however, trapped by that villainous scoundrel at the bottom of a stainless steel vault 2 by 2 by 6 meters deep. Suddenly water begins pouring into the vault, filling the small space to a depth of 1.5 meters. Almost immediately, a pair of opposite walls start to slowly and inexorably move towards Agent Castillo. After a few minutes of observation, she calculates that the walls are moving at a rate of 5 cm per minute. Checking her watch, Carmen Serita determines that the power will shut down in 15 minutes, stopping the walls dead in their tracks. Will Carmen Serita Castillo be able to float to the top of the rising column of water, reach the door, and escape? Or will she be crushed by the doors or doomed to a watery end? How high will the water level reach when the power goes off?

Dr. Aquatic and His Fiendish Plan

2.* The mad and evil scientist, Dr. Aquatic, has hatched a fiendish plan to drain all the oceans into 15 cylindrical holes in the earth. He has calculated that the volume of water in the oceans is about 330,000,000 cubic miles (3.3×10^8 cubic miles). He has determined that the holes should be 2,800 miles deep. At this depth the holes will reach tremendous temperatures and the water filling them will turn into steam and create a waterless dead planet surrounded by a steam room atmosphere. (According to his analyst, Dr. Aquatic never learned to swim causing a neurotic paranoia about water. This mental state makes him feel dirty and creates a need for daily steambaths.) Your job is to find where Dr. Aquatic is doing his dirty work so he can be stopped. With the aid of powerful telescopes mounted on satellites, you can search the ocean floors. What are you looking for? What must the radius of the holes be if they are to hold all the water from the oceans? Use 22/7 for π. Give the answer to the nearest mile.

Lois Lovely Saves Machoman

3.* Tex Ruthlus has recently received a lead box containing femenite. This material emits femenite rays extremely allergic to Machoman. Tex has developed a devious device called a Femenite Derandomizer. It is a ray gun type device that, when turned on, injects a megavolt of electricity into the femenite. This causes the femenite to decay into a powerful beam of femenite rays. The eventual target, of course, is Machoman. Machoman needs to know the volume of the femenite bar in the possession of terrible Tex Ruthlus. Once it has received a one megavolt charge, femenite is known to decay at the rate of one cubic centimeter per minute. Machoman's girlfriend, Lois Lovely, has obtained a diagram of the bar and a copy of the plans for the Femenite Derandomizer. The diagram shows that the bar is a regular hexagonal prism 20 centimeters long. Each edge of the hexagonal base is 4 centimeters. Machoman was never as fast as a speeding bullet when it came to math, so it is up to you to save him. Calculate the total volume in cubic centimeters of the bar of femenite. What is the volume? From this you can determine how many minutes the ray gun can operate before all the femenite decays. How many minutes does Machoman have from the time Tex Ruthlus turns on the device until all the femenite decays? Hurry, I think Tex Ruthlus is approaching! (Use 1.7 for $\sqrt{3}$.)

Boleen Ali and the Treasure of King Caitiff

4.* Young Princess Boleen Ali of Dristan completed a long and hazardous journey. After years of fighting demons and fire-breathing dragons, she was finally approaching her quest, the Mountain of Doom and the cave containing the fabulous treasure of King Caitiff. As Ali was about to enter the fabled cave, there was a tremendous flash of red light and smoke. Out from behind the smoke stepped an old man. "Advance no farther! Behold, I am Algebar, the protector of the treasure of Caitiff. You have traveled far, but you must pass one last test before you can lay claim to the treasure." Algebar placed his outstretched hands palms up and shouted, "Pythagoras!" Two puffs of smoke arose. As the smoke cleared, a brilliant gem appeared in each hand. One, a ruby, was in the shape of a cube; the other, a diamond, was a regular octahedron. "If each edge of this ruby is three-fourths as long as each edge of this diamond, which is larger?" questioned Algebar. "Answer correctly and the treasure is yours. Make a mistake and I will turn you into a precious jewel yourself and add you to the growing treasure of Caitiff. The princess answered correctly and claimed the fabulous treasure. If you had been in her spot, would you have been able to figure out which gem had the larger volume? Which gem is larger?

Bandalf the Elder

5.* Lord Darman, friend and ally of King Bandalf the Elder, has just been placed in a trance by his arch enemy Thorn. In order to release Lord Darman from the trance, Bandalf must solve this puzzle:

> *I am a shining prism of ivory. Six-eight-twelve my measure be. Pierced through am I by non-intersecting prisms three.*
>
> *The longest being triangular on its face. The shortest being rectangular on its base. The third, isosceles trapezoidal in its place.*
>
> *The triangle measures three by four by five. The rectangle exactly two by five. The trapezoid is the strangest of them all. Its sides measure two, four, $\sqrt{2}$ twice, quite irrational.*
> *What part of me is missing, for that is our quest? The ratio of what is missing to that of what is left.*

Perplexed and anguished, Bandalf journeys to the ancient kingdom of Persia to seek help. There grows the Tree of Knowledge where the great Simurgh nests. The Simurgh is a bird of wisdom and, according to the ancient tales, has metallic feathers, a peacock's tail, and a small silver head. Rulers, sages, and scholars travel to Simurgh to gain the wisdom of the ages.

Simurgh gives Bandalf a diagram and says, "Now it shall be clear, for none of the piercing prisms intersect. Thus, calculate the missing volume and the present volume, and their simple ratio is the answer you seek." What is the magic ratio?

11 Similarity

Circle Limit III,
M. C. Escher, 1959

Figures that have the same shape are similar. In this chapter you will discover some of the basic properties of similarity, and then you will use them to solve problems. You will be asked to find the heights of trees, flagpoles, and buildings by measuring shadows and using similar right triangles. Similarity is important in industry; it is used in the fields of film, photography, optics, architecture, and integrated circuits. Solving similarity proportions is a useful skill for the fields of chemistry, physics, and medicine.

Lesson 11.1

Ratio and Proportion

The amount a person uses his imagination is inversely proportional to the amount of punishment he will receive for using it.

– Anonymous

The study of similar geometric figures involves ratios and proportions. You may be a little rusty working ratios and proportions. Let's review. What is ratio?

*A **ratio** is an expression that compares two quantities by division.*

If a and b are two numbers, then the ratio of a to b is written a/b. The ratio of a to b is also written $a{:}b$ or *a is to b*.

Example A

Find the ratio of shaded to unshaded area.

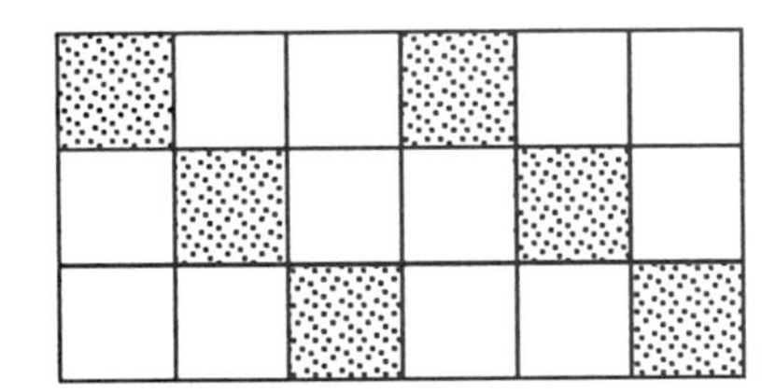

$$\frac{\text{Area}_{\text{shaded}}}{\text{Area}_{\text{unshaded}}} = \frac{6}{12} = \frac{1}{2}$$

Example B

Find the ratio of ✪ to ✧.

The ratio of ✪ to ✧ is $\frac{7}{4}$.

EXERCISE SET A

1. Find the ratio of ❖ to ☆.

2.* Find the ratio of shaded area to the area of the whole figure.

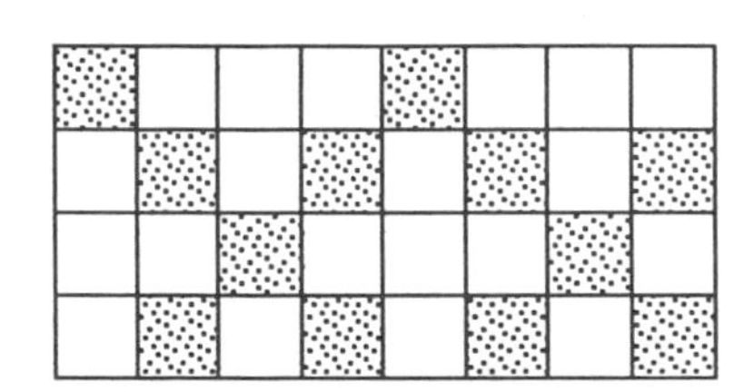

3. Use the diagram below to find these ratios: AC/CD, CD/BD, and BD/BC.

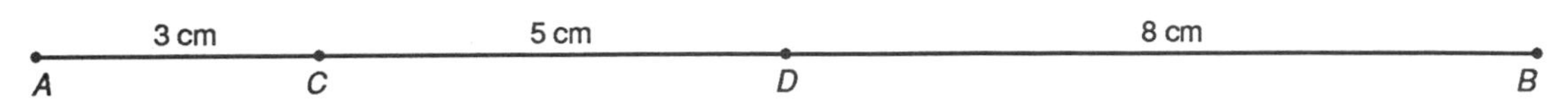

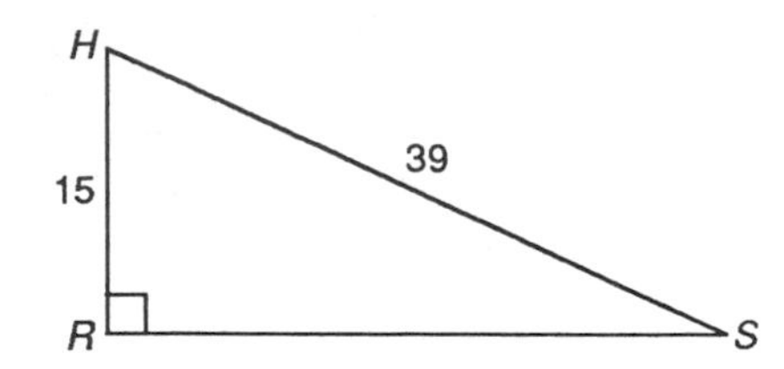

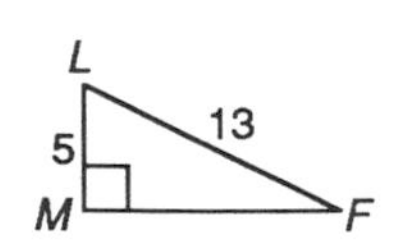

4. Find the ratio of the perimeter of triangle RSH to the perimeter of triangle MFL.

5. Find the ratio of the area of triangle RSH to the area of triangle MFL.

When two ratios are equal, you have a proportion. $\frac{3}{4} = \frac{6}{8}$ is a proportion.

*A **proportion** is a statement of equality between two ratios.*

You may remember how to solve for a variable in an equation with fractions. If you have forgotten, one approach is to cross multiply (if $a/b = c/d$, then $ad = bc$). If one fraction is a multiple of the other, you may use a more direct method. Proportions are used to solve problems that involve comparing similar objects or situations. Let's look at a few examples.

Example A

Solve for x in $\frac{26}{50} = \frac{x}{75}$.

Before you cross multiply, ask yourself, "Can I reduce fractions?" You can.

Rewrite $\frac{26}{50}$ as $\frac{13}{25}$. $(\frac{13}{25} = \frac{x}{75})$

Next, before you cross multiply, check to see if one numerator (or denominator) is a multiple of the numerator (or denominator) in the other fraction. In this problem, since $25 \bullet 3 = 75$, x must be $13 \bullet 3$ or 39.

$x = 39$

Example B

If you earn $380 working for two weeks, how much will you earn in 15 weeks?

$$\frac{380}{2} = \frac{x}{15}$$

$$2x = (380)(15)$$

$$x = (380)(15)/2$$

$$x = 2850$$

You will earn $2850 in 15 weeks.

EXERCISE SET B

Find the missing number in each proportion.

1. $\frac{7}{21} = \frac{a}{18}$

2. $\frac{10}{b} = \frac{15}{24}$

3. $\frac{20}{13} = \frac{60}{c}$

4. $\frac{4}{5} = \frac{x}{7}$

5.* $\frac{2}{y} = \frac{y}{32}$

6. $\frac{10}{10 + z} = \frac{35}{56}$

EXERCISE SET C

Use a proportion to solve each problem.

1. A car travels 106 miles on 4 gallons of gas. How far can it be expected to travel on a full tank of 12 gallons?

2. If a 425-pound Landrover weighs 68 pounds on the moon, how much will a 150-pound woman weigh on the moon? How much would you weigh on the moon?

3.* A pitcher gave up 34 runs in 106 innings of baseball. What is his earned run average (ERA)? In other words, how many runs would he give up in nine innings? (Give your answer accurate to two decimal places.)

4. A recipe for six dozen cookies calls for 2 1/2 cups of flour. How many cups of flour are needed for 10 dozen cookies?

5. The floor plan of a house is drawn to the scale of 1/4" = 1'. The master bedroom measures 3" by 3 3/4" on the blueprints. What is the actual size of the room?

6. Altar and Zenor are visiting ambassadors from Titan, one of the small moons of Saturn. On Titan the atmosphere is so light that the Titans have evolved multiple antennae to pick up sound waves. Altar has revealed to the Biological Research Division that the sum of the lengths of the antennae on a Titan is a direct measure of that Titan's age. Altar has antennae with lengths of 8 cm, 10 cm, 13 cm, 16 cm, 14 cm, and 12 cm. Zenor has an age of 130 and her seven antennae have an average length of 17 cm. What is the age of Ambassador Altar?

Altar and Zenor

The Case of the Belgian Stamp Murder

7. Hemlock Bones, the famous consulting detective, is studying the evidence in the murder of Sir Osborne Chatsworth III. The evidence is a set of photos of the murder victim's bedroom taken by the police photographer. One photo shows the nightstand that sat next to Sir Osborne's bed. On the nightstand is a half-empty teacup, a magnifying glass, and a stamped envelope. Hemlock recognizes the stamp as an 1889 Belgian six-Franc commemorative. There were two types made that year — slightly different in size. One is quite valuable, the other practically worthless. Hemlock suspects that Thaddeus O'Malley may be involved in the crime since Thaddeus recently sold the valuable 1889 Belgian six-Franc commemorative to Chatsworth. Hemlock reasons that the murderer killed Sir Osborne, removed the valuable stamp from the envelope and replaced it with the other Belgian six-Franc! If Hemlock can establish that the stamp on the envelope is an 1889 Belgian six-Franc commemorative but *not* the valuable one, the murderer's devious plan will be uncovered. Hemlock must figure out the actual size of the postage stamp in the photo to determine whether or not his analysis is correct. The valuable stamp is 4 cm square, the worthless one is 3 cm square. The envelope is obviously the standard legal size, 24 cm in length. The length of the envelope in the photo is 1.6 cm. The stamp in the photo measures .2 cm on each edge. Is the stamp on the envelope the rare one or is it the worthless one? Is Thaddeus O'Malley our murderer? What is the actual size of the postage stamp on the envelope?

Special Project

The Golden Ratio

German psychologist Gustav Fechner in 1876 made psychological studies with a rectangle having special proportions. He found that many people (about 75%) selected this specially proportioned rectangle as the most pleasing of a group of rectangular shapes. Let's try a version of Fechner's experiment.

Step 1: Make a tally chart like the one at right.

Step 2: Ask ten or more people which rectangle in each of the groups is the most pleasing or best looking rectangular shape. Record their choices for each group in your chart.

Step 3: Compare your results with the results of others. Did most people select the same rectangle as the most pleasing shape?

Person's initials	Group 1	Group 2
1. —?—	–?–	–?–
2. —?—	–?–	–?–
3. —?—	–?–	–?–
⋮	⋮	⋮
Rectangle voted best looking?	–?–	–?–

Group 1

b.

a.

c.

Group 2

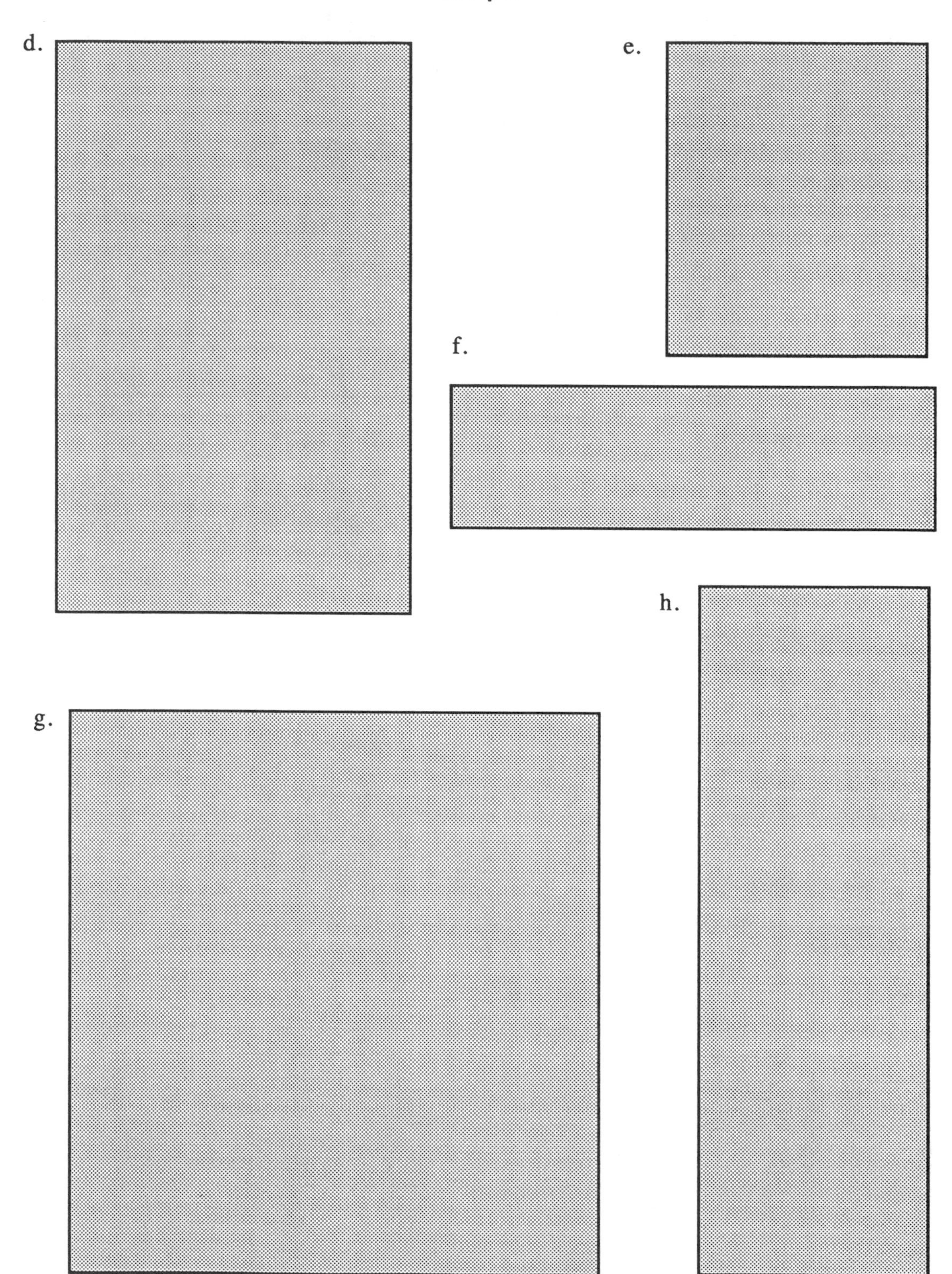

What was special about these special rectangles? Clearly it is not their size. It was their proportions. The rectangles c and d were probably the rectangles chosen as having the most pleasing shapes. Measure the lengths of the sides of these rectangles. Calculate the ratio of the length of the longer side to the length of the shorter side for each rectangles.

Did you get the same result each time? Was it approximately 1.6? This ratio approximates the famous Golden Ratio of the ancient Greeks. These special rectangles are called Golden Rectangles because the ratio of the length of the longer side to the length of the shorter side is the Golden Ratio.

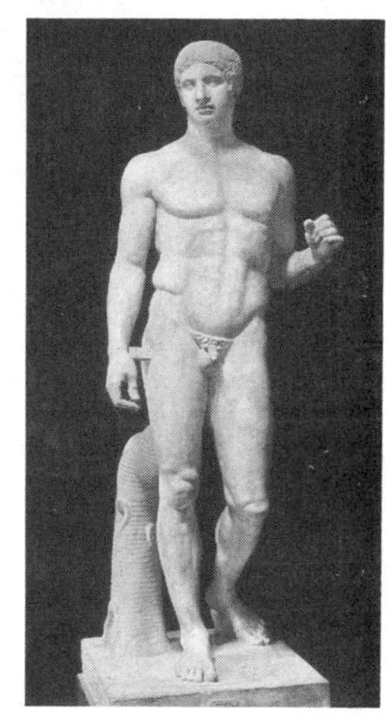

Doryphoros

Golden Rectangles can be found in the shape of playing cards, windows, book covers, file cards, ancient buildings, and modern skyscrapers. Many artists, including Renaissance artist Leonardo da Vinci and modern artist Piet Mondrian, have incorporated the Golden Rectangle into their works because of its aesthetic appeal. It is believed by some researchers that classical Greek sculptures of the human body were proportioned so that the ratio of the total height to the height of the navel was the Golden Ratio. Polycleitos (c. 450–440 B.C.) used the Golden Ratio in creating the proportions for his *Doryphoros*, the Spearbearer.

The ancient Greeks considered the Golden Rectangle to be the most aesthetically pleasing of all rectangular shapes. It was used many times in the design of the famous Greek temple, the Parthenon.

American researcher, Jay Hambridge, established that indeed the Golden Ratio can be found not only in Greek temples and sculpture, but also in the proportions of the human skeleton. The ratio of the total height to the height of the navel is a close approximation to the Golden Ratio. Other writers and researchers have claimed that ratios of many other parts of the human body are also in the Golden Ratio. In other words, we probably all have some proportions close to the Greek ideal somewhere in our bodies.

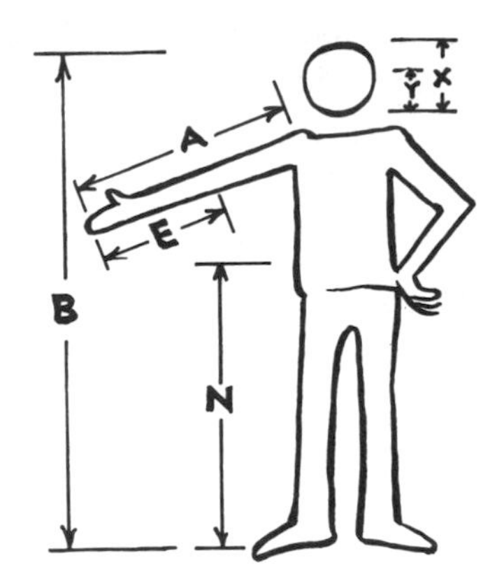

Let's see if the Golden Ratio is somewhere in each of us. Form groups of four or five. Make a table like the one on the bottom of this page. Include the name of each person in your group.

Step 1: Measure the height (B) and the navel height (N) of each member of your group. Calculate the ratios B/N. Record them in your table.

Step 2: Measure the length (F) of an index finger and the distance (K) from the finger tip to the big knuckle of each member of your group. Calculate the ratios F/K. Record them in your table.

Step 3: Measure the length (L) of a leg and the distance (H) from the hip to the kneecap of everyone in your group. Calculate and record the ratios L/H.

Step 4: Measure the length (A) of an arm and the distance (E) from the finger tips to the elbow of everyone in your group. Calculate and record the ratios A/E.

Step 5: Select another pair of lengths (X and Y) on the body that you suspect may be in the golden ratio. Measure these lengths. Calculate the ratios (large to small) and record them.

Person's name	B/N	F/K	L/H	A/E	X/Y
1. —?—	–?–	–?–	–?–	–?–	–?–
2. —?—	–?–	–?–	–?–	–?–	–?–
⋮	⋮	⋮	⋮	⋮	⋮

The Golden Ratio can be approximated with the help of a familiar pattern of numbers. Can you determine the next number in this pattern:

1, 1, 2, 3, 5, 8, 13, 21, 34, 55, 89, 144, 233, –?–

In Chapter 1 you worked with this sequence of numbers. It is called the Fibonacci sequence. The numbers in the Fibonacci sequence occur in many branches of mathematics. They also appear in nature and in art. To determine a decimal approximation of the Golden Ratio, you can calculate the ratio of pairs of consecutive Fibonacci numbers. Use a calculator to estimate the ratios of the following pairs of consecutive Fibonacci numbers as decimals accurate to four decimal places.

1. $34/21 \approx$ –?–
2. $144/89 \approx$ –?–
3. $377/233 \approx$ –?–
4. $610/377 \approx$ –?–
5. $987/610 \approx$ –?–
6. $1597/987 \approx$ –?–
7. What is the Golden Ratio accurate to four decimal places? You should be able to find it using the pattern from the first six problems.

So far you found an approximate value for the Golden Ratio accurate to four places. It was about 1.6180. What *is* the Golden Ratio?

> *If a line segment is divided into two lengths such that the ratio of the segment's entire length to the longer length is equal to the ratio of the longer length to the shorter length, then the segment has been divided into the* ***Golden Ratio****.*

What is the actual value of the Golden Ratio and how is it found? Let's use algebra to find the actual value of the Golden Ratio. Let ø be the Golden Ratio and let a and b be the lengths of two parts of a segment that has been divided into the Golden Ratio.

a b

$$ø = \frac{a}{b} \text{ where } a > b$$

$$\frac{a+b}{a} = \frac{a}{b} \quad \text{(According to the definition of the Golden Ratio)}$$

$$\frac{a}{a} + \frac{b}{a} = \frac{a}{b}$$

$$1 + \frac{b}{a} = \frac{a}{b}$$

$$1 + \frac{1}{ø} = ø \quad \left(\text{If } ø = \frac{a}{b}, \text{ then } \frac{1}{ø} = \frac{b}{a}\right)$$

$$ø + 1 = ø^2 \quad \text{(Multiplying both sides of the last equation by ø)}$$

$$ø^2 - ø - 1 = 0$$

Solve for ø using the quadratic formula: $ø = \dfrac{-(-1) \pm \sqrt{(-1)^2 - 4(1)(-1)}}{2} = \dfrac{1 \pm \sqrt{5}}{2}$.

The positive root is the Golden Ratio. A calculator will verify that $\dfrac{1 + \sqrt{5}}{2} \approx 1.6180$.

Given a line segment, $\overline{GL}$, how can you find a point O between the endpoints G and L that divides the segment into the Golden Ratio by construction?

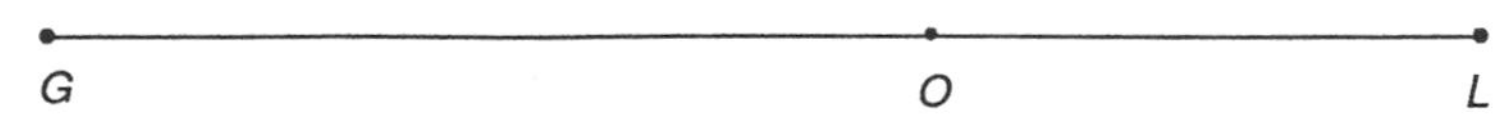

That is, how could you locate O such that $\dfrac{GL}{GO} = \dfrac{GO}{OL}$?

Let's take a step-by-step look at the construction. Start with a segment called $\overline{GL}$.

Step 1

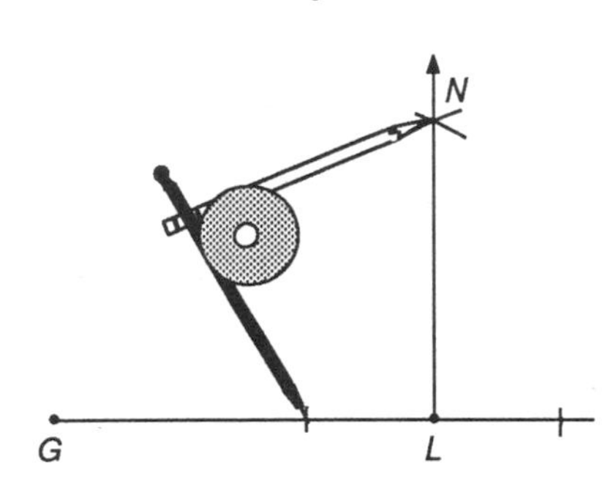

Construct $\overrightarrow{LN}$ perpendicular to the endpoint L of $\overline{GL}$.

Step 2

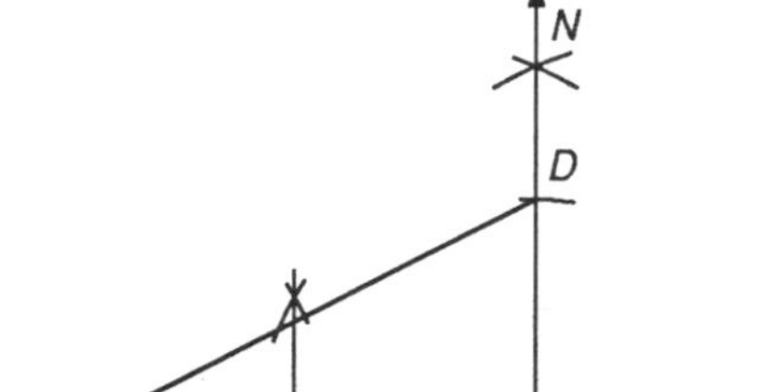

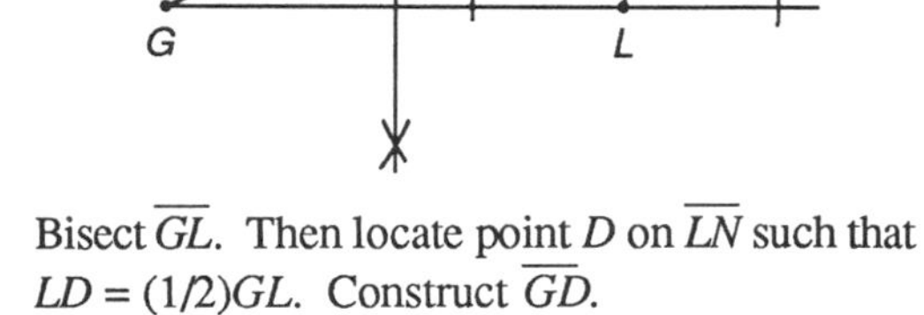

Bisect $\overline{GL}$. Then locate point D on $\overline{LN}$ such that $LD = (1/2)GL$. Construct $\overline{GD}$.

Step 3

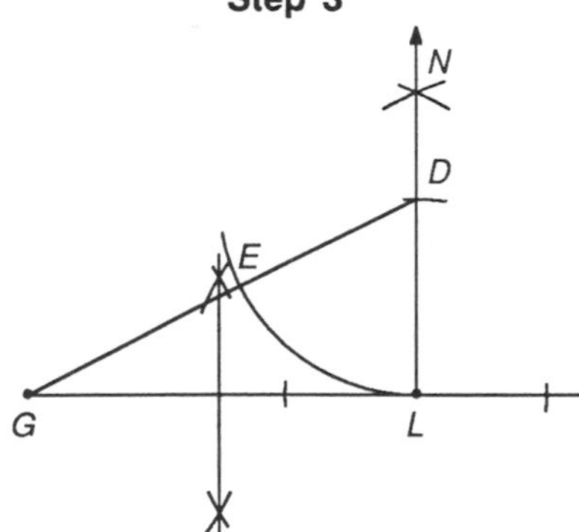

Locate a point E on $\overline{GD}$ such that $DE = DL$.

Step 4

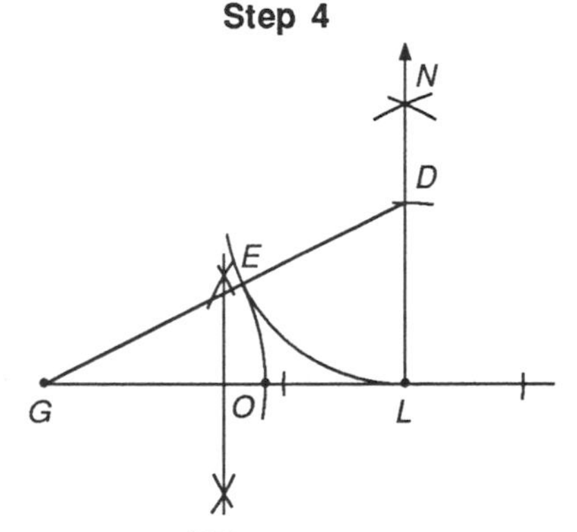

Locate point O on $\overline{GL}$ such that $GO = GE$. O is the point that divides $\overline{GL}$ into the Golden Ratio.

1. Construct a line segment and divide it into the Golden Ratio.
2. Use the construction from Problem 1 to construct a Golden Rectangle.
3. A Golden Triangle is an isosceles triangle in which the ratio of the length of a leg to the length of the base is the Golden Ratio. Use the construction from Problem 1 to construct a Golden Triangle. Each of the five tips of a pentagram is a Golden Triangle.

Lesson 11.2
Similarity

Small Things and Great.
He that lets the small things bind him
Leaves the great undone behind him.

– Piet Hein

Similarity plays an important part in the construction of large objects such as airplanes and small objects such as integrated circuits. Before building an airplane, aeronautical engineers design scale drawings of the plane, use the scale drawings to build scale models, and then run tests with the scale models. To fabricate integrated circuits, electrical engineers create a large scale map of the integrated circuit using a computer. They then reduce the circuit design and transfer it onto minute silicon chips.

EXERCISE SET A

Identify any figure on the right that is a reduction or an enlargement of one on the left.

1.

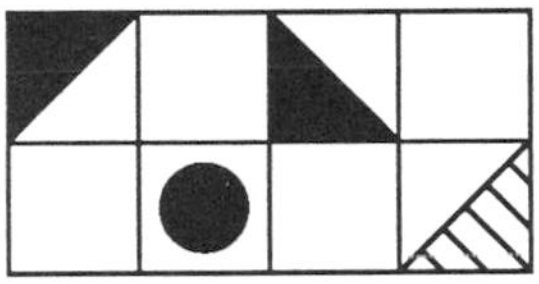

2.

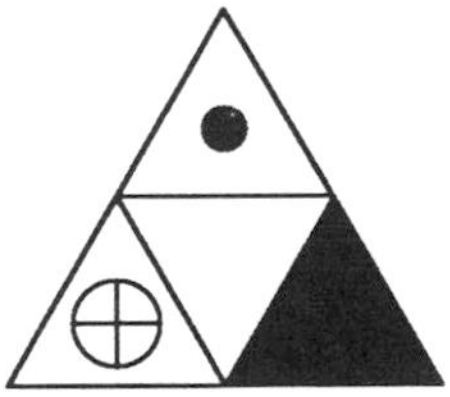

3.

4.

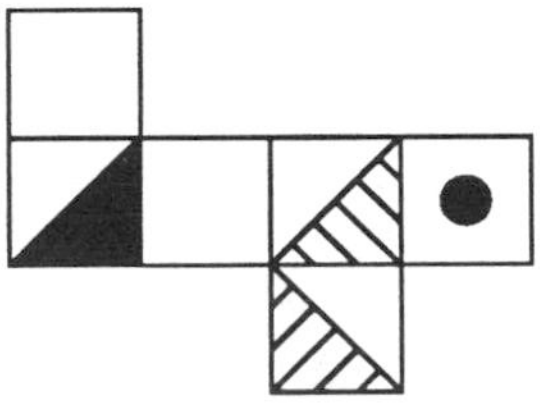

a.

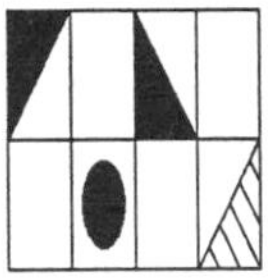

b.

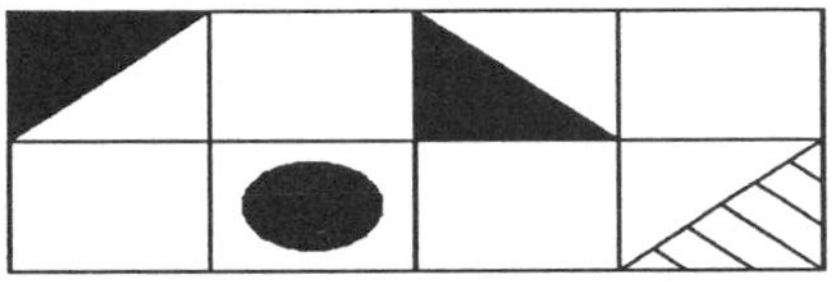

c.

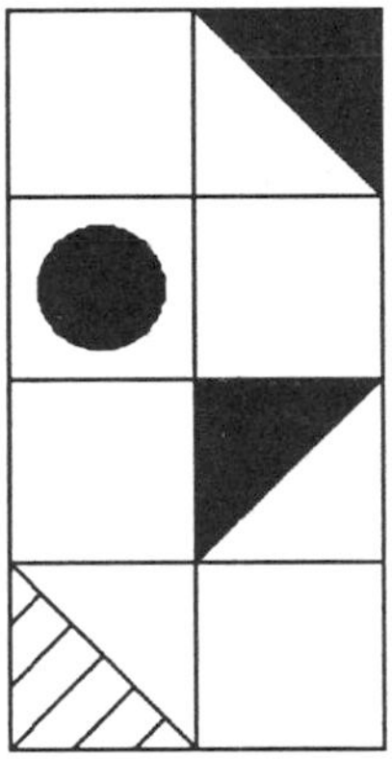

d.

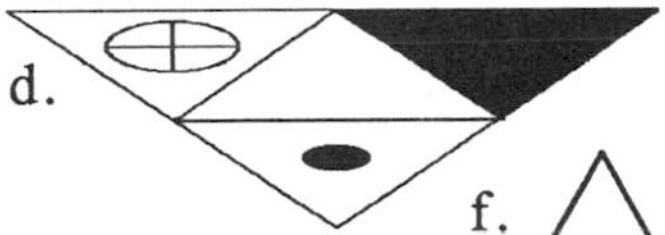

e.

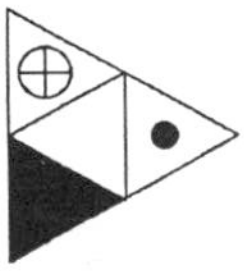

f.

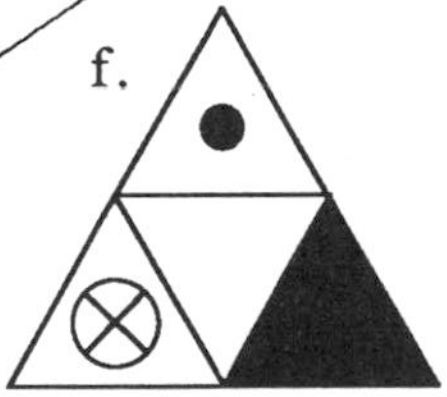

g.

h.

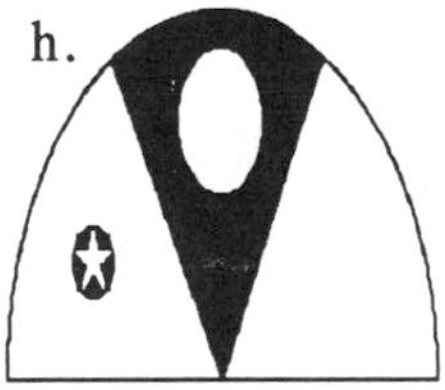

i.

j.

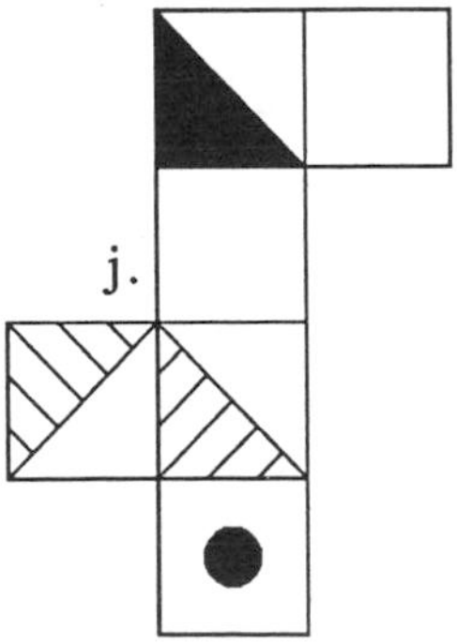

k.

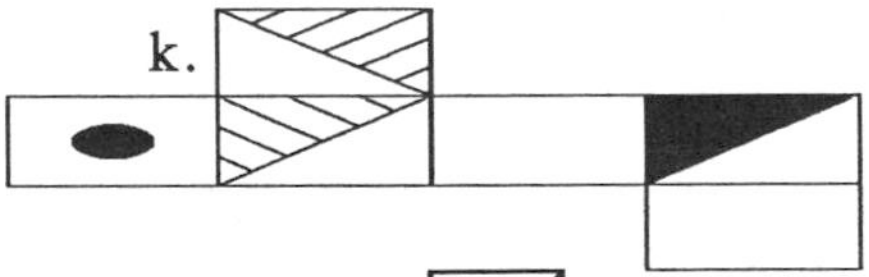

l.

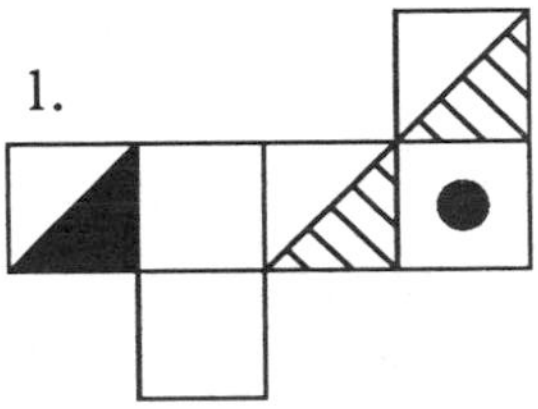

Figures that have the same shape and the same size are congruent figures. Figures that have the same shape but not necessarily the same size are **similar figures**. This, however, is not a precise definition for similarity. What does it mean to be the same shape?

A person looking at reflections in different fun house mirrors sees different images. Do we want to say that the images are similar to the original? They certainly have a lot of features in common, but they are not similar in a mathematical sense. Similar shapes can be thought of as enlargements or reductions where no irregular distortions have taken place. If you can place one figure on a photocopy machine and enlarge or reduce it to fit over another figure, then the two original figures are similar. Are all rectangles similar? They certainly have many common characteristics, but they are not all similar because some could not be enlarged or reduced to fit over others. What about other geometric figures? All squares are similar to each other and all circles are similar to each other, but all triangles are not similar to each other. In this lesson you will arrive at a mathematical definition for similar polygons.

EXERCISE SET B

Sketch a figure similar but not congruent to each figure below. Use graph paper.

1.

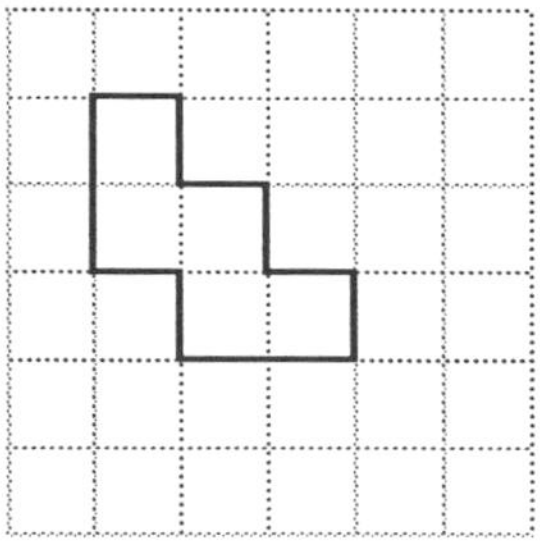

2.

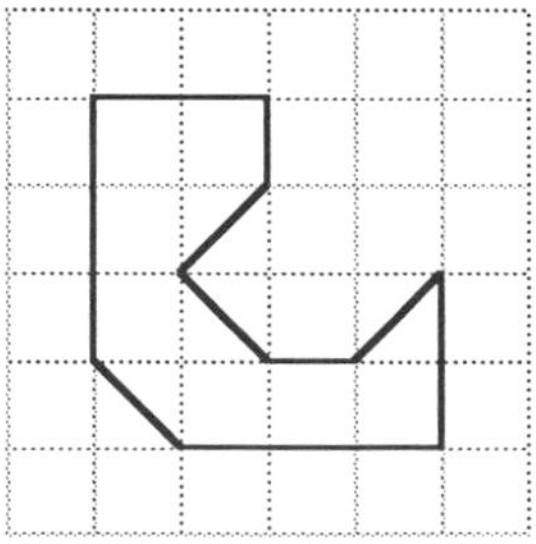

3.

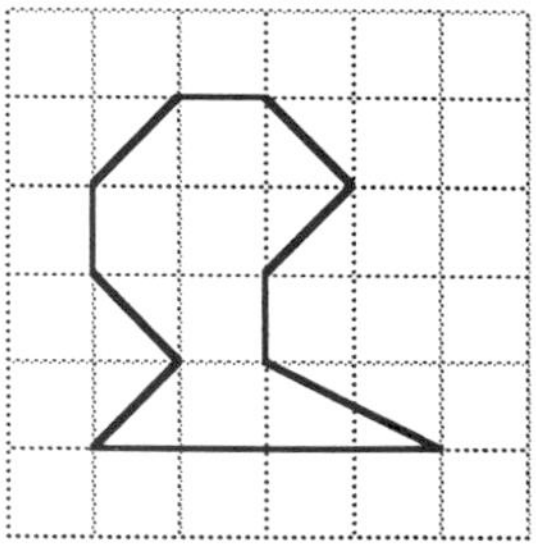

A movie projector projects images onto a screen that are larger than and similar to the images on the film. Some photocopy machines can make enlargements and reductions and both copies are similar to the original.

The two illustrations below are a photocopy enlargement and a reduction of the same drawing. The two photocopies are of a drawing of an impossible three-dimensional solid.

Let's use the photocopy enlargement and reduction below to see what makes polygons similar. Even though the impossible figures appear to be three-dimensional, please try to see each figure as just a two-dimensional figure.

Investigation 11.2.1 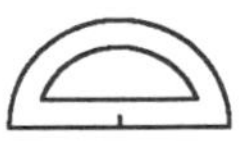 

A hexagon has been outlined in both copies. Measure the angles indicated in both hexagons. Measure the corresponding segments in both hexagons. Find the ratios of the lengths of corresponding sides as indicated.

$m\angle ABC \approx$ –?– $\quad m\angle PQR \approx$ –?–

$m\angle BCD \approx$ –?– $\quad m\angle QRS \approx$ –?–

$AB \approx$ –?– $\quad BC \approx$ –?–

$CD \approx$ –?– $\quad EF \approx$ –?–

$PQ \approx$ –?– $\quad QR \approx$ –?–

$RS \approx$ –?– $\quad TU \approx$ –?–

$\frac{AB}{PQ} \approx$ –?– $\quad \frac{BC}{QR} \approx$ –?–

$\frac{CD}{RS} \approx$ –?– $\quad \frac{EF}{TU} \approx$ –?–

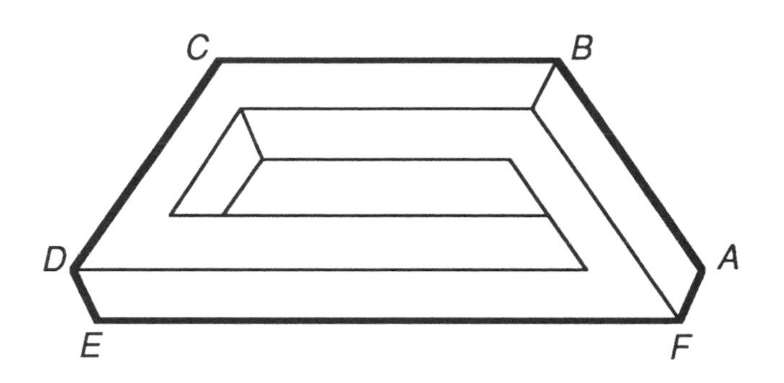

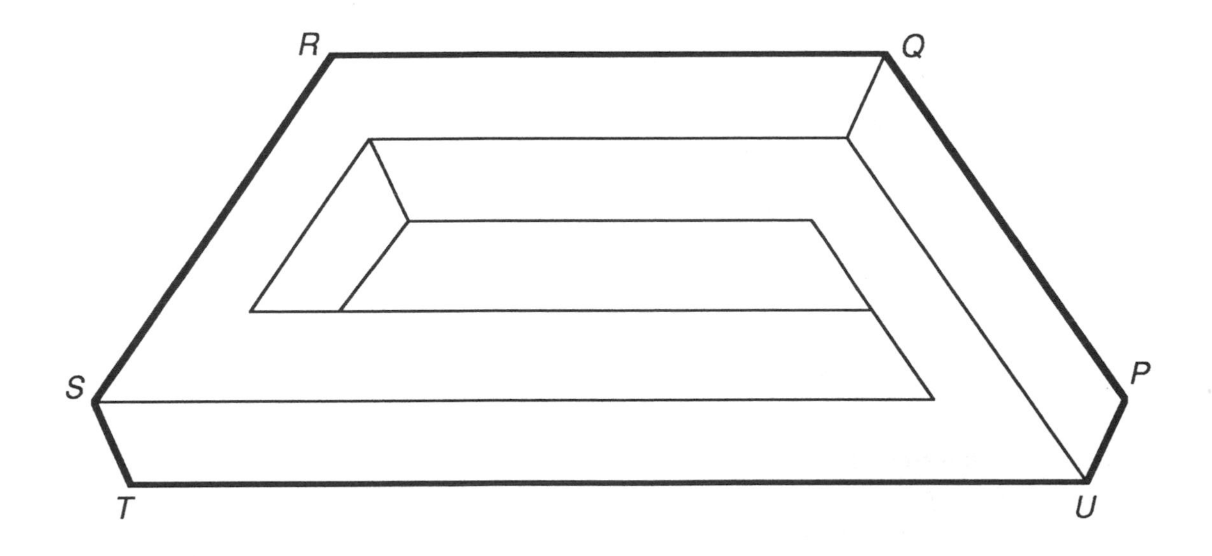

How do the corresponding angles of similar polygons compare? How do the corresponding sides of similar polygons compare? From your observations you should be able to state that if two polygons are similar, then their corresponding angles are congruent and their corresponding sides are proportional. This is reversible. That is, if you constructed two polygons that have corresponding angles congruent and corresponding sides proportional, then one polygon would be an enlargement or reduction of the other. One polygon would be similar to the other. Let's state a more mathematical definition for similar polygons.

> *Two polygons are* ***similar polygons*** *if and only if the corresponding angles are congruent and the corresponding sides are proportional.*

The symbol for *is similar to* is ~. You use it the same way that you use the symbol for congruence. If the two quadrilaterals *CORN* and *MAIZ* are similar, you write *CORN* ~ *MAIZ*. Just as in statements of congruence, the order of the letters tells you which segments and which angles in the two polygons correspond.

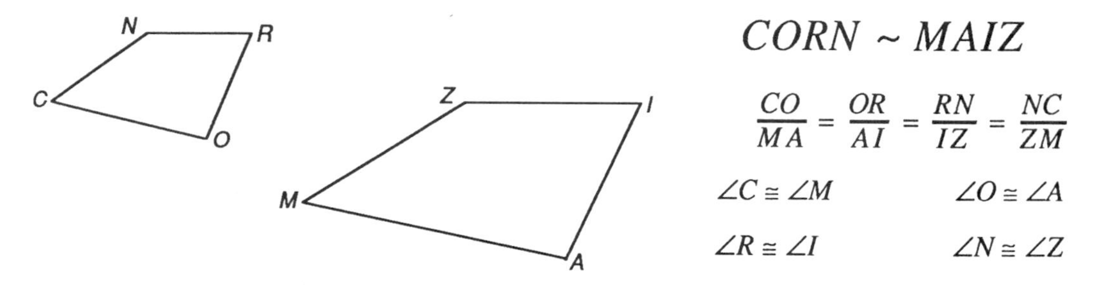

In this lesson you discovered that two polygons are similar if and only if their corresponding angles are congruent and if the corresponding sides are proportional. Do we need both conditions to guarantee that the two polygons are similar? In other words, if you know that two polygons have only their corresponding angles congruent, can you conclude that they have to be similar? Or, if two polygons have their corresponding sides proportional, do they necessarily have to be similar?

Polygons with Corresponding Angles Congruent

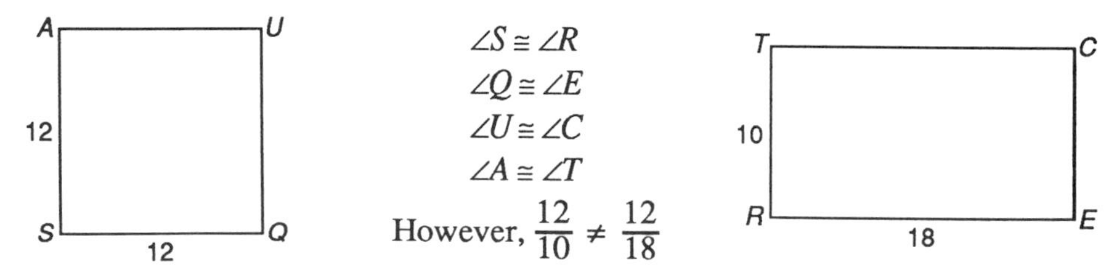

Square *SQUA* and rectangle *RECT* have their corresponding angles congruent, but their corresponding sides are not proportional. Clearly, the two polygons are not similar. Therefore, you cannot determine whether or not two polygons are similar just because the corresponding angles of the two polygons are congruent.

Polygons with Corresponding Sides Proportional

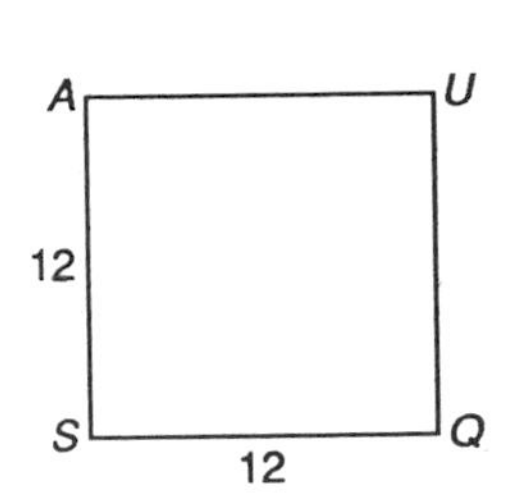

$$\frac{12}{18} = \frac{12}{18}$$

However, $\angle S \not\cong \angle R$

Square *SQUA* and rhombus *RHOM* have their corresponding sides proportional, but their corresponding angles are not congruent. Clearly, the two polygons are not similar. Therefore, you cannot determine whether or not two polygons are similar just because the corresponding sides of the two polygons are proportional.

You have just discovered from the two counter-examples above that you cannot determine whether or not two polygons are similar just because the corresponding sides of the two polygons are proportional *or* the corresponding angles are congruent.

From the definition of similar polygons, you know that if two polygons are similar, then their corresponding angles are congruent and their corresponding sides are proportional. You can use the definition to find missing parts of similar polygons.

Example A

Find the measure of the side labeled x and the measure of the angle labeled y in the similar polygons below.

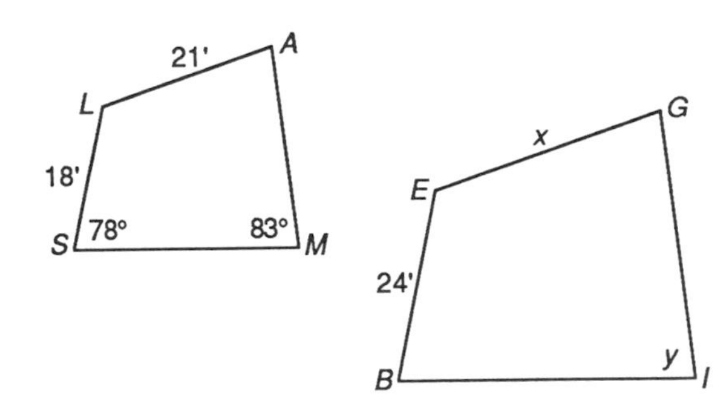

$SMAL \sim BIGE$

$$\frac{18}{24} = \frac{21}{x}$$

$$\frac{3}{4} = \frac{21}{x}$$

$$3x = (4)(21)$$

$$x = 28$$

So the side labeled x is 28'.

$\angle M \cong \angle I$

So the angle labeled y is 83°.

From the definition of similar polygons you know that if two polygons have their corresponding angles congruent and their corresponding sides proportional, then the two polygons are similar. You can use the definition to determine whether two polygons are similar if you know the measures of their angles and the lengths of their sides. Look at the next example.

Example B

Determine whether or not the polygons below are similar.

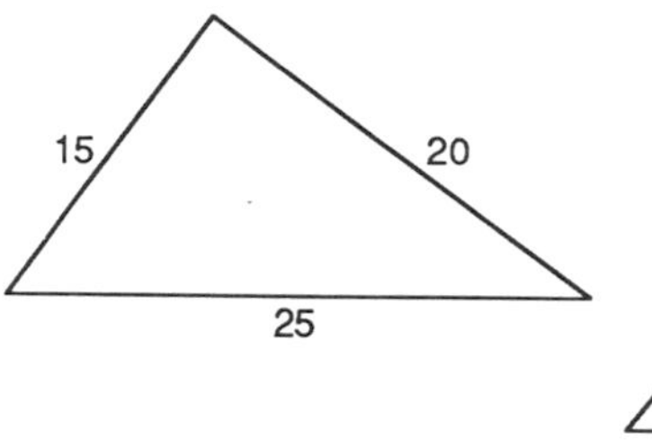

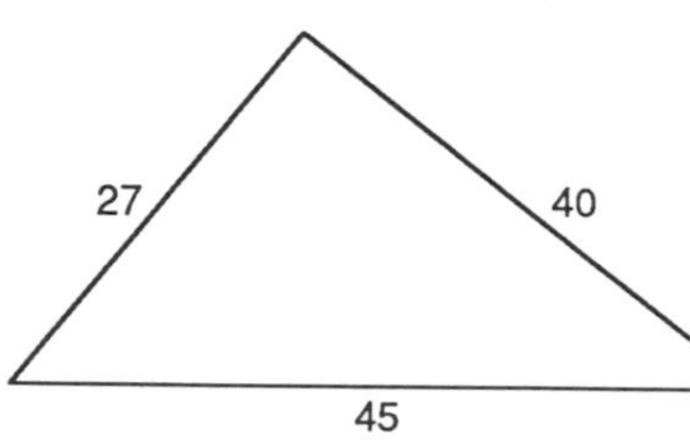

$\frac{15}{27} = \frac{5}{9}$ and $\frac{25}{45} = \frac{5}{9}$

But $\frac{20}{40} = \frac{1}{2}$

Therefore the triangles are not similar.

EXERCISE SET C

Use the definition of similar polygons to solve each problem. All measurements are in centimeters.

1.* *THINK* ~ *LARGE*
Find *AL, RA, RG, KN.*

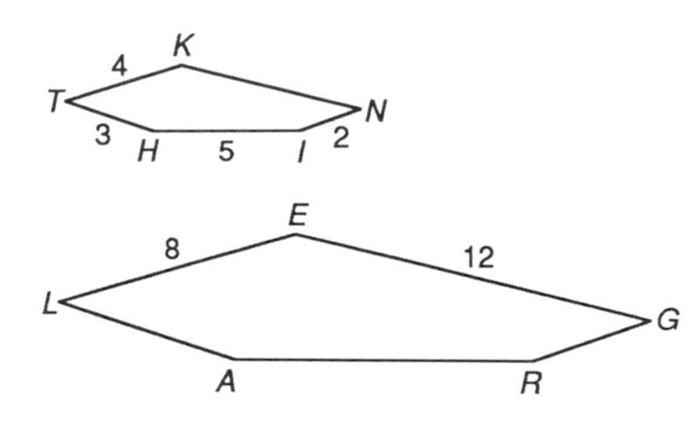

2. Are the polygons similar?

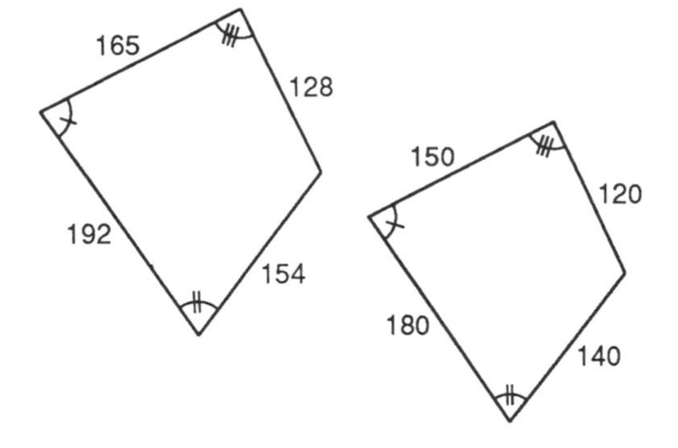

3. *SPIDER* ~ *HNYCMB*
Find *NY, YC, CM, MB.*

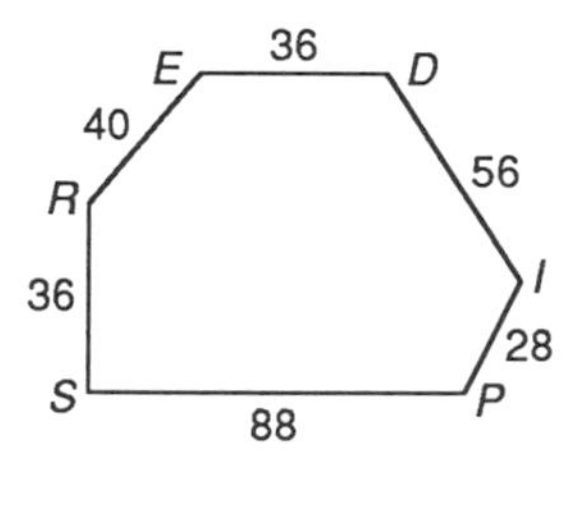

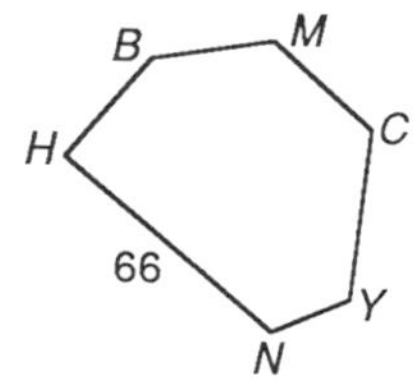

4.* Are the polygons similar?

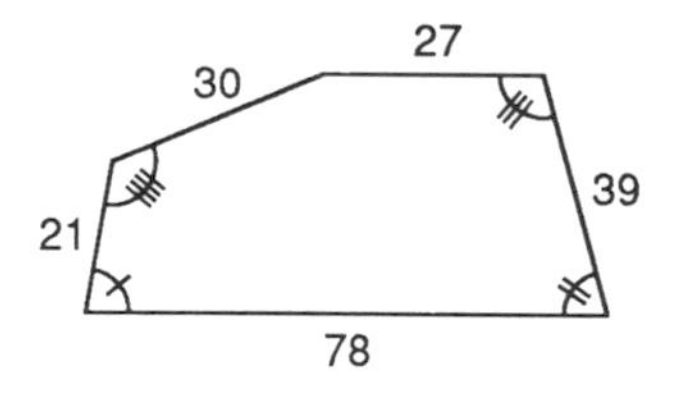

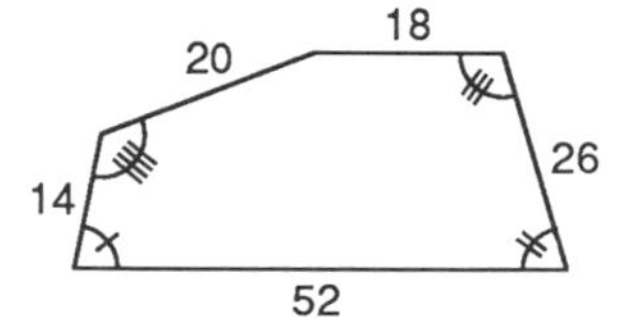

5. $\Delta ACE \sim \Delta IKS$

$x = -?-$

$y = -?-$

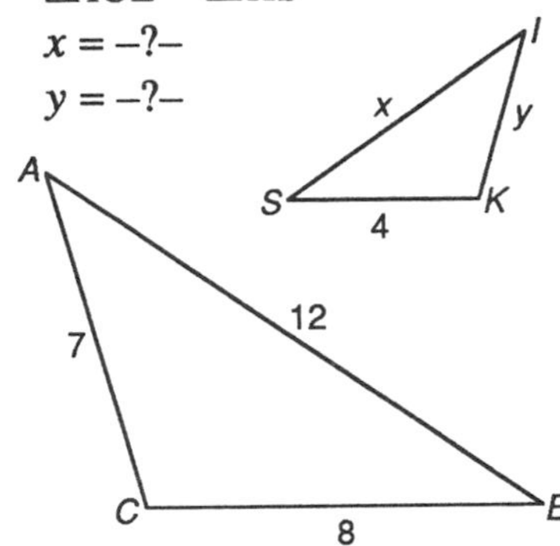

6. $\Delta RAM \sim \Delta XAE$

$z = -?-$

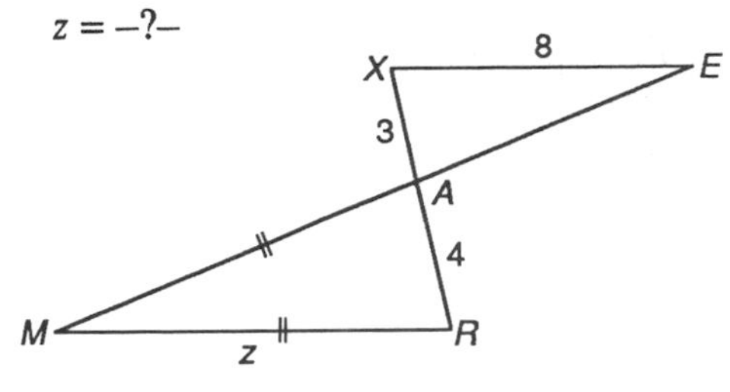

7.* $\overline{DE} \parallel \overline{BC}$

Are corresponding angles congruent?

Are corresponding sides proportional?

Is $\Delta AED \sim \Delta ABC$?

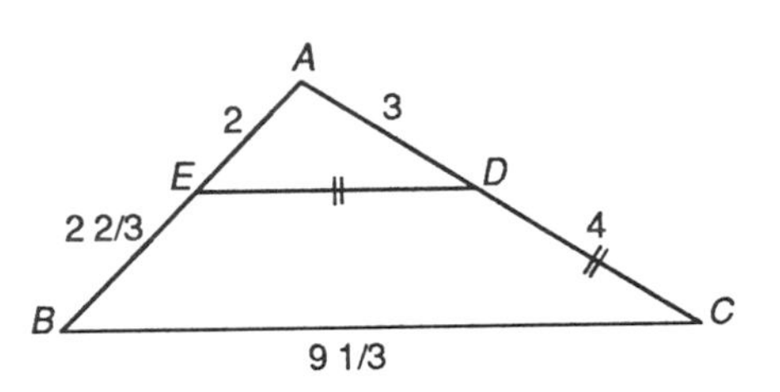

8. $\Delta ABC \sim \Delta DBA$

$m = -?-$

$n = -?-$

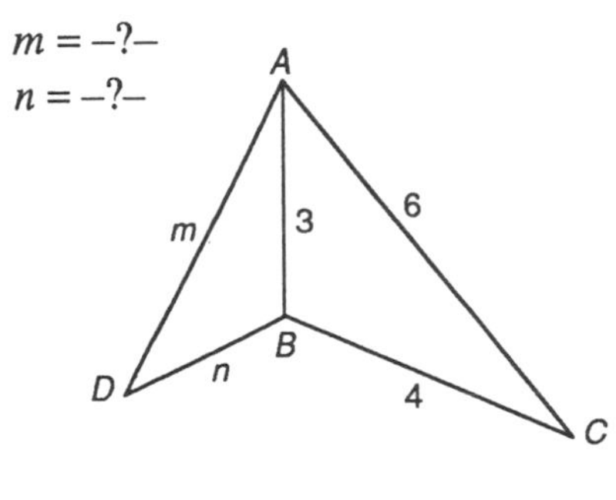

9. $\Delta ABC \sim \Delta DEF$

Use algebra to show that:

If $\frac{a}{d} = \frac{c}{f}$, then $\frac{a}{c} = \frac{d}{f}$.

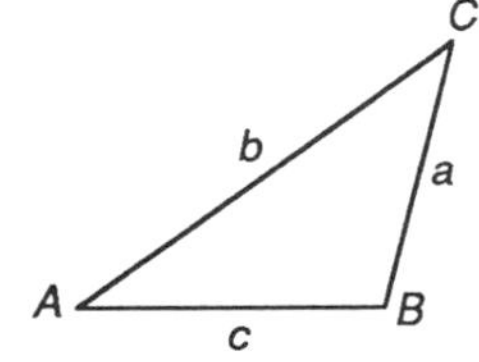

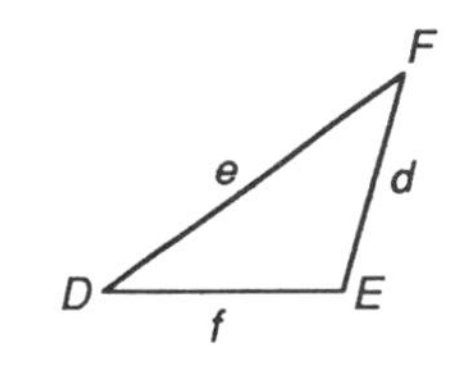

An image on movie film and the image projected onto the screen are similar as long as the movie is projected at right angles to the screen. If the projector is moved half the distance to the screen, each dimension of the image is cut in half. If the projector is moved away from the screen three times the original distance, the dimensions on the new image are all three times the original size.

Let's use this idea to construct polygons similar to an original polygon.

EXERCISE SET D

1. On a sheet of unlined paper, perform the steps below to construct a quadrilateral similar to but larger than an original quadrilateral.

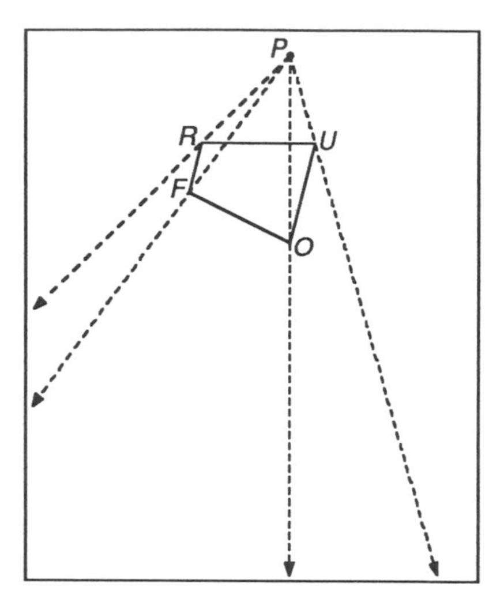

Step 1: Construct a quadrilateral *FOUR* in the upper third of your paper and place a point *P* above it as shown in the diagram. (The quadrilateral is like an image on movie film. The point you select is the projector's light source.)

Step 2: From point *P* construct four rays: $\overrightarrow{PF}$, $\overrightarrow{PO}$, $\overrightarrow{PU}$, $\overrightarrow{PR}$. Extend the rays to the end of the paper. (The rays are like the light rays of the projector lamp.)

Step 3: With your compass measure the distances: *PF*, *PO*, *PU*, *PR*. Transfer the distances twice each so that you find points *X*, *Y*, *Z*, and *W* such that: *P*, *F*, and *X* are collinear and $FX = 2(PF)$; *P*, *O*, and *Y* are collinear and $OY = 2(PO)$; *P*, *U*, and *Z* are collinear and $UZ = 2(PU)$; *P*, *R*, and *W* are collinear and $RW = 2(PR)$.

Step 4: You have now located four points, *X*, *Y*, *Z*, and *W*, that are each three times as far from point *P* as the original four points of the quadrilateral. Construct quadrilateral *XYZW*. (Like the images on the film and the screen, quadrilaterals *FOUR* and *XYZW* are similar.)

2. On a sheet of unlined paper, perform the steps below to construct a pentagon similar to but smaller than an original pentagon.

Step 1: Construct a pentagon *FIVER* at the bottom third of your paper and place a point *P* above it in the top third of your paper as shown in the diagram.

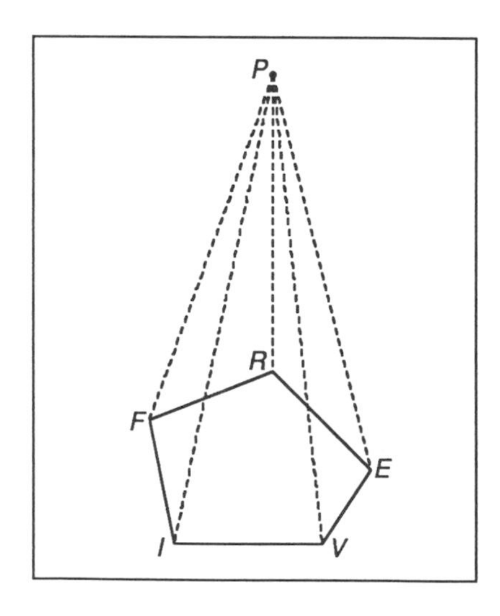

Step 2: Construct the five segments: $\overline{PF}$, $\overline{PI}$, $\overline{PV}$, $\overline{PE}$, $\overline{PR}$. Construct the midpoint of each of these five segments.

Step 3: Label the midpoints of $\overline{PF}$, $\overline{PI}$, $\overline{PV}$, $\overline{PE}$, and $\overline{PR}$ as points *C*, *Y*, *N*, *K*, and *O* respectively. Construct polygon *CYNKO*. (Like the images on the film and the screen, pentagons *FIVER* and *CYNKO* are similar.)

Special Project

Making a Mural

Every child is an artist. The problem is how to remain an artist after he (she) grows up.

– Picasso

Mural artists use similarity to help them create their large art works. A muralist begins a mural by making a small drawing. Then the muralist places a grid of squares over it. The muralist divides the surface on which the mural will appear into a similar but larger grid of squares. The artist next draws the lines and shapes of the original drawing square-by-square in the corresponding positions of the large squares of the mural surface. Finally, the artist colors in the regions to complete the mural.

The design in the small grid of squares below left is similar to the design in the large grid of squares below right. The enlargement was made by matching points in the original drawing to the corresponding points in the larger grid. For example, point A in the grid on the left is in the same position as point A' is in the grid on the right.

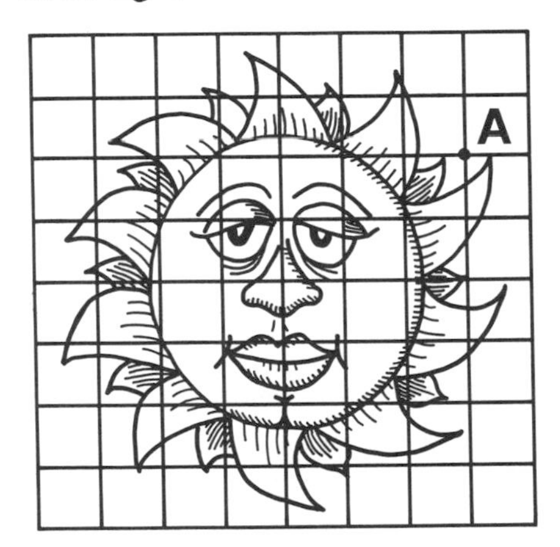

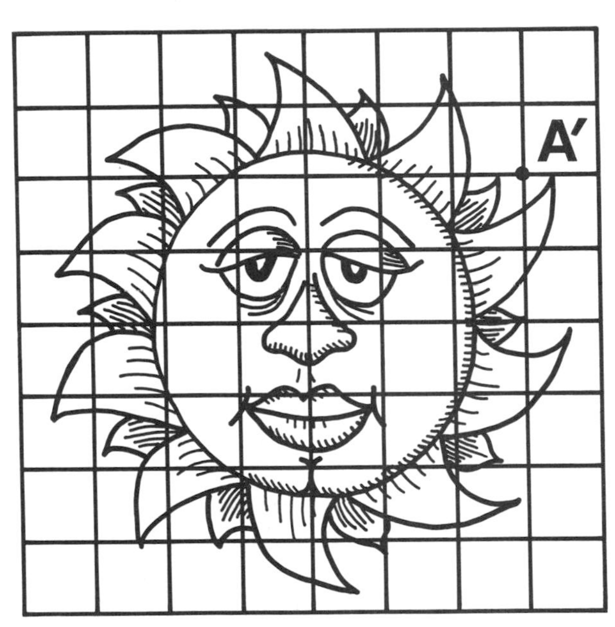

To create your own mural you will need the following materials:

- Photograph or small drawing to enlarge
- Large surface for your mural (large sheet of butcher paper)
- Drawing equipment (ruler, meter stick, marking pens, colored pencils, etc.)

Begin by constructing a grid of squares on a photocopy of your photograph or drawing. The more squares, the more accurately your mural will depict the original. Divide the surface on which you are creating the mural into a similar grid of larger squares. Do all this lightly in pencil. To create the mural, carefully match the lines and curves of the drawing in the small squares to the corresponding positions in the large squares. Finally, color in the appropriate regions.

Lesson 11.3

Similar Triangles

Earlier in the year you found four shortcuts (SSS, SAS, ASA, and SAA) for determining that two triangles are congruent. In the last lesson we were unable to find simple shortcuts for determining that quadrilaterals are similar. How about for triangles? Are there shortcuts for determining that two triangles are similar like the shortcuts you found for determining triangles congruent?

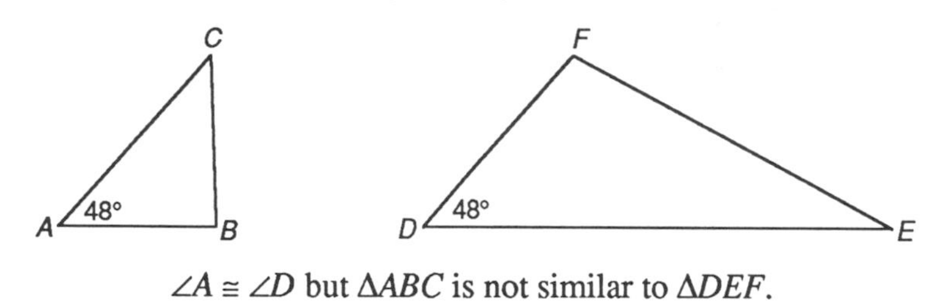

$\angle A \cong \angle D$ but ΔABC is not similar to ΔDEF.

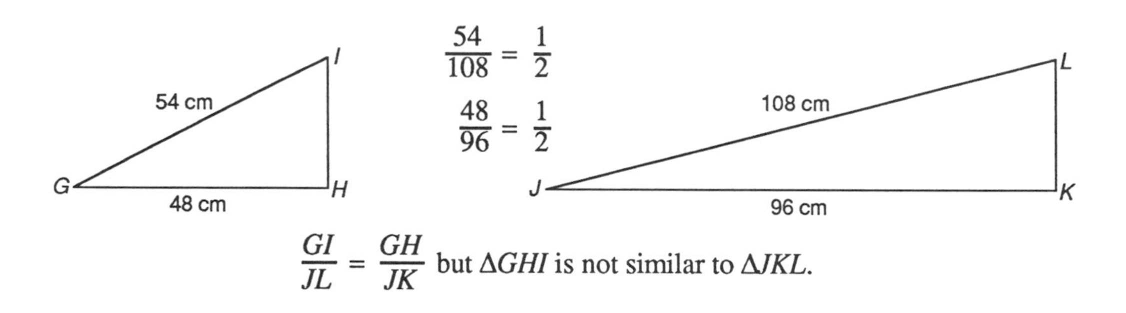

$\frac{GI}{JL} = \frac{GH}{JK}$ but ΔGHI is not similar to ΔJKL.

If there are shortcuts, the two examples above illustrate some limitations. If there is a shortcut using only angles, you need to investigate at least two angles in one triangle congruent to the two corresponding angles in the other triangle to see if they force the two triangles to be similar. And, if there is a shortcut using just sides, the example above demonstrates that two sides are not sufficient; you need to investigate all three sides in one triangle proportional to the corresponding three sides in the other triangle to see if they force the two triangles to be similar. Let's see if we can find shortcuts for determining that two triangles are similar.

The two investigations in this lesson work best if you work in groups of four or five. Each group member should start with a different triangle and perform the steps outlined for the investigation. Share your results with the other members of your group and together make a conjecture based on the investigation.

Investigation 11.3.1

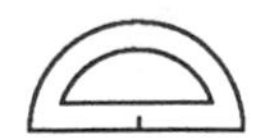

If three sides of one triangle are proportional to the three sides of another triangle, must the two triangles be similar? On a clean sheet of unlined paper perform the following steps:

Step 1: Construct a triangle.

Step 2: Construct a second triangle with the length of the sides some multiple of the original triangle (twice, three times, or four times larger or smaller).

Step 3: With a protractor, compare the corresponding angles of the two triangles.

Compare your results with the results of others near you. You constructed two triangles with all three pairs of sides in the same ratio. If this forced the angles to be congruent, then both conditions for similar polygons hold and the two triangles are similar. You should be ready to state a conjecture.

If the three sides of one triangle are proportional to the three sides of another triangle, then the two triangles are —?—. ***(SSS Similarity Conjecture)***

Investigation 11.3.2

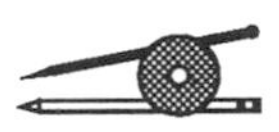

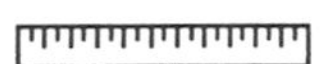

If two angles of one triangle are congruent to two angles of another triangle (which forces the third pair of angles to be congruent), must the two triangles be similar? On a clean sheet of unlined paper perform the following steps:

Step 1: Construct a triangle ABC.

Step 2: Construct a second triangle DEF with $\angle D \cong \angle A$ and $\angle E \cong \angle B$ (which forces $\angle F \cong \angle C$ since the measures of three angles must add to 180).

Step 3: Carefully measure the lengths of the sides of both triangles. With the aid of a calculator, calculate and compare the ratios of the corresponding sides. Is $\frac{AB}{DE} \approx \frac{AC}{DF} \approx \frac{BC}{EF}$?

Compare your results with the results of others near you. You constructed two triangles with two pairs of angles congruent (which forced the third pair of angles also to be congruent). If this forced the ratios of the corresponding sides of the two triangles to be equal, then both conditions for similar polygons hold and it follows that the two triangles are similar. You should be ready to state a conjecture.

C-73

If —?— angles of one triangle are congruent to —?— angles of another triangle, then —?—. ***(AA Similarity Conjecture)***

EXERCISE SET

Use your new conjectures to solve each problem. All measurements are in centimeters.

1. $g = $ –?–

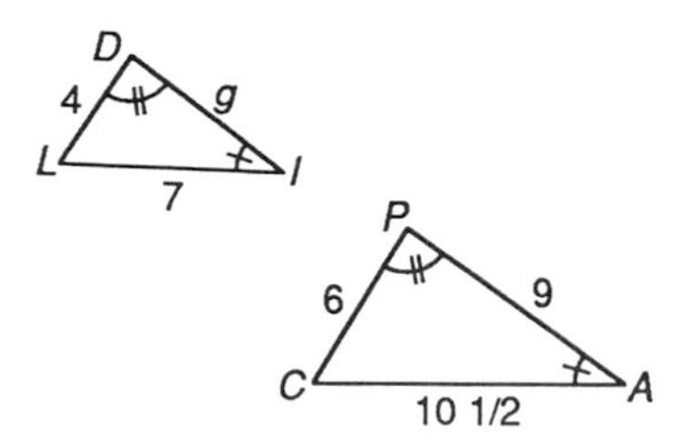

2. $h = $ –?–
$k = $ –?–

3.* $m = $ –?–

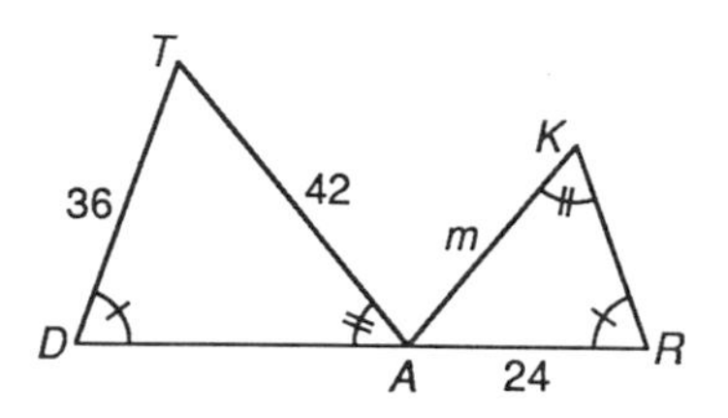

4. $n = $ –?–
$s = $ –?–

5. Is $\Delta AUL \sim \Delta MST$?

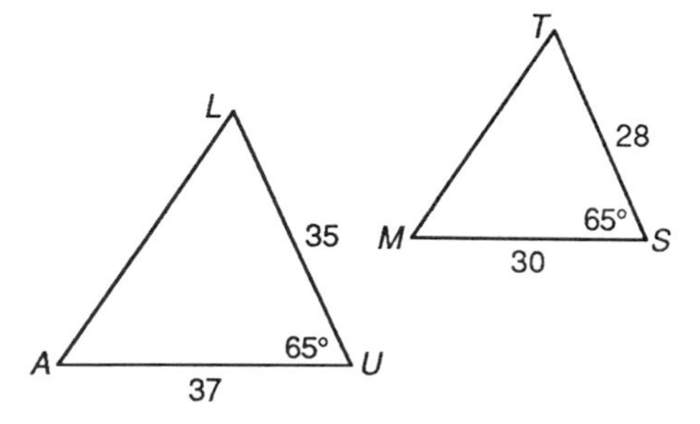

6.* Is $\Delta PHY \sim \Delta YHT$?

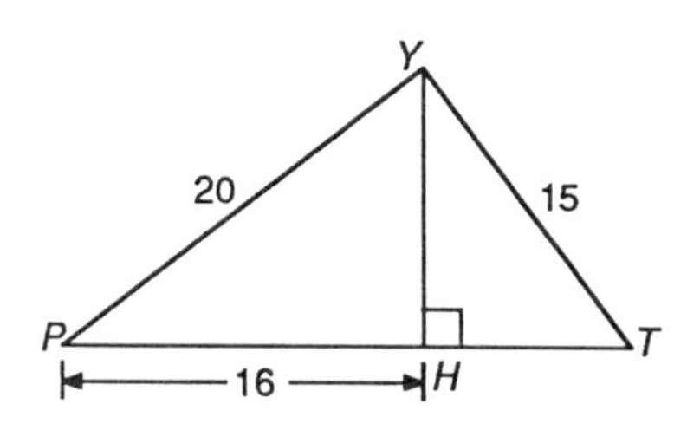

7. $\overline{TA} \parallel \overline{UR}$
Is $\angle QTA \cong \angle TUR$?
Is $\angle QAT \cong \angle ARU$?
Why is $\Delta QTA \sim \Delta QUR$?
$e = $ –?–

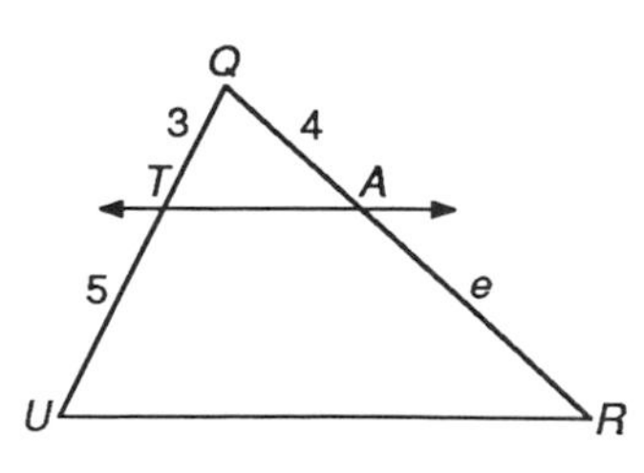

8. $\overline{OT} \parallel \overline{NR} \parallel \overline{VE}$
$f = $ –?–
$j = $ –?–

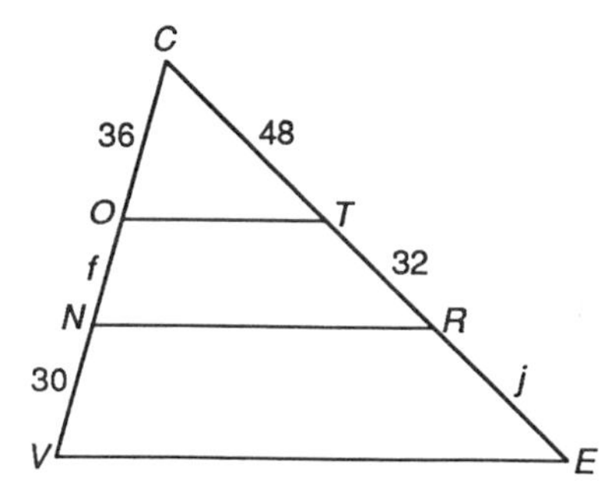

9.* Is $\angle TFA \cong \angle GHA$?
Is $\angle FTA \cong \angle HGA$?
Why is $\Delta FAT \sim \Delta HAG$?
p = –?–
q = –?–

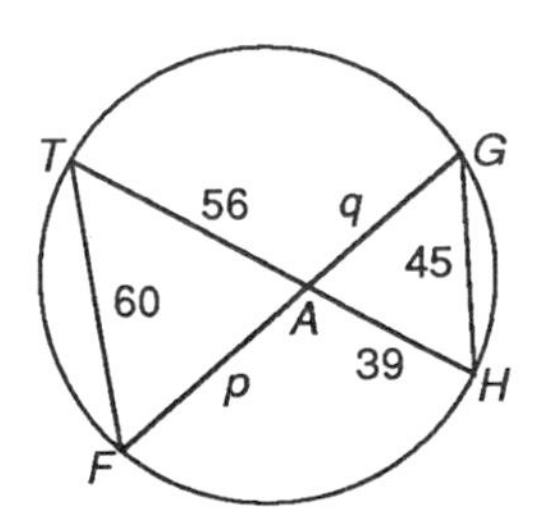

10. *TRAP* is a trapezoid.
Is $\angle RPA \cong \angle TRP$?
Is $\angle PAT \cong \angle RTA$?
Why is $\Delta PAZ \sim \Delta RTZ$?
r = –?–
s = –?–

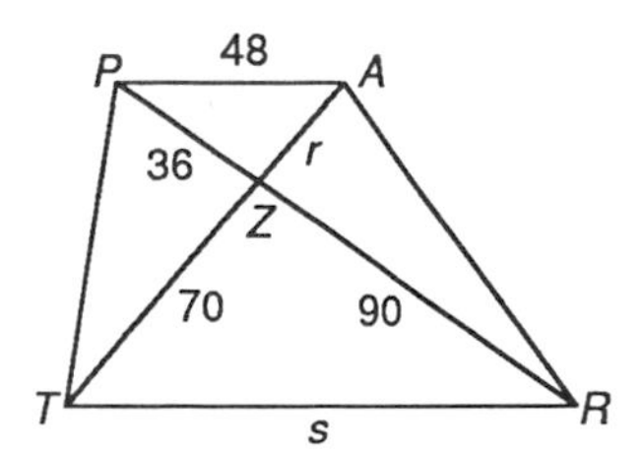

Improving Visual Thinking Skills

The Spider and the Fly

Henry E. Dudeney (1857–1930) is one of the greatest inventors of puzzles in the English-speaking world. Many of the puzzles that challenge people today can be traced back to him. His spider-and-fly problem first appeared in an English newspaper. It has challenged puzzle enthusiasts around the world for over three-quarters of a century.

Inside a rectangular room, measuring 30' in length and 12' in width and height, a spider is at a point on the middle of one of the end walls, 1 foot from the ceiling, as shown at A; and a fly is on the opposite wall, 1 foot from the floor in the center, as shown at B. What is the shortest distance that the spider must crawl in order to reach the fly, which remains stationary? Of course the spider never drops or uses its web, but crawls fairly.

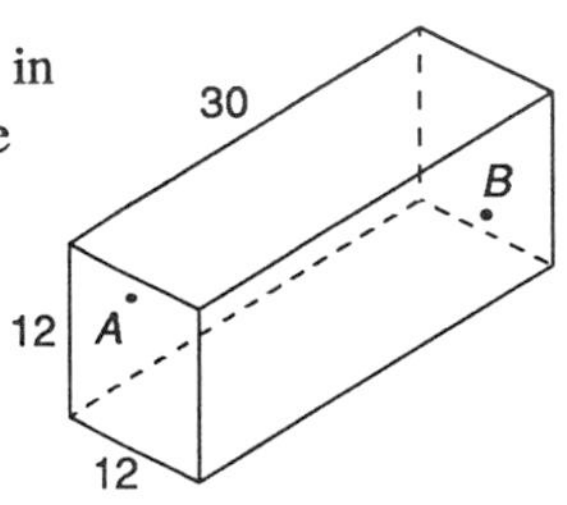

Lesson 11.4

Solving Problems with Similar Triangles

EXERCISE SET

1. If a triangle with sides of 5, 12, and 13 centimeters is similar to a triangle whose longest side is 39 centimeters, then what is the perimeter of the larger triangle?

2. What is the ratio of the perimeter of the small triangle to the perimeter of the large triangle in Exercise 1? What is the ratio of their areas (small to large)?

3.* Similar triangles can be used to find distances that are difficult to measure directly. For example, the distance across the canyon shown below can be calculated by sighting a rock on the opposite side at point R. Next, points G and D can be selected so that $\overline{GD}$ is perpendicular to $\overline{RG}$. Next, a convenient distance ND (with $\overline{ND}$ perpendicular to $\overline{DG}$) can be measured off. Next, point A, the intersection of $\overline{RN}$ and $\overline{GD}$, can be located. Since $\angle D$ and $\angle G$ are congruent and $\angle DAN$ and $\angle GAR$ are congruent, then $\Delta DAN \sim \Delta GAR$. Since the triangles are similar, the distance across the canyon can be determined. If GA is 120 meters, DA is 60 meters, and ND is 50 meters, find GR, the distance across the canyon.

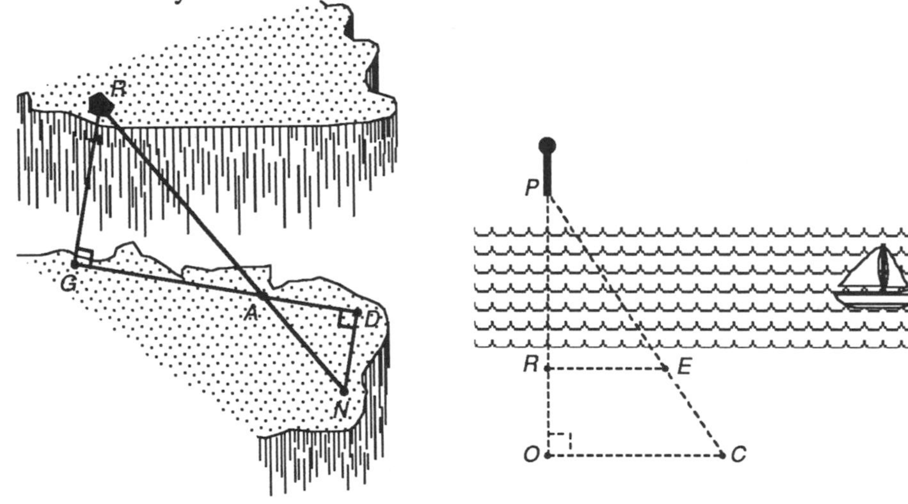

4.* The distance across the river above can be calculated by sighting a pole on the opposite bank at point P. Next, points R and O are aligned on the near bank so that P, R, and O are collinear. Next, a convenient distance OC with $\overline{OC} \perp \overline{PO}$ can be measured. Point E can be located by sighting the intersection of $\overline{CP}$ with $\overline{RE}$ ($\overline{RE} \perp \overline{PO}$). Since $\Delta PRE \sim \Delta POC$, the distance across the river, PR, can be calculated. If RO is 45 meters, OC is 90 meters, and RE is 60 meters, find the distance across the river.

5.* A pin-hole box camera is a very simple device. Place unexposed film at one end of a box (a shoe box will work nicely) and make a pin-hole at the opposite end. When you let light in through the pin-hole, an inverted image is produced on the film. Suppose you are taking a picture of a painting 30 cm wide by 45 cm high with a pin-hole box camera that has a depth of 20 cm. How far from the front of the box should you place the object if you wish to make an image 2 cm by 3 cm? Draw a diagram showing the box camera, film, and painting to be photographed.

In the last lesson you discovered two shortcuts for demonstrating similarity between triangles. The SSS Similarity Conjecture says: *If the three sides of one triangle are proportional to the three sides in another triangle, then the two triangles are similar.* The AA Similarity Conjecture says: *If the two angles of one triangle are congruent to two angles of another triangle, then the two triangles are similar.* There remain two possible cases to investigate. (Since AA is a shortcut, then ASA, AAS, and AAA are automatically shortcuts. Since SSS is a shortcut, this leaves only SAS and SSA as possible shortcuts that use only three parts of a triangle.) One is a shortcut for similarity between triangles and the other is not. In this next investigation you will determine which one is and which one is not.

Investigation 11.4.1

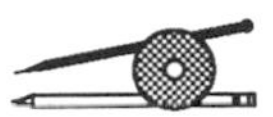

For each case below, demonstrate that the case does or does not force two triangles to be similar.

Case 1: Is SAS a shortcut for similarity? (Try to construct two different triangles that are not similar but have two pairs of sides proportional and the pair of included angles equal in measure.)

Case 2: Is SSA a shortcut for similarity? (Try to construct two different triangles that are not similar but have two pairs of sides proportional and a pair of corresponding non-included angles equal in measure. See Lesson 5.3 on SSA congruence.)

Compare your results with the results of others near you. You should be ready to state a conjecture.

C-74 *If two sides of one triangle are proportional to two sides of another triangle and —?— in one triangle is congruent to the —?— in the other triangle, then the two triangles are/are not similar.* ***(—?— Similarity Conjecture)***

Lesson 11.5

Indirect Measurement with Similar Triangles

Similar triangles can be used to calculate the height of objects that you are unable to measure directly. Suppose you and a friend wanted to find the height of a lamppost on a sunny afternoon. The lamppost is probably too tall to measure directly. The light rays, vertical lamppost, and its shadow, however, form a right triangle that is similar to the right triangle formed by the light rays, your vertical friend, and your friend's shadow. How tall is the lamppost? Since the triangles are similar:

$$\frac{\text{your friend's height}}{\text{your friend's shadow}} = \frac{\text{lamppost's height}}{\text{lamppost's shadow}}$$

You can easily measure the shadows and your friend. The height of the lamppost is the only value in the proportion that you don't know. Calculate it.

Example

If a person is 5 feet tall and casts a shadow of 6 feet at the same time that a lamppost casts an 18-foot shadow, what is the height of the lamppost?

$$\frac{5}{6} = \frac{x}{18}$$

$$6x = (5)(18)$$

$$x = (5)(18)/6$$

$$x = 15$$

The lamppost is about 15 feet tall.

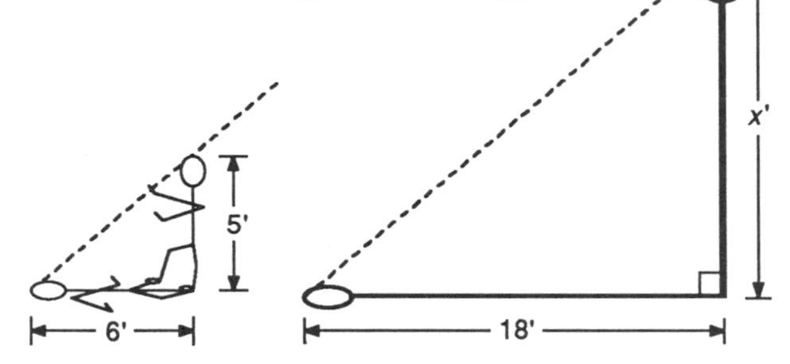

You can also use mirrors and similar triangles to measure indirectly the height of objects. This method will work on overcast days when there are no shadows. It will also work when the sun is directly overhead and there are no shadows. Suppose you needed to find the height of a flagpole. Place a mirror with cross hairs (an X) drawn on it flat on the ground between yourself and the flagpole. Look into the mirror and walk forward and back until the top of the flagpole lines up with the cross hairs on the mirror. Since the light rays from the top of the flagpole to the mirror and back up to your eye form equal angles (just like the incoming and outgoing angles on a pool table), similar right triangles are formed.

$$\frac{\text{your height}}{\text{distance from you to mirror}} = \frac{\text{flagpole's height}}{\text{distance from flagpole to mirror}}$$

You can use your height, the distance from you to the mirror, and the distance from the mirror to the flagpole to calculate the flagpole's height.

Example

If your eye is 168 centimeters above the ground and you are 114 centimeters from the mirror and the mirror is 570 centimeters from the flagpole, how tall is the flagpole?

$$\frac{168}{114} = \frac{x}{570}$$

$$114x = (168)(570)$$

$$x = (168)(570)/114$$

$$x = 840$$

The flagpole is about 840 centimeters tall.

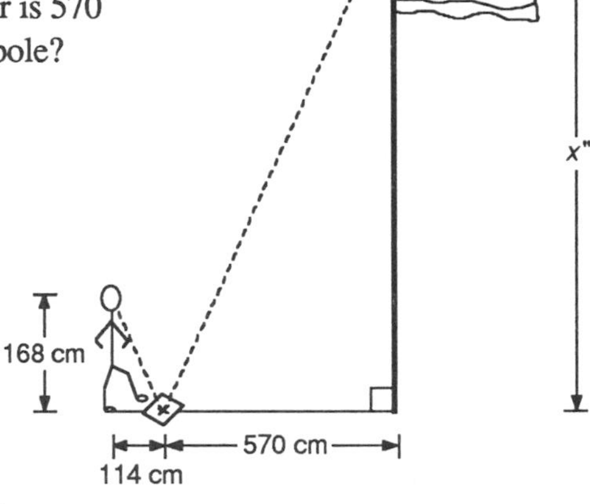

In the *Special Project: The Shadow Knows*, you will get a chance to use the indirect methods described above to determine the heights of objects that you are unable to measure directly. But for now, just practice on the problems below.

EXERCISE SET

1. If a 4-meter flagpole casts a shadow of 6 meters at the same time that a nearby building casts a 24-meter shadow, how tall is the building?

2.* If five-foot-tall Madeleine casts an 84-inch shadow, then how tall is her friend if he has a shadow at the same time which is one foot shorter than hers?

3. A rope from the tip of a flagpole reaches all the way down to the end of the flagpole shadow, a distance of 10 meters. The length of the shadow is 6 meters. How tall is the nearby football goal post if it has a shadow of 4 meters?

4.* Juanita, who is 1.82 meters tall, wishes to find the height of a tree in her backyard. She walks 12.20 meters from the base of the tree along the shadow of the tree until her head is in a position were the tip of her shadow exactly overlaps the end of the tree top's shadow. She is now 6.10 meters from the end of the shadows. How tall is the tree?

5. One overcast day, Private Eye Samantha Diamond needed to calculate the height of a window in a nearby building. Since there were no shadows available, she decided to use mirrors. Sam positioned a mirror on the ground between herself and the building in such a way that when she looked into the mirror while standing upright, she saw into the window. If the mirror was 122 centimeters from her feet and 7.32 meters from the base of the building and the eye of Sam was 1.82 meters above the ground, how high up on the building was the window located?

The Adventures of Robin Hood

6.* Late one afternoon, while being chased by the Sheriff of Nottingham into an unfamiliar part of Sherwood forest, Robin Hood and Little John found themselves trapped between a wide chasm and the approaching evil sheriff. Fortunately, the sign from the old collapsed bridge was still standing, for it gave Robin the information necessary to plan his escape. The sign said the chasm was 36 feet across. A large tree grew near the chasm. It was the only tree within 50 yards of the chasm. Robin quickly paced off the distance from the cliff edge to the tree and found that it was 24 feet. He noticed that the shadow cast by the tree stretched directly across the chasm and that the tip of the shadow just reached the opposite edge of the chasm. Robin hastily measured the shadow created by his 55-inch frame and found it to be 77 inches. Using this information, Robin calculated the height of the tree. What was the height of the tree to the nearest foot? If Robin and Little John were to chop down the tree, would it be long enough to reach across the chasm? Was Robin Hood able to elude the Sheriff of Nottingham?

Special Project

The Shadow Knows

Every shadow points to the sun.

– Ralph Waldo Emerson

In this special project you are going to use the indirect methods outlined in the previous lesson to determine the height of two different size objects that you are unable to measure directly (school building, football goal post, flagpole, tall tree, etc.). Use both methods on each object. This exercise works best in groups of four or five. You may wish to divide up the tasks. The tasks include being responsible for taking measurements, recording the measurements, performing the calculations, keeping track of the equipment, and being the person measured. You will need the following:

- Measuring tape or meter sticks
- Notebook for recording your measurements
- Mirror with cross hairs (Use a straightedge and grease pencil or tape to make two lines intersecting at right angles in the middle of your mirror.)
- Sunshine

Don't try this project at midday because the sun will be directly overhead and won't cast shadows. Before you begin, make tables for your measurements similar to the tables below.

Table for Measurements by Shadow Method

Name of object to be measured	Height of observer	Shadow length of observer	Shadow length of object	Calculated height of object
1. —?—	–?–	–?–	–?–	–?–
2. —?—	–?–	–?–	–?–	–?–

Table for Measurements by Mirror Method

Name of object to be measured	Height of observer's eye	Distance from observer to mirror	Distance from object to mirror	Calculated height of object
1. —?—	–?–	–?–	–?–	–?–
2. —?—	–?–	–?–	–?–	–?–

Follow the procedure below to determine indirectly the height of two objects.

Step 1: Begin by measuring one person's height (we'll call this person the observer), the height to the observer's eye, and the length of the observer's shadow. Record the measurements in your tables.

Step 2: Locate two tall objects with heights that would be difficult to measure directly (school building, football goal post, flagpole, tall tree, etc.). Measure the length of their shadows. Enter the measurements into the appropriate table.

Step 3: Place the mirror flat on the ground between the observer and the object to be measured. Move the observer forward or back until the observer sees the top of the object over the cross hairs of the mirror. Measure the distance from the observer to the mirror and the distance from the mirror to the base of the object. Enter the measurements into the appropriate table.

Step 4: Perform all the calculations for both methods to approximate the height of the two objects.

Step 5: If the calculated heights for an object from the two methods differ significantly, you should remeasure and recalculate. If they still differ significantly, then you must decide if one method is more accurate than the other or average the answers for your calculated height.

Improving Visual Thinking Skills

Build a Two Piece Puzzle

Construct two copies of the figure on the top right. To construct each copy:

- Construct a regular hexagon.
- Construct an equilateral triangle on two edges as shown.
- Construct a square on the edge between the two equilateral triangles as shown.

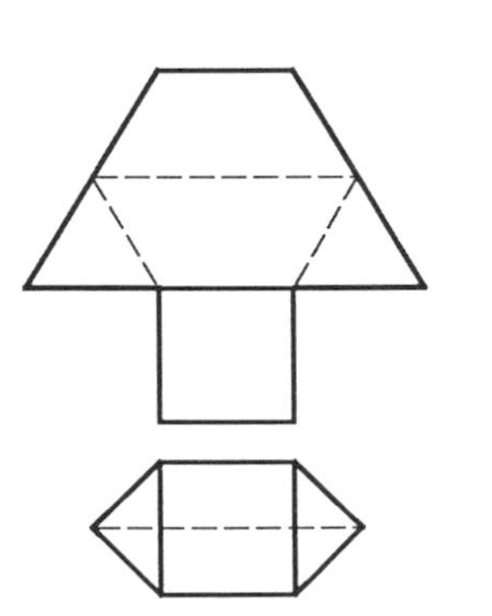

Cut out each copy and fold them along the dashed lines into two identical solids like you see in the figure at the bottom right. Tape the edges. Now arrange your two solids to form a regular tetrahedron.

Lesson 11.6

Corresponding Parts of Similar Triangles

In this lesson you are going to see if there is a relationship between corresponding parts (other than sides) in similar triangles. Is there a relationship between corresponding altitudes in similar triangles, between corresponding medians in similar triangles, or between corresponding angle bisectors in similar triangles? (There must be since we are asking, right?) Let's investigate. This investigation works best if you work in groups, sharing the construction work.

Investigation 11.6.1 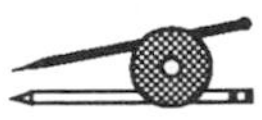 

On a sheet of unlined paper, perform the following steps:

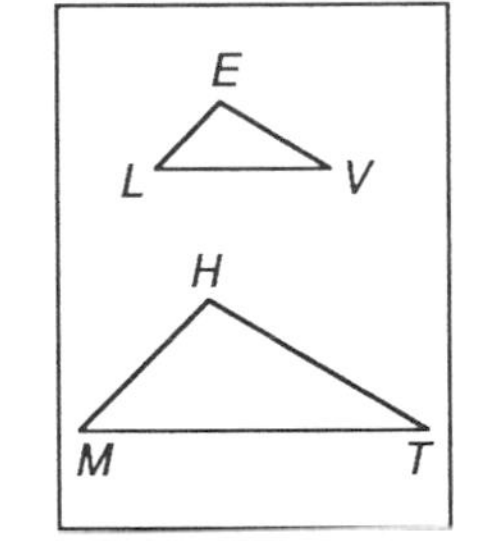

Step 1: Construct ΔLVE.

Step 2: Construct a second triangle ΔMTH with the length of the sides some multiple of the original triangle (twice, three times, or four times larger or smaller) so that $\Delta LVE \sim \Delta MTH$.

Step 3: Construct a pair of corresponding altitudes, $\overline{EO}$ in ΔLVE and $\overline{HA}$ in ΔMTH. Measure the lengths EO, HA, LV, and MT. Is $\frac{EO}{HA} \approx \frac{LV}{MT}$?

Step 4: Construct a pair of corresponding medians, median $\overline{VR}$ in ΔLVE and median $\overline{TS}$ in ΔMTH. Measure their lengths. Is $\frac{VR}{TS} \approx \frac{LV}{MT}$?

Step 5: Construct a pair of corresponding angle bisectors, $\overline{LD}$ in ΔLVE and $\overline{MC}$ in ΔMTH. Measure their lengths. Is $\frac{LD}{MC} \approx \frac{LV}{MT}$?

Compare your results with the results of others near you. Each person in your group constructed a different pair of similar triangles. Did you get different results? Are the corresponding angle bisectors, medians, and altitudes proportional to the corresponding sides in similar triangles? You should be ready to make a conjecture.

If two triangles are similar, then the corresponding —?—, corresponding —?—, and corresponding —?— are —?— to the corresponding sides. ***(Proportional Parts Conjecture)***

There is another proportion relationship involving angle bisectors. Recall when you first saw an angle bisector in a triangle. Perhaps you thought that the angle bisector not only divided the angle into two equal parts, but that it also divided the opposite side into two equal parts. A quick check shows that this is not always true.

$\overline{OC}$ is an angle bisector.
Point C is the midpoint of $\overline{IS}$.

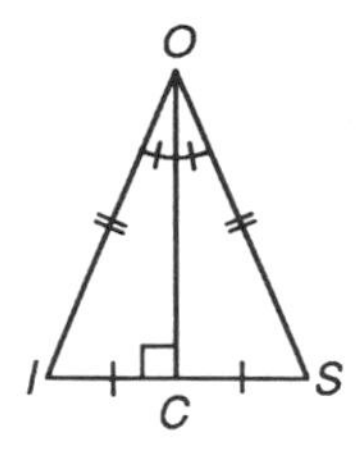

For Isosceles triangles - Yes!

$\overline{RT}$ is an angle bisector.
Point H is the midpoint of $\overline{OE}$.

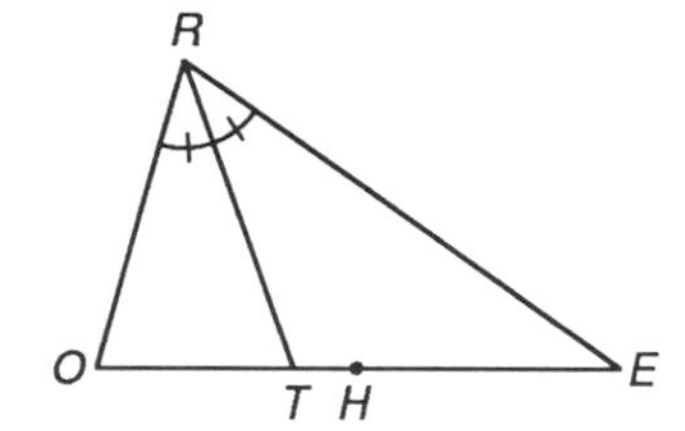

But not for other triangles.

The angle bisector does, however, divide the opposite side into two lengths related by a special ratio. Let's see if you can discover this ratio. Again, you should work in groups.

Investigation 11.6.2 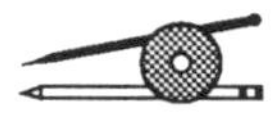

On a sheet of unlined paper, perform the following steps:

Step 1: Construct an angle. Label it $\angle A$. On one ray mark off 8 units (use centimeters or 8 markings with a compass). Label point C so that AC is 8 units. Locate point B on the other ray so that AB is 16 units.

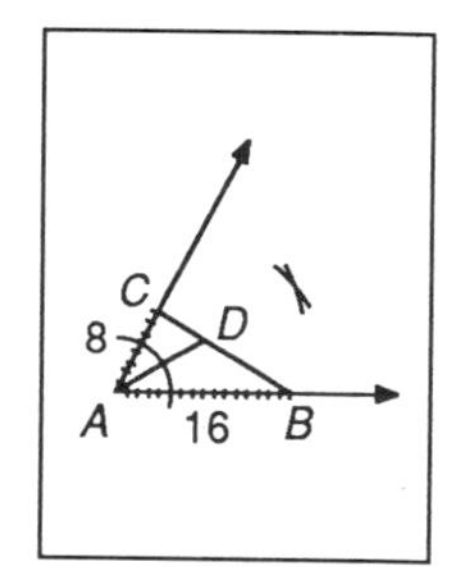

Step 2: Construct $\overline{BC}$, the third side of ΔABC. Construct the angle bisector of $\angle A$. Label as point D the intersection of the bisector with side $\overline{BC}$.

Step 3: With your compass, compare CD and BD. How many times does $\overline{CD}$ fit on $\overline{BD}$?

Step 4: Calculate and compare the ratios CA/BA and CD/BD.

Repeat steps 1 to 4 with $AC = 6$ units and $AB = 18$ units. How do the ratios compare this time? Try this one more time. Repeat Steps 1 to 2, this time choosing any two lengths for sides $\overline{AB}$ and $\overline{AC}$. With your ruler, measure AB, AC, CD, and BD. Calculate the ratios CA/BA and CD/BD. Are they close? Compare your results with the results of others in your group. You should be ready to state a conjecture.

C-76 *The angle bisector in a triangle divides the opposite side into two segments whose lengths are in the same ratio as —?—.*

You already know how to divide a segment into two equal parts with the construction of a perpendicular bisector. By repeatedly constructing perpendicular bisectors you can divide a segment into a few other ratios. For example, you could divide a segment into two parts in the ratio of 3 to 5 by dividing the segment into eight parts using repeated constructions of perpendicular bisectors. But how would you divide a segment into a ratio of 2 to 3? Conjecture 76 gives you a way to do this. But how? Let's investigate.

Investigation 11.6.3 

Devise a method to divide a line segment AB into a ratio of 2 to 3. Your method should make use of Conjecture 76. To get you started, here are the first steps:

Step 1: Construct a segment of any length and label it $\overline{AB}$.

Step 2: Construct a second segment. Call its length x.

Step 3: Construct two new segments, one with length $2x$ and the other with length $3x$. (Any two of the three lengths $2x$, $3x$, and AB must have a sum greater than the third length. The three segments should make a triangle.)

Step 4: Construct a triangle with lengths $2x$, $3x$, and AB. You're on your own from here!

Once you have discovered a method to divide a segment into a ratio of 2 to 3, generalize your method and develop a method for dividing any segment into any given ratio. Write a description of your method beneath the investigative work in your notebook.

EXERCISE SET A

Use your new conjectures to solve each problem. All measurements are in centimeters.

1.* $\Delta AGE \sim \Delta WOE$
$h = $ –?–

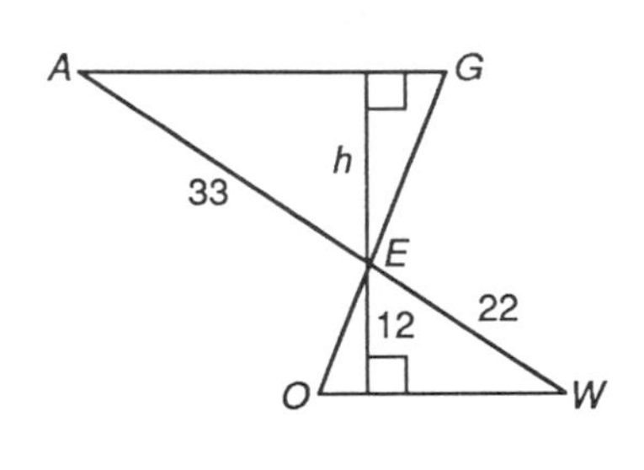

2. $\Delta NOH \sim \Delta ARK$
$x = $ –?–

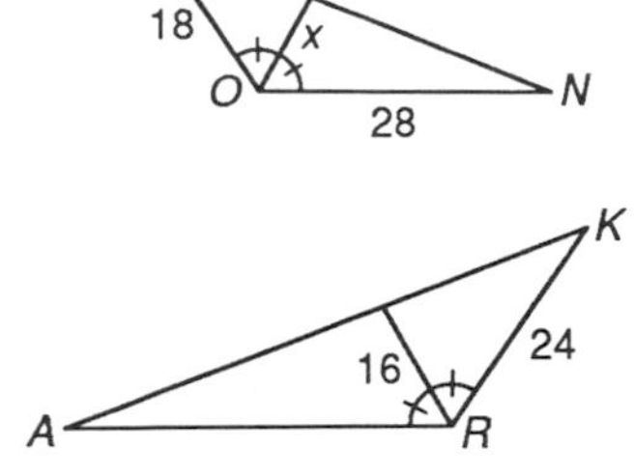

3.* $\Delta PIE \sim \Delta OIS$
O is the midpoint of PI.
$SL = $ –?–
$SO = $ –?–

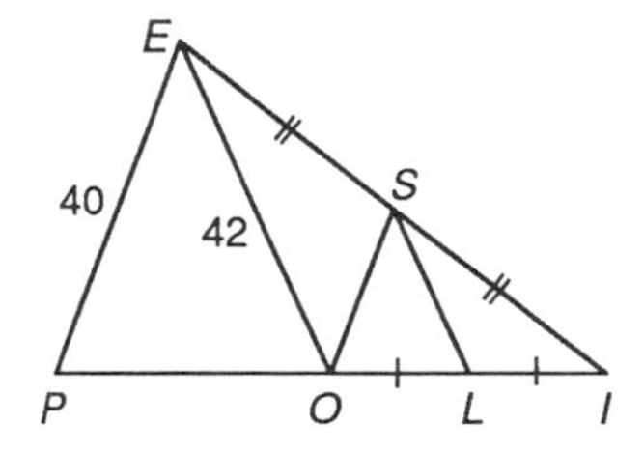

4. $\Delta CAP \sim \Delta DAY$
$FD =$ –?–

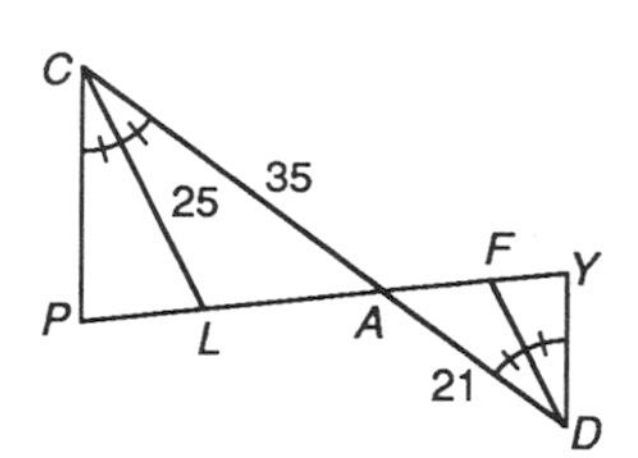

5.* $\Delta HAT \sim \Delta CLD$
$x =$ –?–

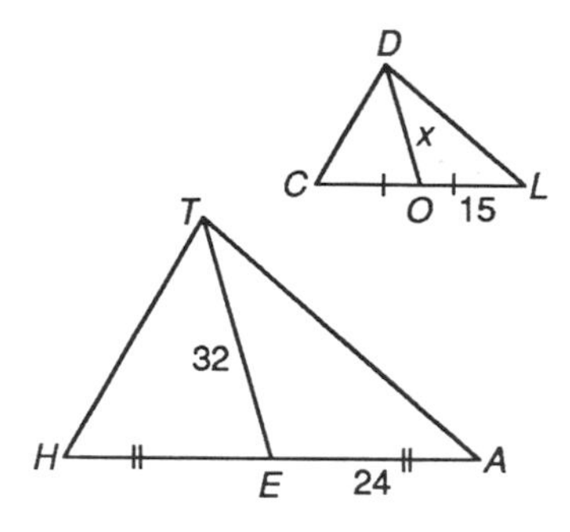

6. $\Delta ARM \sim \Delta LEG$
Area $_{\Delta ARM} =$ –?–
Area $_{\Delta LEG} =$ –?–

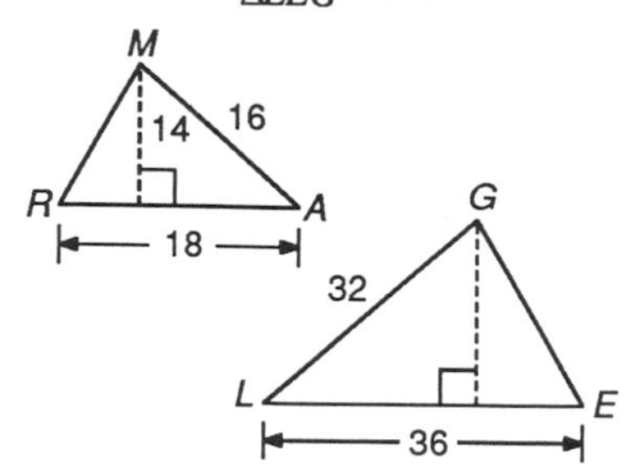

7.* $a/b =$ –?–
$a/p =$ –?–

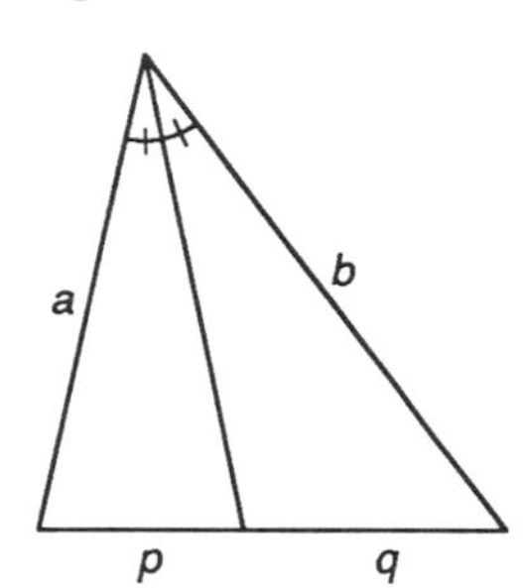

8. $v =$ –?–

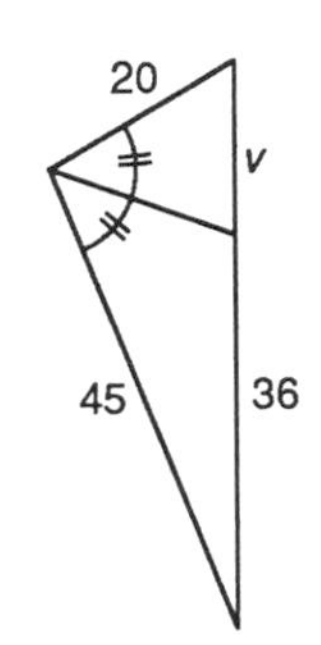

9. $y =$ –?–

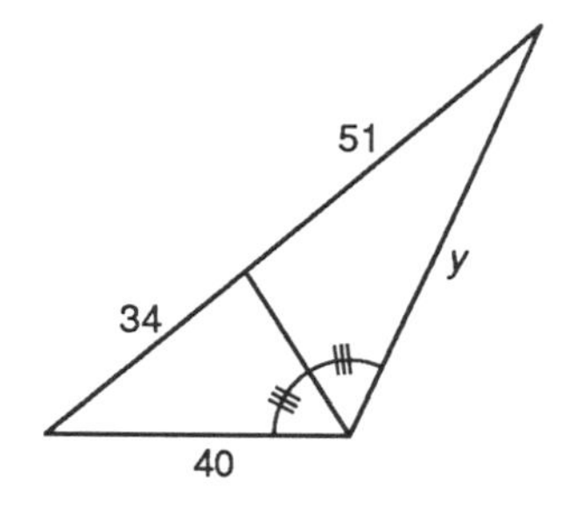

10.* $x =$ –?–

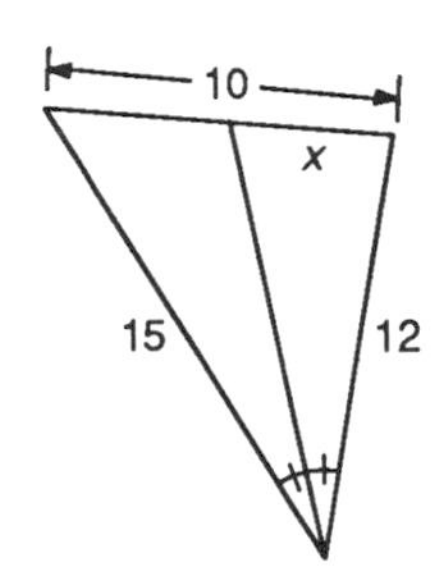

11.* $k =$ –?–

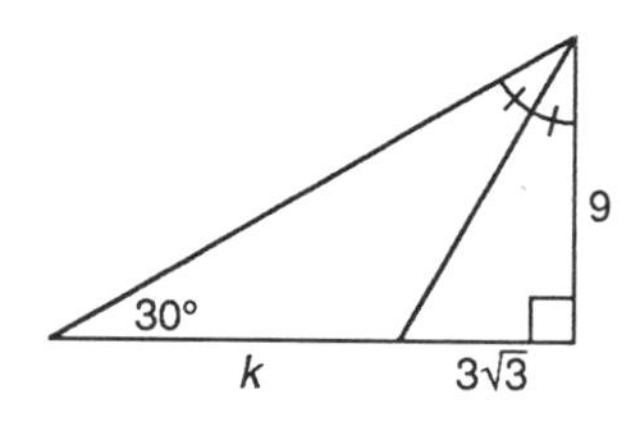

12.* $x =$ –?–
$y =$ –?–
$z =$ –?–

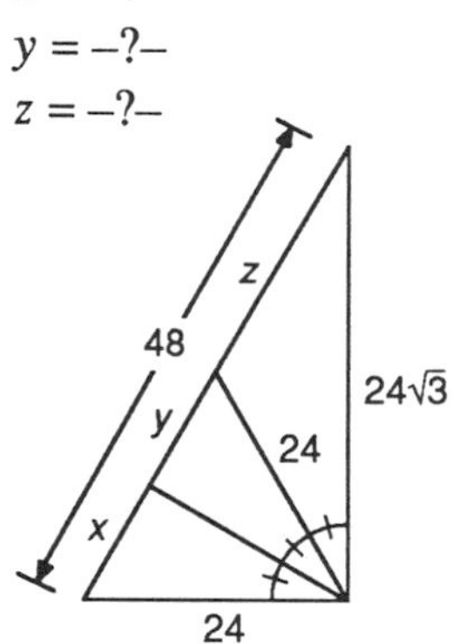

13.* Aunt Teak has passed away and left a very valuable plot of land to her two nephews, Chip and Dale. The plot of land is in the shape of an isosceles right triangle with the equal legs of length 8 kilometers. The will states that the land is to be divided into two parts by bisecting one of the two equal angles. This gives two plots of land of unequal area. Whichever of the two nephews is first to calculate the ratio of the two plots of land, shall have first choice in selecting his inheritance. What is the ratio of the larger area to the smaller area?

EXERCISE SET B

1.* Construct $\overline{CD}$. Then find a point P that divides it into the ratio of 3:4.

2. Construct $\overline{EF}$. Then find a point Q that divides it into the ratio of 4:5.

3.* Construct $\overline{GH}$. Find two points R and S that divide it into the ratio of 2:3:4.

4. Claudia Lewis and Luella Clark are free-lance space explorers and entrepreneurs looking for adventure and quick riches around the moons of Jupiter. They have been searching the unexplored regions of Ganymede and Io for the rare Ammonia-Tellurium ice crystals called *Jupiter's Needles*. These rare crystals are in the shape of long thin rods and are extremely valuable. Claudia has invested 70% of the money for this exploration and, thus, any rod they discover is to be split into the ratio of 7 to 3. If they discover the rod shown below, locate the point where it should be divided. (Trace or copy the length shown onto another sheet of paper and construct the point that divides the length into the ratio of 7:3.)

Improving Reasoning Skills

Container Problem II

You have a small, graduated, cylindrical measuring glass with a maximum capacity of 250 ml. All the graduation marks are worn off except for the 150-ml and 50-ml marks. You also have a second large container that is unmarked. It is not difficult to put exactly 350 ml into the large container by filling the graduated cylinder twice to the 150-ml mark and once to the 50-ml mark. It is also possible to put exactly 350 ml into the large container in only two measurings. How?

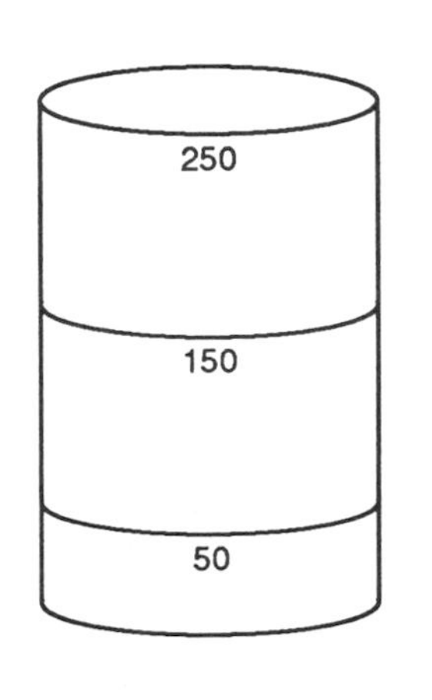

Lesson 11.7

Proportions with Area

Similarity is useful in estimating the surface areas of large regions. Suppose an urban planner has created designs for a city park. If the park on the plans has an area of 600 cm^2 and the plans are drawn to the scale of 1:1,000, that is, each dimension is actually 1,000 times larger than on the plans, what would be the actual area of the park? Would you have figured 60,000 m^2 (600,000,000 cm^2)? In this lesson you will discover the relationship between the areas of similar polygons and the relationship between the areas of similar circles.

Investigation 11.7.1

Find each ratio below. The figures in each problem are similar.

1. $\frac{RE}{AN} = $ –?–

   $\frac{\text{Area } RECT}{\text{Area } ANGL} = $ –?–

2. $\frac{MS}{BI} = $ –?–

   $\frac{\text{Area } \Delta SML}{\text{Area } \Delta BIG} = $ –?–

3. $\frac{r}{R} = $ –?–

   $\frac{\text{Area circle } O}{\text{Area circle } Q} = $ –?–

   $r = 2$

   $R = 3$

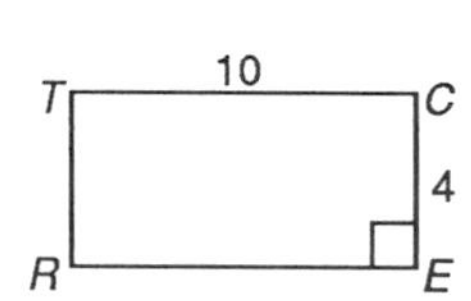

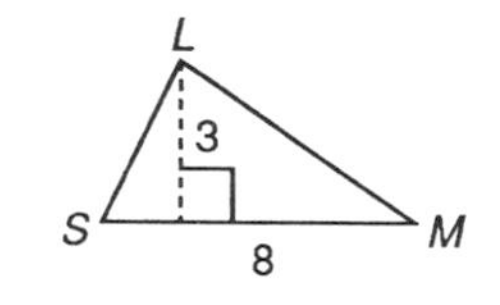

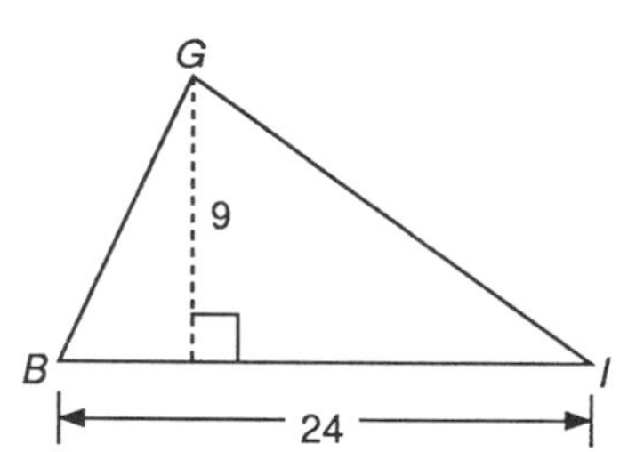

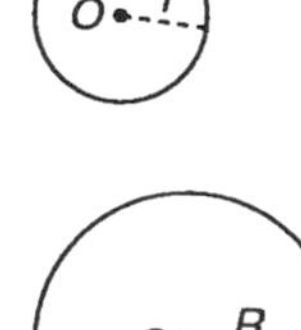

How did the ratios of areas compare with the ratios of corresponding sides or radii in each pair of similar figures? You may be ready to state a conjecture about this relationship, but first let's use algebra to look at the question.

In the last lesson you discovered that if two triangles are similar, then their corresponding altitudes are proportional to the corresponding sides. If the corresponding bases of two similar triangles are in the ratio of a/b, then their corresponding altitudes are in the ratio of a/b. Therefore, their areas must be in the ratio of $[(1/2)(a^2)]/[(1/2)(b^2)]$ or $(a/b)^2$. Thus, if the ratio of corresponding parts of triangles are in the ratio of a/b, then their areas are in the ratio of $(a/b)^2$.

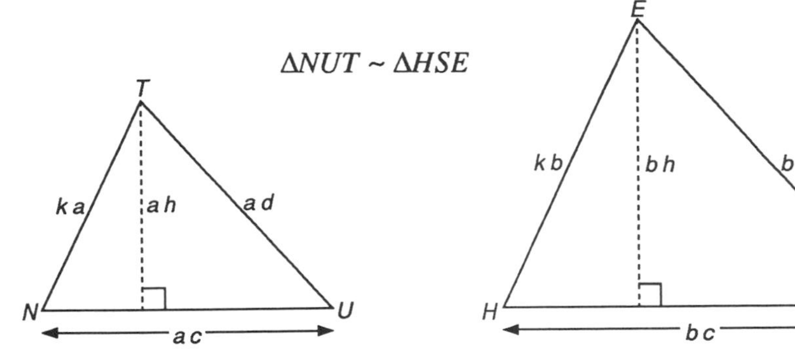

Ratio of corresponding parts: $\frac{ka}{kb} = \frac{ad}{bd} = \frac{ac}{bc} = \frac{ah}{bh} = \frac{a}{b}$

Ratio of areas: $\frac{\text{Area}_{NUT}}{\text{Area}_{HSE}} = \frac{(1/2)(ac)(ah)}{(1/2)(bc)(bh)} = \frac{(1/2)a^2ch}{(1/2)b^2ch} = \frac{a^2}{b^2} = \left(\frac{a}{b}\right)^2$

Is this true for parallelograms as well as triangles? That is, if two parallelograms are similar, is the ratio of corresponding altitudes the same as the ratio of corresponding sides? Clearly, in every parallelogram you can draw a perpendicular from an obtuse angle to the opposite side to form a right triangle. If parallelogram *PARE* is similar to parallelogram *LOGM* (see the diagrams below) and $\overline{ET} \perp \overline{PA}$ and $\overline{MS} \perp \overline{LO}$, then $\Delta PTE \sim \Delta LSM$ by the AA Similarity Conjecture. Therefore, if two parallelograms are similar, then the ratio of corresponding altitudes is equal to the ratio of corresponding sides. If the ratio of the corresponding sides of two similar parallelograms is a/b, then the ratio of the corresponding altitudes is also a/b. Therefore, the ratio of areas for two similar parallelograms is a^2/b^2 or $(a/b)^2$.

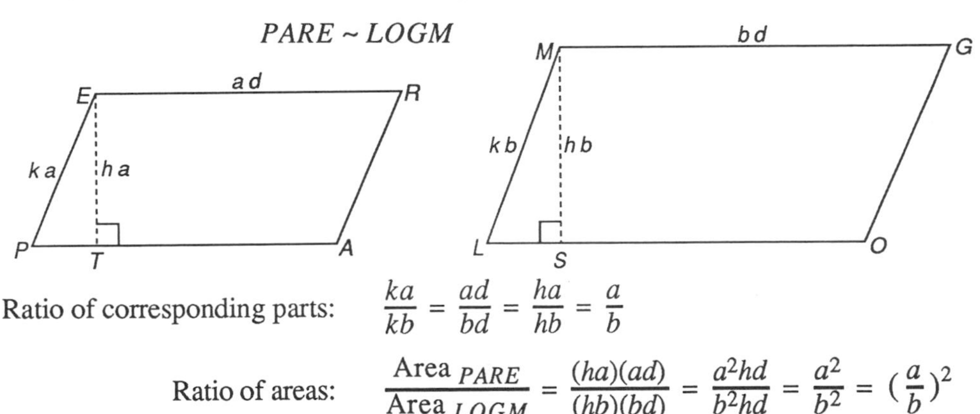

Ratio of corresponding parts: $\frac{ka}{kb} = \frac{ad}{bd} = \frac{ha}{hb} = \frac{a}{b}$

Ratio of areas: $\frac{\text{Area}_{PARE}}{\text{Area}_{LOGM}} = \frac{(ha)(ad)}{(hb)(bd)} = \frac{a^2hd}{b^2hd} = \frac{a^2}{b^2} = \left(\frac{a}{b}\right)^2$

Is this also true for circles? All circles are similar. To compare two circles we compare their radii. If circle *P* has radius *p* and circle *Q* has radius *q*, then the ratio of their radii is p/q and the ratio of their areas is $\pi p^2/\pi q^2$ which simplifies to p^2/q^2 or $(p/q)^2$. Therefore, the ratio of the areas is equal to the square of the ratio of the radii.

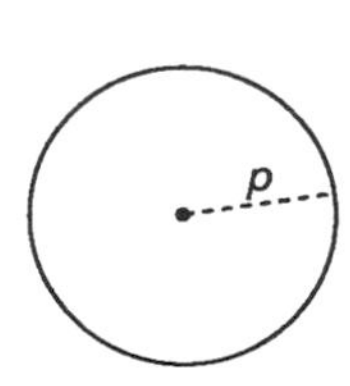

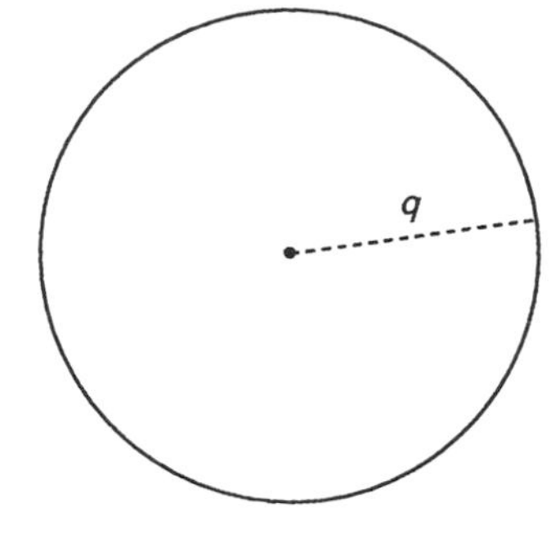

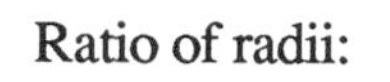

Ratio of radii: $\frac{p}{q}$

Ratio of areas: $\frac{\text{Area}_{\text{circle } P}}{\text{Area}_{\text{circle } Q}} = \frac{\pi p^2}{\pi q^2} = \frac{p^2}{q^2} = \left(\frac{p}{q}\right)^2$

Can this relationship be generalized to all similar figures? Sure! Since all polygons can be divided into triangles, the relationship seems to follow for all similar polygons. How does the ratio of corresponding sides or radii compare with the ratio of areas of similar polygons and circles? State this as your next conjecture.

C-77 *If two similar polygons (or circles) have corresponding sides (or radii) in the ratio of m/n, then their areas are in the ratio of —?—.* ***(Proportional Area Conjecture)***

EXERCISE SET

1.* $\Delta CAT \sim \Delta MSE$

$\text{Area}_{\Delta CAT} = 72 \text{ cm}^2$

$\text{Area}_{\Delta MSE} = $ –?–

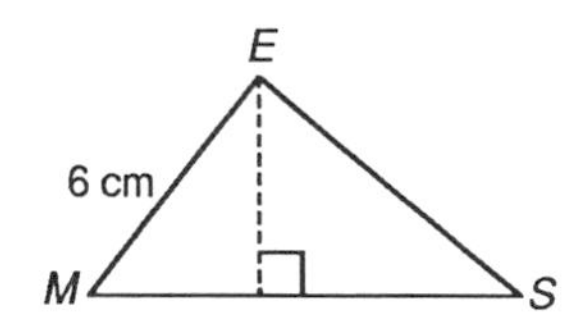

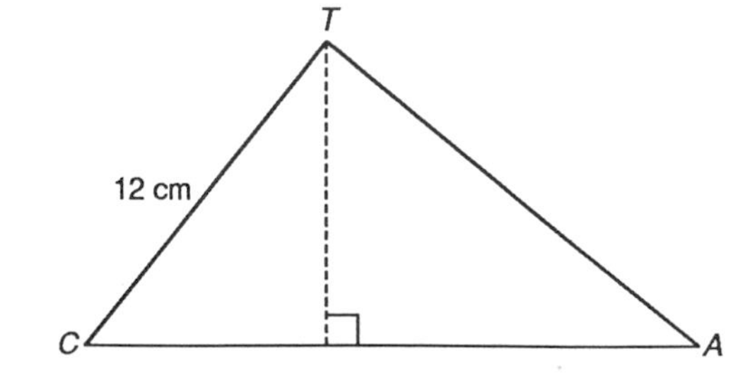

2. Rectangle $RECT \sim$ Rectangle $ANGL$

$\frac{\text{Area}_{RECT}}{\text{Area}_{ANGL}} = \frac{9}{16}$

$TR = $ –?–

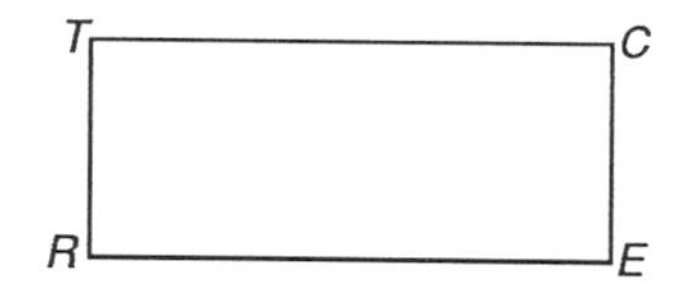

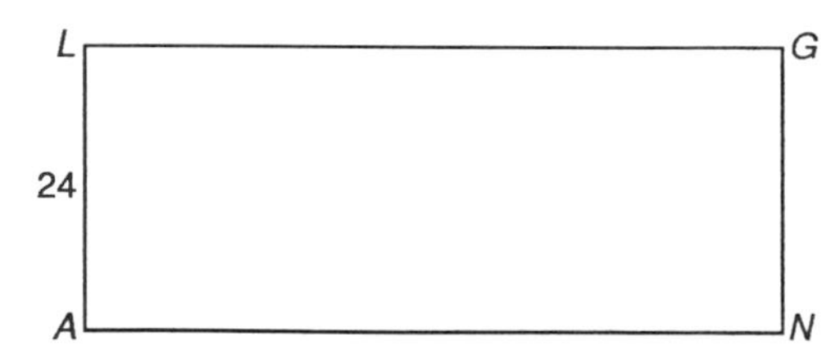

3.* *TRAP* ~ *ZOID*

$$\frac{\text{Area}_{ZOID}}{\text{Area}_{TRAP}} = \frac{16}{25}$$

$x =$ –?–

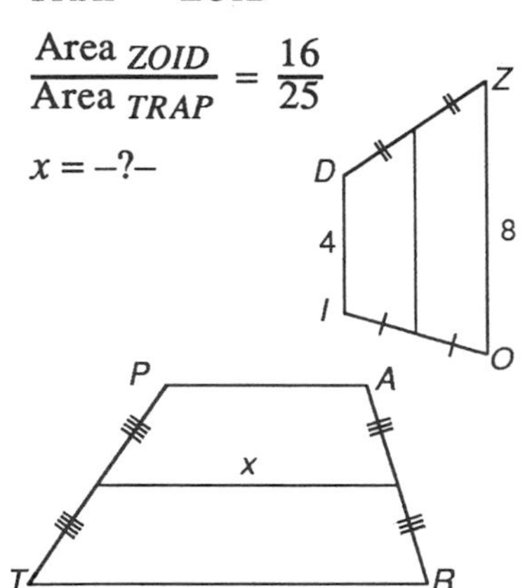

4. Circle *R* ~ Circle *S*

$$\frac{r}{s} = \frac{3}{5}$$

Area $_{\text{semicircle } S} = 75\pi \text{ cm}^2$

Area $_{\text{semicircle } R} =$ –?–

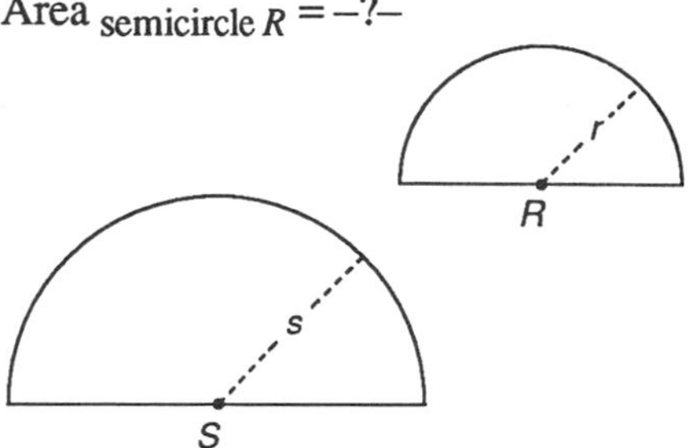

5. The corresponding diagonals of two similar trapezoids are in the ratio of 1:7. What is the ratio of their areas?

6. The ratio of the perimeters of two similar parallelograms is 3:7. What is the ratio of their areas?

7. The ratio of the areas of two similar trapezoids is 1:9. What is the ratio of their altitudes?

8. The areas of two circles are in the ratio of 25:16. What is the ratio of their radii?

9.* Two cubes have edges in the ratio of m/n. What is the ratio of their surface areas?

10.* The Jones family paid $150 to a painting contractor to stain their 12 ft. by 15 ft. back deck. Their neighbors, the Smiths, have a similar deck 16 ft. by 20 ft. If the Smiths wish to "keep up with the Joneses," what is a proportional price the Smith family should expect to pay to have their deck stained by the contractor?

Improving Reasoning Skills

Calculator Cunning

Using the calculator shown on the right, what is the largest number you can form by pressing the 1, 2, and 3 buttons exactly once each and by pressing y^x at most once? You cannot press any other buttons.

Lesson 11.8

Proportions with Volume

Similarity is useful in estimating the volume of large objects. For example, if a scale model of a piece of machinery weighs 35 pounds (the weight is calculated from the volume), how much will the actual machinery weigh if all three dimensions are five times larger and it is made of the same material? Would you have figured 4375 pounds?

Similar solids are solids that have the same shape but not necessarily the same size. All cubes are similar, but not all prisms are similar. All spheres are similar, but not all cylinders are similar. Two polyhedrons are similar if all their corresponding faces are similar and the lengths of their corresponding edges are proportional. Two right cylinders (or right cones) are similar if the ratio of their radii equals the ratio of their corresponding heights.

EXERCISE SET A

Is each pair of solids similar or not similar? All measurements are in centimeters.

1. Figures are right rectangular prisms.

2. Figures are right cylinders.

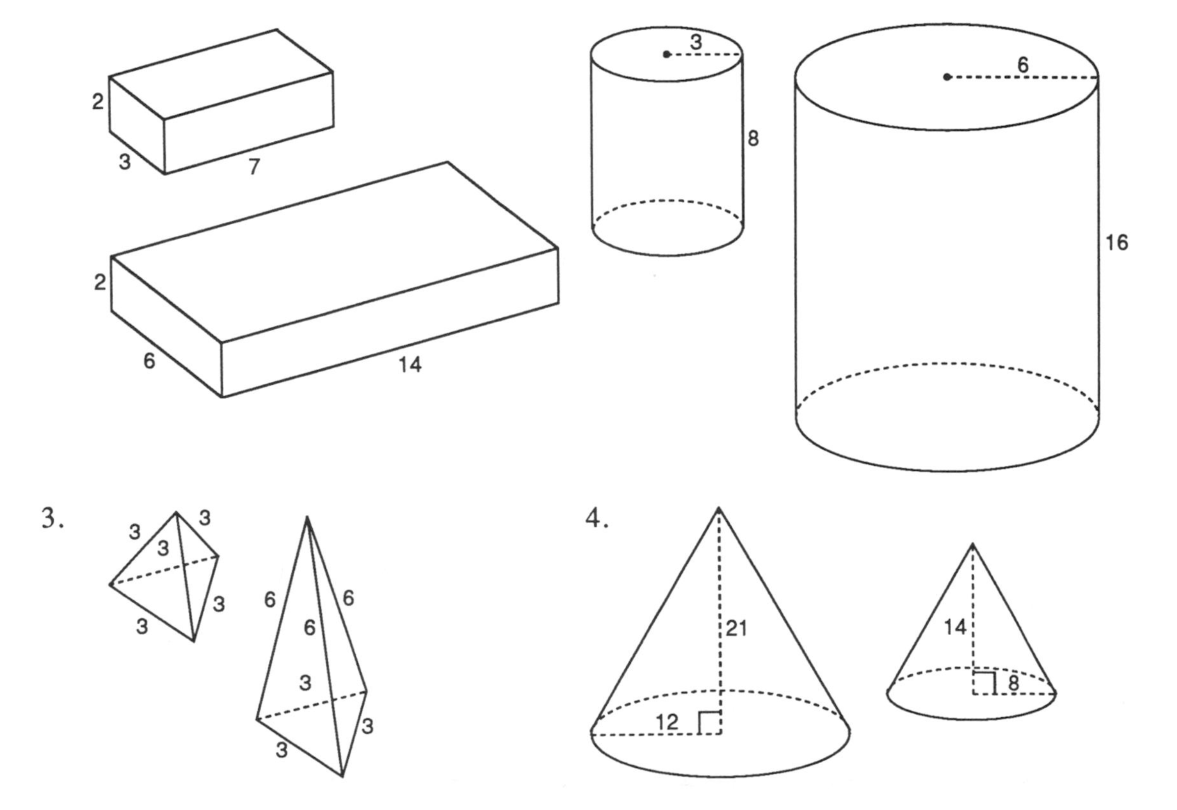

How does the ratio of lengths of corresponding edges of similar solids compare with the ratio of their volumes? Perform the following investigation to find out.

Investigation 11.8.1 

Find the ratio of corresponding lengths and the ratio of volumes in each problem below. All measurements are in centimeters.

1. Both figures are cubes.

   $\frac{x}{y} = $ –?–

   $\frac{\text{Volume}_{\text{small cube}}}{\text{Volume}_{\text{large cube}}} = $ –?–

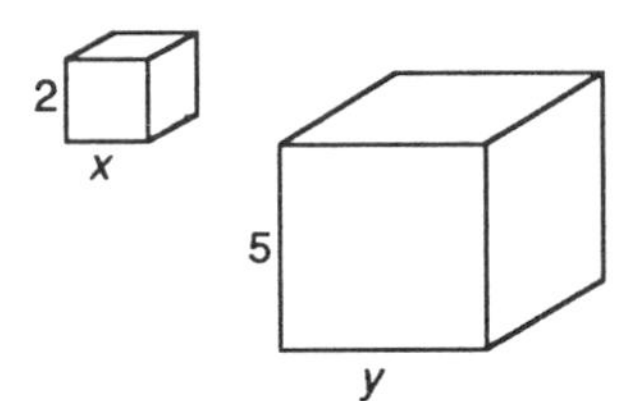

2. Both figures are spheres.

   $\frac{x}{y} = \frac{3}{4}$

   $\frac{\text{Volume}_{\text{small sphere}}}{\text{Volume}_{\text{large sphere}}} = $ –?–

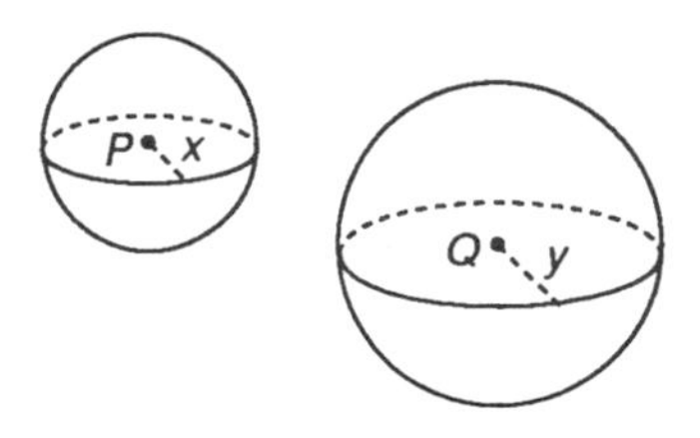

3. The rectangular prisms are similar.

   $\frac{x}{y} = $ –?–

   $\frac{\text{Volume}_{\text{small prism}}}{\text{Volume}_{\text{large prism}}} = $ –?–

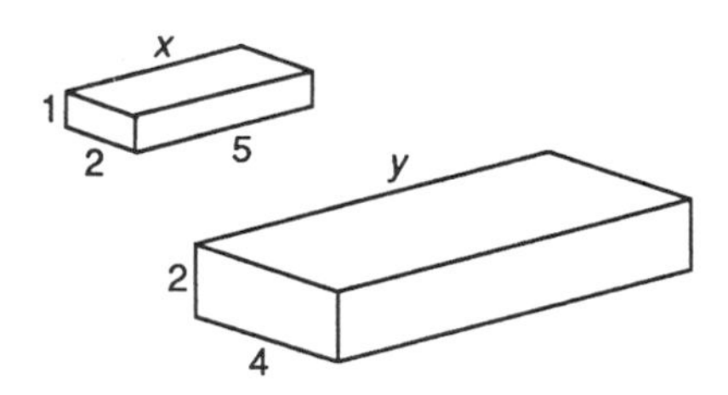

4. The cylinders are similar.

   $\frac{x}{y} = $ –?–

   $\frac{\text{Volume}_{\text{small cylinder}}}{\text{Volume}_{\text{large cylinder}}} = $ –?–

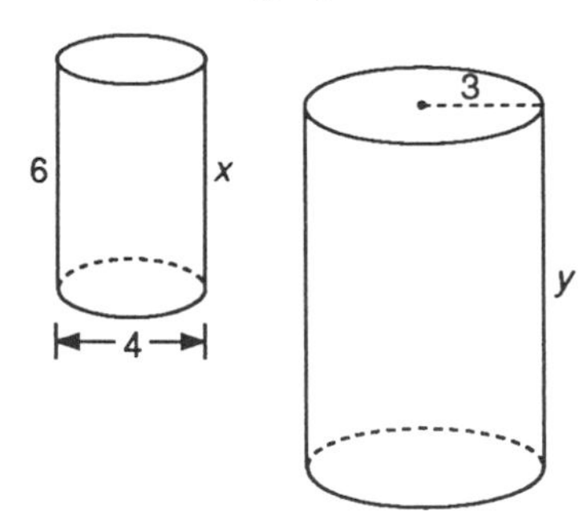

Are you ready to make a conjecture? In each problem in the investigation there is a relationship between the ratio of volumes and the ratio of corresponding dimensions. State this relationship as your next conjecture.

If two similar solids have corresponding dimensions in the ratio of m/n, then their volumes are in the ratio of —?—. ***(Proportional Volume Conjecture)***

This conjecture can be demonstrated as shown below.

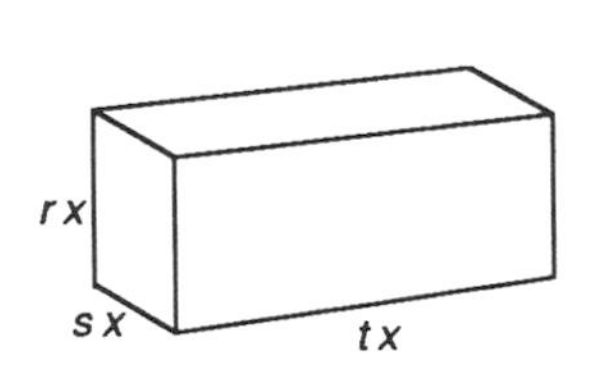

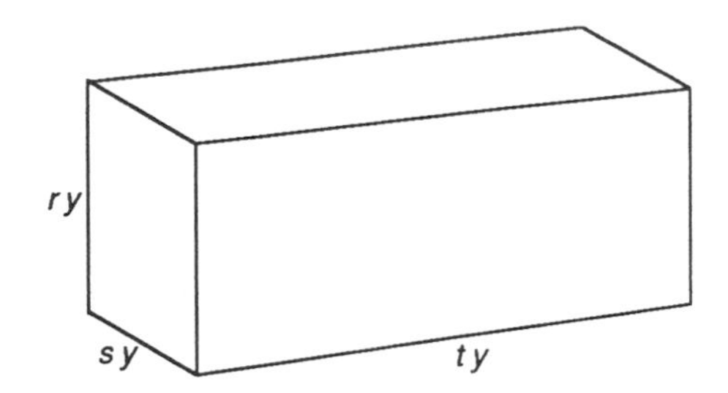

Ratio of corresponding dimensions: $\frac{rx}{ry} = \frac{sx}{sy} = \frac{tx}{ty} = \frac{x}{y}$

Ratio of volumes: $\frac{\text{Volume}_{\text{small prism}}}{\text{Volume}_{\text{large prism}}} = \frac{(rx)(sx)(tx)}{(ry)(sy)(ty)} = \frac{rstx^3}{rsty^3} = \frac{x^3}{y^3} = (\frac{x}{y})^3$

EXERCISE SET B

Use the Proportional Volume Conjecture to solve each problem.

1.* The corresponding edges of two similar triangular prisms are in the ratio of 5:3. What is the ratio of their volumes?

2. The volumes of two similar pentagonal prisms are in the ratio of 8:125. What is the ratio of their heights?

3. The ratio of the volumes of two similar circular cylinders is 27:64. What is the ratio of the diameters of their similar bases?

4. The right trapezoidal prisms are similar.
   $\text{Volume}_{\text{small prism}} = 324 \text{ cm}^3$
   $\frac{\text{Area}_{\text{base of small prism}}}{\text{Area}_{\text{base of large prism}}} = \frac{9}{25}$
   $\text{Volume}_{\text{large prism}} =$ –?–

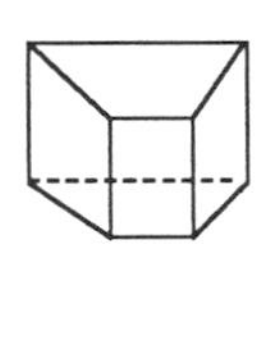

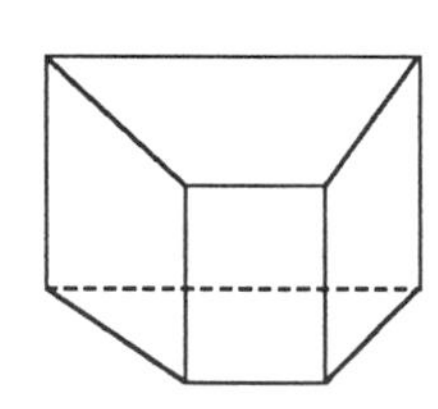

5.* The right cylinders are similar.
   $\text{Volume}_{\text{large cylinder}} = 4608\pi$ cu. ft.
   $H =$ –?–

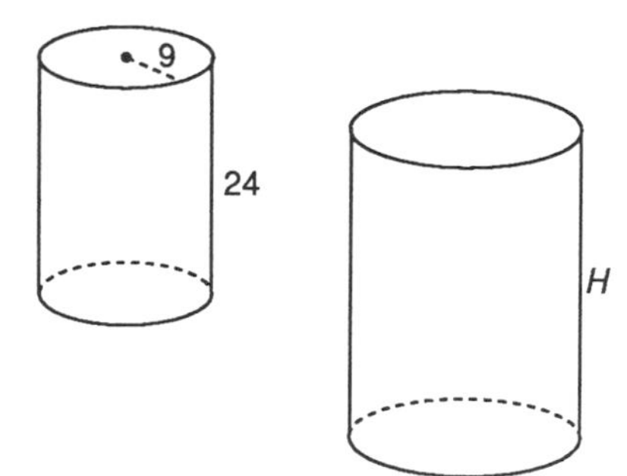

6.* The surface areas of two cubes are in the ratio of 49:81. What is the ratio of their volumes?

7.* The ratio of the weights of two spherical steel balls is 8:27. What is the ratio of the diameters of the two steel balls?

8. After building a chest for his toys, Charlie learns that Lucy has built one with dimensions twice as large. If it takes one-half of a gallon of paint to cover the surface of Charlie's toy chest, how many gallons of paint would be needed to paint Lucy's toy chest? How many times more volume will Lucy's toy chest hold than Charlie's?

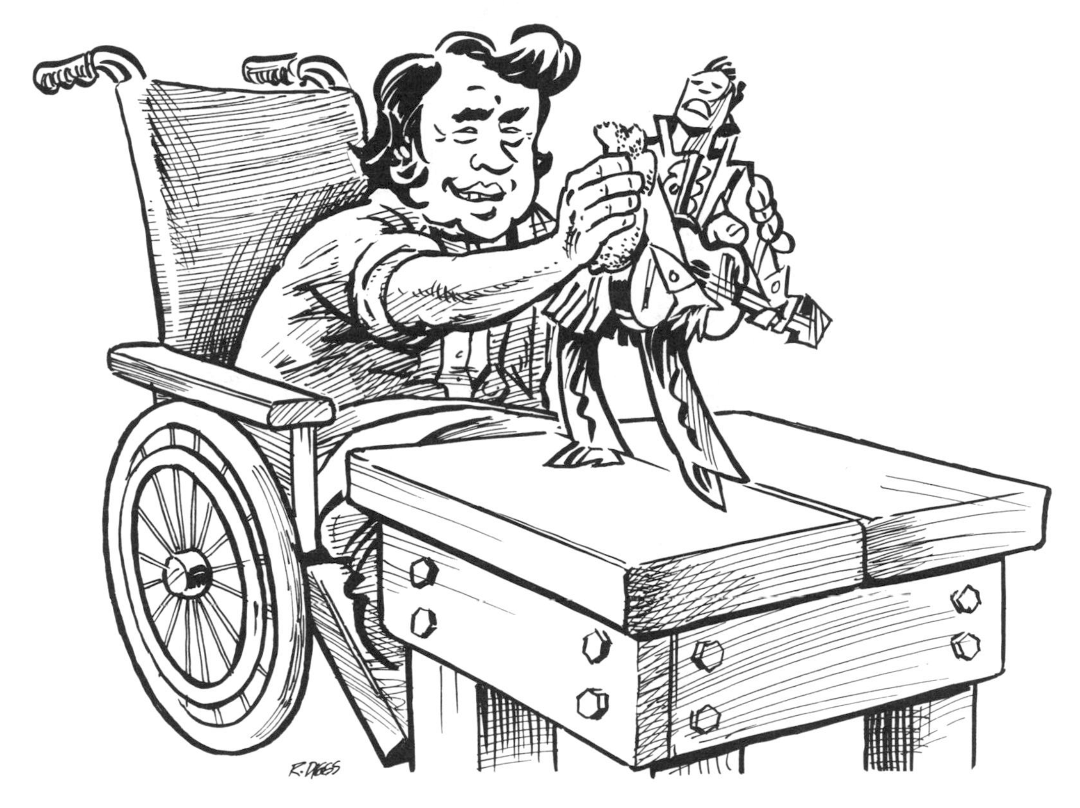

Mert the Metal Sculptor

9. Mert the metal sculptor has just created a solid steel statue in miniature of his singing idol Melvis Bresley. The statue weighs 38 lbs. He plans to make the full-scale statue four times larger in each dimension. How much will it weigh if it is also to be made of solid steel?

10.* ZAP Electronics has just installed an air conditioning unit in the small warehouse section of their VCR plant. The energy needed to operate an air conditioning unit is a function of the volume or space that is being air conditioned. The energy to operate the air conditioning system costs ZAP Electronics about $125 per day. They are considering installing a similar unit in their main storage warehouse. The dimensions of the main storage facility are two and a half times larger in each dimension than the VCR warehouse. What will be the daily operating cost of the system for their main storage warehouse?

Schulhoff Tam
High school student

The Theory of Dragons

11.* The noted Medieval scholar, Professor Otto Rosseforp, is the author of *Dragon Physics*, a scholarly work outlining his theory of dragon flight. In his paper, the professor claims that mature dragons were able to fly because of the presence of gasses in their intestinal tracts. The professor explains that when a dragon ate certain foreign "foods," methane and ammonia gasses were produced in the digestive process. This gas collected in the lower intestinal cavity and provided the lift for dragon flight. When the dragon's stomach became too upset or it wished to land, the methane gas was ignited by the sparking of metals (old swords stuck in its teeth), and the gasses were burned off. This was the cause of the reports of "fire-breathing dragons." After years of studying reports of large and small dragons in medieval manuscripts, the professor has come to the conclusion that a small dragon 6 m in length had the stomach capacity to eat 440 kg of "food." 440 kg was equivalent to nine fair damsels or four large knights (in armor). This 440 kg of "food" was capable of giving a 6 m dragon a flight distance of 50 km. If the stomach capacity of a dragon was proportional to the cube of its length, then what was the food capacity of a dragon of length 18 m? If the distance a dragon was capable of traveling on one meal was directly proportional to its stomach capacity, then how far could a dragon of 18 m length travel on a full dinner?

Lesson 11.9

Proportional Segments by Parallel Lines

In the diagram below, $\overleftrightarrow{MT} \parallel \overline{LU}$. Does ΔLUV appear to be similar to ΔMTV? It does. Let's see if we can support this observation with a paragraph proof.

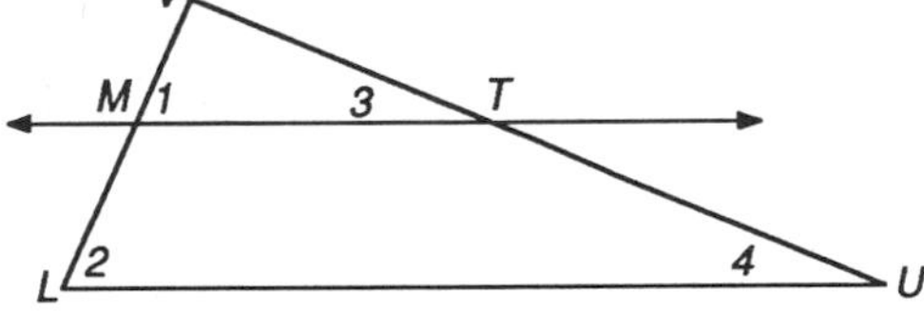

Given: ΔLUV with $\overleftrightarrow{MT} \parallel \overline{LU}$

Show: $\Delta LUV \sim \Delta MTV$

Paragraph Proof:

If $\overleftrightarrow{MT} \parallel \overline{LU}$, then $\angle 1 \cong \angle 2$ and $\angle 3 \cong \angle 4$ by the Corresponding Angles Conjecture.

If $\angle 1 \cong \angle 2$ and $\angle 3 \cong \angle 4$, then $\Delta LUV \sim \Delta MTV$ by the AA Similarity Conjecture.

If the two triangles are similar, then the corresponding sides are proportional.

In the diagram above, $\dfrac{LV}{MV} = \dfrac{VU}{VT} = \dfrac{LU}{MT}$.

EXERCISE SET A

All measurements are in centimeters.

1.* $l \parallel \overline{AB}$

$x = $ –?–

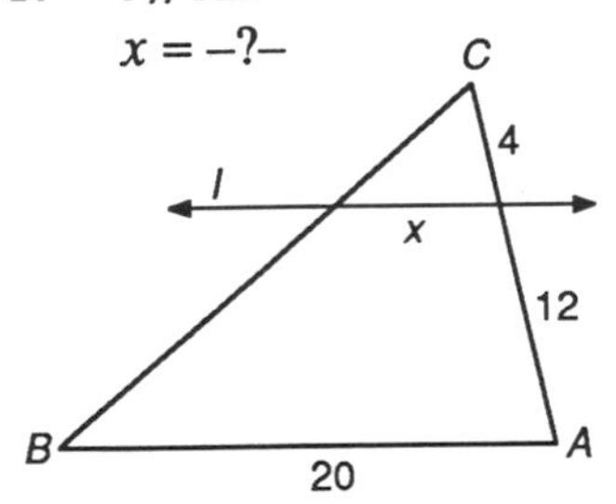

2. $m \parallel \overline{DE}$

$x = $ –?–

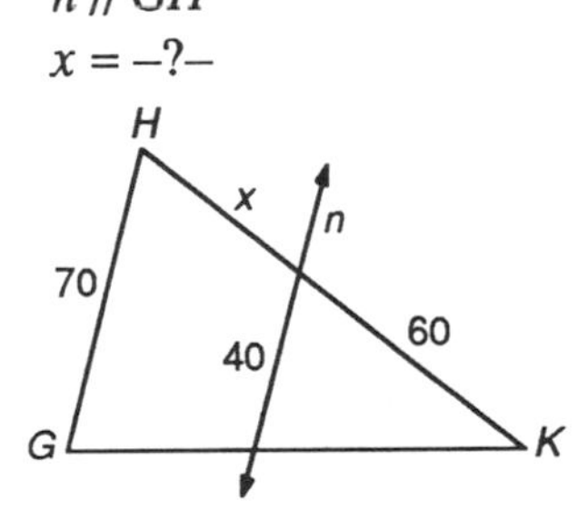

3. $n \parallel \overline{GH}$

$x = $ –?–

In each of the problems find x, and then find the ratios indicated.

4.* $\overleftrightarrow{EC} \parallel \overline{AB}$

$x = $ –?–

$\dfrac{DE}{AE} = $ –?– $\quad \dfrac{DC}{BC} = $ –?–

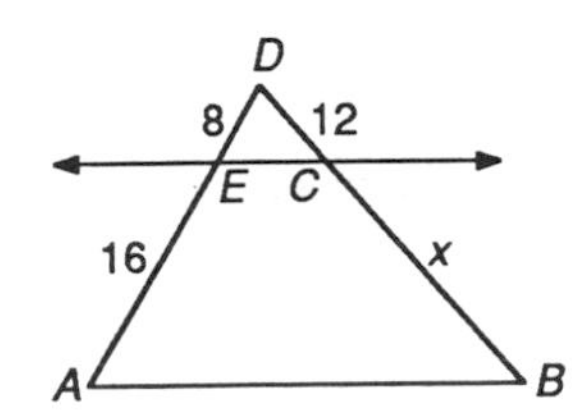

5. $\overleftrightarrow{KH} \parallel \overline{FG}$

$x = $ –?–

$\dfrac{JK}{KF} = $ –?– $\quad \dfrac{JH}{HG} = $ –?–

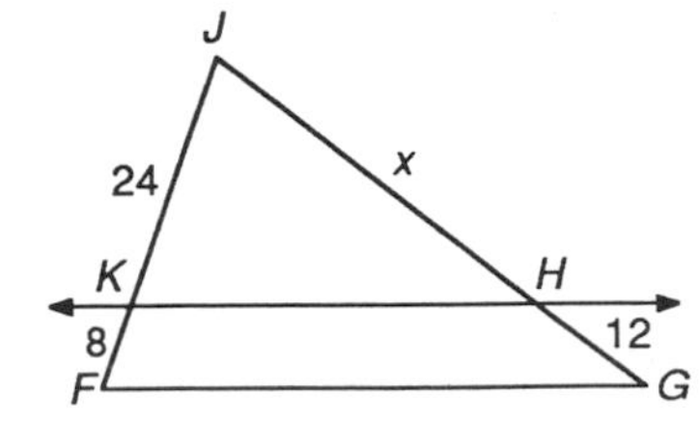

6. $\overleftrightarrow{QN} \parallel \overline{LM}$

$x = $ –?–

$\dfrac{PQ}{QL} = $ –?– $\quad \dfrac{PN}{MN} = $ –?–

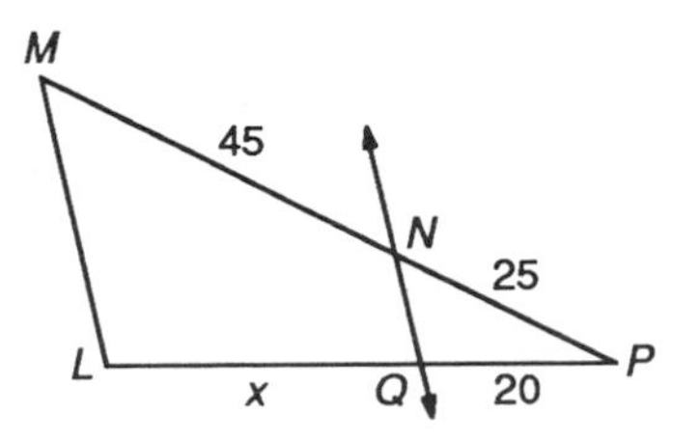

7.* What do you notice about the ratios of the lengths of the segments that have been cut by the parallel lines? Complete this conjecture: *If a line parallel to one side of a triangle passes through the other two sides, then it divides the other two sides —?—.*

8.* The conjecture in Exercise 7 can be demonstrated by algebra. If $l \parallel \overleftrightarrow{MN}$, then, from the proof at the beginning of this lesson, it follows that $\frac{a+b}{a} = \frac{c+d}{c}$.

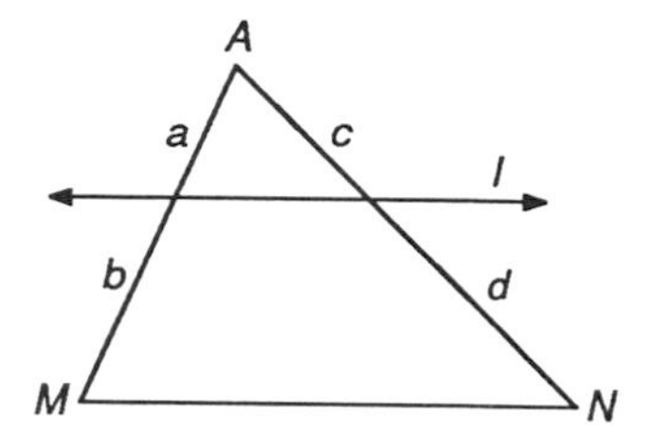

Show by algebra that if $\frac{a+b}{a} = \frac{c+d}{c}$, then $\frac{a}{b} = \frac{c}{d}$.

In Exercise A7 you discovered: *If a line parallel to one side of a triangle passes through the other two sides, then it divides the other two sides proportionally.* In Exercise A8 you demonstrated this conjecture. Is the converse true? That is, if a line divides two sides of a triangle proportionally, is it parallel to the third side?

Investigation 11.9.1

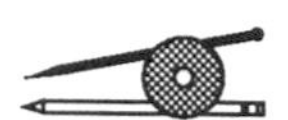
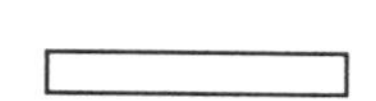
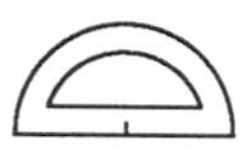

On a clean sheet of unlined paper perform the following steps:

Step 1: Construct an acute angle P.

Step 2: Beginning at P, mark off lengths of 8 cm and 10 cm on one ray. Label the points A and B.

Step 3: Mark off lengths of 12 cm and 15 cm on the other ray. Label the points C and D. Notice that $8/10 = 12/15$.

Step 4: Construct $\overline{AC}$ and $\overline{BD}$. Notice that $\overline{AC}$ divides $\overline{PB}$ and $\overline{PD}$ into segments such that $PA/AB = PC/CD$.

Step 5: With a protractor, measure $\angle PAC$ and $\angle PBD$.

Repeat the steps, but this time mark off lengths of 6 cm and 8 cm on one ray and 12 cm and 16 cm on the other ray. Notice that $6/8 = 12/16$. Compare your results with the results of others near you. If $\angle PAC \cong \angle PBD$, then $\overline{AC} \parallel \overline{BD}$. You should be ready to state a conjecture.

If a line parallel to one side of a triangle passes through the other two sides, then it divides them —?—. Conversely, if a line cuts two sides of a triangle proportionally, then it is —?— to the third side. (**Parallel Proportionality Conjecture**)

You can restate the Parallel Proportionality Conjecture using a diagram as follows:

If $l \parallel \overline{MN}$, then $\frac{a}{b} = \frac{c}{d}$.

If $\frac{a}{b} = \frac{c}{d}$, then $l \parallel \overline{MN}$.

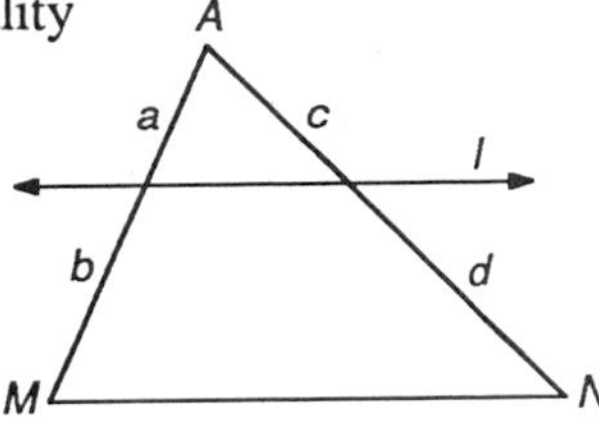

EXERCISE SET B

All measurements are in centimeters.

1.* $l \parallel \overline{TR}$
$a =$ –?–

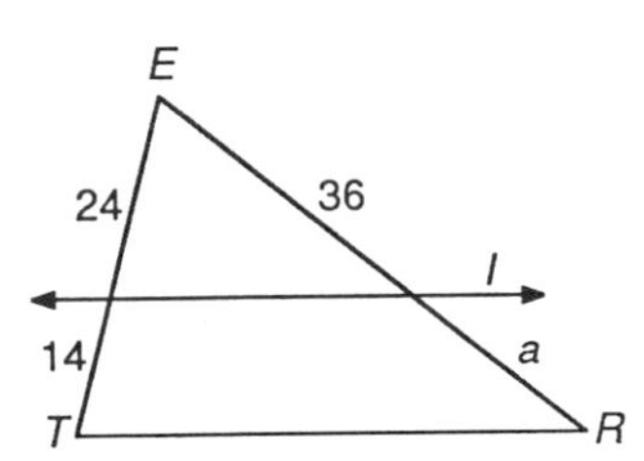

2. $l \parallel \overline{IN}$
$e =$ –?–

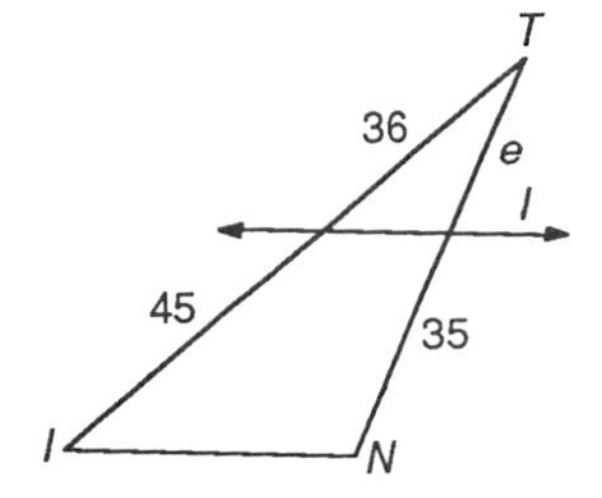

3. Is $n \parallel \overline{MA}$?

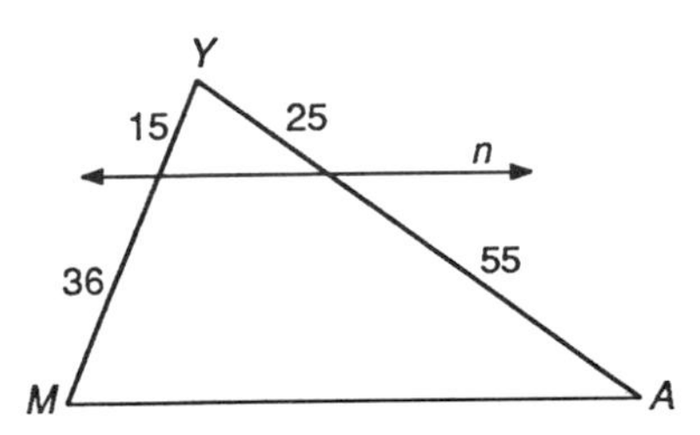

4. Is $s \parallel \overline{BE}$?

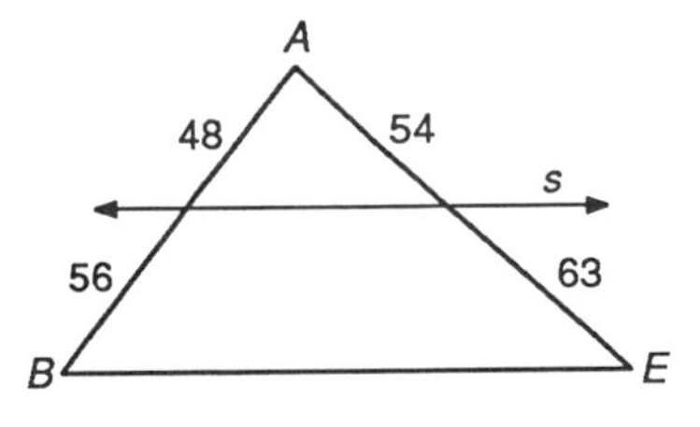

5.* $l \parallel \overline{GO}$
$g =$ –?–

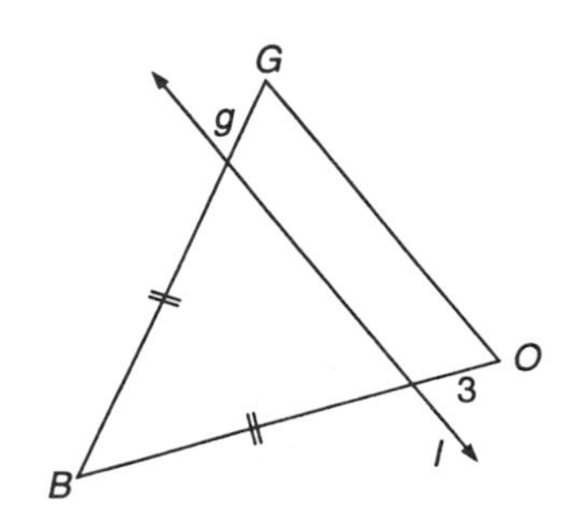

6. $r \parallel s \parallel \overline{RI}$
$m =$ –?– $n =$ –?–

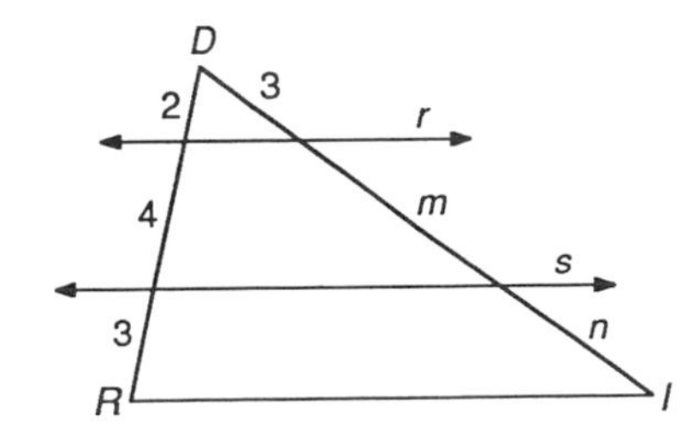

Lesson 11.10

Dividing a Segment into *n* Equal Parts

In the last lesson you discovered the Parallel Proportionality Conjecture. It states:

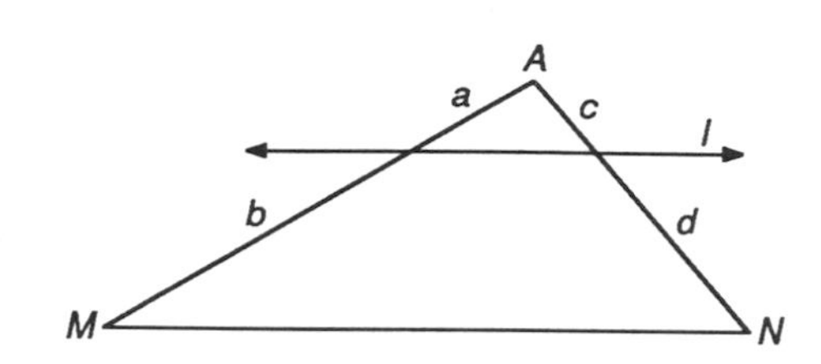

If $l \parallel \overline{MN}$, then $\frac{a}{b} = \frac{c}{d}$.

Conversely, if $\frac{a}{b} = \frac{c}{d}$, then $l \parallel \overline{MN}$.

What if more than one line passes through the two sides of a triangle parallel to the third side? Are all the ratios of lengths of corresponding segments equal?

Example

$\overline{IT} \parallel \overline{NO} \parallel \overline{PR}$

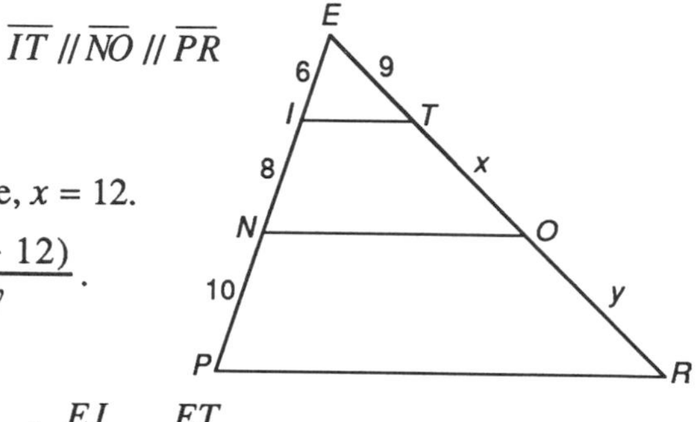

Find x. Find y.

Since $\overline{IT} \parallel \overline{NO}$ in ΔNOE, $\frac{6}{8} = \frac{9}{x}$. Therefore, $x = 12$.

Since $\overline{NO} \parallel \overline{PR}$ in ΔPRE, $\frac{(6+8)}{10} = \frac{(9+12)}{y}$.

Therefore, $\frac{14}{10} = \frac{21}{y}$ and $y = 15$.

Not only is $\frac{EI}{NI} = \frac{ET}{TO}$ but also $\frac{NI}{NP} = \frac{TO}{RO}$ and $\frac{EI}{NP} = \frac{ET}{RO}$.

It appears to be true, at least for this example, that if two lines passing through two sides of a triangle are parallel to the third side of the triangle, then the corresponding segments on those two sides are proportional. Let's investigate a few more problems like the example above.

Investigation 11.10.1

Use the Parallel Proportionality Conjecture to answer each question.

1. $\overline{FT} \parallel \overline{LA} \parallel \overline{GR}$

   $x = $ –?– $y = $ –?–

   Is $\frac{FL}{LG} = \frac{TA}{AR}$?

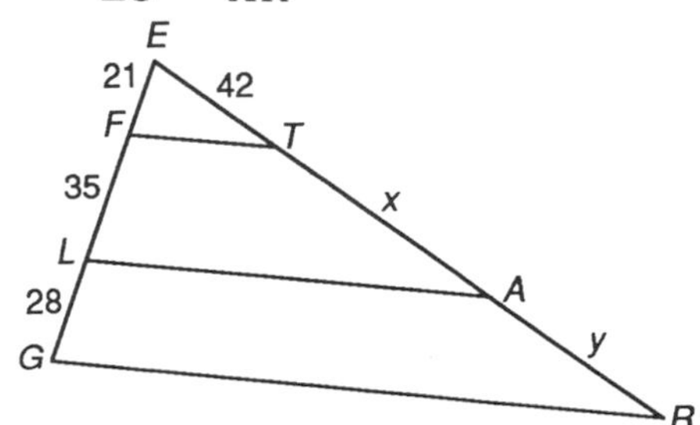

2. $\overline{OE} \parallel \overline{IP} \parallel \overline{DA} \parallel \overline{TR}$

   $a = $ –?– $b = $ –?– $c = $ –?–

   Is $\frac{OI}{ID} = \frac{EP}{PA}$? Is $\frac{ID}{DT} = \frac{PA}{AR}$?

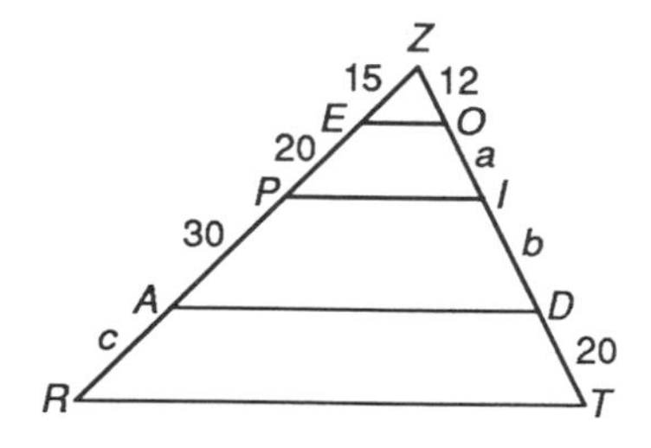

Compare your results from the two problems with the results of others near you. Does it appear that your last conjecture can be extended to hold even if two or more lines pass through the two sides of a triangle parallel to the third side? Complete the conjecture below.

C-80 *If two or more lines pass through two sides of a triangle parallel to the third side, then they divide the two sides —?—.*

This conjecture can be demonstrated by a paragraph proof.

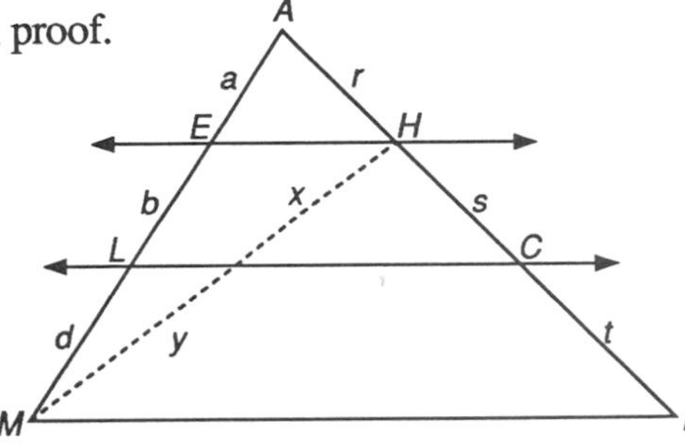

Conjecture: *If* $\overleftrightarrow{EH} \parallel \overleftrightarrow{LC} \parallel \overline{MI}$, *then* $\frac{b}{d} = \frac{s}{t}$.

Given: $\overleftrightarrow{EH} \parallel \overleftrightarrow{LC} \parallel \overline{MI}$

Show: $\frac{b}{d} = \frac{s}{t}$

Paragraph Proof:

Since you need to connect b, d, s, and t into one equation, you construct $\overline{MH}$. This gives you ΔHEM with $\overline{LC} \parallel \overline{EH}$ and ΔMIH with $\overline{LC} \parallel \overline{MI}$

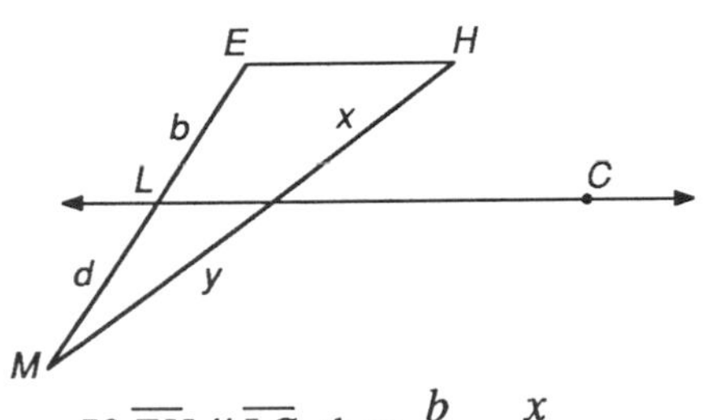

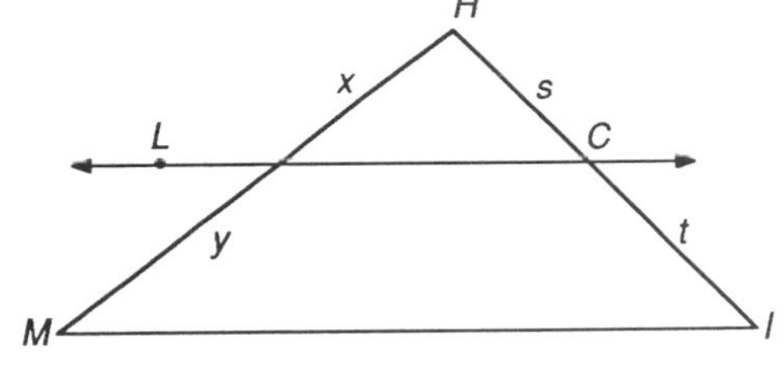

If $\overline{EH} \parallel \overline{LC}$, then $\frac{b}{d} = \frac{x}{y}$.

If $\overline{LC} \parallel \overline{MI}$, then $\frac{x}{y} = \frac{s}{t}$.

Therefore, since $\frac{b}{d} = \frac{x}{y}$ and $\frac{x}{y} = \frac{s}{t}$, then $\frac{b}{d} = \frac{s}{t}$.

Is the converse of this conjecture true? What is the converse?

Converse: If two lines divide two sides of a triangle proportionally, then the two lines are parallel to the third side.

It's not hard to find a counter-example to this converse. It is possible to draw two lines through two sides of a triangle so that the corresponding parts on the two sides are proportional, but the two lines are not parallel. The diagram on the right is a counter-example demonstrating that the converse is not true.

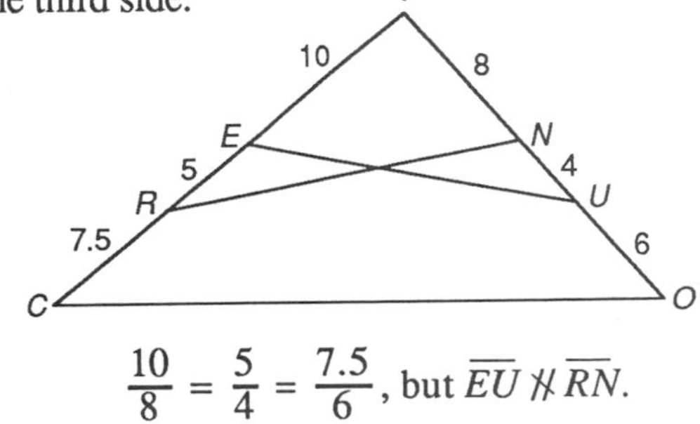

$\frac{10}{8} = \frac{5}{4} = \frac{7.5}{6}$, but $\overline{EU} \not\parallel \overline{RN}$.

EXERCISE SET A

Use your new conjecture to solve each problem. All measurements are in centimeters.

1. $\overline{AB} \parallel m \parallel n$

   $x =$ –?–

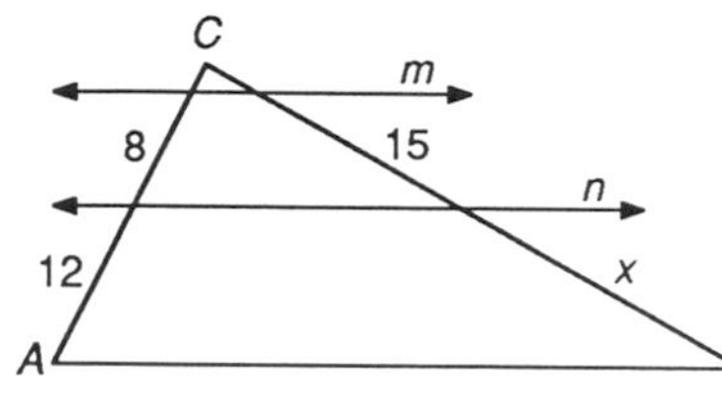

2. $\overline{DE} \parallel p \parallel q$

   $x =$ –?–

   $y =$ –?–

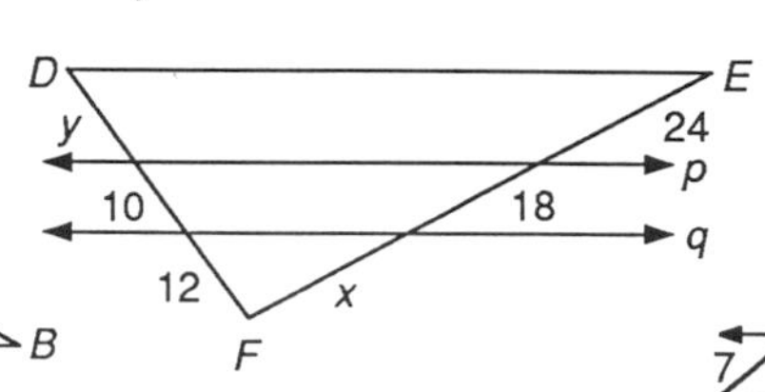

3. $\overline{GH} \parallel r \parallel s \parallel t$

   $x =$ –?–

   $y =$ –?–

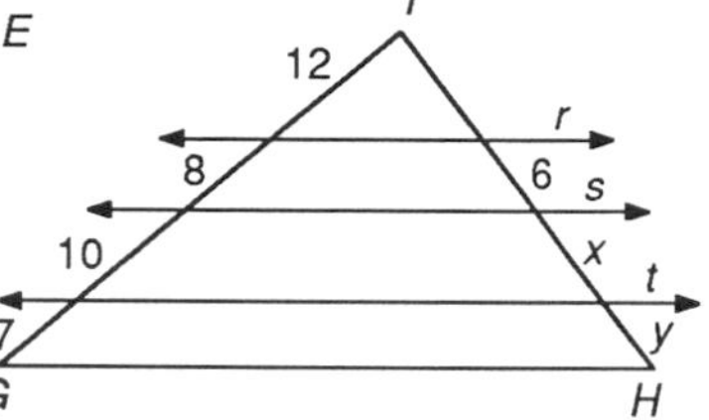

4.* $\overleftrightarrow{XY} \parallel \overline{AB}$

   $YB =$ –?–

   $AB =$ –?–

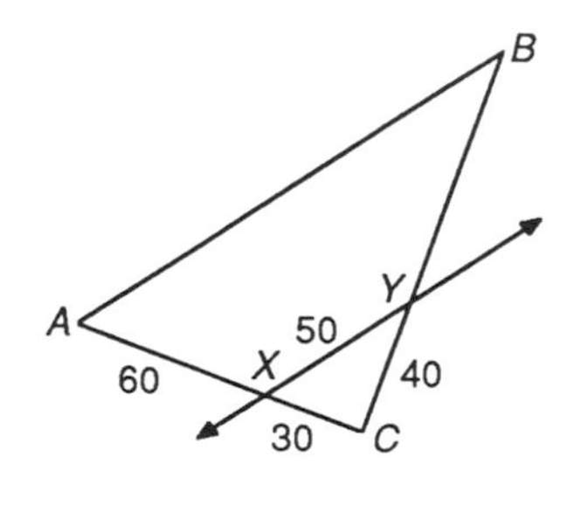

5.* $\dfrac{CM}{MA} =$ –?–

   $\dfrac{CN}{NB} =$ –?–

   Is $\overline{MN} \parallel \overline{AB}$?

   $\dfrac{MN}{AB} =$ –?–

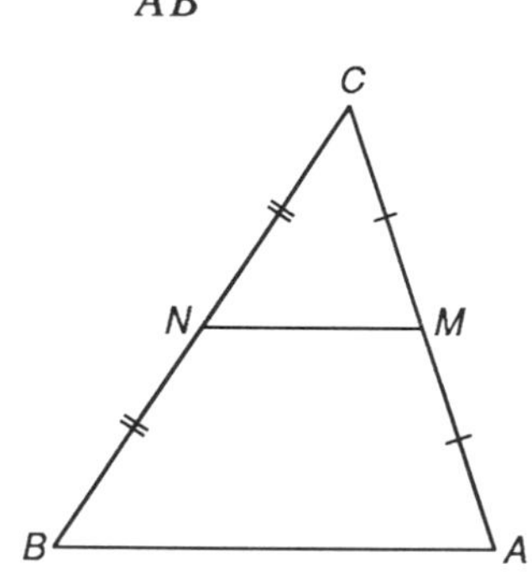

6. 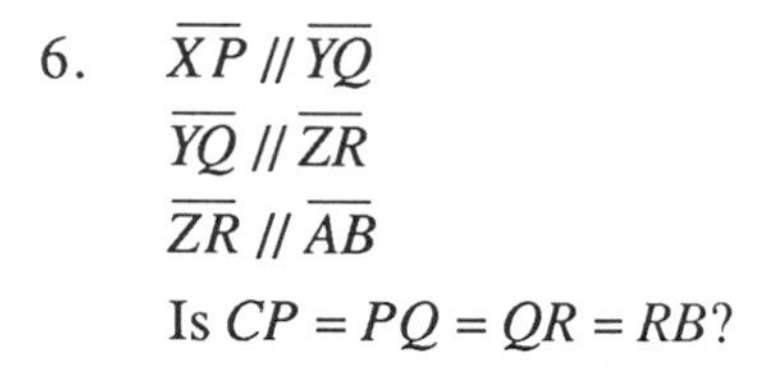

   $\overline{XP} \parallel \overline{YQ}$

   $\overline{YQ} \parallel \overline{ZR}$

   $\overline{ZR} \parallel \overline{AB}$

   Is $CP = PQ = QR = RB$?

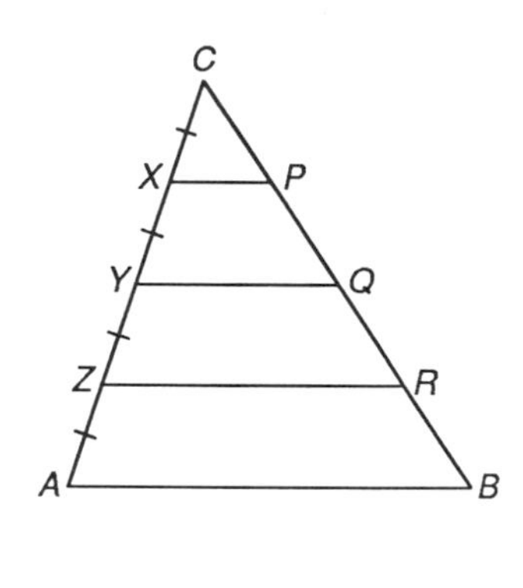

7. Is $m \parallel \overline{AB}$?

   Is $n \parallel \overline{AB}$?

   Is $m \parallel n$?

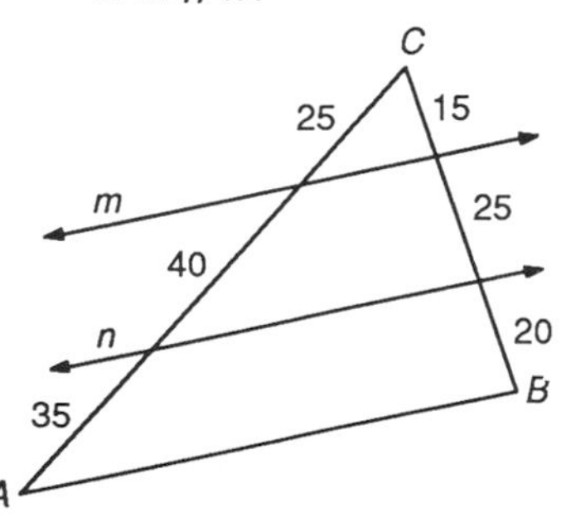

8. 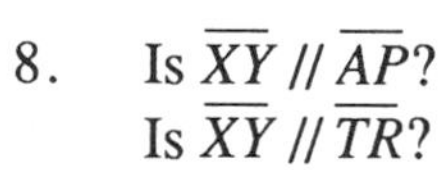

   Is $\overline{XY} \parallel \overline{AP}$?

   Is $\overline{XY} \parallel \overline{TR}$?

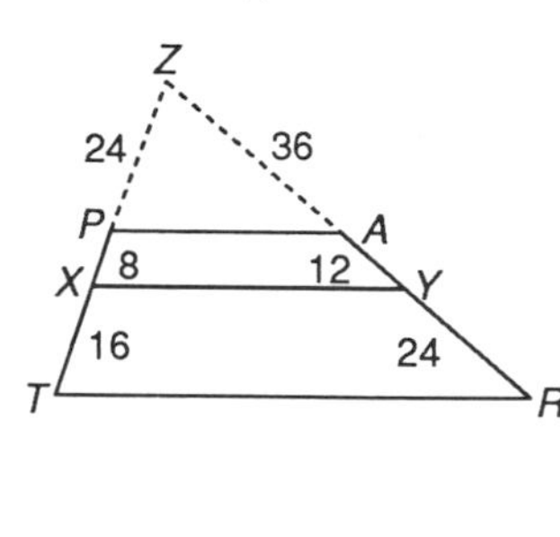

9.* *TOAD* is a trapezoid.

   Is $k \parallel \overline{AD}$?

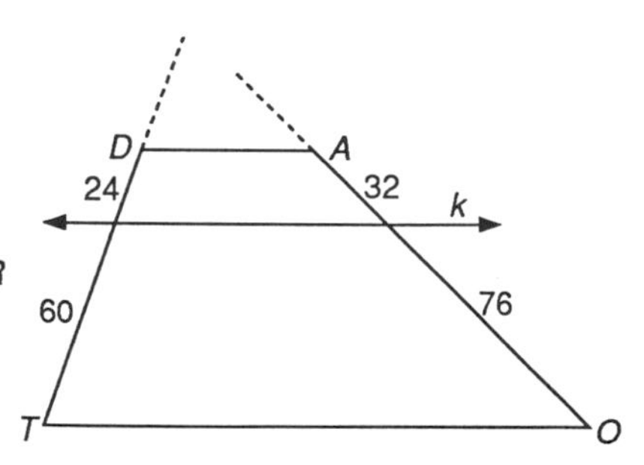

Your new conjecture is very useful. You already know how to use the perpendicular bisector construction to divide a segment into two, four, or eight equal parts. Now you can use your new conjecture to divide a segment into *any* number of equal parts.

Example

Construct segment AB. Then divide it into three equal parts.

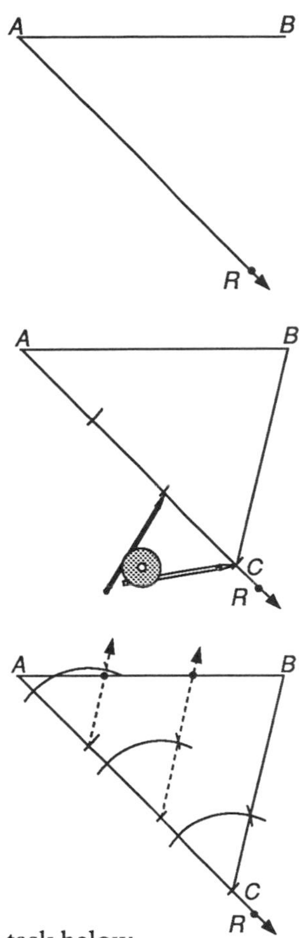

Step 1: Construct segment AB. From one endpoint of segment AB, say point A, construct a ray AR forming an angle BAR of about 45°.

Step 2: On ray AR mark off three equal lengths with your compass. From the endpoint C of the third segment, construct the segment BC to endpoint B. You now have triangle BAC with side $\overline{AC}$ divided into three equal lengths.

Step 3: Next, through the two points on the ray AC, construct rays parallel to side $\overline{BC}$ so that the parallel rays pass through the given segment AB. The two parallel rays intersect segment AB at two points which divide the segment into three equal parts.

EXERCISE SET B

Use your compass and straightedge to complete each task below.

1. Construct $\overline{EF}$. Then divide it into five equal parts.

2.* Construct $\overline{IJ}$. Then construct a regular hexagon with IJ as the perimeter.

3.* Construct $\overline{KL}$. Then find a point P that divides KL into the ratio of 2:3.

Special Project

The Golden Ratio II

Geometry has two great treasures: one is the theorem of Pythagoras; the other is the division of a line into extreme and mean ratio. The first we may compare to a measure of gold; the second we may name a precious jewel.

– Johann Kepler (1571–1630)

In an earlier special project you learned an approximate value for the Golden Ratio (1.6180), the actual value ($\frac{1+\sqrt{5}}{2}$), how to construct the Golden Ratio, and the definition of the Golden Ratio:

> *If a line segment is divided into two lengths such that the ratio of the segment's entire length to the longer length is equal to the ratio of the longer length to the shorter length, then the segment has been divided into the* ***Golden Ratio****.*

If the ratio of the length of the longer side to the length of the shorter side of a rectangle is the Golden Ratio, then the rectangle is a Golden Rectangle. In this special project you will learn how to construct a Golden Rectangle and how to construct a spiral within the rectangle.

$$\frac{a}{b} = \frac{1+\sqrt{5}}{2}$$

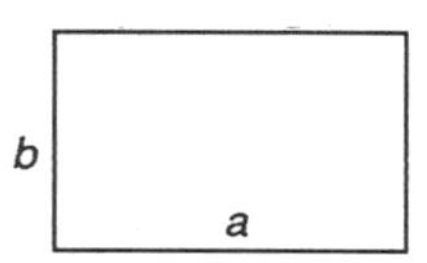

Golden Rectangle

How do you construct a Golden Rectangle?

Step 1

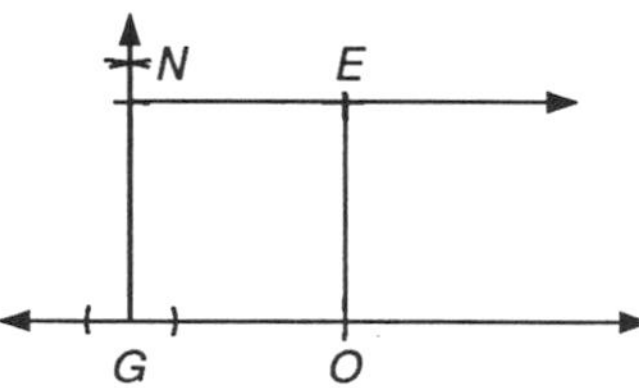

Construct a square. Label it *GOEN*. Extend $\overline{GO}$. Extend $\overline{NE}$.

Step 2

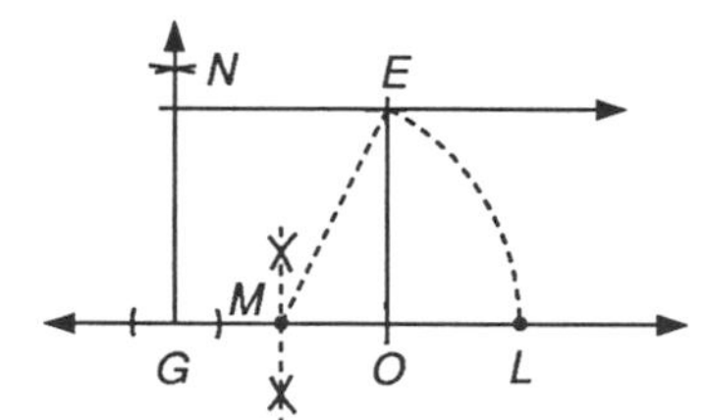

Bisect $\overline{GO}$. Label the midpoint *M*. With *ME* as your radius and *M* as center, construct an arc intersecting $\overleftrightarrow{GO}$ at point *L*.

Step 3

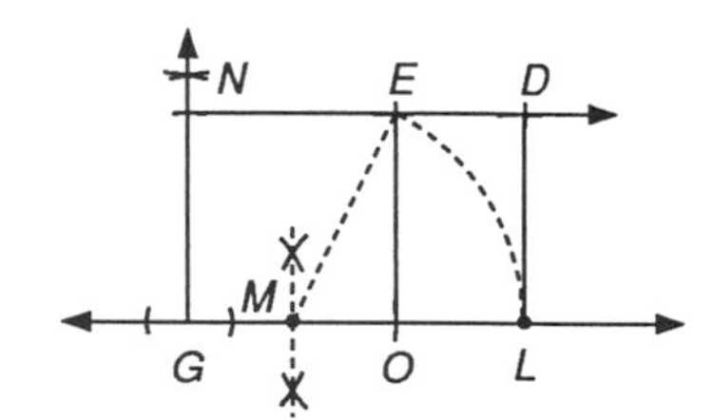

Construct the rectangle *OLDE*. *GLDN* is a Golden Rectangle.

1. Use the Pythagorean Theorem to show that the ratio of the length of the longer side to the length of the shorter side is equal to the Golden Ratio. Let $OE = 2x$, then $MO = x$. Calculate ME. But $GL = x + ME$. Calculate GL/LD.

   Show that: $\dfrac{GL}{LD} = \dfrac{1 + \sqrt{5}}{2}$.

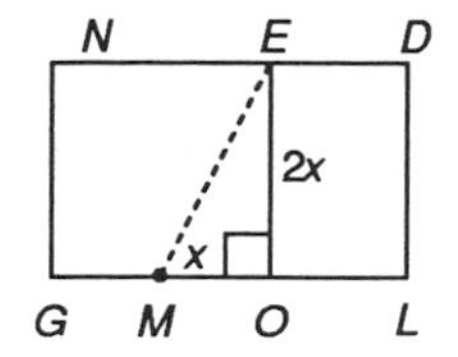

2. Construct as large a Golden Rectangle as possible on a full sheet of paper.

Did you notice that $\overline{OE}$ divides the Golden Rectangle into a square and another rectangle? Does the small rectangle appear to be similar to the original? In fact the smaller rectangle is also a Golden Rectangle.

3. If you start with a Golden Rectangle, you can divide it into a square and another Golden Rectangle. If you repeat this process with each succeeding smaller rectangle, you get the pattern shown at right. Try it. Use the Golden Rectangle constructed in Problem 2. Use construction tools to divide your rectangle into a square and new, smaller Golden Rectangle. Do this again with the new, smaller Golden Rectangle. Repeat this process two more times.

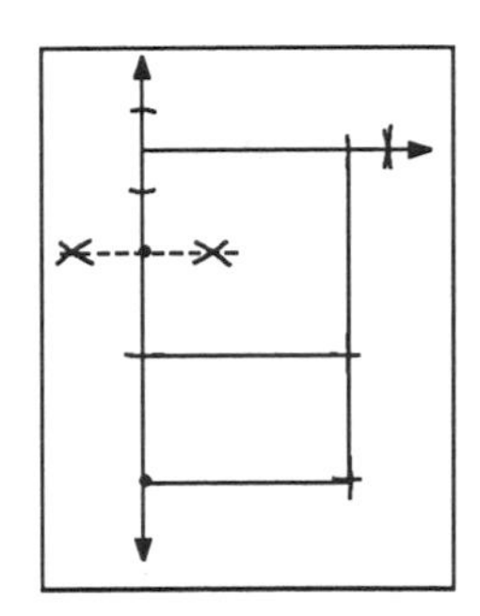

4. This can also be done in reverse. Construct as small a Golden Rectangle as possible. Use your construction tools to add on a square to its left (as shown in the illustration) using the long side of the rectangle as the side of the square. The square and rectangle combine to form a larger Golden Rectangle. Then add on another square above the new rectangle, creating still another larger Golden Rectangle. Do this two more times.

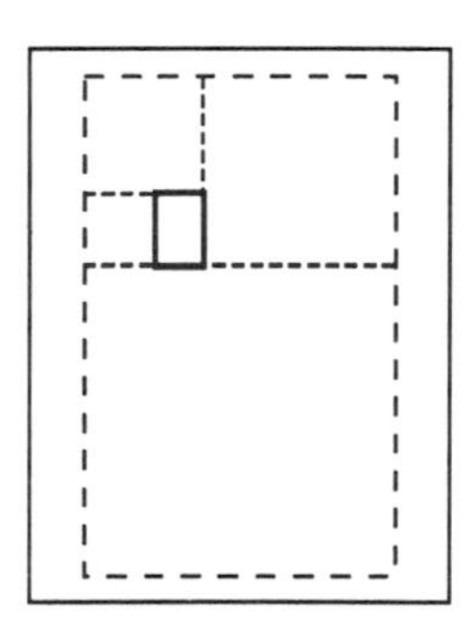

If arcs are drawn in the Golden Rectangle, as shown, you get a very graceful curve that is related to the beautiful spiral of a nautilus seashell. This curve has many names, due in part to its many different but related properties. René Descartes (1596–1650) called it the Equiangular Spiral, and Edmund Halley (1656–1742) called it the Proportional Spiral. Jakob Bernoulli (1654–1705) named it the Logarithmic Spiral and asked that it be engraved on his tombstone.

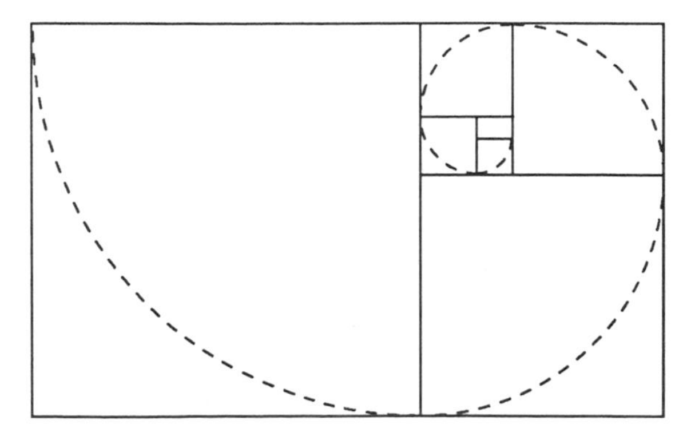

5. Use the Golden Rectangles from Problem 3 to construct an approximation to the Logarithmic Spiral. You can do this by constructing a 90° arc in each square.

6. Use the Golden Rectangles from Problem 4 to construct another approximation to the Logarithmic Spiral. You can do this by placing a point in the center of each square and then sketch in a spiral connecting the center of each square.

The type of growth exhibited by the shell of the nautilus is called "gnomonic growth" by Sir D'Arcy Thompson in his book *Growth and Form*. As an animal grows by gnomonic growth, the size of the animal's shell increases but the shape remains unchanged. The nautilus shell grows longer and wider to make room for the growing animal within, but it grows at one end only. Each new section is increased in size so that the overall shape remains similar. The repeating process of adding on a new square to a Golden Rectangle to get a new, larger Golden Rectangle is analogous to the nautilus adding a new chamber to its shell to accommodate its growth.

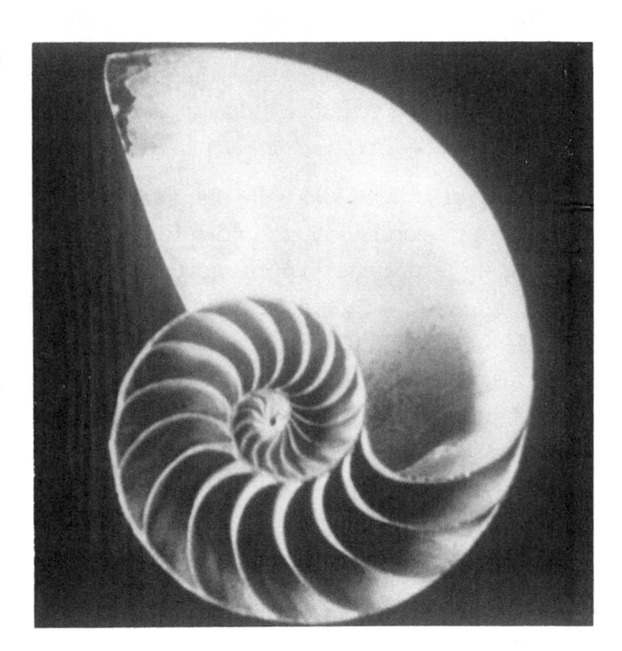

Improving Reasoning Skills

Coin Swap IV

Find five light coins (dimes) and five dark coins (pennies) and arrange them on a grid of eleven squares as shown. Your task is to switch the position of the five light and five dark coins in exactly 35 moves. A coin can slide into an empty square next to it or it can jump over one coin into an empty space. Record your solution by listing in order which color coin is moved. For example, your list might begin DLDLDDLL . . . .

How many moves would it take to switch the position of six light and six dark coins?

Lesson 11.11
Chapter Review

EXERCISE SET A

Identify each statement as true or false.

1. If the three sides of one triangle are proportional to the three sides of another triangle, then the two triangles are similar.

2. If two angles of one triangle are congruent to two angles of another triangle, then the two triangles are similar.

3. If two sides of one triangle are proportional to the sides of another triangle, then the two triangles are similar.

4. If the four angles of one quadrilateral are congruent to the four corresponding angles of another quadrilateral, then the two quadrilaterals are similar.

5. An angle bisector in a triangle divides the opposite side into two segments whose lengths are in the same ratio as the corresponding adjacent sides.

6. If two triangles are similar, then the corresponding altitudes, corresponding medians, and corresponding angle bisectors are proportional to the corresponding sides.

7. If two similar polygons (or circles) have corresponding sides (or radii) in the ratio of m/n, then their areas are in the ratio of m/n.

8. If two similar solids have corresponding dimensions in the ratio of m/n, then their volumes are in the ratio of m/n.

9. If a line parallel to one side of a triangle passes through the other two sides, then it divides them proportionally.

10. If a line cuts two sides of a triangle proportionally, then it is parallel to the third side.

11. If two or more lines pass through two sides of a triangle parallel to the third side, then they divide the two sides equally.

12. Six statements in this true false review are false.

EXERCISE SET B

Solve each problem below.

1. $\frac{x}{15} = \frac{8}{5}$

2. $\frac{4}{11} = \frac{24}{x}$

3. $\frac{4}{x} = \frac{x}{9}$

4. $\frac{x}{x + 3} = \frac{34}{40}$

In Exercises 5 to 8, measurements are in centimeters.

5.* $\Delta ABC \sim \Delta DBA$

$x = \text{–?–}$ $y = \text{–?–}$

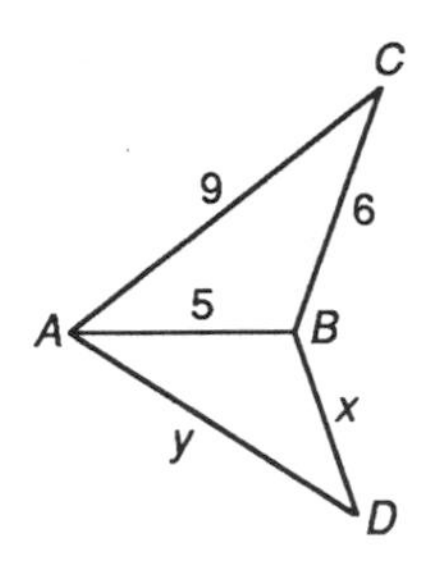

6. $ABCDE \sim FGHIJ$

$w = \text{–?–}$ $x = \text{–?–}$ $y = \text{–?–}$ $z = \text{–?–}$

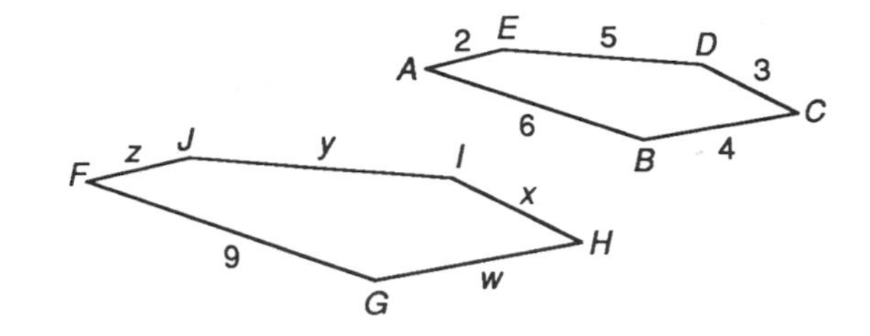

7. $k \parallel l \parallel m \parallel n$

$w = \text{–?–}$

$x = \text{–?–}$

$y = \text{–?–}$

$z = \text{–?–}$

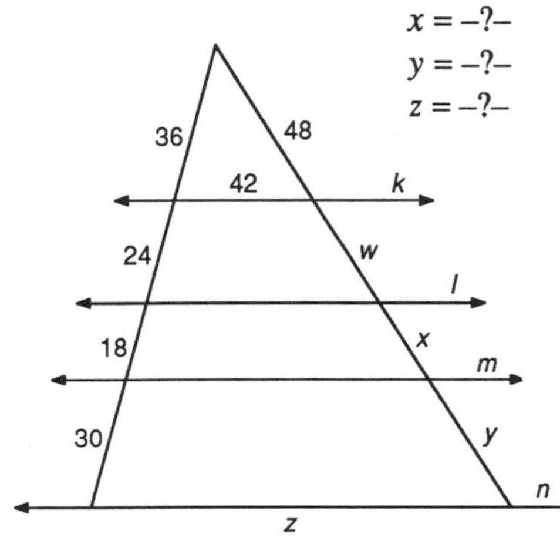

8.* The dimensions of the smaller cylinder are two-thirds of the larger. The volume of the larger cylinder is 2160π cm^3. Find the volume of the smaller cylinder.

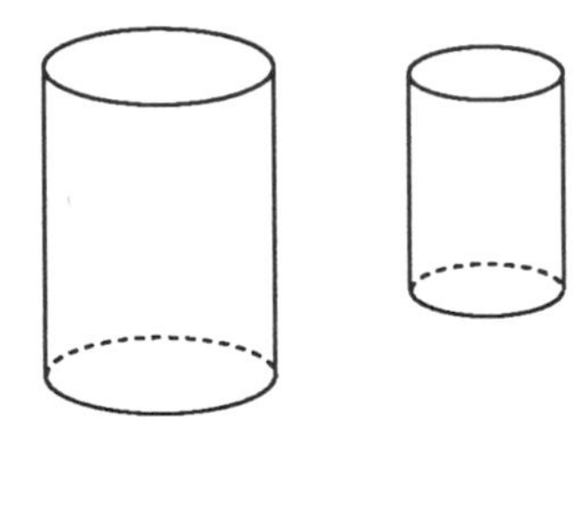

9. A man 5' 8" tall wishes to find the height of an oak tree in his front yard. He walks along the shadow of the tree until his head is in a position where the end of his shadow exactly overlaps the end of the treetop's shadow. He is now 11' 3" from the foot of the tree and 8' 6" from the end of the shadows. How tall is the oak tree?

10. Edith-the-Flagpole-Sitter is on top of her circular platform which is 50' above the ground. Edith is sitting in the lotus position in the very center of her platform which places her eye 2.5' above the platform. If the platform has a circumference of 22', find the area on the ground beneath her that is invisible to her eye. (Use 22/7 for π.)

The Ling Ring Sisters Circus

11.* The Ling Ring Sisters Circus has arrived in town. Ling Mai is the star of the show. She does a juggling act atop a stool on top of a rotating ball that spins on the top of a 20-m pole. The diameter of the ball is 4 m and Ling Mai's eye is 2 m above the ball. The circus manager, Ho Ping, needs to know the radius of the circular region beneath the ball where spectators would be unable to see eye to eye with Ling Mai. Find the radius to the nearest tenth meter for Ho Ping so that he can begin to put in the seats for the show. (Use 1.7 for $\sqrt{3}$.)

12.* The design on the right was a favorite of the Greek mathematician Archimedes. He liked it so much that he requested that it be placed on his tombstone! You are probably thinking, "What is so special about it?"
Let's look and see. The design is composed of a square, an isosceles triangle, and a circle. The circle and isosceles triangle are inscribed in the square. Calculate the ratio of the area of the square, the area of the circle, and the area of the isosceles triangle. Copy and complete the statement of proportionality between the three:

Area $_{\text{square}}$ to Area $_{\text{circle}}$ to Area $_{\text{triangle}}$ is –?– to –?– to –?–.

When each of the figures (square, circle, and isosceles triangle) is revolved about the vertical axis of symmetry, it generates a solid of revolution (cylinder, sphere, and cone). Calculate the volume of the cylinder, the sphere, and the cone. Copy and complete the statement of proportionality between the three:

Volume $_{\text{cylinder}}$ to Volume $_{\text{sphere}}$ to Volume $_{\text{cone}}$ is –?– to –?– to –?–.

Cooperative Problem Solving

Similarity in Space

1. The Hydroponics Farm Module produces all the fruits and vegetables for the Lunar Research Facility. The chief agricultural engineer, Figuroa Newton, has been instructed to design a second hydroponics facility for a larger lunar colony to be built next year. The new lunar colony will house nine times as many people as the original Lunar Research Facility. Fig Newton knows that the amount of food produced in a hydroponics farm module is proportional to the floor area of the facility. One of the major cost factors, however, is the volume of atmosphere that must be maintained in the hemispherical dome of a hydroponics farm. The radius of the hemispherical dome on the Research Facility's hydroponics farm is 12 meters. What is the volume of atmosphere that Fig will have to maintain in the new hemispherical dome of the hydroponics farm for the new lunar colony?

2. Professor Bernie Bernoulli and Chief Engineer Mia Mitsubishi are planning a pizza party at the Lunar Research Facility. The party is to celebrate the Columbus Day landing of the crew of Italian astronauts. From previous experience, Bernie and Mia know that one pizza with a 48 cm diameter will serve three people. How many pizzas with 72 cm diameters are needed to serve the party of twenty-seven people?

3. While on a lunar trek on the back side of the moon, Professor A. Brayne has come upon a strange monolith. The monolith is sitting vertically in the middle of an unexplored lunar ravine. The professor believes the monolith may have been built by an alien civilization. To determine the height of the mysterious object, the professor uses her 36 cm long lunar core sampler device (LCS). She holds the LCS vertically out in front of her 64 cm away from her eye and moves towards the monolith until she sees the top of the monolith just over the top of her LCS and views the base of the monolith just beneath the bottom of her LCS. She paces off the distance from that point to the base of the monolith and finds that it is 28 meters. How tall is the monolith?

4. At the lunar colony, the students of the Luke Skywalker Elementary School are busy designing a mural that will be a scale drawing of the Earth, Moon, inner planets, and Sun. The wall on which they plan to put up the mural is 38 meters long by four meters tall. What is the approximate distances on the mural between the Sun, Mercury, Venus, and Earth? (Use the table below.) What is the approximate distance on the mural between the Earth and Moon if the actual distance between the two is about 384,000 kilometers. For aesthetic reasons they plan to place the center of the Sun one meter in from the left edge and Earth one meter in from the right edge. The diameter of the Sun is about 1,392,000 kilometers. What should be the diameter of the Sun in the mural if it is to be in proportion with the rest of the mural? Sketch your mural on graph paper. Let each square represent one square meter. Locate and label the Sun, Mercury, Venus, and Earth.

Planet	Average Distance from Sun
Mercury	58,000,000 km
Venus	108,000,000 km
Earth	150,000,000 km

12 Trigonometry

Balcony,
M. C. Escher, 1945

Right triangle trigonometry is the study of the relationships between the sides and angles of similar right triangles. In this chapter you will discover some of the properties and applications of trigonometry. You will use trigonometry to calculate distances that are difficult or impossible to measure directly. Trigonometry is a branch of mathematics that you will study in great depth in your next mathematics course. It is a useful subject.

Lesson 12.1

Trigonometric Ratios

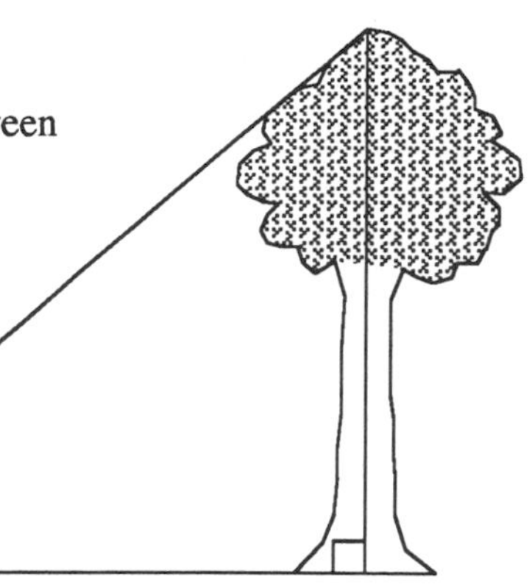

Right triangle trigonometry is the study of the relationships between the sides and angles of right triangles. In this lesson you are going to discover some of these relationships.

In the previous chapter you used mirrors and shadows to measure heights indirectly. Trigonometry gives you an additional method for doing this. For example, you can use trigonometry to calculate the height of a tree by measuring the angle of elevation and the distance from the vertex of the angle to the tree.

How did right triangle trigonometry develop? Early mathematicians and astronomers calculated the equivalent of the ratios of the sides for different right triangles and put their results into tables. They discovered that whenever the length of the shorter leg of a right triangle divided by the length of the longer leg was close to some specific fraction, the angle opposite the shorter leg was close to the same size. This was true no matter how big or small the triangle was.

For example, in every right triangle in which the length of the shorter leg divided by the length of the longer leg is close to the fraction 3/5, the angle opposite the shorter leg is close to 31°. What is x?

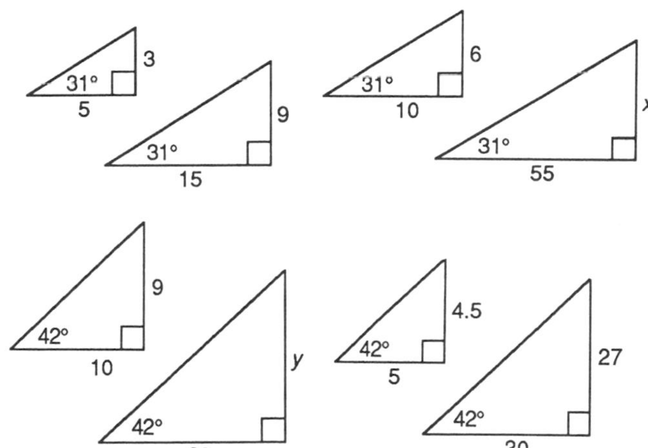

In every right triangle in which the length of the shorter leg divided by the length of the longer leg is always close to the fraction 9/10, the angle opposite the shorter leg is always close to 42°. What is y?

Early mathematicians found ratios like this for various integer degree measures and placed them in tables. By the late Middle Ages mathematicians converted the tables of fractions to tables of decimal fractions. These tables were very useful. Using these trigonometric tables, mathematicians and astronomers were able to calculate distances that they were unable to measure directly.

This, of course, can be supported from what you know about similar triangles. If two right triangles each have an acute angle of the same measure, then the triangles are similar (AA Similarity Conjecture). If two triangles are similar, then the ratios of their corresponding sides are proportional. For example, in the similar right triangles below right, the following proportions are true:

$$\frac{AB}{DE} = \frac{BC}{EF} \qquad \frac{AB}{GH} = \frac{BC}{HI} \qquad \frac{AB}{JK} = \frac{BC}{KL}$$

But, with the help of algebra, this can also be stated another way — the ratio of a pair of sides in one triangle is equal to the ratio of the corresponding sides in each of the other similar triangles. Or;

$$\frac{AB}{BC} = \frac{DE}{EF} = \frac{GH}{HI} = \frac{JK}{KL}$$

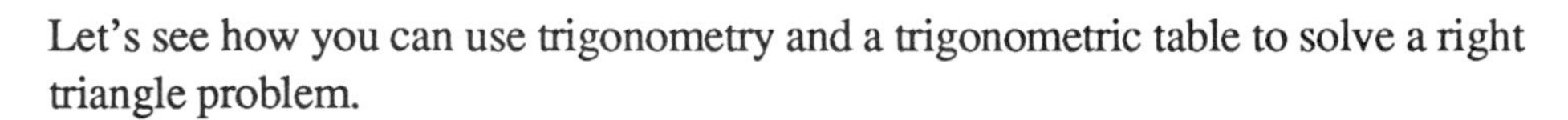

Let's see how you can use trigonometry and a trigonometric table to solve a right triangle problem.

Example

Find the height of a tree (HT of right ΔHAT) if the angle of elevation ($\angle A$) is 40° and the distance to the tree (HA) is 36 meters.

If you look in a trigonometry table, you will find that in a right triangle the ratio of the side opposite a 40° angle divided by the side adjacent to a 40° angle is about 0.84. Therefore:

$$\frac{HT}{HA} \approx 0.84$$

$$\frac{HT}{36} \approx 0.84$$

$$HT \approx (36)(.84)$$

$$HT \approx 30.24$$

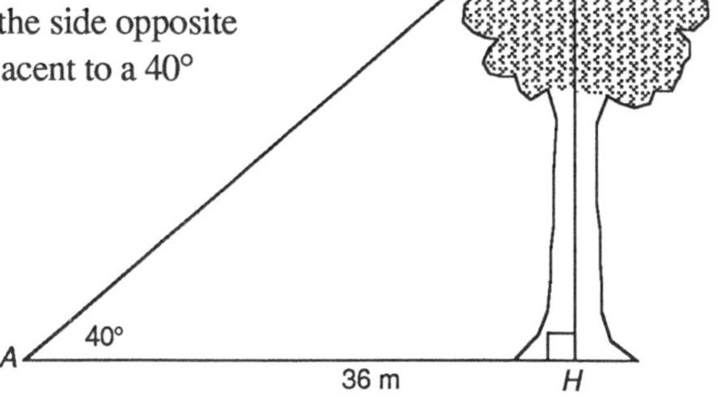

The height of the tree (HT) is approximately 30 meters.

The ratio of the length of the opposite side divided by the length of the adjacent side in a right triangle eventually came to be called the **tangent** of the angle. Early mathematicians also calculated tables of values for the other ratios of sides and eventually gave these ratios names as well. There are six possible ratios that can be formed by taking two sides at a time from a set of three sides. The three ratios that you will be working with are defined below.

For any right triangle *ABC* with an acute angle *A*:

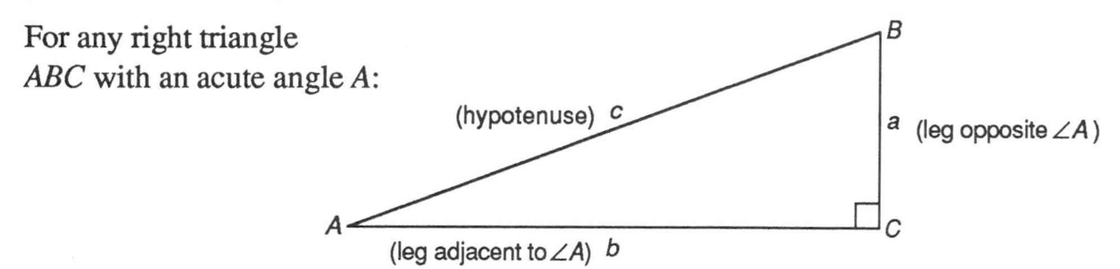

$$\text{tangent } \angle A = \frac{\text{length of leg opposite } \angle A}{\text{length of leg adjacent to } \angle A} \qquad \tan A = \frac{a}{b}$$

$$\text{sine } \angle A = \frac{\text{length of leg opposite } \angle A}{\text{length of hypotenuse}} \qquad \sin A = \frac{a}{c}$$

$$\text{cosine } \angle A = \frac{\text{length of leg adjacent to } \angle A}{\text{length of hypotenuse}} \qquad \cos A = \frac{b}{c}$$

Example

Find sin P, cos P, tan P, sin T, cos T, and tan T.

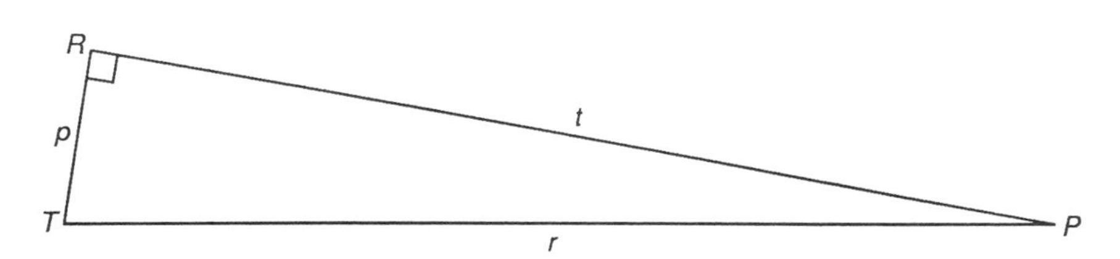

$$\sin P = \frac{\text{length of leg opposite } \angle P}{\text{length of hypotenuse}} = \frac{p}{r} \qquad \sin T = \frac{\text{length of leg opposite } \angle T}{\text{length of hypotenuse}} = \frac{t}{r}$$

$$\cos P = \frac{\text{length of leg adjacent to } \angle P}{\text{length of hypotenuse}} = \frac{t}{r} \qquad \cos T = \frac{\text{length of leg adjacent to } \angle T}{\text{length of hypotenuse}} = \frac{p}{r}$$

$$\tan P = \frac{\text{length of leg opposite } \angle P}{\text{length of leg adjacent to } \angle P} = \frac{p}{t} \qquad \tan T = \frac{\text{length of leg opposite } \angle T}{\text{length of leg adjacent to } \angle T} = \frac{t}{p}$$

EXERCISE SET

Use the definitions of the three trigonometric ratios to complete each statement.

1.* $\sin A =$ –?–
$\cos A =$ –?–
$\tan A =$ –?–

2. $\sin B =$ –?–
$\cos B =$ –?–
$\tan B =$ –?–

3. $\sin C =$ –?–
$\cos C =$ –?–
$\tan C =$ –?–

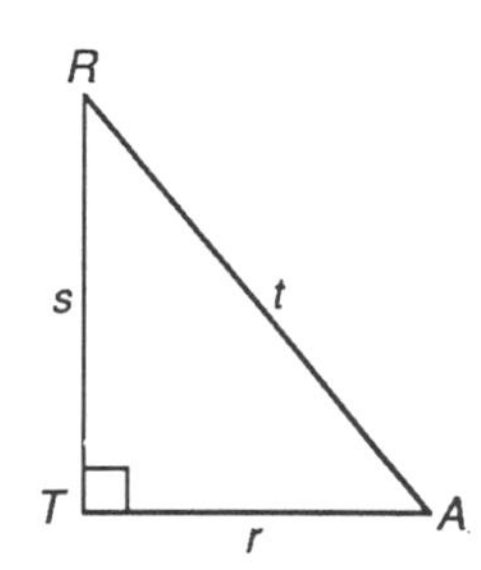

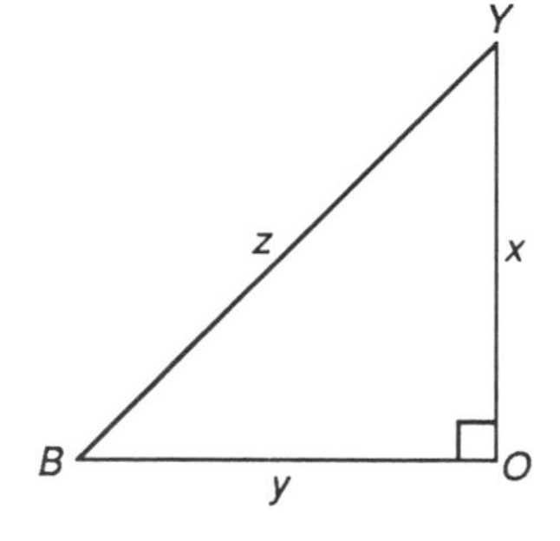

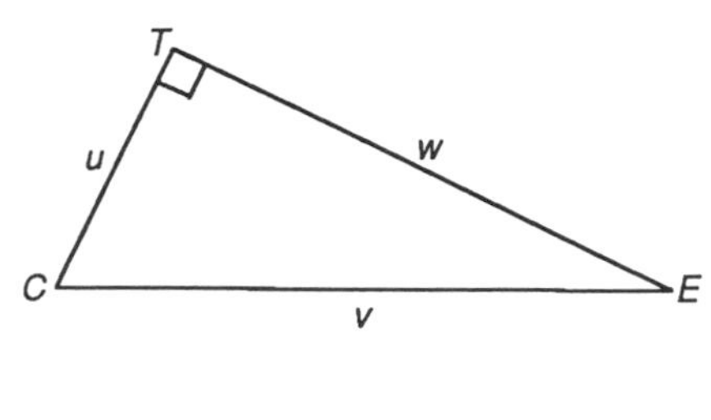

4. $\sin A =$ –?– $\sin B =$ –?–
$\cos A =$ –?– $\cos B =$ –?–
$\tan A =$ –?– $\tan B =$ –?–

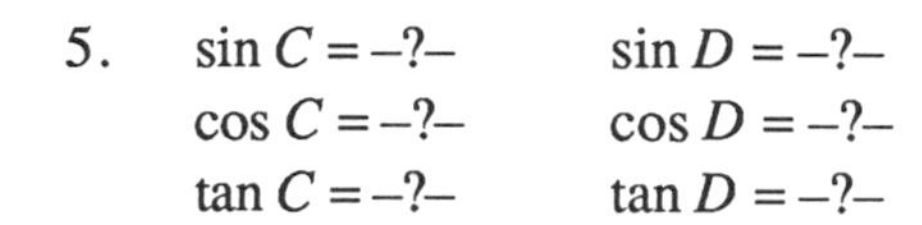

5. $\sin C =$ –?– $\sin D =$ –?–
$\cos C =$ –?– $\cos D =$ –?–
$\tan C =$ –?– $\tan D =$ –?–

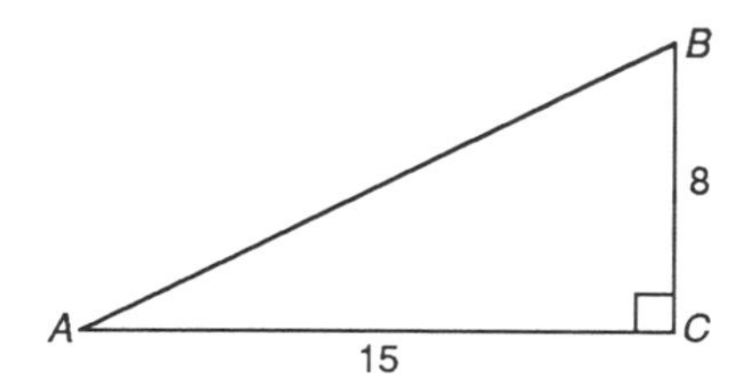

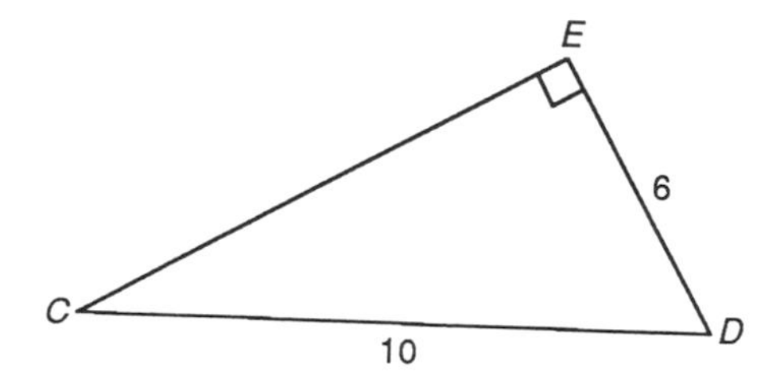

6. $\sin E =$ –?– $\sin F =$ –?–
$\cos E =$ –?– $\cos F =$ –?–
$\tan E =$ –?– $\tan F =$ –?–

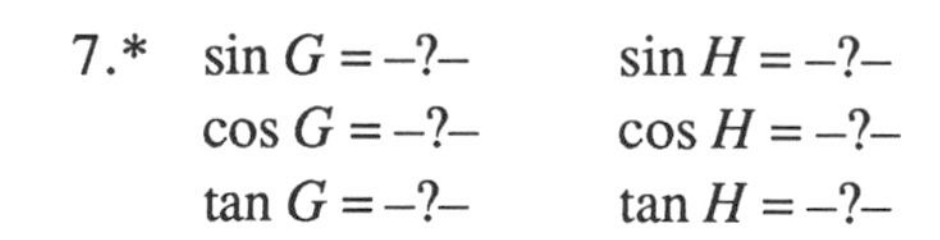

7.* $\sin G =$ –?– $\sin H =$ –?–
$\cos G =$ –?– $\cos H =$ –?–
$\tan G =$ –?– $\tan H =$ –?–

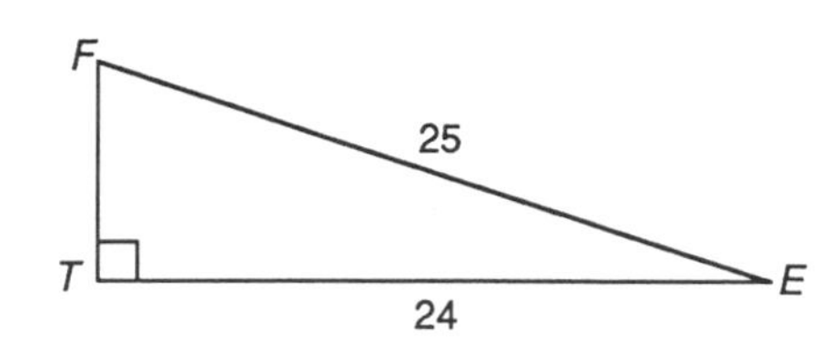

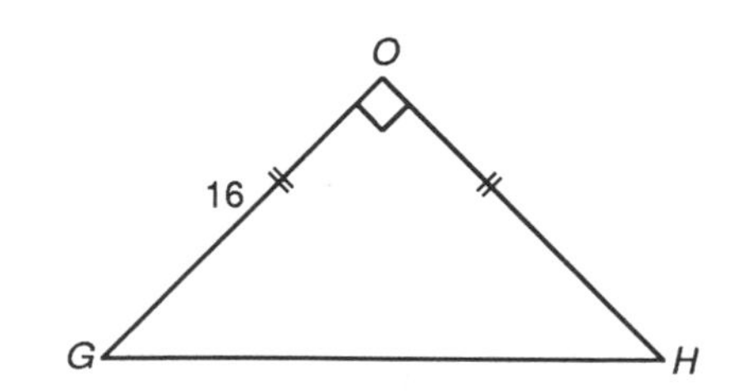

Lesson 12.2

Trigonometric Tables

Research is what I am doing when
I don't know what I'm doing.

– Werner Von Braun

With a lot of patience and plenty of calculations, early mathematicians made tables of values for the trigonometric ratios for many different sized angles. In this investigation you too will make a small trigonometric table. You will measure the lengths of sides of a number of right triangles and then calculate the sine, cosine, and tangent for angles ranging from 5° to 85° in increments of 5°.

Investigation 12.2.1

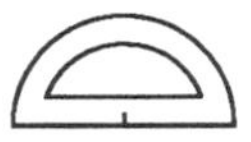

Step 1: Set up a table of values like the one shown below.

$m\angle A$	sin A	cos A	tan A	$m\angle C$	sin C	cos C	tan C
5	–?–	–?–	–?–	85	–?–	–?–	–?–
10	–?–	–?–	–?–	80	–?–	–?–	–?–
15	–?–	–?–	–?–	75	–?–	–?–	–?–
20	–?–	–?–	–?–	70	–?–	–?–	–?–
25	–?–	–?–	–?–	65	–?–	–?–	–?–
30	–?–	–?–	–?–	60	–?–	–?–	–?–
35	–?–	–?–	–?–	55	–?–	–?–	–?–
40	–?–	–?–	–?–	50	–?–	–?–	–?–
45	–?–	–?–	–?–	45	–?–	–?–	–?–

Step 2: On a sheet of graph paper or scratch paper, use your protractor to make a right ΔABC as large as possible with $m\angle B = 90$, $m\angle A = 5$, and $m\angle C = 85$.

Step 3: Measure sides $\overline{AB}$, $\overline{AC}$, and $\overline{BC}$ with your ruler to the nearest millimeter.

Step 4: Use a calculator to calculate the ratios BC/AB (tan A), AB/BC (tan C), BC/AC (sin A and cos C), AB/AC (cos A and sin C).

Step 5: Compare your results with the results of others in your group. Calculate the average of each ratio found by the members of your group. Enter the average value for each ratio in your table accurate to two places. Now you have the values for the ratios of sides for triangles with angles of 5° and 85°.

Step 6: Repeat steps 2 to 5 with the rest of the angle measures in the table.

What is the value of a table of trigonometric values? In Chapter 11, you used similar triangles to calculate the heights of objects that you were unable to measure directly. To do this you measured three lengths. Applying trigonometry, it is also possible to determine heights indirectly — but by using only two measurements. Once you have a table of trigonometric values, you can find the approximate lengths of sides of right triangles given the measures of any acute angle and any side.

Example A

Find the height of a right triangle to the nearest meter if one acute angle measures 35° and the adjacent side measures 24 m.

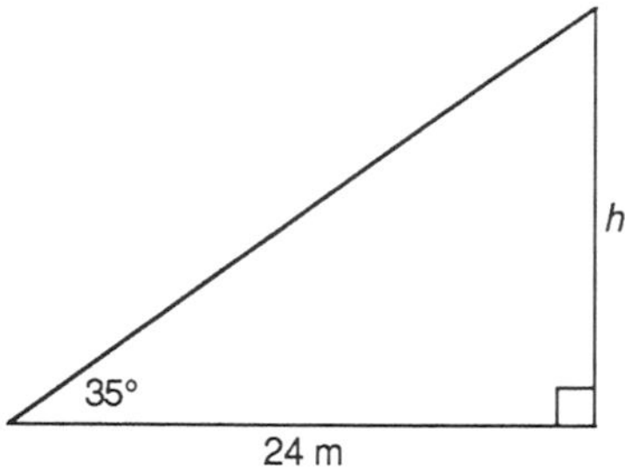

The relationship that connects the length of the side opposite the 35° angle (h) and the length of the side adjacent to the 35° angle (24) is the tangent relationship. Therefore:

h
35°
24 m

$$\frac{h}{24} = \tan 35°$$

$$h = (24)(\tan 35°)$$

From your table you know that $\tan 35° \approx .70$.

$$h \approx (24)(.70)$$

$$\approx 16.8$$

The height is approximately 17 m.

Example B

Find the length of the hypotenuse of a right triangle to the nearest foot if one of the acute angles measures 20 and the opposite side measures 410 feet.

The relationship that connects the lengths of the opposite side (410) and the hypotenuse (x) is the sine relationship. Therefore:

x
410'
20°

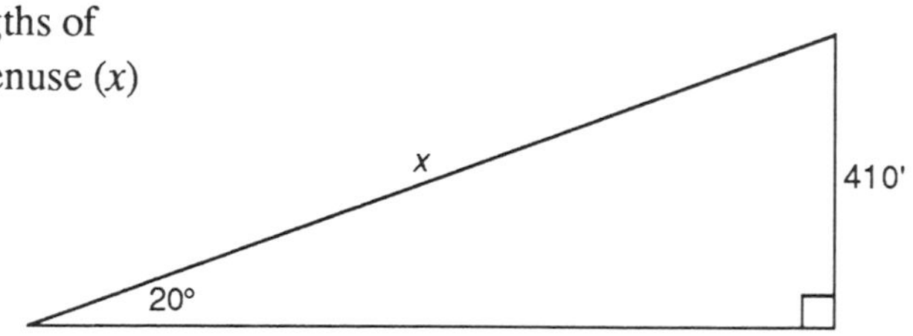

$$\sin 20° = \frac{410}{x}$$

$$x(\sin 20°) = 410$$

$$x = \frac{410}{\sin 20°}$$

From the table you know that $\sin 20° \approx .342$.

$$x \approx \frac{410}{.342}$$

$$x \approx 1198.8$$

The length of the hypotenuse is about 1199 feet.

EXERCISE SET A

Use your table of trigonometric values to approximate each length. Express each answer accurate to the nearest whole unit.

1.* $a \approx$ –?–

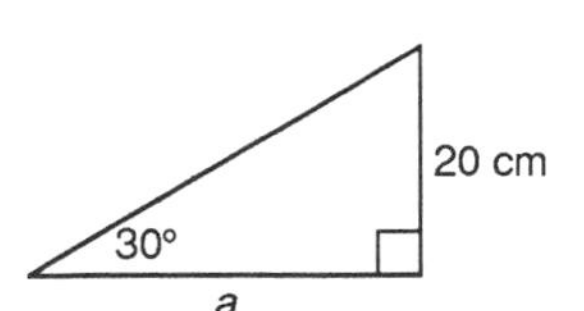

2. $b \approx$ –?–

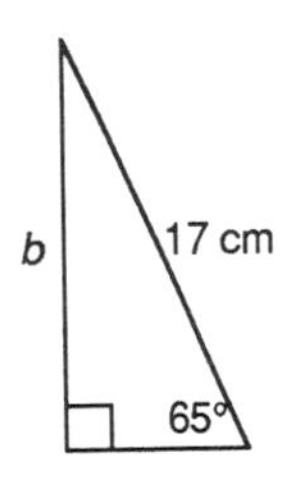

3. $c \approx$ –?–

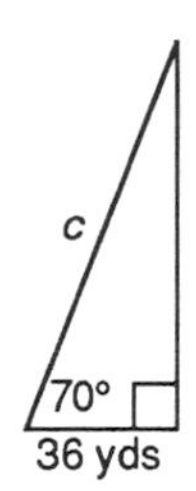

4. $d \approx$ –?–

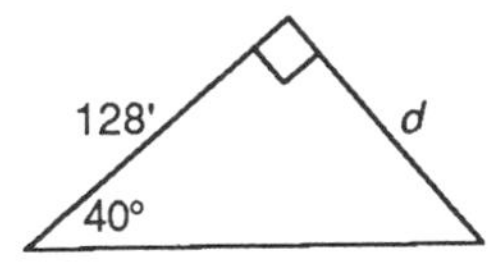

5. $e \approx$ –?–

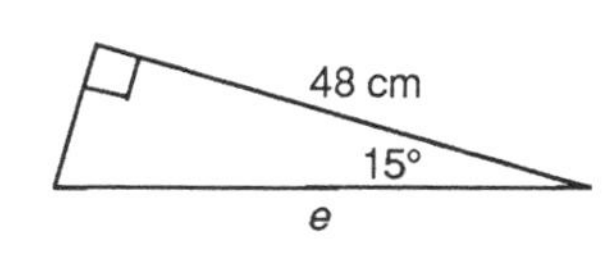

6. $f \approx$ –?–

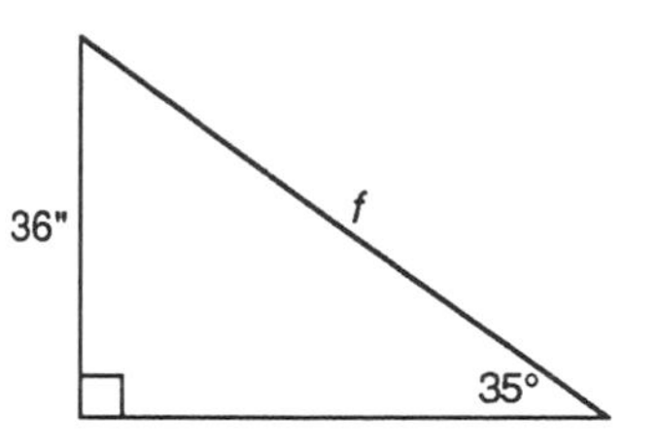

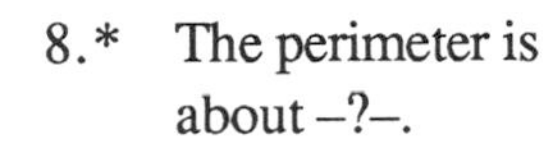

7. The radius is 12 inches. $g \approx$ –?–

8.* The perimeter is about –?–.

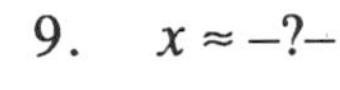

9. $x \approx$ –?–

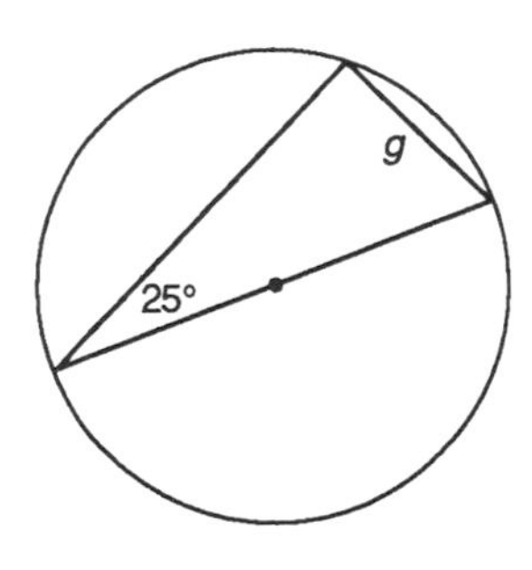

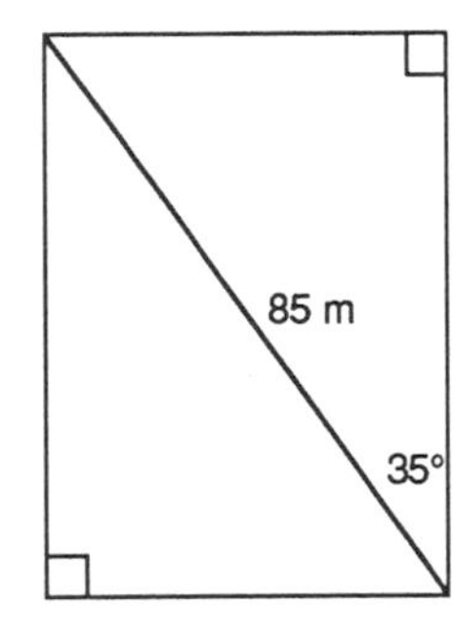

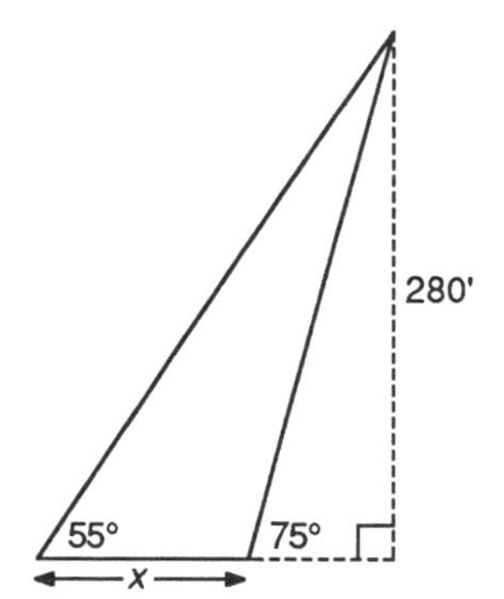

EXERCISE SET B

1.* How does the sine of angle A change as the measure of A goes from 5° to 85°?

2. As the angles get closer and closer to 0°, the values of the sine ratio get closer and closer to what value?

3. As the angles get closer and closer to 90°, the values of the sine ratio get closer and closer to what value?

4. How does the cosine of angle A change as the measure of A goes from 5° to 85°?

5. As the angles get closer and closer to 0°, the values of the cosine ratio get closer and closer to what value?

6. As the angles get closer and closer to 90°, the values of the cosine ratio get closer and closer to what value?

7. What is the measure of acute angle A if the sine of A equals the cosine of A?

8. How does the tangent of angle A change as the measure of A goes from 5° to 85°?

9. As the angles get closer and closer to 0°, the values of the tangent ratio get closer and closer to what value?

10. As the angles get closer and closer to 90°, the values of the tangent ratio get closer and closer to what value?

11. Angles A and C are the two acute angles of a right triangle. What is the measure of angle A if the sine of A equals the sine of C?

12. Angles A and C are the two acute angles of a right triangle. What is the measure of angle A if the sine of A equals the cosine of C?

Improving Visual Thinking Skills

Alpha-Ominos

The alphabet can be divided into three physical types: "ups," "downs," and "middle roaders." Ups include: *b*, *d*, *f*, *h*, and *k*. Downs include: *g*, *j*, and *p*. Middle roaders include: *c*, *i*, *m*, *n*, *r*, *s*, *v*, and *w*.

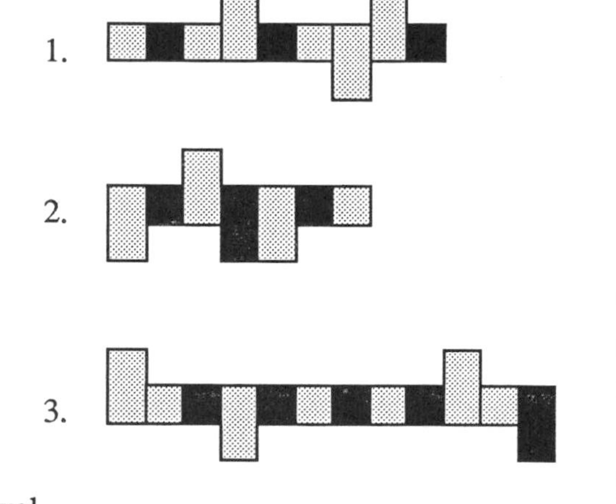

In the puzzle at right, letters in familiar geometric terms have been replaced by rectangles (for ups and downs) and squares (for middle roaders). Consonants have been shaded lightly and vowels are solid black. For the purpose of this puzzle, *y* is always a vowel and all letters are in lowercase. Use visual thinking to determine what familiar geometric terms are represented by each pattern of rectangles and squares.

Lesson 12.3

Trigonometry with Calculators

The table of trigonometric ratios you created in the last lesson was a small one. Tables of the trigonometric ratios have been used for centuries. Some occupy entire volumes. Today trigonometric tables have been replaced by scientific calculators with sin, cos, and tan buttons. You can find a trigonometric ratio faster, easier, and more accurately with a calculator. If you can, use a scientific calculator rather than a trigonometric table for the lessons in the rest of this chapter. If you do not have a scientific calculator, you need to locate a table of trigonometric values.

Example

Find sin 43°.

Using a Calculator

Step 1: Enter 43 (the number of degrees).

Step 2: Press the sin button.

Step 3: Read the display.
It gives sin 43° ≈ .68199836.

Using a Table

Step 1: Locate 43° in the angle measure column.

Step 2: Go across the 43° row to the decimal value in the sine column.

Step 3: Read the value in the sine column and 43° row. It gives sin 43° ≈ .6820.

Table of Trigonometric Ratios

Angle	Sine	Cosine	Tangent
41°	.6561	.7547	.8693
42°	.6691	.7431	.9004
43°	.6820	.7314	.9325
44°	.6947	.7193	.9657
45°	.7071	.7071	1.0000

EXERCISE SET A

Use a trigonometric table or scientific calculator to determine the values accurate to four decimal places.

1.* sin 37° ≈ –?–

2. cos 29° ≈ –?–

3. tan 8° ≈ –?–

4. sin 67° ≈ –?–

5. tan 79° ≈ –?–

6. cos 45° ≈ –?–

With the help of a scientific calculator or trigonometric table, it is possible to determine the size of either acute angle in a right triangle if you know the length of any two sides of the right triangle. Let's look at an example.

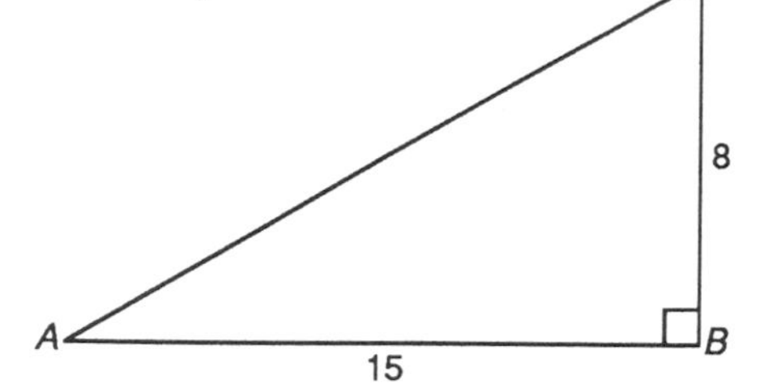

Example

Right ΔABC has legs of length 8 and 15 inches. What is the $m\angle A$, the angle opposite the side of length 8 inches and adjacent to the side of length 15 inches?

Using a Calculator

Step 1: Find the trigonometric relationship that ties together the lengths of the two sides and the angle. The relationship is $\tan A = 8/15$.

Step 2: Calculate the decimal approximation of 8/15. (You should get .53333333 on your display.) Therefore, $\tan A \approx .53333333$.

Step 3: Since you wish to find the angle such that the tangent of that angle has a value of .53333333, you will, in effect, reverse the steps you performed in the previous exercise. Press the inverse button (on some calculators it's labeled as the 2nd function button), and then the tan button. This gives you the measure of $\angle A$ such that $\tan A \approx .53333333$. 28.072487 should appear on your display. Round this value off to the nearest degree. You should find that the measure of $\angle A$ is about 28°.

Using a Table

Step 1: Find the trigonometric relationship that ties together the lengths of the two sides and the angle. The relationship is $\tan A = 8/15$.

Step 2: Calculate the decimal approximation of 8/15 ($8/15 \approx .5333$). Therefore, $\tan A \approx .5333$.

Step 3: Since you wish to find the angle such that the tangent of that angle has a value of .5333, you will, in effect, reverse the steps you performed in the previous exercise. Look up and down the tangent column on your table for the decimal value closest to .5333, and then follow that row over to the angle measure column to find the angle. You should find that $\angle A$ is about 28°.

EXERCISE SET B

Find the measure of each angle to the nearest degree.

1.* $\sin A = .5$
2. $\cos B = .6$
3. $\tan C = .5773$
4. $\sin D = .7071$
5. $\tan E = 1.7321$
6. $\cos F = .5$

If you have a tool for measuring angles and a scientific calculator, you can use trigonometry to determine heights indirectly. There are a variety of devices that can be made for measuring an angle (see the special project following this lesson). In each problem below, an angle measuring device is used to determine a distance indirectly.

7.* Igor's pet bat Natasha is flying at the end of a 50-foot leash. Using an angle measuring device, Igor spots Natasha at an angle of 55 degrees up from the horizontal. To the nearest foot, how high is Natasha flying if the leash is taut and anchored to the ground?

8. Archaeologist Ertha Diggs is using an angle measuring device to determine the height of an ancient temple. When she views the top of the remains of the ancient temple through her angle measuring device, she records an angle of 37 degrees up from the horizontal. She is standing 130 meters from the center of the temple's base and her eye is 1.5 meters above the ground. How tall is the temple to the nearest tenth of a meter?

9.* Meteorologist Wendy Storm is using an angle measuring device to determine the height of a weather balloon. When she views the weather balloon through her measuring device (which is sighted one meter above the level ground), she measures an angle on the device of 44 degrees up from the horizontal. The radio signal from the balloon tells her that the balloon is 1400 meters from her measuring device. To the nearest meter, how high is the balloon? To the nearest meter, how far is it to a position directly beneath the balloon?

Special Project

Indirect Measurement

If I have seen farther than others it is because I have stood on the shoulders of giants.

– Isaac Newton

In this special project you will use trigonometric methods (using a scientific calculator and an angle measuring device) to determine the heights of two different sized objects that you are unable to measure directly (school building, football goal post, flagpole, tall tree, etc.). This project works best in groups of four or five. Divide up the tasks. Some of the tasks include being responsible for taking measurements, recording the measurements, performing the calculations, keeping track of the equipment, and being the person measured. You will need the following equipment:

- Measuring tape or meter sticks
- Notebook for recording your measurements
- Calculator or trigonometric tables
- Device for measuring angles (Three sample angle measuring devices are described below. Build one of them or make one of your own design.)

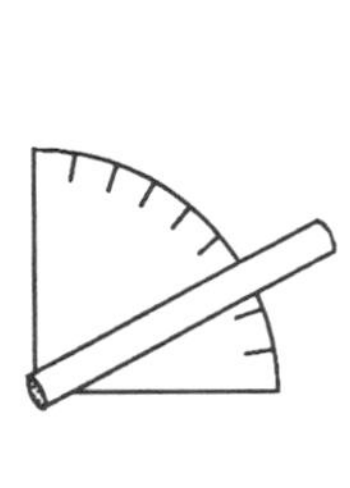

Device 1

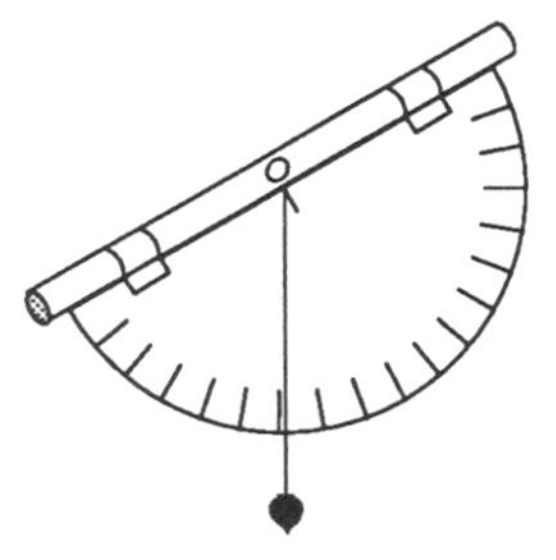

Device 2

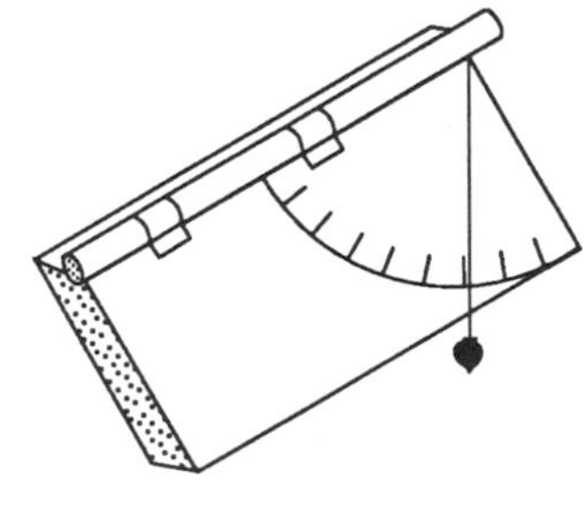

Device 3

The simplest of the devices for measuring angles uses half a large protractor (like half your teacher's demonstration chalkboard protractor) with a viewing tube attached. The viewing tube (it could be a large straw) is attached to the vertex point (O) of the protractor in such a way that it can rotate at this point. You sight the top of the object through the rotating viewing tube and measure the angle at the point (C) where the center of the tube crosses the markings on the protractor. As long as the "zero-edge" ($\overline{OB}$) of the protractor is kept horizontal, the angle formed by the horizontal edge and the viewing straw ($\angle BOC$) is the desired angle.

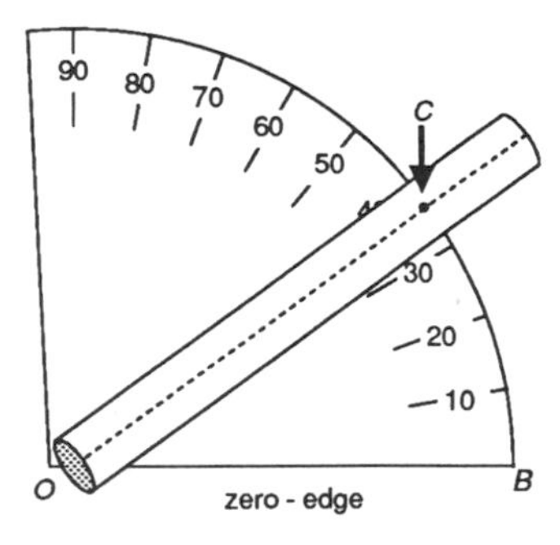

The second angle measuring device uses a similar large protractor with viewing tube. However, in this device the tube is attached to the "zero-edge" ($\overline{AB}$) and a plumb line is attached to the vertex point (O). Hold the device in such a way that when the top of the object is sighted through the viewing tube, the plumb line crosses the angle measurements on the protractor (C) forming an acute angle ($\angle AOC$). This angle is the complement of the desired angle.

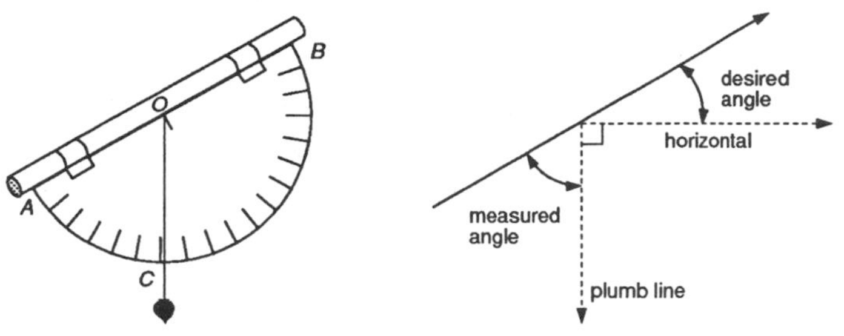

The third angle measuring device is very clever. This device is sometimes used by the U.S. Forestry Service to measure the heights of trees. Hold a rectangular solid (block of wood or cardboard) with a rectangular face labeled $ABCD$ in a position so that the top of the object is sighted along the top edge ($\overline{AB}$). Suspend a plumb line from the top corner (B). The angle between the plumb line and the end edge ($\overline{BC}$) is equal to the desired angle.

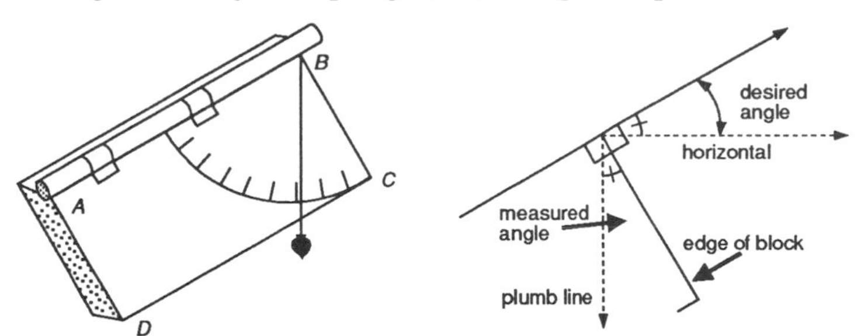

Before you begin, make a table of your measurements similar to the table below.

Table for Measurements by Trigonometry Method

Name of object to be measured	Angle of elevation	Height of observer's eye	Distance from observer to object	Calculated height of object
1. —?—	–?–	–?–	–?–	–?–
2. —?—	–?–	–?–	–?–	–?–

Follow the procedure below to determine indirectly the heights of the two objects.

Step 1: Begin by measuring one person's eye height (we'll call this person the observer). Record the measurement in your table.

Step 2: Locate two tall objects with heights that would be difficult to measure directly (school building, football goal post, flagpole, tall tree, etc.).

Step 3: With your angle measuring device measure the observer's viewing angle from the horizontal to the top of each object to be measured. Measure the distance from the observer to the base of each object. Enter the measurements into the appropriate table.

Step 4: Perform all the calculations to approximate the heights of the two objects.

Lesson 12.4

Problem Solving with Right Triangles

Many problems in right triangle trigonometry involve finding the height of a tall object you are unable to measure directly. To solve a problem of this type you might need to measure an angle from the horizontal to the top of the object you are trying to measure

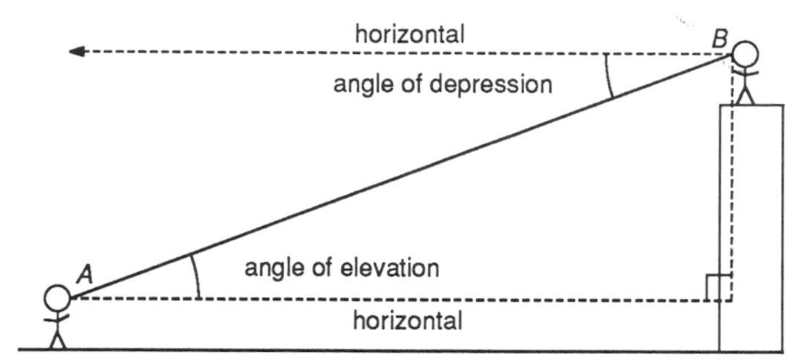

If a person is looking up, then the angle from the horizontal to the line of sight is called the **angle of elevation.** If a person is looking down, then the angle from the horizontal down to the line of sight is called the **angle of depression.** Notice that the angle of elevation from viewer A looking up to point B is the same as the angle of depression for a viewer at point B looking down at point A (AIA Conjecture!). Let's look at a sample problem with a solution that uses the angle of elevation to find a distance indirectly.

Example

The angle of elevation from a sailboat to the top of a 121 foot lighthouse on the shore is 16°. To the nearest foot, how far is the sailboat from shore ?

$$\tan 16^\circ = \frac{121}{d}$$

$$d(\tan 16^\circ) = 121$$

$$d = \frac{121}{\tan 16^\circ}$$

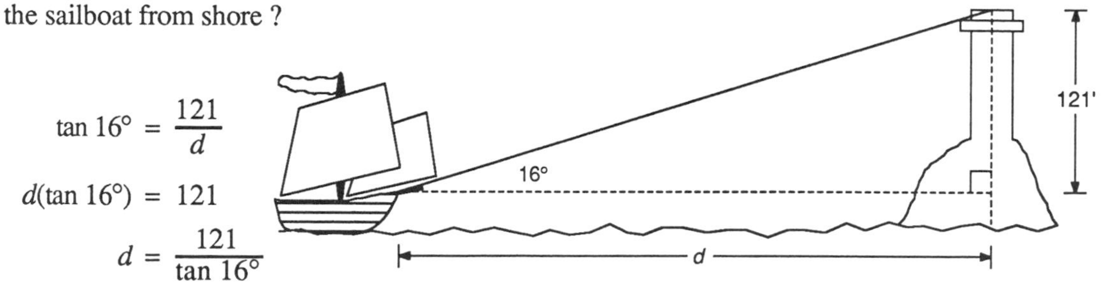

To find d, push the following sequence of buttons on your calculator: 121 [÷] 16 [tan] [=]. Your display should show 421.97715.

The sailboat is approximately 422 feet from shore.

EXERCISE SET

Calculate each distance or angle.

1.* The angle of elevation from a ship to the top of a 42-meter lighthouse on the shore is 33 degrees. How far is the ship from the shore to the nearest meter?

2. A lighthouse 55 meters above sea level spots a distress signal from a sailboat. The angle of depression to the sailboat is 21 degrees. How far away is the sailboat from the base of the lighthouse to the nearest meter?

3. Ben is flying a kite with 125 meters of kite string out. His kite string makes an angle of 39 degrees with the level ground. How high is his kite to the nearest meter?

4. Chip Woodman, the foreman at the paper plant, must make an estimate of the volume of a conical wood chip pile. The distance from the tip of the cone down to the edge of the base (the slant height) is 304 feet and forms an angle of 54 degrees with the ground. What is the height of the cone to the nearest foot? What is the area of the base of the cone to the nearest thousand square feet? What is the volume of the cone to the nearest hundred thousand cubic feet? Use 3.14 for π.

5.* Ertha Diggs has uncovered the remains of a square-based Egyptian pyramid. The base is intact and measures 130 meters on a side. The top portion of the pyramid eroded away over the centuries but what remains of each face of the pyramid forms an angle of 65 degrees with the ground (angle of ascent). What was the original height of the pyramid to the nearest meter?

6. A lighthouse is observed by a ship's officer on watch at an angle of 42° to the path of the ship. At the next sighting the lighthouse is observed at an angle of 90° to the path of the ship. The distance traveled between sightings is 1800 m. To the nearest meter, how far away is the ship to the lighthouse at this second sighting?

7.* It is believed that Galileo used the "Leaning Tower of Pisa" to conduct his experiments on the laws of gravity. When he dropped objects from the top of the 55 meter tower (measured length of tower, not height), they landed 4.8 meters from the base of the tower. To the nearest degree, what is the angle ϕ that the tower leans off from the vertical?

8. One of the most impressive of the Mayan pyramids is El Castillo in the Yucatan of Mexico. The pyramid has a platform on the top and a flight of 91 steps on each of the four sides. (Four flights of 91 steps make 364 steps in all. With the top platform adding a level, there are 365 levels to represent the 365 days of the Mayan year.) What is the height to the nearest centimeter of the top platform if each of the 91 steps is 30 cm deep by 26 cm high? What is the angle of ascent to the nearest degree?

The Flight of the Albatross

9.* Sidney Yendis III is the son of NASA scientist and billionaire eccentric Sidney Yendis II. Eager to follow in his father's footsteps, young Sidney launched his sister's pet tribble into space with his homemade rocket, *Albatross*. He was attempting to set a world record for tribble flight. The record was 412 feet. He selected the backyard football field as his launch site. The launch was perfect. The *Albatross* soared straight up. Sidney's father observed the launch through his telescope from the comfort of the announcers' booth 30 feet above the playing field. He measured the angle of depression to the launch-pad and found it to be 21 degrees. He measured the angle of elevation at the moment that the *Albatross* reached its maximum altitude and found it to be 78 degrees. Did his sister's pet tribble set a record? What was the height reached by the *Albatross*?

Computer Activity

Fractals with Logo

Suppose you were to photograph a coastline on the North American continent from the vantage point of outer space. How would it look? Perhaps you've seen photos of the earth taken from space craft. The coast appears as a winding boundary where land meets water. Some portions of the coast seem smooth and straight. Other stretches appear jagged and highly irregular. How do you think the appearance of the coastline would change if you were to photograph it periodically as you descended from space? Certainly, the closer you came to the continent, the more detail you would see in the pictures. Portions of the coast that at first seemed smooth and straight might begin to look less and less regular. Bays and peninsulas too small to see from space might appear along stretches of coast that had seemed straight from a higher viewpoint. The coast might appear "self-similar." That is, a small stretch viewed from close-in might resemble a larger piece viewed from farther out.

In recent years, mathematicians have become fascinated with a broad class of "self-similar" objects. A French mathematician, Benoit Mandelbrot, has named one group of these objects "fractals." A coastline is shaped like a fractal. It is a highly irregular fractal. Other fractals reveal more pattern. Fractal geometry is a new and exciting field of mathematical inquiry and research. Fractal designs, generated by high-powered supercomputers, reveal tantalizing intricacy and beauty. Here three views of the Mandelbrot Set, a fractal named after Mandelbrot. The second and third views are "close-ups" of a portion of the fractal from the previous view. Can you see why fractals are called "self-similar"?

The Mandelbrot Set
Photos courtesy of Art Matrix

1.

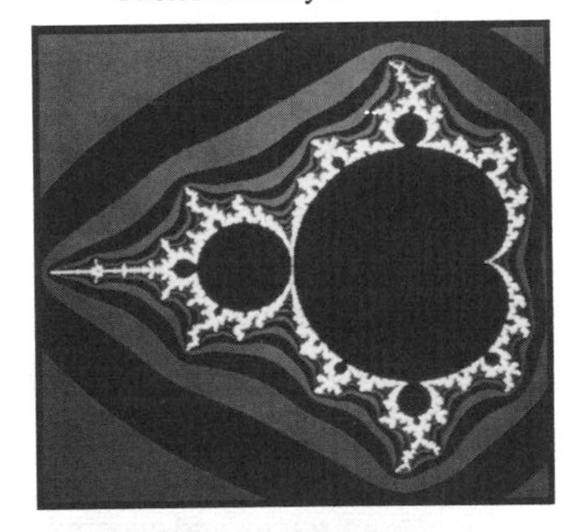

2.

3.

In this text you have seen how geometry can be used to describe many real objects that surround us. Yet when you look closely at natural objects, you do not see objects composed of perfect squares, perfect circles, or even perfect lines. Strange as it may seem, mathematicians hope that the study of fractals will yield a new language that can more accurately describe the shapes of clouds, mountain ranges, and coastlines. Fractal geometry provides a way to solve problems involving systemic irregularity and similarity under changes of scale. Fractals might give order to chaos.

In this computer activity you will use Logo to examine some simple fractal curves — fractals that resemble snowflakes. These snowflakes were first produced and investigated by a Swedish mathematician Helge von Koch in 1904 and are widely known as Koch curves. Mandelbrot suggests that Koch curves are a suggestive beginning model of a coastline. For this activity you will need:

- Computer and disk with Logo computer language
- Resource person familiar enough with Logo to help you get started
- Disk with the file of procedures called FRACTALS
  (The listings of the procedures are in *Appendix B: Logo Procedures*.)

Start up your computer using the disk with Logo on it. Load the file called FRACTALS from your procedures disk. The FRACTALS file includes two procedures: SNOWFLAKE and DIV.SIDE. They work together to produce snowflake fractals.

1. SNOWFLAKE requires two inputs: the level of the fractal (:L) and the length of a side of the base polygon (:S). Try SNOWFLAKE with inputs of 1 and 100. Without clearing the screen, try SNOWFLAKE 2 100. Continue with snowflakes of levels up to 8. Because the DIV.SIDE procedure is recursive (it uses itself as a subprocedure), the time the turtle takes to draw a snowflake fractal increases drastically (geometrically) with an increase in the level of the fractal. You may have to wait a long time for the turtle if you choose a level that is even moderately large.

2. Clear your screen and experiment with SNOWFLAKE using side-length inputs greater than 100. By increasing the side-length you can increase the level of the fractal and still see what is going on. This permits you to see greater detail, however, your entire snowflake will not necessarily fit on the screen in one piece. If your version of Logo has a WINDOW command (or an equivalent), you may want to use it to turn the screen wrap off. Make a snowflake fractal with level 8. Can you see how the snowflake is "self-similar"?

The snowflake fractal produced by the SNOWFLAKE procedure has a very curious property. As the snowflake grows, its area increases, but it never increases beyond a certain size. The additional growth caused by drawing the snowflake to higher levels adds a negligible amount to the total area. Stated in mathematical terms, as the level of the snowflake increases, the area of the snowflake approaches a fixed limit. The perimeter of the snowflake behaves quite differently. It also increases as the level of the snowflake increases. The perimeter, however, unlike the area, grows without limit. Stated in mathematical terms, its limit is infinite.

You can see the difference in the behavior of the area and perimeter if you watch the screen as the turtle draws snowflakes of higher and higher levels. You can see that the area of the snowflake never exceeds the area of the hexagon formed by connecting the tips of the level 2 snowflake. The area of the hexagon is an upper bound or limit on the area. (It's not the least upper bound, however.) The perimeter of the snowflake can not be easily measured. However, the length of time that it takes the turtle to work its way around each level of snowflake is a rough approximation of the perimeter. (The turtle proceeds forward at a roughly even pace.) As you can easily see, it takes the turtle longer and longer to successive levels. If you choose a level sufficiently large, you can wait as long as you want.

The sharp contrast between the behavior of the area and perimeter is one of the interesting aspects of snowflake fractals.

If you are familiar enough with Logo to write a Logo procedure, you can create procedures which will create other Koch curves. Mandelbrot calls these curves "Koch islands." Here's a Logo programming challenge.

3. Write a new procedure called KOCHISLAND which starts with a square (level 1). Grow your Koch island by replacing each unit segment in the square with a zig-zag path composed of eight smaller units each one-fourth as long as a side of the original square (level 2). The shape of the zig-zag path is shown on the right. To form each successive level, replace each of the eight units in the zig-zag path of the previous level with a similar, smaller zig-zag path. Koch islands of levels 1, 2, and 3 are pictured below. Write KOCHISLAND so that it uses recursion and accepts two inputs: the level of the fractal (:L) and the length of a side of the base square (:S).

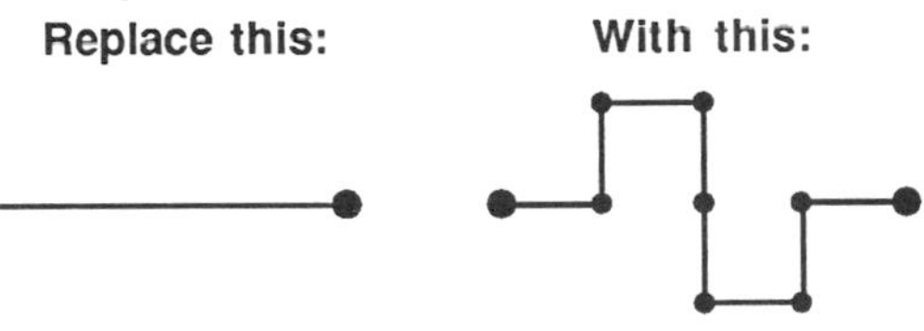

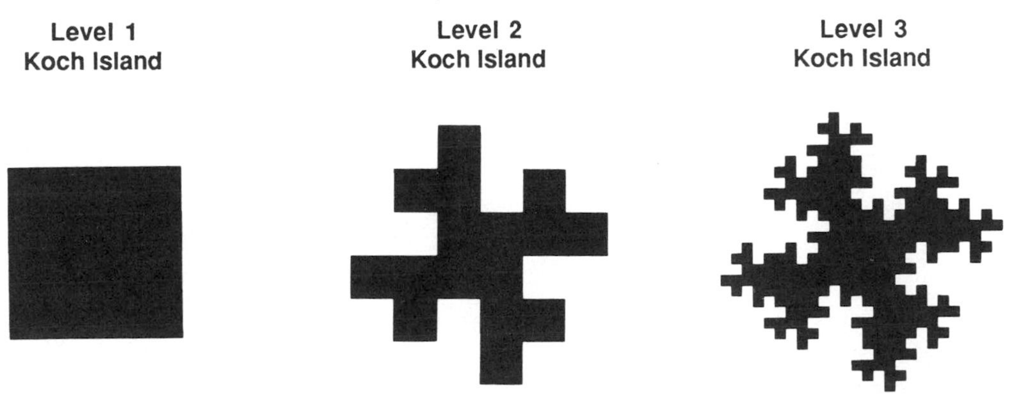

Lesson 12.5

The Law of Sines

Rules are for the obedience of fools and the guidance of wise men.

– David Ogilvy

So far you have only used trigonometry to solve problems involving right triangles. Trigonometry can also be used with triangles other than right triangles. For example, if you know the measures of two angles and one side of a triangle, (either ASA or SAA) you can find the other two sides with the help of a trigonometric property called the Law of Sines. The Law of Sines for acute triangles can be discovered while using trigonometry to find areas of acute triangles. Let's look at an example.

Example

Find the area of ΔABC if $AB = 150$ meters, $BC = 100$ meters, and $m\angle B = 40$.

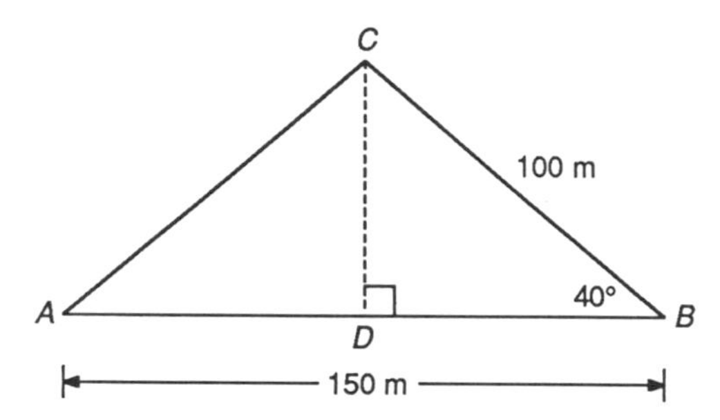

$$\sin 40° = \frac{CD}{100}$$

$$CD = (100)(\sin 40°)$$

$$\text{Area}_{\Delta ABC} \approx (.5)(150)(CD)$$

$$\text{Area}_{\Delta ABC} \approx (.5)(150)[(100)(\sin 40°)]$$

To find the area, push the following sequence of buttons on your calculator:
.5 [x] 150 [x] 100 [x] 40 [sin] [=] . Your display should show 4820.9071.

The area is about 4821 m^2.

EXERCISE SET A

Find the area of each triangle.

1.

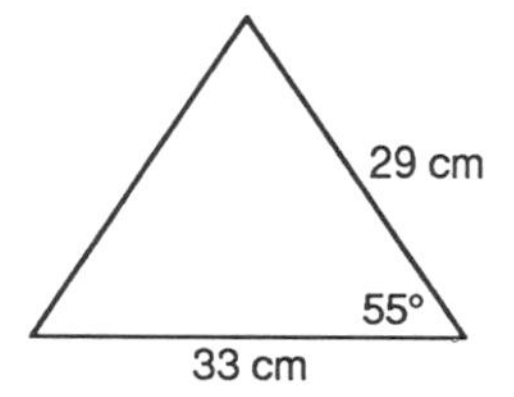

2.

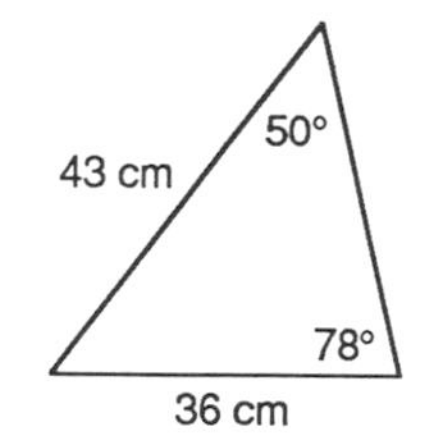

3.

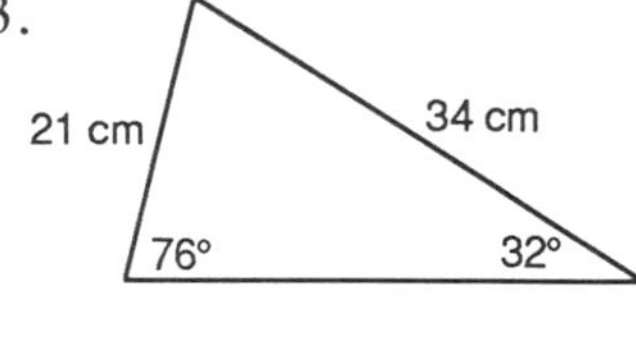

Use the diagram on the right for Exercises 4 to 9.

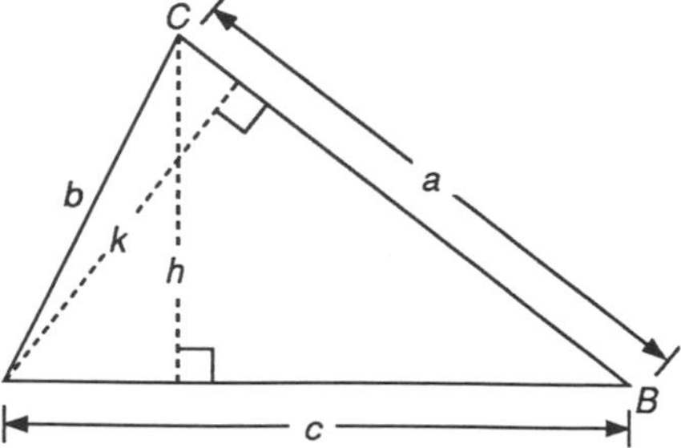

4.* Find altitude h in terms of a and the sine of an angle.

5. Find altitude h in terms of b and the sine of an angle.

6.* Show $\dfrac{\sin A}{a} = \dfrac{\sin B}{b}$

7.* Find altitude k in terms of c and the sine of an angle.

8. Find altitude k in terms of b and the sine of an angle.

9. Show $\frac{\sin B}{b} = \frac{\sin C}{c}$.

In Exercises A6 and A9 you showed that $\frac{\sin A}{a} = \frac{\sin B}{b}$ and $\frac{\sin B}{b} = \frac{\sin C}{c}$.

These two statements can be combined into one called the Law of Sines.

> *The **Law of Sines** says that for any triangle with angles of measures A, B, and C, and sides of lengths a, b, and c (a opposite A, b opposite B, and c opposite C),* $\frac{\sin A}{a} = \frac{\sin B}{b} = \frac{\sin C}{c}$.

This is a very useful equation. It says that the sine of an angle in a triangle divided by the length of the side opposite that angle will equal the sine of another angle in the triangle divided by the length of the side opposite that angle. If you know the right combination of parts (ASA or SAA), you can use the Law of Sines to find lengths of sides or measures of angles of triangles that are not right triangles.

Example

Find the length b of side $\overline{AC}$ in acute ΔABC with $BC = 350$ cm, $m\angle A = 59°$, $m\angle B = 38°$.

Start with the Law of Sines.

$$\frac{\sin A}{a} = \frac{\sin B}{b}$$

Now solve for b.

$$b \sin A = a \sin B$$

$$b = \frac{a \sin B}{\sin A}$$

$$b = \frac{(350)(\sin 38°)}{\sin 59°}$$

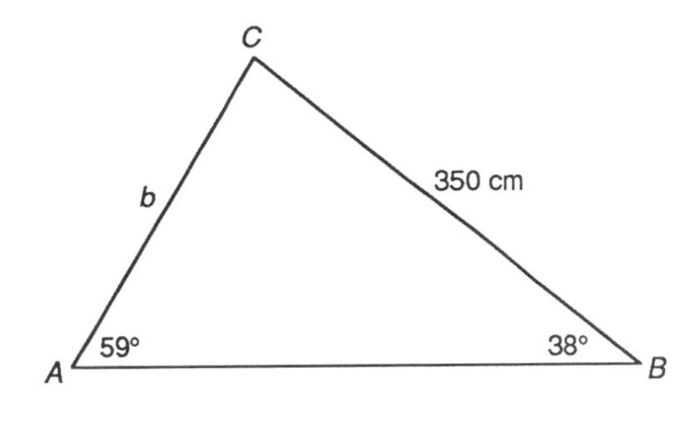

To find b, push the following sequence of buttons on your calculator:
350 [x] 38 [sin] [÷] 59 [sin] [=]. Your display should show 251.38793.

The length b of side $\overline{AC}$ is approximately 251 cm.

EXERCISE SET B

Find the length of each side indicated to the nearest centimeter.

1.* $w = -?-$ 2. $x = -?-$ 3. $y = -?-$ 4. $z = -?-$

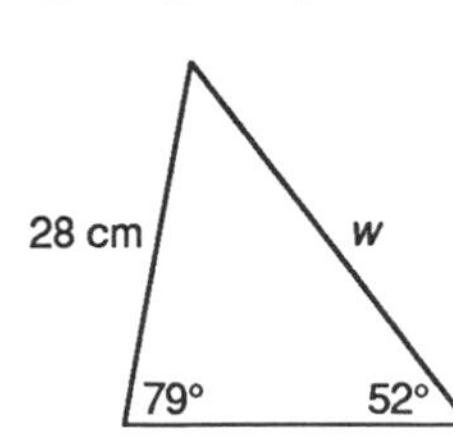

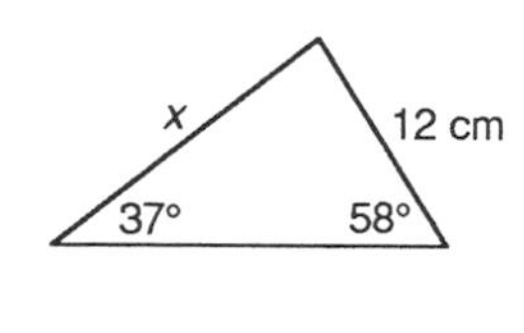

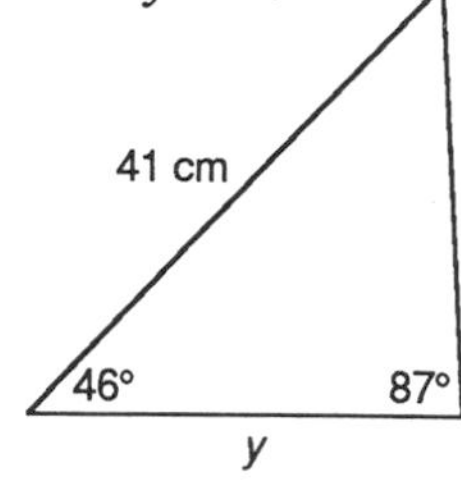

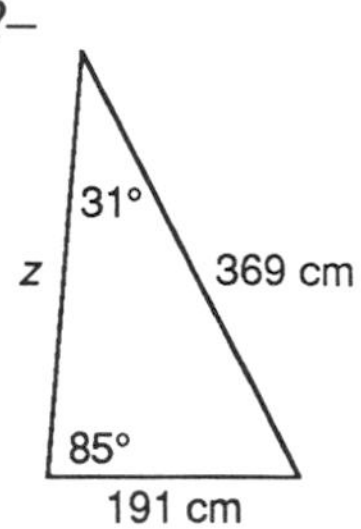

The Law of Sines can also be used to find the measure of an angle.

Example

Find the measure of acute angle B in ΔABC
with $BC = 250$ cm, $AC = 150$ cm, $m\angle A = 69°$.

Start with the Law of Sines.

$$\frac{\sin A}{a} = \frac{\sin B}{b}$$

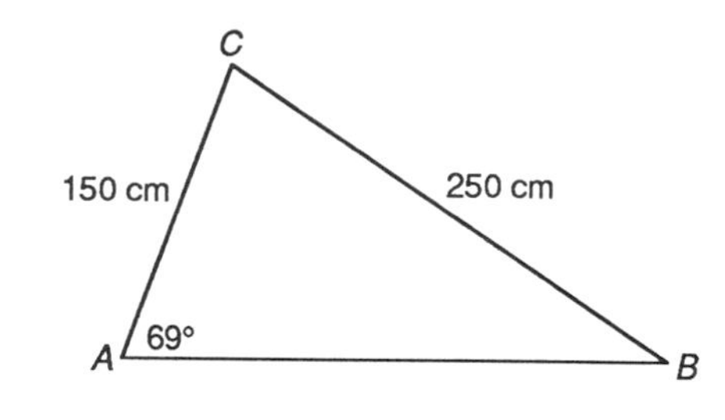

Now solve for sin B.

$$a \sin B = b \sin A$$

$$\sin B = \frac{b \sin A}{a}$$

$$\sin B = \frac{(150)(\sin 69°)}{250}$$

To find the measure of angle B, push the following sequence of buttons on your calculator:
150 [x] 69 [sin] [÷] 250 [=] [inv] [sin] . Your display should show 34.066051.

The measure of angle B is approximately 34°.

EXERCISE SET C

In Exercises 1 to 3, find the measure of each angle indicated to the nearest degree.

1.* $m\angle A = -?-$ 2. $m\angle B = -?-$ 3. $m\angle C = -?-$

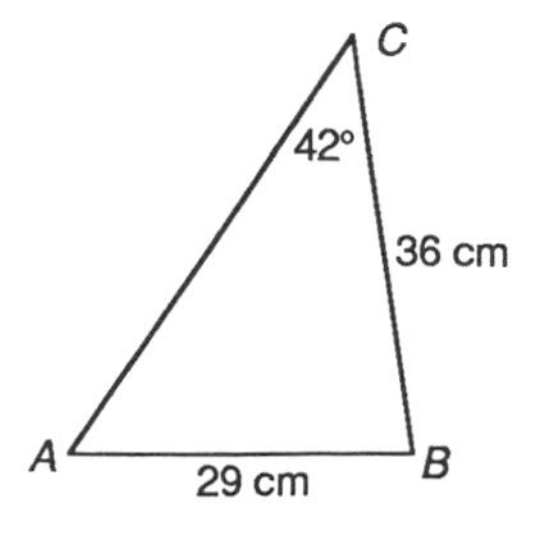

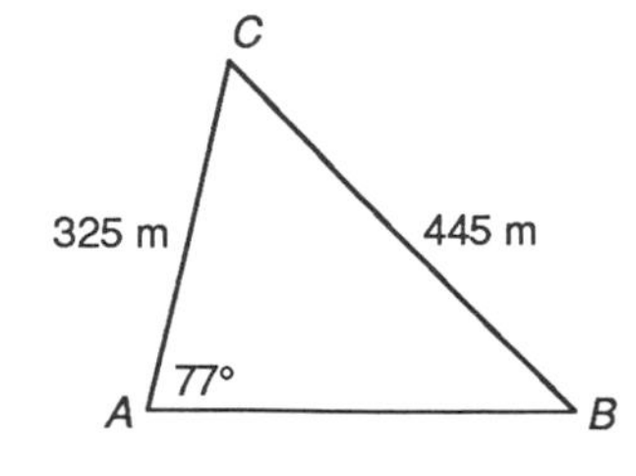

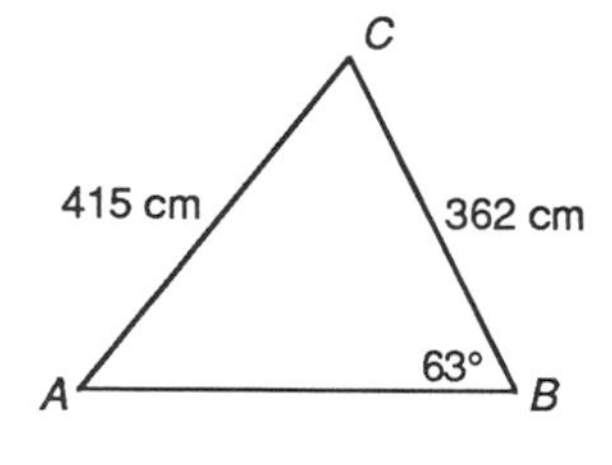

Lesson 12.6

The Law of Cosines

Since Chapter 9 on the Pythagorean Theorem you've solved many problems with the famous formula $c^2 = a^2 + b^2$. It is perhaps your most important geometric conjecture. But its use is limited to right triangles. Another useful trigonometric property, the Law of Cosines, does for all triangles what the Pythagorean Theorem does for right triangles.

The Law of Cosines was probably first developed when a mathematician asked, "What happens to the Pythagorean formula if the right angle were to get smaller or larger?" Huh? Take a look at the diagrams below.

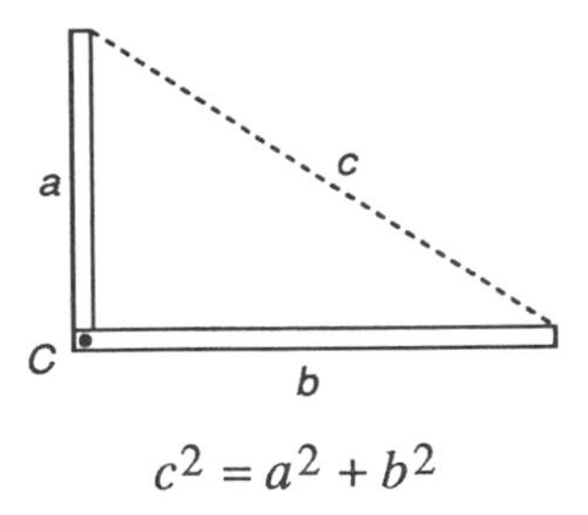

$c^2 = a^2 + b^2$

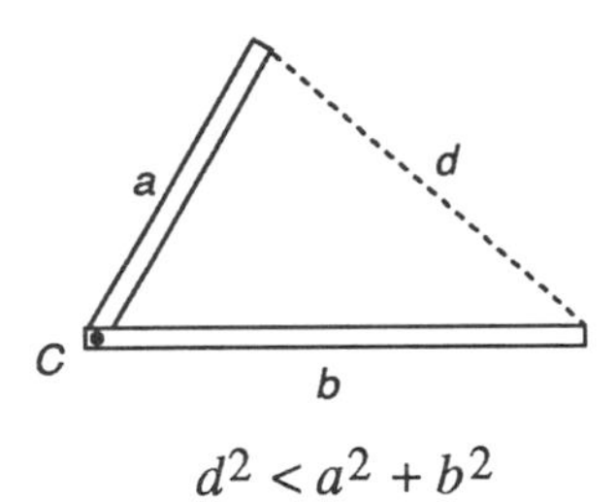

$d^2 < a^2 + b^2$

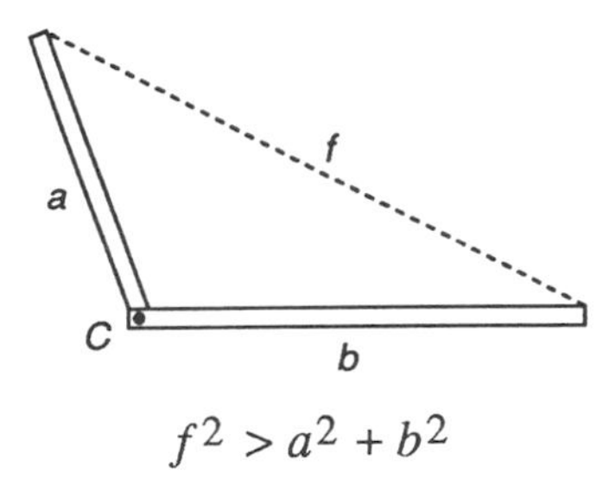

$f^2 > a^2 + b^2$

If two sticks with lengths a and b were connected at one pair of endpoints forming a right angle ($\angle C$), the distance c between the other endpoints could be calculated from $c^2 = a^2 + b^2$.

If the sticks, still attached at the same endpoints, were moved so that the angle between them ($\angle C$) was now an acute angle, the distance d between the endpoints would be less than before. That is, $d^2 < a^2 + b^2$ or $d^2 = a^2 + b^2 -$ something.

If the sticks were pushed apart so that the angle between them ($\angle C$) was now an obtuse angle, the distance f between the endpoints would be greater than before. That is, $f^2 > a^2 + b^2$ or $f^2 = a^2 + b^2 +$ something.

This observation led to a mathematical search for that "something." The "something" turned out to be $2ab \cos C$. The Pythagorean Theorem was thus generalized to all triangles. This property is called the Law of Cosines.

> *The* ***Law of Cosines*** *says that for any triangle with sides of lengths a, b, and c and with C the angle opposite the side with length c,*
> $c^2 = a^2 + b^2 - 2ab \cos C$.

This, like the Law of Sines, is a very useful equation. For example, if you know two sides and the included angle of a triangle (SAS), you can use the Law of Cosines to find the third side. If you know all three sides (SSS), you can use the Law of Cosines to find the measure of one of its angles.

The problems in this book will be restricted to finding the sine, cosine, or tangent of angles less than 90°; the trigonometric ratios for angles greater than 90° must be redefined. This is done in a full trigonometry course.

Example

Find the length r of side $\overline{CT}$ in acute ΔCRT with $RT = 45$ cm, $CR = 52$ cm, and $m\angle R = 36$.

To find r, use the Law of Cosines; $c^2 = a^2 + b^2 - 2ab \cos C$.

In this problem, the Law of Cosines becomes

$$r^2 = c^2 + t^2 - 2ct \cos R.$$

Substitute the known values.

$$r^2 = (45)^2 + (52)^2 - 2(45)(52)(\cos 36°)$$

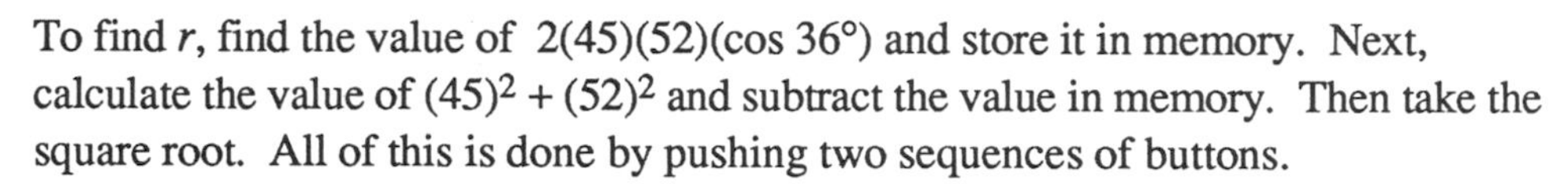

To find r, find the value of $2(45)(52)(\cos 36°)$ and store it in memory. Next, calculate the value of $(45)^2 + (52)^2$ and subtract the value in memory. Then take the square root. All of this is done by pushing two sequences of buttons.

First, push the following sequence:
2 [x] 45 [x] 52 [x] 36 [cos] [=] [memory] . Your display should show 3786.1995.

Now, clear the screen but not the memory and push the second sequence (the symbol [x^2] represents your squaring button):
45 [x^2] [+] 52 [x^2] [=] [–] [memory recall] = [$\sqrt{\ }$] . You should get 30.705056.

The length r of side $\overline{CT}$ is about 31 cm.

EXERCISE SET A

Find the length of each side to the nearest centimeter.

1.* $w = $ –?–

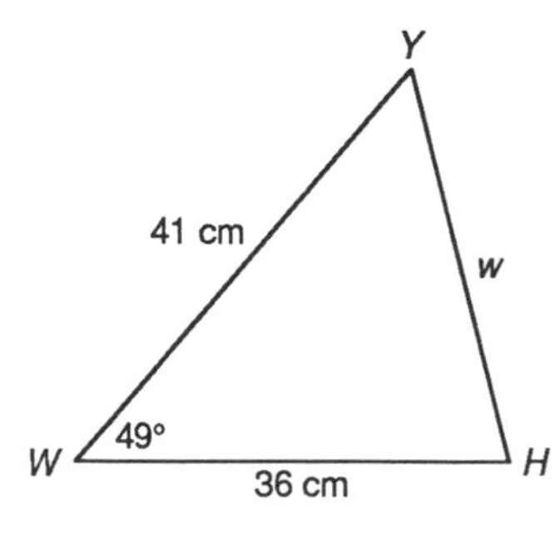

2. $x = $ –?–

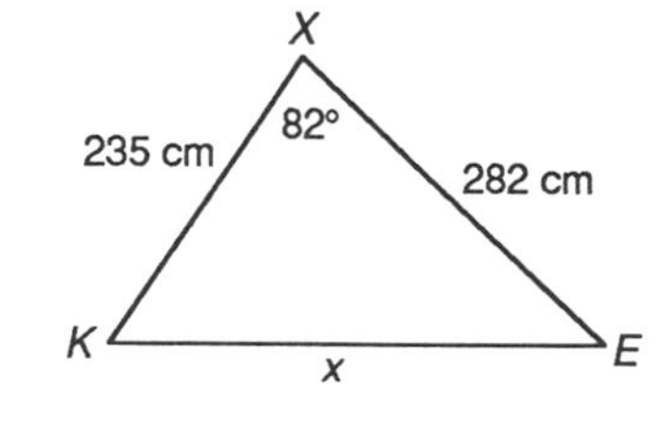

3. $y = $ –?–

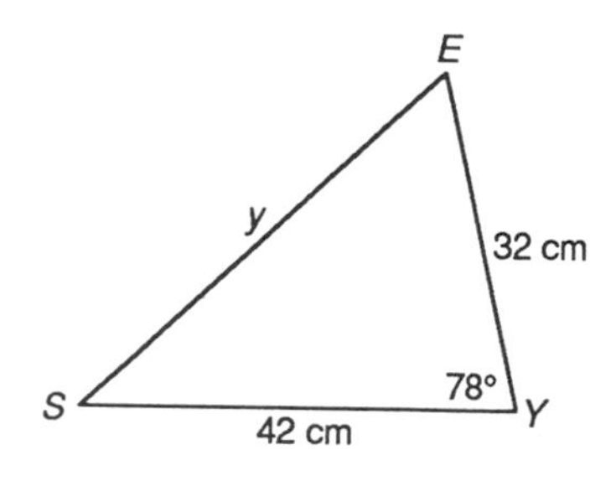

If you are given all three sides of a triangle (SSS), the Law of Cosines can be used to find the measure of an angle.

Example

Find the measure of angle Q in acute ΔQED with $ED = 250$ cm, $QD = 175$ cm, and $QE = 225$ cm.

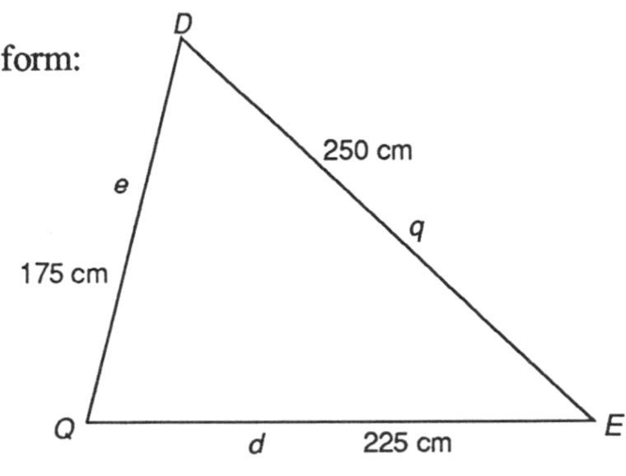

Use the Law of Cosines. To find $m\angle Q$, start with the form:

$$q^2 = e^2 + d^2 - 2ed \cos Q$$

Solve for cos Q.

$$\cos Q = \frac{q^2 - e^2 - d^2}{-2ed}$$

Substitute the known values.

$$\cos Q = \frac{(250)^2 - (175)^2 - (225)^2}{-2(175)(225)}$$

To find $m\angle Q$, push the following sequence of buttons on your calculator:
250 [x^2] [–] 175 [x^2] [–] 225 [x^2] [=] [÷] -2 [÷] 175 [÷] 225 [=] [inv] [cos] [=] .

Your display should show 76.225853.

The measure of angle Q is about 76°.

EXERCISE SET B

Find the measure of each angle indicated to the nearest degree.

1.* $m\angle A = –?–$

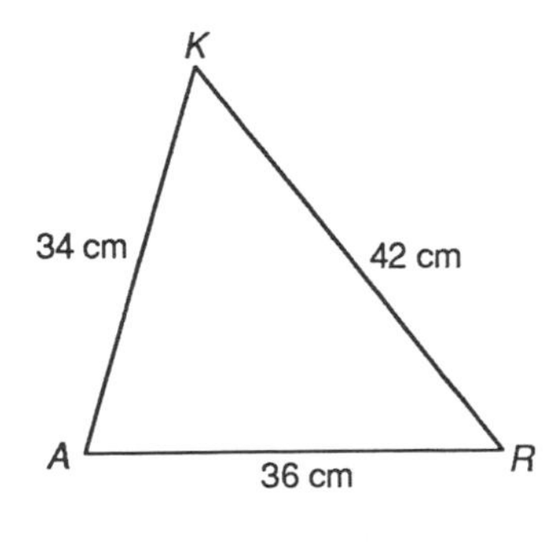

2. $m\angle B = –?–$

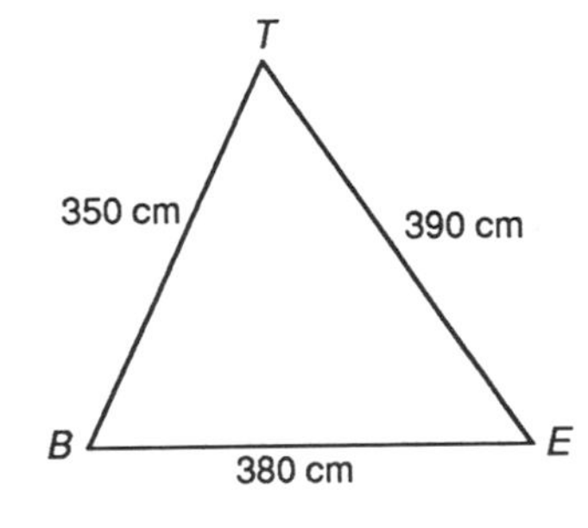

3. $m\angle C = –?–$

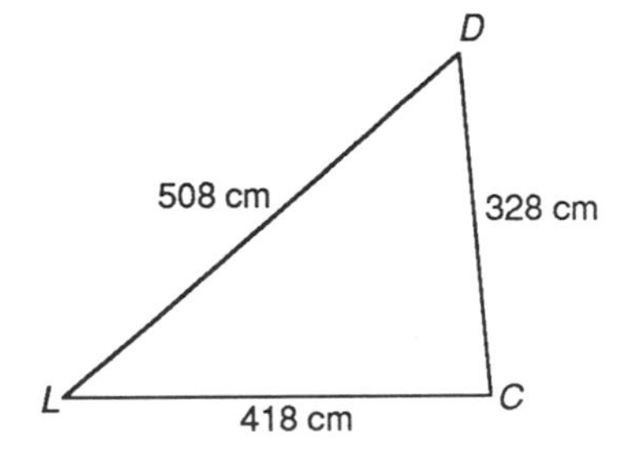

Lesson 12.7

Problem Solving with Trigonometry

There are many practical applications of trigonometry. To solve right triangle trigonometric problems you need the definitions for sine, cosine, and tangent. To solve problems involving acute triangles, the Law of Cosines or the Law of Sines may be helpful.

EXERCISE SET

1.* The steps to the front entrance of a public building rise a total of one meter. The steps are to be torn out and replaced by a ramp for wheelchairs. By a city ordinance, the angle of inclination for a wheelchair ramp cannot be greater than 9 degrees. What is the minimum distance from the entrance that the ramp must begin? Express your answer to the nearest tenth of a meter.

2. A lighthouse is due east from a coast guard patrol boat. The coast guard station is 20 km due north of the lighthouse. The coast guard radar officer aboard ship measures the angle between the lighthouse and coast guard station to be 23°. How far is the ship from the coast guard station to the nearest kilometer?

3. A surveyor needs to calculate the distance from the dock on Nomansan Island (C) to the dock on the mainland (A). She locates a point B up the shoreline so that $\overline{AB}$ is perpendicular to $\overline{AC}$. If AB measures 150 meters and $m\angle ABC = 58$, what is the distance between the two docks to the nearest meter?

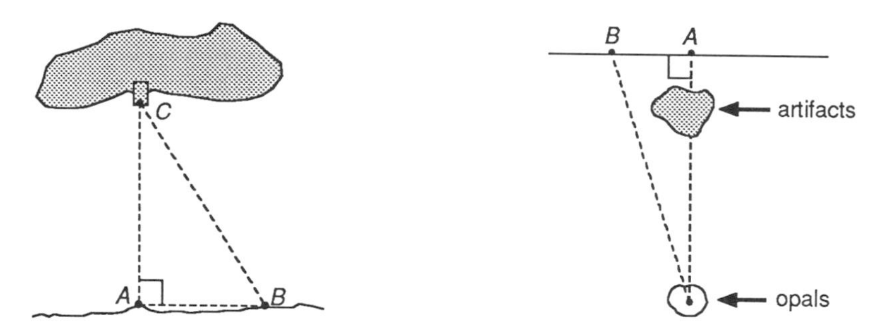

4. A pocket of opal matrix is known to be 24 meters below point A. The Outback Mining Company, however, is unable to dig straight down because of sacred aboriginal artifacts buried a few meters below the surface. The company has been given permission to dig at point B, 8 meters over from point A. At what angle to the level surface must the mining crew dig to reach the opal matrix? What distance must they dig?

5.* Find the measure to the nearest degree of the smallest angle in a triangle having sides of 4 m, 7 m, and 8 m.

6.* Find the area of a triangular plot of land with sides measuring 36 feet by 42 feet by 38 feet.

7.* Find the area of a regular octagon inscribed in a circle with a radius of 12 cm.

8. A tall evergreen tree has been damaged in a strong wind. The top of the tree is cracked and bent over — touching the ground as if the trunk were hinged. The tip of the tree touches the ground 20 feet 6 inches from the base of the tree and forms an angle of 38 degrees with the ground. Forest Ranger Willow Green must determine the original height of the damaged tree. Help her. What was the original height of the tree to the nearest foot?

9. A tree is growing vertically on a hillside that is inclined at an angle of 16° to the horizontal. The tree casts a shadow 18 m long up from the slope. If the angle of elevation of the sun is 68°, how tall is the tree?

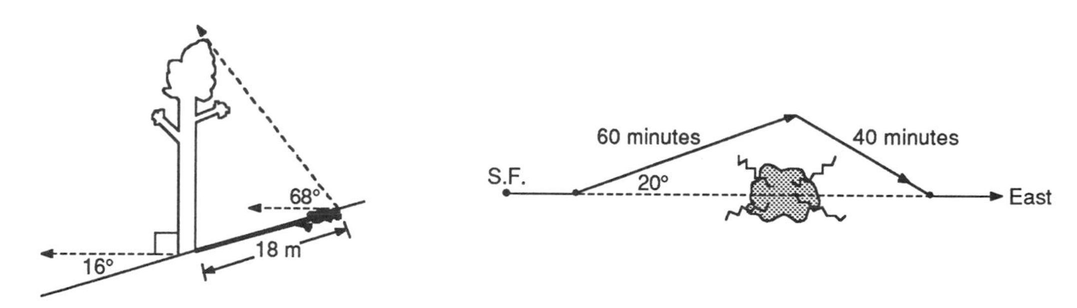

10. Captain Buck "Ace" Malloy is flying commercial flight 1123 out of S.F. heading due east at 725 km/hr when he spots an electrical storm straight ahead. He turns the jet 20° to the north to avoid the storm and continues in this direction for one hour. Then he makes a second correction back towards his original flight path. Forty minutes after his second correction he enters his original flight path at an acute angle and his craft is back on course. How much time did "Ace" lose from his original flight plan in order to avoid the storm?

11.* Army Private Olivia Drabb observes a fighter plane known to be flying at a level elevation of 6.3 km moving directly away from her. At one point she observes the plane at an angle of elevation of 49° while one minute later the angle of elevation is 15°. How fast is the plane flying in km/hr?

12. Farmer Goldie McDonald is planning to lay a water pipe through a small hill. She needs to determine the length of pipe needed. From a point off to the side of the hill on level ground, she runs one rope to the pipe's entry point and a second rope to the exit point. The first rope measures 14.5 meters and the second is 11.2 meters. The two ropes meet at an angle of 58 degrees. What is the length of pipe needed? At what angle with the first rope should the pipe be drilled so that it comes out at the correct exit point?

Lesson 12.8

Chapter Review

EXERCISE SET A

Use the definitions of the three trigonometric ratios to complete each statement.

1. $\sin A = $ –?–
   $\cos A = $ –?–
   $\tan A = $ –?–

2. $\sin B = $ –?–
   $\cos B = $ –?–
   $\tan B = $ –?–

3. $\sin C = $ –?–
   $\cos C = $ –?–
   $\tan C = $ –?–

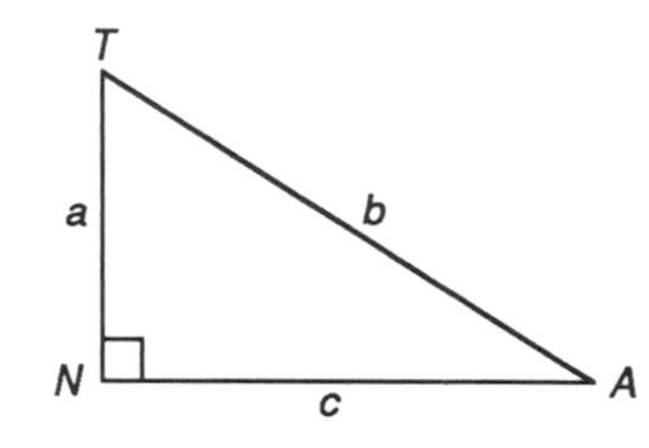

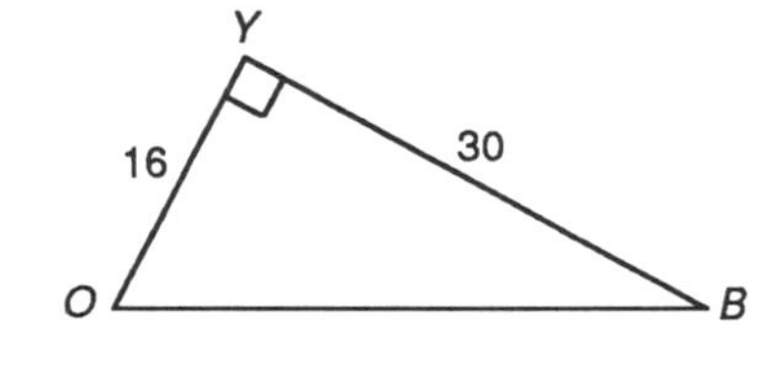

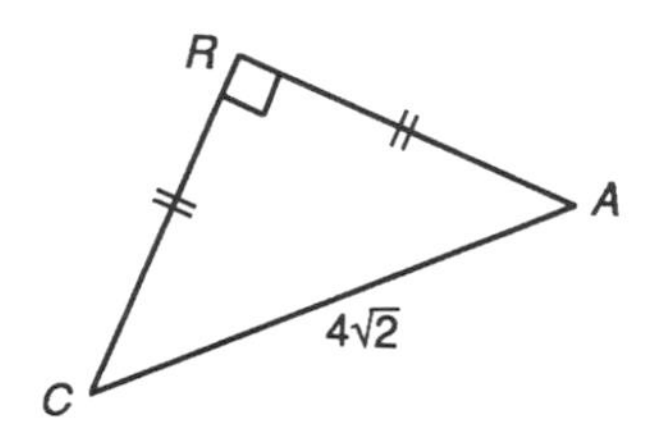

EXERCISE SET B

Use a scientific calculator (or trigonometric table) to determine the values accurate to four decimal places.

1. $\sin 57° \approx$ –?–
2. $\cos 9° \approx$ –?–
3. $\tan 88° \approx$ –?–

Find the measure of each angle to the nearest degree.

4. $\sin A = .5447$
5. $\cos B = .0696$
6. $\tan C = 2.9043$

EXERCISE SET C

Calculate each distance or angle.

1. The angle of elevation from a ship to the top of a 32 meter lighthouse on the shore is 23 degrees. How far is the ship from the shore to the nearest meter?

2. A lighthouse is due east from a sailboat. The berth of the sailboat is 30 km due north of the lighthouse. The captain aboard the sailboat measures the angle between the lighthouse and the sailboat berth and finds it to be 35°. To the nearest kilometer, how far is the the sailboat from the berth ?

3. Benny is flying a kite with 200 m of kite string out. His kite string makes an angle of 54° with the level ground. How high is his kite to the nearest meter?

4. A tree is observed on the opposite bank of a canal. The canal is known to be 115 feet wide. The angle of elevation from the observer to tip of the tree is 24 degrees and is measured on a line 4 feet above the base of the tree. How tall is the tree to the nearest foot?

5. Bulwinkle is a summer intern for the U.S. Forestry Service and has been instructed to measure the heights of the taller trees in his sector of the forest. Using his angle measuring device on the first tree, he measures the angle to be about 69°. He walks off 72 paces to the base of the tree. If each pace is two feet, how tall is the tree to the nearest foot?

6. Air traffic controller Seymour Flytz must quickly calculate the height of an incoming jet. To do this, he records the jet's angle of elevation to be 6°. If the jet signals that its land distance is 44 km from the control tower, calculate the height of the plane for Seymour. (Ignore the height of the tower.)

7. A Coast Guard patrol boat has spotted a hovering helicopter dropping a package near the Florida shoreline. Officer Sandy Shore measures the angle of elevation to the helicopter to be 15° and the distance to the helicopter to be 6800 m. To the nearest meter, how far is Sandy's Coast Guard boat from the point directly beneath the helicopter at the moment the package was dropped?

8. Pat Carpenter is constructing a new house. The house is 32 feet wide. She needs to cut rafters that will rise at an angle of 36 degrees and meet above the center line of the house. If each rafter is to overhang the side of the house by two feet, how long, to the nearest foot, should she make each rafter?

EXERCISE SET D

Use the Law of Sines or the Law of Cosines to solve each problem. Express each length to the nearest centimeter and each angle to the nearest degree.

1. $w = -?-$

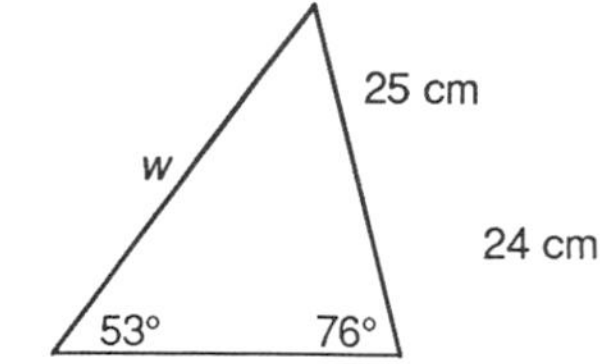

2. $x = -?-$

55°
40 cm
24 cm
x

3. $m\angle A = -?-$

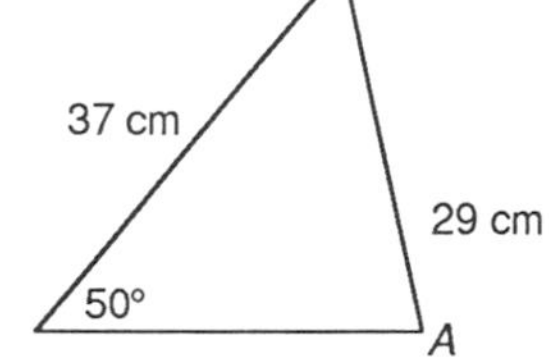

4. $m\angle B = -?-$

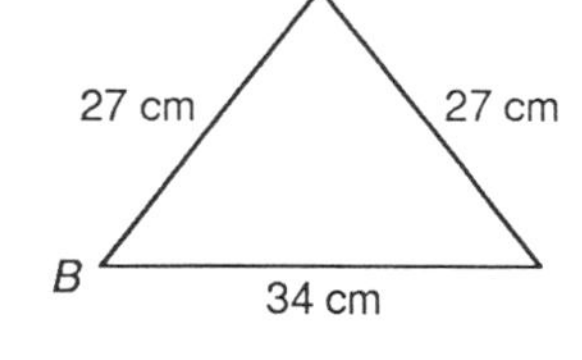

EXERCISE SET E

1. Draw a diagram and present an algebraic argument demonstrating why the angle measured with Device 2 (see *Special Project: Indirect Measurement*) is the complement of the desired angle.

2. Draw a diagram and present an algebraic argument demonstrating why the angle measured on Device 3 (see *Special Project: Indirect Measurement*) is the desired angle.

3. Exercises A4 to A9 of Lesson 12.5 can be restated as a proof of the Law of Sines. Prove the Law of Sines for acute triangles.

4.* Prove the Law of Cosines for acute triangles.

Cooperative Problem Solving

Problem Solving at Mare Imbrium

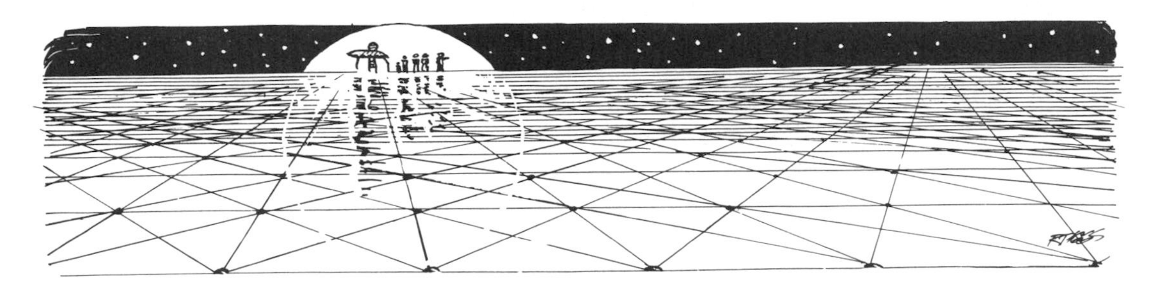

You and your cooperative problem solving team have been taken by lunar shuttle to the Mare Imbrium Recreational Facility (MIRF). Mare Imbrium is the largest flat region of the lunar surface and has been used for decades as a location for school field trips. MIRF is located equidistant (350 km) from the four craters Plato, Aristillus, Archimedes, and Timocharis. Your math and astronomy instructors have given you four days to complete four problems. After four days, you must return to the lunar colony Galileo. You should be able to complete the problems within a day and a half. For the rest of your stay, you and your CPS team may take advantage of the moon's largest and most complete recreational facility.

1.* Your first task is to determine the area of the Mare Imbrium between the centers of the four craters Plato, Aristillus, Archimedes, and Timocharis. Moving counterclockwise with a lunar transit, your CPS team sweeps the angles between the crater centers . The angle from Plato to MIRF to Aristillus is 115°. The angle from Aristillus to MIRF to Archimedes is 45°. The angle from Archimedes to MIRF to Timocharis is 60°. The angle from Timocharis to MIRF to Plato is 140°. What is the area of the quadrilateral connecting the centers of the four craters?

2.* Your second task is to determine the length of a tunnel to be drilled through a lunar peak at the center of a crater. From a point on the crater floor, you and your CPS team run ropes to the two ends of the tunnel. Measuring the ropes, you find that the distance along the first rope is 24.5 m and along the second is 21.2 m. The two ropes meet at an angle of 68°. What will be the length of the tunnel? At what angle with the first rope should the tunnel be drilled so that it comes out at the correct exit point?

3. Your third task is to determine the thickness of a vein of bauxite located 30 kilometers southeast of MIRF. On the surface, the vein appears as a 36-meter wide strip that is over 100 meters long. The distance across the vein at the surface, however, is not the thickness of the vein. Lunar geologists know that the vein recedes from the surface at an angle. It is your task to determine that angle and from it calculate the actual thickness of the vein.

You and your CPS team walk away from the vein along a path perpendicular to the edge of the vein. After walking a short distance, you stop and measure the distance straight down to the bauxite. Your team finds that the vein at this point is eight meters below the surface. After walking ten meters along the same path, you stop and again measure straight down to the vein. Your team finds that the vein at this point is 12 meters below the surface. What is the thickness of the vein? (In other words, what is the perpendicular distance between the two parallel planes bounding the bauxite?)

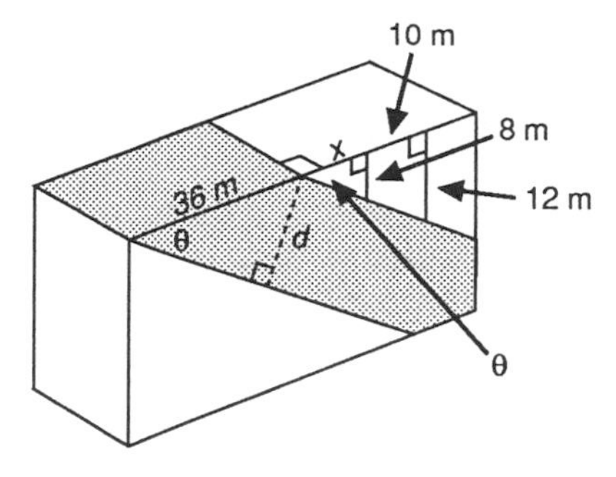

4. Your fourth task is to determine the identity of a piece of clear, hard material uncovered by lunar historians near one of the early moon landing sites. Geologists have narrowed your search down to four possiblities: quartz, glass, zircon, or diamond. Use Snell's Law to determines the identity of the material. **Snell's Law** expresses the relationship between the angles formed by a light ray moving from one medium to another. Light bends as it passes into more dense material because it slows down. By measuring the angles formed by light bending as it moves from the vacuum of the lunar surface through the clear hard material, you can determine the identity of the material.

The figure on the right shows the angles formed by a light ray moving from one medium to another. $\angle 1$ is the angle of incidence. $\angle 2$ is the angle of refraction. They are related by the following formula (Snell's Law):

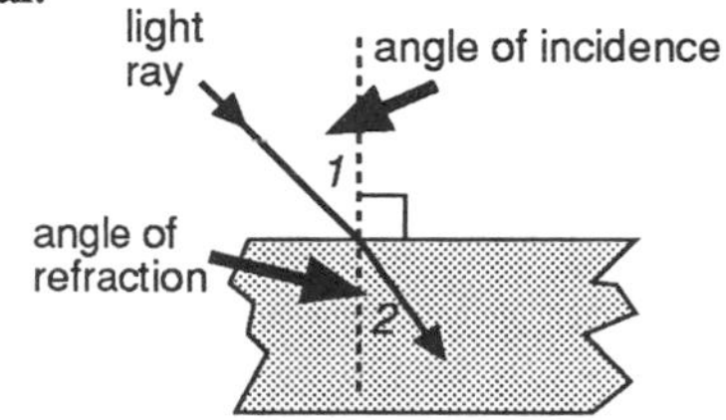

$$\sin \angle 1 = k \sin \angle 2$$

The constant k depends on the nature of the two mediums. The table below gives the value of Snell's constant for light going from a vacuum or air into quartz, glass, zircon, and diamond.

Substance entered from air or vacuum	Quartz	Glass	Zircon	Diamond
Snell's index of refraction	1.46	1.5	1.9	2.42

You and your CPS team take a number of measurements of different angles of incidence and the corresponding angles of refraction for the glass-like material. Your results are shown in the table below. Use Snell's Law to identify the mysterious material. Is it glass, quartz, zircon, or diamond? (Unfortunately, you suspect that one of the measurements was taken incorrectly. Therefore, you may have to test more than one pair of angles.)

Angle of incidence ($\angle 1$)	15°	20°	25°	30°	40	50°	60°
Angle of refraction ($\angle 2$)	6.1°	10.4°	12.9°	15.3°	19.8°	23.8°	27.1°

13 Deductive Reasoning

Belvedere, M. C. Escher, 1958

You used inductive reasoning throughout this text to discover many basic properties of geometry. In this chapter you will learn about the other type of reasoning, deductive reasoning (commonly called logical reasoning). You will learn to translate logical arguments to symbolic form and then use rules of logic to determine whether or not the argument is valid.

Lesson 13.1

"That's Logical!"

I refuse to join any club that would have me for a member.
– Groucho Marx

"That's logical!" You've probably heard that expression many times. What do we mean when we say someone is thinking logically? One dictionary defines *logical* as "capable of reasoning or using reason in an orderly fashion that brings out fundamental points."

EXERCISE SET A

Use logical reasoning to solve each problem below.

1.* Beatrice is older than Catherine, and Michelle is younger than Catherine. Is Beatrice older or younger than Michelle?

2. Every cheerleader at Washington High School is a junior. Mark is a cheerleader at Washington High School. Is Mark a junior at Washington High School?

3. Every student at Brightmore High School who takes Auto Shop also takes Driver's Education. Roberto and Sharon are both students at Brightmore High School. Roberto is not taking Driver's Education. Sharon is taking Auto Shop. Is Roberto taking Auto Shop? Is Sharon taking Driver's Education?

4. Boris and Natasha each own a very unique pet. One owns a monkey and the other owns an alligator. The alligator owner would like to own a golden retriever but she is allergic to animal hair. Who owns the monkey?

5. Edith, Ernie, and Eva have careers as economist, electrician, and engineer, but not necessarily in that order. The economist does consulting work for Eva's business. Ernie hired the electrician to rewire his new kitchen. Edith earns less than the engineer but more than Ernie. Match the names with the occupations.

"Prove it!" That's another expression you've probably heard many times. It is also an expression that is used by someone concerned with logical thinking. In daily life proving something often means that you can present some authority or fact to support a point. In the example below, an eyewitness is cited as one person's proof.

Example A

"All right, Butch, where were you on the night of January 11?" asked Inspector Vida. "Inspector," replied Butch in an injured tone, "I can prove I was with ma dear ol' mom. Call and ask her."

Sometimes to prove a point you may cite accepted rules to support a conclusion. This occurs often in sports.

Example B

Samantha Slugger has two strikes against her. She attempts to bunt and fouls the ball. The umpire calls her out. Samantha complains, "Prove I'm out!" The ump states the official league rule on attempted bunts with two strikes.

EXERCISE SET B

1.* Lori declares, "I live at 2332 Oak Street." Audrey replies, "I don't believe you, prove it." How might Lori prove it?

2. David tells Manny, "I am over six feet tall." "No, you aren't, and I can prove it," states Manny. How might Manny prove it?

3. Ann proudly tells Amy, "I can spell parallelogram correctly." "Prove it!" replied Amy. How might Ann prove it?

4. Trucker Tucker was pulled over by Highway Patrol Officer Brandenburg for speeding. "But officer, I wasn't speeding," pleaded Tucker, "the other vehicles were just going very slow." "You were speeding and I can prove it," growled Officer Brandenburg. In court, how will the officer prove Tucker was speeding?

5. Jock says he can run the 100-yard dash faster than anyone else in the world. How many people does Jock have to beat in order to prove his statement? How many people have to beat Jock to disprove his statement?

Lesson 13.2
Sherlock Holmes

There are two types of reasoning in mathematics: inductive and deductive reasoning. You have been using inductive reasoning by observing patterns and making conjectures about your observations. This is the creative, investigative form of reasoning that mathematicians use most often. In the coming chapters, you will take a look at the other form of reasoning, deductive reasoning, to see if your discoveries are logically consistent.

> ***Deductive reasoning*** *(or logical reasoning) is the process of demonstrating that if certain statements are accepted as true, then other statements can be shown to follow from them.*

You use inductive reasoning to make new discoveries. You can use deductive reasoning to show that your discoveries are logically consistent with each other.

Illustration by Sidney Paget, from the original publication in *The Strand Magazine*, 1892

What better way to begin our study of deductive reasoning than with the master of deduction, Sherlock Holmes. Sit back and read the excerpt below from the Sherlock Holmes tale, *The Adventure of the Dancing Men*, by Sir Arthur Conan Doyle.

Holmes had been seated for some hours in silence with his long, thin back curved over a chemical vessel in which he was brewing a particularly malodorous product. His head was sunk upon his breast, and he looked from my point of view like a strange, lank bird, with dull gray plumage and a black top-knot.

"So, Watson," said he, suddenly, "you do not propose to invest in South African securities?"

I gave a start of astonishment. Accustomed as I was to Holmes' curious faculties, this sudden intrusion into my most intimate thoughts was utterly inexplicable.

"How on earth do you know that?" I asked.

He wheeled round upon his stool, with a steaming test-tube in his hand, and a gleam of amusement in his deep-set eyes.

"Now, Watson, confess yourself utterly taken aback," said he.

"I am."

"I ought to make you sign a paper to that effect."

"Why?"

"Because in five minutes you will say that it is all so absurdly simple."

"I am sure that I shall say nothing of the kind."

"You see, my dear Watson," he propped his test-tube in the rack, and began to lecture with the air of a professor addressing his class, "it is not really difficult to construct a series of inferences, each dependent upon its predecessor and each simple in itself. If, after doing so, one simply knocks out all the central inferences and presents one's audience with the starting point and the conclusion, one may produce a startling, though possibly a meretricious, effect. Now, it was not really difficult, by inspection of the groove between your left forefinger and thumb, to feel sure that you did not propose to invest your small capital in the gold fields."

"I see no connection."

"Very likely not; but I can quickly show you a connection. Here are the missing links of the very simple chain: 1. You had chalk between your left finger and thumb when you returned from the club last night. 2. You put chalk there when you play billiards, to steady the cue. 3. You never play billiards except with Thurston. 4. You told me four weeks ago that Thurston had an option on some South African property which would expire in a month, and which he desired you to share with him. 5. Your checkbook is locked in my drawer, and you have not asked for the key. 6. You do not propose to invest your money in this manner."

"How absurdly simple!" I cried.

"Quite so!" said he, a little nettled. "Every problem becomes very childish when once it is explained to you.

Let's take a closer look at the logical reasoning of Sherlock Holmes. Mr. Holmes used two of the basic forms of logical reasoning. Below is an example of one.

Example

Watson had chalk between his fingers upon returning from the club.
If Watson had chalk on his fingers, then he was playing billiards.
Therefore, Watson was playing billiards.

EXERCISE SET A

Now you play Sherlock Holmes. Complete the logical argument in each problem below by providing the conclusion.

1. If Watson was playing billiards, then he was playing with Thurston.
   Watson was playing billiards.
   Therefore, —?—.
2. If Professor Moriarty was at the train station, then he is the kidnapper.
   The professor was at the train station.
   Therefore, —?—.
3. If ΔABC is isosceles, then the base angles are congruent.
   ΔABC is isosceles.
   Therefore, —?—.
4.* If Sharon studies, then she does well on her tests.
   If Sharon does well on her tests, then she gets good grades on her report card.
   Sharon studies.
   Therefore, —?— (two steps).

In the deduction below, Mr. Holmes uses a second basic form of logical reasoning.

Example

If Watson wishes to invest his money in South African securities with Thurston, then he would have his checkbook when playing billiards with Thurston.

Watson did not have his checkbook when he played billiards with Thurston.

Therefore, Watson does not wish to invest his money in South African securities with Thurston.

EXERCISE SET B

Now, you complete the logical argument.

1. If the object in the box is a rose, then the object is a flower.
   The object is not a flower.
   Therefore, the object in the box is —?—.

2.* If ΔABC is equilateral, then $m\angle A = 60$.
$m\angle A \neq 60$.
Therefore, —?—.

3. If Colonel Moran is the murderer, then he has powder burns on his shirt.
The Colonel does not have powder burns on his shirt.
Therefore, —?—.

4. If Jill is not busy on the computer, then Dana will be able to use the computer.
Dana will not be able to use the computer.
Therefore, —?—.

Let's look at another Sherlock Holmes adventure, Sir Arthur Conan Doyle's *The Hound of the Baskervilles*. Pay close attention to the conclusions that Mr. Holmes arrives at from the facts he gathers observing a walking stick.

> Mr. Sherlock Holmes, who was usually very late in the mornings, save upon those not infrequent occasions when he was up all night, was seated at the breakfast table. I stood upon the hearth-rug and picked up the stick which our visitor had left behind him the night before. It was a fine, thick piece of wood, bulbous-headed, of the sort which is known as a "Penang lawyer." Just under the head was a broad silver band, nearly an inch across. "To James Mortimer, M.R.C.S., from his friends of the C.C.H ," was engraved upon it, with the date "1884." It was just such a stick as the old-fashioned family practitioner used to carry — dignified, solid, and reassuring.
>
> "Well, Watson, what do you make of it?"
>
> Holmes was sitting with his back to me, and I had given him no sign of my occupation.
>
> "How did you know what I was doing? I believe you have eyes in the back of your head."
>
> "I have, at least, a well-polished, silver-plated coffee-pot in front of me," said he. "But tell me, Watson, what do you make of our visitor's stick? Since we have been so unfortunate as to miss him and have no notion of his errand, this accidental souvenir becomes of importance. Let me hear you reconstruct the man by an examination of it."
>
> "I think," said I, following as far as I could the methods of my companion, "that Dr. Mortimer is a successful, elderly medical man, well esteemed, since those who know him give him this mark of their appreciation."
>
> "Good!" said Holmes. "Excellent!"
>
> "I think also that the probability is in favour of his being a country practitioner who does a great deal of his visiting on foot."
>
> "Why so?"
>
> "Because this stick, though originally a very handsome one, has been so knocked about that I can hardly imagine a town practitioner carrying it. The thick iron ferrule is worn down, so it is evident that he has done a great amount of walking with it."
>
> "Perfectly sound!" said Holmes.
>
> "And then again, there is the friends of the C.C.H. I should guess that to be the

Something Hunt, the local hunt to whose members he has possibly given some surgical assistance, and which has made him a small presentation in return."

"Really, Watson, you excel yourself," said Holmes, pushing back his chair and lighting a cigarette. "I am bound to say that in all the accounts which you have been so good as to give of my own small achievements, you have habitually underrated your own abilities. It may be that you are not yourself luminous, but you are a conductor of light. Some people without possessing genius have a remarkable power of stimulating it. I confess, my dear fellow, that I am very much in your debt."

He had never said as much before, and I must admit that his words gave me a keen pleasure, for I had often been piqued by his indifference to my admiration and to the attempts which I had made to give publicity to his methods. I was proud, too, to think that I had so far mastered his system as to apply it in a way which earned his approval. He now took the stick from my hands and examined it for a few minutes with his naked eyes. Then with an expression of interest he laid down his cigarette, and, carrying the cane to the window, he looked over it again with a convex lens.

"Interesting, though elementary," said he as he returned to his favorite corner of the settee. "There are certainly one or two indications upon the stick. It gives us the basis for several deductions.

"I am afraid, my dear Watson, that most of your conclusions were erroneous. When I said that you stimulated me I meant, to be frank, that in noting your fallacies I was occasionally guided towards the truth. Not that you are entirely wrong in this instance. The man is certainly a country practitioner. And he walks a good deal."

Illustration by Sidney Paget, from the original publication in *The Strand Magazine*, August 1901

"Then I was right."

"To that extent."

"But that was all."

"No, no, my dear Watson, not all — by no means all. I would suggest, for example, that a presentation to a doctor is more likely to come from a hospital than from a hunt, and that when initials 'C.C.' are placed before that hospital the words 'Charing Cross' very naturally suggest themselves."

"You may be right."

"The probability lies in that direction. And if we take this as a working hypothesis we have a fresh basis from which to start our construction of this unknown visitor."

"Well, then, supposing the 'C.C.H.' does stand for 'Charing Cross Hospital,' what further inferences may we draw?"

"Do none suggest themselves? You know my methods. Apply them!"

"I can only think of the obvious conclusion that the man has practised in town before going to the country."

"I think that we might venture a little farther than this. Look at it in this light. On what occasion would it be most probable that such a presentation would be made? When would his friends unite to give him a pledge of their good will? Obviously at the moment when Dr. Mortimer withdrew from the service of the hospital in order to start in practice for himself. We know that there has been a presentation. We believe there has been a change from a town hospital to a country practice. Is it, then, stretching our inference too far to say that the presentation was on the occasion of the change?"

"It certainly seems probable."

"Now, you will observe that he could not have been on the staff of the hospital, since only a man well-established in a London practice could hold such a position, and such a one would not drift into the country. What was he, then? If he was in the hospital and yet not on the staff he could only have been a house-surgeon or a house-physician — little more than a senior student. And he left five years ago — the date is on the stick. So your grave, middle-aged family practitioner vanishes into thin air, my dear Watson, and there emerges a young fellow under thirty, amiable, unambitious, absent-minded, and the possessor of a favorite dog, which I should describe roughly as being larger than a terrier and smaller than a mastiff."

I laughed incredulously as Sherlock Holmes leaned back in his settee and blew little wavering rings of smoke up to the ceiling.

"As to the latter part, I have no means of checking you," said I, "but at least it is not difficult to find out a few particulars about the man's age and professional career." From my small medical shelf I took down the Medical Directory and turned up the name. There were several Mortimers, but only one who could be our visitor. I read his record aloud.

> "Mortimer, James, M.R.C.S., 1882, Grimpen, Dartmoor, Devon, House surgeon, from 1882 to 1884, at Charing Cross Hospital. Winner of the Jackson prize for Comparative Pathology, with essay entitled "Is Disease a Reversion?" Corresponding member of the Swedish Pathological Society. Author of "Some Freaks of Atavism" (Lancet, 1882). "Do We Progress?" (Journal of Psychology, March, 1883). Medical Officer for the parishes of Grimpen, Thorsley, and High Barrow."

"No mention of the local hunt, Watson," said Holmes with a mischievous smile, "but a country doctor, as you very astutely observed. I said, if I remember right, amiable, unambitious, and absent-minded. It is my experience that it is only an amiable man in this world who receives testimonials, only an unambitious one who abandons a London career for the country, and only an absent-minded one who leaves his stick and not his visiting-card after waiting an hour in your room."

"And the dog?"

"Has been in the habit of carrying this stick behind his master. Being a heavy stick the dog has held it tightly by the middle, and the marks of his teeth are very plainly visible. The dog's jaw as shown in the space between these marks, is too broad in my opinion for a terrier and not broad enough for a mastiff."

EXERCISE SET C

Mr. Holmes studied the walking stick carefully and came up with many surprising conclusions. He concluded that the owner of the walking stick was a country practitioner who walked a great deal. He had left Charing Cross Hospital five years before and he was absent-minded and owned a dog. Complete the logical argument in each deduction below. Return to the story if necessary.

1. If the stick had been knocked about so much, then the stick must have belonged to a country practitioner.
   The stick had been knocked about.
   Therefore, —?—.

2.* If the iron tip of the walking stick was worn down, then —?—.
   The iron tip of the walking stick was worn down.
   Therefore, —?—.

3. If the date on the stick was 1884, then he left the hospital staff five years ago.
   The date on the stick was 1884.
   Therefore, —?—.

4. If —?—, then —?—.
   The owner of the walking stick forgot it.
   Therefore, the man was absent-minded.

5. If the initials on the stick were C.C.H., then the owner worked at Charing Cross Hospital.
   —?—.
   Therefore, —?—.

6. If —?—, then —?—.
   —?—.
   Therefore, the owner of the stick owned a dog.

Illustration by Sidney Paget, from the original publication in *The Strand Magazine*,1891

Lesson 13.3

Forms of Valid Reasoning

Readers are plentiful,
thinkers are rare.

– Harriet Martineau
(1802–1876)

In the previous lesson Sherlock Holmes demonstrated some very amazing deductive reasoning (logic). Mr. Holmes used forms of valid reasoning in drawing his conclusions. In this lesson you will learn to translate logical arguments from English into symbols. Then you too will learn to use two of the accepted forms of reasoning found in logical arguments.

A **logical argument** consists of a set of premises and a conclusion. Each given statement is a **premise**, and the statement that is arrived at through reasoning is called the **conclusion**. An argument is **valid** if the conclusion has been arrived at through accepted forms of reasoning. If all the premises of an argument are true and the argument is valid, then and only then must the conclusion be true.

When you translate an argument to symbolic form, you use capital letters (P, Q, R, S, etc.) to stand for simple statements that are either true or false. When you write IF P THEN Q, you are writing a **conditional statement**. For example, if P stands for the sentence *Watson had chalk between his fingers* and Q stands for the sentence *Watson was playing billiards*, then the conditional statement IF P THEN Q translates to *If Watson had chalk between his fingers, then Watson was playing billiards.*

Example

LET P: I study.
LET Q: I get good grades.
IF P THEN Q: If I study, then I get good grades.

EXERCISE SET A

LET P: I get a job.
LET Q: I will earn money.
LET R: I will go to the movies.
LET S: I will spend my money.

Use the legend above to translate the following from symbols to English.

1.* IF P THEN Q

2. IF Q THEN R

3. IF P THEN R

4. IF R THEN S

LET P: Today is Wednesday. LET Q: Tomorrow is Thursday.

LET R: Friday is coming. LET S: Yesterday was Tuesday.

Use the legend above to translate the following from English to symbols.

5.* If today is Wednesday, then tomorrow is Thursday.

6. If tomorrow is Thursday, then Friday is coming.

7. If yesterday was Tuesday, then tomorrow is Thursday.

8. If yesterday was Tuesday, then Friday is coming.

One of the accepted forms of reasoning used by Sherlock Holmes is Modus Ponens.

> *According to **Modus Ponens** (MP), if you accept IF P THEN Q as true and you accept P as true, then you must logically accept Q as true.*

Holmes' use of Modus Ponens is shown below. The symbolic form is on the right.

English Argument

If Watson had chalk between his fingers, then he was playing billiards. Watson had chalk between his fingers. Therefore, Watson was playing billiards.

Translation

LET P: Watson had chalk between his fingers.

LET Q: Watson was playing billiards.

IF P THEN Q
P
THEREFORE Q

Let's look at another example of Modus Ponens.

English Argument

Sonya had the flu. If Sonya had the flu, then Sonya gave the flu to Melissa. Therefore, Sonya gave the flu to Melissa.

Translation

LET S: Sonya had the flu.

LET M: Sonya gave the flu to Melissa.

S
IF S THEN M
THEREFORE M

EXERCISE SET B

In each problem below, if the two premises fit the logically valid reasoning pattern of Modus Ponens, state the conclusion in English. Then create a legend and translate the complete argument into symbolic form. If the statements do not fit the logically valid reasoning pattern of Modus Ponens, simply write *no valid conclusion.*

1.* If Sherlock Holmes gathers all the clues, then he will be victorious.
Sherlock Holmes gathers all the clues.

2. If Golum is carnivorous, then Golum is dangerous.
   Golum is dangerous.

3. Triangle *ABC* is isosceles.
   If a triangle is isosceles, then its base angles are congruent.

4. If Conchita doesn't feed her pet lizard El Brujo, then El Brujo will eat the mailman. Conchita doesn't feed El Brujo.

5. If Merlo the Magnificent performs magic on Wednesday, then he will receive his royalty check from the Magician's Guild.
   Merlo the Magnificent performs magic on Thursday.

6. If $\overline{ED}$ is a midsegment in ΔABC, then $\overline{ED}$ is parallel to a side of ΔABC.
   $\overline{ED}$ is a midsegment in ΔABC.

The **negation** of a sentence is made by appropriately placing the word *not*, or by preceding the sentence with the phrase *It is not the case that.* For example, the negation of *Paula will receive her inheritance* can be expressed two ways: *Paula will not receive her inheritance*, or *It is not the case that Paula will receive her inheritance.* To express the negation of a sentence that already contains a negation, either remove the *not*, or precede the sentence with *It is not the case that.* For example, the negation of *It is not raining* can be expressed two ways: *It is raining*, or *It is not the case that it is not raining.* The two expressions are equivalent and this property is called **double negation**. If *P* stands for the sentence *Mike won the sweepstakes*, then NOT *P* stands for the sentence *Mike did not win the sweepstakes.*

Another valid form of reasoning used by Sherlock Holmes is called Modus Tollens.

> *According to **Modus Tollens** (MT), if you accept IF P THEN Q as true and you accept NOT Q as true, then you must logically accept NOT P as true.*

Holmes' use of Modus Tollens is illustrated in the next example.

English Argument

If Watson wishes to invest his money in South African securities with Thurston, then he would have had his checkbook with him when playing billiards with Thurston. Watson did not have his checkbook when playing billiards with Thurston. Therefore, Watson does not wish to invest his money in South African securities with Thurston.

Translation

LET *S*: Watson wishes to invest in South African securities with Thurston.

LET *T*: Watson has his checkbook when playing billiards with Thurston.

IF *S* THEN *T*
NOT *T*
THEREFORE NOT *S*

Let's look at another example of Modus Tollens.

English Argument	***Translation***
I do not get wet when I go outside. If it is raining, then I get wet when I go outside. Therefore, it is not raining.	LET W: I get wet when I go outside. LET R: It is raining. NOT W IF R THEN W THEREFORE NOT R

EXERCISE SET C

In each problem below, if the two premises fit the logically valid reasoning patterns of Modus Tollens or Modus Ponens, state which reasoning pattern and state the conclusion in English. Then translate the complete argument into symbolic form. If the statements do not fit the logically valid reasoning patterns of Modus Tollens or Modus Ponens, simply write *no valid conclusion.*

1. If $2x + 3 = 17$, then $x = 7$. But $x \neq 7$.

2. If Fontaine goes out with David, then she will have a good time Saturday night. Fontaine will not have a good time Saturday night.

3.* If Paul divorces Veronica, then he will not receive his inheritance from Uncle Lake. But Paul will receive his inheritance from Uncle Lake.

4. If the sun rises in the west, then morning shadows point east. Morning shadows do not point east.

5.* If Linda takes the bus, then she will be late for her job interview. Linda does not take the bus.

6. If Hemlock Bones decodes the secret message, then Agent Otto will not be captured. Agent Otto will not be captured.

7. If $\angle A$ and $\angle B$ are vertical angles, then they are congruent. $\angle A$ and $\angle B$ are not vertical angles.

8. If $ABCD$ is a rectangle, then the diagonals are congruent. The diagonals of $ABCD$ are not congruent.

Improving Visual Thinking Skills

Painted Faces III

Unit cubes are assembled to form a larger cube, and then some of the faces of this larger cube are painted. After the paint dries, the larger cube is disassembled into the unit cubes and it is found that 60 of these have no paint on any of their faces. How many faces of the larger cube were painted?

Lesson 13.4

The Symbols of Logic

Logician's use symbols to shorten logical arguments making them easier to understand:

IF P THEN Q is written $P \rightarrow Q$ *NOT R* is written $\sim R$ *THEREFORE* is written $\therefore$

Each statement below has been translated from English to symbolic form.

English Statement	***Symbolic Translation***
If the evening sky is red, then tomorrow will be a beautiful day.	LET P: The evening sky is red. LET Q: Tomorrow will be a beautiful day. $P \rightarrow Q$
If yesterday was Tuesday, then tomorrow will not be Friday.	LET Y: Yesterday was Tuesday. LET T: Tomorrow will be Friday. $Y \rightarrow \sim T$
It is not true that, if I fail the test, then I will not graduate.	LET F: I fail the test. LET G: I will graduate. $\sim(F \rightarrow \sim G)$

A logical argument is translated from English to symbolic form for you below.

English Argument	***Symbolic Translation***
If the integer is divisible by six, then it is divisible by three. The integer is not divisible by three. Therefore, it is not divisible by six.	LET P: The integer is divisible by six. LET Q: The integer is divisible by three. $P \rightarrow Q$ $\sim Q$ $\therefore \sim P$

EXERCISE SET

In Exercises 1 to 6, shorten the statement by using logician's shorthand.

1.* IF P THEN NOT Q

2. IF NOT Q THEN R

3. IF Q THEN NOT R

4. IF NOT Q THEN NOT P

5.* IT IS NOT TRUE THAT IF R THEN S

6. IT IS NOT TRUE THAT IF NOT P THEN R

In Exercises 7 to 9, translate the conditional sentence from English to symbolic form. Use the letter substitutions indicated below.

LET P: World peace is in jeopardy.
LET R: The destruction of civilization could result.

LET Q: International tensions are reduced.
LET S: The buildup of nuclear weapons continues.

7. If international tensions are not reduced, then world peace is in jeopardy

8. If the buildup of nuclear weapons continues, then the destruction of civilization could result.

9. It is not true that if the buildup of nuclear weapons does not continue, then international tensions are reduced.

Translate the symbolic sentences in Exercises 10 to 15 into English sentences. Use the same letter substitutions from above.

10. $P \rightarrow R$ 11. $S \rightarrow {\sim}Q$ 12.* ${\sim}Q \rightarrow R$

13. ${\sim}P \rightarrow {\sim}S$ 14.* ${\sim}({\sim}P \rightarrow {\sim}Q)$ 15. ${\sim}({\sim}S \rightarrow {\sim}R)$

In Exercise 16 to 21, identify the symbolic argument as Modus Ponens (MP) or Modus Tollens (MT). If the argument is not valid, write *no valid conclusion*.

16. $T \rightarrow R$
T
$\therefore R$

17.* $S \rightarrow {\sim}Q$
Q
$\therefore {\sim}S$

18. ${\sim}Q \rightarrow R$
${\sim}Q$
$\therefore R$

19.* $T \rightarrow R$
R
$\therefore {\sim}T$

20. $P \rightarrow Q$
${\sim}P$
$\therefore {\sim}Q$

21. ${\sim}R \rightarrow S$
${\sim}S$
$\therefore R$

In Exercise 22 to 27, if the two premises fit the logically valid reasoning patterns of Modus Tollens or Modus Ponens, state the conclusion in English. Set up a legend showing the sentence that each letter represents, and then translate the complete argument into symbolic form. If the statements do not fit the logically valid reasoning patterns of MT or MP, simply write *no valid conclusion*.

22. If the treasure is discovered, then pirate Rufus T. Ruffian will walk his own plank. The treasure is discovered.

23. If you use Shining Smile toothpaste, then you will be successful. You do not use Shining Smile toothpaste.

24. If Chrissy is an athlete, then she scores high on spatial visualization tests. Chrissy scores high on spatial visualization tests.

25. If Professor Moriarty wrote a paper about the Binomial Theorem, then he is familiar with Pascal's Triangle. Professor Moriarty is not familiar with Pascal's Triangle.

26. If $\overline{AC}$ is the longest side in ΔABC, then $\angle B$ is the largest angle in ΔABC. $\angle B$ is not the largest angle in ΔABC.

27. *If* $\angle B$ is the largest angle in ΔABC, then the measure of $\angle B$ is greater than 60. $\angle B$ is the largest angle in ΔABC.

Lesson 13.5

The Law of Syllogism

What is thinking? I should have thought I would have known.

– Karl Gerstner

So far you have learned two forms of valid reasoning, Modus Ponens and Modus Tollens. Now you will learn a third form of valid reasoning, the Law of Syllogism.

What conclusion can you draw from the two conditional statements below?

If I study, then I'll do well on the test tomorrow. If I do well on the test tomorrow, then I'll get a good grade in the class.

If you accept the two statements as true, then you must accept the truth of the conclusion, *If I study, then I'll get a good grade in the class.*

This valid form of reasoning is called the the Law of Syllogism. The Law of Syllogism draws a conditional conclusion from two conditional statements.

> *According to the* ***Law of Syllogism*** *(LS), if you accept* IF *P* THEN *Q as true and if you accept* IF *Q* THEN *R as true, then you must logically accept* IF *P* THEN *R as true.*

The argument below is an example of the Law of Syllogism.

English Argument

If I eat a pizza after midnight, then I will have nightmares. If I have nightmares, then I will get very little sleep. Therefore, if I eat pizza after midnight, then I will get very little sleep.

Symbolic Translation

LET P: I eat pizza after midnight.
LET N: I will have nightmares.
LET S: I will get very little sleep.

$P \rightarrow N$
$N \rightarrow S$
$\therefore P \rightarrow S$

Let's look at another example of the Law of Syllogism.

English Argument

If I earn money, then I will buy a computer. If I get a job, then I will earn money. Therefore, if I get a job, then I will buy a computer.

Symbolic Translation

LET P: I will earn money.
LET Q: I will buy a computer.
LET R: I get a job.

$P \rightarrow Q$
$R \rightarrow P$
$\therefore R \rightarrow Q$

Often a logical argument contains more than one logically valid reasoning pattern. Each example below uses two logically valid reasoning patterns.

English Argument

If I study all night, then I will miss my nighttime soap. If Jeannine comes over to study, then I study all night. Jeannine comes over to study. Therefore, I will miss my nighttime soap.

Symbolic Translation

LET P: I study all night.
LET Q: I will miss my nighttime soap.
LET R: Jeannine comes over to study.

$P \rightarrow Q$
$R \rightarrow P$
R
$\therefore Q$

You can show that this argument is valid in two logical steps:

$R \rightarrow P$
R
$\therefore P$ by Modus Ponens

$P \rightarrow Q$
P
$\therefore Q$ by Modus Ponens

English Argument

If the consecutive sides of a parallelogram are congruent, then the parallelogram is a rhombus. If the parallelogram is a rhombus, then the diagonals are perpendicular bisectors of each other. The diagonals are not perpendicular bisectors of each other. Therefore, the consecutive sides of the parallelogram are not congruent.

Symbolic Translation

LET P: The consecutive sides of a parallelogram are congruent.
LET Q: The parallelogram is a rhombus.
LET R: The diagonals are perpendicular bisectors of each other.

$P \rightarrow Q$
$Q \rightarrow R$
$\sim R$
$\therefore \sim P$

You can show that this argument is valid in two logical steps:

$P \rightarrow Q$
$Q \rightarrow R$
$\therefore P \rightarrow R$ by the Law of Syllogism

$P \rightarrow R$
$\sim R$
$\therefore \sim P$ by Modus Tollens

EXERCISE SET

In each problem below, if two of the premises fit the logically valid reasoning patterns of the Law of Syllogism, Modus Tollens, or Modus Ponens, state the conclusion in English. Set up a legend showing the sentence that each letter represents, and then translate the complete argument into symbolic form. If the statements do not fit the logically valid reasoning pattern of LS, MT, or MP, simply write *no valid conclusion.*

1.* If the treasure is discovered, then pirate Rufus T. Ruffian will walk his own plank. If Iris Noble finds the treasure map, then the treasure is discovered.

2. If Aunt Teak sells her 1964 Mustang, then used car dealer Nick Dixon will make a fortune on the deal. If used car dealer Nick Dixon makes a fortune on the deal, then he will fly to Hawaii.

3. If you use Shining Smile toothpaste, then you will be successful. If you use Shining Smile toothpaste, then you may become a TV star.

4. If a young woman is an athlete, then she probably scores high on spatial visualization tests. If she probably scores high on spatial visualization tests, then she will probably do well in geometry.

5. If you are open-minded, you will listen to both sides of a story. If you listen to both sides of a story, then you will make a more intelligent decision. If you make a more intelligent decision, then you will have a more successful career.

6. If two base angles of a triangle are congruent, then the triangle has two sides congruent. If two sides of a triangle are congruent, then the triangle is isosceles. The triangle is not isosceles.

7. If $\overline{EF}$ is a segment connecting the midpoints of the non-parallel sides $\overline{AD}$ and $\overline{BC}$ in trapezoid $ABCD$, then $\overline{EF}$ is a midsegment of trapezoid $ABCD$. If $\overline{EF}$ is a midsegment of trapezoid $ABCD$, then $\overline{EF}$ is parallel to side $\overline{AB}$. If $\overline{EF}$ is parallel to side $\overline{AB}$, then $ABFE$ is a trapezoid.

In Exercises 9 to 14, identify the symbolic argument as Modus Ponens (MP), Modus Tollens (MT), the Law of Syllogism (LS), or some combination. If the symbolic argument is not valid, write *no valid conclusion*.

8. $P \rightarrow R$
   $R \rightarrow S$
   $\therefore P \rightarrow S$

9. $S \rightarrow \sim Q$
   S
   $\therefore \sim Q$

10. $\sim Q \rightarrow \sim R$
    Q
    $\therefore R$

11. $T \rightarrow R$
    $P \rightarrow R$
    $\therefore T \rightarrow P$

12. $T \rightarrow S$
    T
    $P \rightarrow \sim S$
    $\therefore \sim P$

13. $\sim R \rightarrow S$
    $S \rightarrow P$
    $P \rightarrow \sim Q$
    $\therefore \sim R \rightarrow \sim Q$

Improving Reasoning Skills

Bagels III

In this puzzle you are to determine a 3-digit number (no digit repeated) by making "educated guesses." After each guess, you will be given a clue about your guess. The clues:

bagels: no digit is correct
pico: one digit is correct but in the wrong position
fermi: one digit is correct and in the correct position

In each of the problems below, a number of guesses have been made with the clue for each guess shown to the right. From the given set of clues, determine the 3-digit number. If there is more than one solution, find them all.

1.

2 3 4	*pico*
5 6 7	*pico*
8 9 1	*fermi*
6 4 1	*bagels*
8 2 5	*pico*
? ? ?	

2.

5 1 4	*pico*
9 6 7	*pico*
6 3 1	*pico*
3 9 2	*pico fermi*
8 0 7	*bagels*
3 5 9	*pico pico*
? ? ?	

Lesson 13.6

The Law of the Contrapositive

> *"You should say what you mean," the March Hare went on. "I do," Alice hastily replied; "At least – at least I mean what I say – that's the same thing, you know." "Not the same thing a bit!" Said the Hatter. "Why, you might just as well say that 'I see what I eat' is the same thing as 'I eat what I see'!"*
>
> *– Alice in Wonderland by Lewis Carroll*

Every conditional statement has three other conditionals associated with it. They are the converse, the inverse, and the contrapositive of the conditional statement.

To create the **converse** of a conditional statement, you simply reverse the two parts of the statement. To create the **inverse**, you negate the two parts. To create the **contrapositive**, you reverse *and* negate the two parts .

Statement:	$P \to Q$
Converse:	$Q \to P$
Inverse:	$\sim P \to \sim Q$
Contrapositive:	$\sim Q \to \sim P$

In logic, conditional statements are either true or false. If a statement is true, is its converse necessarily true? How about its inverse? Contrapositive? Let's look at one example of a true conditional and its converse, inverse, and contrapositive.

Example

Statement: If two angles are vertical angles, then they are congruent. ($P \to Q$)
The statement is *true*.

Converse: If two angles are congruent, then they are vertical angles. ($Q \to P$)
The converse is *false*.

Inverse: If two angles are not vertical angles, then they are not congruent. ($\sim P \to \sim Q$)
The inverse is *false*.

Contrapositive: If two angles are not congruent, then they are not vertical angles. ($\sim Q \to \sim P$)
The contrapositive is *true*.

EXERCISE SET A

In Exercises 1 to 4, write the converse, inverse, and contrapositive for each conditional statement. Then identify each statement, converse, inverse, and contrapositive as true or false.

1.* If it is a rose, then it is a flower.

2. If you're out of chocolate cake, you're out of dessert.

3. If the triangle is isosceles, then the base angles are congruent.

4. If ΔBOY is congruent to ΔGRL, then $BO = GR$.

In each exercise above, the contrapositive has the same truth value as the original conditional. This leads to our fourth form of logical reasoning, the Law of the Contrapositive.

> *The **Law of the Contrapositive** (LC) says that if a conditional statement is true, then its contrapositive is also true and conversely.*

Symbolically,	And conversely,
$P \rightarrow Q$	$\sim Q \rightarrow \sim P$
$\therefore \sim Q \rightarrow \sim P$	$\therefore P \rightarrow Q$

The Law of Contrapositive just says that any conditional and its contrapositive are logically the same.

From the exercises it should be clear that both a statement and its converse need not be true. A statement and its contrapositive always have the same truth value (LC). If one is true, then the other is true. If one is false, then the other is false. The same holds for the inverse and converse. They are either both true or both false.

So far, you have learned four basic forms of valid reasoning.

Four Forms of Valid Reasoning

$P \rightarrow Q$	$P \rightarrow Q$	$P \rightarrow Q$	$P \rightarrow Q$
P	$\sim Q$	$Q \rightarrow R$	$\therefore \sim Q \rightarrow \sim P$
$\therefore Q$	$\therefore \sim P$	$\therefore P \rightarrow R$	
by MP	by MT	by LS	by LC

EXERCISE SET B

Determine whether or not each logical argument is valid. If the argument is valid, tell what logical reasoning pattern or patterns are involved (MP, MT, LS, LC). If the argument is not valid, write *no valid conclusion.*

1.	2.	3.	4.*
$P \rightarrow Q$	$R \rightarrow S$	$P \rightarrow R$	$\sim P \rightarrow S$
P	S	$\sim R$	$\sim P$
$\therefore Q$	$\therefore R$	$\therefore \sim P$	$\therefore S$

5. $P \to Q$
$\sim P$
$\therefore \sim Q$

6. $R \to P$
$P \to S$
$\therefore R \to S$

7. $S \to P$
$R \to P$
$\therefore S \to R$

8.* $\sim(P \to Q)$
$\therefore \sim P \to \sim Q$

9.* $P \to Q$
$Q \to S$
$S \to T$
$\therefore P \to T$

10. $P \to \sim Q$
Q
$\therefore \sim P$

11. $\sim Q \to \sim R$
R
$\therefore Q$

12. $\sim S \to P$
$R \to \sim S$
$\therefore R \to P$

13. $P \to Q$
$\sim R \to \sim Q$
$\therefore P \to R$

14.* $\sim T \to \sim Q$
$S \to Q$
$\therefore S \to T$

15. R
$R \to (P \to T)$
$\therefore P \to T$

16. $(P \to Q) \to R$
$P \to Q$
$R \to S$
$\therefore S$

Improving Reasoning Skills

The Perils of Power and Passion

In the soap opera *The Perils of Power and Passion*, the character Richard Lanceworth is married to socialite Rita Lanceworth. Rita is played by actress Cybil Taylor. The heavy, Boris Dropdeadski, played by stage actor Winston Smythe, is romantically linked to the industrial spy Carlotta Vendetta. Captain I. M. Tuff of the S. F. police vice squad, played by ex-pro running back Foxy Brown, is about to marry Broadway dancer Laverne Seymour. In real life the actor Rocky Stone is married to the actress Carole Esteban. Carole Esteban, who plays the love interest of Boris in the soap, is a close friend of Diane Wilson who plays the dancer Laverne. All six of the actors and actresses in the soap are married to each other in real life, yet none of them plays a character romantically linked with whom he or she is married to in real life. Match up the names of each actor and actress who are married to each other in real life.

Lesson 13.7

Direct Proofs

There are three basic approaches to proving logical arguments. They are direct proofs, indirect proofs, and conditional proofs. In this lesson you will look at direct proofs. In a direct proof of a logical argument, you begin by stating the given information or premises, and then use valid patterns of reasoning, such as MP, MT, LS, and LC, to arrive directly at the conclusion.

Below is the symbolic form of a logical argument. The form is called a two-column proof. In a two-column proof, each statement in the logical argument is written in the left hand column. The reason for each statement is written directly across in the right hand column. To show that the argument is valid you use a direct proof.

Example of a Direct Proof

		Proof:	
Premises:	$P \rightarrow Q$	1. $P \rightarrow Q$	1. Premise
	$R \rightarrow P$	2. $\sim Q$	2. Premise
	$\sim Q$	3. $\sim P$	3. From lines 1 and 2 using MT
Conclusion:	$\sim R$	4. $R \rightarrow P$	4. Premise
		5. $\therefore \sim R$	5. From lines 3 and 4 using MT

Since the proof arrived at $\sim R$ by using the premises and logical reasoning, the argument is valid. You should be convinced that, if $P \rightarrow Q$, $R \rightarrow P$, and $\sim Q$ are true, then $\sim R$ must also be true (a proof is a convincing logical argument). This proof could have been written in more than one way. In proofs, there is usually more than one possible sequence of steps.

EXERCISE SET A

Copy the proof below, including the list of premises and conclusion. Provide the reason for each step in the proof of the logical argument.

		Proof:	
Premises:	P	1. $P \rightarrow Q$	1. —?—
	$P \rightarrow Q$	2. $Q \rightarrow R$	2. —?—
	$Q \rightarrow R$	3. $P \rightarrow R$	3.* —?—
	$R \rightarrow S$	4. $R \rightarrow S$	4. —?—
Conclusion:	S	5. $P \rightarrow S$	5. —?—
		6. P	6. —?—
		7. $\therefore S$	7. —?—

EXERCISE SET B

Copy the proof below, including the list of premises and conclusion. Provide the reason for each step in the proof of the logical argument.

Premises:		*Proof:*	
	$P \to Q$		
	$Q \to \sim R$	1. $Q \to \sim R$	1. —?—
	R	2. R	2. —?—
		3. $\sim Q$	3. —?—
Conclusion:	$\sim P$	4. $P \to Q$	4. —?—
		5. $\therefore \sim P$	5. —?—

EXERCISE SET C

Copy the proof below, including the list of premises and conclusion. Provide the reason for each step in the proof.

Premises:		*Proof:*	
	$P \to R$		
	$T \to S$	1. $T \to S$	1. —?—
	$\sim T \to P$	2. $\sim S$	2. —?—
	$\sim S$	3. $\sim T$	3. —?—
		4. $\sim T \to P$	4. —?—
Conclusion:	R	5. P	5.* —?—
		6. $P \to R$	6. —?—
		7. $\therefore R$	7. —?—

EXERCISE SET D

There can be more than one way to prove an argument. The argument in Exercise C can also be proved in the way shown below. Copy the proof below, including the list of premises and conclusion. Provide the reason for each step.

Premises:		*Proof:*	
	$P \to R$		
	$T \to S$	1. $T \to S$	1. —?—
	$\sim T \to P$	2. $\sim S \to \sim T$	2.* —?—
	$\sim S$	3. $\sim T \to P$	3. —?—
		4. $\sim S \to P$	4. —?—
Conclusion:	R	5. $P \to R$	5. —?—
		6. $\sim S \to R$	6. —?—
		7. $\sim S$	7. —?—
		8. $\therefore R$	8. —?—

EXERCISE SET E

Copy the proof below, including the list of premises and conclusion. Provide the missing steps and reasons in the proof below.

Premises: $S \to Q$
$P \to S$
$\sim R \to P$
$\sim Q$

Conclusion: R

Proof:

	Statement		Reason
1.	$P \to S$	1.	Premise
2.	$S \to Q$	2.	Premise
3.	$P \to Q$	3.	From lines 1 and 2 using —?—
4.	$\sim Q$	4.	—?—
5.	—?—	5.	—?—
6.	—?—	6.	—?—
7.	$\therefore$ —?—	7.	—?—

EXERCISE SET F

1. Provide the steps and reasons necessary to prove the following logical argument.

   Premises: $P \to Q$
   $\sim P \to R$
   $\sim R$

   Conclusion: Q

2. The following argument in English has been translated into a symbolic argument. Prove that the argument is valid.

English Argument	***Symbolic Translation***
If you listen to both sides of a story, then you will make an intelligent decision. If you are open-minded, then you will listen to both sides of a story. You did not make an intelligent decision. Therefore, you are not open-minded.	LET P: You listen to both sides of a story. LET Q: You will make an intelligent decision. LET R: You are open-minded. $P \to Q$ $R \to P$ $\sim Q$ $\therefore \sim R$

In Exercises 3 and 4, translate the English argument into symbolic form and then prove that the argument is valid.

3.* If all wealthy people are happy, then money can buy happiness. If money can buy happiness, then love doesn't exist. But love exists. Therefore, not all wealthy people are happy.

4. If 60 out of 104 majors at the local university require calculus, then calculus is an important class to take in college. If calculus is an important class to take in college, then high school students should take all the math necessary to enter calculus. In fact, 60 out of 104 majors at the local university require calculus. Therefore, high school students should take all the math necessary to enter calculus.

Lesson 13.8

Conditional Proofs

One of the most common methods of proof in mathematics is the conditional proof. You use a conditional proof to prove that a $P \rightarrow Q$ type of statement follows logically from a set of premises. In a conditional proof you begin by making an assumption. You assume that the first part of the conditional statement, called the **antecedent**, is true. Then you use logical reasoning to demonstrate that the second part of the conditional statement, called the **consequent**, must also be true. If you succeed in doing this, you have demonstrated that *If P is true, then Q must be true.* In other words, a conditional proof shows that the antecedent implies the consequent.

Example of a Conditional Proof

Premises: $P \rightarrow R$

$S \rightarrow \sim R$

Conclusion: $P \rightarrow \sim S$

Proof:

1. P	1. Assume the antecedent
2. $P \rightarrow R$	2. Premise
3. R	3. From lines 1 and 2 using MP
4. $S \rightarrow \sim R$	4. Premise
5. $\sim S$	5. From lines 3 and 4 using MT

Assuming P is true, the truth of $\sim S$ is established. $\therefore P \rightarrow \sim S$

The logical argument above was proved using a conditional proof format. The same argument could also have been proved by a direct proof format. Many logical arguments can be proved using more than one type of proof. With practice you will be able to tell which method you should use on a particular logical argument.

EXERCISE SET A

Copy the proof below, including the list of premises and conclusion. Provide the reason for each step in the proof.

Premises: $P \rightarrow R$

$S \rightarrow \sim R$

$\sim P \rightarrow Q$

Conclusion: $\sim Q \rightarrow \sim S$

Proof:

1. $\sim Q$	1. —?—
2. $\sim P \rightarrow Q$	2. —?—
3. P	3. From lines 1 and 2 using —?—
4. $P \rightarrow R$	4. —?—
5. R	5. From lines 3 and 4 using —?—
6. $S \rightarrow \sim R$	6. —?—
7. $\sim S$	7.* From lines 5 and 6 using —?—

Assuming —?— is true, the truth of —?— is established.
$\therefore \sim Q \rightarrow \sim S$

EXERCISE SET B

Copy the proof below, including the list of premises and conclusion. Provide the reason for each step in the proof of the logical argument.

Premises: $\sim R \rightarrow \sim Q$
$T \rightarrow \sim R$
$S \rightarrow T$

Conclusion: $S \rightarrow \sim Q$

Proof:

Statement	Reason
1. S	1. —?—
2. $S \rightarrow T$	2. —?—
3. T	3. From lines 1 and 2 using —?—
4. $T \rightarrow \sim R$	4. —?—
5. $\sim R$	5. From lines 3 and 4 using —?—
6. $\sim R \rightarrow \sim Q$	6. —?—
7. $\sim Q$	7. From lines 5 and 6 using —?—

Assuming S is true, the truth of $\sim Q$ is established. $\therefore$ —?—

EXERCISE SET C

Copy the proof below, including the list of premises and conclusion. Provide the reason for each step in the proof of the logical argument.

Premises: $\sim T \rightarrow S$
$T \rightarrow (R \rightarrow Q)$
$P \rightarrow \sim S$

Conclusion: $P \rightarrow (R \rightarrow Q)$

Proof:

Statement	Reason
1. P	1. —?—
2. $P \rightarrow \sim S$	2. —?—
3. $\sim S$	3. —?—
4. $\sim T \rightarrow S$	4. —?—
5. T	5. From lines 3 and 4 using —?—
6. $T \rightarrow (R \rightarrow Q)$	6. —?—
7. $R \rightarrow Q$	7. From lines 5 and 6 using —?—

Assuming P is true, the truth of —?— is established.
$\therefore P \rightarrow (R \rightarrow Q)$

EXERCISE SET D

Copy the proof below, including the list of premises and conclusion. Provide the reason for each step in the proof of the logical argument.

Premises: P
$Q \rightarrow R$

Conclusion: $(P \rightarrow Q) \rightarrow R$

Proof:

Statement	Reason
1. $P \rightarrow Q$	1. —?—
2. P	2. —?—
3. Q	3. —?—
4. $Q \rightarrow R$	4. —?—
5. R	5. —?—

Assuming $P \rightarrow Q$ is true, the truth of R is established.
$\therefore$ —?—

EXERCISE SET E

In Exercises 1 and 2, provide the steps and reasons necessary to prove the following logical arguments.

1. *Premises:* $P \to (Q \to R)$
   $R \to S$
   Q
   Conclusion: $P \to S$

2. *Premises:* $(P \to Q) \to R$
   $\sim(P \to Q) \to T$
   $R \to S$
   Conclusion: $\sim S \to T$

In Exercises 3 and 4, translate the argument into symbolic terms and then prove that the argument is valid. Some arguments may contain statements that do not logically support the conclusion. Be on guard for unnecessary statements.

3. If Evette is innocent, then Alfa is telling the truth. If Romeo is telling the truth, then Alfa is not. If Romeo is not telling the truth, then he has something to gain. Romeo has nothing to gain. Therefore, if Romeo has nothing to gain, then Evette is not innocent.

4.* If the consecutive sides of a parallelogram are congruent, then the parallelogram is a rhombus. If the parallelogram is a rhombus, then the diagonals are perpendicular bisectors of each other. If a parallelogram is a rhombus, then the diagonals bisect the angles. The diagonals are not perpendicular bisectors of each other in parallelogram *ABCD*. Therefore, the consecutive sides of parallelogram *ABCD* are not congruent.

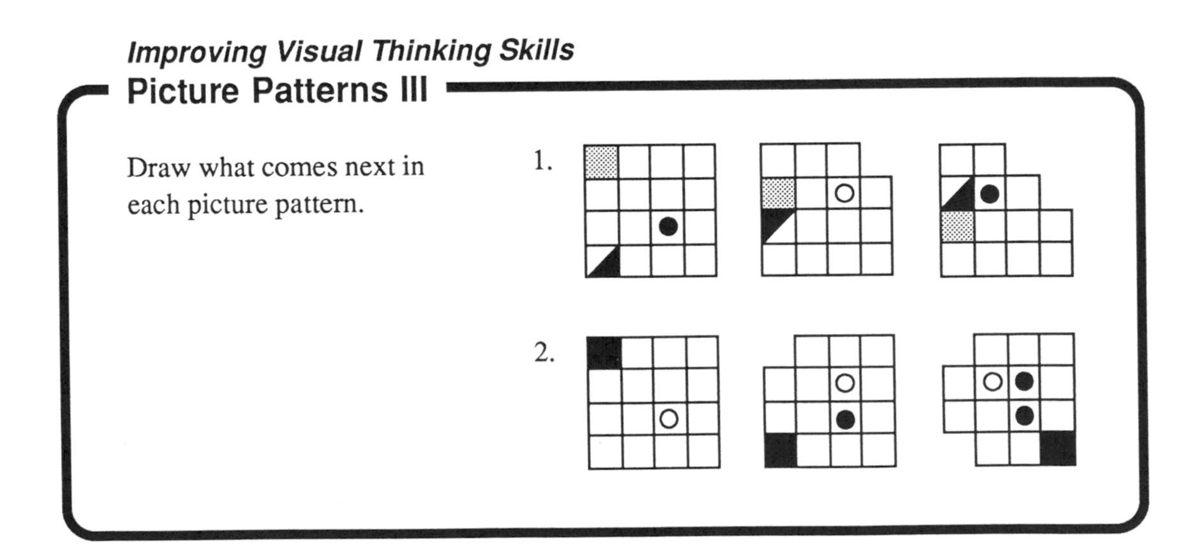

Lesson 13.9

Indirect Proofs

How often have I said to you that when you have eliminated the impossible, whatever remains, however improbable, must be the truth?

– Sherlock Holmes in The Sign of Four, Sir Arthur Conan Doyle

The third type of proof is an indirect proof. An indirect proof is a very clever, almost sneaky, approach to proving something. In an indirect proof, you first recognize all the possibilities that can be true. Next, you eliminate all but one possibility by showing that the others contradict some given fact or accepted idea. Therefore, you are forced to accept the one remaining possibility as proven to be true. The mystery story below is an example of an indirect proof.

> Five people alone on a tropical island have a loud argument after a monopoly game. When four of them return to the scene of the argument later in the day, they discover the fifth person seriously injured with an arrow in his back. The hero of the story, Sheerluck, a cousin of Sherlock Holmes, knowing that he didn't commit the crime, eliminates himself as a suspect. He eliminates his fiance as a suspect because she was with him the entire time. He eliminates Colonel Moran because the Colonel recently injured both arms and would therefore be unable to shoot with a bow and arrow. Sheerluck also eliminates the wounded man as a suspect since an arrow in the back could not be self inflicted. Since there is only one other person on the island, Sheerluck concludes that the guilty party is the fifth person, Sir Charles Mortimer.

Indirect reasoning can be a clever strategy in answering a multiple-choice question if you are unsure of the correct answer.

Example

In which year was a United States president born?

a. 1492 b. 1676 c. 1809 d. 1969

Assuming that you cannot select the correct choice immediately, let's look at the four choices. If you can eliminate three of them, the remaining choice must be the correct answer. You can eliminate 1492 and 1676 because the person would be a hundred years old or older in 1776, the year the U.S. became a country. You can also eliminate 1969 because anyone born in 1969 would still not be old enough to be president. The minimum age is thirty-five. Therefore, since you have only four choices, and you have eliminated three of them, the remaining choice, 1809, must be the correct one. Therefore, a United States President was born in 1809. Do you know who he was?

EXERCISE SET A

For each multiple choice question, determine the correct answer by indirect reasoning. Explain why each incorrect choice was eliminated.

1. What is the capital of Mali?
   a. Paris
   b. Tucson
   c. London
   d. Bamako
2. Which Italian scientist used the new invention, the telescope, to discover the moons of Jupiter?
   a. Sir Edmund Halley
   b. Julius Caesar
   c. Galileo Galilei
   d. Madonna
3. Which person twice won a Nobel Prize?
   a. Ringo Starr
   b. Madame Curie
   c. Joan of Arc
   d. Leonardo da Vinci

In mathematics and logic, if you wish to show that a statement is true, begin by assuming that it is not true and then demonstrate that this assumption leads us to contradict a true statement. For example, if you are given a set of premises and are asked to show that some conclusion, say P, is true, begin the proof by assuming that the opposite of P, namely $\sim P$, is true. Then show that this assumption leads to a contradiction of an earlier statement. If $\sim P$ led to a contradiction, it must be false and P must be true.

Let's look at an example of an indirect proof.

Example of an Indirect Proof

Premises: $R \to S$
$\sim R \to \sim P$
P

Conclusion: S

Proof:

	Statement		Reason
1.	$\sim S$	1.	Assume opposite of conclusion
2.	$R \to S$	2.	Premise
3.	$\sim R$	3.	From lines 1 and 2 using MT
4.	$\sim R \to \sim P$	4.	Premise
5.	$\sim P$	5.	From lines 3 and 4 using MP
6.	P	6.	Premise

But lines 5 and 6 contradict each other.
Therefore, $\sim S$, the assumption, is false. $\therefore S$

EXERCISE SET B

Copy the proof below, including the list of premises and conclusion. Provide the reason for each step in the proof of the logical argument.

Premises: $P \to Q$
$R \to P$
$\sim Q$

Conclusion: $\sim R$

Proof:

Statement	Reason
1. R	1. Assume the opposite of the —?—
2. $R \to P$	2. —?—
3. P	3. From lines 1 and 2 using —?—
4. $P \to Q$	4. —?—
5. Q	5. From lines –?– and –?– using MP
6. $\sim Q$	6. —?—

But lines 5 and 6 contradict each other.
Therefore, R, the assumption, is false. $\therefore \sim R$

EXERCISE SET C

Copy the proof below, including the list of premises and conclusion. Provide the reason for each step in the proof of the logical argument.

Premises: $P \to (Q \to R)$
$Q \to \sim R$
Q

Conclusion: $\sim P$

Proof:

Statement	Reason
1. P	1. Assume the —?— of the —?—
2. $P \to (Q \to R)$	2. —?—
3. $Q \to R$	3.* —?—
4. Q	4. —?—
5. R	5. —?—
6. $Q \to \sim R$	6. —?—
7. $\sim R$	7. From lines –?– and –?– using MP

But lines 5 and 7 contradict each other.
Therefore, P, the assumption, is false. $\therefore \sim P$

EXERCISE SET D

Copy the proof below, including the list of premises and conclusion. Provide the missing steps and reasons in the proof of the logical argument.

Premises: $(P \to S) \to \sim Q$
$\sim Q \to R$
$\sim R$

Conclusion: $\sim(\sim S \to \sim P)$

Proof:

Statement	Reason
1. $\sim S \to \sim P$	1. Assume the —?— of the —?—
2. —?—	2. —?—
3. —?—	3. —?—
4. —?—	4. —?—
5. —?—	5. —?—
6. —?—	6. —?—
7. —?—	7. —?—

But lines –?– and –?– contradict each other.
Therefore, $\sim S \to \sim P$, the assumption, is false. $\therefore \sim(\sim S \to \sim P)$

EXERCISE SET E

It's your turn. In Exercises 1 and 2, provide the steps and reasons to prove the logical arguments.

1. *Premises:* $(R \to S) \to P$
   $T \to Q$
   $\sim T \to \sim P$
   $\sim Q$
   Conclusion: $\sim(R \to S)$

2. *Premises:* $P \to (R \to Q)$
   P
   $\sim(\sim Q \to S)$
   Conclusion: $\sim(\sim R \to S)$

In Exercises 3 and 4, translate the arguments and then prove that they are valid.

3. If Clark is a mathemagician, then Lois is his assistant. If Clark is a mathemagician, then everyone at the theater has a good time. If everyone at the theater has a good time, then Lois is not his assistant. Therefore, Clark is not a mathemagician.

4.* If King Kong Curly defeats Rocking Rosco, then if Bull Moose Morgan vanquishes Wild Bill Higgins, King Kong Curly will be the World Champion of the *All*Star*Wrestling*Association* (ASWA). If King Kong Curly becomes the World Champion of ASWA, then Curly will retire from wrestling and devote himself to bonsai gardening. Bull Moose Morgan will vanquish Wild Bill Higgins. King Kong Curly will not retire from wrestling and devote himself to bonsai gardening. Therefore, King Kong Curly will not defeat Rocking Rosco.

Lesson 13.10

Chapter Review

EXERCISE SET A

Match.

1.	IF-THEN statements	a.	Modus Ponens
2.	P in a $P \to Q$ statement	b.	Modus Tollens
3.	Q in a $P \to Q$ statement	c.	Law of Syllogism
4.	Converse of $P \to Q$	d.	Conditional statements
5.	Inverse of $P \to Q$	e.	Conclusion
6.	Contrapositive of $P \to Q$	f.	Premises
7.	If $P \to Q$ is true and P is true,	g.	$Q \to P$
	then Q is true	h.	$\sim P \to \sim Q$
8.	If $P \to Q$ is true and $\sim Q$ is true,	i.	$\sim Q \to \sim P$
	then $\sim P$ is true	j.	Antecedent
9.	If $P \to Q$ is true and $Q \to R$ is	k.	Consequent
	true, then $P \to R$ is true	l.	Contradiction
10.	Given statements in an argument	m.	Sherlock Holmes
11.	Final statement in an argument	n.	Dr. Watson
12.	Master of Deduction	o.	Professor Moriarty

EXERCISE SET B

In Exercises 1 to 4, translate each English statement to symbolic notation.

LET P: Smoking causes cancer. LET Q: It is dangerous to smoke.

LET R: Smoking is glamorous. LET S: Only fools smoke.

1. If smoking causes cancer, then only fools smoke.
2. If smoking is glamorous, then smoking doesn't cause cancer.
3. If it is not dangerous to smoke, then it is not true that only fools smoke.
4. It is not true that if smoking is glamorous, then smoking doesn't cause cancer.

In Exercises 5 to 8, translate each symbolic statement into English.

LET P: Students are smart.

LET Q: Students will study.

LET R: Teachers are human.

LET S: They make mistakes.

5. $P \to Q$

6. $\sim S \to \sim R$

7. Converse of $\sim P \to S$

8. Inverse of $S \to P$

EXERCISE SET C

Copy the proof below, including the list of premises and conclusion. Provide the missing steps and reasons in the proof.

Premises: $(\sim Q \to P) \to \sim R$

$T \to S$

$\sim R \to \sim S$

T

Conclusion: $\sim(\sim P \to Q)$

Proof:

Statements	Reasons
1. $\sim P \to Q$	1. Assume the opposite of the conclusion
2. $\sim Q \to P$	2. From line 1 using LC
3. —?—	3. —?—
4. —?—	4. —?—
5. —?—	5. —?—
6. —?—	6. —?—
7. —?—	7. —?—
8. —?—	8. —?—
9. —?—	9. —?—

But lines –?– and –?– contradict each other.
Therefore, $\sim P \to Q$, the assumption, is false. $\therefore \sim(\sim P \to Q)$

EXERCISE SET D

In Exercises 1 and 2, provide the steps and reasons to prove the logical arguments.

1. *Premises:* $\sim P \to (Q \to \sim W)$

   $\sim S \to Q$

   $\sim T$

   $P \to T$

   Conclusion: $W \to S$

2. *Premises:* $(P \to Q) \to (R \to S)$

   $(\sim S \to \sim R) \to T$

   Conclusion: $\sim T \to \sim(P \to Q)$

3. Translate the argument below into symbolic form.

   If Sherlock Holmes is successful, then Professor Moriarty will be apprehended. If Dr. Watson doesn't slip up, then Sherlock Holmes will locate the missing clue. If Sherlock Holmes locates the missing clue, then he is successful. Dr. Watson doesn't slip up. Therefore, Professor Moriarty will be apprehended.

4. Prove that the argument in Exercise 3 is valid.

Computer Activity

Tessellating with Logo

In this computer activity you will use the Logo computer language to explore tessellations. This computer activity will take more time to complete and require a more extensive knowledge of the Logo than the earlier computer activities. You'll probably want to work on it over an extended period of time. For this activity, you will need:

- Computer and disk with Logo computer language
- Resource person familiar enough with Logo to help you get started and to show you how to edit Logo procedures
- Disk with the file of procedures called ESCHER.TESS (The listings of the procedures are in *Appendix B: Logo Procedures*.)

Start up your computer and load the Logo file ESCHER.TESS from your procedures disk. The file includes four primary superprocedures, variations of them, and examples of Escher-like transformations of regular polygons to help you create tessellations.

- RPOLY :N :S requires two inputs: the number of sides (:N) and the side-length (:S). It draws a regular polygon.
- TRANSPOLY :N :S :L requires three inputs: the number of sides and side-length of the regular polygon, and the level (:L). It makes a translation to the third vertex of the original polygon and draws a new image of the polygon. TRANSPOLY is recursive (it uses itself as a subprocedure) and continues the translation process to a level of recursion indicated by the third input (:L). It then returns the turtle to its beginning position.
- TESSELLATE :N :S :L1 :L2 requires four inputs: the number of sides and side-length of the regular polygon, the level of recursion for TRANSPOLY (:L1), and its own level of recursion (:L2). The two levels of recursion determine the dimensions (number of polygons along edges) of the tessellation. TESSELLATE will tessellate a section of the plane by translating the result of TRANSPOLY.
- WHOLETESS :N :S :L1 :L2 requires the same inputs as TESSELLATE and will tessellate the whole plane (to the levels specified by :L1 and :L2) by rotating the result of TESSELLATE through the interior angle of the regular polygon.

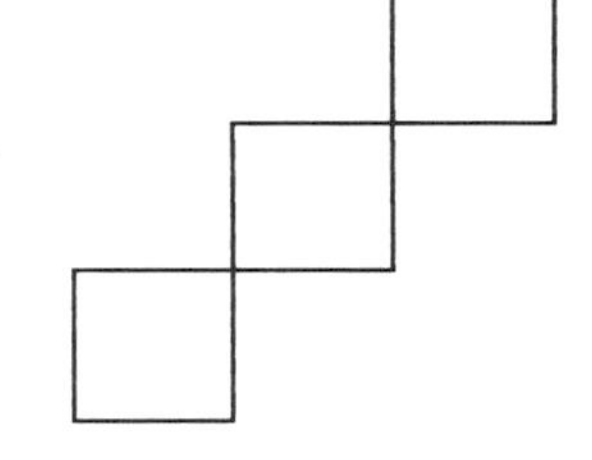

TRANSPOLY 4 40 3

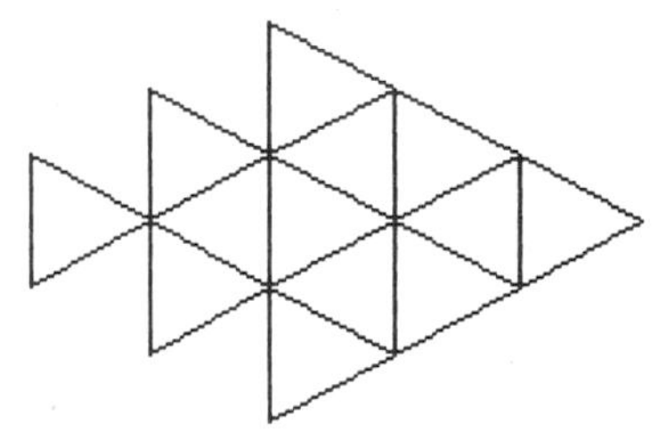

TESSELLATE 3 40 3 3

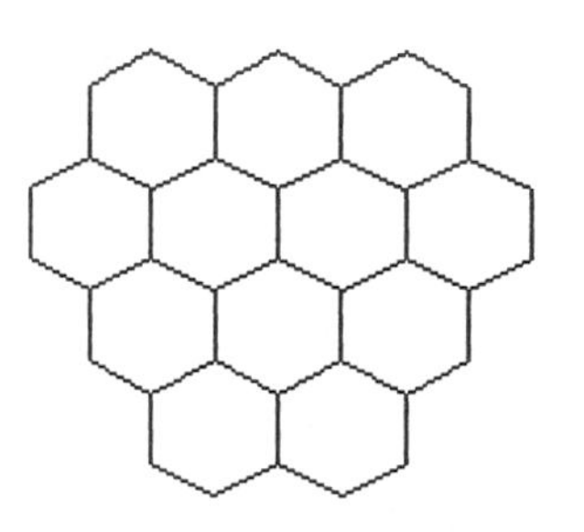

WHOLETESS 6 20 2 2

1. Experiment with the four superprocedures. Draw some regular polygons with RPOLY. Use TRANSPOLY with appropriate inputs to explore different levels of translation of different regular polygons. Use TESSELLATE with the number 2 as the last input to study the second order translation (the translation of TRANSPOLY). Try different regular polygons. Use TESSELLATE with higher levels of recursion to determine which regular polygons will tessellate (completely cover the plane without leaving gaps and without overlapping).

2. Verify your predictions with WHOLETESS. Try using WHOLETESS with five-sided and eight-sided regular polygons. Use side-lengths in the range 20 to 40 turtle steps. Use levels in the range 3 to 6.

Changing the Shape of the Regular Polygon

It is possible to create very beautiful and intriguing designs based on tessellations of shapes that are distortions of the regular polygons which tessellate. The Dutch artist M. C. Escher created many famous works of art based on regular tessellations of the plane. For the distorted polygon to tessellate, certain rules must be followed which preserve certain symmetries of the shape. The Logo procedures you've explored work because of the symmetries of the polygons: opposite sides and consecutive sides are congruent, and they have rotational symmetry. Any distortions of the polygons which preserve these properties will also tessellate using the procedures.

Point-Symmetric Distortions of Each Side

A curve is point-symmetric if when rotated about its midpoint 180 degrees, it coincides with itself. Any distortions of each side of a regular polygon based on a point-symmetric curve (or path) will create figures which tessellate in the same manner as the parent polygon.

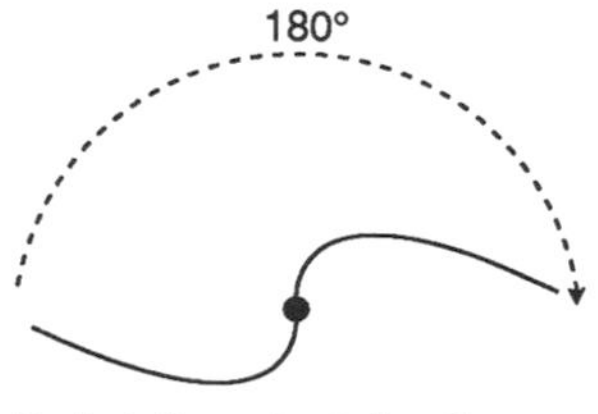

Point-Symmetric Curve

In each of the four primary superprocedures listed on the previous page, one procedure (RPOLY) is used to create the regular polygons, and one command (FD) in that procedure draws each side of the polygon. Thus, in order to create a point-symmetric distortion of each side of a regular polygon, all you have to do is replace the FD command in RPOLY with a new command which draws a point-symmetric path and results in an equivalent forward displacement of the turtle.

3. The procedures CFD and SFD in the ESCHER.TESS file require the same input as FD and move the turtle the same distance forward; however, they take the turtle on a point-symmetric curved path. Edit RPOLY and replace the FD command with SFD. Experiment with WHOLETESS and admire the results! Use inputs which result in tessellations. Edit RPOLY again and replace SFD with CFD. Experiment again with WHOLETESS.

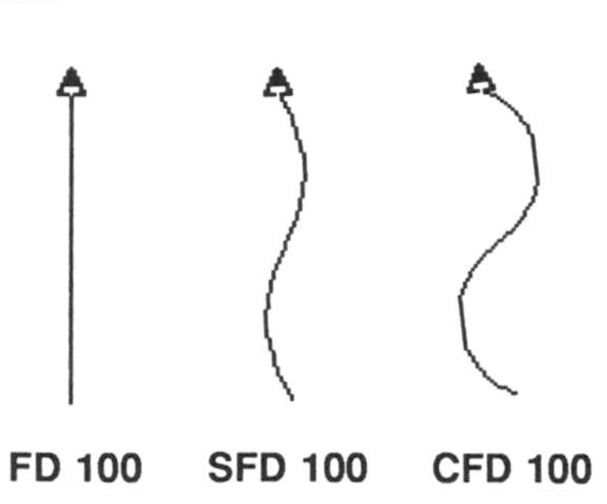

The ability to create completely new designs by changing just one command is an example of Logo's power. The SIMFD procedure will help you create point-symmetric pathways. SIMFD will act just like the FD command except that the turtle will travel on a point-symmetric path instead of a straight line. SIMFD uses a subprocedure called PATH which you must define according to certain rules. PATH need not be point-symmetric; SIMFD creates a point-symmetric curve using PATH.

- PATH must take one input which determines how far it will move in the current direction of the turtle. (Use :D as this input.)
- The endpoint of the path must coincide with the endpoint of a straight line drawn with the FD command using the same input as PATH. (In other words, FD :D and PATH :D must leave the turtle in the same location.)
- The turtle must return to its beginning position and be heading the same direction as when it started. (Use PU and PD with turtle moves.)
- The path must not cross itself.

By redesigning and redefining PATH, you can redesign the tessellations created by the procedures that use PATH. Different paths will create different tessellations.

You can use grid paper to design a path. Make a dashed segment of arbitrary length, say ten units, on the grid. This is your unit length. Draw a path which starts at the beginning of your dashed segment and connects points on your grid, ending up at the endpoint. You may move to either or both sides from your dashed segment, but do not move left or right by more than your unit length.

Once you've designed a path on the grid, you can work out the angles through which the turtle will have to turn to trace your path and the distances the turtle will have to move *as a fraction of your unit length.* Then you can translate the path into turtle moves to define a PATH procedure. (If your unit length :D was ten grid units, then each move of one grid unit would translate to a turtle move of :D / 10 turtle steps.)

4.* Complete the Logo procedure to translate the grid path below into a PATH procedure.

```
TO PATH :D
RT 90 FD :D / 10
LT 90 FD (:D / 10) * 5
:
PU BK :D PD
END
```

5.* Translate the grid path below into a PATH procedure. The unit length (:D) for the path is 10 grid units. The triangles are equilateral.

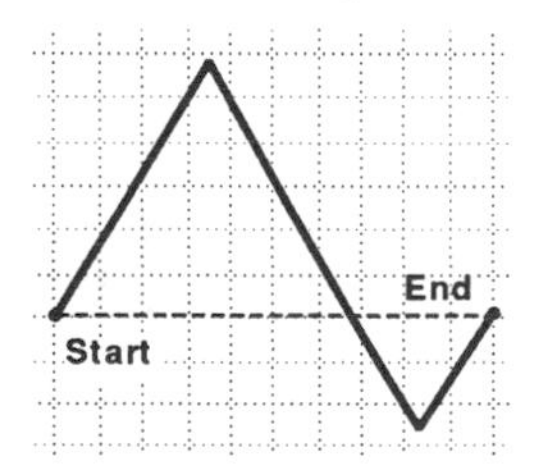

6.* Draw the path created by the following procedure on a grid:

```
TO PATH :D
LT 45 FD :D / SQRT 2
LT 45 BK :D / 2
RT 90 FD :D / 2
PU BK :D PD
END
```

Note: SQRT outputs the square root of its input.

7. Now comes the fun part! Use grid paper to design your own path. Follow the rules. Translate your path into a PATH procedure. Try a simple rectangular path or a path based on equilateral triangles to start. When you think you have defined your PATH procedure, try it using the command PATH 100. Does it follow the rules? Try PATH 100 followed by FD 100 to check.

8. After you have defined your PATH, change the FD command in RPOLY to SIMFD. Try RPOLY with different numbers of sides to see the effects of your new path. Use TESSELLATE to check that your new shape will tessellate through translation. Use WHOLETESS to check that it will tessellate through rotation.

9. Experiment with other paths. You can build a collection of paths by defining them with slightly different names like PATHA, PATHB, etc. To use these, simply change the PATH command in SIMFD by adding the suffix you used (for instance, change PATH to PATHA). Make a variety of tessellating designs. Be careful — tessellating with Logo can become addictive!

10. Procedures which draw arcs of circles have been supplied in ESCHER.TESS so that you can create curved paths. Procedures PATH2, PATH3, and PATH4 are examples of curved paths. To use them, simply change the PATH command in SIMFD by adding the numerical suffix. Try these curved paths.

11. If you have access to a printer, print out your favorite designs. (Ask your computer resource person how to print in your version of Logo.) Fill in details on your shapes to create faces, animals, etc. If you are familiar with Logo and your version of Logo has a FILL command, use turtle moves and FILL to fill parts of your design with solid shading or color. The design on the right was created using the SFD command instead of the FD command in RPOLY and WHOLETESS 3 40 2 2. FILL was used to fill alternate shapes.

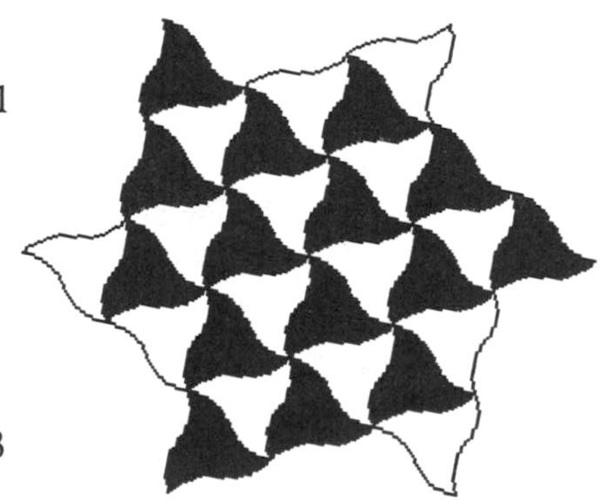

Using a Shape Name as Input to Your Tessellation Procedures

The procedures TRANS :SHAPE :N :S :L, TESS.SHAPE :SHAPE :N :S :L1 :L2, and WHOLE.SHAPE :SHAPE :N :S :L1 :L2 work just like TRANSPOLY, TESSELLATE, and WHOLETESS except that you need to tell these new procedures which shape to use. They will work with any shape that you have defined based on point-symmetric distortions of the sides of a regular polygon. Your shape *must* use two inputs: number of sides and side-length in that order. These new procedures are useful because they make it easy for you to save and use new shapes that you have created instead of changing the FD command in RPOLY, or the PATH procedure in SIMFD each time. You can create a family of point-symmetric poly shapes based on different paths that you create, and then use these new procedures to tessellate them.

12. SPOLY and CPOLY use SFD and CFD respectively. To tessellate with SPOLY, simply use "SPOLY as the first input to TESS.SHAPE or WHOLE.SHAPE. (You must quote your shape name when using it as input.) Try TESS.SHAPE "SPOLY 3 50 3 3 and WHOLE.SHAPE "SPOLY 3 50 3 3. Try TESS.SHAPE and WHOLE.SHAPE with CPOLY.

Distortions of Regular Polygons Based on Translations of Opposite Sides

These new procedures also give you an opportunity to experiment with distortions of sides of regular polygons which are not point-symmetric. They can be used with distortions of squares and hexagons where a distorted side (not necessarily point-symmetric) is translated to the opposite side. The procedures BSQ, BHEX, RSQ, and RHEX are created from a square and a regular hexagon by translation of a distorted side to the opposite side of the polygon. These new shapes will fit together when translated, but what happens when they are rotated?

13. Try TESS.SHAPE "BSQ 4 30 3 3. Try WHOLE.SHAPE "BSQ 4 30 3 3. Are there "holes" in this pattern? Try the same thing with RSQ.

14. Try TESS.SHAPE "BHEX 6 30 3 3 and WHOLE.SHAPE "BHEX 6 30 3 3. Are there any holes in this pattern? Do the same things with RHEX.

15. Why do squares and hexagons appear to behave differently even though we used the same paths for each side of the transformed polygons?

16. You can create your own translation transformations of squares and hexagons by changing the BFD and RFD procedures. Note that these come in pairs. What is the relationship between BFD1 and BFD2? How can this relationship create a translation of one side to the opposite when these procedures are used inside BSQ and BHEX? Do the resulting shapes have reflective symmetry? If so, draw the lines of symmetry on printouts of each shape.

Rotation Tessellations Based on Regular Polygons

The procedures TESSELLATE, WHOLETESS, TESS.SHAPE, and WHOLE.SHAPE all use translation of a shape to a point on the original shape to create the tessellations. Escher created some tessellations which were based on the rotation of consecutive sides of the regular polygon, rather than translation to the opposite side or point-symmetric distortions of each side. One of Escher's most famous tessellating designs is his *Study of Regular Division of the Plane with Reptiles* (Lesson 7.8). The reptile was created from a regular hexagon by distorting a side and then rotating this side about the common vertex to form the consecutive side. Thus, three separate distortions created three pairs of consecutive sides of the hexagon. The procedure REPTILE :S will create a shape very similar to Escher's reptile. It requires only one input — the side-length of its parent hexagon.

17. Clear the screen and type REPTILE 50. Then change the pen color (ask your resource person for help if you're not sure how to do this) and draw the parent hexagon by typing RPOLY 6 50. Now locate the vertices about which the distorted sides were rotated. Can you see how the rotated sides match in pairs?

18.* Rotate REPTILE through 120 degrees (the interior angle of a hexagon) using the command: REPEAT 3[REPTILE 45 RT 120]. You'll get three interlocking reptiles like those on the right. To see the regular hexagons on which the reptiles were based, leave the reptiles on the screen, change the pen color, and give the command: REPEAT 3[RPOLY 6 45 RT 120]. (Check RPOLY first and make sure it uses the original FD command and not CFD, SFD, or SIMFD.) The group of three reptiles can tessellate by translation. Can you see where to move the turtle so that another REPEAT 3[REPTILE 45 RT 120] command will draw three new reptiles that fit with the first three? You can see if you are correct by moving the turtle to the new position using the turtle move commands (FD, BK, RT, and LT) and then issuing another REPEAT 3[REPTILE 45 RT 120].

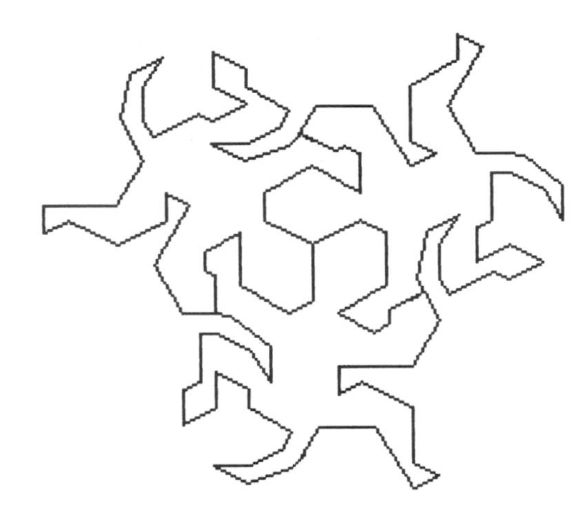

REPEAT 3[REPTILE 45 RT 120]

19. The procedures ROT.HEX :SHAPE :S :L, ROTATE.6 :SHAPE :S :L1 :L2, and WHOLE.HEX :SHAPE :S :L1 :L2 will tessellate any shape formed by rotation of consecutive sides of a regular hexagon. Try the procedures with the REPTILE. Try ROT.HEX "REPTILE 30 2 and WHOLE.HEX "REPTILE 20 3 3, for example. You might want to print and decorate your results.

The procedure HEX.SIDES :S will create a figure based on rotation of consecutive sides of a hexagon. You can create three pairs of sides in the same way that you created paths with the PATH procedure. You must call the three side procedures HSIDE1, HSIDE2, and HSIDE3. Each must take a single input which defines the length of the side of the parent hexagon. Each must create a path which moves the turtle to a position as if it had moved directly forward the length of the side of the parent hexagon. The turtle must return to the beginning of the path and be heading in its original direction. (See the earlier instructions for creating PATH.) Examples of HSIDE procedures are included in the ESCHER.TESS file.

20. Use HEX.SIDES :S with WHOLE.HEX to create a tessellating design by typing WHOLE.HEX "HEX.SIDES 30 3 3. Now edit the HSIDE procedures to create your own tessellating designs.

Another type of transformation which Escher used on equilateral triangles combined the rotation of one side through the common vertex to form the consecutive side with a point-symmetric distortion of the third side. The procedure BIRD.TRI :S was formed using these transformations. TSIDE1 :S is the rotated side and SIMFD1 :S is the point-symmetric side created with PATH1.

21. Will BIRD.TRI tessellate using TESS.SHAPE or WHOLE.SHAPE? Find out by using BIRD.TRIN which creates the same design as BIRD.TRI but requires two inputs, the first being 3 (the number of sides of a triangle).

22. Will BIRD.TRI tessellate with ROT.HEX, ROTATE.6 or WHOLE.HEX? Try it and find out. Explain what happens.

23. The procedures ROT.TRI :SHAPE :S :L, ROTATE.3 :SHAPE :S :L1 :L2, and WHOLE.TRI :SHAPE :S :L1 :L2 can be used with any equilateral triangle based shape formed using the above transformations. Try the procedures with the BIRD.TRI. For example, try ROT.TRI "BIRD.TRI 65 1 and WHOLE.TRI "BIRD.TRI 40 1 1. You might want to print and decorate your results.

ROT.TRI "BIRD.TRI 65 1

24. The procedure TRI.SIDES :S has been designed to help you create shapes that will tessellate with WHOLE.TRI. To use TRI.SIDES you need to define a PATH procedure for the point-symmetric side and a TSIDE procedure for the rotated side. Follow the same instructions as for your HSIDE procedures. (You could just change the name of one of your HSIDE procedures to TSIDE in the Logo editor. You will not lose the HSIDE procedure when you do this.) The ESCHER.TESS file contains examples of TSIDE and PATH procedures. (You might want to check and see that your ESCHER.TESS file contains the original forms of these procedures since you worked with them and may have changed them. If they have been changed, you can look at the listing in *Appendix B* and re-edit them to restore them to their original forms.) The figure on the right was created using these and the command WHOLE.TRI "TRI.SIDES 40 1 1.

WHOLE.TRI "TRI.SIDES 40 1 1

25. Experiment with your own transformations of regular polygons and use the tessellating procedures to create your own unique designs.

Cooperative Problem Solving

Logic in Space

1. Three mysterious black boxes recently appeared in the cargo bay of the space station *Entropy*. The three boxes are labeled "widgets," "gadgets," and "widgets and gadgets." A note was included, obviously by some prankster on the recreational space station *Empyrean*, which read, "Each label is incorrect. But it is possible to determine the correct label for each box by selecting only one object from one of the boxes. How?" The note was signed — Space Happy. How indeed?

2. On the lunar space port Galileo there are three married couples. Leroy, Martin, and Nikko are married to Linda, Michelle, and Nancy, but not necessarily in that order. Nancy, who is Leroy's cousin, has no children. Nikko and Linda are neighbors and Nikko's son plays on the same team as one of Linda's children. Leroy has three children, all girls. Match up the couples. Who is married to whom?

3. The six astronaut crew of the space station *Entropy* recently received four small visitors from Titan. The six astronauts and the four Titans are scheduled to attend the Lunar Space Conference at the lunar space port Galileo. The only two operating lunar shuttles, however, can carry just two Titans or one astronaut on any one trip between the station and moon surface. How many one way shuttle trips must be made to convey all ten conventioneers to the Lunar Space Conference?

4. When all shuttles are in service, it becomes quite busy at the docking stations of the space station *Entropy* and its sister station, the recreational space port *Empyrean.* This active schedule works out well for the seven close friends and space cadets of the graduating class of 2043. Adrianne is stationed on *Entropy* while Brenda, Carl, Dianna, and Eduardo are stationed on *Empyrean.* Their other two cadet classmates, Fred and Gianna, are stationed at the lunar research facility Galileo II. Brenda's job brings her to *Entropy* every second day. Carl's schedule brings him to *Entropy* every third day. Dianna's work requires her to check in at *Entropy* every fourth day. Eduardo's duties send him to the station every fifth day. Cadets Fred and Gianna have projects that require trips up to *Entropy* every sixth and seventh day respectively. Today they have all, to their surprise, arrived on the same day. They have agreed to celebrate the next time they are all together here at *Entropy.* How many days from today will this happen?

5. Lunar explorer Eugene O. Regan, based at the lunar research facility Galileo II, is planning an 800 kilometer trip across the lunar surface to the U.N. Research Facility. Eugene has a logistics problem since the lunar rover at the base can carry only enough fuel to travel a distance of 500 kilometers. Of course, there are no fuel stations between the two bases so he must build up a series of fueling points along the route to store fuel. The lunar rover fuel is packed in 100-kilometer units. How many full loads of fuel will Eugene need to complete his trip to the U.N. Research Facility?

14 Geometric Proofs

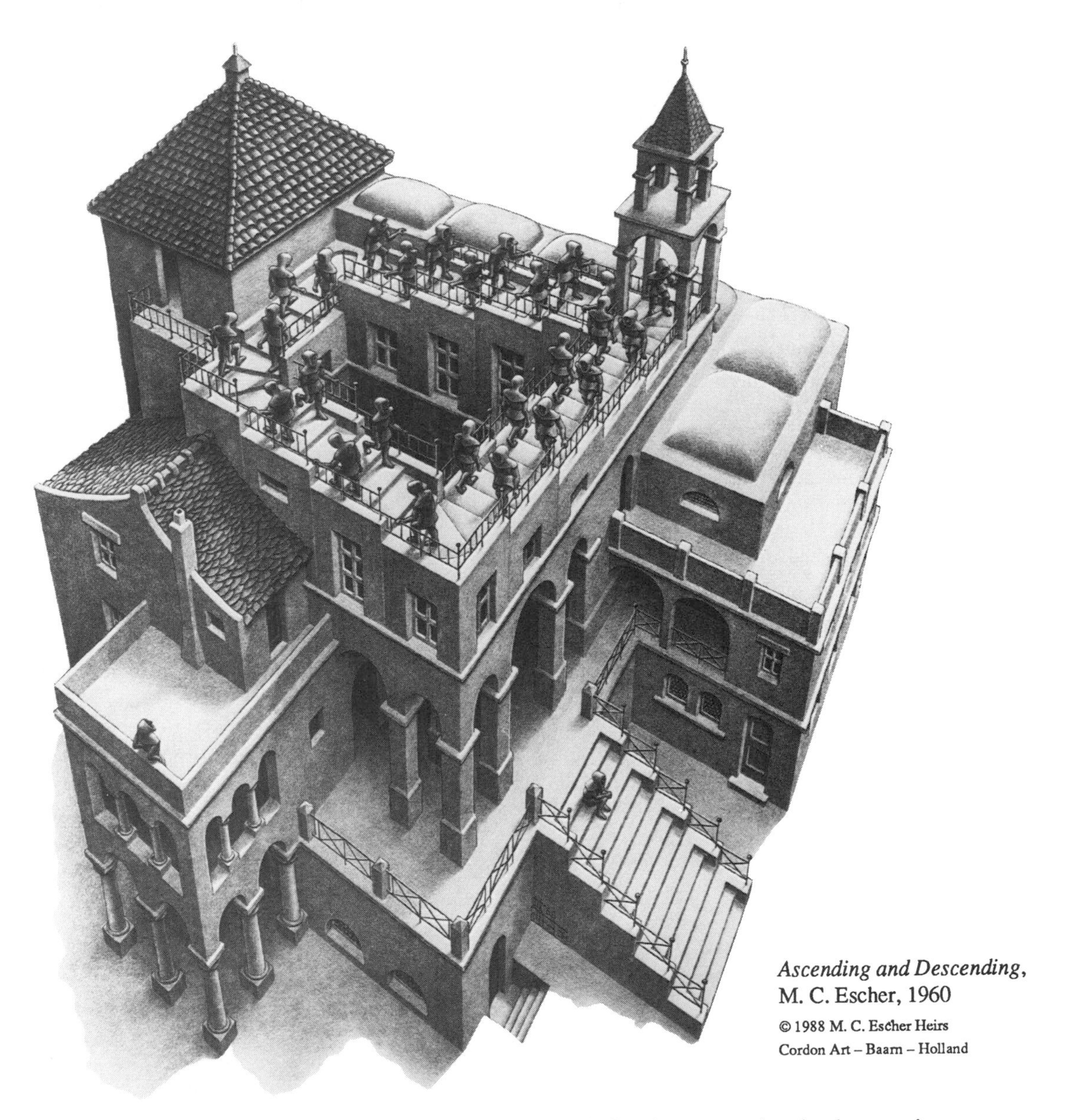

Ascending and Descending, M. C. Escher, 1960

In this chapter you will use what you learned about deductive reasoning in the previous chapter to prove many of your earlier conjectures. You will use flow charts to aid you in planning your proofs. You will prove geometric conjectures by direct proof or indirect proof. You will see how some discoveries are logically related to each other.

Lesson 14.1

Premises of Geometry

Geometry began as a collection of "rules-of-thumb" developed over thousands of years by the Babylonians and Egyptians. Some of the rules were used to compute simple areas and volumes. The rules were used in surveying to reestablish land boundaries after floods. They were practical instructions for building canals and tombs. They were the result of centuries of trial and error.

By 600 B.C., however, there was a new development. Thales of Miletus, a Greek mathematician and one of the "Seven Wise Men of Antiquity," like other mathematicians before him, made a number of geometric conjectures. More importantly, however, Thales supported his discoveries with logical reasoning. Over the next 300 years, the process of supporting mathematical conjectures with logical arguments became more and more refined. Other Greek mathematicians began showing that separate chains of logical reasoning could be linked together. In his famous work about geometry and number theory, *The Elements*, Euclid established a single chain of deductive arguments for all of geometry. Euclid started from a collection of statements that he regarded as obviously true (**postulates**). Then he systematically demonstrated that one after another geometric discovery could be shown to follow logically from his postulates and his previously verified conjectures (**theorems**). In doing this, Euclid created a deductive system. A **deductive system** consists of a set of premises and the laws of logic.

Up to now, you have been learning geometry inductively, the way early civilizations learned geometry. You observed geometric figures and made conjectures about them. For example, you discovered that the sum of the measures of the three angles of every triangle is 180. You did not measure the three angles of every triangle. Instead, you studied enough triangles to become convinced that the conjecture is true.

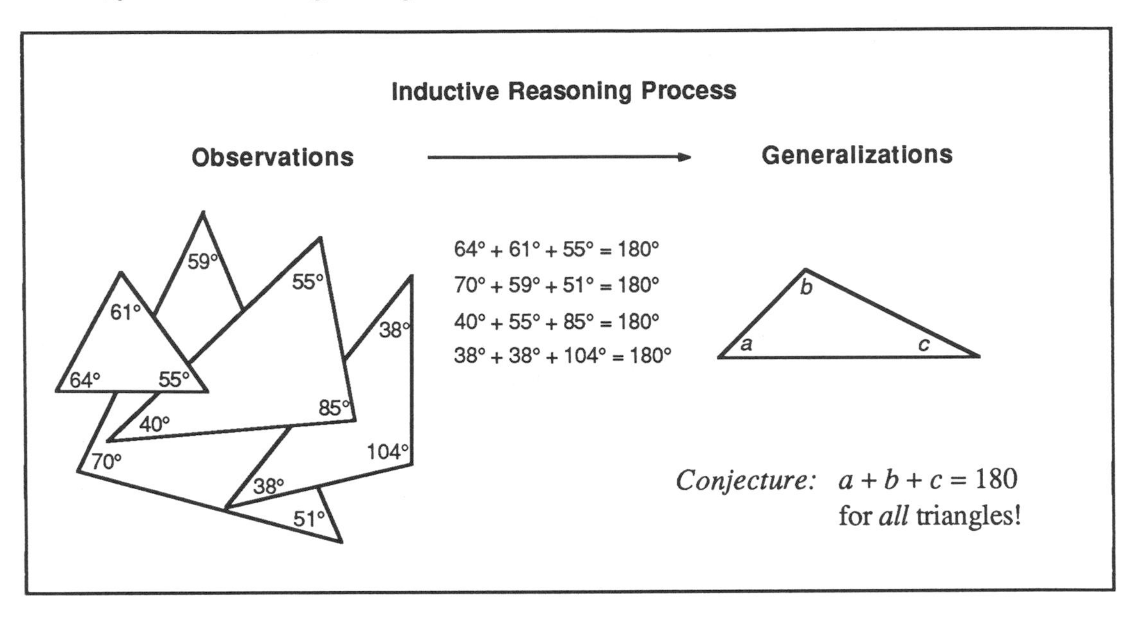

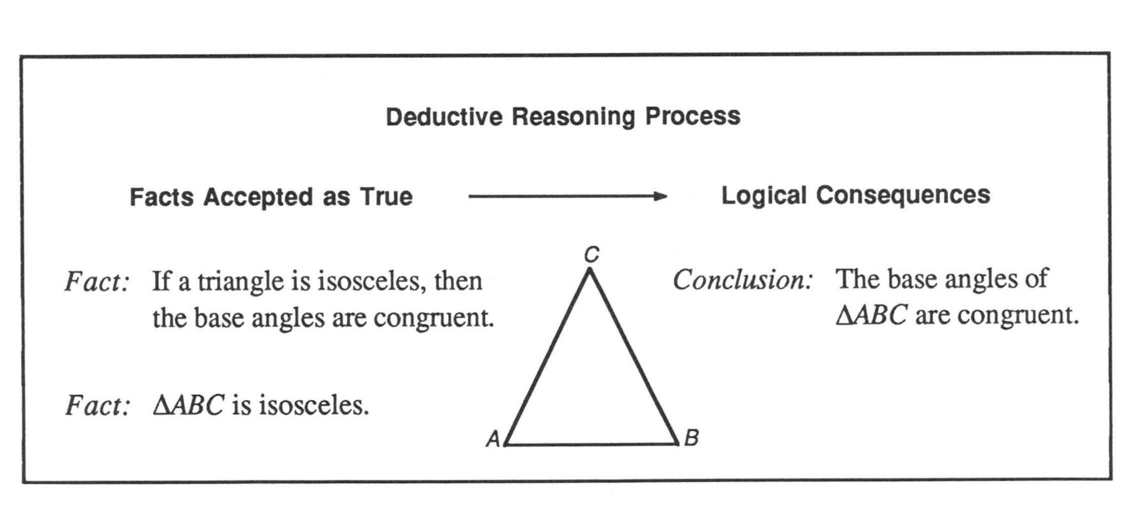

In this chapter you are going to look at geometry as Euclid did. You are going to create, as Euclid did, a deductive system based on a collection of premises. These premises will include definitions and undefined terms; properties of algebra, equality, and congruence; and postulates. From these premises you will prove some of your early conjectures. Proven conjectures will become theorems, and from the theorems you'll prove other conjectures.

The process for geometric proofs is the same as in logic proofs, except that now there are different types of premises.

Premises for Geometric Arguments

1. Definitions and undefined terms
2. Properties of algebra, equality, and congruence
3. Postulates of geometry
4. Previously accepted or proven geometric conjectures (theorems)

You should already be familiar with the first type of premise in the list, the undefined terms (point, line, and plane) and definitions. In fact, you should have a complete list of the definitions in your notebook.

Let's take a closer look at the second group of premises, the properties of algebra, equality, and congruence.

Properties of Algebra

Here are three important properties of algebra.

*The **Commutative Property of Addition** says that* $a + b = b + a$.
*The **Commutative Property of Multiplication** says that* $ab = ba$.

*The **Associative Property of Addition** says that* $(a + b) + c = a + (b + c)$.
*The **Associative Property of Multiplication** says that* $(ab)c = a(bc)$.

*The **Distributive Property** says that* $a(b + c) = ab + ac$.

Properties of Equality

Whether or not you're familiar with the names, you've used properties of equality.

*The **Reflexive Property of Equality** says that* $a = a$. *In other words, any number is equal to itself. (This property is also called the **Identity Property**.)*

*The **Transitive Property of Equality** says that if* $a = b$ *and* $b = c$, *then* $a = c$. *This property often takes the form of the **Substitution Property** which says that if* $a = b$, *then a may be replaced by b in an algebraic expression. For example, if* $x + a = c$ *and* $a = b$, *then* $x + b = c$.

*The **Symmetric Property of Equality** says that if* $a = b$, *then* $b = a$.

*The **Addition Property of Equality** says that if* $a = b$, *then* $a + c = b + c$. *(Also, if* $a = b$ *and* $c = d$, *then* $a + c = b + d$.*)*

*The **Subtraction Property of Equality** says that if* $a = b$, *then* $a - c = b - c$. *(Also, if* $a = b$ *and* $c = d$, *then* $a - c = b - d$.*)*

*The **Multiplication Property of Equality** says that if $a = b$, then $ac = bc$. (Also, if $a = b$ and $c = d$, then $ac = bd$.)*

*The **Division Property of Equality** says that if $a = b$, then $a/c = b/c$ provided that $c \neq 0$. (Also, if $a = b$ and $c = d$, then $a/c = b/d$ provided that $c \neq 0$ and $d \neq 0$.)*

You use the properties of algebra and equality to solve algebraic equations. The solution of an equation is really an algebraic proof. Each step can be supported by a property. The Addition Property of Equality, for example, permits you to add the same number to both sides of an equation to get an equivalent equation.

Example A

Equation: $5x - 12 = 3(x + 2)$

Solution:

$5x - 12 = 3(x + 2)$	Given
$5x - 12 = 3x + 6$	Distributive Property
$5x = 3x + 18$	Addition Property of Equality
$2x = 18$	Subtraction Property of Equality
$x = 9$	Division Property of Equality

Example B

Conjecture: *If* $ax + b = c$, *then* $x = \dfrac{c - b}{a}$ *provided* $a \neq 0$

Proof:

$ax + b = c$	Given (You assume the antecedent is true in a conditional proof.)
$ax = c - b$	Subtraction Property of Equality
$x = \dfrac{c - b}{a}$	Division Property of Equality

Why are the properties of algebra and equality important in geometry? Since the lengths of segments and the measures of angles involve numbers, you will occasionally need to use these properties in geometric proofs. Whenever you use a property of algebra or a property of equality in a geometric proof, refer to it by name.

Definition of Congruence and Properties of Congruence

While you use equality to express a relation between numbers, you use congruence to express a relation between geometric figures. Recall from the definition of congruent segments and angles that if $AB = CD$, then $\overline{AB} \cong \overline{CD}$ and that if $m\angle A = m\angle B$, then $\angle A \cong \angle B$. The properties of congruence come from extending the first three properties of equality to congruence relationships.

The Reflexive Property of Congruence is a property of congruence used frequently in geometric arguments.

> ***The Reflexive Property of Congruence*** *says that Figure A $\cong$ Figure A. In other words, any geometric figure is congruent to itself.*

Example

Given: Isosceles ΔABC with $\overline{AB} \cong \overline{AC}$ and angle bisector $\overline{AD}$

You can state that $\overline{AD}$ in ΔADB is congruent to $\overline{AD}$ in ΔADC because the Reflexive Property of Congruence says $\overline{AD} \cong \overline{AD}$.

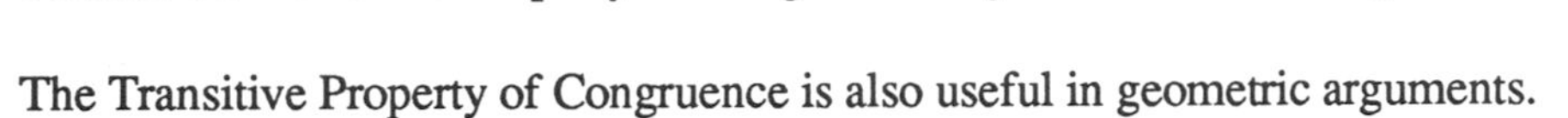

The Transitive Property of Congruence is also useful in geometric arguments.

> ***The Transitive Property of Congruence*** *says that if Figure A $\cong$ Figure B and Figure B $\cong$ Figure C, then Figure A $\cong$ Figure C.*

Example

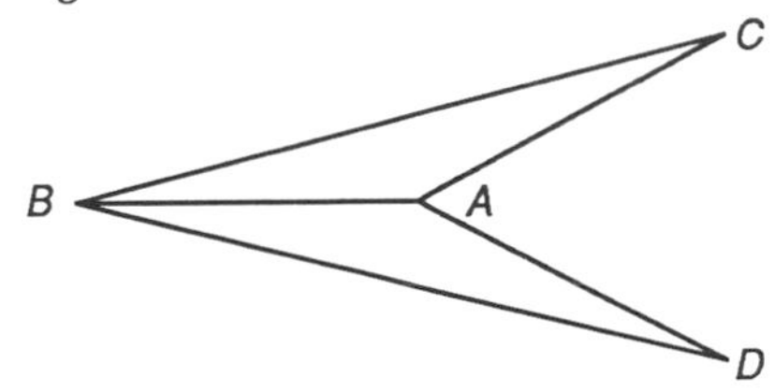

Given: $\overline{AC} \cong \overline{AB}$ and $\overline{AB} \cong \overline{AD}$

You can state that $\overline{AC} \cong \overline{AD}$ because of the Transitive Property of Congruence.

The Symmetric Property of Congruence at first appears a bit silly, if not outright useless, but you use it all the time, probably without being aware of it.

> ***The Symmetric Property of Congruence*** *says that if Figure A $\cong$ Figure B, then Figure B $\cong$ Figure A.*

You use the Symmetric Property of Equality to rewrite the equation $3 = x$ as $x = 3$. The Symmetric Property of Congruence allows you to switch expressions on the right and left hand sides of a congruence symbol. It allows you to turn $\overline{AD} \cong \overline{AB}$ around to read $\overline{AB} \cong \overline{AD}$.

EXERCISE SET

1.* When you state $\overline{AC} \cong \overline{AC}$, you are using what property?

2. *If $\overline{AC} \cong \overline{BD}$ and $\overline{BD} \cong \overline{HK}$, then $\overline{AC} \cong \overline{HK}$* is supported by which property?

3. *If $x + 120 = 180$, then $x = 60$* is supported by which property?

4. *If $2(x + 14) = 36$, then $x + 14 = 18$* is supported by which property?

5. *If $\overline{AC} \cong \overline{BD}$, then $\overline{BD} \cong \overline{AC}$* is supported by which property?

In Exercises 6 to 9, use the properties of equality and algebra to provide reasons for the steps in solving each algebraic equation or in proving each algebraic argument.

6.* *Equation:* $7x - 22 = 4(x + 2)$

Solution:

$7x - 22 = 4(x + 2)$	Given
$7x - 22 = 4x + 8$	—?— Property
$3x - 22 = 8$	—?— Property of Equality
$3x = 30$	—?— Property of Equality
$x = 10$	—?— Property of Equality

7. *Equation:* $\frac{5(x - 12)}{4} = 3(2x - 7)$

Solution:

$\frac{5(x - 12)}{4} = 3(2x - 7)$	Given
$5(x - 12) = 12(2x - 7)$	—?— Property of Equality
$5x - 60 = 24x - 84$	—?— Property
$5x = 24x - 24$	—?— Property of Equality
$-19x = -24$	—?— Property of Equality
$x = \frac{24}{19}$	—?— Property of Equality

8.* *Conjecture:* *If* $\frac{x}{m} - c = d$, *then* $x = m(c + d)$ *provided* $m \neq 0$

Proof:

$\frac{x}{m} - c = d$	—?—
$\frac{x}{m} = d + c$	—?—
$x = m(d + c)$	—?—
$x = m(c + d)$	—?—

9.* *Conjecture:* *If* $a + b = 180$ *and* $c + d = 180$ *and* $a = c$, *then* $b = d$

Proof:

$a + b = 180$	—?—
$c + d = 180$	—?—
$a + b = c + d$	—?—
$a = c$	—?—
$b = d$	—?—

In Exercises 10 to 15, copy the statement/reason table in each exercise. To the right of each statement, name the definition, property of algebra, or property of congruence that supports it.

10.* *Given:* $ABCD$ is a parallelogram

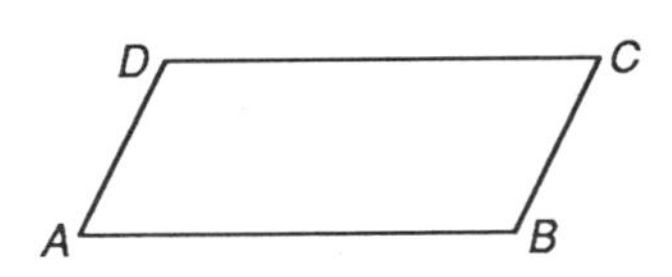

Statement:	*Reason:*
$\overline{AB} \parallel \overline{CD}$	*a.* —?—

11. *Given:* $\overline{CD}$ is a median

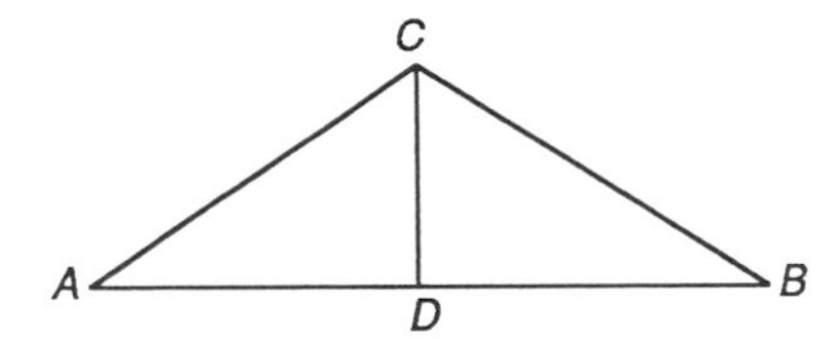

Statement:	*Reason:*
D is a midpoint of $\overline{AB}$	*a.* —?—
$AD = BD$	*b.* —?—

12. *Given:* $\overleftrightarrow{CM}$ is $\perp$ bisector of $\overline{AB}$

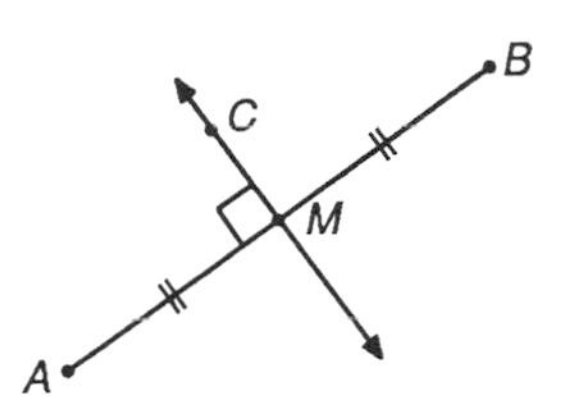

Statement:	*Reason:*
$AM = MB$	*a.* —?—
$\overline{AB} \perp \overleftrightarrow{CM}$	*b.* —?—

13. *Given:* $AP = PC$
$PC = CD$

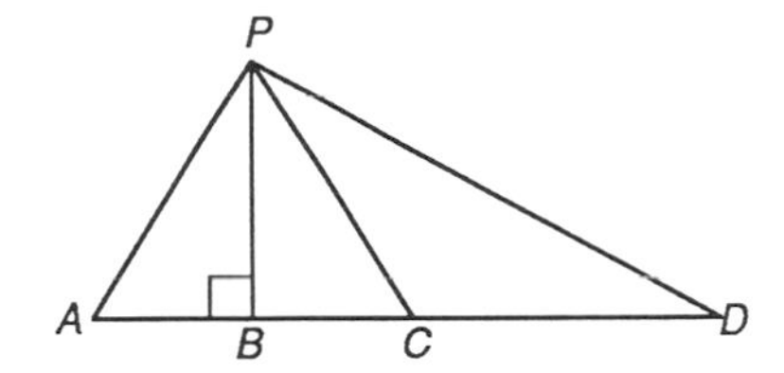

Statement:	*Reason:*
$AP = CD$	*a.* —?—

14. *Given:* $\overline{AO}$ and $\overline{BO}$ are radii

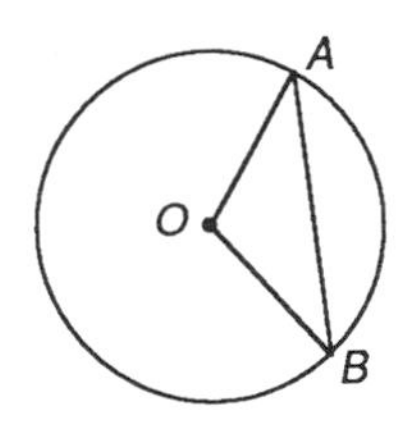

Statement:	*Reason:*
$AO = OB$	*a.* —?—
ΔAOB is isosceles	*b.* —?—

15. *Given:* $AB = BE$
B is the midpoint of $\overline{AC}$ and $\overline{DE}$

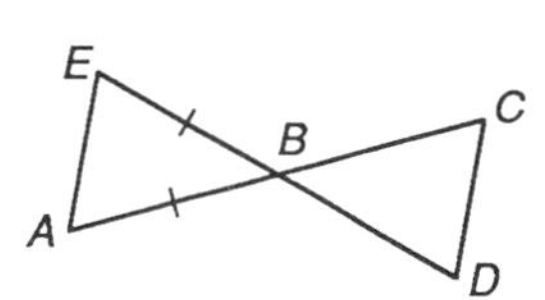

Statement:	*Reason:*
$AB = BC$	*a.* —?—
$BD = BE$	*b.* —?—
$BC = BE$	*c.* —?—

Lesson 14.2

Postulates of Geometry

Geometry is the art of correct reasoning on incorrect figures.

– George Polya

In a deductive system, you demonstrate that statements are logical consequences of previously accepted or proven statements. Since this chain of reasoning must begin somewhere, you must begin with some statements that you accept without proof. These properties of geometry are called the **postulates** of the geometric system. They form the third group of premises for geometric arguments. Some of the conjectures you made earlier will become postulates and accepted as true. Others you will derive using your deductive system. They will become **theorems**.

The postulates of Euclid's geometry were based on the geometric constructions that are possible with compass and straightedge. The first group of postulates of the geometry presented in this book are also based on geometric constructions. When you discovered how to perform basic geometric constructions, you were assuming some of these "obvious" truths. What are they?

P-1 *Exactly one line can be constructed through any two points. (**Line Postulate**)*

P-2 *The intersection of two lines is exactly one point. (**Line Intersection Postulate**)*

P-3 *Exactly one midpoint can be constructed in any line segment. (**Midpoint Postulate**)*

P-4 *Exactly one angle bisector can be constructed in any angle. (**Angle Bisector Postulate**)*

P-5 *Through a point not on a given line, exactly one line can be constructed parallel to the given line. (**Parallel Postulate**)*

P-6 *Through a point not on a given line, exactly one line can be constructed perpendicular to the given line. (**Perpendicular Postulate**)*

P-7 *If B is on $\overline{AC}$ and between A and C, then $AB + BC = AC$.* ***(Segment Addition Postulate)***

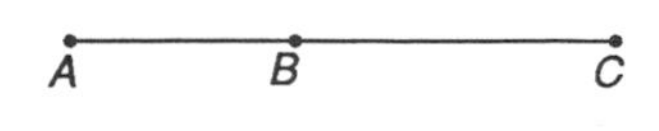

P-8 *If point D lies in the interior of $\angle ABC$, then $m\angle ABD + m\angle DBC = m\angle ABC$.* ***(Angle Addition Postulate)***

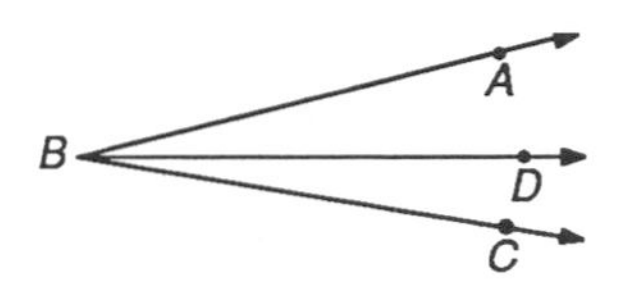

The second group of postulates of the geometry presented in this book are based on some of the most basic and useful of your geometric conjectures. What are they?

P-9 *If two angles are a linear pair, then they are supplementary. (**Linear Pair Conjecture**, now called **Linear Pair Postulate**)*

P-10 *If two parallel lines are cut by a transversal, then the alternate interior angles are congruent. Conversely, if two lines are cut by a transversal forming congruent alternate interior angles, then the lines are parallel. (**AIA Conjecture**, now called **AIA Postulate**)*

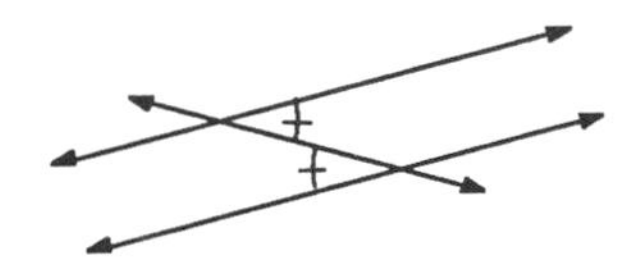

P-11 *If the three sides of one triangle are congruent to three sides of another triangle, then the two triangles are congruent. (**SSS Congruence Conjecture**, now called **SSS Congruence Postulate**)*

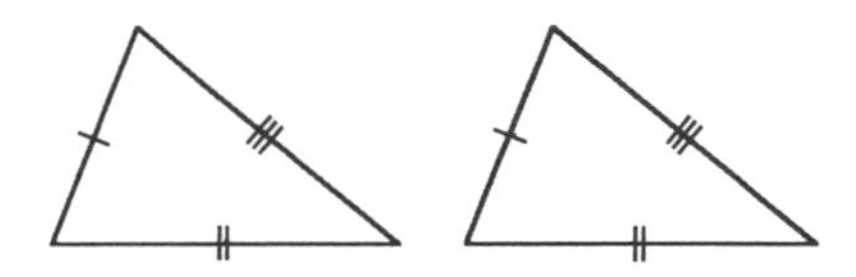

P-12 *If two sides and the angle between them in one triangle are congruent to the two sides and the angle between them in another triangle, then the two triangles are congruent.* ***(SAS Congruence Postulate)***

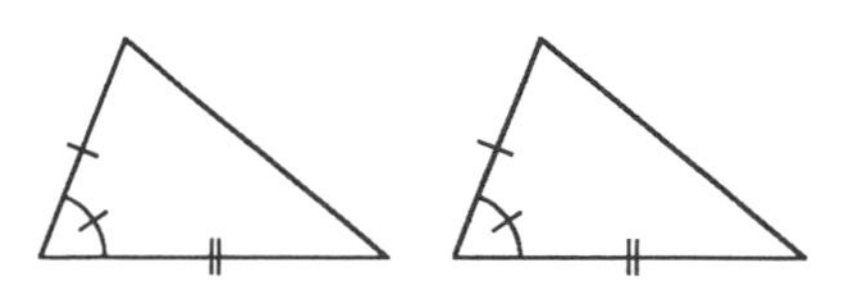

P-13 *If two angles and the side between them in one triangle are congruent to two angles and the side between them in another triangle, then the two triangles are congruent.* ***(ASA Congruence Postulate)***

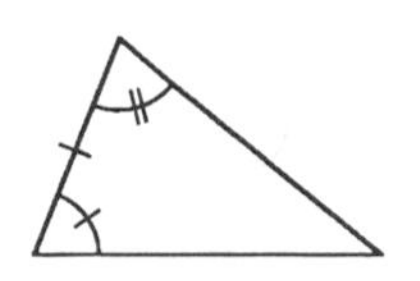

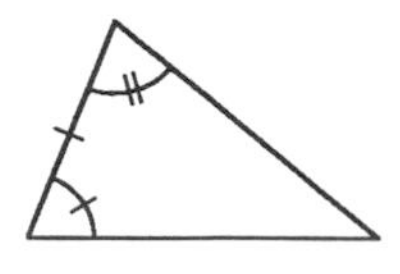

EXERCISE SET

In Exercises 1 to 6, identify each statement as true or false. Then state which definition, property of algebra, property of congruence, or postulate supports your answer.

1. If M is the midpoint of $\overline{AB}$, then $AM = BM$.

2.* If M is the midpoint of $\overline{CD}$ and N is the midpoint of $\overline{CD}$, then M and N are the same point.

3. If $\overrightarrow{AB}$ bisects $\angle CAD$, then $\angle CAB \cong \angle DAB$.

4. If $\overrightarrow{AB}$ bisects $\angle CAD$ and $\overrightarrow{AF}$ bisects $\angle CAD$, then $\overrightarrow{AB}$ and $\overrightarrow{AF}$ are the same ray.

5. Line l and line m intersect at points A and B.

6. Line l passes through points A and B and line m passes through points A and B, but line l and line m are not the same line.

Name the definition, property of algebra, property of congruence, or postulate that supports each statement in Exercises 7 to 12.

7. *Given:* $\angle 1 \cong \angle 2$

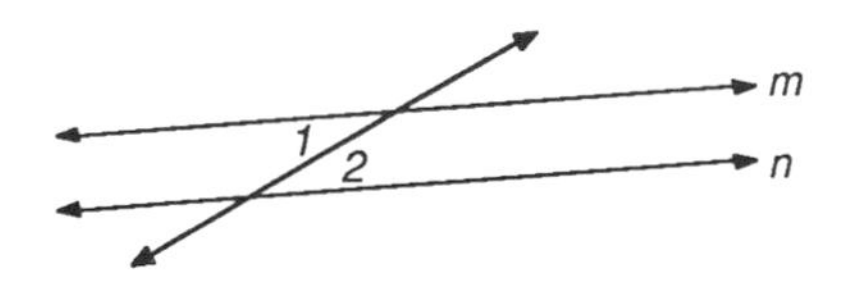

Statement:	*Reason:*
$m \parallel n$	a. —?—

8. *Given:* $\Delta ABC \cong \Delta DEF$

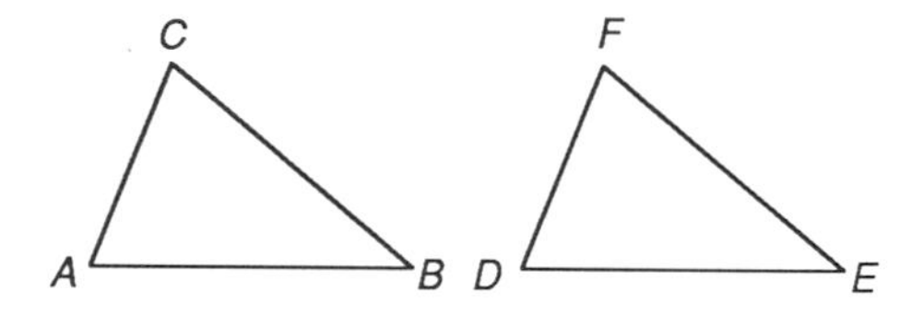

Statement:	*Reason:*
$\overline{AB} \cong \overline{ED}$	a. —?—

9. *Given:* $\overline{AC} \cong \overline{BD}$; $\overline{AD} \cong \overline{BC}$

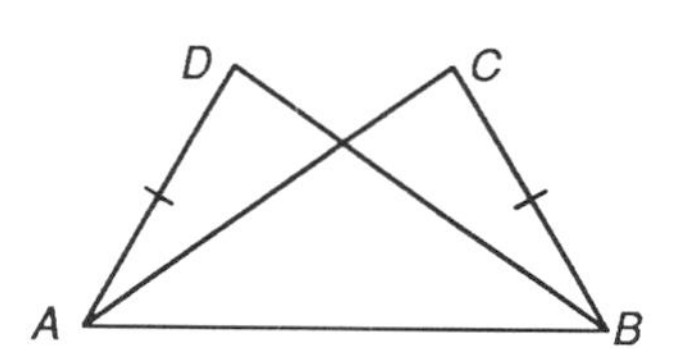

Statement:	*Reason:*
$\overline{AB} \cong \overline{AB}$	a. —?—
$\Delta ABC \cong \Delta BAD$	b.* —?—
$\angle D \cong \angle C$	c. —?—

10.* *Given:* $\overline{AB} \parallel \overline{CD}$

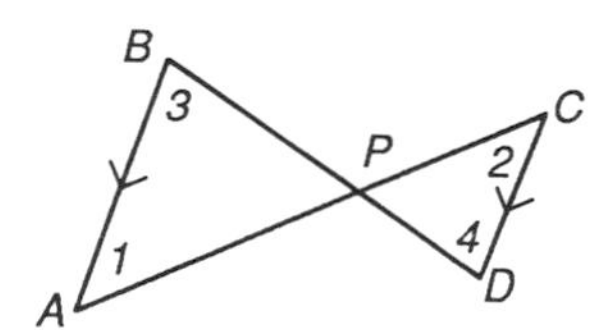

Statement:	*Reason:*
$\angle 1 \cong \angle 2$	a. —?—
$\angle 3 \cong \angle 4$	

11. *Given:* Isosceles ΔABC ($\overline{AB} \cong \overline{BC}$)

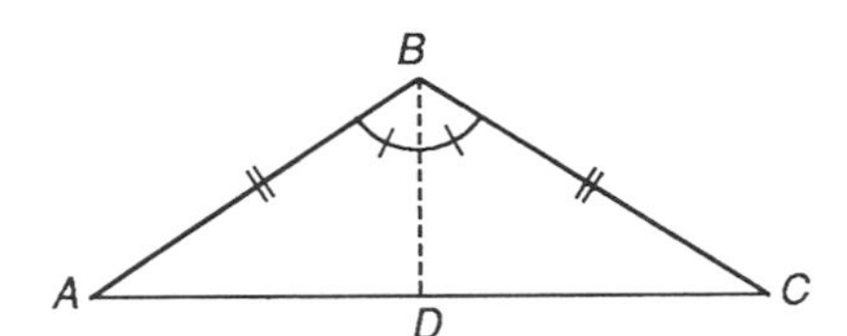

Statement:	*Reason:*
Construct angle bisector $\overline{BD}$	*a.** —?—
$\overline{BD} \cong \overline{BD}$	*b.* —?—

12. *Given:* ΔABC

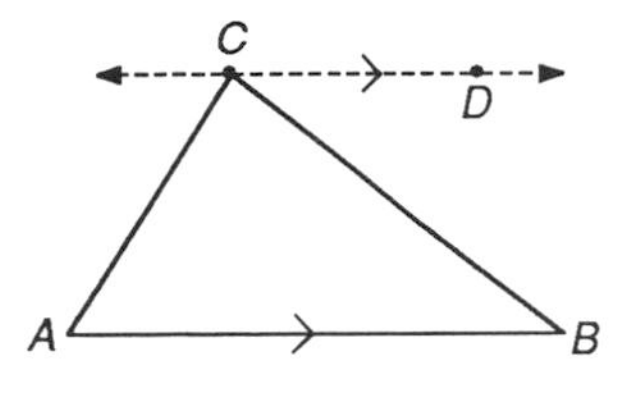

Statement:	*Reason:*
Construct $\overleftrightarrow{CD}$ with $\overline{CD} \parallel \overline{AB}$	*a.* —?—
$\angle DCB \cong \angle ABC$	*b.* —?—

Improving Visual Thinking Skills

Organic Chemistry

The element carbon bonds with more elements than all the other elements combined. One group of carbon compounds (called hydrocarbons) consists of combinations of carbon (C), hydrogen (H), and oxygen (O) which combine according to the following five rules.

- The elements C, H, and O bond together by single bonds (diagrammed by a single dash, –), double bonds (diagrammed by a double dash, =), or triple bonds (diagrammed by a triple dash, ≡).
- All carbon atoms (C) must be linked to each other in a chain (so that it is possible to pass from carbon atom to carbon atom across bonds without having to pass over other elements).
- Each C must have exactly four dashes bonding it with adjacent atoms.
- Each H must have exactly one dash bonding it with an adjacent atom.
- Each O must have exactly two dashes bonding it with adjacent atoms

The diagram that shows the bonding for the compound methane (CHHHH or CH_4) is shown at right. Use the rules above to diagram the bonding of the organic compounds below.

```
     H
     |
H —  C — H
     |
     H
```

1. CCHHHH (C_2H_4)   2. CCHHHHHH (C_2H_6)   3. CH_4O

4. The 14 atoms in the compound C_4H_{10} can be combined in a number of ways but there are two fundamentally different solutions. Find them.

5. The 12 atoms in the compound C_4H_8 can be combined in a number of ways but there are five fundamentally different solutions. Find them all.

Lesson 14.3
Geometric Proofs

All geometric reasoning is in the last result, circular.
– Bertrand Russell

A proof in geometry consists of a sequence of statements, each supported by a reason, that starts with a given set of premises and leads to a valid conclusion. Each statement follows from one or more of the previous statements. A reason for a statement can come from the set of given premises or from one of the four types of other premises: definitions; postulates; properties of algebra, equality, or congruence; or from previously proven theorems. A geometric proof can be demonstrated in the same manner as a logic proof.

Example A

Conjecture: *The angle bisector of the vertex angle of an isosceles triangle is also a median to the base.*

Given: Isosceles ΔABC with $\overline{AC} \cong \overline{BC}$ and with $\overline{CD}$ an angle bisector of vertex angle C

Show: $\overline{CD}$ is a median to the base

Two-Column Proof:

Statement	Reason
1. $\overline{AC} \cong \overline{BC}$	1. Given (premise)
2. $\overline{CD}$ an angle bisector of $\angle C$	2. Given (premise)
3. If $\overline{CD}$ an angle bisector of $\angle C$, $\angle 1 \cong \angle 2$	3. Definition of angle bisector
4. $\angle 1 \cong \angle 2$	4. From lines 2 and 3 using MP
5. $\overline{CD} \cong \overline{CD}$	5. Reflexive Property of Congruence
6. If $\overline{AC} \cong \overline{BC}$, $\angle 1 \cong \angle 2$, and $\overline{CD} \cong \overline{CD}$, then $\Delta ADC \cong \Delta BDC$	6. SAS Congruence Postulate
7. $\Delta ADC \cong \Delta BDC$	7. From lines 1, 4, 5, and 6 using MP
8. If $\Delta ADC \cong \Delta BDC$, then $\overline{AD} \cong \overline{BD}$	8. Definition of congruent triangles; that is, corresponding parts of congruent triangles are congruent (CPCTC)
9. $\overline{AD} \cong \overline{BD}$	9. From lines 7 and 8 using MP
10. If $\overline{AD} \cong \overline{BD}$, then D is a midpoint of $\overline{AB}$	10. Definition of midpoint
11. D is a midpoint of $\overline{AB}$	11. From lines 9 and 10 using MP
12. If D is a midpoint, then $\overline{CD}$ is a median	12. Definition of median
13. Therefore, $\overline{CD}$ is a median	13. From lines 11 and 12 using MP

Because geometric proofs are usually longer and involve more premises than logic proofs, it is customary to write them in shorter form. For example, you use Modus Ponens but no longer mention it directly. In example B, the proof is demonstrated again in a shorter form.

Example B

Conjecture: *The angle bisector of the vertex angle of an isosceles triangle is also a median to the base.*

Given: Isosceles ΔABC with $\overline{AC} \cong \overline{BC}$ and with $\overline{CD}$ an angle bisector of vertex angle C

Show: $\overline{CD}$ is a median to the base

Two-Column Proof:

	Statement		Reason
1.	$\overline{AC} \cong \overline{BC}$	1.	Given (premise)
2.	$\overline{CD}$ an angle bisector of $\angle C$	2.	Given (premise)
3.	$\angle 1 \cong \angle 2$	3.	Definition of angle bisector
4.	$\overline{CD} \cong \overline{CD}$	4.	Reflexive Property of Congruence
5.	$\Delta ADC \cong \Delta BDC$	5.	SAS Congruence Postulate
6.	$\overline{AD} \cong \overline{BD}$	6.	Definition of congruent triangles (CPCTC)
7.	D is a midpoint of $\overline{AB}$	7.	Definition of midpoint
8.	Therefore, $\overline{CD}$ is a median	8.	Definition of median

Once proved, you can call the conjecture above a theorem. As a theorem, it becomes a premise for geometric arguments that you can use to prove other conjectures.

Theorem: *The angle bisector of the vertex angle of an isosceles triangle is also a median to the base.*

As you work through this chapter, you will need to keep track of which conjectures from earlier chapters have been postulated (made into postulates) and which have been proven as theorems. You have a list of the postulates in Lesson 14.2. Start a theorem list. Whenever you prove a geometric statement (such as the one above), add it to your list of theorems. You will need to refer to this list often, so keep it updated and handy.

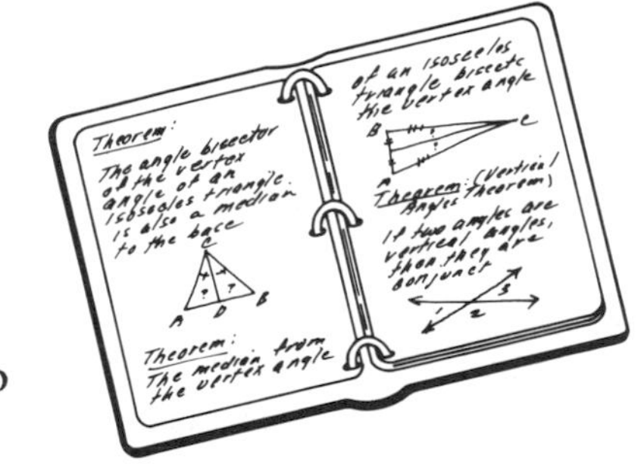

The form of proof above is called a **two-column proof**.

In this chapter you will write proofs of many statements and conjectures. You may have difficulty with two-column proofs through no fault of your own. This is because two-column proofs display logical arguments as a linear sequence of statements. That is, line 2 of the proof seems to follow from line 1; line 3 seems to follow from line 2; line 4 seems to follow from line 3; and so on. However, the logic in proofs is rarely this straightforward.

Proofs require planning. One excellent planning strategy is "thinking backwards." If you know where you are headed, but are unsure where to start, start at the end of the problem and work your way one step at a time back to the beginning.

The fire fighter below asks her assistant to turn on one of the water hydrants. But which one? A mistake could mean disaster (a nozzle flying around loose under all that pressure). Delay could waste valuable time. Which hydrant should the fire fighter's assistant turn on? Hurry! The fire is growing!

Did you "think backwards" to solve the puzzle? Good. You'll also find it a useful strategy as you write proofs.

To help plan a proof and visualize the flow of reasoning, you can make a flowchart. A **flowchart** is a diagram that shows a step-by-step procedure through a complicated system. (Euclid didn't use flowcharts; the concept is adapted from the world of computers.) As you think a proof through backwards, you draw a flowchart backwards to show the steps in your thinking.

Planning a proof can be a difficult task, especially the first time you do it. Let's watch Tut and Nut work together on their first proof. They'll use backward thinking to make their plan. They'll start with the conclusion and reason back to the given information. Here is their task:

Given: ΔABC is isosceles ($\overline{AC} \cong \overline{BC}$);
ΔABD is isosceles ($\overline{AD} \cong \overline{BD}$)

Show: $\overline{CD}$ bisects $\angle ACB$

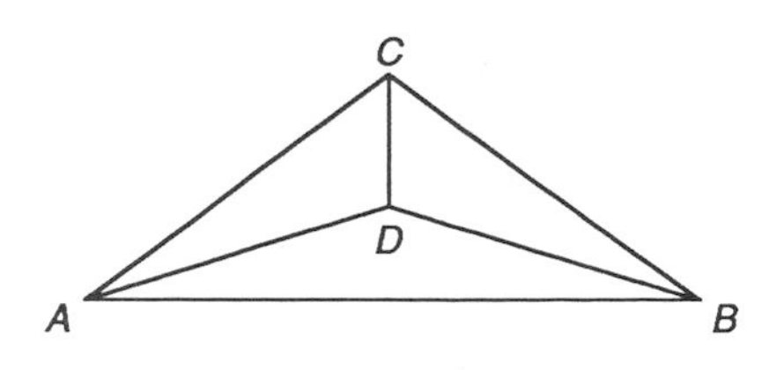

Tut draws a diagram and marks the given information on it. After puzzling over their diagram, he turns to his friend, Nut, for help.

Tut: We need to show $\overline{CD}$ bisects $\angle ACB$.

Nut: How can we show that a ray bisects an angle?

Tut: By showing that the two angles formed are congruent.

Nut: Good. Now how can we show that two angles are congruent?

Tut: One way is to show that they are corresponding parts of congruent triangles.

Nut: Do you see two triangles that might be congruent?

Tut: Yes, ΔADC and ΔBDC.

Nut: How can we show triangles congruent?

Tut: By SSS, SAS, or ASA.

Nut: Which of these fits the given situation? Look at the information you marked on your diagram. What did you indicate?

Tut: $\overline{AC} \cong \overline{BC}$ and $\overline{AD} \cong \overline{BD}$.

Nut: How did we know that?

Tut: Because it was given that ΔABC is isosceles ($\overline{AC} \cong \overline{BC}$) and that ΔABD is isosceles ($\overline{AD} \cong \overline{BD}$).

Nut: Good, but that's not enough. We need three things to show triangles congruent. Are any other parts congruent?

Tut: Yes, $\overline{CD} \cong \overline{CD}$ by the Reflexive Property.

Nut: Good, Tut. Mark this on the diagram too. Now we can see that the triangles are congruent by SSS. We're ready to write our proof.

Tut: Let's draw a flow chart. Let's see, we can show that $\overline{CD}$ bisects $\angle ACB$ if $\angle ACD \cong \angle BCD$. We can show $\angle ACD \cong \angle BCD$ if $\Delta ADC \cong \Delta BDC$. From what was given we know that ΔABC is isosceles ($\overline{AC} \cong \overline{BC}$) and ΔABD is isosceles ($\overline{AD} \cong \overline{BD}$). Marking the diagram, I notice that since $\overline{CD} \cong \overline{CD}$, I'll be able to show the two triangles congruent by SSS. Let's see. By reasoning backwards, we have figured out that since $\overline{AC} \cong \overline{BC}$, $\overline{AD} \cong \overline{BD}$, and $\overline{CD} \cong \overline{CD}$, then $\Delta ADC \cong \Delta BDC$. And since $\Delta ADC \cong \Delta BDC$, then $\angle ACD \cong \angle BCD$. And since $\angle ACD \cong \angle BCD$, then $\overline{CD}$ is an angle bisector of $\angle ACB$. Hey. We can do it!

Tut's and Nut's completed flowchart is shown below.

Flowchart:

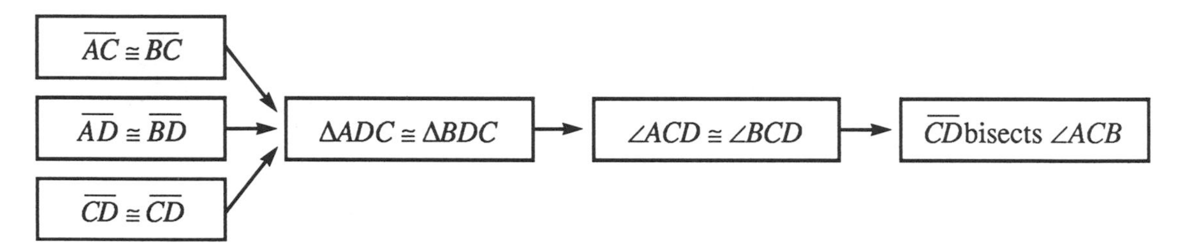

Notice how the flowchart starts with boxes containing the given information and ends with what Tut is trying to demonstrate. The arrows indicate the flow of the logical argument. From the flowchart above, Tut and Nut can write a two-column proof.

Two-Column Proof:

	Statement		Reason
1.	$\overline{AC} \cong \overline{BC}$	1.	Given
2.	$\overline{AD} \cong \overline{BD}$	2.	Given
3.	$\overline{CD} \cong \overline{CD}$	3.	Reflexive Property of Congruence
4.	$\Delta ADC \cong \Delta BDC$	4.	SSS Congruence Postulate
5.	$\angle ACD \cong \angle BCD$	5.	CPCTC
6.	$\therefore \overline{CD}$ bisects $\angle ACB$	6.	Definition of angle bisector

Each statement in the two-column proof comes from a box in the flowchart. The first boxes in the flowchart are the early statements in the two-column proof. The last box becomes the last statement in the two-column proof.

EXERCISE SET A

Copy the empty flowchart. Use the plan to complete it.

Given: $\angle CDA \cong \angle CDB$; $\angle ACD \cong \angle BCD$

Show: $\Delta ADC \cong \Delta BDC$

Plan:
- Let's think this through backwards. I can show $\Delta ADC \cong \Delta BDC$ by SSS, SAS, or ASA. Which one?
- From what is given, I know that $\angle CDA \cong \angle CDB$ and $\angle ACD \cong \angle BCD$.
- Any sides congruent? Yes, $\overline{CD}$ is in both triangles, and therefore, $\overline{CD} \cong \overline{CD}$.
- I can show the two triangles congruent by ASA.

Flowchart:

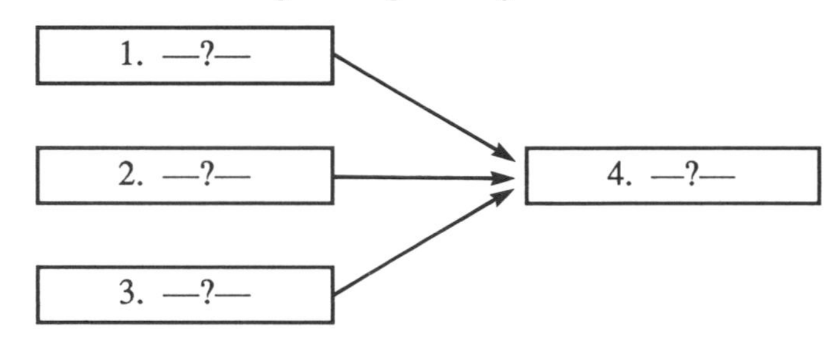

By adding the reason for each statement underneath each box in your flowchart, you can make the flowchart into a complete **flowchart proof**. Provide the reason for each step in the flowchart to create a flowchart proof.

Flowchart Proof:

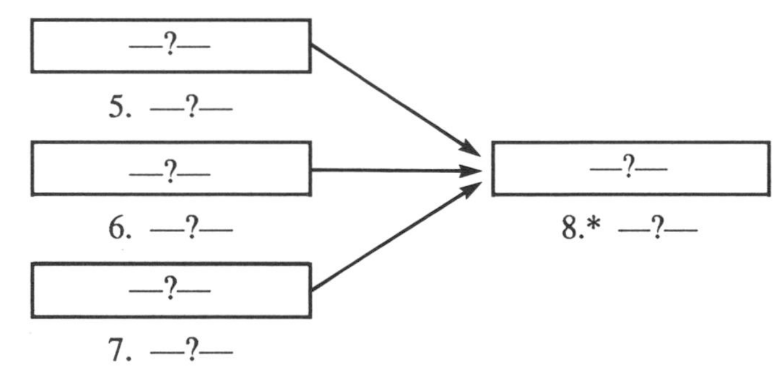

EXERCISE SET B

Now provide the missing statements and reasons to write a two-column proof from the flowchart proof in Exercise Set A.

Two-Column Proof:

1. $\angle CDA \cong \angle CDB$	1. —?—
2. —?—	2. —?—
3. —?—	3. —?—
4. $\therefore \Delta ADC \cong \Delta BDC$	4. —?—

EXERCISE SET C

Now it's your turn. Use the plan below for help.

Given: Quadrilateral *KITE* with consecutive sides $\overline{KI} \cong \overline{KE}$ and $\overline{IT} \cong \overline{TE}$

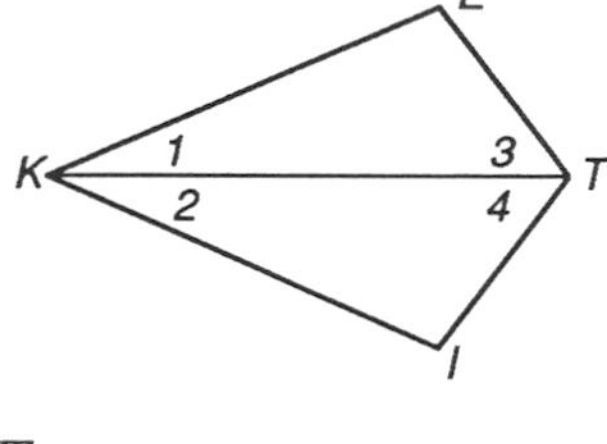

Show: Diagonal $\overline{KT}$ bisects both $\angle EKI$ and $\angle ITE$

Plan:

- I can show $\overline{KT}$ bisects $\angle EKI$ if $\angle 1 \cong \angle 2$.
- I can show $\overline{KT}$ bisects $\angle ITE$ if $\angle 3 \cong \angle 4$.
- I can show $\angle 1 \cong \angle 2$ and $\angle 3 \cong \angle 4$ if $\Delta EKT \cong \Delta IKT$.
- But, since $\overline{KI} \cong \overline{KE}$, $\overline{IT} \cong \overline{TE}$, and $\overline{KT} \cong \overline{KT}$, then $\Delta EKT \cong \Delta IKT$ by SSS.

1. Complete the flowchart. It may fit better if you rotate your paper 90°.

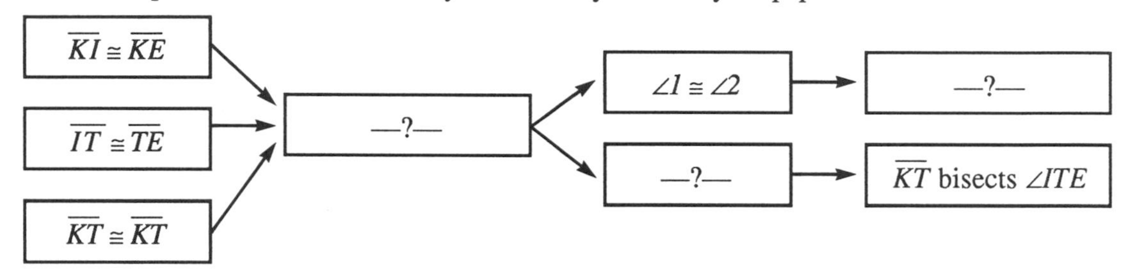

2. Write a two-column proof.

EXERCISE SET D

The proof of the conjecture below is very similar to the proof demonstrated at the beginning of the lesson. Use the proof from example B as a guide.

Conjecture: *The median from the vertex angle of an isosceles triangle bisects the vertex angle.*

Given: Isosceles ΔABC with $\overline{AC} \cong \overline{BC}$ and with $\overline{CD}$ a median to the base

Show: $\overline{CD}$ is an angle bisector of $\angle ACB$

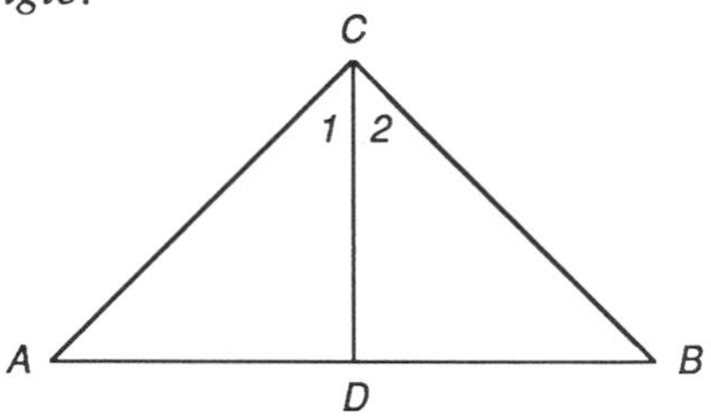

1. Plan a proof and write a flowchart.

2.* Write a two-column proof.

When you're finished you can add your new theorem to your list of theorems.

Theorem: *The median from the vertex angle of an isosceles triangle bisects the vertex angle.*

Improving Reasoning Skills

Crossnumber Puzzle I

Copy the crossnumber grid at right. In this puzzle, the answers you know will help you answer problems you don't know. Did Leonardo da Vinci really create a proof for the Pythagorean Theorem? You'll soon find out. Each clue consists of a statement followed by two numbers. If the statement is true, enter the first number of the pair. If the statement is false, enter the second number.

1				2	3
4		5	6		7
	8		9	10	
	11	12		13	
14		15			16
17				18	

Across

1. The hypotenuse is always the longest side in a right triangle. (98, 91)
2. The consecutive angles of a rhombus are supplementary. (76, 91)
4. There are exactly 9 diagonals in a hexagon. (5, 9)
5. 20 lines in a plane intersect in a maximum of 210 points. (68, 43)
7. Pythagoras taught his followers never to poke a fire with iron because a flame was the symbol of truth. (2, 5)
8. The diagonals of a rectangle are perpendicular to each other. (7, 1)
9. The exact angle formed by the hands of a clock at 1:20 is 80°. (35, 54).
11. The acute angles of a right triangle are supplementary. (96, 79)
13. SSS and SAS are two ways to show two triangles congruent. (2, 5)
14. Over the entrance to Plato's Academy was inscribed the following, "Let no one unversed in geometry enter here." (4, 8)
15. The area of a rhombus is equal to half the product of the two diagonals. (68, 72)
16. You use the formula $A = bh$ to find the area of a triangle. (4, 9)
17. If two angles of a triangle are congruent, then the triangle is equilateral. (87, 75)
18. The mathematician, Blaise Pascal, is credited with the invention of the one-wheeled wheelbarrow. (42, 57)

Down

1. Abraham Lincoln learned logic by studying Euclid's *Elements*. (95, 79)
3. Leonardo da Vinci created a proof of the Pythagorean Theorem. (62, 15)
6. The perpendicular bisector of a chord passes through the center of the circle. (33, 85)
8. In a 30-60 right triangle, the side opposite the 30° angle is half the length of the hypotenuse. (17, 79)
10. The measure of each exterior angle of a regular octagon is 45°. (52, 45)
12. The *z* in the abbreviation *oz.* for ounce comes from the Spanish coin "onza de oro" which contained one ounce of gold. (96, 67)
14. The great mathematician, Carl Friedrich Gauss, at the age of 3 corrected his father's arithmetic. (47, 88)
16. If the corresponding angles of two triangles are congruent, then the two triangles are congruent. (47, 92)

Lesson 14.4

Short Proofs

Do not worry about your difficulties in mathematics. I can assure you mine are still greater.

– Albert Einstein

Let's start with a puzzle. Copy the 5 by 5 puzzle grid below. You are to start at square 1 and move to square 100. You can move to an adjacent square, horizontally, vertically, or diagonally, whenever you can add, subtract, multiply, or divide the number in the square you occupy by 2 or 5 to get the number in the adjacent square. For example, if you happen to be in square 11, you could move to square 9 by subtracting 2, or to square 55 by multiplying by 5. When you find the one path from 1 to 100, show it with arrows.

1	5	10	20	30
2	3	22	6	28
4	27	8	14	19
20	17	11	55	95
18	9	50	57	100

Notice that in this puzzle you may start with different moves. You could start with $1 \rightarrow 5$, or $1 \rightarrow 3$, or $1 \rightarrow 2$. Did you try working backwards? If you start from square 100 and work backwards, the problem becomes much easier.

In this lesson you will prove one of your first conjectures. Watch how thinking backwards helps to plan. Watch how the conjecture translates into the given and show information.

EXERCISE SET A

Provide the reasons for the steps in the two-column proof of the Vertical Angles Conjecture. To help with the proof, read the plan and follow the flowchart.

Conjecture: *If two angles are vertical angles, then they are congruent.* ***(Vertical Angles Conjecture)***

Given: Lines *m* and *n* intersecting to form vertical angles *1* and 2

m
n
1
2
3

Show: $\angle 1 \cong \angle 2$

Plan:

- Thinking backwards, I realize that I need to show $\angle 1 \cong \angle 2$. Looking over the postulates, the only one that looks like it may be useful is the Linear Pair Postulate. From the Linear Pair Postulate, I know $\angle 1$ and $\angle 3$ are supplementary and $\angle 3$ and $\angle 2$ are supplementary.

- If they are supplementary, I know $m\angle 1 + m\angle 3 = 180$ and $m\angle 3 + m\angle 2 = 180$. If the sum of the measures of each pair of angles is 180, then $m\angle 1 + m\angle 3 = m\angle 3 + m\angle 2$.
- If $m\angle 1 + m\angle 3 = m\angle 3 + m\angle 2$, then subtracting $m\angle 3$ from both sides gives $m\angle 1 = m\angle 2$. If $m\angle 1 = m\angle 2$, then $\angle 1 \cong \angle 2$.

Flowchart:

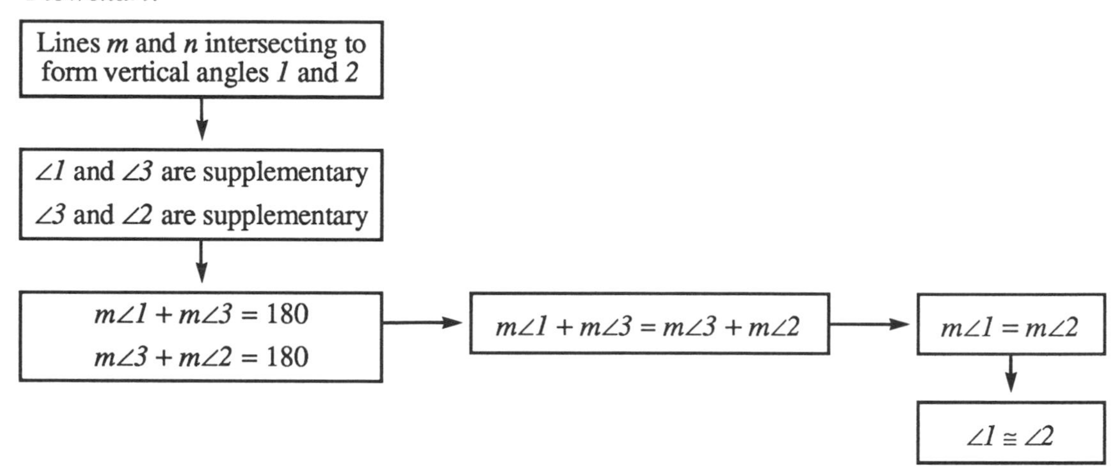

Two Column Proof:

Statement	Reason
1. Lines *m* and *n* intersecting to form vertical angles *1* and *2*	1. —?—
2. $\angle 1$ and $\angle 3$ are supplementary; $\angle 3$ and $\angle 2$ are supplementary	2. Linear Pair Postulate
3. $m\angle 1 + m\angle 3 = 180$; $m\angle 3 + m\angle 2 = 180$	3. —?—
4. $m\angle 1 + m\angle 3 = m\angle 3 + m\angle 2$	4. —?— Property
5. $m\angle 1 = m\angle 2$	5.* —?— Property of Equality
6. $\therefore \angle 1 \cong \angle 2$	6. Definition of congruence

Now that you have proved the Vertical Angles Conjecture, you can call it the Vertical Angles Theorem. As a theorem, it becomes a premise that you can use in proving other conjectures. Add the Vertical Angles Theorem to the list of theorems in your notebook.

Theorem: *If two angles are vertical angles, then they are congruent.* ***(Vertical Angles Theorem)***

EXERCISE SET B

Provide the reasons for the steps in the two-column proof. To help with the proof, read the plan and follow the flowchart. One of the reasons in the proof is the Vertical Angles Theorem that you proved in Exercise Set A.

Given: $\overline{AB}$ and $\overline{CD}$ intersect at M so that M is the midpoint of both $\overline{AB}$ and $\overline{CD}$

Show: $\overline{AC} \cong \overline{BD}$

Plan:

- I need to show $\overline{AC} \cong \overline{BD}$. One possible way is to show they are corresponding parts of congruent triangles.
- What two triangles might be congruent? ΔAMC and ΔBMD. I can show triangles congruent by SSS, SAS, or ASA. Which one?
- From what is given I know that $\overline{AM} \cong \overline{BM}$, $\overline{CM} \cong \overline{DM}$. But I need three things to show triangles congruent. Are any other parts congruent? Yes, $\angle AMC \cong \angle BMD$ because $\angle AMC$ and $\angle BMD$ are vertical angles. Marking the diagram I notice that I'll be able to show the two triangles congruent by SAS.
- Reasoning backwards, since $\overline{AM} \cong \overline{BM}$, $\overline{CM} \cong \overline{DM}$ and $\angle AMC \cong \angle BMD$, then $\Delta AMC \cong \Delta BMD$. Since $\Delta AMC \cong \Delta BMD$, then $\overline{AC} \cong \overline{BD}$.

Flowchart:

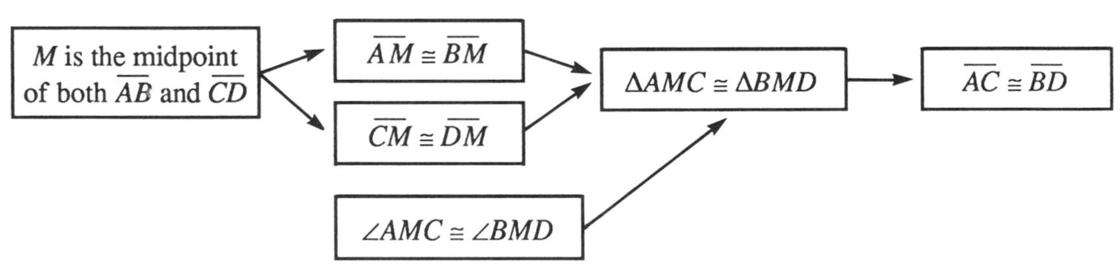

Two-Column Proof:

	Statement		Reason
1.	M is the midpoint of both $\overline{AB}$ and $\overline{CD}$	1.	—?—
2.	$\overline{AM} \cong \overline{BM}$	2.*	—?—
3.	$\overline{CM} \cong \overline{DM}$	3.	—?—
4.	$\angle AMC \cong \angle BMD$	4.*	—?—
5.	$\Delta AMC \cong \Delta BMD$	5.	—?—
6.	$\therefore \overline{AC} \cong \overline{BD}$	6.	—?—

With proofs involving algebraic operations, it is sometimes convenient to let lower case letters represent measures of angles and line segments because then there are fewer symbols.

Example

Instead of writing $m\angle BAC + m\angle ABC + m\angle BCA = 180$,
let $a = m\angle BAC$, $b = m\angle ABC$, and $c = m\angle BCA$.
Then write $a + b + c = 180$.

Instead of writing $m\overline{BA} + m\overline{AC} = m\overline{BC}$,
let $x = m\overline{BA}$, $y = m\overline{AC}$, and $z = m\overline{BC}$.
Then write $x + y = z$.

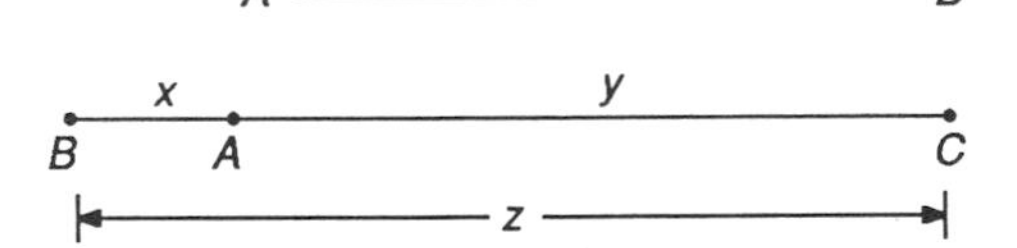

EXERCISE SET C

Provide the missing reasons in the proof below.

Conjecture: *Supplements of congruent angles (or the same angle) are congruent.*

Given: $\angle A$ and $\angle C$ are supplementary;
$\angle B$ and $\angle D$ are supplementary;
$\angle A \cong \angle B$

Show: $\angle C \cong \angle D$

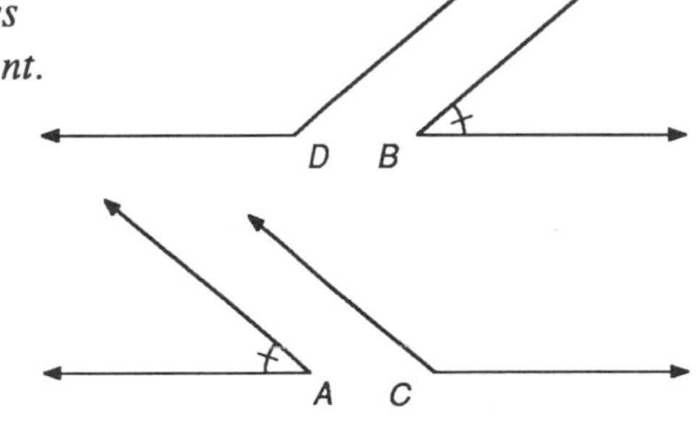

Two-Column Proof:

Let $a = m\angle A$, $b = m\angle B$, $c = m\angle C$, and $d = m\angle D$.

1. $\angle A$ and $\angle C$ are supplementary; $\angle B$ and $\angle D$ are supplementary	1. Given
2. $a + c = 180$; $b + d = 180$	2. —?—
3. $a + c = b + d$	3. —?— Property
4. $\angle A \cong \angle B$	4. —?—
5. $a = b$	5. Definition of congruence
6. $c = d$	6. Subtraction Property of Equality
7. $\angle C \cong \angle D$	7. —?—

Add the theorem to the list of theorems in your notebook.

Theorem: *Supplements of congruent angles (or the same angle) are congruent.*

Lesson 14.5
Linking Proofs

In this lesson you , like Euclid, will start with postulates and prove a theorem. Then you'll use theorems that you've proved to prove other theorems. The concept map shows how the four theorems you'll prove in this lesson are logically linked together.

Logical Relationship of Postulates and Theorems in this Lesson

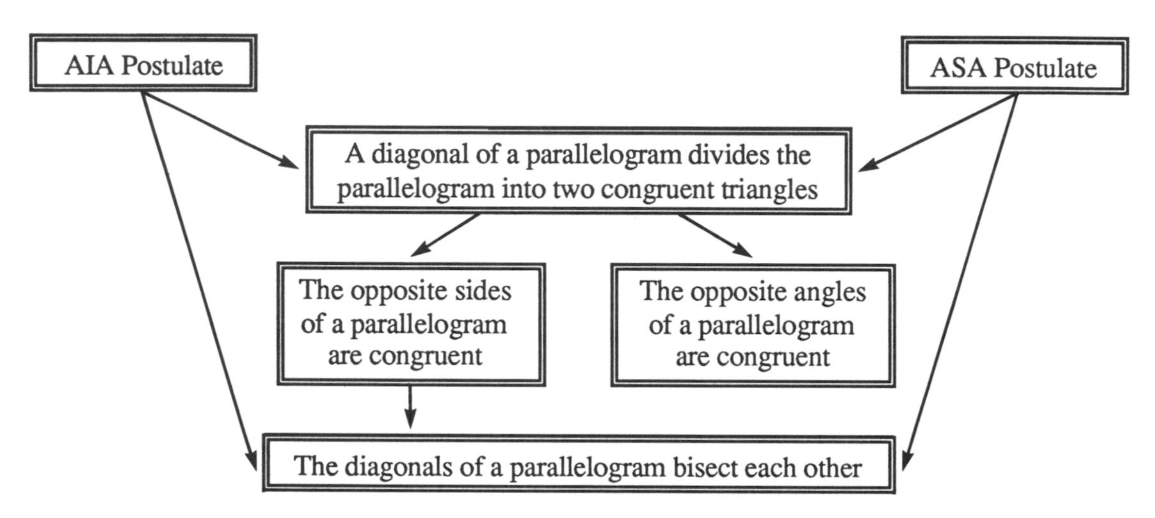

EXERCISE SET A

Watch how the conjecture is translated into the given and show information. Copy the flowchart. Using the plan, fill in each box. Then write the reason for each step below its box. This completes a flowchart proof of the conjecture. Finally, provide the statements and reasons for a two-column proof.

Conjecture: *A diagonal of a parallelogram divides the parallelogram into two congruent triangles.*

Given: Parallelogram $ABCD$ with diagonal $\overline{AC}$

Show: $\Delta ABC \cong \Delta CDA$

Plan:

- Thinking backwards, I realize that I can show $\Delta ABC \cong \Delta CDA$ by either SSS, SAS, or ASA. Which one?
- From what is given, I know that $ABCD$ is a parallelogram. Therefore, $\overline{AB} \parallel \overline{CD}$ and $\overline{AD} \parallel \overline{BC}$.
- Since $\overline{AB} \parallel \overline{CD}$, $\angle 1 \cong \angle 2$. Since $\overline{AD} \parallel \overline{BC}$, $\angle 3 \cong \angle 4$. (AIA Postulate)
- Are any other parts congruent in the two triangles? $\overline{AC} \cong \overline{AC}$, so the triangles are congruent by the ASA Congruence Postulate.

Flowchart Proof:

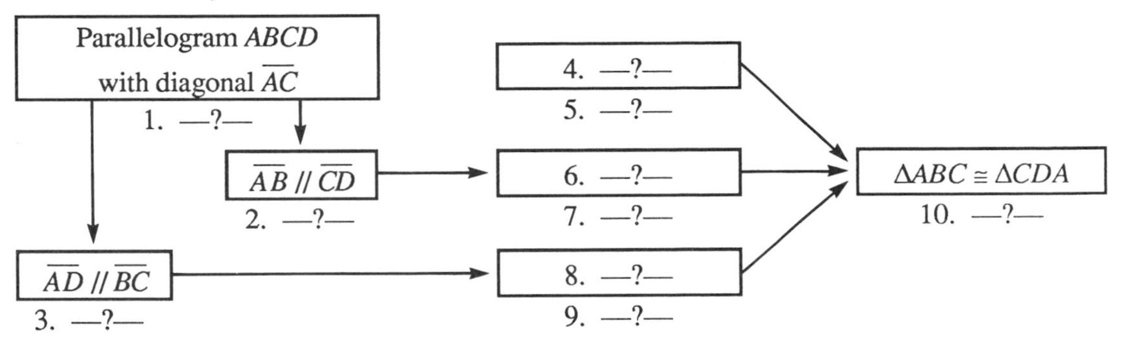

11. Provide the statements and reasons for a two-column proof of the conjecture.

Now that you have proved the conjecture, you can call it a theorem. As a theorem, it becomes a premise that you can use to prove other conjectures. Add this theorem to your list.

Theorem: *A diagonal of a parallelogram divides the parallelogram into two congruent triangles.*

EXERCISE SET B

Copy the flowchart. Using the plan, fill in each box. Then write the reason for each step below its box. You can use the theorem proved in Exercise Set A as a reason for one of the steps in your proof. This completes a flowchart proof of the conjecture. Finally, provide the statements and reasons for a two-column proof.

Conjecture: *The opposite sides of a parallelogram are congruent. (C-24)*

Given: Parallelogram $SANE$ with diagonal $\overline{AE}$

Show: $\overline{SA} \cong \overline{NE}$ and $\overline{SE} \cong \overline{NA}$

Plan:

- I can show $\overline{SA} \cong \overline{NE}$ and $\overline{SE} \cong \overline{NA}$ by CPCTC if $\Delta SAE \cong \Delta NEA$.
- But I can show $\Delta SAE \cong \Delta NEA$ because I know that $SANE$ is a parallelogram. Thus, $\Delta SAE \cong \Delta NEA$ by the theorem proved in Exercise Set A.

Flowchart Proof:

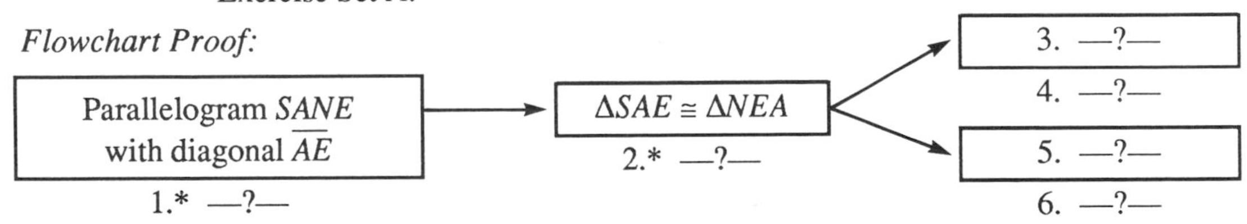

7. Provide the statements and reasons for a two-column proof of the conjecture.

Once the proof is complete, you can rename Conjecture 24 as Theorem 24 and use it to prove other conjectures. Add this theorem to your list of theorems.

Theorem: *The opposite sides of a parallelogram are congruent. (T-24)*

EXERCISE SET C

Copy the flowchart for the proof below. Then complete the missing steps and write the reason for each step below its box to create a flowchart proof. The plan will help.

Conjecture: *The opposite angles of a parallelogram are congruent. (C-22)*

Given: Parallelogram $SANE$ with diagonals $\overline{AE}$ and $\overline{SN}$

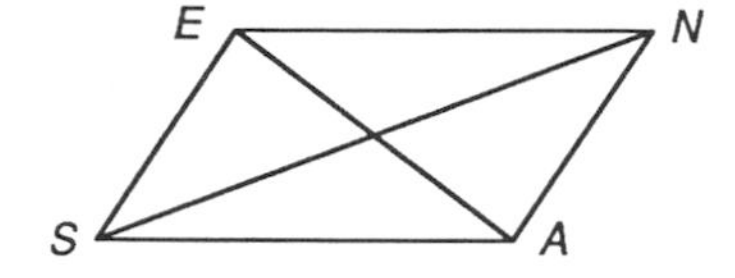

Show: $\angle SAN \cong \angle NES$ and $\angle ESA \cong \angle ANE$

Plan:

- I can show $\angle SAN \cong \angle NES$ by CPCTC if $\Delta SAN \cong \Delta NES$.
- I can show $\angle ESA \cong \angle ANE$ by CPCTC if $\Delta SAE \cong \Delta NEA$.
- But I can show $\Delta SAE \cong \Delta NEA$ because $SANE$ is a parallelogram and, therefore, $\Delta SAE \cong \Delta NEA$ by the theorem in Exercise Set A.
- Likewise, I can show $\Delta SAN \cong \Delta NES$ by the same reasoning.

Flowchart Proof:

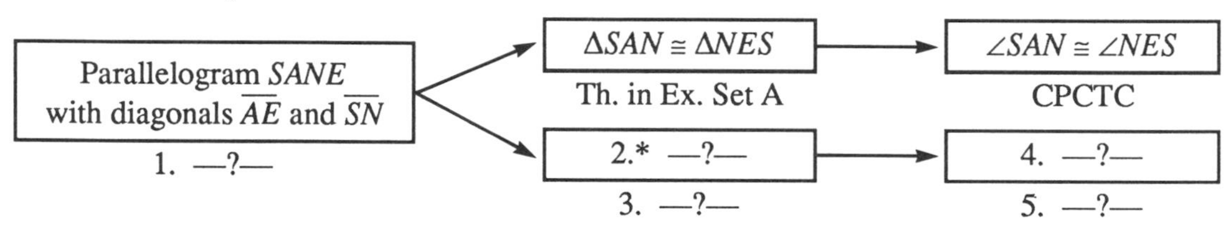

Once proved, Conjecture 22 can be called Theorem 22. Add it to your theorem list.

Theorem: *The opposite angles of a parallelogram are congruent. (T-22)*

EXERCISE SET D

1.* Create a flowchart for the proof of the conjecture below.

Conjecture: *The diagonals of a parallelogram bisect each other. (C-25)*

Given: Parallelogram $SHAM$ with diagonals $\overline{SA}$ and $\overline{MH}$ intersecting at E

Show: Diagonals $\overline{SA}$ and $\overline{MH}$ bisect each other

Plan:

- I can show $\overline{SA}$ bisects $\overline{MH}$ if $\overline{HE} \cong \overline{ME}$ and I can show $\overline{MH}$ bisects $\overline{SA}$ if $\overline{SE} \cong \overline{AE}$.
- I can show $\overline{HE} \cong \overline{ME}$ and $\overline{SE} \cong \overline{AE}$ if I can show $\Delta SHE \cong \Delta AME$.
- But, I can show $\Delta SHE \cong \Delta AME$ by the ASA Congruence Postulate since $\overline{HS} \cong \overline{MA}$ (opposite sides of a parallelogram are congruent), $\angle HSA \cong \angle MAS$ and $\angle SHM \cong \angle AMH$ by the AIA Postulate.

2. Write the reason for each step in the flowchart below its box to create a flowchart proof, or write a two-column proof.

Once proved, Conjecture 25 can be called Theorem 25. Add this theorem to your list.

Theorem: *The diagonals of a parallelogram bisect each other. (T-25)*

Improving Reasoning Skills

Crossnumber Puzzle II

Copy the crossnumber grid at right. In this puzzle, the answers you know will help you answer problems you don't know. Did Napoleon really study geometry? You'll soon find out. Each clue consists of a statement followed by two numbers. If the statement is true, enter the first number of the pair. If the statement is false, enter the second number.

Across

1. Every quadrilateral is a trapezoid. (47, 14)
2. If two angles are vertical angles, then they are congruent. (91, 67)
4. The sum of the measures of the inferior angles of a pentagon is 360°. (5, 6)
5. The diagonals of a rectangle are congruent. (25, 48)
7. Each angle of a regular octagon has a measure of 160°. (7, 3)
8. Euclid wrote the *Elements.* (6, 5)
9. Pythagoras believed that "number" ruled the universe. (49, 27).
11. Lewis Carroll was a mathematician who wrote *Alice in Wonderland.* (64, 24)
13. The midsegment in a trapezoid is parallel to the bases. (8, 9)
14. The symbol *AB* stands for the line through points *A* and *B*. (5, 1)
15. An isosceles right triangle is impossible. (00, 10)
16. An obtuse right triangle is impossible. (0, 4)
17. Each base angle of an isosceles right triangle measures 45°. (12, 87)
18. Much of Victor Vasarely's art is geometrical in form. (11, 28)

Down

1. Leonardo da Vinci invented calculus. (45, 16)
3. Napoleon studied geometry and found a new proof for the famous Pythagorean Theorem. (13, 77)
6. Albert Einstein used non-Euclidian geometry to arrive at his theories of relativity. (54, 89)
8. The angle bisectors in a trapezoid are concurrent. (52, 66)
10. The diagonals of a rhombus are congruent. (79, 98)
12. The incenter of a triangle is also the center of mass. (40, 41)
14. A Roman soldier killed Archimedes while Archimedes was working on a geometry diagram. (11, 58)
16. Sir Arthur Conan Doyle is the author of the tales of Sherlock Holmes, the master of deduction. (01, 38)

Lesson 14.6

Proving the CA Conjecture

If two parallel lines are cut by a transversal, then the corresponding angles are congruent. Conversely, if two lines are cut by a transversal forming congruent corresponding angles, then the lines are parallel. ***(CA Conjecture)***

EXERCISE SET A

Translate each conjecture into a diagram labeled with what is given and what you must show. Then create a proof of the conjecture. The first two exercises are the two halves of the Corresponding Angles Conjecture.

1.* *Conjecture:* *If two parallel lines are cut by a transversal, then the corresponding angles are congruent. (Part of CA Conjecture)*

Given: $l_1 \parallel l_2$ with l_3 a transversal forming corresponding angles $\angle 1$ and $\angle 2$

Show: $\angle 1 \cong \angle 2$

2.* *Conjecture:* *If two lines are cut by a transversal forming congruent corresponding angles, then the lines are parallel. (Part of CA Conjecture)*

Given: l_1 and l_2 with l_3 a transversal forming congruent corresponding angles and with $\angle 1 \cong \angle 2$

Show: $l_1 \parallel l_2$

Once the proofs are complete in Exercises 1 and 2, you have proved both parts of the CA Conjecture. Now you can call it the CA Theorem. Add it to your theorem list.

Theorem: *If two parallel lines are cut by a transversal, then the corresponding angles are congruent. Conversely, if two lines are cut by a transversal forming congruent corresponding angles, then the lines are parallel.* ***(CA Theorem)***

3.* *Conjecture:* *If two lines are parallel to a third line in the plane, then they are parallel to each other.*

Given: $l_1 \parallel l_2$ and $l_1 \parallel l_3$ with l_4 a transversal through all three lines

Show: $l_2 \parallel l_3$

The concept map below shows one way of logically relating the postulates and theorems that were used in Exercises 1 to 3. In later lessons you will be asked to produce similar concept maps.

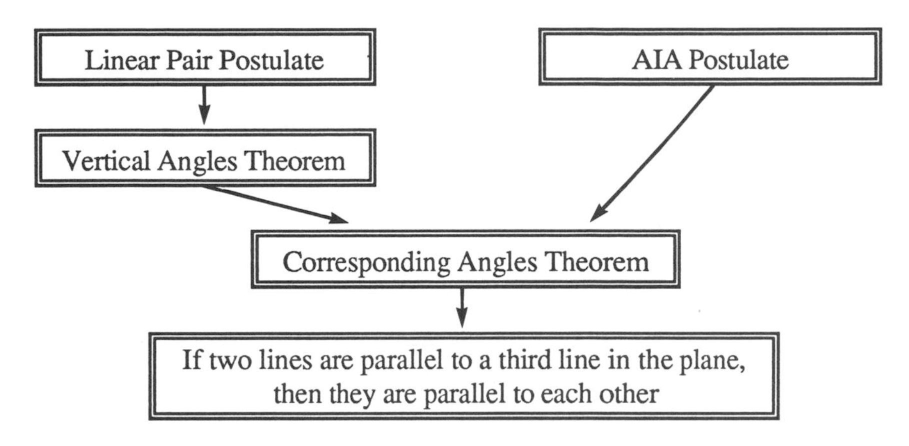

EXERCISE SET B

Use the concept map above or refer back to your proofs to answer each question.

1. Upon what postulates did the proof of the CA Conjecture depend?
2. What theorems were used in the proof of the CA Conjecture?
3. What theorems were used in the proof of the conjecture in Exercise 3?

Improving Visual Thinking Skills

Triangle Cards

Visualize six cards shaped like equilateral triangles with sides measuring 6 cm, 5 cm, 4 cm, and so on down to 1 cm. The triangles with odd perimeters are black and the other triangles are white. Begin by placing the largest triangle down on the table. Then place on it the triangle with the 5-cm edge, not centered, but placed into one corner (A) of the first triangle. Next, place the triangle with the 4-cm edge onto the 5-cm triangle, not centered, but into the corner (B) counterclockwise from corner A. Next, place the 3-cm triangle onto the 4-cm triangle, not centered, but into the corner (C) counterclockwise from corner B. Continue in this way with the remaining triangles with the positions rotating inward counterclockwise. Draw a top view of what the cards would look like when the process is finished.

Lesson 14.7

Proving Conjectures about Parallel Lines

Mistakes are part of the dues one pays for a full life.

– Sophia Loren

EXERCISE SET A

Translate each conjecture into a diagram labeled with what is given and what you must show. Then create a proof of the conjecture.

1. *Conjecture:* *If two lines in the same plane are perpendicular to a third line, then they are parallel to each other.*

   Given: $l_1 \perp l_2$ and $l_1 \perp l_3$

   Show: $l_2 \parallel l_3$

2.* *Conjecture:* *If two parallel lines are intersected by a transversal, then the interior angles on the same side of the transversal are supplementary.*

   Given: $l_1 \parallel l_2$

   Show: $\angle 1$ and $\angle 2$ are supplementary

3.* *Conjecture:* *The consecutive angles of a parallelogram are supplementary. (C-23)*

   Given: Parallelogram *SANE*

   Show: $\angle ESA$ and $\angle NAS$ are supplementary

4.* *Conjecture:* *Parallel lines are everywhere equidistant. (If two lines are parallel, then the distances from all points on one line to the other line are equal.)*

   Given: $l_1 \parallel l_2$ with points *A* and *B* on l_1 and points *C* and *D* on l_2 such that $\overline{AC} \perp l_2$ and $\overline{BD} \perp l_2$

   Show: $AC = BD$

Once the proofs of the conjectures are completed, they can be called theorems. Add these theorems to the list of theorems. You will need them as premises for future proofs.

EXERCISE SET B

1. When completed, the concept map below shows one way of logically relating the postulates and theorems that were used in this lesson to prove that the consecutive angles of a parallelogram are supplementary. Copy and complete the concept map by drawing in the missing arrows.

Logical Relationship of Postulates and Theorems Needed to Prove that Consecutive Angles of a Parallelogram Are Supplementary

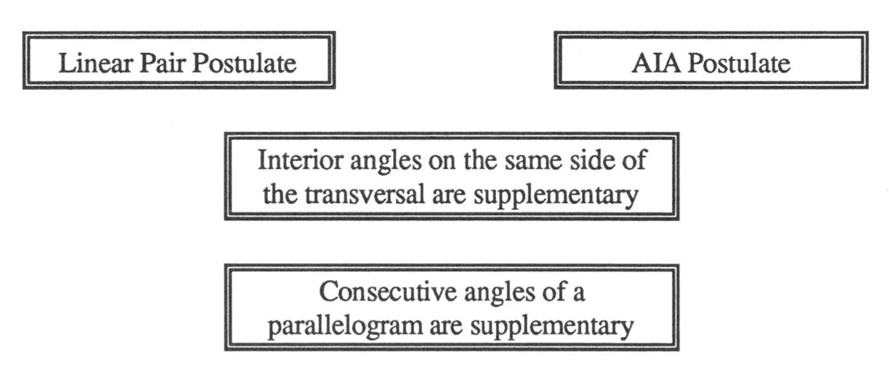

2. Copy and fill in the boxes to complete the concept map to show one way of logically relating the postulates and theorems that were used to prove that parallel lines are everywhere equidistant.

Logical Relationship of Postulates and Theorems Needed to Prove that Parallel Lines Are Everywhere Equidistant

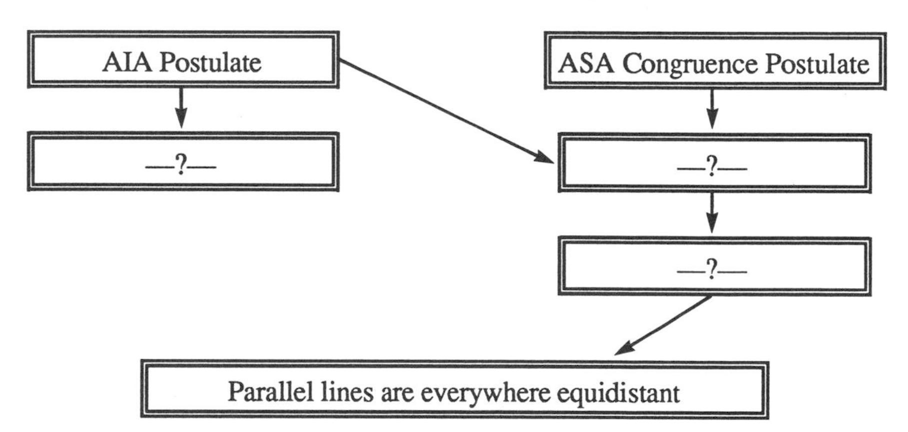

Lesson 14.8

Proofs with Overlapping Figures

Sometimes triangles you are trying to prove congruent overlap in a diagram. To help see the triangles more clearly, redraw them so they don't overlap. For example, if you wish to determine whether or not $\Delta ABE \cong \Delta BAD$ in the diagram below left, you can redraw the two triangles separated from the original diagram as shown below right.

Original Diagram

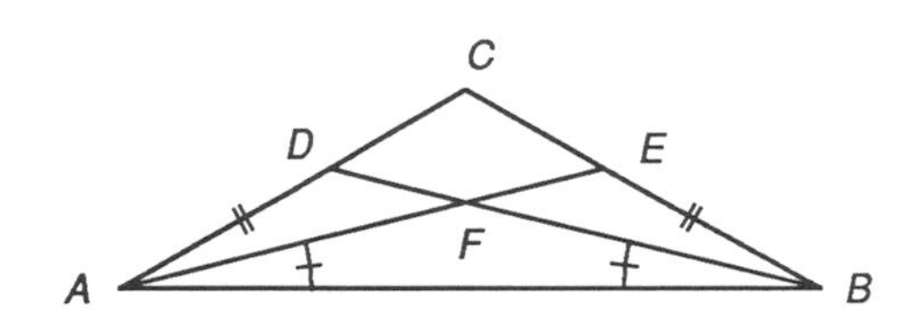

Separated Triangles

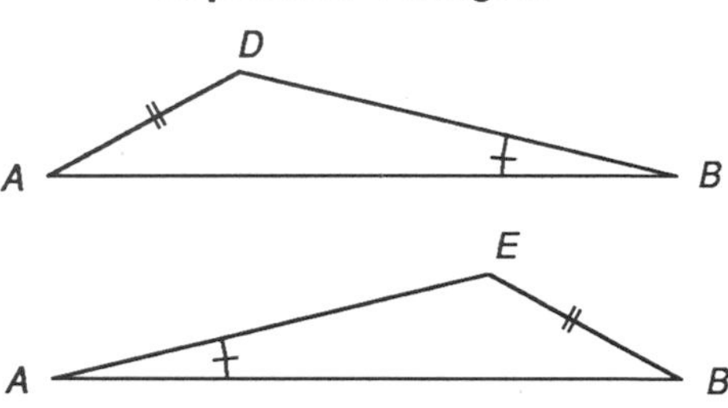

EXERCISE SET A

Write a proof for each given and show below. The triangles you must prove congruent overlap. Redraw the triangles to make writing a proof easier. You might want to make a flowchart.

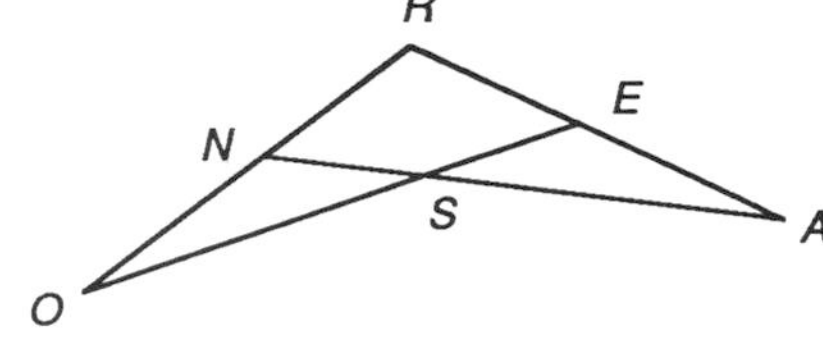

1.* *Given:* $\angle O \cong \angle A$; $\overline{RO} \cong \overline{RA}$

Show: $\Delta REO \cong \Delta RNA$

2. *Given:* ΔTRA and ΔTRP with $\overline{TA} \cong \overline{RP}$ and $\overline{TP} \cong \overline{AR}$

Show: $\angle TPR \cong \angle RAT$

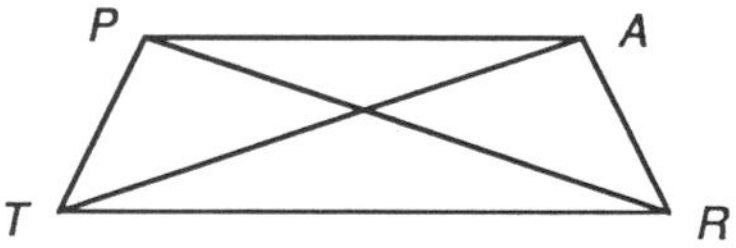

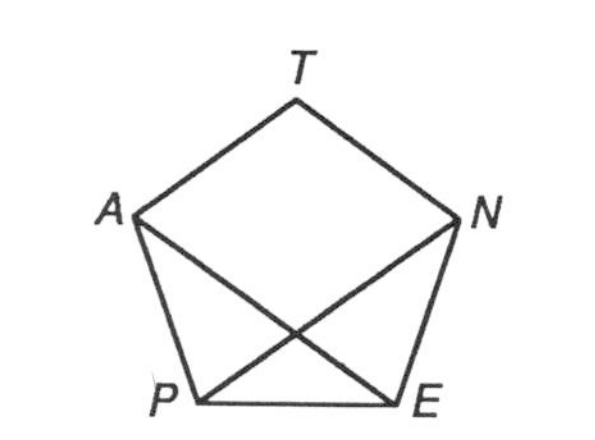

3. *Given:* Regular pentagon $PENTA$

Show: $\Delta PEA \cong \Delta EPN$

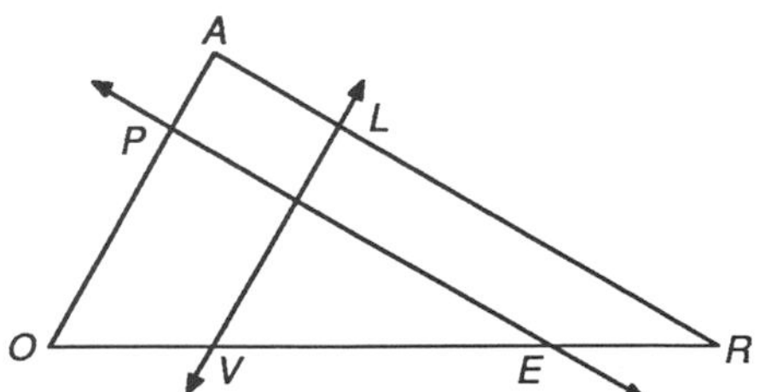

4.* *Given:* ΔORA with $\overleftrightarrow{PE} \parallel \overline{AR}$ and $\overleftrightarrow{LV} \parallel \overline{AO}$ and $\overline{OV} \cong \overline{ER}$

Show: $\overline{LV} \cong \overline{PO}$

EXERCISE SET B

Prove each conjecture. Once the proofs of the conjectures are complete, add these theorems to your list of theorems. You will need them as premises for later proofs. In Exercise 1, use the plan for help.

1. *Conjecture:* *The diagonals of a rectangle are congruent. (C-29)*

   Given: Rectangle $WREK$ with diagonals $\overline{WE}$ and $\overline{RK}$

   Show: $\overline{WE} \cong \overline{RK}$

   Plan:
   - I can show $\overline{WE} \cong \overline{RK}$ if they are corresponding sides of congruent triangles. By redrawing the two overlapping triangles, ΔWRK and ΔRWE, I see that, if I can show they are congruent, then $\overline{WE} \cong \overline{RK}$ by CPCTC.
   - But, I can show $\Delta WRK \cong \Delta RWE$ since $\overline{WK} \cong \overline{RE}$ (opposite sides of a parallelogram are congruent), $\overline{WR} \cong \overline{WR}$ (Reflexive Property of Congruence), and $\angle KWR \cong \angle ERW$ (from the definition of a rectangle).

Once the proof is complete, Conjecture 29 can be called Theorem 29.

2.* *Conjecture:* *If the diagonals of a parallelogram are congruent, then the parallelogram is a rectangle.*

   Given: Parallelogram $PARE$ with diagonals $\overline{PR} \cong \overline{AE}$

   Show: $PARE$ is a rectangle

3.* *Conjecture:* *The medians to the congruent sides of an isosceles triangle are congruent.*

   Given: Isosceles triangle TRN with $\overline{NT} \cong \overline{NR}$ and with medians $\overline{TA}$ and $\overline{RS}$

   Show: $\overline{TA} \cong \overline{RS}$

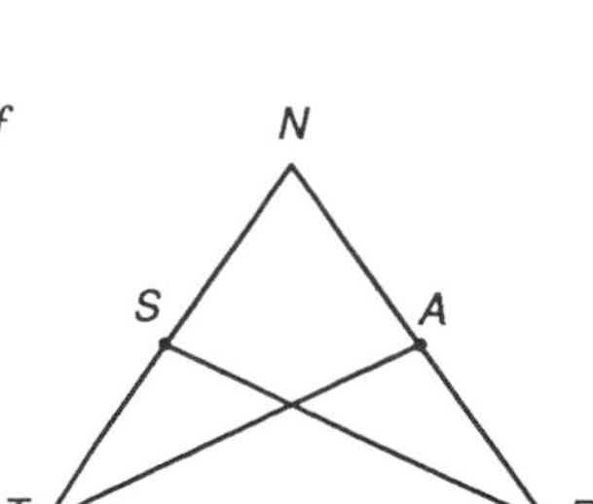

EXERCISE SET C

When completed, the concept map below shows one way of logically relating the postulates and theorems that were used to prove the theorems in Exercise Set B.

1. Copy and complete the concept map by filling in the empty boxes and drawing in the missing arrows.

Logical Relationship of Postulates and Theorems in this Lesson

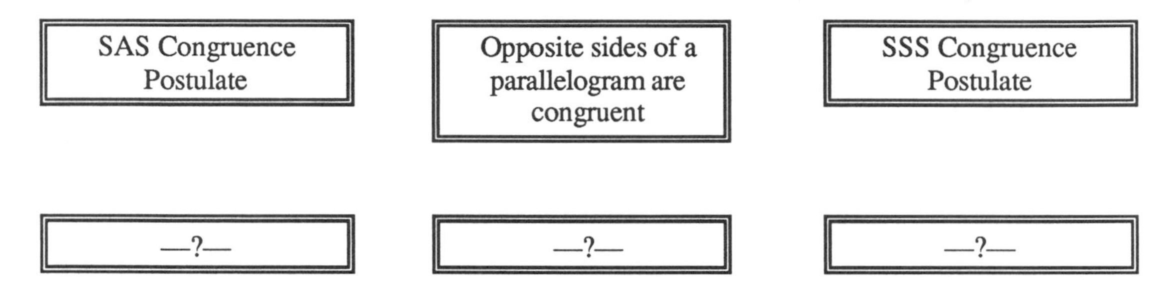

2.* Can you prove the conjecture in Exercise B1 by using ASA instead of SAS?

3. Can you prove the conjecture in Exercise B2 by using SAS instead of SSS?

4. Can you use overlapping triangles *TRS* and *RTA* to prove the conjecture in Exercise B3?

Improving Reasoning Skills

How Did the Farmer Get to the Other Side?

A farmer was taking her daughter's large pet rabbit, a basket of prize winning baby carrots, and her small, but hungry, rabbit-chasing dog to town. She came to a river and realized that she had a problem. The little boat she found tied to the pier was only big enough to carry herself and one of three possessions. She couldn't leave her dog on the bank with her little rabbit (the dog would frighten the poor rabbit) and she couldn't leave the rabbit alone with the carrots (the rabbit would eat all the carrots). She had to figure out how to cross the river safely with one possession at a time while insuring that the dog and rabbit were not left together, and that the rabbit and carrots were not left together during the crossing. How could she safely move back and forth across the river to get her three possessions to the other side?

Lesson 14.9

Starting from the Beginning

In the last six lessons you have proved many conjectures. In each case, however, the proof process was started for you. You were given specific information and asked to show specific information based on a labeled diagram. In this lesson you will have to identify for yourself what is given and what you must show in order to prove a conjecture and you will have to create your own labeled diagram.

Many conjectures of geometry are stated as conditionals. For example, *If a triangle is isosceles, then the base angles are congruent* is a conditional statement. To prove that a conditional statement is true, you assume, as you did in logic proofs, that the first part of the conditional is true, and then you logically demonstrate the truth of the second part of the conditional. You demonstrate that the first part (the antecedent) implies the second part (the consequent). The antecedent is what you assume to be true in the proof; it is the *given* information. The consequent is the part you logically demonstrate in the proof; it is what you want to *show*.

If a triangle is isosceles, then the base angles are congruent.
(Given) (Show)

Once you have identified what is given and what you must show in the statement, you create a diagram that illustrates the statement. Label the parts of the diagram. Then use the labeled diagram to restate what is given and what you must show. You may have to refine your diagram in this process.

Example A

Conjecture: *If two lines are parallel to a third line, then they are parallel to each other.*

Task 1: Identify what is given and what you must show.

Given: Two lines, each parallel to a third line

Show: The two lines are parallel to each other

Task 2: Draw and label a diagram.

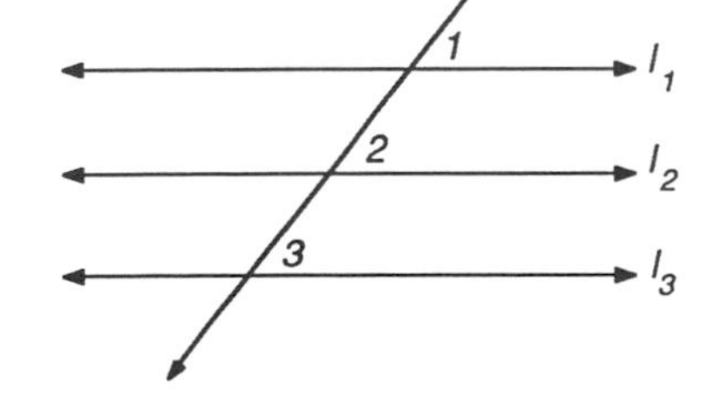

Task 3: Restate what is given and what you must show in terms of the diagram.

Given: $l_1 \parallel l_2$ and $l_1 \parallel l_3$ with l_4 a transversal through all three lines

Show: $l_2 \parallel l_3$

Often a conjecture is not stated in the form of a conditional. When this occurs, it is more difficult to recognize what is given and what you are trying to show.

Example B

Conjecture: *The diagonals of a rhombus are perpendicular bisectors of each other.*

Task 1: Identify what is given and what you must show.

Given: A rhombus with both diagonals

Show: The diagonals are perpendicular bisectors of each other

Task 2: Draw and label a diagram.

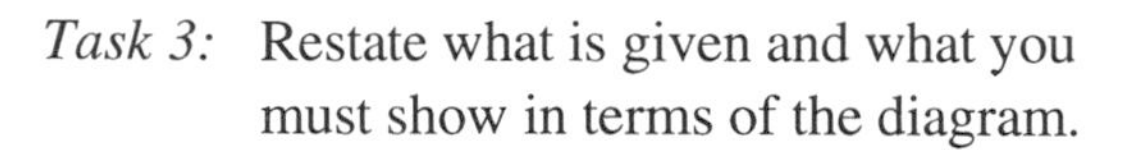

Task 3: Restate what is given and what you must show in terms of the diagram.

Given: Rhombus $RHOM$ with diagonals $\overline{RO}$ and $\overline{MH}$ intersecting at B

Show: $\overline{RO}$ and $\overline{MH}$ are perpendicular bisectors of each other

Once you've created a diagram to illustrate your conjecture, and know where to start and where to go, then you are ready to plan your proof. Once you have your plan, you can draw a flowchart and write the proof. Here's the complete process:

Task 1: Identify what is given and what you must show in the conditional statement.

Task 2: Draw and label a diagram to go with the given information.

Task 3: Restate what is given and what you must show in terms of your diagram.

Task 4: Plan a proof. Organize your reasoning mentally or on paper.

Task 5: Draw a flowchart.

Task 6: Write a proof (flowchart proof, two-column proof, or paragraph proof).

Example C

Conjecture: *If both pairs of opposite sides of a quadrilateral are congruent, then the quadrilateral is a parallelogram. (Converse of Theorem 24)*

Task 1: Identify what is given and what you must show.

Given: Opposite sides of a quadrilateral are congruent

Show: The quadrilateral is a parallelogram

Task 2: Draw and label a diagram.

Task 3: Restate what is given and what you must show in terms of the diagram.

Given: Quadrilateral *PARE* with $\overline{PE} \cong \overline{RA}$ and $\overline{PA} \cong \overline{RE}$

Show: *PARE* is a parallelogram

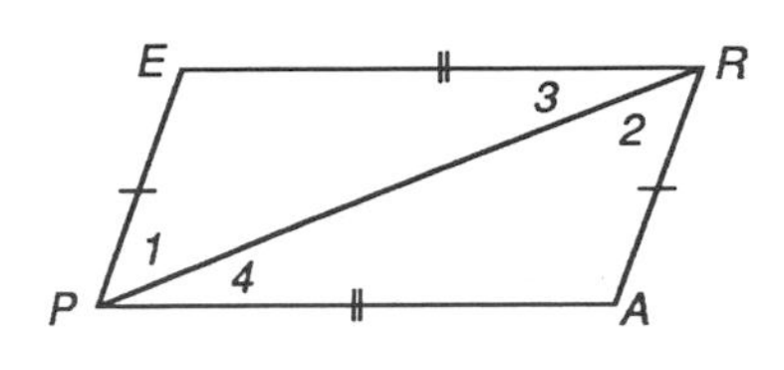

Task 4: Plan a proof.

Plan:

- I can show *PARE* is a parallelogram if the opposite sides are parallel.
- I draw in diagonal $\overline{PR}$ to get two pair of alternate interior angles and, hopefully, two triangles that are congruent.
- If I can show $\angle 1 \cong \angle 2$, then $\overline{PE} \parallel \overline{AR}$ by AIA Postulate.
- If I can show $\angle 3 \cong \angle 4$, then $\overline{PA} \parallel \overline{ER}$ by AIA Postulate.
- If I can show $\triangle PAR \cong \triangle REP$, then $\angle 1 \cong \angle 2$ and $\angle 3 \cong \angle 4$ by CPCTC.
- How can I show the two triangles congruent? By looking at what is given, I see that $\overline{PA} \cong \overline{RE}$ and $\overline{PE} \cong \overline{RA}$. Also, by the Reflexive Property of Congruence, $\overline{PR} \cong \overline{PR}$.
- Therefore, I can show $\triangle PAR \cong \triangle REP$ by SSS.

Task 5: Draw a flowchart.

Flowchart:

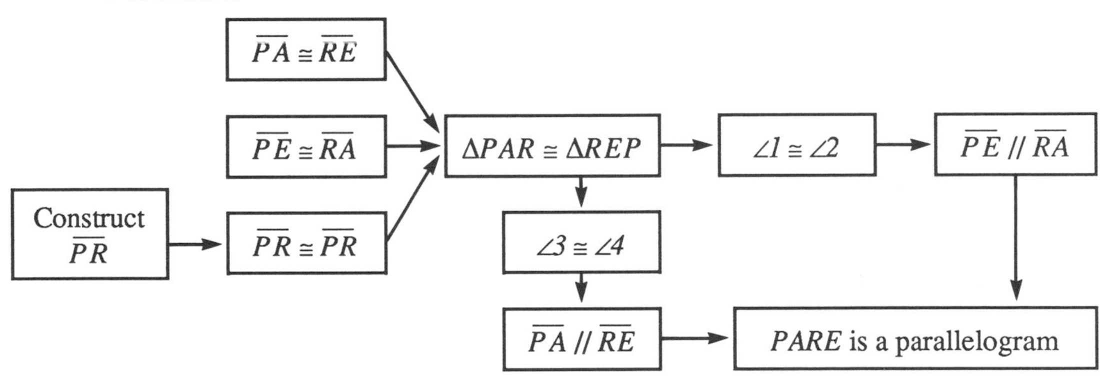

Task 6: Write a proof.

Two-Column Proof:

1. Construct $\overline{PR}$	1. Line Postulate
2. $\overline{PA} \cong \overline{RE}$	2. Given
3. $\overline{PE} \cong \overline{RA}$	3. Given
4. $\overline{PR} \cong \overline{PR}$	4. Reflexive Property of Congruence
5. $\triangle PAR \cong \triangle REP$	5. SSS Congruence Postulate
6. $\angle 1 \cong \angle 2$; $\angle 3 \cong \angle 4$	6. CPCTC
7. $\overline{PE} \parallel \overline{RA}$; $\overline{PA} \parallel \overline{RE}$	7. AIA Postulate
8. *PARE* is a parallelogram	8. Definition of a parallelogram

EXERCISE SET A

Complete the first three tasks of the proof process for each conditional statement.

1.* If a diagonal of a quadrilateral divides the quadrilateral into two congruent triangles, then the quadrilateral is a parallelogram.

2. If both pairs of opposite sides of a quadrilateral are congruent, then the quadrilateral is a parallelogram

3. If both pairs of opposite angles of a quadrilateral are congruent, then the quadrilateral is a parallelogram

4. If the consecutive angles of a quadrilateral are supplementary, then the quadrilateral is a parallelogram.

5. The sum of the measures of the three angles of a triangle is 180.

6. The base angles of an isosceles triangle are congruent.

EXERCISE SET B

Now for some real fun. For each conjecture, first determine whether it is true or false. If false, draw a counter-example. If true, prove it by performing all six tasks in the proof process and add the statement to your theorem list.

1. If two sides of a quadrilateral are parallel and congruent, then the other two sides are parallel.

2. If one pair of opposite sides of a quadrilateral are congruent, then they are parallel.

3. If the diagonals of a quadrilateral are congruent, then the quadrilateral is a parallelogram.

4.* If the diagonals of a quadrilateral bisect each other, then the quadrilateral is a parallelogram.

Improving Reasoning Skills
Abbreviated Equations

Each equation below contains the first letters of words that will make the equation correct. For example: *12 = M. in a Y.* would be *12 = Months in a Year.* Find the missing words in each equation below.

1. 45 = D. in each A.A. of an I.R.T.
2. 7 = S. on a H.
3. 90 = D. in each A. of a R.
4. 5 = D. in a P.
5. 360 = D. in the I.A. of a Q.
6. 180 = D. in a S.

Lesson 14.10

Proofs Using Auxiliary Lines

Sometimes a proof requires the help of additional lines or segments. These helping lines are called **auxiliary lines**. Your construction postulates justify the use of auxiliary lines. For example, when a parallel line needs to be constructed, the Parallel Postulate (*Through a point not on a given line, exactly one line can be constructed parallel to the given line*) can be used as a reason. The trick to using auxiliary lines in proofs is knowing when and where to draw them. This knowledge comes after a lot of experience with proofs and experimentation with the figures. The proof of the Triangle Sum Conjecture is a nice, short proof once you construct an auxiliary line.

EXERCISE SET A

1. The first five tasks in the proof of the Triangle Sum Conjecture have been completed below. You perform the last task. Copy the flowchart and provide reasons beneath each box to create a flowchart proof or write a two-column proof. Once finished, the conjecture can be called the Triangle Sum Theorem. Add it to your list.

Conjecture: *The sum of the measures of the three angles of every triangle is 180.* ***(Triangle Sum Conjecture)***

Given: $\triangle ABC$

Show: $m\angle 1 + m\angle 2 + m\angle 3 = 180$

Plan:

- A nice way to prove this conjecture is to construct a line through one vertex parallel to the opposite side. (The Parallel Postulate allows me to construct it.)

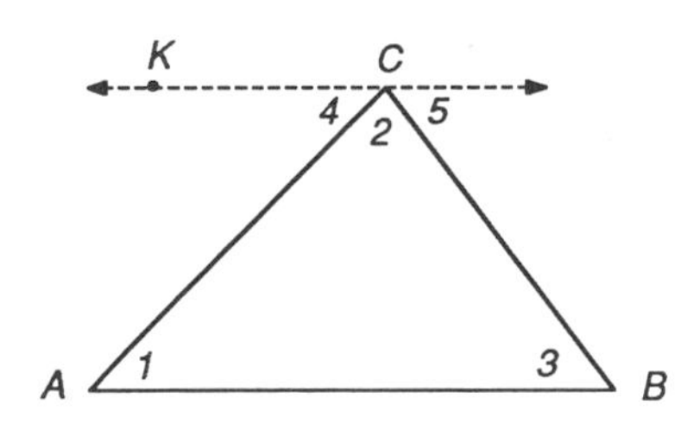

- I must show $m\angle 1 + m\angle 2 + m\angle 3 = 180$. I know I can show $m\angle KCB + m\angle 5 = 180$ since they form a linear pair of angles.
- But $m\angle KCB = m\angle 4 + m\angle 2$ by the Angle Addition Postulate.
- Therefore, I can substitute $m\angle KCB = m\angle 4 + m\angle 2$ into the equation $m\angle KCB + m\angle 5 = 180$ to get $m\angle 4 + m\angle 2 + m\angle 5 = 180$.
- But I can show $m\angle 1 = m\angle 4$ and $m\angle 3 = m\angle 5$ by the AIA Postulate.
- Therefore, I can substitute $m\angle 1$ for $m\angle 4$ and $m\angle 3$ for $m\angle 5$ into the equation $m\angle 4 + m\angle 2 + m\angle 5 = 180$ to get $m\angle 1 + m\angle 2 + m\angle 3 = 180$.

Flowchart:

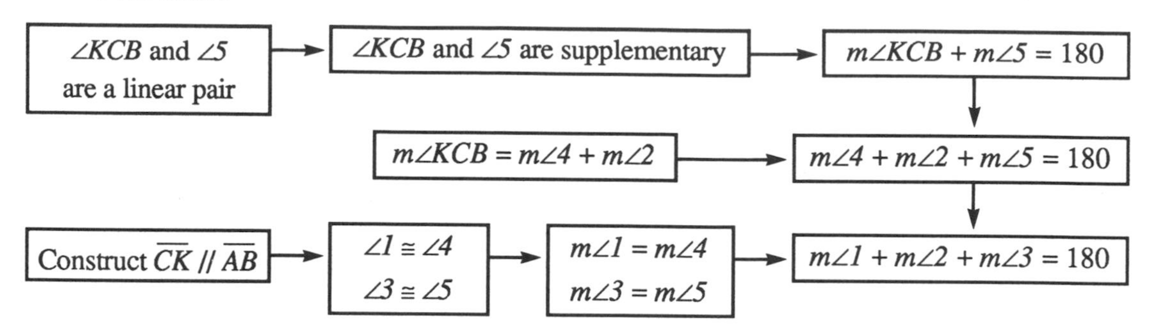

An auxiliary line is also helpful in proving the Isosceles Triangle Conjecture. Quite often, when attempting to prove two segments (or angles) congruent, you try to show that they are corresponding sides (or angles) of congruent triangles. This leads you to construct a segment in the isosceles triangle dividing it into two congruent triangles. But what kind of segment? It turns out that either a median or an angle bisector does the trick. Let's take a look.

2. Write a proof of the Isosceles Triangle Conjecture. Copy the flowchart and provide reasons to create a flowchart proof or write a two-column proof. Once you're finished, the conjecture can be called the Isosceles Triangle Theorem. Add it to your list.

Conjecture: *If a triangle is isosceles, then the base angles are congruent.* ***(Isosceles Triangle Conjecture)***

Given: Isosceles $\triangle ABC$ with $\overline{AC} \cong \overline{BC}$

Show: $\angle CAB \cong \angle CBA$

Plan:

- In order to prove $\angle CAB \cong \angle CBA$, I need to prove two triangles congruent. Therefore, I construct the angle bisector $\overline{CD}$ of the vertex angle. (The Angle Bisector Postulate allows me to construct it.) This gives me $\triangle ADC$ and $\triangle BDC$. I can show $\overline{AC} \cong \overline{BC}$ if I can show $\triangle ADC \cong \triangle BDC$.

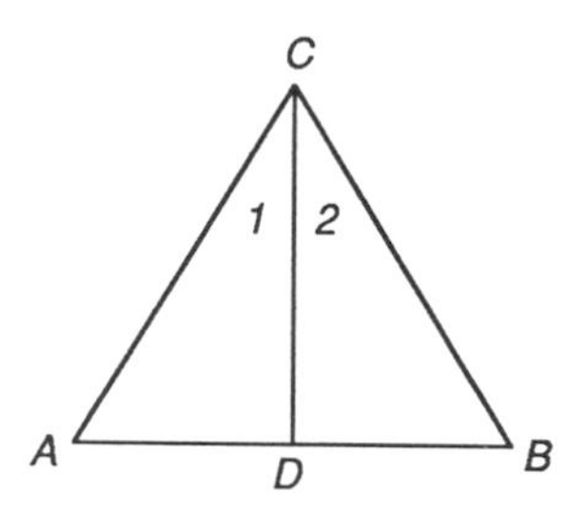

- I can show $\triangle ADC \cong \triangle BDC$ by SSS, SAS, or ASA. Which one?
- Looking at the given information, I know that $\overline{AC} \cong \overline{BC}$ (isosceles triangle), $\angle 1 \cong \angle 2$ ($\overline{CD}$ is an angle bisector), and $\overline{CD} \cong \overline{CD}$ (Reflexive Property of Congruence). Marking the diagram, I see that the triangles are congruent by the SAS Congruence Postulate.

Flowchart:

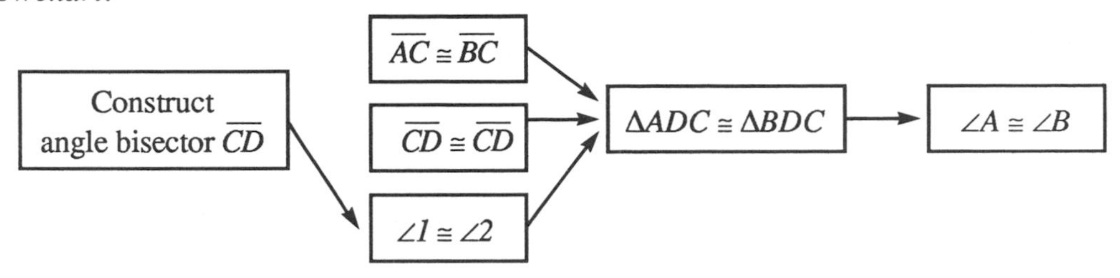

Earlier you discovered a number of angle conjectures. The order in which you discovered them is not the order in which one logically follows from another. Some have become postulates; others you will prove in this lesson. The concept map shows one order in which the conjectures in this lesson can be proved.

Logical Relationship of Postulates and Theorems in this Lesson

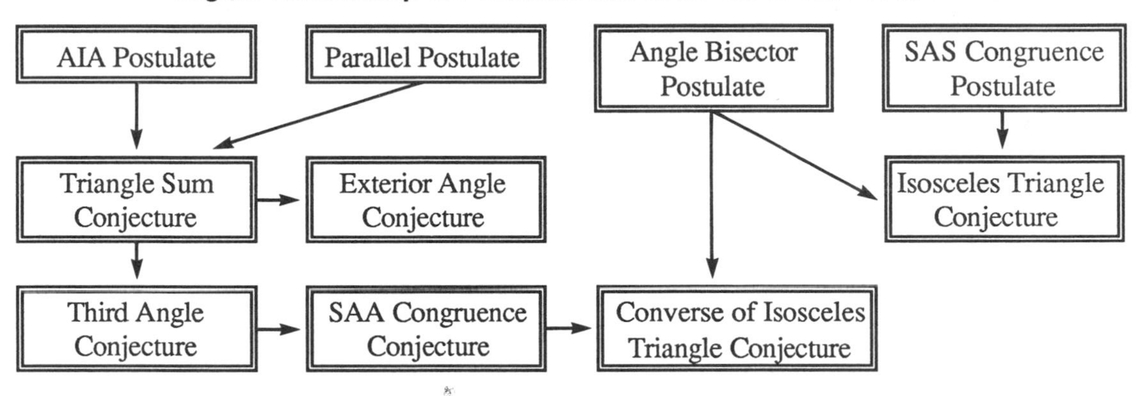

EXERCISE SET B

Perform all six steps of the proof process to prove each conjecture below. When you're finished, add them as theorems to your theorem list. You're going to need them to prove other conjectures.

1. If two angles of one triangle are congruent to two angles of another triangle, then the remaining two angles are congruent. (**Third Angle Conjecture**)

2. The measure of an exterior angle of a triangle equals the sum of the measures of its two remote interior angles. (**Exterior Angle Conjecture**)

3. If two angles and a side that is not between them in one triangle are congruent to two angles and a side that is not between them in another triangle, then the two triangles are congruent. (**SAA Congruence Conjecture**)

4.* If a triangle has two congruent angles, then the triangle is isosceles. (**Converse of the Isosceles Triangle Conjecture**)

Lesson 14.11

Indirect Geometric Proofs

We don't need no education.

– Another Brick in the Wall
by Pink Floyd

An indirect proof in geometry is a little less structured than an indirect logic proof. Do you recall the procedure for an indirect proof in logic?

Procedure for an Indirect Proof

1. Begin by assuming the opposite of what you wish to show is true.
2. Then, using logical reasoning, try to reach a contradiction.
3. Once you have reached a contradiction, point out that since you argued logically, the assumption must be responsible for the contradiction and, therefore, the desired conclusion must be true.

Example A

Conjecture: *If $m\angle N \neq m\angle O$ in ΔNOT, then $NT \neq OT$.*

Given: ΔNOT with $m\angle N \neq m\angle O$

Show: $NT \neq OT$

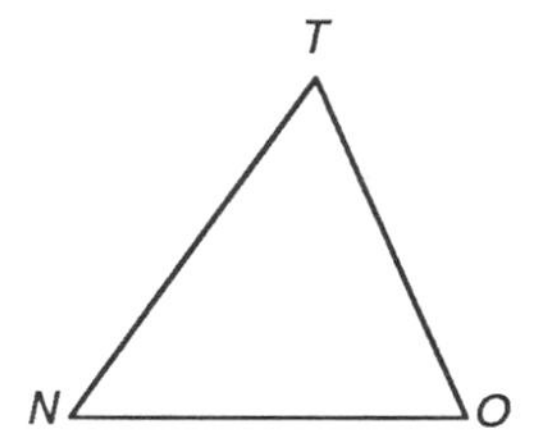

Paragraph Proof:

Assume $NT = OT$. If $NT = OT$, then $m\angle N = m\angle O$ by the Isosceles Triangle Conjecture. But this contradicts the given fact that $m\angle N \neq m\angle O$. Therefore, the assumption ($NT = OT$) is false and so the opposite ($NT \neq OT$) is true.

Example B

Conjecture: *The diagonals of a trapezoid do not bisect each other.*

Given: Trapezoid $ZOID$ with bases $\overline{ZO} \parallel \overline{ID}$ and diagonals $\overline{DO}$ and $\overline{IZ}$ intersecting at Y

Show: Diagonals of trapezoid $ZOID$ do not bisect each other (i.e., $DY \neq OY$ and $ZY \neq IY$)

Paragraph Proof:

Assume the diagonals of trapezoid *ZOID* *do* bisect each other. In Lesson 14.9, Exercise B4, you proved that if the diagonals of a quadrilateral bisect each other, the quadrilateral is a parallelogram. So *ZOID* is a parallelogram. Thus *ZOID* has two pairs of opposite sides parallel. But since it is a trapezoid, it has exactly one pair of parallel sides. This is contradictory. So the assumption is false and the conjecture is true.

EXERCISE SET A

Complete the reasoning in the proof.

Conjecture: *No triangle has two right angles.*

Given: ΔABC

Show: No two angles are right angles

Two-Column Proof:

1.	Assume ΔABC has two right angles (Assume $m\angle A = 90$ and $m\angle B = 90$)	1.	—?—
2.	$m\angle A + m\angle B + m\angle C = 180$	2.	—?—
3.	$90 + 90 + m\angle C = 180$	3.	—?—
4.	$m\angle C =$ –?–	4.	—?—

But if $m\angle C = 0$, then the two sides $\overline{AC}$ and $\overline{BC}$ overlap. This contradicts the given information. So the assumption is false. Therefore, no triangle has two right angles.

EXERCISE SET B

Answer each question to complete the indirect proof.

Conjecture: *The bases of a trapezoid have unequal lengths.*

Given: Trapezoid $ZOID$ with bases $\overline{ZO} \parallel \overline{ID}$

Show: $ZO \neq ID$

Paragraph Proof:

1. Assume $ZOID$ is a trapezoid with $ZO = ID$. If $\overline{ZO}$ and $\overline{ID}$ are bases of the trapezoid, then $\overline{ZO} \parallel \overline{ID}$. Why?

2.* If $\overline{ZO} \parallel \overline{ID}$ and $ZO = ID$, then $\overline{ZD} \parallel \overline{IO}$. Why?

3. But if $\overline{ZO} \parallel \overline{ID}$ and $\overline{ZD} \parallel \overline{IO}$, then $ZOID$ is not a trapezoid. Why not?

But this is a contradiction. Therefore, our assumption is false and $ZO \neq ID$.

EXERCISE SET C

Answer each question to complete the indirect proof.

Conjecture: *A tangent is perpendicular to the radius drawn to the point of tangency.*

Given: Circle O with tangent $\overleftrightarrow{AT}$ tangent at point A and radius $\overline{AO}$

Show: $\overline{AO} \perp \overleftrightarrow{AT}$

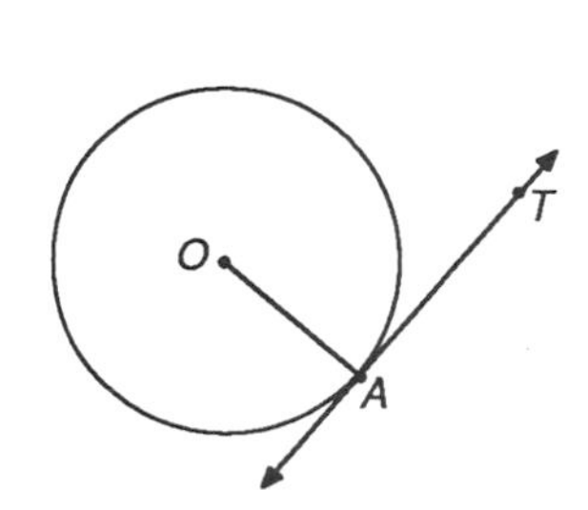

Paragraph Proof:

1. Assume $\overline{AO}$ is not perpendicular to $\overleftrightarrow{AT}$. Construct a perpendicular from O to $\overleftrightarrow{AT}$ and label the intersection point B ($\overline{OB} \perp \overleftrightarrow{AT}$). Which postulate tells you this is possible?
2. Select a point C on $\overline{AT}$ so that B is the midpoint of $\overline{AC}$. Which postulate tells you this is possible?
3. Next construct $\overline{OC}$. Which postulate tells you this is possible?
4. $\angle ABO \cong \angle CBO$ What reason(s) tells you this?
5.* $\overline{AB} \cong \overline{BC}$. What definition tells you this?
6. $\overline{OB} \cong \overline{OB}$. What property of congruence tells you this is possible?
7. Therefore, $\Delta ABO \cong \Delta CBO$. Which shortcut congruence conjecture tells you the triangles are congruent?
8. If $\Delta ABO \cong \Delta CBO$, then $\overline{AO} \cong \overline{CO}$. Why?

Thus, C must be a point of the circle (since a circle is the set of *all* points in the plane at a given distance from the center and A and C are both the same distance from the center). Therefore, $\overleftrightarrow{AT}$ intersects the circle in *two* points (A and C) and thus $\overleftrightarrow{AT}$ is not a tangent. But this contradicts the given information. So the assumption that $\overline{AO}$ is not perpendicular to $\overleftrightarrow{AT}$ is false and $\overline{AO} \perp \overleftrightarrow{AT}$.

Improving Visual Thinking Skills

Design III

Use your geometry tools to create a larger version of the design on the right.

Then decorate your design.

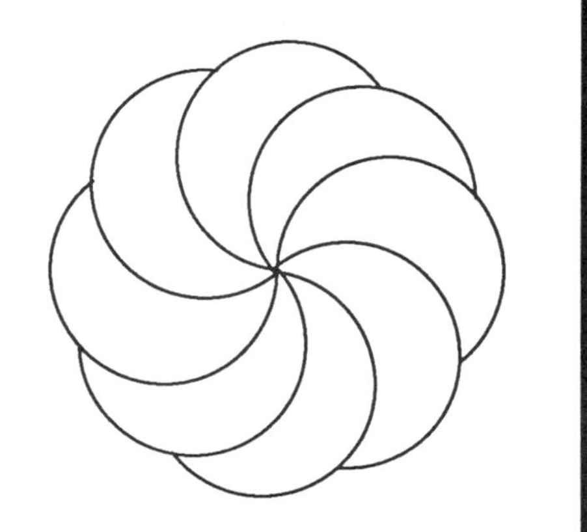

Lesson 14.12

Chapter Review

EXERCISE SET A

Identify each statement as true or false. Consult your theorem list.

1. The angle bisector of the vertex angle of an isosceles triangle is also a median.
2. The median from the vertex angle of an isosceles triangle is also an angle bisector of the vertex angle.
3. If two angles are congruent, then they are vertical angles.
4. Supplements of the same angle or congruent angles are supplementary.
5. A diagonal of a parallelogram divides the parallelogram into two isosceles triangles.
6. The diagonals of a parallelogram are congruent.
7. If two lines are parallel to a third line, then they are parallel to each other.
8. If two lines in the same plane are perpendicular to a third line, then they are perpendicular to each other.
9. If two parallel lines are intersected by a transversal, then the interior angles on the same side of the transversal are congruent.
10. The consecutive angles of a parallelogram are supplementary.
11. If the diagonals of a parallelogram are congruent, then it is a rectangle.
12. The medians to the congruent sides of an isosceles triangle are congruent.
13. If both pairs of opposite sides of a quadrilateral are congruent, then the quadrilateral is a parallelogram.
14. If two sides of a quadrilateral are parallel and congruent, then the other two sides are parallel.
15. If the diagonals of a quadrilateral bisect each other, then the quadrilateral is a parallelogram.
16. If two angles of one triangle are congruent to two angles of another triangle, then the third pair of angles are congruent.
17. A tangent is perpendicular to the radius drawn to the point of tangency.

EXERCISE SET B

1. When you say $\overline{AC} \cong \overline{AC}$, you are using which property of congruence?

2.* Which postulate supports the statement: *If M is the midpoint of* $\overline{CD}$, *then* $CM + MD = CD$?

3. What do you call a diagram that shows a step-by-step procedure through a complicated system?

4. If the triangles you must prove congruent overlap, what should you do to make the proof easier?

5. If $\Delta ABC \cong \Delta DEF$, then $\overline{AC} \cong \overline{DF}$. What is the reason you would give for this conclusion in a proof?

6. Sometimes a proof requires the help of additional lines or segments. What are these helping lines called?

7. How would you state the following as a conditional: *The base angles of an isosceles triangle are congruent?*

8. What is the procedure in an indirect proof?

EXERCISE SET C

1.* Create a flowchart for the proof below. Then write the reason for each step below its box to create a flowchart proof, or write a two-column proof.

Given: Isosceles ΔWHY with $\overline{WY} \cong \overline{HY}$; N, O, and T are midpoints of $\overline{WH}$, $\overline{HY}$, and $\overline{WY}$ respectively

Show: ΔNOT is isosceles

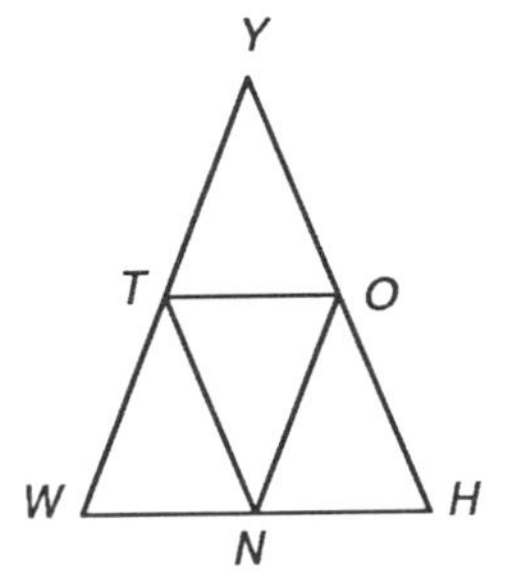

2. Prove the conjecture. When you're finished, add it to your theorem list.

Conjecture: *The sum of the measures of the four angles of a quadrilateral is 360.*

Given: Quadrilateral $ABCD$

Show: $m\angle 1 + m\angle 2 + m\angle 3 + m\angle 4 = 360$

3.* Prove the conjecture. When you're finished, add it to your theorem list.

Conjecture: *If both pairs of opposite angles of a quadrilateral are congruent, then the quadrilateral is a parallelogram.*

Given: Quadrilateral $ABCD$ with $\angle DAB \cong \angle BCD$ and $\angle ADC \cong \angle ABC$

Show: Quadrilateral $ABCD$ is a parallelogram

4.* Prove the following conjecture.

Conjecture: *The base angles of an isosceles trapezoid are congruent.*

Given: Isosceles trapezoid $TRAP$ with $\overline{TP} \cong \overline{AR}$

Show: $\angle T \cong \angle R$ and $\angle P \cong \angle A$

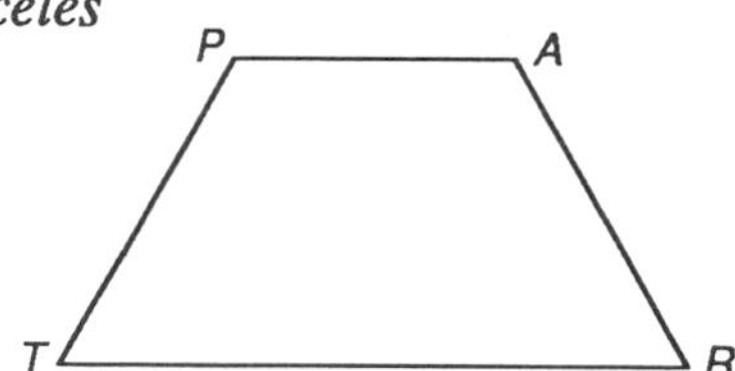

EXERCISE SET D

In each exercise, select one of the three conjectures. Investigate it to determine whether or not it is true. If it is false, demonstrate a counter-example. If it is true, prove it.

1. a. The angle bisectors of the base angles of an isosceles triangle are congruent.

   b. If the diagonals of a parallelogram bisect the angles, then the parallelogram is a square.

   c. The medians of an equilateral triangle are congruent.

2. a. The altitudes to the congruent sides of an isosceles triangle are congruent.

   b.* If the diagonals of a quadrilateral bisect each other, then the quadrilateral is a parallelogram.

   c. The angle bisectors of an equilateral triangle are congruent.

Cooperative Problem Solving

The Centauri Challenge

For the past sixty years, astronomers of the lunar colony have been communicating with an intelligent life-form in the Centauri star-system. Because of the vast distances, communication is very slow. It takes twenty years for a message to be sent and a response received. The most recent communication included *The Centauri Challenge. The Centauri Challenge* was sent from logic students of the Centauri star system to students at the lunar colony. The challenge is shown below.

The Centauri Challenge

Centauri is a formal system of strings of the letters *P*, *Q*, *R*, *S* with four rules governing their behavior. Using some combination of the four rules, it is possible to change one string of letters into a different string. The four rules are:

Rule 1: Any two adjacent letters in a string can change places with each other. (*PQ* >> *QP*)

Rule 2: If a string ends in the same two letters, then you may substitute a *Q* in place of those two letters. (*RSS* >> *RQ*)

Rule 3: If a string begins in the same two letters, then you may add an *S* in front of these two letters. (*PPR* >> *SPPR*)

Rule 4: If a string of letters starts and finishes with the same letter, then you may substitute an *R* in place of all the letters between the first and last letters. (*PQRSP* >> *PRP*)

A theorem in this system is a string followed by a >> followed by a string. For example, *PQQRSS* >> *QRQ* says, if given string *PQQRSS*, then by using the rules you can arrive at *QRQ*. An example of the proof of this theorem is shown below.

Example

Show:	*PQQRSS* >> *QRQ*	
Proof:	*PQQRSS*	Given
	PQQRQ	By Rule 2
	QPQRQ	By Rule 1
	QRQ	By Rule 4
	Therefore, *PQQRSS* >> *QRQ*.	

Challenge 1:

Use the rules of Centauri to prove the theorems below.

1. *PQPRQ* >> *RQ*
2. *PQRSSQR* >> *RQ*
3. *PSSRS* >> *RQ*
4. *PSRQQRSQPSSS* >> *RQ*
5. *PQQQQP* >> *RQ*
6. *QQQQQQ* >> *SRQ*

Challenge 2:

Use the rules of Centauri to answer the questions below.

1. Can you find a string of five or more letters that *cannot* be reduced to *RQ*? If yes, produce it. If not, prove that you cannot.
2. One of the rules of Centauri can be removed without losing any of the first five theorems proved in the first challenge. Which one of the rules can be removed? Why must this rule be used in the sixth theorem? Create another theorem that must use this rule.

15 Geometric Proofs II

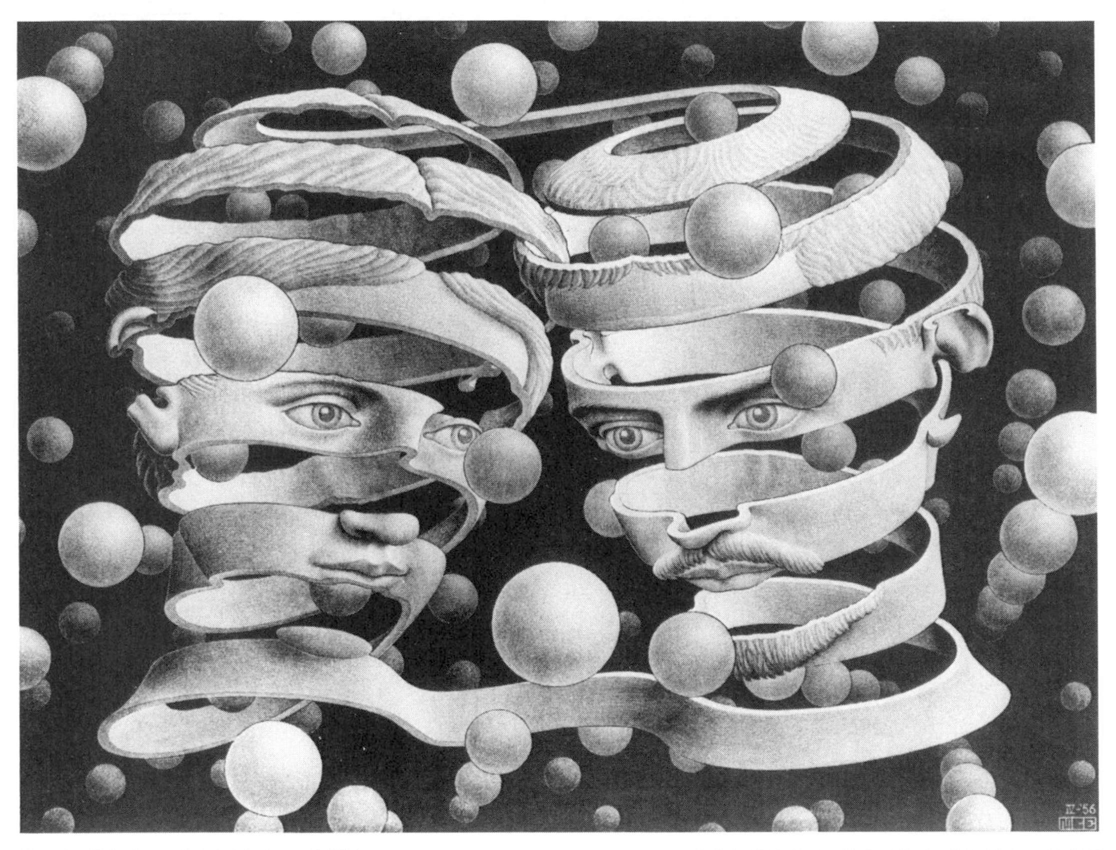

Bond of Union, M. C. Escher, 1956

In this final chapter you will be "playing the whole ball game." You will perform geometric investigations, make conjectures, and then prove or disprove your conjectures. This is the complete process that a working mathematician goes through. Discoveries by themselves are questionable until they have been verified by logical argument. However, without discoveries, there would be nothing to prove.

Lesson 15.1

Proving Your Own Conjectures

Nothing is more dangerous than an idea when it is the only one you have.

– Emile Chartier

In this lesson you will perform investigations leading to conjectures that can be proved using theorems and postulates from Chapter 14. There are five investigations leading to new conjectures. The first investigate-conjecture-prove problem is completed as an example.

Example

Investigate: Construct a linear pair of angles. Construct the angle bisector of each.

Conjecture: What do you observe about the angle bisectors? State a conjecture about the bisectors of a linear pair of angles.

Prove: Prove your conjecture.

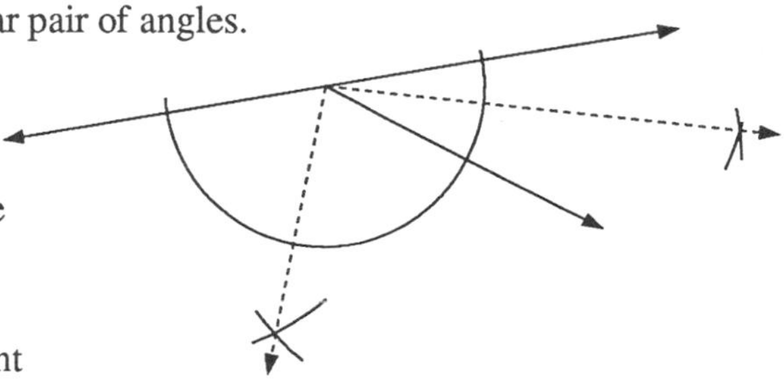

After completing the investigation, you might wind up with a drawing similar to the one pictured at right. The angle bisectors appear perpendicular. Is this always true? Check with your protractor. You don't want to make a conjecture based on only one case, so try again beginning with other linear pairs of angles. After going through the construction several more times, you may satisfy yourself that your observation seems to hold for every linear pair of angles. You're ready to make a conjecture.

Conjecture: *The bisectors of a linear pair of angles are perpendicular.*

How do you prove the conjecture? Start by making a diagram. You may leave off the construction marks. Label your diagram. State what you are given and what you must show in terms of your diagram. Then plan and proceed with your proof.

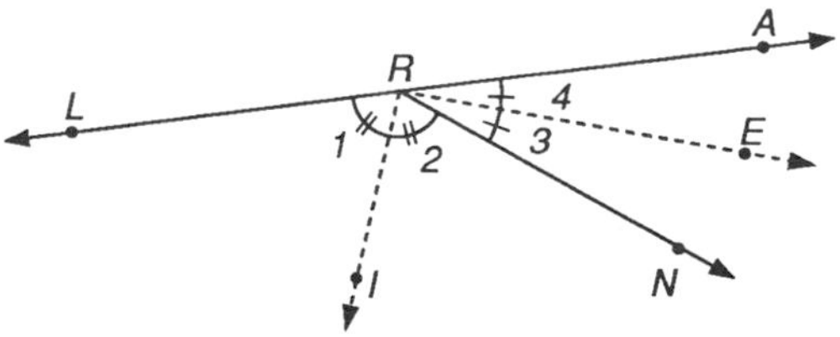

A R L 4 1 3 E 2 I N

Given: $\angle LRN$ and $\angle NRA$ are a linear pair with $\overrightarrow{RI}$ and $\overrightarrow{RE}$ their respective bisectors

Show: $\overrightarrow{RI} \perp \overrightarrow{RE}$

Plan:

- I wish to prove $\overrightarrow{RI} \perp \overrightarrow{RE}$. I can do this if I can prove that $m\angle IRE = 90$. I can show $m\angle IRE = 90$ if I can show $m\angle 2 + m\angle 3 = 90$ since $m\angle IRE = m\angle 2 + m\angle 3$.

- I know by the Angle Addition Postulate that $m\angle 1 + m\angle 2 = m\angle LRN$ and $m\angle 3 + m\angle 4 = m\angle NRA$. By the Linear Pair Postulate and the definition of supplementary angles, $m\angle LRN + m\angle NRA = 180$. Thus, $m\angle 1 + m\angle 2 + m\angle 3 + m\angle 4 = 180$ by the Substitution Property.
- I know that $m\angle 1 = m\angle 2$ and $m\angle 3 = m\angle 4$ by the definition of an angle bisector. Therefore, I can show $m\angle 2 + m\angle 2 + m\angle 3 + m\angle 3 = 180$.
- If I can show that $m\angle 2 + m\angle 2 + m\angle 3 + m\angle 3 = 180$, then using algebra I can show $2m\angle 2 + 2m\angle 3 = 180$ or $2(m\angle 2 + m\angle 3) = 180$ or $m\angle 2 + m\angle 3 = 90$ and, thus, $m\angle IRE = 90$ and $\overrightarrow{RI} \perp \overrightarrow{RE}$.

Flowchart:

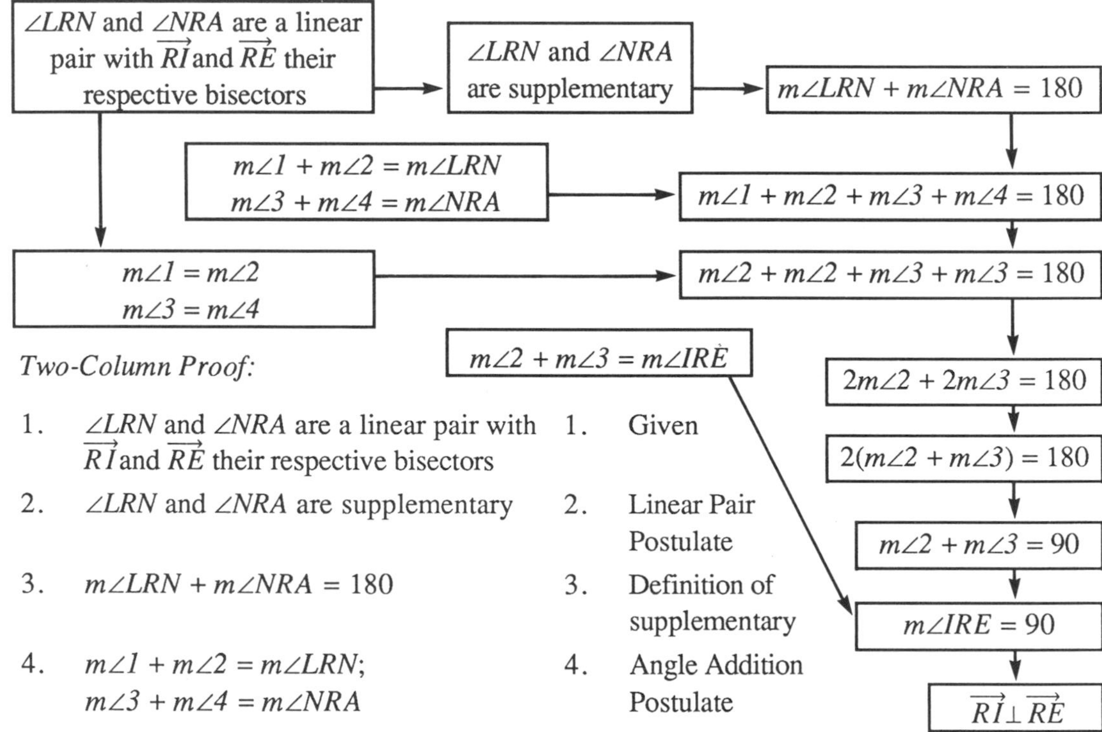

Two-Column Proof:

	Statement		Reason
1.	$\angle LRN$ and $\angle NRA$ are a linear pair with $\overrightarrow{RI}$ and $\overrightarrow{RE}$ their respective bisectors	1.	Given
2.	$\angle LRN$ and $\angle NRA$ are supplementary	2.	Linear Pair Postulate
3.	$m\angle LRN + m\angle NRA = 180$	3.	Definition of supplementary
4.	$m\angle 1 + m\angle 2 = m\angle LRN$; $m\angle 3 + m\angle 4 = m\angle NRA$	4.	Angle Addition Postulate
5.	$m\angle 1 + m\angle 2 + m\angle 3 + m\angle 4 = 180$	5.	Substitution Property
6.	$m\angle 1 = m\angle 2$; $m\angle 3 = m\angle 4$	6.	Definition of angle bisector
7.	$m\angle 2 + m\angle 2 + m\angle 3 + m\angle 3 = 180$	7.	Substitution Property
8.	$2m\angle 2 + 2m\angle 3 = 180$	8.	Addition
9.	$2(m\angle 2 + m\angle 3) = 180$	9.	Distributive Property
10.	$m\angle 2 + m\angle 3 = 90$	10.	Division Property of Equality
11.	$m\angle 2 + m\angle 3 = m\angle IRE$	11.	Angle Addition Postulate
12.	$m\angle IRE = 90$	12.	Substitution Property
13.	$\overrightarrow{RI} \perp \overrightarrow{RE}$	13.	Definition of perpendicular

Now it is your turn. The four investigations and proofs in the exercise set should be worked in groups of four. Each student in the group should:

- Perform one of the investigations.
- Make the conjecture for that investigation.
- Plan and write a proof.
- Share your conjecture and proof with the others in the group.

EXERCISE SET

Perform each investigation. Then make a conjecture. Finally, prove your conjecture. Once you have proved your conjectures, they may be added to your theorem list.

1. *Investigate:* Use both edges of your straightedge to create a pair of parallel lines. Draw a transversal through the pair of parallel lines. Construct the bisectors of a pair of alternate interior angles.

   Conjecture: What do you observe about the angle bisectors? State a conjecture about the bisectors of a pair of alternate interior angles.

   Prove: Prove your conjecture.

2. *Investigate:* Use both edges of a straightedge to create a pair of parallel lines. With the same straightedge, draw a second pair of parallel lines intersecting the first pair to form a parallelogram.

   Conjecture: What special type of parallelogram is it? State a conjecture about the parallelogram formed by two pairs of parallel lines that are the same distance apart.

   Prove: Prove your conjecture.

3.* *Investigate:* Use both edges of a straightedge to create a pair of parallel lines. With your compass, construct an isosceles trapezoid. Construct the diagonals.

   Conjecture: What do you observe about the diagonals? State your conjecture.

   Prove: Prove your conjecture.

4.* *Investigate:* A kite is a quadrilateral with two pairs of consecutive sides congruent. Construct three kites on your paper. Construct the diagonals in each.

   Conjecture: What do you observe about the diagonals in each kite? State your observations as a conjecture.

   Prove: Prove your conjecture.

Lesson 15.2

Proving Circle Conjectures

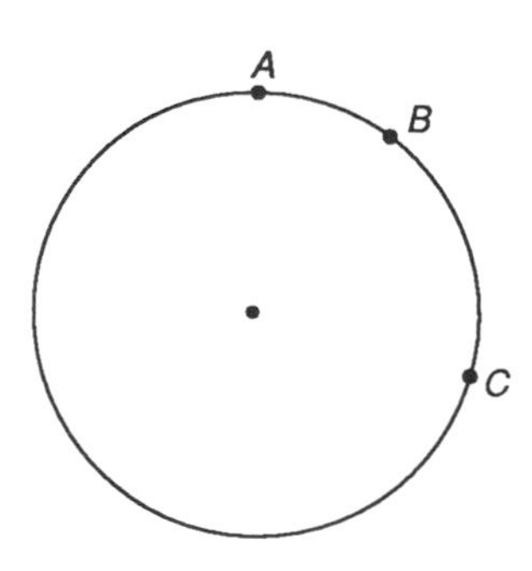

In Chapter 6 you discovered a number of circle properties. Now you will prove some of them. To prove circle conjectures we need to add the Arc Addition Postulate to our list of postulates.

P-14 *If B is on $\overset{\frown}{AC}$ and between A and C, then* $m\overset{\frown}{AB} + m\overset{\frown}{BC} = m\overset{\frown}{AC}$. ***(Arc Addition Postulate)***

Let's start with the proof of the Inscribed Angle Conjecture (*The measure of an inscribed angle in a circle equals one-half the measure of its intercepted arc).* The proof of the Inscribed Angle Conjecture is interesting because you first prove the conjecture for a special condition or case, and then you use that case to extend the conjecture to all conditions.

When you inscribe an angle in a circle, exactly one of the cases below must be true.

Case 1

The center is on the angle.

Case 2

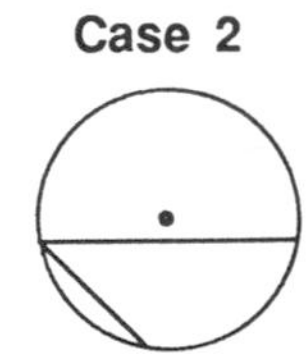

The center is outside the angle.

Case 3

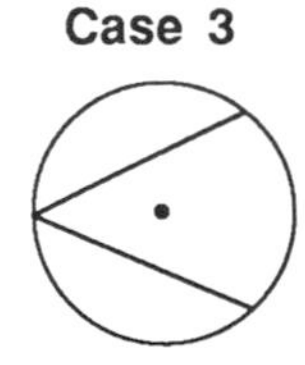

The center is inside the angle.

To prove the Inscribed Angle Conjecture, we will demonstrate that it is true when one of the sides of the angle passes through the center of the circle (case 1). Then we will use this special case to prove two other possible cases. When finished, the conjecture will be proved for all possible cases and you can call it the Inscribed Angle Theorem.

Conjecture: *The measure of an inscribed angle in a circle equals one-half the measure of its intercepted arc when a side of the angle passes through the center of the circle. (Case 1)*

Given: Circle O with $\angle MDR$ on diameter $\overline{DR}$

Show: $m\angle MDR = \frac{1}{2} m\overset{\frown}{MR}$

Plan:

- Since all radii are congruent, ΔDOM is isosceles and $x = z$.
- But $x + z = y$ by the Exterior Angle Theorem.
- So $x + x = y$ which gives $2x = y$. Therefore, $x = (1/2)\,y$.
- But $m\angle MDR = x$ and $m\overset{\frown}{MR} = y$. Therefore, $m\angle MDR = (1/2)\, m\overset{\frown}{MR}$.

Two-Column Proof of Case 1:

Let $x = m\angle MDR$, $z = m\angle OMD$, and $y = m\angle ROM$.

	Statement		Reason
1.	$\overline{DO} \cong \overline{OM}$	1.	All radii of a circle are congruent (definition of a circle)
2.	ΔDOM is isosceles	2.	Definition of isosceles triangle
3.	$\angle MDR \cong \angle OMD \quad (x = z)$	3.	Isosceles Triangle Theorem
4.	$x + z = y$	4.	Exterior Angle Theorem
5.	$x + x = y \quad (2x = y)$	5.	Substitution Property
6.	$x = \frac{1}{2}y$	6.	Division Property of Equality
7.	$m\angle MDR = \frac{1}{2}m\angle ROM$	7.	Substitution Property
8.	$m\angle ROM = m\widehat{MR}$	8.	The measure of a minor arc is equal to the measure of its determined central angle
9.	$m\angle MDR = \frac{1}{2}m\widehat{MR}$	9.	Substitution Property

A paragraph proof is less structured than the two-column proof or flowchart proof. We will use case 1 to write a paragraph proof for case 2 and case 3.

Conjecture: *The measure of an inscribed angle in a circle equals one-half the measure of its intercepted arc when the center of the circle is outside the angle. (Case 2)*

Given: Circle O with inscribed angle $\angle MDK$ on one side of diameter $\overline{DR}$

Show: $m\angle MDK = \frac{1}{2}m\widehat{MK}$

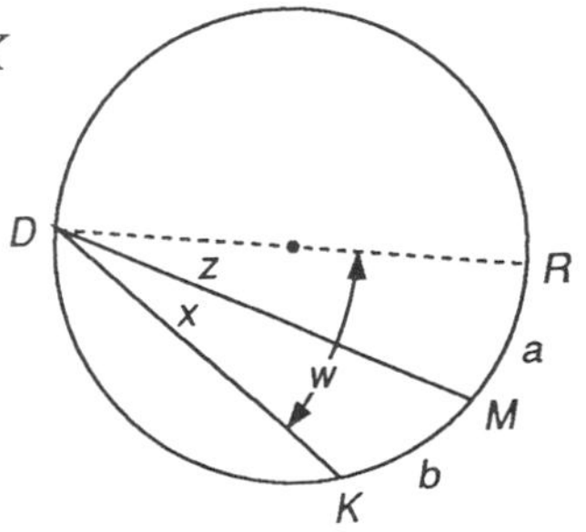

Paragraph Proof of Case 2:

Let $z = m\angle MDR$ and $x = m\angle MDK$.
Let $w = x + z = m\angle MDR + m\angle MDK = m\angle KDR$.
Let $a = m\widehat{MR}$ and $b = m\widehat{MK}$.
Thus $a + b = m\widehat{MR} + m\widehat{MK} = m\widehat{KR}$ (Arc Addition Postulate).

Stated in terms of x and b, we wish to show that $x = \frac{b}{2}$.

From case 1 we know that $w = \frac{(a+b)}{2}$ and $z = \frac{a}{2}$.

Since $w = x + z$ (Angle Addition Postulate),
then $x = w - z$ (Subtraction Property of Equality).

Therefore, $x = w - z = \frac{(a+b)}{2} - \frac{a}{2} = \frac{a}{2} + \frac{b}{2} - \frac{a}{2} = \frac{b}{2}$, or $m\angle MDK = \frac{1}{2}m\widehat{MK}$.

EXERCISE SET A

Copy and complete the statements in the paragraph proof of case 3 below.

Conjecture: *The measure of an inscribed angle in a circle equals one-half the measure of its intercepted arc when the center of the circle is inside the angle. (Case 3)*

Given: Circle O with inscribed angle $\angle MDK$ on both sides of diameter $\overline{DR}$

Show: $m\angle MDK = \frac{1}{2} m\widehat{MK}$

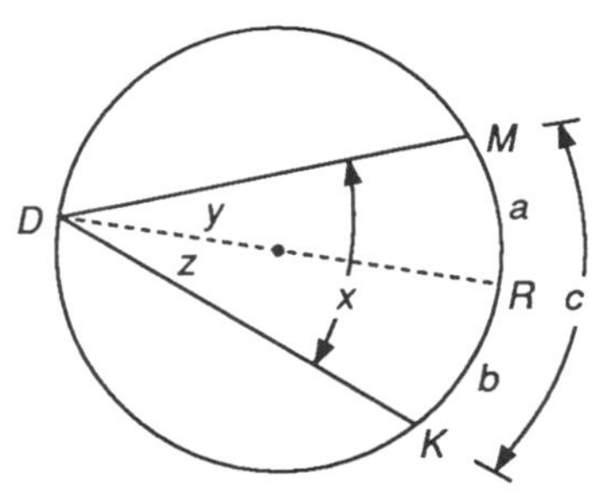

Paragraph Proof of Case 3:

Let $z = m\angle KDR$ and $y = m\angle MDR$.
Let $x = y + z = m\angle KDR + m\angle MDR = m\angle MDK$.
Let $a = m\widehat{MR}$, $b = m\widehat{RK}$.
Let $c = a + b = m\widehat{MR} + m\widehat{RK} = m\widehat{MK}$.

1. We wish to show that $m\angle MDK = \frac{1}{2} m\widehat{MK}$ or that $x = \frac{-?-}{2}$.
2. From case 1 we know that $y = \frac{-?-}{2}$ and $z = \frac{-?-}{2}$.
3. Therefore, $x = y + z = \frac{-?-}{2} + \frac{-?-}{2} = \frac{(-?- + -?-)}{2} = \frac{-?-}{2}$.
4. But this is the same as —?—.

Once you have proved the three cases above, you have proved the Inscribed Angle Conjecture. It can now be renamed as the Inscribed Angle Theorem and it can be added to your list of theorems. You will need the Inscribed Angle Theorem to prove theorems in the next lesson.

When a conjecture can be proved as an immediate consequence of another theorem, the new theorem is a ***corollary*** *of the original theorem.*

EXERCISE SET B

The four conjectures in this exercise set can be proved almost immediately using the Inscribed Angle Theorem. They are corollaries of the Inscribed Angle Theorem. Prove each conjecture. As corollaries they can be added to your theorem list.

1. Angles inscribed in the same arc are congruent. (C-45)
2. Every angle inscribed in a semicircle is a right angle. (C-46)
3. The opposite angles of a quadrilateral inscribed in a circle are supplementary. (C-47)
4. Parallel lines intercept congruent arcs on a circle. (C-48)

Lesson 15.3

Proving New Circle Conjectures

This lesson should be worked in groups of four divided into two pairs. A pair of students should:

- Perform one of the two investigations.
- Make the new circle conjecture for that investigation.
- Complete the proof of the conjecture in Exercise Set A or Exercise Set B.

Then the four group members should share what they have learned in the lesson with the others and together work the problems in Exercise Set C.

In your first investigation you will discover a formula for calculating the measure of an angle formed by two chords intersecting within a circle. To do this you need to recall three of the theorems you recently proved.

> *The sum of the measures of the three angles of every triangle is 180.* ***(Triangle Sum Theorem)***
>
> *The measure of an exterior angle of a triangle equals the sum of the measures of the two remote interior angles.* ***(Exterior Angle Theorem)***
>
> *The measure of an inscribed angle in a circle equals one-half the measure of its intercepted arc.* ***(Inscribed Angle Theorem)***

Investigation 15.3.1 

Use the conjectures above to determine the values indicated.

1. $a =$ –?– $x =$ –?– $b =$ –?–

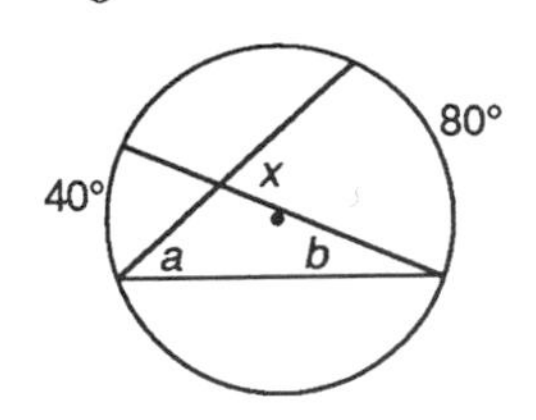

2. $c =$ –?– $y =$ –?– $d =$ –?–

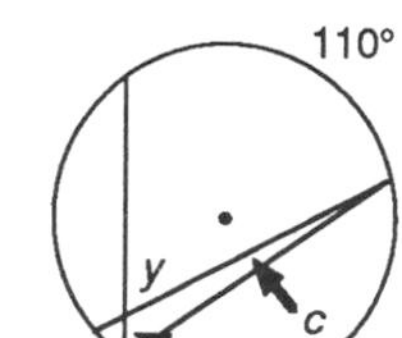
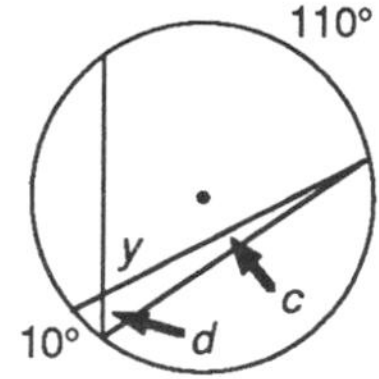

3. $e =$ –?– $z =$ –?– $f =$ –?–

4. $g =$ –?– $w =$ –?– $h =$ –?–

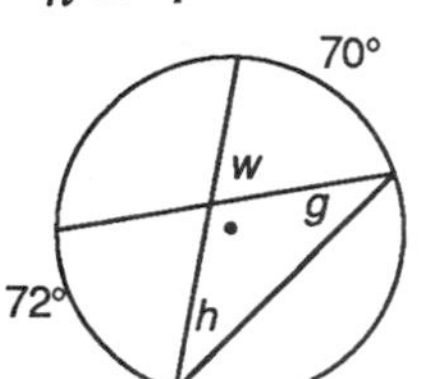

5. $x =$ –?–

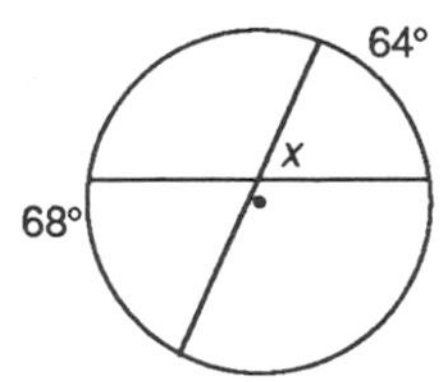

6. $a =$ –?–

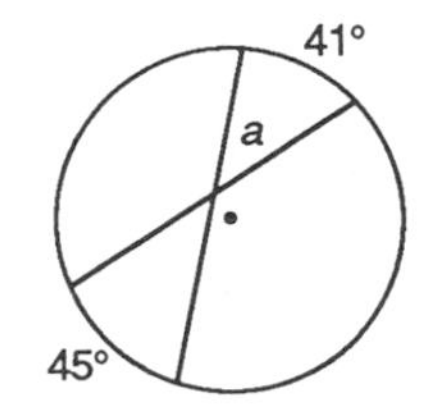

Do you notice a relationship between an angle formed by intersecting chords and the intercepted arcs? Compare your observations with the observations of students near you. State your findings as your next conjecture.

The measure of an angle formed by the intersection of two chords in a circle is equal to —?—.
(Intersecting Chords Conjecture)

In the next investigation, you are looking for a formula to calculate the measure of an angle formed by two secants intersecting outside the circle.

Investigation 15.3.2

Use the Inscribed Angle and Exterior Angle Theorems to calculate the measure of the angle formed by the two secants in each problem.

1. $a = $ –?–

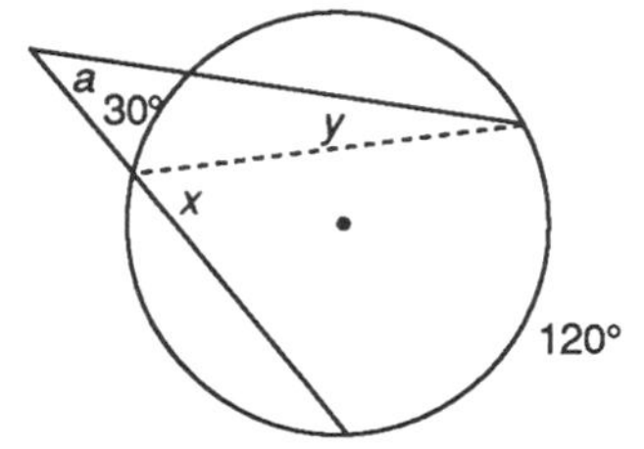

2. $b = $ –?–

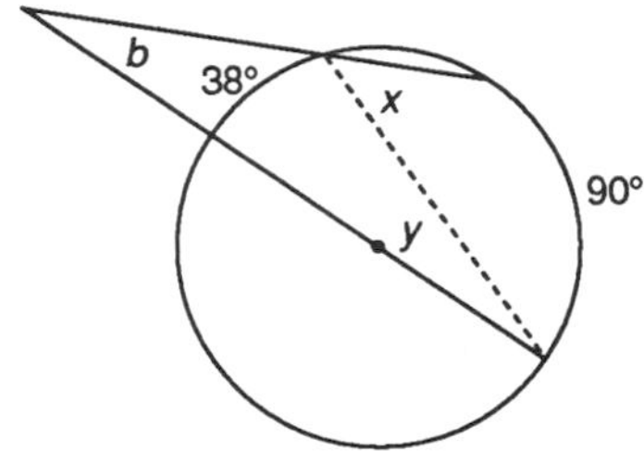

3. $c = $ –?–

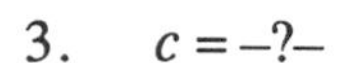

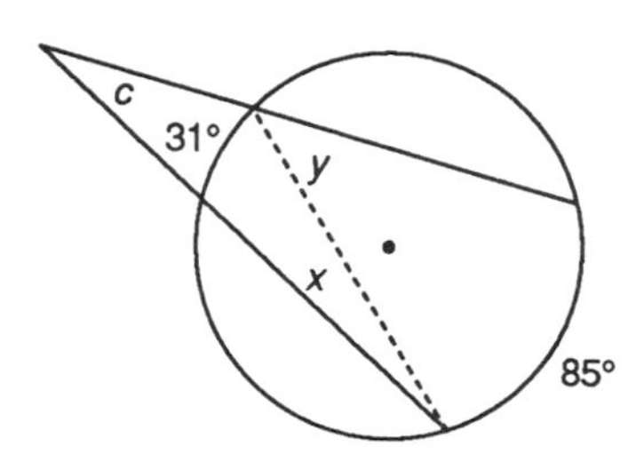

4. $d = $ –?–

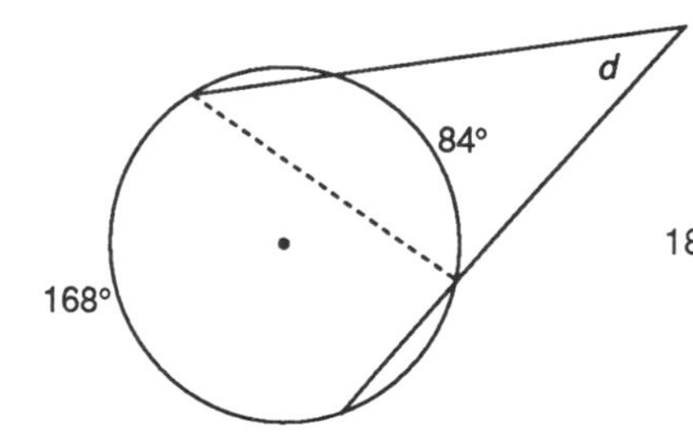

5. $e = $ –?–

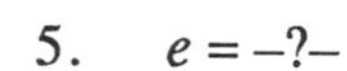

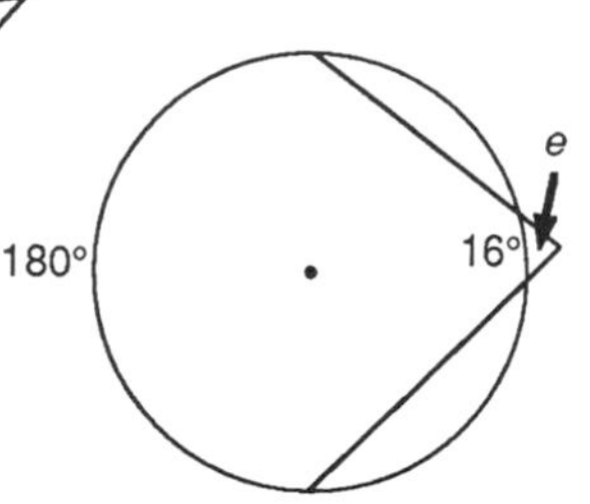

6. $a = $ –?–

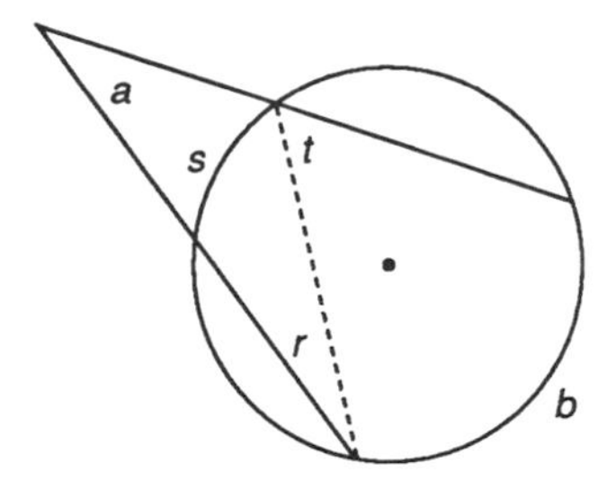

Do you notice a relationship between the angle formed by the two intersecting secants and the two intercepted arcs? Compare your results with the results of students near you. State your observations as your next conjecture.

The measure of an angle formed by two secants intersecting outside a circle is equal to —?—.
(Intersecting Secants Conjecture)

EXERCISE SET A

Use the Inscribed Angle and Exterior Angle Theorems to prove the Intersecting Chords Conjecture. Fill in the blanks in the logical argument.

Conjecture: *The measure of an angle formed by the intersection of two chords in a circle is equal to —?—. **(Intersecting Chords Conjecture)***

Given: Circle O with chords $\overline{PQ}$ and $\overline{RT}$ intersecting at V

Show: $m\angle RVP = \frac{1}{2}(m\widehat{PR} + m\widehat{QT})$

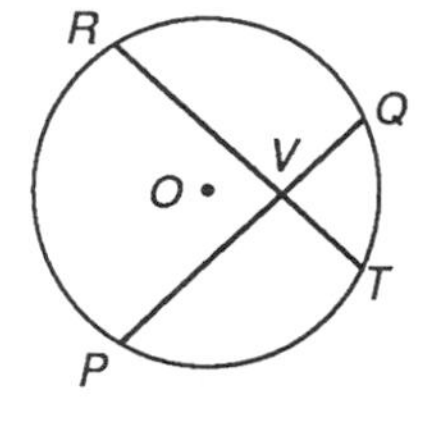

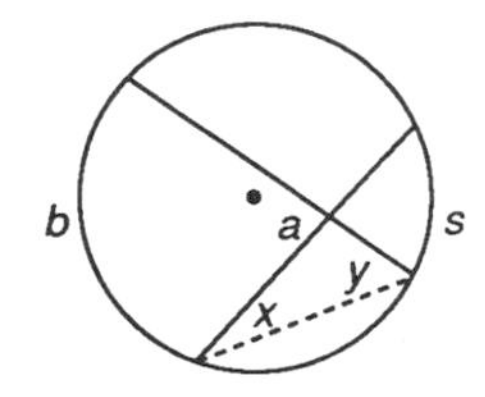

Two-Column Proof:

Let $a = m\angle RVP$, $b = m\widehat{PR}$, $s = m\widehat{QT}$, $x = m\angle QPT$, and $y = m\angle RTP$.

1. Construct $\overline{PT}$	1. —?—
2. $y = \frac{1}{2}b$; $x = \frac{1}{2}s$	2. —?—
3. $a = y + x$	3. —?—
4. $a = \frac{1}{2}b + \frac{1}{2}s$	4. —?— Property
5. $a = \frac{1}{2}(b + s)$	5. —?— Property
6. $m\angle RVP = \frac{1}{2}(m\widehat{PR} + m\widehat{QT})$	6. —?— Property

Now that you have proved the conjecture, you can call it a theorem. As a theorem, it becomes a premise that you can use to prove other conjectures. Add this theorem to your list.

Theorem: *The measure of an angle formed by the intersection of two chords in a circle is equal to —?—. **(Intersecting Chords Theorem)***

EXERCISE SET B

You should be able to use the Inscribed Angle and Exterior Angle Theorems, some algebra, and logical reasoning to prove the Intersecting Secants Conjecture. Fill in the blanks in the logical argument.

Conjecture: *The measure of an angle formed by two secants intersecting outside a circle is equal to one-half the —?— of the measures of the two intercepted arcs. **(Intersecting Secants Conjecture)***

Given: Circle O with secants $\overline{PQ}$ and $\overline{RQ}$ intersecting outside the circle at Q

Show: $m\angle Q = \frac{1}{2}(m\widehat{PR} - m\widehat{TU})$

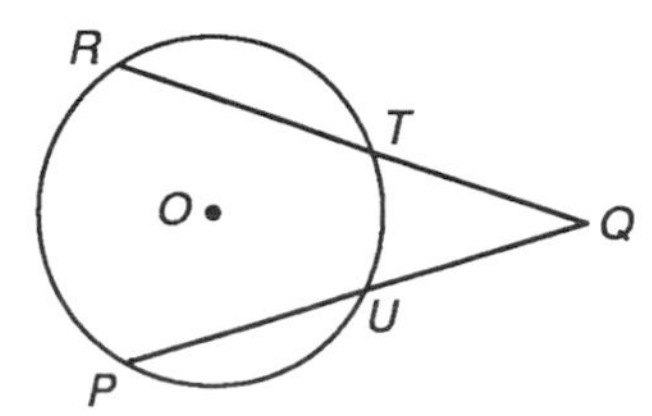

Two-Column Proof:

Let $a = m\angle Q$, $b = m\widehat{PR}$, $s = m\widehat{TU}$, $x = m\angle TPU$, and $y = m\angle RTP$.

	Statement		Reason
1.	Construct $\overline{PT}$	1.	—?—
2.	$y = \frac{1}{2}b;\ x = \frac{1}{2}s$	2.	—?—
3.	$y = a + x$	3.	—?—
4.	$a =$ —?—	4.	Subtraction and Symmetric Properties
5.	$a = \frac{1}{2}(\text{–?–}) - \frac{1}{2}(\text{–?–})$	5.	—?— Property
6.	$a = \frac{1}{2}(\text{—?—})$	6.	—?— Property
7.	$m\angle Q = \frac{1}{2}(m\widehat{PR} - m\widehat{TU})$	7.	—?— Property

Now that you have proved the conjecture, you can call it a theorem. As a theorem, it becomes a premise that you can use to prove other conjectures. Add this theorem to your list.

Theorem: *The measure of an angle formed by two secants intersecting outside a circle is equal to one-half the —?— of the measures of the two intercepted arcs.* ***(Intersecting Secants Theorem)***

EXERCISE SET C

To help you remember the two formulas, think: *s* for small arc, *b* for big arc, and *a* for angle. Notice that when the chords cross, they form "an addition sign" and when the chords do not cross, they form "subtraction signs."

$$a = \frac{(b + s)}{2}$$

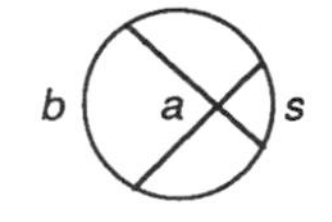

$$a = \frac{(b - s)}{2}$$

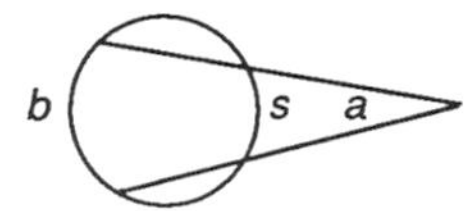

Use your two new conjectures to solve each problem.

1. $a =$ –?–

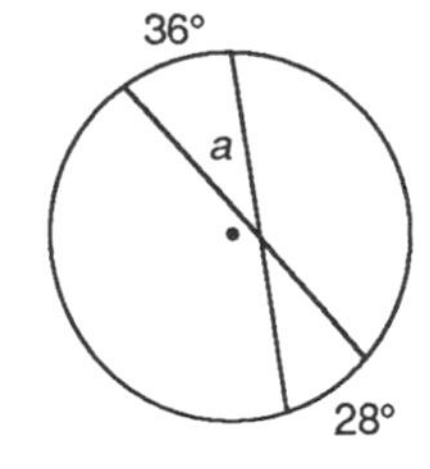

2.* $b =$ –?–

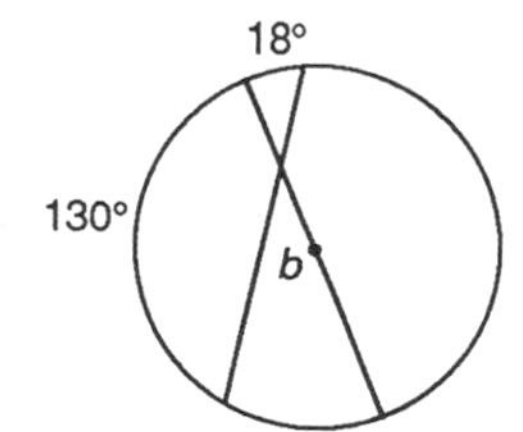

3.* $c =$ –?–

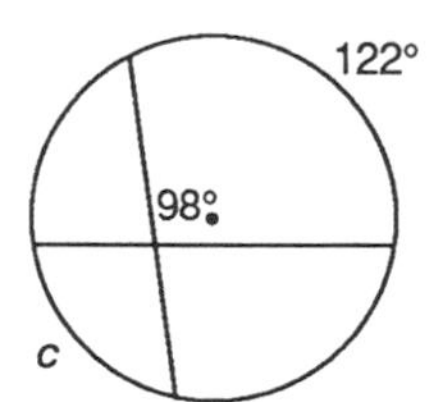

4. $d =$ –?–

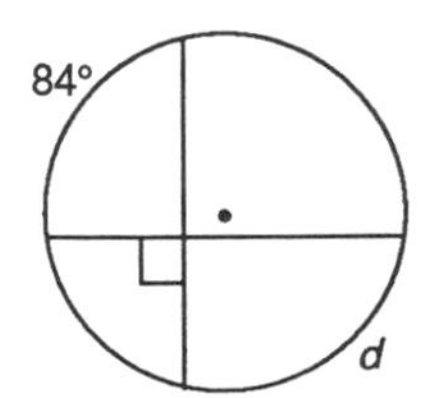

5. $e =$ –?–

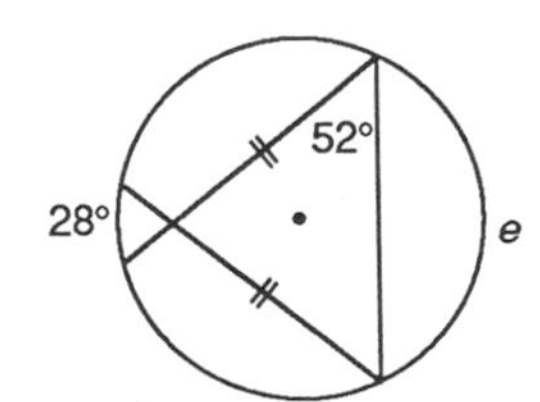

6.* $f =$ –?– $\overleftrightarrow{PQ} \parallel \overline{RS}$

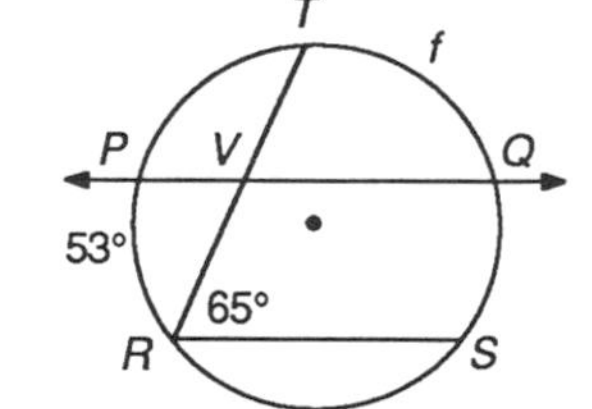

7. $a = $ –?–

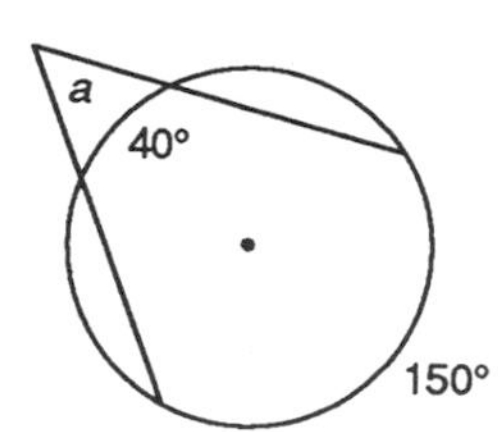

8. $b = $ –?–

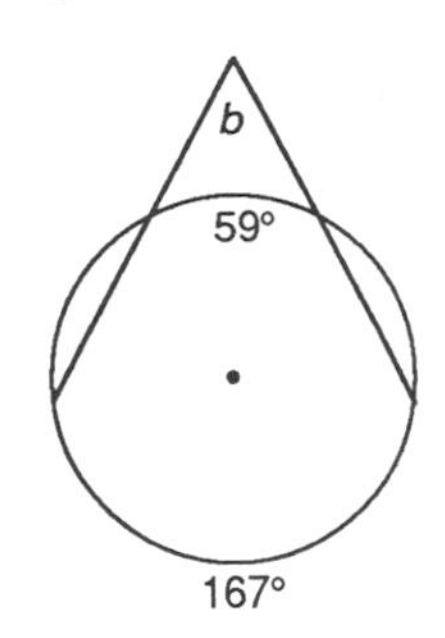

9.* $c = $ –?–

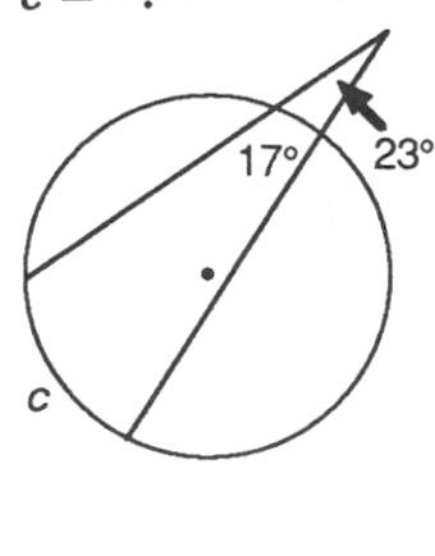

10. $d = $ –?–

11. $e = $ –?–

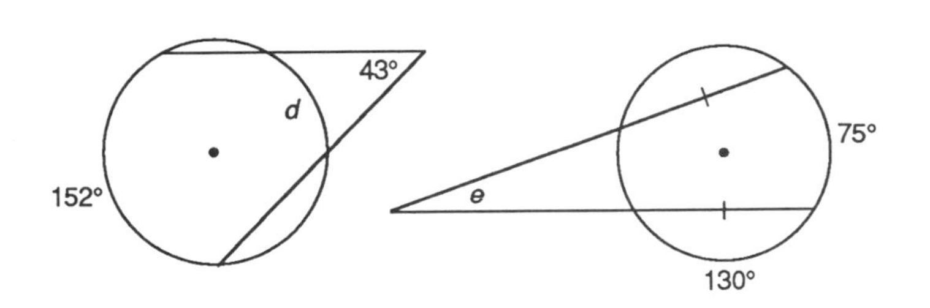

12.* $f = $ –?– $\overline{AB} \parallel \overline{CD}$

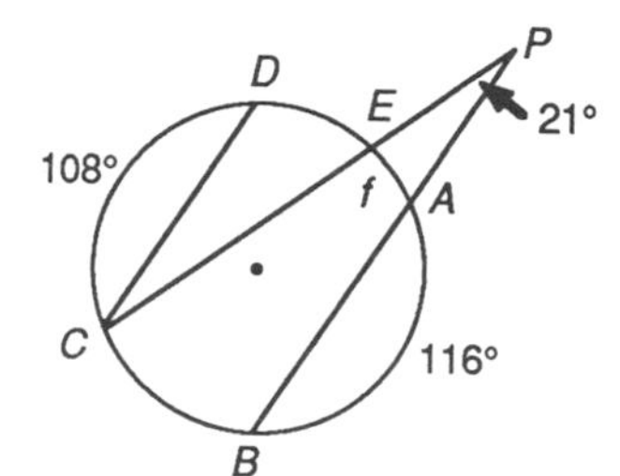

Improving Visual Thinking Skills

Container Problem III

You have a small, graduated, cylindrical measuring glass with a maximum capacity of 250 ml. All the graduation marks are worn off except for the 150-ml and 50-ml marks. You also have a second large container that is unmarked. It is possible to put exactly 125 ml into the large container. What is the fewest measurings necessary to do this? How?

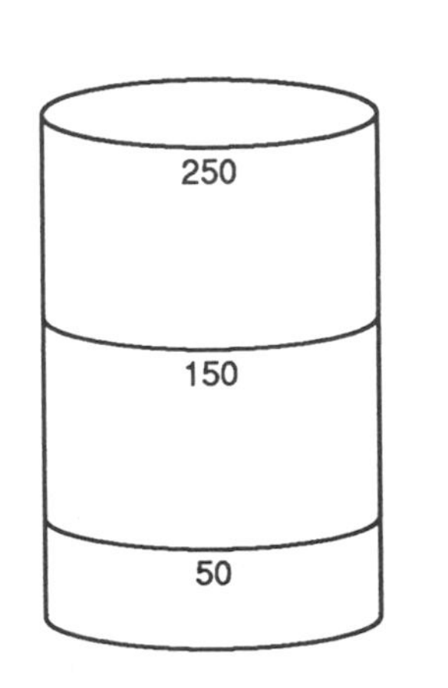

Lesson 15.4

Proving Similarity Conjectures

To prove similarity conjectures we need to add to our list of postulates. In the similarity chapter you discovered some shortcuts for showing that two triangles are similar. We are going to rename the SSS, AA, and SAS Similarity Conjectures as postulates and use them in our proofs. They are listed below.

P-15 *If the three sides of one triangle are proportional to the three sides of another triangle, then the two triangles are similar.* ***(SSS Similarity Postulate)***

$$\frac{nx}{x} = \frac{ny}{y} = \frac{nz}{z}$$

x z y nx nz ny

P-16 *If two angles of one triangle are congruent to two angles of another triangle, then the two triangles are similar.* ***(AA Similarity Postulate)***

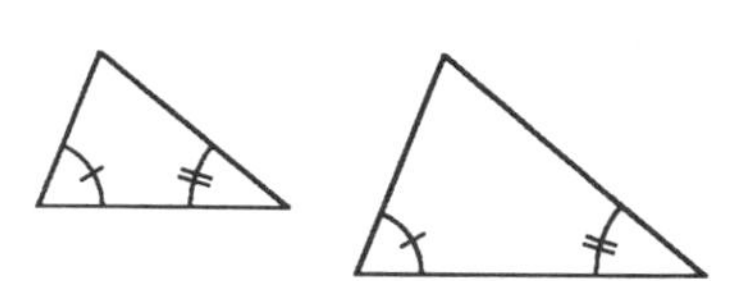

P-17 *If two sides of one triangle are proportional to two sides of another triangle, and the angle between the sides in one triangle is congruent to the angle between the sides in the other triangle, then the two triangles are similar.* ***(SAS Similarity Postulate)***

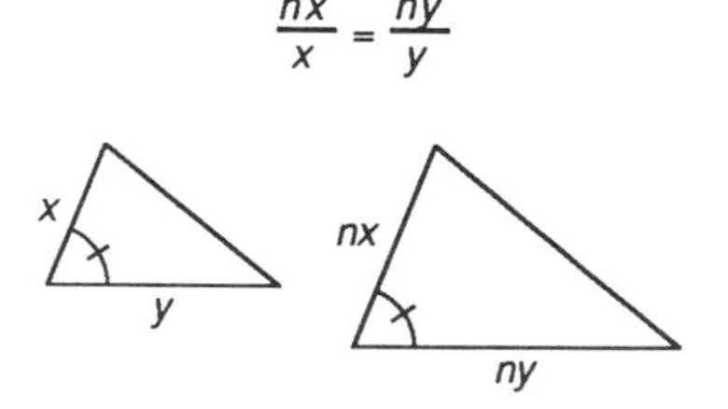

In this lesson we'll look at two of your earlier conjectures that can be proved with the help of your similarity postulates.

If a line parallel to one side of a triangle passes through the other two sides, then it divides them proportionally. Conversely, if a line cuts two sides of a triangle proportionally, then it is parallel to the third side. ***(Parallel Proportionality Conjecture)***

If two triangles are similar, then the corresponding altitudes, corresponding medians, and corresponding angle bisectors are proportional to the corresponding sides.
(Proportional Parts Conjecture)

EXERCISE SET A

To prove the Parallel Proportionality Conjecture, you can use the AA Similarity Postulate. This conjecture is really two conjectures rolled into one, a conjecture and its converse. Let's prove the first part. Copy the flowchart below and provide the reasons for the steps to complete the flowchart proof. You'll prove the second part of the conjecture (the converse of the first part) in the next exercise set. Once the proof is complete, you can call the conjecture the Parallel Proportionality Theorem.

Conjecture: *If a line parallel to one side of a triangle passes through the other two sides, then it divides them proportionally. (First part of Parallel Proportionality Conjecture)*

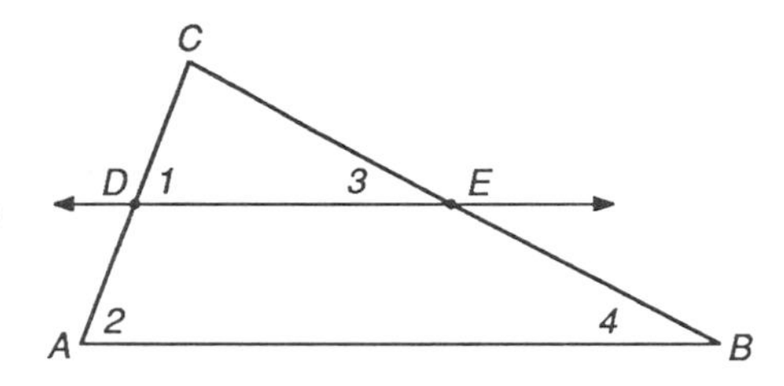

Given: $\triangle ABC$ with $\overleftrightarrow{DE} \parallel \overline{AB}$

Show: $\frac{AD}{CD} = \frac{BE}{CE}$

Plan:

- With some clever algebra I can show that if $\frac{AC}{DC} = \frac{BC}{EC}$, then $\frac{AD}{CD} = \frac{BE}{CE}$.
- I can show $\frac{AC}{DC} = \frac{BC}{EC}$ if I can show $\triangle ABC \sim \triangle DEC$.
- I can show $\triangle ABC \sim \triangle DEC$ if I can show $\angle 1 \cong \angle 2$ and $\angle 3 \cong \angle 4$.
- But, I can show $\angle 1 \cong \angle 2$ and $\angle 3 \cong \angle 4$ by the CA Theorem.
- *Clever algebra:*

$$\frac{AC}{DC} = \frac{BC}{EC}$$

$$\frac{AC}{DC} - \frac{DC}{DC} = \frac{BC}{EC} - \frac{EC}{EC} \quad \text{(subtracting 1 from both sides)}$$

$$\frac{AC - DC}{DC} = \frac{BC - EC}{EC}$$

But, $AC - DC = AD$ and $BC - EC = BE$, and, therefore, $\frac{AD}{CD} = \frac{BE}{CE}$.

Flowchart Proof:

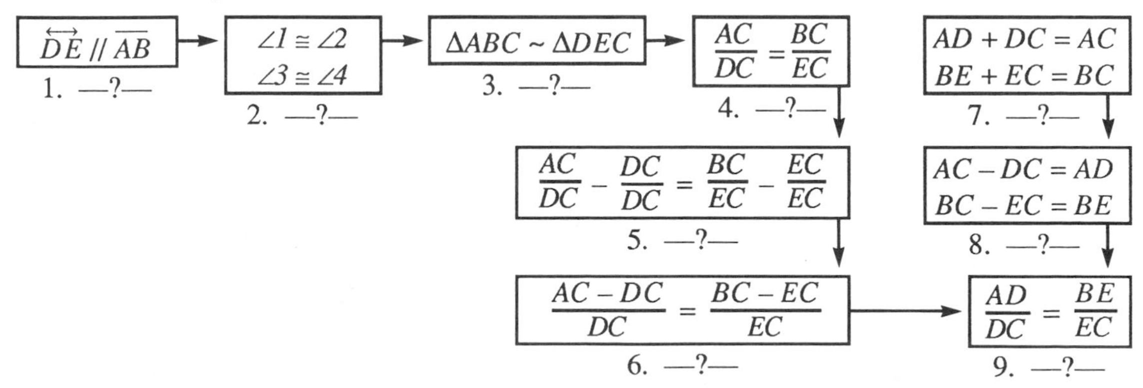

EXERCISE SET B

Prove each conjecture below. Once you have proved the conjectures in this exercise set, the Parallel Proportionality Conjecture and the Proportional Parts Conjecture can be called theorems and added to your list of theorems.

1. If a line cuts two sides of a triangle proportionally, then it is parallel to the third side. (Second part of Parallel Proportionality Conjecture)

2.* If two triangles are similar, then the corresponding altitudes are proportional to the corresponding sides. (Part of Proportional Parts Conjecture)

3. If two triangles are similar, then the corresponding angle bisectors are proportional to the corresponding sides. (Part of Proportional Parts Conjecture)

4. If two triangles are similar, then the corresponding medians are proportional to the corresponding sides. (Part of Proportional Parts Conjecture)

Improving Visual Thinking Skills

Checkerboard Puzzle

1. Four checkers are arranged on the corner of a checkerboad as shown. With exactly three horizontal or vertical jumps, remove all three white checkers, leaving the single black checker. Any checker can jump any other checker. Copy and complete the table to record your solution.

D				
C				
B	○	○		
A	●	○		
	1	2	3	4

From	To
—?—	—?—
—?—	—?—
—?—	—?—

2. Now with exactly seven horizontal or vertical jumps, remove all seven white checkers, leaving the single black checker. Copy and complete the table to record your solution.

D					
C					
B	○	○	○	○	
A	●	○	○	○	
	1	2	3	4	5

From	To
—?—	—?—
—?—	—?—
—?—	—?—
—?—	—?—
—?—	—?—
—?—	—?—
—?—	—?—

Lesson 15.5

Proving New Similarity Conjectures

This lesson should be worked in groups of four. There are three investigations leading to new similarity conjectures. Everyone in the group should work together on the first investigation, make the conjecture, and finally prove that conjecture in Exercise Set A. Then break into pairs and:

- Perform one of the remaining two investigations.
- Make the conjecture for that investigation.
- Prove that conjecture in Exercise Set A.

Then the four group members should share what they have learned in the lesson with the others, study the examples, and finally work the problems in Exercise Set B.

Investigation 15.5.1

Answer the questions for each problem in this investigation.

1. *TRAP* is a trapezoid.
   Why is $\angle ATR \cong \angle PAT$?
   Why is $\angle PRT \cong \angle APR$?
   Why is $\Delta TRE \sim \Delta APE$?

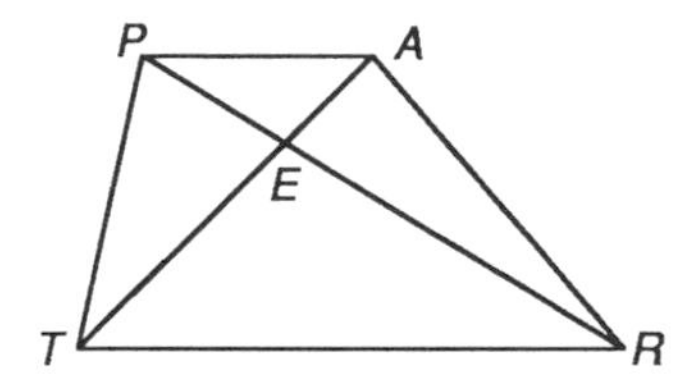

2. *ZOID* is a trapezoid.
   Why is $\Delta ZOE \sim \Delta IDE$?
   $\dfrac{DE}{EO} = \dfrac{IE}{-?-}$

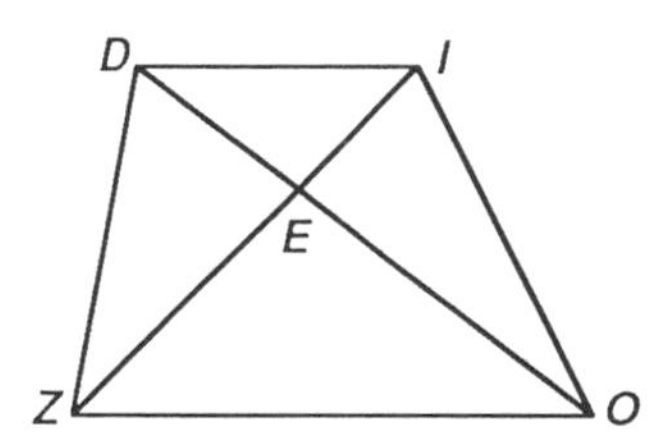

In each problem, similar triangles were formed when you constructed diagonals. Is this always true? Trapezoid *JANE* has diagonals $\overline{EA}$ and $\overline{JN}$ intersecting at point *T*. At least one pair of triangles appears to be similar. If a pair of triangles are similar, then the corresponding sides of the triangles (segments on the diagonals) are proportional. State this as your next conjecture.

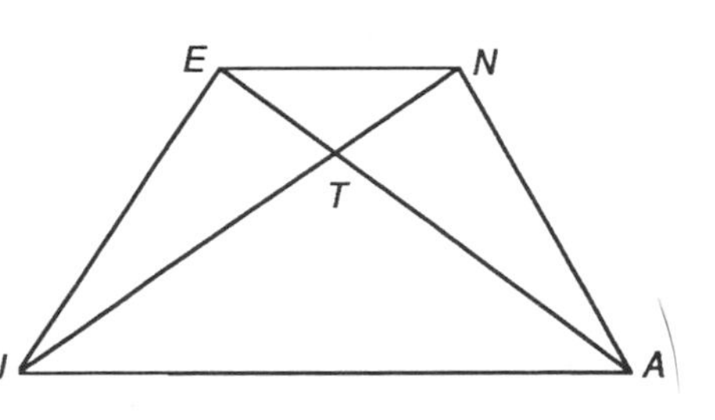

C-83 *The diagonals of a trapezoid divide each other —?—.*

Let's look at this logically. Since $\overline{JA} \parallel \overline{NE}$, $\angle NJA \cong \angle ENJ$ and $\angle EAJ \cong \angle NEA$ by the AIA Postulate. Since the angles are congruent, $\Delta JAT \sim \Delta NET$ by the AA Similarity Postulate. Since the triangles are similar, $\dfrac{ET}{TA} = \dfrac{NT}{TJ}$.

Proportions can also be found in circles. When two chords intersect, there is a nice relationship between the four segments formed. You can discover this relationship with the help of similar triangles.

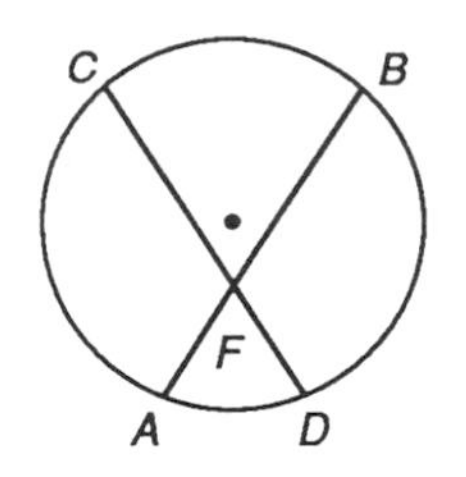

Investigation 15.5.2

When chords $\overline{AC}$ and $\overline{BD}$ are drawn in, you get two triangles that appear to be similar. Is this true? Let's check and see.

Since $\angle CAB$ and $\angle CDB$ are inscribed in the same arc, $\angle CAB \cong \angle CDB$. Likewise, $\angle ACD \cong \angle ABD$ since they are also inscribed in the same arc. Therefore $\Delta AFC \sim \Delta DFB$ by the AA Similarity Postulate. Since the triangles are similar, the corresponding sides are proportional. In order to see the relationship between the four segments formed, complete the statements in each problem below.

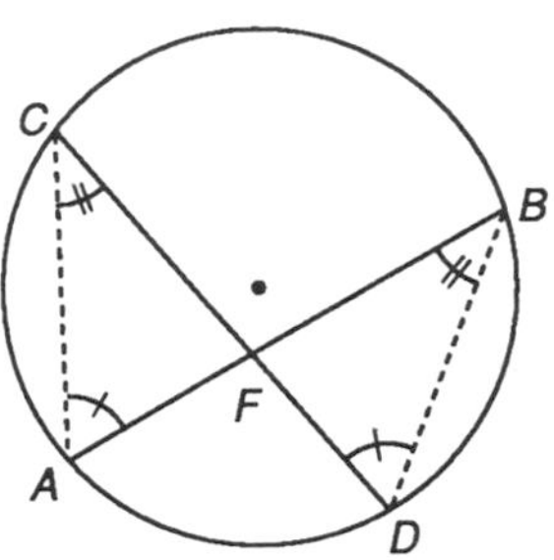

1. $\Delta DEF \sim \Delta$–?–

   $\dfrac{DF}{GF} = \dfrac{EF}{-?-}$

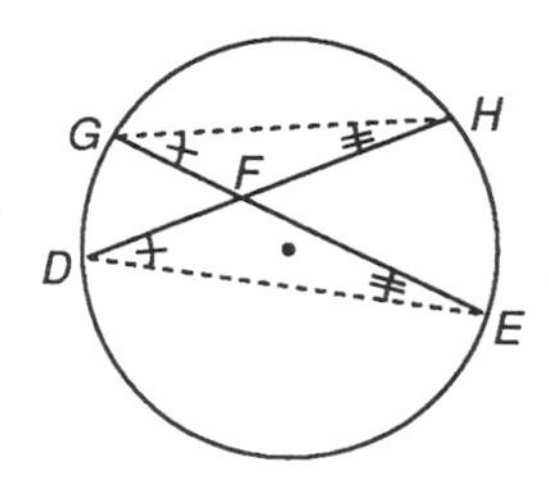

2. $\dfrac{AE}{DE} = \dfrac{-?-}{CE}$

   $(AE)(CE) = (-?-)(-?-)$

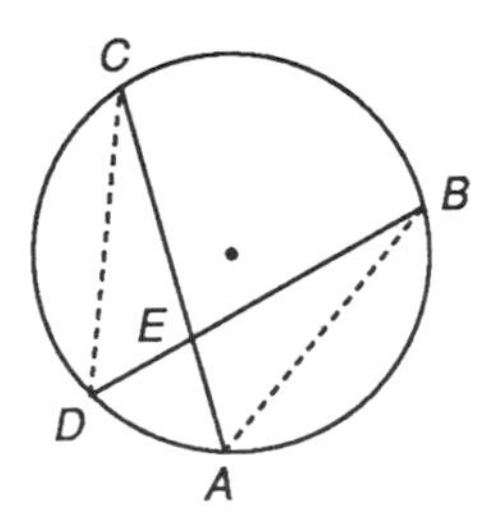

In the investigation, you wrote proportions from similar triangles in circles. In the diagram on the right, $\Delta AFC \sim \Delta DFB$.

Since the triangles are similar, $\dfrac{AF}{DF} = \dfrac{CF}{BF}$.

By cross-multiplying in this proportion you get: $(AF)(BF) = (CF)(DF)$.

This is an easy equation to remember. This equation says that if you multiply the lengths of the segments on one chord and multiply the lengths of the segments on the other chord, the two products will be equal. State this as your next conjecture.

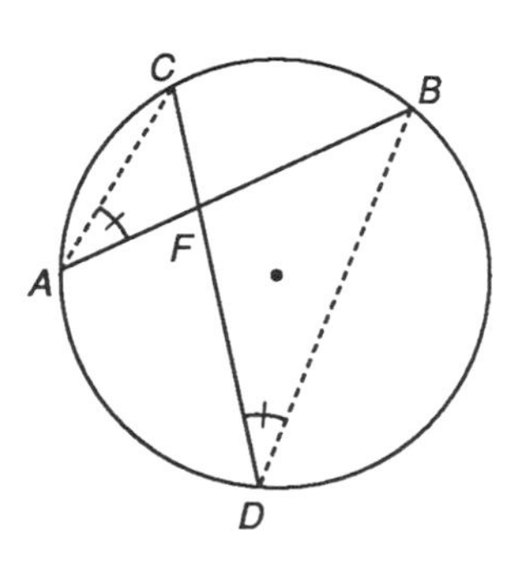

If two chords in a circle intersect, —?—.

There are other conjectures that you can make about the lengths of segments related to circles. Your last conjecture says that if two chords intersect in a circle, the product of the lengths on one chord is equal to the product of the lengths on the other chord. What about the lengths on intersecting secants? Is there a relationship? Since we are asking, there must be, right? In order to see this relationship more clearly, complete the statements in each problem below.

Investigation 15.5.3

1. Since $\angle C \cong \angle A$
   and $\angle S \cong \angle S$,
   then $\Delta CSN \sim \Delta$–?–

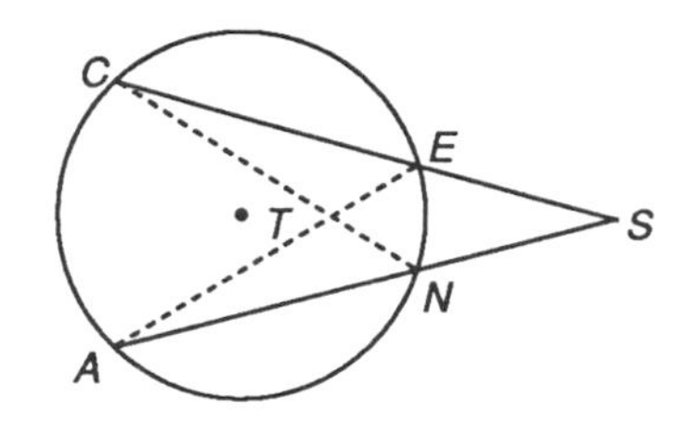

2. $\Delta TRN \sim \Delta$–?–
   $\dfrac{TR}{TN} = \dfrac{TA}{\text{–?–}}$
   $(TR)(TG) = $ (–?–)(–?–)

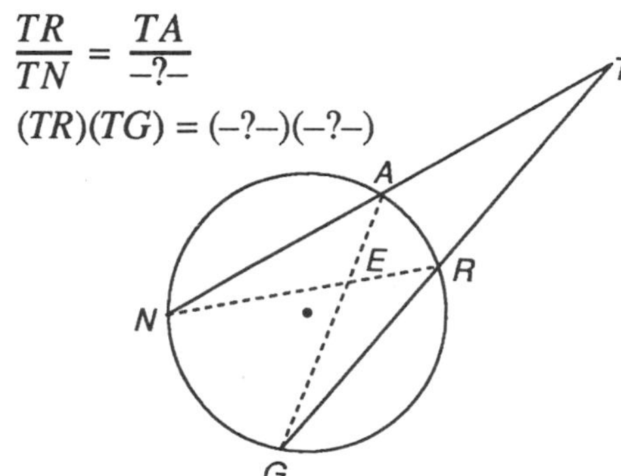

3. $\Delta CER \sim \Delta$–?–
   $\dfrac{CI}{CE} = \dfrac{\text{–?–}}{\text{–?–}}$
   $(CR)(CI) = $ (–?–)(–?–)

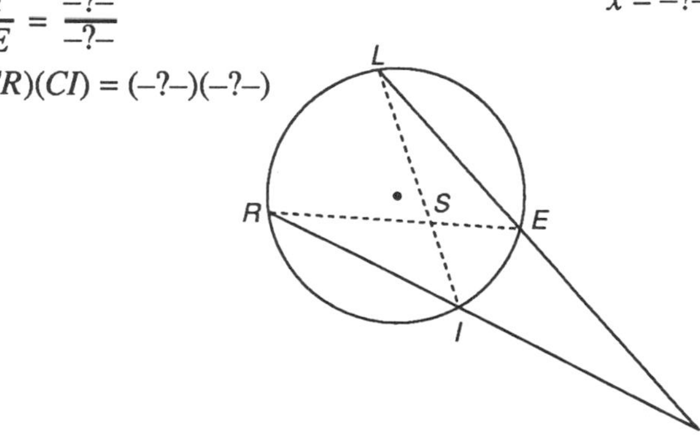

4. $\Delta POA \sim \Delta$–?–
   $x = $ –?–

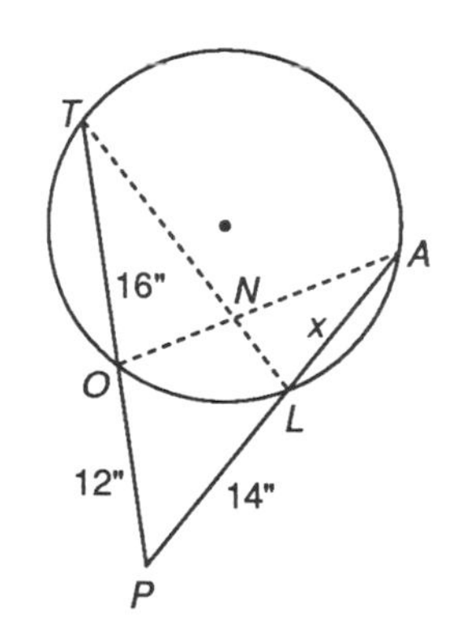

5. Write a statement of proportion between the values w, x, y, and z.

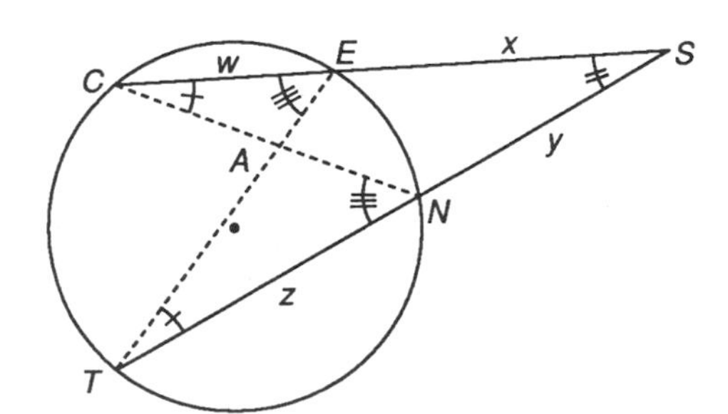

Let's look at Problem 5 logically. When the triangles are separated you can see that $\Delta CSN \sim \Delta TSE$ by the AA Similarity Postulate. If $\Delta CSN \sim \Delta TSE$, then $y/(w + x) = x/(y + z)$. Cross-multiplying, you get $x(w + x) = y(y + z)$. $\overline{SC}$ and $\overline{ST}$ are **secant segments**. $\overline{SE}$ and $\overline{SN}$ are **external parts**. The length $x + w$ is the length of the secant segment and x is the length of the external part of the same secant segment. The length $y + z$ is the length of the other secant segment and y is the length of the external part of that secant segment. The equation $x(w + x) = y(y + z)$ can be stated in the form of a conjecture. Complete the conjecture below.

If two secants intersect outside a circle, the product of the length of one secant segment and the length of its external part is equal to the product of —?—.

EXERCISE SET A

Write a proof for each of the following new conjectures.

1. Conjecture 83
2. Conjecture 84
3. Conjecture 85

Let's look at four examples of how you can use Conjectures 83, 84, and 85 to find the length of a segment on intersecting diagonals, chords, or secants.

Example A

Find w.

$$\frac{w}{25} = \frac{12}{20}$$
$$20w = (12)(25)$$
$$w = (12)(25)/20$$

w is 15 cm.

Example B

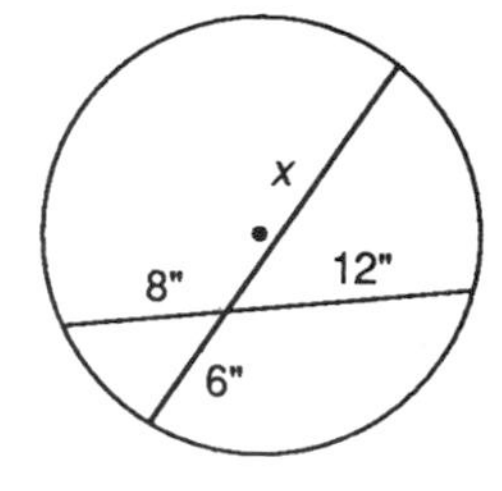

Find x.

$$6x = (8)(12)$$
$$x = (8)(12)/6$$

x is 16 inches.

Example C

Find y.

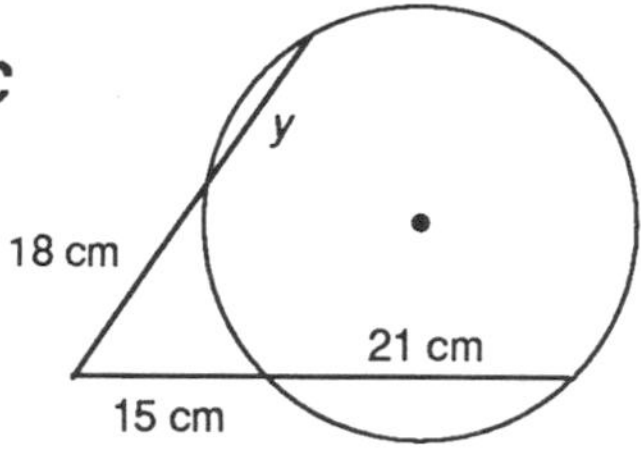

$$18(18 + y) = (15)(36)$$
$$18 + y = (15)(36)/18$$
$$18 + y = 30$$
$$y = 12$$

y is 12 cm.

Example D

Find z.

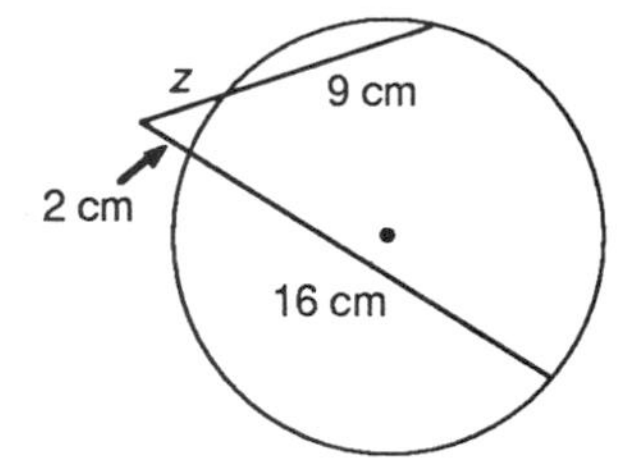

$$z(z + 9) = (2)(2 + 16)$$
$$z^2 + 9z - 36 = 0$$
$$(z + 12)(z - 3) = 0$$
$$z = 3 \text{ or } -12$$

z must be 3 cm because a length cannot be negative.

EXERCISE SET B

Use your new conjectures to solve each problem. The quadrilaterals in Exercises 1 to 3 are trapezoids. All measurements are in centimeters.

1. $x = $ –?–

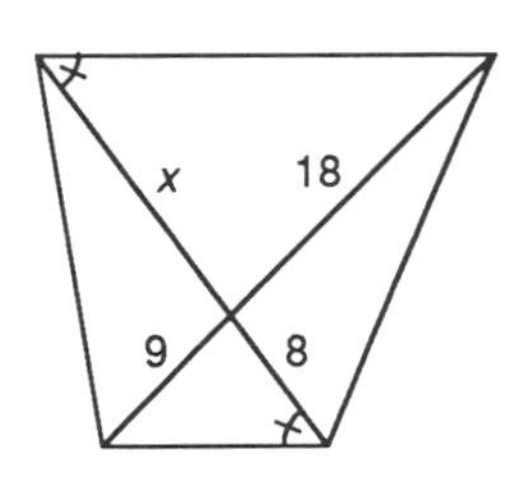

2. $y = $ –?–

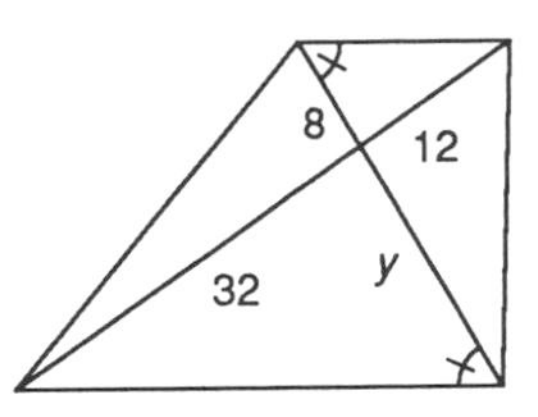

3.* $w = $ –?– $z = $ –?–

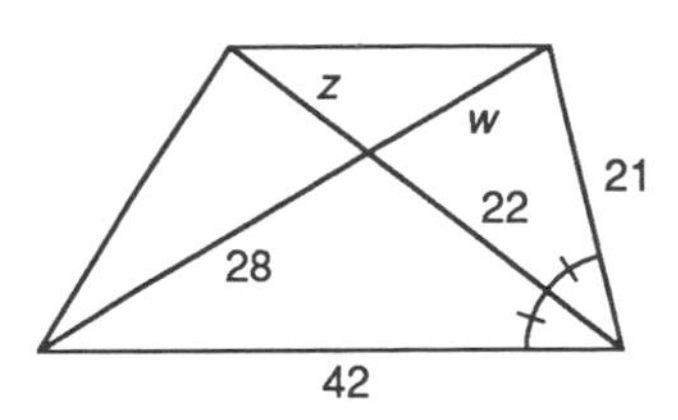

4. $a = $ –?–

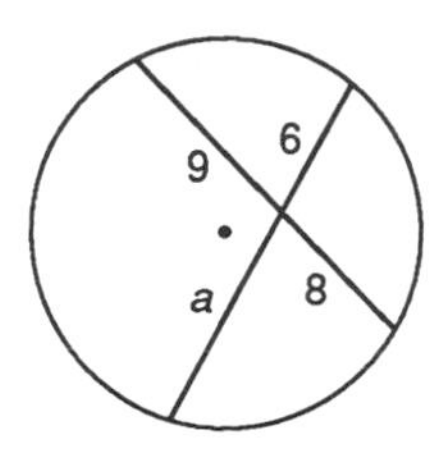

5. $b = $ –?–

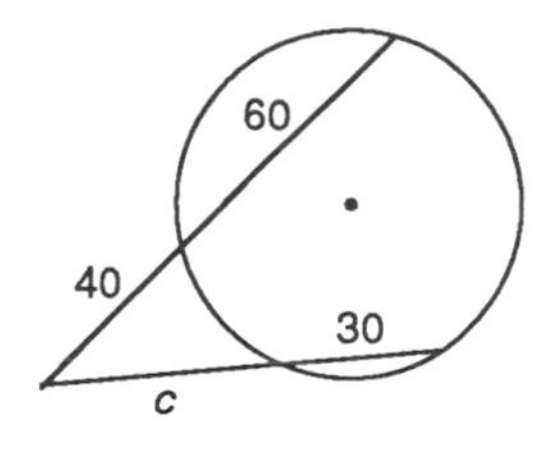

6.* $c = $ –?–

7.* $d = $ –?–

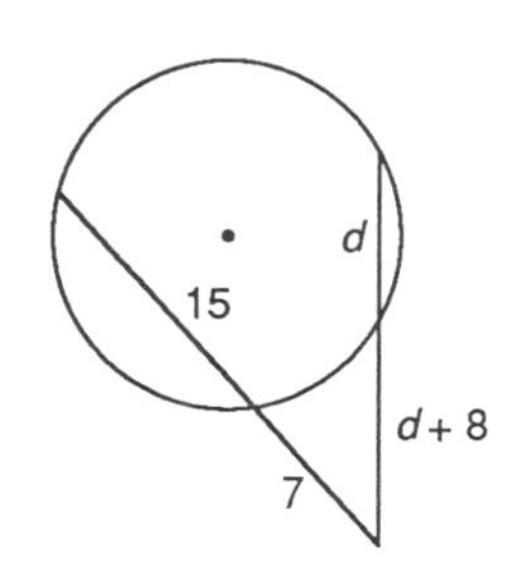

8. Sam and Samantha are twin salamanders that live in a circular pond. One day both were practicing their backstrokes for the special Salamander Swimming Sweepstakes coming in September. On this particular day, it happened that their straight line paths crossed after 36 strokes by Sam and 48 strokes by Samantha. (Their strokes are the same length.) Samantha continued on her same path to the edge in 21 more strokes. How many more strokes did Sam need if he also continued on in a straight path to the edge of the pond?

9. The design of an ironing board uses similar triangles (intersecting diagonals in a trapezoid) to guarantee that the board is always parallel to the floor.

Egbert wishes to design and build an ironing board for his mother (he broke her old one practicing his surfing moves). Egbert found two pipes like those shown in the ironing board diagram on the right. The 100 cm pipe ($\overline{BC}$) already has a hole 30 cm from one end. Where should he drill the hole on the other 120 cm pipe ($\overline{AD}$) so that the two holes can be bolted together to form the center hinge at E?

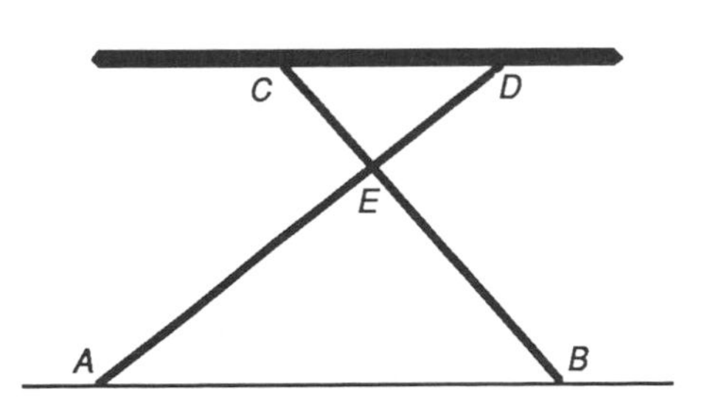

The Puzzle of Pharaoh Tu-Khamon

10.* The Pharaoh Tu-Khamon has just offered Olie Oxenfree a trapezoidal piece of very fertile land along the Nile. The Pharaoh, a lover of puzzles, has included a number of special conditions to his offer. To claim the 500 square cubits of land, Olie must situate the center of his house at the intersection of the two diagonals of the trapezoidal plot of land. The Pharaoh Tu-Khamon also insists that the center of the house be 8 cubits and 12 cubits from the two bases of the trapezoid. Before Olie can begin building, the Pharaoh Tu-Khamon insists that Olie Oxenfree figure out the lengths of the two bases. Olie would love to say, "Thanks, but no thanks," to the Pharaoh, but you just don't say "no" to Tu-Khamon! Help save Olie's neck. What are the lengths of the two bases?

Improving Visual Thinking Skills

Moving Coins

Create a triangle of coins similar to the one on the near right. How can you move exactly three coins so that the triangle is pointing down rather than up? When you have found a solution, draw two triangles of circles similar to those in the diagram. Write letters in each of the empty circles in your drawing to show your solution.

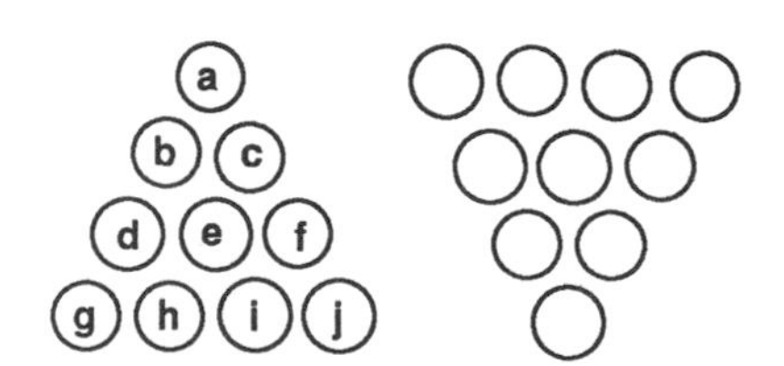

Lesson 15.6

Proving the Pythagorean Theorem

In the last lesson you made and proved similarity conjectures with trapezoids and circles. In this lesson you will investigate and discover another similarity conjecture that can be found in right triangles. Once you have proved this new conjecture, you can use it to prove the Pythagorean Theorem!

Investigation 15.6.1

An altitude has been constructed to the hypotenuse of each right triangle below. This creates two more right triangles within each right triangle. Calculate the measure of the acute angles in each right triangle.

1. $a =$ –?–
   $b =$ –?–
   $c =$ –?–

2. $a =$ –?–
   $b =$ –?–
   $c =$ –?–

3. $a =$ –?–
   $b =$ –?–
   $c =$ –?–

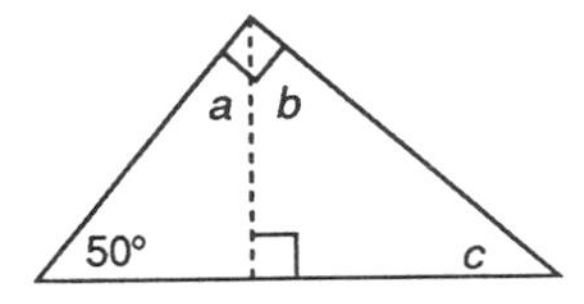

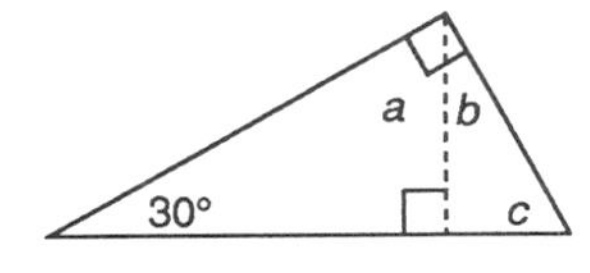

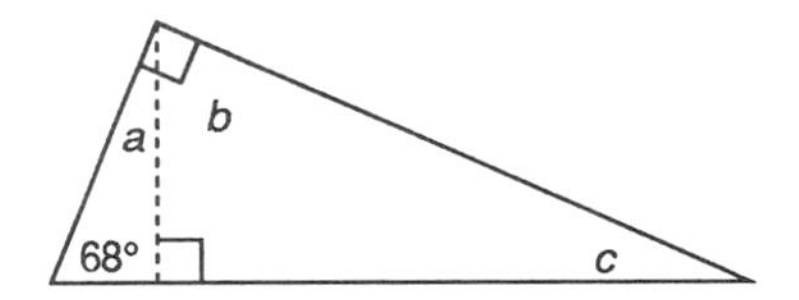

Did you notice that if an altitude is dropped from the right angle to the hypotenuse, it divides the right triangle into two right triangles that are similar to the original right triangle?

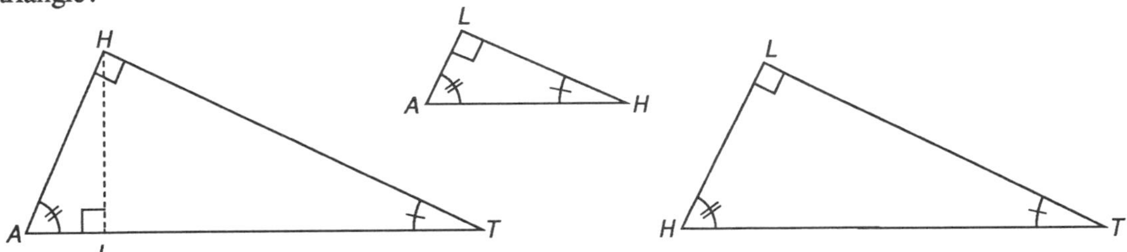

Write the statement of similarity between the three right triangles in each problem.

4. $\Delta ALE \sim \Delta$–?– $\sim \Delta$–?–
5. $\Delta BEK \sim \Delta$–?– $\sim \Delta$–?–
6. $\Delta CHA \sim \Delta$–?– $\sim \Delta$–?–

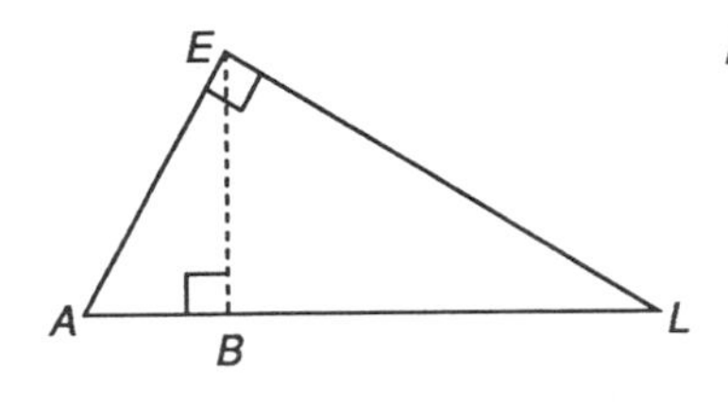

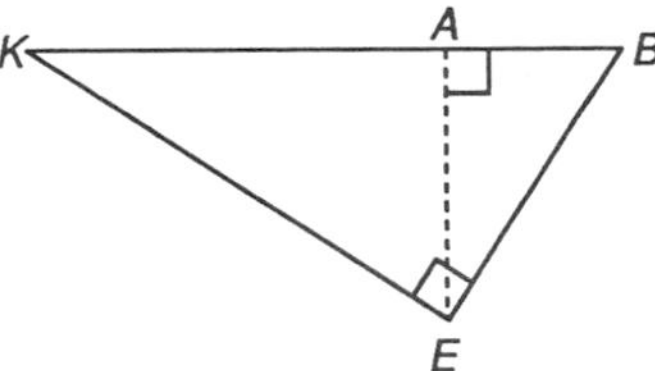

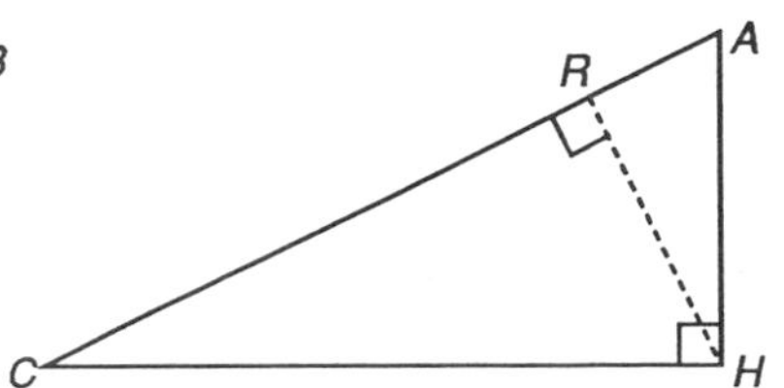

Since there are three similar right triangles when you drop an altitude to the hypotenuse, then there are a number of proportion statements that you can write.

In the diagram below $\frac{x}{h} = \frac{h}{y} = \frac{b}{a}$.

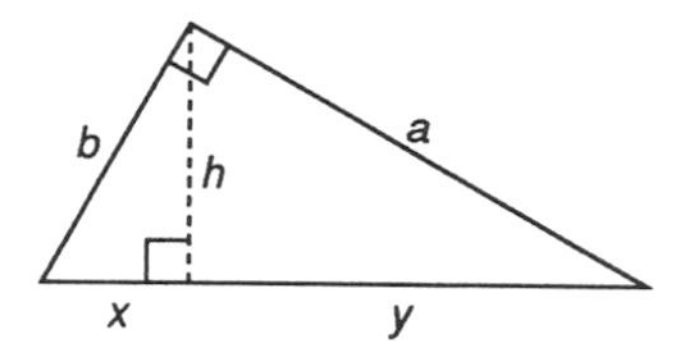

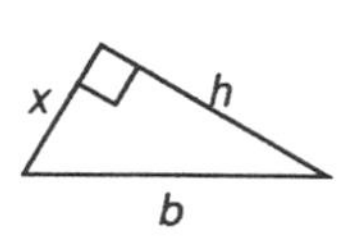

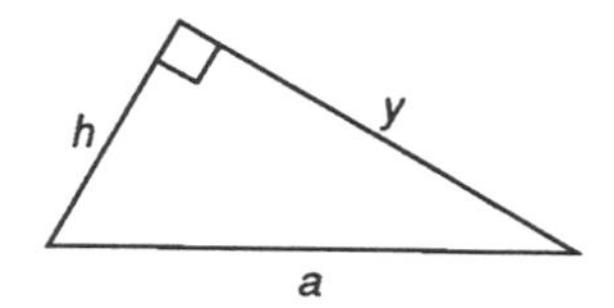

Complete each proportion statement.

7. $\frac{b}{h} = \frac{-?-}{c}$

8. $\frac{c}{h} = \frac{h}{-?-}$

9. $\frac{x}{-?-} = \frac{-?-}{y}$

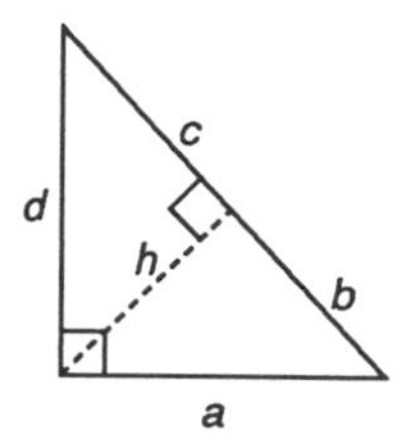

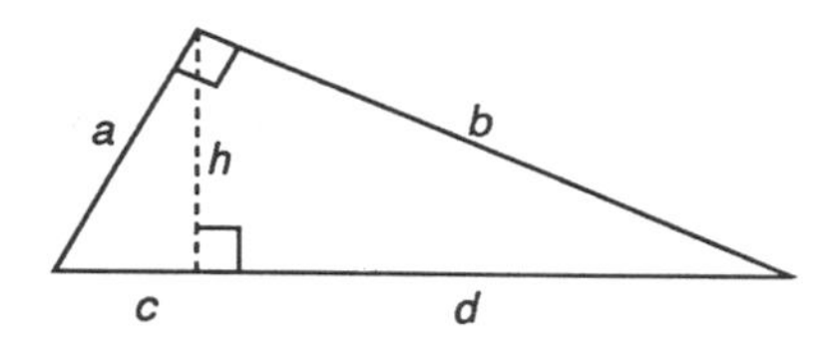

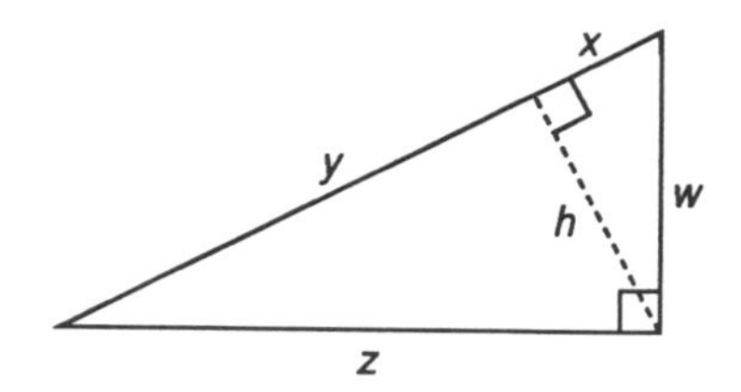

In a proportion, if the numerator of one ratio is the same as the denominator of the other ratio, then that number is called the **mean proportional** of the other two numbers. For example, in the proportion on the right, x is the mean proportional of a and b. Notice that x, the mean proportional of the two terms a and b, is equal to $\pm\sqrt{ab}$.

$$\frac{a}{x} = \frac{x}{b}$$

$$x^2 = ab$$

$$x = \pm\sqrt{ab}$$

In Problems 7 to 9, the altitude to the hypotenuse was the mean proportional of the two segments on the hypotenuse. State this as your next conjecture.

The altitude to the hypotenuse of a right triangle is the —?— to the segments on the hypotenuse.

EXERCISE SET A

Write a proof for each of the following conjectures.

1. Conjecture 86
2.* Pythagorean Theorem
3.* Converse of the Pythagorean Theorem

EXERCISE SET B

1.* $a = –?–$ 2. $b = –?–$ 3. $c = –?–$

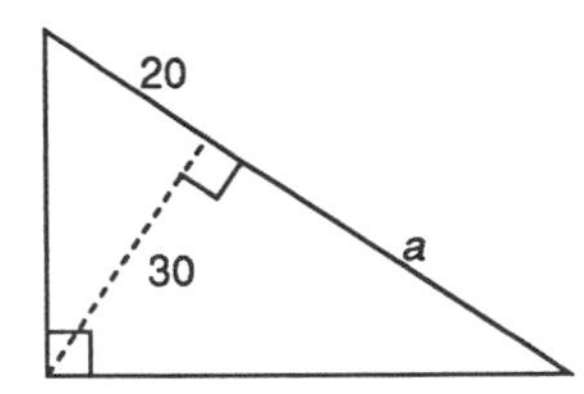

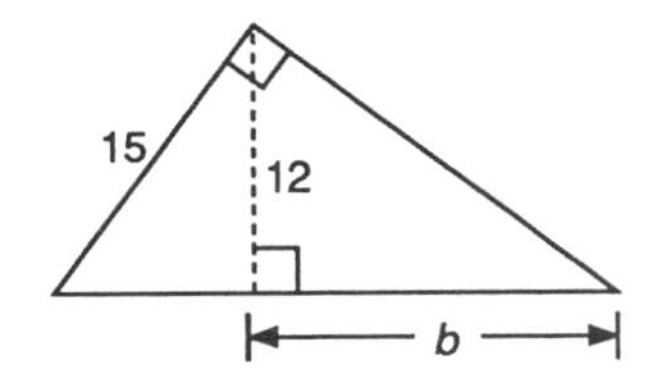

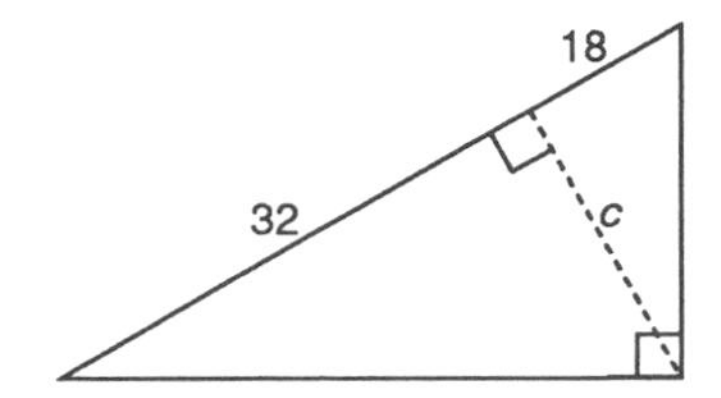

4.* While flying beneath the cloud cover of Saro-Gahtyp Valley, archaeologist and adventurer Montana Smith was able to photograph mysterious geometical markings on the valley floor. These markings had been undiscovered by archaeologists. The markings consist of a series of right triangles with a geometric figure on each side of each right triangle. One right triangle has a semicircle on each side. Another right triangle has a regular pentagon on each side. Still another has an equilateral triangle on each side. The ancient valley culture left no written language, but the diagrams hint at a great geometric discovery by a long lost civilization. What is that discovery? Prove it for one of the special cases shown.

Montana Smith and the Saro-Gahtyp Discovery

Lesson 15.7

Proof with Coordinate Geometry

Conjectures involving parallel, perpendicular, congruent segments, or midpoints can be proved nicely with coordinate geometry.

Do you recall the definition of slope and the conjectures you made in earlier lessons on coordinate geometry? Three conjectures are shown below. We are going to accept these three conjectures as postulates and use them in the coordinate proofs of this lesson.

P-18 *If (x_1, y_1) and (x_2, y_2) are the coordinates of the endpoints of a segment, then the coordinates of the midpoint are $\left(\frac{x_1+x_2}{2}, \frac{y_1+y_2}{2}\right)$.* ***(Coordinate Midpoint Postulate)***

P-19 *In a coordinate plane, two lines are parallel if and only if their slopes are equal.* ***(Parallel Slope Postulate)***

P-20 *In a coordinate plane, two lines are perpendicular if and only if their slopes are negative reciprocals of each other.* ***(Perpendicular Slope Postulate)***

In a coordinate proof, you draw figures on a coordinate plane. Perhaps the trickiest part of coordinate proofs is locating and labeling the coordinates of the vertices of the figures you draw. It helps to keep the coordinates as simple as possible. Let's practice labeling the vertices of figures.

EXERCISE SET A

In Exercises 1 to 6, the diagram demonstrates a convenient position for a polygon. Provide the missing coordinates without adding any new letters.

1. Square
2. Rectangle
3. Trapezoid

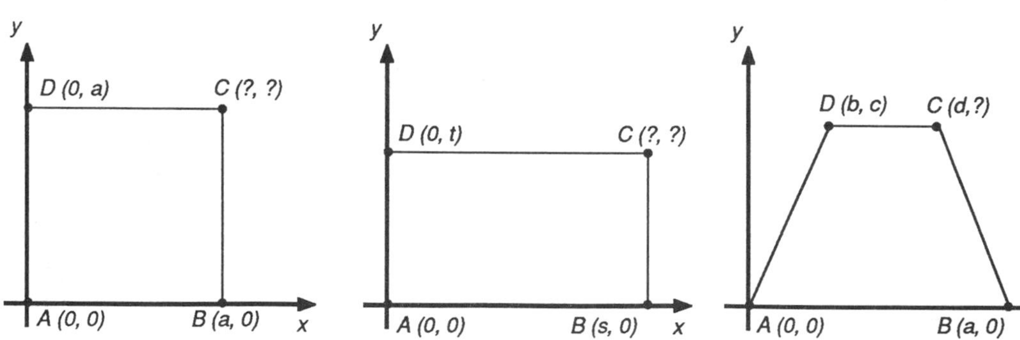

4. Isosceles triangle

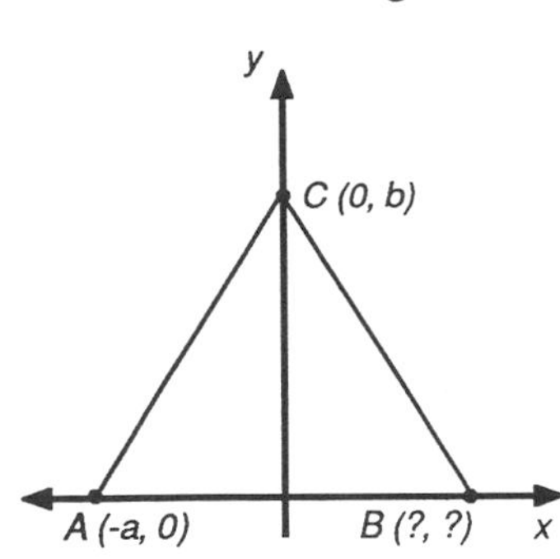

5.* Parallelogram

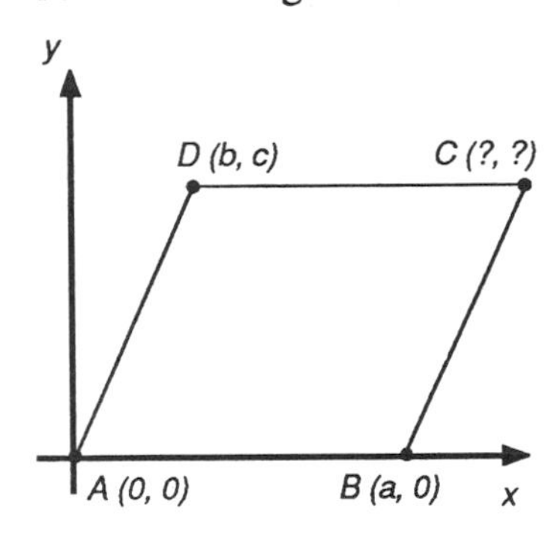

6.* Rhombus

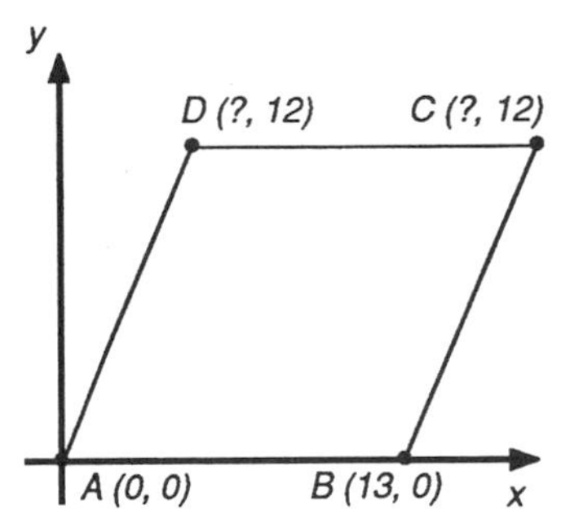

In Exercises 7 to 9, find the slope or length indicated.

7. Find the slope of each diagonal in rhombus *RHOM*.

8. Find the length of each diagonal in isosceles trapezoid *ZOID*.

9. Find the length of the medians to the congruent sides of isosceles ΔTRY.

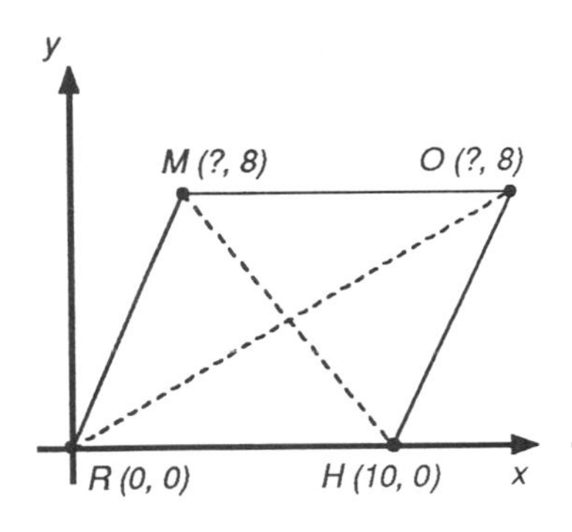

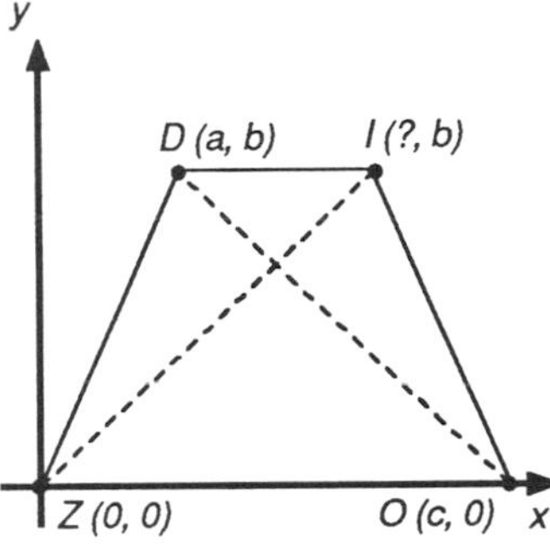

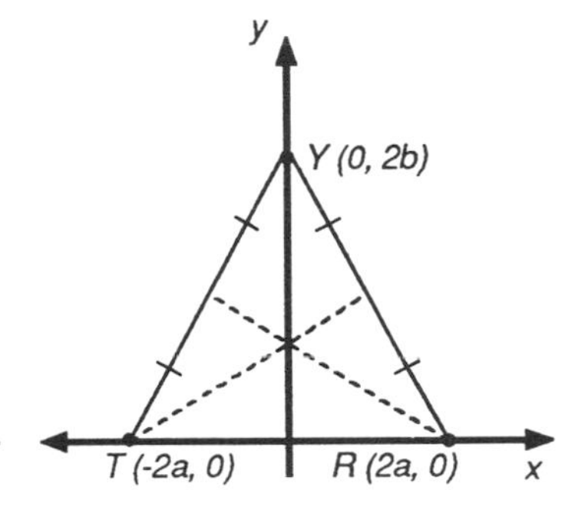

Another very important conjecture you made in an earlier lesson on coordinate geometry is the Distance Formula shown below. It is really the Pythagorean Theorem in coordinate disguise! In fact we will use the Pythagorean Theorem to prove the Distance Formula. Then you will use the Distance Formula in the coordinate proofs of this lesson.

If the coordinates of points A and B are (x_1, y_1) *and* (x_2, y_2) *respectively, then* $AB^2 = (x_2 - x_1)^2 + (y_2 - y_1)^2$ *and* $AB = \sqrt{(x_2 - x_1)^2 + (y_2 - y_1)^2}$. ***(Distance Formula)***

To prove the Distance Formula, you need to recall how you find the distance between two points horizontal to each other and the distance between two points vertical to each other. To find the distance between two points horizontal to each other you subtract the x-coordinates and take the absolute value of the difference. To find the distance between two points vertical to each other you subtract the y-coordinates and take the absolute value of the difference.

Length of a horizontal segment: If $\overline{AB}$ is a horizontal segment with endpoints (x_1, y_1) and (x_2, y_2), then $AB = |x_2 - x_1|$.

Length of a vertical segment: If $\overline{AB}$ is a vertical segment with endpoints (x_1, y_1) and (x_2, y_2), then $AB = |y_2 - y_1|$.

Let's use coordinate methods to prove the Distance Formula.

Conjecture: *If the coordinates of points A and B are (x_1, y_1) and (x_2, y_2), then $AB^2 = (x_2 - x_1)^2 + (y_2 - y_1)^2$ and $AB = \sqrt{(x_2 - x_1)^2 + (y_2 - y_1)^2}$.*

Given: Points A and B with coordinates (x_1, y_1) and (x_2, y_2)

Show: $AB = \sqrt{(x_2 - x_1)^2 + (y_2 - y_1)^2}$

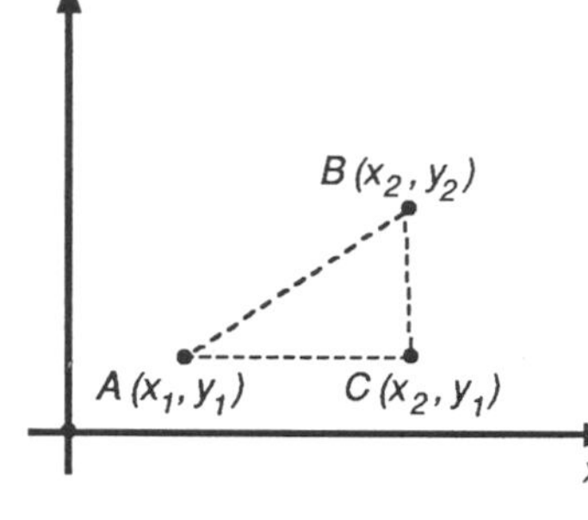

Proof: Locate C with coordinates (x_2, y_1). $\overline{BC}$ is a vertical segment since C has the same x-coordinate as B. $\overline{AC}$ is a horizontal segment since C has the same y-coordinate as A. Since $\overline{AC}$ and $\overline{BC}$ are horizontal and vertical segments, ΔABC is a right triangle. If ΔABC is a right triangle, then $AB^2 = AC^2 + BC^2$. But $AC = |x_2 - x_1|$ and $BC = |y_2 - y_1|$. Therefore, $AB^2 = AC^2 + BC^2 = |x_2 - x_1|^2 + |y_2 - y_1|^2$. Since the square of the absolute value of a number equals the square of the number, $AB^2 = (x_2 - x_1)^2 + (y_2 - y_1)^2$. Taking the positive square root of both sides of the equation, $AB = \sqrt{(x_2 - x_1)^2 + (y_2 - y_1)^2}$.

Now that it has been proved, you can add the Distance Formula to your theorem list.

EXERCISE SET B

1. Copy and complete the coordinate proof of the following conjecture. When you're finished, you can add it to your theorem list.

   Conjecture: *The midpoint of the hypotenuse of a right triangle is the same distance from the three vertices. (C-68)*

   Given: Right ΔRHT with M the midpoint of the hypotenuse. Locate right angle R at the origin, H at $(2a, 0)$, and T at $(0, 2b)$ so that two sides are on the axes.

   Show: $MR = MH = MT$

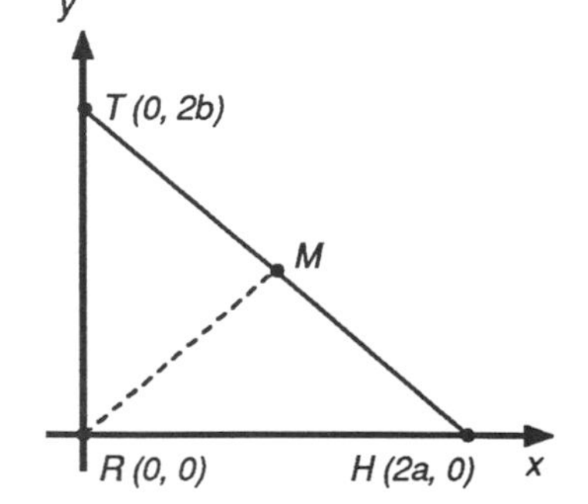

Proof:

By the Coordinate Midpoint Postulate, the coordinates of M are:

$$\left(\frac{2a+0}{2}, \frac{2b+0}{2}\right) \text{ or } \left(\frac{2a}{2}, \frac{2b}{2}\right) \text{ or } (a, b)$$

By the Distance Formula:

$$MR = \sqrt{(a-0)^2 + (b-0)^2} = \sqrt{a^2 + b^2}$$

$$MH = \sqrt{(2a-a)^2 + (\text{—?—})^2} = \sqrt{a^2 + b^2}$$

$$MT = \sqrt{(\text{—?—})^2 + (\text{—?—})^2} = \sqrt{\text{—?—}}$$

Therefore, $M(a, b)$ is the same distance from all three vertices.

You can now rename Conjecture 68 as Theorem 68.

2. Use the plan to write a coordinate proof of the Triangle Midsegment Conjecture. When you're finished, you can add it to your theorem list.

Conjecture: *A midsegment of a triangle is parallel to the third side and one-half the length of the third side.* ***(Triangle Midsegment Conjecture)***

Given: ΔAOK with M and N the midpoints of $\overline{AO}$ and $\overline{OK}$ respectively. Locate vertex O at the origin, vertex A at $(2a, 0)$ so that one side is on an axis, and point K at $(2b, 2c)$.

Show: $MN = \frac{1}{2}AK$

$\overline{MN} \parallel \overline{AK}$

Plan:

- I can use the midpoint formula to find the midpoints M and N. The midpoint of $\overline{AO}$ is $M(a, 0)$. What is the midpoint of $\overline{OK}$?
- I can find MN and AK with the distance formula. Is $MN = \frac{1}{2}AK$?
- I can find the slopes of $\overline{MN}$ and $\overline{AK}$ with the slope formula. Is $\overline{MN} \parallel \overline{AK}$?

3. Use the plan to write a coordinate proof of the Trapezoid Midsegment Conjecture. Then add it as a theorem to your theorem list.

Conjecture: *The midsegment of a trapezoid is parallel to the bases and is equal in length to the average of the lengths of the bases.* ***(Trapezoid Midsegment Conjecture)***

Given:	Trapezoid $TRAP$ with $\overline{TR} \parallel \overline{AP}$ and with E and Z the midpoints of sides $\overline{PT}$ and $\overline{AR}$ respectively. Locate one vertex at the origin with one base on the x-axis.
Show:	$EZ = \frac{1}{2}(TR + AP)$ $\overline{EZ} \parallel \overline{TR} \parallel \overline{AP}$

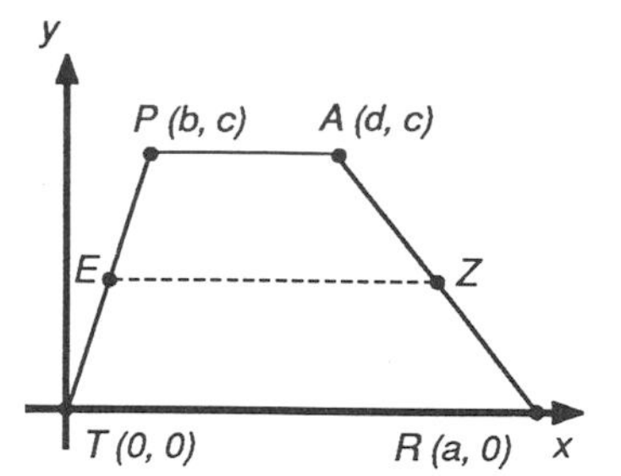

Plan:

- I can find the coordinates of E and Z with the midpoint formula.
- I can use the distance formula to find EZ, TR, and AP.
  Is $EZ = \frac{1}{2}(TR + AP)$?
- I can use the slope formula to find the slopes of $\overline{EZ}$ and $\overline{AP}$.
  Is $\overline{EZ} \parallel \overline{TR} \parallel \overline{AP}$?

Improving Visual Thinking Skills

Zendalf the Wise

Translate the last sentence of this story by Anne Arkey.

> Once upon a time there came to the land of Teknologee a wizard, Zendalf the Wise. The wizard Zendalf discovered that in this land of Teknologee, all the work was done by computers and robots. The robots buwlt the homes and grew the food for the humanowds, whwle the computers managed everythwng wncludwng the communwcatwon systems. Thx humanowds bxcamx vxry dxpxndxnt on the machwnxs and thus thxy grxw vxry wxak wn charactxr. Thxwr yrt ynd culturx dxclwnxd. Thx wwzyrd, wwshwng tz pzwnt zut the fzlly zf thws dxpxndxncx zn mychwnxs, put y spxll zn yll typxwrwtxrs, cyuswng thxm tz slzwly chyngx lxttxrs. Thws cyvsxd y czmplxtx brxykdzwn wn yll czmmvnwcytwzns. Thx hvmynzwds wxrx zncx ygywn fzrcxd tz thwnk fzr thxmsxlvxs.

Lesson 15.8

Triangle Midsegment Theorem Revisited

What is now proved was once only imagined.
– William Blake

The Triangle Midsegment Theorem, proved in the previous lesson, can be used to prove other conjectures. In this lesson you will perform investigations leading to conjectures that can be proved using the Triangle Midsegment Theorem. The first investigate-conjecture-prove problem is completed as an example.

Example

Investigate: Construct a quadrilateral. Construct the midpoint of each side. Connect the four midpoints in order to form another quadrilateral.

Conjecture: What do you observe about the opposite sides in this new quadrilateral? State a conjecture about the quadrilateral formed by connecting the midpoints of a quadrilateral in order?

Prove: Prove your conjecture.

After completing the investigation above, you wind up with a drawing similar to the one pictured at right. The opposite sides in the smaller quadrilateral appear parallel. Is this always true? You don't want to make a conjecture based on only one case, so try again beginning with other quadrilaterals. After going through the construction several more times, you may satisfy yourself that your observations seem to hold for every quadrilateral. You are ready to make a conjecture.

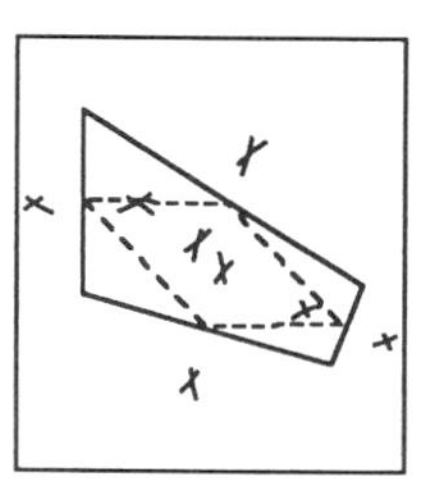

Conjecture: *The quadrilateral formed by connecting in order the midpoints of the sides of a quadrilateral is a parallelogram.*

How do you prove the conjecture? Start by making a diagram. You may leave off the construction marks. Label your diagram. State what you are given and what you must show in terms of your diagram. Then plan and proceed with your proof.

Given: Quadrilateral *QUAD* with inscribed quadrilateral *PREL* where *P*, *R*, *E*, and *L* are the midpoints of $\overline{UA}$, $\overline{AD}$, $\overline{DQ}$, and $\overline{QU}$ respectively

Show: Quadrilateral *PREL* is a parallelogram

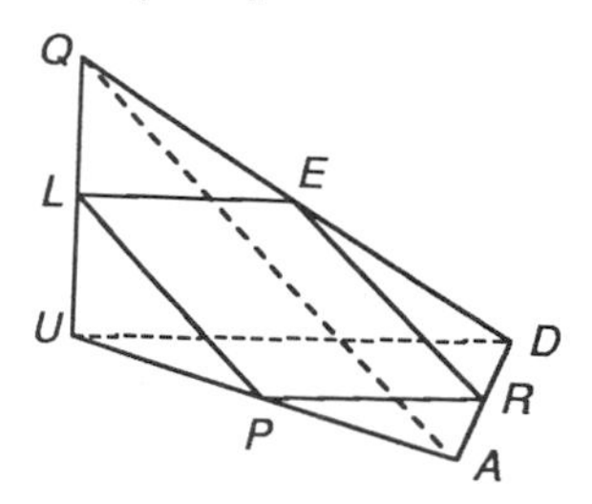

Plan:

- I wish to prove *PREL* is a parallelogram.
  I can do this if I can prove that the opposite sides are parallel.
- In Lesson 14.6, Exercise A3, it was proved that if two lines are parallel to a third line in the plane, then they are parallel to each other.
- Therefore, I construct diagonal $\overline{QA}$ because it appears to be parallel to both $\overline{ER}$ and $\overline{LP}$.
- Therefore, if I can show $\overline{ER} \parallel \overline{QA}$ and $\overline{LP} \parallel \overline{QA}$, then $\overline{ER} \parallel \overline{LP}$.
- $\overline{ER}$ is a line segment connecting the midpoints of two sides of ΔAQD and $\overline{LP}$ is a line segment connecting the midpoints of two sides of ΔAQU. The Triangle Midsegment Theorem says that if a line segment connects the midpoints of two sides of a triangle, then it is parallel to the third side. Therefore, $\overline{ER} \parallel \overline{LP}$.
- By a similar argument, I can construct diagonal $\overline{DU}$ and show that $\overline{LE} \parallel \overline{DU}$ and $\overline{PR} \parallel \overline{DU}$. Therefore, $\overline{LE} \parallel \overline{PR}$.
- Since I can show $\overline{ER} \parallel \overline{LP}$ and $\overline{LE} \parallel \overline{PR}$, then *PREL* is a parallelogram by the definition of a parallelogram.

Flowchart:

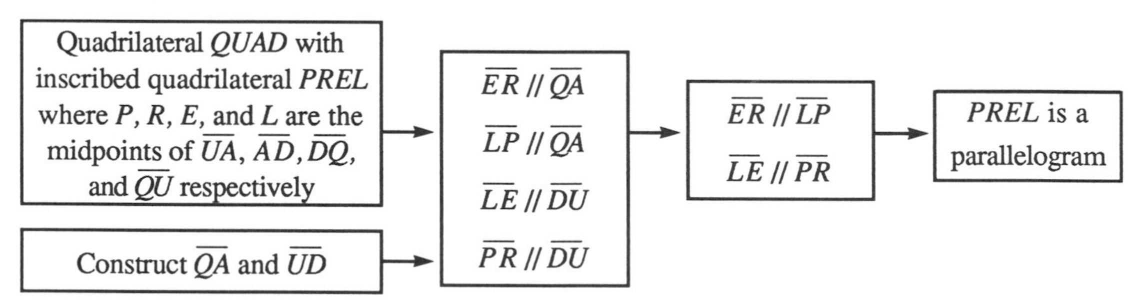

Two-Column Proof:

1. Quadrilateral *QUAD* with inscribed quadrilateral *PREL* where *P*, *R*, *E*, and *L* are the midpoints of $\overline{UA}$, $\overline{AD}$, $\overline{DQ}$ and $\overline{QU}$ respectively.	1. Given
2. Construct $\overline{QA}$ and $\overline{UD}$	2. Line Postulate
3. $\overline{ER} \parallel \overline{QA}$; $\overline{LP} \parallel \overline{QA}$; $\overline{LE} \parallel \overline{DU}$; $\overline{PR} \parallel \overline{DU}$	3. Triangle Midsegment Theorem
4. $\overline{ER} \parallel \overline{LP}$; $\overline{LE} \parallel \overline{PR}$	4. If two lines are parallel to a third line in the plane, then they are parallel to each other. (Lesson 14.6, Exercise A3)
5. *PREL* is a parallelogram	5. Definition of a parallelogram

Now it is your turn. The four investigations and proofs in the exercise set should be worked in groups of four. Each student in the group should:

- Perform one of the investigations and make a conjecture.
- Plan and write a proof (two-column, flowchart, or coordinate proof).
- Share your conjecture and proof with the others in the group.

EXERCISE SET

Perform each investigation. Then make a conjecture. Finally, prove your conjecture. Once you have proved your conjectures, they may be added to your theorem list.

1. *Investigate:* Construct a rectangle. Construct the midpoint of each side. Connect the four midpoints to form another quadrilateral.

   Conjecture: What do you observe about the quadrilateral formed? From the example at the beginning of the lesson, you know that it is a parallelogram, but what type of parallelogram? State a conjecture about the parallelogram formed by connecting the midpoints of a rectangle.

   Prove: Prove your conjecture.

2. *Investigate:* Construct a rhombus. Construct the midpoint of each side. Connect the four midpoints to form another quadrilateral.

   Conjecture: From the example at the beginning of the lesson, you know that it is a parallelogram; but what type of parallelogram? State a conjecture about the parallelogram formed by connecting the midpoints of a rhombus.

   Prove: Prove your conjecture.

3. *Investigate:* Construct a kite (a quadrilateral with two pair of consecutive sides congruent). Construct the midpoint of each side. Connect the four midpoints to form another quadrilateral.

   Conjecture: State a conjecture about the parallelogram formed by connecting the midpoints of a kite.

   Prove: Prove your conjecture.

4. *Investigate:* Construct an isosceles trapezoid. Construct the midpoint of each side. Connect the four midpoints to form another quadrilateral.

   Conjecture: State a conjecture about the parallelogram formed by connecting the midpoints of an isosceles trapezoid.

   Prove: Prove your conjecture.

Lesson 15.9

Non-Euclidean Geometries

Anyone that plays games has probably also changed the rules of a game. Sometimes by changing just one rule, a game becomes completely different. Geometry may be compared to a game — the postulates of geometry are the rules. If you change even one postulate, you may create a new geometry.

The geometry of Euclid, called Euclidean geometry (the geometry discovered in this text) was based on a number of postulates or self-evident truths. A postulate, according to the contemporaries of Euclid, was an obvious truth that could not be derived from other postulates. The first five of Euclid's postulates are listed below.

Postulate 1: *A straight line may be drawn between any two points.*

Postulate 2: *Any segment may be extended indefinitely.*

Postulate 3: *A circle may be drawn with any given point as center and any given radius.*

Postulate 4: *All right angles are equal.*

Postulate 5: *If two straight lines lying in a plane are met by another line, and if the sum of the interior angles on one side is less than two right angles, then the straight lines if extended sufficiently will meet on the side on which the sum of the angles is less than two right angles.*

Euclid's fifth postulate is known as the Parallel Postulate. (That's right! There is no mention of parallel in the fifth postulate?) It has that name because it is logically equivalent to: *Through a given point not on a given line there passes at most one line which is parallel to the given line (Playfair Axiom).*

The first four postulates seem very straightforward. The fifth postulate, however, sounds complicated. Because the fifth postulate does not sound like the others, mathematicians tried for centuries to prove that it could be derived from the others.

Attempting to use indirect proof, they began by assuming that the fifth postulate (or others logically equivalent) was false and tried to reach a logical contradiction. There are two possible assumptions: through a given point not on a given line there passes *more than one line* or there passes *no lines* parallel to the given line.

The German mathematician Carl Friedrich Gauss (1777–1855), the Russian mathematician Nicolai Lobachevsky (1793–1856) and the Hungarian Janos Bolyai (1802–1860) made the first assumption: through a given point

Nicolai Ivanovitch Lobachevsky

not on a given line there passes *more than one line* which is parallel to the given line. This assumption led to a new non-Euclidean geometry that did not contradict any of Euclid's other postulates. It did not contradict any of the theorems that didn't depend on the parallel postulate. This geometry is called **hyperbolic geometry**.

Georg Friedrich Bernhard Riemann (1826–1866) of Germany made the second assumption: through a given point not on a given line there passes *no lines* which are parallel to the given line. This assumption, together with adjustments to other postulates, led to a second non-Euclidean geometry that also did not contradict any of Euclid's other postulates or any of the theorems not dependent on the Parallel Postulate. Riemann's geometry is called **elliptic geometry**.

All the theorems of Euclidean geometry, except those depending directly or indirectly on the Parallel Postulate, are also valid in hyperbolic and elliptic geometries. The most familiar example of a theorem of Euclidean geometry that relies on the Parallel Postulate is the Triangle Sum Theorem: *The sum of the measures of the three angles of a triangle is 180.* The corresponding theorem in hyperbolic geometry states: *The sum of the measures of the three angles of a triangle is less than 180.* In elliptic geometry the theorem states: *The sum of the measures of the three angles of a triangle is greater than 180.*

You can use the surface of a sphere as a model of elliptic geometry. To establish the model, you make two changes to the Euclidean postulates. First, you replace the word *line* with the phrase *great circle.* (A great circle is a circle formed on the surface of a sphere by a plane passing through the center of the sphere. The equator is a great circle on the surface of the earth.) Second, you agree to call each pair of diametrically opposite points (points on opposite ends of a diameter) on the sphere a single point and replace the word *point* with the phrase *point pair.*

Model of Elliptic Geometry

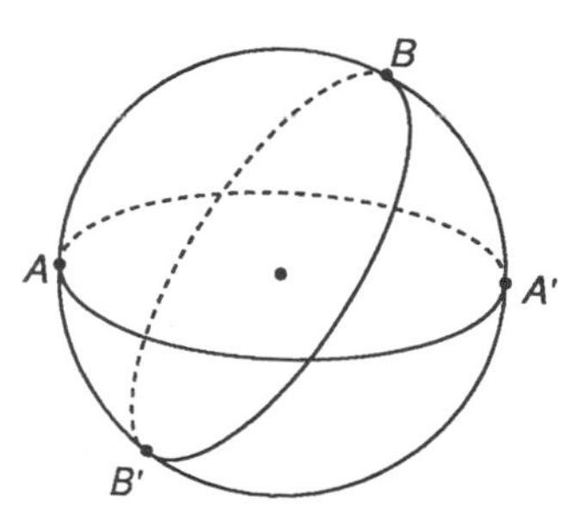

A and A' form a point pair. So do B and B'. More than one great circle can pass through A and A'. However, only one great circle can pass through both A and A' and B and B'.

In Euclidean geometry lines never end and are infinite in length. In elliptic geometry great circles never end, however, their length is finite! You can travel around and around the equator forever. You can also measure it! That part of a great circle between two points of the great circle is called a *geodesic*. A geodesic is the line segment of elliptic geometry.

Recall that Riemann assumed that through a given point not on a given line there passes *no* lines which are parallel to the given line. Simply put, there are no parallel lines in elliptic geometry. The spherical model of elliptic geometry supports this because all great circles intersect.

Let's look at a model of hyperbolic geometry devised by mathematician Felix Klein (1849–1925). The model is all the points in the interior of a circle. The "lines" in this geometry are open-ended chords of the circle (points of the circle are not included in this geometry). To establish the model, you replace the word *line* in Euclid's postulates with the phrase *open-ended chord.* The diagram on the right shows that many "lines" through a common point can be parallel to a given line. "Lines" p, r, and s all pass through point M, yet none of them intersect "line" t.

Model of Hyperbolic Geometry

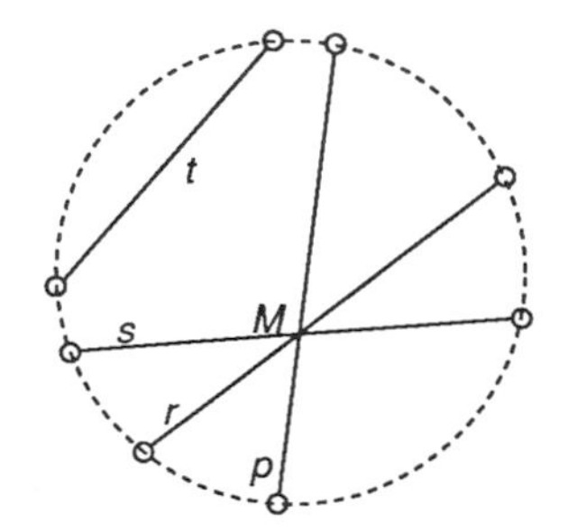

The circle is dotted because the points of the circle are not included in the geometry. The endpoints are open circles because the endpoints are not included.

EXERCISE SET

1. Write the first five postulates of elliptic geometry by replacing the word *line* in Euclid's postulates with the phrase *great circle,* the word *segment* with *geodesic,* and the word *point* with the phrase *point pair.* Use the Playfair Axiom for the Euclid's fifth postulate and change *at most one* to *none.*

2. Write the first five postulates of hyperbolic geometry by replacing the word *line* with the phrase *open-ended chord.* Use the Playfair Axiom for the Euclid's fifth postulate and change *at most one* to *more than one.*

3. Draw a model of hyperbolic geometry. Show an example in your model of two "lines" perpendicular to the same "line" that are parallel.

4. Draw a model of elliptic geometry. Show an example on your model of two "lines" perpendicular to the same "line" that are not parallel.

5. On your model of elliptic geometry, show that two points determine a line. (Remember, that any two diametrically opposite points are considered one point. Why is this necessary?)

6. On your model of hyperbolic geometry, show that two points determine a line.

7. On your model of hyperbolic geometry, show four "lines" passing through the same point with all four "lines" parallel (non-intersecting) to a fifth "line."

8. Draw an "isosceles triangle" on each model. Does the Isosceles Triangle Theorem hold in elliptic geometry? Does it hold in hyperbolic geometry?

9. The sum of the measures of the three angles of a triangle is always greater than 180 in elliptic geometry and always less than 180 in hyperbolic geometry. Draw a "triangle" on both of your models and explain in a logical argument why there are no squares in hyperbolic or elliptic geometry.

Lesson 15.10

Chapter Review

EXERCISE SET A

Complete each statement.

1. The bisectors of a linear pair of angles are —?—.
2. The measure of an angle formed by the intersection of two chords in a circle is equal to —?—. **(Intersecting Chords Theorem)**
3. The measure of an angle formed by two secants intersecting outside a circle is equal to one-half the —?— of the measures of the two intercepted arcs. **(Intersecting Secants Theorem)**
4. If the three sides of one triangle are proportional to the three sides of another triangle, then —?—. **(SSS Similarity Postulate)**
5. If —?— angles of one triangle are congruent to —?— angles of another triangle, then —?—. **(AA Similarity Postulate)**
6. If two sides of one triangle are proportional to two sides of another triangle and —?— in one triangle is congruent to the —?— in the other triangle, —?—. **(—?— Similarity Postulate)**
7. If a line —?— to one side of a triangle passes through the other two sides, then it divides them —?—. Conversely, if a line cuts two sides of a triangle —?—, then it is —?— to the third side. **(Parallel Proportionality Theorem)**
8. If two triangles are similar, then the corresponding —?—, corresponding —?—, and corresponding —?— are —?— to the corresponding sides. **(Proportional Parts Theorem)**
9. The diagonals of a trapezoid divide each other —?—.
10. If two chords in a circle intersect, —?—.
11. If two secants intersect outside a circle, the product of the length of one secant segment and the length of its external part is equal to the product of —?—.
12. The altitude to the hypotenuse of a right triangle is the —?— to the segments on the hypotenuse.
13. The midpoint of the hypotenuse of a right triangle is —?— the three vertices.
14. A midsegment of a triangle is —?— to the third side and —?— the length of the third side.

EXERCISE SET B

1. If $\Delta ABC \cong \Delta DEF$, then $\overline{AC} \cong \overline{DF}$. What reason would you give for this in a proof?
2. Sometimes a proof requires the help of additional constructions. If an angle bisector is needed in the proof, what postulate tells you that you may construct one?
3. How would you state the following as a conditional: *The sum of the measures of the three angles of a triangle is 180*?
4. Sometimes a conjecture can be proved as an immediate consequence of another theorem. What do you call this new theorem?
5. What is the procedure in an indirect proof?
6. What postulate do you use to prove two lines parallel in a coordinate proof?

EXERCISE SET C

1. Create a flowchart for the proof below. Then write the reasons for each step below its box to create a flowchart proof or write a two-column proof.

   Given: Circle O with chords $\overline{PN}, \overline{ET}, \overline{NA}, \overline{TP}, \overline{AE}$

   Show: $m\angle P + m\angle E + m\angle N + m\angle T + m\angle A = 180$

2. Prove the following conjecture.

   Conjecture: *If a right triangle is isosceles, then the base angles measure 45.*

   Given: Isosceles right triangle ABC with $\overline{BC} \cong \overline{AC}$, and with right $\angle ACB$

   Show: $m\angle ABC = 45$

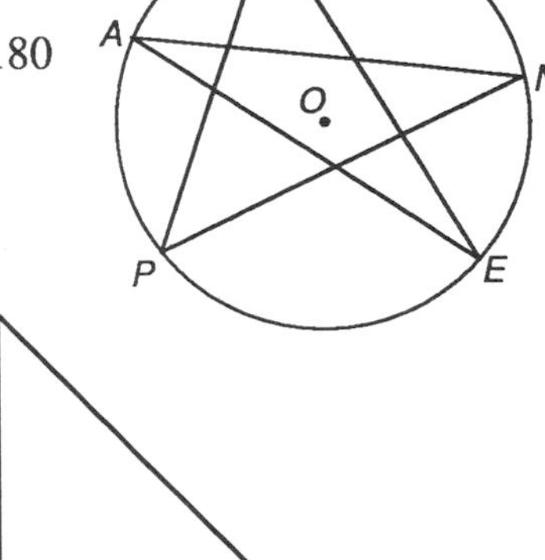

3.* Prove the following conjecture.

   Conjecture: *Tangent segments to a circle from a point outside the circle are congruent.*

   Given: Circle O with tangent segments TA and TB

   Show: $\overline{TA} \cong \overline{TB}$

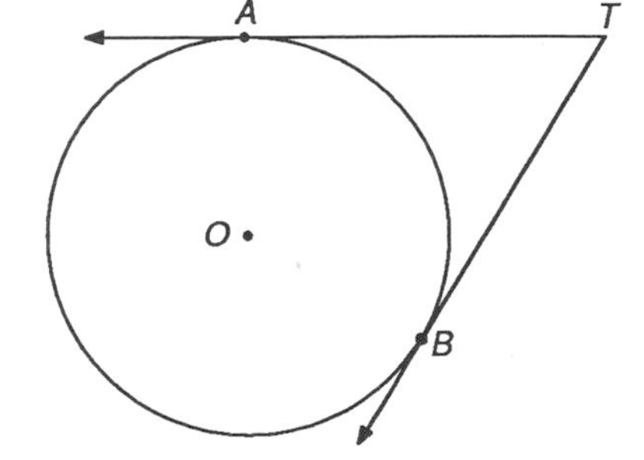

EXERCISE SET D

In each exercise, select one of the three conjectures. Investigate to determine whether or not it is true. If false, demonstrate a counter-example. If true, prove it.

1. a. If the consecutive angles of a quadrilateral are supplementary, then the quadrilateral is a parallelogram.

   b. If the diagonals of a quadrilateral bisect its angles, then the quadrilateral is a rhombus.

   c. If the diagonals of a quadrilateral are perpendicular bisectors of each other, then the quadrilateral is a rhombus.

2. a. If the midpoints of the sides of a quadrilateral are connected in order to form a rectangle then the quadrilateral is a rhombus.

   b. If the diagonals of a quadrilateral are congruent, then the quadrilateral is a rectangle.

   c. The segments joining the midpoints of the opposite sides of a quadrilateral bisect each other.

EXERCISE SET E

Draw concept maps to show how the postulates and theorems in each group are logically related. Refer back to the proofs of the theorems of this chapter.

1.*
- AIA Postulate
- Exterior Angle Theorem
- Isosceles Triangle Theorem
- Inscribed Angle Theorem
- Parallel lines intercept congruent arcs on a circle.

2.*
- SSS Congruence Postulate
- AA Similarity Postulate
- The Pythagorean Theorem
- The Converse of the Pythagorean Theorem
- The altitude to the hypotenuse of a right triangle is the mean proportional to the segments on the hypotenuse.

Improving Reasoning Skills

Think Dinosaur

If the letter in the word *dinosaur* which is three letters after the second vowel is also found in the alphabet before the twentieth letter of the alphabet, then print the word *dinosaur* horizontally and circle it. Otherwise print the word *dinosaur* vertically and cross out the second letter after the first vowel.

Cooperative Problem Solving

Party Time

The lunar colony geometry class has been talking about celebrating the end of the school term with a class party. They feel it is time to use their logical thinking skills and organizational skills for some fun and games! They have decided to have food and entertainment at their party. They want to serve a couple main dishes, desserts, hot and cold beverages, and a variety of salads (pasta salad, green salad, potato salad, and/or fruit salad).

Your job is to help them organize the party. Where do you start? What are the tasks?

- Make a list of all the things that have to be done before the party gets underway (decorations, purchasing, food preparation, drinks and ice, music, dancing, games, clean up).
- Once you have listed all the tasks, you need to put them in chronological order. For example, you can't make the salad until you figure out what type(s) of salad people want, so you need to plan a detailed menu.
- Match up the tasks with people. Who is going to be responsible for what?
- You can't shop for groceries until you know how many people are attending, so you must make a guest list and approximate how much food per person.
- When it comes time to prepare the food, you need to record the approximate time it will take to complete each task. For example, if the party is a barbecue, the time that you wish to start eating determines the time to light the charcoal.

Don't underestimate the complexity of the tasks involved in planning a successful party. Successful entertaining is an art that requires logical thinking and finely tuned organizational skills. Once you have organized the party, why not give it! Enjoy.

Appendix A

Hints for Selected Problems

You will find hints below for problems that are marked with an asterisk (*) in the text. If you turn to a hint before you've tried to solve a problem on your own, SHAME ON YOU! You should go back and make a serious effort to solve the problem without help. If, however, you need additional help to solve a problem, this is the place to look.

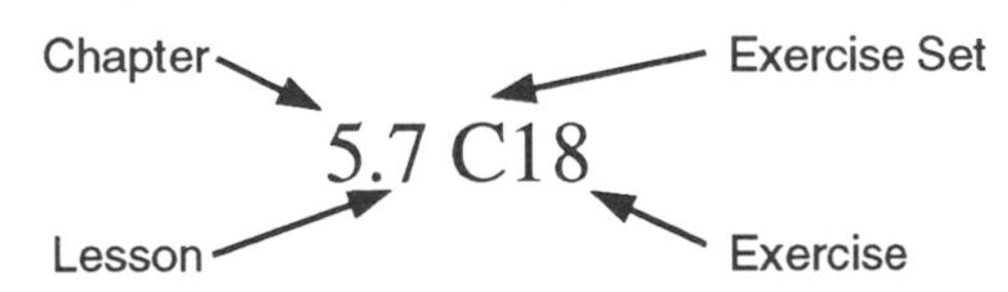

Chapter 0

0.1 3	Perhaps a quilt, basket weaving, or hand embroidered shirt or blouse.
0.4 2	Look at the intersections of the white lines.
0.4 3	How many cubes can you count? 30? Can you hold the book in such a way that you see a different number of cubes? How many?

0.6 2 — One way to find the center of an arc is to "eye-ball-it" or estimate the center. Your eye is a very good estimator. For a more exact mathematical approach, use a compass. The steps are outlined below.

Step 1: Place your compass tip on one endpoint of the arc and swing another arc over thc original arc.

Step 2: Then, without changing the compass setting, place your compass tip on the other end of your original arc and swing another arc that intersects the arc created from the other endpoint.

Step 3: Place a dot where these two arcs cross. Line up the edge of your straightedge with that point of intersecion and the center of the circle. The center of the arc is the point where the edge of your straightedge crosses the arc.

0.8 A4 — You might start with a large rectangle with a smaller rectangle centered within. Then select a horizon line and vanishing point and turn the rectangles into three dimensional boxes.

0.9 6a — The aztecs created a stone calendar for keeping track of the year.

Chapter 1

1.1 B1 — "Oh –?– –?– –?– –?– too!"

1.1 D1 — Each time you change a light bulb you might notice that you turn it clockwise to put it in and counterclockwise to take it out. Therefore, you conjecture that all light bulbs work that way!

1.2 A4 — $1^2, 2^2, 3^2, 4^2, \ldots$

1.2 A5 — Look at the differences. The sequence of differences in this exercise are the same as the sequence of numbers in the last exercise.

1.2 A7 — Change all fractions to the same denominator.

1.2 A8 — The letter pattern skips a letter.

1.2 A11 — $1 + 1 = 2,\ 1 + 2 = 3,\ 2 + 3 = 5,\ 3 + 5 = 8, \ldots$

1.2 A13 — Look at the differences. The differences are the numbers in 1.2 A2.

1.2 A18 — The pattern of the differences is 2, 6, 18, 54, 162, ... How is this pattern generated? 3 times 2 is 6; 3 times 6 is 18; 3 times 18 is 54; and so on.

1.2 A21 — The pattern of the differences is 1, 3, 9, 27, 81, ... How is this pattern generated? $1 = 3^0$; $3 = 3^1$; $9 = 3^2$; $27 = 3^3$; $81 = 3^4$; and so on.

1.2 B1 — The sum of two odd numbers is always an <u>even number.</u>

1.3 A6 — Each new line crosses all of the previously drawn lines.

1.4 A1

Term	1	2	3	4	5	6
Value	1 + 4	2 + 4	3 + 4	4 + 4	5 + 4	6 + 4

1.4 B1

Term	1	2	3	4	5	6
Value	-1 • 1	0 • 2	1 • 3	2 • 4	3 • 5	4 • 6

1.5 A1

Term	1	2	3	4	5	6
Value	0	1	3	6	10	15
Doubled	0	2	6	12	20	30
Factored	0 • 1	1 • 2	2 • 3	3 • 4	4 • 5	5 • 6

1.5 A6

Term	1	2	3	4	5	6
Value	$\frac{2}{3}$	2	4	$\frac{20}{3}$	10	14
Tripled	2	6	12	20	30	42
Factored	1 • 2	2 • 3	3 • 4	4 • 5	5 • 6	6 • 7

1.5 C1 — First *two* odd numbers total 4 or 2^2.
First *three* odd numbers total 9 or 3^2.
First *four* odd numbers total 16 or 4^2.
First *eight* odd numbers total –?– or $(–?–)^2$.
Therefore, the first *thirty* odd numbers total $(–?–)^2$ or –?–.

1.5 D2 — See 1.2 A11 but subtract.

1.5 D5 — The differences are found in 1.2 A3.

1.5 D6 — The differences are the same as the numbers in the last exercise.

1.5 D7 — The differences are the same as the numbers in the last exercise.

1.6 A1

Points	1	2	3	4	5	6
Infinite	2	2	2	2	2	2
Finite	0	1	2	3	4	5
Total	2	3	4	5	6	7

1.6 B1 Treat each person as a point and each conversation as a line segment connecting them. See 1.6 A2.

1.6 B3 Treat each team as a point and each set of four games that a team plays against another team as a line segment connecting them. See 1.6 A2.

1.6 B4 See 1.6 A6.

1.7 A3
$1 = 1 = 1^2$
$1 + 3 = 4 = 2^2$
$1 + 3 + 5 = 9 = 3^2$
$1 + 3 + 5 + 7 = 16 = 4^2$
Therefore, the first 4000 odd numbers total $(–?–)^2$ or –?–.

1.7 A4
The first even number is 2 = 2 = 1 • 2.
The sum of the first *two* even numbers is 2 + 4 = 6 = 2 • 3.
The sum of the first *three* even numbers is 2 + 4 + 6 = 12 = 3 • 4.
The sum of the first *four* even numbers is 2 + 4 + 6 + 8 = 20 = 4 • 5.
Therefore, the first 700 even numbers total (–?–)(–?–) or –?–.

1.7 A7 See 1.7 A1.

1.7 A8 See 1.7 A3.

1.7 A9 1 by 1 Squares

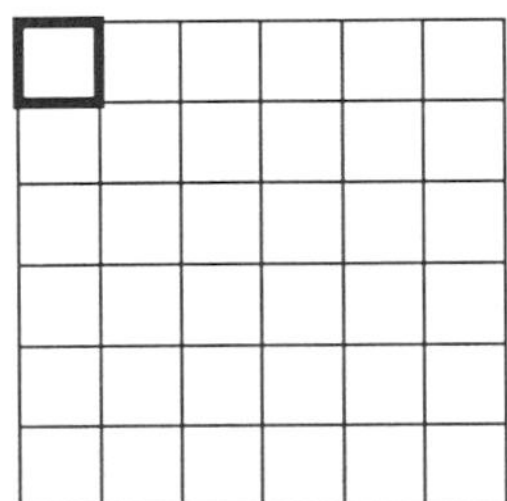

There are 6 • 6 or 36 of the 1 by 1 squares.

2 by 2 Squares

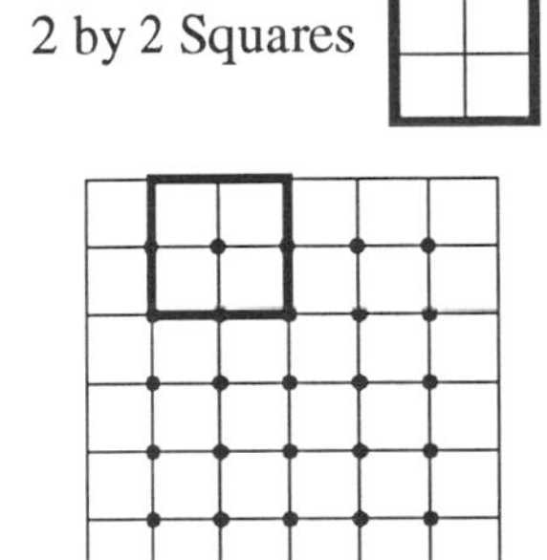

Each dot represents the center of a 2 by 2 square. Therefore, there are 5 • 5 or 25 of the 2 by 2 squares.

Continue to find the number of 3 by 3 squares, 4 by 4 squares, looking for a pattern in their totals.

1.7 A10 Try to solve easier (smaller) problems. How many cubes in a 1 by 1 by 1 cube? One! How many cubes in a 2 by 2 by 2 cube? 1 (whole cube) + 8 (small cubes) or $1^3 + 2^3$. How many cubes in a 3 by 3 by 3 cube? The entire 3 by 3 by 3, one 2 by 2 by 2 in each corner, and 27 of the smallest 1 by 1 by 1 cubes for a total of 1 + 8 + 27 (or $1^3 + 2^3 + 3^3$). How many cubes in a 4 by 4 by 4? Did you find a pattern in your answers?

1.7 B4

Term	1	2	3	4	5	6
Value	0	2	5	9	14	20
Doubled	0	4	10	18	28	40
Factored	0 • 3	1 • 4	2 • 5	3 • 6	4 • 7	5 • 8

1.7 B5

Term	1	2	3	4	5	6
Value	-5	0	15	40	75	120
Doubled	-10	0	30	80	150	240
Factored	5 • -2	5 • 0	5 • 6	5 • 16	5 • 30	5 • 48
	5 • -2	5 • 0	5•1•6	5•2•8	5•3•10	5•4•12
	5•-1•2	5• 0•4	5•1•6	5•2•8	5•3•10	5•4•12

SP 1

No rings; no moves. 1 ring; 1 move. 2 rings; 3 moves. 3 rings; 7 moves. Notice that each number, 1, 3, 7, is one less than a power of two. The *n* will be in the exponent position in the formula for the number of necessary moves for *n* rings. See the factored pattern in the hint for 1.8 A6.

1.8 A1

Notice that 1 and 2 are above 3. 3 and 3 are above 6. 4 and 6 are above 10. 5 and 10 are above 15. See a pattern? Think *add*!

1.8 A6

Number of row	0	1	2	3	4	5	6	...	n
Sum in row	1	2	4	8	16	32	64	...	–?–
Factored	2^0	2^1	2^2	2^3	2^4	2^5	2^6		

1.8 B1

Try adding three numbers.

1.8 B2

Three factors.

1.8 B3

$1 = 1 = 1^2$

$1 + 2 + 1 = 4 = 2^2$

$1 + 2 + 3 + 2 + 1 = 9 = 3^2$

$1 + 2 + 3 + 4 + 3 + 2 + 1 = 16 = 4^2$

1.9 2

See 1.7 A3

1.9 4

Try the alphabet backwards.

1.9 6

The black arm moves around the star counter-clockwise while the top alternates black and white.

1.9 8

The sums of the numbers in the two rows on top are always the top numbers in the front face.

1.9 9

The shading of the semicircles on the bottom alternate left and right while the shading in the top semicircles alternate right to left. The top triangles alternate with and without the black top while the bottom triangles do the same.

1.9 11

Term	1	2	3	4	5	6
Value	0	3	10	21	36	55
Doubled	0	6	20	42	72	110
Factored	–?–	–? –	–?–	6 • 7	–?–	–?–

1.9 13

Term	1	2	3	4	5	6
Value	-4	0	6	14	24	36
Doubled	-8	0	12	28	48	72
Factored	–?–	–? –	–?–	4 • 7	6 • 8	–?–

1.9 15 $1 = 1$
$1 + 3 = 4$
$1 + 3 + 5 = 9$
$1 + 3 + 5 + 7 = 16$
And so on . . .

1.9 17 The maximum number of points will be when the lines are random. See 1.6 A3

1.9 19 Try spelling smaller words first.

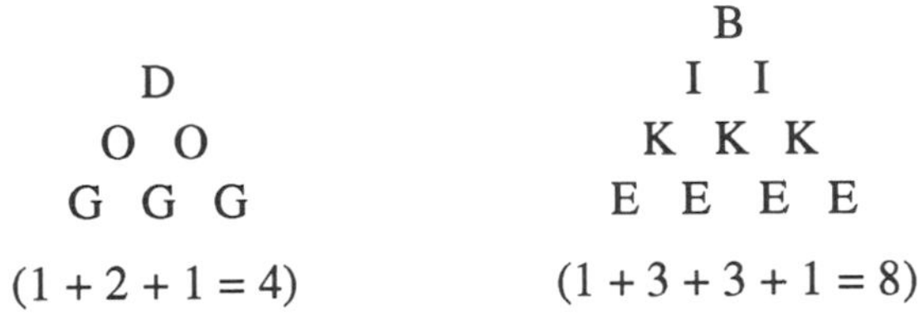

CPS On a sheet of graph paper lay out the grid of streets as described. How many ways from Third and A to Third and B? How many ways from Third and A to Third and C? Continue in this way — finding the number of ways to go from Third and A to each of the intersections. Place your answers on your graph paper on the points of intersection. Look for a pattern in the triangle of numbers?

Chapter 2

2.1 A2 Use only two letters in naming the line. Do not use three letters. $\overleftrightarrow{ART}$ is not acceptable. You may use any two of the three named points. There are six possible ways to name this line.

2.1 C1 Use only two letters in naming the ray. Do not use three letters. $\overrightarrow{ABC}$ is not acceptable. The letter on the left *must* be the initial point A. You may use either one of the two remaining named points. There are two possible ways to name this ray.

2.1 D1 Use three letters in naming the angle. The middle letter *must* be the vertex point A. There are two possible ways to name this angle. In this case, but not always, they spell words.

2.1 D7 There is no ambiguity about angle A or angle C. However, angle D could be one of three different angles. How about angle B?

2.2 A7 $m\angle COB = m\angle COA - m\angle BOA \approx$ –?–

2.2 B3 Place the zero-edge of your protractor against the "cushion" with the center mark at the point where the ball will bounce. Place your straightedge so that it shows the path of the ball leaving the cushion.

2.2 D1

H
6
8
T
O

2.2 D5 $m\overline{AK} = m\overline{RK}$; $m\angle A = m\angle B$

2.3 A1	Widgets are empty inside and have two different looking tails.
2.3 B1	*A **right angle** is an angle whose measure is –?–.*
2.3 B2	*An **acute angle** is an angle whose measure is less than –?–.*
2.4 A7	If $\overleftrightarrow{AB}$ and $\overleftrightarrow{CD}$ intersect at point P, then $\angle APC$ and $\angle BPD$ are a pair of . . . .
2.4 A8	If X, Y, and Z are consecutive collinear points and W is a point not on $\overleftrightarrow{XZ}$, then $\angle XYW$ and –?– form a linear pair of angles.
2.4 B1	In geometry we sometimes use the expression "one and only one." This expression is used to combine two statements. For example, consider the following sentence: *There is one and only one line that can be constructed to pass through two given points.* This sentence says two things. First, it claims there exists a line; *there is one line* that can be constructed to pass through two points. Second, it claims the line is unique; *there is only one line* that can be constructed to pass through two points.
2.4 B5	Think of a pencil as representing a line. Place a point on the pencil. How many pencils can you place perpendicular to the pencil at that point?
2.5 A1	A three-sided polygon is a triangle.
2.5 A11	*FIVER* is one possible name. You find another.
2.5 B4	Can you use "equilateral" or "equiangular" in your definition?
2.6 A1	*A **right triangle** is a triangle with one right angle.*
2.7 A1	*A **trapezoid** is a quadrilateral with exactly one pair of parallel sides.*
2.7 A4	Do not use right angles in your definition.
2.7 A5	You can define a square three ways: as a special rhombus, as a special rectangle, or as a regular polygon.
2.8 A5	What if they are not in the same plane?
2.8 A12	To help you visualize this, hold pencils at right angles into a sheet of cardboard.
2.9 A3	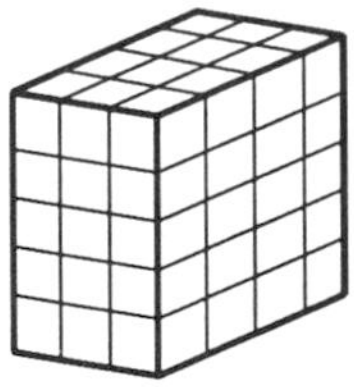
2.9 A5	Picture cutting off a corner.
2.9 B2	It comes in pretty handy when you have to do long division.
2.10 A4	Draw a diagram. Look at the twenty-sixth and twenty-seventh mornings.
2.10 A8	Draw a diagram of a clock face. The big hand will be somewhere between the 3 and 6 while the little hand will be between the 6 and 7.
2.11 C11	The plane intersects all six faces. Copy the drawing on the right, then finish the drawing by putting in the last dashed hidden line on the bottom. Shade in the hexagon. 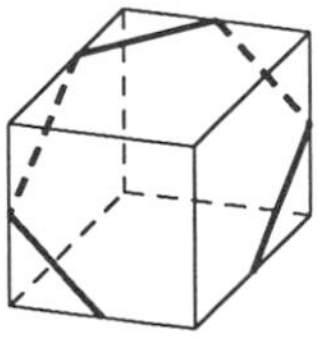

2.11 D10

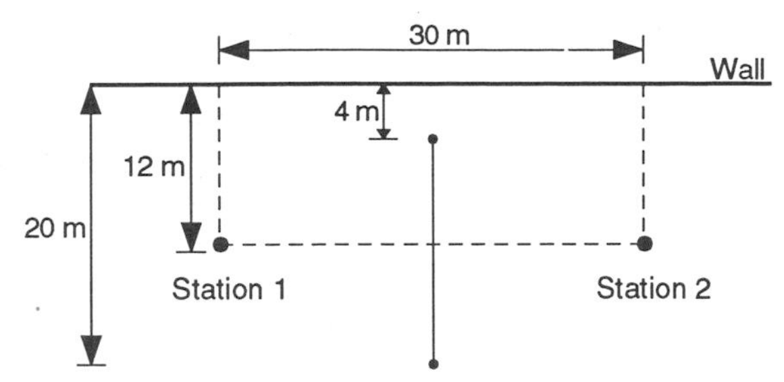

2.11 D13 Don't let this one scare you! It only sounds tough. Draw a diagram (two circles with the same center, one large circle for the outside edge of the record, the other small circle for the label). The record is moving around but what is the needle doing? It is just moving straight across!

Chapter 3

3.1 2

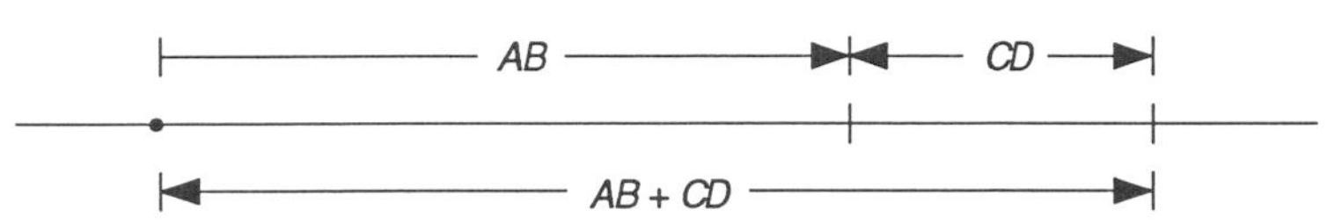

3.1 6 Begin with a construction line and starting point *P*. Swing a large arc with center on *P* passing through your construction line and having a radius the same size as the arc on the original angle. With your compass, measure the distance between the two points where the arc crosses the original angle. With this compass setting, place the needle of your compass on the intersection point of the arc and construction line and swing a second arc intersecting the first arc. Line up this point of intersection and point *P* with your straightedge and draw the second ray of your duplicate angle.

3.1 11 You duplicated a triangle in 3.1 9. Think of a quadrilateral as two triangles stuck together (draw in a diagonal). You do not need to duplicate angles.

3.2 B2 Construct the perpendicular bisector of $\overline{QD}$. Then construct the perpendicular bisector of each half.

3.2 B3 Swing both sets of intersecting arcs on one side of the segment.

3.2 B4 The average is half the sum of the two lengths. Therefore, add the two lengths with your compass and then construct the perpendicular bisector of their sum.

3.2 B5 Construct the perpendicular bisector of $\overline{CD}$ and subtract its length from two AB's.

3.3 A3

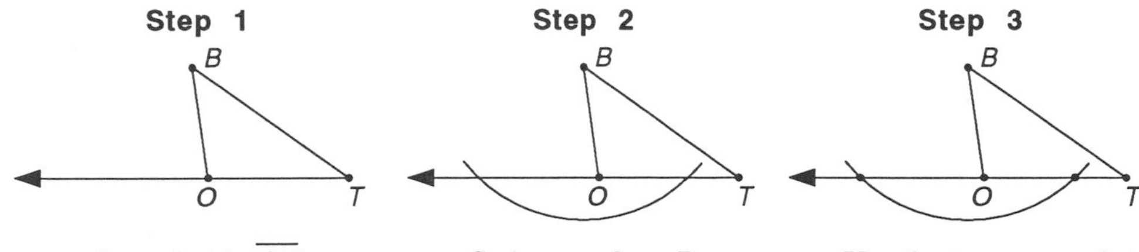

Extend side $\overline{OT}$.

Swing arc from B through $\overrightarrow{TO}$.

Use the two new points to finish.

3.3 A5 Label the point on the line point C. With your compass, mark off two points, one on each side of C, so that C becomes the midpoint of some new line segment. Label segment MN. From the two points, M and N, swing arcs above and below $\overline{MN}$. Now construct the line through the two points where the pairs of arcs intersect. This will be the perpendicular through point C.

3.3 B4 Draw the square first!

3.4 A5 To construct a 60° angle, construct an equilateral triangle.

3.4 B1 Construct an equilateral triangle. Each angle will be 60°. Bisect one angle.

3.4 B7 Construct a $22\frac{1}{2}$° and a 30° angle and add them. You can add angles by constructing them with a common vertex and a common side.

3.4 B9 Construct a perpendicular. Bisect one of the right angles. $90 + 45 = 135$.

3.5 1

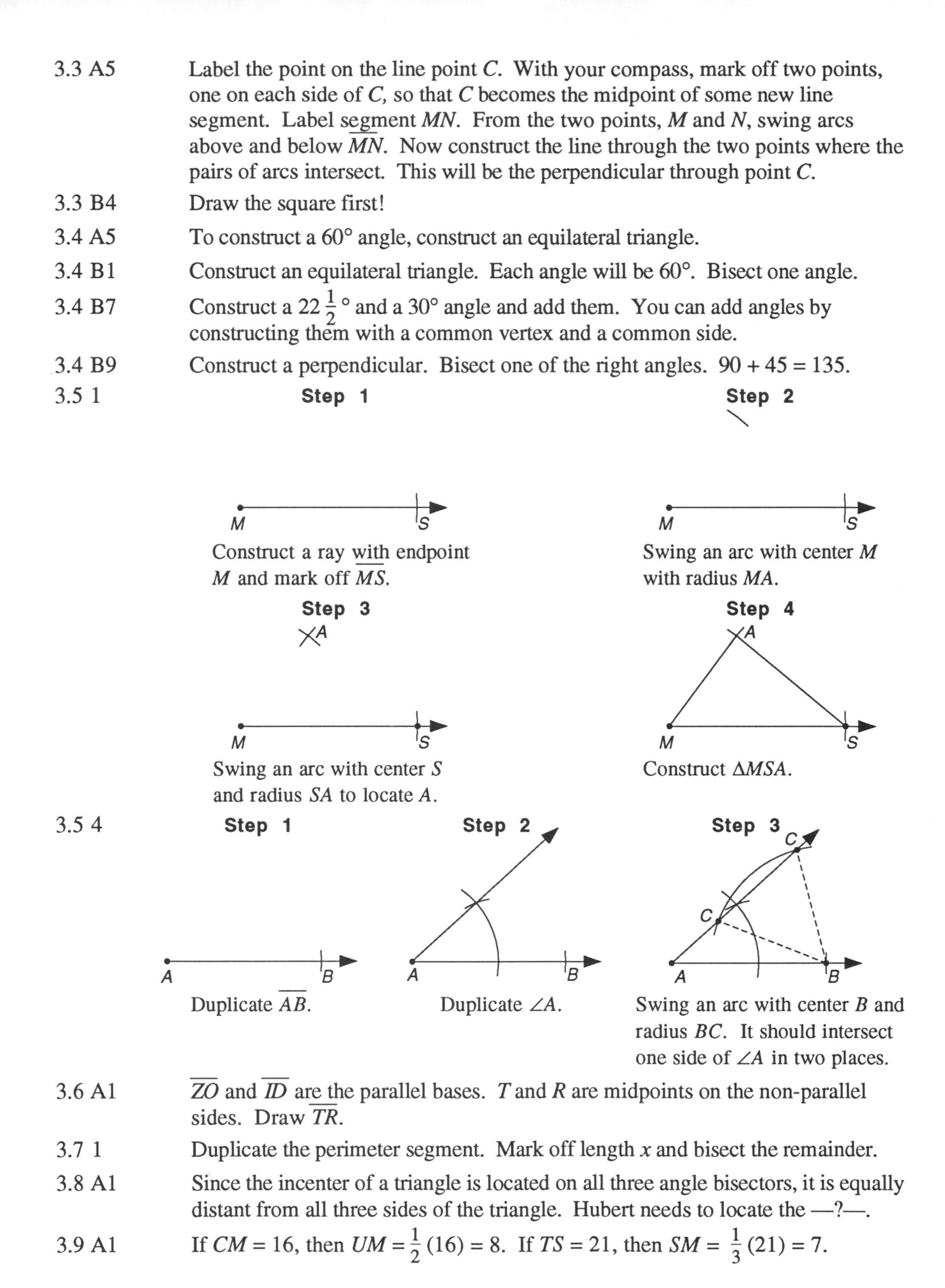

Step 1

Construct a ray with endpoint M and mark off $\overline{MS}$.

Step 2

Swing an arc with center M with radius MA.

Step 3

Swing an arc with center S and radius SA to locate A.

Step 4

Construct ΔMSA.

3.5 4

Step 1

Duplicate $\overline{AB}$.

Step 2

Duplicate $\angle A$.

Step 3

Swing an arc with center B and radius BC. It should intersect one side of $\angle A$ in two places.

3.6 A1 $\overline{ZO}$ and $\overline{ID}$ are the parallel bases. T and R are midpoints on the non-parallel sides. Draw $\overline{TR}$.

3.7 1 Duplicate the perimeter segment. Mark off length x and bisect the remainder.

3.8 A1 Since the incenter of a triangle is located on all three angle bisectors, it is equally distant from all three sides of the triangle. Hubert needs to locate the —?—.

3.9 A1 If $CM = 16$, then $UM = \frac{1}{2}(16) = 8$. If $TS = 21$, then $SM = \frac{1}{3}(21) = 7$.

3.10 C8 Construct the perpendicular bisector of a segment with length z to get $\frac{1}{2}z$.

3.10 C10 Construct $\angle A$ and $\angle B$ at the opposite ends of $\overline{AB}$ so that their rays intersect.

3.10 C14 $\frac{7}{2}x$ is three and one-half x's.

Chapter 4

4.1 3 $a = 60°$; $b = c = 120°$

4.2 A1 $x = 180 - (55 + 52)$

4.2 A3 Look at the diagram on the right:
$a = 180 - 120 = 60$; $b = 180 - 130 = 50$;
$c = 180 - (60 + 50)$; $z = 180 - c$

4.2 A5 Do you see two large triangles that overlap?

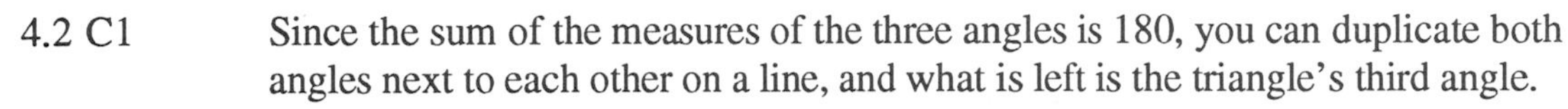

4.2 C1 Since the sum of the measures of the three angles is 180, you can duplicate both angles next to each other on a line, and what is left is the triangle's third angle.

4.2 C2 Find the supplement of $\angle D$. Then bisect the supplement.

4.3 A1 First find the measure of each angle in the equiangular pentagon.

4.3 A3

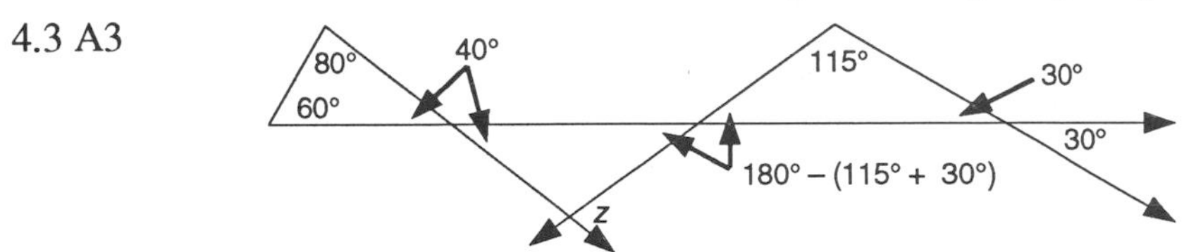

4.3 B3 The sum of the measures of the angles of a polygon with n sides is $(n - 2)180$. Therefore, $(n - 2)180 = 2700$. Solve for n.

4.3 B6 $[(n - 2)180]/n = 170$. To solve for n, first multiply both sides of the equation by n to get rid of the fractional expression on the left.

4.4 A3 If the polygon is a regular octagon, the eight interior angles are equal in measure. Therefore, the exterior angles are equal in measure. If all eight exterior angles are equal in measure and their sum is 360, each measures –?–.

4.4 A4 The number of sides is equal to the number of exterior angles. The number of exterior angles times 24 must equal 360 ($24n = 360$).

4.4 A8 If each interior angle measures 165, then each exterior angle measures 15. How many 15's in 360?

4.4 B6 Sometimes a diagram has unnecessary information. Ignore the dotted line! Add the three exterior angles (100 + 115 + 50) and subtract from 360 to get the measure of z.

SP 3 Draw a network with points standing for rooms and lines representing paths through doors. How many odd points are in the network. Remember, Samantha starts outside the house (odd point) and ends up outside (odd point).

SP 4 Since Elmer must start and finish at point B, it follows that all the points must become even. Make all the points even by adding lines. These lines represent roads Elmer must travel twice. Choose the lines you add carefully so that you add a minimum distance to Elmer's route.

4.5 B1 The measure of the angle opposite the side of length a is $180 - (70 + 35) = 75$. Since $75 > 70 > 35$ then $a >$ –?– $>$ –?–.

4.6 1 $m\angle H + m\angle O = 158$ (since $180 - 22 = 158$); but, since $m\angle H = m\angle O$, each angle measures half of 158.

4.6 7 Find this triangle in the figure:

56°
a

To find a, solve the equation: $a + a + 56 = 180$.

4.7 A6 Since the little triangle is isosceles, the two base angles are equal in measure. Since the lines are parallel, each base angle of the isosceles triangle measures 67. Therefore, $z + 67 + 67 = 180$. Solve for z.

4.8 3 Construct $\angle A$ and $\angle B$ at both ends of $\overline{AB}$. Construct $\overline{BC}$. Construct a line through C parallel to $\overline{AB}$.

4.8 4 Sketch the trapezoid before you construct it. Then mark the given information by drawing thicker lines.

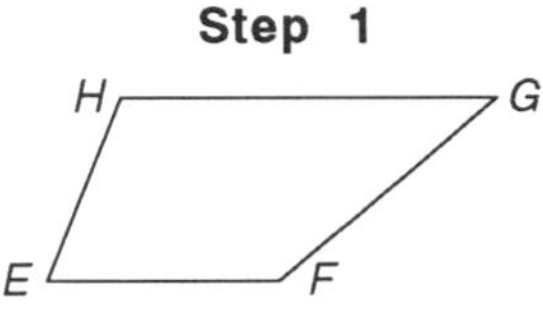

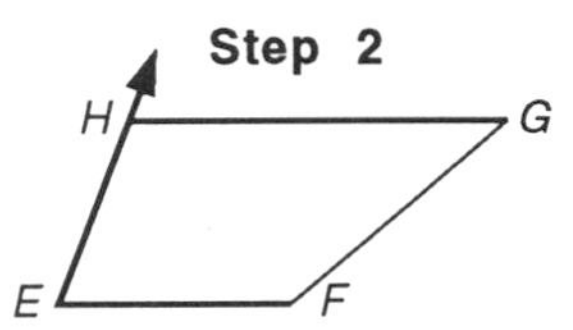

Your sketch tells you to start by constructing $\overline{EF}$. Then construct $\angle E$. Next, mark off $\overline{EH}$ on the ray. Finally, construct $\overline{HG} \parallel \overline{EF}$.

4.9 A1 If $RA = 20$, then $PO = \frac{1}{2}(RA) =$ –?–. If $PR = 8$, the $PT =$ –?–. If $OA = 10$, then $TO =$ –?–. Therefore, the perimeter $PO + OT + PT =$ –?–.

4.9 A3 Each side of the center triangle is parallel to the opposite side of the large triangle. Mark all the equal corresponding angles. This tells you that all four small triangles have the same three angles.

4.10 A4 $x - 4 = 17$. Solve for x. Find the value of $x + 3$. Then find the perimeter.

4.10 A6 If $VF = 36$, then $VN =$ –?–. If $EI = 42$, then $NI =$ –?–. If $EF = 24$, then $VI =$ –?–. Therefore, the perimeter of ΔNVI is $VN + NI + VI$.

4.10 B2 Since you are given the two diagonals, you should ask yourself, "What do I know about the diagonals of a parallelogram?" You know that the diagonals bisect each other. Therefore, start by constructing one of the diagonals ($\overline{MN}$), and then find its midpoint. Find the midpoint of the second diagonal ($\overline{EO}$) and swing a circle with a radius half the length of this diagonal centered on the midpoint of $\overline{MN}$. This is the locus of points where the endpoints E and O must be. Finally, measure side $\overline{MO}$ and, well, you can finish from here.

4.11 A1 It is true for a rectangle, but it is not true for all parallelograms. Draw a long skinny parallelogram. Clearly the two diagonals are not equal in length. This is a counter example. You should try to find a counter-example for each exercise in this exercise set.

4.11 B3 $\angle P$ and $\angle PEA$ are supplementary. Therefore, $48 + (y + 95) = 180$.

4.11 C1 You are given a diagonal of a rhombus. What do you know about the diagonals of a rhombus? They bisect each angle of the rhombus. So, construct the given angle and bisect it. Mark off the diagonal on the angle bisector. What else do you know about the diagonals of a rhombus? They are perpendicular bisectors of each other. So, construct the other diagonal by constructing the perpendicular bisector of the given diagonal.

4.11 C2 You are given a diagonal of a rectangle. What do you know about the diagonals of a rectangle? They are equal in length and they bisect each other.

4.11 C3 The diagonals of a square are equal in length (a square is a rectangle). And the diagonals of a square are perpendicular bisectors of each other (a square is a rhombus).

4.12 A1 (3, 0) and (3, 2) are two possible solutions. You find another.

4.12 A3 There are six possible solutions. (1, 5) and (5, -3) are two possible solutions. Can you find two more?

4.12 A6 You need to find a point D so that $\overline{AD} \parallel \overline{BC}$. How is $\overline{BC}$ slanted? It goes up how much and over how much from B to C?

4.12 B6 One of them is (3, 2).

Chapter 5

5.1 A1 b is too big. In c, the dot is in the wrong position.

5.1 C1 $\Delta DOG \cong \Delta RAT$

5.1 C6 If $\overline{PR} \cong \overline{TK}$ and $TK = 34$, then $PR = 34$.

5.2 1 $\Delta TAN \cong \Delta BOD$

5.2 10 ASS is not a congruence shortcut.

5.2 13 You do not know whether any angles are congruent. You only know that two sides in one triangle are congruent to two sides in the other triangle.

5.3 1 SAS

5.3 13 $\angle CGH \cong \angle NGI$ by the Vertical Angles Conjecture. $\overline{HG} \cong \overline{IG}$ by the Converse of the Isosceles Triangle Conjecture. Why, then, is $\Delta CGH \cong \Delta NGI$?

5.4 A4 SAA

5.4 A7 Cannot be determined (AAS does not correspond to ASA.)

5.4 A9 If you separate the two triangles, it is easier to see the congruence.

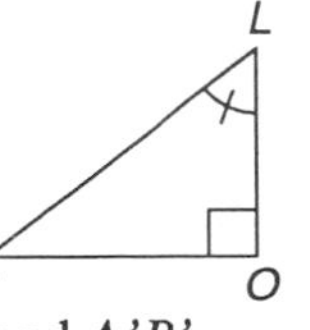

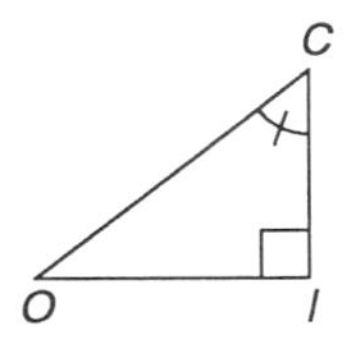

5.4 B4 Start with the two different segments AB and $A'B'$.

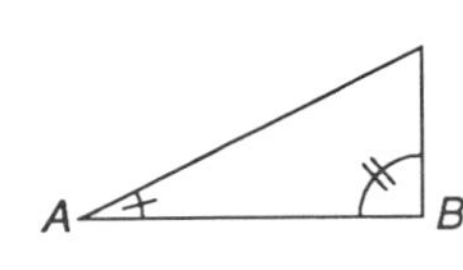

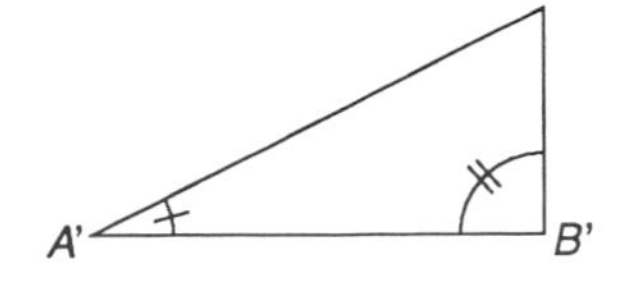

5.4 B5 Start with two congruent segments $\overline{AB} \cong \overline{A'B'}$. Construct $\angle A \cong \angle A'$. Select a length for the third side $\overline{BC} \cong \overline{B'C'}$ so that arcs with radius BC (and $B'C'$) will intersect one side of $\angle A$ (and $\angle A'$) in two different points.

SP 1

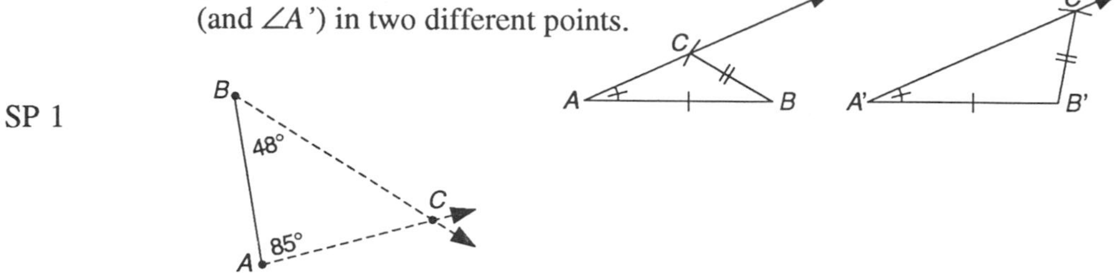

5.5 1 $\Delta RCE \cong \Delta RCA$ by SSS. Therefore, $\angle A \cong \angle E$ by CPCTC.

5.5 3 If $\overline{SA} \parallel \overline{NE}$, then $\angle ENS \cong \angle ASN$ (AIA Conjecture). If $\overline{SA} \parallel \overline{NA}$, then $\angle NSE \cong \angle ANS$ (AIA Conjecture). $SN \cong SN$ (same segment). Therefore, $\Delta SAN \cong \Delta NES$ by —?—.

5.5 5 "Just because they look it," is not a good enough reason!

5.5 8 Construct $\overline{FU}$.

5.5 9 Construct $\overline{UT}$.

5.6 D1 You can show that $\overline{RT} \cong \overline{SA}$ if you can show $\Delta RET \cong \Delta AES$. What is necessary to show that triangles are congruent?

5.6 C6 ASA

CA 6 There is a connection between the first and third inputs.

5.7 A1 If $AC = BC = 18$, then $AB = 48 - 36 = 12$.

If $AB = 12$ and $AD = \frac{1}{2}AB$, then $AD =$ –?–.

5.7 B1 $\overline{AD} \cong \overline{BD}$

5.7 B2 CPCTC

5.8 3 Make a list of all the possible triples formed by taking one length from each of the three sets. For example: (3, 3, 3); (3, 3, 12); (3, 6, 3); (3, 6, 12); and so on. Then find which sets of triples satisfy the triangle inequality conjecture.

5.8 8

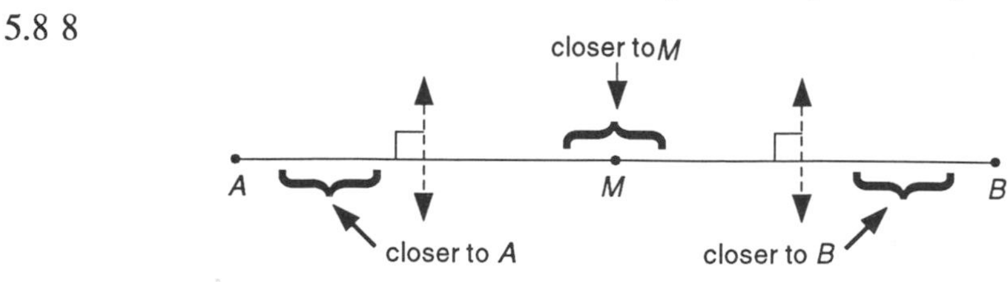

5.8 9 Organize your counting. For example:

ABC ACD ADE AEF BCD BDE BEF
ABD ACE ADF BCE BDF
ABE ACF BCF
ABF

And so on . . .

5.9 C1 SAS

5.9 C4 Since $\overline{LA} \parallel \overline{TR}$, $\angle A \cong \angle T$.

5.9 C7 Nothing is marked to indicate that $\overline{OE}$ is an angle bisector, median, or altitude. Therefore you cannot determine whether or not any triangles are congruent.

5.9 C9 If you separate the two triangles it is easier to see the congruence.

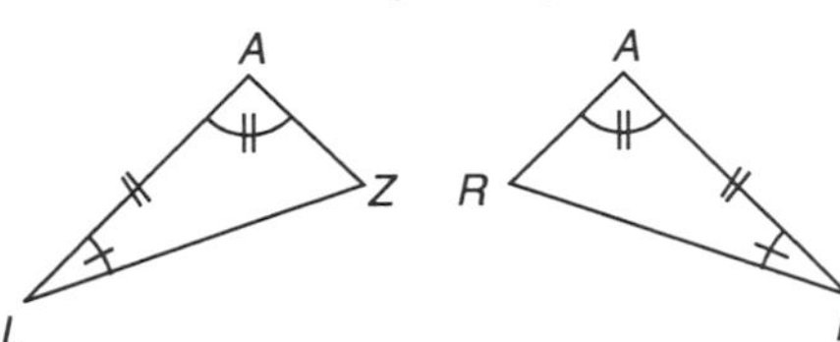

Chapter 6

6.1 A5 *An **inscribed angle** is an angle whose sides are —?— of a circle and whose vertex is a point on the circle.*

6.1 C4 Since $PAQB$ is a rhombus, $\overline{AB}$ and $\overline{PQ}$ are perpendicular bisectors of each other.

6.2 A3 Since the three chords are congruent, their three arcs are congruent. Therefore, $x + x + x + 72 = 360$.

6.3 A1 $130 + 90 + w + 90 = 360$

6.3 A2 Since the triangle is isosceles, $x + x + 70 = 180$.

6.3 B1 From the Tangent Conjecture you know that the tangent is perpendicular to the radius at the point of tangency. Therefore, to construct a tangent through a point on a circle, you perform the following steps:

Step 1: Construct a large circle. Label the center O.

Step 2: Choose a point on the circle and label it P.

Step 3: Construct $\overline{OP}$.

Step 4: Construct the perpendicular to $\overline{OP}$ through P.

6.4 3 $c + 120 = (2)(95)$

6.4 6 Draw the radius to the tangent. You get a right triangle. Find the remaining angle (central angle). The arc (h) has the same measure as the central angle.

6.4 9

$$120 + p + 98 + p = 360$$
$$98 + q + q = 180$$

6.4 11 The measure of each angle is half the intercepted arc. However, all five arcs add to 360.

6.4 12 $a = (1/2)(70)$
$b = (1/2)(80)$
$y = a + b$

6.5 4 $C = \pi D = \pi(5\pi) = 5\pi^2$

6.5 7 If the perimeter of the square is 24, then each side is 6. If each side is 6, then the diameter of the circle is 6. If the diameter is 6, the circumference is –?–.

6.5 12 $C = 2\pi r$
$44 = (2)(22/7)r$

6.6 3 $\text{speed} = \dfrac{\text{distance}}{\text{time}} = \dfrac{\text{circumference}}{12 \text{ hours}}$

$\text{speed} = \dfrac{(2)(22/7)(2000 + 6400) \text{ km}}{12 \text{ hours}}$

6.7 A3 $m\widehat{BIG} = 360 - 150 = 210$
$210/360 = 7/12$
Arc length of $\widehat{BIG}$ is $(7/12)\pi(24)$.

6.7 A6 $m\widehat{SO} = 180 - 100$
$80/360 = 2/9$
Arc length of $\widehat{SO}$ is $(2/9)\pi(18)$.

6.7 A9 If $\overleftrightarrow{PA} \parallel \overleftrightarrow{RE}$, then $m\widehat{AR} = m\widehat{PE}$. If $m\widehat{AR} = m\widehat{PE}$, then $m\widehat{AR} = \frac{1}{2}[360 - (146 + 70)] = \frac{1}{2}[144] = 72$. But, $\frac{72}{360} = \frac{1}{5}$. Therefore, the length of $\widehat{AR}$ is $\frac{1}{5}$ of the circumference. So $40\pi = \frac{1}{5}(2\pi r)$.

6.7 B3 If you draw a radius from the center of the circle that contains the arc to each endpoint of the bridge, you create an equilateral triangle. Therefore, the radius for the arc is 180. Therefore, the arc length is $(1/6)[2\pi(180)]$.

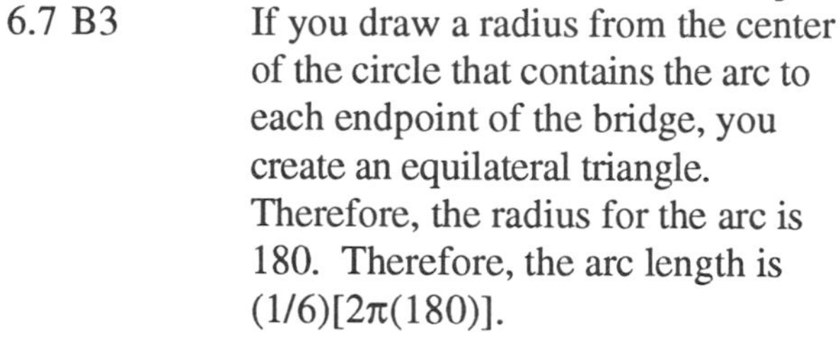

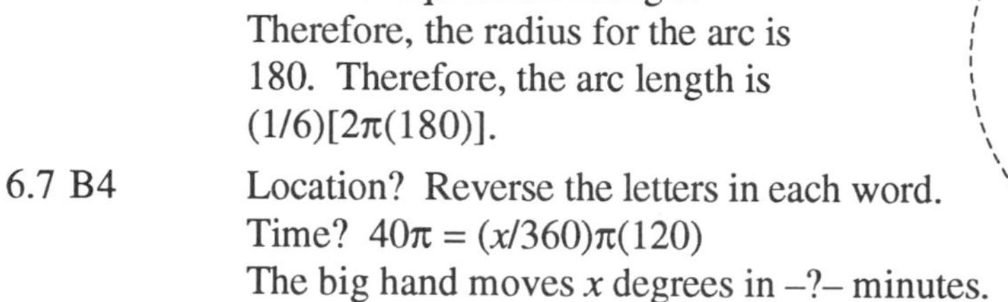

6.7 B4 Location? Reverse the letters in each word.
Time? $40\pi = (x/360)\pi(120)$
The big hand moves x degrees in –?– minutes.

6.8 C2 Draw in the radius to the point of tangency. Find the central angle.

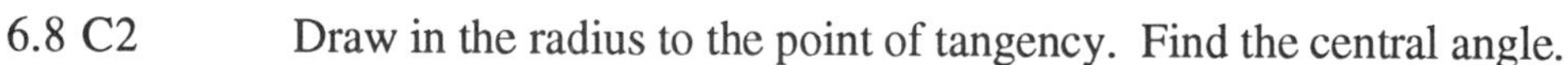

6.8 C5 $a = (1/2)(64)$
$b = (1/2)(60)$
$a + b + e = 180$

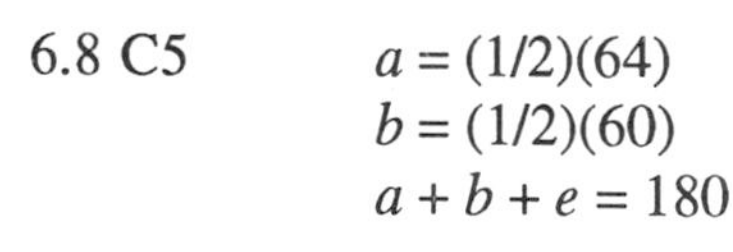

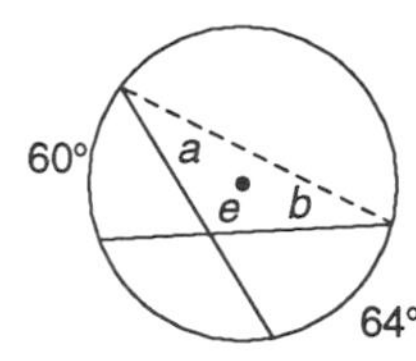

6.8 C6 $a = (1/2)f$
$b = (1/2)(118)$
$a + b + 88 = 180$

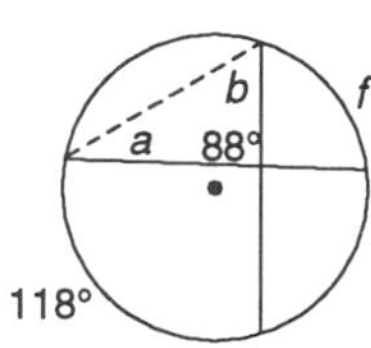

6.8 C10 $C = 2\pi r$
$132 = (2)(22/7)r$

6.8 C11 Arc length is $\frac{100}{360}C$ or $\frac{5}{18}(2\pi)(27)$ or –?–.

6.8 C12 $a = (1/2)(60)$
$b = a + 50$
$c = 2b$

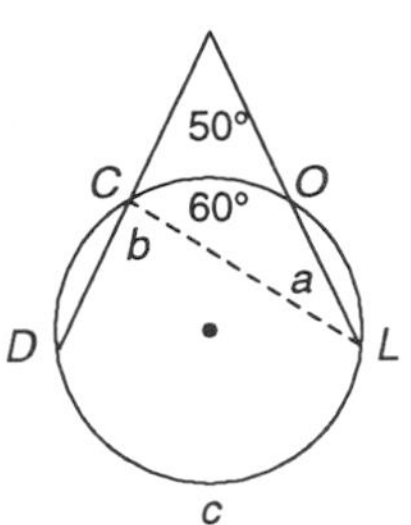

6.8 D1 $C = \pi D = \pi(50) \approx (\frac{22}{7})(50) \approx$ –?–

6.8 D2 Circumference of anchor spool is about $(\frac{22}{7})(1)$.
Length of chain is about $42(\frac{22}{7})(1) \approx$ –?–.

6.8 D4 Since there are 150 knights and each knight needs 2', the circumference of the table should be (150)(2') = 300'.
$C = \pi D$
$300 = \frac{22}{7}D$

6.8 D5 One degree of arc is 1/360 of the circumference. Therefore, one minute of arc is 1/60 of 1/360 of the circumference.

Find one nautical mile on a "polar" great circle:
1 polar nautical mile =
$\frac{1}{60}$ of $\frac{1}{360}$ of the circumference $= \frac{1}{21600} \cdot C = \frac{2\pi R_{polar}}{21600} = \frac{2(3.14)(6357)}{21600}$

Find one nautical mile along an "equatorial" great circle the same way, using the equatorial radius.

CPS One possible basic design. This design lacks many of the details that you will need in your design.

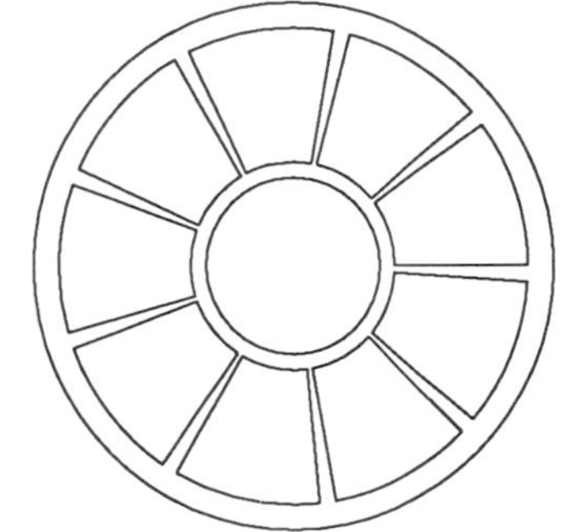

Chapter 7

7.1 4 Use Conjecture 41: *The perpendicular bisector of a chord of a circle passes through the center of the circle.*

7.2 9

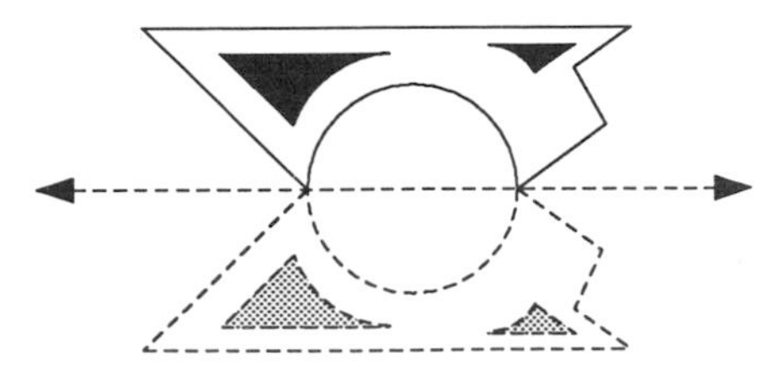

7.2 11 Twice.

7.3 B13 In this chapter you discovered that if a figure is reflected twice about two intersecting reflection lines, then the two isometries is equivalent to one rotation with the angle of rotation equal to twice the angle between the reflection lines. In the case of vertical and horizontal reflection lines, the measure of the angle between them is 90 and thus the two reflections are equivalent to a half-turn (180° rotation). Therefore, if a figure has both horizontal and vertical axes of symmetry, then it must have point symmetry.

7.4 13 Place your pencil tip at point *A*. How many different rotational symmetries at this point? Try this at points *B* and *C*.

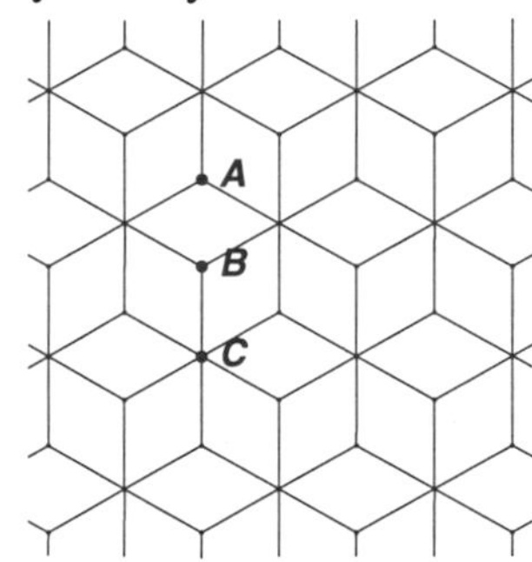

Chapter 8

8.1 A4 Pair off pieces of squares along the boundary that make up a full square.

8.1 B5 $96 = 12b$

8.1 C8 $2508 = 44b$, so the perimeter is $2b + (2)(48)$.

8.2 A3 $A = (1/2)(11)(9) =$ –?–

8.2 A5 $31.5 = (.5)(9)h$

8.2 A9 Every side of the triangle can be a base. For every base of the triangle, there is a corresponding height. Therefore, the area of the triangle can be calculated three different ways: $(1/2)(5)y = (1/2)(12)x = (1/2)(6)(8)$.

8.2 B3 $A = (1/2)(18)(22 + 37) =$ –?–

8.2 B4
$$180 = (1/2)(9)(24 + b)$$
$$360 = (9)(24 + b)$$
$$40 = (24 + b)$$

8.2 C1

b_1 I II h b_2 => b_1 I h + II h

area $\Delta \text{I} = \frac{1}{2}b_1h$ b_2 area $\Delta \text{II} = \frac{1}{2}b_2h$

Therefore, the area of a trapezoid is given by the formula $A = \frac{1}{2}b_1h + \frac{1}{2}b_2h$ or $A = \frac{1}{2}h\,(b_1 + b_2)$.

8.2 C2

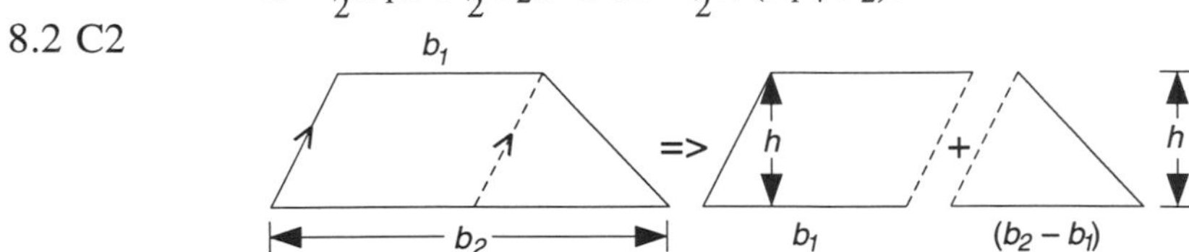

8.2 C3 Extend the non-parallel sides until they intersect, or construct a line parallel to one of the non-parallel sides through the vertex of one of the base angles.

8.3 1 Divide the figure into three rectangles.

8.3 2 Use your compass to find which has the greatest perimeter. To find the area you will have to draw an altitude in each parallelogram.

8.3 3 Divide the figure into two triangles.

8.3 5 You could select one vertex and draw in all the diagonals from that vertex, but there would be six different size triangles (a lot of work!). If you found the center of the circumscribed circle and drew in the radius to each vertex, all the triangles would be identical. Then, to find the area of the dodecagon, just find the area of one triangle and multiple by twelve.

8.3 6 You can use every other vertical line of the paper to create rectangles; then, you can find the sum of the areas of all the rectangles. Or, you could also divide the region into trapezoids and find the sum of the areas of all the trapezoids.

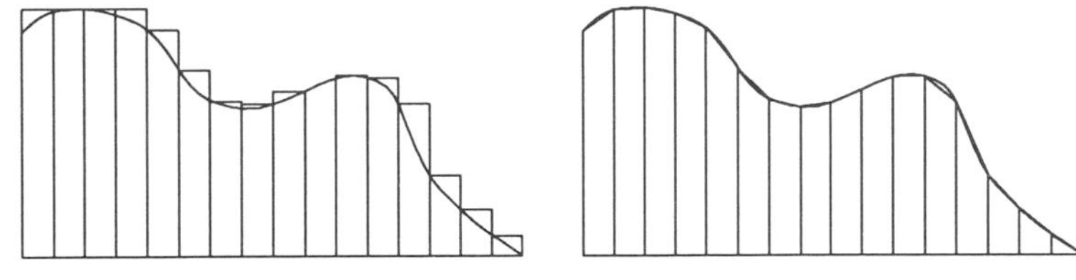

8.4 2 $\text{Area}_{\text{frosting}} = \text{Area}_{\text{top}} + \text{Area}_{\text{four sides}}$
$= [400 \bullet 600] + [(2)(180 \bullet 400) + (2)(180 \bullet 600)]$

The number of liters of frosting needed is $\frac{\text{Area}_{\text{frosting}}}{1200}$.

8.4 3 The total cost is \$20 per square yard. \$20 per square yard is \$20/9 per sq. ft.
$\text{Area}_{\text{carpet}} = 17 \bullet 27 - (6 \bullet 10 + 7 \bullet 9)$

8.4 5 First find the area of all the rectangles in the first floor. Then find the area of the front and back triangles.

8.4 6 You can find the total area of the four sides below the roof line by multiplying the height to the roof by the perimeter of the floor. You can find the total area of the end pieces under the roof by doubling the areas of the trapezoid and triangle. (24)(30 + 40 + 30 + 40) + (2)[(1/2)(12)(30 + 12)] + 2[(1/2)(2.5)(12)] is the painted area. The shingle area is (40)(15 + 6.5 + 6.5 + 15) or –?–.

To find the cost for paint, divide the painted area by 250 and round up to a whole number of gallons; then, multiple the number of gallons by \$25.

To find the cost for shingles, divide the shingle area by 100 and round up to a whole number of bundles; then, multiply the number of bundles by \$65.

8.5 11 Find the area of the large hexagon and subtract the area of the small hexagon. Since it is a regular hexagon, the distance from the center to each vertex is the same as the length of each side.

8.5 12 As the number of sides increases, the perimeter of the polygon approaches the circumference of the circumscribed circle ($P \to C$). As the number of sides increases, the apothem of the regular polygon approaches the radius of the circumscribed circle ($a \to r$). As the number of sides increases, the area of the polygon approaches the area of the circumscribed circle ($\text{Area}_{\text{polygon}} \to \text{Area}_{\text{circle}}$). Therefore, $(1/2)aP \to (1/2)rC$.

8.6 A3 $A = \pi r^2 \approx (22/7)(1/2)^2 \approx (22/7)(1/4) \approx$ –?–

8.6 A7

$$A = \pi r^2$$
$$3\pi = \pi r^2$$
$$3 = r^2$$
$$r = \sqrt{3}$$

8.6 B3 Area of the small circle is $\pi(3^2)$ or 9π.
Area of the big circle is $\pi(6^2)$ or 36π.
36π is –?– times larger than 9π.

8.7 3 The shaded area is 120/360 or 1/3 of the area of the circle.

8.7 6 The shaded area is $(3/4)\pi(8^2) + (1/2)(8)(8)$ or –?–.

8.7 9 Since a sector of 60 degrees (one-sixth of the circle) is removed, the shaded area is $(5/6)[\pi(R^2 - r^2)]$.

8.7 10 $(1/3)\pi r^2 = 12\pi$

8.7 11 The area of each small shaded piece is $(1/4)\pi r^2 - (1/2)r^2$, so the area of the entire shaded region is $(4)[(1/4)\pi r^2 - (1/2)r^2]$. But, there is an easier way . . . .

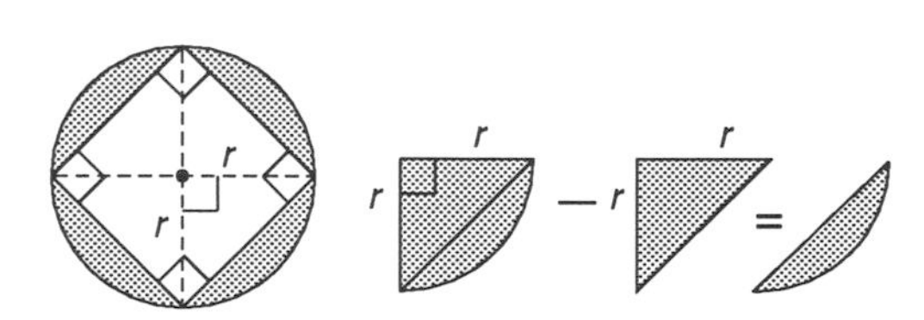

8.7 12 $32\pi = \pi(9^2 - r^2)$

8.7 13 $\frac{x}{360}(\pi 24^2) = 120\pi$

8.7 14 $(\frac{m\angle ABC}{360})(\pi 10^2 - \pi 8^2) = 10\pi$

8.7 21 The area of the square equals the area of four right triangles. Therefore, the shaded area equals the area of the circle less the area of the four right triangles or $\pi 6^2 - (4)(1/2)(6)(6)$ or –?–.

8.7 23 The area of one shaded piece is the area of the square less the area of one quarter of a circle.

CA 7 Try listing your sequences in order under the directional headings E-S-W-N. For instance, the order-6 right spirolateral created by RT.ANY [1 1 2 3 5 8] would list as shown on the right.)

E	S	W	N
1	1	2	3
5	8	1	1
2	3	5	8

8.8 A6

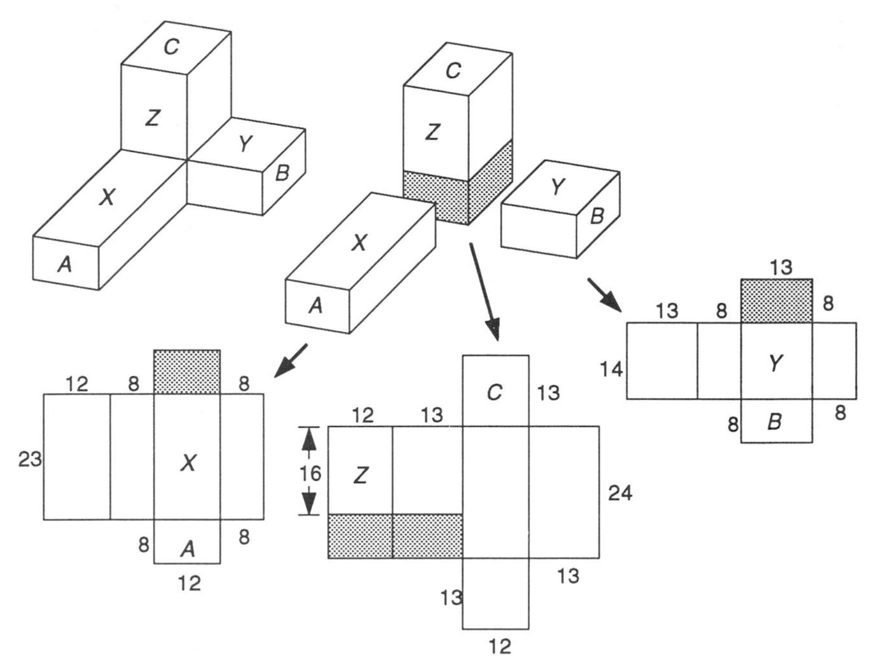

8.8 A7 The total surface area is equal to the area of the regular pentagonal base and the area of the five isosceles triangles.

$$\text{Surface area} = (1/2)nas + (5)[(1/2)bh]$$
$$= (1/2)(5)(11)(16) + (5)[(1/2)(16)(15)]$$
$$= –?–$$

8.8 A8

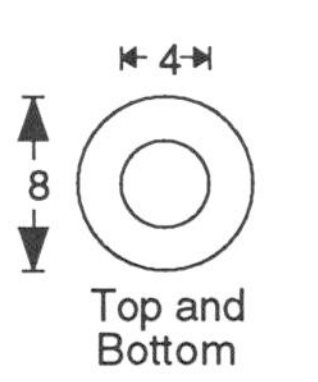

$C = \pi(8)$

9

Outer Surface

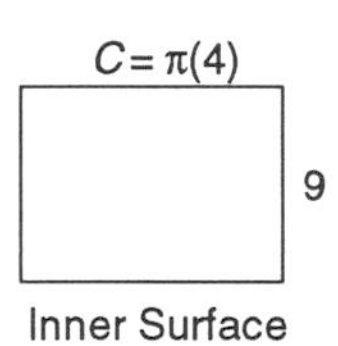

8.8 A9

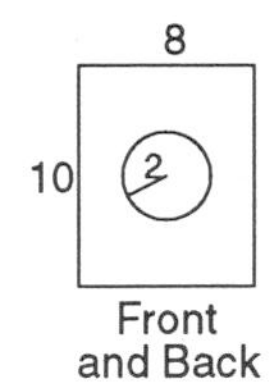

8 4 8 4 10 4 10 4

Sides

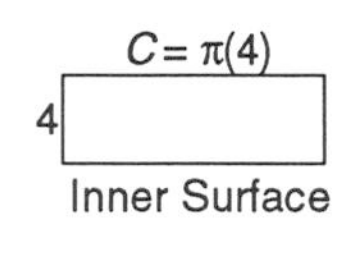

8.8 B2

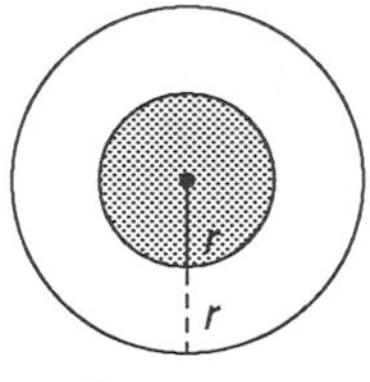

8.8 B3 $\frac{\pi 4^2}{\pi 20^2}$

8.9 A4 — The amount of pizza per dollar is the same whether you calculate it for one pizza or 8 pizzas. The size pizza that gives the most area per dollar can be quickly found by comparing: $\frac{25}{10.5}$; $\frac{36}{12.75}$; $\frac{49}{15.5}$; $\frac{64}{19.5}$. (You don't need π to compare.)

8.9 B3 — The shaded area is (12)(15 + 25) + (8)(20). The area of the vertical sides is (1/2)(12)(17 + 26) + (2)[(1/2)(12)(16)] + (12)(12) + (1/2)(12)(34 + 25).

To find the cost of the shingles, divide the shaded area by 100 and round up to a whole number of bundles; then, multiply the number of boxes by $35.

To find the cost of the wood stain, divide the vertical area by 150 and round up to a whole number of gallons; then, multiply the number of gallons by $15.

8.10 C1 — First we constructed an altitude from the vertex of an obtuse angle to the base. Then we cut off the right triangle and moved it to the opposite side forming a rectangle. Since the area hasn't changed, the area of the parallelogram is the same as the area of the rectangle. Since the area of the rectangle is given by the formula $A = bh$, the area of the parallelogram is given by $A = bh$ also.

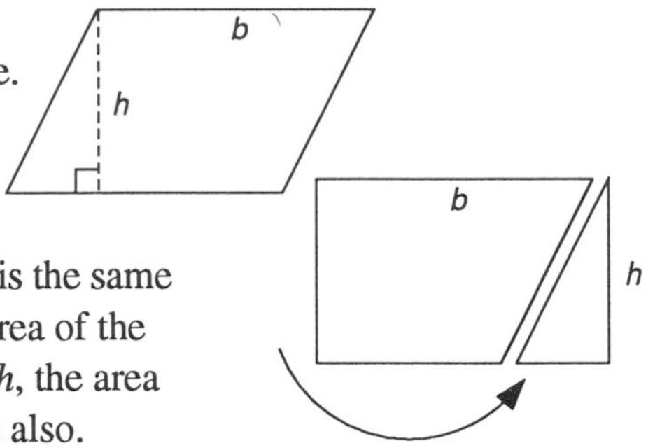

8.10 D8

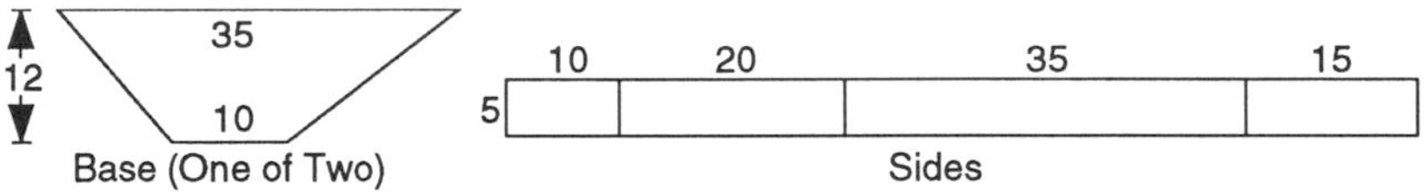

35, 12, 10

Base (One of Two)

10, 20, 35, 15, 5

Sides

8.10 D10 — Copy the diagram. Draw in segments connecting the midpoints of the opposite sides. These two segments divide the shaded area into four parts. Look carefully at those parts. How do they compare to the four non-shaded parts?

8.10 D13 — $A = \pi(5.5)^2 - \pi(4^2)$

8.10 D14 — Use trial and error to reach your solution.

8.10 D15

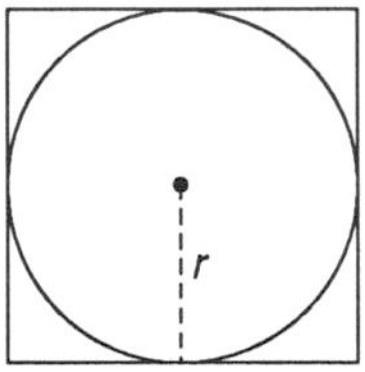

$\text{Area}_{\text{circle}} = \pi r^2$

$\text{Area}_{\text{square}} = (2r)^2 = 4r^2$

$\frac{\text{Area}_{\text{circle}}}{\text{Area}_{\text{square}}} = \text{–?–}$

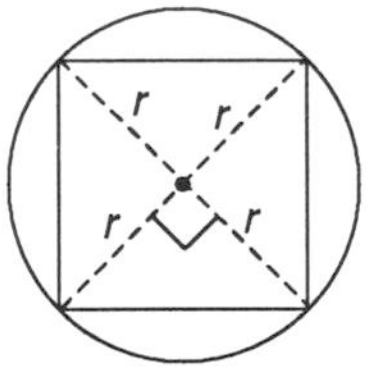

$\text{Area}_{\text{circle}} = \pi r^2$

$\text{Area}_{\text{square}} = (4)(1/2)r^2 = 2r^2$

$\frac{\text{Area}_{\text{square}}}{\text{Area}_{\text{circle}}} = \text{–?–}$

CPS 5 — Ertha's translation tells us how they discovered this formula. They noticed that the sector of an annulus looks like a trapezoid on a ball, therefore they used the formula for the area of a trapezoid to find the area of a sector of an annulus. Therefore, the area formula for the sector of an annulus is $A = (1/2)s(m + n)$ where s is the distance between the two arcs, and m and n are the length of the two arcs. This sounds convincing but it takes clever algebra to logically demonstrate this. To find the area of a sector of an annulus, determine the area of the larger sector less the area of the smaller sector. The area of the larger sector is $(\theta/360)\pi R^2$. The area of the smaller sector is $(\theta/360)\pi r^2$. The arc length $m = (\theta/360)2\pi R$. The arc length $n = (\theta/360)2\pi r$. The rest is up to you.

Chapter 9

9.1 A1 — Simplifying a square root means you are taking the square root of any square factors in the number. Your first task, therefore, is to find the factors. You could break the number down to its prime factors. First (if the number is even) divide by 2, then divide the quotient by 2, and so on until there are no more even factors. Next divide the quotient by 3, and so on until all the factors of 3 are found. Next divide the quotient by 5 (the next prime) and find all factors of 5. Then 7, and so on, until all the prime factors of the original number are found. Next group the like factors in pairs. Each pair of factors creates a square. Finally, take the square root of each square. For example: $12 = 2 \cdot 6 = 2 \cdot 2 \cdot 3 = (2 \cdot 2)(3) = (4)(3)$. Therefore, $\sqrt{12} = \sqrt{4} \cdot \sqrt{3} = 2\sqrt{3}$. Another approach to simplifying a square root is to break the number down to a product of squares instead of primes. This approach is often faster.

9.1 A4 — Look for the largest square factor. $32 = 16 \cdot 2$. So, $\sqrt{32} = \sqrt{16} \cdot \sqrt{2} = 4\sqrt{2}$.

9.1 A14 — Be creative in your factoring. Since 185 ends in 5, it is divisible by 5. $185 = 5 \cdot 37$. Since both factors are prime, there are no squares as factors. Therefore, $\sqrt{185}$ cannot be simplified.

9.1 A15 — Be flexible! 490 ends in a zero, therefore, it is divisible by 10. $490 = 49 \cdot 10$. Aha, 49 is a square. Therefore, $\sqrt{490} = \sqrt{49} \cdot \sqrt{10} = 7\sqrt{10}$. Since there are no squares as factors in 10, you are finished.

9.1 A17 — There are many ways to break large numbers into factors. Since 720 is even, you could start dividing by 4 (the first even square). Or, since it ends in a zero, you could start by dividing by 10 (breaking it into $72 \cdot 10$). This probably makes it easier to see other factors. $720 = 72 \cdot 10 = (8 \cdot 9)(2 \cdot 5)$. You may now regroup to make squares. $720 = 16 \cdot 9 \cdot 5$. Therefore, $\sqrt{720} = \sqrt{16} \cdot \sqrt{9} \cdot \sqrt{5} = 4 \cdot 3\sqrt{5} = 12\sqrt{5}$. (Of course if you had recognized that $720 = 144 \cdot 5$ at the start, you would have been done even faster.)

9.1 A21 — $(3\sqrt{2})^2 = (3\sqrt{2})(3\sqrt{2}) = 9 \cdot 2 = 18$

9.1 A27 — $\sqrt{\dfrac{5}{12}} = \dfrac{\sqrt{5}}{\sqrt{12}} = \dfrac{\sqrt{5}\sqrt{3}}{\sqrt{12}\sqrt{3}} = \dfrac{\sqrt{15}}{\sqrt{36}} = \dfrac{\sqrt{15}}{6}$

Why did you multiply numerator and denominator by $\sqrt{3}$? To make the denominator the square root of a square. That way the denominator can be simplified to a rational number.

9.2 1 $5^2 + 12^2 = 25 + 144 = 169$. Therefore, $c = \sqrt{169} = 13$.

9.2 6 $6^2 + 6^2 = 72$. Therefore, $c = \sqrt{72} = 6\sqrt{2}$.

9.2 8 $7^2 + b^2 = 25^2$. $b^2 = 625 - 49$. Therefore, $b = \sqrt{625 - 49} = \sqrt{576} = 24$.

9.2 11 $s^2 + s^2 = 25$. $2s^2 = 25$. $s^2 = \frac{25}{2}$. Therefore, $s = \sqrt{\frac{25}{2}} = \frac{\sqrt{25}}{\sqrt{2}} = \frac{\sqrt{25 \cdot 2}}{\sqrt{2 \cdot 2}} = \frac{5\sqrt{2}}{2}$.

9.2 12 $(8\sqrt{2})^2 + x^2 = (9\sqrt{3})^2$. $64 \cdot 2 + x^2 = 81 \cdot 3$.
Therefore, $x = \sqrt{243 - 128} = \sqrt{115}$.

9.3 A1 $8^2 + 15^2 = 64 + 225 = 289 = 17^2$. Since 8-15-17 fits in the Pythagorean formula, the triangle is a right triangle.

9.3 B2 Find the other leg: $b^2 + 5^2 = 13^2$. Then $A = \frac{1}{2}(5)(b)$.

9.3 B3 Since 9-12-15 fits in the Pythagorean formula, the lengths form a right triangle. Therefore, the angle is a right angle and the parallelogram is a rectangle.

9.3 B6

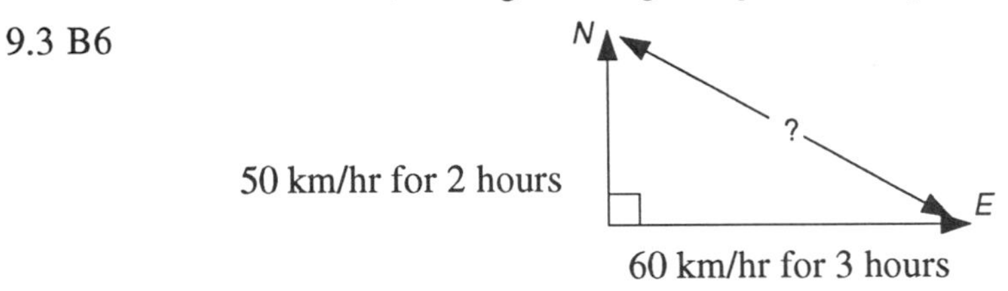

9.4 5 Let s represent the length of the side of the square. Then $s^2 + s^2 = 32^2$. $2s^2 = 1024$. Therefore, $s^2 = 512$. But, the area of the square is s^2. Therefore, the area of the square is 512 square meters.

9.4 12 Let x represent the distance from the crack to the top of the pole. Then $9^2 + 12^2 = x^2$. $x = 15$. Therefore, the original height is $9 + x = 9 + 15 = 24$. The original height of the pole is 24 feet.

9.4 13

9.4 14

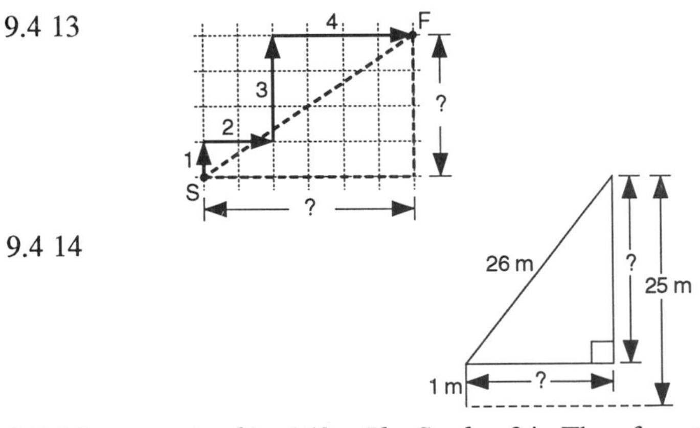

9.5 A5 $A = bh$. $168 = 7b$. So, $b = 24$. Therefore, the triangle is the familiar 7-24-?.

9.5 A8 7-?-25. The third side must be 24. If the diameter of the semicircle is 24, the radius is 12. Therefore, the area is $(1/2)\pi(12^2)$.

9.5 B1 9-12-15

9.5 C1

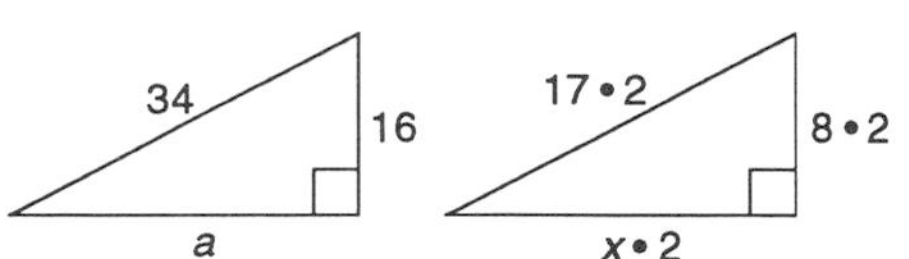

8-x-17
Aha!
8-15-17

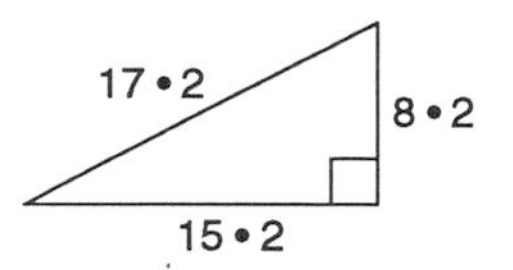

So $a = 30$.

9.5 C5 First find x from the triple: x-24-26. Then find e from the triple 6-e-x.

9.6 3 $(1/2)x^2 = 98$. Therefore, $x^2 = 196$. Let c be the hypotenuse; then, $x^2 + x^2 = c^2$. Therefore, $196 + 196 = c^2$. Use a calculator from here.

9.6 4

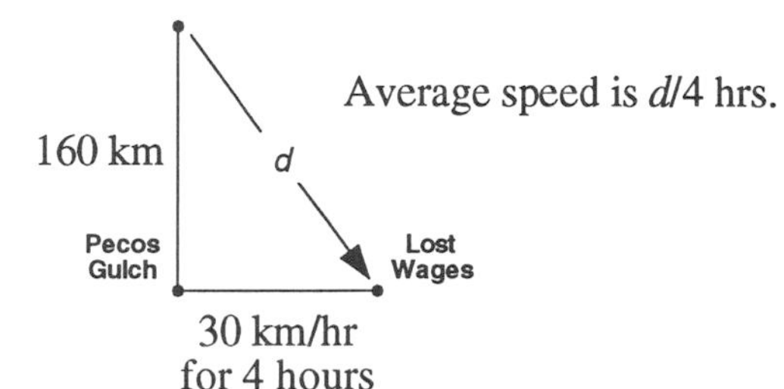

Average speed is d/4 hrs.

9.6 5

$$x^2 + 24^2 = (36 - x)^2$$
$$x^2 + 24^2 = 36^2 - 72x + x^2$$
$$24^2 = 36^2 - 72x$$

Solve for x.

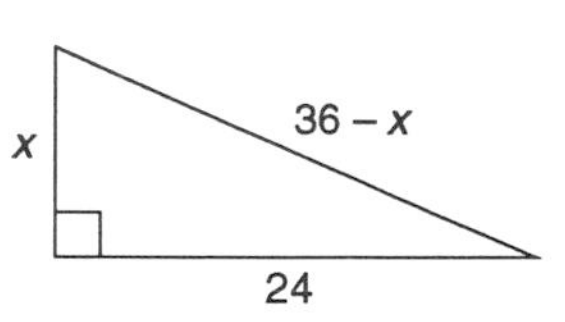

9.6 6 The distance x is called the space diagonal. It is the Pythagorean Theorem in three dimensions. First find y, the length of the diagonal in the bottom of the box. Use $24^2 + 30^2 = y^2$. Then find the space diagonal. Use $18^2 + y^2 = x^2$.

9.6 7

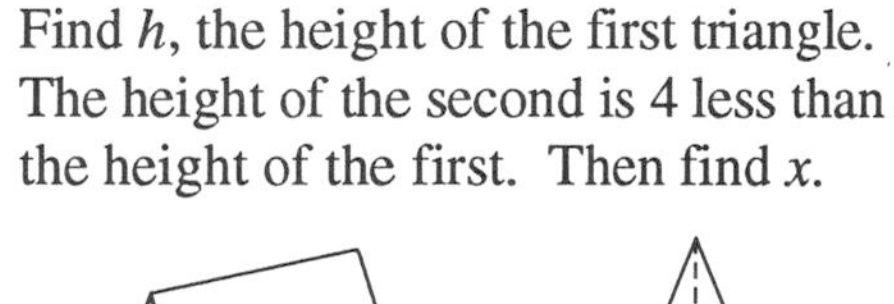

Find h, the height of the first triangle. The height of the second is 4 less than the height of the first. Then find x.

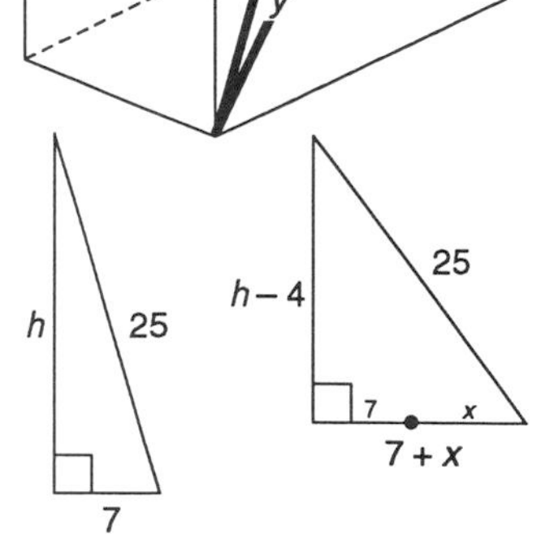

9.6 8

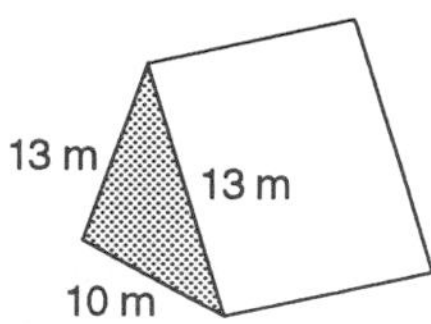

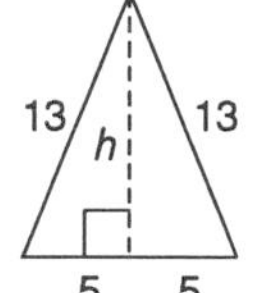

9.7 Investigation 9.7.1

1. $a^2 = 3^2 + 3^2 = 9 \cdot 2$. Therefore, $a = \sqrt{9 \cdot 2} = 3\sqrt{2}$.

9.7 Investigation 9.7.3

7. $12^2 = 6^2 + J^2$. $144 = 36 + J^2$. $108 = J^2$. Therefore, $J = \sqrt{108} = \sqrt{36 \cdot 3} = 6\sqrt{3}$.

9.7 2 $b\sqrt{2} = 13\sqrt{2}$. Therefore, $b = 13$.

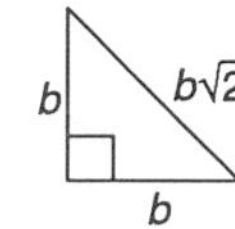

9.7 4 — Let s be the length of a leg of the right triangle. Then $s^2 + s^2 = (8\sqrt{2})^2$. Therefore, $2s^2 = 128$. But, the area of the triangle is $(1/2)s^2$.

9.7 7 — $a = 10$; $b = 5\sqrt{3}$

9.7 11 — $m = \sqrt{3} \bullet \sqrt{2} = \sqrt{6}$; $k = \dfrac{\sqrt{6}}{\sqrt{3}} = \sqrt{2}$

9.7 15

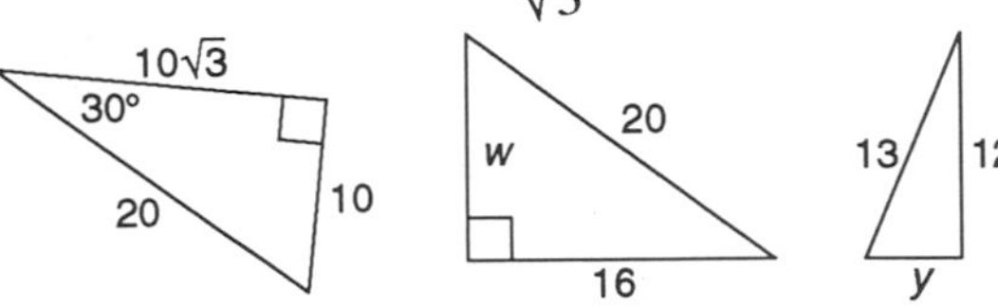

9.7 17 — Make a list of all triples, then find all Pythagorean triples.
(3,4,5), (3,4,9), (3,4,17) . . .
(3,8,5), (3,8,9), (3,8,17) . . .

9.8 A1 — Each median is $6\sqrt{3}$. Therefore, x is $(1/3)(6\sqrt{3})$ or $2\sqrt{3}$ and y is two-thirds of the median.

9.8 A2 — $x = \frac{1}{3}(18) = 6$

9.8 A3

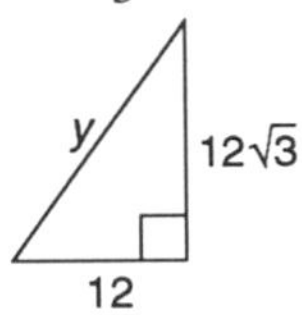

9.8 A4 — x is twice $\sqrt{3}$. One-third of the length of the altitude is $\sqrt{3}$ and, therefore, the height is $3\sqrt{3}$. Half the length of the side of the equilateral triangle is 3 and so y is 6.

9.8 A7 — Since the length of a side is 12, half the length is 6. Therefore, the height is $6\sqrt{3}$. The area is half the base times the height or $(6)(6\sqrt{3})$ or –?–.

9.8 A9 — Since two-thirds of the length of the median is $2\sqrt{3}$, then the length of the median (and altitude) is $(3/2)(2\sqrt{3}) = 3\sqrt{3}$. If the height is $3\sqrt{3}$, then half the length of a side is 3. $A = (3)(3\sqrt{3}) =$ –?–.

9.8 B1 — The medians (altitudes, etc.) of an equilateral triangle divide each other at the incenter/circumcenter/centroid. Therefore, the longer segment on the median (two-thirds as long as the median) is the radius of the circumscribed circle and the shorter segment on the median (one-third as long as the median) is the radius of the inscribed circle. If the length of a side is 6, then half the length is 3 and, thus, the length of the median is $3\sqrt{3}$. Therefore, the radius of the circumscribed circle is $(2/3)(3\sqrt{3})$ or $2\sqrt{3}$ and the radius of the inscribed circle is $(1/3)(3\sqrt{3})$ or $\sqrt{3}$. You're on your own from here.

9.8 B3 — Since the altitude is $9\sqrt{3}$, then the radius of the circumscribed circle is $(2/3)(9\sqrt{3})$ and the radius of the inscribed circle is $(1/3)(9\sqrt{3})$. Since the length of each altitude is $9\sqrt{3}$, then half the length of a side is 9. Therefore, the area of the triangle is –?–.

9.8 B4 — The area of an equilateral triangle is $(1/2)sh$. But, $h = (1/2)s\sqrt{3}$. Therefore, $A = (1/4)s^2\sqrt{3}$. But, $A = 25\sqrt{3}$. Therefore, $(1/4)s^2\sqrt{3} = 25\sqrt{3}$. Solve for s. The perimeter is $3s$.

9.8 B6 A regular hexagon can be divided into six equilateral triangles. The radius of the inscribed circle is the height of one of the six triangles. Therefore, half the length of a side of the equilateral triangle is $3\sqrt{3}$. Thus, the length of each side is $6\sqrt{3}$. The area of the regular hexagon can be found with the formula $A = (1/2)ap$ where a is the apothem (9) and p is the perimeter $((6)(6\sqrt{3}))$.

9.9 A4 

$x^2 = 4^2 + 4^2$

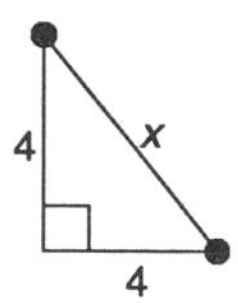

9.9 A7 $d^2 = 5^2 + 12^2$

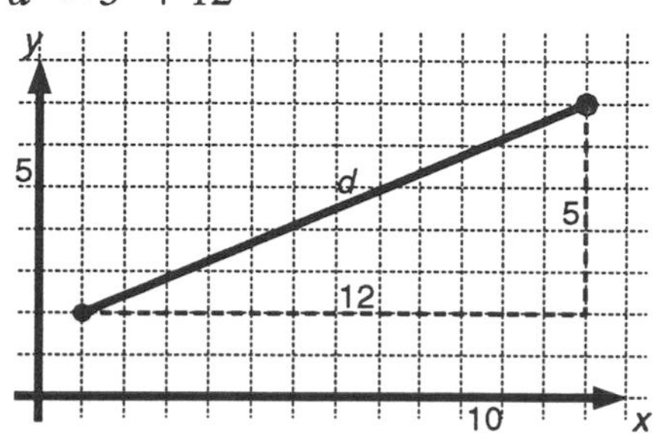

9.9 B1 $d^2 = (13 - 10)^2 + (20 - 16)^2$

9.9 B6 $GH = \sqrt{(18 - 2)^2 + (6 - 2)^2}$

$GI = \sqrt{(12 - 2)^2 + (12 - 6)^2}$

$HI = \sqrt{(18 - 12)^2 + (12 - 2)^2}$

To check if the triangle is a right triangle, see if GH, GI, and HI are Pythagorean triples or compare the slopes to see if two segments are perpendicular.

9.9 C1 Points A, B, G, D, E, and F divide the circle into six equal parts. If C is selected on $\widehat{AB}$, then ΔABC would be obtuse. If C is selected at D or E, then ΔABC would be right triangles. Therefore, if C is selected on $\widehat{BD}$ or $\widehat{AE}$, then ΔABC would be obtuse. If C is selected on $\widehat{DE}$, then ΔABC would be acute.

9.10 A1 $m\angle DOB = 180 - 105$ since two angles of the quadrilateral are right angles.

9.10 A3 When $\overline{TO}$ is constructed, it forms two 30-60 right triangles with the tangent segments as the longer legs. Therefore, the radius of the circle is the shorter leg in each 30-60 right triangle. Therefore, $r = 6$.

9.10 A4 The shaded area is $(2)[(1/2)(8)(8\sqrt{3}) - (1/6)\pi(8^2)]$ or –?–.

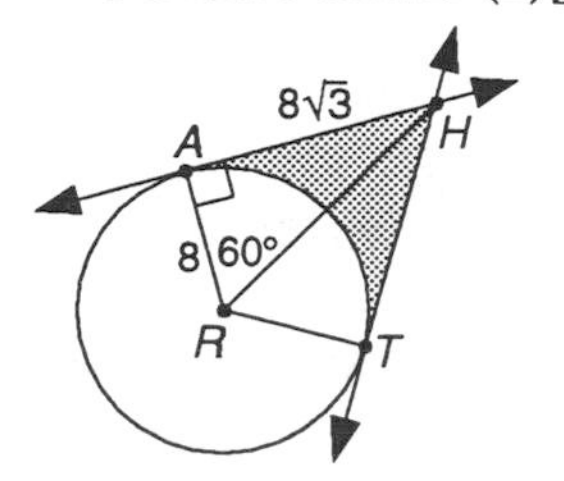

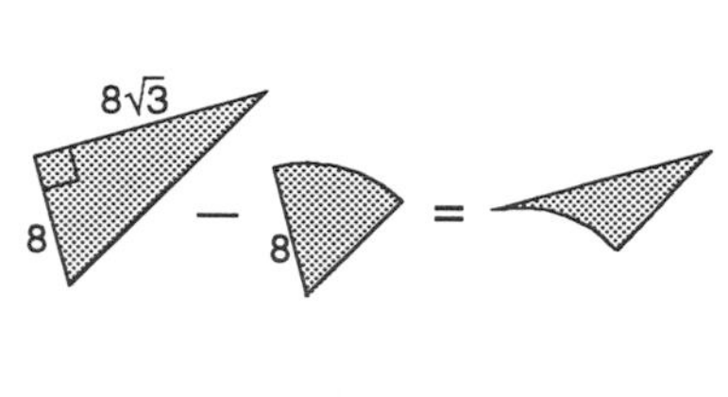

9.10 A6 The shaded area is $(1/3)\pi 8^2 - (1/2)(8\sqrt{3})(4)$ or –?–.

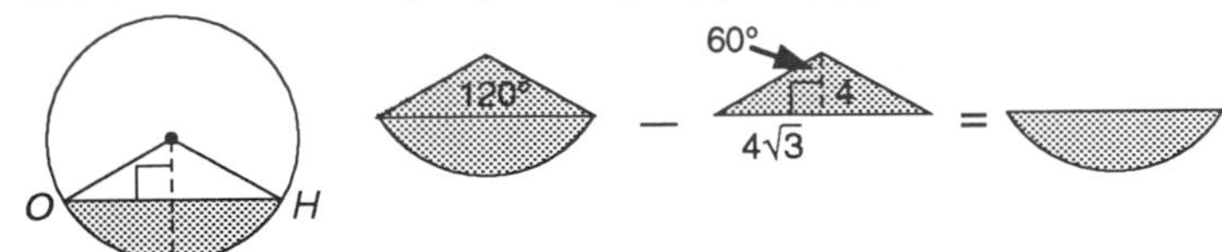

9.10 B1 When $\overline{OT}$ and $\overline{OA}$ are drawn, they form a right triangle with $OA = 15$ (length of hypotenuse) and $OT = 12$ (length of leg). Thus the length of the third side (half the chord) can be calculated.

9.10 B2 Using the same diagram (from 9.10 B1), draw in $\overline{OA}$ with length R and $\overline{OT}$ with length r. The area of the annulus is $\pi R^2 - \pi r^2$ or $\pi(R^2 - r^2)$.
But, $R^2 = r^2 + 18^2$, so, $R^2 - r^2 = 324$.

9.10 B3 The arc length of AC is $(80/360)[2\pi(9)] = (2/9)(18\pi) = 4\pi$.
Therefore, the circumference of the base of the cone is 4π. From this you can determine the radius of the base. The radius of the sector (9) becomes the slant height (the distance from the tip of the cone to the circumference of the base). The radius of the base, the slant height, and the height of the cone form a right triangle. Solve for the height.

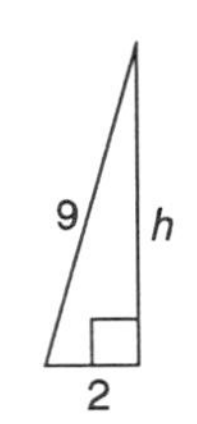

9.10 B4 $C = \pi D$, so, $336 \approx (3.14)D$. $D = x\sqrt{2}$, so, $x = \dfrac{D}{\sqrt{2}}$.

9.10 B5 The belt length is $(2/3)\pi(36) + (2/3)\pi(24) + (2)(18\sqrt{3}) + (2)(12\sqrt{3})$ or –?–.

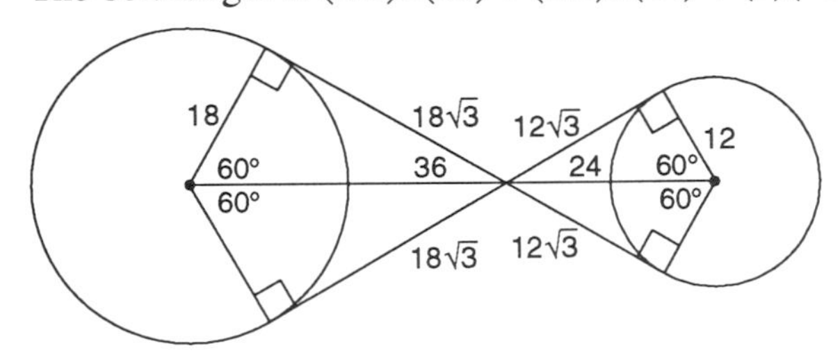

9.11 A1 To construct $a\sqrt{2}$, construct an isosceles right triangle with sides of length a. The length of the hypotenuse will be $a\sqrt{2}$. To construct $a\sqrt{3}$, construct a right triangle using $a\sqrt{2}$ and a as the length of the legs. The hypotenuse will have a length of $a\sqrt{3}$.

9.11 A3 Begin by sketching a picture of what you want when finished with the construction. One approach is to start with the circle and try to construct the angle so that both sides are tangent to the circle. Radii to the points of tangency are congruent and perpendicular to the sides of the angle. The line segment from the center of the circle to the vertex of the angle is the angle bisector. Therefore, the angle at the center of the circle (formed by the angle bisector and a radius) is the complement of half the given angle. Therefore, bisect the given angle and construct the complement of this angle at the center of the circle. Another approach? Construct the given angle; then, bisect it. Next construct a line parallel to one of the sides of the angle at a distance OR from the side. The intersection of this line and the angle bisector is the center of the circle. Do you see why this works? Try both methods.

9.11 A5 Since the midpoint of the hypotenuse is equally distant from the three vertices, the median is half the length of the hypotenuse. Therefore, construct a semicircle with radius OA. Label center point A and the endpoints of the diameter T and M. Point O is on the semicircle. You do the rest.

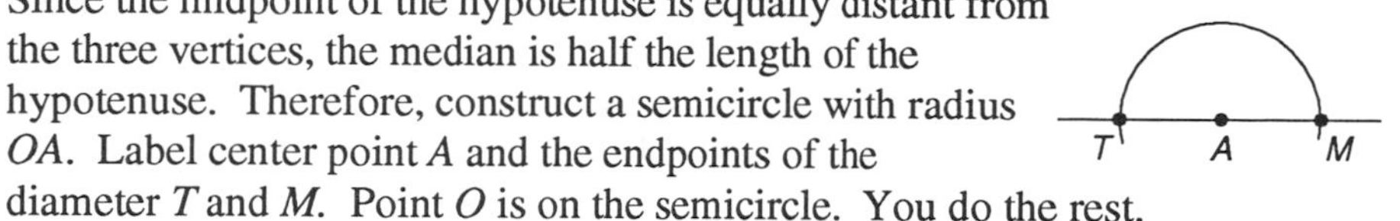

9.12 B9 Half the shaded area is equal to the area of a quadrant (quarter) of the circle less the area of the isosceles right triangle.

9.12 C4 $(45 \bullet 2)^2 + (60 \bullet 2)^2 = d^2$

9.12 C8 If the height is $12\sqrt{3}$, then the radius of the inscribed circle is $(1/3)(12\sqrt{3})$.

9.12 C10 $(x + 18)^2 = 36^2 + x^2$

9.12 C12

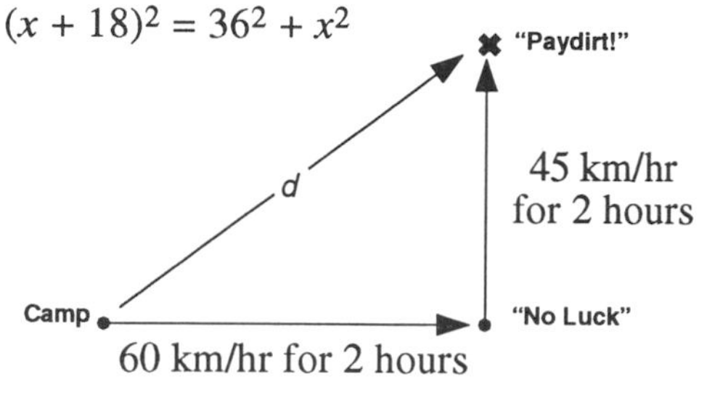

CPS 3 The smallest triangle that would enclose three congruent mutually tangent circles would be the equilateral triangle that is tangent to two circles on each side. The diagram on the right should help.

CPS 4 A drawing of two hallways intersecting at right angles shows that the limiting size is the diagonal ($\overline{AC}$) of the smallest rectangular plane ($ABCD$) that can maneuver the right angle turn. This will happen when $AE = BE$.

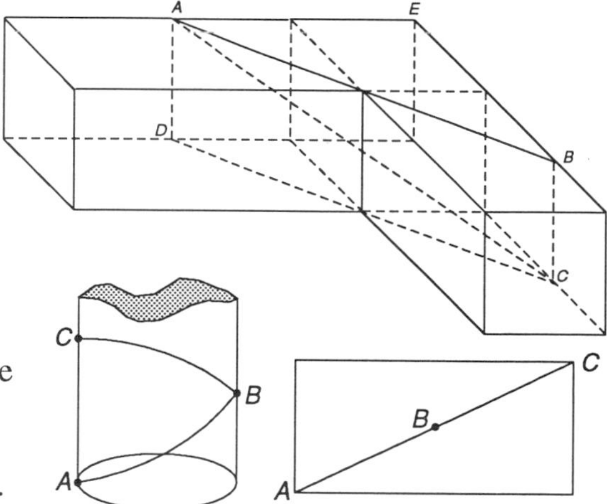

CPS 5 If you unwrap one complete loop (one of the 30), you get the diagonal of a rectangle. The diagonal is one-thirtieth of the total distance. See the diagram.

Chapter 10

10.2 C4 True. You can think of a prism as a stacking up of thin copies of the bases.

10.3 A2 False. It is a sector of a circle.

10.3 D1 A die is one of a pair of dice (as in "seven come eleven!").

10.3 D7

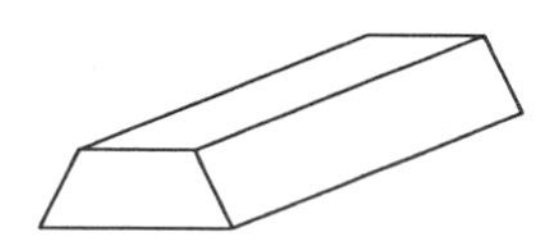

10.3 D11

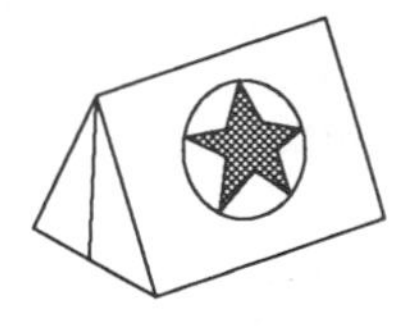

10.4 A2

$$\begin{aligned} V &= BH \\ &= [(1/2)(3)(4)](4) \\ &= 24 \end{aligned}$$

10.4 A3

$$\begin{aligned} V &= BH \\ &= [(1/2)(3)(3 + 5)](9) \\ &= \text{–?–} \end{aligned}$$

10.4 A5

$$\begin{aligned} V &= BH \\ &= [(1/2)(\pi)(3^2)](8) \\ &= \text{–?–} \end{aligned}$$

10.4 A6

90/360, or 1/4 of the cylinder is removed. Therefore, you need to find 3/4 of the volume of the whole cylinder.

10.4 B1

$$V = [\tfrac{1}{2}(6)(8)](20)$$

10.5 A1

$$\begin{aligned} V &= (1/3)BH \\ &= (1/3)[(1/2)(5)(4 + 8)](15) \\ &= (1/3)(30)(15) \\ &= 150 \end{aligned}$$

10.5 A2

$$\begin{aligned} V &= (1/3)BH \\ &= (1/3)(\pi 6^2)(7) \\ &= (1/3)(36\pi)(7) \\ &= 84\pi \end{aligned}$$

10.5 A5

$$\begin{aligned} V &= (1/3)BH \\ &= (1/3)[(1/2)(5)(12)](6) \\ &= 60 \end{aligned}$$

10.5 A6

The figure is a cylinder with a cone removed.

$$\begin{aligned} V &= V_{\text{cylinder}} - V_{\text{cone}} \\ &= \pi(6^2)(16) - (1/3)\pi(6^2)(16) \\ &= (2/3)\pi(6^2)(16) \\ &= \text{–?–} \end{aligned}$$

10.6 A5

The figure is a cylinder with a cylinder removed.

$$\begin{aligned} V &= V_{\text{larger cylinder}} - V_{\text{smaller cylinder}} \\ &= \pi(8y)^2(6y) - \pi(3y)^2(6y) \\ &= \pi(6y)[(8y)^2 - (3y)^2] \\ &= \text{–?–} \end{aligned}$$

10.6 A8

$$\begin{aligned} V &= (1/3)[\pi(6x)^2](8x) \\ &= \text{–?–} \end{aligned}$$

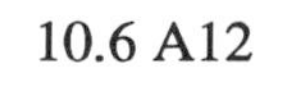

10.6 A12

$V = BH$
$= [\pi(\sqrt{Q})^2]T$
$= QT\pi$ (Cutie pie! Sorry about that.)

10.6 A15

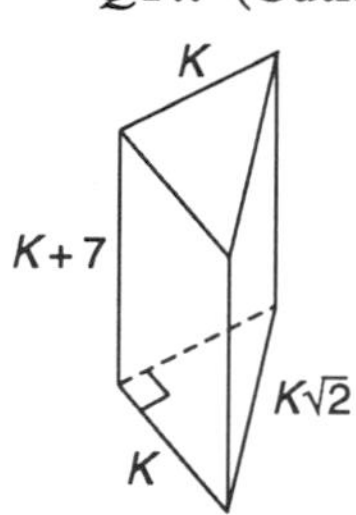

10.6 A18

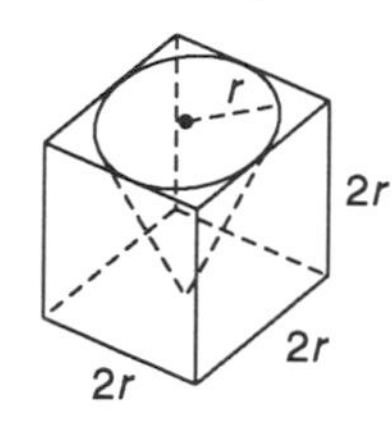

10.6 B1 Each of the small cubes on each of the 8 corners of the rectangular prism has red on three faces.

10.7 A4 Suppose the smaller prism has dimensions 4 by 5 by 6 (volume of 120 cm^3). Then the dimensions of the larger prism is 8 by 10 by 12 (volume of 960 cm^3). The volume is eight times larger. Is this always true? Try it.

10.7 B1

$V = (1/3)BH$
$138\pi = (1/3)(46\pi)H$

Solve for H. (Divide both sides by 46π; then, multiply both sides by 3.)

10.7 B3

$V = (1/3)BH$
$3168 = (1/3)[(1/2)(20 + 28)h](36)$

Solve for h.

10.7 C1 First change 8" to 2/3'. To find the volume multiply (5.5)(6.5)(2/3). To convert volume (cubic feet) to weight (pounds), multiply the volume by 63 pounds/cubic foot.

10.7 C3

$A = (1/2)ap$
$= (1/2)[(3/2)(\sqrt{3})](18) = 27\sqrt{3}/2$
$V = (27\sqrt{3}/2)(3) \approx 68.9$

The number of gallons is about (68.9)(7.5) or –?–.

10.7 C4 The swimming pool is a pentagonal prism resting on one of its lateral faces. The area of the pentagonal base can be found by dividing it into a rectangular region and a trapezoidal region.

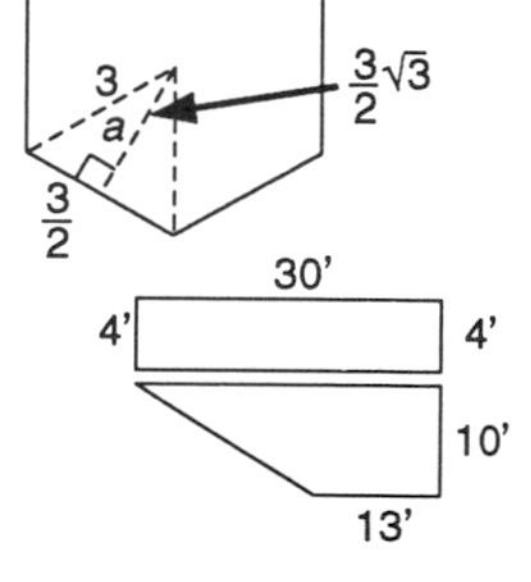

10.8 A4

$V_{\text{displacement}} = (7/8)(V_{\text{block of ice}})$
$(35)(50)(4) = (7/8)(V_{\text{block of ice}})$
$(8/7)(35)(50)(4) = V_{\text{block of ice}}$
$V_{\text{block of ice}} =$ –?–

10.8 B3 The density of the clump of metal equals its mass divided by its volume.

$$.97 = \frac{145.5}{V_{\text{displacement}}}$$

$$V_{\text{displacement}} = \frac{145.5}{.97}$$

But, $V_{\text{displacement}}$ is $(10)(10)H$. So $\frac{145.5}{.97} = (10)(10)H$. Solve for H.

10.9 A1 $r^2 + 12^2 = 15^2$
$r^2 = 15^2 - 12^2$
$A = \pi r^2$

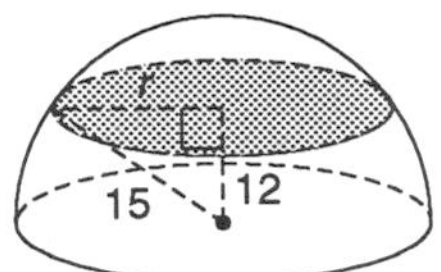

10.9 B4

$$\begin{aligned} V_{\text{capsule}} &= V_{\text{two hemispheres}} + V_{\text{cylinder}} \\ &= 2[(2/3)\pi(6^3)] + [\pi(6^2)(12)] \\ &= (4/3)(216)\pi + (12)(36)\pi \\ &= \text{–?–} \end{aligned}$$

10.9 B6 40/360, or 1/9 of the hemisphere is missing. What fraction is still there?

10.9 C4

$$\begin{aligned} V_{\text{solid}} &= V_{\text{hemisphere}} - V_{\text{cone}} \\ &= (2/3)\pi r^3 - (1/3)\pi r^2(r) \\ &= (2/3)\pi r^3 - (1/3)\pi r^3 \\ &= \text{–?–} \end{aligned}$$

10.10 2

$$V_{\text{container}} = \pi(3^2)(10) = 90\pi$$
$$V_{\text{scoop}} = (4/3)\pi(3/2)^3 = (9/2)\pi$$

The number of scoops is $(90\pi)/[(9/2)\pi]$ or –?–.

10.10 6

$$V_{\text{ball}} = (4/3)\pi(6^3) \approx \text{–?–}$$

The weight of the ball is equal to the product of the volume and the density.

10.10 8

$$\begin{aligned} V_{\text{hollow ball}} &= V_{\text{outer sphere}} - V_{\text{inner sphere}} \\ &\approx (4/3)(22/7)(7^3) - (4/3)(22/7)(r^3) \\ &\approx \text{–?–} \end{aligned}$$

However, the volume of the hollow ball is its weight divided by its density.

$$V_{\text{hollow ball}} = \frac{327.36}{.28}$$

Therefore, $\frac{327.36}{.28} = (4/3)(22/7)(7^3 - r^3)$. Solve for r.

10.10 9 $V_{\text{shell}} = V_{\text{big hemisphere}} - V_{\text{small hemisphere}}$

$$= \frac{2}{3}\pi R^3 - \frac{2}{3}\pi r^3 = \frac{2}{3}\pi(R^3 - r^3)$$

The inner diameter is 5.4 meters, and thus, the inner radius is 2.7 meters. Find the outer radius.

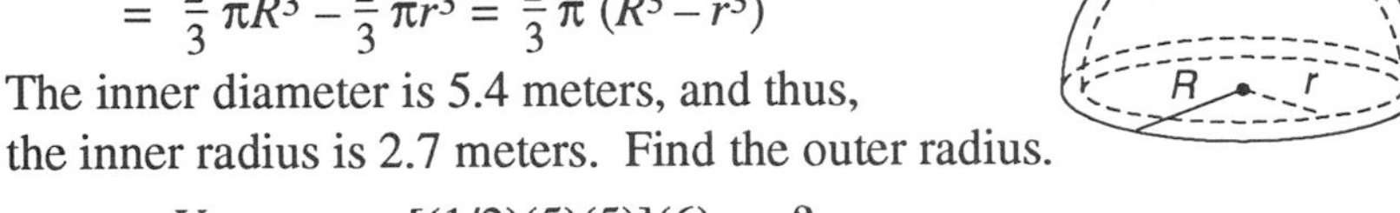

10.11 1

$$V_{\text{pup tent}} = [(1/2)(5)(5)](6) = \text{–?–}$$
$$V_{\text{tepee}} = (1/3)(5)[(1/2)(2\sqrt{3})(24)] = \text{–?–}$$

10.11 3

$$\begin{aligned} V_{\text{ring}} &= V_{\text{larger prism}} - V_{\text{missing prism}} \\ &= [(1/2)(3\sqrt{3}(36)](2) - [(1/2)(2\sqrt{3})(24)](2) \\ &= \text{–?–} \end{aligned}$$

The volume of the hole ($48\sqrt{3}$) does not equal the volume of the ring ($60\sqrt{3}$).

10.11 4

$$\begin{aligned} V_{\text{cone pipe}} &= V_{\text{outer cylinder}} - V_{\text{inner cylinder}} \\ &= \pi(3^2)(160) - \pi(2.5)^2(160) \\ &= \pi(160)[3^2 - (2.5)^2] \\ &\approx (22/7)(160)(2.75) \end{aligned}$$

Therefore, the weight of 200 pipes is $(200)(V_{\text{one pipe}})(\text{density})$ or $(200)(V_{\text{one pipe}})(7.7)$ or –?–.

10.11 6

Since the directions north and east are at right angles, the diameter of the pool can be determined. Once the diameter is determined and the diagonal length across the pool is known, the depth of the pool can be calculated. Knowing the depth and diameter, the volume can be calculated.

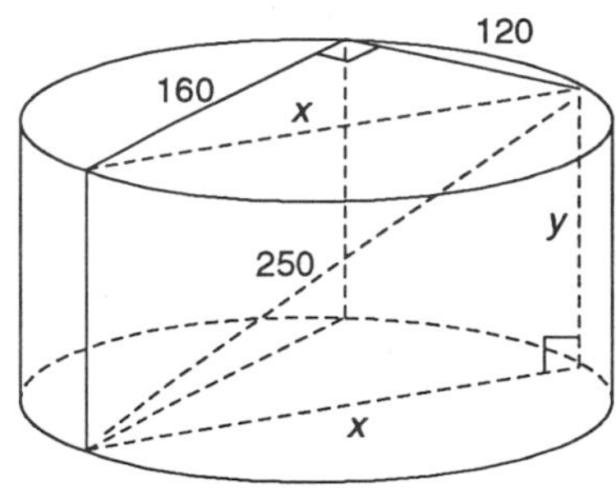

10.12 B5

$V = (2/3)\pi(15^3) = $ –?–

10.12 B6

The area of the six sided base is (12)(12) – (6)(4) or –?–.

10.12 B9

One-fourth of the hemisphere is missing. Therefore, the volume is given by: $V = (\frac{3}{4})[\frac{2}{3}\pi r^3] = \frac{1}{2}\pi r^3$. If $V = 256\pi$, then $256\pi = \frac{1}{2}\pi r^3$.
Solve this equation for r^3, then take the cube root.

10.12 C1

The volume of the object equals the area of the hexagonal base of the glass prism multiplied by the rise in the water level.

$$\begin{aligned} V_{\text{object}} &= [(1/2)(5/2)(\sqrt{3})(30)](4) \\ &= 150\sqrt{3} \end{aligned}$$

The density of the object is its weight divided by its volume.

$$\begin{aligned} \text{Density}_{\text{object}} &= \frac{5457}{(150\sqrt{3})} \\ &\approx \frac{5457}{(150)(1.7)} \\ &\approx \text{–?–} \end{aligned}$$

10.12 C2

Since the liquid level drops 2 cm for each person, then the liquid level must be at least 18 cm below the top of the tub for Tessie and her eight guests.

$$\begin{aligned} V_{\text{sparkling cider}} &= \pi(120^2)(150 - 18) \\ &\approx (3.14)(120^2)(132) \\ &\approx \text{–?–} \end{aligned}$$

CPS 1

The volume of the water is (2)(2)(1.5) or six cubic meters. To find the floor dimensions after two walls have moved 5 cm/minute for fifteen minutes, you must recognize that *two* walls moving towards each other at that rate means a loss of 10 cm/minute for fifteen minutes or a loss of 1.5 meters in one floor dimension. Hence, the new floor dimensions become two meters by a half meter. If the volume is still six cubic meters, how high will the water level reach? If you let H represent the height of the water, then $(2)(.5)H = 6$.

CPS 2 The total volume of ocean water is fifteen times the volume of one hole.

$$V_{\text{ocean water}} = 15(V_{\text{one hole}})$$
$$3.3 \times 10^8 = 15[\pi r^2(2800)]$$
$$3.3 \times 10^8 \approx (4.2 \times 10^4)(22/7)r^2$$

Solve for r^2. Then solve for r.

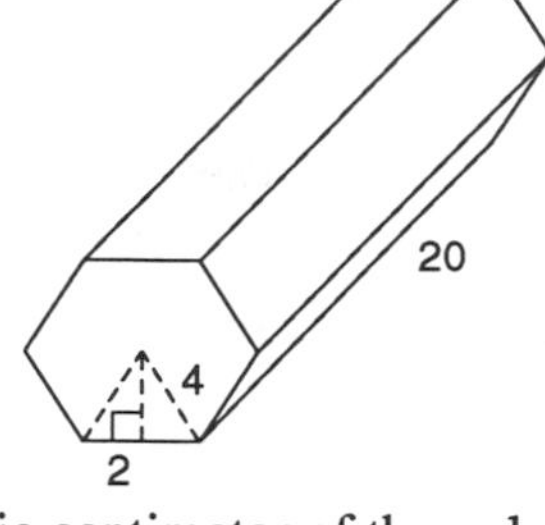

CPS 3 The volume of the hexagonal bar is equal to the area of the regular hexagon times the length (20). To find the area of the regular hexagon, use $A = (1/2)ap$, where $a = 2\sqrt{3}$ and the perimeter p is (6)(4) or 24. The time is one minute for each cubic centimeter of the volume.

CPS 4 The octahedron is two square-based pyramids. Let each edge be one unit. Find the volume of one pyramid and double your answer for the octahedron. The altitude of the pyramid touches the pyramid's base at the intersection of the diagonals. If you know the length of the edge of the pyramid and the length of half a diagonal, you can calculate the height of the pyramid. Once you know the height of the pyramid, you can calculate the volume of the octahedron. Let each edge of the cube be 3/4. Then find the volume of the cube.

CPS 5 The object is a prism 6 by 8 by 12. This prism has three different prisms cut out of it. One cut-out prism is a rectangular prism with dimensions of 2 by 5. Another cut-out prism is a triangular prism with a right triangle base. The third cut-out prism is an isosceles trapezoidal prism. None of the three cut-out prisms intersect. Find the volume of each and subtract from the whole.

Chapter 11

11.1 A2 $\dfrac{\text{shaded area}}{\text{total area}} = \dfrac{12}{32} = $ –?–

11.1 B5 $y^2 = 64$, so, $y = \pm\sqrt{64} = \pm$ –?–

11.1 C3 $\dfrac{x \text{ runs}}{9 \text{ innings}} = \dfrac{34 \text{ runs}}{106 \text{ innings}}$

11.2 C1 $\dfrac{1}{2} = \dfrac{3}{AL}$; $\dfrac{1}{2} = \dfrac{5}{RA}$; etc. Or, $\dfrac{4}{3} = \dfrac{8}{AL}$; $\dfrac{4}{5} = \dfrac{8}{AR}$; etc.

11.2 C4 All the corresponding angles are congruent. Are all these ratios equal:

$\dfrac{21}{14} =$ –?– ; $\dfrac{30}{20} =$ –?– ; $\dfrac{27}{18} =$ –?– ; $\dfrac{39}{26} =$ –?– ; $\dfrac{78}{52} =$ –?–

11.2 C7 Since the lines are parallel, the corresponding angles are congruent. Are all these ratios equal:

$\dfrac{2}{2 + 2\ 2/3} =$ –?– ; $\dfrac{3}{3 + 4} =$ –?– ; $\dfrac{4}{9\ 1/3} =$ –?–

11.3 3 It helps to rotate ΔARK so that you can see which sides correspond.

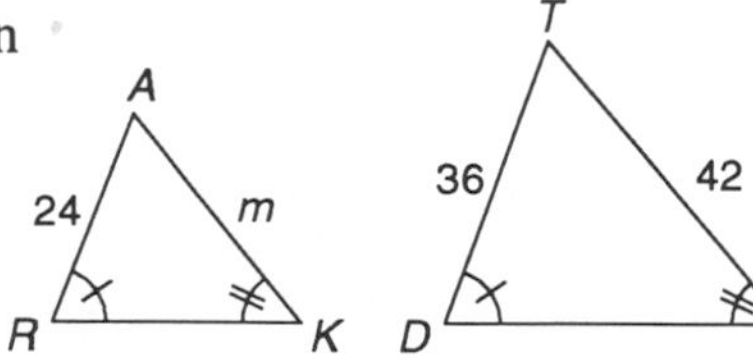

11.3 6 Again, it helps to rotate one of the triangles so that it is easier to see the equal angles in corresponding positions. Since ΔPHY is a right triangle, $20^2 = 16^2 + HY^2$. Once you know HY, you can find HT since $15^2 = HY^2 + HT^2$. Once you have the values for HY and HT, compare the ratios: 20/15, $16/HY$, HY/HT.

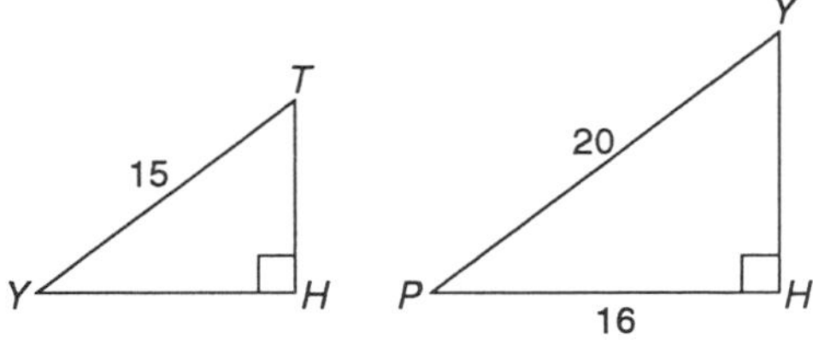

11.3 9 Since $\angle F$ and $\angle H$ are two angles inscribed in the same arc, they are congruent (Conjecture 49). $\angle T$ and $\angle G$ are congruent for the same reason. Therefore, $\Delta FTA \sim \Delta HGA$ by the AA shortcut.

11.4 3 Since $\Delta GAR \sim \Delta DAN$, then $\dfrac{AD}{AG} = \dfrac{ND}{RG}$.

11.4 4 Since $\Delta PRE \sim \Delta POC$, then $\dfrac{PR}{RE} = \dfrac{PO}{OC}$. Let $x = PR$. Then $\dfrac{x}{60} = \dfrac{x + 45}{90}$.

11.4 5

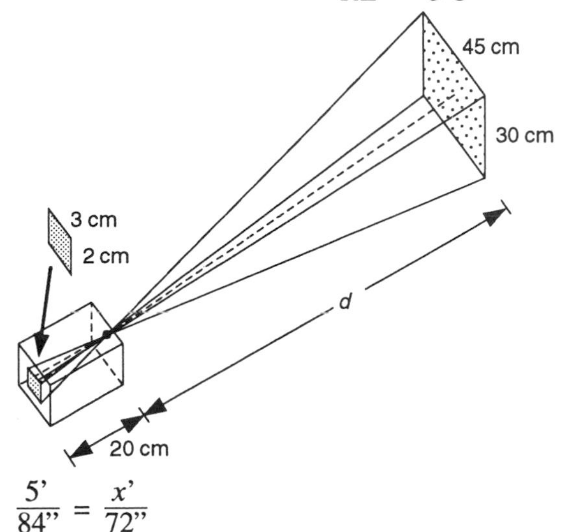

11.5 2 $\dfrac{5'}{84''} = \dfrac{x'}{72''}$

11.5 4

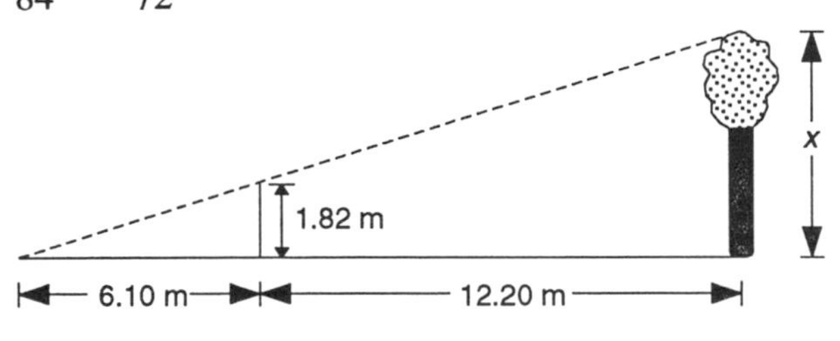

11.5 6

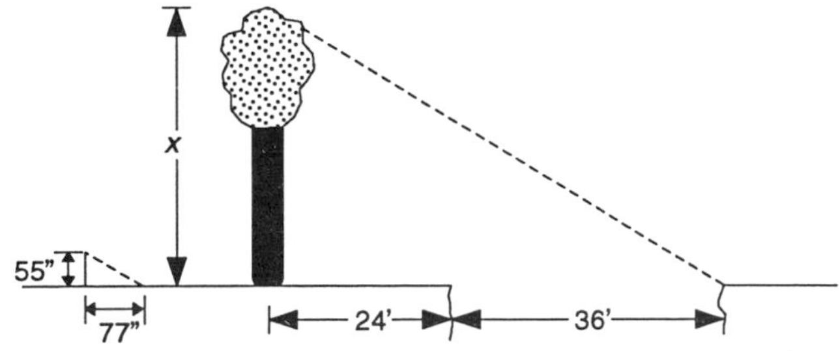

11.6 A1 $\frac{22}{33} = \frac{12}{h}$

11.6 A3 $\frac{IS}{IE} = \frac{SO}{PE}$; $\frac{LI}{OI} = \frac{SL}{OE}$

11.6 A5 $\frac{5}{8} = \frac{x}{32}$

11.6 A7 $\frac{a}{b} = \frac{p}{q}$

11.6 A10 $\frac{12}{15} = \frac{x}{10 - x}$

11.6 A11 $\frac{9}{18} = \frac{3\sqrt{3}}{k}$

11.6 A12 Since $48^2 = 24^2 + (24\sqrt{3})^2$, then ΔABC is a 30-60 right triangle. Since $m\angle A = 60$, then ΔABD is equilateral and, therefore, $AD = 24$. But, since $\frac{24}{24} = \frac{x}{y}$, then $x = y = 12$.

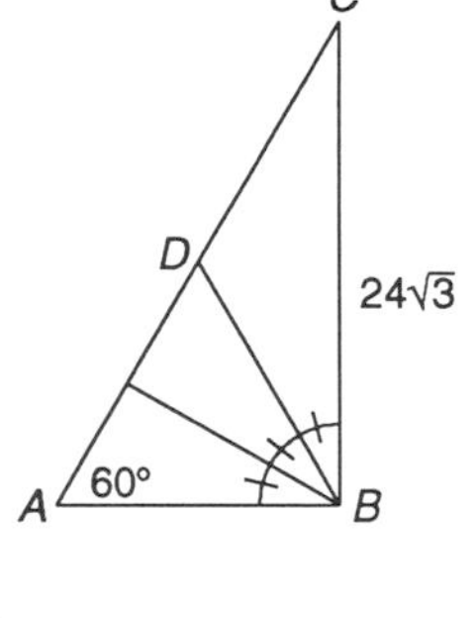

11.6 A13 Area $_{\text{large triangle}} = (1/2)(8)a$

Area $_{\text{small triangle}} = (1/2)(8)b$

So, $\frac{\text{Area}_{\text{large triangle}}}{\text{Area}_{\text{small triangle}}} = \frac{(1/2)(8)a}{(1/2)(8)b} = \frac{a}{b}$

But you know that the ratio of a to b equals the ratio of the corresponding sides with lengths of $8\sqrt{2}$ and 8.

So, $\frac{\text{Area}_{\text{large triangle}}}{\text{Area}_{\text{small triangle}}} = \frac{a}{b} = \frac{8\sqrt{2}}{8} = \sqrt{2}$

11.6 B1 Use the method you discovered in the last investigation. That is, construct two segments in the ratio of 3:4. Use them with $\overline{CD}$ to construct a triangle. Then construct the angle bisector of the angle opposite $\overline{CD}$. The point where the angle bisector intersects $\overline{CD}$ is the point that divides $\overline{CD}$ into two segments in the ratio of 3:4.

11.6 B3 First, find the point S that divides GH into the ratio 5:4.

Second, find the point R that divides GS into the ratio 2:3.

11.7 1 $(\frac{1}{2})^2 = \frac{\text{Area}_{\Delta MSE}}{72}$

11.7 3 Since $\frac{\text{Area}_{ZOID}}{\text{Area}_{TRAP}} = \frac{16}{25}$, then the ratio of the corresponding sides is $\frac{4}{5}$.

$\frac{4}{AP} = \frac{4}{5}$ and $\frac{8}{TR} = \frac{4}{5}$, so $x = \frac{1}{2}(AP + TR)$.

11.7 9

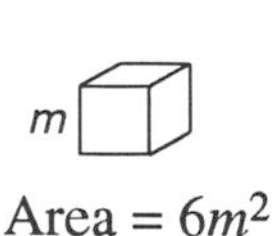

Area = $6m^2$

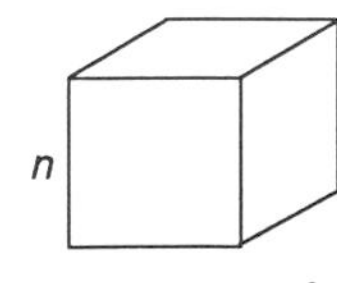

Area = $6n^2$

11.7 10 $\frac{(12)(15)}{(16)(20)} = \frac{150}{x}$

11.8 B1 $(\frac{5}{3})^3 = $ –?–

11.8 B5 $\frac{\text{Volume}_{\text{small cylinder}}}{\text{Volume}_{\text{large cylinder}}} = \frac{\pi(81)(24)}{4608\pi} = \frac{1944\pi}{4608\pi} = \frac{27}{64}$

But, $(\frac{\text{Height}_{\text{small cylinder}}}{\text{Height}_{\text{large cylinder}}})^3 = \frac{\text{Volume}_{\text{small cylinder}}}{\text{Volume}_{\text{large cylinder}}}$. So, $\frac{24^3}{H^3} = \frac{27}{64}$.

11.8 B6 If $\frac{r^2}{s^2} = \frac{49}{81}$, then $\frac{r}{s} = \frac{7}{9}$.

If $\frac{r}{s} = \frac{7}{9}$, then $\frac{r^3}{s^3} = \frac{7^3}{9^3} = $ –?–.

11.8 B7 If $(\frac{r}{R})^3 = \frac{8}{27}$, then $\frac{r}{R} = \frac{2}{3}$. Therefore, $\frac{d}{D} = $ –?–.

11.8 B10

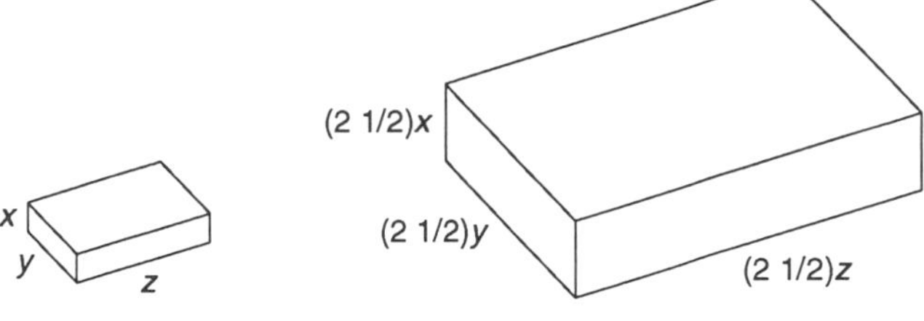

Volume = xyz Volume = $(2\ 1/2)^3 xyz$

The volume of the larger storage facility is $(2\ 1/2)^3$ times larger, so, the cost will bc $(2\ 1/2)^3$ times larger.

$(2\ 1/2)^3 = (\frac{5}{2})^3 = \frac{125}{8}$, so the cost is $(\frac{125}{8})(\$125)$ per day.

11.8 B11 $\frac{\text{Stomach capacity}_{\text{dragon A}}}{\text{Stomach capacity}_{\text{dragon B}}} = \frac{(\text{Length}_{\text{dragon A}})^3}{(\text{Length}_{\text{dragon B}})^3}$. So $\frac{440}{x} = \frac{6^3}{18^3} = (\frac{1}{3})^3$.

11.9 A1 $\frac{4}{4+12} = \frac{x}{20}$

11.9 A4 $\frac{8}{8+16} = \frac{12}{12+x}$ or $\frac{1}{3} = \frac{12}{12+x}$ or $12 + x = 36$. So, $x = 24$.

$\frac{DE}{AE} = \frac{8}{16} = \frac{1}{2}$ and $\frac{DC}{BC} = \frac{12}{24} = \frac{1}{2}$

11.9 A7

A, a, c, l, b, d, M, N

$\frac{a}{b} = \frac{c}{d}$

11.9 A8 If $\frac{a+b}{a} = \frac{c+d}{c}$, then $\frac{a}{a} + \frac{b}{a} = \frac{c}{c} + \frac{d}{c}$.

Then subtract one from each side.

11.9 B1 If $\frac{24}{14} = \frac{36}{a}$, then $\frac{12}{7} = \frac{36}{a}$.

11.9 B5 $\frac{x}{3} = \frac{x}{g}$ and since the numerators are the same, the denominators must be too.

11.10 A4 To find YB, you can use $\frac{40}{YB} = \frac{30}{60}$. But to find AB you **cannot** use $\frac{50}{AB} = \frac{30}{60}$. Why not? Reread conjecture C-80. It doesn't apply to the ratio $\frac{XY}{AB}$. You need to use similar triangles.

11.10 A5 $\frac{CM}{MA} = 1 \qquad \frac{CN}{NB} = 1$

11.10 A9 $\frac{24}{60} = \frac{2}{5};\ \frac{32}{76} = \frac{8}{19};\ \frac{2}{5} \neq \frac{8}{19}$

11.10 B2 Divide $\overline{IJ}$ into six parts. Then use one part as the radius of a circle. Divide the circle into six equal arcs (daisy construction).

11.10 B3 See 11.10 B1.

11.11 B5 Redraw the two triangles so that the corresponding angles are in the same position.

$\frac{6}{5} = \frac{5}{x}$ and $\frac{9}{6} = \frac{y}{5}$

11.11 B8 $\frac{2^3}{3^3} = \frac{V}{2160\pi}$

11.11 B11 $\Delta ABE \sim \Delta ADC$

$\frac{AB}{BE} = \frac{AD}{CD}$

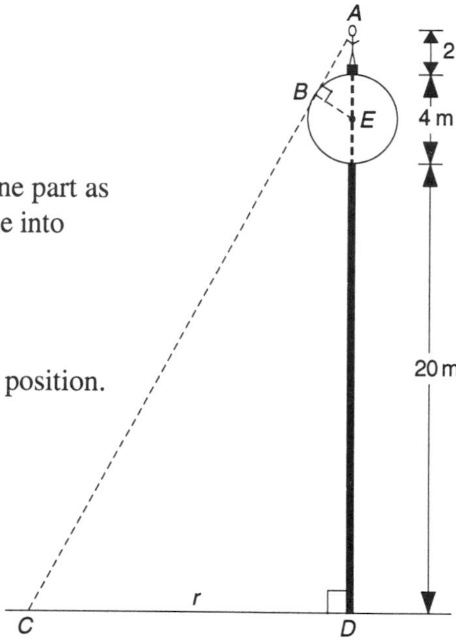

11.11 B12 Since you are concerned with ratios, it doesn't make any difference what lengths you choose. Therefore, let's let the square have a side of length 2. Then the area of the square is 4, the area of the circle is $\pi(1)^2$ or π, and the area of the isosceles triangle is (1/2)(2)(2) or 2. Therefore, the ratio of the three areas is –?– : –?– : –?–.

The volume of the cylinder is $\pi(1)^2(2)$ or 2π, the volume of the sphere is $(4/3)\pi(1)^3$ or $(4/3)\pi$, the volume of the cone is $(1/3)\pi(1)^2(2)$ or $(2/3)\pi$.

The ratio of the three volumes is $2\pi : \frac{4}{3}\pi : \frac{2}{3}\pi$.

This can be reduced to a very nice ratio of three whole numbers.

Chapter 12

12.1 1 $\sin A = \frac{s}{t};\ \cos A = \frac{r}{t};\ \tan A = \frac{s}{r}$

12.1 7 $\sin G = \frac{16}{16\sqrt{2}} = \frac{1}{\sqrt{2}} = \frac{\sqrt{2}}{2} = \sin H;\ \cos G = \frac{16}{16\sqrt{2}} = \frac{1}{\sqrt{2}} = \frac{\sqrt{2}}{2} = \cos H;$
$\tan G = 1 = \tan H$

12.2 A1 $\tan 30° = \frac{20}{a}$. But, $\tan 30° \approx .58$. Therefore, $\frac{20}{a} \approx .58$.

12.2 A8 $\sin 35° = \frac{x}{85}$ and $\cos 35° = \frac{y}{85}$. The perimeter is $2(x + y)$.

12.2 B1 The sine of A increases from almost zero to almost one.

12.3 A1 $\sin 37° \approx .6018$

12.3 B1 Using a calculator

Step 1: Enter .5 (the sine of A).

Step 2: Press the second function or inverse button and then the sin button.

Step 3: Read the display. It gives the measure of A as 30.

12.3 B7 $\sin 55° = \frac{h}{50}$

12.3 B9 $\sin 44° = \frac{h}{1400}$ where h is the height of the balloon.

$\cos 44° = \frac{d}{1400}$ where d is the distance from Wendy to a position directly beneath the balloon.

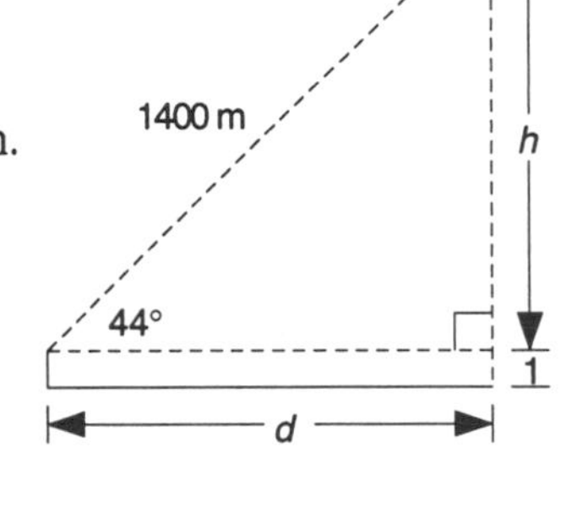

12.4 1 $\tan 33° = \frac{42}{d}$

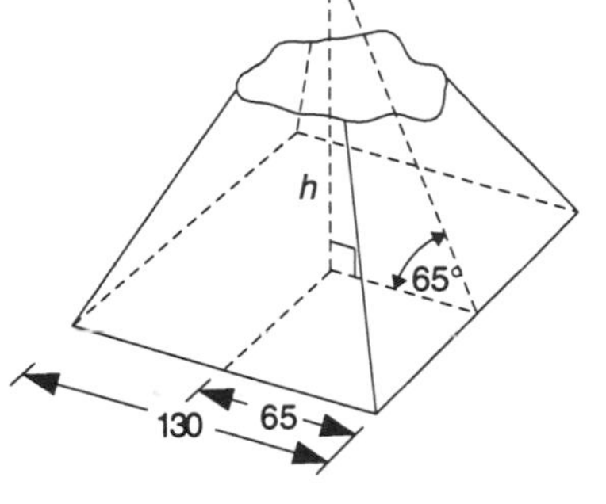

12.4 5 $\tan 65° = \frac{d}{65}$

12.4 7 $\sin ø = \frac{4.8}{55}$

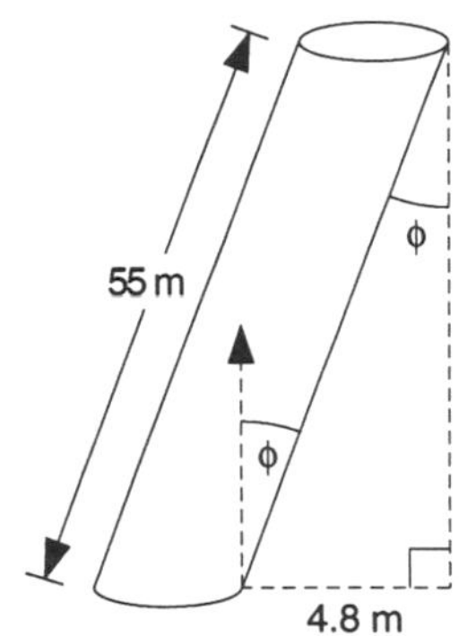

12.4 9 Make a sketch. One approach would be to find the length of a side in each of two different right triangles.

12.5 A4 $\sin B = \frac{h}{a}$

12.5 A6 If $\sin B = \frac{h}{a}$, then $h = a \sin B$.

Likewise, if $\sin A = \frac{h}{b}$, then $h = b \sin A$.

12.5 A7 $\sin B = \frac{k}{c}$

12.5 B1 $\frac{28}{\sin 52°} = \frac{w}{\sin 79°}$

12.5 C1 $\frac{\sin 42°}{29} = \frac{\sin A}{36}$

12.6 A1 $w^2 = 36^2 + 41^2 - (2)(36)(41)(\cos 49°)$

12.6 B1 $42^2 = 34^2 + 36^2 - (2)(36)(34)(\cos A)$

12.7 1 $\tan 9° = \dfrac{1}{x}$

12.7 5 $4^2 = 7^2 + 8^2 - (2)(7)(8)(\cos \phi)$

12.7 6 First, find one of the angles, say ø.

$36^2 = 38^2 + 42^2 - (2)(38)(42)(\cos \phi)$

$\sin \phi = \dfrac{h}{38}$

$\text{Area} = \dfrac{1}{2}(42)(h)$

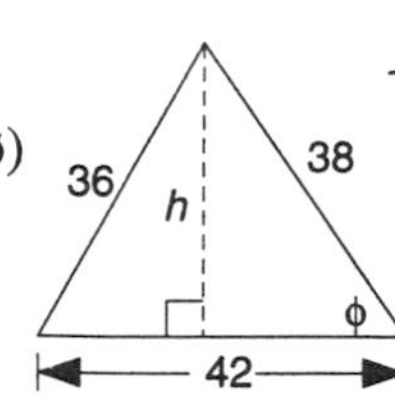

12.7 7 An octagon can be divided into eight isosceles triangles with the vertex angle measuring 45. Therefore, the base angles measure –?–. The altitude h to the base can be calculated by using:

$\sin_{\text{base angle}} = \dfrac{h}{\text{radius}}$

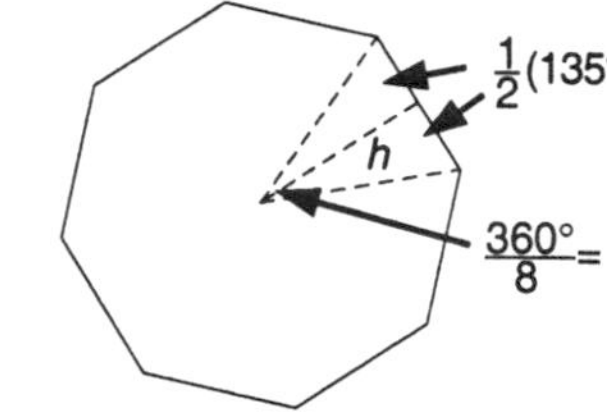

12.7 11 $\tan 49° = \dfrac{(6.3)}{x}$

$\tan 15° = \dfrac{(6.3)}{y}$

Distance traveled is $y - x$.

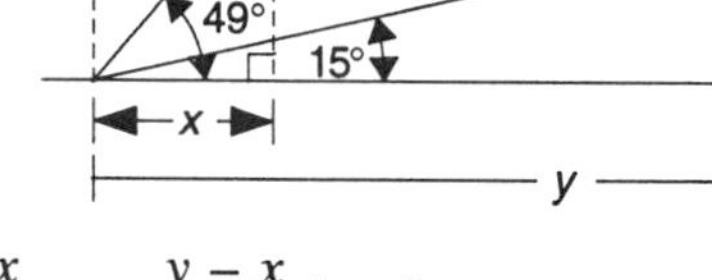

Speed is $\dfrac{\text{distance}}{\text{time}} = \dfrac{y - x}{1 \text{ minute}} = \dfrac{y - x}{\frac{1}{60}}$ km/hr.

12.8 E4 To prove $a^2 = b^2 + c^2 - 2bc \cos A$, begin by constructing an altitude $\overline{CD}$ in ΔABC. In ΔBDC, $a^2 = x^2 + (c - y)^2$ or $a^2 = x^2 + c^2 - 2cy + y^2$ (equation 1). In ΔADC, $b^2 = x^2 + y^2$ (equation 2). Substituting equation 2 into equation 1, you get $a^2 = b^2 + c^2 - 2cy$ (equation 3). Now if you can show that $y = b \cos A$ (equation 3) and substitute equation 4 into equation 3, you're in like Flynn.

CPS 1

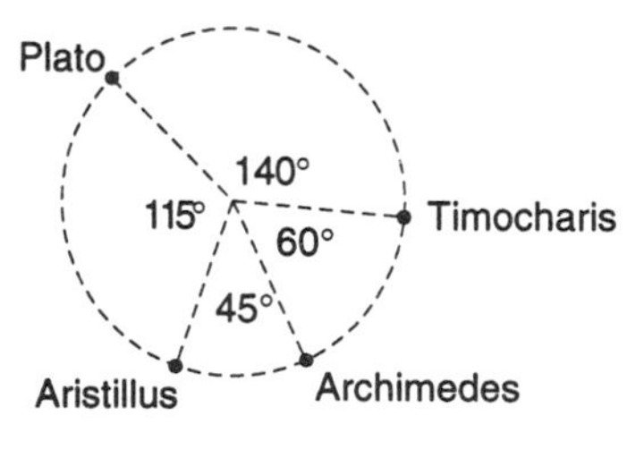

CPS 2

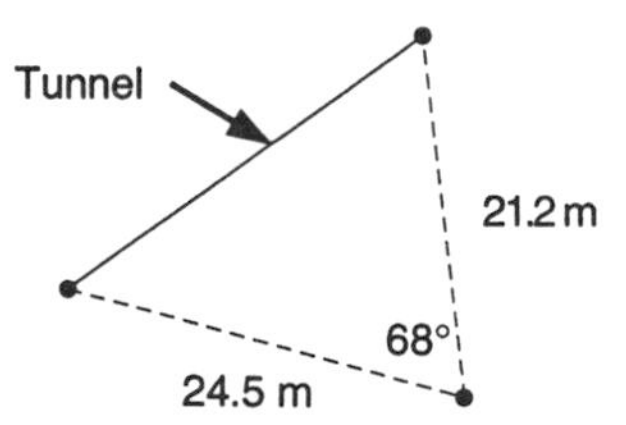

Chapter 13

13.1 A1 — B, C | B, C, M

13.1 B1 — Present some official identification showing the address. Other ideas?

13.2 A4 — If Sharon studies, then she does well on her tests.
Sharon studies.
Therefore, Sharon does well on her tests.

If Sharon does well on her tests, then she gets good grades on her report card.
Sharon does well on her tests.
Therefore, —?—.

13.2 B2 — ΔABC is not equilateral.

13.2 C2 — If the iron tip of the walking stick was worn down, then the owner of the walking stick walks a great deal.
The iron tip of the walking stick was worn down.
Therefore, —?—.

13.3 A1 — If I get a job, then I will earn money.

13.3 A5 — IF P THEN Q

13.3 B1 — He will be victorious.
LET P: Sherlock Holmes gathers all the clues.
LET Q: He will be victorious.
IF P THEN Q
P
THEREFORE Q

13.3 C3 — *If Paul divorces Veronica, then he will not receive his inheritance from Uncle Lake. Paul will receive his inheritance from Uncle Lake.* This last statement is the opposite or negation of *he will not receive his inheritance.* We could write *It is not true that Paul will not receive his inheritance*, or we could write *Paul will receive his inheritance.* These two statements are equivalent by double negation. The logical conclusion is *Paul doesn't divorce Veronica.*

13.3 C5 — No valid conclusion

13.4 1 — $P \rightarrow \sim Q$

13.4 5 — $\sim(R \rightarrow S)$

13.4 12 — If international tensions are not reduced, then —?—.

13.4 14 — It is not true that, if world peace is not in jeopardy, then international tensions are not reduced.

13.4 17 — MT

13.4 19 — No valid conclusion

13.5 1 — Use LS. If Iris Noble finds the treasure map, then pirate Rufus T. Ruffian will walk his own plank.

13.6 A1 — *Converse:* If it is a flower, then it is a rose. (False)
Inverse: If it is not a rose, then it is not a flower. (False)
Contrapositive: If it is not a flower, then it is not a rose. (True)

13.6 B4 — Valid using MP

13.6 B8 — No valid conclusion. Even though this looks like it should be a valid argument, it does not fit any of our four accepted forms of reasoning.

13.6 B9 — Valid using LS twice

13.6 B14 — Valid using LC and LS

13.7 A3 — From lines 1 and 2 using LS

13.7 C5 — From lines 3 and 4 using MP

13.7 D2 — From line 1 using LC

13.7 F3 — LET P: All wealthy people are happy.
LET Q: Money can buy happiness.
LET R: Love exists.
$P \rightarrow Q$
$Q \rightarrow \sim R$
$\sim R$
$\therefore \sim P$
The proof of the argument is left for you.

13.8 A7 — From lines 5 and 6 using MT

13.8 E4 — LET P: The consecutive sides of a parallelogram are congruent.
LET Q: The parallelogram is a rhombus.
LET R: The diagonals are perpendicular bisectors of each other.
LET S: The diagonals bisect the angles.
$P \rightarrow Q$
$Q \rightarrow R$
$Q \rightarrow S$
$\sim R$
$\therefore \sim P$
The proof is left for you. One sentence in the argument is unnecessary.

13.9 C3 — From lines 1 and 2 using MP

13.9 E4 — LET P: King Kong Curly defeats Rocking Rosco.
LET Q: Bull Moose Morgan vanquishes Wild Bill Higgins.
LET R: Curly will be World Champion of *All *Star *Wrestling *Association.
LET S: Curly will retire from wrestling and devote himself to bonsai gardening.

If King Kong Curly defeats Rocking Rosco, then if Bull Moose Morgan vanquishes Wild Bill Higgins, Curly will be World Champion of ASWA.

IF P THEN (IF Q THEN R)
$P \rightarrow (Q \rightarrow R)$
Now you translate the rest.

CA 4

```
TO PATH :D
RT 90 FD :D / 10
LT 90 FD (:D / 10) * 5
LT 90 FD (:D / 10) * 5
RT 90 FD (:D / 10) * 5
RT 90 FD (:D / 10) * 4
LT 90
PU BK :D PD
END
```

CA 5

```
TO PATH :D
LT 60 FD (:D / 10) * 7
RT 120 FD (:D / 10) * 7
FD (:D / 10) * 3
LT 120 FD (:D / 10) * 3
RT 60
PU BK :D PD
END
```

CA 6

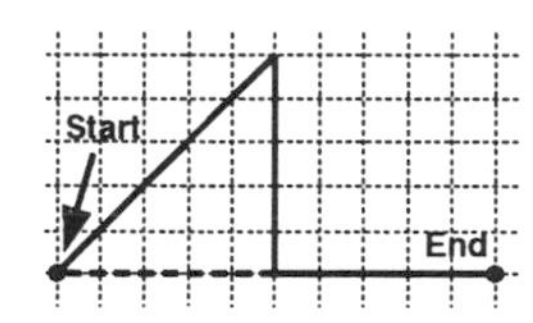

CA 18 Try FD with multiples of 40.

Chapter 14

14.1 1 It's also called the Identity Property.

14.1 6 Distributive Property; Subtraction Property of Equality; —?—; —?—

14.1 8 —?—; —?—; Multiplication Property of Equality; —?—

14.1 9 —?—; —?—; Substitution Property; —?—; —?—

14.1 10 $\overline{AB} \parallel \overline{CD}$ a. Definition of a parallelogram

14.2 2 True. The Midpoint Postulate says that a segment has exactly one midpoint.

14.2 9b SSS Congruence Postulate

14.2 10 $\angle 1 \cong \angle 2$; $\angle 3 \cong \angle 4$ a. AIA Postulate

14.2 11a Angle Bisector Postulate

14.3 A8 ASA Congruence Postulate

14.3 D2 One possible proof:

Statement	Reason
1. $\overline{AC} \cong \overline{BC}$	1. —?— (premise)
2. $\overline{CD}$ is a —?— to the base	2. Given (premise)
3. D is a —?—.	3. Definition of median
4. $\overline{AD} \cong \overline{BD}$	4. Definition of —?—
5. $\overline{CD} \cong \overline{CD}$	5. —?— Property of Congruence

6.	$\Delta ADC \cong \Delta BDC$	6. —?— Congruence Postulate
7.	$\angle 1 \cong \angle 2$	7. —?—
8.	Therefore, $\overline{CD}$ is an —?—	8. Definition of angle bisector

14.4 A5 Subtraction Property of Equality

14.4 B2 Definition of midpoint

14.4 B4 Vertical Angles Theorem

14.5 B1 Given

14.5 B2 Theorem from Exercise Set A: *A diagonal of a parallelogram divides the parallelogram into two congruent triangles.*

14.5 C2 $\Delta SAE \cong \Delta NEA$

14.5 D1

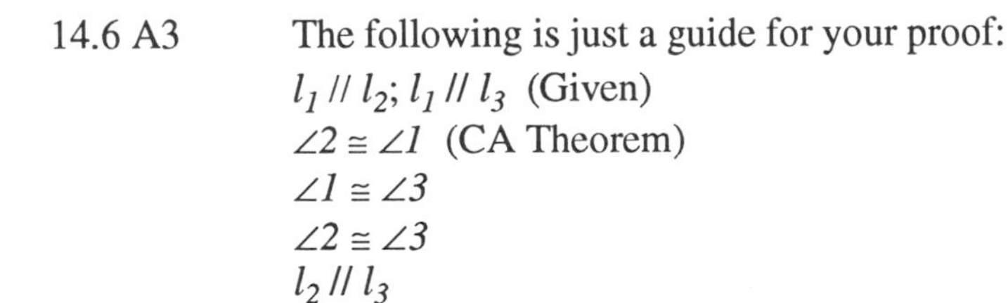

14.6 A1 The following is just a guide for your proof:
$l_1 \parallel l_2$ (Given)
$\angle 1 \cong \angle 3$ (AIA Postulate)
$\angle 3 \cong \angle 2$ (Vertical Angles Theorem)
$\angle 1 \cong \angle 2$ (Transitive Property of Congruence)

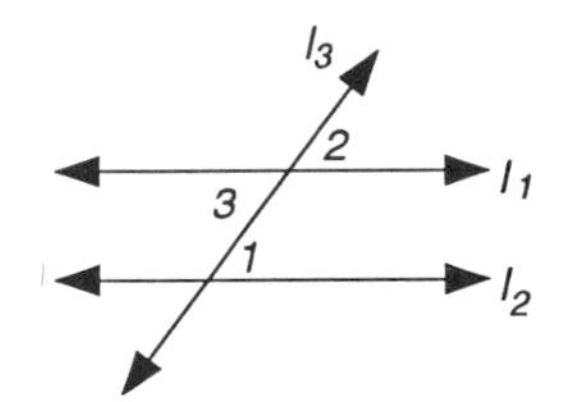

14.6 A2 The following is just a guide for your proof:
$\angle 1 \cong \angle 2$ (Given)
$\angle 2 \cong \angle 3$ (Vertical Angles Theorem)
$\angle 1 \cong \angle 3$ (Transitive Property of Congruence)
$l_1 \parallel l_2$ (AIA Postulate)

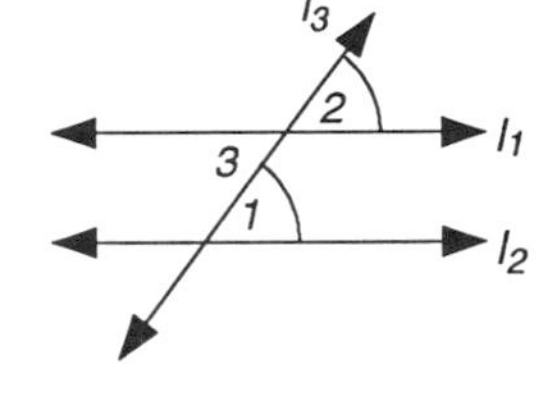

14.6 A3 The following is just a guide for your proof:
$l_1 \parallel l_2$; $l_1 \parallel l_3$ (Given)
$\angle 2 \cong \angle 1$ (CA Theorem)
$\angle 1 \cong \angle 3$
$\angle 2 \cong \angle 3$
$l_2 \parallel l_3$

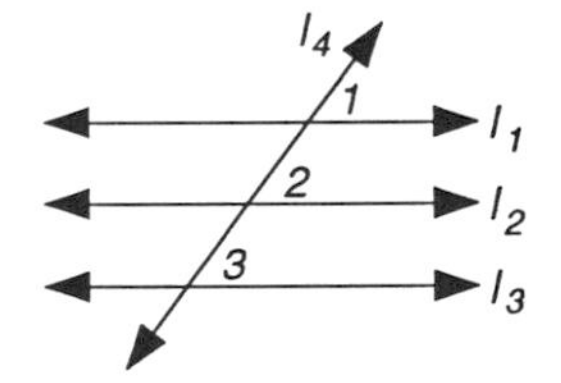

14.7 A2 The following is just a guide for your proof:
$\angle 1 \cong \angle 3$
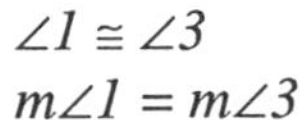
$m\angle 1 = m\angle 3$
$\angle 3$ and $\angle 2$ are supplementary
$m\angle 3 + m\angle 2 = 180$
$m\angle 1 + m\angle 2 = 180$
$\angle 1$ and $\angle 2$ are supplements

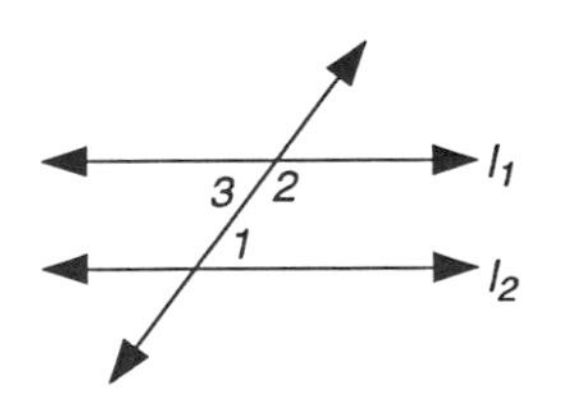

14.7 A3 You could just repeat the proof from 14.7 A2 since the conjecture in 14.7 A3 is just a special case of the theorem proved in 14.7 A2. Or, you could write a brief paragraph proof arguing that the proof of the conjecture in 14.7 A3 follows from the theorem proved in 14.7 A2. The theorem in 14.7 A3 is a corollary of the theorem in 14.7 A2.

14.7 A4 The following is just a guide for your proof:

If $\overline{AC} \perp l_2$ and $\overline{BD} \perp l_2$, then $\overline{AC} \parallel \overline{BD}$ (by 14.7 A1).

If $\overline{AC} \parallel \overline{BD}$ and $\overline{AB} \parallel \overline{CD}$, then $ABCD$ is a parallelogram (by definition of a parallelogram).

If $ABCD$ is a parallelogram, then $\overline{AC} \cong \overline{BD}$ (opposite sides of a parallelogram are congruent).

14.8 A1

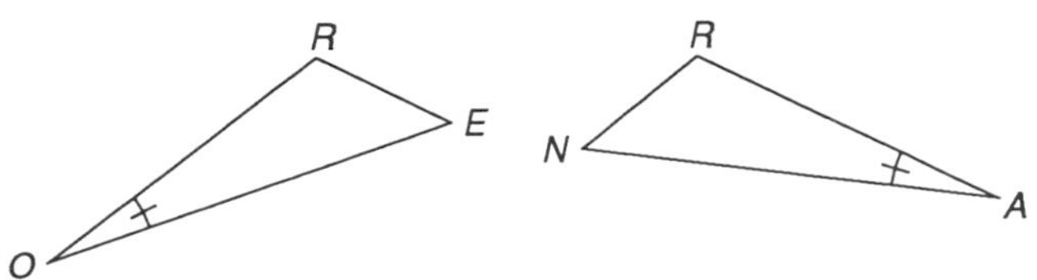

14.8 A4 Show $\Delta OEP \cong \Delta VRL$ by ASA.

14.8 B2 The following is a guide for your proof:

$\overline{PA} \cong \overline{PA}$ (Reflexive Property)

$\overline{AE} \cong \overline{PR}$ (Given)

$\overline{PE} \cong \overline{AR}$ (The opposite sides of a parallelogram are congruent.)

You finish.

14.8 B3 The following is just a guide for your proof:

$\angle N \cong \angle N$

$\overline{NT} \cong \overline{NR}$

$NT = TS + SN$

$NR = NA + AR$

$TS = SN$

$NA = AR$

$NT = SN + SN = 2(SN)$

$NR = NA + NA = 2(NA)$

$2(SN) = 2(NA)$

$SN = NA$

$\overline{SN} \cong \overline{NA}$

$\Delta SNR \cong \Delta ANT$

$\overline{TA} \cong \overline{RS}$

14.8 C2 Label the intersection of the diagonals X. Use Theorem 25 (*The diagonals of a parallelogram bisect each other*) to show that $\Delta KWX \cong \Delta ERX$.

14.9 A1 *Given:* Quadrilateral $QUAD$ with diagonal $\overline{QA}$ forming ΔQUA and ΔADQ with $\Delta QUA \cong \Delta ADQ$

Show: $QUAD$ is a parallelogram

14.9 B4 *Given:* Quadrilateral $ABCD$ where diagonals $\overline{AC}$ and $\overline{BD}$ bisect each other at E

Show: $ABCD$ is a parallelogram

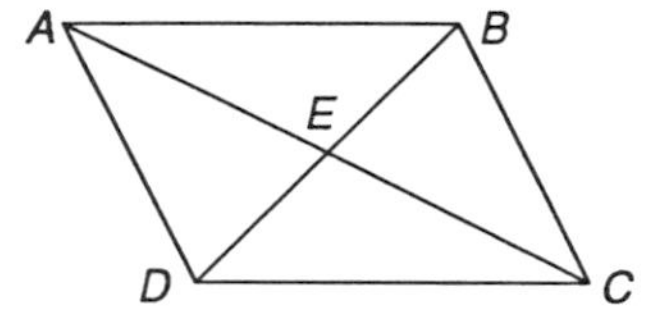

Plan:
- I can prove $ABCD$ is a parallelogram if I can show that a pair of opposite sides are both congruent and parallel.
- I can show $\overline{AB}$ and $\overline{CD}$ are congruent and parallel if I can show $\Delta ABE \cong \Delta CDE$.
- Since the diagonals bisect each other, that gives me two pairs of congruent sides.
- Since vertical angles are congruent, $\angle DEC \cong \angle BEA$.
- Therefore, I can prove $\Delta ABE \cong \Delta CDE$ by SAS.

14.10 B4 The following is just a guide for your proof:
Construct angle bisector $\overline{CD}$
$\angle ACD \cong \angle BCD$
$\angle A \cong \angle B$
$\overline{CD} \cong \overline{CD}$
$\Delta ADC \cong \Delta BDC$
$\overline{AC} \cong \overline{BC}$

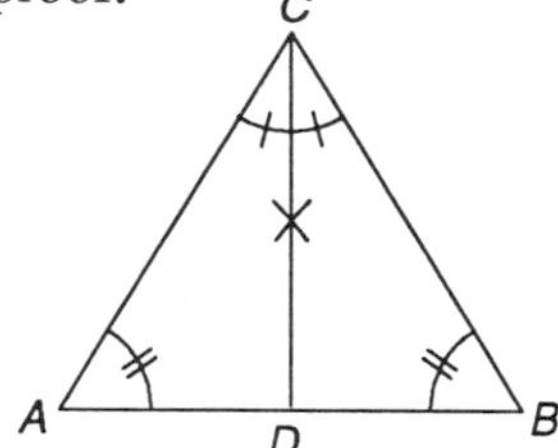

14.11 B2 In an earlier exercise (14.9 B1) you proved: *If two sides of a quadrilateral are parallel and congruent, then the other two sides are parallel.*

14.11 C5 Midpoint

14.12 B2 It's not the Midpoint Postulate.

14.12 C1 *Plan:*
- I can show ΔNOT is isosceles if I can show $\overline{NT} \cong \overline{NO}$.
- I can show $\overline{NT} \cong \overline{NO}$ if $\Delta WNT \cong \Delta HNO$.
- I can show the triangles congruent by SSS, SAS, or ASA. Which?
- Looking at the given, I see that $\overline{WN} \cong \overline{HN}$ since N is the midpoint of $\overline{WH}$ and $\angle W \cong \angle H$ since ΔWHY is isosceles. Since $\overline{WY} \cong \overline{HY}$ and T and O are their midpoints, it can be shown that $\overline{WT} \cong \overline{HO}$.
- Therefore, I will be able to show the triangles congruent by SAS.

14.12 C3 Use the theorem proved in the previous exercise.

14.12 C4 Construct a line through P parallel to $\overline{AR}$ (by Parallel Postulate). Label the point of intersection on $\overline{TR}$ as point Z. Since $\overline{ZR} \parallel \overline{PA}$ and $\overline{PZ} \parallel \overline{AR}$, then $PARZ$ is a parallelogram. If $PARZ$ is a parallelogram, then $\overline{PZ} \cong \overline{AR}$. But, $\overline{AR} \cong \overline{PT}$. Therefore, $\overline{PZ} \cong \overline{PT}$. If $\overline{PZ} \cong \overline{PT}$, then $\angle 2 \cong \angle 3$. But, $\angle 3 \cong \angle 4$ (Vertical Angles Theorem) and $\angle 4 \cong \angle 1$ (AIA Postulate), therefore, $\angle 2 \cong \angle 1$ (Transitive Property of Congruence). This proves the bottom pair of base angles are congruent.

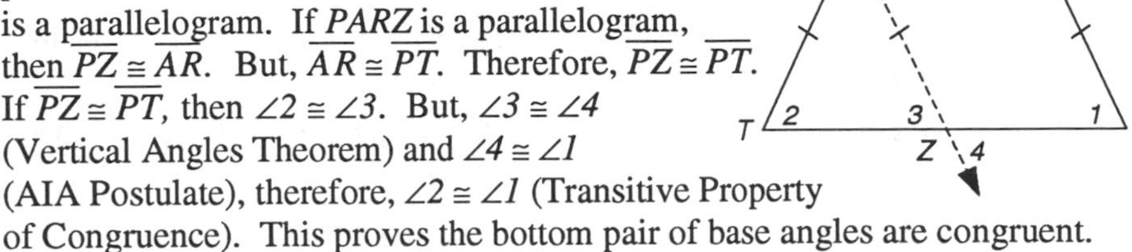

Next you prove that the top pair are congruent. Since $\angle 1 \cong \angle 2$, then $\angle 7 \cong \angle 8$ since supplements of congruent angles are congruent ($\angle 2$ and $\angle 7$ are supplements and $\angle 1$ and $\angle 8$ are supplements).

14.12 D2b True. Triangles congruent (by SAS). Therefore, opposite sides congruent (by CPCTC). Therefore, the quadrilateral is a parallelogram (if the opposite sides of a quadrilateral are congruent then the quadrilateral is a parallelogram).

Chapter 15

15.1 3

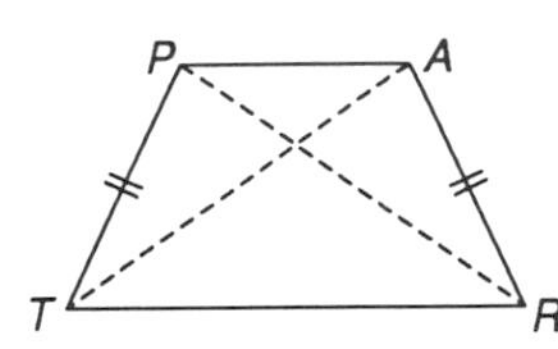

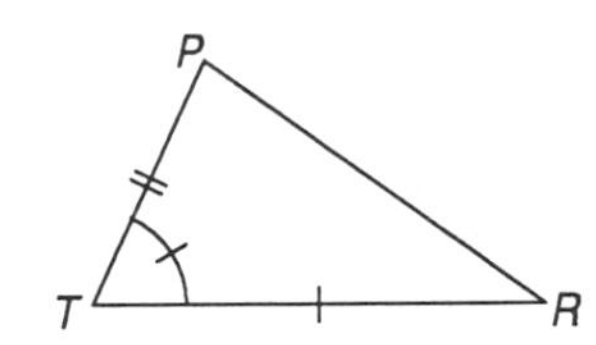

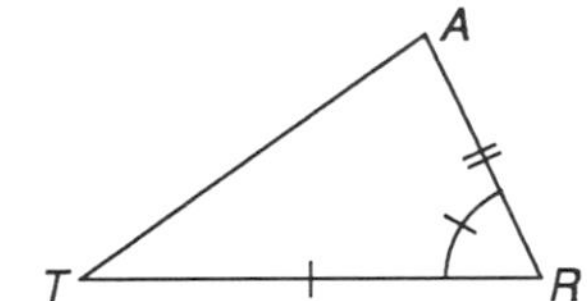

15.1 4 There are a number of conjectures that you can make in this investigation. The diagonals don't bisect all the angles. Both diagonals are not perpendicular bisectors of the other, but one is.

15.3 C2 $b = \frac{1}{2}(18 + 50)$

15.3 C3 $98 = \frac{1}{2}(c + 122)$

15.3 C6 $65 = \frac{1}{2}(f + 53)$

15.3 C9 $23 = \frac{1}{2}(c - 17)$

15.3 C12 $\angle C \cong \angle P$, therefore, $m\widehat{DE} = 42$. Since $\overline{CD} \parallel \overline{AB}$, $m\widehat{AD} = m\widehat{CB}$. Therefore, $m\widehat{AD}$ equals half of what is left after subtracting the sum of 108 and 116 from 360.

$$m\widehat{CB} = \frac{1}{2}(360 - 224)$$
$$42 = m\widehat{AD} - f$$

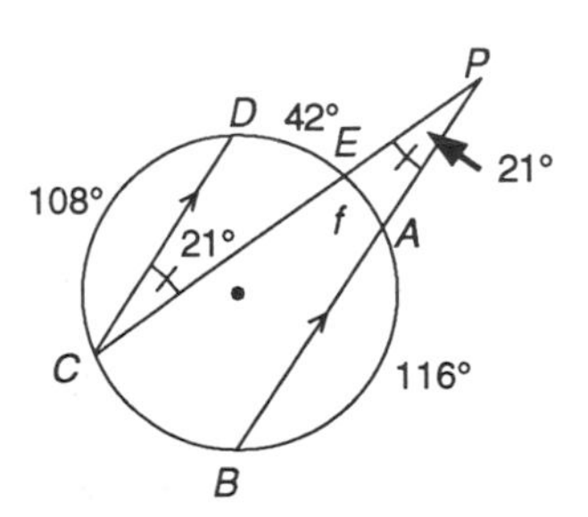

15.4 B2 Given $\Delta SAL \sim \Delta BGE$, show $\frac{LM}{EI} = \frac{LA}{GE}$.

Plan: I can show $\frac{LM}{EI} = \frac{LA}{GE}$ if I can show $\Delta LMA \sim \Delta EIG$.

I can show two triangles are similar if I can show two angles in one triangle are congruent to two angles in another triangle.

But $\angle A \cong \angle G$ (since $\Delta SAL \sim \Delta BGE$), and $\angle LMA \cong \angle EIG$ because both are right angles.

15.5 B3 Recall the conjecture: *An angle bisector divides the opposite side into two segments whose lengths are in the same ratio as the ratio of the corresponding adjacent sides.*

You can use it to find length w ($\frac{21}{42} = \frac{w}{28}$). Use w to find z ($\frac{z}{22} = \frac{w}{28}$).

15.5 B6 $c(c + 30) = (40)(100)$

15.5 B7

$$(d + 8)(2d + 8) = (7)(22)$$
$$2d^2 + 24d + 64 = 154$$
$$2d^2 + 24d - 90 = 0$$
$$d^2 + 12d - 45 = 0$$

Factor and solve for d.

15.5 B10 Let a and b be the two bases of the trapezoid.

$$A = \frac{1}{2}(h)(a + b)$$
$$500 = \frac{1}{2}(8 + 12)(a + b)$$

Solve for $a + b$. But, $\frac{a}{b} = \frac{8}{12}$ or $\frac{a}{b} = \frac{2}{3}$.

Since you know the ratio of a to b and their sum, you can solve for a and b.

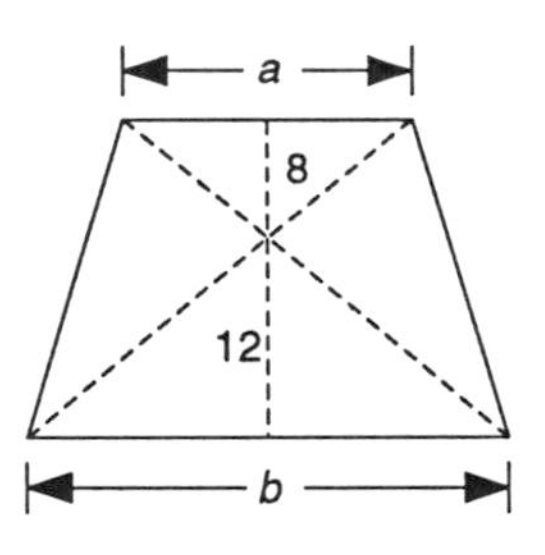

15.6 A2 *Conjecture:* *In a right triangle, if a and b are the lengths of the legs and c is the length of the hypotenuse, then $c^2 = a^2 + b^2$.*

Given: Right triangle ABC with right $\angle ACB$

Show: $c^2 = a^2 + b^2$

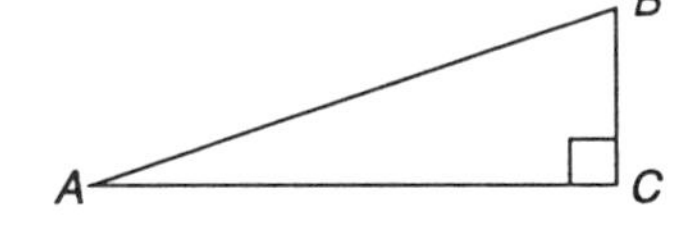

Plan:

- Construct an altitude $\overline{CD}$ to the hypotenuse. Label the height h. Label the lengths of the two parts on the hypotenuse x and y.
- Since $\Delta BDC \sim \Delta BCA$ (by the AA similarity postulate), then $\frac{y}{a} = \frac{a}{c}$ (corresponding sides of similar triangles are proportional). Cross multiplying you get: $a^2 = cy$.
- Similarly, since $\Delta CDA \sim \Delta BCA$, $\frac{x}{b} = \frac{b}{c}$ and thus: $b^2 = xc$.
- If $a^2 = cy$, $b^2 = xc$, and $c = x+y$, use algebra to get $c^2 = a^2 + b^2$.

15.6 A3 *Conjecture:* *If a, b, and c are the lengths of the sides of ΔABC and $c^2 = a^2 + b^2$, then ΔABC is a right triangle.*

Given: a, b, and c are the lengths of the sides of ΔABC and $c^2 = a^2 + b^2$

Show: ΔABC is a right triangle

Plan:

- Construct a right ΔDEF with legs of lengths a and b and hypotenuse of length x.
- Then $x^2 = a^2 + b^2$ by the Pythagorean Theorem.
- It is given that $c^2 = a^2 + b^2$. Therefore, $x^2 = c^2$ or $x = c$.
- If $x = c$, then $\Delta DEF \cong \Delta ABC$.

15.6 B1 $\frac{20}{30} = \frac{30}{a}$

15.6 B4 The discovery is a generalization of the Pythagorean Theorem: *If similar figures are constructed on the sides of a right triangle so that corresponding sides of the similar figures are the sides of the right triangle, then the —?— of the areas of two similar figures constructed on the legs is equal to the area of a ––?— figure constructed —?—.* To prove the discovery for one of the special cases, find the area of each figure formed on the sides of one of the right triangles below in terms of a, b, and c. Does your discovery hold for this special case?

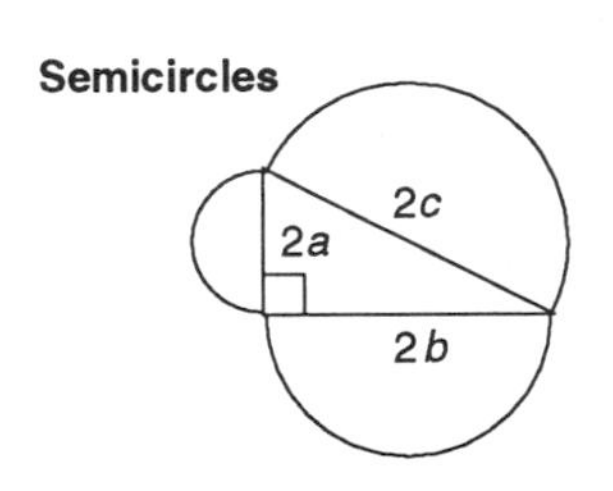

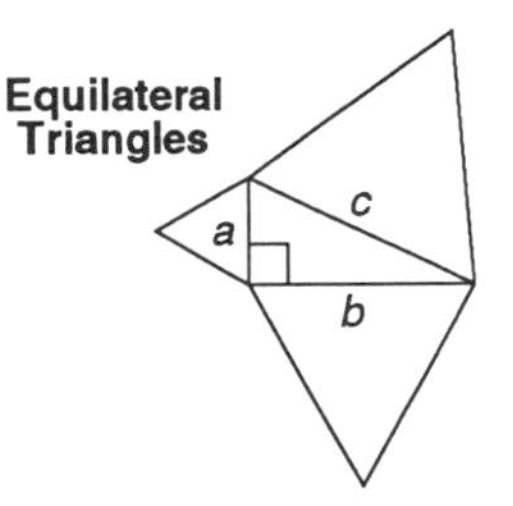

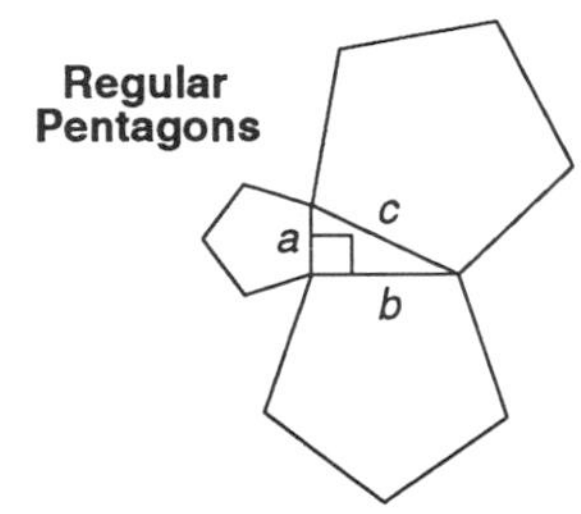

15.7 A5

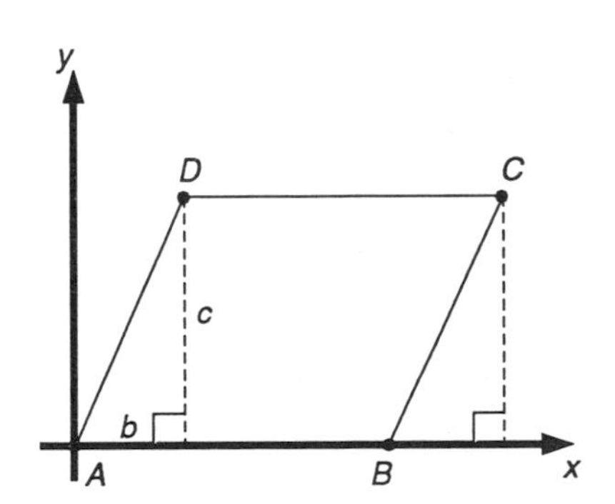

15.7 A6 — Look for a special right triangle.

15.10 C3 — Construct $\overline{AO}$, $\overline{BO}$, and $\overline{OT}$. Use the Pythagorean Theorem.
$OT^2 = OB^2 + BT^2$
$OT^2 = OA^2 + AT^2$
Therefore, $OB^2 + BT^2 = OA^2 + AT^2$.
But $OB = OA$, therefore $OB^2 = OA^2$. You finish it.

15.10 E1 — The proof of the conjecture about parallel lines and congruent arcs used the AIA Postulate and the Inscribed Angle Theorem. The proof of the Inscribed Angle Theorem is in Lesson 15.2.

15.10 E2 — The proof of the Converse of the Pythagorean Theorem used the Pythagorean Theorem and the SSS Congruence Postulate.

Appendix B

Logo Procedures for Computer Activities

Below are the listings of all the Logo procedures used in the five Logo computer activities. Use the listings to make a *Discovering Geometry* Logo procedures disk. You can enter the procedures activity by activity, or you can enter them all at once.

The procedures will run without modification if you are using LCSI™ Logo II or Apple® Logo II. They may need modifications to run with other versions of Logo. The modifications needed for LCSI LogoWriter® (either the Apple II or IBM versions), IBM® Logo, Terrapin™ Logo for the Apple, and Logo PLUS™ by Terrapin are outlined below. You will have to consult your Logo manuals if the procedures do not run as written with other versions of Logo.

LCSI and LogoWriter are trademarks of Logo Computer Systems Inc. Terrapin and Logo PLUS are trademarks of Terrapin, Inc. Apple is a trademark of Apple Computer, Inc. IBM is a trademark of International Business Machines Corp.

If you are using LCSI LogoWriter (Apple II or IBM versions):

Change all the CS commands to CG. Change TOOT to TONE in RT.S3, SPIRO, and ANY.SPIRO. Change EMPTYP to EMPTY? in ANY.SUBL. Change IF in the FWD procedure to IFELSE. Substitute the procedure at right for RESPONSE1. Eliminate WINDOW in WHOLETESS, WHOLE.SHAPE, WHOLE.HEX, and WHOLE.TRI.

```
TO RESPONSE1
CC
TYPE [This spirolateral does not
     close.]
TYPE [You must use the logo "break" key
     sequence]
TYPE [to stop this procedure.]
SHOW "
END
```

If you are using IBM Logo:

Change TOOT to TONE in RT.S3, SPIRO, and ANY.SPIRO. Remove the SETBG 6 command in RT.ANY. Remove the PI procedure from ESCHER.TESS because PI is a primitive in IBM Logo.

If you are using Terrapin Logo for the Apple or Logo PLUS by Terrapin:

Change all the CS commands to DRAW. Change the syntax of the IF...[...] commands to IF...THEN.... Change the INT command in ARCR and ARCL to INTEGER. Change EMPTYP to EMPTY? in ANY.SUBL. Change SETBG to BG in ANY.SPIRO. Change ARCTAN 0.75 to ATAN 3 4 in PATH1. In MOVETO, change SETPOS LIST :X :Y to SETXY :X :Y. In SIDE1 and SIDE2, change the MAKE "START POS command to MAKE "X XCOR MAKE "Y YCOR and change SETPOS :START to SETXY :X :Y. Eliminate WINDOW in WHOLETESS, WHOLE.SHAPE, WHOLE.HEX, and WHOLE.TRI. If you are using Terrapin Logo for the Apple, eliminate the TOOT command in RT.S3, SPIRO, and ANY.SPIRO. If you are using Logo PLUS by Terrapin, change the TOOT commands to NOTE. If you are using Terrapin Logo for the Apple, add the following WAIT procedure to the SPIROS file since there is no WAIT primitive.

```
TO WAIT :X
REPEAT 100 * :X []
END
```

Exploring Parallelograms with Logo

```
TO PARA :SIDE1 :ANGLE :SIDE2
REPEAT 2 [FD :SIDE1 RT :ANGLE FD :SIDE2 RT 180 - :ANGLE]
END
```

Regular Polygons and Star Polygons with Logo

Organize the four procedures below into a Logo file called POLYGONS.

```
TO SIDEPOLY :N
REPEAT :N [FD 30 RT (360 / :N)]
END

TO REGPOLY :N :S
REPEAT :N [FD :S RT (360 / :N)]
END

TO ANYPOLY :N :S :A
REPEAT :N [FD :S RT :A]
END

TO STARPOLY :N :S :M
REPEAT :N [FD :S RT (:M * 360 / :N)]
END
```

Logo Spirolaterals

Organize the procedures below into a Logo file called SPIROS.

```
TO SCALE :N
MAKE "SCALE :N
END

TO RT.S3 :FIRST :SECOND :THIRD
REPEAT 4 [FWD :FIRST RT 90 FWD :SECOND
   RT 90 FWD :THIRD ST WAIT 25 TOOT 440 5
   RT 90 HT]
END

TO FWD :DISTANCE
HT
IF :DISTANCE < 0 [REPEAT - :DISTANCE
   [FD - :SCALE TIC.MARK]] [REPEAT
   :DISTANCE [FD :SCALE TIC.MARK]]
ST
END

TO TIC.MARK
RT 90 FD 1.5 BK 1.5 LT 90
END

TO MOVETO :X :Y
PU SETPOS LIST :X :Y PD
END

TO MOVE :UNIT
PU FD :UNIT * :SCALE PD
END

TO RT.S :ORDER
SPIRO :ORDER 90
END

TO SPIRO :ORDER :ANGLE
DEGREE :ORDER :ANGLE
ST
REPEAT :DEGREE [SA 1 :ORDER :ANGLE TOOT
   220 5 WAIT 20]
HT
IF :DEGREE = 1 [SPIRO :ORDER :ANGLE]
END

TO DEGREE :ORDER :ANGLE
MAKE "DEGREE 360 / GCD ROUND (:ORDER *
   :ANGLE) 360
IF :DEGREE = 1 [RESPONSE1]
END

TO GCD :A :B
IF :B = 0 [OP :A]
OP GCD :B REMAINDER :A :B
END

TO RESPONSE1
PR [This spirolateral does not close.]
PR [You must use the logo "break" key
   sequence]
PR [to stop this procedure.]
END

TO SA :N :ORDER :A
IF :N > :ORDER [STOP]
FD :SCALE * :N RT :A
SA :N + 1 :ORDER :A
END

TO RT.ANY :LIST
ANY.SPIRO :LIST 90
END

TO ANY.SPIRO :LIST :ANGLE
DEGREE COUNT :LIST :ANGLE
SETBG 6 ST
REPEAT :DEGREE [ANY.SUBL :LIST :ANGLE
   TOOT 220 5 WAIT 20]
HT
IF :DEGREE = 1 [ANY.SPIRO :LIST :ANGLE]
END

TO ANY.SUBL :LIST :ANGLE
IF EMPTYP :LIST [STOP]
FWD FIRST :LIST RT :ANGLE
ANY.SUBL BF :LIST :ANGLE
END

TO TESS.CROSS
CS
REPEAT 3 [RT.ANY [1 -1 1] MOVE 1 RT 90
   MOVE 1 LT 90 MOVE 1]
PU HOME
RT 90 MOVE 2 LT 90 MOVE -1 PD
REPEAT 3 [RT.ANY [1 -1 1] MOVE 1 RT 90
   MOVE 1 LT 90 MOVE 1]
END
```

This command sets the initial value of the scale at 10. Include it in the SPIROS file.

```
MAKE "STARTUP [SCALE 10]
```

Fractals with Logo

Organize these two procedures into a Logo file called FRACTALS.

```
TO SNOWFLAKE :L :S
REPEAT 3 [DIV.SIDE :L :S RT 120]
END
```

```
TO DIV.SIDE :L :S
IF :L = 1 [FD :S STOP]
DIV.SIDE :L - 1 :S / 3 LT 60
DIV.SIDE :L - 1 :S / 3 RT 120
DIV.SIDE :L - 1 :S / 3 LT 60
DIV.SIDE :L - 1 :S / 3
END
```

Tessellating with Logo

Organize the procedures that follow into a Logo file called ESCHER.TESS.

The primary superprocedures and their subprocedures:

```
TO RPOLY :N :S
REPEAT :N [FD :S RT 360 / :N]
END

TO TRANSPOLY :N :S :L
IF :L < 1 [STOP]
RPOLY :N :S
UP.OVER :N :S
TRANSPOLY :N :S :L - 1
BK.DOWN :N :S
END

TO UP.OVER :N :S
PU FD :S RT 360 / :N
FD :S LT 360 / :N PD
END

TO BK.DOWN :N :S
PU RT 360 / :N BK :S
LT 360 / :N BK :S PD
END
```

```
TO TESSELLATE :N :S :L1 :L2
IF :L2 < 1 [STOP]
TRANSPOLY :N :S :L1
BK.DOWN - :N :S
TESSELLATE :N :S :L1 :L2 - 1
UP.OVER - :N :S
END

TO WHOLETESS :N :S :L1 :L2
WINDOW CS
REPEAT 360 / INT.ANGLE :N [TESSELLATE :N
  :S :L1 :L2 RT INT.ANGLE :N]
END

TO INT.ANGLE :N
OP 180 - (360 / :N)
END
```

Transformations of RPOLY based on point-symmetric curved sides:

```
TO SPOLY :N :S
REPEAT :N [SFD :S RT 360 / :N]
END

TO SFD :S
LT 30
ARCR :S / 2 60
ARCL :S / 2 60
RT 30
END
```

```
TO CPOLY :N :S
REPEAT :N [CFD :S RT 360 / :N]
END

TO CFD :S
LT 60
ARCR :S / (2 * SQRT 3) 120
ARCL :S / (2 * SQRT 3) 120
RT 60
END
```

Arc procedures for drawing radius based arcs of given angle:

```
TO ARCR :R :A
MAKE "TIMES INT (:A / 6) + 1
MAKE "TURN :A / (:TIMES * 2)
REPEAT :TIMES [RT :TURN FD ((ARC :R) * :A
  / :TIMES) RT :TURN]
END

TO ARCL :R :A
MAKE "TIMES INT (:A / 6) + 1
MAKE "TURN :A / (:TIMES * 2)
REPEAT :TIMES [LT :TURN FD ((ARC :R) * :A
  / :TIMES) LT :TURN]
END
```

```
TO ARC :RADIUS
OP (CIRC :RADIUS) / 360
END

TO CIRC :RADIUS
OP PI * :RADIUS * 2
END

TO PI
OP 3.14159
END
```

A point-symmetric FD move which may be used in RPOLY instead of FD:

```
TO SIMFD :S
PU FD :S / 2 PD
REPEAT 2 [PATH :S / 2 RT 180]
PU FD :S / 2 PD
END
```

Example PATHs which can be used by SIMFD:

```
TO PATH :D
LT 45 FD :D / SQRT 2
LT 45 BK :D / 2
RT 90 FD :D / 2
PU BK :D PD
END

TO PATH1 :D
LT 90
FD 3 * :D / 8
RT 90 FD :D / 2
RT ARCTAN 0.75 FD 5 * :D / 8
LT ARCTAN 0.75 PU BK :D PD
END

TO PATH2 :D
LT 30 ARCR :D 60
LT 30 PU BK :D PD
END

TO PATH3 :D
LT 60 ARCR :D / SQRT 3 120
LT 60 PU BK :D PD
END

TO PATH4 :D
LT 90 ARCR :D / 2 180
LT 90 PU BK :D PD
END
```

General procedures to tessellate any shape based on a regular polygon:

```
TO TRANS :SHAPE :N :S :L
IF :L < 1 [STOP]
RUN (SE :SHAPE :N :S)
UP.OVER :N :S
TRANS :SHAPE :N :S :L - 1
BK.DOWN :N :S
END

TO TESS.SHAPE :SHAPE :N :S :L1 :L2
IF :L2 < 1 [STOP]
TRANS :SHAPE :N :S :L1
BK.DOWN - :N :S
TESS.SHAPE :SHAPE :N :S :L1 :L2 - 1
UP.OVER - :N :S
END

TO WHOLE.SHAPE :SHAPE :N :S :L1 :L2
WINDOW CS
REPEAT 360 / INT.ANGLE :N [TESS.SHAPE
   :SHAPE :N :S :L1 :L2 RT INT.ANGLE :N]
END
```

Non point-symmetric shapes to try with TESS.SHAPE and WHOLE.SHAPE:

```
TO BSQ :N :S
REPEAT 2 [BFD1 :S RT 90]
REPEAT 2 [BFD2 :S RT 90]
END

TO RSQ :N :S
REPEAT 2 [RFD1 :S RT 90]
REPEAT 2 [RFD2 :S RT 90]
END

TO BHEX :N :S
REPEAT 3 [BFD1 :S RT 360 / 6 BFD2 :S RT 360 / 6]
END

TO RHEX :N :S
REPEAT 3 [RFD1 :S RT 360 / 6 RFD2 :S RT 360 / 6]
END

TO BFD1 :S
FD :S / 3
RT 60 FD :S / 3
LT 120 FD :S / 3
RT 60 FD :S / 3
END

TO BFD2 :S
FD :S / 3
LT 60 FD :S / 3
RT 120 FD :S / 3
LT 60 FD :S / 3
END

TO RFD1 :S
FD :S / 3
RT 90 FD :S / 5
LT 90 FD :S / 3
LT 90 FD :S / 5
RT 90 FD :S / 3
END

TO RFD2 :S
FD :S / 3
LT 90 FD :S / 5
RT 90 FD :S / 3
RT 90 FD :S / 5
LT 90 FD :S / 3
END
```

Escher's REPTILE based on rotations of consecutive sides of a regular hexagon:

```
TO REPTILE :S
REPEAT 2 [SIDE1 :S RT 120]
RT 120
PU REPEAT 2 [FD :S RT 60] PD
REPEAT 2 [SIDE2 :S RT 120]
RT 120
PU REPEAT 2 [FD :S RT 60] PD
REPEAT 2 [SIDE3 :S RT 120]
RT 120
PU REPEAT 2 [FD :S RT 60] PD
END
```

```
TO REPTILEN :N :S
REPEAT 2 [SIDE1 :S RT 120]
RT 120
PU REPEAT 2 [FD :S RT 60] PD
REPEAT 2 [SIDE2 :S RT 120]
RT 120
PU REPEAT 2 [FD :S RT 60] PD
REPEAT 2 [SIDE3 :S RT 120]
RT 120
PU REPEAT 2 [FD :S RT 60] PD
END
```

The three sides of the hexagon which are rotated to form Escher's REPTILE:

```
TO SIDE1 :S
LT 60 FD :S / 2
RT 60 FD :S / 4
RT 60 FD :S / 2
RT 60 FD :S / 2
LT 120 FD :S * 13 / 36
LT 60 FD :S / 6
LT 60 FD :S / 9
RT 60 FD :S / 3
BK :S / 9 RT 60
PU BK :S PD
END
```

```
TO SIDE2 :S
MAKE "START POS
LT 60 FD :S / 2
LT 60 FD :S / 4
RT 120 FD :S / 4
RT 90 FD (:S * 3 * SQRT 3) / 8
LT 60 FD :S / 2
LT 60 FD 7 * :S / 24 RT 30
PU SETPOS :START PD
END
```

```
TO SIDE3 :S
MAKE "START POS
LT 120 FD :S / 3
RT 120 FD :S / 3
RT 60 FD :S / 6
LT 60 FD 4 *:S / 9
RT 90 FD :S / 6
RT 30 FD :S / 3
RT 30 FD :S / 3
LT 150 FD :S / 3
LT 50 FD 4 * :S / 9
LT 30 FD :S / 7 RT 80
PU SETPOS :START PD
END
```

Rotation tessellations of non point-symmetric hexagon based shapes:

```
TO ROT.HEX :SHAPE :S :L
IF :L < 1 [STOP]
REPEAT 3 [RUN SE :SHAPE :S RT 120]
PU FD :S * 3 PD
ROT.HEX :SHAPE :S :L - 1
PU BK :S * 3 PD
END
```

```
TO ROTATE.6 :SHAPE :S :L1 :L2
IF :L2 < 1 [STOP]
ROT.HEX :SHAPE :S :L1
RT 120
PU FD :S * 3 PD
LT 120
ROTATE.6 :SHAPE :S :L1 :L2 - 1
RT 120
PU BK :S * 3 PD
LT 120
END
```

```
TO WHOLE.HEX :SHAPE :S :L1 :L2
CS WINDOW
UP.OVER 6 :S
REPEAT 3 [ROTATE.6 :SHAPE :S :L1 :L2 PU
  BK :S * 3 RT 120 PD]
BK.DOWN 6 :S
END
```

```
TO HEX.SIDES :S
REPEAT 2 [HSIDE1 :S RT 120]
RT 120
PU REPEAT 2 [FD :S RT 60] PD
REPEAT 2 [HSIDE2 :S RT 120]
RT 120
PU REPEAT 2 [FD :S RT 60] PD
REPEAT 2 [HSIDE3 :S RT 120]
RT 120
PU REPEAT 2 [FD :S RT 60] PD
END
```

```
TO HSIDE1 :D
LT 45 FD :D / SQRT 2
LT 45 BK :D / 2
RT 90 FD :D / 2
PU BK :D PD
END

TO HSIDE2 :D
FD :D / 3
LT 90 FD :D / 6
RT 90 FD :D / 3
RT 90 FD :D / 6
LT 90 FD :D / 3
PU BK :D PD
END

TO HSIDE3 :D
RT 90 FD :D / 4
LT 90 FD :D / 2
LT 90 FD :D / 4
RT 90 FD :D / 2
PU BK :D PD
END
```

Rotation tessellations of non point-symmetric triangle based shapes:

```
TO BIRD.TRI :S
TSIDE1 :S PU BK :S
PD RT 60 TSIDE1 :S
LT 120 SIMFD1 :S
RT 60 PU BK :S PD
END

TO BIRD.TRIN :N :S
TSIDE1 :S PU BK :S
PD RT 60 TSIDE1 :S
LT 120 SIMFD1 :S
RT 60 PU BK :S PD
END

TO TSIDE1 :S
LT 30 FD 2 * :S / (3 * SQRT 3)
RT 90 FD :S / 3
LT 120 FD :S / 3
RT 90 FD 2 * :S / (3 * SQRT 3)
LT 30
END

TO SIMFD1 :S
PU FD :S / 2 PD
REPEAT 2 [PATH1 :S / 2 RT 180]
PU FD :S / 2 PD
END

TO ROT.TRI :SHAPE :S :L
IF :L < 1 [STOP]
REPEAT 6 [RUN SE :SHAPE :S RT 60]
PU RT 60 FD :S LT 60 FD :S PD
ROT.TRI :SHAPE :S :L - 1
PU BK :S RT 60 BK :S LT 60 PD
END

TO ROTATE.3 :SHAPE :S :L1 :L2
IF :L2 < 1 [STOP]
ROT.TRI :SHAPE :S :L1
PU RT 60 FD :S RT 60 FD :S LT 120 PD
ROTATE.3 :SHAPE :S :L1 :L2 - 1
PU RT 120 BK :S LT 60 BK :S LT 60 PD
END

TO WHOLE.TRI :SHAPE :S :L1 :L2
CS WINDOW
PU RT 60 FD :S RT 60 FD :S LT 120 PD
REPEAT 6 [ROTATE.3 :SHAPE :S :L1 :L2 PU BK
:S RT 60 BK :S PD]
PU RT 120 BK :S LT 60 BK :S LT 60 PD
ROT.TRI :SHAPE :S 1
END

TO TRI.SIDES :S
TSIDE :S RT 60 TSIDE :S
PU FD :S PD
LT 120 SIMFD :S
RT 60 PU BK :S PD
END

TO TSIDE :D
FD :D / 3
LT 60 FD :D / 3
RT 120 FD :D / 3
LT 60 FD :D / 3
PU BK :D PD
END
```

Index

Acknowledgements

Peter Rasmussen, Donald Rasmussen, Lore Rasmussen, Dana Curtin, Madeleine Mulgrew, Sherry Fraser, Diane Barrows, Scott Kim, Jerry Legé, Cindy Sima, Hong Jen Tsui, Arianne Leong, Mike Bobadilla, Paul Woodward, Marc Roth, Dave Stonerod, and many excellent speakers on geometry at the CMC-N Asilomar Conference and many valuable articles on geometry in *The Mathematics Teacher* and the magazine *Games*.

Credits

Graphic Design	Eleanor Henderson
Computer Graphics	Ann Rothenbuhler, Eleanor Henderson, Felicia Woytak
Illustrations	R. Diggs, Jerry Simpfenderfer, Marylin Hill, Eleanor Henderson
Cover Design	John Odam
Photo Coordination	Linda Graham
Index	Shirley Manley

Photo Credits

1, 304	*Snow Crystals*, Bentley and Humphreys, Reprinted by Dover Publications, 1962
14, 15, 22, 477	Alinari/Art Resource, New York
18	Courtesy of the Trustees of the British Museum
23	Metropolitan Museum of Art, Harris Brisbane Dick Fund, 1941
24	*Perspective*, Jan Vredeman de Vries, Reprinted by Dover Publications, 1968
64, 191, 204	Library of Congress
261	Wide World Photos, New York
324, 330, 332, 335	Collection Haags Gemeentemuseum – The Hague
331	National Gallery of Art, Washington D.C, Gift of Mr. C. V. S. Roosevelt
422	Fitzwilliam Museum, Cambridge, England
477	Alison Frantz, Princeton, New Jersey
524	National Council of Teachers of Mathematics
547	Art Matrix
688	Picture Collection, The Branch Libraries, The New York Public Library